Study and Solutions Guide

PRECALCULUS FUNCTIONS AND GRAPHS: A GRAPHING APPROACH

THIRD EDITION

and

PRECALCULUS WITH LIMITS: A GRAPHING APPROACH

THIRD EDITION

Larson/Hostetler/Edwards

Bruce H. Edwards
University of Florida
Gainesville, Florida

Dianna L. Zook
Indiana University—
Purdue University at
Fort Wayne, Indiana

HOUGHTON MIFFLIN COMPANY **Boston** **New York**

Editor-in-Chief: Jack Shira
Managing Editor: Cathy Cantin
Development Manager: Maureen Ross
Associate Editor: Laura Wheel
Assistant Editor: Carolyn Johnson
Supervising Editor: Karen Carter
Project Editor: Patty Bergin
Art Supervisor: Gary Crespo
Marketing Manager: Michael Busnach
Senior Manufacturing Coordinator: Sally Culler
Composition and Art: Meridian Creative Group

Printed in the United States of America

ISBN: 0-618-07410-4

23456789-CRS-04 03 02 01

CONTENTS

TO THE STUDENT

The *Study and Solutions Guide for Precalculus Functions and Graphs: A Graphing Approach*, Third Edition, and *Precalculus with Limits: A Graphing Approach*, Third Edition, is a supplement to the textbook by Ron Larson, Robert P. Hostetler, and Bruce H. Edwards.

As mathematics instructors, we often have students come to us with questions about assigned homework. When we ask to see their work, the reply often is "I didn't know where to start." The purpose of the *Study Guide* is to provide brief summaries of the topics covered in the textbook and enough detailed solutions to problems so that you will be able to work the remaining exercises.

A special thanks to Meridian Creative Group for typing this guide. Also, we would like to thank our spouses, Consuelo Edwards and Edward Schlindwein, for their support.

If you have any corrections or suggestions for improving this *Study Guide*, we would appreciate hearing from you.

Good luck with your study of precalculus.

Bruce H. Edwards
358 Little Hall
University of Florida
Gainesville, FL 32611
Be@math.ufl.edu

Dianna L. Zook
Indiana University
Purdue University
Fort Wayne, IN 46805
Zook@ipfw.edu

STUDY STRATEGIES

- Attend all classes and come prepared. Have your homework completed. Bring the text, paper, pen or pencil, and a calculator (scientific or graphing) to each class.
- Read the section in the text that is to be covered before class. Make notes about any questions that you have and, if not answered during the lecture, ask them at the appropriate time.
- Participate in class. As mentioned above, ask questions — and answer them.
- Take notes on all definitions, concepts, rules, formulas, and examples. After class, read your notes and fill in any gaps, or make notations of any questions that you have.
- DO THE HOMEWORK!!! You learn mathematics by doing it yourself. Allow at least two hours outside of each class for homework. Do not fall behind.
- Seek help when needed. Visit your instructor during office hours and come prepared with specific questions; check with your school's tutoring service; find a study partner in class; check additional books in the library for more examples – just do something before the problem becomes insurmountable.
- Do not cram for exams. Each chapter in the text contains a chapter review and a chapter test and this *Study Guide* contains a practice test at the end of each chapter. (The answers are at the end of Part I.) Work these problems a few days before the exam and review any areas of weakness.

CHAPTER P
Prerequisites

CHAPTER P
Prerequisites

Section P.1 Graphical Representation of Data

- You should be able to plot points.
- You should know that the distance between (x_1, y_1) and (x_2, y_2) in the plane is

  $$d = \sqrt{(x_2 - x_1)^2 + (y_2 - y_1)^2}.$$
- You should know that the midpoint of the line segment joining (x_1, y_1) and (x_2, y_2) is

  $$\left(\frac{x_1 + x_2}{2}, \frac{y_1 + y_2}{2}\right).$$
- You should know the equation of a circle: $(x - h)^2 + (y - k)^2 = r^2$.
- You should be able to construct scatter plots, bar graphs and line graphs for a set of data.

Solutions to Odd-Numbered Exercises

1.

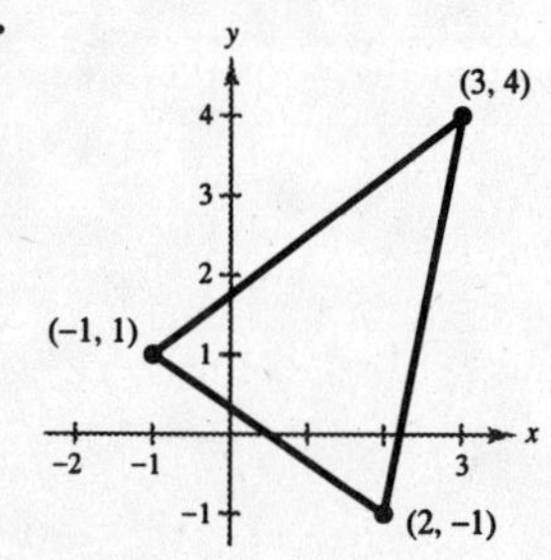

3.

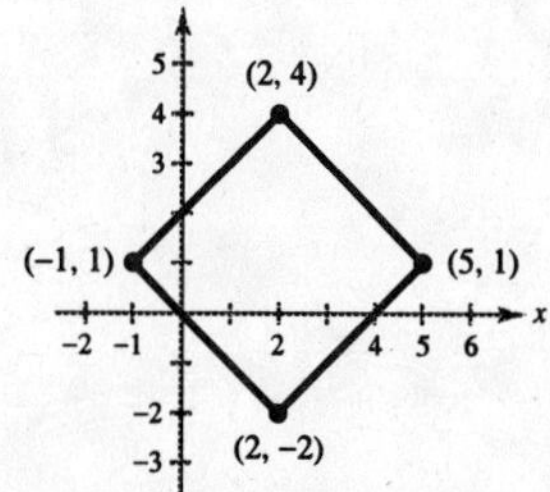

5. A: $(2, 6)$, B: $(-6, -2)$, C: $(4, -4)$, D: $(-3, 2)$

7. A:$(0, 5)$, B:$(-3, -6)$, C:$(1, -4.5)$, D:$(-4, 2)$

9. $(-3, 4)$

11. $(-5, -5)$

13. $x > 0 \implies$ The point lies in Quadrant I or in Quadrant IV.

$y < 0 \implies$ The point lies in Quadrant III or in Quadrant IV.

$x > 0$ and $y < 0 \implies (x, y)$ lies in Quadrant IV.

15. $x = -4 \implies x$ is negative $\implies$ The point lies in Quadrant II or Quadrant III.

$y > 0 \implies$ The point lies in Quadrant I or Quadrant II.

$x = -4$ and $y > 0 \implies (x, y)$ lies in Quadrant II.

17. $y < -5 \implies y$ is negative $\implies$ The point lies in either Quadrant III or Quadrant IV.

19. Since $(x, -y)$ is in Quadrant II, we know that $x < 0$ and $-y > 0$. If $-y > 0$, then $y < 0$.

$x < 0 \implies$ The point lies in Quadrant II or in Quadrant III.

$y < 0 \implies$ The point lies in Quadrant III or in Quadrant IV.

$x < 0$ and $y < 0 \implies (x, y)$ lies in Quadrant III.

21. If $xy > 0$, then either x and y are both positive, or both negative. Hence, (x, y) lies in either Quadrant I or Quadrant III.

23. The x-coordinates are increased by 2, and the y-coordinates are increased by 5: $(0, 1)$, $(4, 2)$, $(1, 4)$.

25. $d = |5 - (-3)| = 8$

27. $d = |-3 - 2| = |-5| = 5$

29. $d = \sqrt{(3 - (-2))^2 + (-6 - 6)^2} = \sqrt{5^2 + (-12)^2} = \sqrt{25 + 144} = \sqrt{169} = 13.$

31. (a)

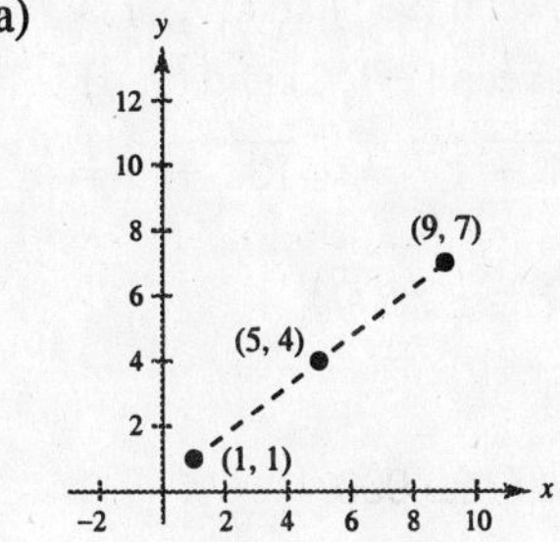

(b) $d = \sqrt{(9 - 1)^2 + (7 - 1)^2}$

$= \sqrt{64 + 36} = 10$

(c) $\left(\dfrac{9 + 1}{2}, \dfrac{7 + 1}{2}\right) = (5, 4)$

33. (a)

(−4, 10)

$(0, \frac{5}{2})$

(4, −5)

(b) $d = \sqrt{(4 + 4)^2 + (-5 - 10)^2}$

$= \sqrt{64 + 225} = 17$

(c) $\left(\dfrac{4 - 4}{2}, \dfrac{-5 + 10}{2}\right) = \left(0, \dfrac{5}{2}\right)$

35. (a)

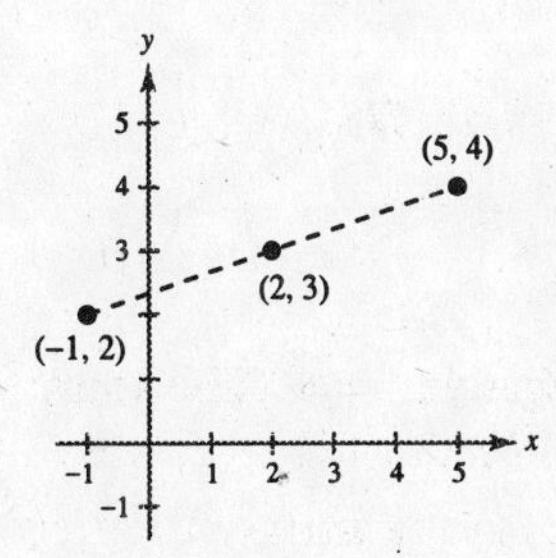

(b) $d = \sqrt{(5 + 1)^2 + (4 - 2)^2}$

$= \sqrt{36 + 4} = \sqrt{40} = 2\sqrt{10}$

(c) $\left(\dfrac{-1 + 5}{2}, \dfrac{2 + 4}{2}\right) = (2, 3)$

37. (a)

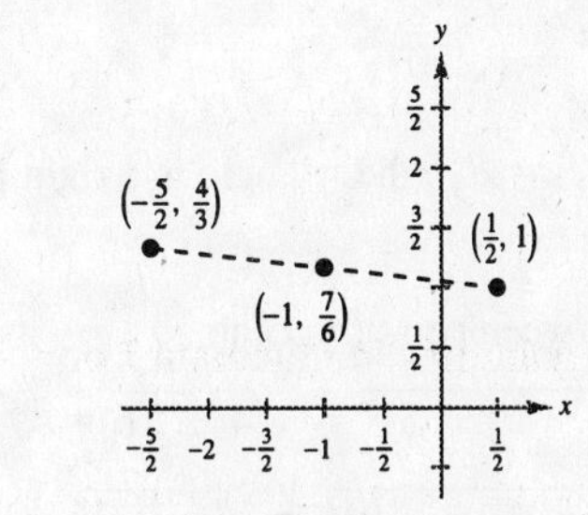

(b) $d = \sqrt{\left(\dfrac{1}{2} + \dfrac{5}{2}\right)^2 + \left(1 - \dfrac{4}{3}\right)^2}$

$= \sqrt{9 + \dfrac{1}{9}} = \dfrac{\sqrt{82}}{3}$

(c) $\left(\dfrac{-\frac{5}{2} + \frac{1}{2}}{2}, \dfrac{\frac{4}{3} + 1}{2}\right) = \left(-1, \dfrac{7}{6}\right)$

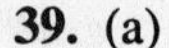

39. (a)

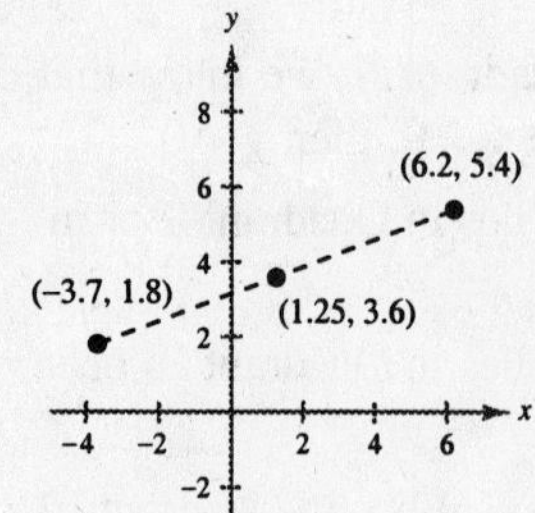

(b) $d = \sqrt{(6.2 + 3.7)^2 + (5.4 - 1.8)^2}$

$= \sqrt{98.01 + 12.96}$

$= \sqrt{110.97}$

(c) $\left(\frac{6.2 - 3.7}{2}, \frac{5.4 + 1.8}{2}\right) = (1.25, 3.6)$

41. (a)

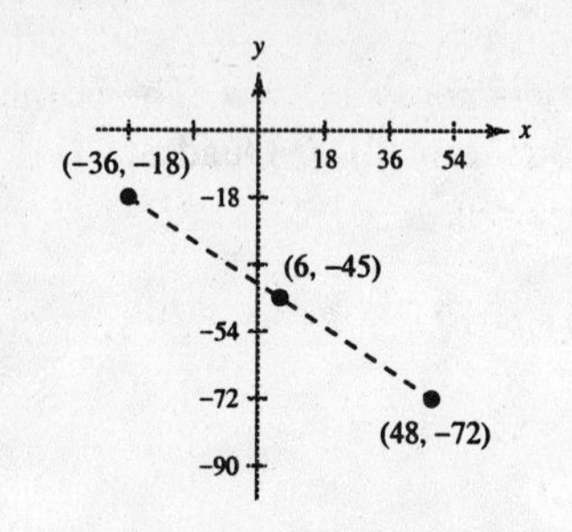

(b) $d = \sqrt{(48 + 36)^2 + (-72 + 18)^2}$

$= \sqrt{7056 + 2916}$

$= \sqrt{9972} = 6\sqrt{277}$

(c) $\left(\frac{-36 + 48}{2}, \frac{-18 - 72}{2}\right) = (6, -45)$

43. (a) The distance between (0, 2) and (4, 2) is 4.
The distance between (4, 2) and (4, 5) is 3.
The distance between (0, 2) and (4, 5) is
$\sqrt{(4 - 0)^2 + (5 - 2)^2} = \sqrt{16 + 9} = \sqrt{25} = 5.$

(b) $4^2 + 3^2 = 16 + 9 = 25 = 5^2$

45. (a) The distance between $(-1, 1)$ and $(9, 1)$ is 10.
The distance between (9, 1) and (9, 4) is 3.
The distance between $(-1, 1)$ and $(9, 4)$ is
$\sqrt{(9 - (-1))^2 + (4 - 1)^2} = \sqrt{100 + 9} = \sqrt{109}.$

(b) $10^2 + 3^2 = 109 = \left(\sqrt{109}\right)^2$

47. $\left(\frac{1996 + 2000}{2}, \frac{\$520{,}000 + \$740{,}000}{2}\right) = (1998, \$630{,}000)$. The sales in 1998 are \$630,000.

49. Find distances between pairs of points.

$d_1 = \sqrt{(4 - 2)^2 + (0 - 1)^2} = \sqrt{5}$

$d_2 = \sqrt{(4 + 1)^2 + (0 + 5)^2} = \sqrt{50}$

$d_3 = \sqrt{(2 + 1)^2 + (1 + 5)^2} = \sqrt{45}$

$\left(\sqrt{5}\right)^2 + \left(\sqrt{45}\right)^2 = \left(\sqrt{50}\right)^2$

Because $d_1^2 + d_2^2 = d_3^2$, the triangle is a right triangle.

51. Find distances between pairs of points.

$d_1 = \sqrt{(0 - 2)^2 + (9 - 5)^2} = \sqrt{4 + 16} = \sqrt{20} = 2\sqrt{5}$

$d_2 = \sqrt{(-2 - 0)^2 + (0 - 9)^2} = \sqrt{4 + 81} = \sqrt{85}$

$d_3 = \sqrt{(0 - (-2))^2 + (-4 - 0)^2} = \sqrt{4 + 16} = \sqrt{20} = 2\sqrt{5}$

$d_4 = \sqrt{(0 - 2)^2 + (-4 - 5)^2} = \sqrt{4 + 81} = \sqrt{85}$

Opposite sides have equal lengths of $2\sqrt{5}$ and $\sqrt{85}$, so the figure is a parallelogram.

53. Since $x_m = \dfrac{x_1 + x_2}{2}$ and $y_m = \dfrac{y_1 + y_2}{2}$ we have:

$$2x_m = x_1 + x_2 \qquad 2y_m = y_1 + y_2$$

$$2x_m - x_1 = x_2 \qquad 2y_m - y_1 = y_2$$

Thus, $(x_2, y_2) = (2x_m - x_1, 2y_m - y_1)$.

(a) $(x_2, y_2) = (2x_m - x_1, 2y_m - y_1) = (2(4) - 1, 2(-1) - (-2)) = (7, 0)$

(b) $(x_2, y_2) = (2x_m - x_1, 2y_m - y_1) = (2(2) - (-5), 2(4) - 11) = (9, -3)$

55. $(x - 0)^2 + (y - 0)^2 = 3^2$

$x^2 + y^2 = 9$

57. $(x - 2)^2 + (y + 1)^2 = 4^2$

$(x - 2)^2 + (y + 1)^2 = 16$

59. $(x + 1)^2 + (y - 2)^2 = r^2$

$(0 + 1)^2 + (0 - 2)^2 = r^2 \implies r^2 = 5$

$(x + 1)^2 + (y - 2)^2 = 5$

61. $r = \dfrac{1}{2}\sqrt{(6 - 0)^2 + (8 - 0)^2} = \dfrac{1}{2}\sqrt{100} = 5$

Center $= \left(\dfrac{0 + 6}{2}, \dfrac{0 + 8}{2}\right) = (3, 4)$

$(x - 3)^2 + (y - 4)^2 = 25$

63. Center: $(0, 0)$

Radius $= 2$

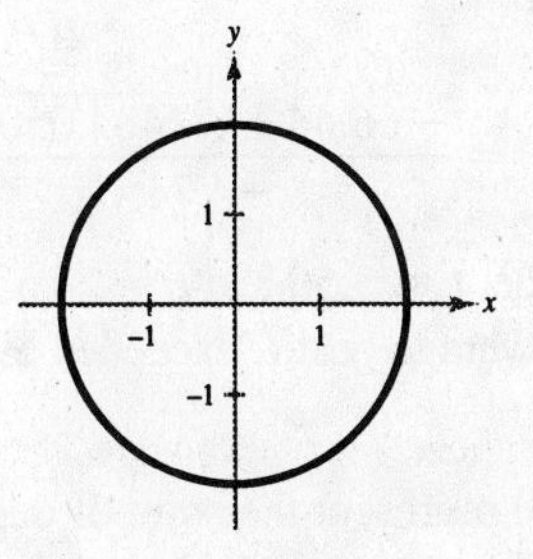

65. Center: $(1, -3)$

Radius $= 2$

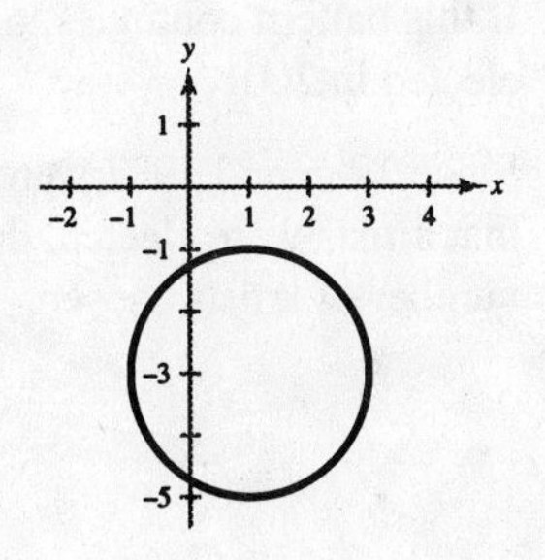

67. Center: $\left(\frac{1}{2}, \frac{1}{2}\right)$

Radius $= \frac{3}{2}$

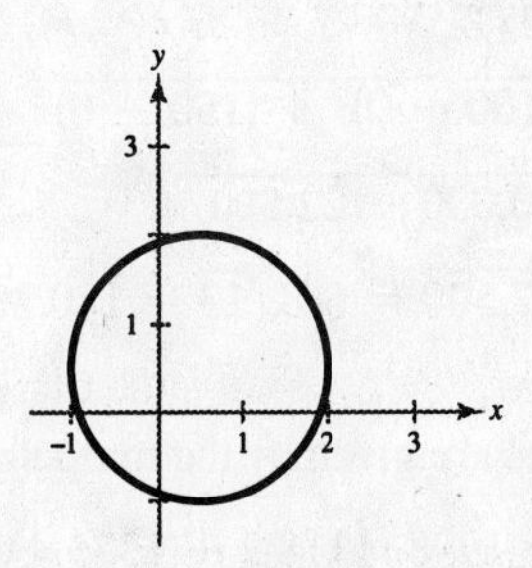

69.

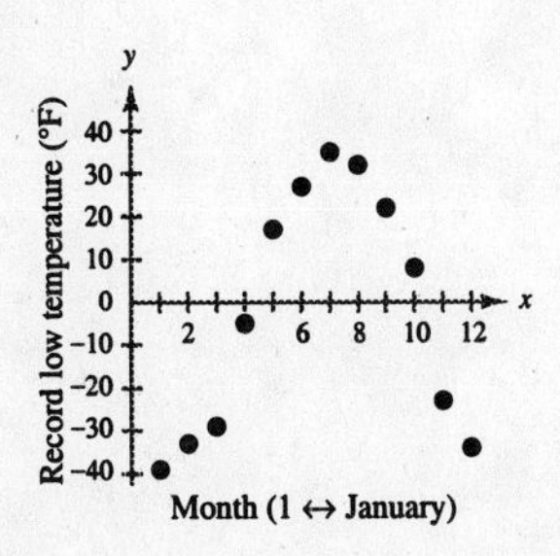

71. The highest price was approximately \$1.66, which occured in 1996.

73. $\dfrac{1600 - 600}{600}(100) \approx 166.67\%$ from 1987 to 1999.

75. The point (65, 83) represents an entrance exam score of 65.

77. Corn: $\frac{45}{240}(100) = 18.75\% \approx 19\%$

Soybeans: $\frac{20}{60}(100) = 33.33\%$

Wheat: $\frac{35}{70}(100) = 50.0\%$

(Answers will vary.)

79.

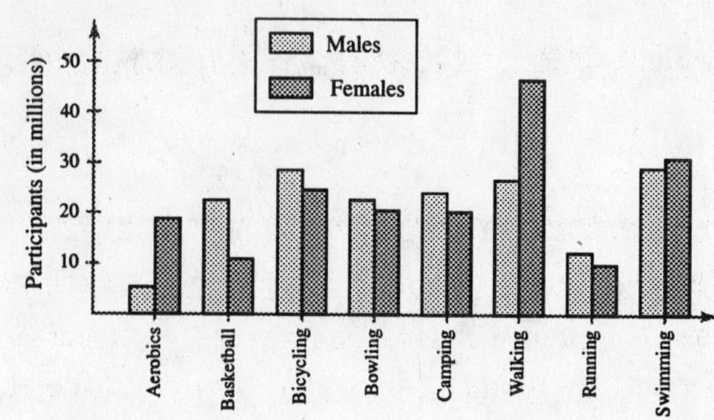

81. (a) The savings decreased from 8.2% to 3.9%. The decrease is $\frac{8.2 - 3.9}{8.2} = 0.52$ or 52%.

(b) No. The trend limits the amount of funds available for capital improvements and investments

83. (a) Solve the equation $C = 900$:

$$-2.37t^2 + 66.44t + 696.39 = 900$$

$$-2.37t^2 + 66.44t - 203.61 = 0$$

By the Quadratic Formula,

$$t = \frac{-66.44 \pm \sqrt{(66.44)^2 - 4(-2.37)(-203.61)}}{2(-2.37)}$$

$$= \frac{-66.44 \pm \sqrt{2484.0508}}{-4.74}$$

Hence, $t \approx 3.50$ and $t \approx 24.53$. Since 24.53 is not in the domain of C, the average cost C exceeded \$900 per day when $t > 3.5$ or the middle of 1993.

(b)

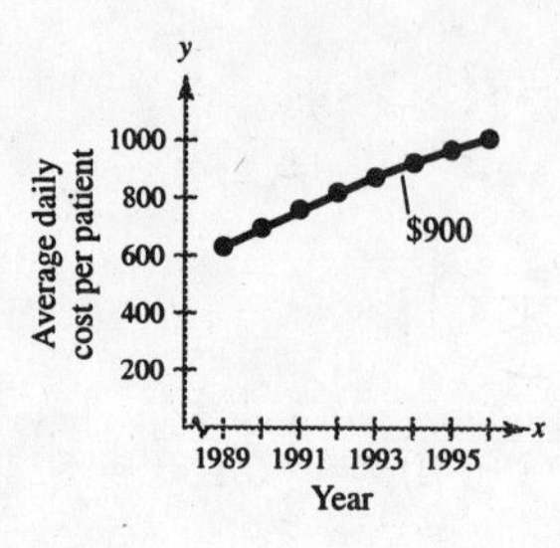

85. Let (0, 0) represent the point of departure and let (100, 150) represent the destination. Then the distance is given by

$$d = \sqrt{(100 - 0)^2 + (150 - 0)^2}$$

$$= \sqrt{10{,}000 + 22{,}500}$$

$$= \sqrt{32{,}500} = 50\sqrt{13} \approx 180.28 \text{ km}.$$

87. (a) It appears that the number of artists elected alternates between 6 and 8 per year in the 1990s. If this pattern continues, 6, 7 or 8 would be elected in 2001.

(b) Since 1986 and 1987 were the first two years that artists were elected, there was a larger number of artists chosen.

89. 1997 sales are given by the midpoint:

$$\left(\frac{1996 + 1998}{2}, \frac{1118.7 + 1371.4}{2}\right) = (1997, 1245.05)$$

The 1997 sales were approximately \$1245 million.

91. $d_1 = \sqrt{(2 - (-8))^2 + (11 - 4)^2} = \sqrt{10^2 + 7^2} = \sqrt{149}$

$d_2 = \sqrt{(-5 - (-8))^2 + (1 - 4)^2} = \sqrt{3^2 + 3^2} = 3\sqrt{2}$

$d_3 = \sqrt{(-5 - 2)^2 + (1 - 11)^2} = \sqrt{49 + 100} = \sqrt{149}$

Since $d_1 = d_3$, the triangle is isosceles. True.

93. On the x-axis, $y = 0$
On the y-axis, $x = 0$

Section P.2 Graphs of Equations

- You should be able to use the point-plotting method of graphing.
- You should be able to find x- and y-intercepts.
  (a) To find the x-intercepts, let $y = 0$ and solve for x.
  (b) To find the y-intercepts, let $x = 0$ and solve for y.
- You should know how to graph an equation with a graphing utility. You should be able to determine an appropriate viewing rectangle.
- You should be able to use the zoom and trace features of a graphing utility.

Solutions to Odd-Numbered Exercises

1. $y = \sqrt{x + 4}$

(a) $(0, 2)$: $2 \stackrel{?}{=} \sqrt{0 + 4}$

$2 = 2$ ✓

Yes, the point *is* on the graph.

(b) $(5, 3)$: $3 \stackrel{?}{=} \sqrt{5 + 4}$

$3 = \sqrt{9}$ ✓

Yes, the point *is* on the graph.

3. $y = 4 - |x - 2|$

(a) $(1, 5)$: $5 \stackrel{?}{=} 4 - |1 - 2|$

$5 \neq 4 - 1$

No, the point *is not* on the graph.

(b) $(1.2, 3.2)$: $3.2 \stackrel{?}{=} 4 - |1.2 - 2|$

$3.2 \stackrel{?}{=} 4 - |-.8|$

$3.2 \stackrel{?}{=} 4 - .8$

$3.2 \stackrel{?}{=} 3.2$ ✓

Yes, the point *is* on the graph.

5. $2x - y - 3 = 0$

(a) $(1, 2)$: $2(1) - (2) - 3 \stackrel{?}{=} 0$

$-3 \neq 0$

No, the point *is not* on the graph.

(b) $(1, -1)$: $2(1) - (-1) - 3 \stackrel{?}{=} 0$

$2 + 1 - 3 = 0$ ✓

Yes, the point *is* on the graph.

7. $x^2y - x^2 + 4y = 0$

(a) $\left(1, \frac{1}{5}\right)$: $(1)^2\left(\frac{1}{5}\right) - (1)^2 + 4\left(\frac{1}{5}\right) \stackrel{?}{=} 0$

$\frac{1}{5} - 1 + \frac{4}{5} = 0$ ✓

Yes, the point *is* on the graph.

(b) $\left(2, \frac{1}{2}\right)$: $(2)^2\left(\frac{1}{2}\right) - (2)^2 + 4\left(\frac{1}{2}\right) \stackrel{?}{=} 0$

$2 - 4 + 2 = 0$ ✓

Yes, the point *is* on the graph.

9. $y = -2x + 3$

x	-1	0	1	$\frac{3}{2}$	2
y	5	3	1	0	-1

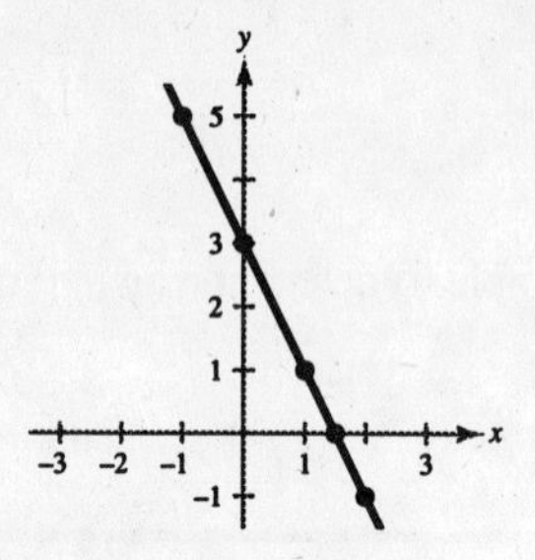

11. $y = x^2 - 2x$

x	-1	0	1	2	3
y	3	0	-1	0	3

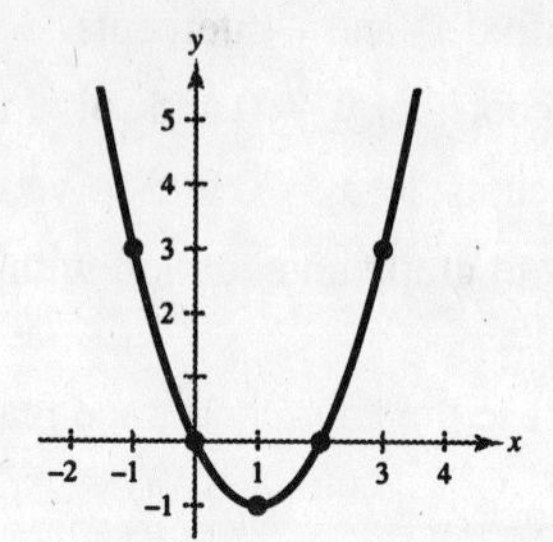

13. $y = 3 - |x - 2|$

x	0	1	2	3	4
y	1	2	3	2	1

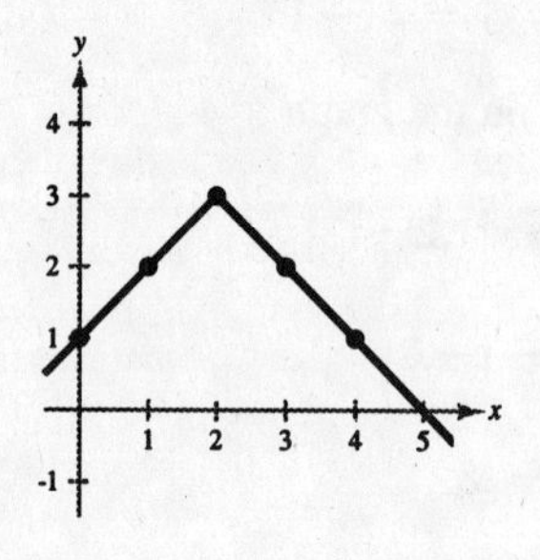

15. (a) $y = \frac{1}{4}x - 3$

x	-2	-1	0	1	2
y	$-\frac{7}{2}$	$-\frac{13}{4}$	-3	$-\frac{11}{4}$	$-\frac{5}{2}$

(b)

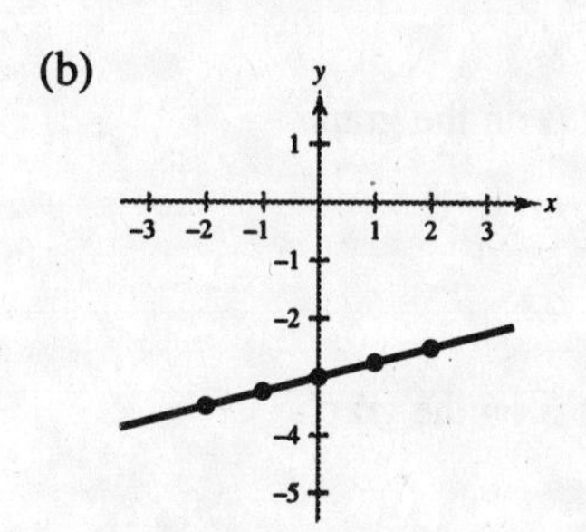

(c) $y = -\frac{1}{4}x - 3$

x	-2	-1	0	1	2
y	$-\frac{5}{2}$	$-\frac{11}{4}$	-3	$-\frac{13}{4}$	$-\frac{7}{2}$

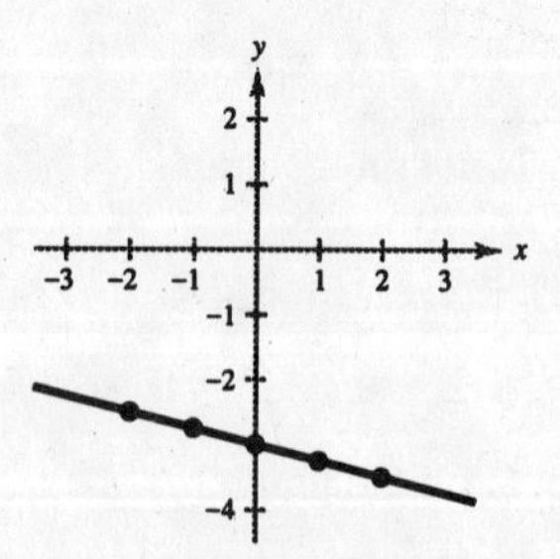

Both graphs are lines. The first graph rises to the right, whereas the second falls. Both pass through $(0, -3)$.

17. $y = 1 - x$ has intercepts $(1, 0)$ and $(0, 1)$. Matches graph (d).

19. $y = \sqrt{9 - x^2}$ has intercepts $(\pm 3, 0)$ and $(0, 3)$. Matches graph (f).

21. $y = x^3 - x + 1$ has a y-intercept of $(0, 1)$ and the points $(1, 1)$ and $(-2, -5)$ are on the graph. Matches (a).

23. $y = -3x + 2$

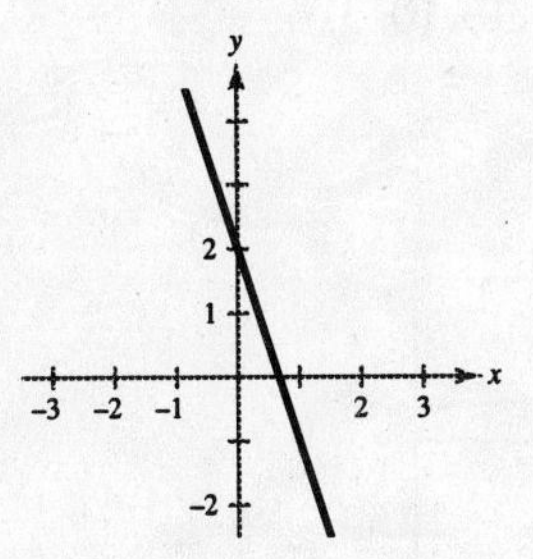

25. $y = 1 - x^2$

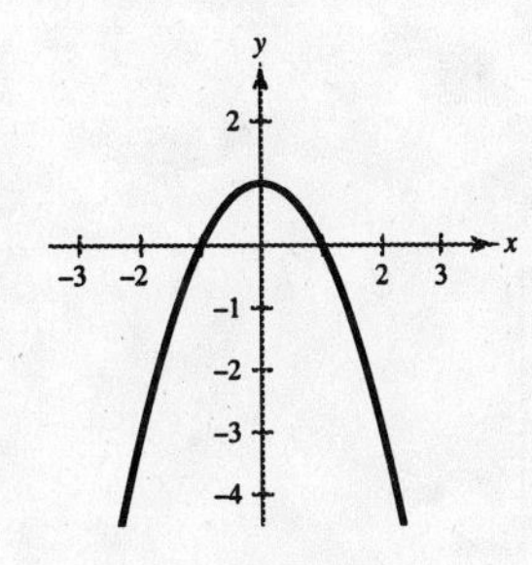

27. $y = x^2 - 3x$

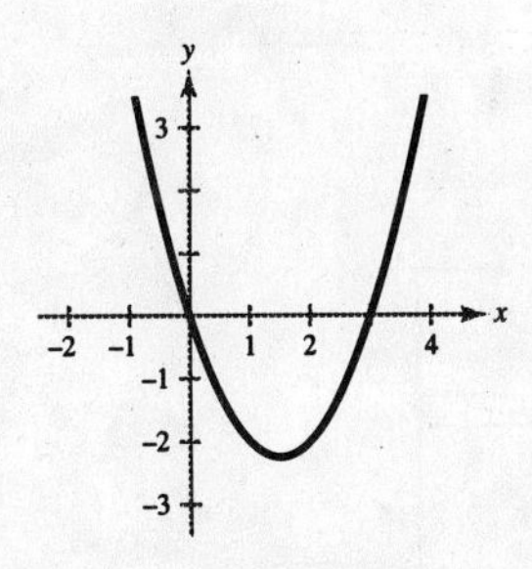

29. $y = x^3 + 2$

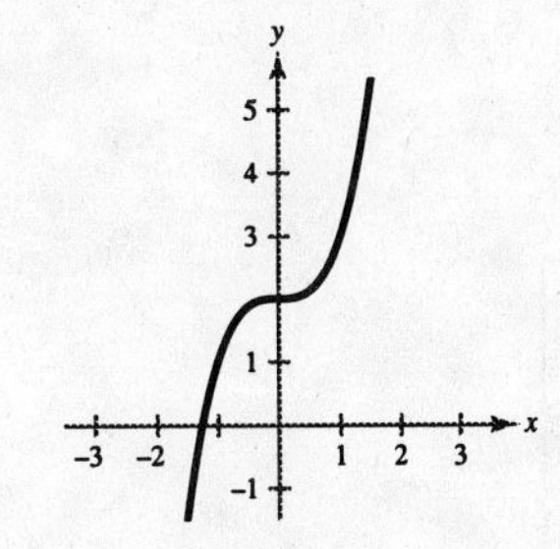

31. $y = \sqrt{x - 3}$

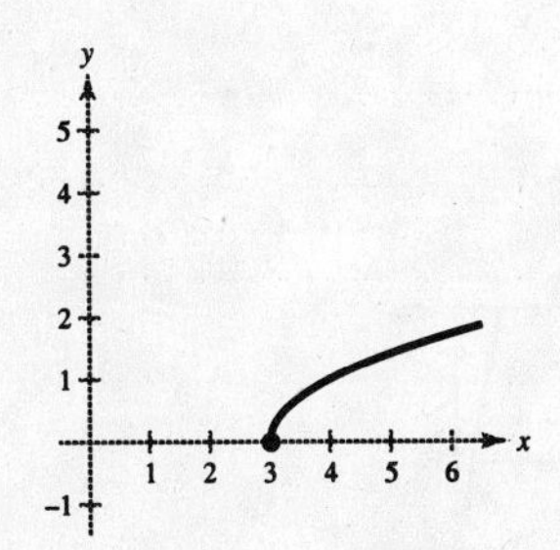

33. $y = |x - 2|$

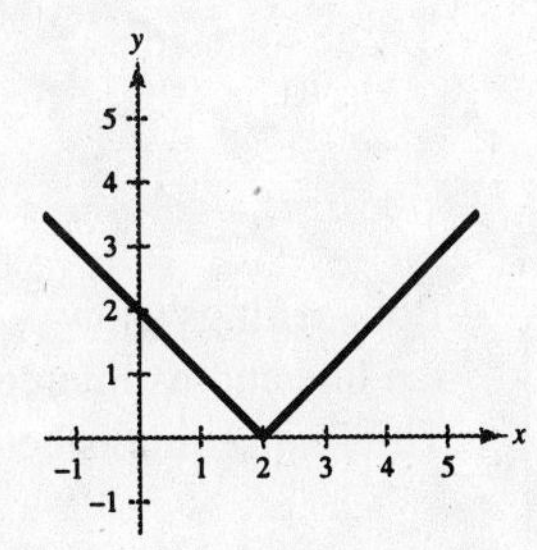

35. $x = y^2 - 1$

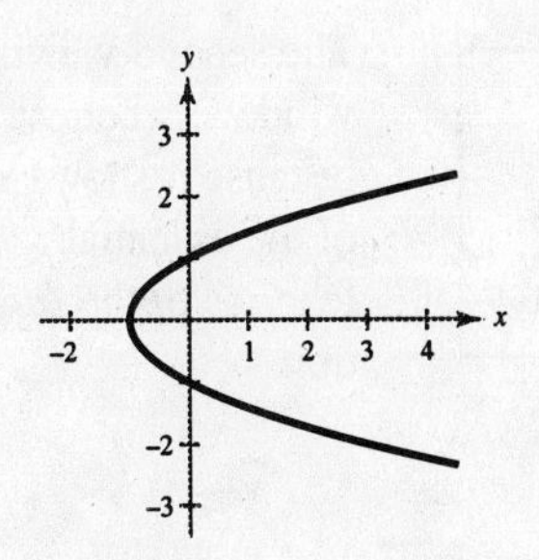

37. $y = x - 5$

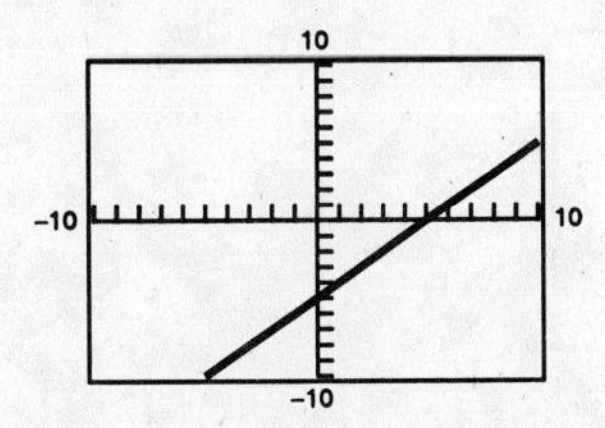

Intercepts: $(0, -5)$, $(5, 0)$

39. $y = 3 - \frac{1}{2}x$

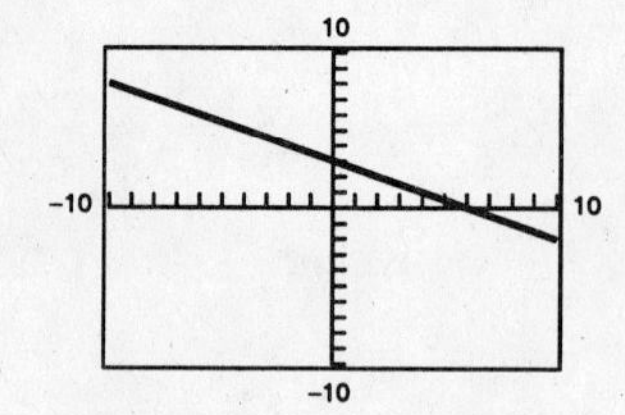

Intercepts: (6,0), (0, 3)

41. $y = x^2 - 4x + 3$

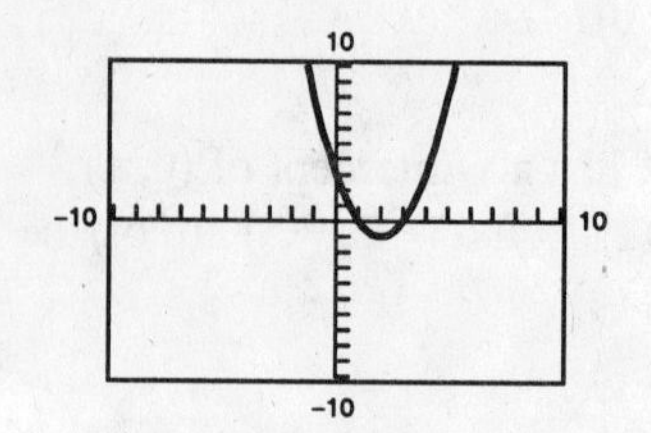

Intercepts: (3, 0), (1, 0), (0, 3)

43. $y = x(x - 2)^2$

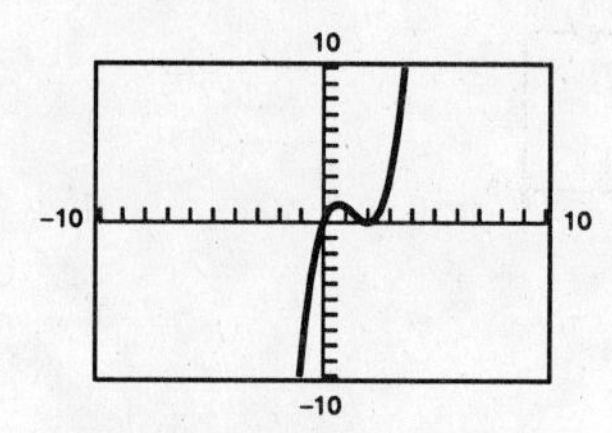

Intercepts: (0, 0), (2, 0)

45. $y = \dfrac{2x}{x - 1}$

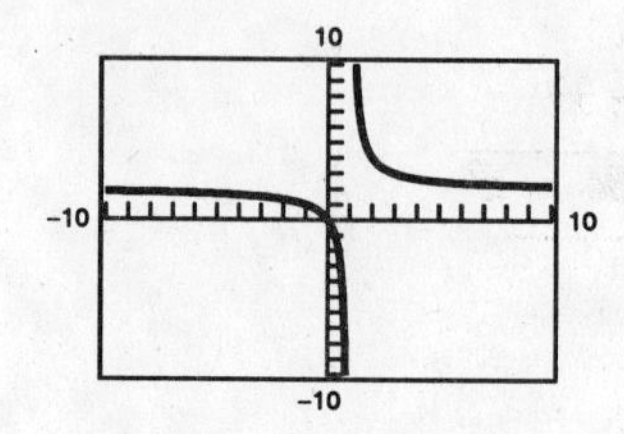

Intercept: (0, 0)

47. $y = x\sqrt{x + 6}$

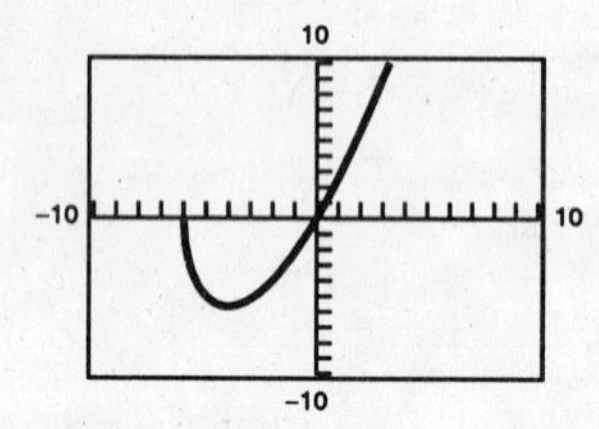

Intercepts: (0, 0), $(-6, 0)$

49. $y = \sqrt[3]{x}$

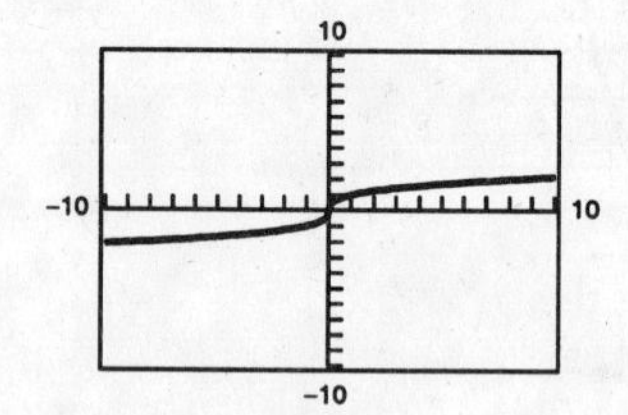

Intercept: (0, 0)

51. $y = \frac{5}{2}x + 5$

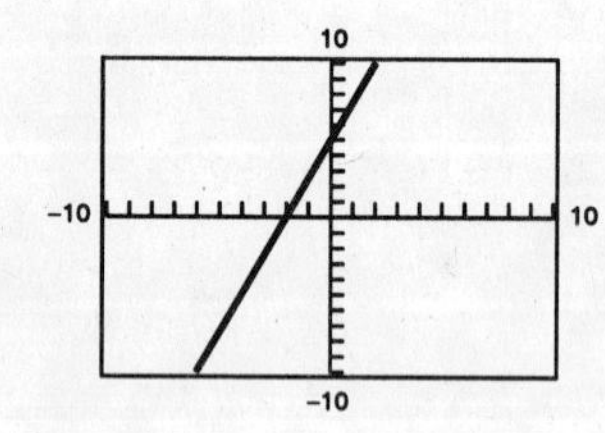

Both settings show the line and its intercept. The first setting is better.

53. $y = -x^2 + 10x - 5$

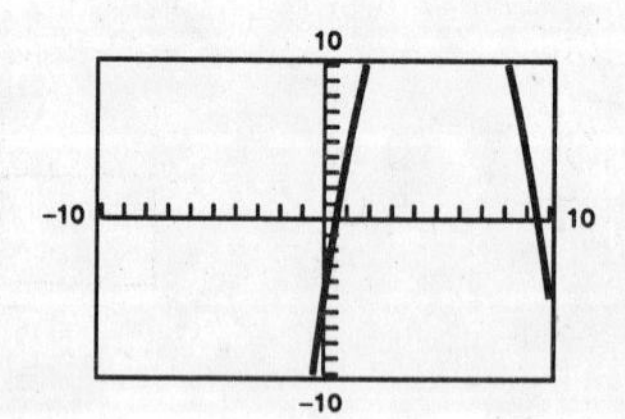

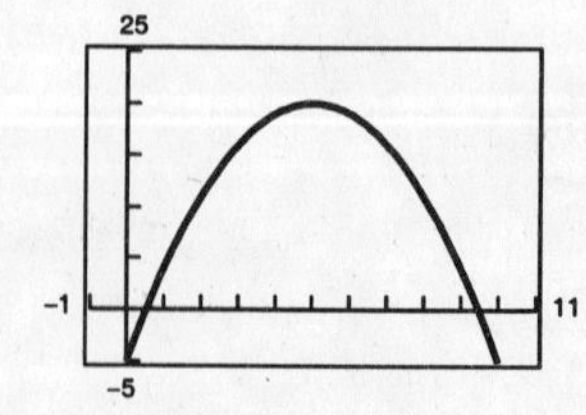

The second viewing window is better because it shows more of the essential features of the function.

55. $y = 4x^2 - 25$

Range/Window

Xmin=-5
Xmax=5
Xscl=1
Ymin=-30
Ymax=10
Yscl=5

57. $y = |x| + |x - 10|$

Range/Window

Xmin=-30
Xmax=30
Xscl=5
Ymin=-10
Ymax=50
Yscl=5

59. $x^2 + y^2 = 64$

$$y^2 = 64 - x^2$$

$$y = \pm\sqrt{64 - x^2}$$

Use: $y_1 = \sqrt{64 - x^2}$

$y_2 = -\sqrt{64 - x^2}$

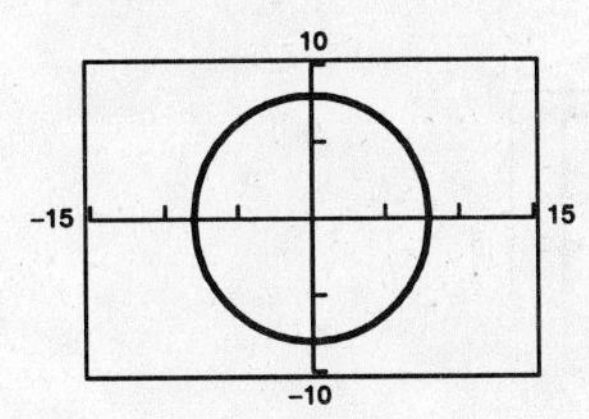

61. $x^2 + y^2 = 49$

$$y^2 = 49 - x^2$$

$$y = \pm\sqrt{49 - x^2}$$

Use: $y_1 = \sqrt{49 - x^2}$

$y_2 = -\sqrt{49 - x^2}$

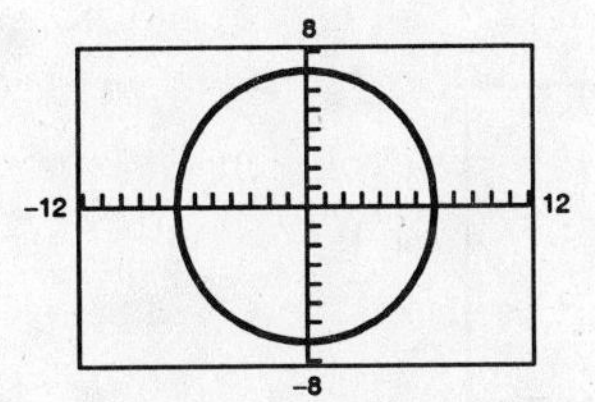

63. $y_1 = \frac{1}{4}(x^2 - 8)$

$y_2 = \frac{1}{4}x^2 - 2$

The graphs are identical.
The Distributive Property is illustrated.

65. $y_1 = \frac{1}{5}[10(x^2 - 1)]$

$y_2 = 2(x^2 - 1)$

The graphs are identical.
The Associative Property of Multiplication is illustrated.

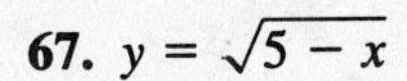

67. $y = \sqrt{5 - x}$

(a) $(2, y) \approx (2, 1.73)$

(b) $(x, 3) = (-4, 3)$

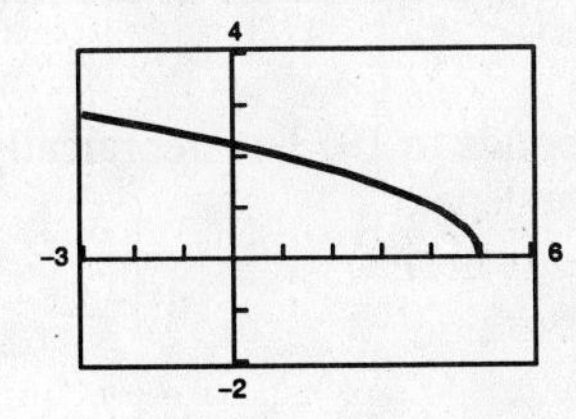

69. $y = x^5 - 5x$

(a) $(-0.5, y) \approx (-0.5, 2.47)$

(b) $(x, -4) = (1, -4)$ or $(x, -4) \approx (-1.65, -4)$

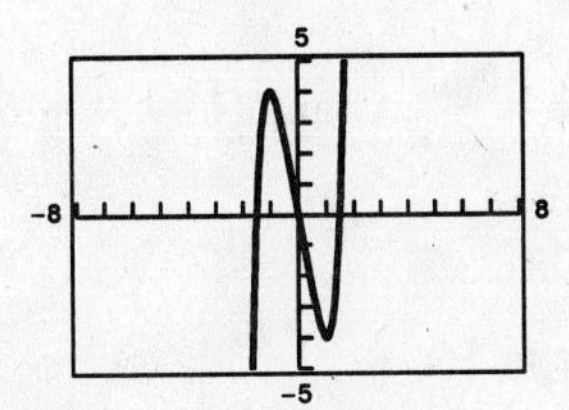

71. (a) $y = 225{,}000 - 20{,}000t\ ,\ 0 \le t \le 8$

Window
$X_{min} = 0$
$X_{max} = 8$
$X_{scl} = 1$
$Y_{min} = 60{,}000$
$Y_{max} = 230{,}000$
$Y_{scl} = 10{,}000$

(b) 230,000 0 8 60,000

(c) When $t = 5.8$, $y = 109{,}000$. Algebraically, $225{,}000 - 20{,}000(5.8) = \$109{,}000$.

(d) When $t = 2.35$, $y = 178{,}000$. Algebraically, $225{,}000 - 20{,}000(2.35) = \$178{,}000$.

73. (a)

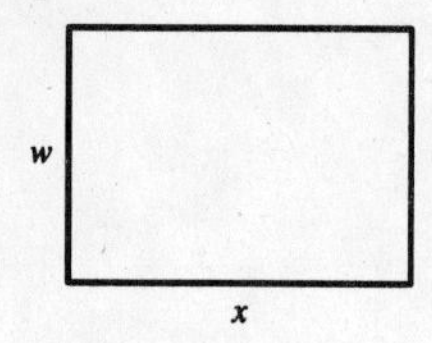

(b) Perimeter: $12 = 2x + 2w$

$$12 = 2(x + w)$$
$$6 = x + w$$

Thus, $w = 6 - x$.

Area: $xw = x(6 - x) \Rightarrow A = x(6 - x)$

(c) 10 0 0 10

(d) When $w = 4.9$, $x = 1.1$ and Area $= 5.39$ meters.
Algebraically, Area $= xw = (1.1)(4.9) = 5.39$ meters.

(e) The maximum area corresponds to the highest point on the graph, which appears to be (3, 9). Thus, $x = 3$ and $w = 3$, and the rectangle is a square.

75. (a) The y-intercept (0, 66.93) indicates the model's estimate of the life expectancy in 1950 ($t = 0$).

(b) $y = 73.2$ when $t = 23.40$, which corresponds to 1973. Algebraically,

$$\frac{66.93 + t}{1 + 0.01t} = 73.2$$

$$66.93 + t = 73.2 + 0.732t$$
$$0.268t = 6.27$$
$$t \approx 23.4$$

(c) 1948 corresponds to $t = -2$. Graphically, $y = 66.26$ when $t = -2$. Algebraically,

$$\frac{66.93 + (-2)}{1 + 0.01(-2)} = \frac{64.93}{0.98} = 66.26 \text{ years}$$

(d) 2005 corresponds to $t = 55$:

$$\frac{66.93 + 55}{1 + 0.01(55)} = \frac{121.93}{1.55} = 78.66 \text{ years}$$

77. (a)

x	10	20	30	40	50	60	70	80	90	100
y	107.3	26.6	11.6	6.4	3.9	2.6	1.8	1.3	0.96	0.71

(b) From the table, $x \approx 45$ when $y = 4.8$.

Algebraically,

$$\frac{10{,}770}{x^2} - 0.37 = 4.8$$

$$\frac{10{,}770}{x^2} = 5.17$$

$$10{,}770 = 5.17x^2$$

$$2083.17 = x^2$$

$$x = 45.6 \text{ mils}$$

(c) When $x = 85.5$, $y = 1.10$ ohms.

(d) As the diameter increases, the resistance decreases.

79. False. The line $x = 0$ has an infinite number of x-intercepts.

81. Answers will vary.

Section P.3 Lines in the Plane

You should know the following important facts about lines.

- The graph of $y = mx + b$ is a straight line. It is called a linear equation.
- The slope of the line through (x_1, y_1) and (x_2, y_2) is

$$m = \frac{y_2 - y_1}{x_2 - x_1}.$$

- (a) If $m > 0$, the line rises from left to right.
  (b) If $m = 0$, the line is horizontal.
  (c) If $m < 0$, the line falls from left to right.
  (d) If m is undefined, the line is vertical.
- Equations of Lines
  (a) Slope-Intercept: $y = mx + b$
  (b) Point-Slope: $y - y_1 = m(x - x_1)$
  (c) Two-Point: $y - y_1 = \frac{y_2 - y_1}{x_2 - x_1}(x - x_1)$
  (d) General: $Ax + By + C = 0$
  (e) Vertical: $x = a$
  (f) Horizontal: $y = b$
- Given two distinct nonvertical lines

$$L_1: y = m_1x + b_1 \text{ and } L_2: y = m_2x + b_2$$

  (a) L_1 is parallel to L_2 if and only if $m_1 = m_2$ and $b_1 \neq b_2$.
  (b) L_1 is perpendicular to L_2 if and only if $m_1 = -1/m_2$.

Solutions to Odd-Numbered Exercises

1. (a) $m = \frac{2}{3}$. Since the slope is positive, the line rises. Matches L_2.

(b) m is undefined. The line is vertical. Matches L_3.

(c) $m = -2$. The line falls. Matches L_1.

3.

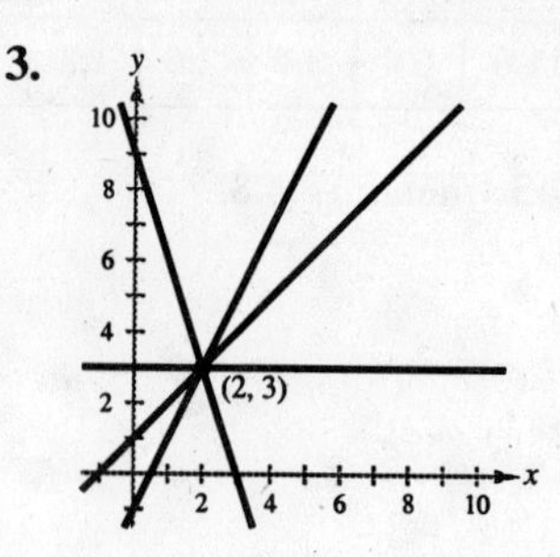

5. $\text{Slope} = \dfrac{\text{rise}}{\text{run}} = \dfrac{3}{2}$

7. $\text{Slope} = \dfrac{\text{rise}}{\text{run}} = \dfrac{0}{1} = 0$

9. $\text{Slope} = \dfrac{\text{rise}}{\text{run}} = \dfrac{-8}{2} = -4$

11. $\text{slope} = \dfrac{0-(-10)}{-4-0} = \dfrac{10}{-4} = -\dfrac{5}{2}$

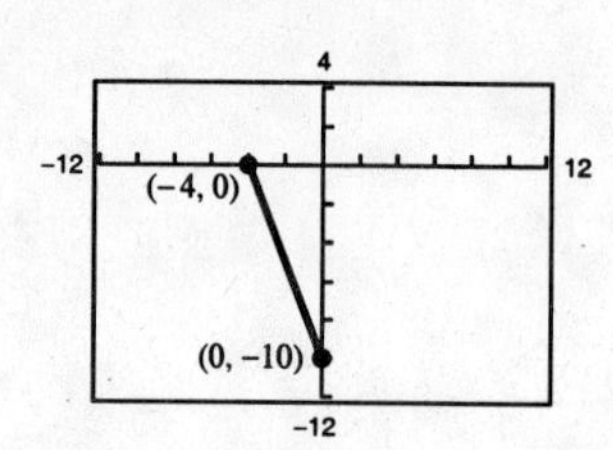

13.

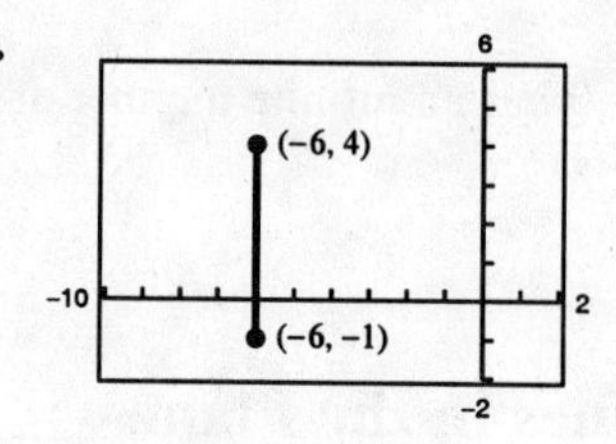

Slope is undefined.

15. Since $m = 0$, y does not change. Three points are $(0, 1)$, $(3, 1)$, and $(-1, 1)$.

17. Since $m = 2$, y increases 2 for every unit increase in x. Three points are $(-4, 6)$, $(-3, 8)$, $(-2, 10)$.

19. Since $m = \frac{1}{2}$, y increases 1 for every increase of 2 in x. Three points are $(9, -1)$, $(11, 0)$, $(13, 1)$.

21. $m_{L_1} = \dfrac{9+1}{5-0} = 2$

$m_{L_2} = \dfrac{1-3}{4-0} = -\dfrac{1}{2} = -\dfrac{1}{m_{L_1}}$

L_1 and L_2 are perpendicular.

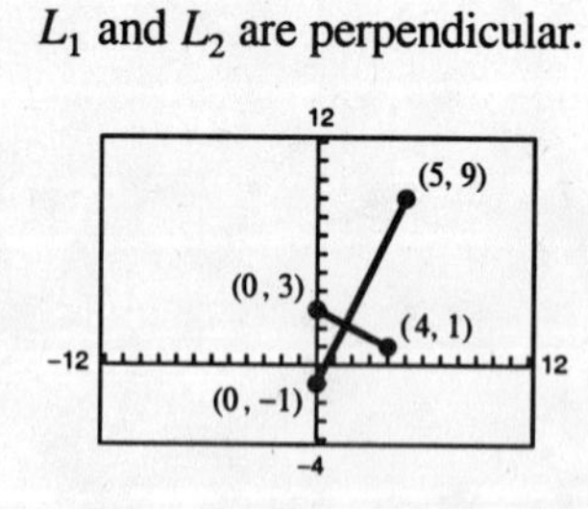

23. $m_{L_1} = \dfrac{0-6}{-6-3} = \dfrac{2}{3}$

$m_{L_2} = \dfrac{\frac{7}{3}+1}{5-0} = \dfrac{2}{3} = m_{L_1}$

L_1 and L_2 are parallel.

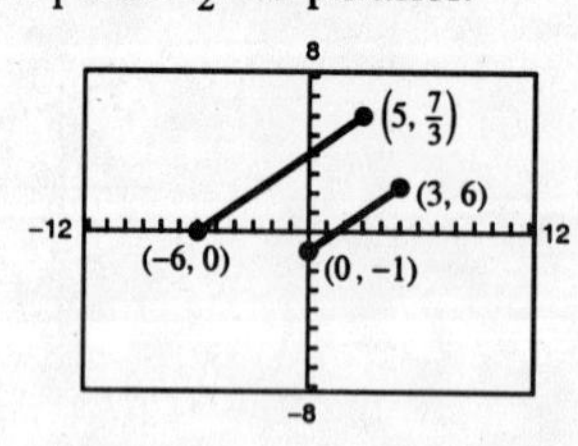

25. $5x - y + 3 = 0$

$y = 5x + 3$

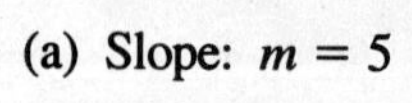

(a) Slope: $m = 5$

y-intercept: $(0, 3)$

(b)

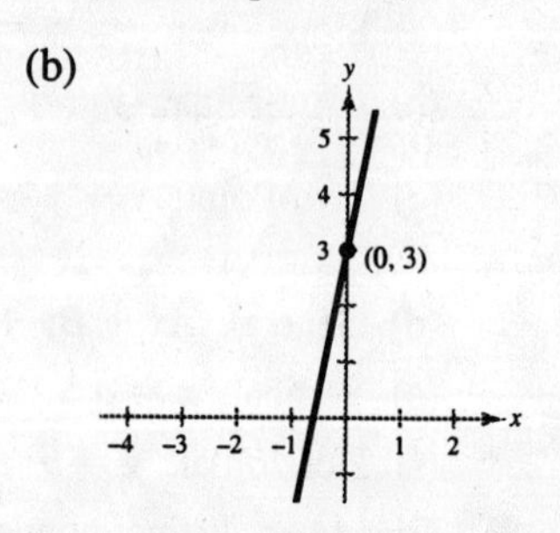

27. $5x - 2 = 0$

$x = \frac{2}{5}$

(a) Slope: undefined

No y-intercept

(b)

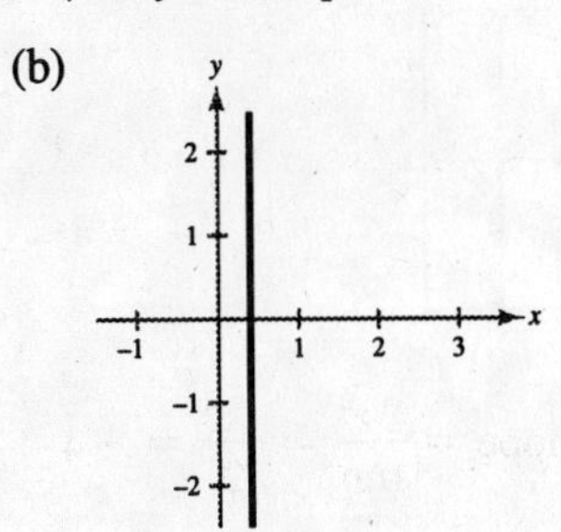

29. $3y + 5 = 0$

(a) $y = -\frac{5}{3}$

Slope: $m = 0$

y-intercept: $\left(0, -\frac{5}{3}\right)$

(b)

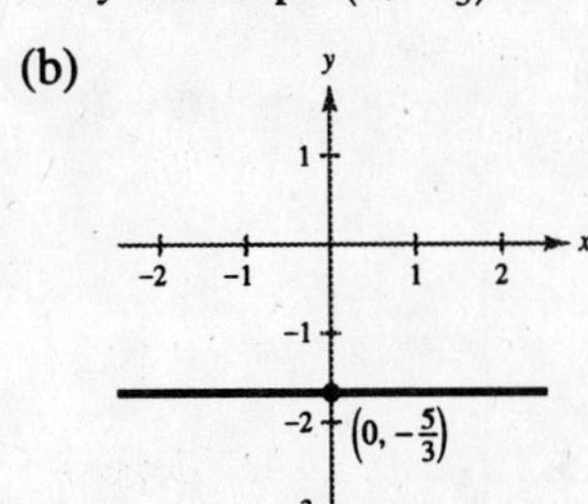

31. $7x + 6y - 30 = 0$

$y = -\frac{7}{6}x + 5$

(a) Slope: $m = -\frac{7}{6}$

y-intercept: $(0, 5)$

(b)

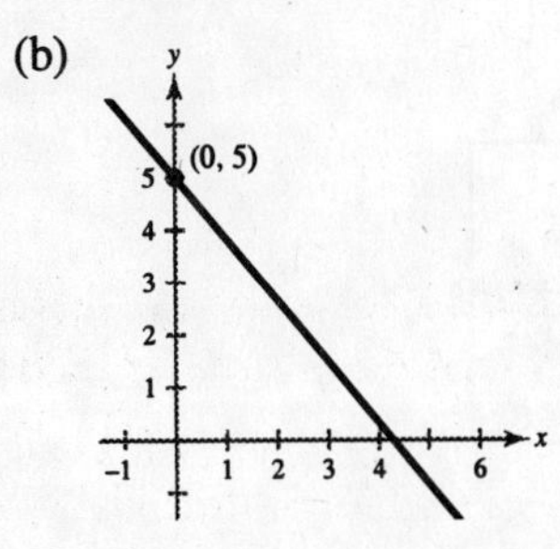

33. $y + 2 = 3(x - 0)$

$y = 3x - 2 \implies 3x - y - 2 = 0$

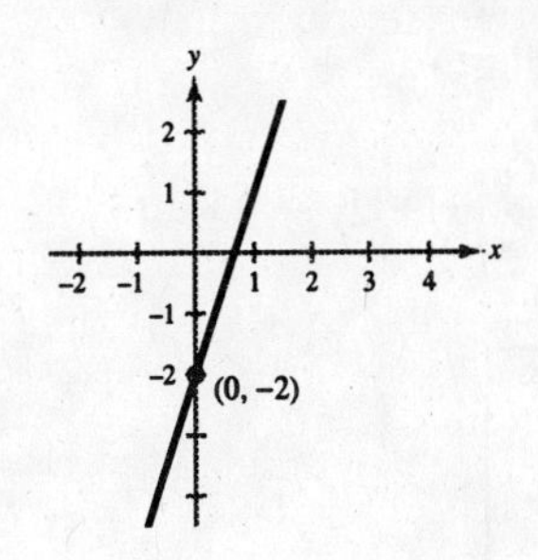

35. $y - 6 = -2(x + 3)$

$y = -2x \implies 2x + y = 0$

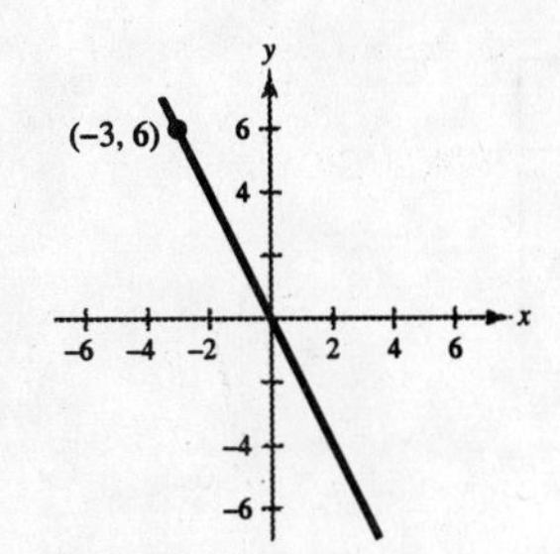

37. $y - 0 = -\frac{1}{3}(x - 4)$

$y = -\frac{1}{3}x + \frac{4}{3} \implies x + 3y - 4 = 0$

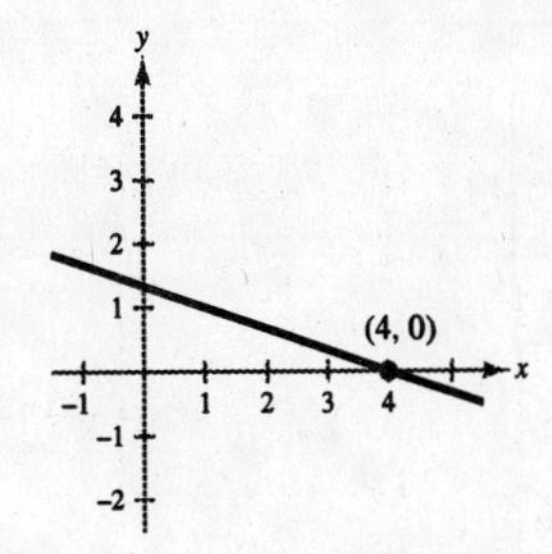

39. $x = 6$

$x - 6 = 0$

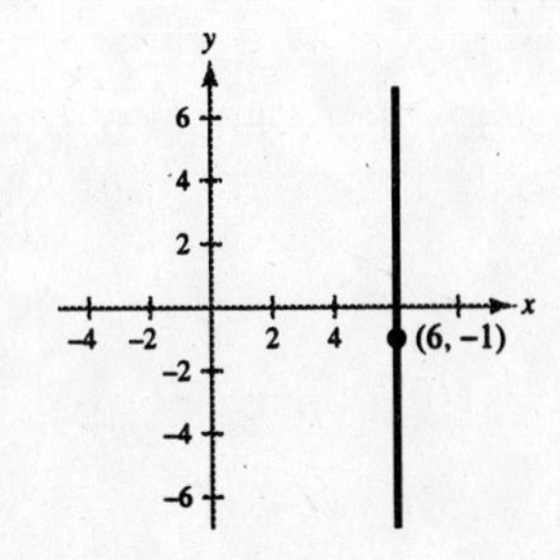

41. $y - \frac{3}{2} = -3\left(x + \frac{1}{2}\right)$

$y = -3x$

$3x + y = 0$

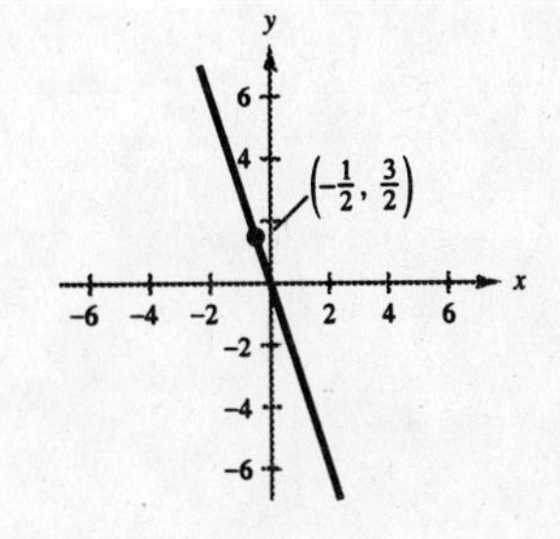

43. $y + 1 = \dfrac{5 + 1}{-5 - 5}(x - 5)$

$y = -\dfrac{3}{5}(x - 5) - 1$

$y = -\dfrac{3}{5}x + 2 \implies 3x + 5y - 10 = 0$

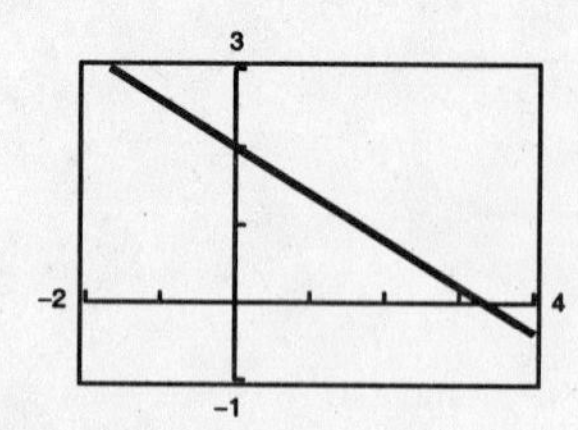

45. Since both points have $x = -8$, the slope is undefined.

$x = -8 \implies x + 8 = 0$

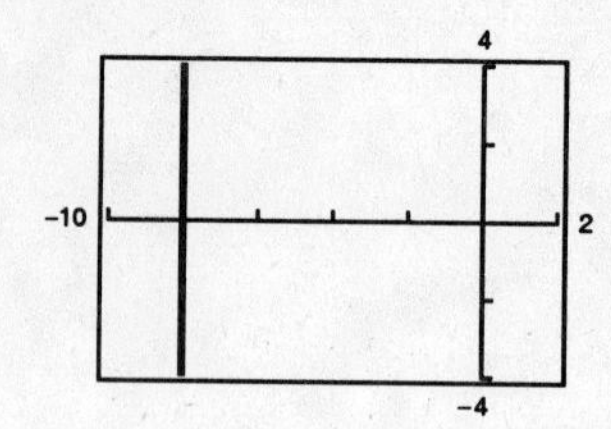

47. $y - \dfrac{1}{2} = \dfrac{\frac{5}{4} - \frac{1}{2}}{\frac{1}{2} - 2}(x - 2)$

$y = -\dfrac{1}{2}(x - 2) + \dfrac{1}{2}$

$y = -\dfrac{1}{2}x + \dfrac{3}{2} \implies x + 2y - 3 = 0$

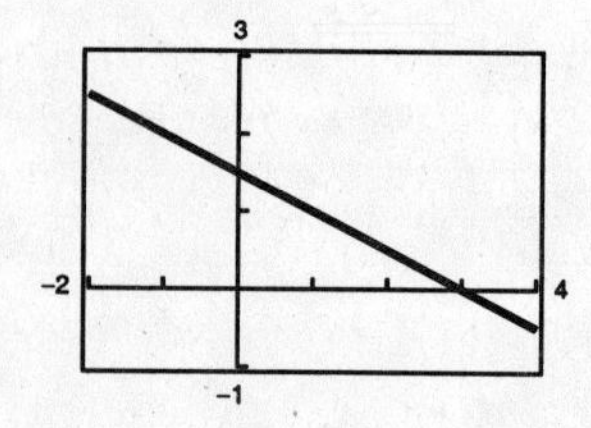

49. $y + \frac{3}{5} = \dfrac{-\frac{9}{5} + \frac{3}{5}}{\frac{9}{10} + \frac{1}{10}}\left(x + \frac{1}{10}\right)$

$y + \frac{3}{5} = -\frac{6}{5}\left(x + \frac{1}{10}\right)$

$y = -\frac{6}{5}x - \frac{18}{25}$

$30x + 25y + 18 = 0$

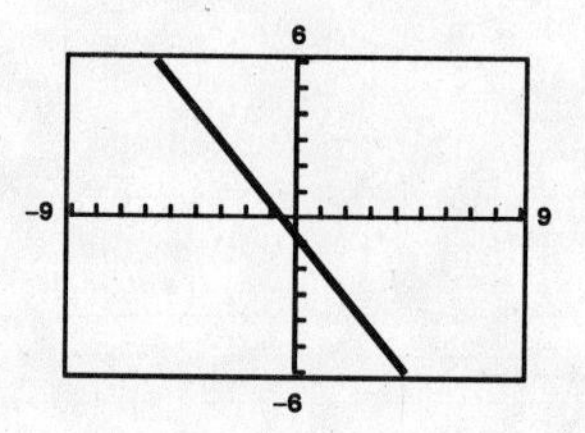

51. $y - 0.6 = \dfrac{-0.6 - 0.6}{-2 - 1}(x - 1)$

$y = 0.4(x - 1) + 0.6$

$y = 0.4x + 0.2 \implies 2x - 5y + 1 = 0$

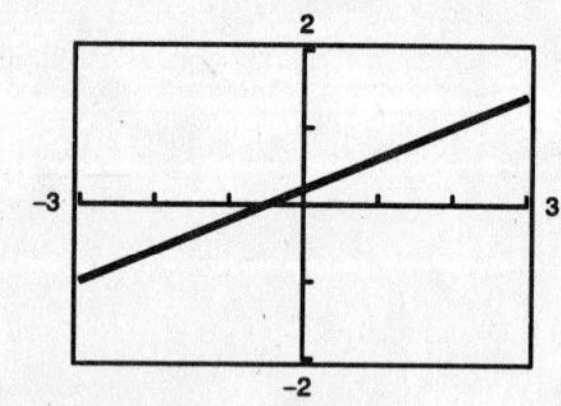

53. $\dfrac{x}{5} + \dfrac{y}{-3} = 1$

$-3x + 5y + 15 = 0$

$a = 5$ and $b = -3$ are the x- and y-intercepts.

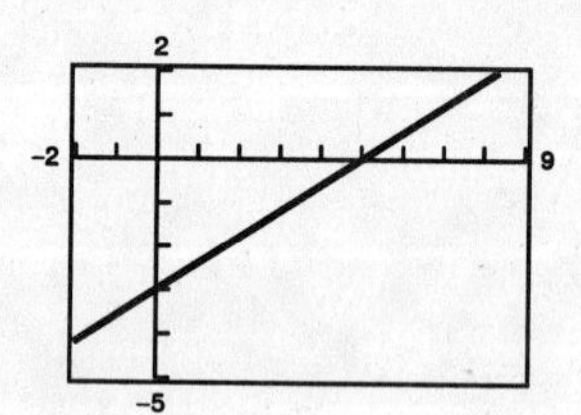

55. $\dfrac{x}{2} + \dfrac{y}{3} = 1$

$3x + 2y - 6 = 0$

57. $\dfrac{x}{-\frac{1}{6}} + \dfrac{y}{-\frac{2}{3}} = 1$

$-6x - \frac{3}{2}y = 1$

$12x + 3y + 2 = 0$

59. $y = 0.5x - 3$

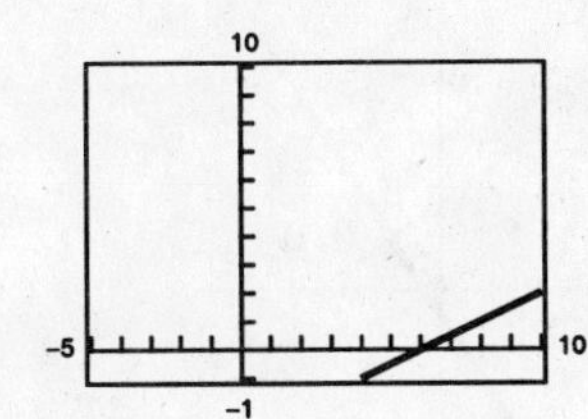

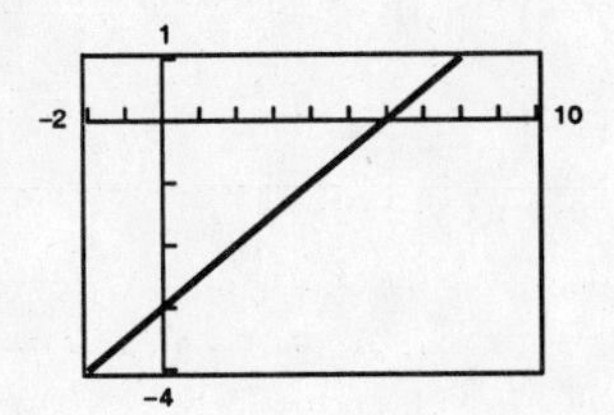

The second setting shows the x and y intercepts more clearly.

61. (a) $y = 2x$ (b) $y = -2x$ (c) $y = \frac{1}{2}x$

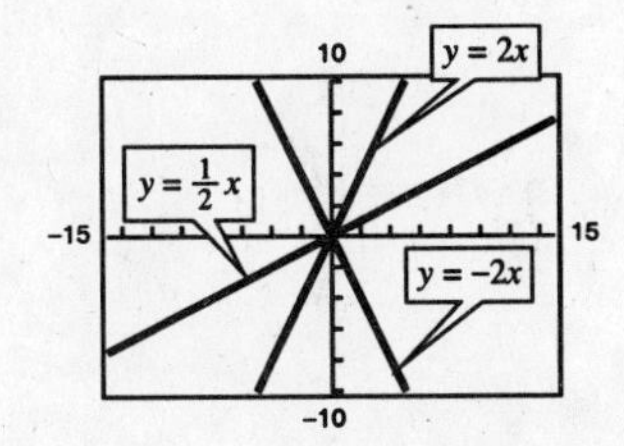

(b) and (c) are perpendicular.

63. (a) $y = -\frac{1}{2}x$ (b) $y = -\frac{1}{2}x + 3$ (c) $y = 2x - 4$

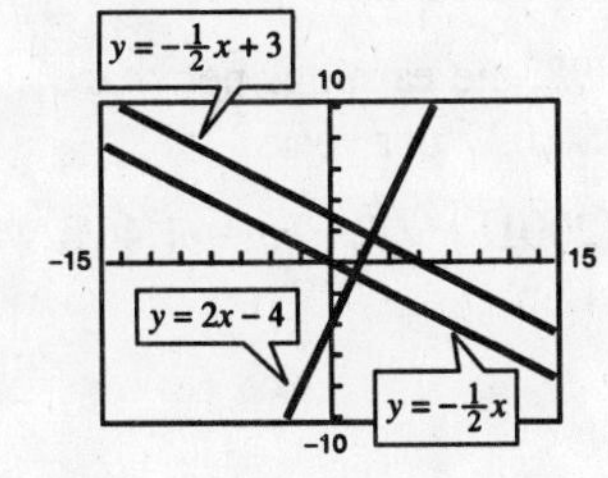

(a) and (b) are parallel.

(c) is perpendicular to (a) and (b).

65. $4x - 2y = 3$

$$y = 2x - \frac{3}{2}$$

Slope: $m = 2$

(a) $y - 1 = 2(x - 2)$

$$y = 2x - 3 \implies 2x - y - 3 = 0$$

(b) $y - 1 = -\frac{1}{2}(x - 2)$

$$y = -\frac{1}{2}x + 2 \implies x + 2y - 4 = 0$$

67. $3x + 4y = 7$

$$y = -\tfrac{3}{4}x + \tfrac{7}{4}$$

slope: $m = -\frac{3}{4}$

(a) $y - \frac{7}{8} = -\frac{3}{4}\left(x + \frac{2}{3}\right)$

$$y = -\tfrac{3}{4}x + \tfrac{3}{8}$$

$$6x + 8y - 3 = 0$$

(b) $y - \frac{7}{8} = \frac{4}{3}\left(x + \frac{2}{3}\right)$

$$y = \tfrac{4}{3}x + \tfrac{127}{72}$$

$$96x - 72y + 127 = 0$$

69. $x - y = 4$

$$y = x - 4$$

slope: $m = 1$

(a) $y - 6.8 = 1(x - 2.5)$

$$y = x + 4.3$$

$$10x - 10y + 43 = 0$$

(b) $y - 6.8 = -1(x - 2.5)$

$$y = -x + 9.3$$

$$10x + 10y - 93 = 0$$

71. Set the distance between $(4, -1)$ and (x, y) equal to the distance between $(-2, 3)$ and (x, y).

$$\sqrt{(x-4)^2 + [y-(-1)]^2} = \sqrt{[x-(-2)]^2 + (y-3)^2}$$

$$(x-4)^2 + (y+1)^2 = (x+2)^2 + (y-3)^2$$

$$x^2 - 8x + 16 + y^2 + 2y + 1 = x^2 + 4x + 4 + y^2 - 6y + 9$$

$$-8x + 2y + 17 = 4x - 6y + 13$$

$$0 = 12x - 8y - 4$$

$$0 = 4(3x - 2y - 1)$$

$$0 = 3x - 2y - 1$$

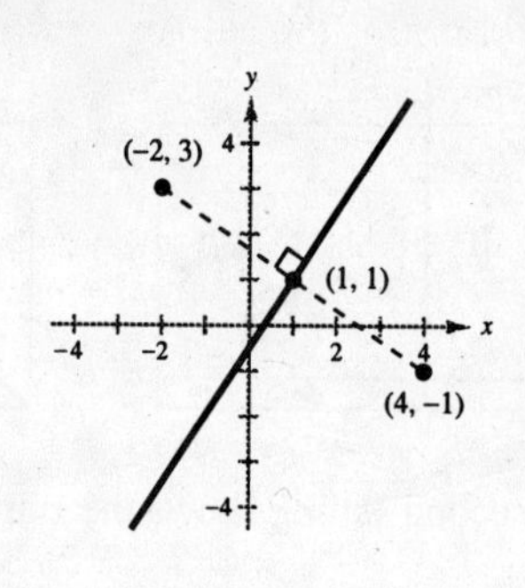

This line is the perpendicular bisector of the line segment connecting $(4, -1)$ and $(-2, 3)$.

73. (a) $m = 135$. The sales are increasing 135 units per year.

(b) $m = 0$. There is no change in sales.

(c) $m = -40$. The sales are decreasing 40 units per year.

75. (a)

Years	Slope
1988–1989	$0.87 - 0.98 = -0.11$
1989–1990	$1.04 - 0.87 = 0.17$
1990–1991	$1.26 - 1.04 = 0.22$
1991–1992	$1.38 - 1.26 = 0.12$
1992–1993	$1.47 - 1.38 = 0.09$
1993–1994	$1.58 - 1.47 = 0.11$
1994–1995	$1.74 - 1.58 = 0.16$
1995–1996	$1.48 - 1.74 = -0.26$
1996–1997	$1.70 - 1.48 = 0.22$
1997–1998	$1.35 - 1.70 = -0.35$

Greatest increase: 1990–1991 and 1996–1997

Greatest decrease: 1997–1998

(b) $(1, 0.98), (11, 1.35)$: $y - 0.98 = \dfrac{1.35 - 0.98}{11 - 1}(x - 1)$

$$y = 0.037(x - 1) + 0.98$$

$$y = 0.037x + 0.943$$

(c) Between 1988 and 1998, the earnings per share increased at a rate of \$0.037 per year.

(d) For 2001, $y = 0.037(2001) - 72.576 = 1.461$, which is a reasonable prediction.

77. $\text{Slope} = \dfrac{\text{Rise}}{\text{Run}}$

$$-\frac{12}{100} = -\frac{2000}{y}$$

$$-12y = -200{,}000$$

$$y = 16{,}666\tfrac{2}{3} \text{ feet} \approx 3.16 \text{ miles}$$

79. $(1, 2540)$ $\quad m = 125$

$$V - 2540 = 125(t - 1)$$

$$V - 2540 = 125t - 125$$

$$V = 125t + 2415$$

81. $(1, 20400)$ $\quad m = -2000$

$$V - 20400 = -2000(t - 1)$$

$$V - 20400 = -2000t + 2000$$

$$V = -2000t + 22400$$

83. The slope is $m = -10$. This represents the decrease in the amount of the loan each week. Matches graph (b).

85. The slope is $m = 0.25$. This represents the increase in travel cost for each mile driven. Matches graph (a).

87. Using the points (0, 32) and (100, 212), we have

$$m = \frac{212 - 32}{100 - 0} = \frac{180}{100} = \frac{9}{5}$$

$$F - 32 = \frac{9}{5}(C - 0)$$

$$F = \frac{9}{5}C + 32.$$

89. Using the points (1998, 28500) and (2000, 32900), we have

$$m = \frac{32900 - 28500}{2000 - 1998} = \frac{4400}{2} = 2200$$

$$S - 28500 = 2200(t - 1998)$$

$$S = 2200t - 4{,}367{,}100$$

When $t = 2003$, $S = 2200(2003) - 4{,}367{,}100$
$= 39{,}500$

91. (a) Using the points (0, 875) and (5, 0), where the first coordinate represents the year t and the second coordinate represents the value V, we have

$$m = \frac{0 - 875}{5 - 0} = -175$$

$V = -175t + 875,\ 0 \le t \le 5.$

t	0	1	2	3	4	5
V	875	700	525	350	175	0

(b)

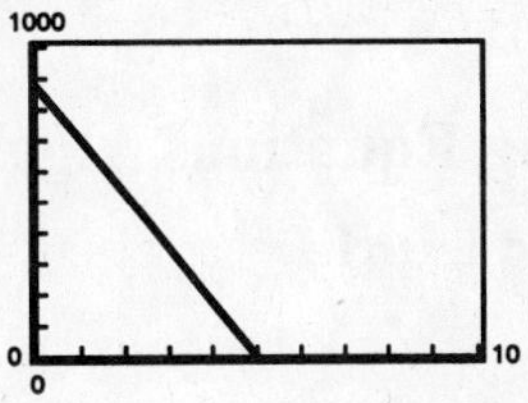

(c) $t = 0$: $V = -175(0) + 875 = 875$

$t = 1$: $V = -175(1) + 875 = 700$

$t = 2$: $V = -175(2) + 875 = 525$

$t = 3$: $V = -175(3) + 875 = 350$

$t = 4$: $V = -175(4) + 875 = 175$

$t = 5$: $V = -175(5) + 875 = 0$

93. (a) $C = 36{,}500 + 5.25t + 11.50t$

$= 16.75t + 36{,}500$

(b) $R = 27t$

(c) $P = R - C$

$= 27t - (16.75t + 36{,}500)$

$= 10.25t - 36{,}500$

(d) $0 = 10.25t - 36{,}500$

$36{,}500 = 10.25t$

$t \approx 3561$ hours

95. (a) $y = 92.84t + 487.82$ (answers will vary)

(b) For 2000, $t = 12$ and $y = 1{,}601$ corresponding to \$1,601,000.

(c) The slope is the average increase per year.

97. False.

The equation of the line joining $(10, -3)$ and $(2, -9)$ is

$$y + 3 = \frac{-9 + 3}{2 - 10}(x - 10)$$

$$y + 3 = \frac{3}{4}(x - 10)$$

$$y = \frac{3}{4}x - \frac{21}{2}$$

For $x = -12$, $y = \frac{3}{4}(-12) - \frac{21}{2}$

$$= -19.5$$

$$\neq \frac{-37}{2}$$

$$= -18.5$$

99. The line with slope -4 is steeper.

101. No, the slopes of perpendicular lines are negative reciprocals of each other.

Section P.4 Solving Equations Algebraically and Graphically

Solutions to Odd-Numbered Exercises

- You should know how to solve linear equations.
  $ax + b = 0$
- An identity is an equation whose solution consists of every real number in its domain.
- To solve an equation you can:
  (a) Add or subtract the same quantity from both sides.
  (b) Multiply or divide both sides by the same nonzero quantity.
- To solve an equation that can be simplified to a linear equation:
  (a) Remove all symbols of grouping and all fractions.
  (b) Combine like terms.
  (c) Solve by algebra.
  (d) Check the answer.
- A "solution" that does not satisfy the original equation is called an extraneous solution.
- You should be able to solve equations graphically.
- You should be able to solve a quadratic equation by factoring, if possible.
- You should be able to solve a quadratic equation of the form $u^2 = d$ by extracting square roots.
- You should be able to solve a quadratic equation by completing the square.
- You should know and be able to use the Quadratic Formula: For $ax^2 + bx + c = 0, a \neq 0$,

$$x = \frac{-b \pm \sqrt{b^2 - 4ac}}{2a}.$$

- You should be able to solve polynomials of higher degree by factoring.
- For equations involving radicals or fractional powers, raise both sides to the same power.
- For equations with fractions, multiply both sides by the least common denominator to clear the fractions.
- For equations involving absolute value, remember that the expression inside the absolute value can be positive or negative.
- Always check for extraneous solutions.

1. $\dfrac{5}{2x} - \dfrac{4}{x} = 3$

(a) $\dfrac{5}{2(-1/2)} - \dfrac{4}{(-1/2)} \stackrel{?}{=} 3$

$3 = 3$

$x = -\frac{1}{2}$ *is* a solution.

(b) $\dfrac{5}{2(4)} - \dfrac{4}{4} \stackrel{?}{=} 3$

$-\dfrac{3}{8} \neq 3$

$x = 4$ *is not* a solution.

(c) $\dfrac{5}{2(0)} - \dfrac{4}{0}$ is undefined.

$x = 0$ *is not* a solution.

(d) $\dfrac{5}{2(1/4)} - \dfrac{4}{1/4} \stackrel{?}{=} 3$

$-6 \neq 3$

$x = \frac{1}{4}$ *is not* a solution.

3. $3 + \dfrac{1}{x+2} = 4$

(a) $3 + \dfrac{1}{(-1)+2} \stackrel{?}{=} 4$

$4 = 4$

$x = -1$ is a solution

(b) $3 + \dfrac{1}{(-2)+2} = 3 + \dfrac{1}{0}$ is undefined

$x = -2$ is not a solution

(c) $3 + \dfrac{1}{0+2} \stackrel{?}{=} 4$

$\dfrac{7}{2} \neq 4$

$x = 0$ is not a solution

(d) $3 + \dfrac{1}{5+2} \stackrel{?}{=} 4$

$\dfrac{22}{7} = 4$

$x = 5$ is not a solution

5. $\dfrac{\sqrt{x+4}}{6} + 3 = 4$

(a) $\dfrac{\sqrt{-3+4}}{6} + 3 \stackrel{?}{=} 4$

$\dfrac{19}{6} \neq 4$

$x = -3$ is not a solution

(b) $\dfrac{\sqrt{0+4}}{6} + 3 \stackrel{?}{=} 4$

$\dfrac{10}{3} \neq 4$

$x = 0$ is not a solution

(c) $\dfrac{\sqrt{21+4}}{6} + 3 \stackrel{?}{=} 4$

$\dfrac{23}{6} \neq 4$

$x = 21$ is not a solution

(d) $\dfrac{\sqrt{32+4}}{6} + 3 \stackrel{?}{=} 4$

$4 = 4$

$x = 32$ is a solution

7. $2(x - 1) = 2x - 2$ is an *identity* by the Distributive Property. It is true for all real values of x.

9. $x^2 - 8x + 5 = (x - 4)^2 - 11$ is an *identity* since $(x - 4)^2 - 11 = x^2 - 8x + 16 - 11 = x^2 - 8x + 5$.

11. $3 + \dfrac{1}{x+1} = \dfrac{4x}{x+1}$ is *conditional.* There are real values of x for which the equation is not true.

13. Method 1: $\frac{3x}{8} - \frac{4x}{3} = 4$

$$\frac{9x - 32x}{24} = 4$$

$$-23x = 96$$

$$x = -\frac{96}{23}$$

Method 2: Graph $y_1 = \frac{3x}{8} - \frac{4x}{3}$ and $y_2 = 4$ in the same viewing window. These lines intersect at $x \approx -4.1739 \approx -\frac{96}{23}$

15. $\frac{5x}{4} + \frac{1}{2} = x - \frac{1}{2}$

$$4\left(\frac{5x}{4}\right) + 4\left(\frac{1}{2}\right) = 4(x) - 4\left(\frac{1}{2}\right)$$

$$5x + 2 = 4x - 2$$

$$x = -4$$

17. $\frac{3}{2}(z + 5) - \frac{1}{4}(z + 24) = 0$

$$4\left(\tfrac{3}{2}\right)(z + 5) - 4\left(\tfrac{1}{4}\right)(z + 24) = 4(0)$$

$$6(z + 5) - (z + 24) = 0$$

$$6z + 30 - z - 24 = 0$$

$$5z = -6$$

$$z = -\tfrac{6}{5}$$

19. $\frac{100 - 4u}{3} = \frac{5u + 6}{4} + 6$

$$12\left(\frac{100 - 4u}{3}\right) = 12\left(\frac{5u + 6}{4}\right) + 12(6)$$

$$4(100 - 4u) = 3(5u + 6) + 72$$

$$400 - 16u = 15u + 18 + 72$$

$$-31u = -310$$

$$u = 10$$

21. $\frac{5x - 4}{5x + 4} = \frac{2}{3}$

$$3(5x - 4) = 2(5x + 4)$$

$$15x - 12 = 10x + 8$$

$$5x = 20$$

$$x = 4$$

23. $\frac{1}{x - 3} + \frac{1}{x + 3} = \frac{10}{x^2 - 9}$

$$\frac{(x + 3) + (x - 3)}{x^2 - 9} = \frac{10}{x^2 - 9}$$

$$2x = 10$$

$$x = 5$$

25. $\frac{7}{2x + 1} - \frac{8x}{2x - 1} = -4$

$$7(2x - 1) - 8x(2x + 1) = -4(2x + 1)(2x - 1)$$

$$14x - 7 - 16x^2 - 8x = -16x^2 + 4$$

$$6x = 11$$

$$x = \frac{11}{6}$$

27. $\frac{1}{x} + \frac{2}{x - 5} = 0$

$$1(x - 5) + 2x = 0$$

$$3x - 5 = 0$$

$$3x = 5$$

$$x = \frac{5}{3}$$

29. $\frac{3}{x(x - 3)} + \frac{4}{x} = \frac{1}{x - 3}$

$$3 + 4(x - 3) = x$$

$$3 + 4x - 12 = x$$

$$3x = 9$$

$$x = 3$$

A check reveals that $x = 3$ is an extraneous solution, so there is no solution.

31. $y = x - 5$

Let $y = 0$: $0 = x - 5 \implies x = 5 \implies (5, 0)$ x-intercept

Let $x = 0$: $y = 0 - 5 \implies y = -5 \implies (0, -5)$ y-intercept

33. $y = x^2 + x - 2$

Let $y = 0$: $(x^2 + x - 2) = (x + 2)(x - 1) = 0 \implies x = -2, 1 \implies (-2, 0), (1, 0)$ x-intercepts

Let $x = 0$: $y = 0^2 + 0 - 2 = -2 \implies (0, -2)$ y-intercept

35. $y = x\sqrt{x + 2}$

Let $y = 0$: $0 = x\sqrt{x + 2} \implies x = 0, -2 \implies (0, 0), (-2, 0)$ x-intercepts

Let $x = 0$: $y = 0\sqrt{0 + 2} = 0 \implies (0, 0)$ y-intercept

37. $y = |x - 2| - 4$

Let $y = 0$: $|x - 2| - 4 = 0 \implies |x - 2| = 4 \implies x = -2, 6 \implies (-2, 0), (6, 0)$ x-intercepts

Let $x = 0$: $|0 - 2| - 4 = |-2| - 4 = 2 - 4 = -2 = y \implies (0, -2)$ y-intercept

39. $xy - 2y - x + 1 = 0$

Let $y = 0$: $-x + 1 = 0 \implies x = 1 \implies (1, 0)$ x-intercept

Let $x = 0$: $-2y + 1 = 0 \implies y = \frac{1}{2} \implies \left(0, \frac{1}{2}\right)$ y-intercept

41. $y = 12 - 4x$

$12 - 4x = 0$

$12 = 4x$

$3 = x$

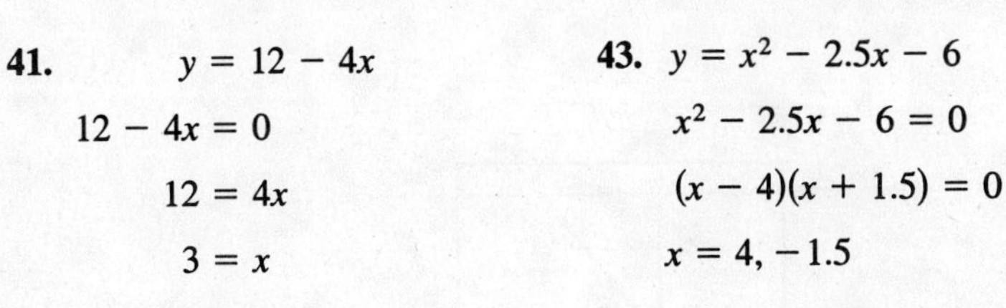

43. $y = x^2 - 2.5x - 6$

$x^2 - 2.5x - 6 = 0$

$(x - 4)(x + 1.5) = 0$

$x = 4, -1.5$

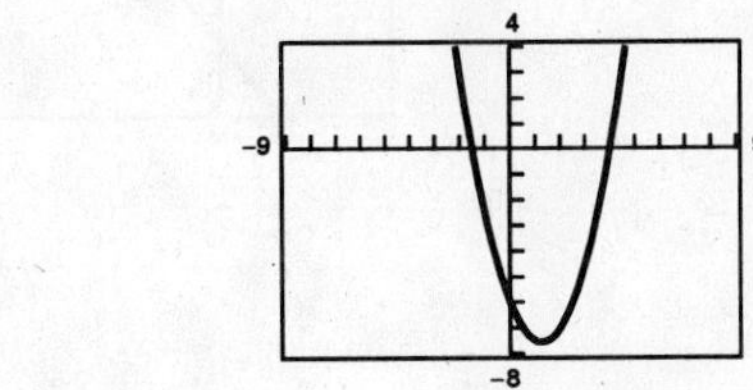

45. $y = \dfrac{x + 2}{3} - \dfrac{x - 1}{5} - 1$

$\dfrac{x + 2}{3} - \dfrac{x - 1}{5} - 1 = 0$

$5(x + 2) - 3(x - 1) - 15 = 0$

$2x = 2$

$x = 1$

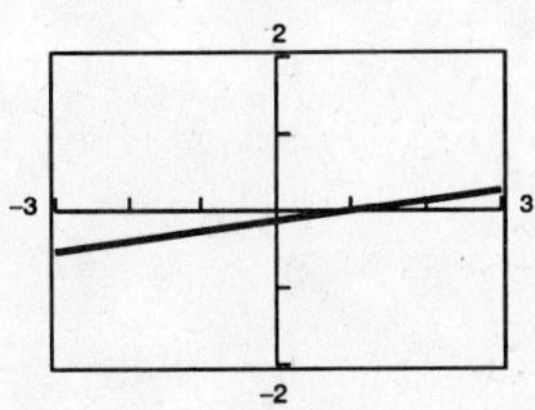

47.

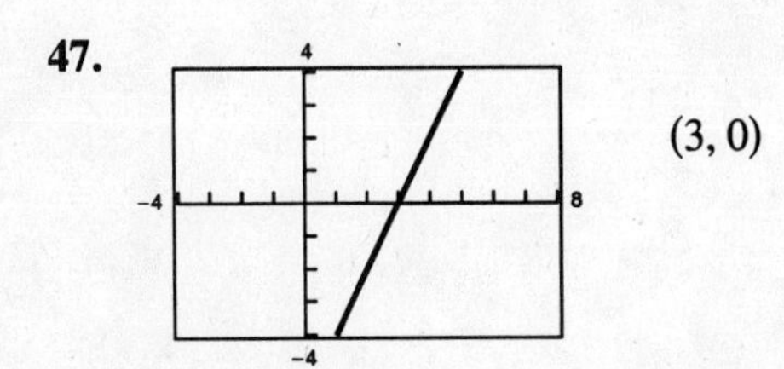

$(3, 0)$

$y = 0 = 2(x - 1) - 4 = 2x - 2 - 4 = 2x - 6 \implies 2x = 6 \implies x = 3$

49.

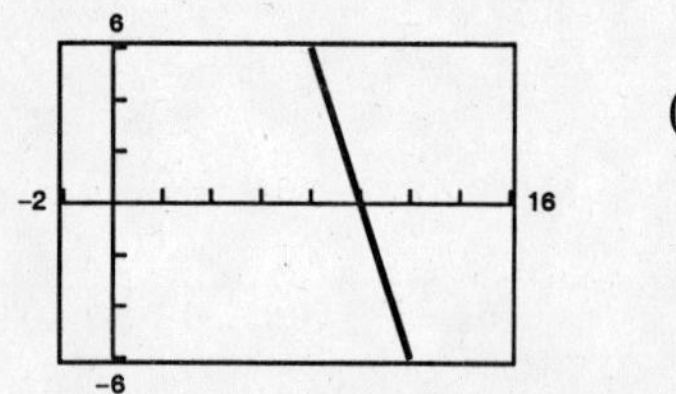

(10, 0)

$y = 0 = 20 - (3x - 10) = 20 - 3x + 10 = 30 - 3x \implies 3x = 30 \implies x = 10$

51.

$$27 - 4x = 12$$
$$-4x = -15$$
$$x = \tfrac{15}{4}$$
$$27 - 4x - 12 = 0$$
$$y = 15 - 4x = 0$$
$$x = 3.75 = \tfrac{15}{4}$$

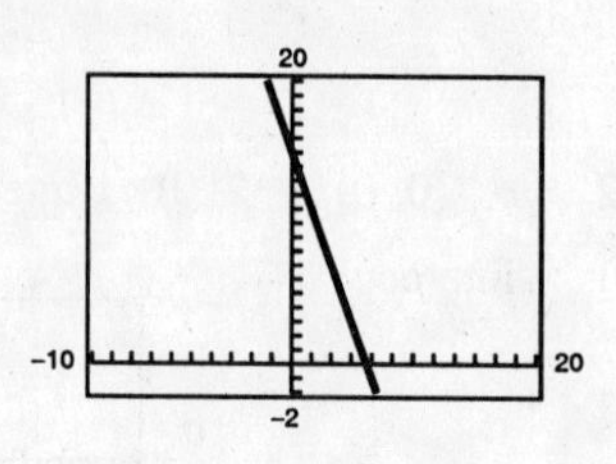

53.

$$25(x - 3) = 12(x + 2) - 10$$
$$25x - 75 = 12x + 24 - 10$$
$$13x - 89 = 0$$
$$x = \tfrac{89}{13}$$
$$y = 25(x - 3) - 12(x + 2) + 10 = 0$$
$$x = 6.846$$

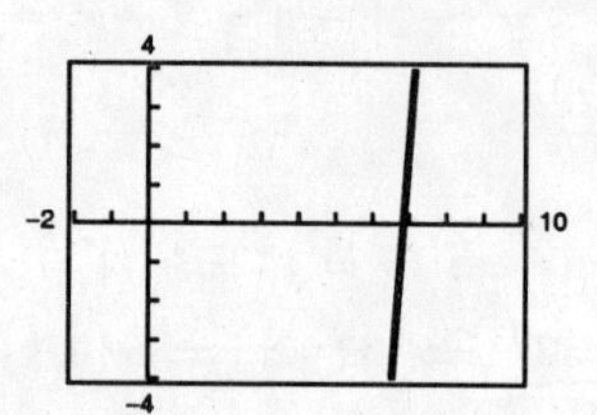

55.

$$\frac{3x}{2} + \frac{1}{4}(x - 2) = 10$$
$$\frac{6x}{4} + \frac{x}{4} = 10 + \frac{1}{2}$$
$$\frac{7x}{4} = \frac{21}{2}$$
$$x = 6$$
$$y = \frac{3x}{2} + \frac{1}{4}(x - 2) - 10 = 0$$
$$x = 6.0$$

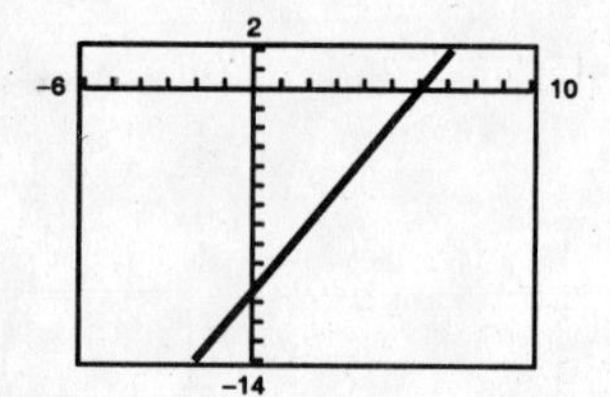

57. $\dfrac{2x}{3} = 10 - \dfrac{24}{x}$

$\dfrac{2x}{3}(3x) = 10(3x) - \dfrac{24}{x}(3x)$

$2x^2 = 30x - 72$

$2x^2 - 30x + 72 = 0$

$x^2 - 15x + 36 = 0$

$(x - 3)(x - 12) = 0$

$x = 3, 12$

$y = \dfrac{2x}{3} - 10 + \dfrac{24}{x}$

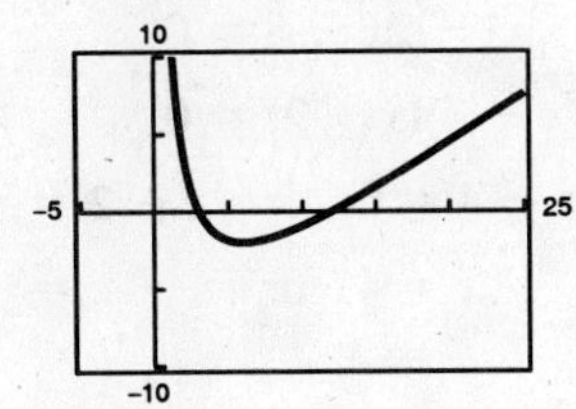

$x = 3, 12$

59. $\dfrac{3}{x+2} - \dfrac{4}{x-2} = 5$

$3(x - 2) - 4(x + 2) = 5(x + 2)(x - 2)$

$3x - 6 - 4x - 8 = 5(x^2 - 4)$

$0 = 5x^2 + x - 6$

$0 = (x - 1)(5x + 6)$

$x = 1, -\dfrac{6}{5}$

$y = \dfrac{3}{x+2} - \dfrac{4}{x-2} - 5 = 0$

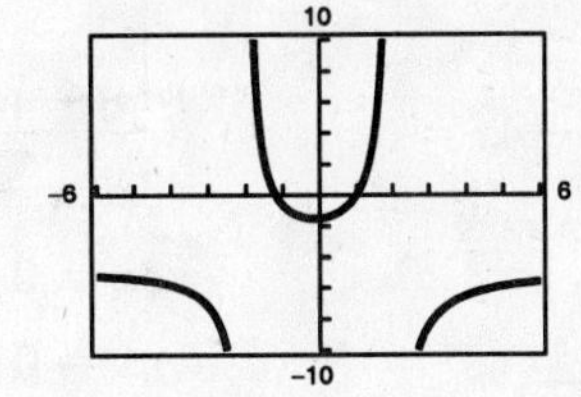

$x = 1.0, -1.2$

61. $3(x + 3) = 5(1 - x) - 1$

$3x + 9 = 5 - 5x - 1$

$8x = -5$

$x = -\frac{5}{8}$

$y = 3(x + 3) - 5(1 - x) + 1 = 0$

$x = -0.625$

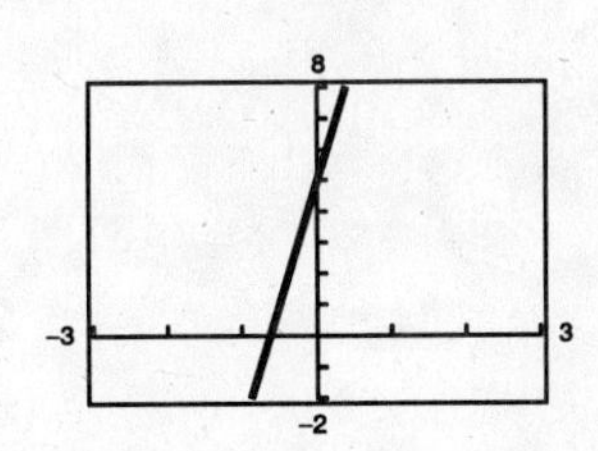

63. $2x^3 - x^2 - 18x + 9 = 0$

$x = -3.0,\ 0.5,\ 3.0$

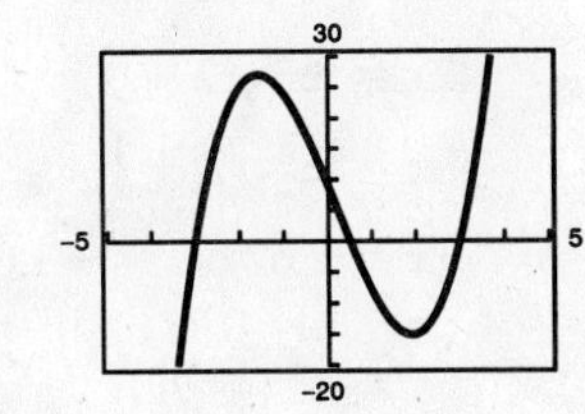

65. $x^4 = 2x^3 + 1$

$x^4 - 2x^3 - 1 = 0$

$x = -0.717,\ 2.107$

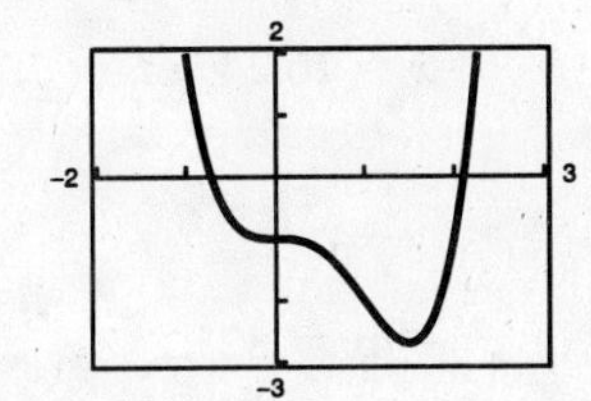

67. $\dfrac{2}{x+2} = 3$

$\dfrac{2}{x+2} - 3 = 0$

$x = -1.333$

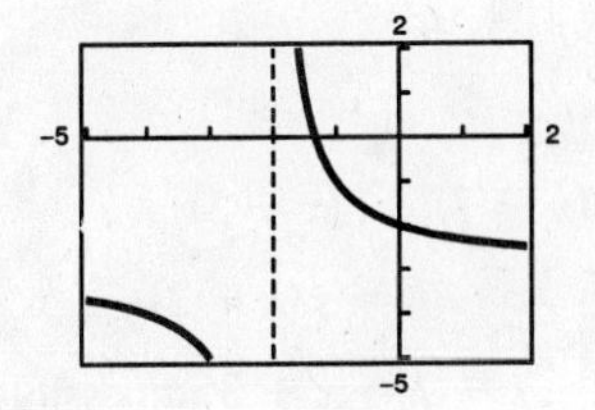

69. $|x - 3| = 4$

$|x - 3| - 4 = 0$

$x = -1, 7$

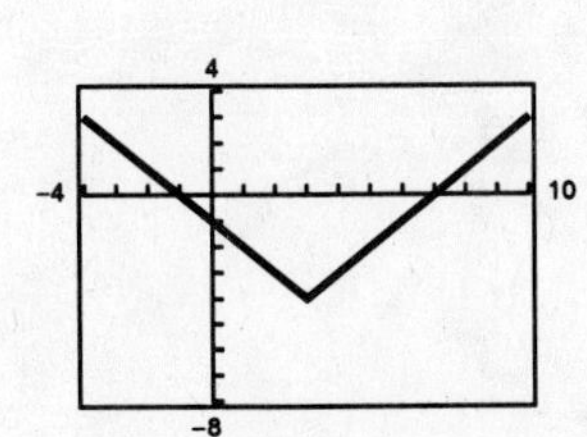

71. $y = 2 - x$

$y = 2x - 1$

$2 - x = 2x - 1$

$3 = 3x$

$x = 1, y = 2 - 1 = 1$

$(x, y) = (1, 1)$

73. $x - y = -4 \Rightarrow y = x + 4$

$x^2 - y = -2 \Rightarrow y = x^2 + 2$

$x^2 + 2 = x + 4$

$x^2 - x - 2 = 0$

$(x - 2)(x + 1) = 0$

$x = 2, y = 6$

$x = -1, y = 3$

$(2, 6), (-1, 3)$

75. $y = 9 - 2x$

$y = x - 3$

$(4, 1)$

77. $y = 4 - x^2$

$y = 2x - 1$

$(x, y) = (1.449, 1.898),$

$(-3.449, -7.899)$

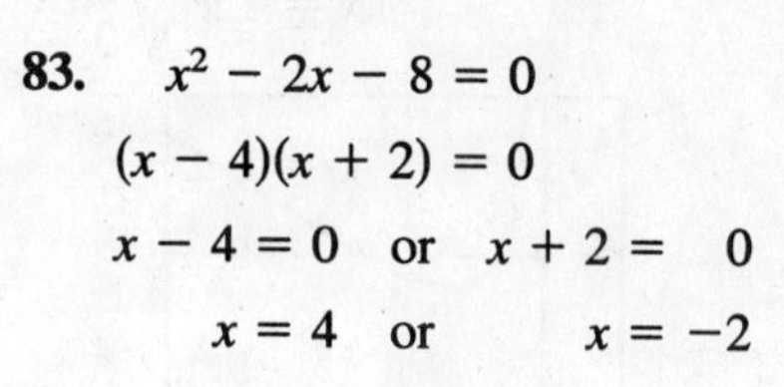

79. $y = 2x^2$

$y = x^4 - 2x^2$

$(x, y) = (0, 0),\ (2, 8),\ (-2, 8)$

81. $6x^2 + 3x = 0$

$3x(2x + 1) = 0$

$3x = 0 \quad \text{or} \quad 2x + 1 = 0$

$x = 0 \quad \text{or} \quad x = -\frac{1}{2}$

83. $x^2 - 2x - 8 = 0$

$(x - 4)(x + 2) = 0$

$x - 4 = 0 \quad \text{or} \quad x + 2 = 0$

$x = 4 \quad \text{or} \quad x = -2$

85. $3 + 5x - 2x^2 = 0$

$(3 - x)(1 + 2x) = 0$

$3 - x = 0 \quad \text{or} \quad 1 + 2x = 0$

$x = 3 \quad \text{or} \quad x = -\frac{1}{2}$

87. $x^2 + 4x = 12$

$x^2 + 4x - 12 = 0$

$(x + 6)(x - 2) = 0$

$x + 6 = 0 \quad \text{or} \quad x - 2 = 0$

$x = -6 \quad \text{or} \quad x = 2$

89. $x^2 = 7$

$x = \pm\sqrt{7}$

$\approx \pm 2.65$

91. $(x - 12)^2 = 18$

$x - 12 = \pm 3\sqrt{2}$

$x = 12 \pm 3\sqrt{2}$

$x \approx 16.24 \quad \text{or} \quad \approx 7.76$

93. $(2x - 1)^2 = 18$

$2x - 1 = \pm\sqrt{18} = \pm 3\sqrt{2}$

$2x = \pm 3\sqrt{2} + 1$

$x = \pm\frac{3}{2}\sqrt{2} + \frac{1}{2}$

$x \approx 2.62 \text{ or } \approx -1.62$

95. $(x - 7)^2 = (x + 3)^2$

$x - 7 = \pm(x + 3)$

$x - 7 = x + 3$ impossible

$x - 7 = -(x + 3) \Rightarrow 2x = 4$

$\Rightarrow x = 2$

97. $x^2 + 4x = 32$

$x^2 + 4x + 4 = 32 + 4$

$(x + 2)^2 = 36$

$x + 2 = \pm 6$

$x = -2 \pm 6$

$x = -8, 4$

99. $x^2 + 6x + 2 = 0$

$x^2 + 6x = -2$

$x^2 + 6x + 3^2 = -2 + 3^2$

$(x + 3)^2 = 7$

$x + 3 = \pm\sqrt{7}$

$x = -3 \pm \sqrt{7}$

101. $9x^2 - 18x + 3 = 0$

$$x^2 - 2x + \frac{1}{3} = 0$$
$$x^2 - 2x = -\frac{1}{3}$$
$$x^2 - 2x + 1^2 = -\frac{1}{3} + 1^2$$
$$(x-1)^2 = \frac{2}{3}$$
$$x - 1 = \pm\sqrt{\frac{2}{3}}$$
$$x = 1 \pm \sqrt{\frac{2}{3}}$$
$$x = 1 \pm \frac{\sqrt{6}}{3}$$

103. $-x^2 + 2x + 2 = 0$

$$x = \frac{-b \pm \sqrt{b^2 - 4ac}}{2a}$$
$$= \frac{-2 \pm \sqrt{2^2 - 4(-1)(2)}}{2(-1)}$$
$$= \frac{-2 \pm 2\sqrt{3}}{-2} = 1 \pm \sqrt{3}$$

105. $x^2 + 8x - 4 = 0$

$$x = \frac{-b \pm \sqrt{b^2 - 4ac}}{2a}$$
$$= \frac{-8 \pm \sqrt{8^2 - 4(1)(-4)}}{2(1)}$$
$$= \frac{-8 \pm 4\sqrt{5}}{2}$$
$$= -4 \pm 2\sqrt{5}$$

107. $4x^2 + 16x + 15 = 0$

$$x = \frac{-b \pm \sqrt{b^2 - 4ac}}{2a}$$
$$= \frac{-16 \pm \sqrt{16^2 - 4(4)(15)}}{2(4)}$$
$$= \frac{-16 + \sqrt{16}}{8}$$
$$= -2 \pm \frac{1}{2} = -\frac{3}{2}, -\frac{5}{2}.$$

109. $x^2 - 2x - 1 = 0$

$$x^2 - 2x = 1$$
$$x^2 - 2x + 1^2 = 1 + 1^2$$
$$(x-1)^2 = 2$$
$$x - 1 = \pm\sqrt{2}$$
$$x = 1 \pm \sqrt{2}$$

111. $(x+3)^3 = 81$

$$x + 3 = \pm 9$$
$$x + 3 = 9 \quad \text{or} \quad x + 3 = -9$$
$$x = 6 \quad \text{or} \quad x = -12$$

113. $x^2 - x - \frac{11}{4} = 0$

$$x^2 - x + \frac{1}{4} = \frac{11}{4} + \frac{1}{4}$$
$$\left(x - \frac{1}{2}\right)^2 = 3$$
$$x - \frac{1}{2} = \pm\sqrt{3}$$
$$x = \frac{1}{2} \pm \sqrt{3}$$
$$x = \frac{1}{2} + \sqrt{3}, \frac{1}{2} - \sqrt{3}$$

115. $4x^4 - 18x^2 = 0$

$$2x^2(2x^2 - 9) = 0$$
$$2x^2 = 0 \implies x = 0$$
$$2x^2 - 9 = 0 \implies x = \pm\frac{3\sqrt{2}}{2}$$

117. $x^4 - 81 = 0$

$(x^2 + 9)(x + 3)(x - 3) = 0$

$x^2 + 9 = 0$ No real solution.

$x + 3 = 0 \Rightarrow x = -3$

$x - 3 = 0 \Rightarrow x = 3$

119. $5x^3 + 30x^2 + 45x = 0$

$5x(x^2 + 6x + 9) = 0$

$5x(x + 3)^2 = 0$

$5x = 0 \Rightarrow x = 0$

$x + 3 = 0 \Rightarrow x = -3$

121. $x^3 - 3x^2 - x + 3 = 0$

$x^2(x - 3) - (x - 3) = 0$

$(x - 3)(x^2 - 1) = 0$

$(x - 3)(x + 1)(x - 1) = 0$

$x - 3 = 0 \Rightarrow x = 3$

$x + 1 = 0 \Rightarrow x = -1$

$x - 1 = 0 \Rightarrow x = 1$

123. $x^4 - 4x^2 + 3 = 0$

$(x^2 - 3)(x^2 - 1) = 0$

$(x + \sqrt{3})(x - \sqrt{3})(x + 1)(x - 1) = 0$

$x + \sqrt{3} = 0 \Rightarrow x = -\sqrt{3}$

$x - \sqrt{3} = 0 \Rightarrow x = \sqrt{3}$

$x + 1 = 0 \Rightarrow x = -1$

$x - 1 = 0 \Rightarrow x = 1$

125. $4x^4 - 65x^2 + 16 = 0$

$(4x^2 - 1)(x^2 - 16) = 0$

$(2x + 1)(2x - 1)(x + 4)(x - 4) = 0$

$2x + 1 = 0 \Rightarrow x = -\frac{1}{2}$

$2x - 1 = 0 \Rightarrow x = \frac{1}{2}$

$x + 4 = 0 \Rightarrow x = -4$

$x - 4 = 0 \Rightarrow x = 4$

127. $\frac{1}{t^2} + \frac{8}{t} + 15 = 0$

$1 + 8t + 15t^2 = 0$

$(1 + 3t)(1 + 5t) = 0$

$1 + 3t = 0 \Rightarrow t = -\frac{1}{3}$

$1 + 5t = 0 \Rightarrow t = -\frac{1}{5}$

129. $2x + 9\sqrt{x} - 5 = 0$

$(2\sqrt{x} - 1)(\sqrt{x} + 5) = 0$

$\sqrt{x} = \frac{1}{2} \Rightarrow x = \frac{1}{4}$

($\sqrt{x} = -5$ is not possible.)

Note: You can see graphically that there is only one solution.

131. $\sqrt{x - 10} - 4 = 0$

$\sqrt{x - 10} = 4$

$x - 10 = 16$

$x = 26$

133. $\sqrt{x + 1} - 3x = 1$

$\sqrt{x + 1} = 3x + 1$

$x + 1 = 9x^2 + 6x + 1$

$0 = 9x^2 + 5x$

$0 = x(9x + 5)$

$x = 0$

$9x + 5 = 0 \Rightarrow x = -\frac{5}{9}$, extraneous

135. $\sqrt{x} - \sqrt{x - 5} = 1$

$\sqrt{x} = 1 + \sqrt{x - 5}$

$(\sqrt{x})^2 = (1 + \sqrt{x - 5})^2$

$x = 1 + 2\sqrt{x - 5} + x - 5$

$4 = 2\sqrt{x - 5}$

$2 = \sqrt{x - 5}$

$4 = x - 5$

$9 = x$

137. $(x - 5)^{2/3} = 16$

$x - 5 = \pm 16^{3/2}$

$x - 5 = \pm 64$

$x = 69, \ -59$

139. $3x(x-1)^{1/2} + 2(x-1)^{3/2} = 0$

$(x-1)^{1/2}[3x + 2(x-1)] = 0$

$(x-1)^{1/2}(5x-2) = 0$

$(x-1)^{1/2} = 0 \Rightarrow x - 1 = 0 \Rightarrow x = 1$

$5x - 2 = 0 \Rightarrow x = \frac{2}{5}$ which is extraneous.

141. $\dfrac{20-x}{x} = x$

$20 - x = x^2$

$0 = x^2 + x - 20$

$0 = (x+5)(x-4)$

$x + 5 = 0 \Rightarrow x = -5$

$x - 4 = 0 \Rightarrow x = 4$

143. $\dfrac{1}{x} - \dfrac{1}{x+1} = 3$

$x(x+1)\dfrac{1}{x} - x(x+1)\dfrac{1}{x+1} = x(x+1)(3)$

$x + 1 - x = 3x(x+1)$

$1 = 3x^2 + 3x$

$0 = 3x^2 + 3x - 1;\ a = 3,\ b = 3,\ c = -1$

$x = \dfrac{-3 \pm \sqrt{(3)^2 - 4(3)(-1)}}{2(3)} = \dfrac{-3 \pm \sqrt{21}}{6}$

145. $x = \dfrac{3}{x} + \dfrac{1}{2}$

$(2x)(x) = (2x)\left(\dfrac{3}{x}\right) + (2x)\left(\dfrac{1}{2}\right)$

$2x^2 = 6 + x$

$2x^2 - x - 6 = 0$

$(2x+3)(x-2) = 0$

$2x + 3 = 0 \Rightarrow x = -\dfrac{3}{2}$

$x - 2 = 0 \Rightarrow x = 2$

147. $|2x - 1| = 5$

$2x - 1 = 5 \Rightarrow x = 3$

$-(2x-1) = 5 \Rightarrow x = -2$

149. $|x| = x^2 + x - 3$

$x = x^2 + x - 3$ OR $-x = x^2 + x - 3$

$x^2 - 3 = 0$

$x = \pm\sqrt{3}$

$x^2 + 2x - 3 = 0$

$(x-1)(x+3) = 0$

$x - 1 = 0 \Rightarrow x = 1$

$x + 3 = 0 \Rightarrow x = -3$

Only $x = \sqrt{3}$, and $x = -3$ are solutions to the original equation. $x = -\sqrt{3}$ and $x = 1$ are extraneous. Note that the graph of $y = x^2 + x - 3 - |x|$ has two x-intercepts.

151. $y = x^3 - 2x^2 - 3x$

(a)

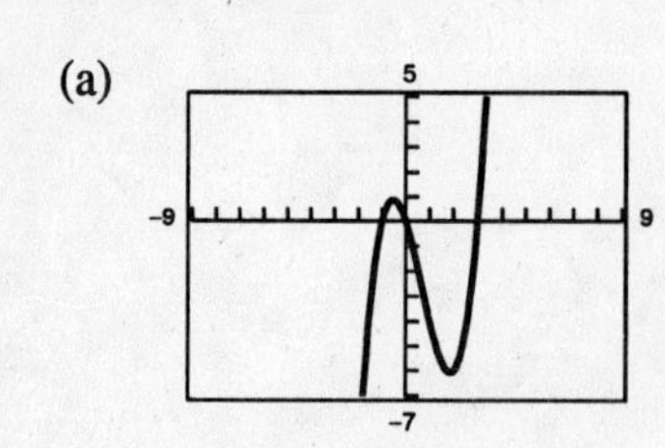

(b) x-intercepts: $(-1, 0), (0, 0), (3, 0)$

(c) $0 = x^3 - 2x^2 - 3x$

$0 = x(x + 1)(x - 3)$

$x = 0$

$x + 1 = 0 \Rightarrow x = -1$

$x - 3 = 0 \Rightarrow x = 3$

153. $y = \sqrt{11x - 30} - x$

(a)

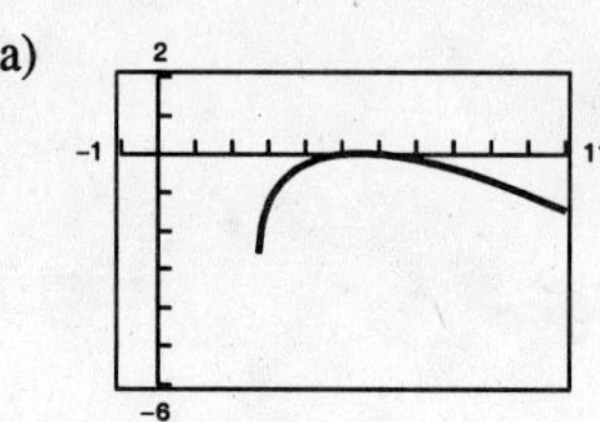

(b) x-intercepts: $(5, 0), (6, 0)$

(c) $0 = \sqrt{11x - 30} - x$

$x = \sqrt{11x - 30}$

$x^2 = 11x - 30$

$x^2 - 11x + 30 = 0$

$(x - 5)(x - 6) = 0$

$x - 5 = 0 \Rightarrow x = 5$

$x - 6 = 0 \Rightarrow x = 6$

155. $y = \dfrac{1}{x} - \dfrac{4}{x - 1} - 1$

(a)

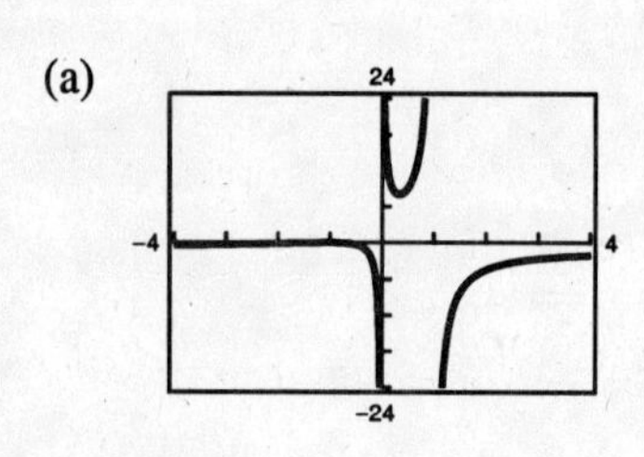

(b) x-intercept: $(-1, 0)$

(c) $0 = \dfrac{1}{x} - \dfrac{4}{x - 1} - 1$

$0 = (x - 1) - 4x - x(x - 1)$

$0 = x - 1 - 4x - x^2 + x$

$0 = -x^2 - 2x - 1$

$0 = x^2 + 2x + 1$

$0 = (x + 1)^2$

$x + 1 = 0 \Rightarrow x = -1$

157. $y = |x + 1| - 2$

(a)

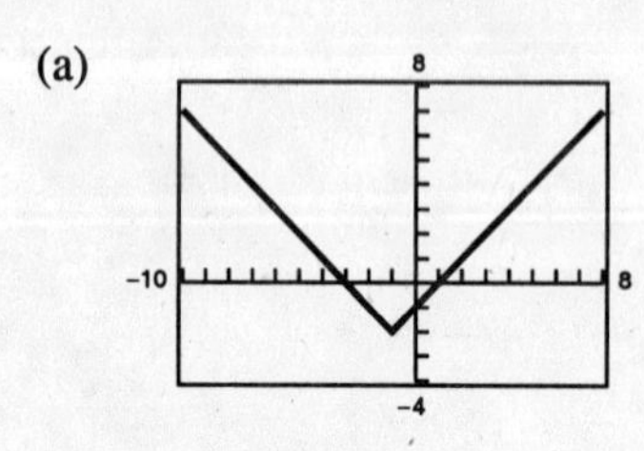

(b) x-intercept: $(1, 0), (-3, 0)$

(c) $0 = |x + 1| - 2$

$2 = |x + 1|$

$x + 1 = 2$ or $-(x + 1) = 2$

$x = 1$ or $-x - 1 = 2$

$-x = 3$

$x = -3$

159. (a) The point of intersection represents the moment when the per capita utilizations of nectarines and cucumbers are equal.

(b) $-0.37t + 6.88 = 0.27t + 4.42$

$$2.46 = 0.64t$$

$$t = 3.84375 \text{ or during 1993}$$

(c)

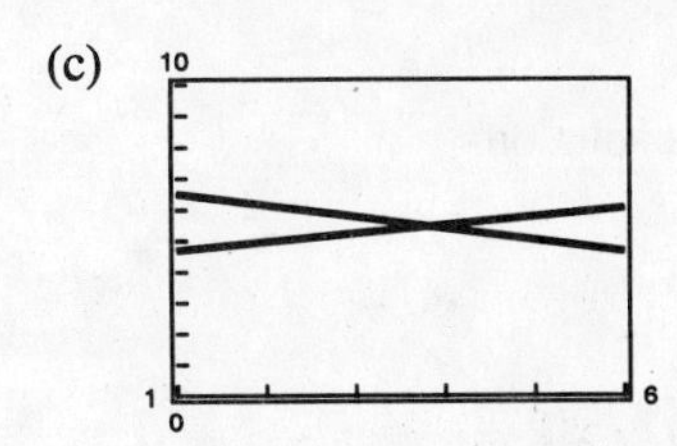

Intersection point: $(3.84375, 5.45781)$

161. $p = 40 - \sqrt{0.0001x + 1}, 0 \leq x$

(a)

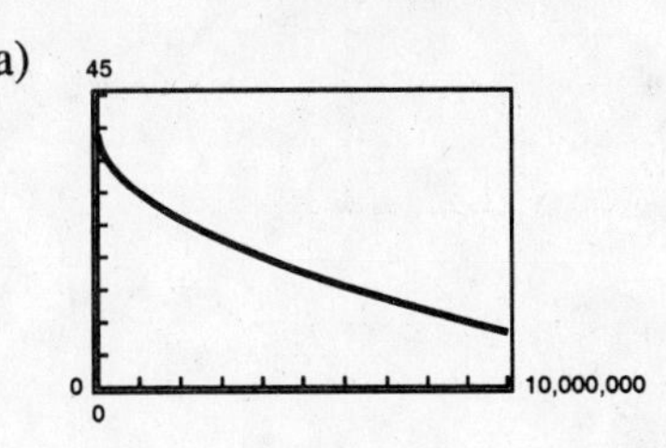

(b) If $p = 12.95$, then $x \approx 7{,}307{,}025$ books

163. False, they could have an infinite number of intersections.

Section P.5 Solving Inequalities Algebraically and Graphically

- You should know the properties of inequalities.
  (a) Transitive: $a < b$ and $b < c$ implies $a < c$.
  (b) Addition: $a < b$ and $c < d$ implies $a + c < b + d$.
  (c) Adding or Subtracting a Constant: $a \pm c < b \pm c$ if $a < b$.
  (d) Multiplying or Dividing by a Constant: For $a < b$,
    1. If $c > 0$, then $ac < bc$ and $\frac{a}{c} < \frac{b}{c}$.
    2. If $c < 0$,then $ac > bc$ and $\frac{a}{c} > \frac{b}{c}$.
- You should know that
  $$|x| = \begin{cases} x & \text{if } x \geq 0 \\ -x & \text{if } x < 0 \end{cases}.$$
- You should be able to solve absolute value inequalities.
  (a) $|x| < a$ if and only if $-a < x < a$.
  (b) $|x| > a$ if and only if $x < -a$ or $x > a$.
- You should be able to solve polynomial inequalities.
  (a) Find the critical numbers.
    1. Values that make the expression zero
    2. Values that make the expression undefined
  (b) Test one value in each interval on the real number line resulting from the critical numbers.
  (c) Determine the solution intervals.
- You should be able to solve rational and other types of inequalities.

Solutions to Odd-Numbered Problems

1. $x < 3$

Matches (d).

3. $-3 < x \le 4$

Matches (c).

5. (a) $x = 3$

$5(3) - 12 \overset{?}{>} 0$

$3 > 0$

Yes, $x = 3$ is a solution.

(b) $x = -3$

$5(-3) - 12 \overset{?}{>} 0$

$-27 \not> 0$

No, $x = -3$ is not a solution.

(c) $x = \frac{5}{2}$

$5\left(\frac{5}{2}\right) - 12 \overset{?}{>} 0$

$\frac{1}{2} > 0$

Yes, $x = \frac{5}{2}$ is a solution.

(d) $x = \frac{3}{2}$

$5\left(\frac{3}{2}\right) - 12 \overset{?}{>} 0$

$-\frac{9}{2} \not> 0$

No, $x = \frac{3}{2}$ is not a solution.

7. $-1 < \dfrac{3 - x}{2} \le 1$

(a) $x = 0$

$-1 \overset{?}{<} \dfrac{3 - 0}{2} \overset{?}{\le} 1$

$-1 \overset{?}{<} \dfrac{3}{2} \overset{?}{\le} 1$

No, $x = 0$ is not a solution.

(b) $x = \sqrt{5}$

$-1 \overset{?}{<} \dfrac{3 - \sqrt{5}}{2} \overset{?}{\le} 1$

$-1 \overset{?}{<} 0.382 \overset{?}{\le} 1$

Yes, $x = \sqrt{5}$ is a solution.

(c) $x = 1$

$-1 \overset{?}{<} \dfrac{3 - 1}{2} \overset{?}{\le} 1$

$-1 \overset{?}{<} 1 \overset{?}{\le} 1$

Yes, $x = 1$ is a solution.

(d) $x = 5$

$-1 \overset{?}{<} \dfrac{3 - 5}{2} \le 1$

$-1 \overset{?}{<} -1 \overset{?}{\le} 1$

No, $x = 5$ is not a solution.

9.

$-10x < 40$

$-\frac{1}{10}(-10) > -\frac{1}{10}(40)$

$x > -4$

-4 -3 -2 -1 x

11. $4(x + 1) < 2x + 3$

$4x + 4 < 2x + 3$

$2x < -1$

$x < -\frac{1}{2}$

$-\frac{1}{2}$

-2 -1 0 1 x

13. $1 < 2x + 3 < 9$

$-2 < 2x < 6$

$-1 < x < 3$

-1 0 1 2 3 x

15. $-8 \le 1 - 3(x - 2) < 13$

$-8 \le 1 - 3x + 6 < 13$

$-8 \le -3x + 7 < 13$

$-15 \le -3x < 6$

$5 \ge x > -2 \Rightarrow -2 < x \le 5$

-2 -1 0 1 2 3 4 5 x

17. $-4 < \dfrac{2x - 3}{3} < 4$

$-12 < 2x - 3 < 12$

$-9 < 2x < 15$

$-\dfrac{9}{2} < x < \dfrac{15}{2}$

$-\frac{9}{2}$ $\frac{15}{2}$

-6 -4 -2 0 2 4 6 8 x

19. $6x > 12$

$x > 2$

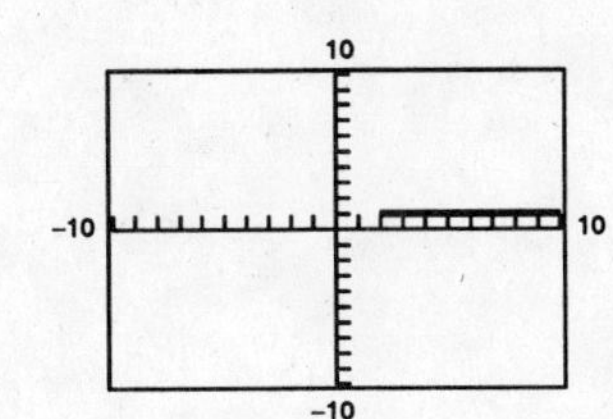

21. $5 - 2x \geq 1$

$-2x \geq -4$

$x \leq 2$

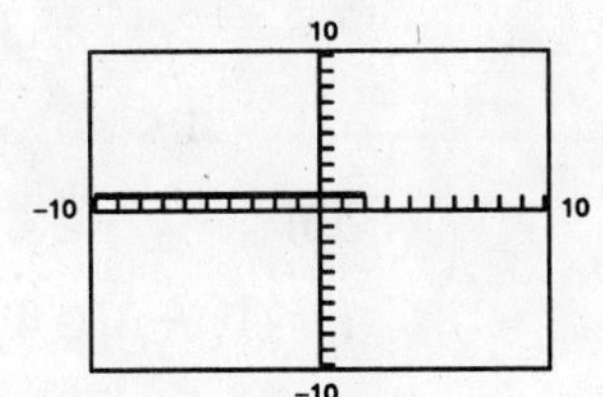

23. $-9 < 6x - 1 < 1$

$-8 < 6x < 2$

$-\frac{4}{3} < x < \frac{1}{3}$

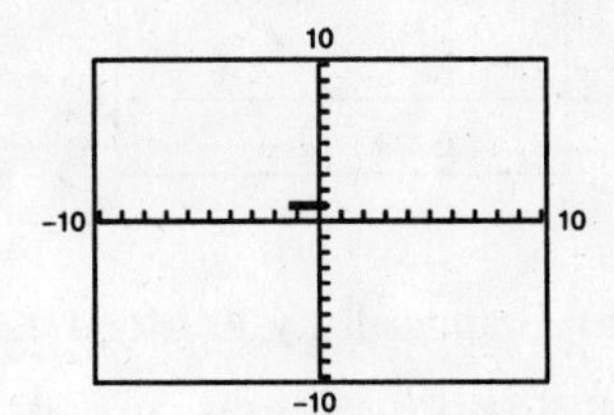

25. Using the graph, (a) $y \geq 1$ for $x \geq 2$ and (b) $y \leq 0$ for $x \leq \frac{3}{2}$.

Algebraically, (a) $y \geq 1$ (b) $y \leq 0$

$2x - 3 \geq 1$ $\quad$ $2x - 3 \leq 0$

$2x \geq 4$ $\quad$ $2x \leq 3$

$x \geq 2$ $\quad$ $x \leq \frac{3}{2}$

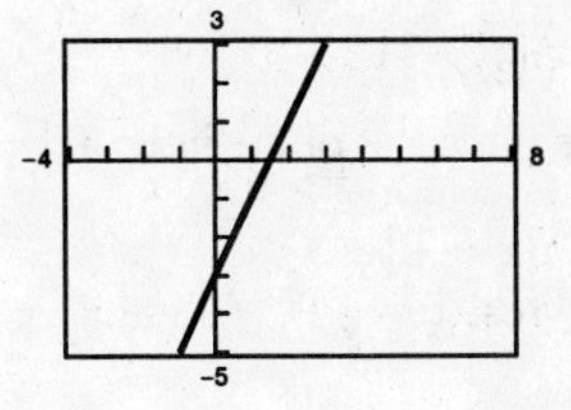

27. Using the graph, (a) $0 \leq y \leq 3$ for $-2 \leq x \leq 4$ and (b) $y \geq 0$ for $x \leq 4$

Algebraically, (a) $0 \leq y \leq 3$ (b) $y \geq 0$

$0 \leq -\frac{1}{2}x + 2 \leq 3$ $\quad$ $-\frac{1}{2}x + 2 \geq 0$

$-2 \leq -\frac{1}{2}x \leq 1$ $\quad$ $2 \geq \frac{1}{2}x$

$4 \geq x \geq -2$ $\quad$ $4 \geq x$

29. $|5x| > 10$

$5x < -10$ or $5x > 10$

$x < -2$ or $x > 2$

31. $|x - 7| < 6$

$-6 < x - 7 < 6$

$1 < x < 13$

33. $|x + 14| + 3 > 17$

$|x + 14| > 14$

$x + 14 < -14$ or $x + 14 > 14$

$x < -28$ or $x > 0$

35. $|1 - 2x| < 5$

$-5 < 1 - 2x < 5$

$-6 < -2x < 4$

$3 > x > -2$

$-2 < x < 3$

37. $y = |x - 3|$

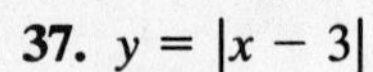

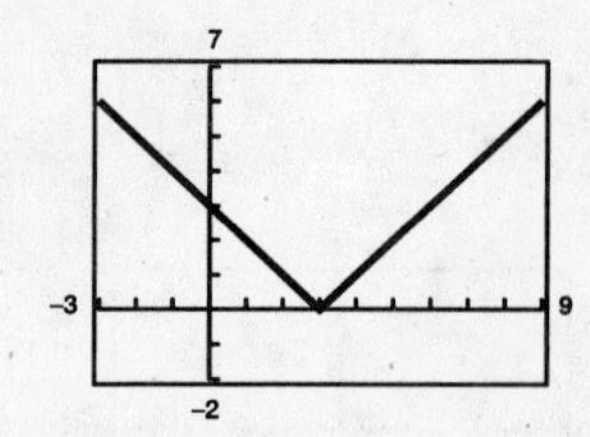

(a) Graphically, $y \le 2$ for $1 \le x \le 5$

and (b) $y \ge 4$ for $x \le -1$ or $x \ge 7$

Algebraically,

(a)
$$y \le 2$$
$$|x - 3| \le 2$$
$$-2 \le x - 3 \le 2$$
$$1 \le x \le 5$$

(b)
$$y \ge 4$$
$$|x - 3| \ge 4$$
$$x - 3 \le -4 \quad \text{or} \quad x - 3 \ge 4$$
$$x \le -1 \qquad\qquad x \ge 7$$

39. The midpoint of the interval $[-3, 3]$ is 0. The interval represents all real numbers x no more than 3 units from 0.

$$|x - 0| \le 3$$
$$|x| \le 3$$

41. The graph shows all real numbers at least 3 units from 7.

$$|x - 7| \ge 3$$

43. All real numbers within 10 units of 12

$$|x - 12| \le 10$$

45.
$$(x + 2)^2 < 25$$
$$x^2 + 4x + 4 < 25$$
$$x^2 + 4x - 21 < 0$$
$$(x + 7)(x - 3) < 0$$

Critical numbers: $x = -7, x = 3$

Test intervals: $(-\infty, -7), (-7, 3), (3, \infty)$

Test: Is $(x + 7)(x - 3) < 0$?

Solution set: $(-7, 3)$

47.
$$x^2 + 4x + 4 \ge 9$$
$$x^2 + 4x - 5 \ge 0$$
$$(x + 5)(x - 1) \ge 0$$

Critical numbers: $x = -5, x = 1$

Test intervals: $(-\infty, -5), (-5, 1), (1, \infty)$

Test: Is $(x + 5)(x - 1) \ge 0$?

Solution set: $(-\infty, -5] \cup [1, \infty)$

49.
$$x^3 - 4x \ge 0$$
$$x(x + 2)(x - 2) \ge 0$$

Critical number: $x = 0, x = \pm 2$

Test intervals: $(-\infty, -2), (-2, 0), (0, 2), (2, \infty)$

Test: Is $x(x + 2)(x - 2) \ge 0$?

Solution set: $[-2, 0] \cup [2, \infty)$

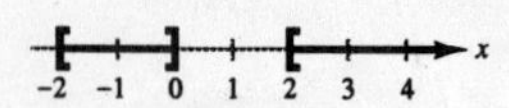

51. $y = -x^2 + 2x + 3$

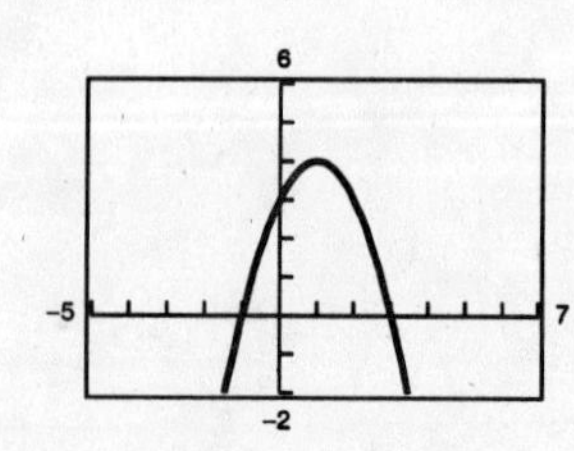

(a) $y \le 0$ when $x \le -1$ or $x \ge 3$.

(b) $y \ge 3$ when $0 \le x \le 2$.

53. $y = \frac{1}{8}x^3 - \frac{1}{2}x$

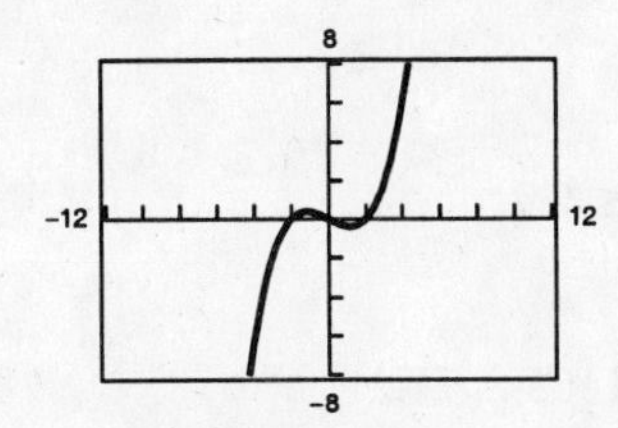

(a) $y \geq 0$ when $-2 \leq x \leq 0, 2 \leq x$.

(b) $y \leq 6$ when $x \leq 4$.

55. $\frac{1}{x} - x > 0$

$\frac{1 - x^2}{x} > 0$

Critical numbers: $x = 0, x = \pm 1$

Test intervals: $(-\infty - 1), (-1, 0), (0, 1), (1, \infty)$

Test: Is $\frac{1 - x^2}{x} > 0$?

Solution set: $(-\infty, -1) \cup (0, 1)$

57. $\frac{x + 6}{x + 1} - 2 < 0$

$\frac{x + 6 - 2(x + 1)}{x + 1} < 0$

$\frac{4 - x}{x + 1} < 0$

Critical numbers: $x = -1, x = 4$

Test intervals: $(-\infty, -1), (-1, 4), (4, \infty)$

Test: Is $\frac{4 - x}{x + 1} < 0$?

Solution set: $(-\infty, -1) \cup (4, \infty)$

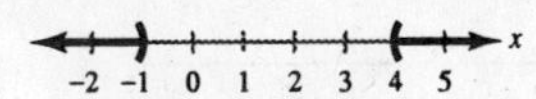

59. $y = \frac{3x}{x - 2}$

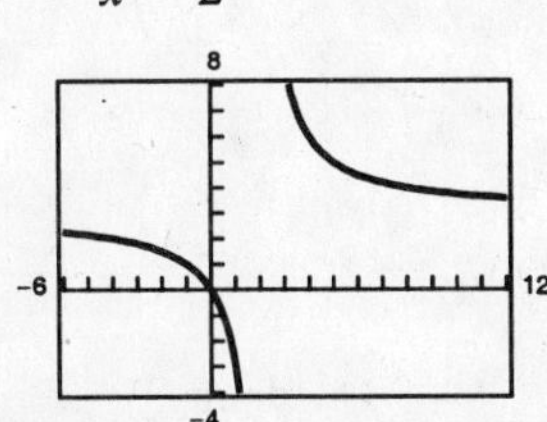

(a) $y \leq 0$ when $0 \leq x < 2$.

(b) $y \geq 6$ when $2 < x \leq 4$.

61. $y = \frac{2x^2}{x^2 + 4}$

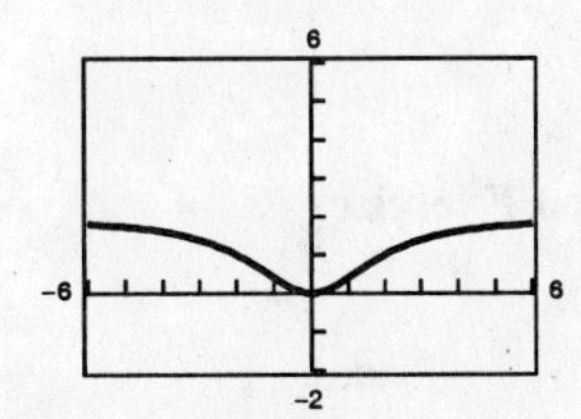

(a) $y \geq 1$ when $x \leq -2$ or $x \geq 2$.

This can also be expressed as $|x| \geq 2$.

(b) $y \leq 2$ for all real numbers x.

This can also be expressed as $-\infty < x < \infty$.

63. $\sqrt{x - 5}$

Need $x - 5 \geq 0$

$x \geq 5$

Domain: $[5, \infty)$

65. $\sqrt[3]{6 - x}$

Domain: all real x

67. $\sqrt[4]{6x + 15}$

Need $6x + 15 \geq 0$

$6x > -15$

$x \geq -\frac{5}{2}$

Domain: $\left[-\frac{5}{2}, \infty\right)$

69. (a), (b)

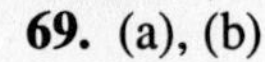

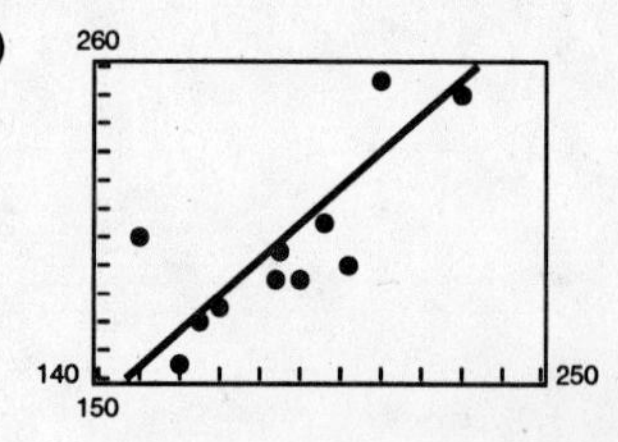

(c) For $y \geq 200$, $x \geq 186.23$ pounds

(d) The model is not accurate. The data is not linear. Other factors include muscle strength, height, physical condition, etc.

71. $\left|\frac{h - 68.5}{2.7}\right| \leq 1$

$|h - 68.5| \leq 2.7$

$-2.7 \leq h - 68.5 \leq 2.7$

$65.8 \leq h \leq 71.2$

h lies in the interval $[65.8, 71.2]$

73. (a) If $t = 2$, $u \approx 330$ vibrations per second.

(b) If $u = 600$, $t \approx 3.6$ mm.

(c) If $200 \leq u \leq 400$, then $1.2 \leq t \leq 2.4$.

(d) If $t < 3$, then $u < 500$ vibrations per second.

75. False. If $-10 \leq x \leq 8$, then $10 \geq -x$ and $-x \geq -8$.

77. (a) The polynomial is zero at $x = a$ and $x = b$.

(b)

Interval	*Sign of* $(x - a)$	*Sign of* $(x - b)$	*Sign of product*
$(-\infty, a)$	$-$	$-$	$+$
(a, b)	$+$	$-$	$-$
(b, ∞)	$+$	$+$	$+$

(c) The zeros of a polynomial are the only places where a polynomial can change signs.

Review for Chapter P

Solutions to Odd-Numbered Exercises

1. Quadrant IV

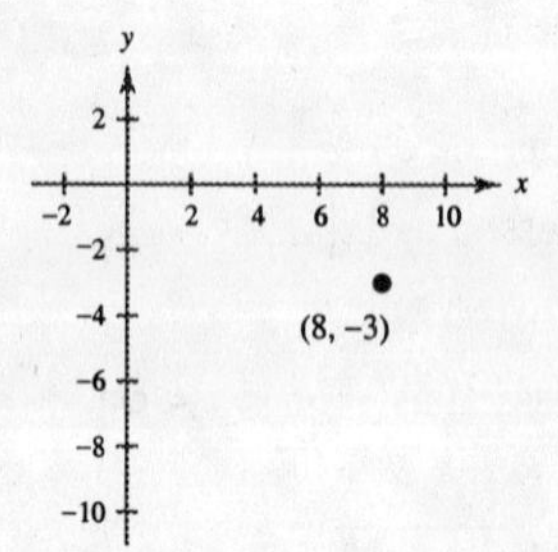

3. Quadrant II

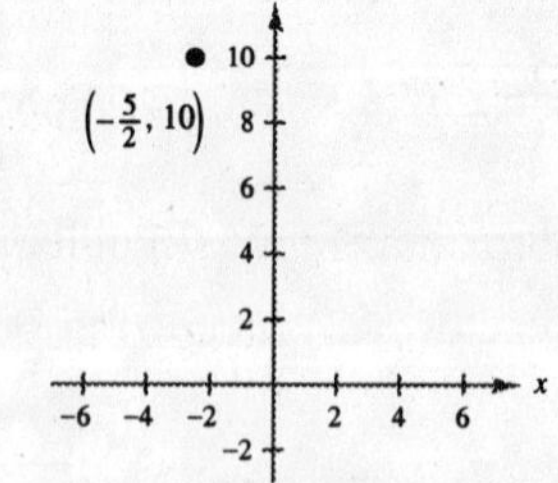

5. $x > 0$, $y = -2 \rightarrow$ Quadrant IV

7. (a)

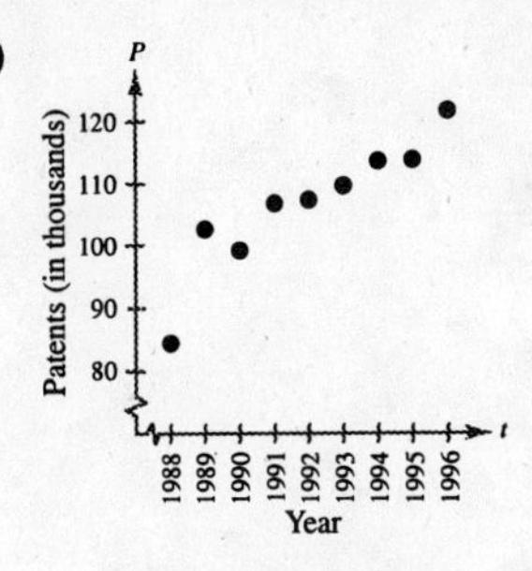

(b) The number of patents increases from 1990 to 1996.

9. $(-3, 8), \ (1, 5)$

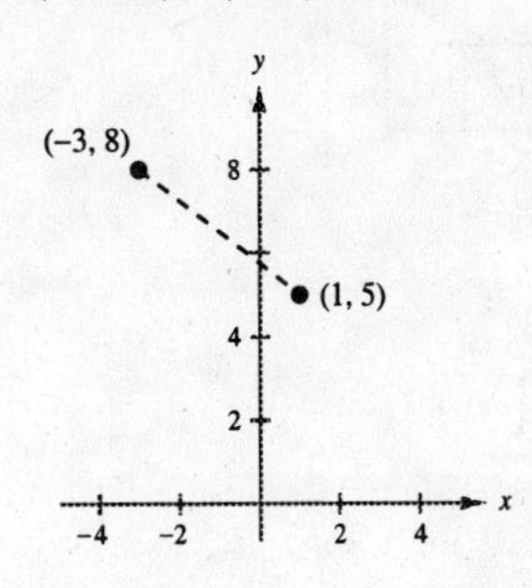

$$d = \sqrt{(1-(-3))^2 + (5-8)^2}$$
$$= \sqrt{4^2 + 3^2} = \sqrt{25} = 5$$

11. $d_1 = \sqrt{(22-3)^2 + (5-2)^2} = \sqrt{19^2 + 3^2} = \sqrt{370}$

$d_2 = \sqrt{(22-11)^2 + (5-13)^2} = \sqrt{11^2 + 8^2} = \sqrt{185}$

$d_3 = \sqrt{(11-3)^2 + (13-2)^2} = \sqrt{64 + 121} = \sqrt{185}$

$d_2^{\,2} + d_3^{\,2} = d_1^{\,2} = 370$

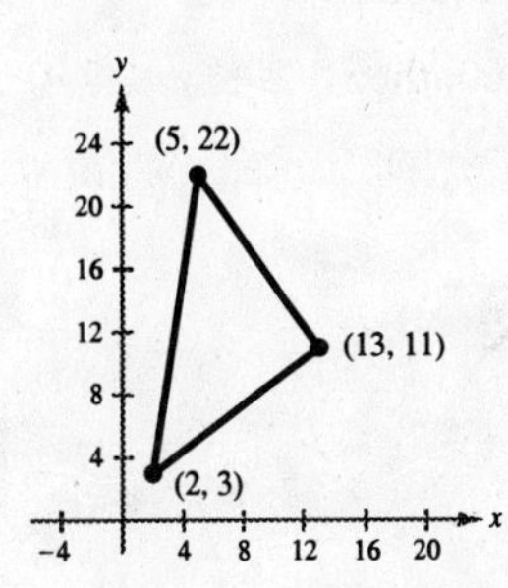

13.

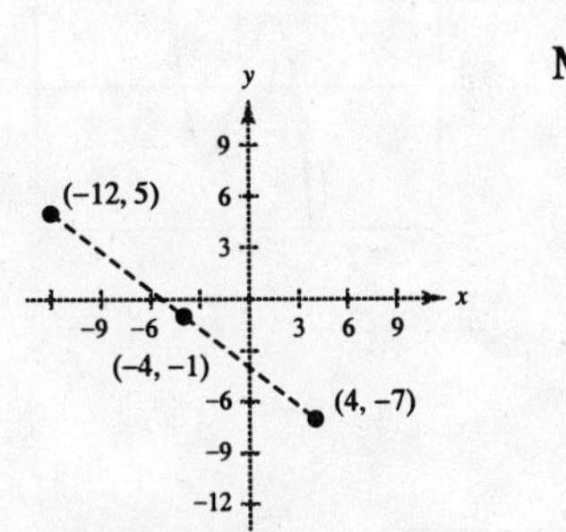

Midpoint: $\left(\dfrac{-12+4}{2}, \dfrac{5-7}{2}\right) = (-4, -1)$

15. (a) Using the midpoint to estimate 1997 revenues,

$$\left(\frac{1996 + 1998}{2}, \frac{329.5 + 375.2}{2}\right) = (1997, 352.35)$$

Revenues were approximately 352.35 million in 1997

(b) The estimate is fairly accurate: Error: $352.35 - 349.4 = 2.95$ million

17. Center: $\left(\dfrac{-4+1-}{2}, \dfrac{6-2}{2}\right) = (3, 2)$

Radius: $\frac{1}{2}\sqrt{(10+4)^2 + (-2-6)^2} = \frac{1}{2}\sqrt{14^2 + 8^2} = \sqrt{65}$

Circle: $(x-3)^2 + (y-2)^2 = 65$

19.

x	-1	0	1	2	3
y	4	0	-2	-2	0

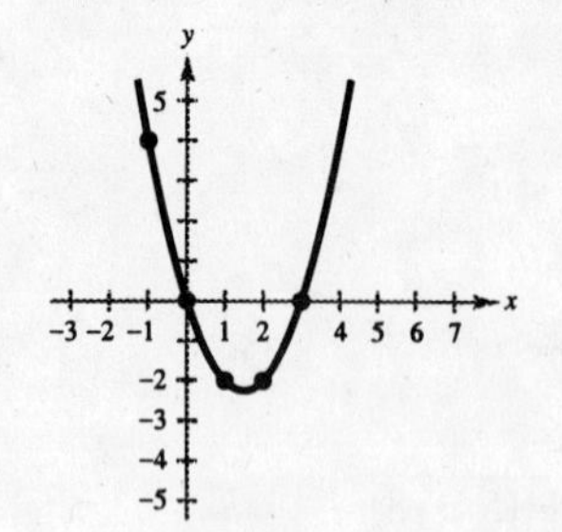

21.

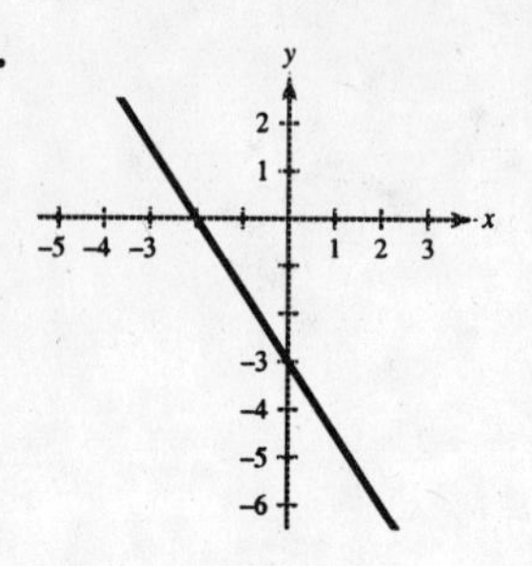

23.

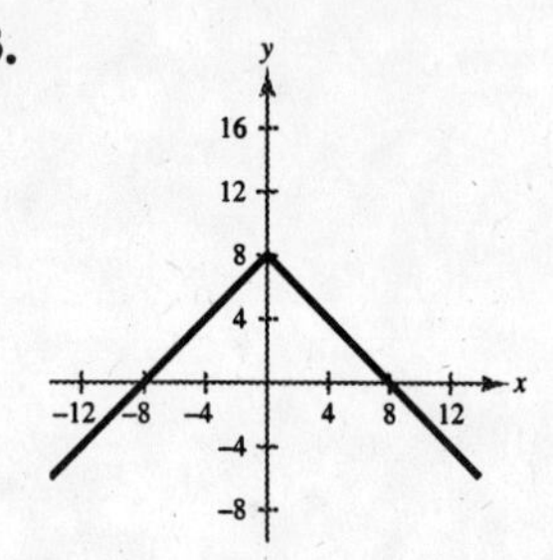

25.

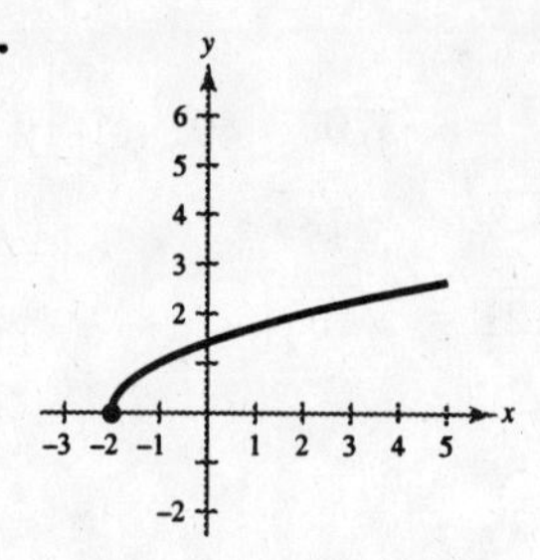

27.

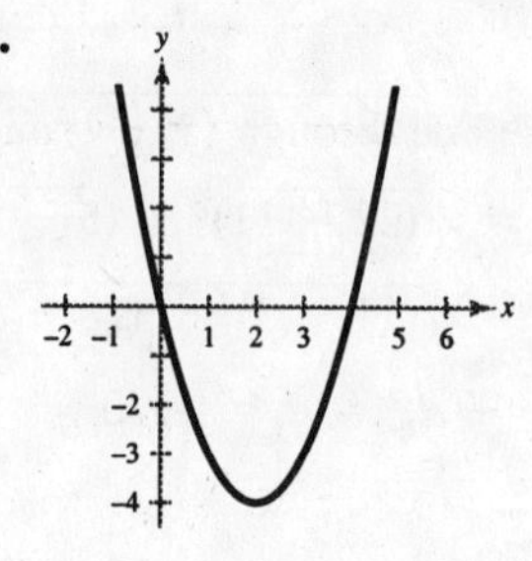

29.

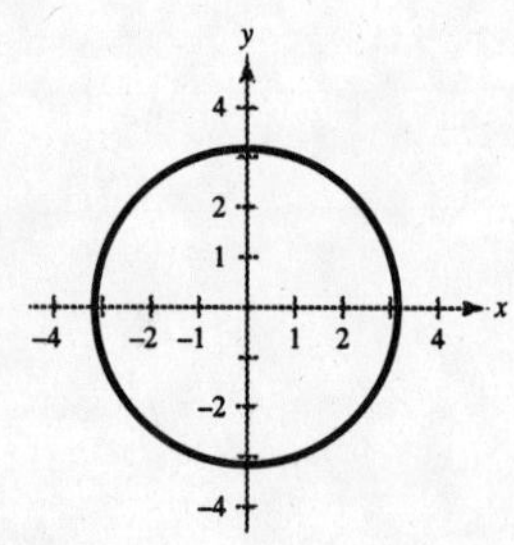

31.

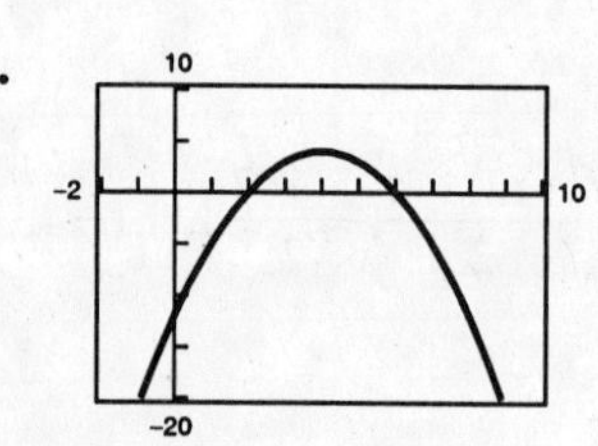

Intercepts:
$(6, 0), (2, 0), (0, -12)$

33.

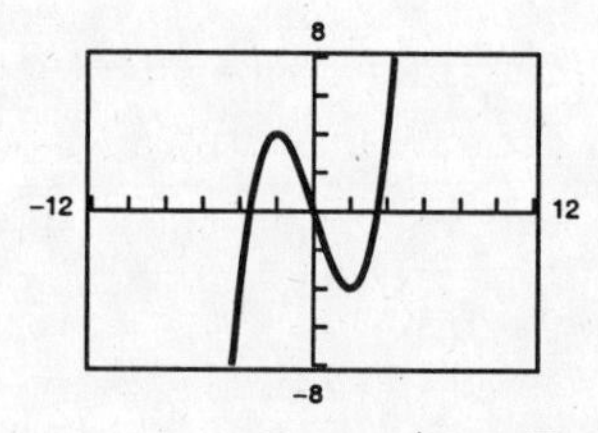

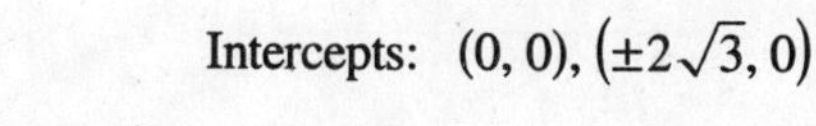

Intercepts: $(0, 0), (\pm 2\sqrt{3}, 0)$

35.

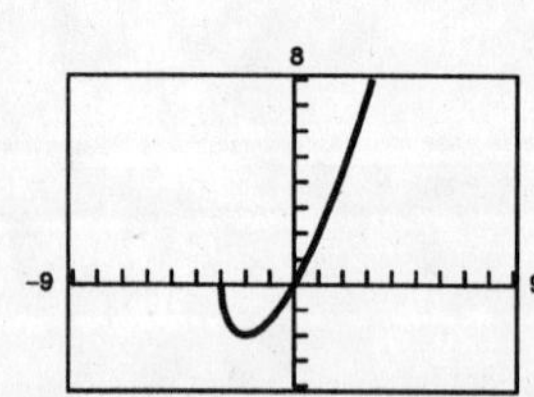

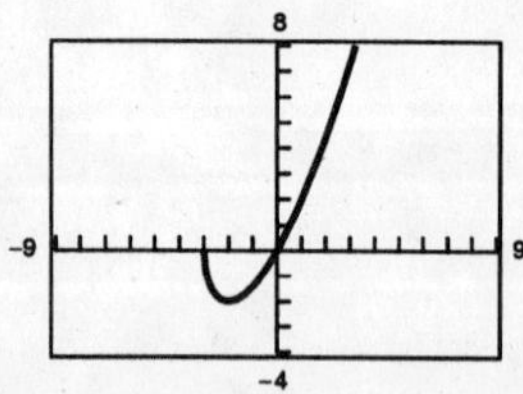

Intercepts: $(0, 0), (-3, 0)$

37.

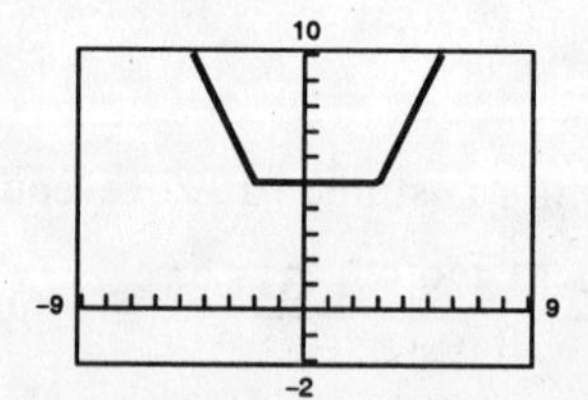

Intercept: $(0, 5)$

39. $m = \dfrac{2 - 2}{8 - (-3)} = \dfrac{0}{11} = 0$

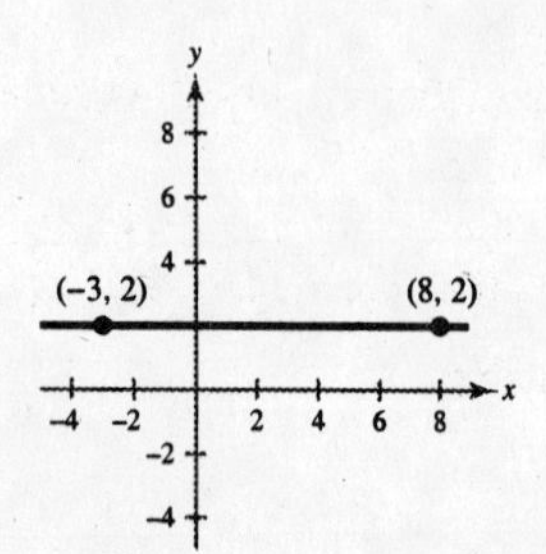

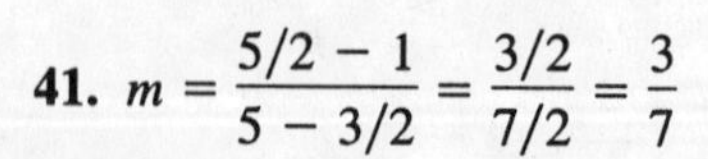

41. $m = \dfrac{5/2 - 1}{5 - 3/2} = \dfrac{3/2}{7/2} = \dfrac{3}{7}$

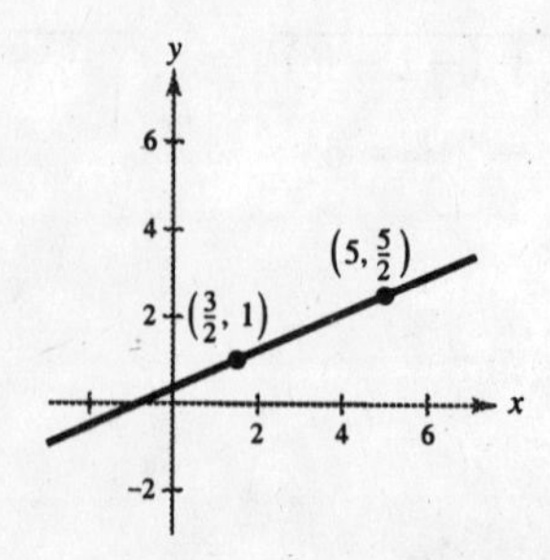

43. $(-4.5, 6), (2.1, 3)$

$$m = \frac{3 - 6}{2.1 - (-4.5)} = \frac{-3}{6.6} = -\frac{30}{66} = -\frac{5}{11}$$

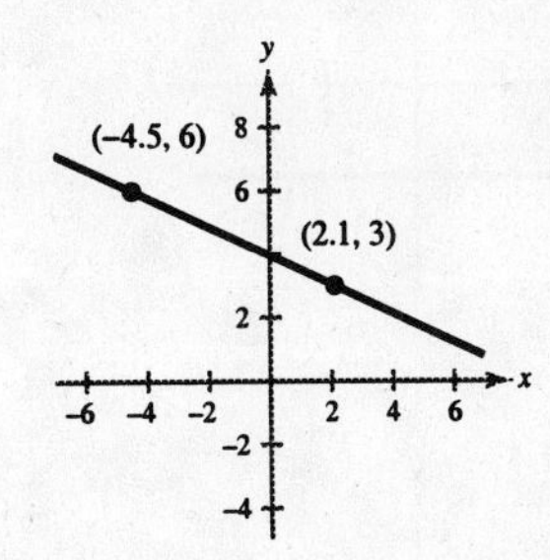

45. $(-2, 5), (0, t), (1, 1)$ are collinear.

$$\frac{t - 5}{0 - (-2)} = \frac{1 - 5}{1 - (-2)}$$

$$\frac{t - 5}{2} = \frac{-4}{3}$$

$$3(t - 5) = -8$$

$$3t - 15 = -8$$

$$3t = 7$$

$$t = \frac{7}{3}$$

47. The line through $(1, -4)$ and $(5, 10)$ is:

$$y + 4 = \frac{10 + 4}{5 - 1}(x - 1)$$

$$y + 4 = \frac{7}{2}(x - 1)$$

$$2y + 8 = 7(x - 1)$$

$$2y + 8 = 7x - 7$$

$$7x - 2y = 15$$

For $(t, 3)$ to be on this line also, it must satisfy the equation $7x - 2y = 15$.

$$7(t) - 2(3) = 15$$

$$7t = 21$$

Thus, $t = 3$.

49. (a) $y + 1 = \frac{1}{4}(x - 2)$

$$4y + 4 = x - 2$$

$$4y - x = -6$$

(b) Three additional points:

$(2 + 4, -1 + 1) = (6, 0)$

$(6 + 4, 0 + 1) = (10, 1)$

$(10 + 4, 1 + 1) = (14, 2)$

(other answers possible)

51. (a) $y + 5 = \frac{3}{2}(x - 0)$

$$2y + 10 = 3x$$

$$2y - 3x = -10$$

(b) Three additional points:

$(0 + 2, -5 + 3) = (2, -2)$

$(2 + 2, -2 + 3) = (4, 1)$

$(4 + 2, 1 + 3) = (6, 4)$

(other answers possible)

53. (a) $y + 5 = -1\left(x - \frac{1}{5}\right)$

$$y + 5 = -x + \frac{1}{5}$$

$$5y + 25 = -5x + 1$$

$$5x + 5y = -24$$

(b) Three additional points:

$\left(\frac{1}{5} + 1, -5 - 1\right) = \left(\frac{6}{5}, -6\right)$

$\left(\frac{6}{5} + 1, -6 - 1\right) = \left(\frac{11}{5}, -7\right)$

$\left(\frac{11}{5} + 1, -7 - 1\right) = \left(\frac{16}{5}, -8\right)$

(other answers possible)

55. (a) $y - 6 = 0(x + 2)$

$$y - 6 = 0$$

$$y = 6$$

(b) Three additional points:

$(0, 6), (1, 6), (2, 6)$

(other answers possible)

57. (a) m is undefined means that the line is vertical.

$$x = 10$$

(b) Three additional points: $(10, 0), (10, 1), (10, 2)$

(other answers possible)

59. (a) $y + 1 = \dfrac{-1+1}{4-2}(x-2) = 0(x-2) = 0 \Rightarrow y = -1$ (slope = 0) (b)

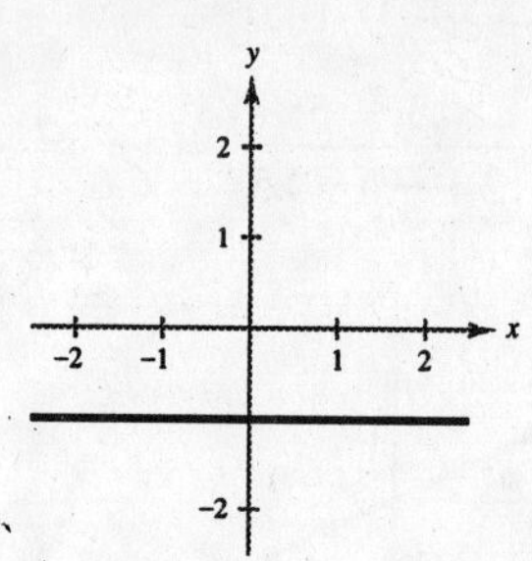

61. (a) $y - 1 = \dfrac{6-1}{14-2}(x-2) = \dfrac{5}{12}(x-2) = \dfrac{5}{12}x - \dfrac{5}{6} \Rightarrow y = \dfrac{5}{12}x + \dfrac{1}{6}$ (b)

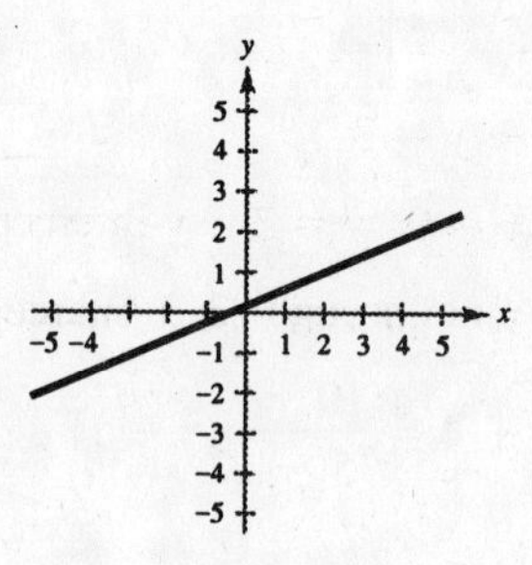

63. (a) $y - 0 = \dfrac{2-0}{6-(-1)}(x+1) = \dfrac{2}{7}(x+1) = \dfrac{2}{7}x + \dfrac{2}{7} \Rightarrow y = \dfrac{2}{7}x + \dfrac{2}{7}$ (b)

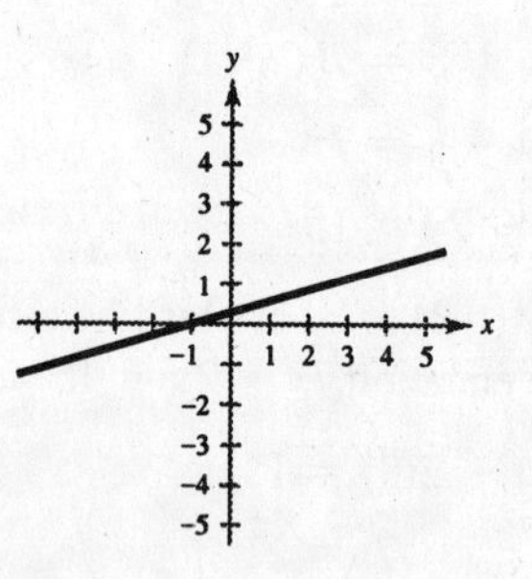

65. $5x - 4y = 8 \Rightarrow y = \frac{5}{4}x - 2$ and $m = \frac{5}{4}$

(a) Parallel slope: $m = \frac{5}{4}$

$$y - (-2) = \tfrac{5}{4}(x - 3)$$
$$4y + 8 = 5x - 15$$
$$0 = 5x - 4y - 23$$

(b) Perpendicular slope: $m = -\frac{4}{5}$

$$y - (-2) = -\tfrac{4}{5}(x - 3)$$
$$5y + 10 = -4x + 12$$
$$4x + 5y - 2 = 0$$

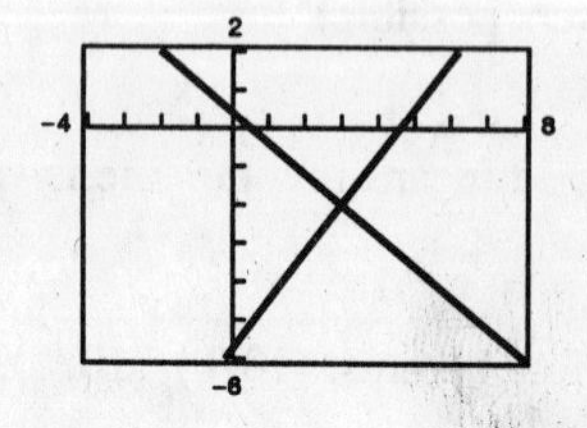

67. $x = 4$ is a vertical line; the slope is not defined.

(a) Parallel line: $x = -6$

(b) Perpendicular slope: $m = 0$

Perpendicular line:

$y - 2 = 0(x + 6) = 0 \Rightarrow y = 2$

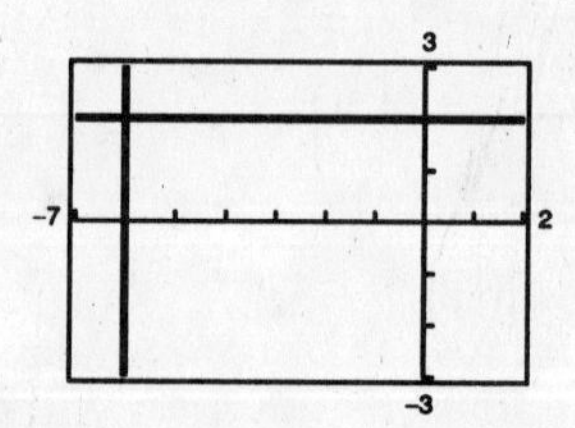

69. $14 + \dfrac{2}{x-1} = 10$

$$\frac{2}{x-1} = -4$$

$$2 = -4(x-1)$$

$$2 = -4x + 4$$

$$4x = 2$$

$$x = \frac{1}{2}$$

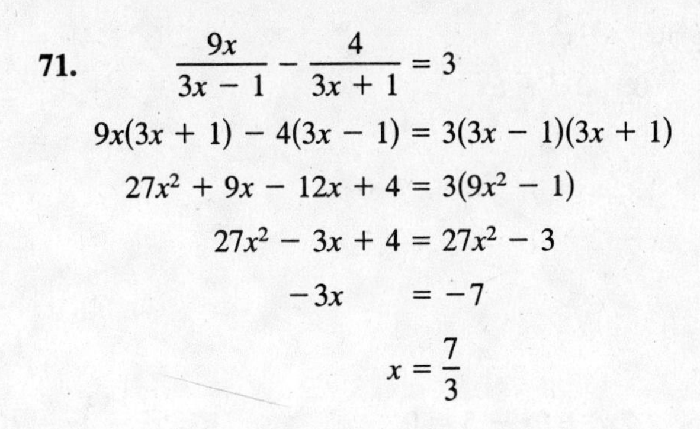

71. $\dfrac{9x}{3x-1} - \dfrac{4}{3x+1} = 3$

$$9x(3x+1) - 4(3x-1) = 3(3x-1)(3x+1)$$

$$27x^2 + 9x - 12x + 4 = 3(9x^2 - 1)$$

$$27x^2 - 3x + 4 = 27x^2 - 3$$

$$-3x = -7$$

$$x = \frac{7}{3}$$

73. $-x + y = 3$

Let $x = 0$: $y = 3$. y-intercept: $(0, 3)$

Let $y = 0$: $x = -3$. x-intercept: $(-3, 0)$

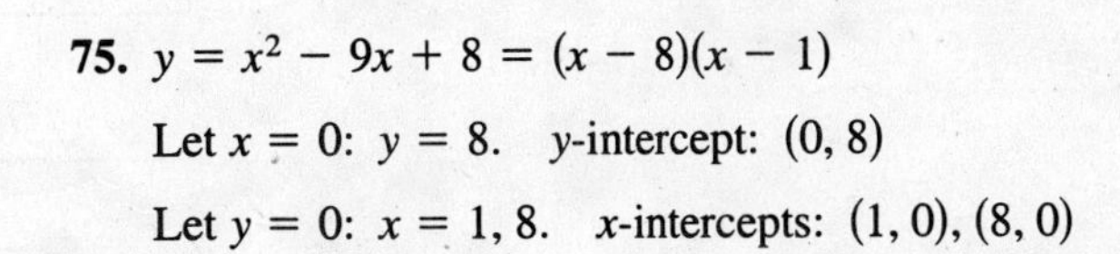

75. $y = x^2 - 9x + 8 = (x-8)(x-1)$

Let $x = 0$: $y = 8$. y-intercept: $(0, 8)$

Let $y = 0$: $x = 1, 8$. x-intercepts: $(1, 0), (8, 0)$

77. $y = -|x+5| - 2$

y-intercept: $(0, -7)$

No x-intercepts

79.

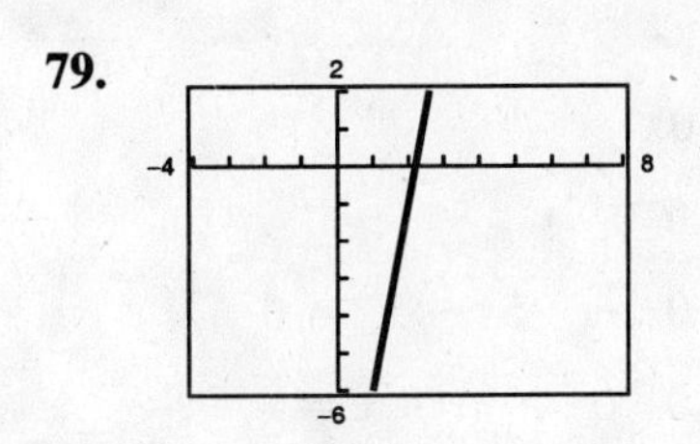

Solution: $x = 2.2$

81.

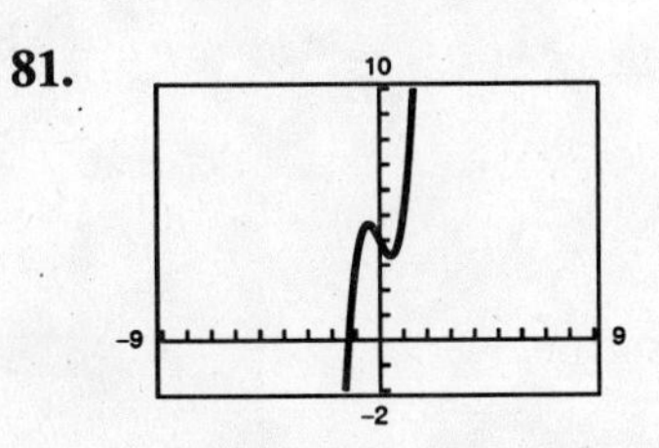

Solution: $x = -1.301$

83.

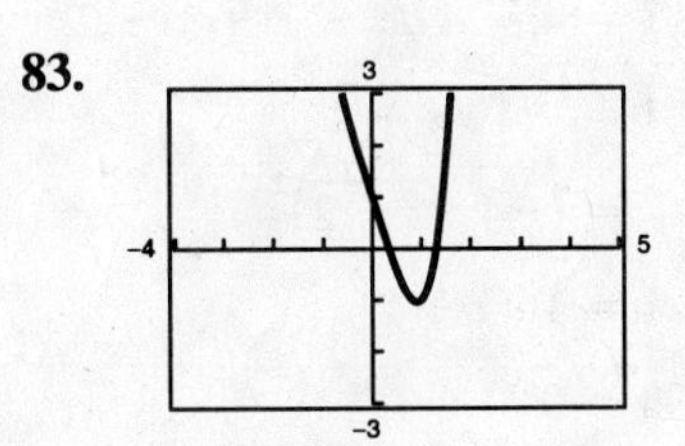

Solutions: $x = 1.307, x = 0.338$

85. $3x + 5y = -7$

$-x - 2y = 3$

From second equation, $x = -2y - 3$. Then

$3(-2y - 3) + 5y = 7$

$$-y - 9 = -7$$

$$y = -2 \text{ and } x = -2(-2) - 3 = 1$$

Intersection point $(1, -2)$

87. $x^2 + 2y = 14$

$3x + 4y = 1$

From equation 2, $y = \frac{1}{4}(1 - 3x)$. Then

$$x^2 + 2\left(\tfrac{1}{4}\right)(1 - 3x) = 14$$

$$x^2 + \tfrac{1}{2} - \tfrac{3}{2}x = 14$$

$$2x^2 - 3x - 27 = 0$$

$$(2x - 9)(x + 3) = 0$$

$$x = \tfrac{9}{2} \Rightarrow y = \tfrac{1}{4}\left(1 - 3\left(\tfrac{9}{2}\right)\right) = -\tfrac{25}{8}$$

$$x = -3 \Rightarrow y = \tfrac{1}{4}\left(1 - 3(-3)\right) = \tfrac{5}{2}$$

Intersection points: $\left(-3, \frac{5}{2}\right), \left(\frac{9}{2}, -\frac{25}{8}\right)$

89. $6x = 3x^2$

$0 = 3x^2 - 6x$

$0 = 3x(x - 2)$

$3x = 0 \Rightarrow x = 0$

$x - 2 = 0 \Rightarrow x = 2$

91. $(x + 4)^2 = 18$

$x + 4 = \pm\sqrt{18}$

$x = -4 \pm 3\sqrt{2}$

93. $x^2 - 12x + 30 = 0$

$x^2 - 12x = -30$

$x^2 - 12x + 36 = -30 + 36$

$(x - 6)^2 = 6$

$x - 6 = \pm\sqrt{6}$

$x = 6 \pm \sqrt{6}$

95. $2x^2 + 9x - 5 = 0$

$(2x - 1)(x + 5) = 0$

$x = \frac{1}{2}, -5$

97. $x^2 - 4x - 10 = 0$

$$x = \frac{4 \pm \sqrt{(-4)^2 - 4(-10)}}{2}$$

$$= \frac{4 \pm \sqrt{56}}{2}$$

$$= 2 \pm \sqrt{14}$$

99. $3x^3 - 26x^2 + 16x = 0$

$x(3x^2 - 26x + 16) = 0$

$x(3x - 2)(x - 8) = 0$

$x = 0, \frac{2}{3}, 8$

101. $5x^4 - 12x^3 = 0$

$x^3(5x - 12) = 0$

$x^3 = 0$ or $5x - 12 = 0$

$x = 0$ or $x = \frac{12}{5}$

103. $\sqrt{x + 4} = 3$

$\left(\sqrt{x + 4}\right)^2 = (3)^2$

$x + 4 = 9$

$x = 5$

105. $\sqrt{2x + 3} + \sqrt{x - 2} = 2$

$\left(\sqrt{2x + 3}\right)^2 = \left(2 - \sqrt{x - 2}\right)^2$

$2x + 3 = 4 - 4\sqrt{x - 2} + x - 2$

$x + 1 = -4\sqrt{x - 2}$

$(x + 1)^2 = \left(-4\sqrt{x - 2}\right)^2$

$x^2 + 2x + 1 = 16(x - 2)$

$x^2 - 14x + 33 = 0$

$(x - 3)(x - 11) = 0$

$x = 3$, extraneous or $x = 11$, extraneous

No solution. (You can verify that the graph of $y = \sqrt{2x + 3} + \sqrt{x - 2} - 2$ lies above the x-axis.)

107. $(x - 1)^{2/3} - 25 = 0$

$(x - 1)^{2/3} = 25$

$(x - 1)^2 = 25^3$

$x - 1 = \pm\sqrt{25^3}$

$x = 1 \pm 125$

$x = 126$ or $x = -124$

109. $3\left(1 - \dfrac{1}{5t}\right) = 0$

$1 - \dfrac{1}{5t} = 0$

$1 = \dfrac{1}{5t}$

$5t = 1$

$t = \dfrac{1}{5}$

111. $\dfrac{4}{(x-4)^2} = 1$

$4 = (x-4)^2$

$\pm 2 = x - 4$

$4 \pm 2 = x$

$x = 6 \quad \text{or} \quad x = 2$

113. $|x - 5| = 10$

$x - 5 = -10 \quad \text{or} \quad x - 5 = 10$

$x = -5 \qquad\qquad x = 15$

115. $|x^2 - 3| = 2x$

$x^2 - 3 = 2x$ OR $x^2 - 3 = -2x$

$x^2 - 2x - 3 = 0 \qquad x^2 + 2x - 3 = 0$

$(x-3)(x+1) = 0 \qquad (x+3)(x-1) = 0$

$x = 3 \text{ or } x = -1 \qquad x = -3 \text{ or } x = 1$

The only solutions to the original equation are $x = 3$ or $x = 1$. ($x = -3$ and $x = -1$ are extraneous.)

117. $8x - 3 < 6x + 15$

$2x < 18$

$x < 9$

119. $-2 < -x + 7 \le 10$

$-9 < -x \le 3$

$9 > x \ge -3$

$-3 \le x < 9$

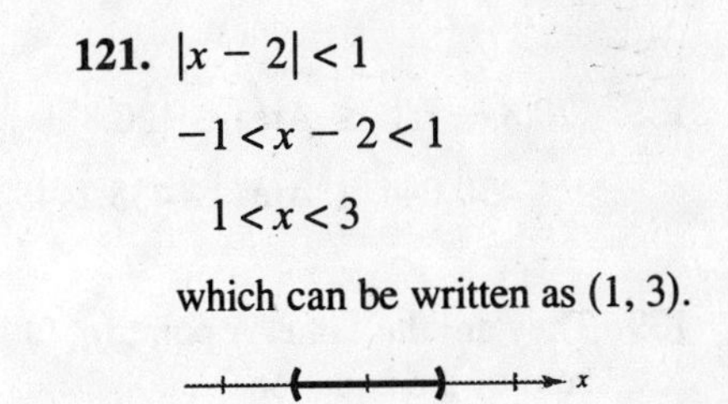

121. $|x - 2| < 1$

$-1 < x - 2 < 1$

$1 < x < 3$

which can be written as $(1, 3)$.

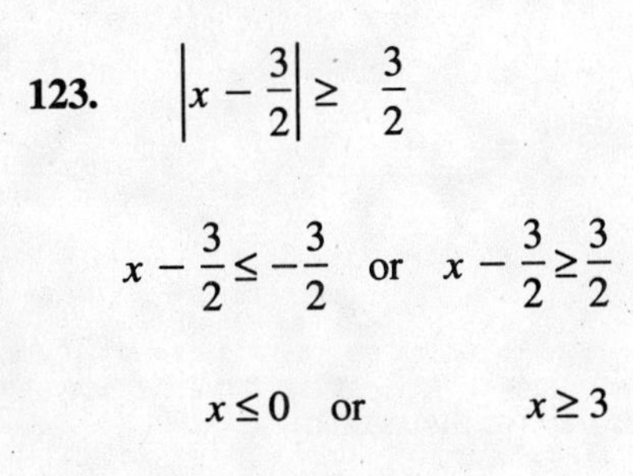

123. $\left|x - \dfrac{3}{2}\right| \ge \dfrac{3}{2}$

$x - \dfrac{3}{2} \le -\dfrac{3}{2} \quad \text{or} \quad x - \dfrac{3}{2} \ge \dfrac{3}{2}$

$x \le 0 \quad \text{or} \quad x \ge 3$

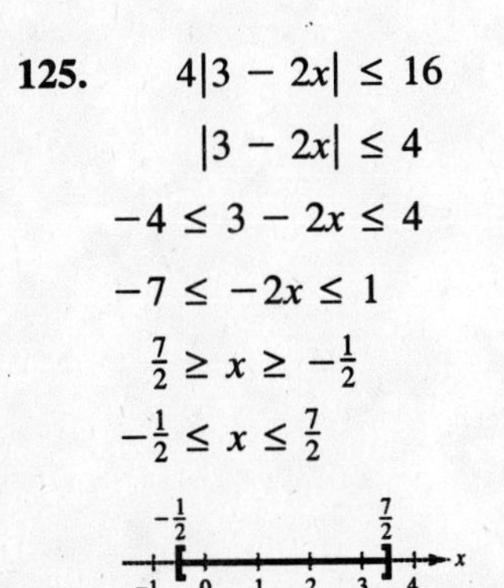

125. $4|3 - 2x| \le 16$

$|3 - 2x| \le 4$

$-4 \le 3 - 2x \le 4$

$-7 \le -2x \le 1$

$\frac{7}{2} \ge x \ge -\frac{1}{2}$

$-\frac{1}{2} \le x \le \frac{7}{2}$

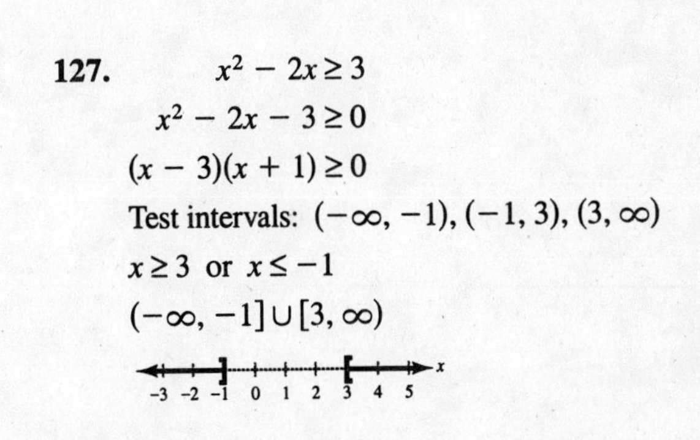

127. $x^2 - 2x \ge 3$

$x^2 - 2x - 3 \ge 0$

$(x-3)(x+1) \ge 0$

Test intervals: $(-\infty, -1), (-1, 3), (3, \infty)$

$x \ge 3 \text{ or } x \le -1$

$(-\infty, -1] \cup [3, \infty)$

129. $x^3 - 16x \geq 0$

$x(x-4)(x+4) \geq 0$

Critical numbers: 0, 4, −4. Testing the four intervals, we obtain $-4 \leq x \leq 0$ or $x \geq 4$

-4 -2 0 2 4 6 x

131. $\dfrac{x-5}{3-x} < 0$

Critical numbers: $x = 5, x = 3$

Test intervals: $(-\infty, 3), (3, 5), (5, \infty)$

Test: Is $\dfrac{x-5}{3-x} < 0$?

Solution set: $(-\infty, 3) \cup (5, \infty)$

2 3 4 5 6 x

133. $\dfrac{3x+8}{x-3} - 4 \leq 0$

$$\frac{3x + 8 - 4(x-3)}{x-3} \leq 0$$

$$\frac{20 - x}{x-3} \leq 0$$

$$\frac{x-20}{x-3} \geq 0$$

3

0 5 10 15 20 25 x

Critical numbers: $x = 3, 20$. Testing the three intervals, we obtain $x \geq 20$ or $x < 3$.

135. $\left(20.8 - \frac{1}{16}\right)^2 \leq \text{Area} \leq \left(20.8 + \frac{1}{16}\right)^2$

$430.044 \leq \text{Area} \leq 435.244$ square inches

137. True. For example, $x^2 + y^2 = 1$

139. They are the same. A point $(a, 0)$ is an x-intercept if it is a solution point of the equation. In other words, a is a zero of the equation.

CHAPTER 1
Functions and Their Graphs

CHAPTER 1
Functions and Their Graphs

Section 1.1 Functions

- Given a set or an equation, you should be able to determine if it represents a function.
- Given a function, you should be able to do the following.
  (a) Find the domain.
  (b) Evaluate it at specific values.

Solutions to Odd-Numbered Exercises

1. Yes, it does represent a function. Each domain value is matched with only one range value.

3. No, it does not represent a function. The domain values are each matched with three range values.

5. Yes, it does represent a function. Each input value is matched with only one output value.

7. No, it does not represent a function. The input values of 10 and 7 are each matched with two output values.

9. (a) Each element of A is matched with exactly one element of B, so it does represent a function.

(b) The element 1 in A is matched with two elements, -2 and 1 of B, so it does not represent a function.

(c) Each element of A is matched with exactly one element of B, so it does represent a function.

(d) The element 2 of A is not matched to any element of B, so it does not represent a function.

11. Each are functions. For each year there corresponds one and only one circulation.

13. $x^2 + y^2 = 4 \implies y = \pm\sqrt{4 - x^2}$

Thus, y *is not* a function of x. For instance, the values $y = 2$ and -2 both correspond to $x = 0$.

15. $x^2 + y = 4 \implies y = 4 - x^2$

Thus, y *is* a function of x.

17. $2x + 3y = 4 \implies y = \frac{1}{3}(4 - 2x)$

Thus, y *is* a function of x.

19. $y^2 = x^2 - 1 \implies y = \pm\sqrt{x^2 - 1}$

Thus, y *is not* a function of x. For instance, the values $y = \sqrt{3}$ and $-\sqrt{3}$ both correspond to $x = 2$.

21. $y = |4 - x|$ This is a function of x.

23. $x = 4$ does not represent y as a function of x. All values of y correspond to $x = 4$.

25. $f(x) = \dfrac{1}{x + 1}$

(a) $f(4) = \dfrac{1}{(4) + 1} = \dfrac{1}{5}$

(b) $f(0) = \dfrac{1}{(0) + 1} = 1$

(c) $f(4t) = \dfrac{1}{(4t) + 1} = \dfrac{1}{4t + 1}$

(d) $f(x + c) = \dfrac{1}{(x + c) + 1} = \dfrac{1}{x + c + 1}$

27. $f(x) = 2x - 3$

(a) $f(1) = 2(1) - 3 = -1$

(b) $f(-3) = 2(-3) - 3 = -9$

(c) $f(x - 1) = 2(x - 1) - 3 = 2x - 5$

29. $h(t) = t^2 - 2t$

(a) $h(2) = 2^2 - 2(2) = 0$

(b) $h(1.5) = (1.5)^2 - 2(1.5) = -0.75$

(c) $h(x + 2) = (x + 2)^2 - 2(x + 2) = x^2 + 2x$

31. $f(y) = 3 - \sqrt{y}$

(a) $f(4) = 3 - \sqrt{4} = 1$

(b) $f(0.25) = 3 - \sqrt{0.25} = 2.5$

(c) $f(4x^2) = 3 - \sqrt{4x^2} = 3 - 2|x|$

33 $q(x) = \dfrac{1}{x^2 - 9}$

(a) $q(0) = \dfrac{1}{0^2 - 9} = -\dfrac{1}{9}$

(b) $q(3) = \dfrac{1}{3^2 - 9}$ is undefined.

(c) $q(y + 3) = \dfrac{1}{(y + 3)^2 - 9} = \dfrac{1}{y^2 + 6y}$

35. $f(x) = \dfrac{|x|}{x}$

(a) $f(2) = \dfrac{|2|}{2} = 1$

(b) $f(-2) = \dfrac{|-2|}{-2} = -1$

(c) $f(x^2) = \dfrac{|x^2|}{x^2} = 1,\ x \neq 0$

37. $f(x) = \begin{cases} 2x + 1, & x < 0 \\ 2x + 2, & x \geq 0 \end{cases}$

(a) $f(-1) = 2(-1) + 1 = -1$

(b) $f(0) = 2(0) + 2 = 2$

(c) $f(2) = 2(2) + 2 = 6$

39. $f(x) = x^2 - 3$

x	-2	-1	0	1	2
$f(x)$	1	-2	-3	-2	1

41. $h(t) = \frac{1}{2}|t + 3|$

t	-5	-4	-3	-2	-1
$h(t)$	1	$\frac{1}{2}$	0	$\frac{1}{2}$	1

43. $f(x) = \begin{cases} -\frac{1}{2}x + 4, & x \leq 0 \\ (x - 2)^2, & x > 0 \end{cases}$

x	-2	-1	0	1	2
$f(x)$	5	$\frac{9}{2}$	4	1	0

45. $15 - 3x = 0$

$3x = 15$

$x = 5$

47. $f(x) = \dfrac{3x - 4}{5} = 0$

$3x - 4 = 0$

$3x = 4$

$x = \frac{4}{3}$

49. $x^2 - 9 = 0$

$x^2 = 9$

$x = \pm 3$

51. $f(x) = \sqrt{x^2 - 16} = 0$

$x^2 - 16 = 0$

$x^2 = 16$

$x = \pm 4$

53.
$$f(x) = g(x)$$
$$x^2 = x + 2$$
$$x^2 - x - 2 = 0$$
$$(x + 1)(x - 2) = 0$$
$$x = -1 \text{ or } x = 2$$

55.
$$f(x) = g(x)$$
$$\sqrt{3x} + 1 = x + 1$$
$$\sqrt{3x} = x$$
$$3x = x^2$$
$$0 = x^2 - 3x$$
$$0 = x(x - 3)$$
$$x = 0 \text{ or } x = 3$$

57. $f(x) = 5x^2 + 2x - 1$

Since $f(x)$ is a polynomial, the domain is all real numbers x.

59. $h(t) = \dfrac{4}{t}$

Domain: All real numbers except $t = 0$

61. $g(y) = \sqrt{y - 10}$

Domain: $y - 10 \geq 0$

$y \geq 10$

63. $f(x) = \sqrt[4]{1 - x^2}$

Domain: $1 - x^2 \geq 0$

$x^2 - 1 \leq 0$

$-1 \leq x \leq 1$

65. $g(x) = \dfrac{1}{x} - \dfrac{1}{x + 2}$

Domain: All real numbers except

$x = 0,\ x = -2.$

67. $f(s) = \dfrac{\sqrt{s - 1}}{s - 4}$

Domain: $s - 1 \geq 0$ and $s - 4 \neq 0$. That is, all real numbers $s \geq 1, s \neq 4$.

69. $f(x) = \dfrac{\sqrt[3]{x - 4}}{x}$. Domain: all $x \neq 0$.

71. $f(x) = x^2$

$\{(-2, 4), (-1, 1), (0, 0), (1, 1), (2, 4)\}$

73. $f(x) = \sqrt{x + 2}$

$\{(-2, 0), (-1, 1), (0, \sqrt{2}), (1, \sqrt{3}), (2, 2)\}$

75. By plotting the points, we have a parabola, so $g(x) = cx^2$. Since $(-4, -32)$ is on the graph, we have $-32 = c(-4)^2 \implies c = -2$. Thus, $g(x) = -2x^2$.

77. Since the function is undefined at 0, we have $r(x) = \dfrac{c}{x}$. Since $(-8, -4)$ is on the graph, we have $-4 = \dfrac{c}{-8} \Rightarrow c = 32$. Thus, $r(x) = \dfrac{32}{x}$.

79. $f(x) = 2x$

$f(x + c) = 2(x + c) = 2x + 2c$

$f(x + c) - f(x) = (2x + 2c) - 2x = 2c$

$\dfrac{f(x + c) - f(x)}{c} = \dfrac{2c}{c} = 2, c \neq 0$

81. $f(x) = x^2 - x + 1$

$$f(2 + h) = (2 + h)^2 - (2 + h) + 1$$
$$= 4 + 4h + h^2 - 2 - h + 1$$
$$= h^2 + 3h + 3$$

$f(2) = (2)^2 - 2 + 1 = 3$

$f(2 + h) - f(2) = h^2 + 3h$

$\dfrac{f(2 + h) - f(2)}{h} = h + 3,\ h \neq 0$

83. $f(x) = x^3$

$f(x + c) = (x + c)^3 = x^3 + 3x^2c + 3xc^2 + c^3$

$$\frac{f(x + c) - f(x)}{c} = \frac{(x^3 + 3x^2c + 3xc^2 + c^3) - x^3}{c}$$

$$= \frac{c(3x^2 + 3xc + c^2)}{c}$$

$$= 3x^2 + 3xc + c^2, \quad c \neq 0$$

85. $f(t) = \dfrac{1}{t}$

$f(1) = 1$

$$\frac{f(t) - f(1)}{t - 1} = \frac{\frac{1}{t} - 1}{t - 1} = \frac{1 - t}{t(t - 1)} = -\frac{1}{t}, t \neq 1$$

87. $A = \pi r^2, \quad C = 2\pi r$

$$r = \frac{C}{2\pi}$$

$$A = \pi\left(\frac{C}{2\pi}\right)^2 = \frac{C^2}{4\pi}$$

89. Area $= A = \frac{1}{2}bh = \frac{1}{2}(s)(s) = \dfrac{s^2}{2}$

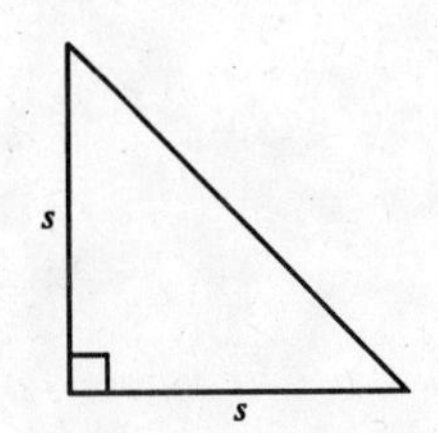

91. (a)

Height, x	Width	Volume, V
1	$24 - 2(1)$	$1[24 - 2(1)]^2 = 484$
2	$24 - 2(2)$	$2[24 - 2(2)]^2 = 800$
3	$24 - 2(3)$	$3[24 - 2(3)]^2 = 972$
4	$24 - 2(4)$	$4[24 - 2(4)]^2 = 1024$
5	$24 - 2(5)$	$5[24 - 2(5)]^2 = 980$
6	$24 - 2(6)$	$6[24 - 2(6)]^2 = 864$

The volume is maximum when $x = 4$.

(b)

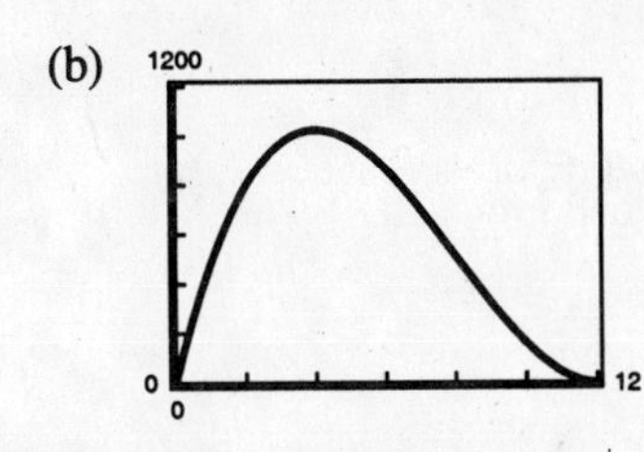

$V = x(24 - 2x)^2$

Domain: $0 < x < 12$

(c) $V(9) = 324$; $V(10) = 160$

(d) $V(9) = 9(24 - 2(9))^2 = 9(36) = 324$

$V(10) = 10(24 - 2(10))^2 = 10(16) = 160$

93. $A = \frac{1}{2}(\text{base})(\text{height}) = \frac{1}{2}xy.$

Since $(0, y)$, $(2, 1)$ and $(x, 0)$ all lie on the same line, the slopes between any pair of points are equal.

$$\frac{1-y}{2-0} = \frac{1-0}{2-x}$$

$$1 - y = \frac{2}{2-x}$$

$$y = 1 - \frac{2}{2-x} = \frac{x}{x-2}$$

Therefore, $A = \frac{1}{2}xy = \frac{1}{2}x\left(\frac{x}{x-2}\right) = \frac{x^2}{2x-4}$

The domain is $x > 2$, since $A > 0$.

95. (a) $V = (\text{length})(\text{width})(\text{height}) = yx^2$

But, $y + 4x = 108$, or $y = 108 - 4x$.

Thus, $V = (108 - 4x)x^2$.

(b) Since $y = 108 - 4x > 0$

$$4x < 108$$

$$x < 27$$

Domain: $0 < x < 27$

(c) 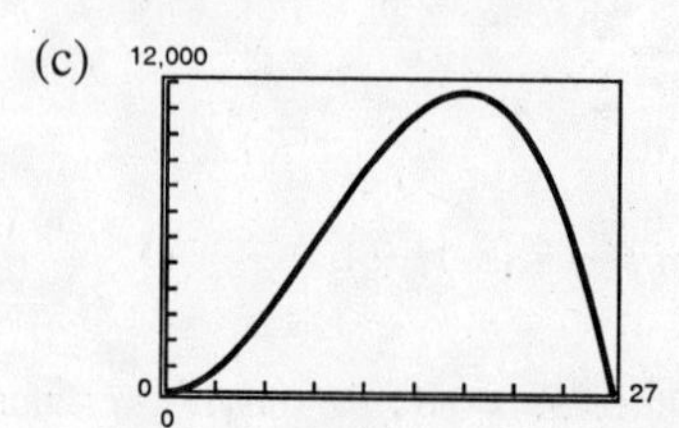

(d) The highest point on the graph occurs at $x = 18$. The dimensions that maximize the volume are $18 \times 18 \times 36$ inches.

97. (a) Cost = variable costs + fixed costs

$$C = 12.30x + 98{,}000$$

(b) Revenue = price per unit $\times$ number of units

$$R = 17.98x$$

(c) Profit = Revenue − Cost

$$P = 17.98x - (12.30x + 98{,}000)$$

$$P = 5.68x - 98{,}000$$

99. (a) $R = (\text{rate})(\text{number of people})$

$$= [8 - 0.05(n - 80)]n$$

$$= (12 - 0.05n)n = \frac{240n - n^2}{20}$$

(b)

n	90	100	110	120	130	140	150
$R(n)$	675	700	715	720	715	700	675

The revenue increases, and then decreases. The maximum revenue occurs when $n = 120$.

(c) 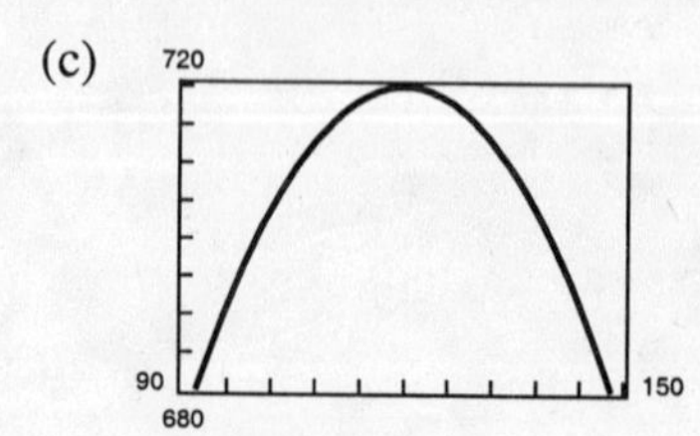

The maximum occurs at $n = 120$.

101. (a)

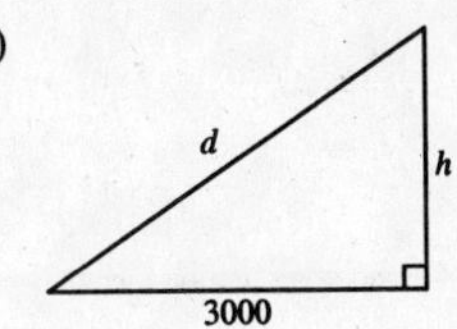

(b) $3000^2 + h^2 = d^2$

$$h^2 = d^2 - 3000^2$$

$$h = \sqrt{d^2 - 3000^2},\ d \geq 3000$$

(c)

20,000

3000 0 20,000

(d) When $d = 10{,}000$, $h \approx 9539.4$ feet.

Algebraically, $h = \sqrt{10{,}000^2 - 3000^2} = \sqrt{91{,}000{,}000}$

$$\approx 9539.4 \text{ feet}$$

103. False. The range of $f(x)$ is $[-1, \infty)$.

105. No. The element 3 in A has two images in B, u and v.

107. An advantage of function notation is that it gives a name to the relationship so it can easily be referenced. When evaluating a function, you see both the input and output values.

109.

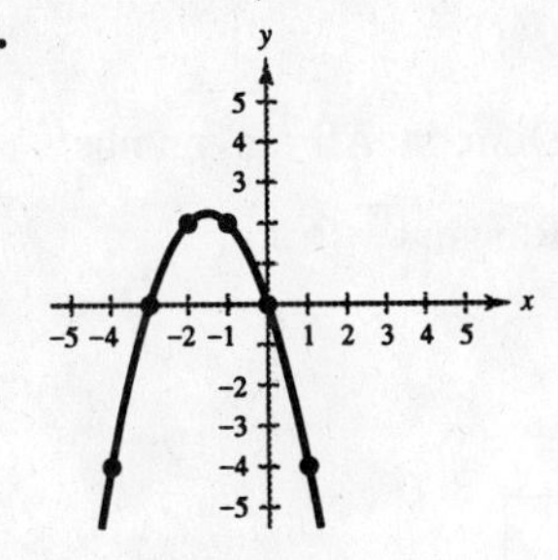

111.

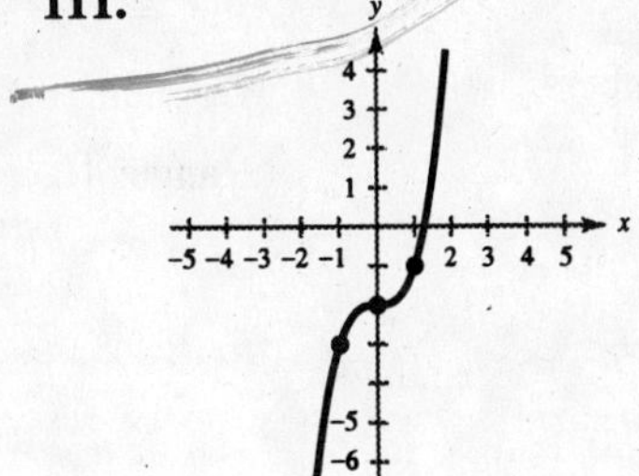

113. Center: $(-8, -5)$

Radius: $\dfrac{3}{4}$

$$(x - (-8))^2 + (y - (-5))^2 = \left(\frac{3}{4}\right)^2$$

$$(x + 8)^2 + (y + 5)^2 = \frac{9}{16}$$

115. Endpoints of dimeter: $(6, -5), (-2, 7)$

Center: $\left(\dfrac{6 - 2}{2}, \dfrac{-5 + 7}{2}\right) = (2, 1)$

Radius: $\sqrt{(6 - 2)^2 + (-5 - 1)^2} = \sqrt{16 + 36}$

$$= \sqrt{52}$$

$$(x - 2)^2 + (y - 1)^2 = 52$$

Section 1.2 Graphs of Functions

- You should be able to determine the domain and range of a function from its graph.
- You should be able to use the vertical line test for functions.
- You should be able to determine when a function is constant, increasing, or decreasing.
- You should be able to find relative maximum and minimum values of a function.
- You should know that f is
  (a) Odd if $f(-x) = -f(x)$.
  (b) Even if $f(-x) = f(x)$.

Solutions to Odd-Numbered Exercises

1. $g(x) = 1 - x^2$

Domain: All real numbers

Range: $(-\infty, 1]$

3. $f(x) = \sqrt{x^2 - 1}$

Domain: $(-\infty, -1] \cup [1, \infty)$

Range: $[0, \infty)$

5. $f(x) = \frac{1}{2}|x - 2|$

Domain: All real numbers

Range: $[0, \infty]$

7. $f(x) = 2x^2 + 3$

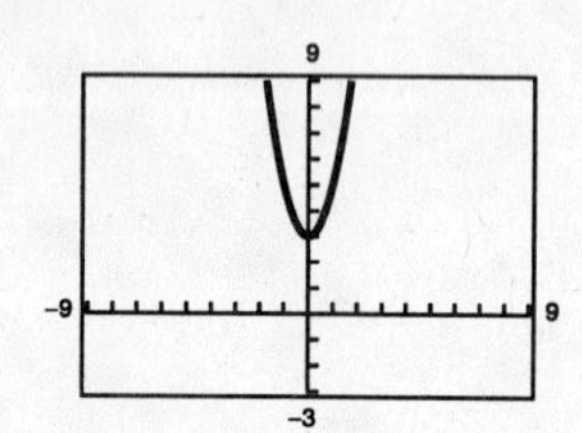

Domain: All real numbers

Range: $[3, \infty]$

9. $f(x) = \sqrt{x - 1}$

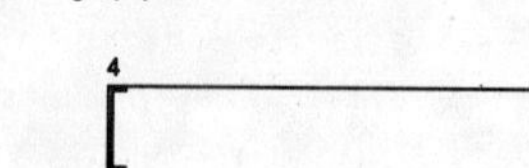
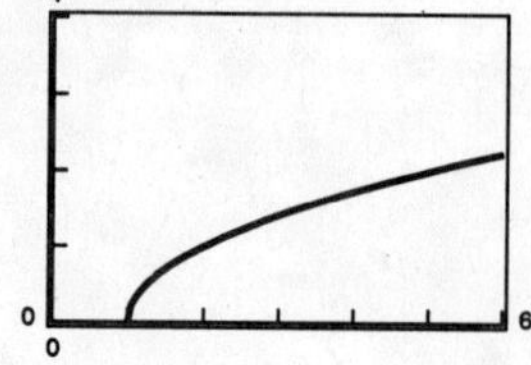

Domain: $x - 1 \geq 0 \Rightarrow x \geq 1$ or $[1, \infty)$

Range: $[0, \infty)$

11. $f(x) = |x + 3|$

Domain: All real numbers

Range: $[0, \infty)$

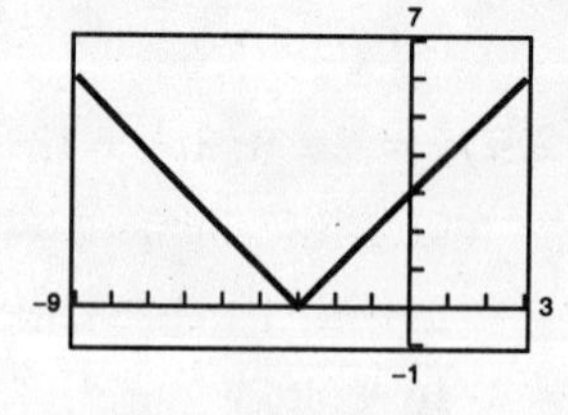

13. $y = \frac{1}{2}x^2$

A vertical line intersects the graph just once, so y is a function of x.

15. $x - y^2 = 1 \quad \Rightarrow \quad y = \pm\sqrt{x - 1}$

y is not a function of x. Graph

$y_1 = \sqrt{x - 1}$ and $y_2 = -\sqrt{x - 1}$.

17. $x^2 = 2xy - 1$

A vertical line intersects the graph just once, so y is a function of x. Solve for y and graph

$$y = \frac{x^2 + 1}{2x}$$

19. $f(x) = \frac{3}{2}x$

(a) f is increasing on $(-\infty, \infty)$.

(b) Since $f(-x) = -f(x)$, f is odd.

21. $f(x) = x^3 - 3x^2 + 2$

(a) f is increasing on $(-\infty, 0)$ and $(2, \infty)$.

f is decreasing on $(0, 2)$.

(b) $f(-x) \neq -f(x)$

$f(-x) \neq f(x)$

f is neither odd nor even.

23. $f(x) = 3x^4 - 6x^2$

(a)

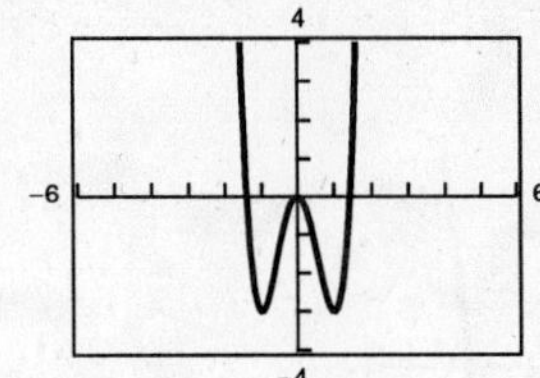

(b) Increasing on $(-1, 0)$ and $(1, \infty)$

Decreasing on $(-\infty, -1)$ and $(0, 1)$

(c) Since $f(-x) = f(x)$, f is even.

25. $f(x) = x^{2/3}$

(a)

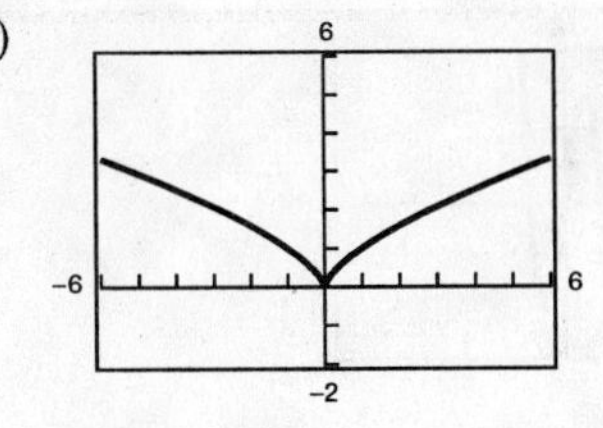

(b) Increasing on $(0, \infty)$

Decreasing on $(-\infty, 0)$

(c) $f(-x) = (-x)^{2/3} = x^{2/3} = f(x) \Rightarrow$
The function is even.

27. $f(x) = x\sqrt{x + 3}$

(a)

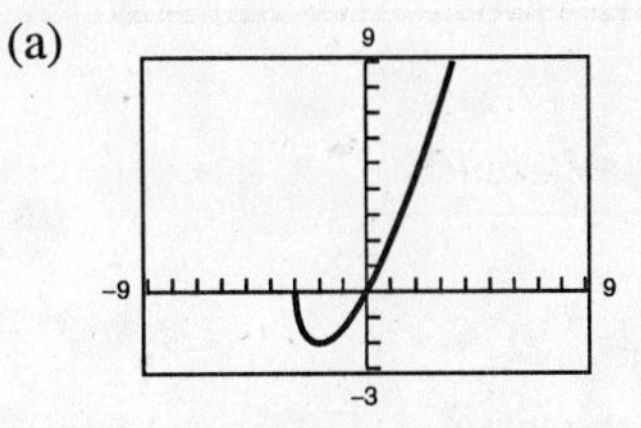

(b) Increasing on $(-2, \infty)$

Decreasing on $(-3, -2)$

(c) $f(-x) \neq -f(x)$

$f(-x) \neq f(x)$

f is neither odd nor even.

29. $f(x) = |x + 1| + |x - 1|$

(a)

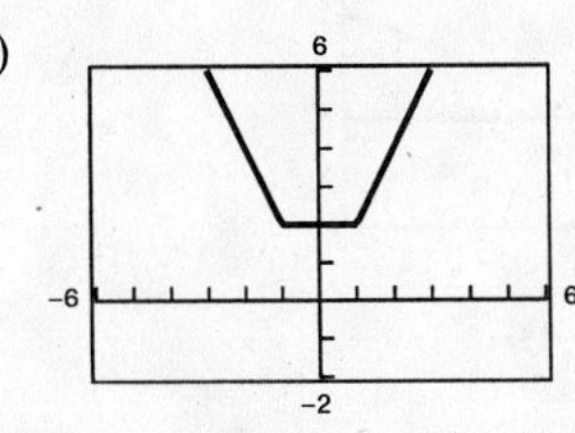

(b) Increasing on $(1, \infty)$, constant on $(-1, 1)$, decreasing on $(-\infty, -1)$

(c) $f(-x) = |-x + 1| + |-x - 1|$

$= |(-1)(x - 1)| + |(-1)(x + 1)|$

$= |x - 1| + |x + 1| = f(x)$

$\Rightarrow$ The function is even.

31. $f(x) = x^2 - 6x$

Relative minimum: $(3, -9)$

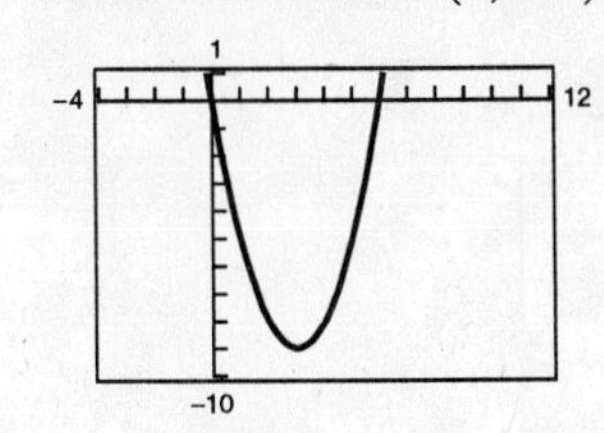

33. $y = 2x^3 + 3x^2 - 12x$

Relative minimum: $(1, -7)$

Relative maximum: $(-2, 20)$

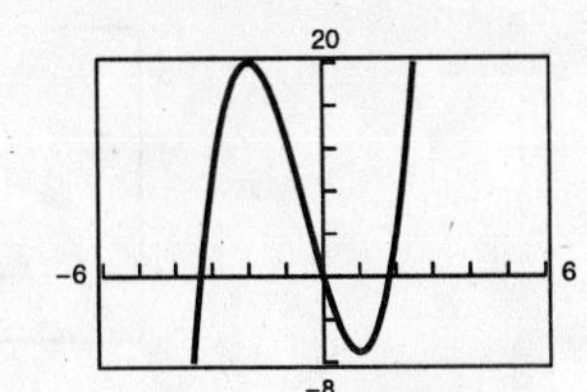

35. $h(x) = (x - 1)\sqrt{x}$

Relative minimum: $(0.33, -0.38)$

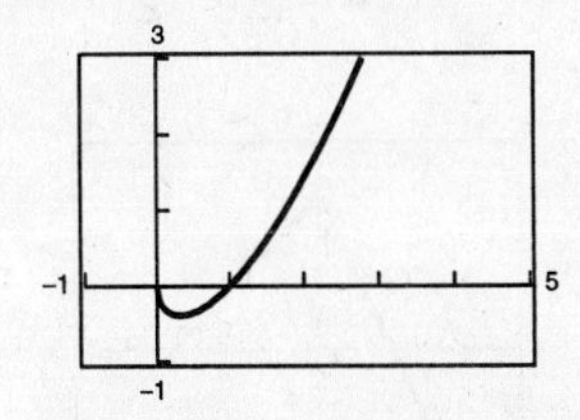

($(0, 0)$ is not a relative maximum because it occurs at the endpoint of the domain $[0, \infty)$.)

37. $f(x) = x^2 - 4x - 5$

(a)

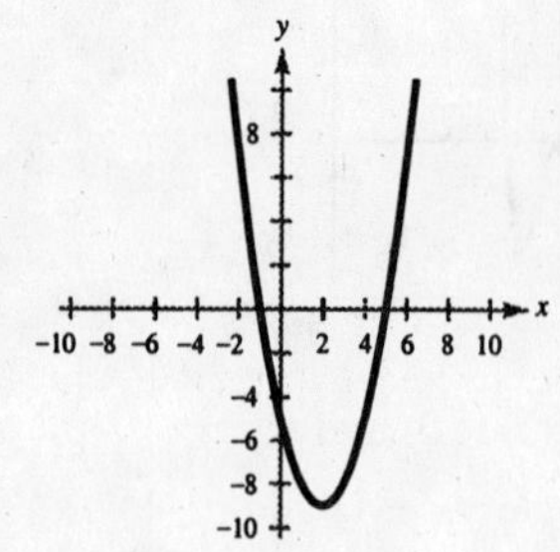

Minimum: $(2, -9)$

(b)

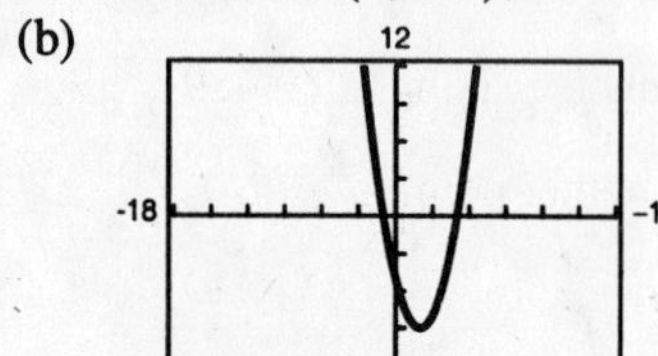

Minimum: $(2, -9)$

(c) Answers are the same

39. $f(x) = x^3 - 8x$

(a)

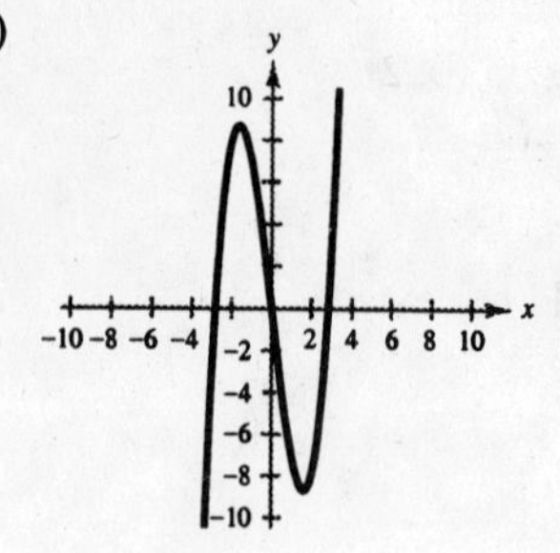

Maximum: Approximately $(-2, 9)$

Minimum: Approximately $(2, -9)$

(b)

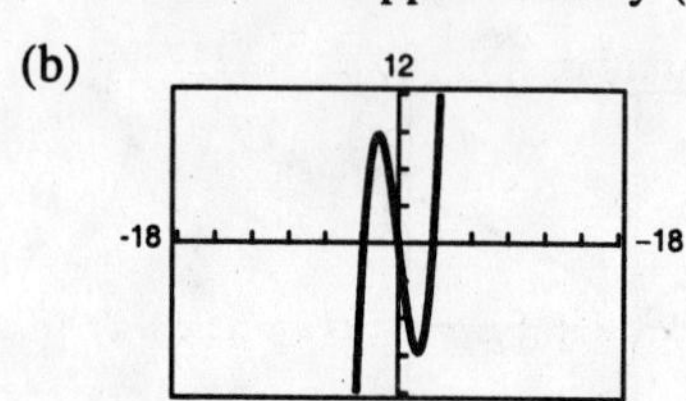

Maximum: $(-1.63, 8.71)$

Minimum: $(1.63, -8.71)$

(c) The answers are similar.

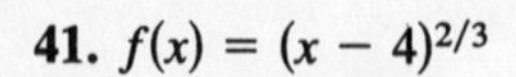

41. $f(x) = (x - 4)^{2/3}$

(a)

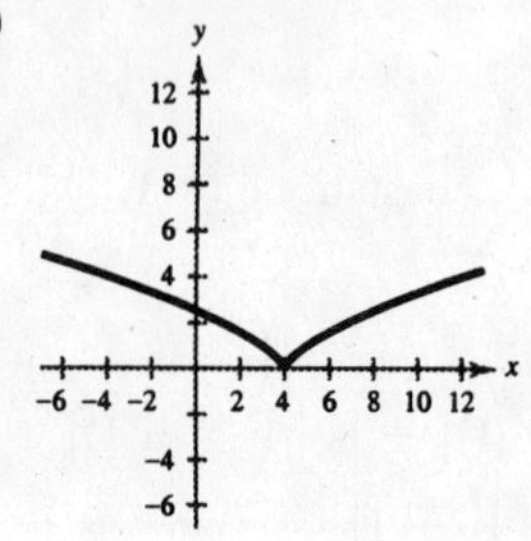

Minimum: $(4, 0)$

(c) The answers are the same.

(b)

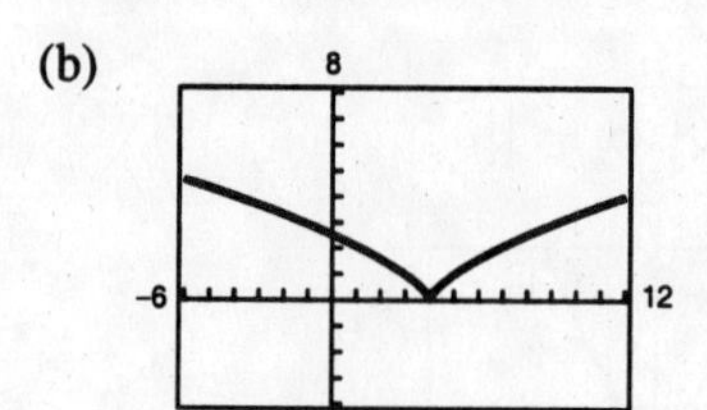

Minimum: $(4, 0)$

43. $f(x) = \begin{cases} 2x + 3, & x < 0 \\ 3 - x, & x \geq 0 \end{cases}$

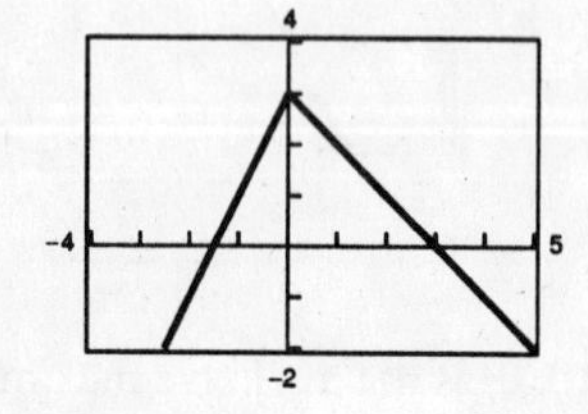

45. $f(x) = \begin{cases} \sqrt{x + 4} & x < 0 \\ \sqrt{4 - x} & x \geq 0 \end{cases}$

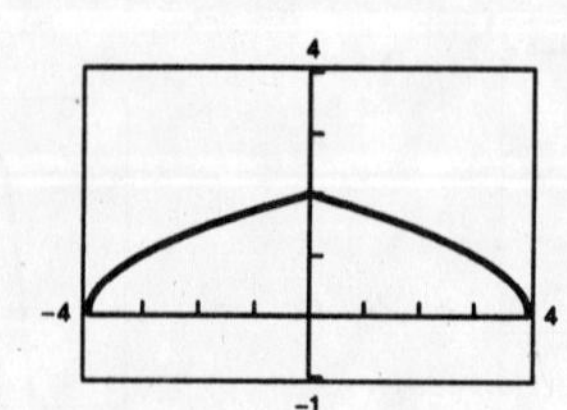

47. $f(-t) = (-t)^2 + 2(-t) - 3$

$= t^2 - 2t - 3$

$\neq f(t) \neq -f(t)$

f is neither even nor odd.

49. $g(-x) = (-x)^3 - 5(-x)$

$= -x^3 + 5x$

$= -g(x)$

g is odd.

51. $f(-x) = (-x)\sqrt{1 - (-x)^2} = -x\sqrt{1 - x^2} = -f(x)$

The function is odd.

53. $g(-s) = 4(-s)^{2/3} = 4s^{2/3} = g(s)$

The function is even.

55. $\left(-\frac{3}{2}, 4\right)$

(a) If f is even, another point is $\left(\frac{3}{2}, 4\right)$.

(b) If f is odd, another point is $\left(\frac{3}{2}, -4\right)$.

57. $(4, 9)$

(a) If f is even, another point is $(-4, 9)$.

(b) If f is odd, another point is $(-4, -9)$.

59. $(x, -y)$

(a) If f is even, another point is $(-x, -y)$.

(b) If f is odd, another point is $(-x, y)$.

61. $f(x) = 5$, even

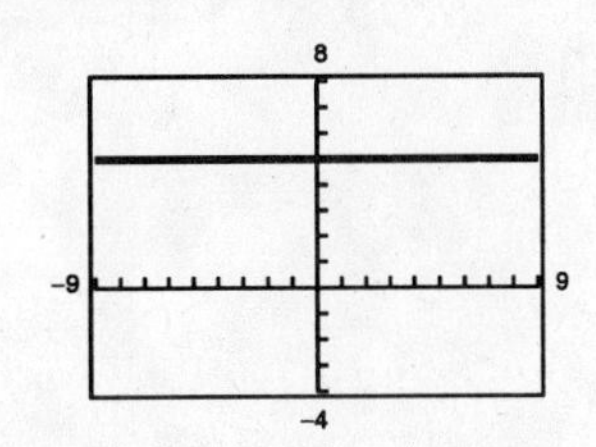

63. $f(x) = 3x - 2$, neither even nor odd

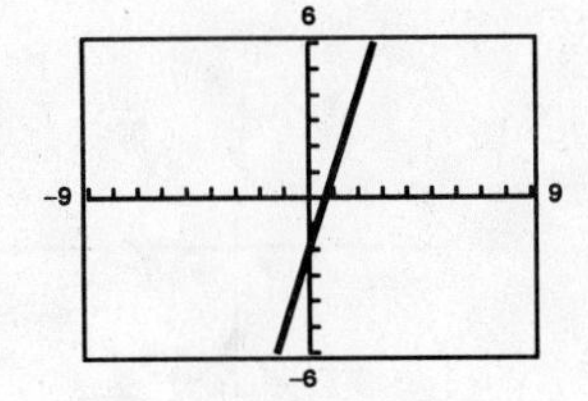

65. $h(x) = x^2 - 4$, even

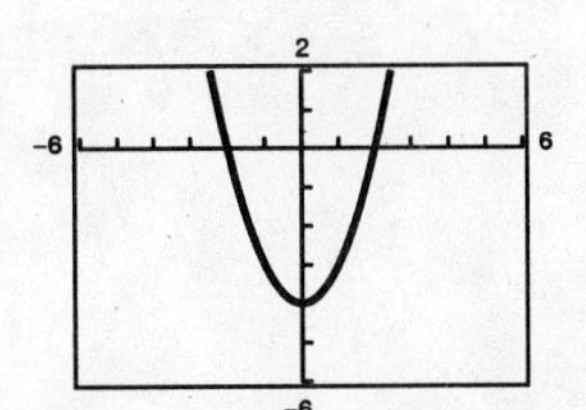

67. $f(x) = \sqrt{1 - x}$, neither even nor odd

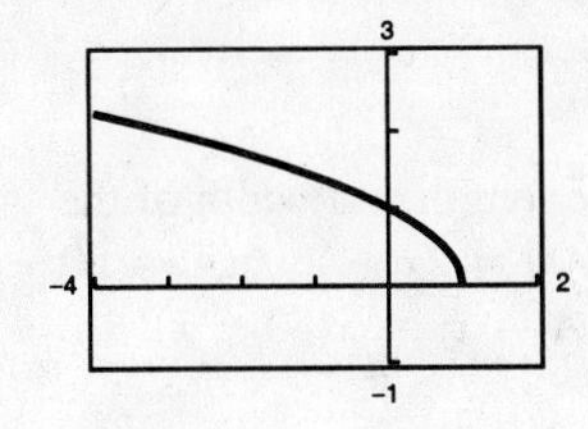

69. $f(x) = |x + 2|$, neither even nor odd

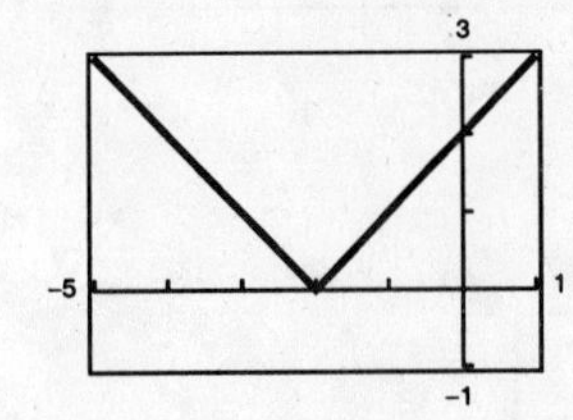

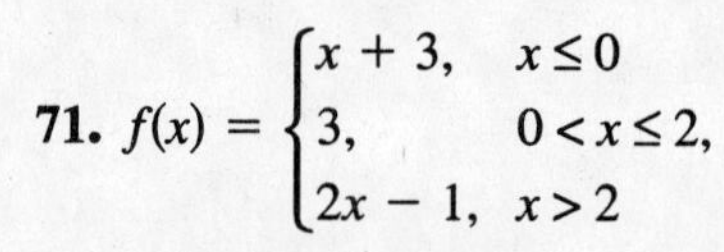

71. $f(x) = \begin{cases} x + 3, & x \le 0 \\ 3, & 0 < x \le 2, \\ 2x - 1, & x > 2 \end{cases}$

Neither even nor odd

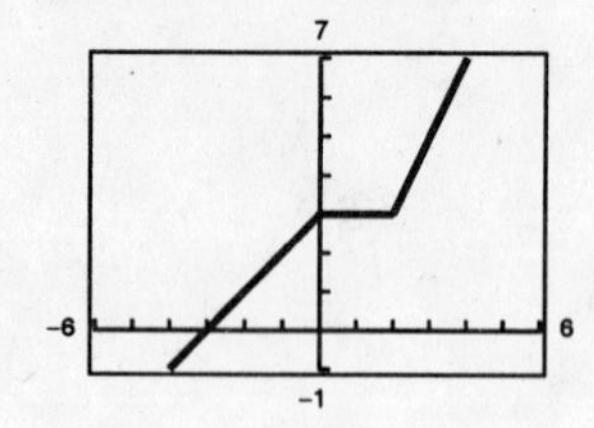

73. $f(x) = 4 - x \ge 0$

$4 \ge x$

$(-\infty, 4]$

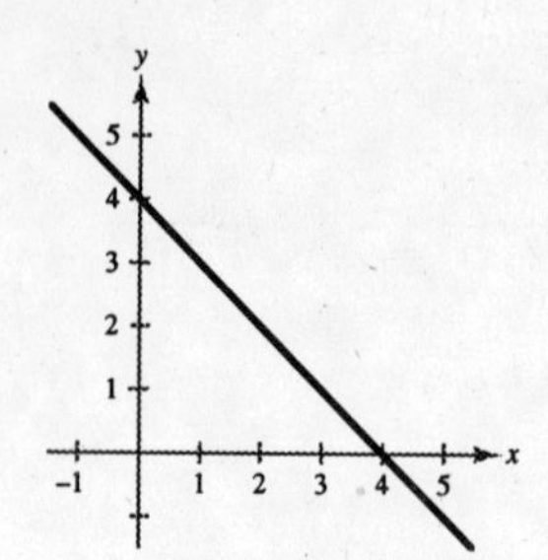

75. $f(x) = x^2 - 9 \ge 0$

$x^2 \ge 9$

$x \ge 3$ or $x \le -3$

$[3, \infty)$ or $(-\infty, -3]$

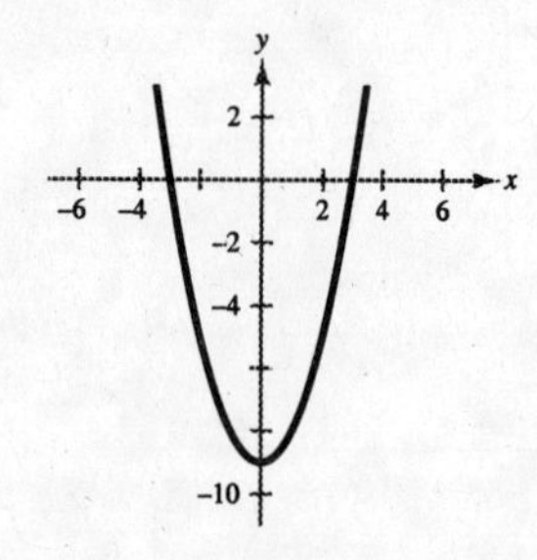

77. $f(x) = 1 - x^4 \ge 0$

$1 \ge x^4$

$-1 \le x \le 1$

$[-1, 1]$

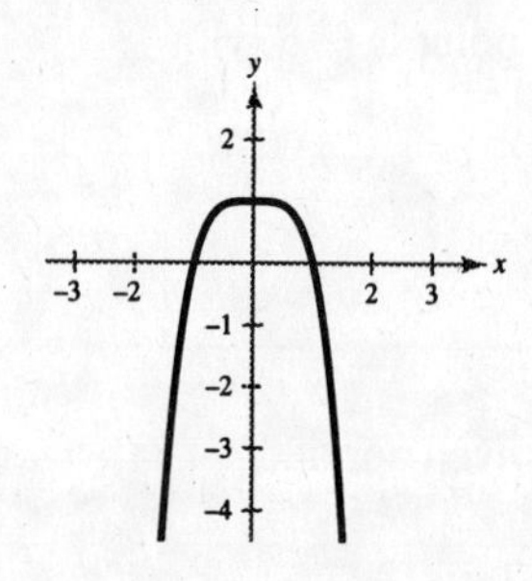

79. $f(x) = \sqrt{x + 2} \ge 0$

$x + 2 \ge 0$

$x \ge -2$

$[-2, \infty)$

81. $f(x) = -(1 + |x|)$

$f(x)$ is never greater than 0.

$f(x) < 0$ for all x.

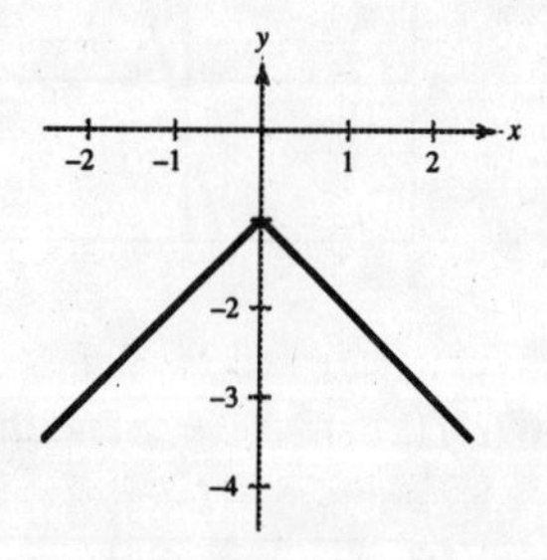

83. $s(x) = 2\left(\frac{1}{4}x - \left[\!\left[\frac{1}{4}x\right]\!\right]\right)$

Domain: $(-\infty, \infty)$

Range: $[0, 2)$

Sawtooth pattern

85. (a) Let x and y be the length and width of the rectangle. Then $100 = 2x + 2y$ or $y = 50 - x$. Thus, the area is $A = xy = x(50 - x)$.

(b)

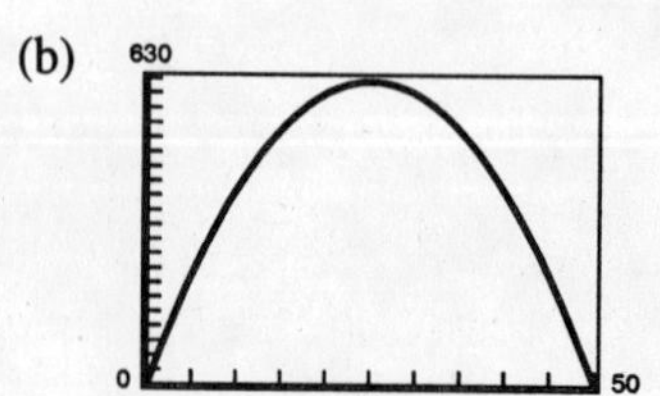

(c) The maximum area is $625\ \text{m}^2$ when $x = y = 25$ m. That is, the rectangle is a square.

87. (a) The second model is correct. For instance,

$$C_2\left(\tfrac{1}{2}\right) = 1.05 - 0.38[[-(\tfrac{1}{2} - 1)]]$$
$$= 1.05 - 0.38\left[\left[\tfrac{1}{2}\right]\right] = 1.05.$$

(b)

The cost of an 18-minute 45-second call is

$$C_2\left(18\tfrac{45}{60}\right)$$
$$= C_2(18.75) = 1.05 - 0.38[[-(18.75 - 1)]]$$
$$= 1.05 - 0.38[[-17.75]] = 1.05 - 0.38(-18)$$
$$= 1.05 + 0.38(18) = \$7.89$$

89. $h = \text{top} - \text{bottom}$

$$= (-x^2 + 4x - 1) - 2$$
$$= -x^2 + 4x - 3,\ 1 \le x \le 3$$

91. $h = \text{top} - \text{bottom}$

$$= (4x - x^2) - 2x$$
$$= 2x - x^2,\ 0 \le x \le 2$$

93. $L = \text{right} - \text{left}$

$$= \tfrac{1}{2}y^2 - 0$$
$$= \tfrac{1}{2}y^2,\ 0 \le y \le 4$$

95. (a) $y = 1.473x^3 + 16.411x^2 + 31.242x - 95.195$

(b) Domain: $[0, 7]$

(c)

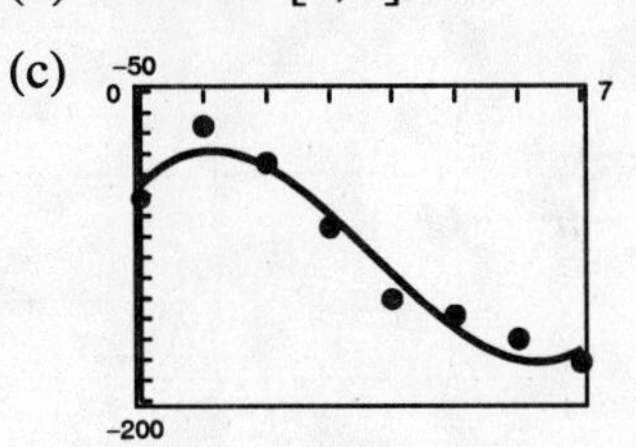

(d) Most accurate: 1992 [error $= -84.5 - (-86.57) = 2.071$]

Least accurate: 1991 [error $= -66.7 - (-78.89) = 12.191$]

(e) Yes, the deficit would decrease as x increases because of the positive x^3 coefficient.

97. False. The domain of $f(x) = \sqrt{x^2}$ is the set of all real numbers.

99. $f(x) = a_{2n+1}x^{2n+1} + a_{2n-1}x^{2n-1} + \cdots + a_3x^3 + a_1x$

$$f(-x) = a_{2n+1}(-x)^{2n+1} + a_{2n-1}(-x)^{2n-1} + \cdots + a_3(-x)^3 + a_1(-x)$$
$$= -a_{2n+1}x^{2n+1} - a_{2n-1}x^{2n-1} - \cdots - a_3x^3 - a_1x = -f(x)$$

Therefore, $f(x)$ is odd.

101. f is an even function.

(a) $g(x) = -f(x)$ is even because
$g(-x) = -f(-x) = -f(x) = g(x)$.

(b) $g(x) = f(-x)$ is even because
$g(-x) = f(-(-x)) = f(x) = f(-x) = g(x)$.

(c) $g(x) = f(x) - 2$ is even because
$g(-x) = f(-x) - 2 = f(x) - 2 = g(x)$.

(d) $g(x) = -f(x - 2)$ is neither even nor odd because
$g(-x) = -f(-x-2) = -f(x+2) \neq g(x)$ nor $-g(x)$.

103. No, $x^2 + y^2 = 25$ does not represent x as a function of y. For instance, $(-3, 4)$ and $(3, 4)$ both lie on the graph.

105. (a) $d = \sqrt{(-5 - 3)^2 + (0 - 6)^2} = \sqrt{64 + 36} = \sqrt{100} = 10$

(b) midpoint $= \left(\dfrac{-5 + 3}{2}, \dfrac{0 + 6}{2}\right) = (-1, 3)$

107. (a) $d = \sqrt{\left(-6 - \dfrac{3}{4}\right)^2 + \left(\dfrac{2}{3} - \dfrac{1}{6}\right)^2} = \sqrt{\left(\dfrac{-27}{4}\right)^2 + \left(\dfrac{1}{2}\right)^2} = \dfrac{\sqrt{733}}{4}$

(b) midpoint $= \left(\dfrac{-6 + \frac{3}{4}}{2}, \dfrac{\frac{2}{3} + \frac{1}{6}}{2}\right) = \left(\dfrac{-21}{8}, \dfrac{5}{12}\right)$

109. $f(x) = -x^2 - x + 3$

(a) $f(4) = -(4)^2 - 4 + 3 = -17$

(b) $f(-2) = -(-2)^2 - (-2) + 3 = 1$

(c) $f(x - 2) = -(x - 2)^2 - (x - 2) + 3$

$= -(x^2 - 4x + 4) - x + 2 + 3$

$= -x^2 + 3x + 1$

111. $f(x) = -\frac{1}{2}x|x + 1|$

(a) $f(-4) = -\frac{1}{2}(-4)|-4 + 1| = 2(3) = 6$

(b) $f(10) = -\frac{1}{2}(10)|10 + 1| = -5(11) = -55$

(c) $f\left(-\frac{2}{3}\right) = -\frac{1}{2}\left(-\frac{2}{3}\right)\left|-\frac{2}{3} + 1\right| = \frac{1}{3}\left(\frac{1}{3}\right) = \frac{1}{9}$

113. $f(x) = 5 + 6x - x^2$

$f(6 + h) = 5 + 6(6 + h) - (6 + h)^2 = 5 + 36 + 6h - (36 + 12h + h^2) = -h^2 - 6h + 5$

$f(6) = 5 + 6(6) - 6^2 = 5$

$\dfrac{f(x + 6) - f(6)}{h} = \dfrac{(-h^2 - 6h + 5) - 5}{h} = \dfrac{h(-h - 6)}{h} = -h - 6, h \neq 0$

Section 1.3 Shifting, Reflecting, and Stretching Graphs

- You should know the graphs of the most commonly used functions in algebra, and be able to reproduce them on your graphing utility.

  (a) Constant function: $f(x) = c$

  (b) Identity function: $f(x) = x$

  (c) Absolute value function: $f(x) = |x|$

  (d) Square root function: $f(x) = \sqrt{x}$

  (e) Squaring function: $f(x) = x^2$

  (f) Cubing function: $f(x) = x^3$

—CONTINUED—

—CONTINUED—

- You should know how the graph of a function is changed by vertical and horizontal shifts.
- You should know how the graph of a function is changed by reflection.
- You should know how the graph of a function is changed by nonrigid transformations, like stretches and shrinks.
- You should know how the graph of a function is changed by a sequence of transformations.

Solutions to Odd-Numbered Exercises

1.

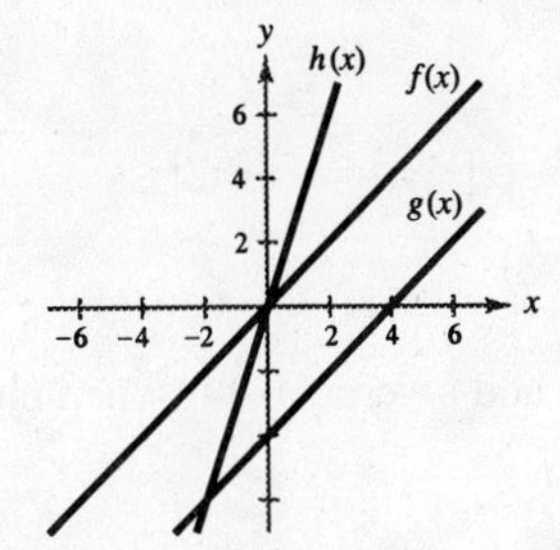

3.

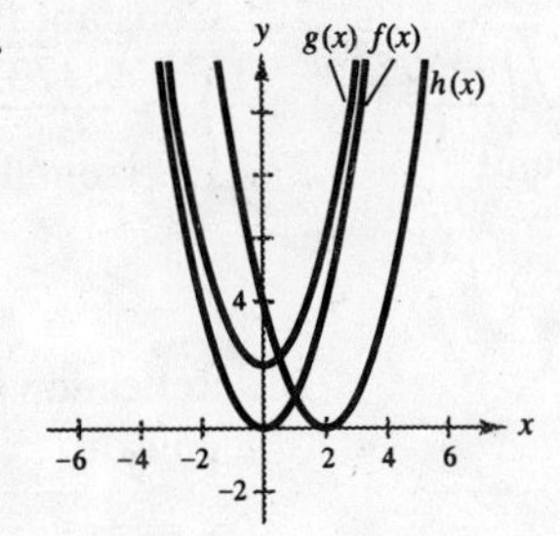

5.

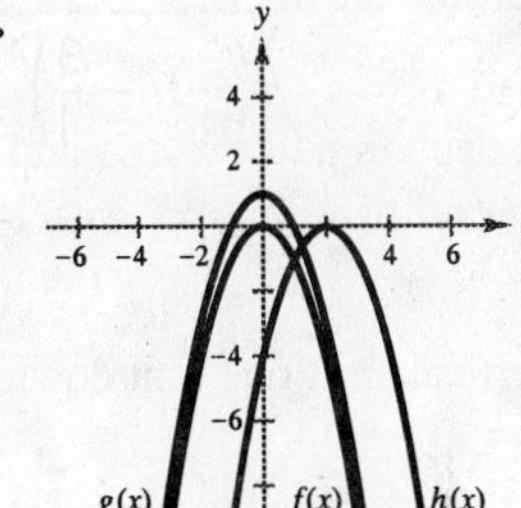

7.

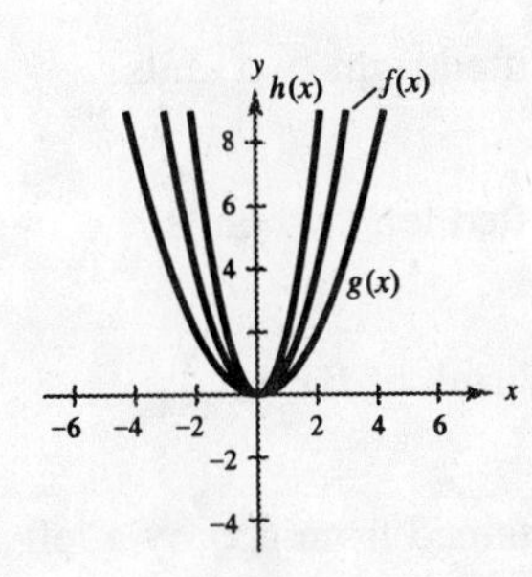

9.

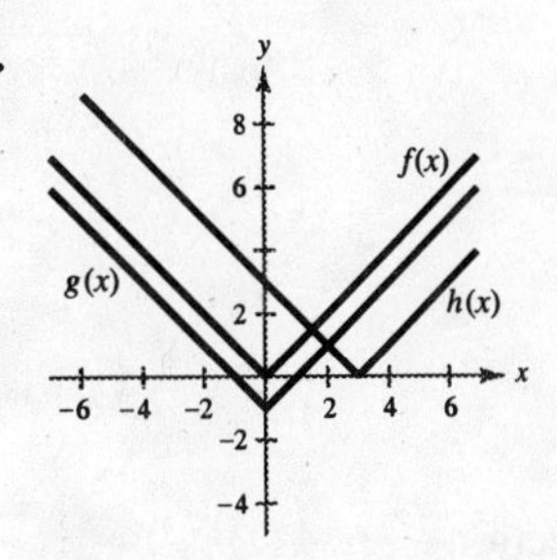

11.

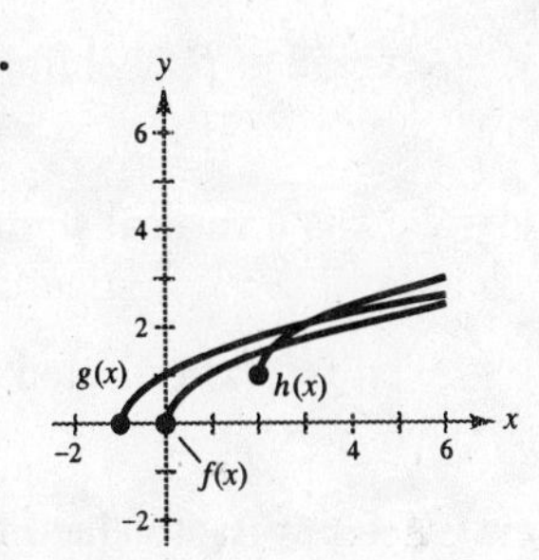

13. (a) $y = f(x) + 2$

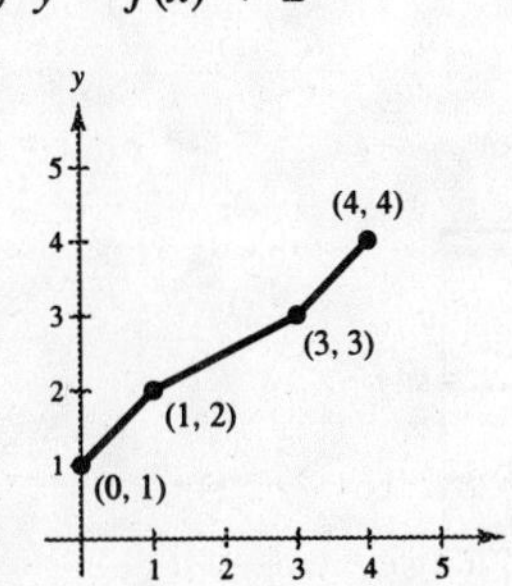

(b) $y = -f(x)$

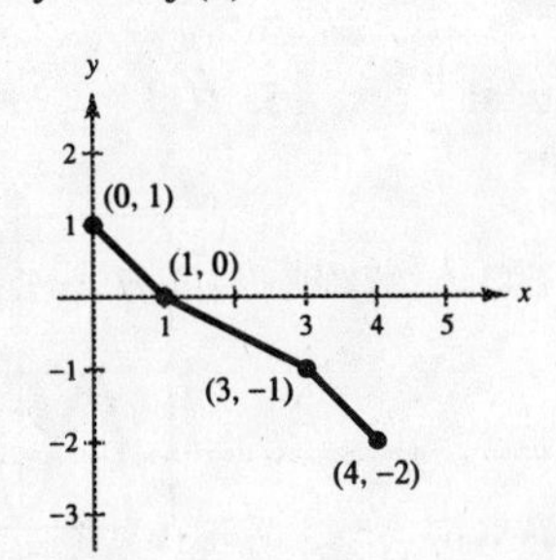

(c) $y = f(x - 2)$

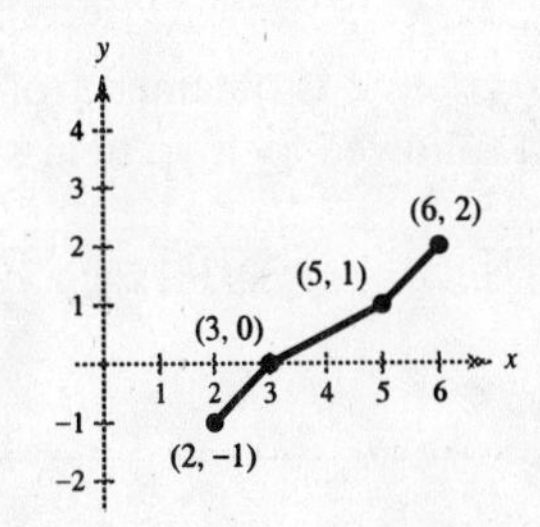

—CONTINUED—

—CONTINUED—

(d) $y = f(x + 3)$

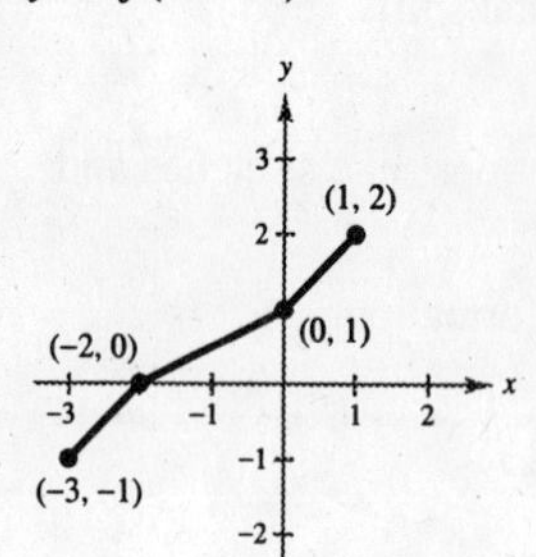

(e) $y = 2f(x)$

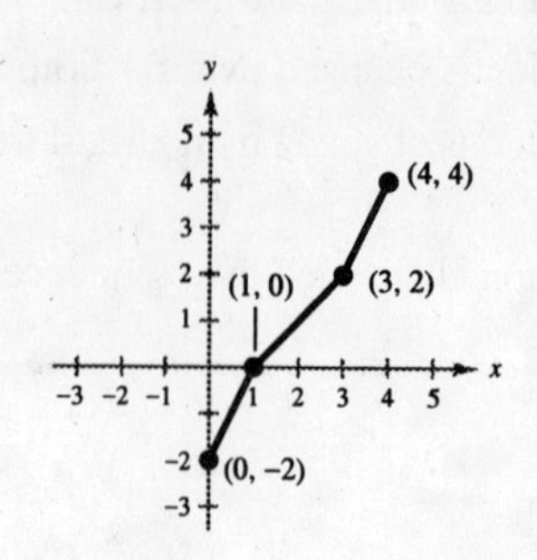

(f) $y = f(-x)$

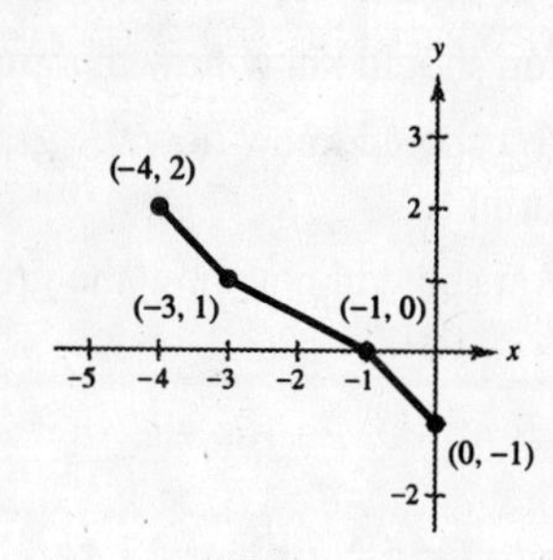

15. Vertical shrink of $y = x : y = \frac{1}{2}x$

17. Constant function: $y = 7$

19. Reflection in the x-axis and a vertical shift one unit upward of $y = \sqrt{x}$: $g = 1 - \sqrt{x}$

21. Horizontal shift of $y = |x| : y = |x + 2|$

23. Vertical shift one unit downward of $y = x^2$
$y = x^2 - 1$

25. Reflection in the x-axis and a vertical shift one unit upward

$y = 1 - x^3$

27. $y = \sqrt{x} + 2$ is $f(x)$ shifted up two units.

29. $y = \sqrt{x - 2}$ is $f(x)$ shifted right two units.

31. $y = 2\sqrt{x}$ is a vertical stretch of $f(x)$ by 2.

33. $y = |x + 2|$ is $f(x)$ shifted left two units.

35. $y = -|x|$ is $f(x)$ reflected in the x-axis.

37. $y = \frac{1}{3}|x|$ is a vertical shrink of $f(x)$.

39. $g(x) = 4 - x^3$ is obtained from $f(x)$ by a reflection in the x-axis followed by a vertical shift upward of four units.

41. $h(x) = \frac{1}{4}(x + 2)^3$ is obtained from $f(x)$ by a left shift of two units and a vertical shrink by a factor of $\frac{1}{4}$.

43. $p(x) = \frac{1}{3}x^3 + 2$ is obtained from $f(x)$ by a vertical shrink, followed by a vertical shift of two units upward.

45. $f(x) = x^3 - 3x^2$

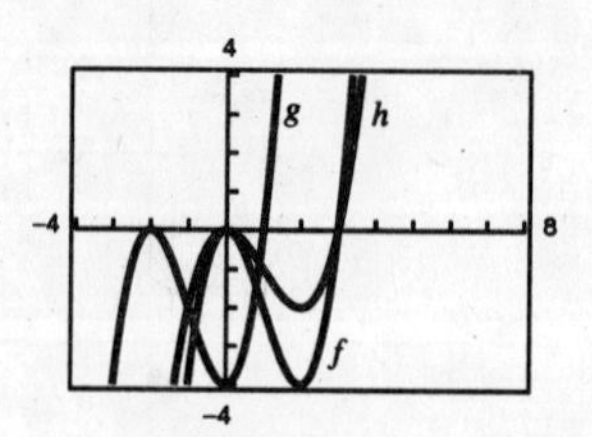

$g(x) = f(x + 2) = (x + 2)^3 - 3(x + 2)^2$ is a horizontal shift 2 units to left

$h(x) = \frac{1}{2}f(x) = \frac{1}{2}(x^3 - 3x^2)$ is a vertical shrink.

47. $f(x) = x^3 - 3x^2$

$g(x) = -\frac{1}{3}f(x) = -\frac{1}{3}(x^3 - 3x^2)$ reflection in the x-axis and vertical shrink

$h(x) = f(-x) = (-x)^3 - 3(-x)^2$ reflection in the y-axis

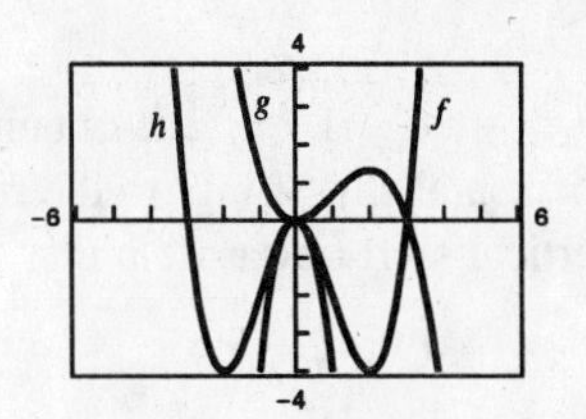

49. The graph of g is obtained from that of f by first negating f, and then shifting vertically one unit upward: $g(x) = -x^3 + 3x^2 + 1$.

51. (a) $f(x) = x^2$

(b) $g(x) = 12 - x^2$ is obtained from f by a reflection in the x-axis followed by a vertical shift upward 12 units.

(c)

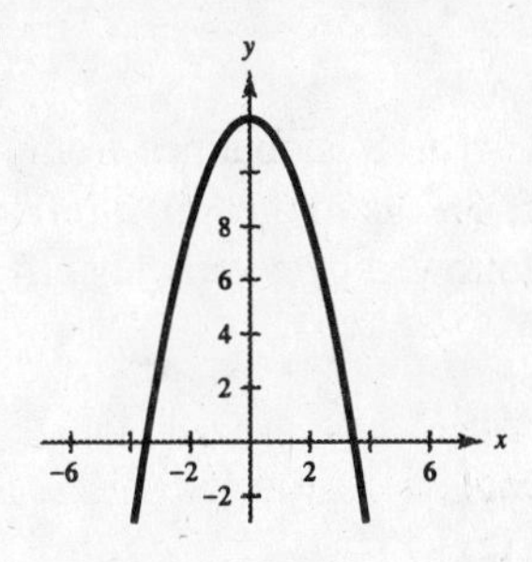

(d) $g(x) = 12 - f(x)$

53. (a) $f(x) = x^2$

(b) $g(x) = 2 - (x + 5)^2$ is obtained from f by a horizontal shift to the left 5 units, a reflection in the x-axis, and a vertical shift upward 2 units.

(c)

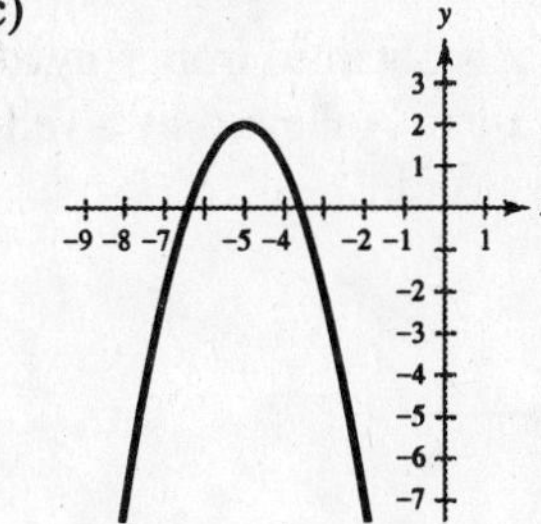

(d) $g(x) = 2 - f(x + 5)$

55. (a) $f(x) = x^2$

(b) $g(x) = 3 + 2(x - 4)^2$ is obtained from f by a horizontal shift 4 units to the right, a vertical stretch of 2, and a vertical shift upward 3 units.

(c)

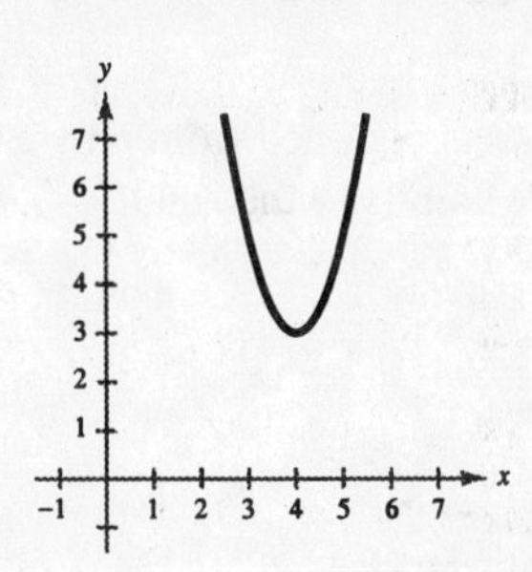

(d) $g(x) = 3 + 2f(x - 4)$

57. (a) $f(x) = x^3$

(b) $g(x) = x^3 + 7$ is obtained from f by a vertical shift upward 7 units.

(c)

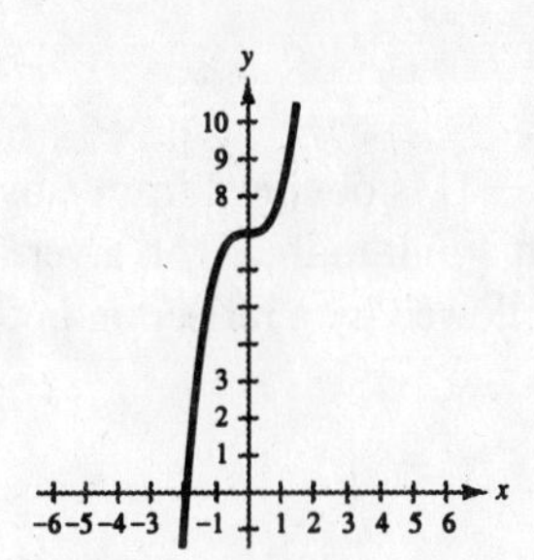

(d) $g(x) = f(x) + 7$

59. (a) $f(x) = x^3$

(b) $g(x) = (x - 1)^3 + 2$ is obtained from f by a horizontal shift 1 unit to the right, and a vertical shift upward 2 units.

(c)

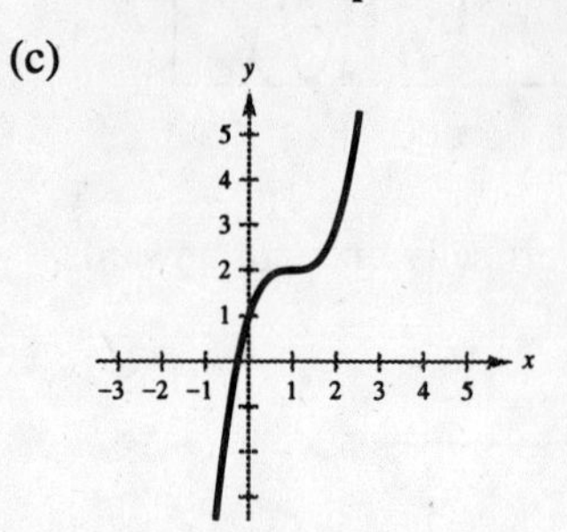

(d) $g(x) = f(x - 1) + 2$

61. (a) $f(x) = x^3$

(b) $g(x) = 3(x - 2)^3$ is obtained from f by a horizontal shift 2 units to the right followed by a vertical stretch of 3.

(c)

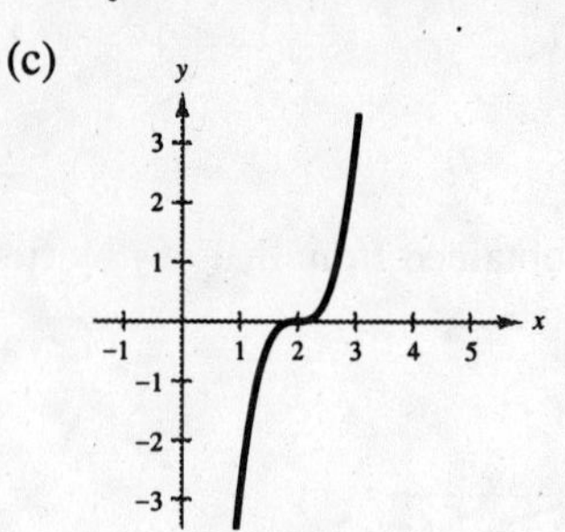

(d) $g(x) = 3f(x - 2)$

63. (a) $f(x) = |x|$

(b) $g(x) = -|x| - 2$ is obtained from f by a reflection in the x-axis, followed by a vertical shift 2 units downward.

(c)

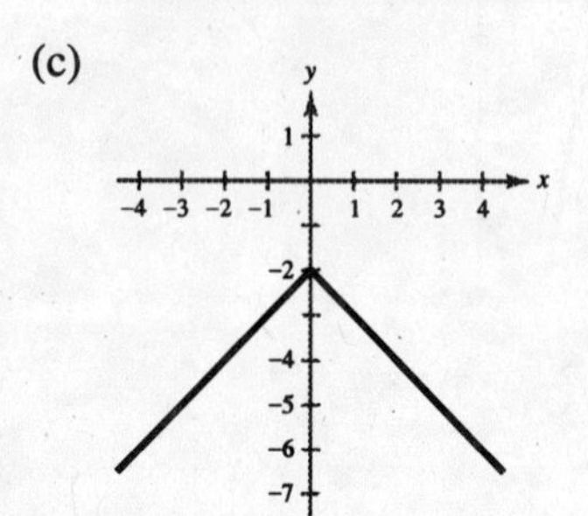

(d) $g(x) = -f(x) - 2$

65. (a) $f(x) = |x|$

(b) $g(x) = -|x + 4| + 8$ is obtained from f by a horizontal shift 4 units to the left, a reflection in the x-axis, followed by a vertical shift 8 units upward.

(c)

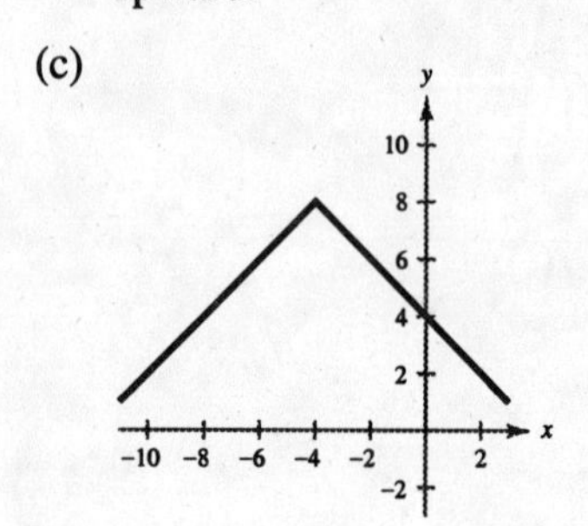

(d) $g(x) = -f(x + 4) + 8$

67. (a) $f(x) = |x|$

(b) $g(x) = -2|x - 1|$ is obtained from f by a horizontal shift 1 unit to the right, a vertical stretch of 2, followed by a reflection in the x-axis.

(c)

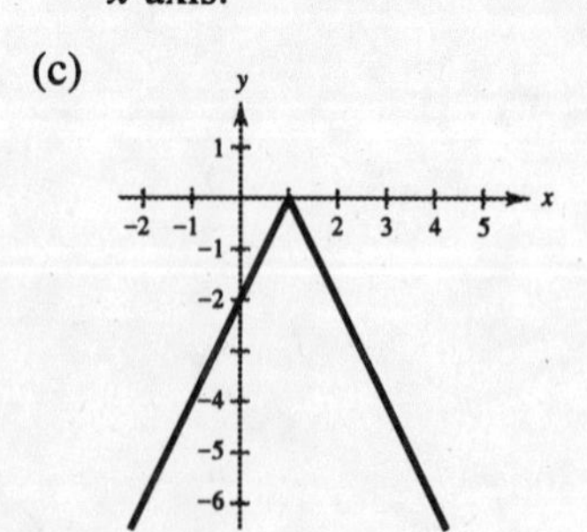

(d) $g(x) = -2f(x - 1)$

69. (a) $f(x) = \sqrt{x}$

(b) $g(x) = \sqrt{x - 9}$ is obtained from f by a horizontal shift 9 units to the right.

(c)

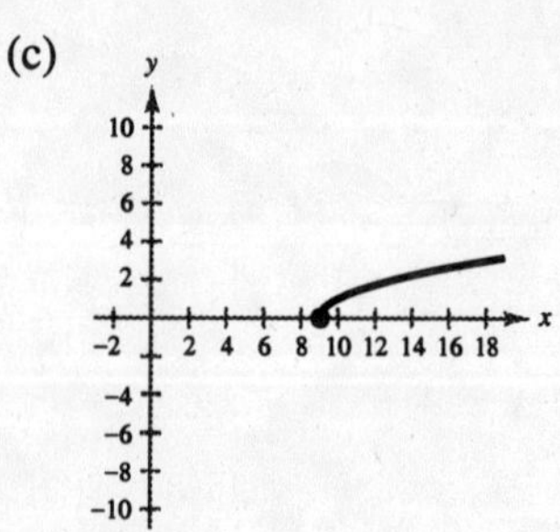

(d) $g(x) = f(x - 9)$

71. (a) $f(x) = \sqrt{x}$

(b) $g(x) = \sqrt{7 - x} - 2$ is obtained from f by a reflection in the y-axis, a horizontal shift 7 to the right, followed by a vertical shift 2 units downward. Equivalently, use a horizontal shift left 7 units, a reflection in the y-axis, and a vertical shift down 2 units.

(c)

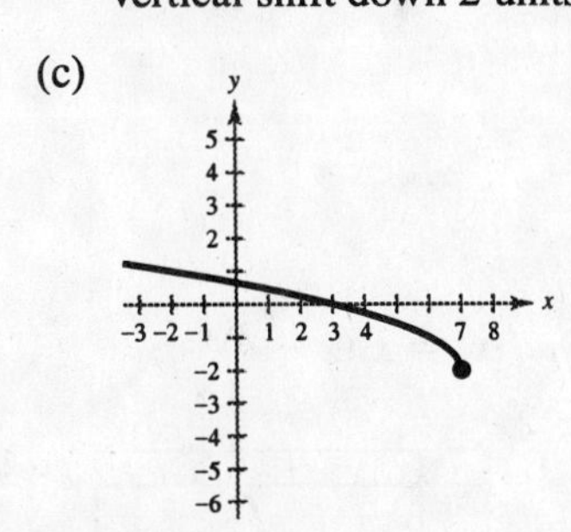

(d) $g(x) = f(7 - x) - 2$

73. (a) $f(x) = \sqrt{x}$

(b) $g(x) = 4\sqrt{x - 1}$ is obtained from f by a horizontal shift 1 unit to the right, followed by a vertical stretch of 4 units.

(c)

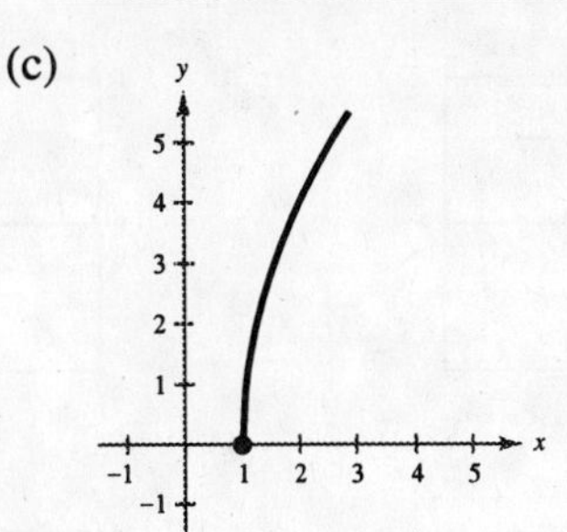

(d) $g(x) = 4f(x - 1)$

75. (a) $P(x) = 80 + 20x - 0.5x^2, 0 \leq x \leq 20$

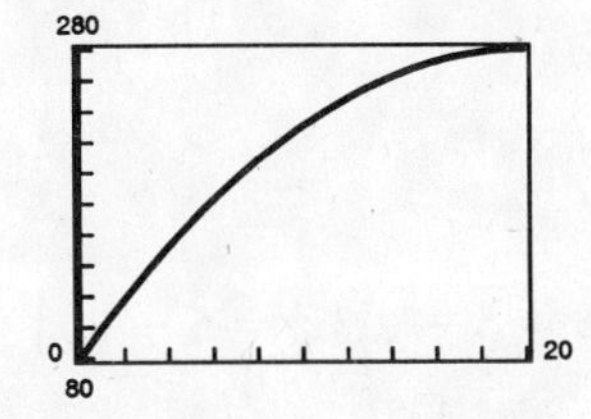

(b) $P(x)$ is shifted downward by a vertical shift of -2500.

$P(x) = -2420 + 20x - 0.5x^2, 0 \leq x \leq 20$

(c) $P(x)$ is changed by a *horizontal stretch.*

$$P(x) = 80 + 20\left(\frac{x}{100}\right) - 0.5\left(\frac{x}{100}\right)^2$$

$$= 80 + 0.2x - 0.00005x^2$$

77. $F(t) = 20.46 + 0.04t^2, 0 \leq t \leq 16,$
$t = 0$ corresponds to 1980

(a) F is obtained from $f(t) = t^2$ by a vertical shrink of 0.04 followed by a vertical shift 20.46 units upward.

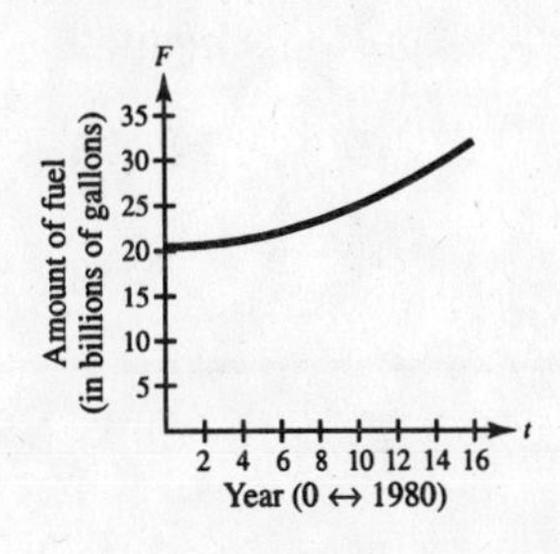

(b) $G(t) = F(t + 10) = 20.46 + 0.04(t + 10)^2,$
$-10 \leq t \leq 6.$

$G(0) = F(10)$ corresponds to 1990.

79. (a) For each time t there corresponds one and only one temperature T.

(b) $T(4) \approx 60°$, $T(15) \approx 72°$

(c) All the temperature changes would be one hour later.

(d) The temperature would be decreased by one degree.

81. False. $f(x) = x^2$ is transformed to $g(x) = -[(x - 6)^2 + 3]$. But, $g(-1) = -52 \neq 28$.

83. $y = x^7$ will look like $y = x^5$, but flatter in $-1 < x < 1$, and steeper for $x < -1$ and $x > 1$.

$y = x^8$ will look like $y = x^6$, but flatter in $-1 < x < 1$, and steeper for $x < -1$ and $x > 1$.

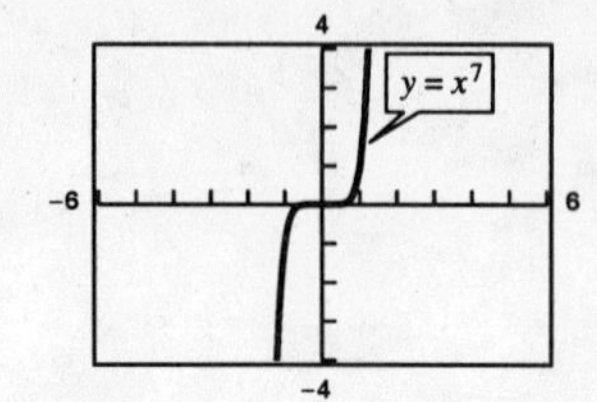

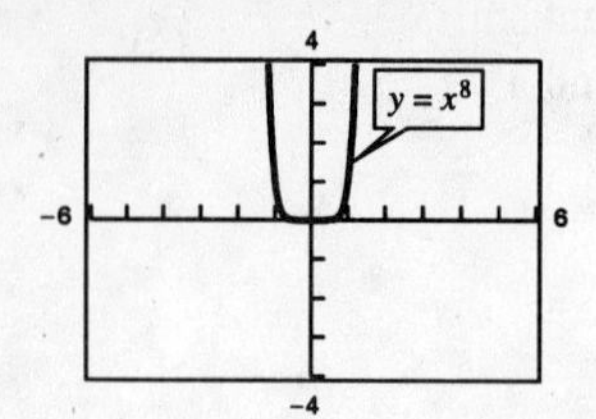

85. $y = (x + 1)^2$

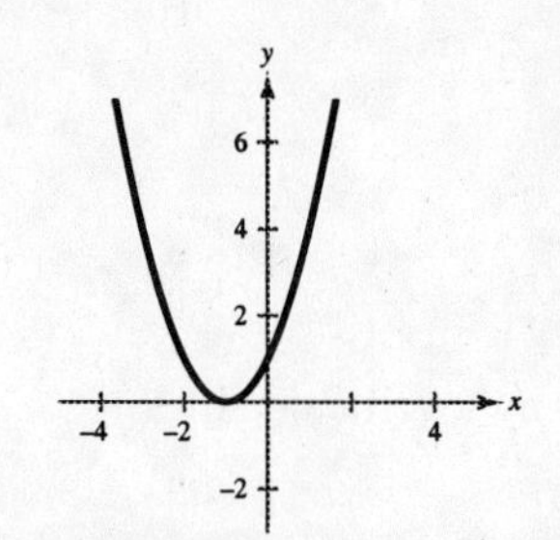

87. $f(x) = x^3(x - 6)^2$

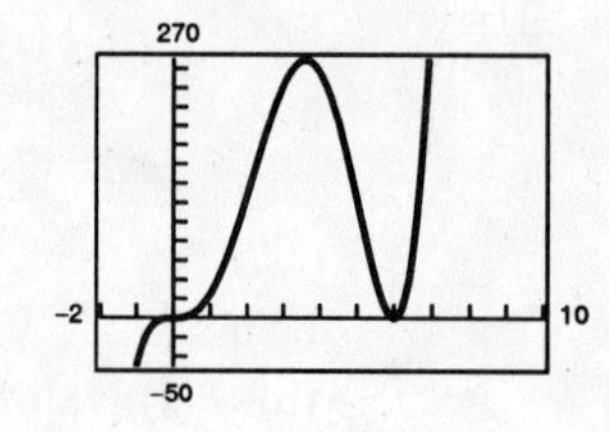

89. $f(x) = x^3(x - 6)^3$

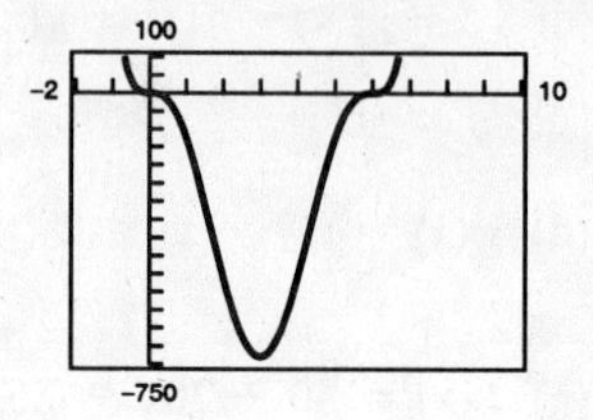

91. $x - 5 \geq 0$ and $x \neq 7 \Rightarrow$ Domain: $x \geq 5, x \neq 7$

93. Domain: all x

95. $y = 9 - 4x$

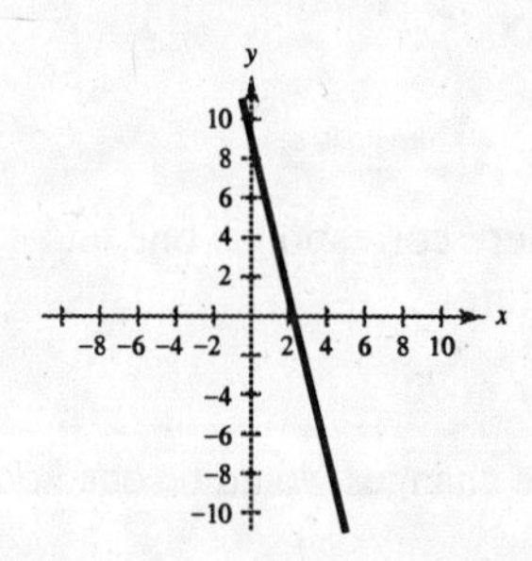

97. $y = -x^3 - 3$

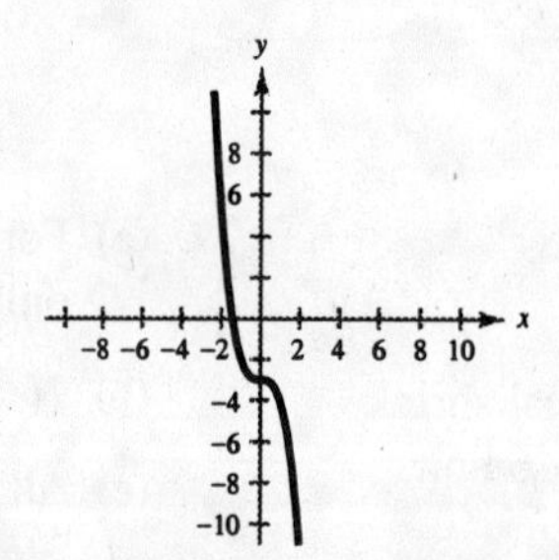

99. $y = 5 - 2|3x| = 5 - 6|x|$

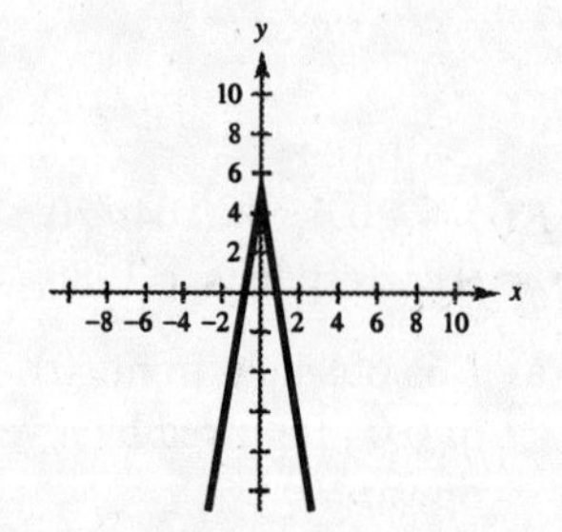

Section 1.4 Combinations of Functions

■ Given two functions, f and g, you should be able to form the following functions (if defined):

1. Sum: $(f + g)(x) = f(x) + g(x)$
2. Difference: $(f - g)(x) = f(x) - g(x)$
3. Product: $(fg)(x) = f(x)g(x)$
4. Quotient: $(f/g)(x) = f(x)/g(x), g(x) \neq 0$
5. Composition of f with g: $(f \circ g)(x) = f(g(x))$
6. Composition of g with f: $(g \circ f)(x) = g(f(x))$

Solutions to Odd-Numbered Exercises

1. $f(x) = x + 1, g(x) = x - 1$

(a) $(f + g)(x) = f(x) + g(x) = (x + 1) + (x - 1) = 2x$

(b) $(f - g)(x) = f(x) - g(x) = (x + 1) - (x - 1) = 2$

(c) $(fg)(x) = f(x) \cdot g(x) = (x + 1)(x - 1) = x^2 - 1$

(d) $\left(\frac{f}{g}\right)(x) = \frac{f(x)}{g(x)} = \frac{x+1}{x-1}, \; x \neq 1$

(e) Domain: all $x \neq 1$.

3. $f(x) = x^2, g(x) = 1 - x$

(a) $(f + g)(x) = f(x) + g(x) = x^2 + (1 - x) = x^2 - x + 1$

(b) $(f - g)(x) = f(x) - g(x) = x^2 - (1 - x) = x^2 + x - 1$

(c) $(fg)(x) = f(x) \cdot g(x) = x^2(1 - x) = x^2 - x^3$

(d) $\left(\frac{f}{g}\right)(x) = \frac{f(x)}{g(x)} = \frac{x^2}{1-x}, \; x \neq 1$

(e) Domain: all $x \neq 1$.

5. $f(x) = x^2 + 5, g(x) = \sqrt{1 - x}$

(a) $(f + g)(x) = f(x) + g(x) = (x^2 + 5) + \sqrt{1 - x}$

(b) $(f - g)(x) = f(x) - g(x) = (x^2 + 5) - \sqrt{1 - x}$

(c) $(fg)(x) = f(x) \cdot g(x) = (x^2 + 5)\sqrt{1 - x}$

(d) $\left(\frac{f}{g}\right)(x) = \frac{f(x)}{g(x)} = \frac{x^2+5}{\sqrt{1-x}}, \; x < 1$

(e) Domain: $x < 1$.

7. $f(x) = \frac{1}{x}, g(x) = \frac{1}{x^2}$

(a) $(f + g)(x) = f(x) + g(x) = \frac{1}{x} + \frac{1}{x^2} = \frac{x+1}{x^2}$

(b) $(f - g)(x) = f(x) - g(x) = \frac{1}{x} - \frac{1}{x^2} = \frac{x-1}{x^2}$

(c) $(fg)(x) = f(x) \cdot g(x) = \frac{1}{x}\left(\frac{1}{x^2}\right) = \frac{1}{x^3}$

(d) $\left(\frac{f}{g}\right)(x) = \frac{f(x)}{g(x)} = \frac{1/x}{1/x^2} = \frac{x^2}{x} = x, \; x \neq 0$

(e) Domain: $x \neq 0$.

9. $(f + g)(3) = f(3) + g(3) = (3^2 + 1) + (3 - 4) = 9$

11. $(f - g)(0) = f(0) - g(0) = [0^2 + 1] - (0 - 4) = 5$

13. $(fg)(4) = f(4)g(4) = (4^2 + 1)(4 - 4) = 0$

15. $\left(\dfrac{f}{g}\right)(5) = \dfrac{f(5)}{g(5)} = \dfrac{5^2 + 1}{5 - 4} = 26$

17. $(f - g)(2t) = f(2t) - g(2t) = [(2t)^2 + 1] - (2t - 4) = 4t^2 - 2t + 5$

19. $(fg)(-5t) = f(-5t)g(-5t) = [(-5t)^2 + 1][(-5t) - 4]$

$= (25t^2 + 1)(-5t - 4) = -125t^3 - 100t^2 - 5t - 4$

21. $\left(\dfrac{f}{g}\right)(-t) = \dfrac{f(-t)}{g(-t)} = \dfrac{(-t)^2 + 1}{-t - 4} = \dfrac{t^2 + 1}{-t - 4}, t \neq -4$

23.

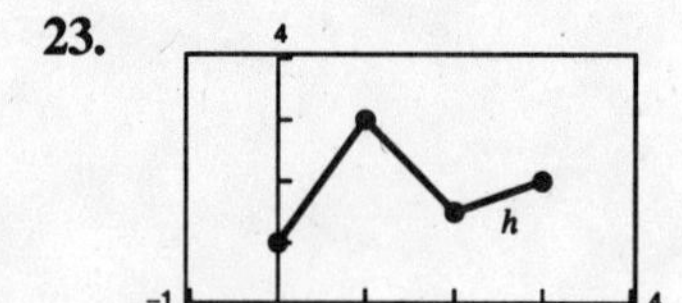

25.

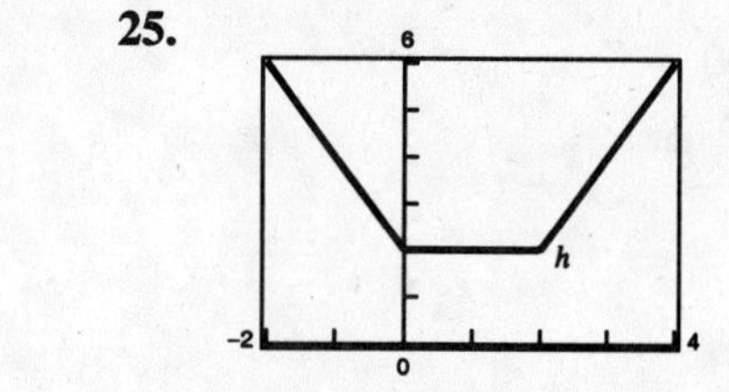

27. $f(x) = \frac{1}{2}x, g(x) = x - 1, (f + g)(x) = \frac{3}{2}x - 1$

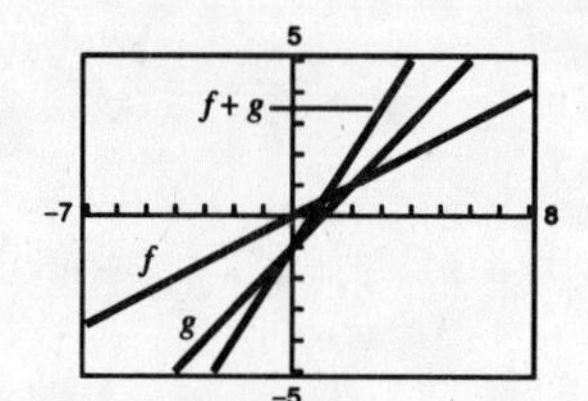

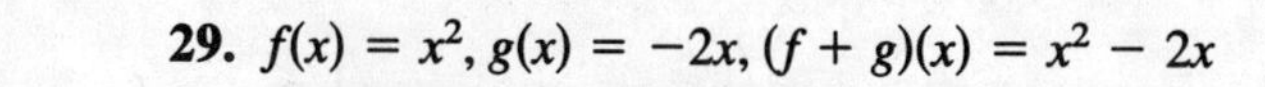

29. $f(x) = x^2, g(x) = -2x, (f + g)(x) = x^2 - 2x$

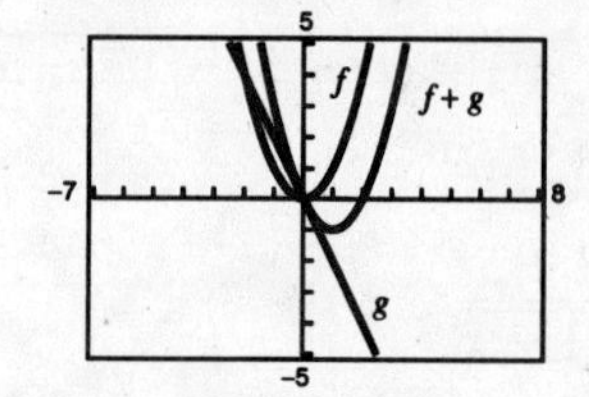

31. $f(x) = 3x, g(x) = -\dfrac{x^3}{10}, (f + g)(x) = 3x - \dfrac{x^3}{10}$

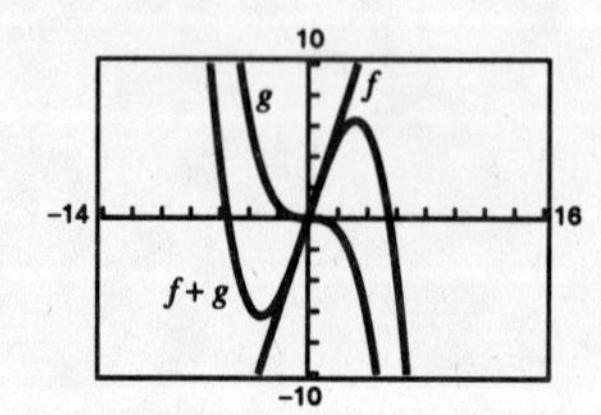

For $0 \leq x \leq 2, f(x)$ contributes more to the magnitude.
For $x > 6, g(x)$ contributes more to the magnitude.

33. $f(x) = 3x + 2, g(x) = -\sqrt{x + 5}, (f + g)(x) = 3x + 2 - \sqrt{x + 5}$

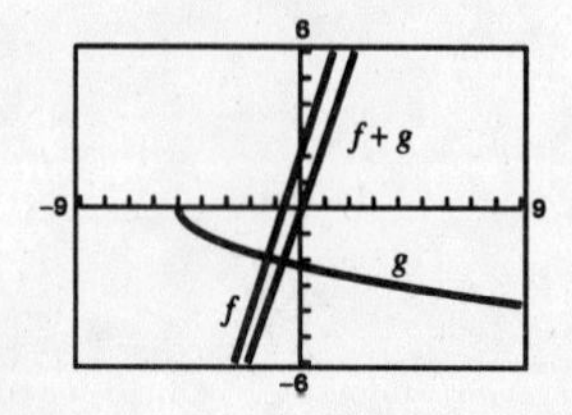

$f(x) = 3x + 2$ contributes more to the magnitude in both intervals.

35. $f(x) = x^2, g(x) = x - 1$

(a) $(f \circ g)(x) = f(g(x)) = f(x - 1) = (x - 1)^2$

(b) $(g \circ f)(x) = g(f(x)) = g(x^2) = x^2 - 1$

37. $f(x) = 3x + 5, g(x) = 5 - x$

(a) $(f \circ g)(x) = f(g(x)) = f(5 - x) = 3(5 - x) + 5 = 20 - 3x$

(b) $(g \circ f)(x) = g(f(x)) = g(3x + 5) = 5 - (3x + 5) = -3x$

39. (a) $(f \circ g)(x) = f(g(x)) = f(x^2) = \sqrt{x^2 + 4}$

$(g \circ f)(x) = g(f(x)) = g(\sqrt{x + 4}) = (\sqrt{x + 4})^2$

$= x + 4, x \geq -4$

(b)

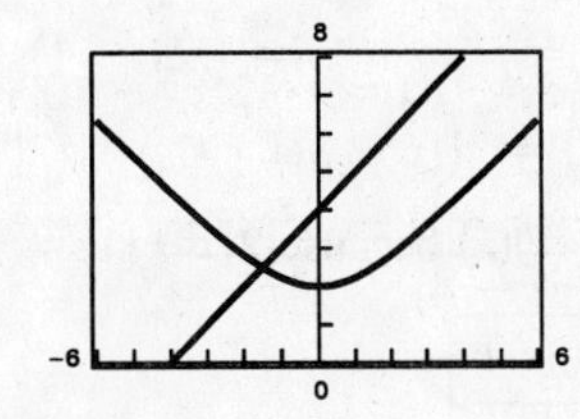

They are not equal.

41. (a) $(f \circ g)(x) = f(g(x)) = f(3x + 1)$

$= \frac{1}{3}(3x + 1) - 3 = x - \frac{8}{3}$

(b) $(g \circ f)(x) = g(f(x)) = g(\frac{1}{3}x - 3)$

$= 3(\frac{1}{3}x - 3) + 1 = x - 8$

(b)

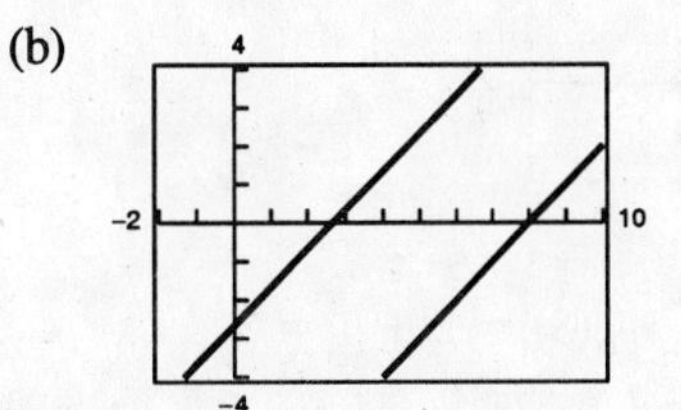

They are not equal.

43. (a) $(f \circ g)(x) = f(g(x)) = f(x^6) = (x^6)^{2/3} = x^4$

$(g \circ f)(x) = g(f(x)) = g(x^{2/3}) = (x^{2/3})^6 = x^4$

(b)

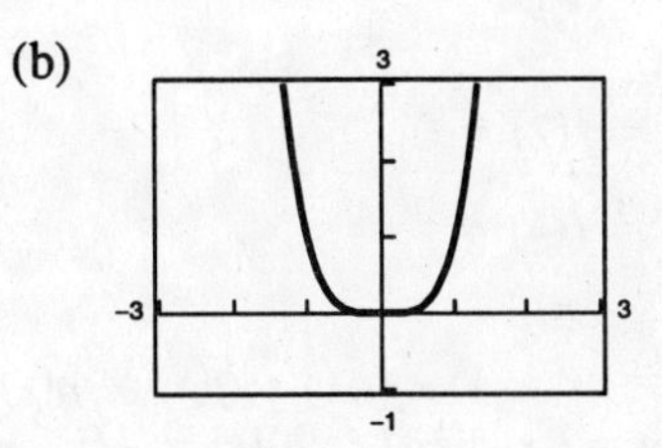

They are equal.

45. (a) $(f \circ g)(x) = f(g(x)) = f(4 - x) = 5(4 - x) + 4 = 24 - 5x$

$(g \circ f)(x) = g(f(x)) = g(5x + 4) = 4 - (5x + 4) = -5x$

(b) No, $(f \circ g)(x) \neq (g \circ f)(x)$ because $24 - 5x \neq -5x$.

(c)

x	$f(g(x))$	$g(f(x))$
0	24	0
1	19	−5
2	14	−10
3	9	−15

47. (a) $(f \circ g)(x) = f(g(x)) = f(x^2 - 5) = \sqrt{(x^2 - 5) + 6} = \sqrt{x^2 + 1}$

$(g \circ f)(x) = g(f(x)) = g(\sqrt{x + 6}) = (\sqrt{x + 6})^2 - 5 = (x + 6) - 5 = x + 1, x \geq -6$

(b) No, $(f \circ g)(x) \neq (g \circ f)(x)$ because $\sqrt{x^2 + 1} \neq x + 1$

(c)

x	$f(g(x))$	$g(f(x))$
0	1	1
-2	$\sqrt{5}$	-1
3	$\sqrt{10}$	4

49. (a) $(f \circ g)(x) = f(g(x)) = f(2x - 1) = |(2x - 1) + 3| = |2x + 2| = 2|x + 1|$

$(g \circ f)(x) = g(f(x)) = g(|x + 3|) = 2|x + 3| - 1$

(b) No, $(f \circ g)(x) \neq (g \circ f)(x)$ because $2|x + 1| \neq 2|x + 3| - 1$

(c)

x	$f(g(x))$	$g(f(x))$
-1	0	3
0	2	5
1	4	7

51. (a) $(f + g)(3) = f(3) + g(3) = 2 + 1 = 3$

(b) $\left(\frac{f}{g}\right)(2) = \frac{f(2)}{g(2)} = \frac{0}{2} = 0$

53. (a) $(f \circ g)(2) = f(g(2)) = f(2) = 0$

(b) $(g \circ f)(2) = g(f(2)) = g(0) = 4$

55. (a) $(f \circ f)(3) = f(f(3)) = f(2) = 0$

(b) $(f \circ f)(4) = f(f(4)) = f(4) = 4$

57. Let $f(x) = x^2$ and $g(x) = 2x + 1$, then $(f \circ g)(x) = h(x)$. This is not a unique solution. For example, if $f(x) = (x + 1)^2$ and $g(x) = 2x$, then $(f \circ g)(x) = h(x)$ as well.

59. Let $f(x) = \sqrt[3]{x}$ and $g(x) = x^2 - 4$, then $(f \circ g)(x) = h(x)$. This answer is not unique. Other possibilities may be:

$f(x) = \sqrt[3]{x - 4}$ and $g(x) = x^2$ or
$f(x) = \sqrt[3]{-x}$ and $g(x) = 4 - x^2$ or
$f(x) = \sqrt[9]{x}$ and $g(x) = (4 - x^2)^3$

61. Let $f(x) = 1/x$ and $g(x) = x + 2$, then $(f \circ g)(x) = h(x)$. Again, this is not a unique solution. Other possibilities may be:

$f(x) = \frac{1}{x + 2}$ and $g(x) = x$

or $f(x) = \frac{1}{x + 1}$ and $g(x) = x + 1$

63. Let $f(x) = x^2 + 2x$ and $g(x) = x + 4$. Then $(f \circ g)(x) = h(x)$. (Answer is not unique.)

65. (a) The domain of $f(x) = \sqrt{x}$ is $x \geq 0$.

(b) The domain of $g(x) = x^2 + 1$ is all real numbers.

(c) $(f \circ g)(x) = f(g(x)) = f(x^2 + 1) = \sqrt{x^2 + 1}$

The domain of $f \circ g$ is all real numbers.

67. (a) The domain of $f(x) = \dfrac{1}{x}$ is all $x \neq 0$.

(b) The domain of $g(x) = x + 3$ is all real numbers.

(c) The domain of $(f \circ g)(x) = f(x + 3) = \dfrac{1}{x + 3}$ is all $x \neq -3$.

69. (a) The domain of $f(x) = \dfrac{2}{|x|}$ is all $x \neq 0$.

(b) The domain of $g(x) = x - 1$ is all real numbers.

(c) The domain of $(f \circ g)(x) = f(x - 1) = \dfrac{2}{|x - 1|}$ is all $x \neq 1$.

71. $f(x) = 3x - 4$

$$\begin{aligned}\frac{f(x+h) - f(x)}{h} &= \frac{[3(x+h) - 4] - (3x - 4)}{h}\\ &= \frac{3x + 3h - 4 - 3x + 4}{h}\\ &= \frac{3h}{h}\\ &= 3,\ h \neq 0\end{aligned}$$

73. $f(x) = 1 - x^2$

$$\begin{aligned}\frac{f(x+h) - f(x)}{h} &= \frac{[1 - (x+h)^2] - [1 - x^2]}{h}\\ &= \frac{1 - x^2 - 2xh - h^2 - 1 + x^2}{h}\\ &= \frac{-2xh - h^2}{h}\\ &= \frac{(-2x - h)h}{h} = -2x - h,\ h \neq 0\end{aligned}$$

75. $f(x) = \dfrac{4}{x}$

$$\begin{aligned}\frac{f(x+h) - f(x)}{h} &= \frac{\dfrac{4}{x+h} - \dfrac{4}{x}}{h} = \frac{\dfrac{4x - 4(x+h)}{x(x+h)}}{\dfrac{h}{1}}\\ &= \frac{4x - 4x - 4h}{x(x+h)} \cdot \frac{1}{h}\\ &= \frac{-4\cancel{h}}{x(x+h)} \cdot \frac{1}{\cancel{h}}\\ &= \frac{-4}{x(x+h)},\ h \neq 0\end{aligned}$$

77. $f(x) = \sqrt{2x + 1}$

$$\begin{aligned}\frac{f(x+h) - f(x)}{h} &= \frac{\sqrt{2(x+h) + 1} - \sqrt{2x + 1}}{h} \cdot \frac{\sqrt{2(x+h) + 1} + \sqrt{2x + 1}}{\sqrt{2(x+h) + 1} + \sqrt{2x + 1}}\\ &= \frac{[2(x+h) + 1] - [2x + 1]}{h\left[\sqrt{2x + 2h + 1} + \sqrt{2x + 1}\right]}\\ &= \frac{2h}{h\left[\sqrt{2x + 2h + 1} + \sqrt{2x + 1}\right]}\\ &= \frac{2}{\sqrt{2x + 2h + 1} + \sqrt{2x + 1}},\ h \neq 0\end{aligned}$$

79. (a) $T(x) = R(x) + B(x) = \frac{3}{4}x + \frac{1}{15}x^2$

(b)

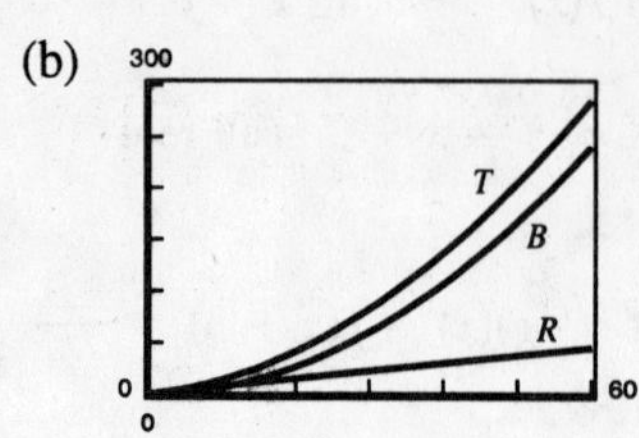

(c) $B(x)$ contributes more to $T(x)$ at higher speeds.

81. $y_1 = -0.587t^2 + 7.661t + 144.905$

$y_2 = 16.579t + 245.064$

$y_3 = 1.836t + 21.921$

83. $(A \circ r)(t)$ gives the area of the circle as a function of time.

$$(A \circ r)(t) = A(r(t))$$
$$= A(0.6t)$$
$$= \pi(0.6t)^2 = 0.36\pi t^2$$

85. (a) $(C \circ x)(t) = C(x(t))$

$$= 60(50t) + 750$$
$$= 3000t + 750$$

$C \circ x$ represents the cost after t production hours.

(b)

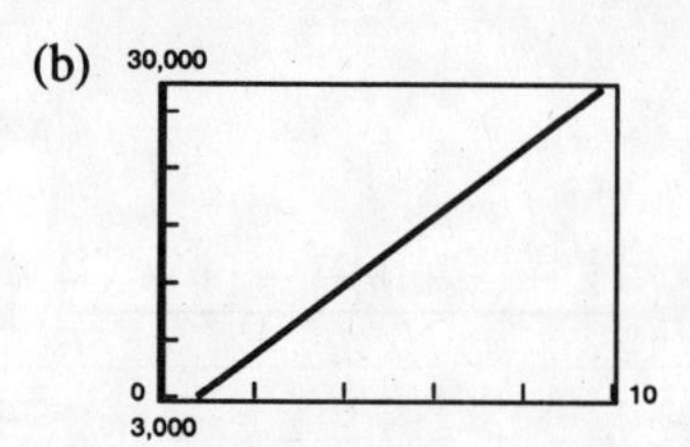

The cost increases to \$15,000 when $t = 4.75$ hours.

87. $g(f(x)) = g(x - 500{,}000) = 0.03(x - 500{,}000)$

represents 3 percent of the amount over \$500,000.

89.

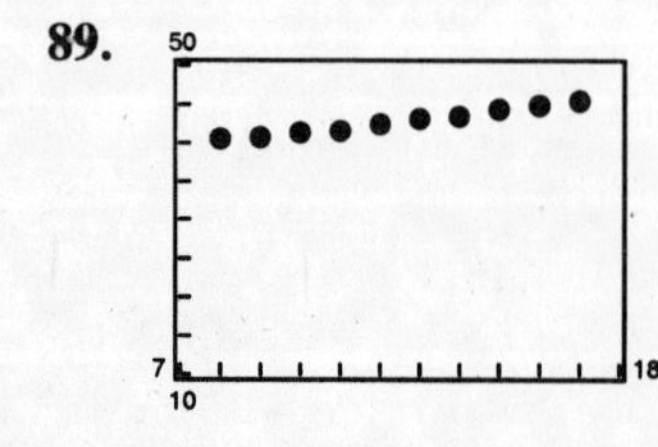

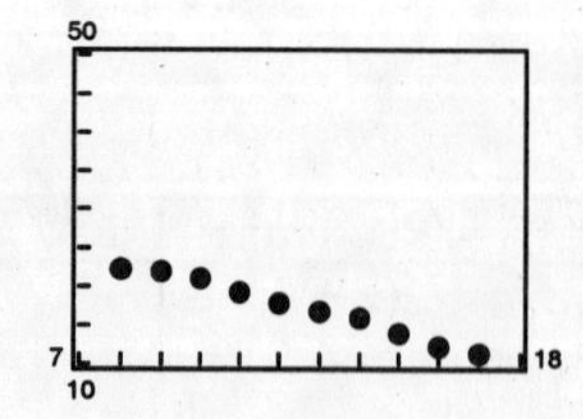

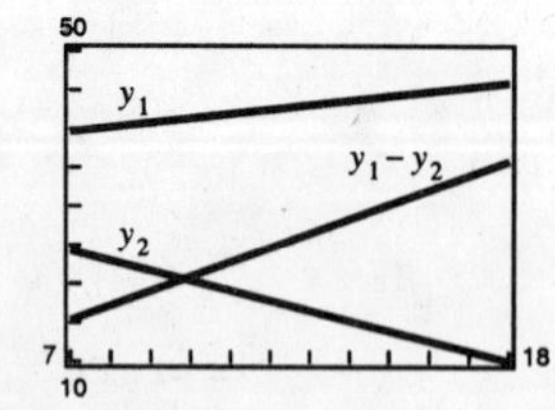

$y_1 = 0.57x + 35.6$

$y_2 = -1.29x + 33.3$

Both data sets appear to be linear.

The difference between circulations is increasing.

91. True. $(f \circ g)(x) = f(g(x))$ is only defined if $g(x)$ is the domain of f.

93. The product of an odd function and an even function is odd. Let $f(x)$ be even, $g(x)$ odd and $h(x) = f(x)g(x)$ their product. Then $h(-x) = f(-x)g(-x) = f(x)(-g(x)) = -f(x)g(x) = -h(x)$. Thus, h is odd.

95. $f(x) = \frac{1}{2}[f(x) + f(-x)] + \frac{1}{2}[f(x) - f(-x)]$

$= \quad g(x) \qquad\qquad + \quad h(x)$

where g is even and h is odd by Exercise 94.

97. $(0, -5), (1, -5), (2, -7)$ (other answers possible)

99. $(\sqrt{24}, 0), (-\sqrt{24}, 0), (0, \sqrt{24})$ (other answers possible)

101. $y - (-2) = \dfrac{8 - (-2)}{-3 - (-4)}(x - (-4))$

$y + 2 = 10(x + 4)$

$y - 10x - 38 = 0$

103. 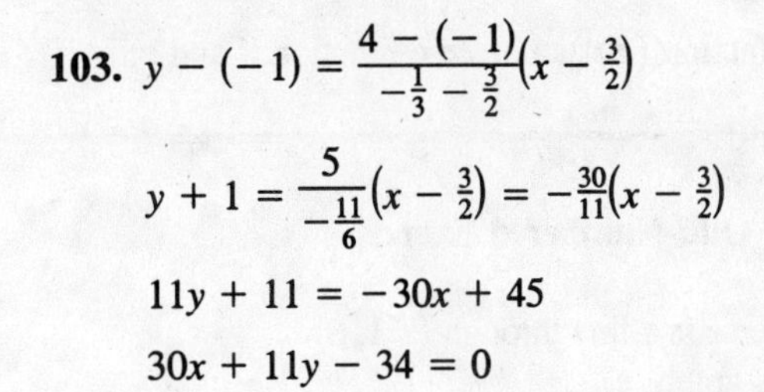

$y - (-1) = \dfrac{4 - (-1)}{-\frac{1}{3} - \frac{3}{2}}\left(x - \frac{3}{2}\right)$

$y + 1 = \dfrac{5}{-\frac{11}{6}}\left(x - \frac{3}{2}\right) = -\frac{30}{11}\left(x - \frac{3}{2}\right)$

$11y + 11 = -30x + 45$

$30x + 11y - 34 = 0$

105. $f(2) = 1$ and $f(-4) = -3$. Thus, for $g(x) = f(x - 4)$, $g(6) = f(2) = 1$ and $g(0) = f(-4) = -3$.

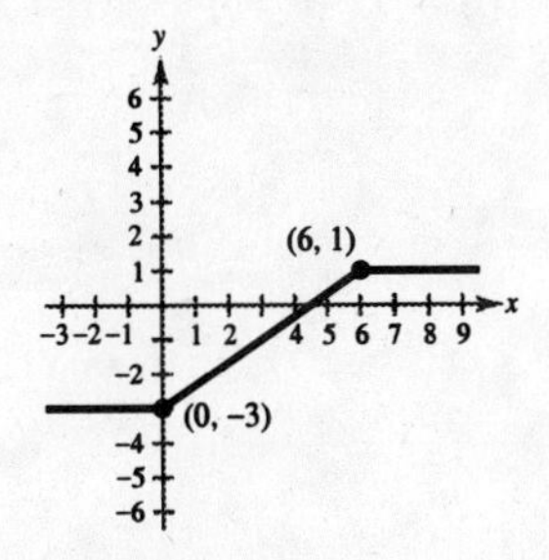

107. $f(2) = 1$ and $f(-4) = -3$. Thus, $g(x) = f(x) + 4$ satisfies $g(2) = 5$ and $g(-4) = 1$.

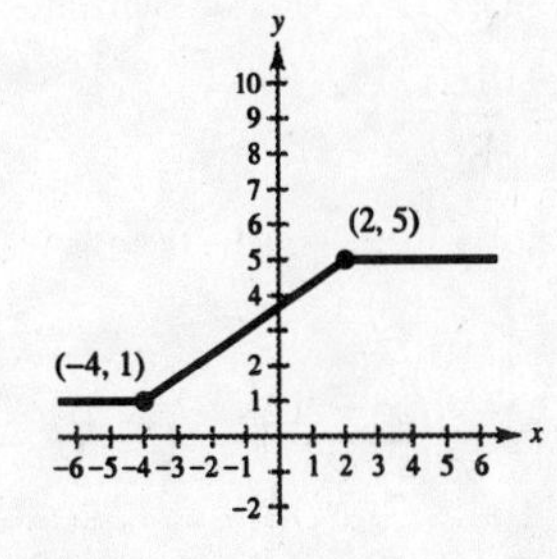

109. $f(2) = 1$ and $f(-4) = -3$. Thus, $g(x) = 2f(x)$ satisfies $g(2) = 2$ and $g(-4) = -6$.

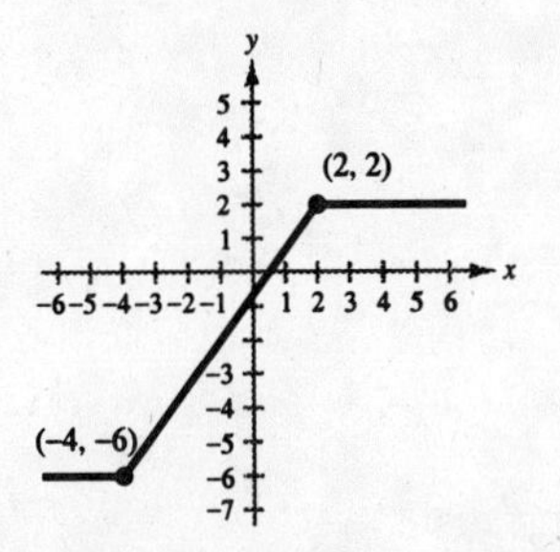

Section 1.5 Inverse Functions

- Two functions f and g are inverses of each other if $f(g(x)) = x$ for every x in the domain of g and $g(f(x)) = x$ for every x in the domain of f.
- Be able to find the inverse of a function, if it exists.
  1. Replace $f(x)$ with y.
  2. Interchange x and y.
  3. Solve for y. If this equation represents y as a function of x, then you have found $f^{-1}(x)$. If this equation does not represent y as a function of x, then f does not have an inverse function.
- A function f has an inverse function if and only if no **horizontal** line crosses the graph of f at more than one point.
- A function f has an inverse function if and only if f is one-to-one.

Solutions to Odd-Numbered Exercises

1. The inverse is a line through $(-1, 0)$.
Matches graph (c).

3. The inverse is half a parabola starting at $(1, 0)$.
Matches graph (a).

5. $f^{-1}(x) = \dfrac{x}{8} = \dfrac{1}{8}x$

$$f(f^{-1}(x)) = f\left(\frac{x}{8}\right) = 8\left(\frac{x}{8}\right) = x$$

$$f^{-1}(f(x)) = f^{-1}(8x) = \frac{8x}{8} = x$$

7. $f^{-1}(x) = x - 10$

$$f(f^{-1}(x)) = f(x - 10) = (x - 10) + 10 = x$$

$$f^{-1}(f(x)) = f^{-1}(x + 10) = (x + 10) - 10 = x$$

9. $f^{-1}(x) = \dfrac{x - 1}{2}$

$$f(f^{-1}(x)) = f\left(\frac{x-1}{2}\right) = 2\left(\frac{x-1}{2}\right) + 1 = (x - 1) + 1 = x$$

$$f^{-1}(f(x)) = f^{-1}(2x + 1) = \frac{(2x + 1) - 1}{2} = \frac{2x}{2} = x$$

11. $f^{-1}(x) = x^3$

$$f(f^{-1}(x)) = f(x^3) = \sqrt[3]{x^3} = x$$

$$f^{-1}(f(x)) = f^{-1}\left(\sqrt[3]{x}\right) = \left(\sqrt[3]{x}\right)^3 = x$$

13. (a) $f(g(x)) = f\left(\dfrac{x}{2}\right) = 2\left(\dfrac{x}{2}\right) = x$

$$g(f(x)) = g(2x) = \frac{2x}{2} = x$$

(b)

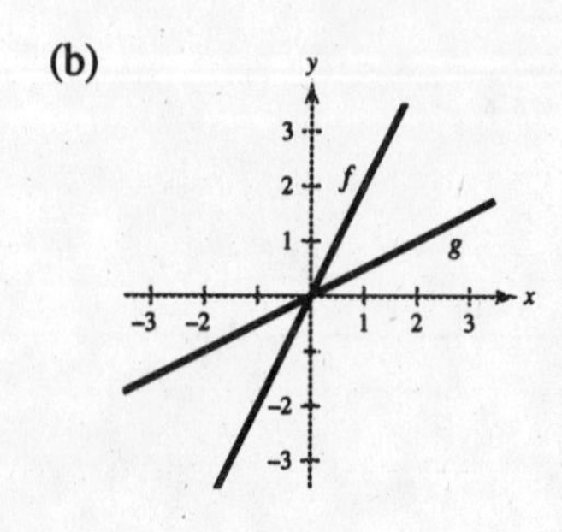

15. (a) 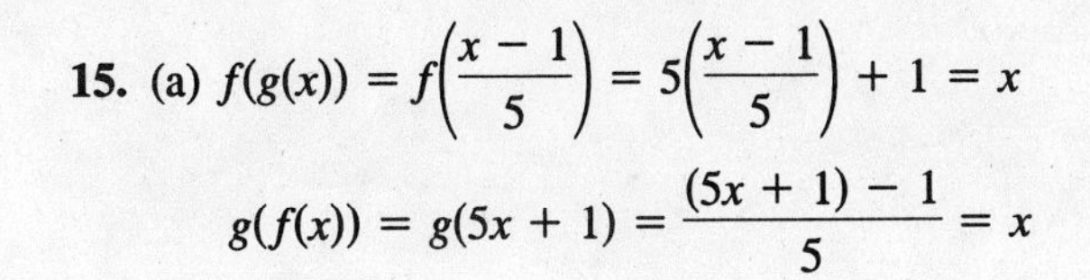

$$f(g(x)) = f\left(\frac{x-1}{5}\right) = 5\left(\frac{x-1}{5}\right) + 1 = x$$

$$g(f(x)) = g(5x+1) = \frac{(5x+1)-1}{5} = x$$

(b)

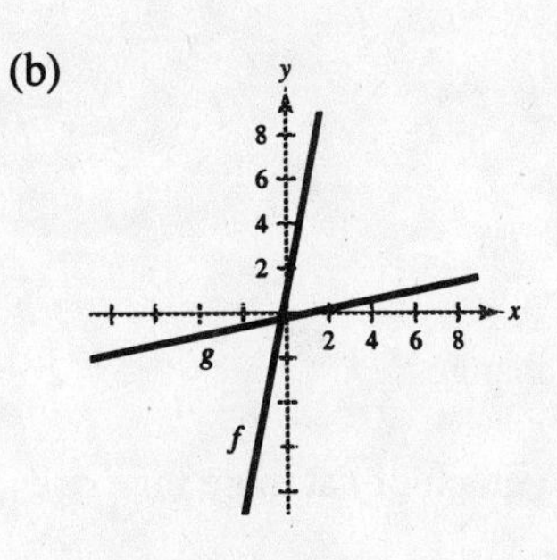

17. (a) $f(g(x)) = f\left(\sqrt[3]{x}\right) = \left(\sqrt[3]{x}\right)^3 = x$

$g(f(x)) = g(x^3) = \sqrt[3]{x^3} = x$

(b)

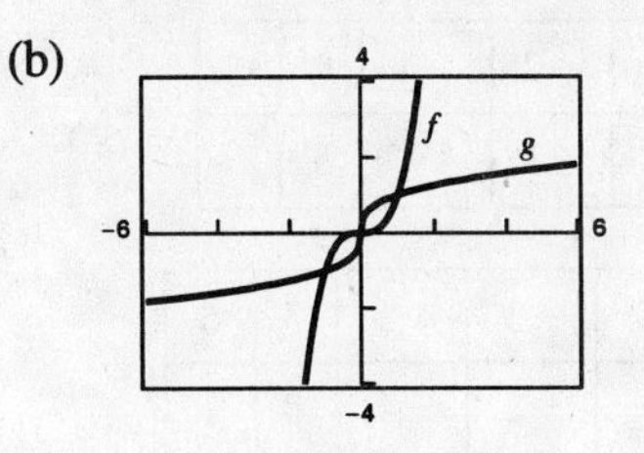

Reflections in the line $y = x$

19. (a) $f(g(x)) = f(x^2 + 4),\ x \geq 0$

$= \sqrt{(x^2 + 4) - 4} = x$

$g(f(x)) = g\left(\sqrt{x-4}\right)$

$= \left(\sqrt{x-4}\right)^2 + 4 = x$

(b)

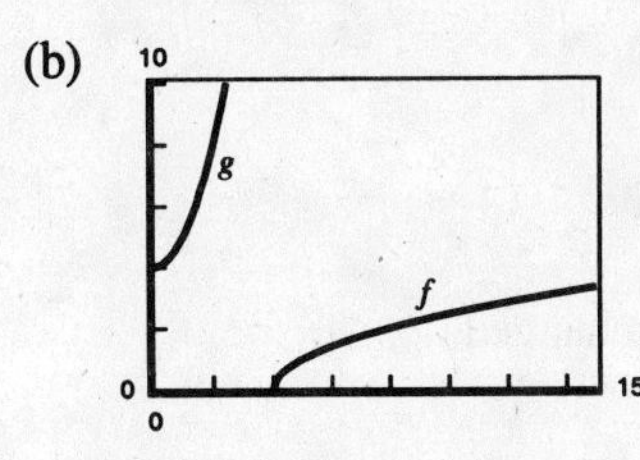

Reflections in the line $y = x$

21. (a) $f(g(x)) = f\left(\sqrt[3]{1-x}\right) = 1 - \left(\sqrt[3]{1-x}\right)^3 = 1 - (1-x) = x$

$g(f(x)) = g(1 - x^3) = \sqrt[3]{1 - (1 - x^3)} = \sqrt[3]{x^3} = x$

(b)

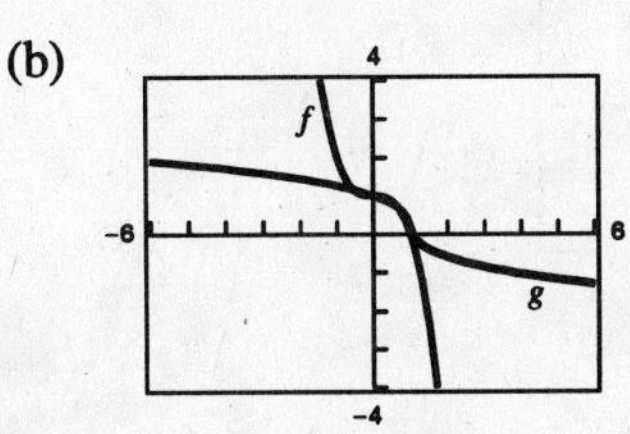

Reflections in the line $y = x$

23. (a) $f(g(x)) = f\left(-\frac{2x+6}{7}\right) = -\frac{7}{2}\left(-\frac{2x+6}{7}\right) - 3 = \frac{2x+6}{2} - 3 = (x+3) - 3 = x$

$g(f(x)) = g\left(-\frac{7}{2}x - 3\right) = -\frac{2\left(-\frac{7}{2}x - 3\right) + 6}{7} = -\frac{-7x - 6 + 6}{7} = \frac{7x}{7} = x$

(b)

x	2	0	-2	-4	-6
$f(x)$	-10	-3	4	11	18

x	-10	-3	4	11	18
$g(x)$	2	0	-2	-4	-6

Note that the entries in the tables are the same except that the rows are interchanged.

25. (a) $f(g(x)) = f(\sqrt[3]{x-5}) = [\sqrt[3]{x-5}]^3 + 5 = (x-5) + 5 = x$

$g(f(x)) = g(x^3 + 5) = \sqrt[3]{(x^3+5) - 5} = \sqrt[3]{x^3} = x$

(b)

x	-3	-2	-1	0	1
$f(x)$	-22	-3	4	5	6

x	-22	-3	4	5	6
$g(x)$	-3	-2	-1	0	1

Note that the entries in the tables are the same except that the rows are interchanged.

27. (a) $f(g(x)) = f(8 + x^2) = -\sqrt{(8 + x^2) - 8} = -\sqrt{x^2} = -(-x) = x$

[Since $x \le 0$, $\sqrt{x^2} = -x$]

$g(f(x)) = g(-\sqrt{x-8}) = 8 + [-\sqrt{x-8}]^2 = 8 + (x-8) = x$

(b)

x	8	9	12	17	24
$f(x)$	0	-1	-2	-3	-4

x	0	-1	-2	-3	-4
$g(x)$	8	9	12	17	24

Note that the entries in the tables are the same except that the rows are interchanged.

29. Since no horizontal line crosses the graph of f at more than one point, f **has** an inverse.

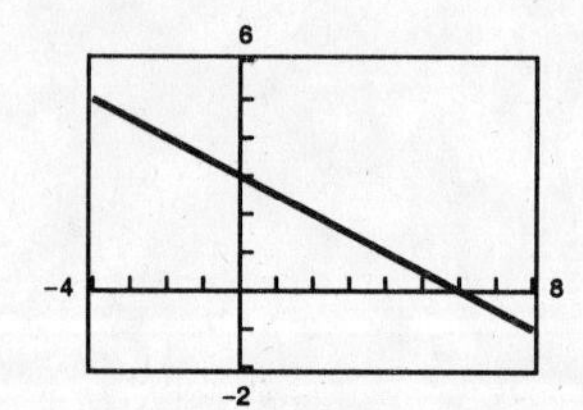

31. Since some horizontal lines cross the graph of f twice, f does **not** have an inverse.

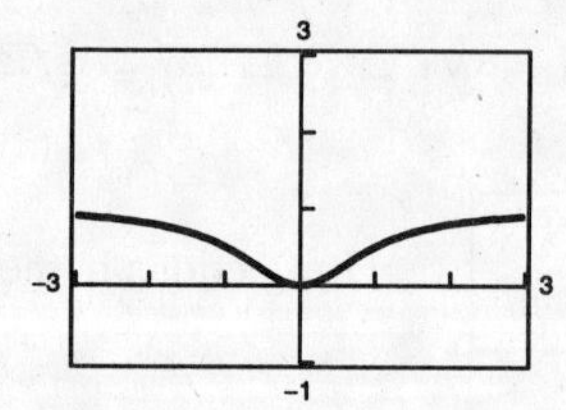

33. No, because some horizontal lines intersect the graph more than once, h does not have an inverse.

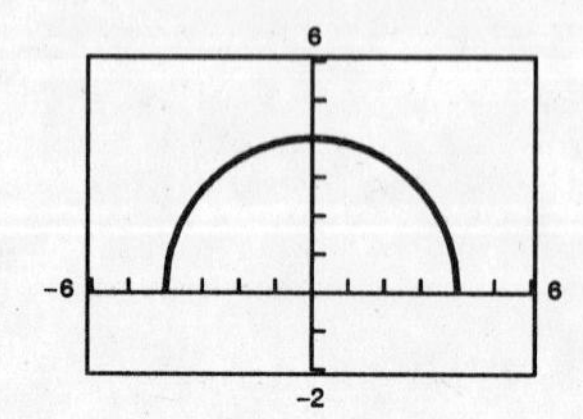

35. Yes, because no horizontal lines intersect the graph at more than one point, f has an inverse.

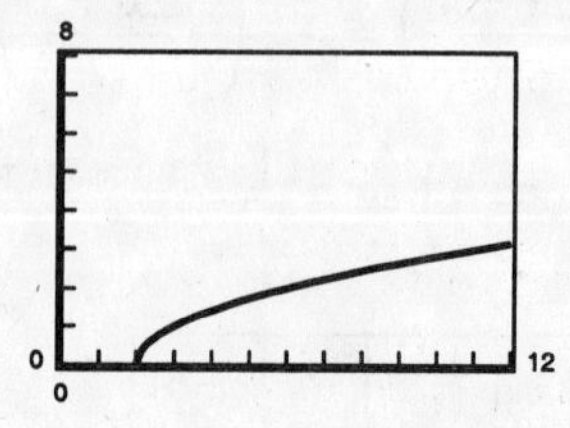

37. f does not pass the horizontal line test, so f has no inverse.

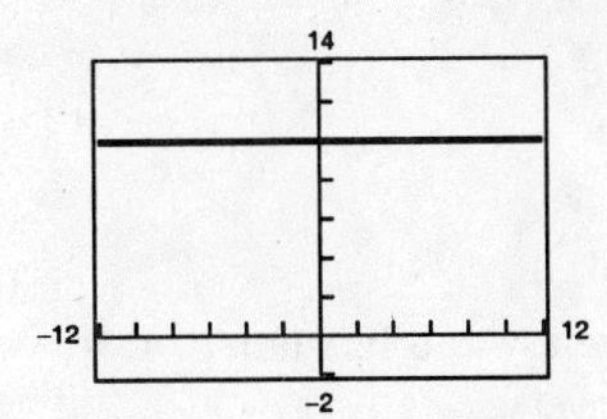

39. g passes the horizontal line test, so g has an inverse.

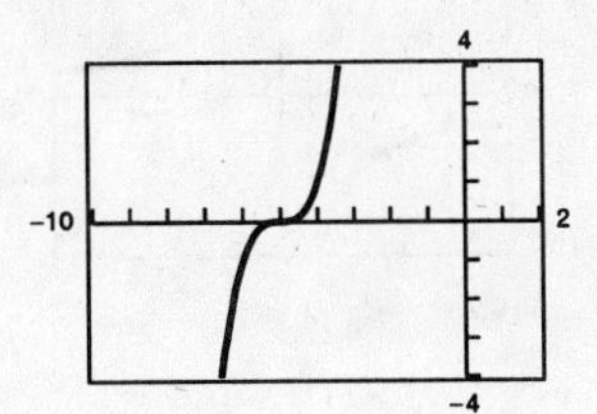

41. $h(x) = |x + 4| - |x - 4|$

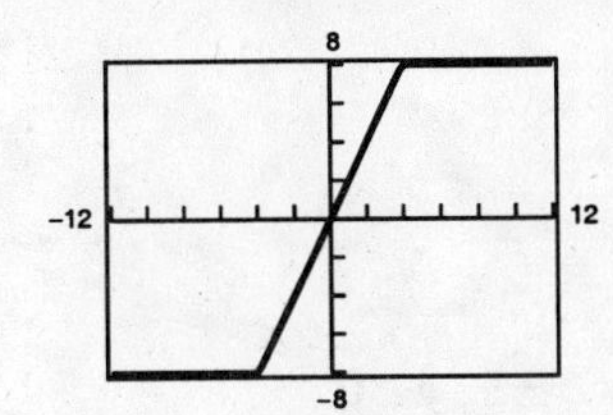

h does not pass the horizontal line test, so h does not have an inverse.

43.
$$f(x) = 2x - 3$$
$$y = 2x - 3$$
$$x = 2y - 3$$
$$y = \frac{x + 3}{2}$$
$$f^{-1}(x) = \frac{x + 3}{2}$$

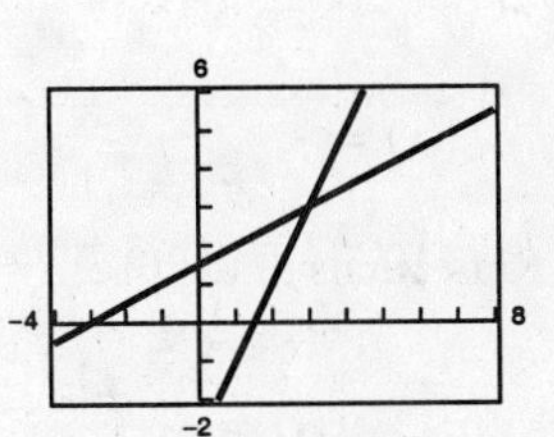

Reflections in the line $y = x$

45.
$$f(x) = x^5$$
$$y = x^5$$
$$x = y^5$$
$$y = \sqrt[5]{x}$$
$$f^{-1}(x) = \sqrt[5]{x}$$

Reflections in the line $y = x$

47.
$$f(x) = \sqrt{x}$$
$$y = \sqrt{x}$$
$$x = \sqrt{y}$$
$$y = x^2$$
$$f^{-1}(x) = x^2,\ x \geq 0$$

Reflections in the line $y = x$

49.
$$f(x) = \sqrt{4 - x^2}, 0 \leq x \leq 2$$
$$y = \sqrt{4 - x^2}$$
$$x = \sqrt{4 - y^2}$$
$$x^2 = 4 - y^2$$
$$y^2 = 4 - x^2$$
$$y = \sqrt{4 - x^2}$$
$$f^{-1}(x) = \sqrt{4 - x^2}, 0 \leq x \leq 2$$

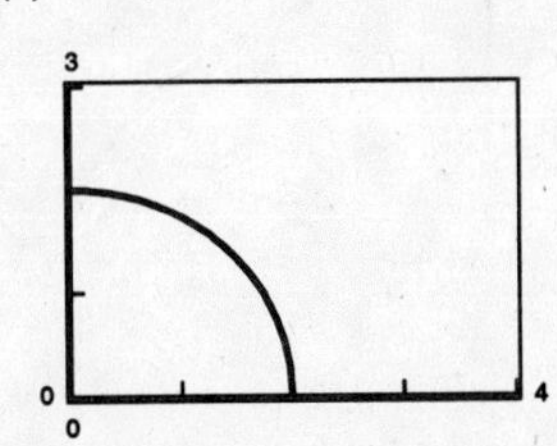

Reflections in the line $y = x$

51.
$$f(x) = \sqrt[3]{x - 1}$$
$$y = \sqrt[3]{x - 1}$$
$$x = \sqrt[3]{y - 1}$$
$$x^3 = y - 1$$
$$y = x^3 + 1$$
$$f^{-1}(x) = x^3 + 1$$

Reflections in the line $y = x$

53. $f(x) = \dfrac{4}{x}$

$y = \dfrac{4}{x}$

$x = \dfrac{4}{y}$

$xy = 4$

$y = \dfrac{4}{x}$

$f^{-1}(x) = \dfrac{4}{x}$

Reflections in the line $y = x$

55. $f(x) = x^4$

$y = x^4$

$x = y^4$

$y = \pm\sqrt[4]{x}$

f is not one-to-one.

This does not represent y as a function of x. f does not have an inverse.

57. $f(x) = \dfrac{3x + 4}{5}$

$y = \dfrac{3x + 4}{5}$

$x = \dfrac{3y + 4}{5}$

$5x = 3y + 4$

$5x - 4 = 3y$

$(5x - 4)/3 = y$

$f^{-1}(x) = \dfrac{5x - 4}{3}$

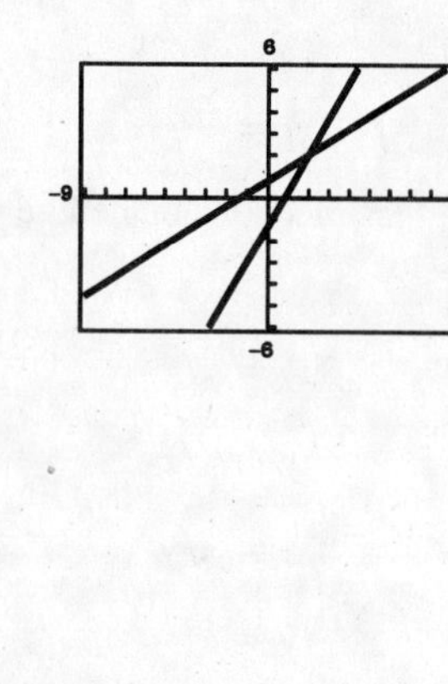

f is one-to-one and has an inverse.

59. $f(x) = (x + 3)^2,\ x \geq -3,\ y \geq 0$

$y = (x + 3)^2,\ x \geq -3,\ y \geq 0$

$x = (y + 3)^2,\ y \geq -3,\ x \geq 0$

$\sqrt{x} = y + 3\ ,\ y \geq -3,\ x \geq 0$

$y = \sqrt{x} - 3,\ x \geq 0,\ y \geq -3$

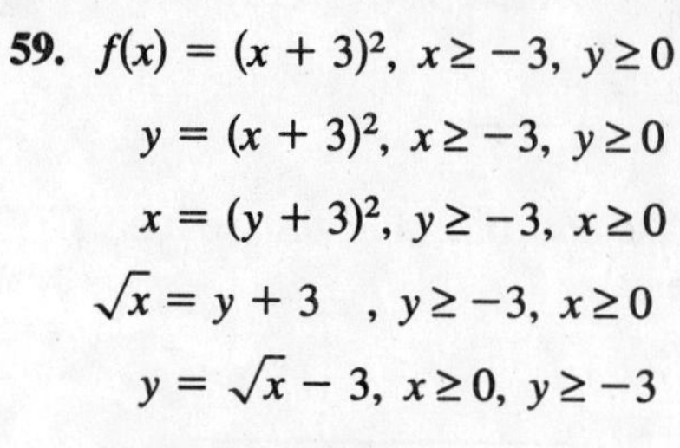

f is one-to-one.

This is a function of x, so f has an inverse.

$f^{-1}(x) = \sqrt{x} - 3,\ x \geq 0$

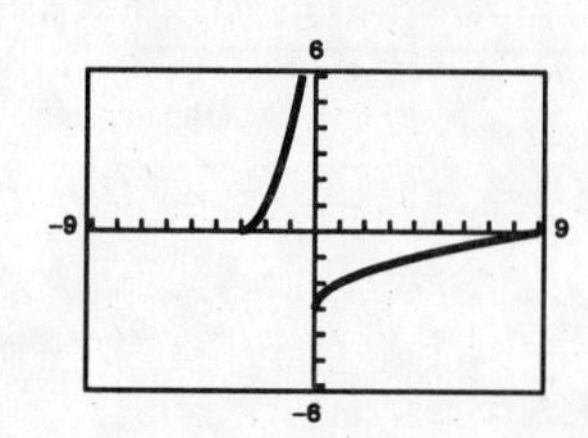

61. $h(x) = \dfrac{4}{x^2}$ is not one-to-one, and does not have an inverse. For example, $h(1) = h(-1) = 4$.

63. $f(x) = \sqrt{2x + 3} \implies x \geq -\dfrac{3}{2},\ y \geq 0$

$y = \sqrt{2x + 3},\ x \geq -\dfrac{3}{2},\ y \geq 0$

$x = \sqrt{2y + 3},\ y \geq -\dfrac{3}{2},\ x \geq 0$

$x^2 = 2y + 3,\ x \geq 0,\ y \geq -\dfrac{3}{2}$

$y = \dfrac{x^2 - 3}{2},\ x \geq 0,\ y \geq -\dfrac{3}{2}$

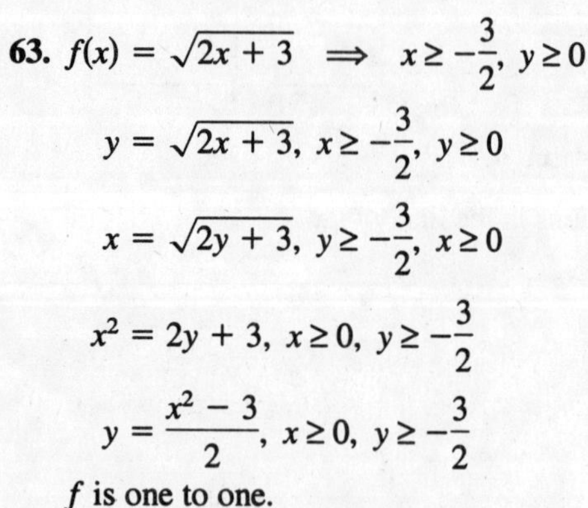

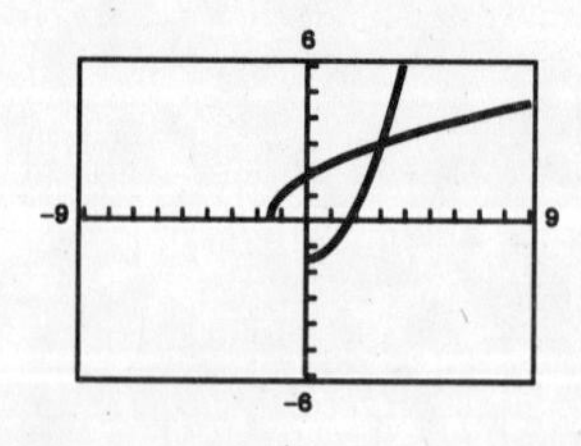

f is one to one.

This is a function of x, so f has an inverse.

$f^{-1}(x) = \dfrac{x^2 - 3}{2},\ x \geq 0$

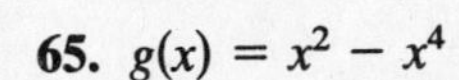

65. $g(x) = x^2 - x^4$

The graph fails the horizontal line test, so g does not have an inverse. g is not one-to-one.

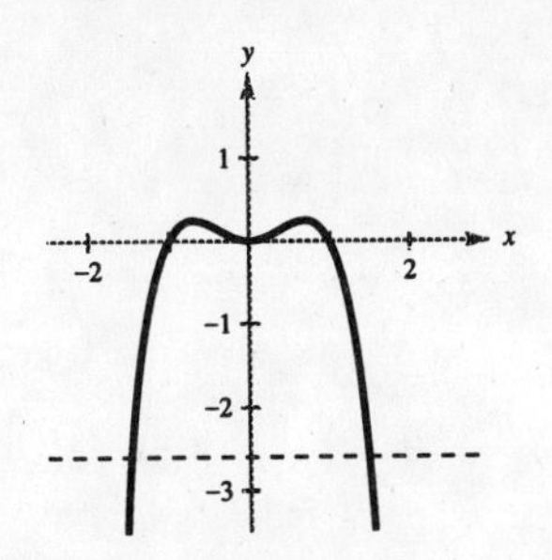

67. $f(x) = ax + b, a \neq 0$

$$y = ax + b$$
$$x = ay + b$$
$$x - b = ay$$
$$y = (x - b)/a$$

f is one-to-one and has an inverse, $f^{-1}(x) = \dfrac{x - b}{a}$.

69. If we let $f(x) = (x - 2)^2$, $x \geq 2$, then f has an inverse. [Note: we could also let $x \leq 2$.]

$$f(x) = (x - 2)^2,\ x \geq 2,\ y \geq 0$$
$$y = (x - 2)^2,\ x \geq 2,\ y \geq 0$$
$$x = (y - 2)^2,\ x \geq 0,\ y \geq 2$$
$$\sqrt{x} = y - 2,\quad x \geq 0,\ y \geq 2$$
$$\sqrt{x} + 2 = y,\quad x \geq 0,\ y \geq 2$$

Thus, $f^{-1}(x) = \sqrt{x} + 2,\ x \geq 0$.

71. If we let $f(x) = |x + 2|$, $x \geq -2$, then f has an inverse. [Note: we could also let $x \leq -2$.]

$$f(x) = |x + 2|,\ x \geq -2$$
$$f(x) = x + 2 \text{ when } x \geq -2$$
$$y = x + 2,\ x \geq -2,\ y \geq 0$$
$$x = y + 2,\ x \geq 0,\ y \geq -2$$
$$x - 2 = y,\quad x \geq 0,\ y \geq -2$$

Thus, $f^{-1}(x) = x - 2,\ x \geq 0$.

73.

x	$f(x)$
-2	-4
-1	-2
1	2
3	3

x	$f^{-1}(x)$
-4	-2
-2	-1
2	1
3	3

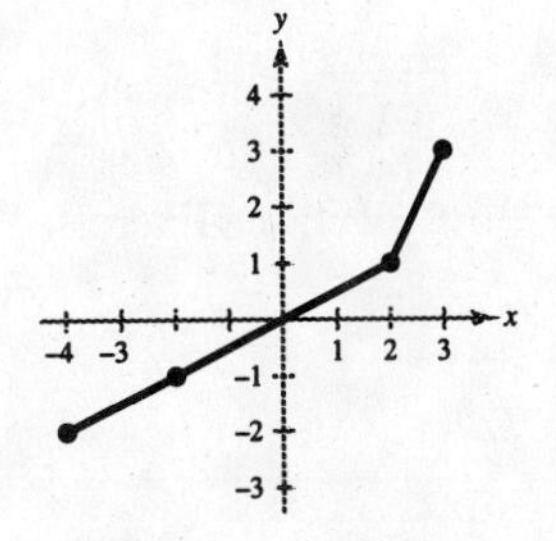

75. $f(x) = x^3 + x + 1$

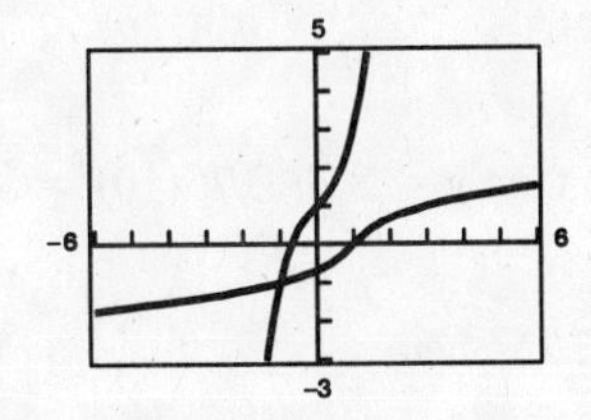

The graph of the inverse relation is an inverse function since it satisfies the vertical line test.

77. $g(x) = \dfrac{3x^2}{x^2 + 1}$

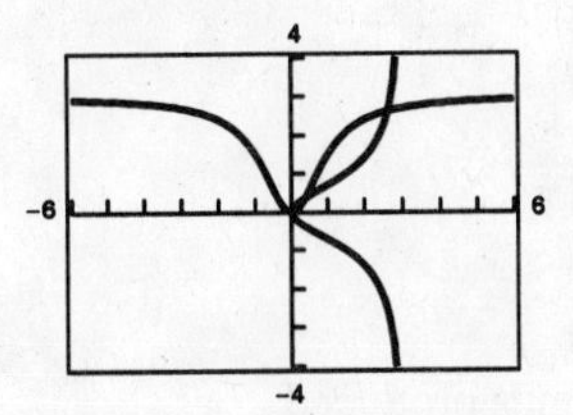

The graph of the inverse relation is not an inverse function since it does not satisfy the vertical line test.

In Exercises 79, 81, and 83, $f(x) = \frac{1}{8}x - 3$, $f^{-1}(x) = 8(x + 3)$, $g(x) = x^3$, $g^{-1}(x) = \sqrt[3]{x}$.

79. $(f^{-1} \circ g^{-1})(1) = f^{-1}(g^{-1}(1)) = f^{-1}(\sqrt[3]{1}) = 8(\sqrt[3]{1} + 3) = 8(1 + 3) = 32$

81. $(f^{-1} \circ f^{-1})(6) = f^{-1}(f^{-1}(6)) = f^{-1}(8[6 + 3]) = f^{-1}(72) = 8(72 + 3) = 600$

83. $(f \circ g)(x) = f(g(x)) = f(x^3) = \frac{1}{8}x^3 - 3$. Now find the inverse of $(f \circ g)(x) = \frac{1}{8}x^3 - 3$:

$$y = \tfrac{1}{8}x^3 - 3$$
$$x = \tfrac{1}{8}y^3 - 3$$
$$x + 3 = \tfrac{1}{8}y^3$$
$$8(x + 3) = y^3$$
$$\sqrt[3]{8(x + 3)} = y$$
$$(f \circ g)^{-1}(x) = 2\sqrt[3]{x + 3}$$

Note: $(f \circ g)^{-1} = g^{-1} \circ f^{-1}$

In Exercises 85 and 87, $f(x) = x + 4$, $f^{-1}(x) = x - 4$, $g(x) = 2x - 5$, $g^{-1}(x) = \dfrac{x + 5}{2}$.

85. $(g^{-1} \circ f^{-1})(x) = g^{-1}(f^{-1}(x)) = g^{-1}(x - 4) = \dfrac{(x - 4) + 5}{2} = \dfrac{x + 1}{2}$

87. $(f \circ g)(x) = f(g(x)) = f(2x - 5) = (2x - 5) + 4 = 2x - 1$ Now find the inverse of $(f \circ g)(x) = 2x - 1$:

$$y = 2x - 1$$
$$x = 2y - 1$$
$$x + 1 = 2y$$
$$y = \frac{x + 1}{2}$$
$$(f \circ g)^{-1}(x) = \frac{x + 1}{2}$$

Note that $(f \circ g)^{-1}(x) = (g^{-1} \circ f^{-1})(x)$; see Exercise 94.

89. (a)
$$y = 8 + 0.75x$$
$$x = 8 + 0.75y$$
$$x - 8 = 0.75y$$
$$\frac{x - 8}{0.75} = y$$
$$y = f^{-1}(x) = \frac{x - 8}{0.75}$$

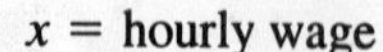

x = hourly wage
y = number of units produced

(b)

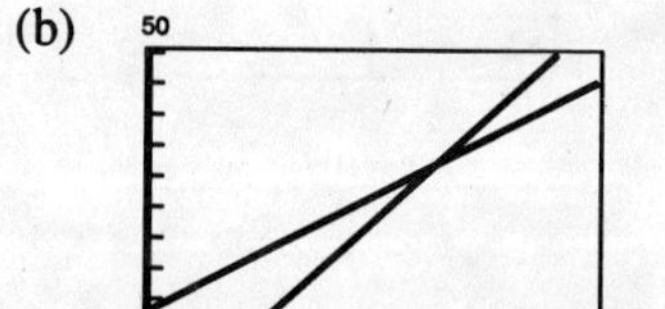

(c) If 10 units are produced, then $y = 8 + 0.75(10) = \$15.50$.

(d) If the hourly wage is \$22.25, then
$$y = \frac{22.25 - 8}{0.75} = 19 \text{ units.}$$

91. (a) Yes, f^{-1} exists because for each value of $f(t)$, there corresponds a unique value of t.

(b) f^{-1} indicates the year t corresponds to the total value of new car sales.

(c) $f^{-1}(456.2) = 5$ (or 1995)

(d) No, in this case the function f would not be one-to-one. $f(4) = f(8) = 430.6$.

93. True. If $(0, b)$ is the y-intercept of f, then $(b, 0)$ is the x-intercept of f^{-1}.

95. If f is one-to-one, then f^{-1} exists. If f is odd, then $f(-x) = -f(x)$. Consider $f(x) = y \leftrightarrow f^{-1}(y) = x$. Then $f^{-1}(-y) = f^{-1}(-f(x)) = f^{-1}(f(-x)) = -x = -f^{-1}(y)$. Thus, f^{-1} is odd.

97. $(f + g)(-(-2)) = (f + g)(2) = f(2) + g(2) = 3 + (-1) = 2$

99. $(fg)(-(-3)) = (fg)(3) = f(3)g(3) = (13)(0) = 0$

101.
$$y = 12x$$
$$x = 12y$$
$$y = \frac{1}{12}x$$
$$f^{-1}(x) = \frac{1}{12}x$$

103.
$$y = x^3 + 7$$
$$x = y^3 + 7$$
$$y^3 = x - 7$$
$$y = \sqrt[3]{x - 7}$$
$$f^{-1}(x) = \sqrt[3]{x - 7}$$

Review Exercises for Chapter 1

Solutions to Odd-Numbered Exercises

1. (a) Not a function. 20 is assigned two different values.

(b) Function

(c) Function

(d) Not a function. No value is assigned to 30.

3.
$$16x - y^4 = 0$$
$$y^4 = 16x$$
$$y = \pm 2\sqrt[4]{x}$$

y is **not** a function of x. Some x-values correspond to two y-values.

For example, $x = 1$ corresponds to $y = 2$ and $y = -2$.

5. $y = \sqrt{1 - x}$

Each x value, $x \le 1$, corresponds to only one y value so y **is** a function of x.

7. $f(x) = x^2 + 1$

(a) $f(2) = 2^2 + 1 = 5$

(b) $f(-4) = (-4)^2 + 1 = 17$

(c) $f(t^2) = (t^2)^2 + 1 = t^4 + 1$

(d) $-f(x) = -(x^2 + 1) = -x^2 - 1$

9. $f(x) = (x - 1)(x + 2)$ is defined for all real numbers.

Domain: $(-\infty, \infty)$

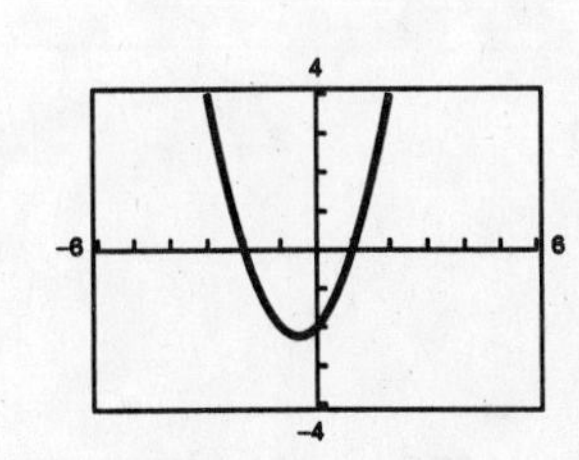

11. $f(x) = \sqrt{25 - x^2}$

Domain: $25 - x^2 \geq 0$

$(5 + x)(5 - x) \geq 0$

Critical numbers: $x = \pm 5$

Test intervals: $(-\infty, -5)$, $(-5, 5)$, $(5, \infty)$

Solution set: $[-5, 5]$

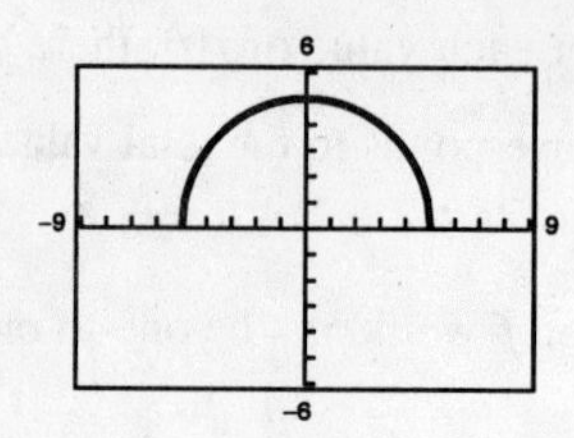

13. $g(s) = \dfrac{5}{3s - 9} = \dfrac{5}{3(s - 3)}$

Domain: All real numbers except $s = 3$

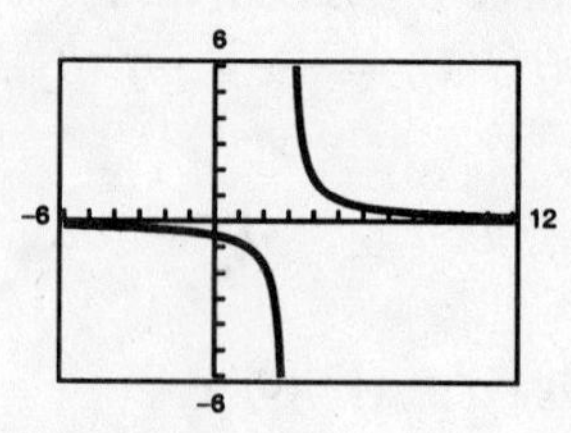

15. (a) $C(x) = 16{,}000 + 5.35x$

(b) $P(x) = R(x) - C(x) = 8.20x - [16{,}000 - 5.35x] = 2.85x - 16{,}000$

17. Domain: all real numbers

Range: all $y \leq 3$

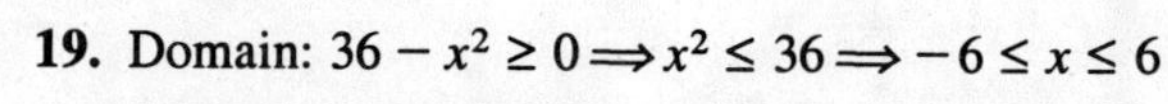

19. Domain: $36 - x^2 \geq 0 \Longrightarrow x^2 \leq 36 \Longrightarrow -6 \leq x \leq 6$

Range: $0 \leq y \leq 6$

21. (a)

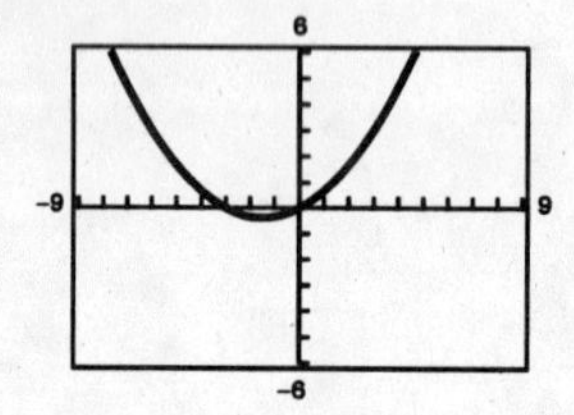

(b) y is a function of x

23. (a) $3x + y^2 = 2$

$y^2 = 2 - 3x$

$y = \pm\sqrt{2 - 3x}$

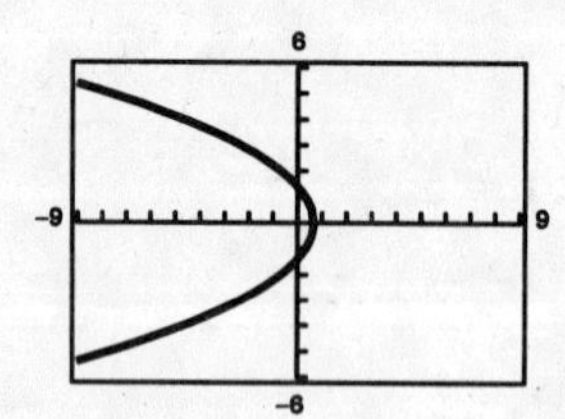

(b) y is not a function of x

25. $f(x) = x^3 - 3x$

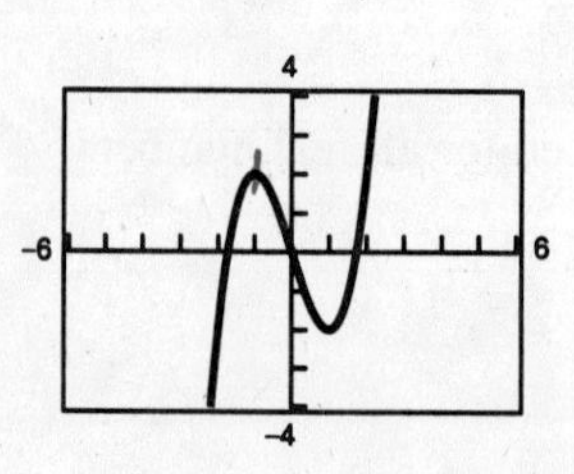

Increasing on $(-\infty, -1)$ and $(1, \infty)$. Decreasing on $(-1, 1)$.

27. $f(x) = x\sqrt{x-6}$

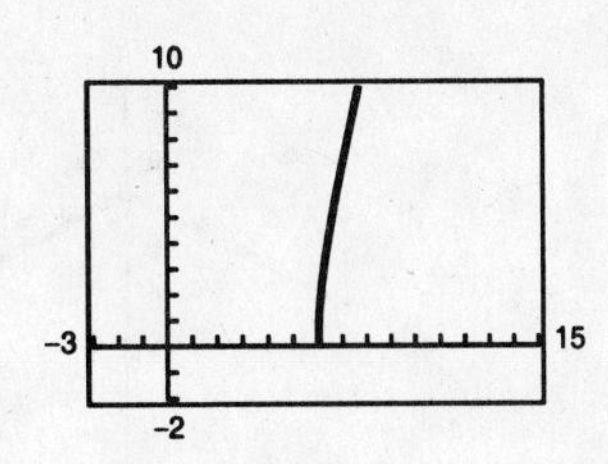

Increasing on $(6, \infty)$

29. $f(x) = (x^2 - 4)^2$. Relative minimums at $(-2, 0)$ and $(2, 0)$. Relative maximum at $(0, 16)$.

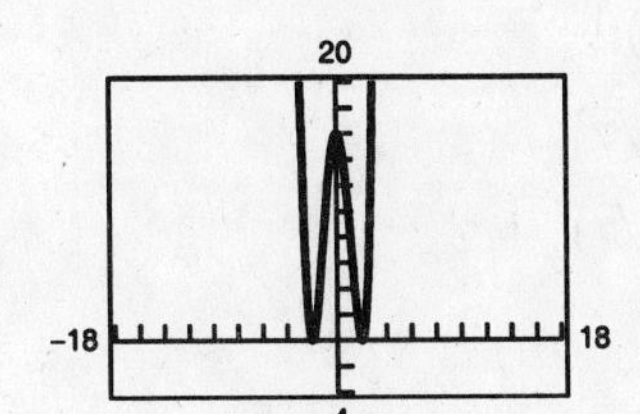

31. $h(x) = 4x^3 - x^4$. Relative maximum $(3, 27)$

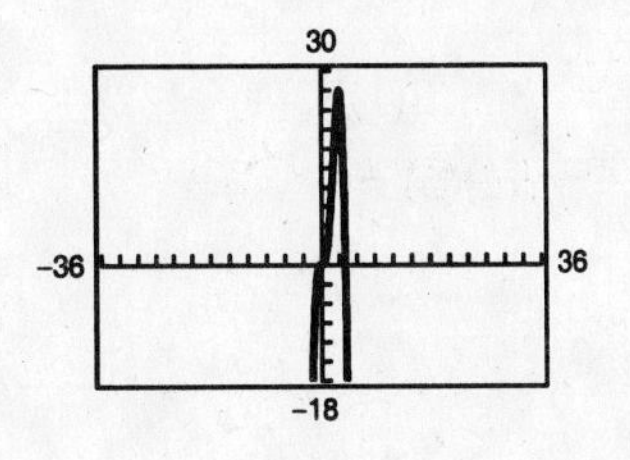

33. $f(x) = \begin{cases} 3x + 5, & x < 0 \\ x - 4, & x \geq 0 \end{cases}$

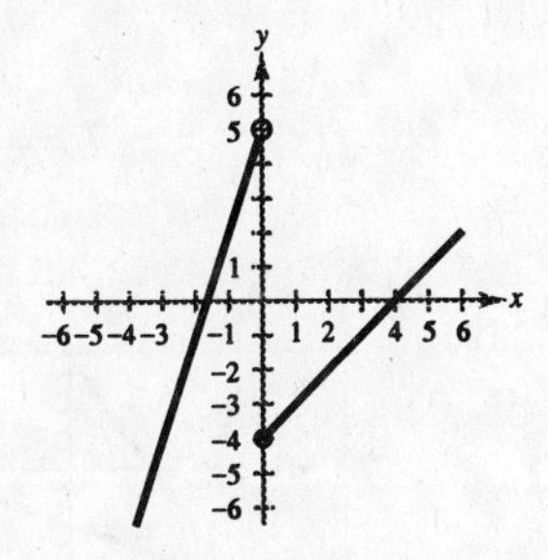

35. $f(-x) = ((-x)^2 - 8)^2 = (x^2 - 8)^2 = f(x)$. f is even.

37. $g(x) = |x| + 3$ is obtained from $f(x) = |x|$ by a vertical shift 3 units upwards. $g(x) = f(x) + 3$

39.

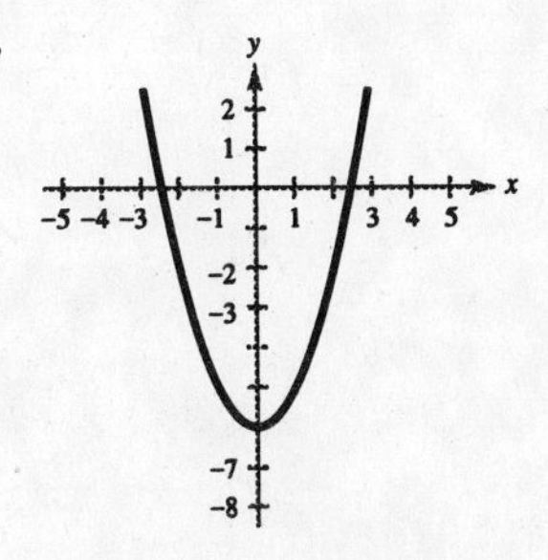

41.

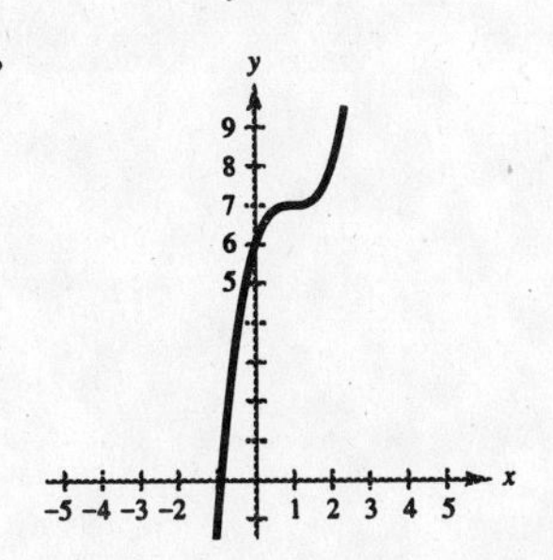

43.

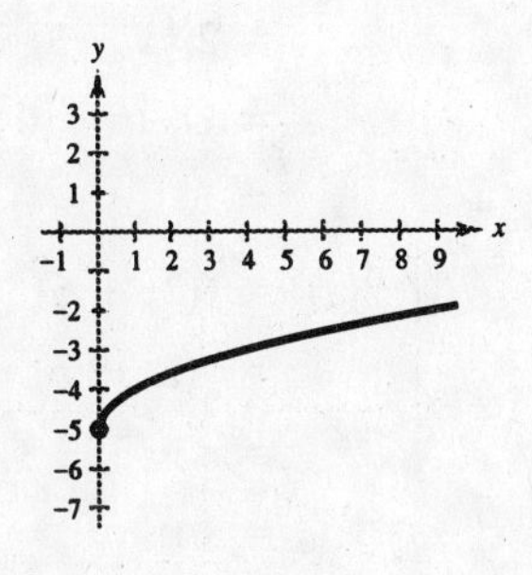

45.

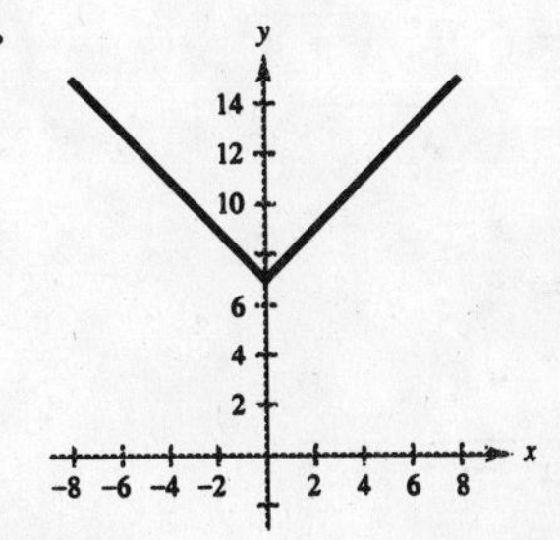

47.

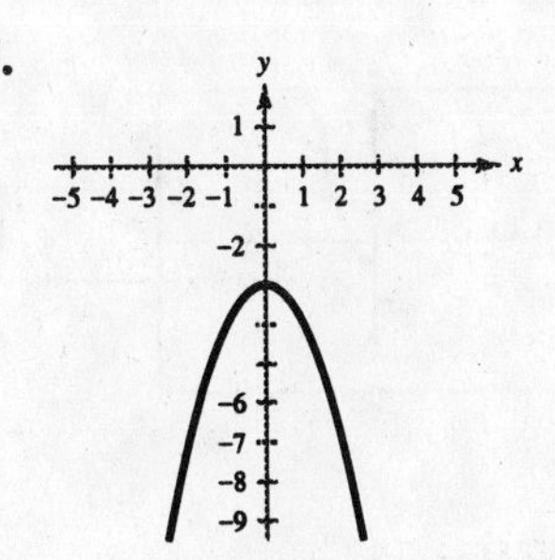

49.

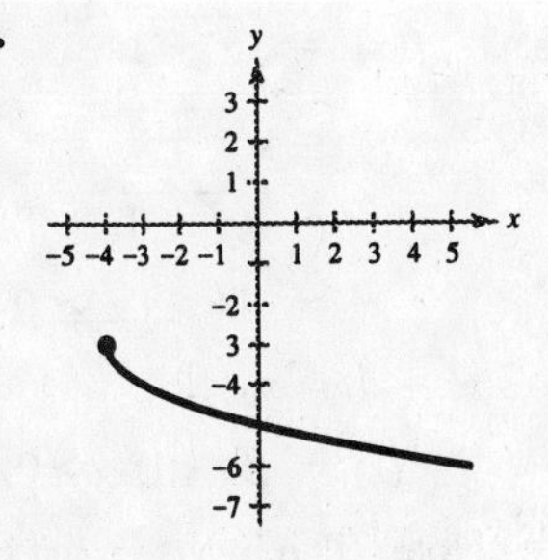

51.

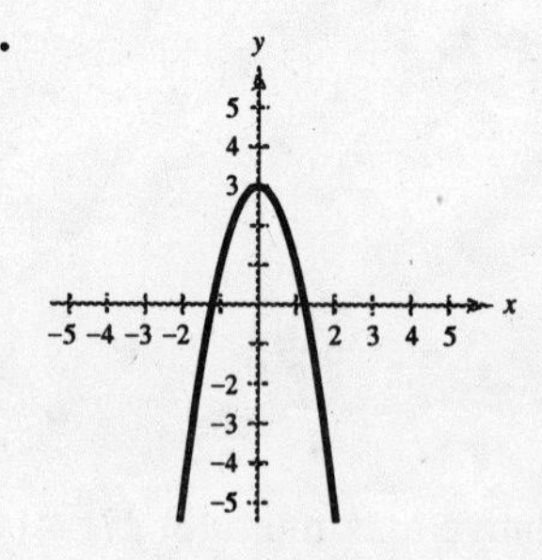

53.

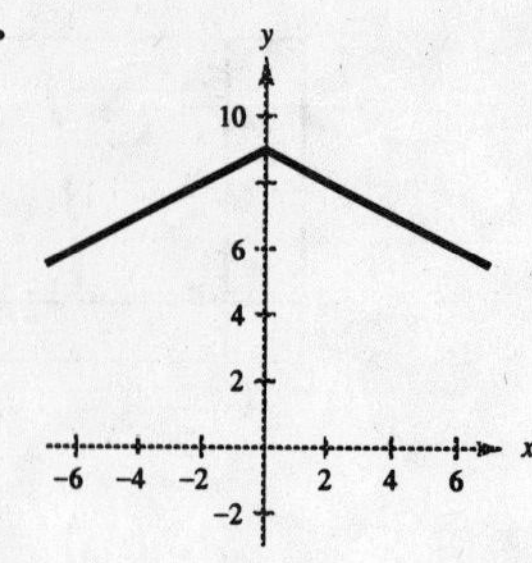

55. $(f - g)(4) = f(4) - g(4)$

$= [3 - 2(4)] - \sqrt{4}$

$= -5 - 2$

$= -7$

57. $(fh)(1) = f(1)h(1) = (3 - 2(1))(3(1)^2 + 2)$

$= (1)(5) = 5$

59. $(h \circ g)(7) = h(g(7))$

$= h(\sqrt{7})$

$= 3(\sqrt{7})^2 + 2$

$= 23$

61. $y_1 = 0.380t^2 + 3.754t + 16.896$

$y_2 = 0.146t^2 + 0.302t + 23.231$

63. $f(x) = 6x \Rightarrow f^{-1}(x) = \frac{1}{6}x$

65. $f(x) = x - 7 \Rightarrow f^{-1}(x) = x + 7$

67. (a)

$f(x) = \frac{1}{2}x - 3$

$y = \frac{1}{2}x - 3$

$x = \frac{1}{2}y - 3$

$x + 3 = \frac{1}{2}y$

$2(x + 3) = y$

$f^{-1}(x) = 2x + 6$

(b)

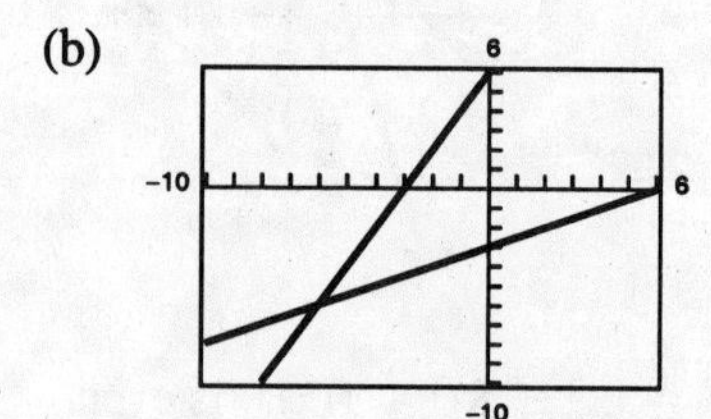

(c) $f^{-1}(f(x)) = f^{-1}\left(\frac{1}{2}x - 3\right)$

$= 2\left(\frac{1}{2}x - 3\right) + 6$

$= x - 6 + 6$

$= x$

$f(f^{-1}(x)) = f(2x + 6)$

$= \frac{1}{2}(2x + 6) - 3$

$= x + 3 - 3$

$= x$

69. (a)

$f(x) = \sqrt{x + 1}$

$y = \sqrt{x + 1}$

$x = \sqrt{y + 1}$

$x^2 = y + 1, \ x \geq 0$

$x^2 - 1 = y$

$f^{-1}(x) = x^2 - 1, \ x \geq 0$

Note: The inverse must have a restricted domain.

(b)

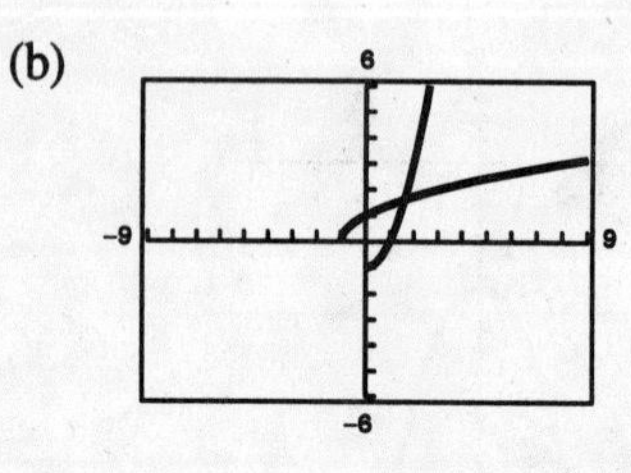

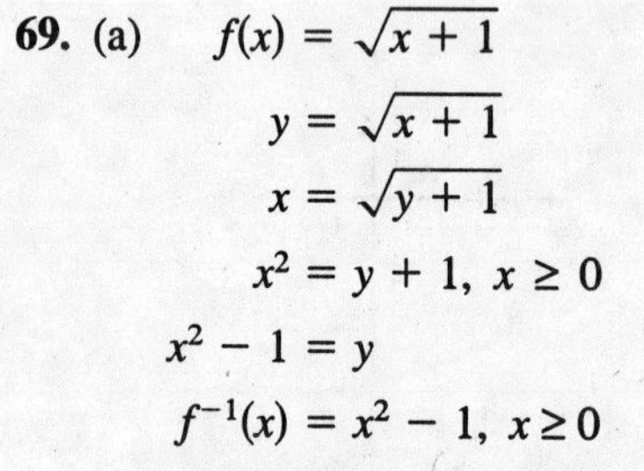

(c) $f^{-1}(f(x)) = f^{-1}\left(\sqrt{x + 1}\right)$

$= \left(\sqrt{x + 1}\right)^2 - 1$

$= x + 1 - 2$

$= x$

$f^{-1} = f(x^2 - 1)$

$= \sqrt{(x^2 - 1) + 1}$

$= \sqrt{x^2} = x$ for $x \geq 0$

71. $f(x) = \dfrac{x}{12}$

$y = \dfrac{x}{12}$

$x = \dfrac{y}{12}$

$12x = y$

$f^{-1}(x) = 12x$

73. $f(x) = 4x^3 - 3$

$y = 4x^3 - 3$

$x = 4y^3 - 3$

$x + 3 = 4y^3$

$\dfrac{x+3}{4} = y^3$

$f^{-1}(x) = \sqrt[3]{\dfrac{x+3}{4}}$

75. $f(x) = \sqrt{x + 10}$

$y = \sqrt{x + 10}, x \geq -10, y \geq 0$

$x = \sqrt{y + 10}\,, y \geq -10, x \geq 0$

$x^2 = y + 10$

$x^2 - 10 = y$

$f^{-1}(x) = x^2 - 10\,, x \geq 0$

77. True. $f^{-1}(x) = x^{1/n}$, n odd

79. The vertical line $x = c$ is not a function because it does not pass the Vertical Line Test. All other lines are functions.

Chapter 1 Practice Test

1. Use a graphing utility to graph the equation $y = 4/x^2 - 5$. Approximate any x-intercepts of the graph.

2. Use a graphing utility to graph the equation $y = |x - 3| + 2$. Approximate any x-intercepts of the graph.

3. Graph $3x - 5y = 15$ by hand.

4. Graph $y = \sqrt{9 - x}$ by hand.

5. Solve $5x + 4 = 7x - 8$.

6. Solve $\frac{x}{3} - 5 = \frac{x}{5} + 1$.

7. Solve $\frac{3x + 1}{6x - 7} = \frac{2}{5}$ graphically and analytically.

8. Solve $(x - 3)^2 + 4 = (x + 1)^2$ graphically and analytically.

9. Find an equation for the line passing through the points $(3, -2)$ and $(4, -5)$. Use a graphing utility to sketch a graph of the line.

10. Find an equation of the line that passes through the point $(-1, 5)$ and has slope -3. Use a graphing utility to sketch a graph of the line.

11. Does the equation $x^4 + y^4 = 16$ represent y as a function of x?

12. Evaluate the function $f(x) = |x - 2|/(x - 2)$ at the points $x = 0, x = 2$, and $x = 4$.

13. Find the domain of the function $f(x) = 5/(x^2 - 16)$.

14. Find the domain of the function $g(t) = \sqrt{4 - t}$.

15. Use a graphing utility to sketch the graph of the function $f(x) = 3 - x^6$ and determine if the function is even, odd, or neither.

16. Use a graphing utility to approximate any relative minimum or maximum values of the function $y = 4 - x + x^3$.

17. Compare the graph of $f(x) = x^3 - 3$ with the graph of $y = x^3$.

18. Compare the graph of $f(x) = \sqrt{x - 6}$ with the graph of $y = \sqrt{x}$.

19. Find $g \circ f$ if $f(x) = \sqrt{x}$ and $g(x) = x^2 - 2$. What is the domain of $g \circ f$?

20. Find f/g if $f(x) = 3x^2$ and $g(x) = 16 - x^4$. What is the domain of f/g?

21. Show that $f(x) = 3x + 1$ and $g(x) = \frac{x - 1}{3}$ are inverse functions algebraically and graphically.

22. Find the inverse of $f(x) = \sqrt{9 - x^2}$, $0 \le x \le 3$. Graph f and f^{-1} in the same viewing rectangle.

CHAPTER 2
Polynomial and Rational Functions

CHAPTER 2
Polynomial and Rational Functions

Section 2.1 Quadratic Functions

You should know the following facts about parabolas.

- $f(x) = ax^2 + bx + c$, $a \neq 0$, is a quadratic function, and its graph is a parabola.
- If $a > 0$, the parabola opens upward and the vertex is the minimum point. If $a < 0$, the parabola opens downward and the vertex is the maximum point.
- The vertex is $(-b/2a, f(-b/2a))$.
- To find the x-intercepts (if any), solve
  $ax^2 + bx + c = 0.$
- The standard form of the equation of a parabola is
  $f(x) = a(x - h)^2 + k$
  where $a \neq 0$.
  (a) The vertex is (h, k).
  (b) The axis is the vertical line $x = h$.

Solutions to Odd-Numbered Exercises

1. $f(x) = (x - 2)^2$ opens upward and has vertex $(2, 0)$. Matches graph (g).

3. $f(x) = x^2 - 2$ opens upward and has vertex $(0, -2)$. Matches graph (b).

5. $f(x) = 4 - (x - 2)^2 = -(x - 2)^2 + 4$ opens downward and has vertex $(2, 4)$. Matches graph (f).

7. $f(x) = x^2 + 3$ opens upward and has vertex $(0, 3)$. Matches graph (e).

9. (a) $y = \frac{1}{2}x^2$

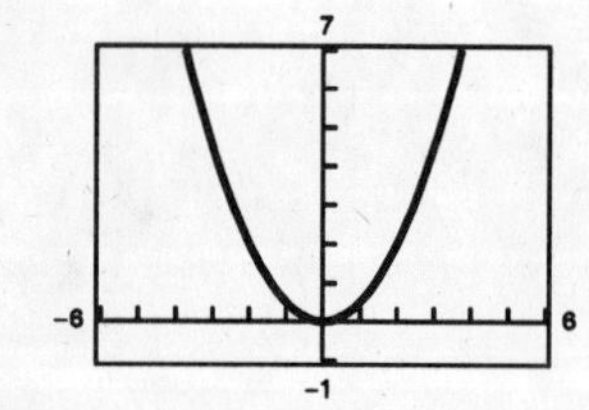

Vertical shrink

(b) $y = -\frac{1}{8}x^2$

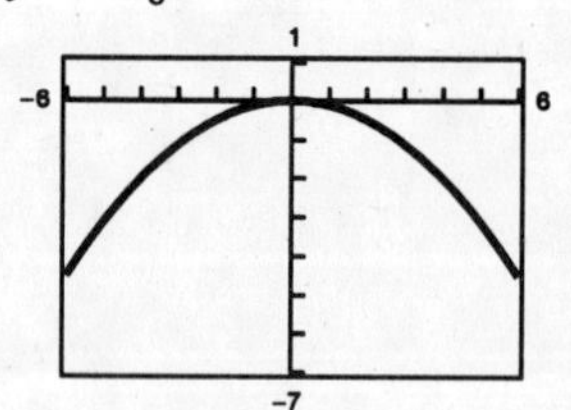

Vertical shrink and reflection in the x-axis

(c) $y = \frac{3}{2}x^2$

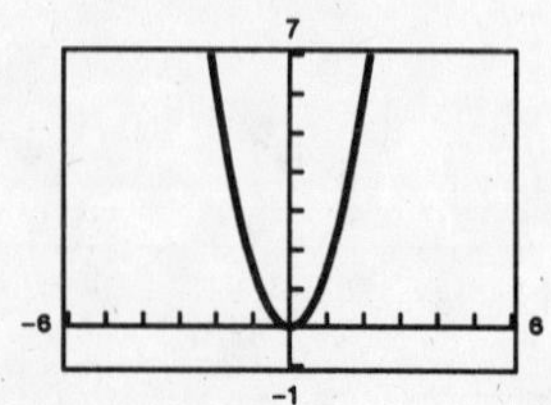

Vertical stretch

(d) $y = -3x^2$

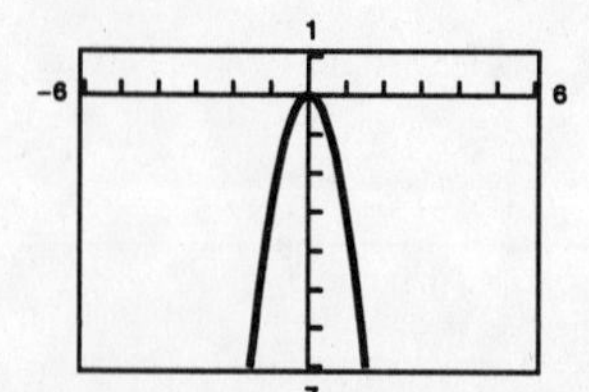

Vertical stretch and reflection in the x-axis

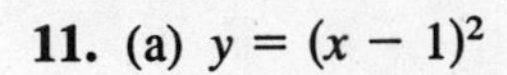

11. (a) $y = (x - 1)^2$

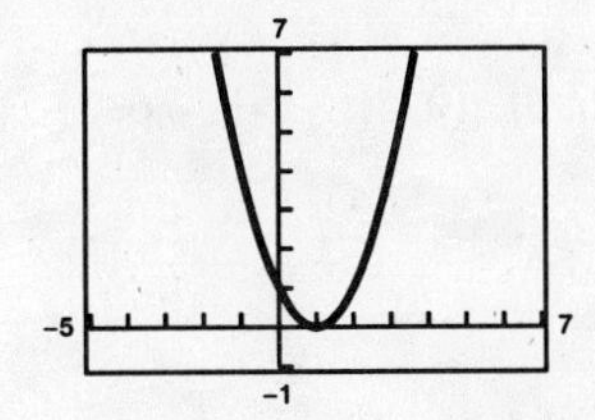

Horizontal shift one unit to the right

(b) $y = (x + 1)^2$

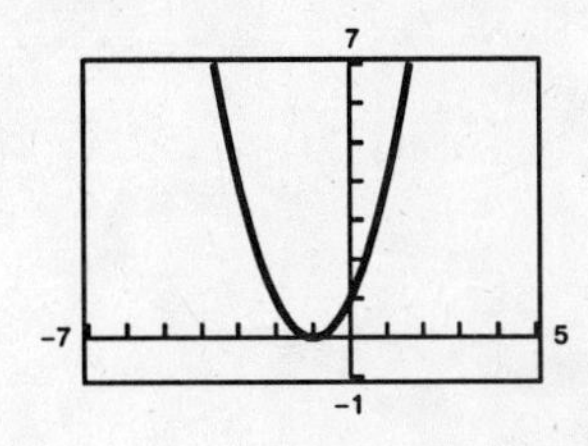

Horizontal shift one unit to the left.

(c) $y = (x - 3)^2$

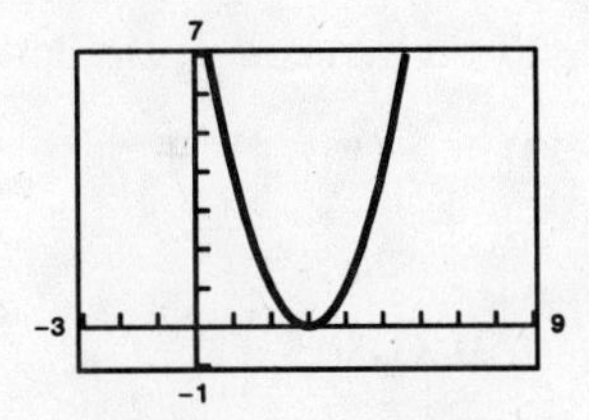

Horizontal shift three units to the right

(d) $y = (x + 3)^2$

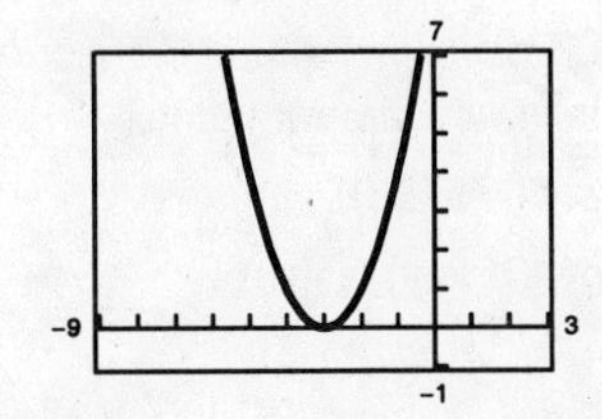

Horizontal shift three units to the left

13. $f(x) = 25 - x^2$

Vertex: $(0, 25)$

Intercepts: $(-5, 0)$, $(0, 25)$, $(5, 0)$

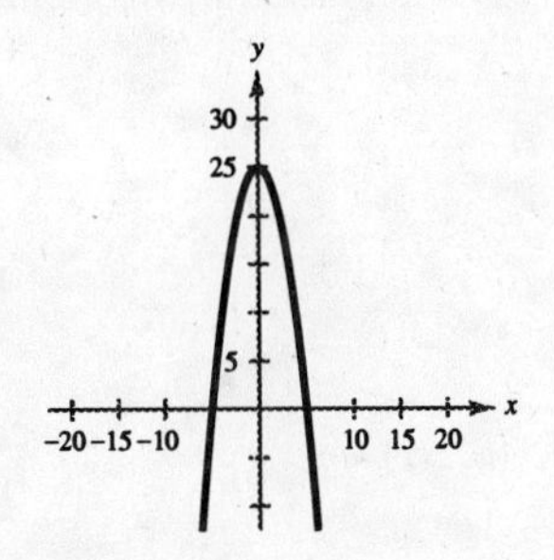

15. $f(x) = \frac{1}{2}x^2 - 4$

Vertex: $(0, -4)$

Intercepts: $(\pm 2\sqrt{2}, 0)$, $(0, -4)$

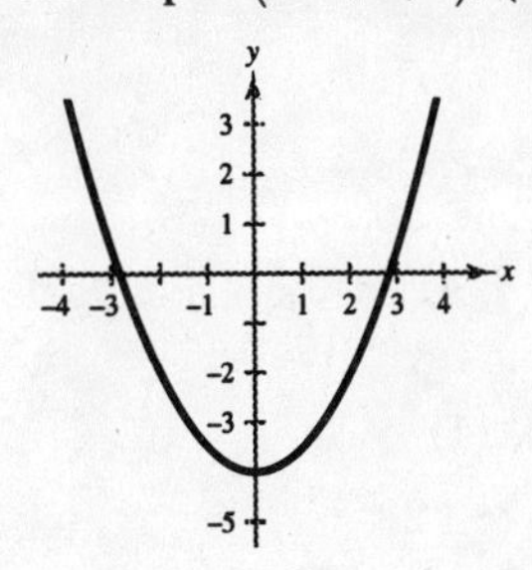

17. $f(x) = (x + 4)^2 - 3$

Vertex: $(-4, -3)$

Intercepts: $(0, 13)$, $(-4 \pm \sqrt{3}, 0)$

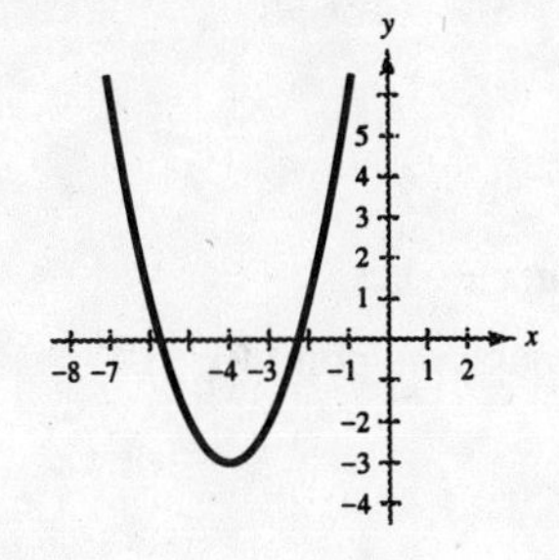

19. $h(x) = x^2 - 8x + 16 = (x - 4)^2$

Vertex: $(4, 0)$

Intercepts: $(0, 16)$, $(4, 0)$

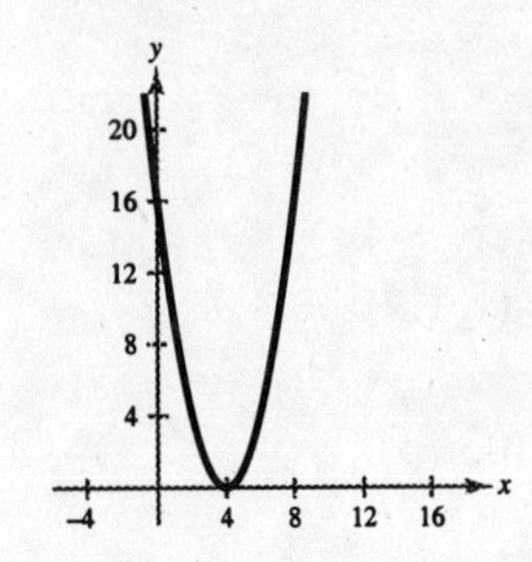

21. $f(x) = x^2 - x + \frac{5}{4} = \left(x - \frac{1}{2}\right)^2 + 1$

Vertex: $\left(\frac{1}{2}, 1\right)$

Intercepts: $\left(0, \frac{5}{4}\right)$

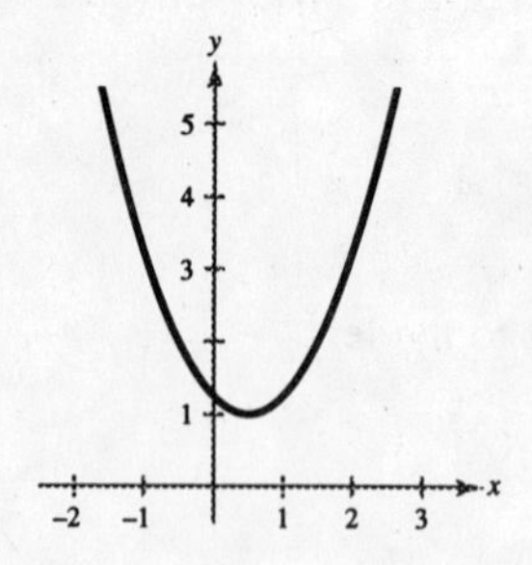

23. $f(x) = -x^2 + 2x + 5 = -(x - 1)^2 + 6$

Vertex: $(1, 6)$

Intercepts: $\left(1 - \sqrt{6}, 0\right)$, $(0, 5)$, $\left(1 + \sqrt{6}, 0\right)$

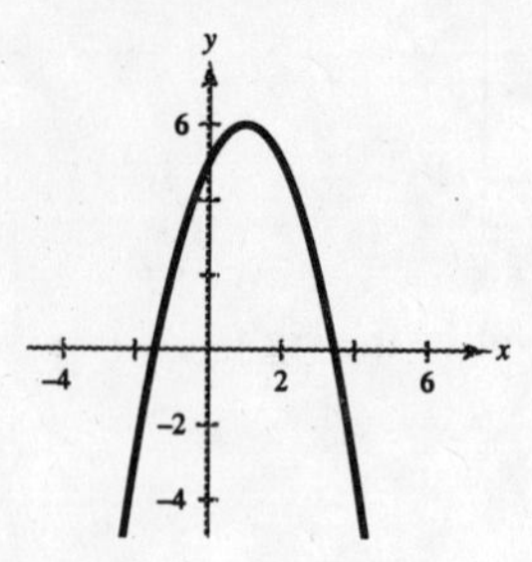

25. $h(x) = 4x^2 - 4x + 21 = 4\left(x - \frac{1}{2}\right)^2 + 20$

Vertex: $\left(\frac{1}{2}, 20\right)$

Intercept: $(0, 21)$

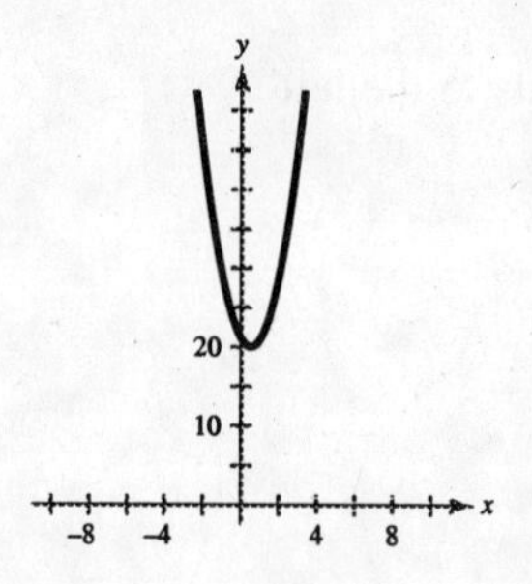

27. $f(x) = -(x^2 + 2x - 3) = -(x + 1)^2 + 4$

Vertex: $(-1, 4)$

Intercepts: $(-3, 0)$, $(0, 3)$, $(1, 0)$

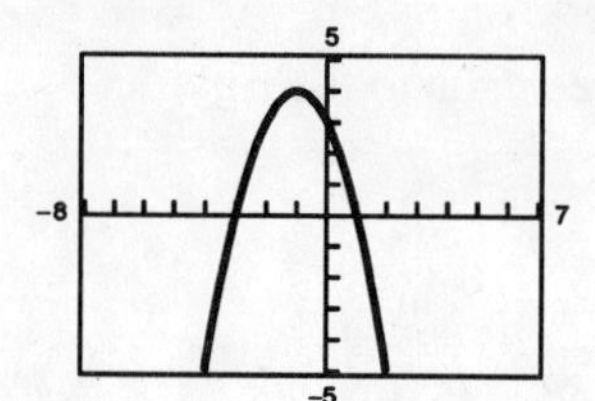

29. $g(x) = x^2 + 8x + 11 = (x + 4)^2 - 5$

Vertex: $(-4, -5)$

Intercepts: $\left(-4 \pm \sqrt{5}, 0\right)$, $(0, 11)$

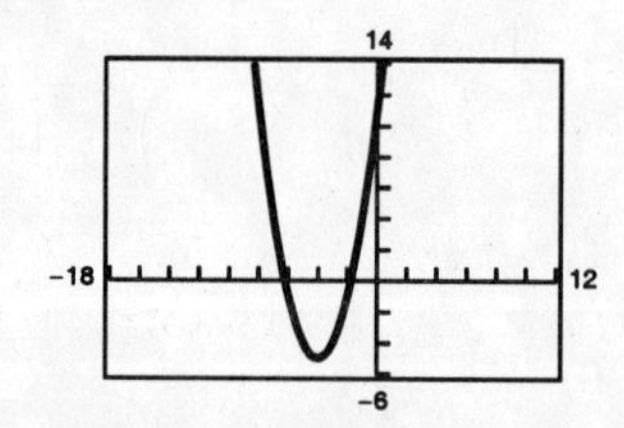

31. $f(x) = 2x^2 - 16x + 31$

$= 2(x - 4)^2 - 1$

Vertex: $(4, -1)$

Intercepts: $\left(4 \pm \frac{1}{2}\sqrt{2}, 0\right)$, $(0, 31)$

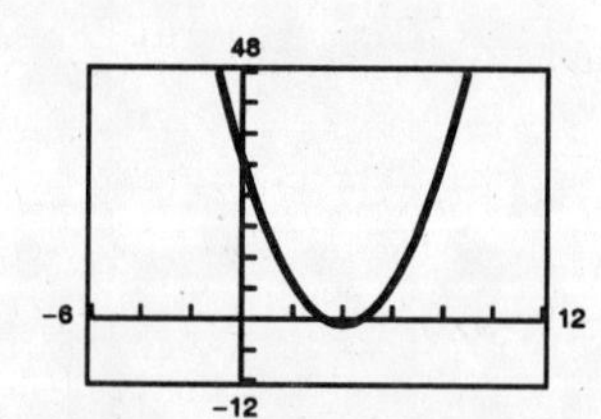

33. $g(x) = \frac{1}{2}(x^2 + 4x - 2) = \frac{1}{2}(x^2 + 4x + 4 - 6)$

$= \frac{1}{2}(x + 2)^2 - 3$

Vertex: $(-2, -3)$

Intercepts: $\left(-2 \pm \sqrt{6}, 0\right)$, $(0, -1)$

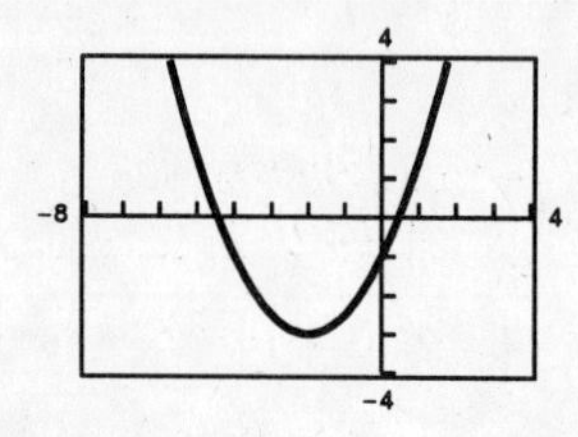

35. $(1, 0)$ is the vertex.

$f(x) = a(x - 1)^2 + 0 = a(x - 1)^2$

Since the graph passes through the point $(0, 1)$ we have:

$1 = a(0 - 1)^2$

$1 = a$

$f(x) = 1(x - 1)^2 = (x - 1)^2$

37. $(-1, 4)$ is the vertex.

$f(x) = a(x + 1)^2 + 4$

Since the graph passes through the point $(1, 0)$ we have

$$0 = a(1 + 1)^2 + 4$$
$$0 = 4a + 4$$
$$-1 = a$$

Thus, $f(x) = -(x + 1)^2 + 4$. Note that $(-3, 0)$ is on the parabola.

39. $(-2, 5)$ is the vertex.

$f(x) = a(x + 2)^2 + 5$

Since the graph passes through the point $(0, 9)$, we have:

$$9 = a(0 + 2)^2 + 5$$
$$4 = 4a$$
$$1 = a$$
$$f(x) = 1(x + 2)^2 + 5 = (x + 2)^2 + 5$$

41. $(3, 4)$ is the vertex.

$f(x) = a(x - 3)^2 + 4$

Since the graph passes through the point $(1, 2)$, we have:

$$2 = a(1 - 3)^2 + 4$$
$$-2 = 4a$$
$$-\tfrac{1}{2} = a$$
$$f(x) = -\tfrac{1}{2}(x - 3)^2 + 4$$

43. $(-2, -2)$ is the vertex

$f(x) = a(x + 2)^2 - 2$

Since the graph passes through $(-1, 0)$,

$$0 = a(-1 + 2)^2 - 2$$
$$0 = a - 2$$
$$2 = a$$

Thus, $f(x) = 2(x + 2)^2 - 2$

45. $\left(\tfrac{5}{2}, -\tfrac{3}{4}\right)$ is the vertex.

$f(x) = a\left(x - \tfrac{5}{2}\right)^2 - \tfrac{3}{4}$

Since the graph passes through $(-2, 4)$,

$$4 = a\left(-2 - \tfrac{5}{2}\right)^2 - \tfrac{3}{4}$$
$$\tfrac{19}{4} = a\left(-\tfrac{9}{2}\right)^2$$
$$19 = 81a$$
$$a = \tfrac{19}{81}$$

Thus, $f(x) = \tfrac{19}{81}\left(x - \tfrac{5}{2}\right)^2 - \tfrac{3}{4}$

47. $y = x^2 - 16$

x-intercepts: $(\pm 4, 0)$

$$0 = x^2 - 16$$
$$x^2 = 16$$
$$x = \pm 4$$

49. $y = x^2 - 4x - 5$

x-intercepts: $(5, 0), (-1, 0)$

$$0 = x^2 - 4x - 5$$
$$0 = (x - 5)(x + 1)$$
$$x = 5 \text{ or } x = -1$$

51. $y = x^2 - 4x$

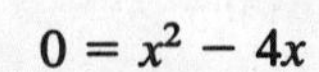

$$0 = x^2 - 4x$$
$$0 = x(x - 4)$$
$$x = 0 \text{ or } x = 4$$

x-intercepts: $(0, 0)$, $(4, 0)$

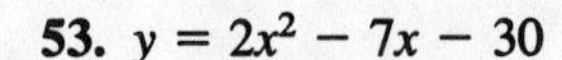

53. $y = 2x^2 - 7x - 30$

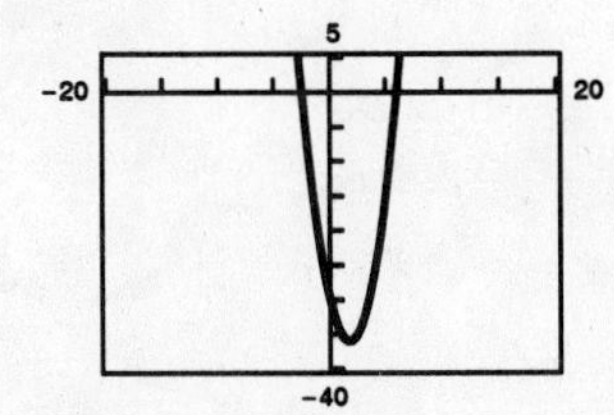

$$0 = 2x^2 - 7x - 30$$
$$0 = (2x + 5)(x - 6)$$
$$x = -\tfrac{5}{2} \text{ or } x = 6$$

x-intercepts:

$\left(-\tfrac{5}{2}, 0\right)$, $(6, 0)$

55. $y = -\tfrac{1}{2}(x^2 - 6x - 7)$

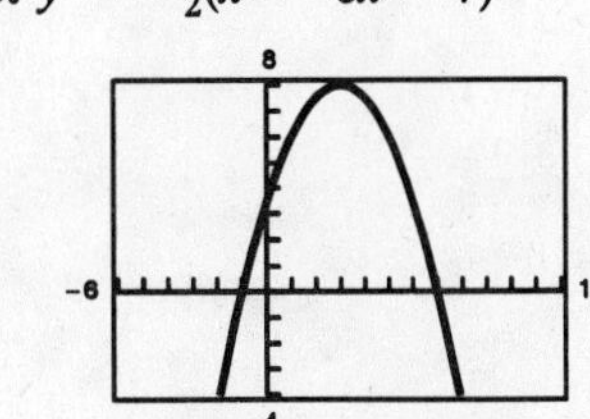

x-intercepts:

$(-1, 0), (7, 0)$

$$0 = -\tfrac{1}{2}(x^2 - 6x - 7)$$
$$0 = x^2 - 6x - 7$$
$$0 = (x + 1)(x - 7)$$
$$x = -1, 7$$

57. $f(x) = [x - (-1)](x - 3)$ opens upward

$= (x + 1)(x - 3)$

$= x^2 - 2x - 3$

$g(x) = -[x - (-1)](x - 3)$ opens downward

$= -(x + 1)(x - 3)$

$= -(x^2 - 2x - 3)$

$= -x^2 + 2x + 3$

Note: $f(x) = a(x + 1)(x - 3)$ has x-intercepts $(-1, 0)$ and $(3, 0)$ for all real numbers $a \neq 0$.

59. $f(x) = [x - (-3)]\left[x - \left(-\frac{1}{2}\right)\right](2)$ opens upward

$= (x + 3)\left(x + \frac{1}{2}\right)(2)$

$= (x + 3)(2x + 1)$

$= 2x^2 + 7x + 3$

$g(x) = -(2x^2 + 7x + 3)$ opens downward

$= -2x^2 - 7x - 3$

Note: $f(x) = a(x + 3)(2x + 1)$ has x-intercepts $(-3, 0)$ and $\left(-\frac{1}{2}, 0\right)$ for all real numbers $a \neq 0$.

61. Let $x =$ the first number and $y =$ the second number. Then the sum is

$$x + y = 110 \implies y = 110 - x.$$

The product is $P(x) = xy = x(110 - x) = 110x - x^2$.

$P(x) = -x^2 + 110x$

$= -(x^2 - 110x + 3025 - 3025)$

$= -[(x - 55)^2 - 3025]$

$= -(x - 55)^2 + 3025$

The maximum value of the product occurs at the vertex of $P(x)$ and is 3025. This happens when $x = y = 55$.

63. Let x be the first number and y be the second number. Then $x + 2y = 24 \implies x = 24 - 2y$. The product is $P = xy = (24 - 2y)y = 24y - 2y^2$. Completing the square,

$P = -2y^2 + 24y$

$= -2(y^2 - 12y + 36) + 72$

$= -2(y - 6)^2 + 72.$

The maximum value of the product P occurs at the vertex of the parabola and equals 72. This happens when $y = 6$ and $x = 24 - 2(6) = 12$.

65. x, y, y, x (rectangle)

$2x + 2y = 100$

$y = 50 - x$

(a) $A(x) = xy = x(50 - x)$
Domain: $0 < x < 50$

(b) 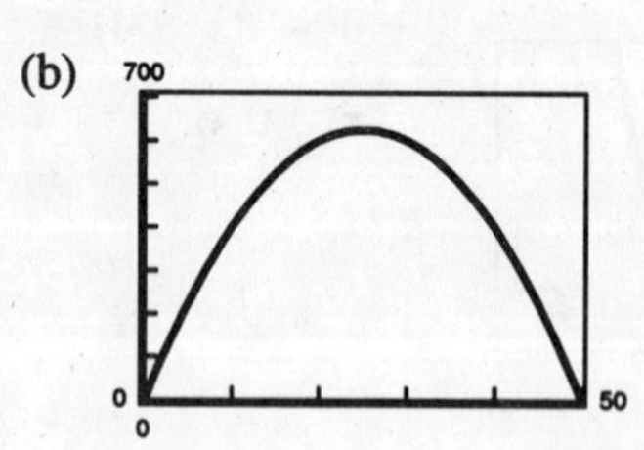

(c) The area is maximum (625 square feet) when $x = y = 25$. The rectangle has dimensions 25 ft × 25 ft. Algebraically, you have:

$A(x) = -(x^2 - 50x)$

$= -(x^2 - 50x + 625) + 625$

$= -(x - 25)^2 + 625$

$A(x)$ is a maximum of 625 when $x = 25$.

67. (a)

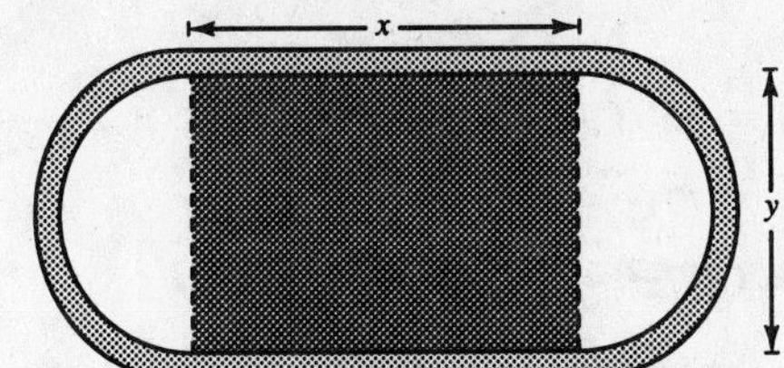

(b) Radius of semicircular ends of track: $r = \frac{1}{2}y$
distance around two semicircular parts of track:

$$d = 2\pi r = 2\pi\left(\frac{1}{2}y\right) = \pi y$$

(c) Distance traveled around track in one lap:

$$d = \pi y + 2x = 200$$
$$\pi y = 200 - 2x$$
$$y = \frac{200 - 2x}{\pi}$$

(d) Area of rectangular region:

$$A = xy = x\left(\frac{200 - 2x}{\pi}\right)$$
$$= \frac{1}{\pi}(200x - 2x^2)$$
$$= -\frac{2}{\pi}(x^2 - 100x)$$
$$= -\frac{2}{\pi}(x^2 - 100x + 2500 - 2500)$$
$$= -\frac{2}{\pi}(x - 50)^2 + \frac{5000}{\pi}$$

The area is maximum when $x = 50$ and
$y = \dfrac{200 - 2(50)}{\pi} = \dfrac{100}{\pi}$.

(e)

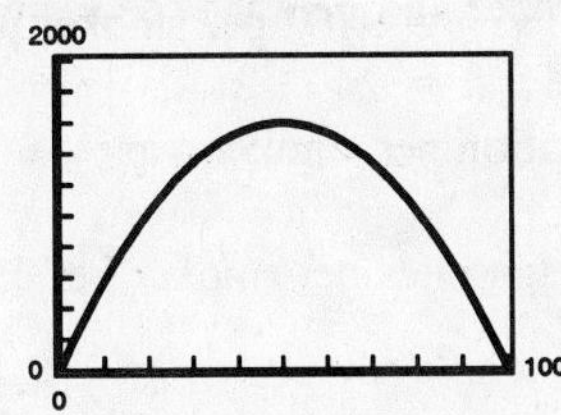

The area is maximum when $x = 50$ and
$y = \dfrac{200 - 2(50)}{\pi} = \dfrac{100}{\pi}$.

69. $C = 800 - 10x + 0.25x^2$

The minimum cost occurs at the vertex.

$$x = -\frac{b}{2a} = -\frac{(-10)}{2(0.25)} = \frac{10}{.5} = 20$$

$C(20) = 700$ is the minimum cost.

Graphically, you could graph
$C = 800 - 10x + 0.25x^2$ in the window
$[0, 40] \times [0, 1000]$ and find the vertex $(20, 700)$.

71. $P = -0.0002x^2 + 140x - 250{,}000$

The vertex of this parabola is at

$$x = -\frac{b}{2a} = -\frac{140}{2(-0.0002)} = \frac{140}{0.0004}$$
$$= 350{,}000 \text{ units}$$

Thus, the maximum profit is attained at a sales level of 350,000 units.

73. $y = -\dfrac{1}{12}x^2 + 2x + 4$

(a)

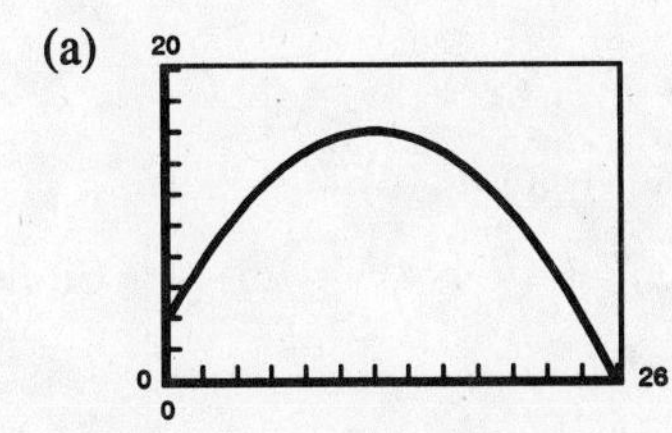

(b) When $x = 0$, $y = 4$ feet.

(c) The vertex occurs at

$$x = -\frac{b}{2a} = -\frac{2}{2(-1/12)} = 12.$$

The maximum height is

$$y = -\frac{1}{12}(12)^2 + 2(12) + 4$$
$$= 16 \text{ feet.}$$

(d) You can solve this part graphically by finding the x-intercept of the graph:
$x \approx 25.856$.
Algebraically,

$$0 = -\frac{1}{12}x^2 + 2x + 4$$
$$0 = x^2 - 24x - 48 \quad \text{(Multiply both sides by } -12.)$$
$$x = \frac{-(-24) \pm \sqrt{(-24)^2 - 4(1)(-48)}}{2(1)}$$
$$= \frac{24 \pm \sqrt{768}}{2} = \frac{24 \pm 16\sqrt{3}}{2} = 12 \pm 8\sqrt{3}$$

Using the positive value for x, we have
$x = 12 + 8\sqrt{3} \approx 25.86$ feet.

75. $V = 0.77x^2 - 1.32x - 9.31, \; 5 \le x \le 40$

(a)

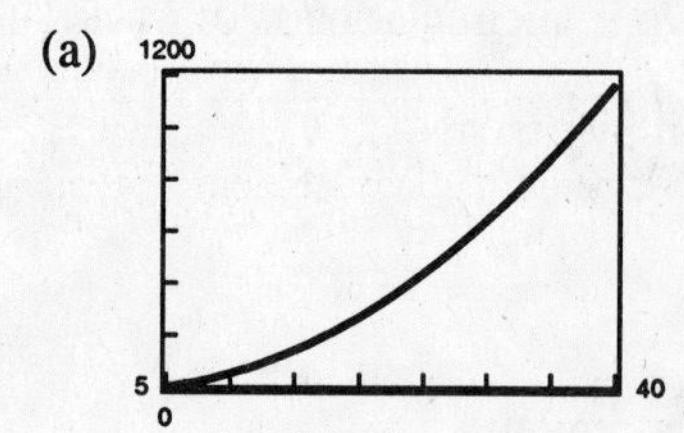

(b) $V(16) = 166.69$ board feet

(c) $500 = 0.77x^2 - 1.32x - 9.31$

$$0 = 0.77x^2 - 1.32x - 509.31$$

Using the Quadratic Formula and selecting the positive value for x, we have $x \approx 26.6$ inches in diameter. Or, use a graphing utility.

77. (a)

4500

0 40

2500

(b) Using a graphing utility, the maximum is approximately 4242 cigarettes at $t = 18.3$, or 1968. Yes, the warnings on cigarette packages seemed to have an effect.

(c) For 1960, $C(10) \approx 4038$ cigarettes per person. The annual consumption per smoker was

$$\frac{4038(116{,}530{,}000)}{48{,}500{,}000} = 9702 \text{ per smoker per year.}$$

The daily consumption per smoker was

$$\frac{9702}{365} \approx 26.6 \text{ cigarettes per smoker per day.}$$

79. True

$$-12x^2 - 1 = 0$$

$$12x^2 = -1 \text{ impossible}$$

81. Model (a) is preferable. $a > 0$ means the parabola opens upward and profits are increasing for t to the right of the vertex,

$$t \ge -\frac{b}{(2a)}.$$

83. $y = 3x - 10 = \frac{1}{4}x + 1$

$$12x - 40 = x + 4$$

$$11x = 44$$

$$x = 4$$

The graphs intersect at (4, 2).

85. $y = x^3 + 2x - 1 = -2x + 15$

$$x^3 + 4x - 16 = 0$$

$$(x - 2)(x^2 + 2x + 8) = 0$$

$$x = 2$$

The graphs intersect at (2, 11).

87. $y^2 = x^2 - 9$

$$y = \pm\sqrt{x^2 - 9}$$

No, y is not a function of x.

89. $x^2 + y^2 - 6x + 8y = 0$

$$(x^2 - 6x + 9) + (y^2 + 8y + 16) = 9 + 16$$

$$(x - 3)^2 + (y + 4)^2 = 25 \qquad \text{Circle}$$

y is not a function of x.

Section 2.2 Polynomial Functions of Higher Degree

- You should know the following basic principles about polynomials.
- $f(x) = a_n x^n + a_{n-1}x^{n-1} + \cdots + a_2x^2 + a_1x + a_0$ is a polynomial function of degree n.
- If f is of odd degree and
  - (a) $a_n > 0$, then
    1. $f(x) \to \infty$ as $x \to \infty$.
    2. $f(x) \to -\infty$ as $x \to -\infty$.
  - (b) $a_n < 0$, then
    1. $f(x) \to -\infty$ as $x \to \infty$.
    2. $f(x) \to \infty$ as $x \to -\infty$.
- If f is of even degree and
  - (a) $a_n > 0$, then
    1. $f(x) \to \infty$ as $x \to \infty$.
    2. $f(x) \to \infty$ as $x \to -\infty$.
  - (b) $a_n < 0$, then
    1. $f(x) \to -\infty$ as $x \to \infty$.
    2. $f(x) \to -\infty$ as $x \to -\infty$.
- The following are equivalent for a polynomial function.
  - (a) $x = a$ is a zero of a function.
  - (b) $x = a$ is a solution of the polynomial equation $f(x) = 0$.
  - (c) $(x - a)$ is a factor of the polynomial.
  - (d) $(a, 0)$ is an x-intercept of the graph of f.
- A polynomial of degree n has at most n distinct zeros.
- If f is a polynomial function such that $a < b$ and $f(a) \neq f(b)$, then f takes on every value between $f(a)$ and $f(b)$ in the interval $[a, b]$.
- If you can find a value where a polynomial is positive and another value where it is negative, then there is at least one real zero between the values.

Solutions to Odd-Numbered Exercises

1. $f(x) = -2x + 3$ is a line with y-intercept $(0, 3)$. Matches graph (f).

3. $f(x) = -2x^2 - 5x$ is a parabola with x-intercepts $(0, 0)$ and $\left(-\frac{5}{2}, 0\right)$ and opens downward. Matches graph (c).

5. $f(x) = -\frac{1}{4}x^4 + 3x^2$ has intercepts $(0, 0)$ and $\left(\pm 2\sqrt{3}, 0\right)$. Matches graph (e).

7. $f(x) = x^4 + 2x^3$ has intercepts $(0, 0)$ and $(-2, 0)$. Matches graph (g).

9. $y = x^3$

(a) $f(x) = (x - 2)^3$

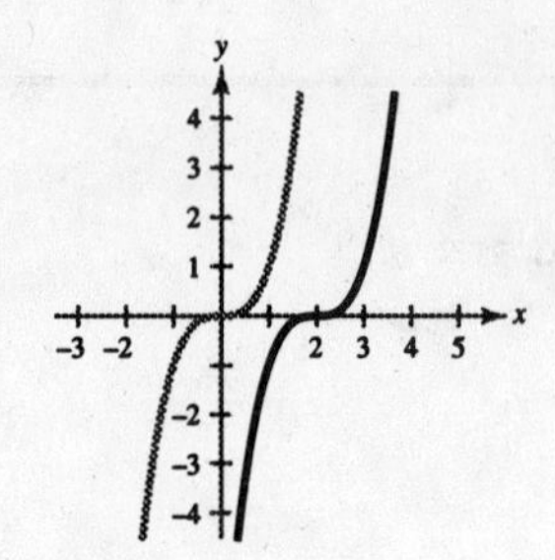

Horizontal shift two units to the right

(b) $f(x) = x^3 - 2$

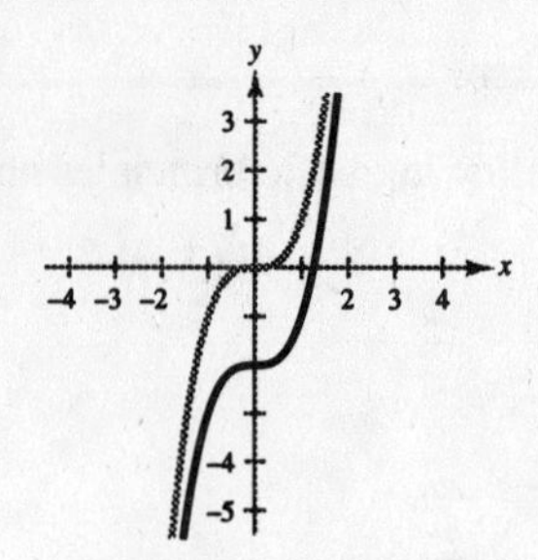

Vertical shift two units downward

(c) $f(x) = -\frac{1}{2}x^3$

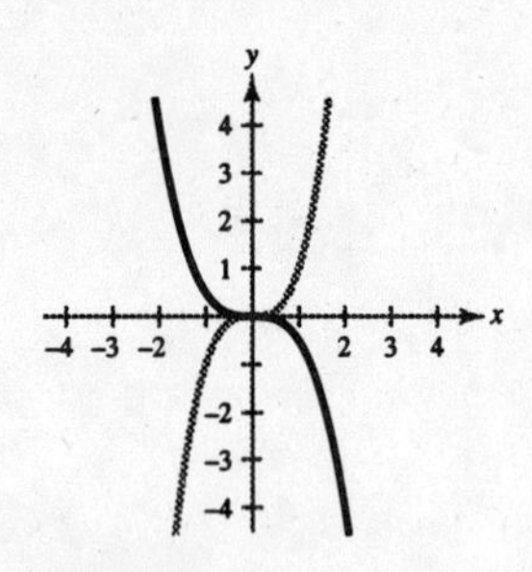

Reflection in the x-axis and a vertical shrink

(d) $f(x) = (x - 2)^3 - 2$

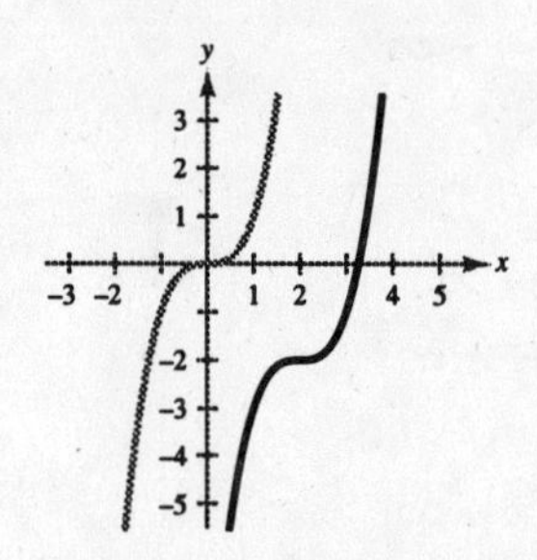

Horizontal shift two units to the right and a vertical shift two units downward

11. $y = x^4$

(a) $f(x) = (x + 5)^4$

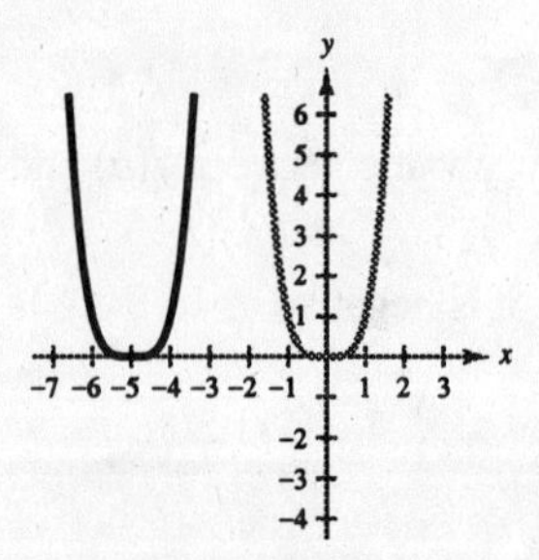

Horizontal shift five units to the left

(b) $f(x) = x^4 - 5$

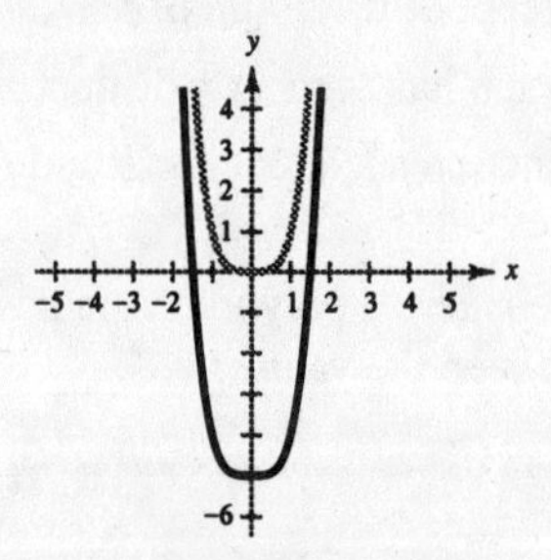

Vertical shift five units downward

(c) $f(x) = 4 - x^4$

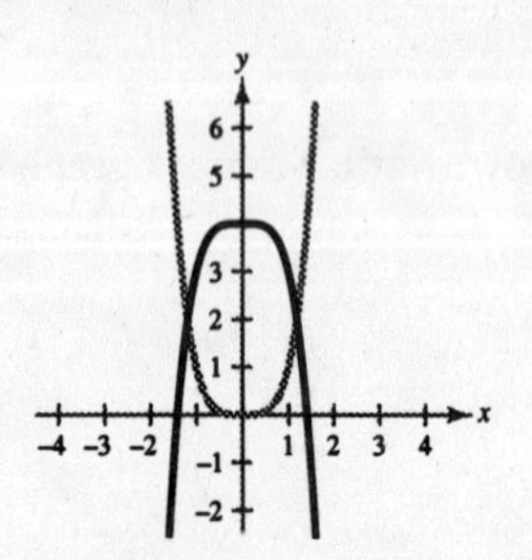

Reflection in the x-axis and then a vertical shift four units upward

(d) $f(x) = \frac{1}{2}(x - 1)^4$

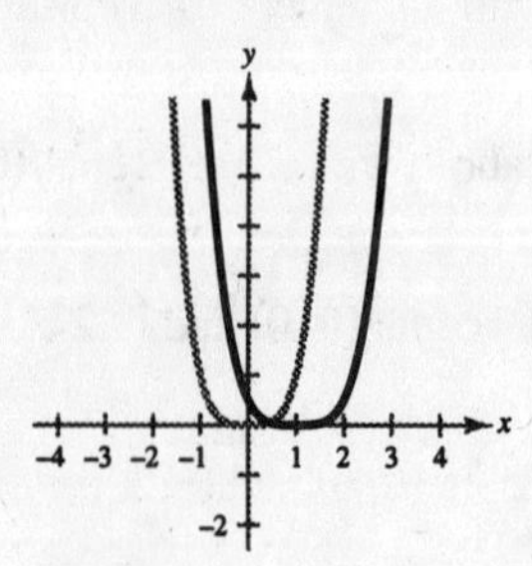

Horizontal shift one unit to the right and a vertical shrink

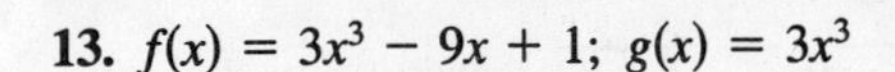

13. $f(x) = 3x^3 - 9x + 1;\ g(x) = 3x^3$

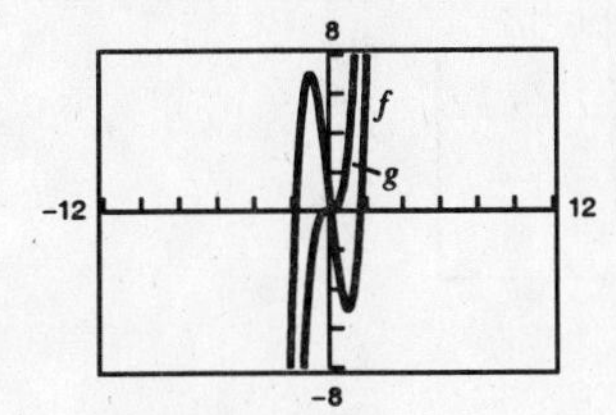

15. $f(x) = -(x^4 - 4x^3 + 16x);\ g(x) = -x^4$

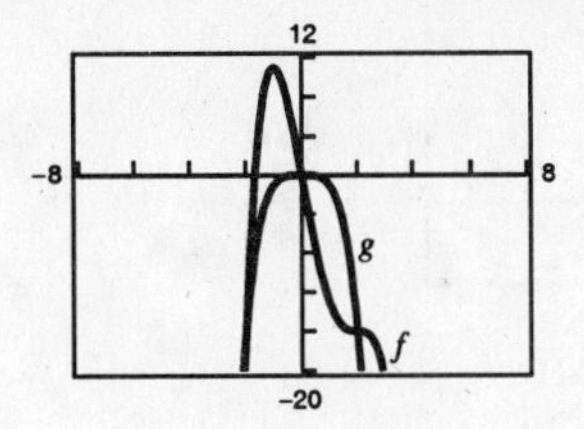

17. $f(x) = 2x^2 - 3x + 1$

Degree: 2

Leading coefficient: 2

The degree is even and the leading coefficient is positive. The graph rises to the left and right.

19. $g(x) = 5 - \frac{7}{2}x - 3x^2$

Degree: 2

Leading coefficient: -3

The degree is even and the leading coefficient is negative. The graph falls to the left and right.

21. $f(x) = -2.1x^5 + 4x^3 - 2$

Degree: 5

Leading coefficient: -2.1

The degree is odd and the leading coefficient is negative. The graph rises to the left and falls to the right.

23. $f(x) = 6 - 2x + 4x^2 - 5x^3$

Degree: 3

Leading coefficient: -5

The degree is odd and the leading coefficient is negative. The graph rises to the left and falls to the right.

25. $h(t) = -\frac{2}{3}(t^2 - 5t + 3)$

Degree: 2

Leading coefficient: $-\frac{2}{3}$

The degree is even and the leading coefficient is negative. The graph falls to the left and right.

27. $f(x) = x^2 - 25$

$= (x + 5)(x - 5)$

$x = \pm 5$

29. $h(t) = t^2 - 6t + 9$

$= (t - 3)^2$

$t = 3$

31. $f(x) = x^2 + x - 2$

$= (x + 2)(x - 1)$

$x = -2, 1$

33. $f(t) = t^3 - 4t^2 + 4t$

$= t(t - 2)^2$

$t = 0, 2$

35. $f(x) = \dfrac{1}{2}x^2 + \dfrac{5}{2}x - \dfrac{3}{2}$

$= \dfrac{1}{2}(x^2 + 5x - 3)$

$x = \dfrac{-5 \pm \sqrt{25 - 4(-3)}}{2} = -\dfrac{5}{2} \pm \dfrac{\sqrt{37}}{2}$

$\approx 0.5414, -5.5414$

37. (a)

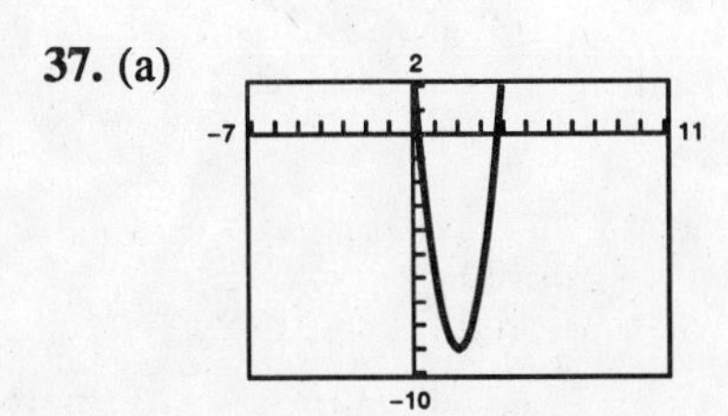

(b) $x \approx 3.732, 0.268$

(c) $f(x) = 3x^2 - 12x + 3$

$= 3(x^2 - 4x + 1)$

$x = \dfrac{4 \pm \sqrt{16 - 4}}{2} = 2 \pm \sqrt{3}$

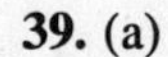

39. (a)

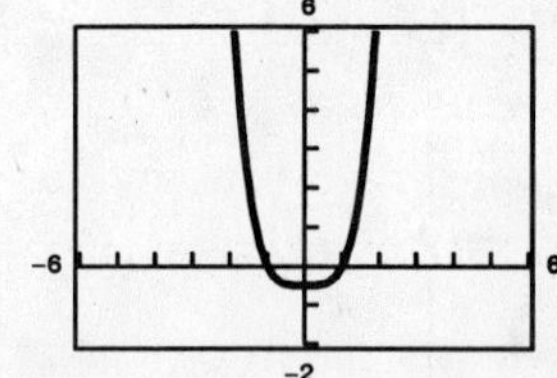

(b) $t = \pm 1$

(c) $g(t) = \frac{1}{2}t^4 - \frac{1}{2}$

$= \frac{1}{2}(t + 1)(t - 1)(t^2 + 1)$

$t = \pm 1$

41. (a)

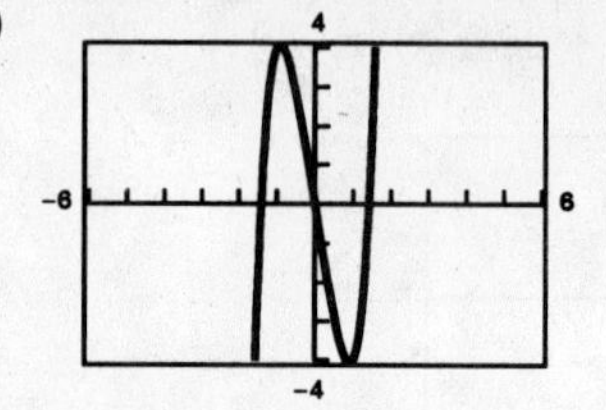

(b) $x = 0, 1.414, -1.414$

(c) $f(x) = x^5 + x^3 - 6x$

$= x(x^4 + x^2 - 6)$

$= x(x^2 + 3)(x^2 - 2)$

$x = 0, \pm\sqrt{2}$

43. (a)

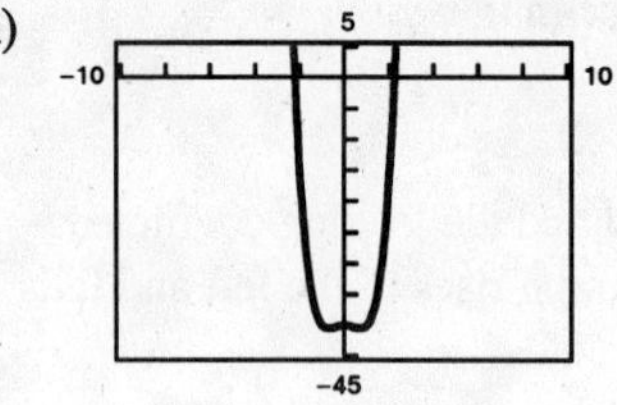

(b) $2.236, -2.236$

(c) $f(x) = 2x^4 - 2x^2 - 40$

$= 2(x^2 + 4)(x + \sqrt{5})(x - \sqrt{5})$

$x = \pm\sqrt{5}$

45. (a)

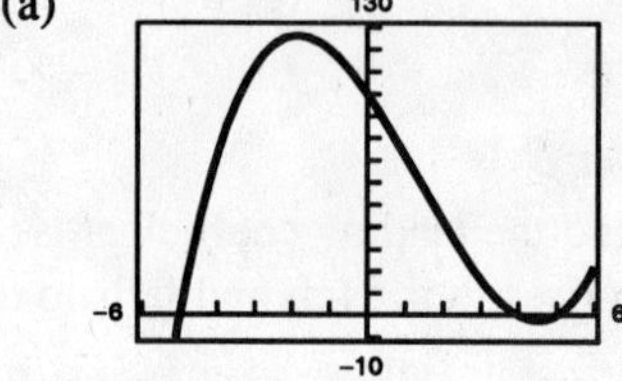

(b) $x = 4, 5, -5$

(c) $f(x) = x^3 - 4x^2 - 25x + 100$

$= x^2(x - 4) - 25(x - 4)$

$= (x^2 - 25)(x - 4)$

$= (x - 5)(x + 5)(x - 4)$

$x = \pm 5, 4$

47. (a)

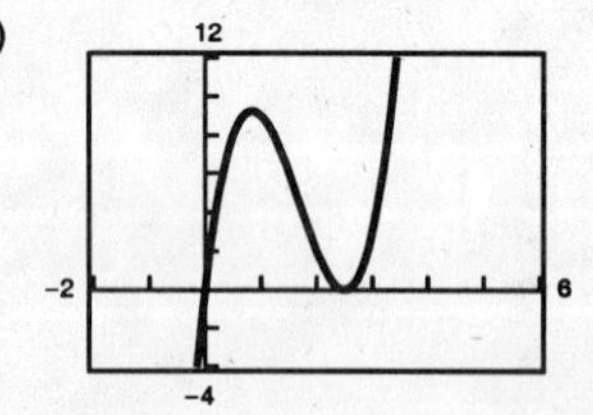

(b) x-intercepts: $(0, 0), \left(\frac{5}{2}, 0\right)$

(c) $y = 4x^3 - 20x^2 + 25x$

$0 = 4x^3 - 20x^2 + 25x$

$0 = x(2x - 5)^2$

$x = 0$ or $x = \frac{5}{2}$

49.

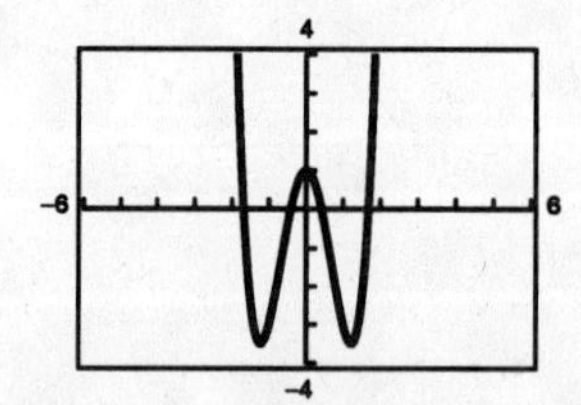

Relative maximum: $(0, 1)$

Relative minimums: $(1.225, -3.5), (-1.225, -3.5)$

51.

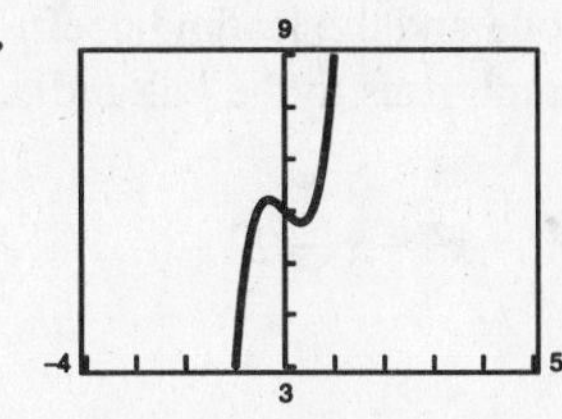

Relative maximum: $(-0.324, 6.218)$

Relative minimum: $(0.324, 5.782)$

53. $f(x) = (x - 0)(x - 12)$

$f(x) = x^2 - 12x$

Note: $f(x) = a(x - 0)(x - 12) = ax(x - 12)$ has zeros 0 and 12 for all nonzero real numbers a.

55. $f(x) = (x - 2)(x - (-6))$

$= (x - 2)(x + 6)$

$= x^2 + 4x - 12$

Note: $f(x) = a(x - 2)(x + 6)$ has zeros 2 and -6 for all nonzero real numbers a.

57. $f(x) = (x - 0)[x - (-4)][x - (-3)]$

$= x(x + 4)(x + 3)$

$= x^3 + 7x^2 + 12x$

Note: $f(x) = ax(x + 4)(x + 3)$ has zeros $0, -4, -3$ for all nonzero real numbers a.

59. $f(x) = (x - 4)(x + 3)(x - 3)(x - 0)$

$= (x - 4)(x^2 - 9)x$

$= x^4 - 4x^3 - 9x^2 + 36x$

Note: $f(x) = a(x^4 - 4x^3 - 9x^2 + 36x)$ has these zeros for all nonzero real numbers a.

61. $f(x) = \left[x - (1 + \sqrt{3})\right]\left[x - (1 - \sqrt{3})\right]$

$= \left[(x - 1) - \sqrt{3}\right]\left[(x - 1) + \sqrt{3}\right]$

$= (x - 1)^2 - \left(\sqrt{3}\right)^2$

$= x^2 - 2x + 1 - 3$

$= x^2 - 2x - 2$

Note: $f(x) = a(x^2 - 2x - 2)$ has these zeros for all nonzero real numbers a.

63. $f(x) = (x - 2)\left[x - (4 + \sqrt{5})\right]\left[x - (4 - \sqrt{5})\right]$

$= (x - 2)\left[(x - 4) - \sqrt{5}\right]\left[(x - 4) + \sqrt{5}\right]$

$= (x - 2)[(x - 4)^2 - 5]$

$= x^3 - 10x^2 + 27x - 22$

Note: $f(x) = a(x - 2)[(x - 4)^2 - 5]$ has these zeros for all nonzero real numbers a.

65. (a) The degree of f is odd and the leading coefficient is 1. The graph falls to the left and rises to the right.

(b) $f(x) = x^3 - 9x = x(x^2 - 9) = x(x - 3)(x + 3)$

zeros: $0, 3, -3$

(c), (d)

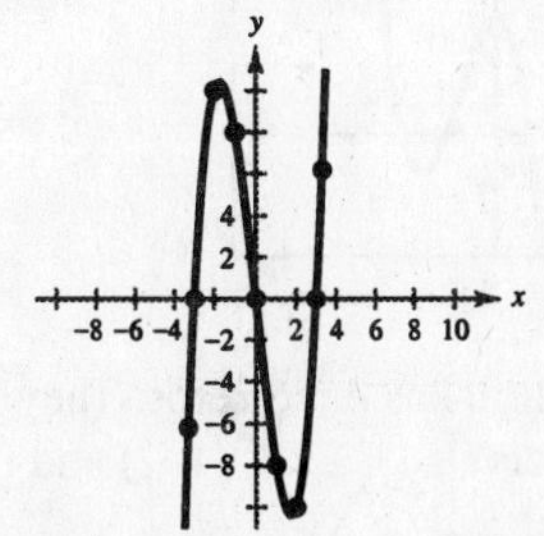

67. (a) The degree of f is even and the leading coefficient is $\frac{1}{4}$. The graph rises to the left and to the right.

(b) $f(t) = \frac{1}{4}(t^2 - 2t + 15)$ has no real zeros.

(c), (d)

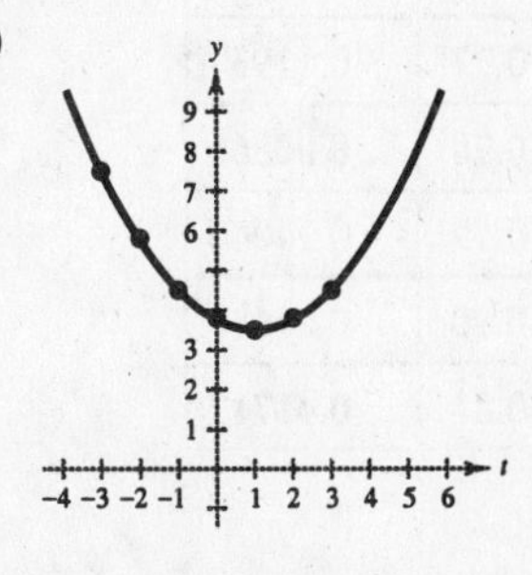

69. (a) The degree of f is odd and the leading coefficient is 1. The graph falls to the left and rises to the right.

(b) $f(x) = x^3 - 3x^2 = x^2(x - 3)$; zeros: 0, 3

(c), (d)

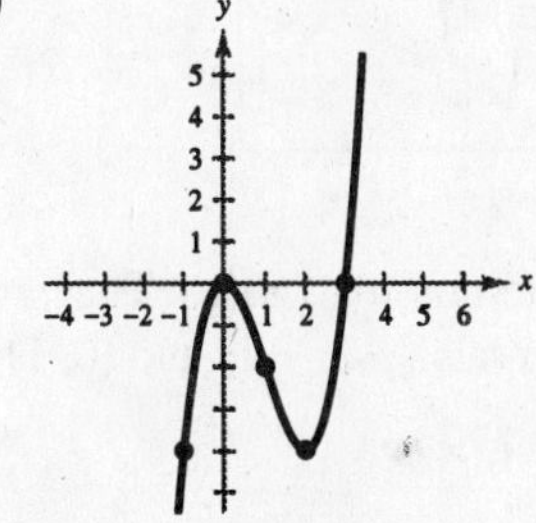

71. (a) The degree of f is odd and the leading coefficient is 3. The graph falls to the left and rises to the right.

(b) $f(x) = 3x^3 - 15x^2 + 18x = 3x(x^2 - 5x + 6)$

$= 3x(x - 2)(x - 3)$

zeros: 0, 2, 3

(c), (d)

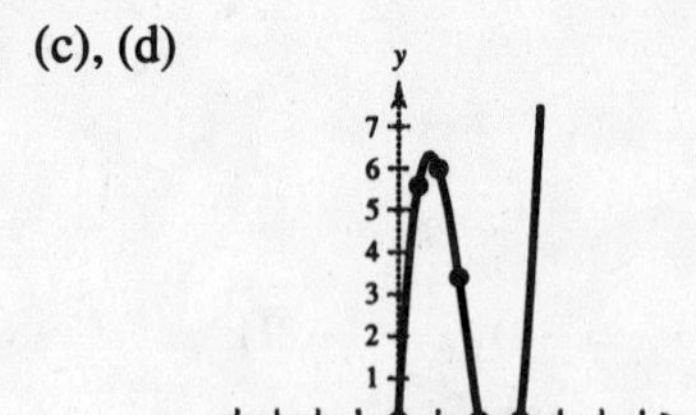

73. (a) The degree of f is odd and the leading coefficient is -1. The graph rises to the left and falls to the right.

(b) $f(x) = -x^3 - 5x^2 = x^2(-x - 5)$

zeros: $0, -5$

(c), (d)

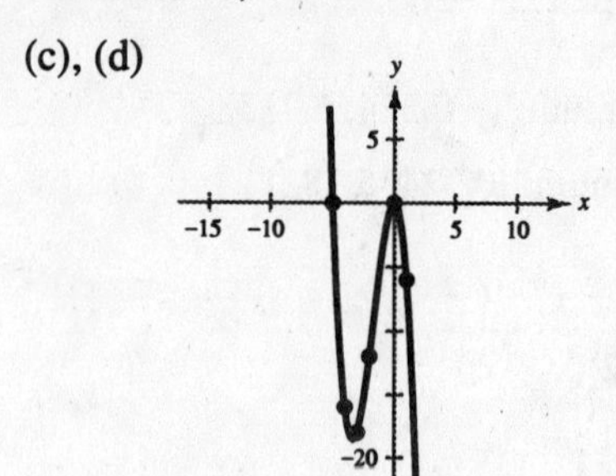

75. (a) The degree of f is odd and the leading coefficient is 1. The graph falls to the left and rises to the right.

(b) $f(x) = x^2(x - 4)$; zeros: 0, 4

(c), (d)

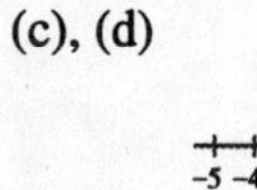

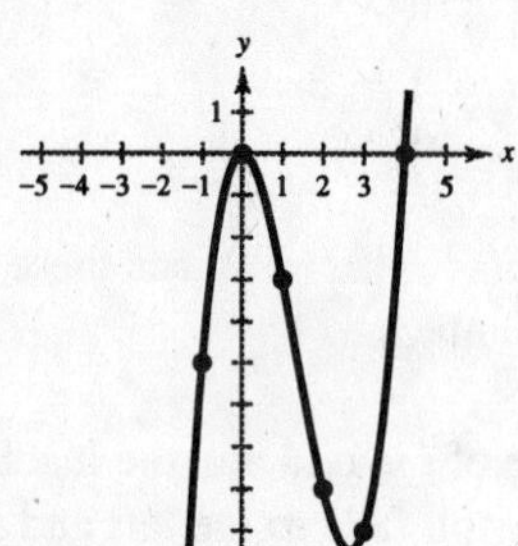

77. (a) The degree of g is even (4) and the leading coefficient is $-\frac{1}{4}$. The graph falls to the left and to the right.

(b) $g(t) = -\frac{1}{4}(t - 2)^2(t + 2)$; zeros: $2, -2$

(c)

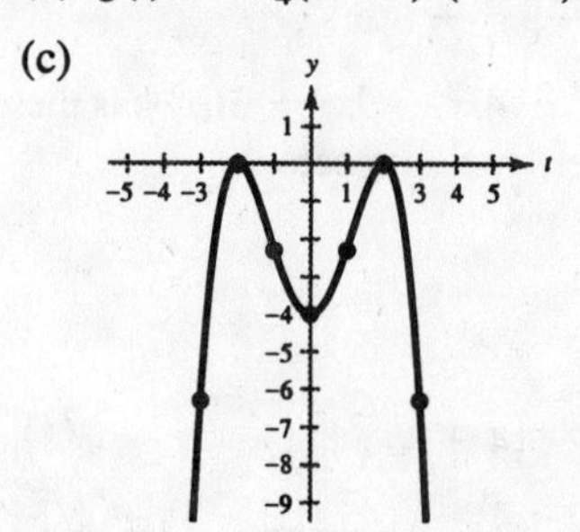

79. $f(x) = x^3 - 3x^2 + 3$

(a)

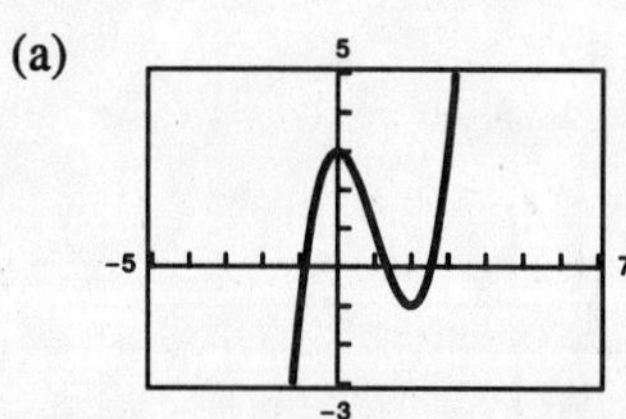

The function has three zeros. They are in the intervals $(-1, 0)$, $(1, 2)$ and $(2, 3)$.

(b) $-0.879, 1.347, 2.532$

(c)

x	Y1
−0.9	−0.159
−0.89	−0.0813
−0.88	−0.0047
−0.87	−0.0708
−0.86	−0.14514
−0.85	−0.21838
−0.84	−0.2905

x	Y1
1.3	0.127
1.31	0.09979
1.32	0.07277
1.33	0.04594
1.34	0.0193
1.35	−0.0071
1.36	−0.0333

x	Y1
2.5	−0.125
2.51	−0.087
2.52	−0.0482
2.53	−0.0084
2.54	0.03226
2.55	0.07388
2.56	0.11642

81. $g(x) = 3x^4 + 4x^3 - 3$

(a)

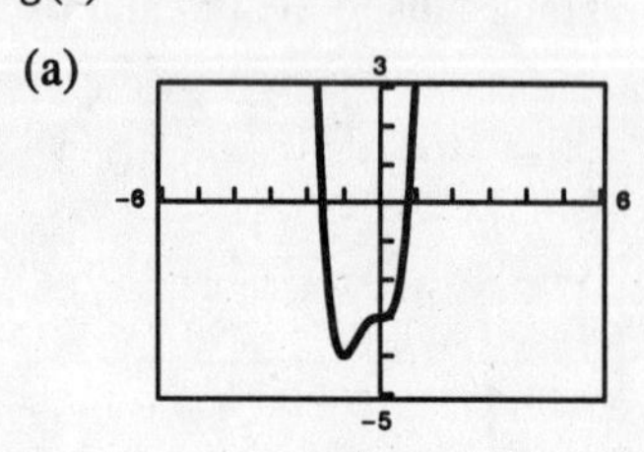

The function has two zeros. They are in the intervals $(-2, -1)$ and $(0, 1)$.

(b) $-1.585, 0.779$

(c)

x	Y1
−1.6	0.2768
−1.59	0.09515
−1.58	−0.0812
−1.57	−0.2524
−1.56	−0.4184
−1.55	−0.5795
−1.54	−0.7356

x	Y1
0.75	−0.3633
0.76	−0.2432
0.77	−0.1193
0.78	0.00866
0.79	0.14066
0.80	0.2768
0.81	0.41717

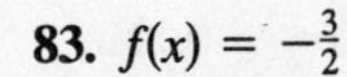

83. $f(x) = -\frac{3}{2}$

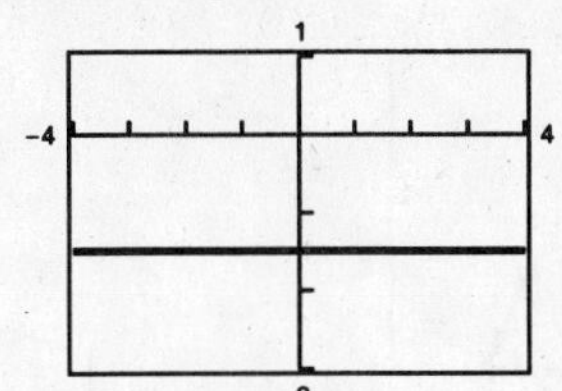

Horizontal line

Xmin=-4
Xmax=4
Xscl=1
Ymin=-3
Ymax=1
Yscl=1

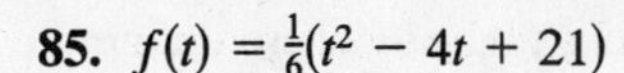

85. $f(t) = \frac{1}{6}(t^2 - 4t + 21)$

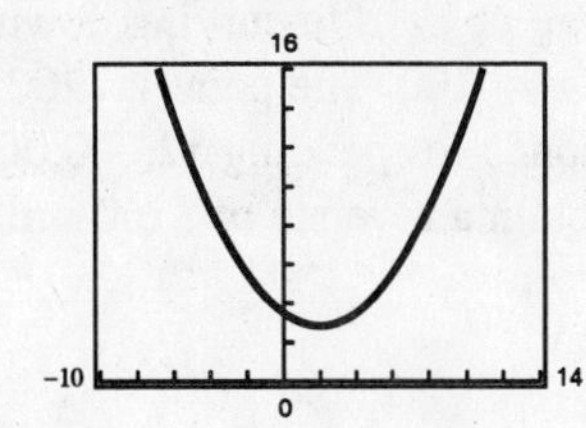

Xmin=-10
Xmax=14
Xscl=2
Ymin=0
Ymax=16
Yscl=2

87.

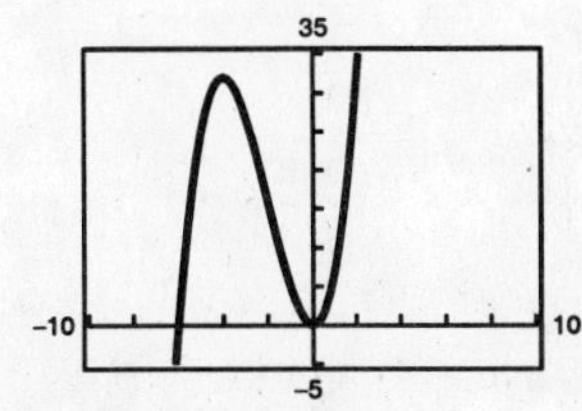

No symmetry

Two x-intercepts

89.

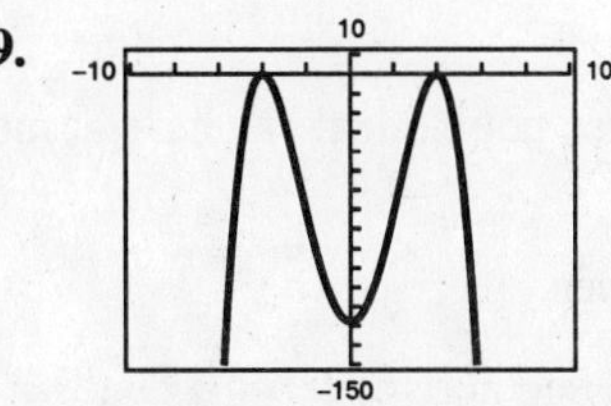

Symmetric about the y-axis

Two x-intercepts

91. $f(x) = x^3 - 4x = x(x + 2)(x - 2)$

Symmetric to origin
Three x-intercepts

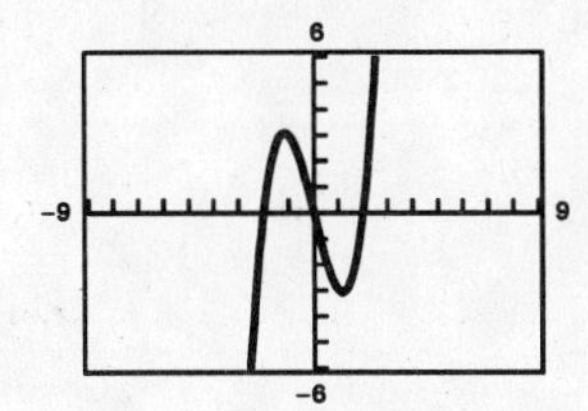

93. $g(x) = \frac{1}{5}(x + 1)^2(x - 3)(2x - 9)$

Three x-intercepts
No symmetry

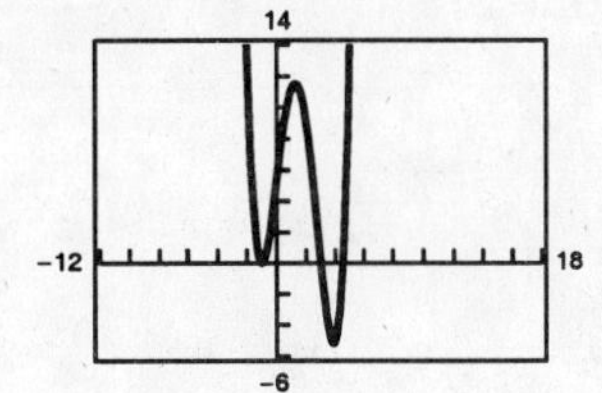

95. (a) and (b)

Height, x	Length and Width	Volume, V
1	$36 - 2(1)$	$1[36 - 2(1)]^2 = 1156$
2	$36 - 2(2)$	$2[36 - 2(2)]^2 = 2048$
3	$36 - 2(3)$	$3[36 - 2(3)]^2 = 2700$
4	$36 - 2(4)$	$4[36 - 2(4)]^2 = 3136$
5	$36 - 2(5)$	$5[36 - 2(5)]^2 = 3380$
6	$36 - 2(6)$	$6[36 - 2(6)]^2 = 3456$
7	$36 - 2(7)$	$7[36 - 2(7)]^2 = 3388$

(c) Volume = length × width × height

Because the box is made from a square, length = width.

Thus:

$$\text{Volume} = (\text{length})^2 \times \text{height}$$
$$= (36 - 2x)^2 x$$

Domain:
$$0 < 36 - 2x < 36$$
$$-36 < -2x < 0$$
$$18 > x > 0$$

(d)

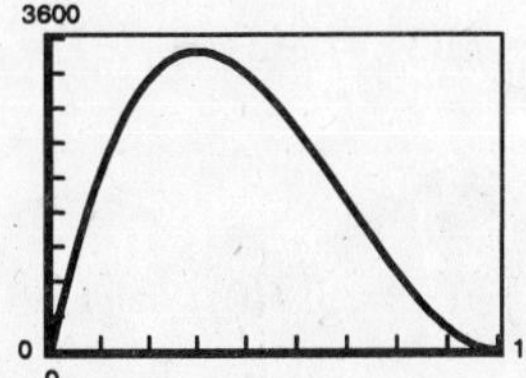

$x = 6$ when $V(x)$ is maximum.

97. The point of diminishing returns (where the graph changes from curving upward to curving downward) occurs when $x = 200$. The point is (200, 160) which corresponds to spending \$2,000,000 on advertising to obtain a revenue of \$160 million.

99. (a) $y = 0.003x^4 - 0.024x^3 + 0.020x^2 + 0.113x$

(b)

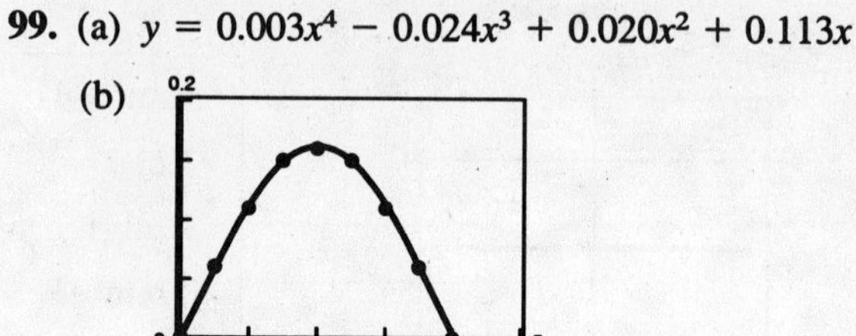

(c) The constant term should be zero. Yes, the model has zero as its constant term.

101. False. A fourth degree polynomial can have at most three turning points.

103. $f(x) = x^4$; $f(x)$ is even.

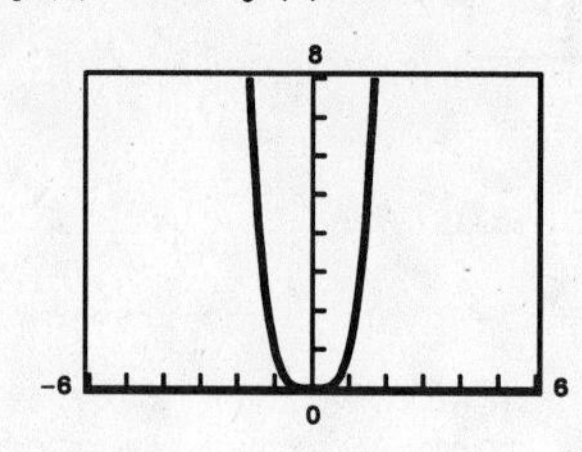

(a) $g(x) = f(x) + 2$

Vertical shift two units upward

$$g(-x) = f(-x) + 2$$
$$= f(x) + 2$$
$$= g(x)$$

Even

(b) $g(x) = f(x + 2)$

Horizontal shift two units to the left

Neither odd nor even

(c) $g(x) = f(-x) = (-x)^4 = x^4$

Reflection in the y-axis
The graph looks the same.
Even

(d) $g(x) = -f(x) = -x^4$

Reflection in the x-axis
Even

(e) $g(x) = f(\frac{1}{2}x) = \frac{1}{16}x^4$

Vertical shrink
Even

(f) $g(x) = \frac{1}{2}f(x) = \frac{1}{2}x^4$

Vertical shrink
Even

(g) $g(x) = f(x^{3/4}) = (x^{3/4}) = x^3$

Odd

(h) $g(x) = (f \circ f)(x) = f(f(x)) = f(x^4) = f(x^4)^4 = x^{16}$

Even

105. $(g - f)(3) = g(3) - f(3) = 8(3)^2 - [14(3) - 3]$
$= 72 - 39 = 33$

107. $\left(\frac{f}{g}\right)(-1.5) = \frac{f(-1.5)}{g(-1.5)} = \frac{-24}{18} = -\frac{4}{3}$

109. $(g \circ f)(0) = g(f(0)) = g(-3) = 8(-3)^2 = 72$

111. $2x^2 - x \geq 1$

$2x^2 - x - 1 \geq 0$

$(2x + 1)(x - 1) \geq 0$

$[2x + 1 \geq 0 \text{ and } x - 1 \geq 0]$ or $[2x + 1 \leq 0 \text{ and } x - 1 \leq 0]$

$\left[x \geq -\frac{1}{2} \text{ and } x \geq 1\right]$ or $\left[x \leq -\frac{1}{2} \text{ and } x \leq 1\right]$

$x \geq 1$ or $x \leq -\frac{1}{2}$

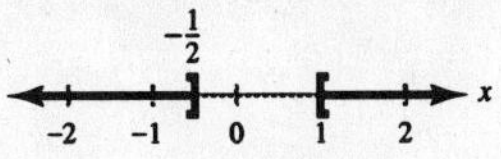

113. $|x + 8| - 1 \geq 15$

$|x + 8| \geq 16$

$x + 8 \geq 16$ or $x + 8 \leq -16$

$x \geq 8$ or $x \leq -24$

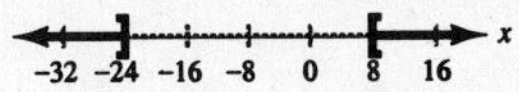

115. Vertex: $(0, -8)$

$f(x) = a(x - 0)^2 - 8 = ax^2 - 8$

Point: $(5, 9) \Rightarrow 9 = a(5)^2 - 8$

$17 = 25a$

$a = \frac{17}{25}$

$f(x) = \frac{17}{25}x^2 - 8$

117. Vertex: $(-5, -2)$

$f(x) = a(x + 5)^2 - 2$

Point: $(0, 3) \Rightarrow 3 = a(0 + 5)^2 - 2$

$5 = 25a$

$a = \frac{1}{5}$

$f(x) = \frac{1}{5}(x + 5)^2 - 2$

Section 2.3 Real Zeros of Polynomial Functions

You should know the following basic techniques and principles of polynomial division.

- The Division Algorithm (Long Division of Polynomials)
- Synthetic Division
- $f(k)$ is equal to the remainder of $f(x)$ divided by $(x - k)$.
- $f(k) = 0$ if and only if $(x - k)$ is a factor of $f(x)$.
- The Rational Zero Test
- The Upper and Lower Bound Rule

Solutions to Odd-Numbered Exercises

1. $y_2 = 4 + \dfrac{4}{x-1}$

$= \dfrac{4(x-1)+4}{x-1}$

$= \dfrac{4x-4+4}{x-1}$

$= \dfrac{4x}{x-1}$

$= y_1$

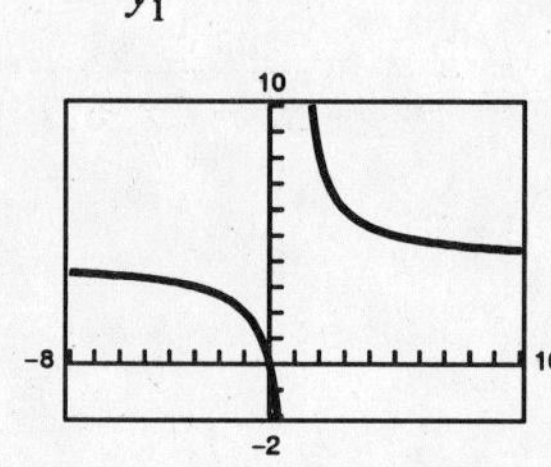

3. $y_2 = x - 2 + \dfrac{4}{x+2}$

$= \dfrac{(x-2)(x+2)+4}{x+2}$

$= \dfrac{x^2-4+4}{x+2}$

$= \dfrac{x^2}{x+2}$

$= y_1$

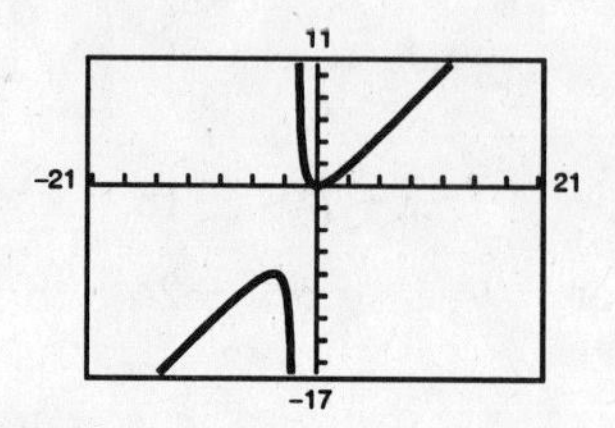

5. $y_2 = x^3 - 4x + \dfrac{4x}{x^2+1}$

$= \dfrac{(x^3-4x)(x^2+1)+4x}{x^2+1}$

$= \dfrac{x^5+x^3-4x^3-4x+4x}{x^2+1}$

$= \dfrac{x^5-3x^3}{x^2+1} = y_1$

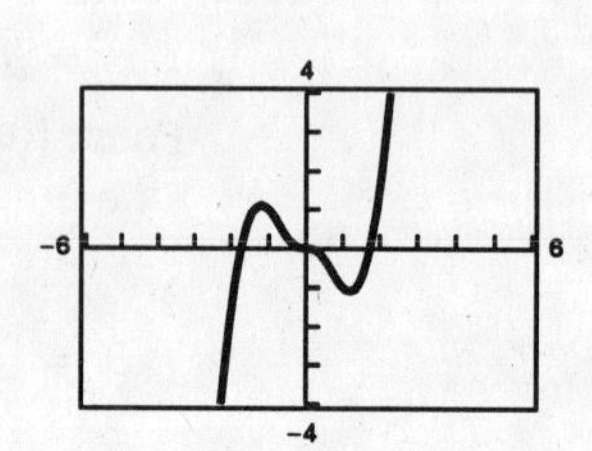

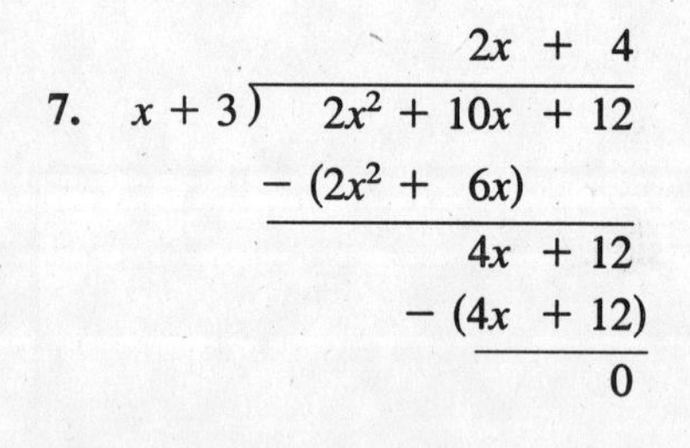

7.

$$\begin{array}{r} 2x + 4 \\ x+3 \overline{)\, 2x^2 + 10x + 12} \\ -(2x^2 + 6x) \\ \hline 4x + 12 \\ -(4x + 12) \\ \hline 0 \end{array}$$

$$\frac{2x^2+10x+12}{x+3} = 2x+4$$

9.

$$\begin{array}{r} x^2 - 3x + 1 \\ 4x+5 \overline{)\, 4x^3 - 7x^2 - 11x + 5} \\ -(4x^3 + 5x^2) \\ \hline -12x^2 - 11x \\ -(-12x^2 - 15x) \\ \hline 4x + 5 \\ -(4x + 5) \\ \hline 0 \end{array}$$

$$\frac{4x^3-7x^2-11x+5}{4x+5} = x^2-3x+1$$

11.
$$\begin{array}{rl} & 7 \\ x+2 \big) & \overline{7x + 3} \\ & -(7x+14) \\ \hline & -11 \end{array}$$

$$\frac{7x+3}{x+2} = 7 - \frac{11}{x+2}$$

13.
$$\begin{array}{rl} & 3x \;+5 \\ 2x^2+0x+1 \big) & \overline{6x^3+10x^2+\;x\;+8} \\ & -(6x^3+\;0x^2+3x) \\ \hline & 10x^2-2x\;+8 \\ & -(10x^2+0x\;+5) \\ \hline & -2x\;+3 \end{array}$$

$$\frac{6x^3+10x^2+x+8}{2x^2+1} = 3x+5-\frac{2x-3}{2x^2+1}$$

15.
$$\begin{array}{rl} & x^2+2x+4 \\ x^2-2x+3 \big) & \overline{x^4+0x^3+3x^2+0x+1} \\ & -(x^4-2x^3+3x^2) \\ \hline & 2x^3+0x^2+0x \\ & -(2x^3-4x^2+6x) \\ \hline & 4x^2-6x+1 \\ & -(4x^2-8x+12) \\ \hline & 2x-11 \end{array} \Rightarrow \frac{x^4+3x^2+1}{x^2-2x+3} = x^2+2x+4+\frac{2x-11}{x^2-2x+3}$$

17.
$$\begin{array}{rl} & 2x \\ x^2-2x+1 \big) & \overline{2x^3-4x^2-15x+5} \\ & -(2x^3-4x^2+\;2x) \\ \hline & -17x+5 \end{array}$$

$$\frac{2x^3-4x^2-15x+5}{(x-1)^2} = 2x - \frac{17x-5}{(x-1)^2}$$

19.
$$\begin{array}{r|rrrr} 4 & 3 & -10 & 12 & -22 \\ & & 12 & 8 & 80 \\ \hline & 3 & 2 & 20 & 58 \end{array}$$

$$\frac{3x^3-10x^2+12x-22}{x-4} = 3x^2+2x+20+\frac{58}{x-4}$$

21.
$$\begin{array}{r|rrrr} 3 & 6 & 7 & -1 & 26 \\ & & 18 & 75 & 222 \\ \hline & 6 & 25 & 74 & 248 \end{array}$$

$$\frac{6x^3+7x^2-x+26}{x-3} = 6x^2+25x+74+\frac{248}{x-3}$$

23.
$$\begin{array}{r|rrrr} 2 & 9 & -18 & -16 & 32 \\ & & 18 & 0 & -32 \\ \hline & 9 & 0 & -16 & 0 \end{array}$$

$$\frac{9x^3-18x^2-16x+32}{x-2} = 9x^2-16$$

25.
$$\begin{array}{r|rrrr} -8 & 1 & 0 & 0 & 512 \\ & & -8 & 64 & -512 \\ \hline & 1 & -8 & 64 & 0 \end{array}$$

$$\frac{x^3+512}{x+8} = x^2-8x+64$$

27.
$$\begin{array}{r|rrrr} -\frac{1}{2} & 4 & 16 & -23 & -15 \\ & & -2 & -7 & 15 \\ \hline & 4 & 14 & -30 & 0 \end{array}$$

$$\frac{4x^3+16x^2-23x-15}{x+\frac{1}{2}} = 4x^2+14x-30$$

29. $f(x) = x^3 - x^2 - 14x + 11,\ k = 4$

$$\begin{array}{r|rrrr} 4 & 1 & -1 & -14 & 11 \\ & & 4 & 12 & -8 \\ \hline & 1 & 3 & -2 & 3 \end{array}$$

$f(x) = (x-4)(x^2+3x-2)+3$

$f(4) = (0)(26)+3 = 3$

31. $f(x) = x^3 + 3x^2 - 2x - 14,\ k = \sqrt{2}$

$$\begin{array}{r|rrrr} \sqrt{2} & 1 & 3 & -2 & -14 \\ & & \sqrt{2} & 2+3\sqrt{2} & 6 \\ \hline & 1 & 3+\sqrt{2} & 3\sqrt{2} & -8 \end{array}$$

$f(x) = \left(x-\sqrt{2}\right)\left[x^2+\left(3+\sqrt{2}\right)x+3\sqrt{2}\right]-8$

$f\left(\sqrt{2}\right) = (0)\left(4+6\sqrt{2}\right)-8 = -8$

33. $$\begin{array}{r|rrrr} 1-\sqrt{3} & 4 & -6 & -12 & -4 \\ & & 4-4\sqrt{3} & 10-2\sqrt{3} & 4 \\ \hline & 4 & -2-4\sqrt{3} & -2-2\sqrt{3} & 0 \end{array}$$

$f(x) = \left(x - 1 + \sqrt{3}\right)\left[4x^2 - \left(2 + 4\sqrt{3}\right)x - \left(2 + 2\sqrt{3}\right)\right]$

$f\left(1 - \sqrt{3}\right) = 0$

35. $f(x) = 4x^3 - 13x + 10$

(a) $$\begin{array}{r|rrrr} 1 & 4 & 0 & -13 & 10 \\ & & 4 & 4 & -9 \\ \hline & 4 & 4 & -9 & \underline{1} = f(1) \end{array}$$

(b) $$\begin{array}{r|rrrr} -2 & 4 & 0 & -13 & 10 \\ & & -8 & 16 & -6 \\ \hline & 4 & -8 & 3 & \underline{4} = f(-2) \end{array}$$

(c) $$\begin{array}{r|rrrr} \frac{1}{2} & 4 & 0 & -13 & 10 \\ & & 2 & 1 & -6 \\ \hline & 4 & 2 & -12 & \underline{4} = f\left(\frac{1}{2}\right) \end{array}$$

(d) $$\begin{array}{r|rrrr} 8 & 4 & 0 & -13 & 10 \\ & & 32 & 256 & 1944 \\ \hline & 4 & 32 & 243 & \underline{1954} = f(8) \end{array}$$

37. $h(x) = 3x^3 + 5x^2 - 10x + 1$

(a) $$\begin{array}{r|rrrr} 3 & 3 & 5 & -10 & 1 \\ & & 9 & 42 & 96 \\ \hline & 3 & 14 & 32 & \underline{97} = f(3) \end{array}$$

(b) $$\begin{array}{r|rrrr} \frac{1}{3} & 3 & 5 & -10 & 1 \\ & & 1 & 2 & -\frac{8}{3} \\ \hline & 3 & 6 & -8 & \underline{-\frac{5}{3}} = f\left(\frac{1}{3}\right) \end{array}$$

(c) $$\begin{array}{r|rrrr} -2 & 3 & 5 & -10 & 1 \\ & & -6 & 2 & 16 \\ \hline & 3 & -1 & -8 & \underline{17} = f(-2) \end{array}$$

(d) $$\begin{array}{r|rrrr} -5 & 3 & 5 & -10 & 1 \\ & & -15 & 50 & -200 \\ \hline & 3 & -10 & 40 & \underline{-199} = f(-5) \end{array}$$

39. $$\begin{array}{r|rrrr} 2 & 1 & 0 & -7 & 6 \\ & & 2 & 4 & -6 \\ \hline & 1 & 2 & -3 & 0 \end{array}$$

$x^3 - 7x + 6 = (x - 2)(x^2 + 2x - 3)$

$= (x - 2)(x + 3)(x - 1)$

Zeros: $2, -3, 1$

41. $$\begin{array}{r|rrrr} \frac{1}{2} & 2 & -15 & 27 & -10 \\ & & 1 & -7 & 10 \\ \hline & 2 & -14 & 20 & 0 \end{array}$$

$2x^3 - 15x^2 + 27x - 10$

$= \left(x - \frac{1}{2}\right)(2x^2 - 14x + 20)$

$= (2x - 1)(x - 2)(x - 5)$

Zeros: $\frac{1}{2}, 2, 5$

43. $$\begin{array}{r|rrrr} -2 & 1 & 2 & -2 & -4 \\ & & -2 & 0 & 4 \\ \hline & 1 & 0 & -2 & 0 \end{array}$$

$x^3 + 2x^2 - 2x - 4 = (x + 2)(x^2 - 2)$

$= (x + 2)\left(x + \sqrt{2}\right)\left(x - \sqrt{2}\right)$

Zeros: $-2, \sqrt{2}, -\sqrt{2}$

45. (a)
$$\begin{array}{r|rrrr} -2 & 2 & 1 & -5 & 2 \\ & & -4 & 6 & -2 \\ \hline & 2 & -3 & 1 & 0 \end{array}$$

$$\begin{array}{r|rrr} 1 & 2 & -3 & 1 \\ & & 2 & -1 \\ \hline & 2 & -1 & 0 \end{array}$$

(b) Remaining factor: $(2x - 1)$

(c) $f(x) = (x + 2)(x - 1)(2x - 1)$

(d) Real zeros: $-2, 1, \frac{1}{2}$

(e)

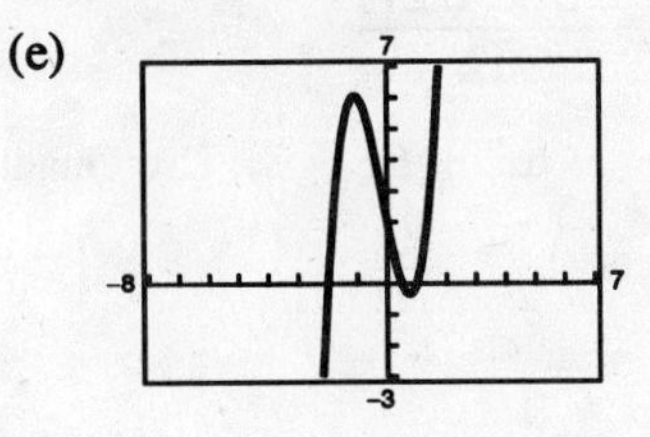

47. (a)
$$\begin{array}{r|rrrrr} 5 & 1 & -4 & -15 & 58 & -40 \\ & & 5 & 5 & -50 & 40 \\ \hline & 1 & 1 & -10 & 8 & 0 \end{array}$$

$$\begin{array}{r|rrrr} -4 & 1 & 1 & -10 & 8 \\ & & -4 & 12 & -8 \\ \hline & 1 & -3 & 2 & 0 \end{array}$$

(b) $x^2 - 3x + 2 = (x - 2)(x - 1)$,

Remaining factors: $(x - 2), (x - 1)$

(c) $f(x) = (x - 5)(x + 4)(x - 2)(x - 1)$

(d) Real zeros: $5, -4, 2, 1$

(e)

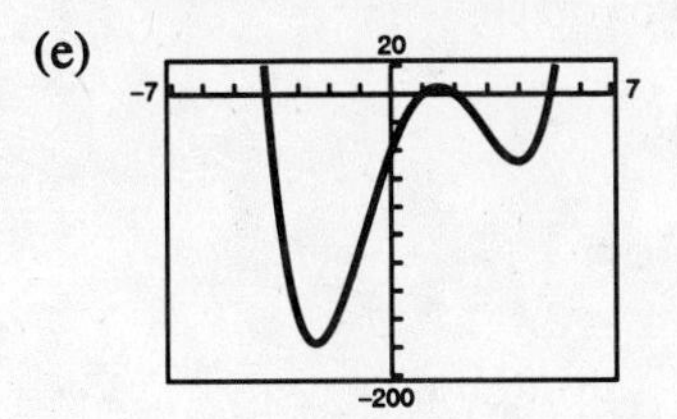

49. (a)
$$\begin{array}{r|rrrr} -\frac{1}{2} & 6 & 41 & -9 & -14 \\ & & -3 & -19 & 14 \\ \hline & 6 & 38 & -28 & 0 \end{array}$$

$$\begin{array}{r|rrr} \frac{2}{3} & 6 & 38 & -28 \\ & & 4 & 28 \\ \hline & 6 & 42 & 0 \end{array}$$

(b) $6x + 42$ Remaining factor $\big(\text{or } 6(x + 7)\big)$

(c) $f(x) = (2x + 1)(3x - 2)(x + 7)$

Note: Use $\frac{1}{6}(6x + 42) = x + 7$

(d) Real zeros: $-\frac{1}{2}, \frac{2}{3}, -7$

(e)

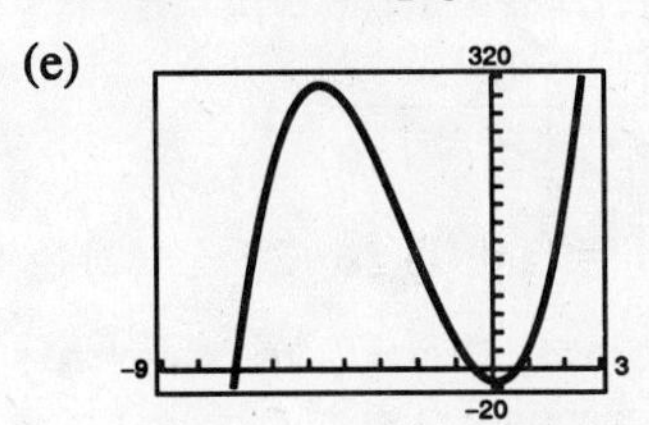

51. $f(x) = x^3 + 3x^2 - x - 3$

Possible rational zeros: $\pm 1, \pm 3$

Zeros shown on graph: $-3, -1, 1$

53. $f(x) = 2x^4 - 17x^3 + 35x^2 + 9x - 45$

Possible rational zeros: $\pm 1, \pm 3, \pm 5, \pm 9, \pm 15, \pm 45,$
$\pm \frac{1}{2}, \pm \frac{3}{2}, \pm \frac{5}{2}, \pm \frac{9}{2}, \pm \frac{15}{2}, \pm \frac{45}{2}$

Zeros shown of graph: $-1, \frac{3}{2}, 3, 5$

55. $f(x) = x^3 + x^2 - 4x - 4$

(a) Possible rational zeros: $\pm 1, \pm 2, \pm 4$

(b)

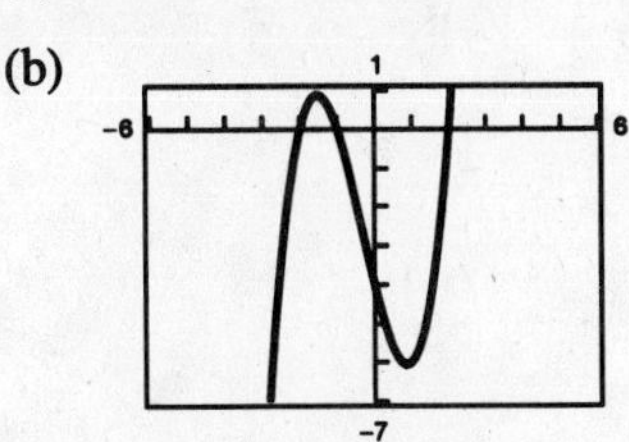

(c) $-2, -1, 2$ on graph

57. $f(x) = -4x^3 + 15x^2 - 8x - 3$

(a) Possible rational zeros: $\pm \frac{1}{4}, \pm \frac{1}{2}, \pm \frac{3}{4}, \pm 1, \pm \frac{3}{2}, \pm 3$

(b)

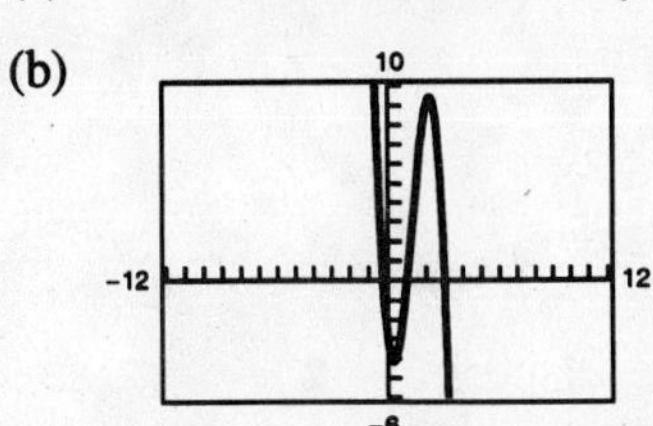

(c) $-\frac{1}{4}, 1, 3$ on graph

59. $f(x) = -2x^4 + 13x^3 - 21x^2 + 2x + 8$

(a) Possible rational zeros:

$\pm\frac{1}{2}, \pm1, \pm2, \pm4, \pm8$

(b)

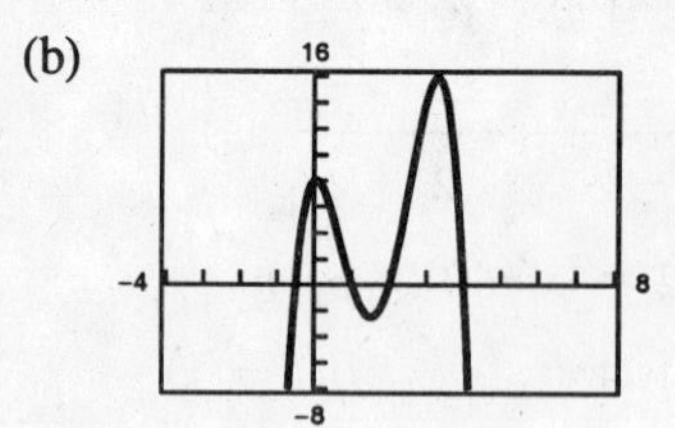

(c) $-\frac{1}{2}$, 1, 2, 4 on graph

61. $f(x) = 6x^3 - x^2 - 13x + 8$

(a) Possible rational zeros:

$\pm\frac{1}{6}, \pm\frac{1}{3}, \pm\frac{1}{2}, \pm\frac{2}{3}, \pm1, \pm\frac{4}{3}, \pm2, \pm\frac{8}{3}, \pm4, \pm8$

(b)

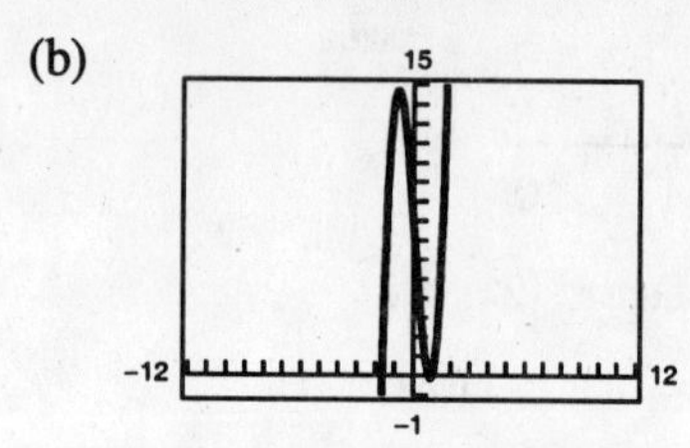

(c) Real zeros: $1, \dfrac{-5 \pm \sqrt{217}}{12}$

$[f(x) = (x - 1)(6x^2 + 5x - 8)$; Use Quadratic Formula$]$

63. $z^4 - z^3 - 2z - 4 = 0$

Possible rational zeros: $\pm1, \pm2, \pm4$

$$\begin{array}{r|rrrrr} -1 & 1 & -1 & 0 & -2 & -4 \\ & & -1 & 2 & -2 & 4 \\ \hline & 1 & -2 & 2 & -4 & 0 \end{array}$$

$$\begin{array}{r|rrrr} 2 & 1 & -2 & 2 & -4 \\ & & 2 & 0 & 4 \\ \hline & 1 & 0 & 2 & 0 \end{array}$$

$z^4 - z^3 - 2z - 4 = (z + 1)(z - 2)(z^2 + 2) = 0$

The only real zeros are -1 and 2. You can verify this by graphing the function $f(z) = z^4 - z^3 - 2z - 4$.

65. $x^4 - 13x^2 - 12x = 0$

$x(x^3 - 13x - 12) = 0$

$x = 0$ is a solution.

Possible rational zeros of $x^3 - 13x - 12 = 0$ are $\pm1, \pm2, \pm3, \pm4, \pm6$ and ±12. Using a graphing utility or synthetic division, you find that the zeros are $0, -1, -3, 4$.

67. $2x^4 - 11x^3 - 6x^2 + 64x + 32 = 0$

Using a graphing utility, you can see that there are three zeros. Using synthetic division, you can verify that these zeros are $-2, -\frac{1}{2}, 4$.

69. $f(x) = x^3 - 2x^2 - 5x + 10$

(a) Zeros: $2, 2.236, -2.236$

(b)
$$\begin{array}{r|rrrr} 2 & 1 & -2 & -5 & 10 \\ & & 2 & 0 & -10 \\ \hline & 1 & 0 & -5 & 0 \end{array} \quad x = 2 \text{ is a zero}$$

$f(x) = (x - 2)(x^2 - 5)$

$= (x - 2)(x + \sqrt{5})(x - \sqrt{5})$

71. $h(t) = t^3 - 2t^2 - 7t + 2$

(a) zeros: $-2, 3.732, 0.268$

(b)
$$\begin{array}{r|rrrr} -2 & 1 & -2 & -7 & 2 \\ & & -2 & 8 & -2 \\ \hline & 1 & -4 & 1 & 0 \end{array} \quad t = -2 \text{ is a zero}$$

$h(t) = (t + 2)(t^2 - 4t + 1)$

$= (t + 2)[t - (\sqrt{3} + 2)][t + (\sqrt{3} - 2)]$

73. $h(x) = x^5 - 7x^4 + 10x^3 + 14x^2 - 24x$

(a) $h(x) = x(x^4 - 7x^3 + 10x^2 + 14x - 24)$

From the calculator we have $x = 0, 3, 4$ and $x \approx \pm 1.414$.

(b)

$$\begin{array}{r|rrrrr} 3 & 1 & -7 & 10 & 14 & -24 \\ & & 3 & -12 & -6 & 24 \\ \hline & 1 & -4 & -2 & 8 & 0 \end{array}$$

$$\begin{array}{r|rrrr} 4 & 1 & -4 & -2 & 8 \\ & & 4 & 0 & -8 \\ \hline & 1 & 0 & -2 & 0 \end{array}$$

$$f(x) = x(x-3)(x-4)(x^2-2)$$
$$= x(x-3)(x-4)(x-\sqrt{2})(x+\sqrt{2})$$

The exact roots are $x = 0,\ 3,\ 4,\ \pm\sqrt{2}$.

75. $f(x) = x^4 - 4x^3 + 15$

(a)

$$\begin{array}{r|rrrrr} 4 & 1 & -4 & 0 & 0 & 15 \\ & & 4 & 0 & 0 & 0 \\ \hline & 1 & 0 & 0 & 0 & 15 \end{array}$$

4 is an upper bound.

(b)

$$\begin{array}{r|rrrrr} -1 & 1 & -4 & 0 & 0 & 15 \\ & & -1 & 5 & -5 & 5 \\ \hline & 1 & -5 & 5 & -5 & 20 \end{array}$$

-1 is a lower bound.

77. $f(x) = x^4 - 4x^3 + 16x - 16$

(a)

$$\begin{array}{r|rrrrr} 5 & 1 & -4 & 0 & 16 & -16 \\ & & 5 & 5 & 25 & 205 \\ \hline & 1 & 1 & 5 & 41 & 189 \end{array}$$

5 is an upper bound.

(b)

$$\begin{array}{r|rrrrr} -3 & 1 & -4 & 0 & 16 & -16 \\ & & -3 & 21 & -63 & 141 \\ \hline & 1 & -7 & 21 & -47 & 125 \end{array}$$

-3 is a lower bound.

79. $P(x) = x^4 - \frac{25}{4}x^2 + 9$

$$= \tfrac{1}{4}(4x^4 - 25x^2 + 36)$$
$$= \tfrac{1}{4}(4x^2 - 9)(x^2 - 4)$$
$$= \tfrac{1}{4}(2x + 3)(2x - 3)(x + 2)(x - 2)$$

The zeros are $\pm\frac{3}{2}$ and ± 2.

81. $f(x) = x^3 - \frac{1}{4}x^2 - x + \frac{1}{4}$

$$= \tfrac{1}{4}(4x^3 - x^2 - 4x + 1)$$
$$= \tfrac{1}{4}[x^2(4x - 1) - 1(4x - 1)]$$
$$= \tfrac{1}{4}(4x - 1)(x^2 - 1)$$
$$= \tfrac{1}{4}(4x - 1)(x + 1)(x - 1)$$

The zeros are $\frac{1}{4}$ and ± 1.

83. $f(x) = x^3 - 1 = (x - 1)(x^2 + x + 1)$

Rational zeros: 1 $(x = 1)$

Irrational zeros: 0

Matches (d).

85. $f(x) = x^3 - x = x(x + 1)(x - 1)$

Rational zeros: 3 $(x = 0, \pm 1)$

Irrational zeros: 0

Matches (b).

87. (a)

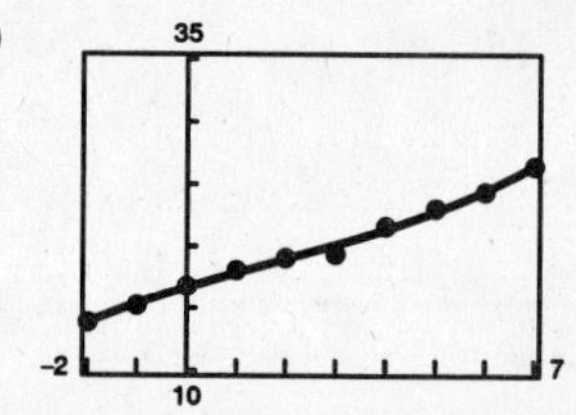

(b) $R = 0.01326t^3 - 0.06765t^2 + 1.2306t + 16.6770$

(c)

t	-2	-1	0	1	2	3	4	5	6	7
R	13.86	15.21	16.78	18.10	19.08	19.39	21.62	23.07	24.41	26.48
Model	13.84	15.37	16.68	17.85	18.97	20.12	21.37	22.80	24.49	26.52

(d)

12	.01326	-0.06765	1.2306	16.6770
		0.15912	1.09764	27.9389
	.01326	0.09147	2.3282	44.6159

$R(12) \approx 44.62$. No. The model will turn sharply upward.

89. (a)

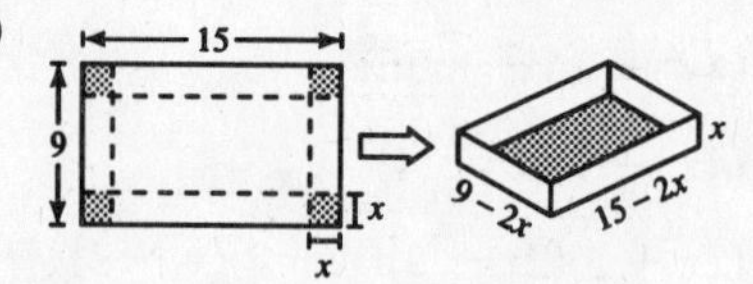

(b) $V = l \cdot w \cdot h = (15 - 2x)(9 - 2x)x$

$= x(9 - 2x)(15 - 2x)$

Since length, width, and height cannot be negative, we have $0 < x < \frac{9}{2}$ for the domain.

(c)

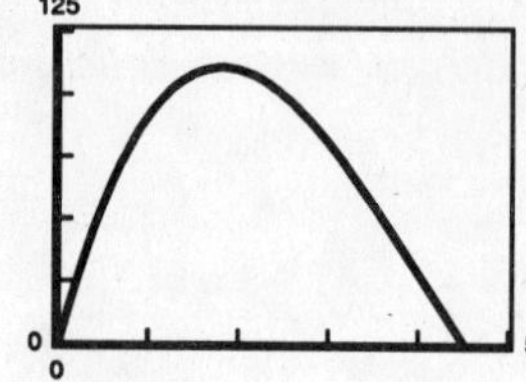

The volume is maximum when $x \approx 1.82$.

The dimensions are: length $= 15 - 2(1.82) = 11.36$

width $= 9 - 2(1.82) = 5.36$

height $= x = 1.82$

1.82 cm × 5.36 cm × 11.36 cm

(d) $56 = x(9 - 2x)(15 - 2x)$

$56 = 135x - 48x^2 + 4x^3$

$0 = 4x^3 - 48x^2 + 135x - 56$

The zeros of this polynomial are $\frac{1}{2}$, $\frac{7}{2}$, and 8.

x cannot equal 8 since it is not in the domain of V. [The length cannot equal -1 and the width cannot equal -7. The product of $(8)(-1)(-7) = 56$ so it showed up as an extraneous solution.]

91. $y = -5.05x^3 + 3857x - 38{,}411.25,\ 13 \le x \le 18$

(a)

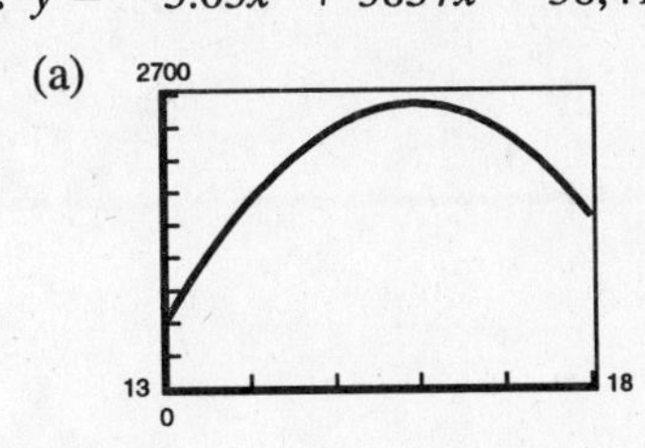

(b) The second air-fuel ration of 16.89 can be obtained by finding the second point where the curves y and $y_1 = 2400$ intersect.

(c) Solve $-5.05x^3 + 3857x - 38{,}411.25 = 2400$ or $-5.05x^3 + 3857x - 40{,}811.25 = 0$. By synthetic division,

$$\begin{array}{r|rrrr} 15 & -5.05 & 0 & 3857 & -40811.25 \\ & & -75.75 & -1136.25 & 40811.25 \\ \hline & -5.05 & -75.75 & 2720.75 & 0 \end{array}$$

(d) The positive zero of the quadratic $-5.05x^2 - 75.75x + 2720.75$ can be found by the Quadratic Formula.

$$x \approx \frac{75.75 - \sqrt{(75.75)^2 - 4(-5.05)(2720.75)}}{2(-5.05)} \approx 16.89$$

93. False, $-\frac{4}{7}$ is a zero of f.

95.
$$\begin{array}{r|rrrrrrr} \frac{1}{2} & 6 & 1 & -92 & 45 & 184 & 4 & -48 \\ & & 3 & 2 & -45 & 0 & 92 & 48 \\ \hline & 6 & 4 & -90 & 0 & 184 & 96 & 0 \end{array}$$

True.

97.
$$\begin{array}{r} x^{2n} - x^n + 3 \\ x^n - 2 \overline{\smash{)}\, x^{3n} - 3x^{2n} + 5x^n - 6} \\ \underline{x^{3n} - 2x^{2n}} \qquad\qquad \\ -x^{2n} + 5x^n \qquad \\ \underline{-x^{2n} + 2x^n} \qquad \\ 3x^n - 6 \\ \underline{3x^n - 6} \\ 0 \end{array}$$

Hence, $\dfrac{x^{3n} - 3x^{2n} + 5x^n - 6}{x^n - 2} = x^{2n} - x^n + 3.$

99. You can check polynomial division by multiplying the quotient by the divisor. This should yield the original dividend if the multiplication was performed correctly.

101. (a) $(f \circ g)(x) = f(x^2 + 1)$
$= 2(x^2 + 1) - 5$
$= 2x^2 - 3$

(b) $(g \circ f)(x) = g(2x - 5)$
$= (2x - 5)^2 + 1$
$= 4x^2 - 20x + 26$

103. (a) $(f \circ g)(x) = f(4x + x^2)$
$= \dfrac{1}{4x + x^2}$

(b) $(g \circ f)(x) = g\left(\dfrac{1}{x}\right) = 4\left(\dfrac{1}{x}\right) + \left(\dfrac{1}{x}\right)^2$
$= \dfrac{4x + 1}{x^2}$

105. $f(x) = (x - 1)(x + 3)(x - 8)$ [answer not unique]
$= x^3 - 6x^2 - 19x + 24$

107. $f(x) = \left[x - (2 + \sqrt{3})\right]\left[x - (2 - \sqrt{3})\right]$
$= \left[(x - 2) - \sqrt{3}\right]\left[(x - 2) + \sqrt{3}\right]$
$= (x - 2)^2 - 3$
$= x^2 - 4x + 4 - 3$
$= x^2 - 4x + 1$

[answer not unique]

Section 2.4 Complex Numbers

- You should know how to work with complex numbers.
- Operations on complex numbers
  (a) Addition: $(a + bi) + (c + di) = (a + c) + (b + d)i$
  (b) Subtraction: $(a + bi) - (c + di) = (a - c) + (b - d)i$
  (c) Multiplication: $(a + bi)(c + di) = (ac - bd) + (ad + bc)i$
  (d) Division: $\dfrac{a + bi}{c + di} = \dfrac{a + bi}{c + di} \cdot \dfrac{c - di}{c - di} = \dfrac{ac + bd}{c^2 + d^2} + \dfrac{bc - ad}{c^2 + d^2}i$
- The complex conjugate of $a + bi$ is $a - bi$:
  $(a + bi)(a - bi) = a^2 + b^2$
- The additive inverse of $a + bi$ is $-a - bi$.
- The multiplicative inverse of $a + bi$ is
  $\dfrac{a - bi}{a^2 + b^2}.$
- $\sqrt{-a} = \sqrt{a}\,i$ for $a > 0$.

Solutions to Odd-Numbered Exercises

1. $a + bi = -9 + 4i$
$a = -9$
$b = 4$

3. $(a - 1) + (b + 3)i = 5 + 8i$
$a - 1 = 5 \Rightarrow a = 6$
$b + 3 = 8 \Rightarrow b = 5$

5. $4 + \sqrt{-25} = 4 + 5i$

7. $12 = 12 + 0i$

9. $-5i + i^2 = -5i - 1 = -1 - 5i$

11. $(\sqrt{-75})^2 = -75$

13. $\sqrt{-0.09} = \sqrt{0.09}\,i = 0.3i$

15. $(4 + i) + (7 - 2i) = 11 - i$

17. $(-1 + \sqrt{-8}) + (8 - \sqrt{-50}) = 7 + 2\sqrt{2}i - 5\sqrt{2}i = 7 - 3\sqrt{2}i$

19. $13i - (14 - 7i) = 13i - 14 + 7i = -14 + 20i$

21. $-\left(\frac{3}{2} + \frac{5}{2}i\right) + \left(\frac{5}{3} + \frac{11}{3}i\right) = -\frac{3}{2} - \frac{5}{2}i + \frac{5}{3} + \frac{11}{3}i$
$= -\frac{9}{6} - \frac{15}{6}i + \frac{10}{6} + \frac{22}{6}i$
$= \frac{1}{6} + \frac{7}{6}i$

23. $(1.6 + 3.2i) + (-5.8 + 4.3i) = -4.2 + 7.5i$

25. $\sqrt{-6} \cdot \sqrt{-2} = (\sqrt{6}i)(\sqrt{2}i) = \sqrt{12}i^2 = (2\sqrt{3})(-1) = -2\sqrt{3}$

27. $(\sqrt{-10})^2 = (\sqrt{10}i)^2 = 10i^2 = -10$

29. $(1 + i)(3 - 2i) = 3 - 2i + 3i - 2i^2$
$= 3 + i + 2$
$= 5 + i$

31. $6i(5 - 2i) = 30i - 12i^2 = 30i + 12 = 12 + 30i$

33. $(\sqrt{14} + \sqrt{10}i)(\sqrt{14} - \sqrt{10}i) = 14 - 10i^2 = 14 + 10 = 24$

35. $(4 + 5i)^2 = 16 + 40i + 25i^2 = 16 + 40i - 25$
$= -9 + 40i$

37. $\sqrt{-6}\sqrt{-6} = \sqrt{6}i\,\sqrt{6}i = 6i^2 = -6$ $\left(\text{Note: } \sqrt{-6}\,\sqrt{-6} \neq \sqrt{(-6)(-6)}\right)$

39. $(4 + 3i)(4 - 3i) = 16 + 9 = 25$

41. $(-6 - \sqrt{5}i)(-6 + \sqrt{5}i) = 36 + 5 = 41$

43. $(22i)(-22i) = 484$

45. $(3 - \sqrt{-2})(3 + \sqrt{-2}) = (3 - \sqrt{2}i)(3 + \sqrt{2}i) = 9 + 2 = 11$

47. $\dfrac{6}{i} = \dfrac{6}{i} \cdot \dfrac{-i}{-i} = \dfrac{-6i}{-i^2} = \dfrac{-6i}{1} = -6i$

49. $\dfrac{4}{4 - 5i} = \dfrac{4}{4 - 5i} \cdot \dfrac{4 + 5i}{4 + 5i} = \dfrac{4(4 + 5i)}{16 + 25} = \dfrac{16 + 20i}{41} = \dfrac{16}{41} + \dfrac{20}{41}i$

51. $\dfrac{2 + i}{2 - i} = \dfrac{2 + i}{2 - i} \cdot \dfrac{2 + i}{2 + i} = \dfrac{4 + 4i + i^2}{4 + 1} = \dfrac{3 + 4i}{5} = \dfrac{3}{5} + \dfrac{4}{5}i$

53. $\dfrac{6 - 7i}{i} = \dfrac{6 - 7i}{i} \cdot \dfrac{-i}{-i} = \dfrac{-6i - 7}{1} = -7 - 6i$

55. $\dfrac{1}{(4 - 5i)^2} = \dfrac{1}{16 - 40i + 25i^2} = \dfrac{1}{-9 - 40i} \cdot \dfrac{-9 + 40i}{-9 + 40i}$
$= \dfrac{-9 + 40i}{81 + 1600} = \dfrac{-9 + 40i}{1681} = -\dfrac{9}{1681} + \dfrac{40}{1681}i$

57. $\dfrac{2}{1 + i} - \dfrac{3}{1 - i} = \dfrac{2(1 - i) - 3(1 + i)}{(1 + i)(1 - i)}$
$= \dfrac{2 - 2i - 3 - 3i}{1 + 1}$
$= \dfrac{-1 - 5i}{2}$
$= -\dfrac{1}{2} - \dfrac{5}{2}i$

59. $-6i^3 + i^2 = -6i^2i + i^2$
$= -6(-1)i + (-1)$
$= 6i - 1$
$= -1 + 6i$

61. $-5i^5 = -5i^2i^2i$
$= -5(-1)(-1)i$
$= -5i$

63. $(\sqrt{-75})^3 = (5\sqrt{3}i)^3 = 5^3(\sqrt{3})^3\, i^3$
$= 125(3\sqrt{3})(-i)$
$= -375\sqrt{3}i$

65. $\dfrac{1}{i^3} = \dfrac{1}{-i} = \dfrac{1}{-i} \cdot \dfrac{i}{i} = \dfrac{1}{-i^2} = \dfrac{i}{1} = i$

67. $4 + 3i$

69. $0 + 6i = 6i$

71. $4 - 5i$

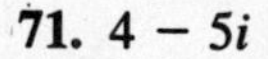

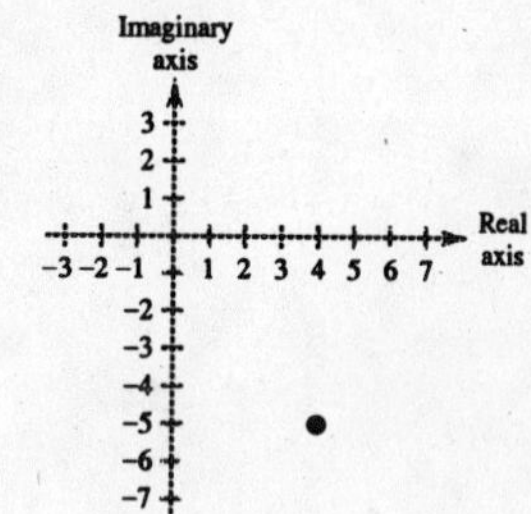

73. -6

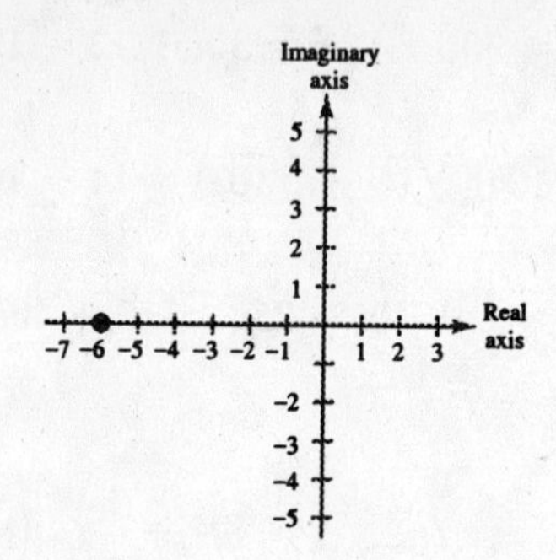

75. The complex number 0 is in the Mandelbrot Set since for $c = 0$, the corresponding Mandelbrot sequence is 0, 0, 0, 0, 0, 0, . . . which is bounded.

77. The complex number $\frac{1}{2}i$, is in the Mandelbrot Set since for $c = \frac{1}{2}i$, the corresponding Mandelbrot sequence is $\frac{1}{2}i, -\frac{1}{4} + \frac{1}{2}i, -\frac{3}{16} + \frac{1}{4}i, -\frac{7}{256} + \frac{13}{32}i, \frac{-10,767}{65,536} + \frac{1957}{4096}i, -\frac{864,513,055}{4,294,967,296} + \frac{46,037,845}{134,217,728}i$

which is bounded. Or in decimal form

$0.5i, -0.25 + 0.5i, -0.1875 + 0.25i, -0.02734 + 0.40625i,$

$-0.164291 + 0.477783i, -0.201285 + 0.343009i.$

79. The complex number 1 is not in the Mandelbrot Set since for $c = 1$, the corresponding Mandelbrot sequence is 1, 2, 5, 26, 677, 458,330 which is unbounded.

81. $(2)^3 = 8$

$$\begin{aligned}(-1 + \sqrt{3}i)^3 &= (-1)^3 + 3(-1)^2(\sqrt{3}i) + 3(-1)(\sqrt{3}i)^2 + (\sqrt{3}i)^3 \\ &= -1 + 3\sqrt{3}i - 9i^2 + 3\sqrt{3}i^3 \\ &= -1 + 3\sqrt{3}i + 9 - 3\sqrt{3}i \\ &= 8\end{aligned}$$

$$\begin{aligned}(-1 - \sqrt{3}i)^3 &= (-1)^3 + 3(-1)^2(-\sqrt{3}i) + 3(-1)(-\sqrt{3}i)^2(\sqrt{3}i)^3 \\ &= -1 - 3\sqrt{3}i - 9i^2 - 3\sqrt{3}i^3 \\ &= -1 - 3\sqrt{3}i + 9 + 3\sqrt{3}i \\ &= 8\end{aligned}$$

The three numbers are cube roots of 8.

83. (a) $z_1 = 5 + 2i$

$z_2 = 3 - 4i$

$$\begin{aligned}\frac{1}{z} = \frac{1}{z_1} + \frac{1}{z_2} &= \frac{1}{5 + 2i} + \frac{1}{3 - 4i} \\ &= \frac{(3 - 4i) + (5 + 2i)}{(5 + 2i)(3 - 4i)} \\ &= \frac{8 - 2i}{23 - 14i}\end{aligned}$$

$$\begin{aligned}z &= \frac{23 - 14i}{8 - 2i}\left(\frac{8 + 2i}{8 + 2i}\right) \\ &= \frac{212 - 66i}{68} \approx 3.118 - 0.971i\end{aligned}$$

(b) $z_1 = 16i + 9$

$z_2 = 20 - 10i$

$$\begin{aligned}\frac{1}{z} = \frac{1}{z_1} + \frac{1}{z_2} &= \frac{1}{9 + 16i} + \frac{1}{20 - 10i} \\ &= \frac{(20 - 10i) + (9 + 16i)}{(9 + 16i)(20 - 10i)} \\ &= \frac{29 + 6i}{340 + 230i}\end{aligned}$$

$$\begin{aligned}z &= \frac{340 + 230i}{29 + 6i}\left(\frac{29 - 6i}{29 - 6i}\right) \\ &= \frac{11240 + 4630i}{877} \approx 12.816 + 5.279i\end{aligned}$$

85. True. $(-i\sqrt{6})^4 - (-i\sqrt{6})^2 + 14 = 36 + 6 + 14 = 56$

87. (a) $i^{40} = (i^4)^{10} = 1^{10} = 1$

(b) $i^{25} = (i^{24})(i) = (i^4)^6(i) = 1(i) = i$

(c) $i^{50} = (i^{48})(i^2) = (i^4)^{12}(-1) = 1(-1) = -1$

(d) $i^{67} = (i^{64})(i^3) = (i^4)^{16}(i^2)(i) = 1(-1)(i) = -i$

89. $(a + bi)(a - bi) = a^2 + abi - abi - b^2i^2 = a^2 + b^2$

which is a real number

91. $2x + 3y = 5$

$3y = -2x + 5$

$y = -\frac{2}{3}x + \frac{5}{3}$ Slope: $-\frac{2}{3}$

(a) Parallel line: $y - 3 = -\frac{2}{3}[x - (-8)]$

$3y - 9 = -2x - 16$

$3y + 2x = -7$

(b) Perpendicular line: $y - 3 = \frac{3}{2}[x - (-8)]$

$2y - 6 = 3x + 24$

$2y - 3x = 30$

93. $y = x^2 + 2x - 8$

Let $y = 0$: $x^2 + 2x - 8 = (x + 4)(x - 2) = 0 \Rightarrow x = -4, 2$. x-intercepts: $(-4, 0)$, $(2, 0)$

Let $x = 0$: $y = -8$. y-intercept: $(0, -8)$

94. $y = |x| - 1$

Let $y = 0$: $|x| = 1 \Rightarrow x = \pm 1$. x-intercepts: $(1, 0)$, $(-1, 0)$

Let $x = 0$: $y = -1$. y-intercept: $(0, -1)$

Section 2.5 The Fundamental Theorem of Algebra

- You should know that if f is a polynomial of degree $n > 0$, then f has at least one zero in the complex number system. (Fundamental Theorem of Algebra)
- You should know that if $a + bi$ is a complex zero of a polynomial f, with real coefficients, then $a - bi$ is also a complex zero of f.
- You should know the difference between a factor that is irreducible over the rationals (such as $x^2 - 7$) and a factor that is irreducible over the reals (such as $x^2 + 9$).

Solutions to Odd-Numbered Exercises

1. $f(x) = x(x - 6)^2 = x(x - 6)(x - 6)$

The three zeros are $x = 0$, $x = 6$, and $x = 6$.

3. $g(x) = (x - 2)(x + 4)^3$

The four zeros are $x = 2$, $x = -4$, $x = -4$, and $x = -4$.

5. $f(x) = (x + 6)(x + i)(x - i)$

The three zeros are $x = -6$, $x = -i$, and $x = i$.

7. $f(x) = (x - 2)(x + 3 - 5i)(x + 3 + 5i)$

The three zeros are $x = 2$, $x = -3 + 5i$, and $x = -3 - 5i$.

9. $f(x) = x^3 - 4x^2 + x - 4 = x^2(x - 4) + 1(x - 4)$
$= (x - 4)(x^2 + 1)$ zeros: $4, \pm i$

The only real zero of $f(x)$ is $x = 4$. This corresponds to the x-intercept of $(4, 0)$ on the graph.

11. $f(x) = x^4 + 4x^2 + 4 = (x^2 + 2)^2$

zeros: $\pm\sqrt{2}i, \pm\sqrt{2}i$

$f(x)$ has no real zeros and the graph of $f(x)$ has no x-intercepts.

13. $h(x) = x^2 - 4x + 1$

h has no rational zeros. By the Quadratic Formula, the zeros are $x = \dfrac{4 \pm \sqrt{16 - 4}}{2} = 2 \pm \sqrt{3}$.

$h(x) = \left[x - (2 + \sqrt{3})\right]\left[x - (2 - \sqrt{3})\right] = \left(x - 2 - \sqrt{3}\right)\left(x - 2 + \sqrt{3}\right)$

15. $f(x) = x^2 - 12x + 26$

f has no rational zeros. By the Quadratic Formula, the zeros are

$$x = \frac{12 \pm \sqrt{(-12)^2 - 4(26)}}{2} = 6 \pm \sqrt{10}.$$

$f(x) = \left[x - (6 + \sqrt{10})\right]\left[x - (6 - \sqrt{10})\right]$
$= \left(x - 6 - \sqrt{10}\right)\left(x - 6 + \sqrt{10}\right)$

17. $f(x) = x^2 + 25$
$= (x + 5i)(x - 5i)$

The zeros of $f(x)$ are $x = \pm 5i$.

19. $f(x) = x^4 - 81$
$= (x^2 - 9)(x^2 + 9)$
$= (x + 3)(x - 3)(x + 3i)(x - 3i)$

The zeros of $f(x)$ are $x = \pm 3$ and $x = \pm 3i$.

21. $f(z) = z^2 - 2z + 2$

f has no rational zeros. By the Quadratic Formula,

the zeros are $z = \dfrac{2 \pm \sqrt{4 - 8}}{2} = 1 \pm i$.

$f(z) = [z - (1 + i)][z - (1 - i)]$
$= (z - 1 - i)(z - 1 + i)$

23. $f(t) = t^3 - 3t^2 - 15t + 125$

Possible rational zeros: $\pm 1, \pm 5, \pm 25, \pm 125$

−5	1	−3	−15	125
		−5	40	−125
	1	−8	25	0

By the Quadratic Formula, the zeros of

$t^2 - 8t + 25$ are $t = \dfrac{8 \pm \sqrt{64 - 100}}{2} = 4 \pm 3i$.

The zeros of $f(t)$ are $t = -5$ and $t = 4 \pm 3i$.

$f(t) = [t - (-5)][t - (4 + 3i)][t - (4 - 3i)]$
$= (t + 5)(t - 4 - 3i)(t - 4 + 3i)$

25. $f(x) = 16x^3 - 20x^2 - 4x + 15$

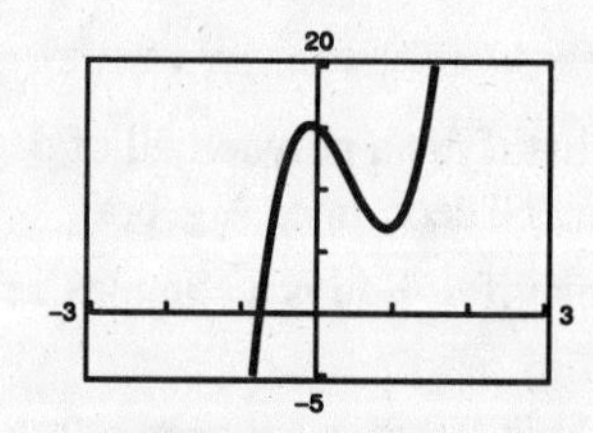

The graph reveals one zero at $x = -\frac{3}{4}$.

$-\frac{3}{4}$	16	−20	−4	15
		−12	24	−15
	16	−32	20	0

By the Quadratic Formula, the zeros of $16x^2 - 32x + 20 = 4(4x^2 - 8x + 5)$ are

$$x = \frac{8 \pm \sqrt{64 - 80}}{8} = 1 \pm \frac{1}{2}i.$$

The zeros of $f(x)$ are $x = -\frac{3}{4}$ and $x = 1 \pm \frac{1}{2}i$.

$$16\left(x + \frac{3}{4}\right)\left(x - 1 + \frac{1}{2}i\right)\left(x - 1 - \frac{1}{2}i\right)$$

27. $f(x) = x^4 + 10x^2 + 9$

$= (x^2 + 1)(x^2 + 9)$

$= (x + i)(x - i)(x + 3i)(x - 3i)$

The zeros of $f(x)$ are $x = \pm i$ and $x = \pm 3i$.

29. $g(x) = x^4 - 4x^3 + 8x^2 - 16x + 16$

Possible rational zeros: $\pm 1, \pm 2, \pm 4, \pm 8, \pm 16$

$$\begin{array}{r|rrrrr} 2 & 1 & -4 & 8 & -16 & 16 \\ & & 2 & -4 & 8 & -16 \\ \hline 2 & 1 & -2 & 4 & -8 & 0 \\ & & 2 & 0 & 8 & \\ \hline & 1 & 0 & 4 & 0 & \end{array}$$

$g(x) = (x - 2)(x - 2)(x^2 + 4)$

$= (x - 2)^2(x + 2i)(x - 2i)$

The zeros of g are 2, 2, and $\pm 2i$.

31. $f(x) = 2x^4 + 5x^3 + 4x^2 + 5x + 2$

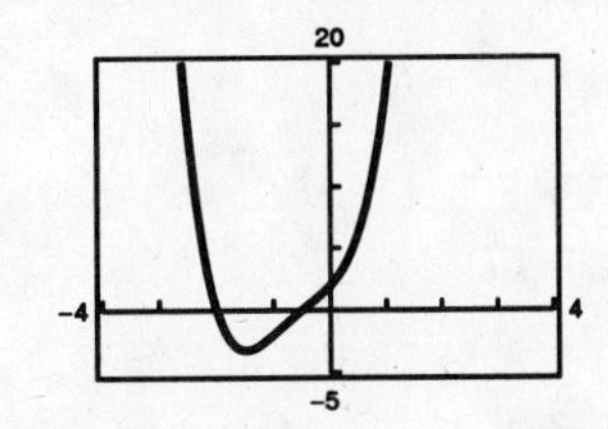

The graph reveals one zero at $x = -2$ and $x = -\frac{1}{2}$.

$$\begin{array}{r|rrrrr} -2 & 2 & 5 & 4 & 5 & 2 \\ & & -4 & -2 & -4 & -2 \\ \hline & 2 & 1 & 2 & 1 & 0 \end{array}$$

$$\begin{array}{r|rrrr} -\frac{1}{2} & 2 & 1 & 2 & 1 \\ & & -1 & 0 & -1 \\ \hline & 2 & 0 & 2 & 0 \end{array}$$

The zeros of $2x^2 + 2 = 2(x^2 + 1)$ are $x = \pm i$.

The zeros of $f(x)$ are $-2, -\frac{1}{2}, \pm i$.

$f(x) = (x + 2)(2x + 1)(x - i)(x + i)$

33. (a) $f(x) = x^2 - 14x + 46$. By the Quadratic Formula

$$x = \frac{14 \pm \sqrt{(-14)^2 - 4(46)}}{2} = 7 \pm \sqrt{3}.$$

The zeros are $7 + \sqrt{3}$ and $7 - \sqrt{3}$.

(b) $f(x) = \left[x - \left(7 + \sqrt{3}\right)\right]\left[x - \left(7 - \sqrt{3}\right)\right]$

$= \left(x - 7 - \sqrt{3}\right)\left(x - 7 + \sqrt{3}\right)$

(c) x-intercepts: $\left(7 + \sqrt{3}, 0\right)$ and $\left(7 - \sqrt{3}, 0\right)$

(d)

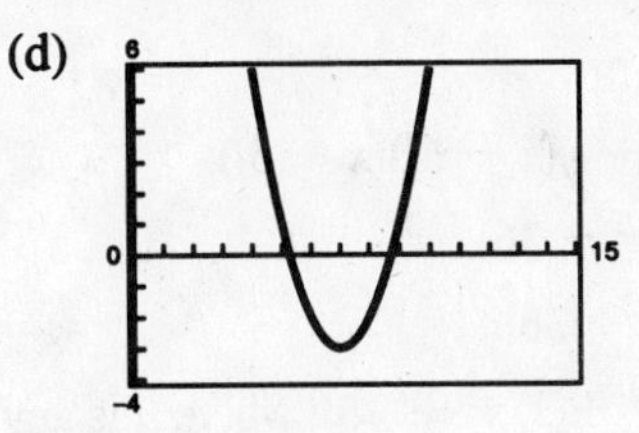

35. (a) $f(x) = x^2 + 14x + 44$. By the Quadratic Formula,

$$x = \frac{-14 \pm \sqrt{14^2 - 4(44)}}{2} = -7 \pm \sqrt{5}$$

The zeros are $-7 + \sqrt{5}$ and $-7 - \sqrt{5}$.

(b) $f(x) = \left[x - \left(-7 + \sqrt{5}\right)\right]\left[x - \left(-7 - \sqrt{5}\right)\right]$

$= \left(x + 7 - \sqrt{5}\right)\left(x + 7 + \sqrt{5}\right)$

(c) x-intercepts: $\left(-7 + \sqrt{5}, 0\right), \left(-7 - \sqrt{5}, 0\right)$

(d)

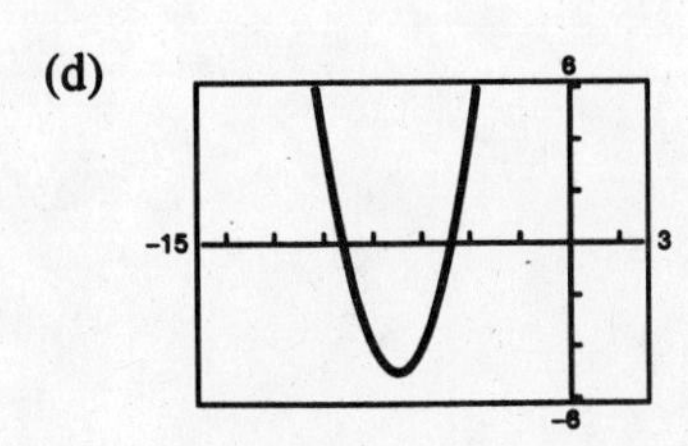

37. (a) $f(x) = x^3 - 11x + 150$
$= (x + 6)(x^2 - 6x + 25).$

Use the Quadratic Formula to find the zeros of $x^2 - 6x + 25$:

$$x = \frac{6 \pm \sqrt{(-6)^2 - 4(25)}}{2} = 3 \pm 4i.$$

The zeros are -6, $3 + 4i$, and $3 - 4i$.

(b) $f(x) = (x + 6)(x - 3 + 4i)(x - 3 - 4i)$

(c) x-intercept: $(-6, 0)$

(d)

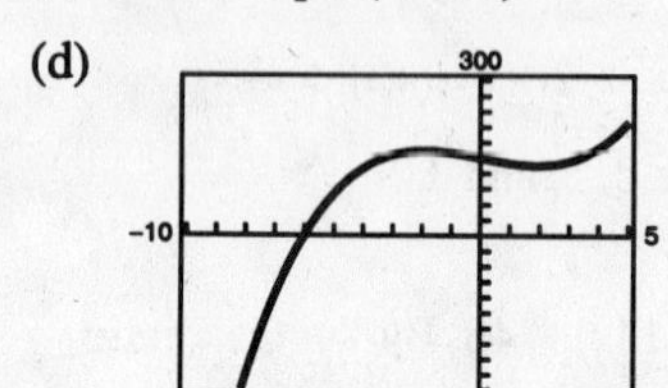

39. (a) $f(x) = x^4 + 25x^2 + 144$
$= (x^2 + 9)(x^2 + 16)$

The zeros are $\pm 3i$, $\pm 4i$.

(b) $f(x) = (x^2 + 9)(x^2 + 16)$
$= (x + 3i)(x - 3i)(x + 4i)(x - 4i)$

(c) No x-intercepts

(d)

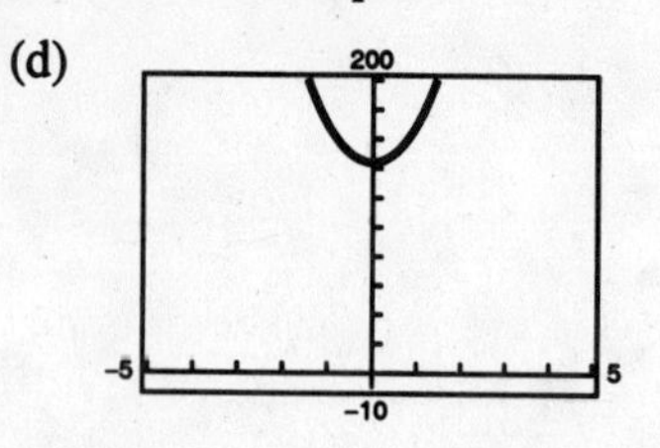

41. $f(x) = (x - 1)(x - 5i)(x + 5i)$
$= (x - 1)(x^2 + 25)$
$= x^3 - x^2 + 25x - 25$

Note: $f(x) = a(x^3 - x^2 + 25x - 25)$, where a is any nonzero real number, has the zero, 1 and $\pm 5i$.

43. $f(x) = (x - 2)(x - 4 - i)(x - 4 + i)$
$= (x - 2)(x^2 - 8x + 17)$
$= x^3 - 10x^2 + 33x - 34$

Note: $f(x) = a(x^3 - 10x^2 + 33x - 34)$, where a is any nonzero real number, has these zeros.

45. $f(x) = (x - i)(x + i)(x - 6i)(x + 6i)$
$= (x^2 + 1)(x^2 + 36)$
$= x^4 + 37x^2 + 36$

Note: $f(x) = a(x^4 + 37x^2 + 36)$, where a is any nonzero real number, has the zeros $\pm i$ and $\pm 6i$.

47. If $1 + \sqrt{3}i$ is a zero, so is its conjugate $1 - \sqrt{3}i$.

$f(x) = (x + 5)^2(x - 1 + \sqrt{3}i)(x - 1 - \sqrt{3}i)$
$= (x^2 + 10x + 25)(x^2 - 2x + 4)$
$= x^4 + 8x^3 + 9x^2 - 10x + 100$

Note: $f(x) = a(x^4 + 8x^3 + 9x^2 - 10x + 100)$, where a is any nonzero real number, has these zeros.

49. $f(x) = x^4 - 6x^2 - 7$

(a) $f(x) = (x^2 - 7)(x^2 + 1)$

(b) $f(x) = \left(x - \sqrt{7}\right)\left(x + \sqrt{7}\right)(x^2 + 1)$

(c) $f(x) = \left(x - \sqrt{7}\right)\left(x + \sqrt{7}\right)(x + i)(x - i)$

51. $f(x) = x^4 - 2x^3 - 3x^2 + 12x - 18$

(a) $f(x) = (x^2 - 6)(x^2 - 2x + 3)$

(b) $f(x) = \left(x + \sqrt{6}\right)\left(x - \sqrt{6}\right)(x^2 - 2x + 3)$

(c) $f(x) =$
$\left(x + \sqrt{6}\right)\left(x - \sqrt{6}\right)\left(x - 1 - \sqrt{2}i\right)\left(x - 1 + \sqrt{2}i\right)$

53. $f(x) = 2x^3 + 3x^2 + 50x + 75$

Since $5i$ is a zero, so is $-5i$.

$$\begin{array}{r|rrrr} 5i & 2 & 3 & 50 & 75 \\ & & 10i & -50+15i & -75 \\ \hline & 2 & 3+10i & 15i & 0 \end{array}$$

$$\begin{array}{r|rrr} -5i & 2 & 3+10i & 15i \\ & & -10i & -15i \\ \hline & 2 & 3 & 0 \end{array}$$

The zero of $2x + 3$ is $x = -\frac{3}{2}$. The zeros of f are $x = -\frac{3}{2}$ and $x = \pm 5i$.

Alternate Solution

Since $x = \pm 5i$ are zeros of $f(x)$, $(x + 5i)(x - 5i) = x^2 + 25$ is a factor of $f(x)$. By long division we have:

$$\begin{array}{r} 2x + 3 \\ x^2 + 0x + 25 \overline{\smash{)}\, 2x^3 + 3x^2 + 50x + 75} \\ \underline{2x^3 + 0x^2 + 50x} \\ 3x^2 + 0x + 75 \\ \underline{3x^2 + 0x + 75} \\ 0 \end{array}$$

Thus, $f(x) = (x^2 + 25)(2x + 3)$ and the zeros of f are $x = \pm 5i$ and $x = -\frac{3}{2}$.

55. $g(x) = x^3 - 7x^2 - x + 87$. Since $5 + 2i$ is a zero, so is $5 - 2i$.

$$\begin{array}{r|rrrr} 5+2i & 1 & -7 & -1 & 87 \\ & & 5+2i & -14+6i & -87 \\ \hline & 1 & -2+2i & -15+6i & 0 \end{array}$$

$$\begin{array}{r|rrr} 5-2i & 1 & -2+2i & -15+6i \\ & & 5-2i & 15-6i \\ \hline & 1 & 3 & 0 \end{array}$$

The zero of $x + 3$ is $x = -3$.

The zeros of f are $-3, 5 \pm 2i$.

57. $h(x) = 3x^3 - 4x^2 + 8x + 8$. Since $1 - \sqrt{3}i$ is a zero, so is $1 + \sqrt{3}i$.

$$\begin{array}{r|rrrr} 1-\sqrt{3}i & 3 & -4 & 8 & 8 \\ & & 3-3\sqrt{3}i & -10-2\sqrt{3}i & -8 \\ \hline & 3 & -1-3\sqrt{3}i & -2-2\sqrt{3}i & 0 \end{array}$$

$$\begin{array}{r|rrr} 1+\sqrt{3}i & 3 & -1-3\sqrt{3}i & -2-2\sqrt{3}i \\ & & 3+3\sqrt{3}i & 2+2\sqrt{3}i \\ \hline & 3 & 2 & 0 \end{array}$$

The zero of $3x + 2$ is $x = -\frac{2}{3}$.

The zeros of h are $x = -\frac{2}{3}, 1 \pm \sqrt{3}i$.

59. $h(x) = 8x^3 - 14x^2 + 18x - 9$. Since $\frac{1}{2}(1 - \sqrt{5}i)$ is a zero, so is $\frac{1}{2}(1 + \sqrt{5}i)$.

$$\begin{array}{r|rrrr} \frac{1}{2}(1-\sqrt{5}i) & 8 & -14 & 18 & -9 \\ & & 4-4\sqrt{5}i & -15+3\sqrt{5}i & 9 \\ \hline & 8 & -10-4\sqrt{5}i & 3+3\sqrt{5}i & 0 \end{array}$$

$$\begin{array}{r|rrr} \frac{1}{2}(1+\sqrt{5}i) & 8 & -10-4\sqrt{5}i & 3+3\sqrt{5}i \\ & & 4+4\sqrt{5}i & -3-3\sqrt{5}i \\ \hline & 8 & -6 & 0 \end{array}$$

The zero of $8x - 6$ is $x = \frac{3}{4}$.

The zeros of h are $x = \frac{3}{4}, \frac{1}{2}(1 \pm \sqrt{5}i)$.

61. (a) The root feature yields the real roots 1 and 2, and the complex roots $-3 \pm 1.414i$.

(b) By synthetic division,

$$\begin{array}{r|rrrrr} 1 & 1 & 3 & -5 & -21 & 22 \\ & & 1 & 4 & -1 & -22 \\ \hline & 1 & 4 & -1 & -22 & 0 \end{array}$$

$$\begin{array}{r|rrrr} 2 & 1 & 4 & -1 & -22 \\ & & 2 & 12 & 22 \\ \hline & 1 & 6 & 11 & 0 \end{array}$$

The complex roots of $x^2 + 6x + 11$ are

$$x = \frac{-6 \pm \sqrt{6^2 - 4(11)}}{2} = -3 \pm \sqrt{2}i.$$

63. (a) The root feature yields the real root 0.75, and the complex roots $0.5 \pm 1.118i$.

(b) By synthetic division,

$$\begin{array}{r|rrrr} \frac{3}{4} & 8 & -14 & 18 & -9 \\ & & 6 & -6 & 9 \\ \hline & 8 & -8 & 12 & 0 \end{array}$$

The complex roots of $8x^2 - 8x + 12$ are

$$x = \frac{8 \pm \sqrt{64 - 4(8)(12)}}{2(8)} = \frac{1}{2} \pm \frac{\sqrt{5}}{2}i.$$

65. $-16t^2 + 48t = 64, \quad 0 \le t \le 3$

$$-16t^2 + 48t - 64 = 0$$

$$t = \frac{-48 \pm \sqrt{1792}i}{-32}$$

Since the roots are imaginary, the ball never will reach a height of 64 feet. You can verify this graphically by observing that $y_1 = -16t^2 + 48t$ and $y_2 = 64$ do not intersect.

67. False, a third degree polynomial must have at least one real zero.

69. $f(x) = x^4 - 4x^2 + k$

(a) f has four real zeros for $0 < k < 4$.

(b) f has two real zeros each of multiplicity 2 for $k = 4$: $f(x) = x^4 - 4x^2 + 4 = (x^2 - 2)^2$.

(c) f has two real zeros and two complex zeros if $k < 0$.

(d) f has four complex zeros if $k > 4$.

71. $f(x) = \left(x - \sqrt{b}i\right)\left(x + \sqrt{b}i\right) = x^2 + b$

73. $f(x) = x^2 - 7x - 8 = \left(x^2 - 7x + \frac{49}{4}\right) - 8 - \frac{49}{4}$

$$= \left(x - \tfrac{7}{2}\right)^2 - \tfrac{81}{4}$$

Vertex: $\left(\frac{7}{2}, -\frac{81}{4}\right)$

$f(x) = (x - 8)(x + 1)$

Intercepts: $(8, 0), (-1, 0), (0, -8)$

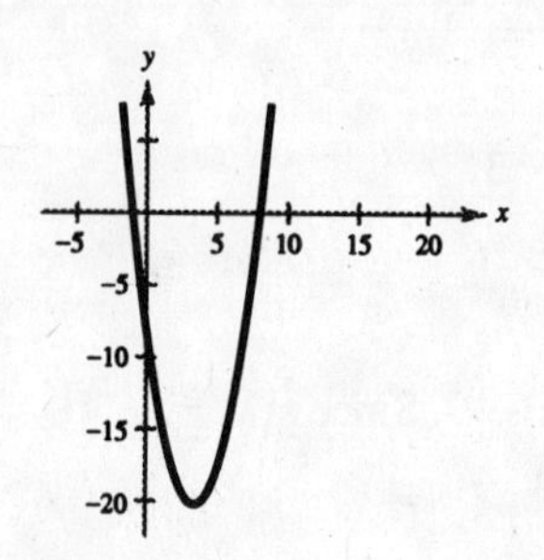

75. $f(x) = 6x^2 + 5x - 6 = (3x - 2)(2x + 3)$

Intercepts: $\left(\frac{2}{3}, 0\right), \left(-\frac{3}{2}, 0\right), (0, -6)$

$f(x) = 6x^2 + 5x - 6$

$$= 6\left(x^2 + \tfrac{5}{6}x + \tfrac{25}{144}\right) - 6 - \tfrac{25}{24}$$

$$= 6\left(x + \tfrac{5}{12}\right)^2 + \tfrac{169}{24}$$

Vertex: $\left(-\frac{5}{12}, -\frac{169}{24}\right)$

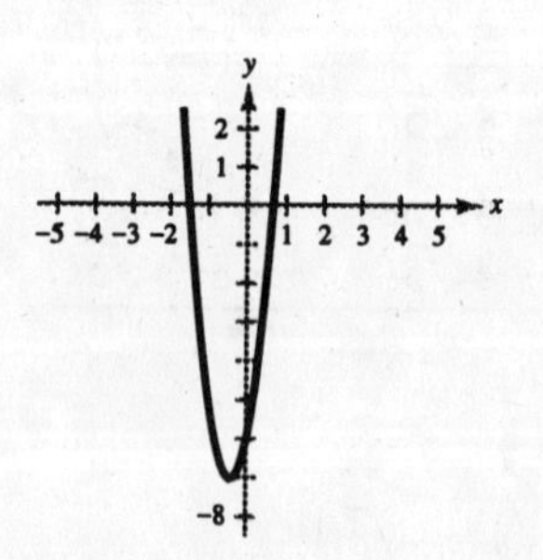

77. $12 - (-7 + 5i) + (-2 + 3i) = 17 - 2i$

79. $(-3 - 8i)^2 = 9 + 48i + (-8i)^2$

$$= 9 + 48i - 64$$

$$= -55 + 48i$$

Section 2.6 Rational Functions and Asymptotes

- You should know the following basic facts about rational functions.
  - (a) A function of the form $f(x) = P(x)/Q(x)$, $Q(x) \neq 0$, where $P(x)$ and $Q(x)$ are polynomials, is called a rational function.
  - (b) The domain of a rational function is the set of all real numbers except those which make the denominator zero.
  - (c) If $f(x) = P(x)/Q(x)$ is in reduced form, and a is a value such that $Q(a) = 0$, then the line $x = a$ is a vertical asymptote of the graph of f. $f(x) \to \infty$ or $f(x) \to -\infty$ as $x \to a$.
  - (d) The line $y = b$ is a horizontal asymptote of the graph of f if $f(x) \to b$ as $x \to \infty$ or $x \to -\infty$.
  - (e) Let $f(x) = \dfrac{P(x)}{Q(x)} = \dfrac{a_nx^n + a_{n-1}x^{n-1} + \cdots + a_1x + a_0}{b_mx^m + b_{m-1}x^{m-1} + \cdots + b_1x + b_0}$ where $P(x)$ and $Q(x)$ have no common factors.
    1. If $n < m$, then the x-axis $(y = 0)$ is a horizontal asymptote.
    2. If $n = m$, then $y = \dfrac{a_n}{b_m}$ is a horizontal asymptote.
    3. If $n > m$, then there are no horizontal asymptotes.

Solutions to Odd-Numbered Exercises

1. $f(x) = \dfrac{1}{x-1}$

(a)

x	$f(x)$	x	$f(x)$	x	$f(x)$	x	$f(x)$
0.5	−2	1.5	2	5	0.25	−5	−0.167
0.9	−10	1.1	10	10	$0.\overline{1}$	−10	−0.0909
0.99	−100	1.01	100	100	$0.\overline{01}$	−100	−0.0099
0.999	−1000	1.001	1000	1000	$0.\overline{001}$	−1000	−0.001

(b) The zero of the denominator is $x = 1$, so $x = 1$ is a vertical asymptote. The degree of the numerator is less than the degree of the denominator so the x-axis, or $y = 0$ is a horizontal asymptote.

(c) The domain is all real numbers except $x = 1$.

3. $f(x) = \dfrac{3x}{|x-1|}$

(a)

x	$f(x)$	x	$f(x)$	x	$f(x)$	x	$f(x)$
0.5	3	1.5	9	5	3.75	−5	−2.5
0.9	27	1.1	33	10	$3.\overline{33}$	−10	−2.727
0.99	297	1.01	303	100	$3.\overline{03}$	−100	−2.970
0.999	2997	1.001	3003	1000	$3.\overline{003}$	−1000	−2.997

(b) The zero of the denominator is $x = 1$, so $x = 1$ is a vertical asymptote. Since $f(x) \to 3$ as $x \to \infty$ and $f(x) \to -3$ as $x \to -\infty$, both $y = 3$ and $y = -3$ are horizontal asymptotes.

(c) The domain is all real numbers except $x = 1$.

5. $f(x) = \dfrac{3x^2}{x^2 - 1}$

(a)

x	$f(x)$
0.5	-1
0.9	-12.79
0.99	-148.79
0.999	-1498

x	$f(x)$
1.5	5.4
1.1	17.29
1.01	152.3
1.001	1502.3

x	$f(x)$
5	3.125
10	$3.\overline{03}$
100	$3.\overline{0003}$
1000	3

x	$f(x)$
-5	3.125
-10	$3.\overline{03}$
-100	$3.\overline{0003}$
-1000	3

(b) The zeros of the denominator are $x = \pm 1$ so both $x = 1$ and $x = -1$ are vertical asymptotes.

Since the degree of the numerator equals the degree of the denominator, $y = \frac{3}{1} = 3$ is a horizontal asymptote.

(c) The domain is all real numbers except $x = \pm 1$.

7. $f(x) = \dfrac{2}{x + 2}$

Vertical asymptote: $x = -2$

Horizontal asymptote: $y = 0$

Matches graph (a)

9. $f(x) = \dfrac{4x + 1}{x}$

Vertical asymptote: $x = 0$

Horizontal asymptote: $y = 4$

Matches graph (c)

11. $f(x) = \dfrac{x - 2}{x - 4}$

Vertical asymptote: $x = 4$

Horizontal asymptote: $y = 1$

Matches graph (b)

13. $f(x) = \dfrac{1}{x^2}$

(a) Domain: all real numbers except $x = 0$

(b) Vertical asymptote: $x = 0$

Horizontal asymptote: $y = 0$

[Degree of $p(x) <$ degree of $q(x)$]

(c)

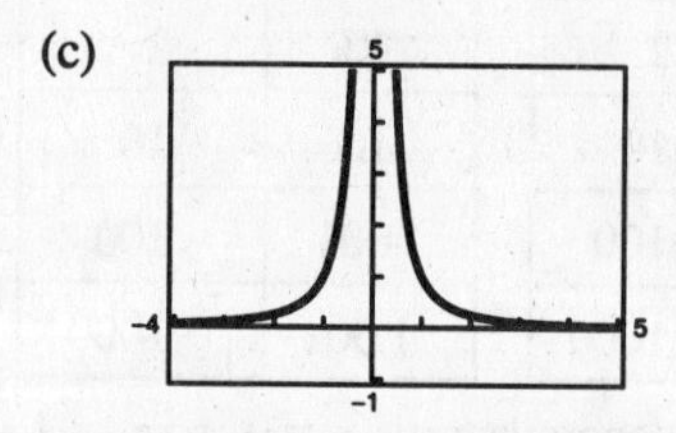

15. $f(x) = \dfrac{3 + x}{3 - x}$

(a) Domain: all real numbers except $x = 3$.

(b) Vertical asymptote: $x = 3$

Horizontal asymptote: $y = -1$

(c)

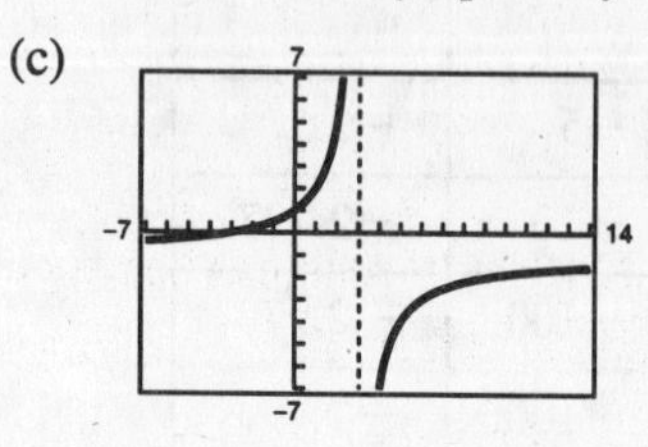

17. $f(x) = \dfrac{2x^3}{x^2 - 1}$

(a) Domain: all real numbers except $x = \pm 1$

(b) Vertical asymptotes: $x = \pm 1$

Horizontal asymptotes: None

[Degree of $p(x) >$ degree of $q(x)$]

(c)

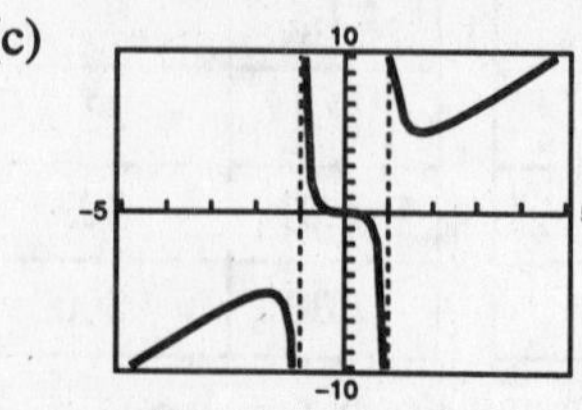

19. $f(x) = \dfrac{x^2 - 4}{x + 2}$, $g(x) = x - 2$

(a) Domain of f: all real numbers except -2
Domain of g: all real numbers

(b) Since $x + 2$ is a common factor of both the numerator and the denominator of $f(x)$, $x = -2$ is not a vertical asymptote of f. f has no vertical asymptotes.

(c)

x	-4	-3	-2.5	-2	-1.5	-1	0
$f(x)$	-6	-5	-4.5	undef.	-3.5	-3	-2
$g(x)$	-6	-5	-4.5	-4	-3.5	-3	-2

(d) f and g differ only where f is undefined.

21. $f(x) = \dfrac{x - 3}{x^2 - 3x}$, $g(x) = \dfrac{1}{x}$

(a) Domain of f: all real number except 0 and 3
Domain of g: all real numbers except 0

(b) Since $x - 3$ is a common factor of both the numerator and the denominator of f, $x = 3$ is not a vertical asymptote of f. The only vertical asymptote is $x = 0$.

(c)

x	-1	-0.5	0	0.5	2	3	4
$f(x)$	-1	-2	undef.	2	$\frac{1}{2}$	undef.	$\frac{1}{4}$
$g(x)$	-1	-2	undef	2	$\frac{1}{2}$	$\frac{1}{3}$	$\frac{1}{4}$

(d) They differ only at $x = 3$, where f is undefined and g is defined.

23. $f(x) = 4 - \dfrac{1}{x}$

(a) As $x \to \pm\infty$, $f(x) \to 4$

(b) As $x \to \infty$, $f(x) \to 4$ but is less than 4

(c) As $x \to -\infty$, $f(x) \to 4$ but is greater than 4

25. $f(x) = \dfrac{2x - 1}{x - 3}$

(a) As $x \to \pm\infty$, $f(x) \to 2$

(b) As $x \to \infty$, $f(x) \to 2$ but is greater than 2

(c) As $x \to -\infty$, $f(x) \to 2$ but is less than 2

27. $f(x) = \dfrac{x^2 - 9}{x + 1} = \dfrac{(x + 3)(x - 3)}{x + 1}$

The zeros of f correspond to the zeros of the numerator and are $x = \pm 3$.

29. $f(x) = 1 - \dfrac{2}{x - 5} = \dfrac{x - 7}{x - 5}$

The zero of f corresponds to the zero of the numerator and is $x = 7$.

31. $C = \dfrac{255p}{100 - p}$, $0 \le p < 100$

(a) $C(10) = \dfrac{255(10)}{100 - 10} \approx 28.33$ million dollars

(b) $C(40) = \dfrac{255(40)}{100 - 40} = 170$ million dollars

(c) $C(75) = \dfrac{255(75)}{100 - 75} = 765$ million dollars

(d)

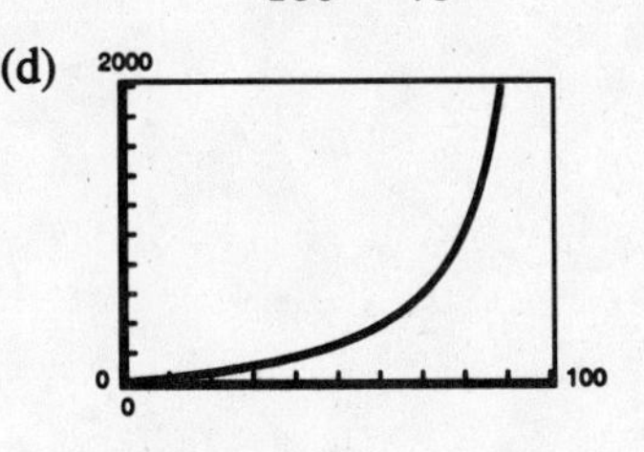

(e) $C \to \infty$ as $x \to 100$. No, it would not be possible to remove 100% of the pollutants.

33. (a)

M	200	400	600	800	1000	1200	1400	1600	1800	2000
t	0.472	0.596	0.710	0.817	0.916	1.009	1.096	1.178	1.255	1.328

The greater the mass, the more time required per oscillation. The model is a good fit to the actual data.

(b) You can find M corresponding to $t = 1.056$ by finding the point of intersection of

$$t = \frac{38M + 16{,}965}{10(M + 500)} \quad \text{and} \quad t = 1.056.$$

If you do this, you obtain $M \approx 1306$ grams.

35. $N = \dfrac{20(5 + 3t)}{1 + 0.04t}, \; 0 \le t$

(a) $N(5) \approx 333$ deer

$N(10) = 500$ deer

$N(25) = 800$ deer

(b) The herd is limited by the horizontal asymptote:

$$N = \frac{60}{0.04} = 1500 \text{ deer}$$

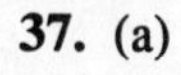

37. (a)

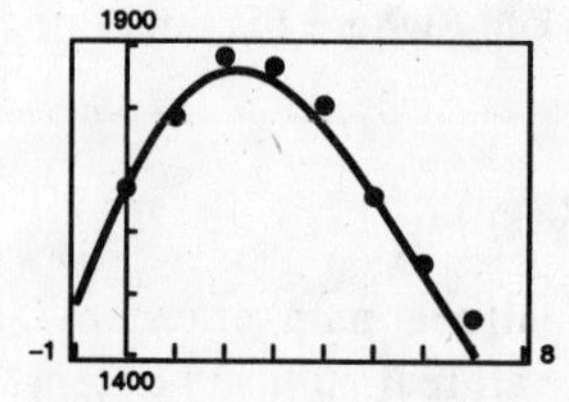

(b) For 2002, $t = 12$ and $M \approx 889$ thousand

(c) No, this model predicts that $M \to 0$ as t increases.

39. True, $f(x) = x^3 - 2x^2 - 5x + 6 = \dfrac{x^3 - 2x^2 - 5x + 6}{1}$ is a rational function.

41. $f(x) = \dfrac{1}{x^2 + 1}$ is one possible answer.

43. $f(x) = \dfrac{-3x^2}{x(2x - 5)} = \dfrac{-3x^2}{2x^2 - 5x}$ is one possible answer.

45.
$$\begin{aligned} x(10 - x) &= 25 \\ 0 &= x^2 - 10x + 25 \\ 0 &= (x - 5)^2 \\ x &= 5 \end{aligned}$$

47.
$$\begin{aligned} t^3 - 50t &= 0 \\ t(t^2 - 50) &= 0 \\ t &= 0, \pm 5\sqrt{2} \end{aligned}$$

49.
$$\begin{aligned} x^4 - 225 &= 0 \\ (x^2 - 15)(x^2 + 15) &= 0 \\ x &= \pm\sqrt{15}, \pm\sqrt{15}\, i \end{aligned}$$

51.
$$\begin{array}{r|rrr} 3 & 1 & -10 & 15 \\ & & 3 & -21 \\ \hline & 1 & -7 & -6 \end{array}$$

$$\frac{x^2 - 10x + 15}{x - 3} = x - 7 + \frac{-6}{x - 3}$$

53.
$$\begin{array}{r|rrr} -6 & 4 & 3 & -10 \\ & & -24 & 126 \\ \hline & 4 & -21 & 116 \end{array}$$

$$\frac{4x^2 + 3x - 10}{x + 6} = 4x - 21 + \frac{116}{x + 6}$$

55. $(x + 2)(x - 6i)(x + 6i) = (x + 2)(x^2 + 36) = x^3 + 2x^2 + 36x + 72$

57.
$$\begin{aligned} (x - 1)(x - (-3 + 2i))(x - (-3 - 2i)) &= (x - 1)((x + 3)^2 + 4) \\ &= (x - 1)(x^2 + 6x + 13) \\ &= x^3 + 5x^2 + 7x - 13 \end{aligned}$$

Section 2.7 Graphs of Rational Functions

- You should be able to graph $f(x) = \dfrac{p(x)}{q(x)}$.
  (a) Find the x- and y-intercepts.
  (b) Find any vertical or horizontal asymptotes.
  (c) Plot additional points.
  (d) If the degree of the numerator is one more than the degree of the denominator, use long division to find the slant asymptote.

Solutions to Odd-Numbered Exercises

1. $g(x) = \dfrac{2}{x} + 1$

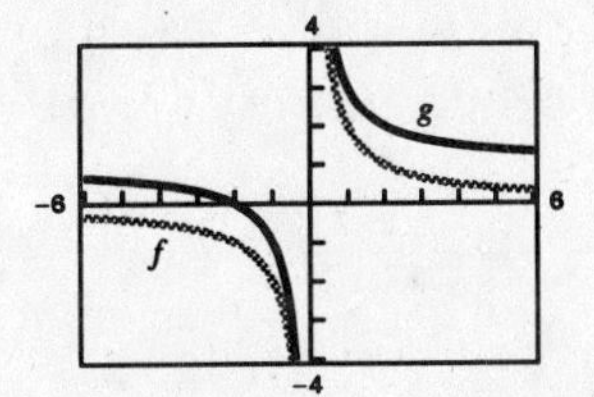

Vertical shift one unit upward

3. $g(x) = -\dfrac{2}{x}$

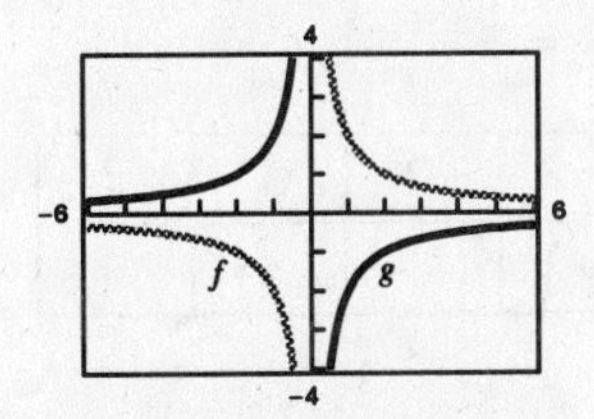

Reflection in the x-axis

5. $g(x) = \dfrac{2}{x^2} - 2$

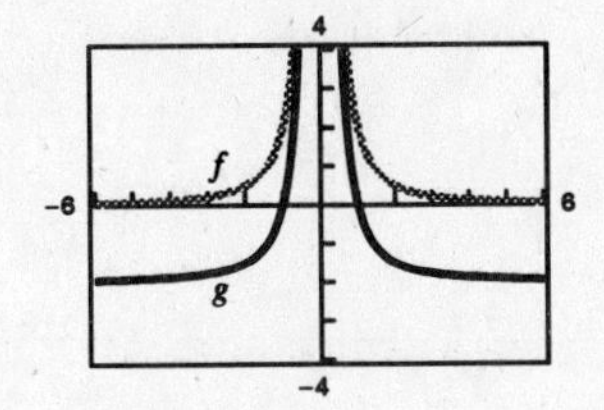

Vertical shift two units downward

7. $g(x) = \dfrac{2}{(x - 2)^2}$

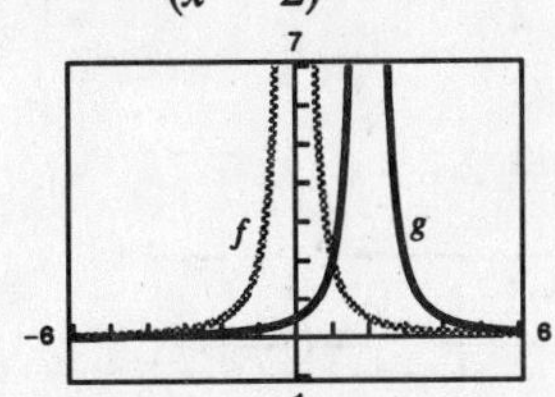

Horizontal shift two units to the right

9. $g(x) = \dfrac{4}{(x + 2)^3}$

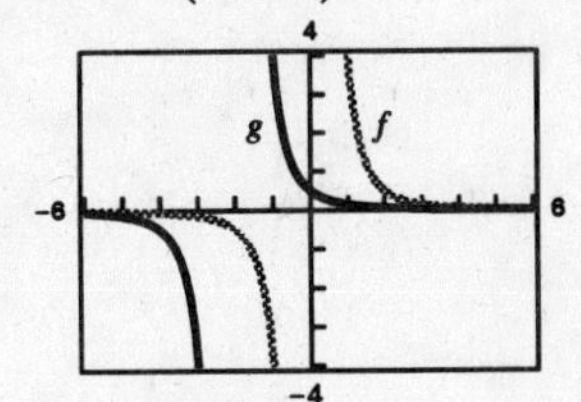

Horizontal shift two units to the left

11. $g(x) = -\dfrac{4}{x^3}$

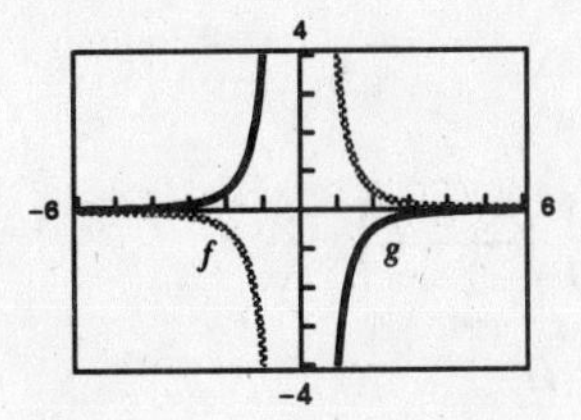

Reflection in the x-axis

13. $f(x) = \dfrac{1}{x+2}$

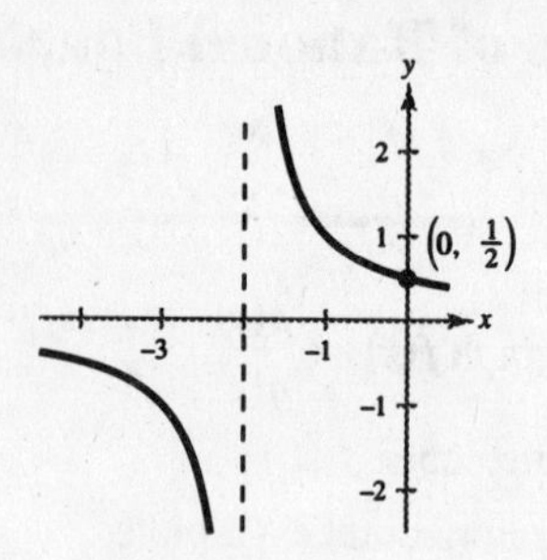

y-intercept: $\left(0, \dfrac{1}{2}\right)$

Vertical asymptote: $x = -2$

Horizontal asymptote: $y = 0$

x	-4	-3	-1	0	1
y	$-\frac{1}{2}$	-1	1	$\frac{1}{2}$	$\frac{1}{3}$

15. $C(x) = \dfrac{5+2x}{1+x} = \dfrac{2x+5}{x+1}$

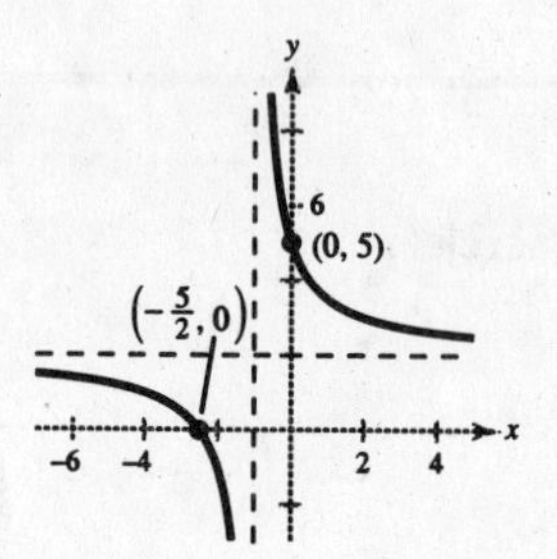

x-intercept: $\left(-\dfrac{5}{2}, 0\right)$

y-intercept: $(0, 5)$

Vertical asymptote: $x = -1$

Horizontal asymptote: $y = 2$

x	-4	-3	-2	0	1	2
$C(x)$	1	$\frac{1}{2}$	-1	5	$\frac{7}{2}$	3

17. $g(x) = \dfrac{1}{x+2} + 2 = \dfrac{2x+5}{x+2}$

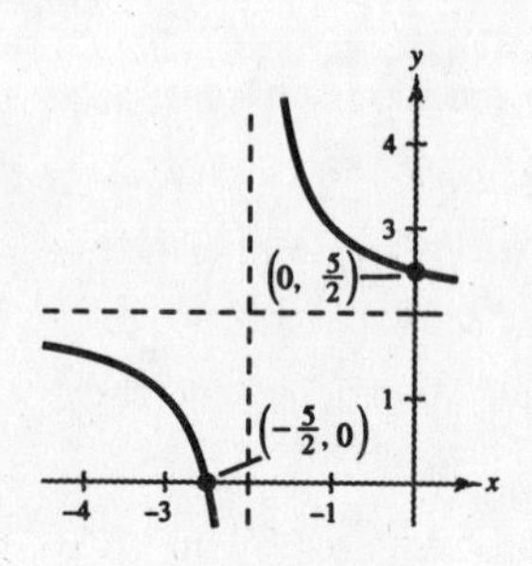

Intercepts: $\left(-\dfrac{5}{2}, 0\right), \left(0, \dfrac{5}{2}\right)$

Vertical asymptote: $x = -2$

Horizontal asymptote: $y = 2$

x	-4	-3	-1	0	1
y	$\frac{3}{2}$	1	3	$\frac{5}{2}$	$\frac{7}{3}$

Note: This is the graph of $f(x) = \dfrac{1}{x+2}$ (Exercise 13) shifted upward two units.

19. $f(x) = 2 - \dfrac{3}{x^2} = \dfrac{2x^2 - 3}{x^2}$

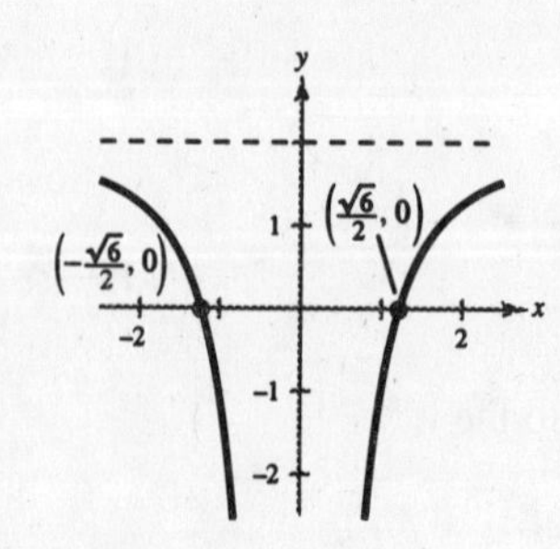

x-intercepts: $\left(-\dfrac{\sqrt{6}}{2}, 0\right), \left(\dfrac{\sqrt{6}}{2}, 0\right)$

Vertical asymptote: $x = 0$

Horizontal asymptote: $y = 2$

y-axis symmetry

x	-2	-1	$-\frac{1}{2}$	$\frac{1}{2}$	1	2
y	$\frac{5}{4}$	-1	-10	-10	-1	$\frac{5}{4}$

21. $f(x) = \dfrac{x^2}{x^2 - 4}$

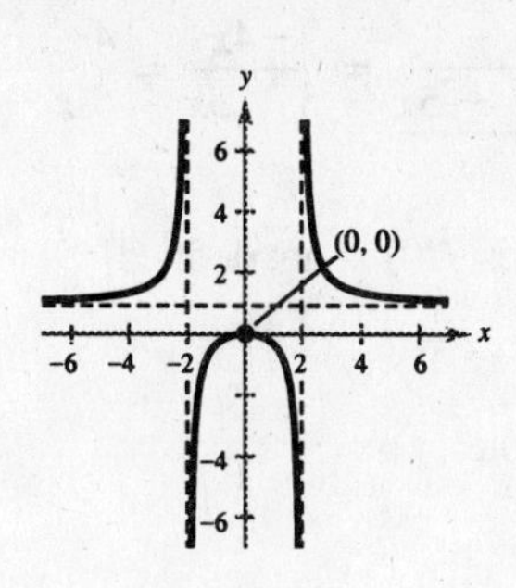

Intercept: $(0, 0)$

Vertical asymptotes: $x = 2,\ x = -2$

Horizontal asymptote: $y = 1$

y-axis symmetry

x	± 4	± 3	± 1	0
y	$\pm\frac{4}{3}$	$\frac{9}{5}$	$-\frac{1}{3}$	0

23. $f(x) = \dfrac{x}{x^2 - 4}$

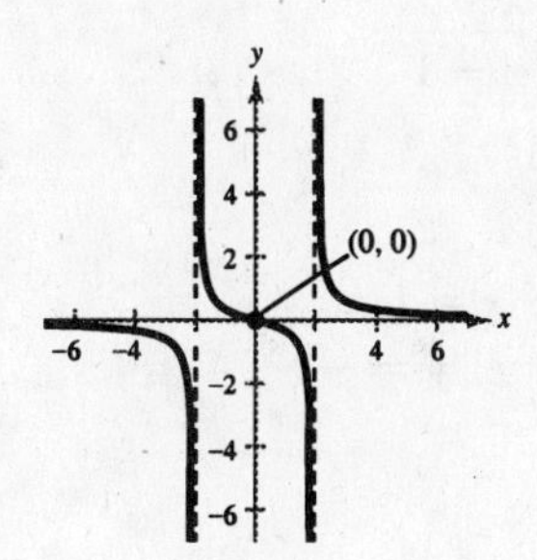

Intercepts: $(0, 0)$

Vertical asymptotes: $x = 2,\ x = -2$

Horizontal asymptote: $y = 0$

Origin symmetry

x	-4	-3	-1	0	1	3	4
y	$-\frac{1}{3}$	$-\frac{3}{5}$	$\frac{1}{3}$	0	$-\frac{1}{3}$	$\frac{3}{5}$	$\frac{1}{3}$

25. $g(x) = \dfrac{4(x + 1)}{x(x - 4)}$

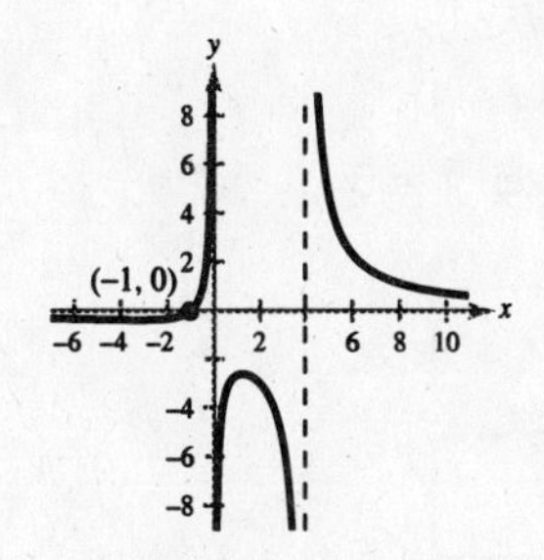

Intercept: $(-1, 0)$

Vertical asymptotes: $x = 0$ and $x = 4$

Horizontal asymptote: $y = 0$

x	-2	-1	1	2	3	5	6
y	$-\frac{1}{3}$	0	$-\frac{8}{3}$	-3	$-\frac{16}{3}$	$\frac{24}{5}$	$\frac{7}{3}$

27. $f(x) = \dfrac{3x}{x^2 - x - 2} = \dfrac{3x}{(x + 1)(x - 2)}$

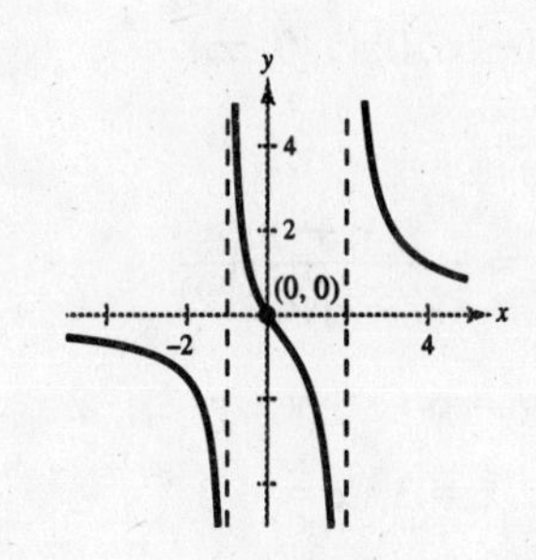

Intercept: $(0, 0)$

Vertical asymptotes: $x = -1, 2$

Horizontal asymptote: $y = 0$

x	-3	0	1	3	4
y	$-\frac{9}{10}$	0	$-\frac{3}{2}$	$\frac{9}{4}$	$\frac{6}{5}$

29. $f(x) = \dfrac{-4}{\dfrac{5}{x} - 5} = \dfrac{-4}{\dfrac{5 - 5x}{x}} = \dfrac{-4x}{5 - 5x} = \dfrac{4}{5}\dfrac{x}{x - 1},\ x \neq 0$

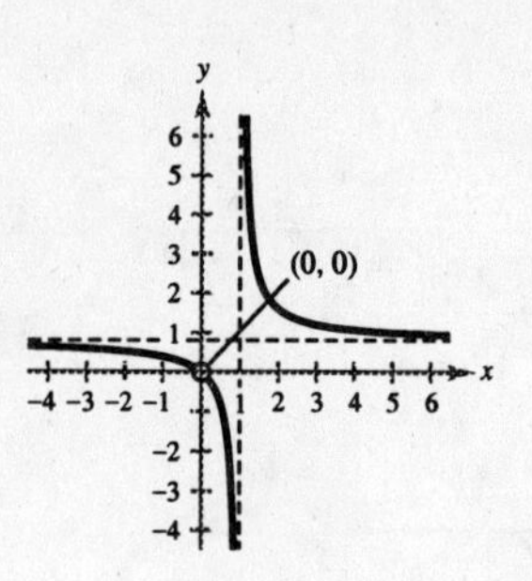

No intercepts [Note: $f(0)$ is not defined]

Vertical asymptote: $x = 1$

Horizontal asymptote: $y = \dfrac{4}{5}$

No symmetry

x	-2	-1	0	1	2	3
y	$\frac{8}{15}$	$\frac{2}{5}$	undef.	undef.	$\frac{8}{5}$	$\frac{6}{5}$

31. $f(x) = \dfrac{2 + x}{1 - x} = -\dfrac{x + 2}{x - 1}$

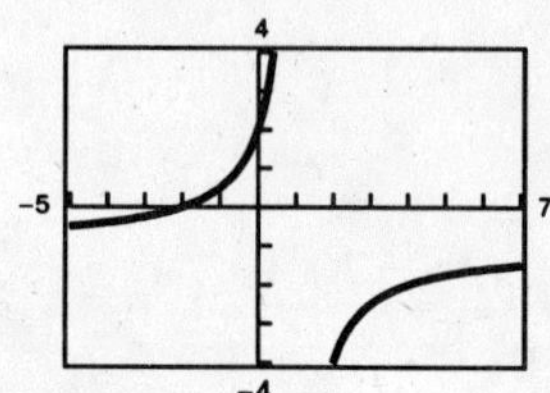
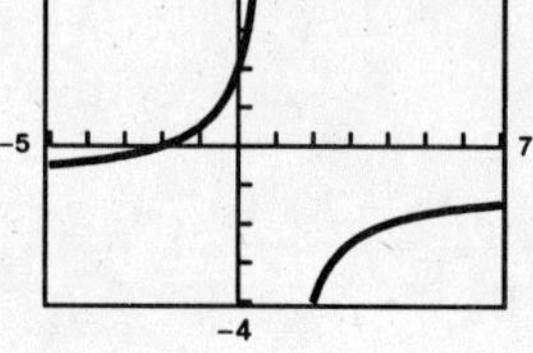

x-intercept: $(-2, 0)$

y-intercept: $(0, 2)$

Vertical asymptote: $x = 1$

Horizontal asymptote: $y = -1$

Domain: $x \neq 1$ or $(-\infty, 1) \cup (1, \infty)$

33. $f(t) = \dfrac{3t + 1}{t}$

t-intercept: $\left(-\dfrac{1}{3}, 0\right)$

Vertical asymptote: $t = 0$

Horizontal asymptote: $y = 3$

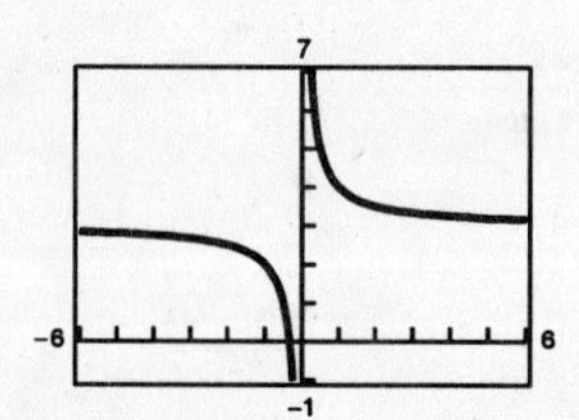

Domain: $t \neq 0$ or $(-\infty, 0) \cup (0, \infty)$

35. $h(t) = \dfrac{4}{t^2 + 1}$

Domain: all real numbers OR $(-\infty, \infty)$

Horizontal asymptote: $y = 0$

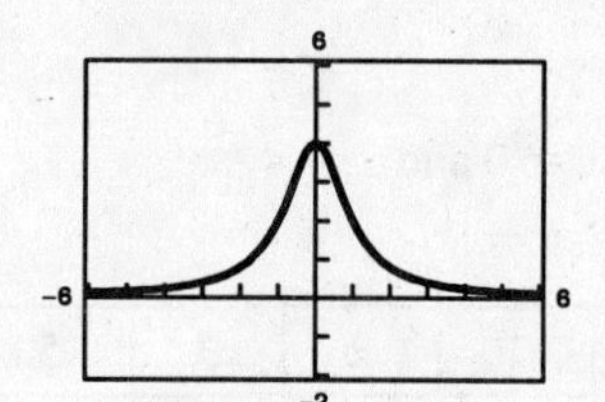

37. $f(x) = \dfrac{x + 1}{x^2 - x - 6} = \dfrac{x + 1}{(x - 3)(x + 2)}$

Domain: all real numbers except $x = 3, -2$

Vertical asymptotes: $x = 3,\ x = -2$

Horizontal asymptote: $y = 0$

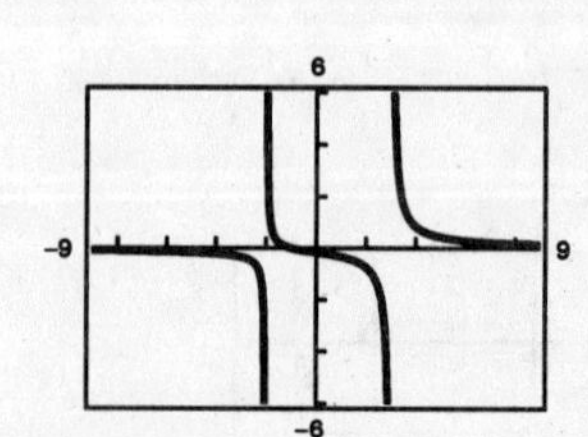

39. $f(x) = \dfrac{20x}{x^2+1} - \dfrac{1}{x} = \dfrac{19x^2-1}{x(x^2+1)}$

Domain: all real numbers except 0,
OR $(-\infty, 0) \cup (0, \infty)$

Vertical asymptote: $x = 0$
Horizontal asymptote: $y = 0$
Origin Symmetry

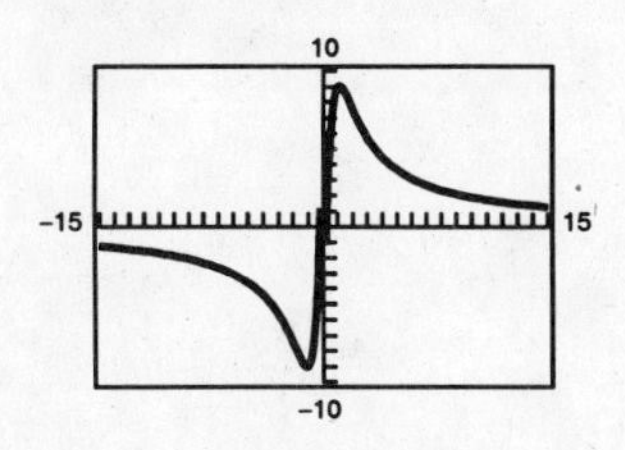

41. $h(x) = \dfrac{6x}{\sqrt{x^2+1}}$

There are two horizontal asymptotes: $y = \pm 6$

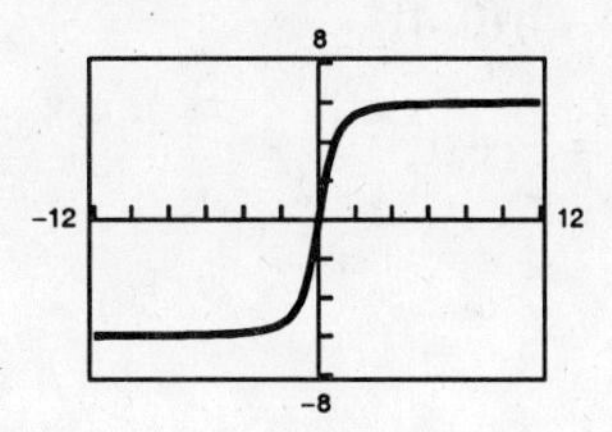

43. $g(x) = \dfrac{4|x-2|}{x+1}$

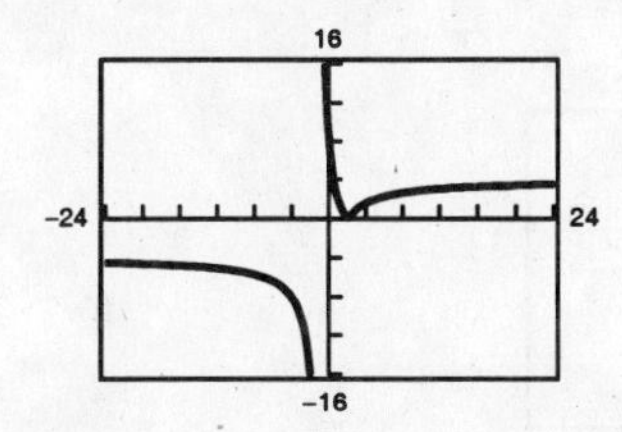

There are two horizontal asymptotes: $y = \pm 4$.

One vertical asymptote: $x = -1$.

45. $f(x) = \dfrac{4(x-1)^2}{x^2-4x+5}$

The graph crosses its horizontal asymptote: $y = 4$

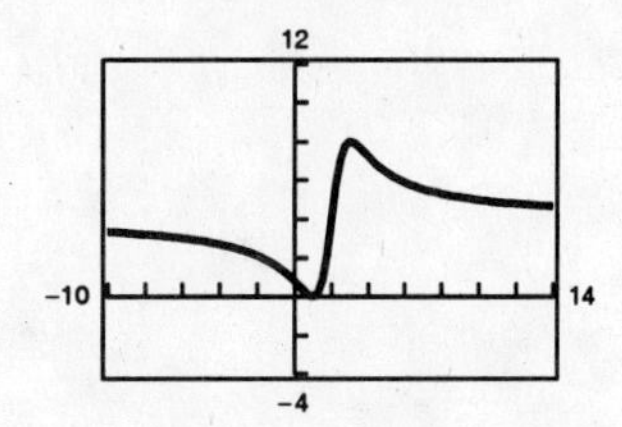

47. $f(x) = \dfrac{2x^2+1}{x} = 2x + \dfrac{1}{x}$

Vertical asymptote: $x = 0$
Slant asymptote: $y = 2x$
Origin symmetry

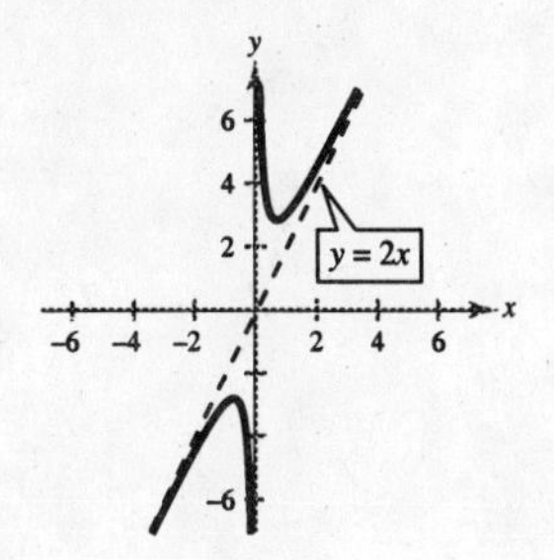

49. $h(x) = \dfrac{x^2}{x-1} = x + 1 + \dfrac{1}{x-1}$

Intercept: $(0, 0)$
Vertical asymptote: $x = 1$
Slant asymptote: $y = x + 1$

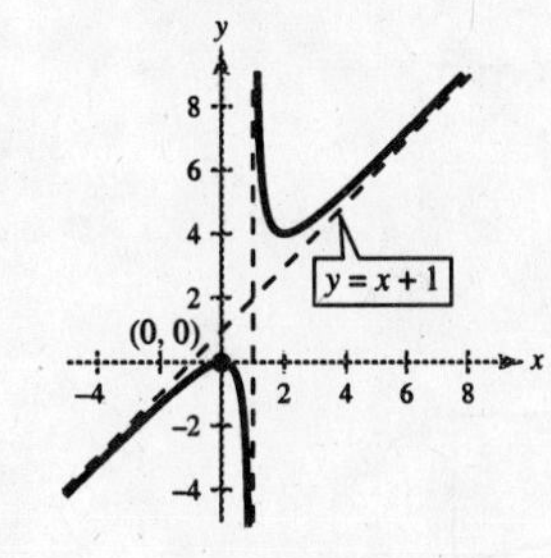

51. $g(x) = \dfrac{x^3}{2x^2-8} = \dfrac{1}{2}x + \dfrac{4x}{2x^2-8}$

Intercept: $(0, 0)$
Vertical asymptotes: $x = \pm 2$

Slant asymptote: $y = \dfrac{1}{2}x$

Origin symmetry

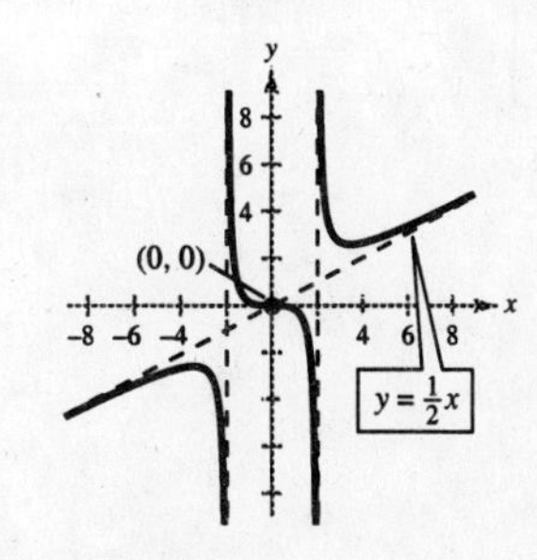

53. $f(x) = \dfrac{x^3 + 2x^2 + 4}{2x^2 + 1} = \dfrac{x}{2} + 1 + \dfrac{3 - \frac{x}{2}}{2x^2 + 1}$

Intercepts: $(-2.594, 0), (0, 4)$

Slant asymptote: $y = \dfrac{x}{2} + 1$

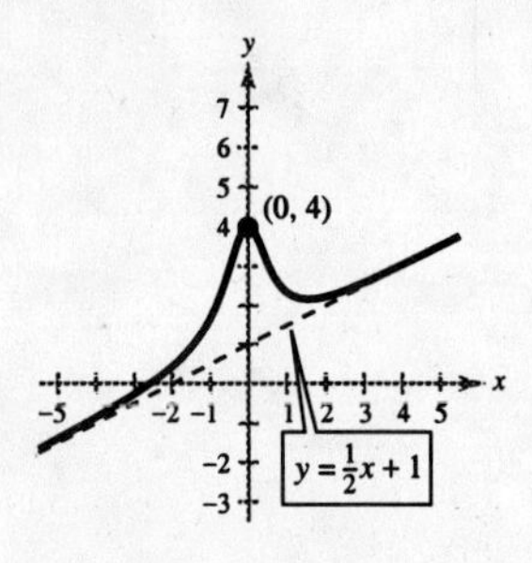

55. (a) x-intercept: $(-1, 0)$

(b) $0 = \dfrac{x + 1}{x - 3}$

$0 = x + 1$

$-1 = x$

57. (a) x-intercepts: $(\pm 1, 0)$

(b) $0 = \dfrac{1}{x} - x$

$x = \dfrac{1}{x}$

$x^2 = 1$

$x = \pm 1$

59. $y = \dfrac{2x^2 + x}{x + 1} = 2x - 1 + \dfrac{1}{x + 1}$

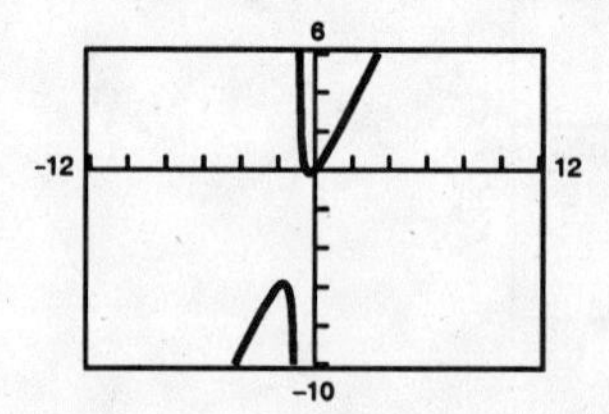

Domain: all real numbers except $x = -1$
Vertical asymptote: $x = -1$
Slant asymptote: $y = 2x - 1$

61. $g(x) = \dfrac{1 + 3x^2 - x^3}{x^2} = \dfrac{1}{x^2} + 3 - x = -x + 3 + \dfrac{1}{x^2}$

Domain:all real numbers except 0
OR $(-\infty, 0) \cup (0, \infty)$
Vertical asymptote: $x = 0$
Slant asymptote: $y = -x + 3$

12

-12 12

-4

63. $y = \dfrac{1}{x + 5} + \dfrac{4}{x}$

(a)

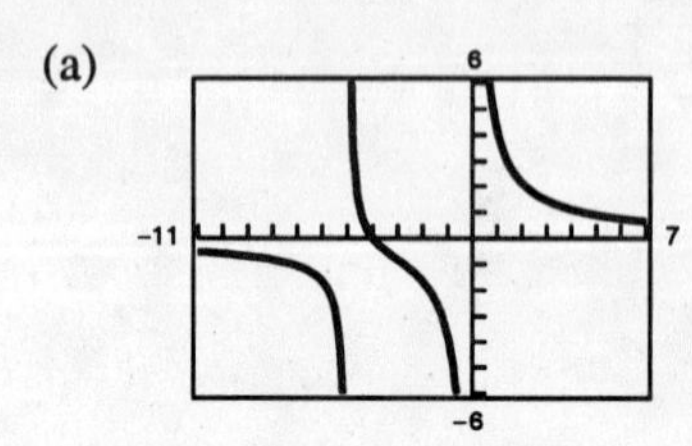

x-intercept: $(-4, 0)$

(b) $0 = \dfrac{1}{x + 5} + \dfrac{4}{x}$

$-\dfrac{4}{x} = \dfrac{1}{x + 5}$

$-4(x + 5) = x$

$-4x - 20 = x$

$-5x = 20$

$x = -4$

65. $y = x - \dfrac{6}{x - 1}$

(a)

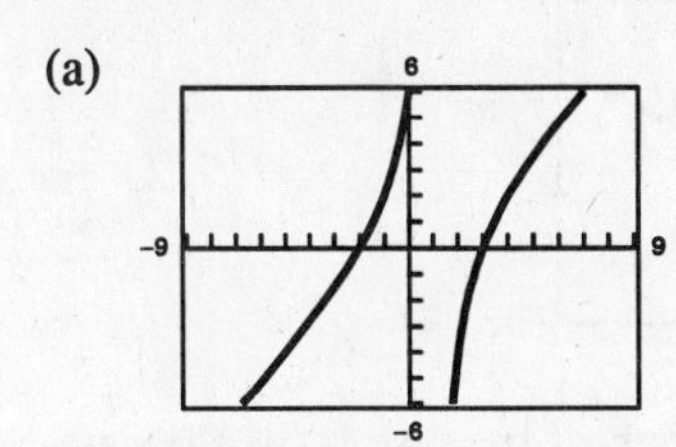

x-intercept: $(-2, 0), (3, 0)$

(b) $$0 = x - \frac{6}{x-1}$$
$$\frac{6}{x-1} = x$$
$$6 = x(x-1)$$
$$0 = x^2 - x - 6$$
$$0 = (x+2)(x-3)$$
$$x = -2, \quad x = 3$$

67. (a) $0.25(10) + 0.75(x) = C(10 + x)$

$$C = \frac{2.5 + 0.75x}{10 + x} \cdot \frac{4}{4}$$
$$C = \frac{10 + 3x}{4(10 + x)}$$
$$= \frac{3x + 10}{4(x + 10)}$$

(b) Domain: $x > 0$ and $x \le 1000 - 10$
Thus, $0 \le x \le 990$ OR $[0, 990]$.

(c)

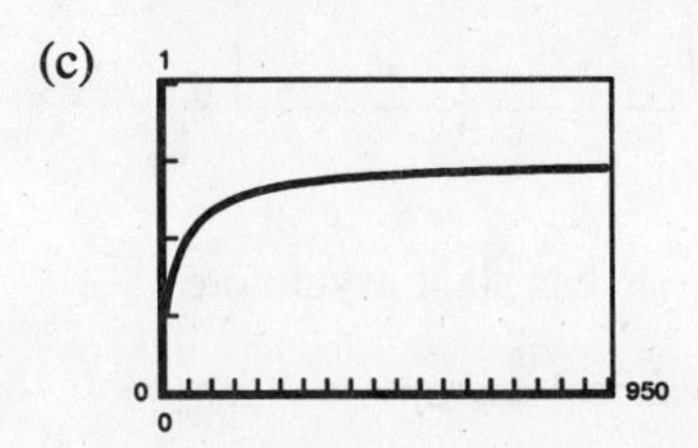

As the tank is filled, the rate that the concentration is increasing slows down. It approaches the horizontal asymptote of $C = \frac{3}{4} = 0.75$. The concentration reaches 74.5% when the tank is full ($x = 990$).

69. (a) $A = xy$ and

$$(x - 2)(y - 4) = 30$$
$$y - 4 = \frac{30}{x - 2}$$
$$y = 4 + \frac{30}{x-2} = \frac{4x + 22}{x - 2}$$

Thus, $A = xy = x\left(\dfrac{4x + 22}{x - 2}\right) = \dfrac{2x(2x + 11)}{x - 2}$.

(b) Domain: Since the margins on the left and right are each 1 inch, $x > 2$, OR $(2, \infty)$.

(c)

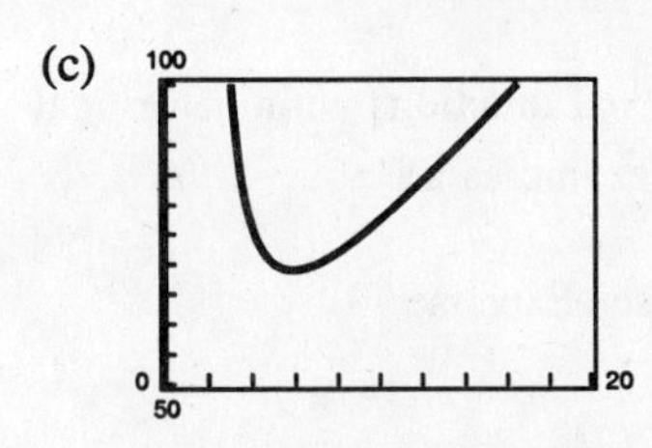

The area is minimum when $x \approx 5.87$ in. and $y \approx 11.75$ in.

71. $C = 100\left(\dfrac{200}{x^2} + \dfrac{x}{x + 30}\right)$, $1 \le x$

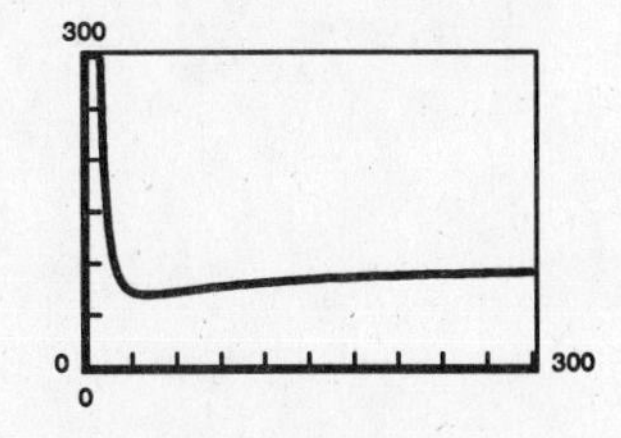

The minimum occurs when $x \approx 40.4 \approx 40$.

73. $C = \dfrac{3t^2 + t}{t^3 + 50}$, $0 \le t$

(a) The horizontal asymptote is the t-axis, or $C = 0$. This indicates that the chemical eventually dissipates.

(b)

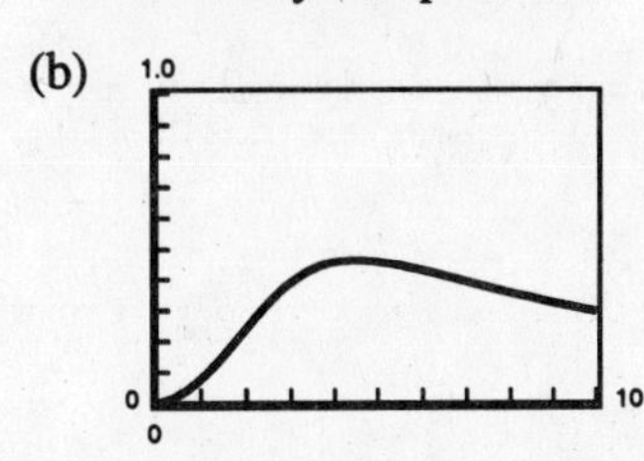

The maximum occurs when $t \approx 4.5$.

(c) $C < 0.345$ when $0 \le t < 2.65$ hours and when $t > 8.32$ hours

75. (a)

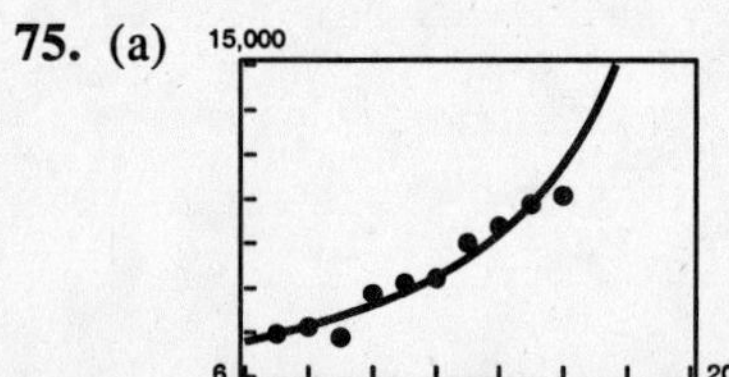

(b) $K = 384.49t + 5937.65$

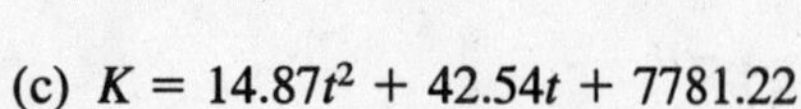

(c) $K = 14.87t^2 + 42.54t + 7781.22$

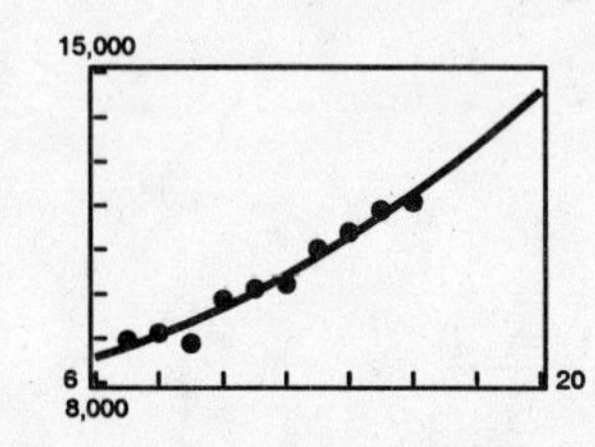

(d) Quadratic model is best because it fits the data well and is easy to use. Answer will vary.

77. False, you will have to lift your pencil to cross the vertical asymptote.

79. $h(x) = \dfrac{6 - 2x}{3 - x} = \dfrac{2(3 - x)}{3 - x}$

Since $h(x)$ is not reduced and $(3 - x)$ is a factor of both the numerator and the denominator, $x = 3$ is not a horizontal asymptote.

There is a hole in the graph at $x = 3$.

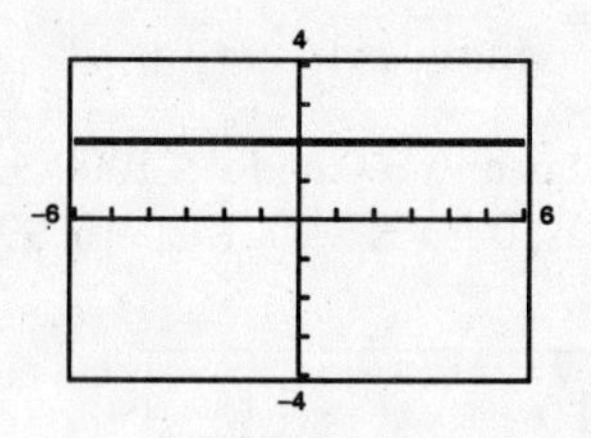

81. No, there are rational functions without vertical asymptotes. Two examples are

$f(x) = \dfrac{6 - 2x}{3 - x}$ (see Exercise 79)

$g(x) = \dfrac{1}{x^2 + 1}$

83. $y = x - 2 + \dfrac{a}{x + 4}$ has slant asymptote $y = x - 2$ and vertical asymptote at $x = -4$.

We determine a so that y has a zero at $x = 3$:

$0 = 3 - 2 + \dfrac{a}{3 + 4} = 1 + \dfrac{a}{7} \Longrightarrow a = -7$

Hence, $y = x - 2 + \dfrac{-7}{x + 4} = \dfrac{x^2 + 2x - 15}{x + 4}$

85. $y = \dfrac{-2(x + 6)}{(x - 3)}$ has vertical asymptote $x = 3$,

horizontal asymptote $y = -2$ and zero at $x = -6$

87. $4x + 5y - 2 = 0 \Longrightarrow y = \dfrac{1}{5}(-4x + 2) = -\dfrac{4}{5}x + \dfrac{2}{5}$ line

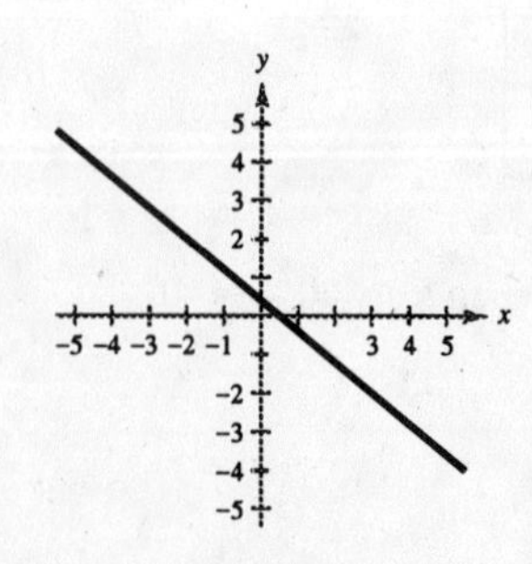

89. $4y - 10 = 0$

$4y = 10$

$y = \frac{5}{2}$ horizontal line

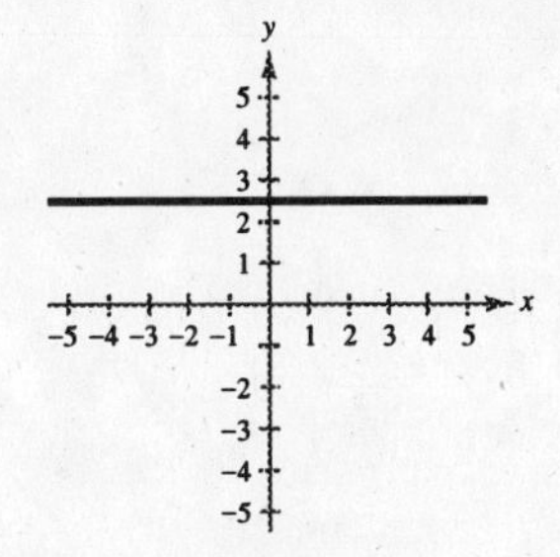

91. $-7 + 8x - 2y = 0$

$-2y = -8x + 7$

$y = 4x - \frac{7}{2}$ line

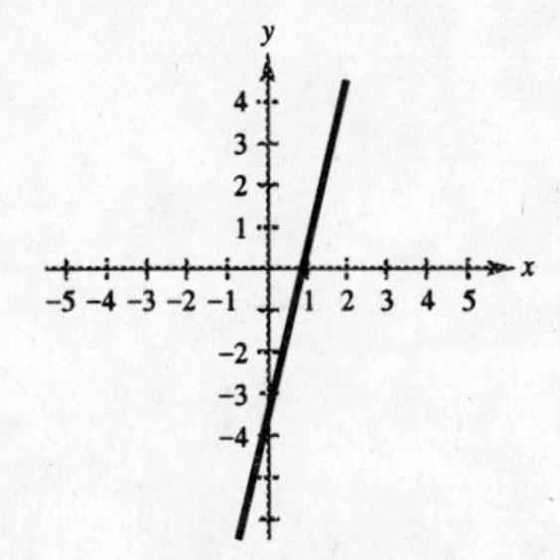

93.

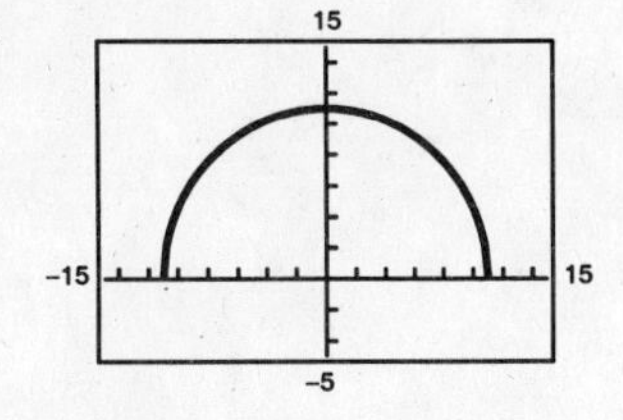

Semicircle

Domain: $-11 \leq x \leq 11$

Range: $0 \leq y \leq 11$

95.

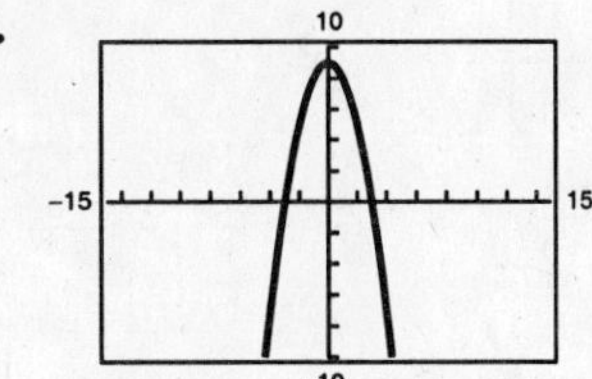

Parabola

Domain: all x

Range: $y \leq 9$

Review Exercises for Chapter 2

Solutions to Odd-Numbered Exercises

1. (a) $y = 2x^2$
Vertical stretch

(b) $y = -2x^2$
Vertical stretch and a reflection in the x-axis

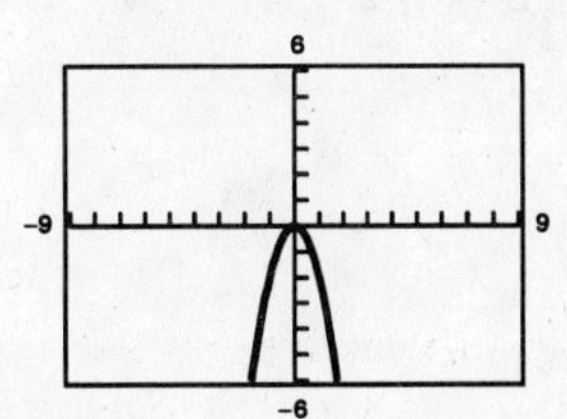

(c) $y = x^2 + 2$
Vertical shift two units upward

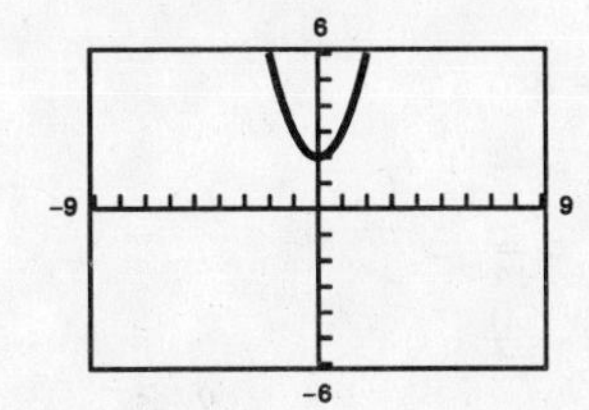

(d) $y = (x + 2)^2$
Horizontal shift two units to the left

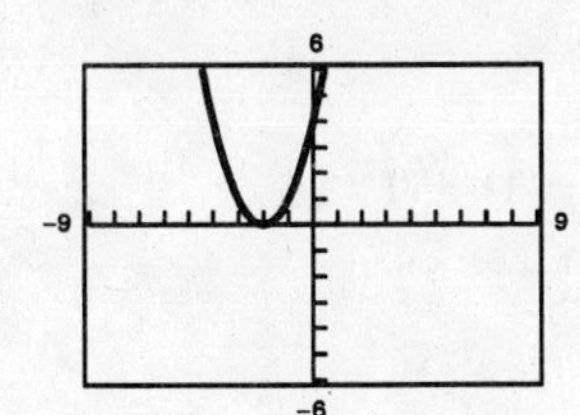

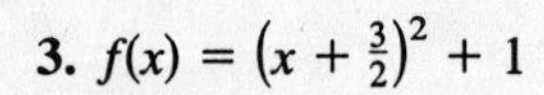

3. $f(x) = \left(x + \frac{3}{2}\right)^2 + 1$

Vertex: $\left(-\frac{3}{2}, 1\right)$

y-intercept: $\left(0, \frac{13}{4}\right)$

No x-intercepts

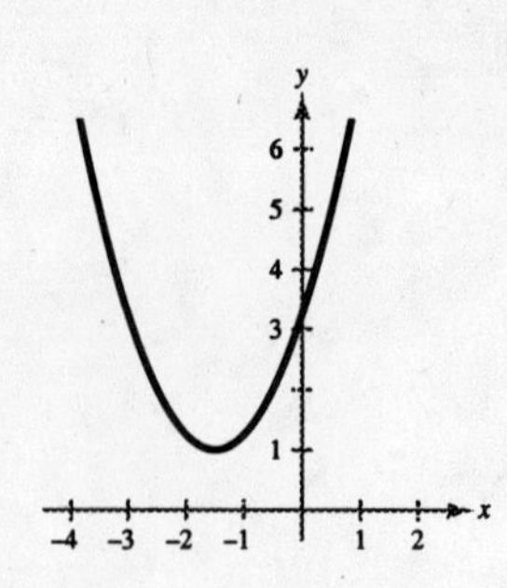

5. $f(x) = \frac{1}{3}(x^2 + 5x - 4)$

$$= \frac{1}{3}\left(x^2 + 5x + \frac{25}{4} - \frac{25}{4} - 4\right)$$

$$= \frac{1}{3}\left[\left(x - \frac{5}{2}\right)^2 - \frac{41}{4}\right]$$

$$= \frac{1}{3}\left(x - \frac{5}{2}\right)^2 - \frac{41}{12}$$

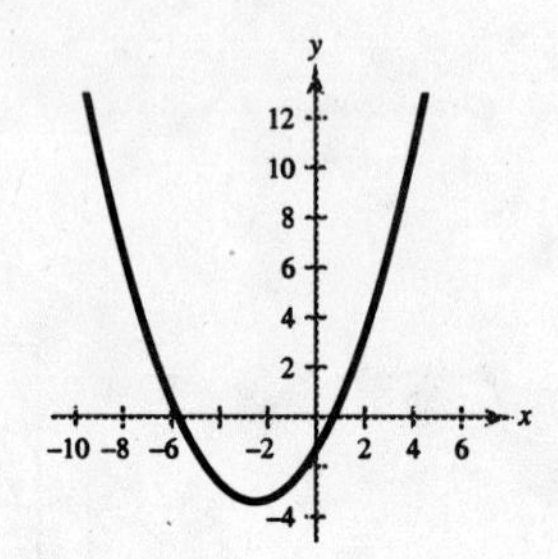

Vertex: $\left(\frac{5}{2}, -\frac{41}{12}\right)$

y-intercept: $\left(0, -\frac{3}{4}\right)$

x-intercepts: $0 = \frac{1}{3}(x^2 + 5x - 4)$

$$0 = x^2 + 5x - 4$$

$$x = \frac{-5 \pm \sqrt{41}}{2} \quad \text{Use the Quadratic Formula.}$$

$$\left(\frac{-5 \pm \sqrt{41}}{2}, 0\right)$$

7. Vertex: $(1, -4) \implies f(x) = a(x - 1)^2 - 4$

Point: $(2, -3) \implies -3 = a(2 - 1)^2 - 4$

$$1 = a$$

Thus, $f(x) = (x - 1)^2 - 4$.

9. $g(x) = x^2 - 2x$

$$= x^2 - 2x + 1 - 1$$

$$= (x - 1)^2 - 1$$

The minimum occurs at the vertex $(1, -1)$.

11. $f(x) = 6x - x^2$

$$= -(x^2 - 6x + 9 - 9)$$

$$= -(x - 3)^2 + 9$$

The maximum occurs at the vertex $(3, 9)$.

13. $f(t) = -2t^2 + 4t + 1$

$$= -2(t^2 - 2t + 1 - 1) + 1$$

$$= -2[(t - 1)^2 - 1] + 1$$

$$= -2(t - 1)^2 + 3$$

The maximum occurs at the vertex $(1, 3)$.

15. $h(x) = x^2 + 5x - 4$

$$= x^2 + 5x + \frac{25}{4} - \frac{25}{4} - 4$$

$$= \left(x + \frac{5}{2}\right)^2 - \frac{25}{4} - \frac{16}{4}$$

$$= \left(x + \frac{5}{2}\right)^2 - \frac{41}{4}$$

The minimum occurs at the vertex $\left(-\frac{5}{2}, -\frac{41}{4}\right)$.

17. (a)

x	y	Area
1	$4 - \frac{1}{2}(1)$	$(1)[4 - \frac{1}{2}(1)] = \frac{7}{2}$
2	$4 - \frac{1}{2}(2)$	$(2)[4 - \frac{1}{2}(2)] = 6$
3	$4 - \frac{1}{2}(3)$	$(3)[4 - \frac{1}{2}(3)] = \frac{15}{2}$
4	$4 - \frac{1}{2}(4)$	$(4)[4 - \frac{1}{2}(4)] = 8$
5	$4 - \frac{1}{2}(5)$	$(5)[4 - \frac{1}{2}(5)] = \frac{15}{2}$
6	$4 - \frac{1}{2}(6)$	$(6)[4 - \frac{1}{2}(6)] = 6$

(b) The dimensions that will produce a maximum area are $x = 4$ and $y = 2$.

(c) $A = xy = x\left(\frac{8 - x}{2}\right)$, since

$x + 2y - 8 = 0 \Rightarrow y = \frac{8 - x}{2}$.

Since the figure is in the first quadrant and x and y must be positive, the domain of

$A = x\left(\frac{8 - x}{2}\right)$ is $0 < x < 8$.

(d)

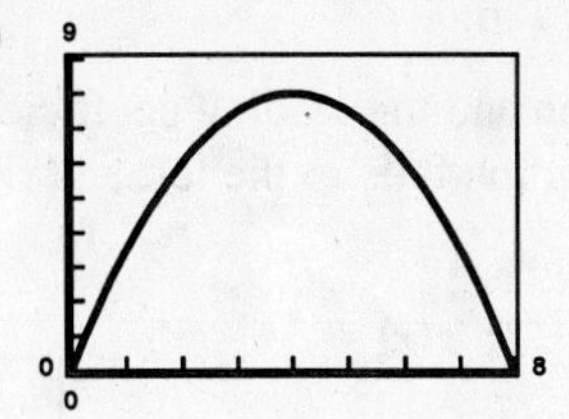

The maximum area of 8 occurs at the vertex when $x = 4$ and $y = \frac{8 - 4}{2} = 2$.

(e) $A = x\left(\frac{8 - x}{2}\right)$

$= \frac{1}{2}(8x - x^2)$

$= -\frac{1}{2}(x^2 - 8x)$

$= -\frac{1}{2}(x^2 - 8x + 16 - 16)$

$= -\frac{1}{2}[(x - 4)^2 - 16]$

$= -\frac{1}{2}(x - 4)^2 + 8$

The maximum area of 8 occurs when $x = 4$ and $y = \frac{8 - 4}{2} = 2$.

19. $y = x^4$

(a)

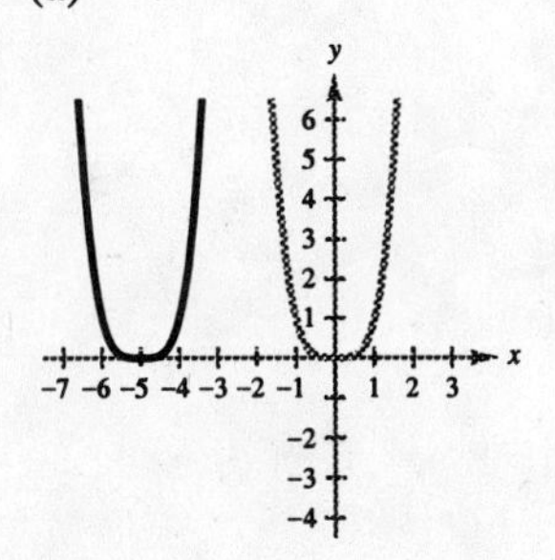

(b)

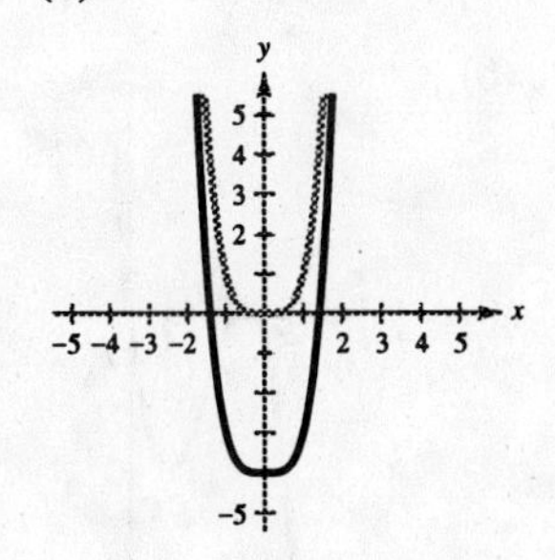

(c)

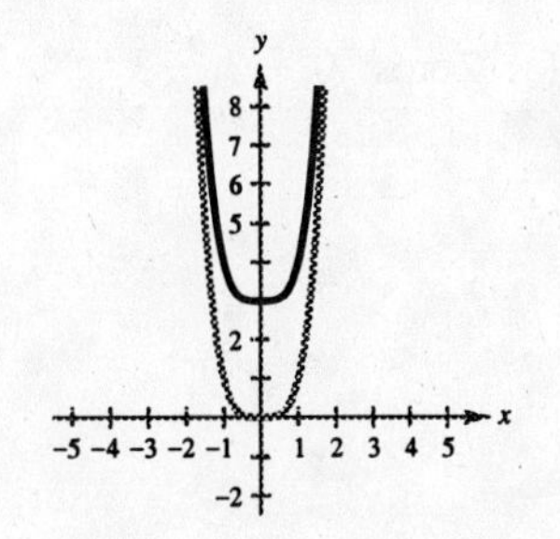

(d)

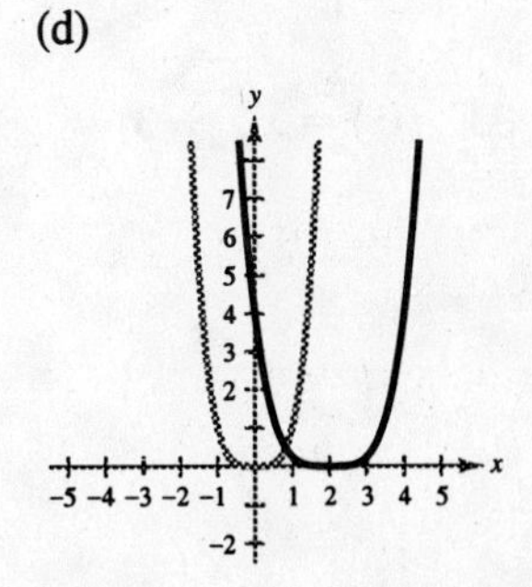

21. $y = x^6$

(a)

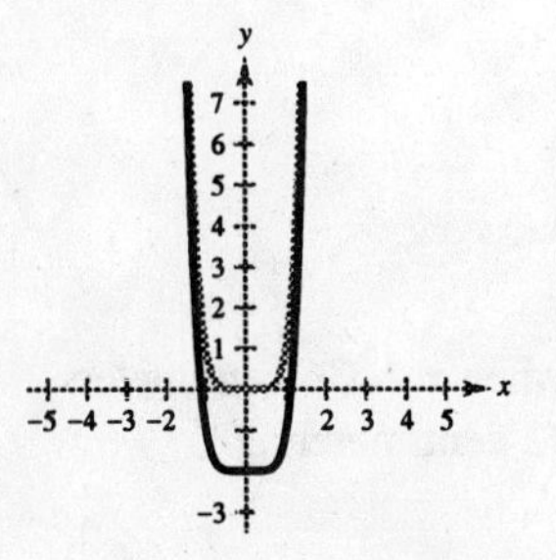

(b)

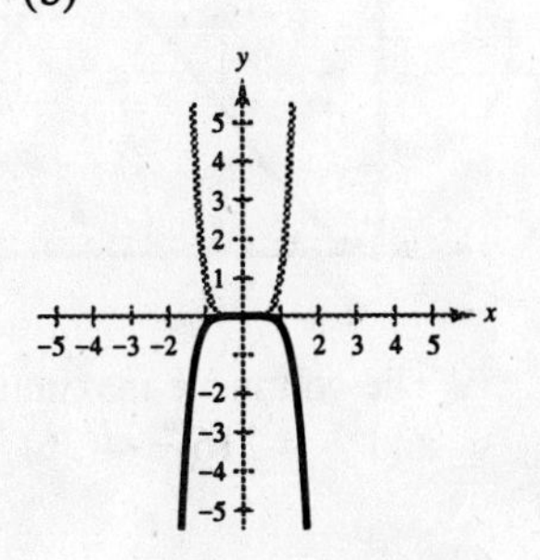

(c)

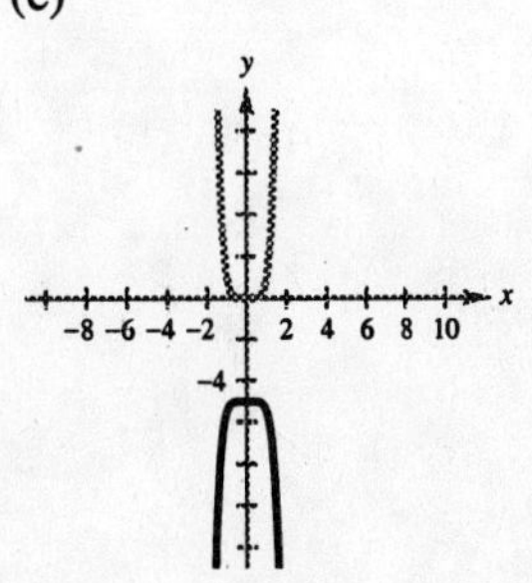

(d)

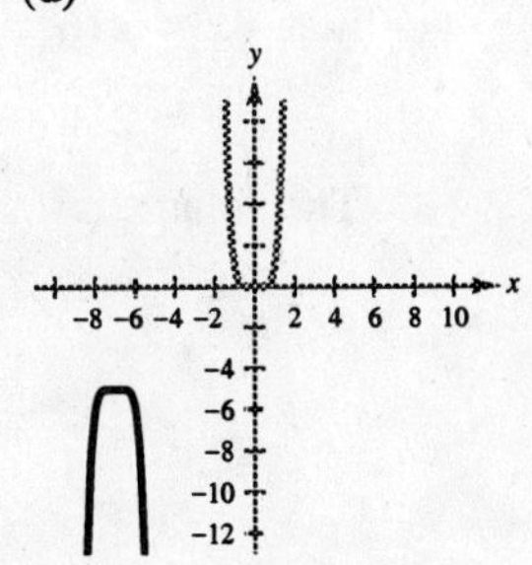

23. $f(x) = -x^2 + 6x + 9$

The degree is even and the leading coefficient is negative. The graph falls to the left and right.

25. $f(x) = \frac{3}{4}(x^4 + 3x^2 + 2)$

The degree is even and the leading coefficient is positive. The graph rises to the left and right.

27. $f(x) = \frac{1}{2}x^3 - 2x + 1;\ g(x) = \frac{1}{2}x^3$

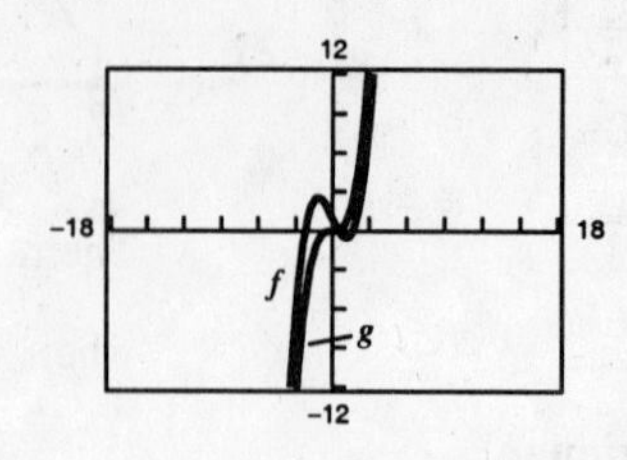

29. $g(x) = x^4 - x^3 - 2x^2$

(a) $0 = x^4 - x^3 - 2x^2$

$0 = x^2(x^2 - x - 2)$

$0 = x^2(x - 2)(x + 1)$

Zeros: $0, 0, 2, -1$

(b)

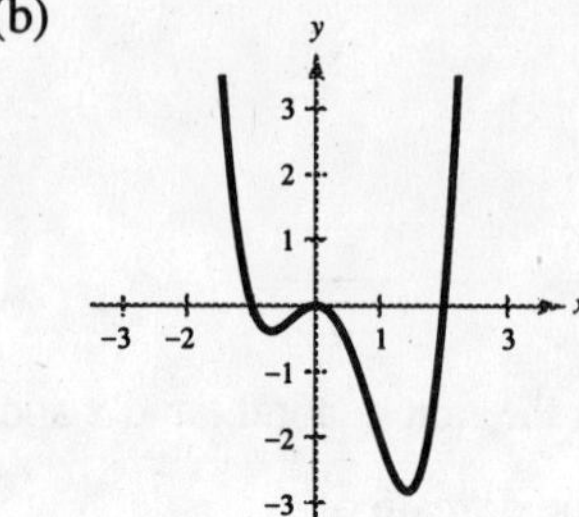

31. $f(t) = t^3 - 3t$

(a) $0 = t^3 - 3t$

$0 = t(t^2 - 3)$

Zeros: $0, \pm\sqrt{3}$

(b)

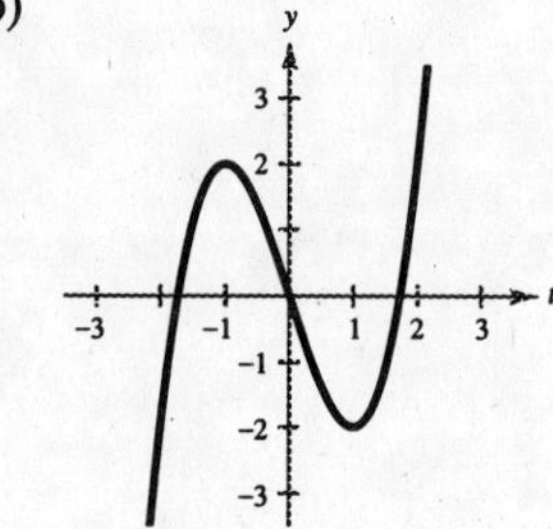

33. $f(x) = x(x + 3)^2$ (a) Zeros: $0, -3, -3$

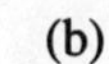
(b)

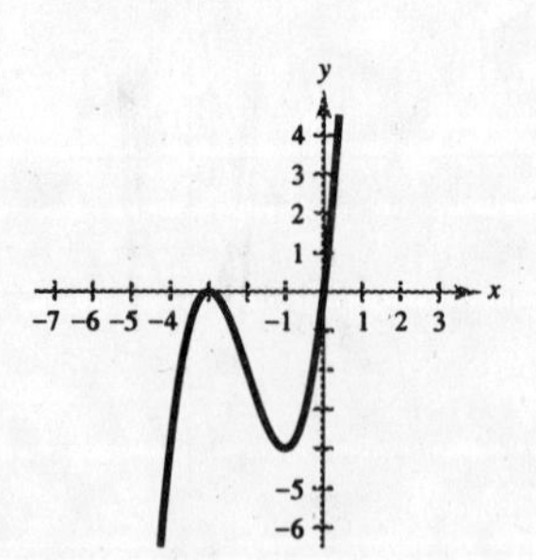

35. (a) The combined length and girth is

$y + 4x = 216$

$y = 216 - 4x.$

The volume is

$V = x^2y = x^2(216 - 4x).$

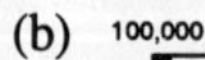
(b) 100,000

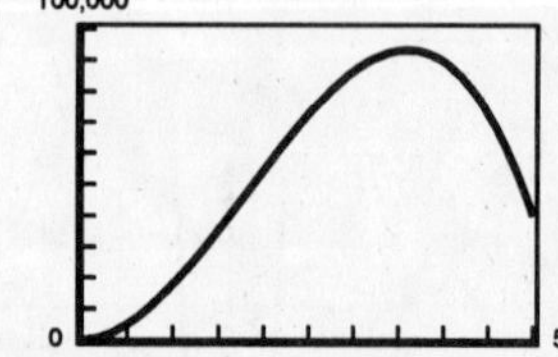

The volume is maximum when $x = 36$ centimeters and $y = 216 - 4(36) = 72$ centimeters.

37. (a) $f(-3) < 0,\ f(-2) > 0 \Rightarrow$ zero in $[-3, -2]$

$f(-1) > 0,\ f(0) < 0 \Rightarrow$ zero in $[-1, 0]$

$f(0) < 0,\ f(1) > 0 \Rightarrow$ zero in $[0, 1]$

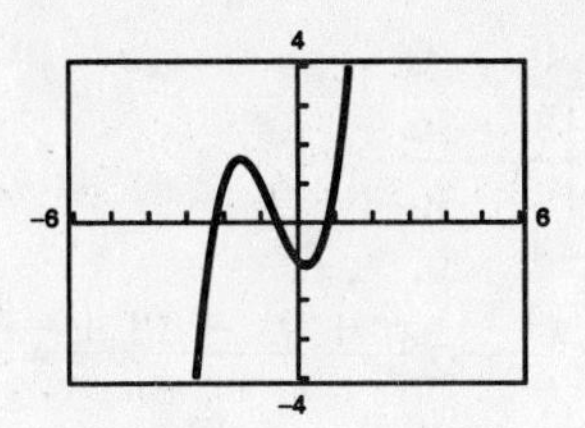

(b) zeros: $-2.247, -0.555, 0.802$

39. (a) $f(-3) > 0,\ f(-2) < 0 \Rightarrow$ zero in $[-3, -2]$

$f(2) < 0,\ f(3) > 0 \Rightarrow$ zero in $[2, 3]$

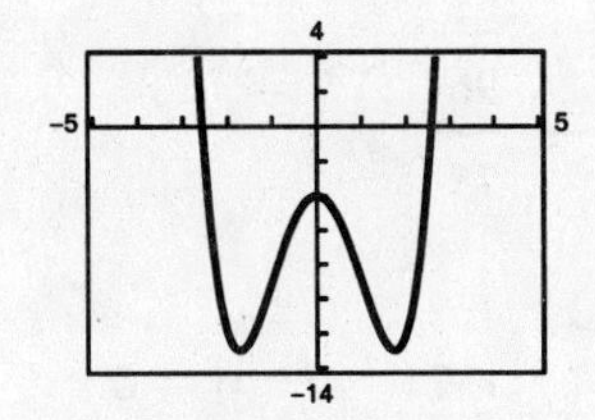

(b) zeros: ± 2.570

41. $y_1 = \dfrac{x^2}{x-2}$

$$\begin{aligned} y_2 &= x + 2 + \frac{4}{x-2} \\ &= \frac{(x+2)(x-2)}{x-2} + \frac{4}{x-2} \\ &= \frac{x^2-4}{x-2} + \frac{4}{x-2} \\ &= \frac{x^2}{x-2} \\ &= y_1 \end{aligned}$$

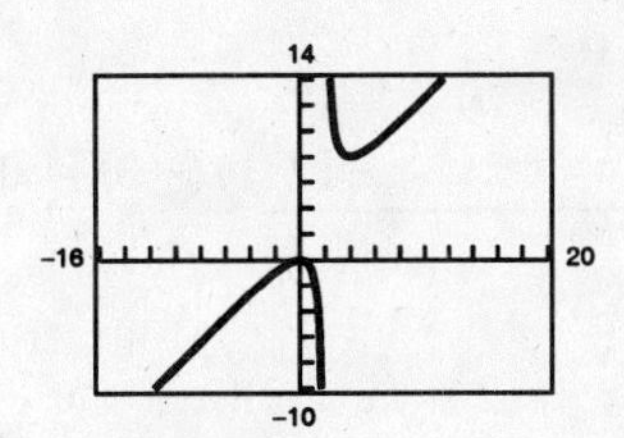

43.

$$\begin{array}{r} 8x + 5 \\ 3x - 2 \overline{)\,24x^2 - x - 8} \\ \underline{24x^2 - 16x} \\ 15x - 8 \\ \underline{15x - 10} \\ 2 \end{array}$$

Thus, $\dfrac{24x^2 - x - 8}{3x - 2} = 8x + 5 + \dfrac{2}{3x - 2}$.

45.

$$\begin{array}{r} x^2 - 2 \\ x^2 - 1 \overline{)\,x^4 - 3x^2 + 2} \\ \underline{x^4 - x^2} \\ -2x^2 + 2 \\ \underline{-2x^2 + 2} \\ 0 \end{array}$$

Thus, $\dfrac{x^4 - 3x^2 + 2}{x^2 - 1} = x^2 - 2 \quad (x \neq \pm 1)$

47.

$$\begin{array}{r} x^2 - x + 1 \\ x^2 + 2x \overline{)\,x^4 + x^3 - x^2 + 2x} \\ \underline{x^4 + 2x^3} \\ -x^3 - x^2 \\ \underline{-x^3 - 2x^2} \\ x^2 + 2x \end{array}$$

Thus, $\dfrac{x^4 + x^3 - x^2 + 2x}{x^2 + 2x} = x^2 - x + 1, \quad (x \neq 0, -2)$.

49.

-2	0.25	-4	0	0	0
		$-\frac{1}{2}$	9	-18	36
	$\frac{1}{4}$	$-\frac{9}{2}$	9	-18	36

Hence, $\dfrac{0.25x^4 - 4x^3}{x+2} = \dfrac{1}{4}x^3 - \dfrac{9}{2}x^2 + 9x - 18 + \dfrac{36}{x+2}$

51.

$\frac{2}{3}$	6	-4	-27	18	0
		4	0	-18	0
	6	0	-27	0	0

Thus, $\dfrac{6x^4 - 4x^3 - 27x^2 + 18x}{x - \left(\frac{2}{3}\right)} = 6x^3 - 27x,\ x \neq \dfrac{2}{3}$

53. (a)

4	2	3	-20	-21
		8	44	96
	2	11	24	75

$f(4) = 75$

(b)

-1	2	3	-20	-21
		-2	-1	21
	2	1	-21	0

$f(-1) = 0$

(c)

$-\frac{7}{2}$	2	3	-20	-21
		-7	14	21
	2	-4	-6	0

$f\left(-\frac{7}{2}\right) = 0$

(d)

0	2	3	-20	-21
		0	0	0
	2	3	-20	-21

$f(0) = -21$

55.

-3	2	5	-11	-20	12
		-6	3	24	-12
	2	-1	-8	4	0

$f(-3) = 0$

$$2x^4 + 5x^3 - 11x^2 - 20x + 12 = (x+3)(2x^3 - x^2 - 8x + 4)$$
$$= (x+3)(2x-1)(x-2)(x+2)$$

Zeros: $-3, \frac{1}{2}, 2, -2$

57. $f(x) = 4x^3 - 11x^2 + 10x - 3$

Possible Rational Zeros: $\pm 1, \pm 3, \pm\frac{1}{2}, \pm\frac{3}{2}, \pm\frac{1}{4}, \pm\frac{3}{4}$. Use a graphing utility to see that $x = 1$ is probably a zero.

1	4	-11	10	-3
		4	-7	3
	4	-7	3	0

$4x^3 - 11x^2 + 10x - 3 = (x-1)(4x^2 - 7x + 3) = (x-1)^2(4x-3)$

Thus, the zeros of f are $x = 1$ (repeated) and $x = \frac{3}{4}$.

59. $f(x) = 6x^3 - 5x^2 + 24x - 20$

Graphing $f(x)$ with a graphing utility suggests that $x = \frac{5}{6}$ is a zero.

$\frac{5}{6}$	6	-5	24	-20
		5	0	20
	6	0	24	0

The quadratic $6x^2 + 24 = 0$ has complex zeros $x = \pm 2i$. Thus, the zeros are $\frac{5}{6}, 2i, -2i$.

61. $f(x) = 6x^4 - 25x^3 + 14x^2 + 27x - 18$

Possible Rational Zeros: $\pm 1, \pm 2, \pm 3, \pm 6, \pm 9, \pm 18, \pm\frac{1}{2}, \pm\frac{3}{2}, \pm\frac{9}{2}, \pm\frac{1}{3}, \pm\frac{2}{3}, \pm\frac{1}{6}$. Use a graphing utility to see that $x = -1$ and $x = 3$ are probably zeros.

$$\begin{array}{r|rrrrr} -1 & 6 & -25 & 14 & 27 & -18 \\ & & -6 & 31 & -45 & 18 \\ \hline & 6 & -31 & 45 & -18 & 0 \end{array}$$

$$\begin{array}{r|rrrr} 3 & 6 & -31 & 45 & -18 \\ & & 18 & -39 & 18 \\ \hline & 6 & -13 & 6 & 0 \end{array}$$

$$\begin{aligned} 6x^4 - 25x^3 + 14x^2 + 27x - 18 &= (x + 1)(x - 3)(6x^2 - 13x + 6) \\ &= (x + 1)(x - 3)(3x - 2)(2x - 3) \end{aligned}$$

Thus, the zeros of f are $x = -1, x = 3, x = \frac{2}{3}$, and $x = \frac{3}{2}$.

63.
$$\begin{array}{r|rrrr} 1 & 4 & -3 & 4 & -3 \\ & & 4 & 1 & 5 \\ \hline & 4 & 1 & 5 & 2 \end{array}$$

All entries positive. $x = 1$ is upper bound.

$$\begin{array}{r|rrrr} -\frac{1}{4} & 4 & -3 & 4 & -3 \\ & & -1 & 1 & -\frac{5}{4} \\ \hline & 4 & -4 & 5 & -\frac{17}{4} \end{array}$$

Alternating signs. $x = -\frac{1}{4}$ is lower bound.

65. $6 + \sqrt{-25} = 6 + 5i$

67. $-2i^2 + 7i = 2 + 7i$

69. $$\begin{aligned} (7 + 5i) + (-4 + 2i) &= (7 - 4) + (5i + 2i) \\ &= 3 + 7i \end{aligned}$$

71. $5i(13 - 8i) = 65i - 40i^2 = 40 + 65i$

73. $(10 - 8i)(2 - 3i) = 20 - 30i - 16i + 24i^2 = -4 - 46i$

75. $$\begin{aligned} (3 + 7i)^2(3 - 7i)^2 &= (9 + 42i - 49) + (9 - 42i - 49) \\ &= -80 \end{aligned}$$

77. $$\begin{aligned} \frac{6 + i}{i} &= \frac{6 + i}{i} \cdot \frac{-i}{-i} = \frac{-6i - i^2}{-i^2} \\ &= \frac{-6i + 1}{1} = 1 - 6i \end{aligned}$$

79. $$\frac{4}{-3i} = \frac{4}{-3i} \cdot \frac{3i}{3i} = \frac{12i}{9} = \frac{4i}{3} = \frac{4}{3}i$$

81.

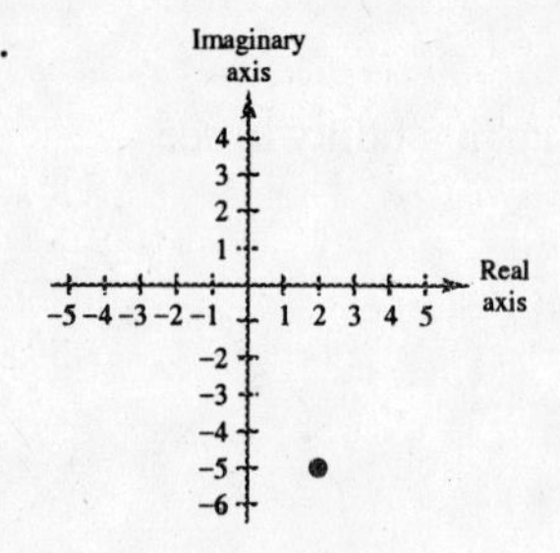

83.

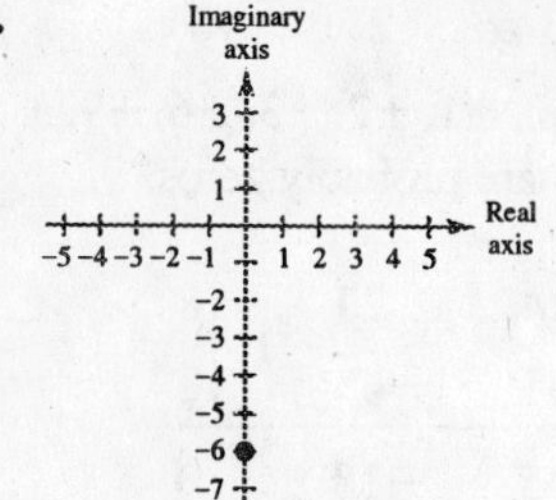

85. $f(x) = x^4 + 2x + 1$

(a)

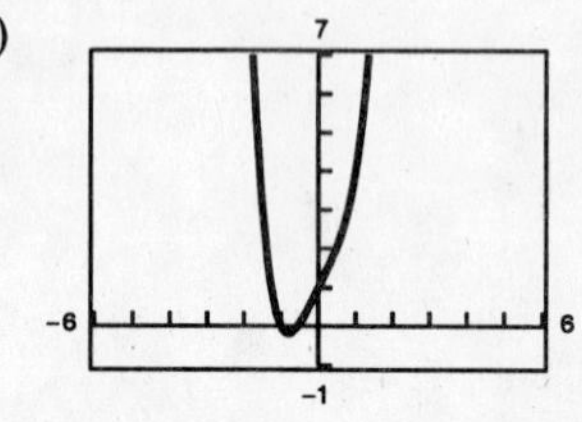

(b) The graph has two x-intercepts, so there are two real zeros.

(c) The zeros are $x = -1$ and $x \approx -0.54$.

87. $h(x) = x^3 - 6x^2 + 12x - 10$

(a)

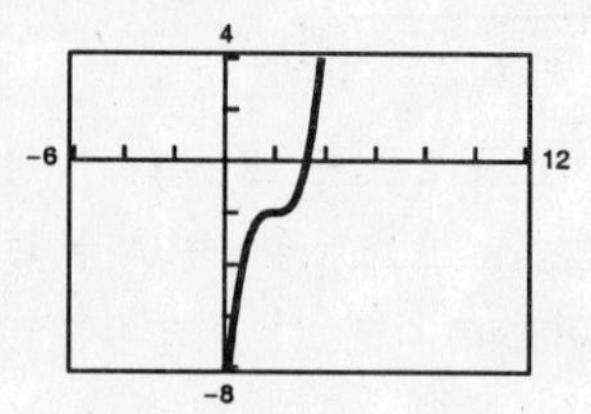

(b) The graph has one x-intercept, so there is one real zero.

(c) $x \approx 3.26$

89. $f(x) = x^3 - 4x^2 + 6x - 4$

$= (x - 2)(x^2 - 2x + 2)$

Use the Quadratic Formula for $x^2 - 2x + 2$

$$x = \frac{2 \pm \sqrt{(-2)^2 - 4(2)}}{2} = \frac{2 \pm \sqrt{-4}}{2} = 1 \pm i$$

zeros: $2, 1 + i, 1 - i$

$f(x) = (x - 2)(x - 1 + i)(x - 1 - i)$

91. $f(x) = x^3 + 6x^2 + 11x + 12$

$= (x + 4)(x^2 + 2x + 3)$

Use the Quadratic Formula for $x^2 + 2x + 3$

$$x = \frac{-2 \pm \sqrt{(2)^2 - 4(3)}}{2} = \frac{-2 \pm \sqrt{-8}}{2} = -1 \pm \sqrt{2}\,i$$

zeros: $-4, -1 + \sqrt{2}\,i, -1 - \sqrt{2}\,i$

$f(x) = (x + 4)\left(x + 1 + \sqrt{2}\,i\right)\left(x + 1 - \sqrt{2}\,i\right)$

93. $f(x) = x^4 + 34x^2 + 225$

$= (x^2 + 25)(x^2 + 9)$

$= (x + 5i)(x - 5i)(x + 3i)(x - 3i)$

zeros: $x = \pm 5i, \pm 3i$

95. $f(x) = (x + 2)(x + 2)(x + 5i)(x - 5i)$

$= (x^2 + 4x + 4)(x^2 + 25)$

$= x^4 + 4x^3 + 29x^2 + 100x + 100$

97. $f(x) = 3\left(x + \frac{2}{3}\right)(x + 1)\left(x - 3 - \sqrt{2}\,i\right)\left(x - 3 + \sqrt{2}\,i\right)$

$= (3x + 2)(x + 1)((x - 3)^2 + 2)$

$= (3x^2 + 5x + 2)(x^2 - 6x + 11)$

$= 3x^4 - 13x^3 + 5x^2 + 43x + 22$

99. $f(x) = x^4 - x^3 - x^2 + 5x - 20$

(a) $f(x) = (x^2 - 5)(x^2 - x + 4)$

(b) $f(x) = \left(x + \sqrt{5}\right)\left(x - \sqrt{5}\right)(x^2 - x + 4)$

(c) $f(x) = \left(x + \sqrt{5}\right)\left(x - \sqrt{5}\right)\left(x - \frac{1}{2} + \frac{\sqrt{15}}{2}i\right)\left(x - \frac{1}{2} - \frac{\sqrt{15}}{2} - 1\right)$

101. Domain: all $x \neq 1$

Horizontal asymptote: $y = -1$

Vertical asymptote: $x = 1$

103. $f(x) = \dfrac{2}{x^2 - 3x - 18} = \dfrac{2}{(x - 6)(x + 3)}$

Domain: all $x \neq 6, -3$

Horizontal asymptote: $y = 0$

Vertical asymptotes: $x = 6, x = -3$

105. $y = -1$ (degree $p(x)$ = degree $q(x)$)

107. $y = \dfrac{4}{2} = 2$ (degree $p(x)$ = degree $q(x)$)

109. $y = 1, y = -1$

111. (a)

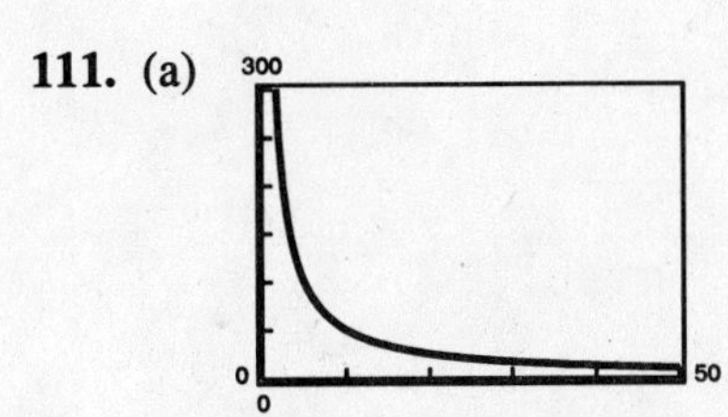

(b) $\overline{C}(50) = 10.50$ dollars

$\overline{C}(100) = 5.50$ dollars

$\overline{C}(1000) = 1.00$ dollar

$\overline{C}(10{,}000) = 0.55$ dollars

(c) As x increases, the average cost approaches its horizontal asymptote, $\overline{C} = 0.5$.

113. $f(x) = \dfrac{2x - 1}{x - 5}$

Intercepts: $\left(0, \frac{1}{5}\right), \left(\frac{1}{2}, 0\right)$

Vertical asymptote: $x = 5$

Horizontal asymptote: $y = 2$

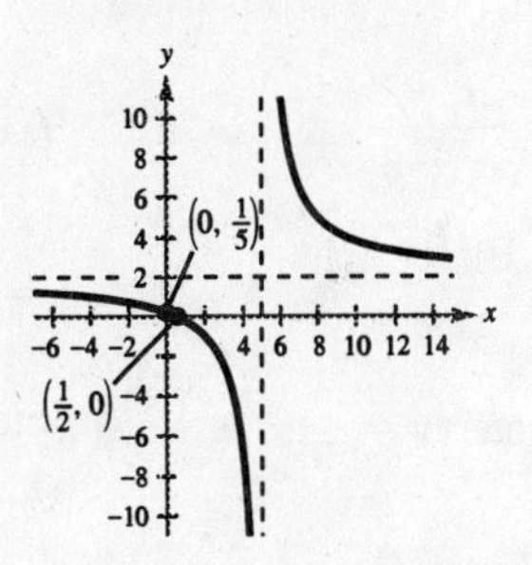

115. $f(x) = \dfrac{2x}{x^2 + 4}$

Intercept: $(0, 0)$

Origin symmetry

Horizontal asymptote: $y = 0$

x	-2	-1	0	1	2
y	$-\frac{1}{2}$	$-\frac{2}{5}$	0	$\frac{2}{5}$	$\frac{1}{2}$

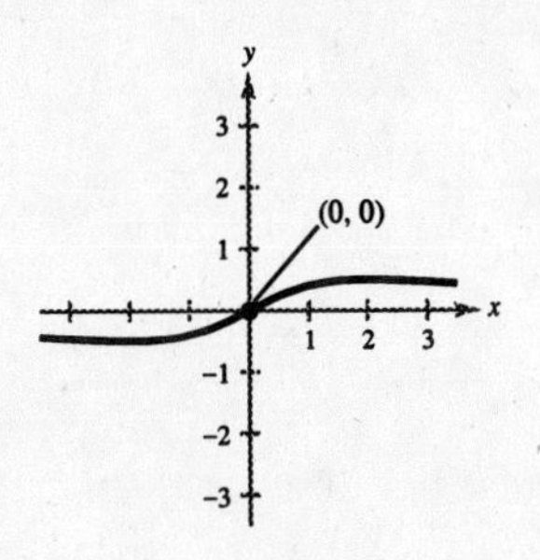

117. $f(x) = \dfrac{2}{(x + 1)^2}$

Intercept: $(0, 2)$

Horizontal asymptote: $y = 0$

Vertical asymptote: $x = -1$

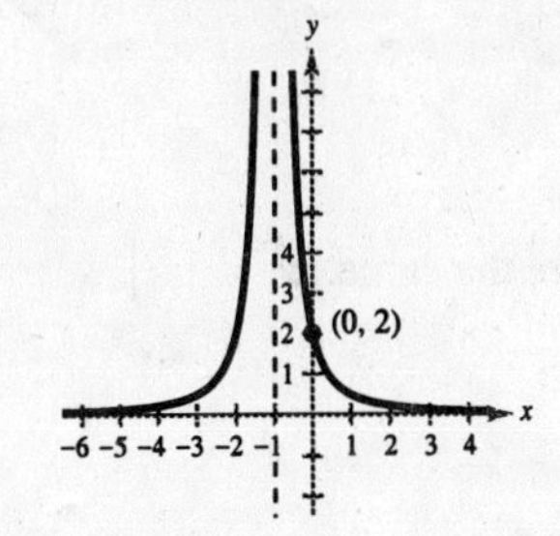

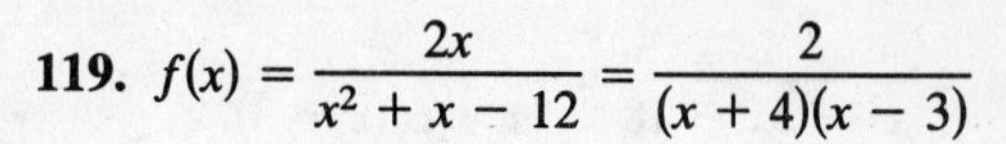

119. $f(x) = \dfrac{2x}{x^2 + x - 12} = \dfrac{2}{(x + 4)(x - 3)}$

Intercept: $(0, 0)$

Vertical asymptotes: $x = -4$, $x = 3$

Horizontal asymptote: $y = 0$

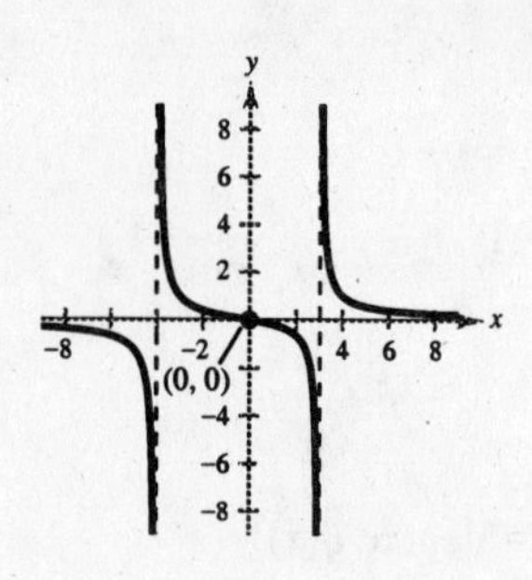

121. $f(x) = \dfrac{2x^3}{x^2 + 1} = 2x - \dfrac{2x}{x^2 + 1}$

Intercept: $(0, 0)$

Origin symmetry

Slant asymptote: $y = 2x$

x	-2	-1	0	1	2
y	$-\frac{16}{5}$	-1	0	1	$\frac{16}{5}$

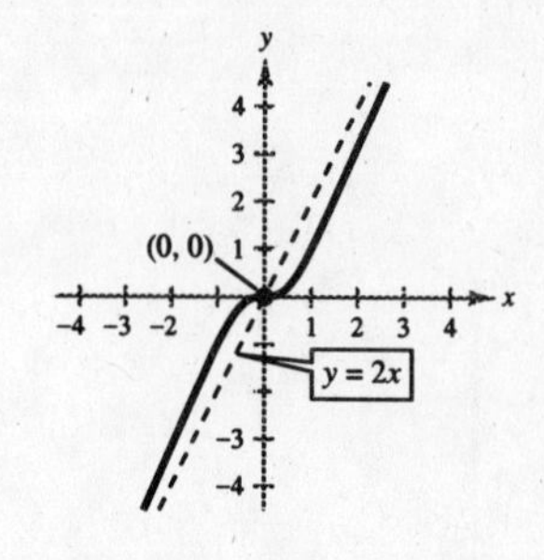

123. $y = \dfrac{1}{x + 3} + 2 = \dfrac{2x + 7}{x + 3}$

Intercepts: $(-3.5, 0)$, $\left(0, 2\frac{1}{3}\right)$

Vertical asymptote: $x = -3$

Horizontal asymptote: $y = 2$

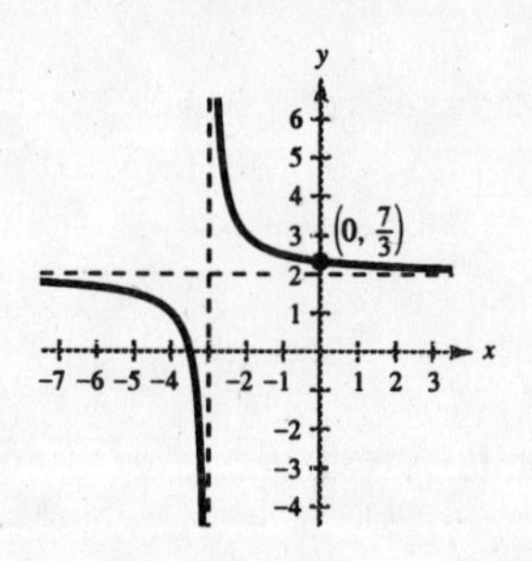

125. $f(x) = \dfrac{x^2 - x + 1}{x - 3}$

$= x + 2 + \dfrac{7}{x - 3}$

Intercept: $\left(0, -\frac{1}{3}\right)$

Vertical asymptote: $x = 3$

Slant asymptote: $y = x + 2$

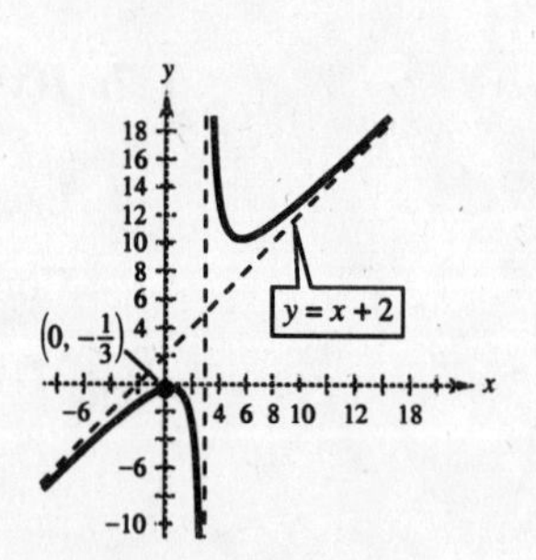

127. (a)

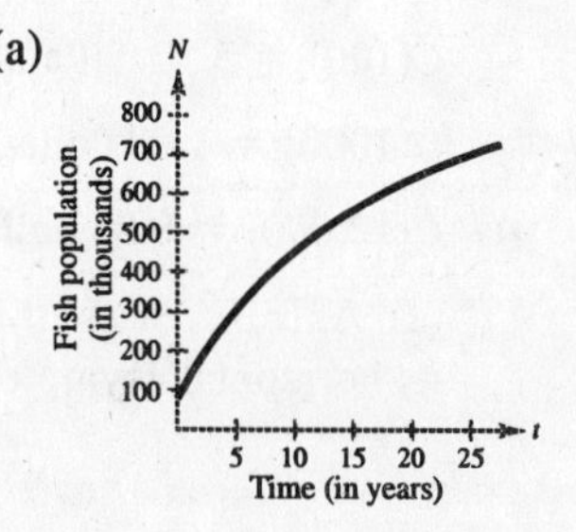

(b) $N(5) = 304{,}000$ fish

$N(10) \approx 453{,}333$ fish

$N(25) \approx 702{,}222$ fish

(c) The limit is

$\frac{60}{0.05} = 1{,}200{,}000$ fish,

the horizontal asymptote.

129. True, the graphs are the same. $x^2 = |x^2|$.

Chapter 2 Practice Test

1. Sketch the graph of $f(x) = x^2 - 6x + 5$ by hand and identify the vertex and the intercepts.

2. Find the number of units x that produce a minimum cost C if $C = 0.01x^2 - 90x + 15{,}000$.

3. Find the quadratic function that has a maximum at $(1, 7)$ and passes through the point $(2, 5)$.

4. Find two quadratic functions that have x-intercepts $(2, 0)$ and $\left(\frac{4}{3}, 0\right)$.

5. Use the leading Coefficient Test to determine the right-hand and left-hand behavior of the graph of the polynomial function $f(x) = -3x^5 + 2x^3 - 17$.

6. Find all the real zeros of $f(x) = x^5 - 5x^3 + 4x$. Verify your answer with a graphing utility.

7. Find the polynomial function with 0, 3, and -2 as zeros.

8. Sketch $f(x) = x^3 - 12x$ by hand.

9. Divide $3x^4 - 7x^2 + 2x - 10$ by $x - 3$ using long division.

10. Divide $x^3 - 11$ by $x^2 + 2x - 1$.

11. Use synthetic division to divide $3x^5 + 13x^4 + 12x - 1$ by $x + 5$.

12. Use synthetic division to find $f(-6)$ when $f(x) = 7x^3 + 40x^2 - 12x + 15$.

13. Find the real zeros of $f(x) = x^3 - 19x - 30$.

14. Find the real zeros of $f(x) = x^4 + x^3 - 8x^2 - 9x - 9$.

15. List all possible rational zeros of the function $f(x) = 6x^3 - 5x^2 + 4x - 15$.

16. Find the rational zeros of the polynomial $f(x) = x^3 - \frac{20}{3}x^2 + 9x - \frac{10}{3}$. product of linear factors.

17. Write $f(x) = x^4 + x^3 + 5x - 10$ as a

18. Find a polynomial with real coefficients that has 2, $3 + i$, and $3 - 2i$ as zeros.

19. Use synthetic division to show that $3i$ is a zero of $f(x) = x^3 + 4x^2 + 9x + 36$.

20. Find a mathematical model for the statement, "z varies directly as the square of x and inversely as the square root of y".

21. Sketch the graph of $f(x) = \dfrac{x - 1}{2x}$ and label all intercepts and asymptotes.

22. Sketch the graph of $f(x) = \dfrac{3x^2 - 4}{x}$ and label all intercepts and asymptotes.

23. Find all the asymptotes of $f(x) = \dfrac{8x^2 - 9}{x^2 + 1}$.

24. Find all the asymptotes of $f(x) = \dfrac{4x^2 - 2x + 7}{x - 1}$.

25. Sketch the graph of $f(x) = \dfrac{x - 5}{(x - 5)^2}$.

CHAPTER 3
Exponential and Logarithmic Functions

CHAPTER 3
Exponential and Logarithmic Functions

Section 3.1 Exponential Functions and Their Graphs

- You should know that a function of the form $y = a^x$, where $a > 0$, $a \neq 1$, is called an exponential function with base a.
- You should be able to graph exponential functions.
- You should be familiar with the number e and the natural exponential function $f(x) = e^x$.
- You should know formulas for compound interest.

  (a) For n compoundings per year: $A = P\left(1 + \frac{r}{n}\right)^{nt}$.

  (b) For continuous compoundings: $A = Pe^{rt}$.

Solutions to Odd-Numbered Exercises

1. $(3.4)^{6.8} \approx 4112.033$

3. $6^{2\pi} \approx 77{,}494.076$

5. $\sqrt[3]{7493} \approx 19.568$

7. $e^{1/2} \approx 1.649$

9. $e^{9.2} \approx 9897.129$

11.
$$\begin{aligned} f(x) &= 3^{x-2} \\ &= 3^x 3^{-2} \\ &= 3^x\left(\frac{1}{3^2}\right) \\ &= \frac{1}{9}(3^x) \\ &= h(x) \end{aligned}$$

Thus, $f(x) \neq g(x)$, but $f(x) = h(x)$. You can confirm your answer graphically by graphing f, g, and h in the same viewing rectangle.

13.
$$\begin{aligned} f(x) &= 16(4^{-x}) \\ &= 4^2(4^{-x}) \\ &= 4^{2-x} \\ &= \left(\frac{1}{4}\right)^{-(2-x)} \\ &= \left(\frac{1}{4}\right)^{x-2} \\ &= g(x) \end{aligned}$$

and

$$\begin{aligned} f(x) &= 16(4^{-x}) \\ &= 16(2^2)^{-x} \\ &= 16(2^{-2x}) \\ &= h(x) \end{aligned}$$

Thus, $f(x) = g(x) = h(x)$. You can confirm your answer graphically by graphing f, g, and h in the same viewing rectangle.

15. $f(x) = 2^x$ rises to the right.

Asymptote: $y = 0$
Intercept: $(0, 1)$
Matches graph (c).

17. $f(x) = 2^{-x}$ falls to the right.

Asymptote: $y = 0$
Intercept: $(0, 1)$
Matches graph (e).

19. $f(x) = 2^x - 4$ rises to the right.

Asymptote: $y = -4$
Intercept: $(0, -3)$
Matches graph (g).

21. $f(x) = -2^{x-2} = -(2^{x-2})$ falls to the right.

Asymptote: $y = 0$
Intercept: $(0, -2^{-2}) = \left(0, -\frac{1}{4}\right)$
Matches graph (a).

23. $f(x) = 3^x$

$g(x) = 3^{x-5} = f(x - 5)$

Horizontal shift five units to the right

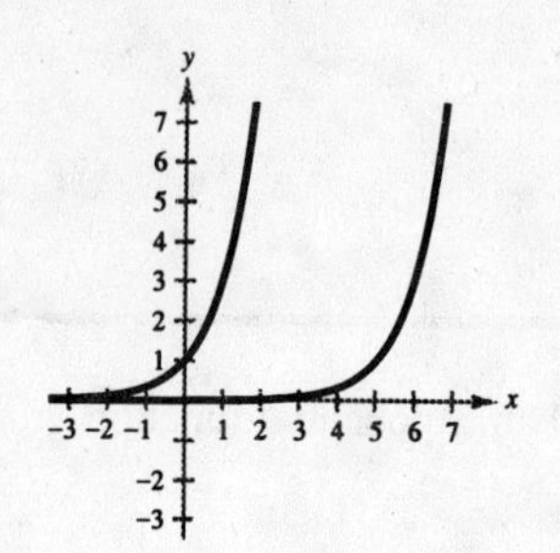

25. $f(x) = \left(\frac{3}{5}\right)^x$

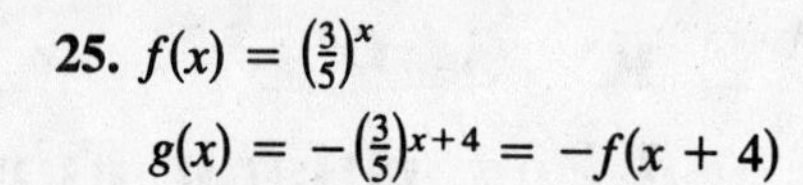

$g(x) = -\left(\frac{3}{5}\right)^{x+4} = -f(x + 4)$

Horizontal shift 4 units to the left, followed by reflection in x-axis.

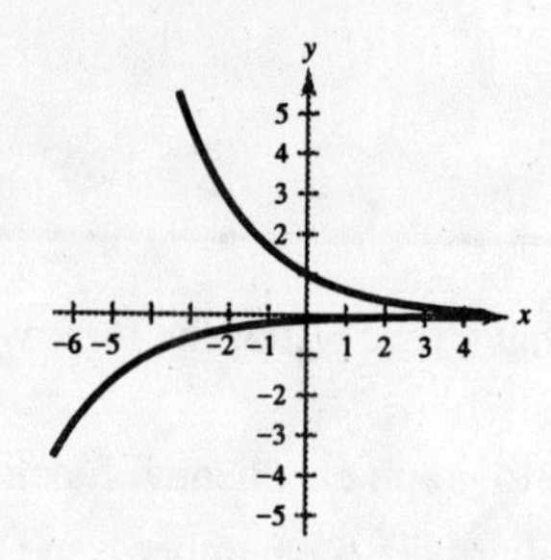

27. $g(x) = 5^x$

x	-2	-1	0	1	2
$g(x)$	$\frac{1}{25}$	$\frac{1}{5}$	1	5	25

(a) Asymptote: $y = 0$
(b) Intercept: $(0, 1)$
(c) Increasing

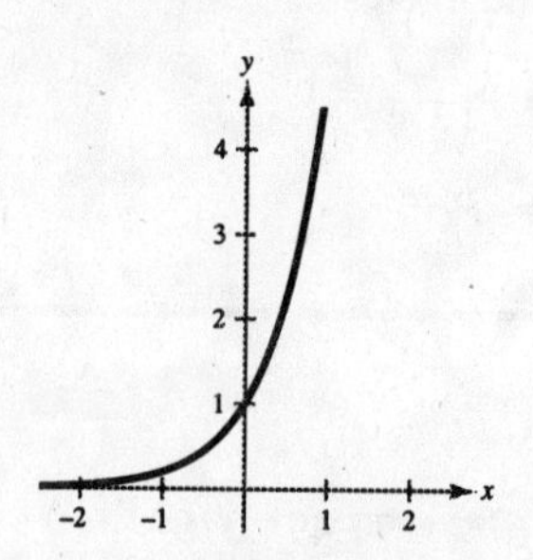

29. $f(x) = \left(\frac{1}{5}\right)^x = 5^{-x}$

x	-2	-1	0	1	2
y	25	5	1	$\frac{1}{5}$	$\frac{1}{25}$

(a) Asymptote: $y = 0$
(b) Intercepts: $(0, 1)$
(c) Decreasing

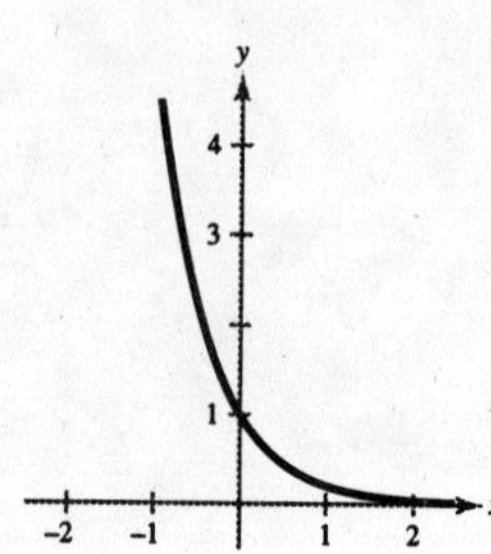

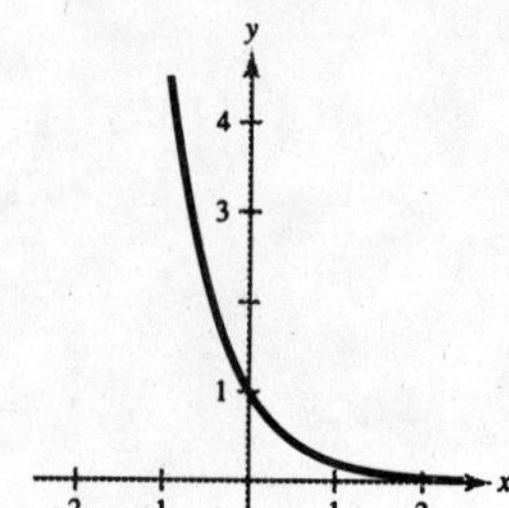

31. $h(x) = 5^{x-2}$

x	-1	0	1	2	3
y	$\frac{1}{125}$	$\frac{1}{25}$	$\frac{1}{5}$	1	5

(a) Asymptote: $y = 0$
(b) Intercepts: $\left(0, \frac{1}{25}\right)$
(c) Increasing

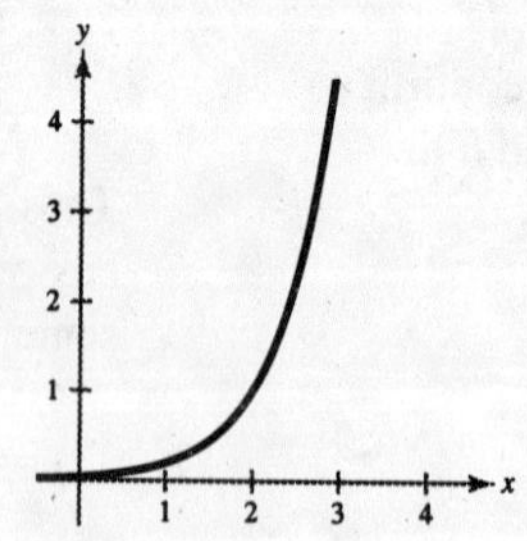

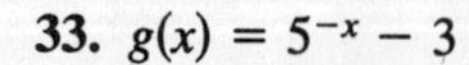

33. $g(x) = 5^{-x} - 3$

x	-1	0	1	2
y	2	-2	$-2\frac{4}{5}$	$-2\frac{24}{25}$

(a) Asymptote: $y = -3$
(b) Intercepts: $(0, -2), (-0.683, 0)$

(c) Decreasing

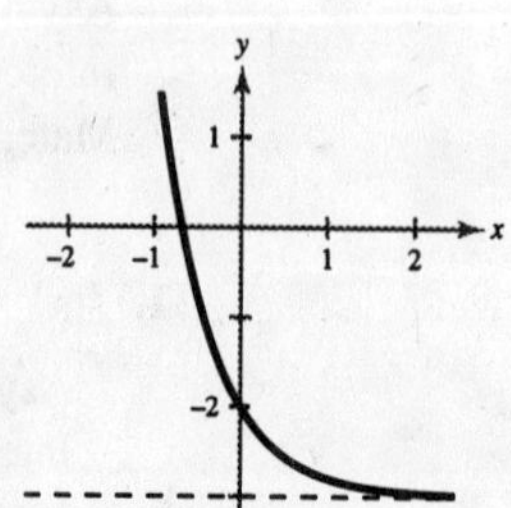

35. $f(x) = \left(\frac{5}{2}\right)^x$

x	-1	0	1	2	3
$f(x)$	0.4	1	2.5	6.25	15.625

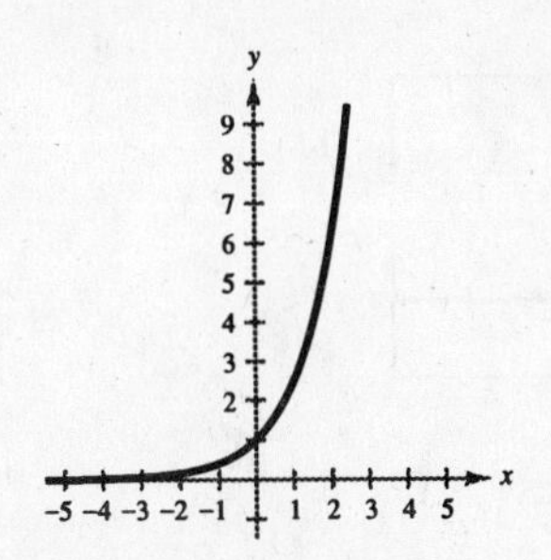

37. $f(x) = 6^x$

x	-1	0	1	2
$f(x)$	0.167	1	6	36

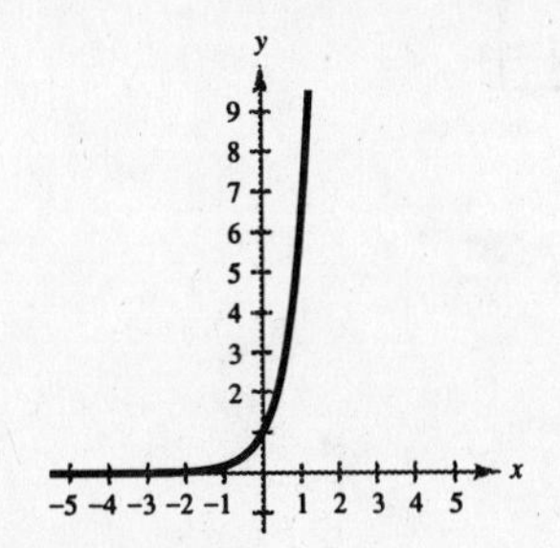

39. $f(x) = 3^{x+2} = 9 \cdot 3^x$

x	-3	-2	0	1
$f(x)$	$\frac{1}{3}$	1	9	27

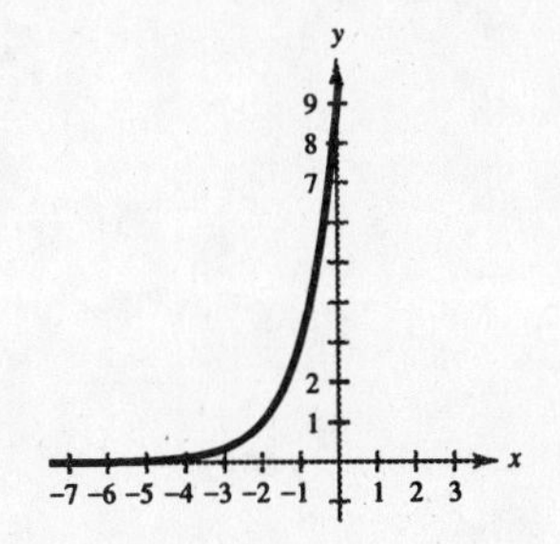

41. $f(x) = 3e^{x+4}$

x	-7	-6	-5	-4	-3
$f(x)$	0.149	0.406	1.104	3	8.155

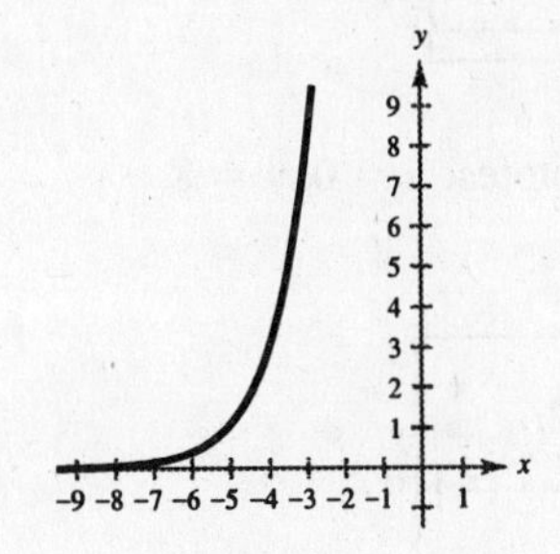

43. $f(x) = 2 + e^{x-5}$

x	2	3	4	5	6	7
$f(x)$	2.05	2.135	2.368	3	4.718	9.389

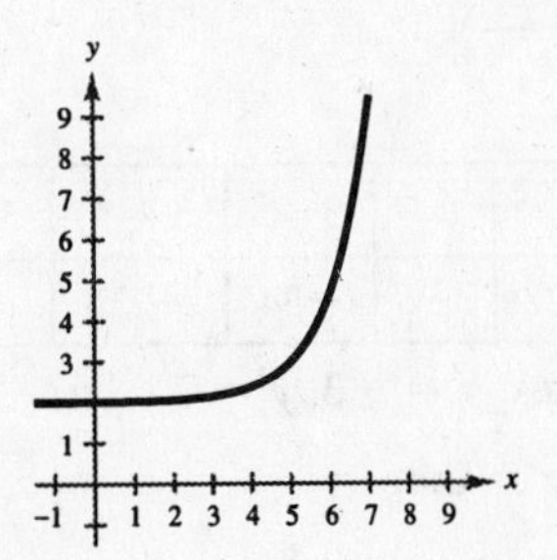

45. $y = 2^{-x^2}$

Asymptote: $y = 0$

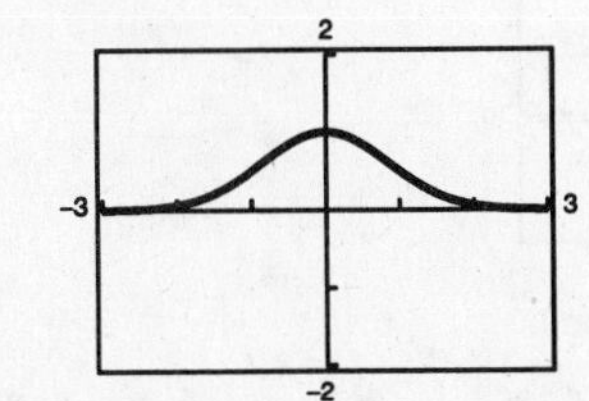

47. $f(x) = 3^{x-2} + 1$

Asymptote: $y = 1$

49. $y = 1.08^{-5x}$

Asymptote: $y = 0$

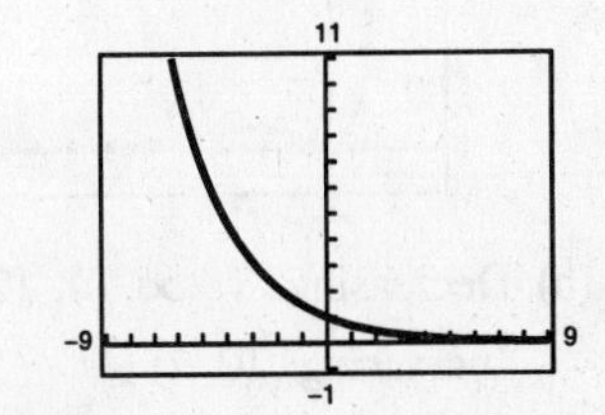

51. $S(t) = 3e^{-0.2t}$

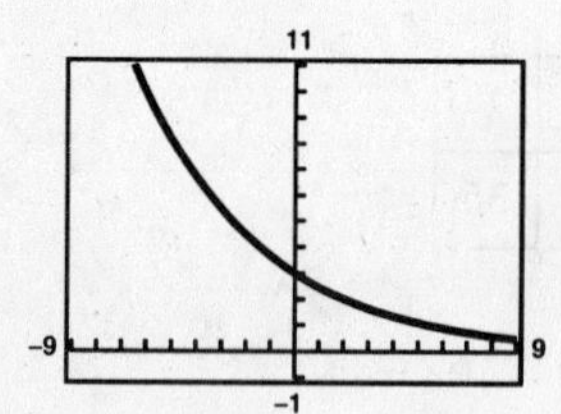

Asymptote: $S(t) = 0$

53. (a)

x	-1	-0.5	0	0.5	1
$f(x)$	$\frac{1}{3}$	0.577	1	1.732	3
$g(x)$	$\frac{1}{4}$	0.5	1	2	4

$4^x < 3^x$ when $x < 0$

(b)

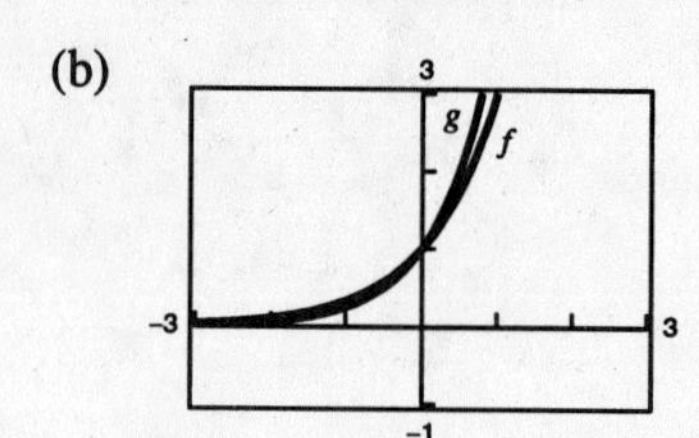

(i) $4^x < 3^x$ when $x > 0$

(ii) $4^x > 3$ when $x > 0$

55. (a)

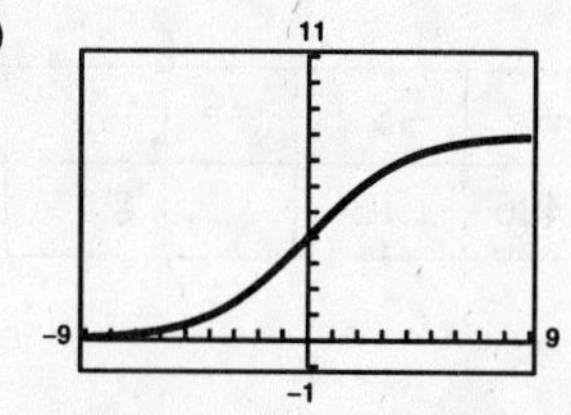

Horizontal asymptotes: $y = 0, y = 8$

(b)

x	-30	-20	-10	0	10	20	30
$f(x)$	≈ 0	≈ 0	0.05	4	7.95	≈ 8	≈ 8

57. (a)

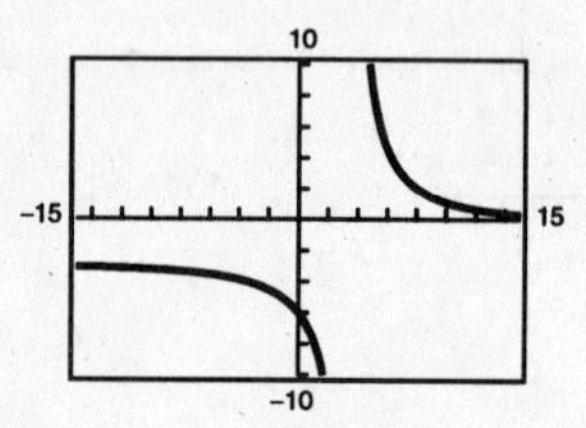

(b)

x	-20	-10	0	3	3.4	3.46	3.47	4	5	10	20
$f(x)$	-3.03	-3.22	-6	-34	-230	-2617	3516	26.6	8.4	1.11	0.11

Horizontal asymptotes: $y = -3, y = 0$

Veritcal asymptote: $x \approx 3.46$

59. $f(x) = x^2e^{-x}$

(a)

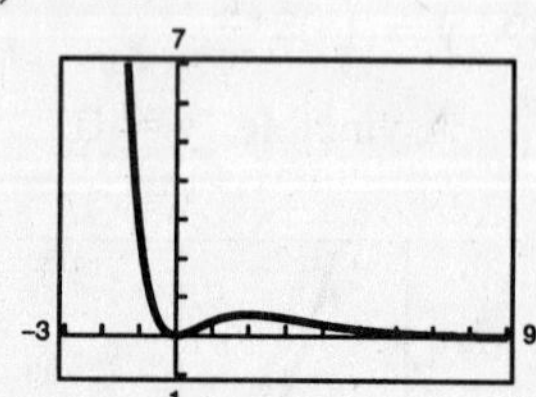

(b) Decreasing: $(-\infty, 0)$, $(2, \infty)$

Increasing: $(0, 2)$

(c) Relative maximum: $(2, 4e^{-2}) \approx (2, 0.541)$

Relative minimum: $(0, 0)$

61. $f(x) = x2^{3-x}$

(a)

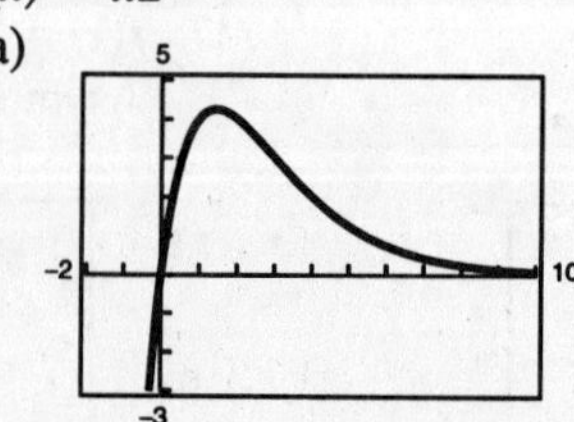

(b) Decreasing: $(1.44, \infty)$

Increasing: $(-\infty, 1.44)$

(c) Relative maximum: $(1.44, 4.25)$

63. $P = 2500, r = 8\% = 0.08, t = 10$

Compounded n times per year: $A = P\left(1 + \frac{r}{n}\right)^{nt} = 2500\left(1 + \frac{.08}{n}\right)^{10n}$

Compounded continuously: $A = Pe^{rt} = 2500e^{(.08)(10)}$

n	1	2	4	12	365	Continuous
A	5397.31	5477.81	5520.10	5549.10	5563.36	5563.85

65. $P = 2500, r = 8\% = 0.08, t = 20$

Compounded n times per year: $A = P\left(1 + \frac{r}{n}\right)^{nt} = 2500\left(1 + \frac{.08}{n}\right)^{20n}$

Compounded continuously: $A = Pe^{rt} = 2500e^{(.08)(20)}$

n	1	2	4	12	365	Continuous
A	11652.39	12002.55	12188.60	12317.01	12380.41	12382.58

67. $P = 12{,}000, r = 8\% = 0.08$, compounded continuously: $A = Pe^{rt} = 12{,}000e^{(.08)t}$

t	1	10	20	30	40	50
A	12,999.44	26,706.49	59,436.39	132,278.12	294,390.36	655,177.80

69. $P = 12{,}000, r = 6.5\% = 0.065, A = P\left(1 + \frac{r}{n}\right)^{nt} = 12{,}000\left(1 + \frac{0.065}{12}\right)^{12t}$

t	1	10	20	30	40	50
A	12,803.66	22,946.21	43,877.36	83,901.58	160,435.23	306,781.64

71. $P = 5000\left(1 - \frac{4}{4 + e^{-0.002x}}\right)$

(a)

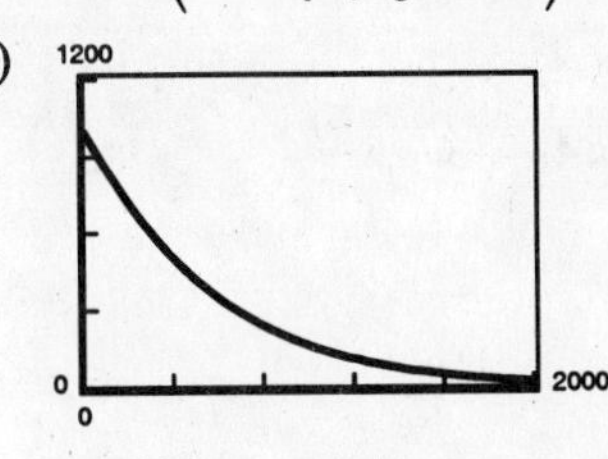

(b) If $x = 500$, $p \approx \$421.12$

(c) For $x = 600$, $p \approx \$350.13$.

73. (a)

2000
0
0
15

(b) $P(0) = 100$

$P(5) \approx 300$

$P(10) \approx 900$

(c) $P(0) = 100e^{0.2197(0)} = 100$

$P(5) = 100e^{0.2197(5)} = 299.966 \approx 300$

$P(10) = 100e^{0.2197(10)} = 899.798 \approx 900$

75. $Q = 25\left(\frac{1}{2}\right)^{t/1620}$

(a) When $t = 0$, $Q = 25\left(\frac{1}{2}\right)^{0/1620} = 25(1) = 25$ grams.

(b) When $t = 1000$, $Q = 25\left(\frac{1}{2}\right)^{1000/1620} \approx 16.30$ grams.

(c)

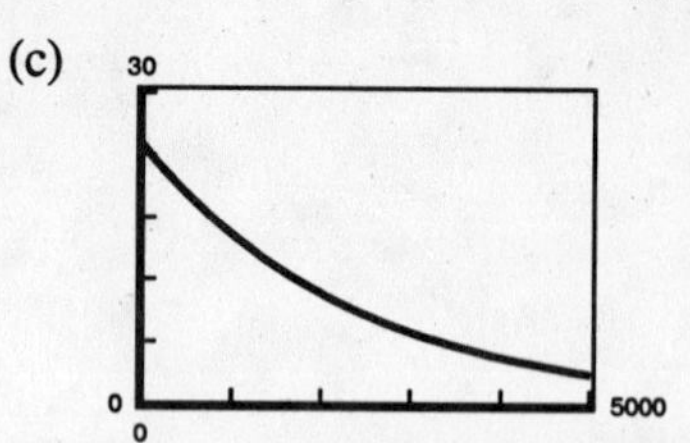

(d) No, $Q \to 0$ as $t \to \infty$, but Q never reaches 0.

77. (a) and (b)

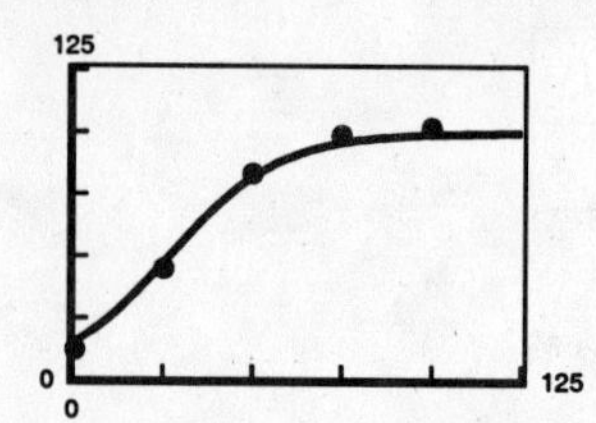

The model fits the data well.

(c)

x	0	25	50	75	100
y	15	47	82	96	99

(d) If $x = 36$, $y \approx 64.7\%$.

(e) If $y = 66.7\%$, $x \approx 37.4$.

79. (a)

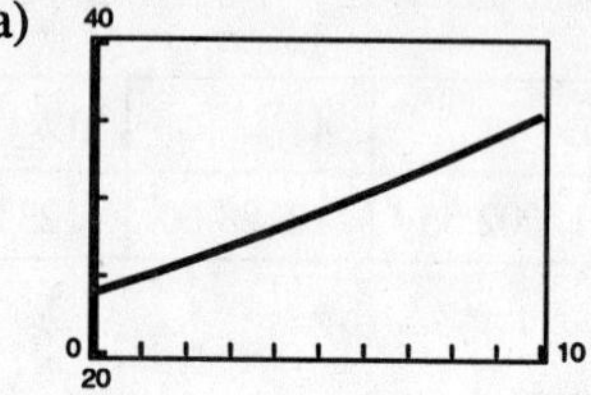

(b) $P(10) \approx 35.45$

(c) $P(10) = 23.95(1.04)^{10} \approx 35.45$

81. True. As $x \to -\infty, f(x) \to 0$

83.

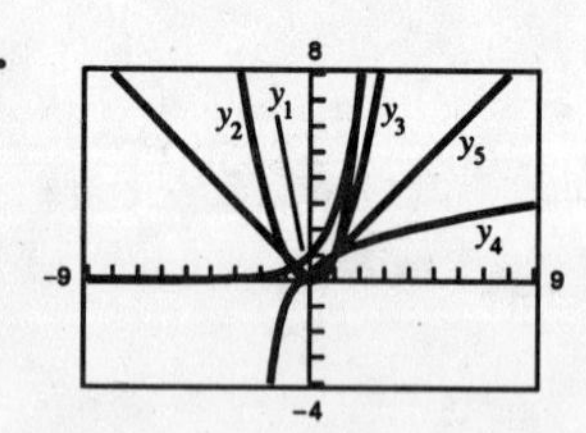

(a) $y_1 = e^x$ increases at the fastest rate.

(b) For any positive interger n, $e^x > x^n$ for x sufficiently large. That is, e^x grows faster than x^n.

(c) A quantity is growing exponentially if its growth rate is of the form $y = ce^{rx}$. This is a faster rate than any polynomial growth rate.

85. Since $\sqrt{2} \approx 1.414$, we know that $1 < \sqrt{2} < 2$.

Thus: $2^1 < 2^{\sqrt{2}} < 2^2$

$2 < 2^{\sqrt{2}} < 4$

87.

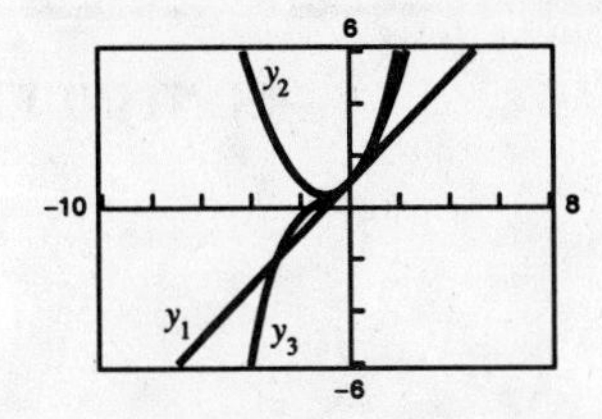

y_3 is the best approximation of $y = e^x$.

89. f is one-to-one, so it has an inverse.

$$f(x) = -\tfrac{2}{3}x + \tfrac{5}{2}$$
$$y = -\tfrac{2}{3}x + \tfrac{5}{2}$$
$$x = -\tfrac{2}{3}y + \tfrac{5}{2}$$
$$x - \tfrac{5}{2} = -\tfrac{2}{3}y$$
$$-\tfrac{3}{2}\left(x - \tfrac{5}{2}\right) = y$$
$$f^{-1}(x) = -\tfrac{3}{2}x + \tfrac{15}{4}$$

91. f is not one-to-one, so it does not have an inverse.

93. $f(x) = \dfrac{x^2 + 3}{x + 1} = x - 1 + \dfrac{4}{x + 1}$

Slant asymptote: $y = x - 1$

Vertical asymptote: $x = -1$

Intercept: $(0, 3)$

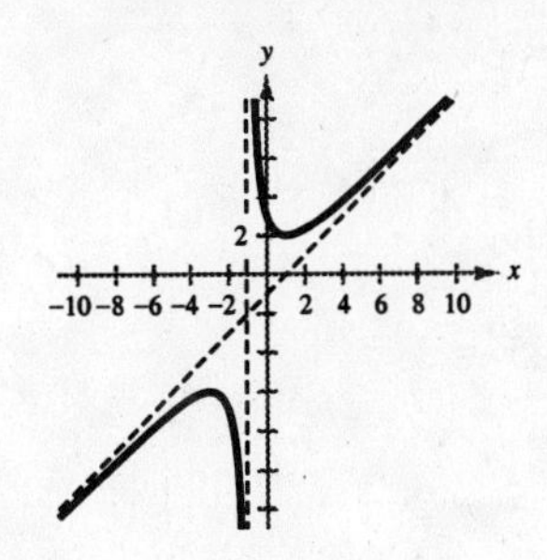

95. $f(x) = \dfrac{2x^3}{(x - 2)^2} = 2x + 8 + \dfrac{24x - 32}{x^2 - 4x + 4}$

Slant asymptote: $y = 2x + 8$

Vertical asymptote: $x = 2$

Intercept: $(0, 0)$

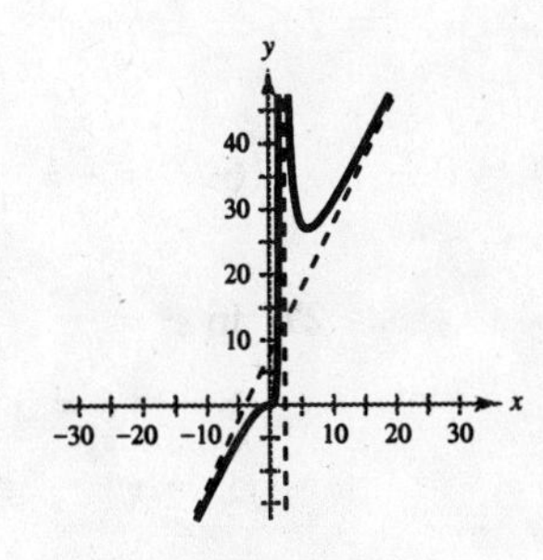

Section 3.2 Logarithmic Functions and Their Graphs

- You should know that a function of the form $y = \log_a x$, where $a > 0$, $a \neq 1$, and $x > 0$, is called a logarithm of x to base a.
- You should be able to convert from logarithmic form to exponential form and vice versa.

  $y = \log_a x \Leftrightarrow a^y = x$
- You should know the following properties of logarithms.

  (a) $\log_a 1 = 0$ since $a^0 = 1$.

  (b) $\log_a a = 1$ since $a^1 = a$.

  (c) $\log_a a^x = x$ since $a^x = a^x$.

  (d) If $\log_a x = \log_a y$, then $x = y$.

—CONTINUED—

—CONTINUED—

- You should know the definition of the natural logarithmic function.

  $\log_e x = \ln x, x > 0$
- You should know the properties of the natural logarithmic function.
  (a) $\ln 1 = 0$ since $e^0 = 1$.
  (b) $\ln e = 1$ since $e^1 = e$.
  (c) $\ln e^x = x$ since $e^x = e^x$.
  (d) If $\ln x = \ln y$, then $x = y$.
- You should be able to graph logarithmic functions.

Solutions to Odd-Numbered Exercises

1. $\log_4 64 = 3 \Rightarrow 4^3 = 64$

3. $\log_7 \frac{1}{49} = -2 \Rightarrow 7^{-2} = \frac{1}{49}$

5. $\log_{32} 4 = \frac{2}{5} \Rightarrow 32^{2/5} = 4$

7. $\ln 1 = 0 \Rightarrow e^0 = 1$

9. $5^3 = 125 \Rightarrow \log_5 125 = 3$

11. $81^{1/4} = 3 \Rightarrow \log_{81} 3 = \frac{1}{4}$

13. $6^{-2} = \frac{1}{36} \Rightarrow \log_6 \frac{1}{36} = -2$

15. $e^3 = 20.0855\ldots \Rightarrow \ln 20.0855\ldots = 3$

17. $e^{2.6} = 13.463\ldots \Rightarrow \ln 13.463\ldots = 2.6$

19. $\log_2 16 = \log_2 2^4 = 4$

21. $\log_{16}\left(\frac{1}{4}\right) = \log_{16}1 - \log_{16}4 = 0 - \log_{16}16^{1/2} = -\frac{1}{2}$

23. $\log_{10} 0.01 = \log_{10} 10^{-2} = -2$

25. $\log_7 x = \log_7 9$

$x = 9$

27. $\ln e^8 = x$

$8 \cdot \ln e = x$

$8 = x$

29. $\log_6 6^2 = x$

$2\log_6 6 = x$

$2 = x$

31. $\log_{10} 345 \approx 2.538$

33. $\log_{10}\left(\frac{4}{5}\right) \approx -0.097$

35. $\ln\left(4 + \sqrt{3}\right) \approx 1.746$

37. $\ln\sqrt{42} \approx 1.869$

39. $6 \log_{10} 14.8 \approx 7.022$

41. $12 \ln 6.4 \approx 22.276$

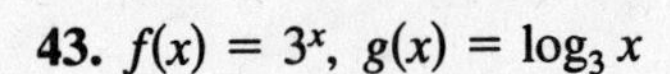

43. $f(x) = 3^x,\ g(x) = \log_3 x$

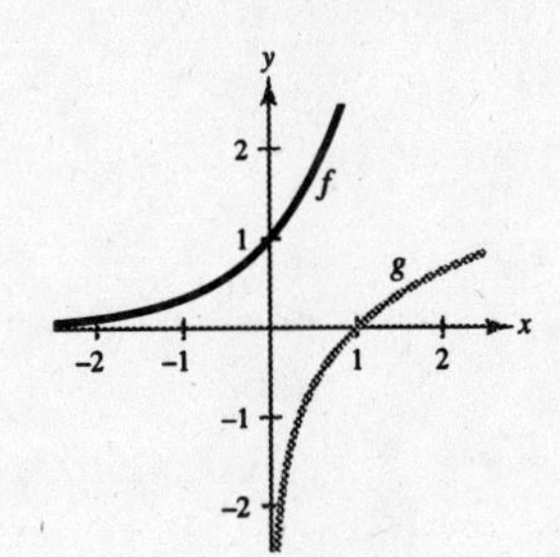

f and g are inverses. Their graphs are reflected about the line $y = x$.

45. $f(x) = e^x,\ g(x) = \ln x$

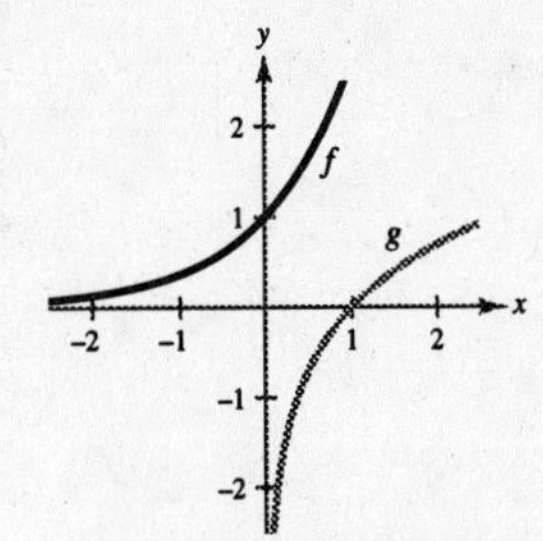

f and g are inverses. Their graphs are reflected about the line $y = x$.

47. $f(x) = \log_3 x + 2$
Asymptote: $x = 0$
Point on graph: $(1, 2)$
Matches graph (c).

49. $f(x) = -\log_3(x + 2)$
Asymptote: $x = -2$
Point on graph: $(-1, 0)$
Matches graph (d).

51. $f(x) = \log_3(1 - x)$
Asymptote: $x = 1$
Point on graph: $(0, 0)$
Matches graph (b).

53. $f(x) = \log_4 x$
Domain: $x > 0 \implies$ The domain is $(0, \infty)$.
Vertical asymptote: $x = 0$
x-intercept: $(1, 0)$
$y = \log_4 x \implies 4^y = x$

x	$\frac{1}{4}$	1	4	2
y	-1	0	1	$\frac{1}{2}$

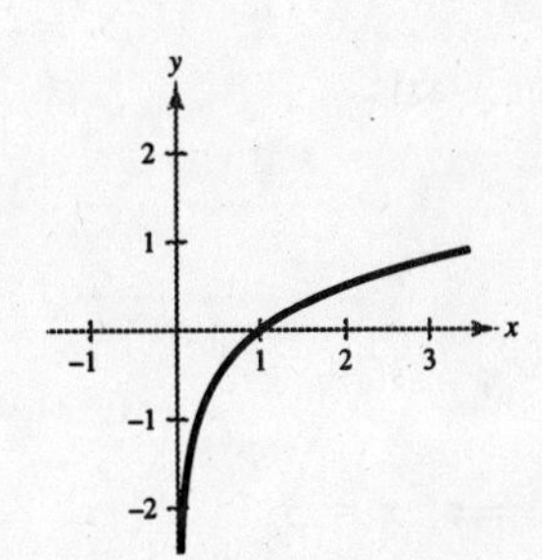

55. $h(x) = \log_4(x - 3)$
Domain: $x - 3 > 0$ or $(3, \infty)$
Vertical asymptote: $x = 3$
Intercept: $(4, 0)$

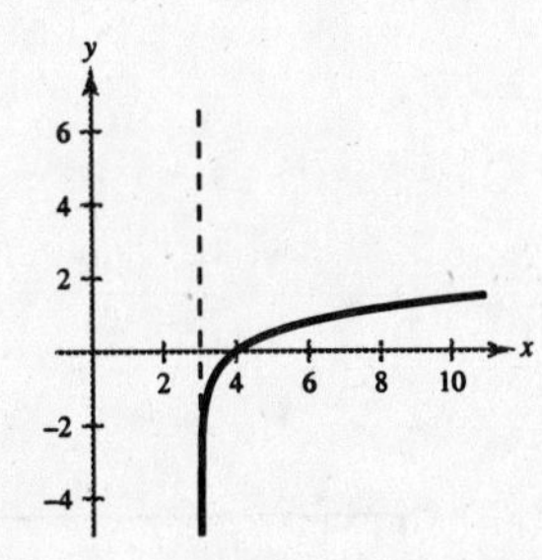

57. $y = -\log_3 x + 2$
Domain: $(0, \infty)$
Vertical asymptote: $x = 0$
x-intercept:
$$\begin{aligned} -\log_3 x + 2 &= 0 \\ 2 &= \log_3 x \\ 3^2 &= x \\ 9 &= x \end{aligned}$$

The x-intercept is $(9, 0)$.
$y = -\log_3 x + 2$
$\log_3 x = 2 - y \implies 3^{2-y} = x$

x	27	9	3	1	$\frac{1}{3}$
y	-1	0	1	2	3

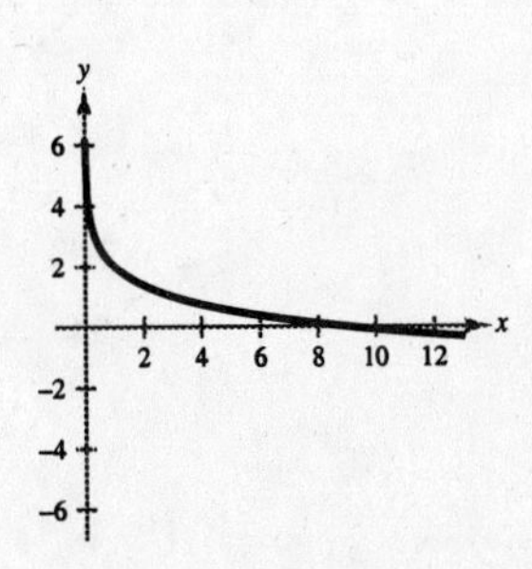

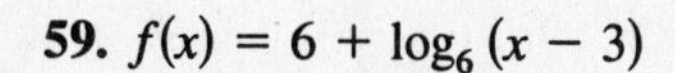

59. $f(x) = 6 + \log_6 (x - 3)$

Domain: $(3, \infty)$

Vertical asymptote: $x = 3$

x-intercept: $\log_6(x - 3) = -6$

$$6^{-6} = x - 3$$

$$x = 3 + 6^{-6} \approx 3$$

x	4	9	$3\frac{1}{6}$
y	6	7	5

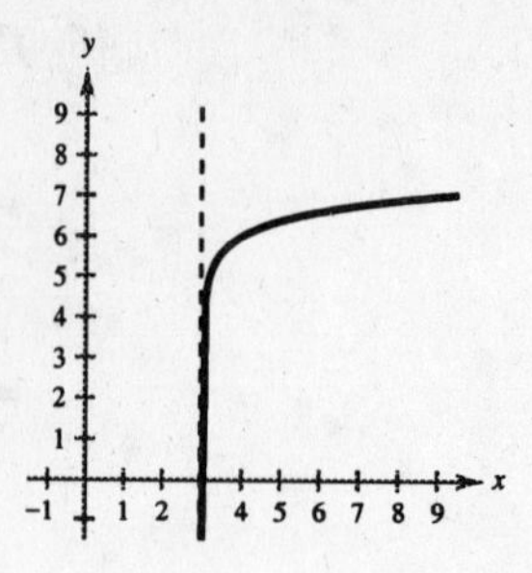

61. $y = \log_{10}\left(\frac{x}{5}\right)$

Domain: $\frac{x}{5} > 0 \Rightarrow x > 0$

The domain is $(0, \infty)$.

Vertical asymptote: $\frac{x}{5} = 0 \Rightarrow x = 0$

The vertical asymptote is the y-axis.

x-intercept: $\log_{10}\left(\frac{x}{5}\right) = 0$

$$\frac{x}{5} = 10^0$$

$$\frac{x}{5} = 1 \Rightarrow x = 5$$

The x-intercept is $(5, 0)$.

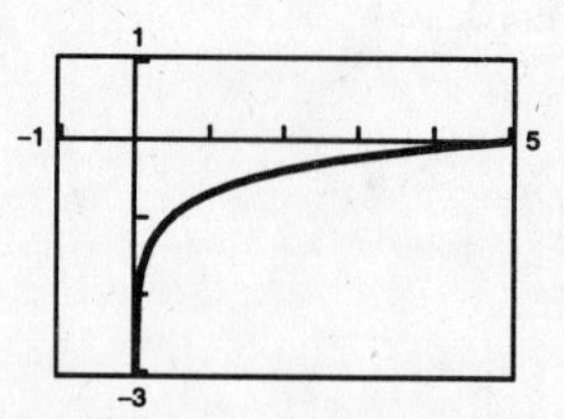

63. $f(x) = \ln(x - 2)$

Domain: $x - 2 > 0 \Rightarrow x > 2$

The domain is $(2, \infty)$.

Vertical asymptote: $x - 2 = 0 \Rightarrow x = 2$

x-intercept: $0 = \ln(x - 2)$

$$e^0 = x - 2$$

$$3 = x$$

The x-intercept is $(3, 0)$.

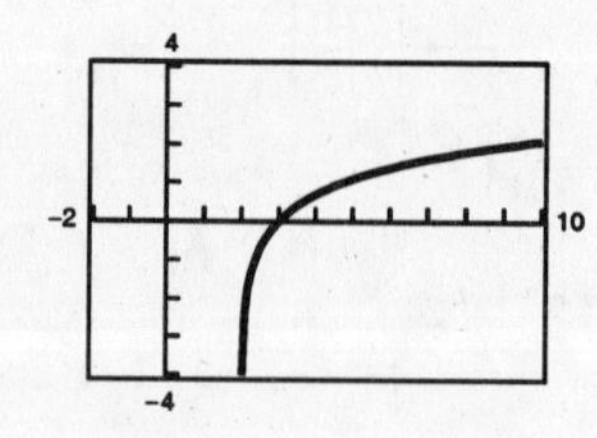

65. $g(x) = \ln(-x)$

Domain: $-x > 0 \implies x < 0$

The domain is $(-\infty, 0)$.

Vertical asymptote: $-x = 0 \implies x = 0$

x-intercept: $0 = \ln(-x)$

$e^0 = -x$

$-1 = x$

The x-intercept is $(-1, 0)$.

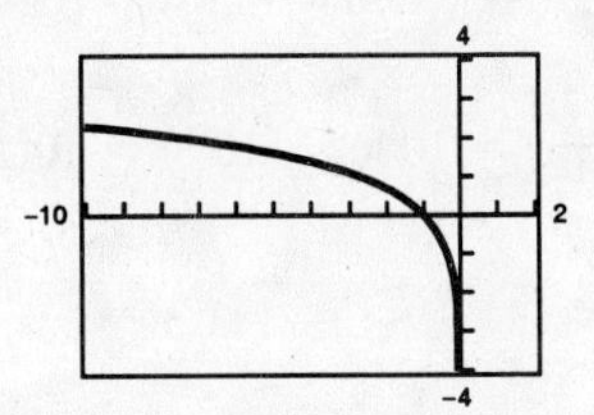

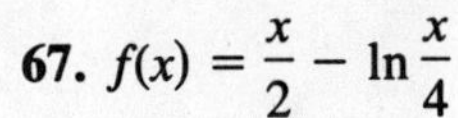

67. $f(x) = \dfrac{x}{2} - \ln\dfrac{x}{4}$

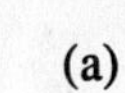

(a)

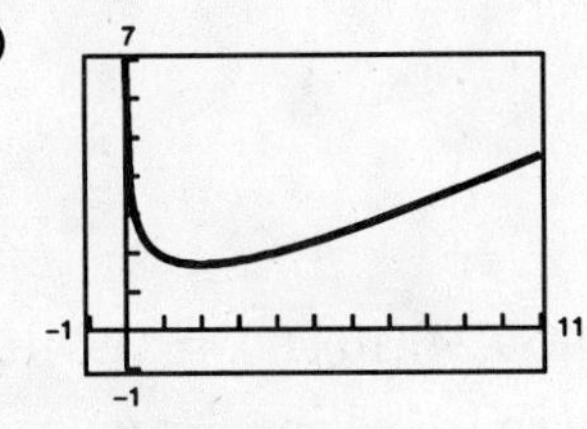

(b) Domain: $(0, \infty)$

(c) Increasing on $(2, \infty)$

Decreasing on $(0, 2)$

(d) Relative minimum: $(2, 1.693)$

69. $h(x) = 4x \ln x$

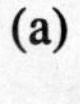

(a)

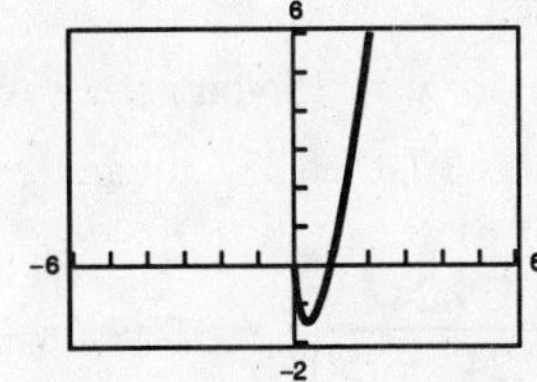

(b) Domain: $(0, \infty)$

(c) Increasing on $(0.368, \infty)$

Decreasing on $(0, 0.368)$

(d) Relative minimum: $(0.368, -1.472)$

71. $t = \dfrac{10 \ln 2}{\ln 67 - \ln 50} \approx 23.68$ years

73. (a) $f(0) = 80 - 17 \log_{10}(0 + 1) = 80$

(b) $f(4) = 80 - 17 \log_{10}(4 + 1) \approx 68.1$

(c) $f(10) = 80 - 17 \log_{10}(10 + 1) \approx 62.3$

(d)

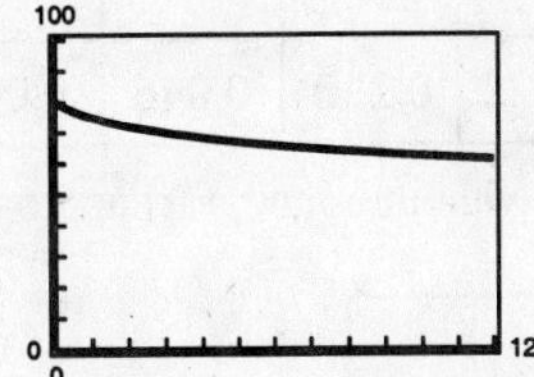

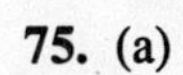

75. (a)

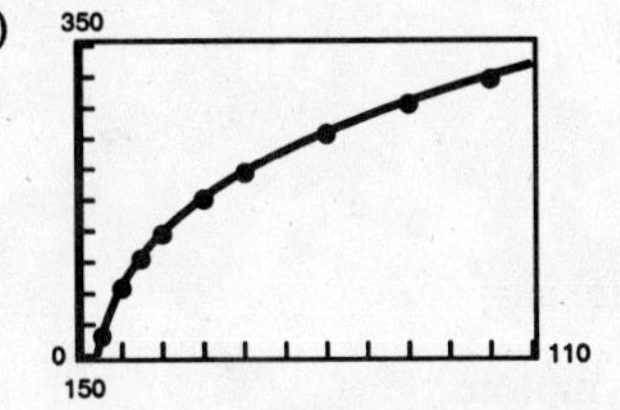

(b) $T > 300°$F when $p > 67.3$ pounds per square inch

[The graph of T and $y = 300$ intersect at $p = 67.3$.]

(c) $T(74) = 306.48°$F

77.

r	0.005	0.010	0.015
t	138.6 yr	69.3 yr	46.2 yr

r	0.020	0.025	0.030
t	34.7 yr	27.7 yr	23.1 yr

The doubling time decreases as r increases.

79. $\beta = 10 \log_{10}\left(\frac{I}{10^{-12}}\right) =$

(a) $I = 1: \beta = 10 \log_{10}\left(\frac{1}{10^{-12}}\right) = 10 \cdot \log_{10}(10^{12}) = 10(12) = 120$ decibels

(b) $I = 10^{-2}: \beta = 10 \log_{10}\left(\frac{10^{-2}}{10^{-12}}\right) = 10 \log_{10}(10^{10}) = 10(10) = 100$ decibels

(c) No, this is a logarithmic scale.

81. $y = 80.4 - 11 \ln x, \quad 100 \le x \le 1500$

(a) $\dfrac{450 \text{ cubic ft per minute}}{30 \text{ children}} = 15$ cubic feet per minute per child

(b) From the graph, for $y = 15$ you get $x \approx 382$ cubic feet.

(c) If ceiling height is 30, then 382 square feet of floor space is needed.

83. $t = 16.625 \ln\left(\frac{1659.24}{1659.24 - 750}\right) \approx 10$ years

85. Total amount $= (1659.24)(10)(12) \approx \$199,108.80$

Interest $= 199,108.80 - 150,000 = \$49,108.80$

87. True. $\log_3(27) = \log_3 3^3 = 3$

89.

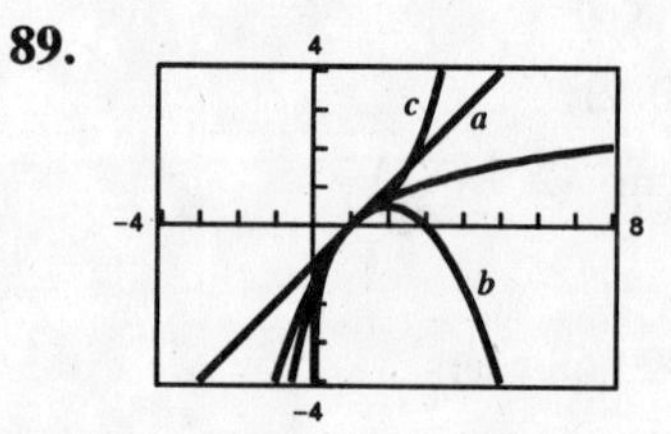

91. (a)

x	1	5	10	10^2	10^4	10^6
$f(x)$	0	0.322	0.230	0.046	0.00092	0.0000138

(b) As x increases without bound, $f(x)$ approaches 0.

(c)

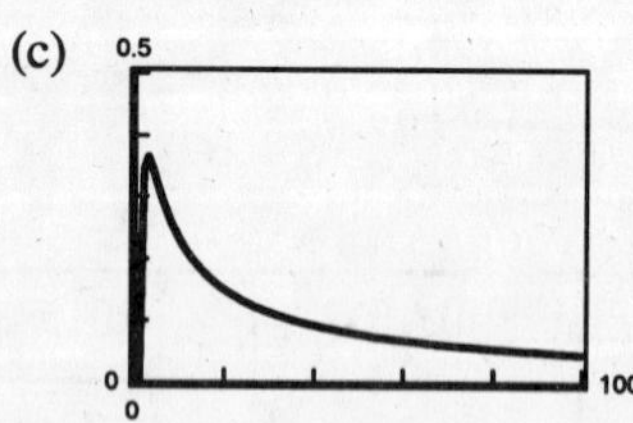

93. Vertical asymptote: $x = 0$

No horizontal asymptote

95. Vertical asymptote: $x = 7$

No horizontal asymptote

97. $e^{7/2} \approx 33.115$

99. $6e^{-8} \approx 0.002$

Section 3.3 Properties of Logarithms

- You should know the following properties of logarithms.

  (a) $\log_a x = \dfrac{\log_b x}{\log_b a}$

  (b) $\log_a (uv) = \log_a u + \log_a v$ $\quad\ln (uv) = \ln u + \ln v$

  (c) $\log_a (u/v) = \log_a u - \log_a v$ $\quad\ln (u/v) = \ln u - \ln v$

  (d) $\log_a u^n = n \log_a u$ $\quad\ln u^n = n \ln u$
- You should be able to rewrite logarithmic expressions using these properties.

Solutions to Odd-Numbered Exercises

1. $f(x) = \log_{10} x$

$g(x) = \dfrac{\ln x}{\ln 10}$

$f(x) = g(x)$

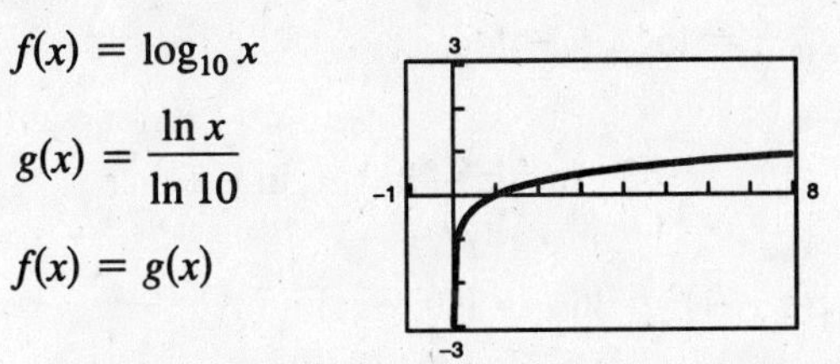

3. $\log_3 7 = \dfrac{\ln 7}{\ln 3} \approx 1.771$

5. $\log_{1/2} 4 = \dfrac{\ln 4}{\ln \frac{1}{2}} = -2$

7. $\log_9(0.8) = \dfrac{\ln(0.8)}{\ln 9} \approx -0.102$

9. $\log_{15} 1460 = \dfrac{\ln 1460}{\ln 15} \approx 2.691$

11. (a) $\log_5 x = \dfrac{\log_{10} x}{\log_{10} 5}$

(b) $\log_5 x = \dfrac{\ln x}{\ln 5}$

13. (a) $\log_{1/5} x = \dfrac{\log_{10} x}{\log_{10} \frac{1}{5}} = \dfrac{\log_{10} x}{-\log_{10} 5}$

(b) $\log_{1/5} x = \dfrac{\ln x}{\ln \frac{1}{5}} = \dfrac{\ln x}{-\ln 5}$

15. (a) $\log_a\left(\dfrac{3}{10}\right) = \dfrac{\log_{10}\left(\frac{3}{10}\right)}{\log_{10} a}$

(b) $\log_a\left(\dfrac{3}{10}\right) = \dfrac{\ln\left(\frac{3}{10}\right)}{\ln a}$

17. (a) $\log_{2.6} x = \dfrac{\log_{10} x}{\log_{10} 2.6}$

(b) $\log_{2.6} x = \dfrac{\ln x}{\ln 2.6}$

19. $f(x) = \log_2 x = \dfrac{\ln x}{\ln 2}$

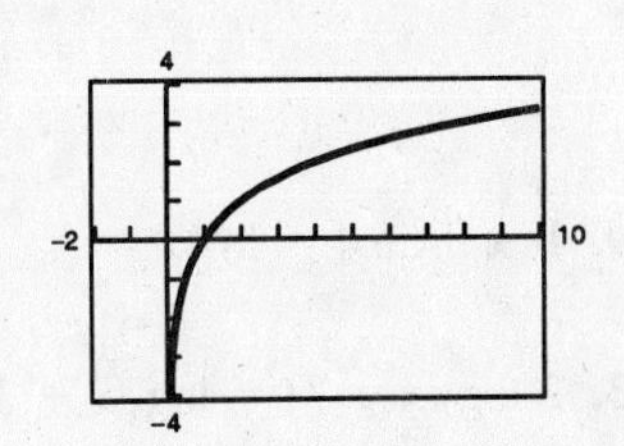

21. $f(x) = \log_{1/2} x = \dfrac{\ln x}{\ln \frac{1}{2}} = -\dfrac{\ln x}{\ln 2}$

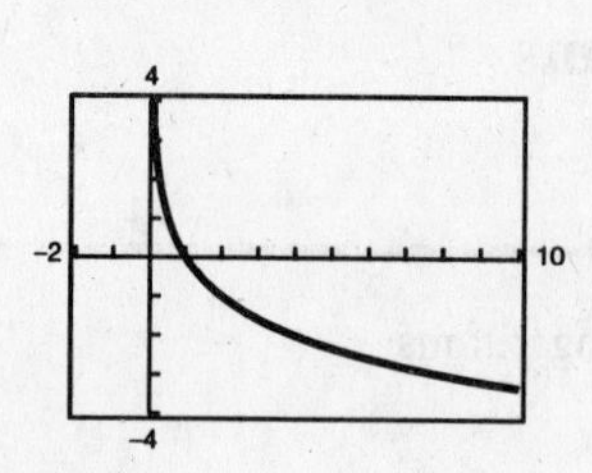

23. $f(x) = \log_{11.8} x = \dfrac{\ln x}{\ln 11.8}$

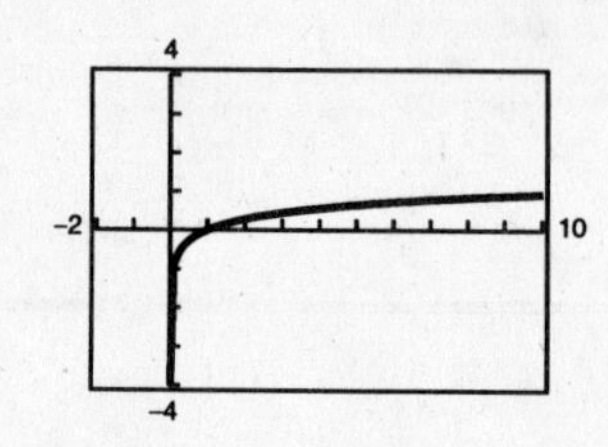

25. $f(x) = \log_3 x^{1/2} = \dfrac{1}{2}\dfrac{\ln x}{\ln 3}$

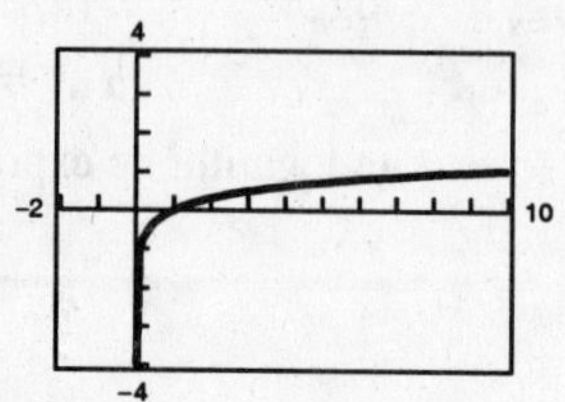

27. $\log_{10} 5x = \log_{10} 5 + \log_{10} x$

29. $\log_{10} \dfrac{5}{x} = \log_{10} 5 - \log_{10} x$

31. $\log_8 x^4 = 4 \log_8 x$

33. $\ln\sqrt{z} = \ln z^{1/2} = \frac{1}{2} \ln z$

35. $\ln xyz = \ln x + \ln y + \ln z$

37. $\ln\sqrt{a-1} = \frac{1}{2} \ln(a-1)$

39. $\ln z(z-1)^2 = \ln z + \ln(z-1)^2$
$= \ln z + 2\ln(z-1)$

41. $\ln \sqrt[3]{\dfrac{x}{y}} = \dfrac{1}{3} \ln \dfrac{x}{y}$
$= \dfrac{1}{3}[\ln x - \ln y]$
$= \dfrac{1}{3} \ln x - \dfrac{1}{3} \ln y$

43. $\ln\left(\dfrac{x^4\sqrt{y}}{z^5}\right) = \ln x^4\sqrt{y} - \ln z^5$
$= \ln x^4 + \ln\sqrt{y} - \ln z^5$
$= 4 \ln x + \dfrac{1}{2} \ln y - 5 \ln z$

45. $\log_b\left(\dfrac{x^2}{y^2z^3}\right) = \log_b x^2 - \log_b y^2z^3$
$= \log_b x^2 - [\log_b y^2 + \log_b z^3]$
$= 2 \log_b x - 2 \log_b y - 3 \log_b z$

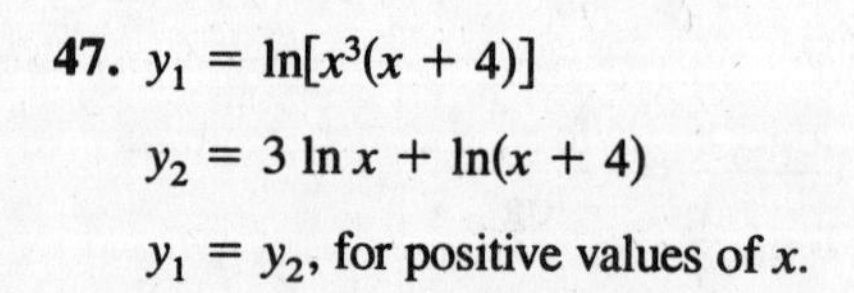

47. $y_1 = \ln[x^3(x+4)]$
$y_2 = 3 \ln x + \ln(x+4)$
$y_1 = y_2$, for positive values of x.

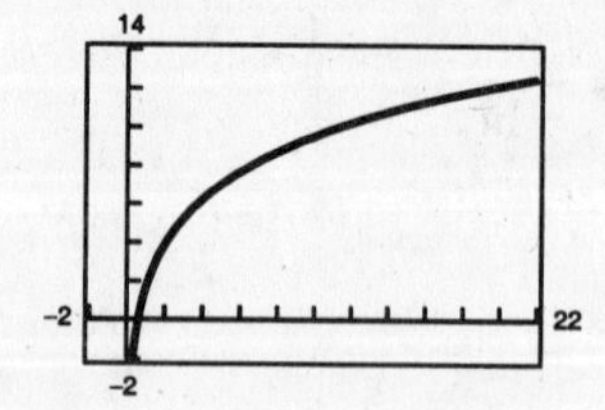

49. $\ln x + \ln 4 = \ln 4x$

51. $\log_4 z - \log_4 y = \log_4 \dfrac{z}{y}$

53. $2 \log_2(x+3) = \log_2(x+3)^2$

55. $\frac{1}{3}\log_3 7x = \log_3(7x)^{1/3} = \log_3 \sqrt[3]{7x}$

57. $\ln x - 3\ln(x+1) = \ln x - \ln(x+1)^3$
$= \ln \dfrac{x}{(x+1)^3}$

59. $\ln(x-2) - \ln(x+2) = \ln\left(\dfrac{x-2}{x+2}\right)$

61. $\ln x - 2[\ln(x+2) + \ln(x-2)] = \ln x - 2\ln[(x+2)(x-2)]$
$= \ln x - 2\ln(x^2 - 4)$
$= \ln x - \ln(x^2 - 4)^2$
$= \ln \dfrac{x}{(x^2 - 4)^2}$

63. $\frac{1}{3}[2\ln(x+3) + \ln x - \ln(x^2 - 1)] = \frac{1}{3}[\ln(x+3)^2 + \ln x - \ln(x^2 - 1)]$
$= \frac{1}{3}[\ln[x(x+3)^2] - \ln(x^2 - 1)]$
$= \frac{1}{3}\ln \dfrac{x(x+3)^2}{x^2 - 1}$
$= \ln \sqrt[3]{\dfrac{x(x+3)^2}{x^2 - 1}}$

65. $\frac{1}{3}[\ln y + 2\ln(y+4)] - \ln(y-1) = \frac{1}{3}[\ln y + \ln(y+4)^2] - \ln(y-1)$
$= \frac{1}{3}\ln[y(y+4)^2] - \ln(y-1)$
$= \ln \sqrt[3]{y(y+4)^2} - \ln(y-1)$
$= \ln \dfrac{\sqrt[3]{y(y+4)^2}}{y-1}$

67. $2\ln 3 - \frac{1}{2}\ln(x^2 + 1) = \ln 3^2 - \ln\sqrt{x^2 + 1}$
$= \ln \dfrac{9}{\sqrt{x^2 + 1}}$

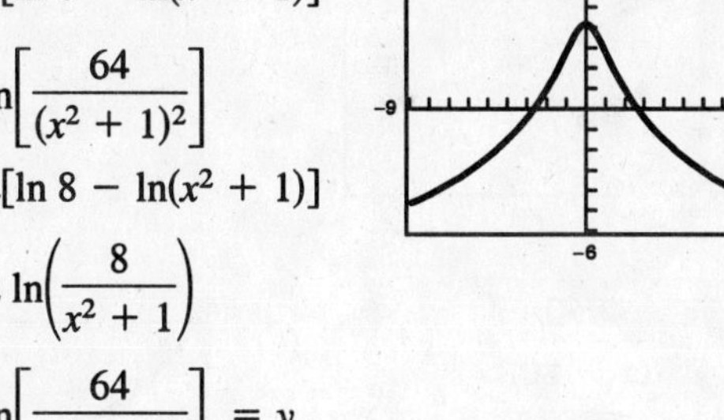

69. $y_1 = 2[\ln 8 - \ln(x^2 + 1)]$

$y_2 = \ln\left[\dfrac{64}{(x^2 + 1)^2}\right]$

$y_1 = 2[\ln 8 - \ln(x^2 + 1)]$
$= 2\ln\left(\dfrac{8}{x^2 + 1}\right)$
$= \ln\left[\dfrac{64}{(x^2 + 1)^2}\right] = y_2$

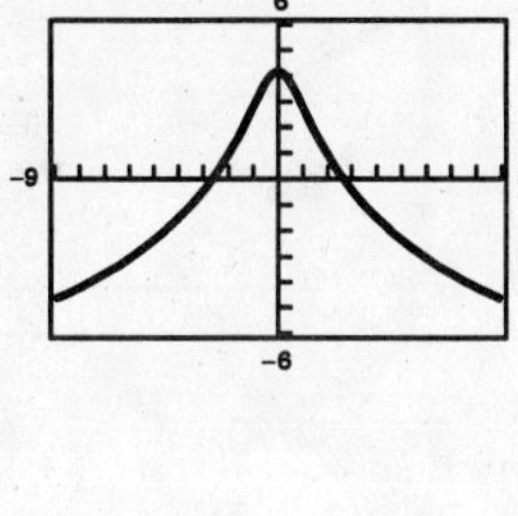

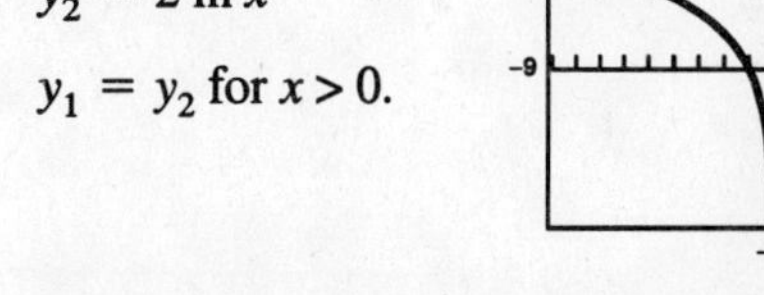

71. $y_1 = \ln x^2$

$y_2 = 2\ln x$

$y_1 = y_2$ for $x > 0$.

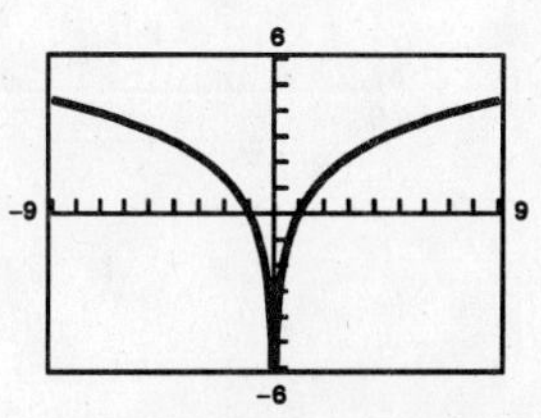

They are not equivalent. The domain of $f(x)$ is all real numbers except 0. The domain of $g(x)$ is $x > 0$.

73. $\log_3 9 = 2\log_3 3 = 2$

75. $\log_4 16^{3.4} = 3.4\log_4(4^2) = 6.8\log_4 4 = 6.8$

77. $\log_2(-4)$ is undefined. -4 is not in the domain of $f(x) = \log_2 x$

79. $\log_5 75 - \log_5 3 = \log_5 \frac{75}{3} = \log_5 25 = \log_5 5^2 = 2$

81. $\ln e^3 - \ln e^7 = 3 - 7 = -4$

83. $\log_{10} 0$ is undefined. 0 is not in the domain of $\log_{10} x$.

85. $\ln e^{8.5} = 8.5$

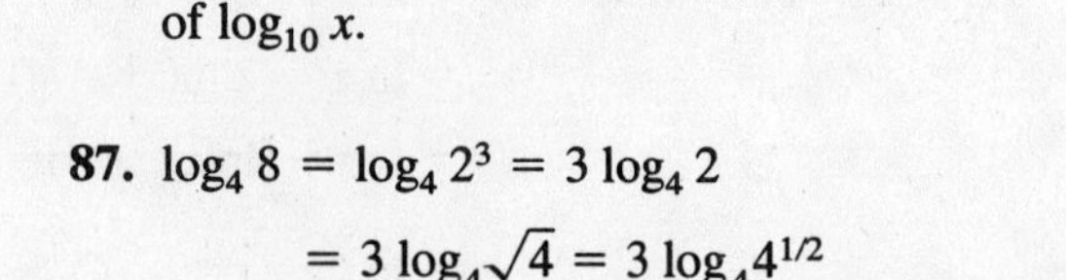

87. $\log_4 8 = \log_4 2^3 = 3 \log_4 2$

$= 3 \log_4 \sqrt{4} = 3 \log_4 4^{1/2}$

89. $\log_7 \sqrt{70} = \frac{1}{2} \log_7 70 = \frac{1}{2} \log_7(10 \cdot 7)$

$= \frac{1}{2} \log_7 10 + \frac{1}{2} \log_7 7 = \frac{1}{2} \log_7 10 + \frac{1}{2}$

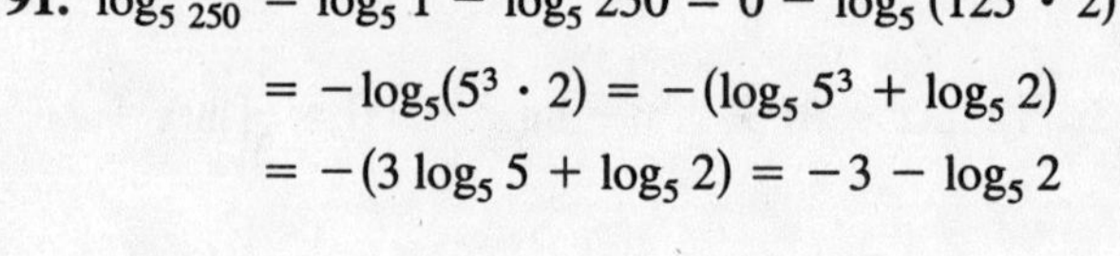

91. $\log_5 \frac{1}{250} = \log_5 1 - \log_5 250 = 0 - \log_5 (125 \cdot 2)$

$= -\log_5(5^3 \cdot 2) = -(\log_5 5^3 + \log_5 2)$

$= -(3 \log_5 5 + \log_5 2) = -3 - \log_5 2$

93. $\ln(5e^6) = \ln 5 + \ln e^6 = \ln 5 + 6 = 6 + \ln 5$

95. (a) $\beta = 10 \cdot \log_{10}\left(\frac{I}{10^{-12}}\right) = 10(\log_{10} I - \log_{10} 10^{-12})$

$= 10[\log_{10} I - (-12) \log_{10} 10]$

$= 10 (\log_{10} I + 12) = 120 + 10 \cdot \log_{10} I$

(b)

I	10^{-4}	10^{-6}	10^{-8}	10^{-10}	10^{-12}	10^{-14}
β	80	60	40	20	0	−20

(c) $\beta(10^{-4}) = 120 + 10 \cdot \log_{10} 10^{-4} = 120 - 40 = 80$

$\beta(10^{-6}) = 120 + 10 \cdot \log_{10} 10^{-6} = 120 - 60 = 60$

$\beta(10^{-8}) = 120 + 10 \cdot \log_{10} 10^{-8} = 120 - 80 = 40$

$\beta(10^{-10}) = 120 + 10 \cdot \log_{10} 10^{-10} = 120 - 100 = 20$

$\beta(10^{-12}) = 120 + 10 \cdot \log_{10} 10^{-12} = 120 - 120 = 0$

$\beta(10^{-14}) = 120 + 10 \cdot \log_{10} 10^{-14} = 120 - 140 = -20$

97. (a)

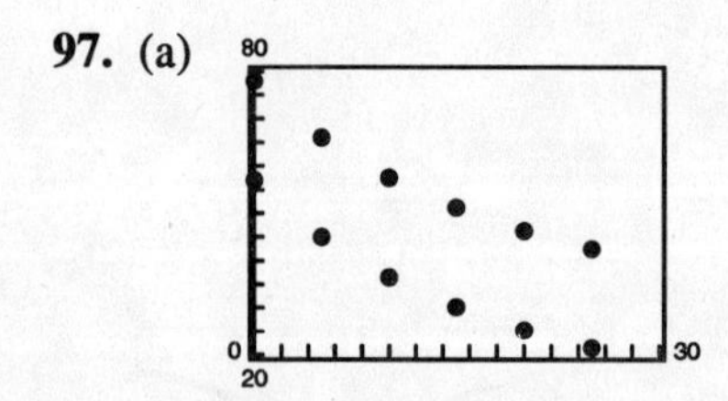

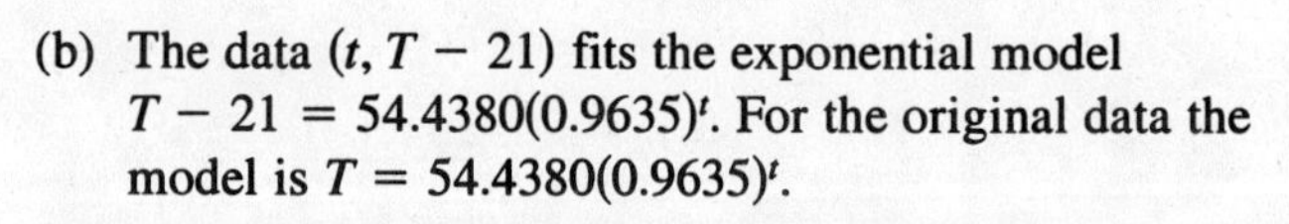

(b) The data $(t, T - 21)$ fits the exponential model $T - 21 = 54.4380(0.9635)^t$. For the original data the model is $T = 54.4380(0.9635)^t$.

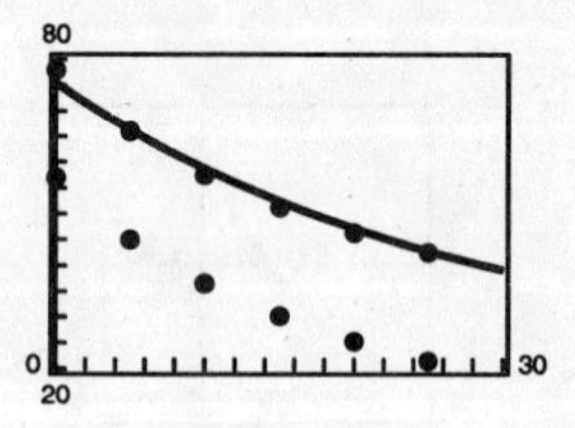

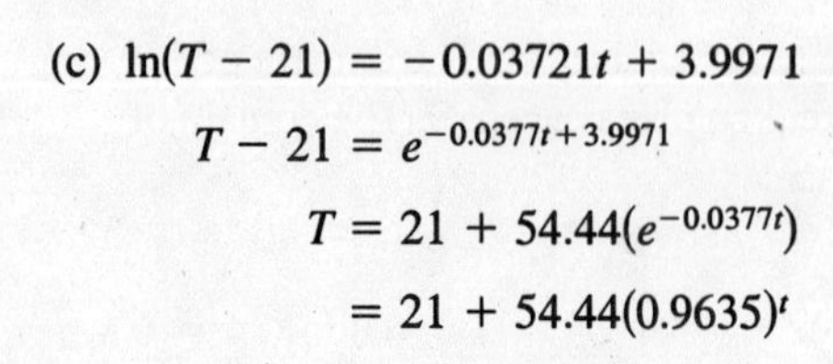

(c) $\ln(T - 21) = -0.03721t + 3.9971$

$T - 21 = e^{-0.0377t + 3.9971}$

$T = 21 + 54.44(e^{-0.0377t})$

$= 21 + 54.44(0.9635)^t$

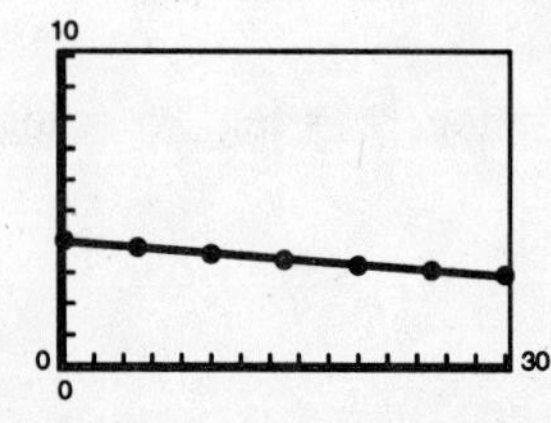

(d) $T = \dfrac{4960}{6t + 80} + 21$

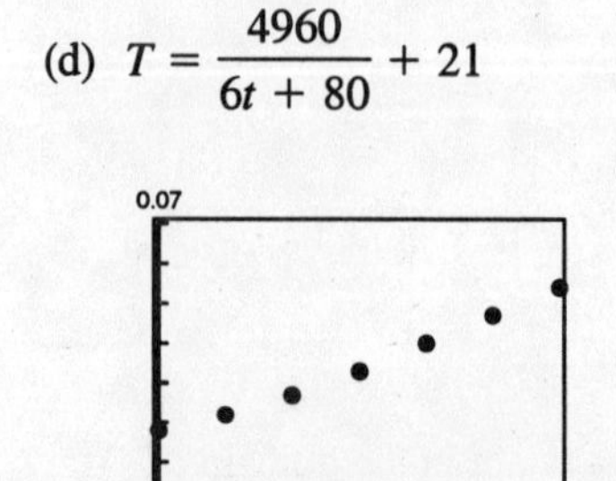

(e)

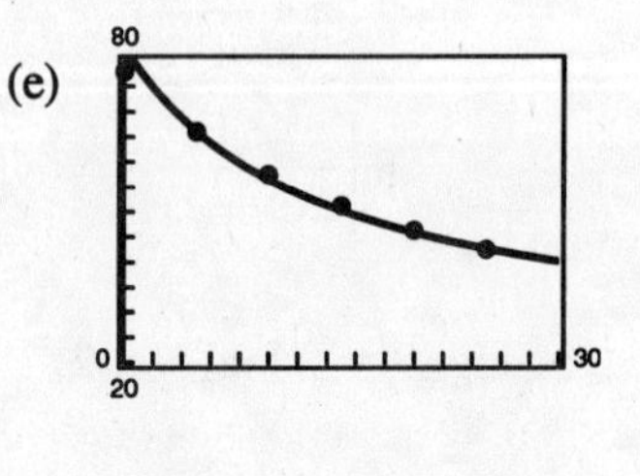

99. $f(x) = \ln x$

False, $f(0) \neq 0$ since 0 is not in the domain of $f(x)$. $f(1) = \ln 1 = 0$

101. False, $f(x) - f(2) = \ln x - \ln 2 = \ln \frac{x}{2} \neq \ln(x - 2)$.

103. False, $f(u) = 2f(v) \implies \ln u = 2 \ln v \implies \ln u = \ln v^2 \implies u = v^2$.

105. $f(x) = \ln \frac{x}{2}$

$g(x) = \frac{\ln x}{\ln 2}$

$h(x) = \ln x - \ln 2$

$f(x) = h(x)$ by Property 2.

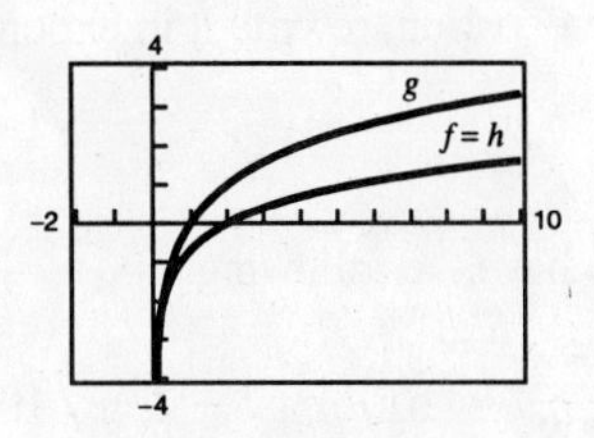

107. Let $x = \log_b u$ and $y = \log_b v$, then $b^x = u$ and $b^y = v$.

$$\frac{u}{v} = \frac{b^x}{b^y} = b^{x-y}$$

$$\log_b\left(\frac{u}{v}\right) = \log_b(b^{x-y}) = x - y = \log_b u - \log_b v$$

109. $f(x) = -\frac{1}{2}(x^2 + 4x)$

Intercepts: $(0, 0), (-4, 0)$

Parabola opening downward.

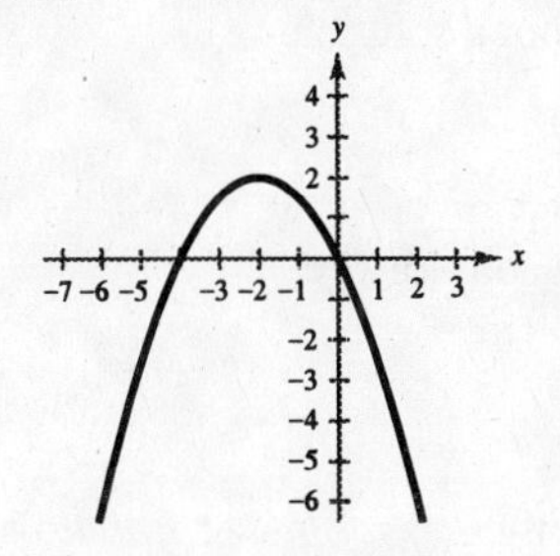

111. $f(x) = x^3 + 2x^2 - 9x - 18 = x^2(x + 2) - 9(x + 2)$

$= (x + 2)(x - 3)(x + 3)$

Zeros: $x = -2, -3, 3$

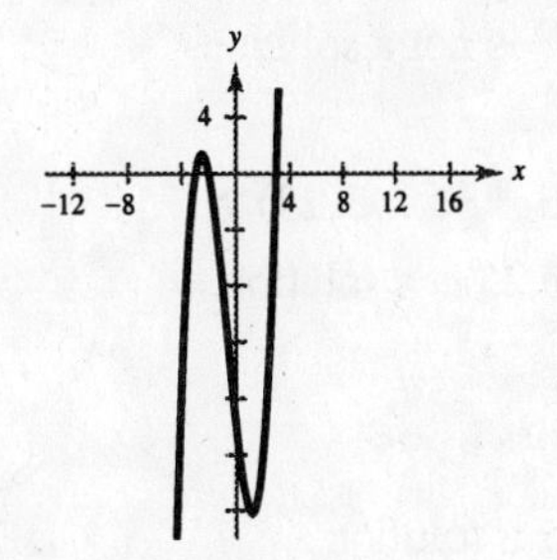

113. $x^2 - 6x + 2 = 0$

$$x = \frac{6 \pm \sqrt{36 - 4(2)}}{2} = 3 \pm \sqrt{7}$$

115. $x^4 - 19x^2 + 48 = 0$

$(x^2 - 16)(x^2 - 3) = 0$

$(x - 4)(x + 4)(x - \sqrt{3})(x + \sqrt{3}) = 0$

$x = \pm 4, \pm 3$

117. $x^3 - 6x^2 - 4x + 24 = 0$

$x^2(x - 6) - 4(x - 6) = 0$

$(x^2 - 4)(x - 6) = 0$

$(x - 2)(x + 2)(x - 6) = 0$

$x = 2, -2, 6$

119. $1.6^{-2\pi} \approx 0.052$

121. $260^{\sqrt{3}} \approx 15{,}235.494$

123. $\log_{10}(220) \approx 2.342$

125. $\ln 2.008 \approx 0.697$

Section 3.4 Solving Exponential and Logarithmic Equations

- To solve an exponential equation, isolate the exponential expression, then take the logarithm of both sides. Then solve for the variable.
  1. $\log_a a^x = x$
  2. $\ln e^x = x$
- To solve a logarithmic equation, rewrite it in exponential form. Then solve for the variable.
  1. $a^{\log_a x} = x$
  2. $e^{\ln x} = x$
- If $a > 0$ and $a \neq 1$ we have the following:
  1. $\log_a x = \log_a y \Rightarrow x = y$
  2. $a^x = a^y \Rightarrow x = y$
- Use your graphing utility to approximate solutions.

Solutions to Odd-Numbered Exercises

1. $4^{2x-7} = 64$

(a) $x = 5$

$4^{2(5)-7} = 4^3 = 64$

Yes, $x = 5$ is a solution.

(b) $x = 2$

$4^{2(2)-7} = 4^{-3} = \frac{1}{64} \neq 64$

No, $x = 2$ is not a solution.

3. $3e^{x+2} = 75$

(a) $x = -2 + e^{25}$

$3e^{(-2+e^{25})+2} = 3e^{e^{25}} \neq 75$

No, $x = -2 + e^{25}$ is not a solution.

(b) $x = -2 + \ln 25$

$3e^{(-2+\ln 25)+2} = 3e^{\ln 25} = 3(25) = 75$

Yes, $x = -2 + \ln 25$ is a solution.

(c) $x \approx 1.2189$

$3e^{1.2189+2} = 3e^{3.2189} \approx 75$

Yes, $x \approx 1.2189$ is a solution.

5. $\log_4(3x) = 3 \Rightarrow 3x = 4^3 \Rightarrow 3x = 64$

(a) $x \approx 20.3560$

$3(20.3560) = 61.0680 \neq 64$

No, $x \approx 20.3560$ is not a solution.

(b) $x = -4$

$3(-4) = -12 \neq 64$

No, $x = -4$ is not a solution.

(c) $x = \frac{64}{3}$

$3\left(\frac{64}{3}\right) = 64$

Yes, $x = \frac{64}{3}$ is a solution.

7. $\ln(x - 1) = 3.8$

(a) $x = 1 + e^{3.8}$

$\ln(1 + e^{3.8} - 1) = \ln e^{3.8} = 3.8$

Yes, $x = 1 + e^{3.8}$ is a solution.

(b) $x \approx 45.7012$

$\ln(45.7012 - 1) = \ln(44.7012) \approx 3.8$

Yes, $x \approx 45.7012$ is a solution.

(c) $x = 1 + \ln 3.8$

$\ln(1 + \ln 3.8 - 1) = \ln(\ln 3.8) \approx 0.289$

No, $x = 1 + \ln 3.8$ is not a solution.

9.

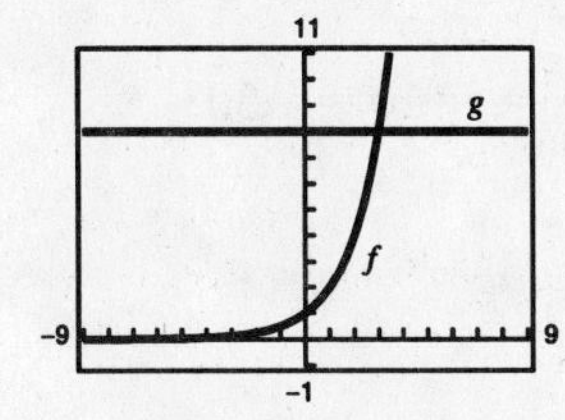

Intersection Point: $(3, 8)$

Algebraically, $2^x = 8$

$2^x = 2^3$

$x = 3 \Rightarrow y = 8 \Rightarrow (3, 8)$

11.

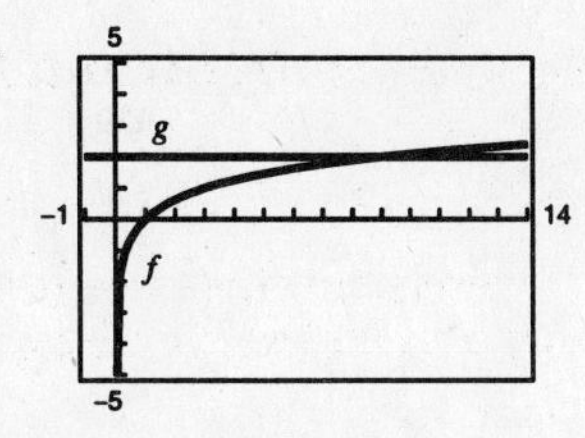

Intersection Point: $(9, 2)$

Algebraically, $\log_3 x = 2$

$x = 3^2 = 9 \Rightarrow y = 2 \Rightarrow (9, 2)$

13.

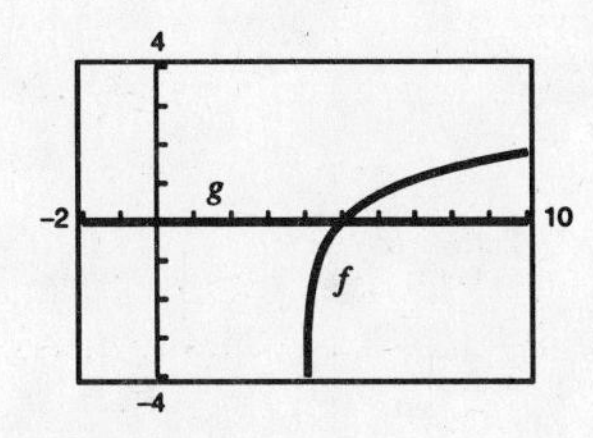

Intersection Point: $(5, 0)$

Algebraically, $\ln(x - 4) = 0$

$x - 4 = e^0 = 1$

$x = 5 \Rightarrow y = 0 \Rightarrow (5, 0)$

15. $4^x = 16$

$4^x = 4^2$

$x = 2$

17. $5^x = 625$

$5^x = 5^4$

$x = 4$

19. $8^x = 4$

$8^x = 8^{2/3}$

$x = \frac{2}{3}$

21. $\left(\frac{1}{4}\right)^x = 64$

$\left(\frac{1}{4}\right)^x = 4^3$

$\left(\frac{1}{4}\right)^x = \left(\frac{1}{4}\right)^{-3}$

$x = -3$

23. $3^{x-1} = 27$

$3^{x-1} = 3^3$

$x - 1 = 3$

$x = 4$

25. $\ln x - \ln 5 = 0$

$\ln x = \ln 5$

$x = 5$

27. $e^x = 4$

$x = \ln 4 \approx 1.386$

29. $\ln x = -7$

$x = e^{-7}$

31. $\log_x 625 = 4$

$x^4 = 625$

$x^4 = 5^4$

$x = 5$

33. $\log_{10} x = -1$

$x = 10^{-1}$

$x = \frac{1}{10}$

35. $\ln(2x - 1) = 0$

$e^0 = 2x - 1$

$1 = 2x - 1$

$2 = 2x$

$1 = x$

37. $\ln e^{x^2} = x^2 \ln e = x^2$

39. $e^{\ln(5x+2)} = 5x + 2$

41. $e^{\ln x^2} = x^2$

43.
$$10^x = 570$$
$$\log_{10} 10^x = \log_{10} 570$$
$$x = \log_{10} 570 \approx 2.756$$

45. $e^x = 10$
$$x = \ln 10 \approx 2.303$$

47. $5^{-t/2} = 0.20 = \dfrac{1}{5}$
$$-\frac{t}{2}\ln 5 = \ln\left(\frac{1}{5}\right)$$
$$-\frac{t}{2}\ln 5 = -\ln 5$$
$$\frac{t}{2} = 1$$
$$t = 2$$

49. $2^{3-x} = 565$
$$(3 - x)\ln 2 = \ln 565$$
$$-x\ln 2 = \ln 565 - 3\ln 2$$
$$x = (3\ln 2 - \ln 565)/\ln 2 \approx -6.142$$

51. $500e^{-x} = 300$
$$e^{-x} = \tfrac{3}{5}$$
$$-x = \ln\tfrac{3}{5}$$
$$x = -\ln\tfrac{3}{5} = \ln\tfrac{5}{3} \approx 0.511$$

53. $7 - 2e^x = 5$
$$-2e^x = -2$$
$$e^x = 1$$
$$x = \ln 1 = 0$$

55. $e^{2x} - 4e^x - 5 = 0$
$$(e^x - 5)(e^x + 1) = 0$$
$$e^x = 5 \text{ or } e^x = -1$$
$$x = \ln 5 \approx 1.609$$
($e^x = -1$ is impossible.)

57. $50(120 - e^{x/2}) = 600$
$$120 - e^{x/2} = 12$$
$$e^{x/2} = 108$$
$$\frac{x}{2} = \ln 108$$
$$x = 2\ln 108 \approx 9.364$$

59. Using the root feature of a graphing utility for
$$y = \left(1 + \frac{0.10}{12}\right)^{12t} - 2 = 0,$$
you obtain $t \approx 6.960$.

61.

x	0.6	0.7	0.8	0.9	1.0
$f(x)$	6.05	8.17	11.02	14.88	20.09

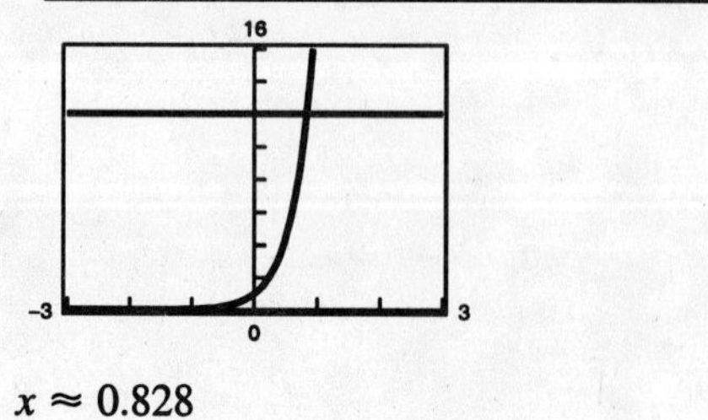

$x \approx 0.828$

63.

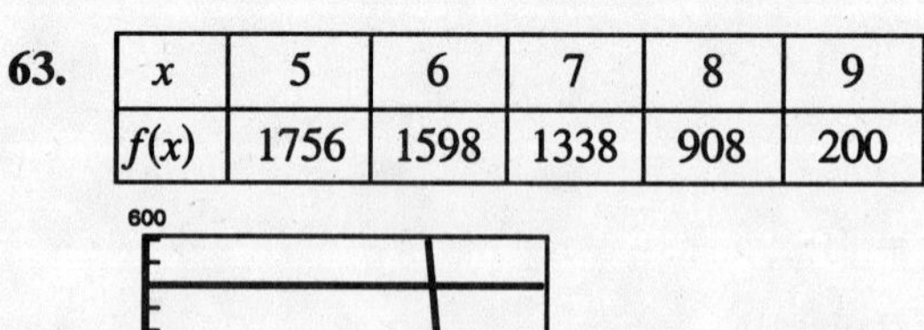

x	5	6	7	8	9
$f(x)$	1756	1598	1338	908	200

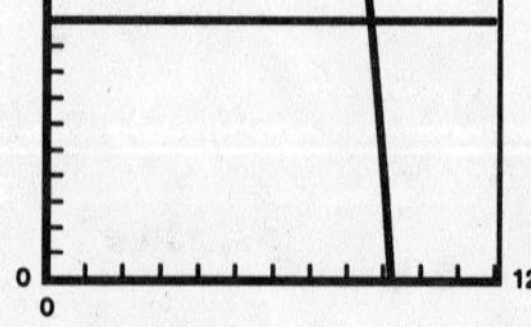

$x \approx 8.635$

65. $2^{3x} = 50$

Graphing $y = 2^{3x} - 50$, you obtain $x \approx 1.881$

67. $2^{-3x} = 0.90$

Graphing $y = 2^{-3x} - 0.90$, you obtain $x \approx 0.051$

69. $5(10^{x-6}) = 7$

Graphing $y = 5(10^{x-6}) - 7$, you obtain $x \approx 6.146$

71. $\left(1 + \frac{0.065}{365}\right)^{365t} = 4 \Longrightarrow t = 21.330$

73. $\frac{3000}{2 + e^{2x}} = 2$

$1500 = 2 + e^{2x}$

$1498 = e^{2x}$

$\ln 1498 = \ln e^{2x}$

$\ln 1498 = 2x$

$\frac{\ln 1498}{2} = x \approx 3.656$

75. $g(x) = 6e^{1-x} - 25$

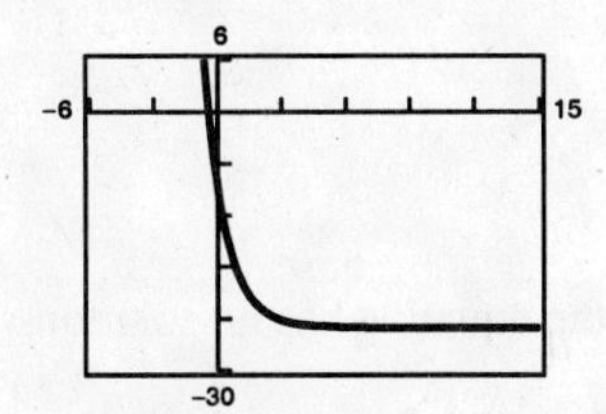

zero at $x = -0.427$

77. $g(t) = e^{0.09t} - 3$

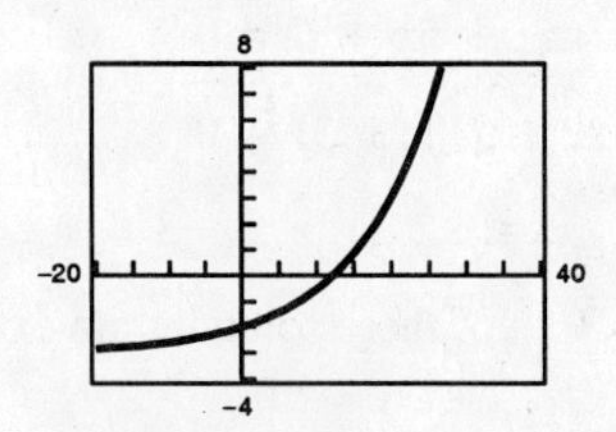

zero at $t = 12.207$

79. $\ln x = -3$

$x = e^{-3} \approx 0.050$

81. $\ln 4x = 2.1$

$4x = e^{2.1}$

$x = \frac{1}{4}e^{2.1}$

≈ 2.042

83. $2\ln 3x = 19$

$\ln 3x = \frac{19}{2} = 9.5$

$3x = e^{9.5}$

$x = \frac{1}{3}e^{9.5} \approx 4453.242$

85. $\log_{10}(z - 3) = 2$

$z - 3 = 10^2$

$z = 10^2 + 3 = 103$

87. $7 \log_4 (0.6x) = 12$

$\log_4(0.6x) = \frac{12}{7}$

$4^{12/7} = 0.6x = \frac{3}{5}x$

$x = \frac{5}{3}4^{12/7} \approx 17.945$

89. $\ln\sqrt{x + 2} = 1$

$\sqrt{x + 2} = e^1$

$x + 2 = e^2$

$x = e^2 - 2 \approx 5.389$

91. $\ln(x + 1)^2 = 2$

$e^{\ln(x+1)^2} = e^2$

$(x + 1)^2 = e^2$

$x + 1 = e$ or $x + 1 = -e$

$x = e - 1 \approx 1.718$

or

$x = -e - 1 \approx -3.718$

93. $\log_4 x - \log_4(x - 1) = \frac{1}{2}$

$\log_4\left(\frac{x}{x - 1}\right) = \frac{1}{2}$

$4^{\log_4(x/x-1)} = 4^{1/2}$

$\frac{x}{x - 1} = 2$

$x = 2(x - 1)$

$x = 2x - 2$

$2 = x$

95.
$$\ln(x+5) = \ln(x-1) - \ln(x+1).$$
$$\ln(x+5) = \ln\left(\frac{x-1}{x+1}\right)$$
$$x+5 = \frac{x-1}{x+1}$$
$$(x+5)(x+1) = x-1$$
$$x^2+6x+5 = x-1$$
$$x^2+5x+6 = 0$$
$$(x+2)(x+3) = 0$$
$$x = -2 \text{ or } x = -3$$

Both of these solutions are extraneous, so the equation has no solution.

97. $\log_{10} 8x - \log_{10}(1+\sqrt{x}) = 2$
$$\log_{10}\frac{8x}{1+\sqrt{x}} = 2$$
$$\frac{8x}{1+\sqrt{x}} = 10^2$$
$$8x = 100 + 100\sqrt{x}$$
$$8x - 100\sqrt{x} - 100 = 0$$
$$2x - 25\sqrt{x} - 25 = 0$$
$$\sqrt{x} = \frac{25 \pm \sqrt{25^2 - 4(2)(-25)}}{4}$$
$$= \frac{25 \pm 5\sqrt{33}}{4}$$

Choosing the positive value, we have $\sqrt{x} \approx 13.431$ and $x \approx 180.384$.

99.

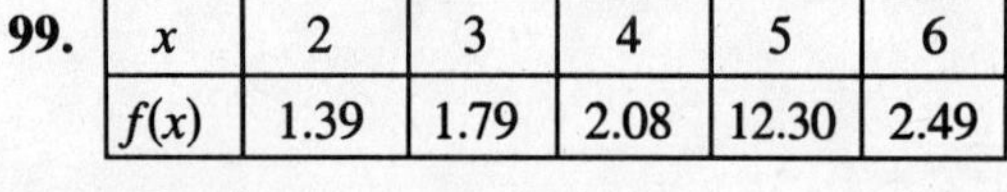

x	2	3	4	5	6
$f(x)$	1.39	1.79	2.08	12.30	2.49

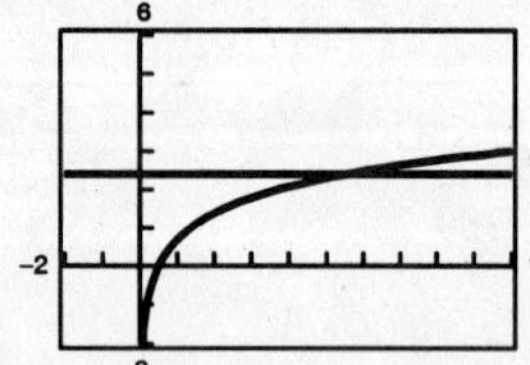

$x \approx 5.512$

101.

x	12	13	14	15	16
$f(x)$	9.79	10.22	10.63	11.00	11.36

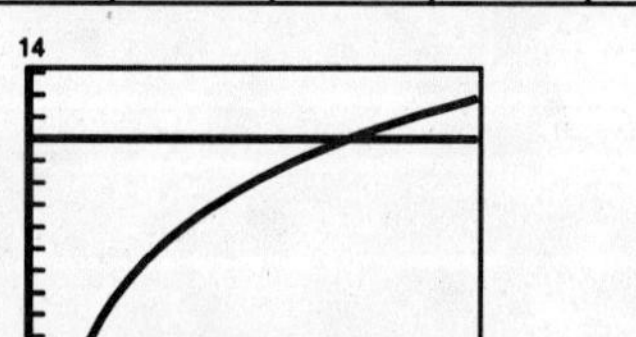

$x \approx 14.988$

103. $\log_{10}(z-4) = 1$

Graphing $y = \log_{10}(z-4) - 1$, you obtain $z = 14$.

105. $3 \ln x = 5$

Graphing $y = 3 \ln x - 5$, you obtain $x \approx 5.294$

107. $\ln x + \ln(x-3) = 1$

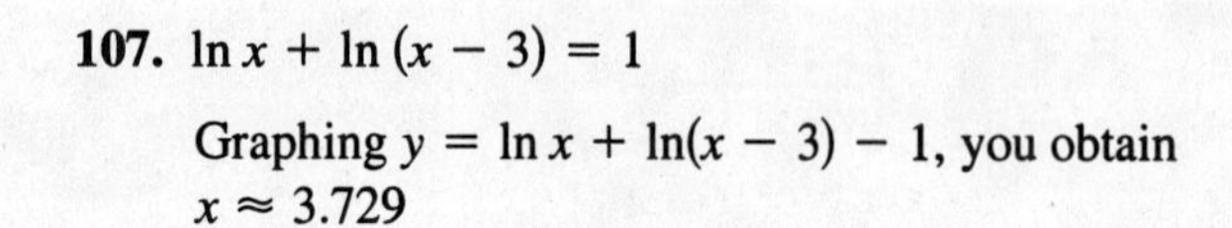

Graphing $y = \ln x + \ln(x-3) - 1$, you obtain $x \approx 3.729$

109. $\ln(x-5) = \ln(x-3) - \ln(x+3)$

Graphing $y = \ln(x-5) - \ln(x-3) + \ln(x+3)$, you obtain $x \approx 5.275$

111. $y_1 = 7$

$y_2 = 2^x$

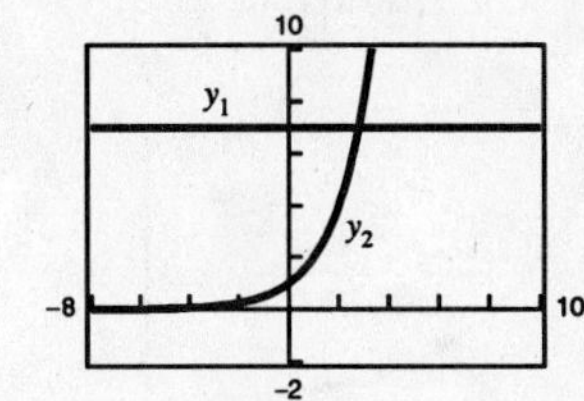

From the graph we have $(x, y) \approx (2.807, 7)$.

113. $y_1 = 8$

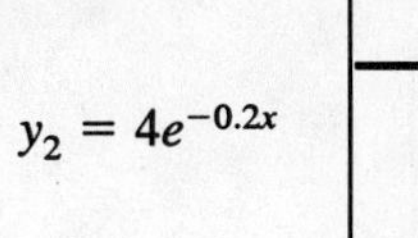

$y_2 = 4e^{-0.2x}$

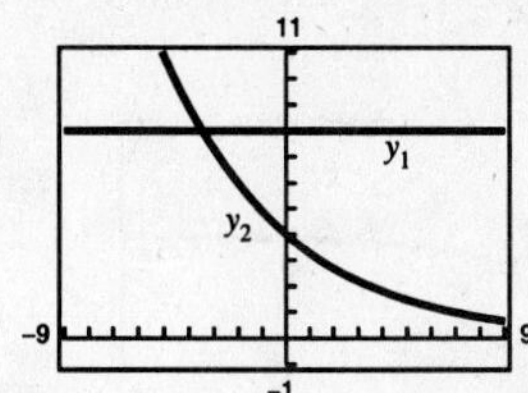

From the graph, we have $(x, y) \approx (-3.466, 8)$.

115. $y_1 = 3$

$y_2 = \ln x$

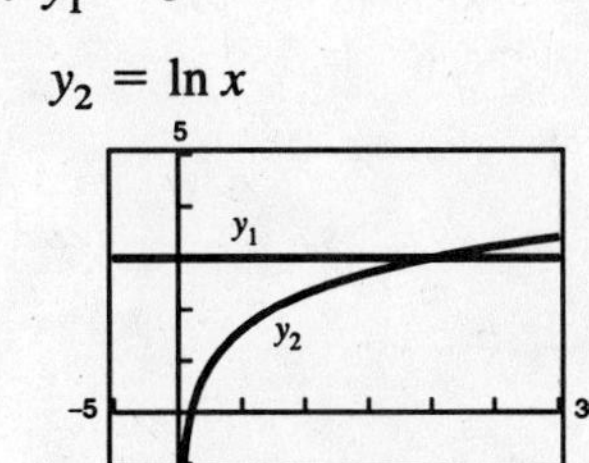

From the graph we have $(x, y) \approx (20.086, 3)$.

117. (a)

$$A = Pe^{rt}$$
$$2000 = 1000e^{0.085t}$$
$$2 = e^{0.085t}$$
$$\ln 2 = 0.085t$$
$$\frac{\ln 2}{0.085} = t$$
$$t \approx 8.2 \text{ years}$$

(b)

$$3000 = 1000e^{0.085t}$$
$$3 = e^{0.085t}$$
$$\ln 3 = 0.085t$$
$$\frac{\ln 3}{0.085} = t$$
$$t \approx 12.9 \text{ years}$$

119. (a)

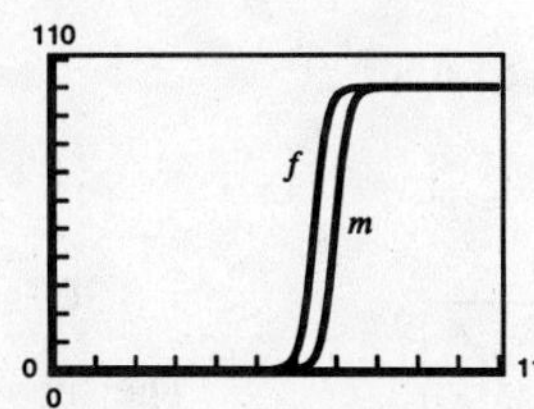

(b) From the graph we see horizontal asymptotes at $y = 0$ and $y = 100$. These represent the lower and upper percent bounds.

(c) Males:

$$50 = \frac{100}{1 + e^{-0.6114(x-69.71)}}$$
$$1 + e^{-0.6114(x-69.71)} = 2$$
$$e^{-0.6114(x-69.71)} = 1$$
$$-0.6114(x - 69.71) = \ln 1$$
$$-0.6114(x - 69.71) = 0$$
$$x = 69.71 \text{ inches}$$

Females:

$$50 = \frac{100}{1 + e^{-0.66607(x-64.51)}}$$
$$1 + e^{-0.66607(x-64.51)} = 2$$
$$e^{-0.66607(x-64.51)} = 1$$
$$-0.66607(x - 64.51) = \ln 1$$
$$-0.66607(x - 64.51) = 0$$
$$x = 64.51 \text{ inches}$$

121. $p = 500 - 0.5(e^{0.004x})$

(a)

$$p = 350$$
$$350 = 500 - 0.5(e^{0.004x})$$
$$300 = e^{0.004x}$$
$$0.004x = \ln 300$$
$$x \approx 1426 \text{ units}$$

(b)

$$p = 300$$
$$300 = 500 - 0.5(e^{0.004x})$$
$$400 = e^{0.004x}$$
$$0.004x = \ln 400$$
$$x \approx 1498 \text{ units}$$

123. $V = 6.7e^{-48.1/t},\ t > 0$

(a)

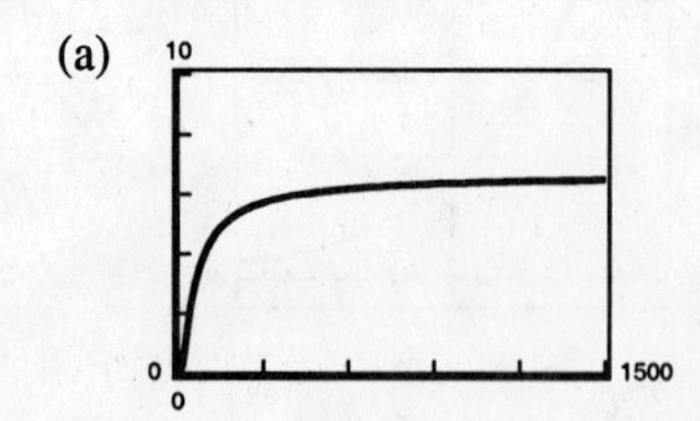

(b) As $t \longrightarrow \infty$, $V \longrightarrow 6.7$.

Horizontal asymptote: $y = 6.7$
The yield will approach
6.7 million cubic feet per acre.

(c) $1.3 = 6.7e^{-48.1/t}$

$$\frac{1.3}{6.7} = e^{-48.1/t}$$

$$\ln\frac{13}{67} = \frac{-48.1}{t}$$

$$t = \frac{-48.1}{\ln\left(\frac{13}{67}\right)} \approx 29.3 \text{ years}$$

125. $T = 20[1 + 7(2^{-h})]$

(a)

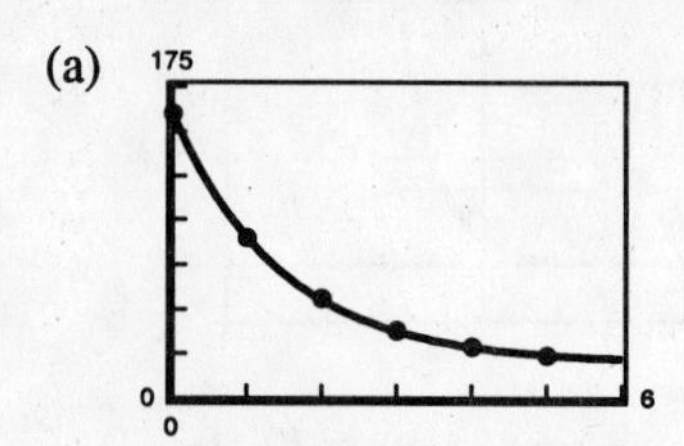

(b) We see a horizontal asymptote at $y = 20$. This represents the room temperature.

(c)
$$100 = 20[1 + 7(2^{-h})]$$
$$5 = 1 + 7(2^{-h})$$
$$4 = 7(2^{-h})$$
$$\frac{4}{7} = 2^{-h}$$
$$\ln\left(\frac{4}{7}\right) = \ln 2^{-h}$$
$$\ln\left(\frac{4}{7}\right) = -h \ln 2$$
$$\frac{\ln(4/7)}{-\ln 2} = h$$
$$h \approx 0.81 \text{ hour}$$

127. (a)

80, 2, 0, 8

$y = 15.17x - 46.15$

$y = 100$ when $x \approx 9.6$, or during 1999.

(b)

$\ln x$	1.0986	1.3863	1.6094	1.7918	1.9459
$\ln y$	2.0541	2.2300	3.0681	3.7658	4.2002

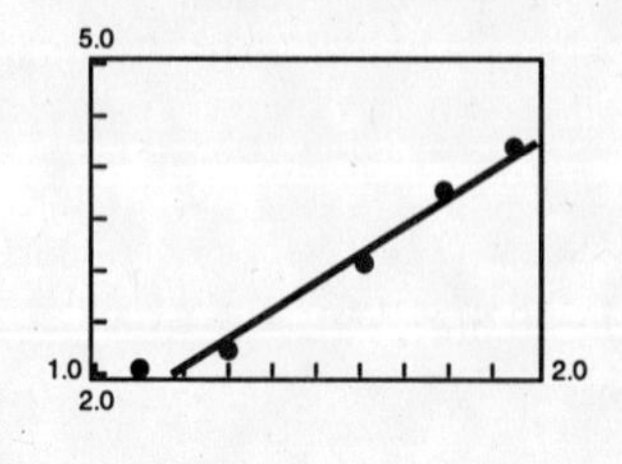

$\ln y = 2.706 \ln x - 1.175$

For $y = 100$, $\ln 100 = 2.706 \ln x - 1.175$

$$5.780 = 2.706 \ln x$$
$$\ln x = 2.136$$
$$x \approx 8.5 \text{ or during 1998}$$

(c) $y = e^{2.706 \ln x - 1.175} = e^{\ln x^{2.706} - 1.175} = e^{-1.175}x^{2.706}$

$= 0.309x^{2.706}$

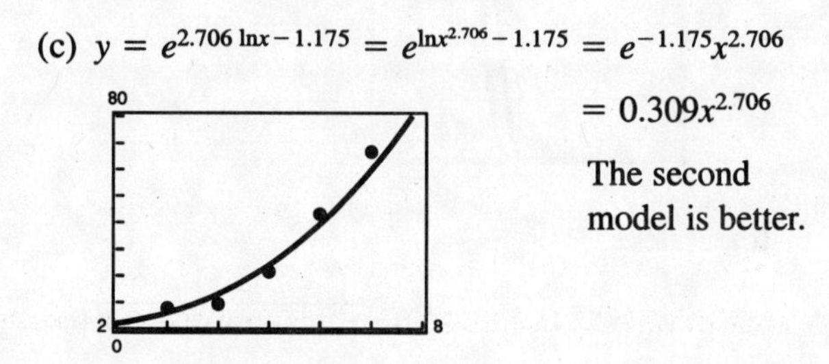

The second model is better.

129. False. A logarithmic equation can have any number of extraneous solutions

131. To find the length of time it takes for an investment P to double to $2P$, solve

$$2P = Pe^{rt}$$
$$2 = e^{rt}$$
$$\ln 2 = rt$$
$$\frac{\ln 2}{r} = t.$$

Thus, you can see that the time is not dependent on the size of the investment, but rather the interest rate.

133. $f(x) = -3^{-x-3} + 5$

135. $f(x) = \left(\frac{1}{2}\right)^{-x} + 3$

$= 2^x + 3$

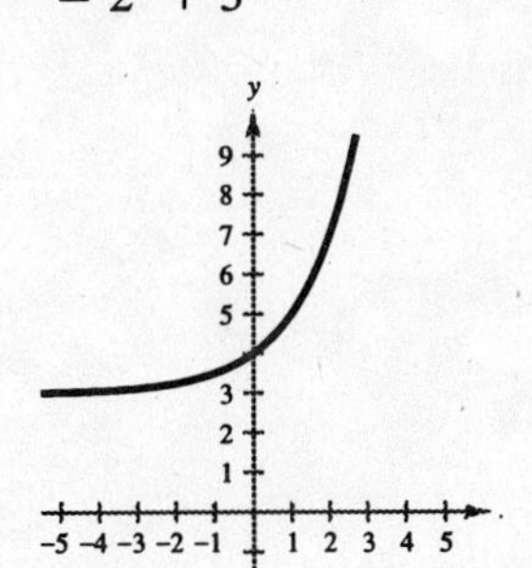

137. $\log_3 22 = \frac{\ln 22}{\ln 3} \approx 2.814$

139. $\log_{21} 140 = \frac{\ln 140}{\ln 21} \approx 1.623$

Section 3.5 Exponential and Logarithmic Models

- You should be able to solve compound interest problems.
  1. $A = P\left(1 + \frac{r}{n}\right)^{nt}$
  2. $A = Pe^{rt}$
- You should be able to solve growth and decay problems.
  (a) Exponential growth if $b > 0$ and $y = ae^{bx}$.
  (b) Exponential decay if $b > 0$ and $y = ae^{-bx}$.
- You should be able to use the Gaussian model
  $y = ae^{-(x-b)^2/c}$.
- You should be able to use the logistics growth model
  $y = \frac{a}{1 + be^{-(x-c)/d}}$.
- You should be able to use the logarithmic models
  $y = \ln(ax + b)$ and $y = \log_{10}(ax + b)$.

Solutions to Odd-Numbered Exercises

1. $y = 2e^{x/4}$

This is an exponential growth model. Matches graph (c).

3. $y = 6 + \log_{10}(x + 2)$

This is a logarithmic model, and contains $(-1, 6)$. Matches graph (b).

5. $y = \ln(x + 1)$

This is a logarithmic model. Matches graph (d).

7. A logarithmic model seems best.

9. A Gaussian model seems best.

11. An exponential model seems best.

13. A Gaussian model seems best.

15. Logarithmic model

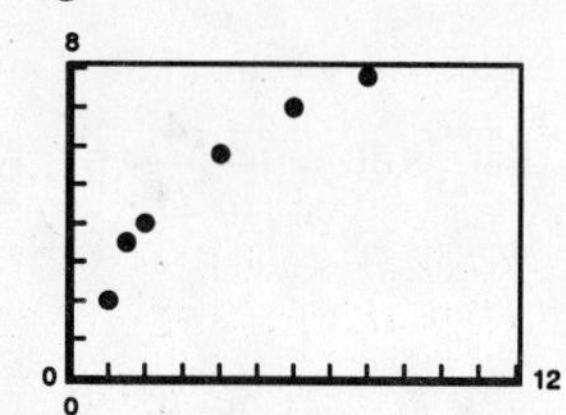

17. Linear model

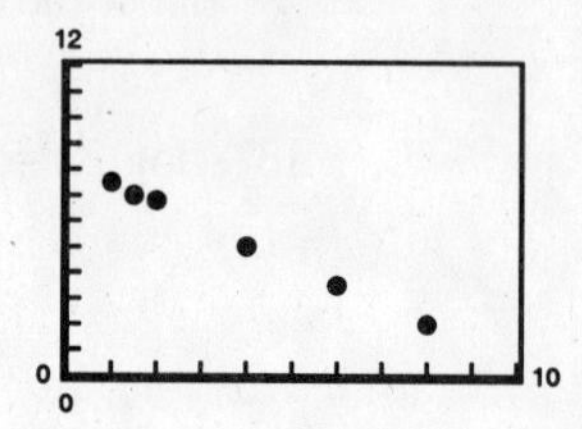

19. Since $A = 1000e^{0.12t}$, the time to double is given by $2000 = 1000e^{0.12t}$ and we have

$$t = \frac{\ln 2}{0.12} \approx 5.78 \text{ years.}$$

Amount after 10 years: $A = 1000e^{1.2} \approx \$3320.12$

21. Since $A = 750e^{rt}$ and $A = 1500$ when $t = 7.75$, we have the following.

$$15000 = 750e^{7.75r}$$

$$r = \frac{\ln 2}{7.75} \approx 0.0894 = 8.94\%$$

Amount after 10 years: $A = 750e^{0.0894(10)} \approx \$1833.67$

23. Since $A = 500e^{rt}$ and $A = 1292.85$ when $t = 10$, we have the following.

$$1292.85 = 500e^{10r}$$

$$r = \frac{\ln(1292.85/500)}{10} \approx 0.0950 = 9.5\%$$

The time to double is given by

$$1000 = 500e^{0.095t}$$

$$t = \frac{\ln 2}{0.095} \approx 7.30 \text{ years.}$$

25. Since $A = Pe^{0.045t}$ and $A = 10{,}000.00$ when $t = 10$, we have the following.

$$10{,}000.00 = Pe^{0.045(10)}$$

$$\frac{10{,}000.00}{e^{0.045(10)}} = P \approx 6376.28$$

The time to double is given by

$$t = \frac{\ln 2}{0.045} \approx 15.40 \text{ years.}$$

27. (a) $3P = Pe^{rt}$

$$3 = e^{rt}$$

$$\ln 3 = rt$$

$$\frac{\ln 3}{r} = t$$

r	2%	4%	6%	8%	10%	12%
$t = \dfrac{\ln 3}{r}$	54.93	27.47	18.31	13.73	10.99	9.16

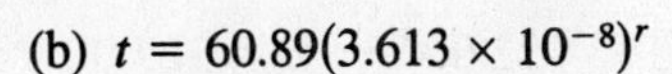

(b) $t = 60.89(3.613 \times 10^{-8})^r$

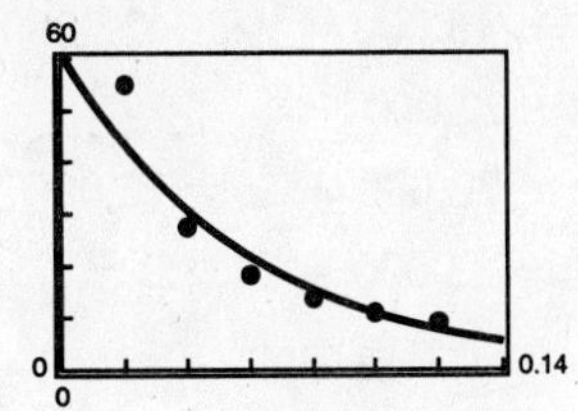

29. Continuous compounding results in faster growth.

$A = 1 + 0.075[\![t]\!]$

and $A = e^{0.07t}$

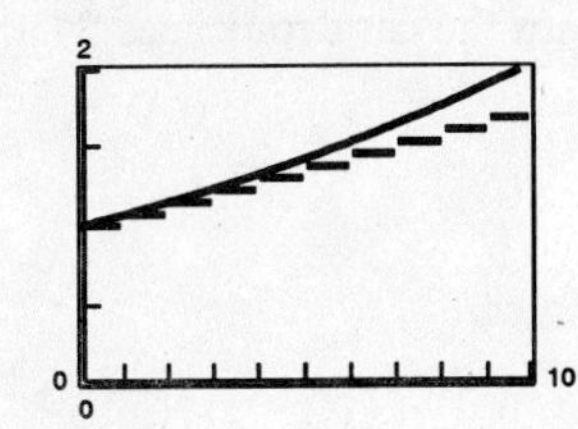

31. $\frac{1}{2}C = Ce^{k(1620)}$

$$k = \frac{\ln 0.5}{1620}$$

Given $C = 10$ grams, after 1000 years, we have

$$y = Ce^{[(\ln 0.5)/1620](1000)}$$

$$C \approx 6.52 \text{ grams.}$$

33. $\frac{1}{2}C = Ce^{k(5730)}$

$$k = \frac{\ln 0.5}{5730}$$

Given $y = 3$ grams, after 1000 years, we have

$$3 = Ce^{[(\ln 0.5)/5730](1000)}$$

$$C \approx 2.66 \text{ grams.}$$

35. $P = 105{,}300e^{0.015t}$

$$150{,}000 = 105{,}300e^{0.015t}$$

$$\ln \tfrac{1500}{1053} = 0.015t$$

$$t \approx 23.59$$

The population will reach 150,000 during 2023. [Note: 2000 + 13.59.]

37. $P = 2500e^{kt}$, $P(0) = 2500$ represents year 2000

For 1945, $t = -55$ and

$$1350 = 2500e^{k(-55)}$$

$$\ln \frac{1350}{2500} = -55k$$

$$k \approx 0.0112$$

For 2010, $t = 10$ and

$$P \approx 2500e^{0.0112(10)} \approx 2796 \text{ people.}$$

39. $N = 100e^{kt}$

$$300 = 100e^{5k}$$

$$k = \frac{\ln 3}{5} \approx 0.2197$$

$$N = 100e^{0.2197t}$$

$$200 = 100e^{0.2197t}$$

$$t = \frac{\ln 2}{0.2197} \approx 3.15 \text{ hours}$$

41. $y = Ce^{kt}$

$$\frac{1}{2}C = Ce^{(1620)k}$$

$$\ln \frac{1}{2} = 1620k$$

$$k = \frac{\ln(1/2)}{1620}$$

When $t = 100$, we have

$$y = Ce^{[\ln(1/2)/1620](100)} \approx 0.958C = 95.8\%C.$$

After 100 years, approximately 95.8% of the radioactive radium will remain.

43. (a) $V = mt + b$; $V(0) = 22{,}000 \Rightarrow b = 22{,}000$

$V(2) = 13{,}000 \Rightarrow 13{,}000 = 2m + 22{,}000 \Rightarrow m = -4500$

$V(t) = -4500t + 22{,}000$

(b) $V = ae^{kt}$; $V(0) = 22{,}000 \Rightarrow a = 22{,}000$

$V(2) = 13{,}000 \Rightarrow 13{,}000 = 22{,}000e^{2k}$

$$\frac{13}{22} = e^{2k}$$

$$\ln\frac{13}{22} = 2k$$

$$k = \frac{1}{2}\ln\frac{13}{22} \approx -0.263$$

$V = 22{,}000e^{-0.263t}$

(c)

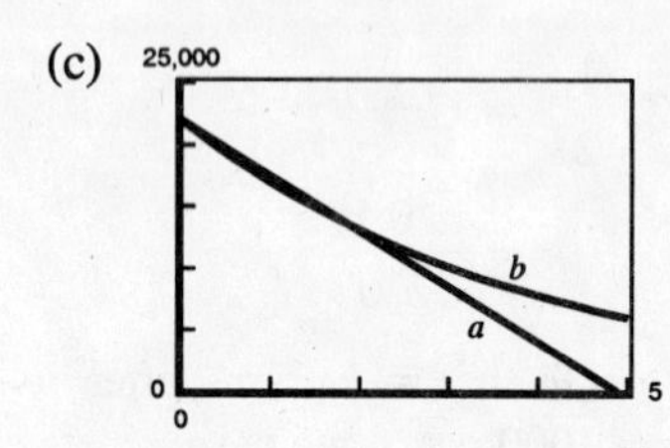

The exponential model depreciates faster in the first two years.

(d) Straight line: $V(1) = \$17{,}500$

$V(3) = \$8500$

Exponential: $V(1) = \$16{,}912$

$V(3) = \$9993$

(e) The negative slope means the car depreciates \$4500 per year.

45. $S(t) = 100(1 - e^{kt})$

(a) $15 = 100(1 - e^{k(1)})$

$-85 = -100e^k$

$k = \ln 0.85$

$k \approx -0.1625$

$S(t) = 100(1 - e^{-0.1625t})$

(b)

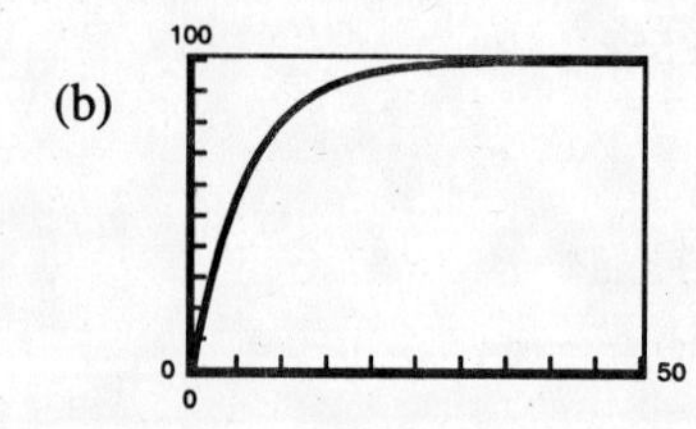

(c) $S(5) = 100(1 - e^{-0.1625(5)})$

$\approx 55.625 = 55{,}625$ units

47. $N = 30(1 - e^{kt})$

(a) $N = 19,\ t = 20$

$19 = 30(1 - e^{20k})$

$20k = \ln\dfrac{11}{30}$

$k \approx -0.050$

$N = 30(1 - e^{-0.050t})$

(b) $N = 25$

$25 = 30(1 - e^{-0.05t})$

$\dfrac{5}{30} = e^{-0.05t}$

$t = -\dfrac{1}{0.05}\ln\dfrac{5}{30} \approx 36$ days

(c) No, this is not a linear function.

49. $R = \log_{10}\left(\dfrac{I}{I_0}\right) = \log_{10} I$, since $I_0 = 1$

(a) $R = \log_{10} 39{,}811{,}000 \approx 7.6$

(b) $R = \log_{10} 12{,}589{,}000 \approx 7.1$

51. $\beta(I) = 10\log_{10}(I/I_0)$, where $I_0 = 10^{-12}$ watts per sq meter

(a) $\beta(10^{-10}) = 10 \cdot \log_{10}\left(\frac{10^{-10}}{10^{-12}}\right) = 10\log_{10}10^2 = 20$ decibels

(b) $\beta(10^{-5}) = 10 \cdot \log_{10}\left(\frac{10^{-5}}{10^{-12}}\right) = 10\log_{10}10^7 = 70$ decibels

(c) $\beta(10^{0}) = 10 \cdot \log_{10}\left(\frac{10^{0}}{10^{-12}}\right) = 10\log_{10}10^{12} = 120$ decibels

53. $\beta = 10\log_{10}\frac{I}{I_0}$

$10^{\beta/10} = \frac{I}{I_0}$

$I = I_0 10^{\beta/10}$

$$\% \text{ decrease} = \frac{I_0 10^{9.3} - I_0 10^{8.0}}{I_0 10^{9.3}} \times 100 \approx 95\%$$

55. $\text{pH} = -\log_{10}[H^+] = -\log_{10}[2.3 \times 10^{-5}] \approx 4.64$

57. $\text{pH} = -\log_{10}[H^+]$

$-\text{pH} = \log_{10}[H^+]$

$10^{-\text{pH}} = [H^+]$

$$\frac{\text{Hydrogen ion concentration of fruit}}{\text{Hydrogen ion concentration of tablet}} = \frac{10^{-2.5}}{10^{-9.5}} = 10^7$$

59. (a) $P = 120{,}000,\ t = 30,\ r = 0.075,\ M = 839.06$

$$u = M - \left(M - \frac{Pr}{12}\right)\left(1 + \frac{r}{12}\right) = 839.06 - (839.06 - 750)(1 + 0.00625)^{12t}$$

$v = (839.06 - 750)(1.00625)^{12t}$

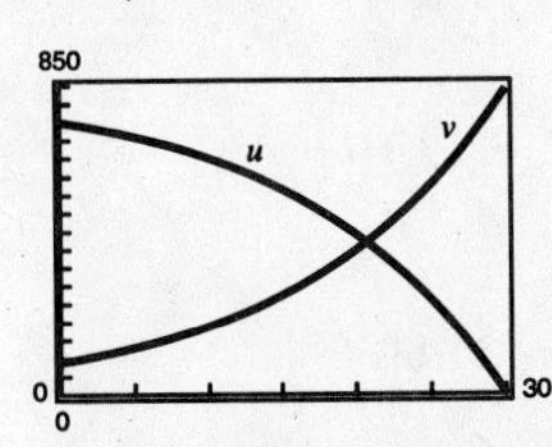

(b) In the early years, the majority of the monthly payment goes toward interest. The interest and principle are equal when $t \approx 20.729 \approx 21$ years.

(c) $P = 120{,}000,\ t = 20,\ r = 0.075,\ M = 966.71$

$u = 966.71 - (966.71 - 750)(1.00625)^{12t}$

$v = (966.71 - 750)(1.00625)^{12t}$

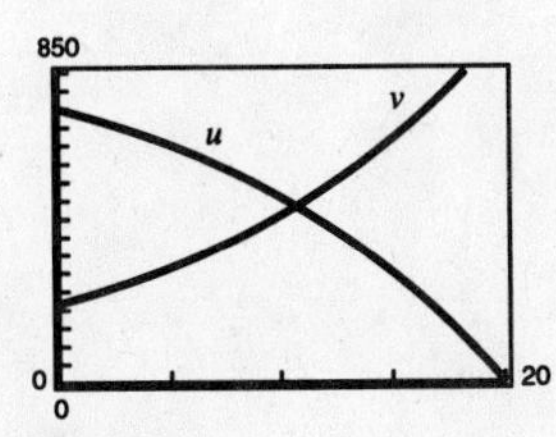

$u = v$ when $t \approx 10.73$ years

61. $y = ae^{bx}$

$1 = ae^{b(0)} \Rightarrow 1 = a$

$10 = e^{b(3)}$

$\ln 10 = 3b$

$\frac{\ln 10}{3} = b \Rightarrow b \approx 0.7675$

Thus, $y = e^{0.7675x}$.

63. $y = ae^{bx}$

$\frac{1}{2} = ae^{b(0)} \Rightarrow a = \frac{1}{2}$

$5 = \frac{1}{2}e^{b(4)}$

$10 = e^{4b}$

$\ln 10 = 4b$

$\frac{\ln 10}{4} = b \Rightarrow b \approx 0.5756$

Thus, $y = \frac{1}{2}e^{0.5756x}$.

65. $t_1 = 40.757 + 0.556s - 15.817 \ln s$

$t_2 = 1.2259 + 0.0023s^2$

(a) Linear Model: $t_3 \approx 0.2729s - 6.0143$
Exponential Model: $t_4 \approx 1.5385e^{1.0291s}$

(b)

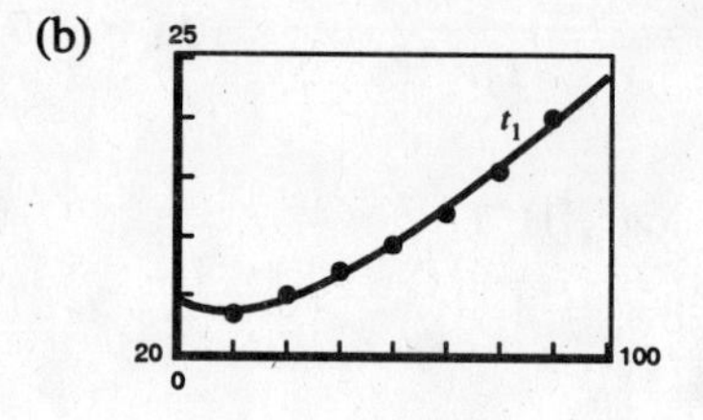

(c)

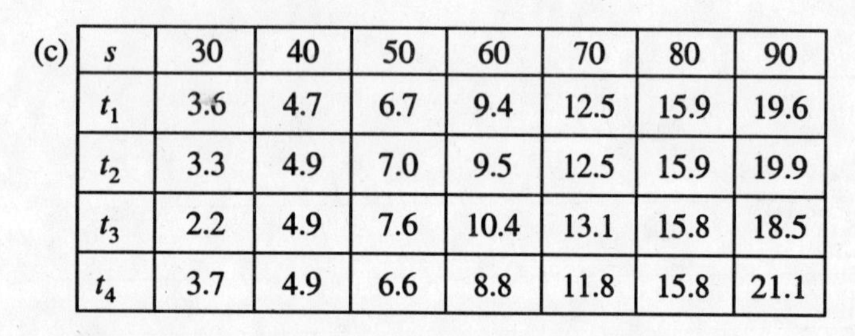

s	30	40	50	60	70	80	90
t_1	3.6	4.7	6.7	9.4	12.5	15.9	19.6
t_2	3.3	4.9	7.0	9.5	12.5	15.9	19.9
t_3	2.2	4.9	7.6	10.4	13.1	15.8	18.5
t_4	3.7	4.9	6.6	8.8	11.8	15.8	21.1

(d) Model t_1: $S_1 = |3.4 - 3.6| + |5 - 4.7| + |7 - 6.7| + |9.3 - 9.4| + |12 - 12.5| + |15.8 - 15.9| + |20 - 19.6| = 1.9$

Model t_2: $S_2 = |3.4 - 3.3| + |5 - 4.9| + |7 - 7| + |9.3 - 9.5| + |12 - 12.5| + |15.8 - 15.9| + |20 - 19.9| = 1.1$

Model t_3: $S_3 = |3.4 - 2.2| + |5 - 4.9| + |7 - 7.6| + |9.3 - 10.4| + |12 - 13.1| + |15.8 - 15.8| + |20 - 18.5| = 5.6$

Model t_4: $S_4 = |3.4 - 3.7| + |5 - 4.9| + |7 - 6.6| + |9.3 - 8.8| + |12 - 11.8| + |15.8 - 15.8| + |20 - 21.1| = 2.6$

t_2, the Quadratic model, is the best fit with the data.

67. $t = -2.5 \ln\left(\frac{T - 70}{98.6 - 70}\right)$

At 9:00 A.M. we have: $t = -2.5 \ln\left(\frac{85.7 - 70}{98.6 - 70}\right) \approx 1.5$ hours.

From this we can conclude that the person died at 7:30 A.M.

69. (a) $y = 0.08245x + 4.45274$

(b) $y = 4.5355(1.01519)^x$

(c) The models are nearly identical.

(d) For 2005, $x = 25$

Linear model: $y = 0.08245(25) + 4.45274 \approx 6.514$ billion

Exponential model: $y = 4.5355(1.01519)^{25} \approx 6.61$ billion

(Answers will vary.)

71. (a) $y = 298.794(1.0851)$

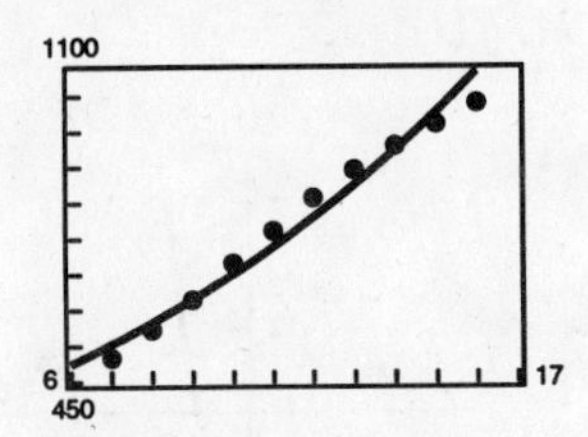

(b) $y = -837.735 + 673.619 \ln(x)$

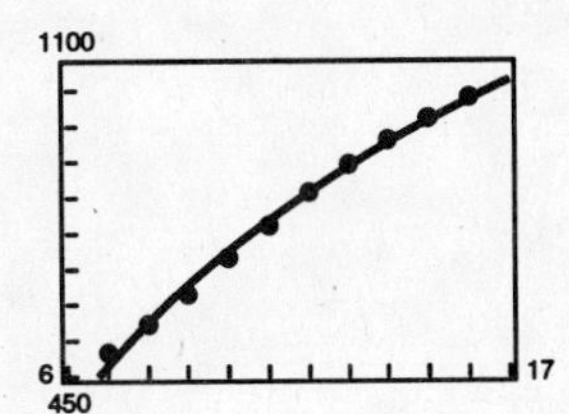

(c) The logarithmic model is more accurate. If the rate of growth of health costs is slowed, then the logarithmic model would be better.

73. (a) $y_1 = -1.81x^3 + 14.58x^2 + 16.39x + 10.00$

$y_2 = 23.07 + 121.08 \ln x$

$y_3 = 38.38(1.4227)^x$

(b)

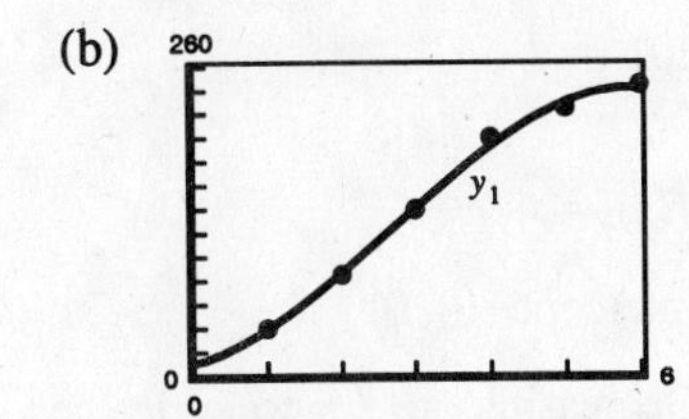

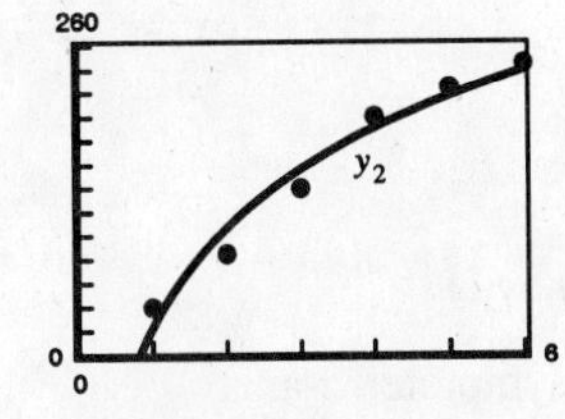

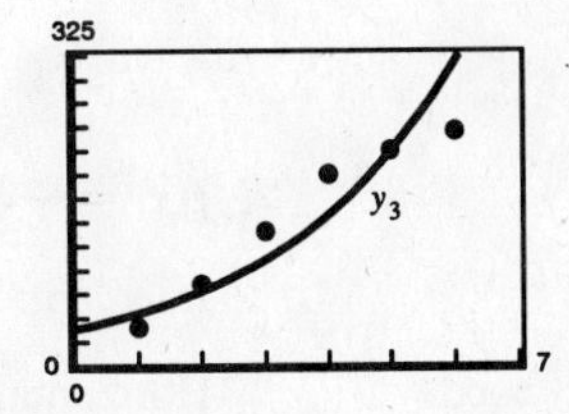

(c) Cubic model

x	y	$y - y_1$	$(y - y_1)^2$	$y - y_2$	$(y - y_2)^2$	$y - y_3$	$(y - y_3)^2$
1	40	0.84	0.71	16.93	286.62	−14.60	213.25
2	85	−1.62	2.62	−22.00	483.84	7.32	53.52
3	140	−1.52	2.31	−16.09	258.89	29.48	869.01
4	200	7.00	49.00	9.08	82.40	42.76	1828.56
5	225	5.20	27.04	7.06	49.83	1.30	1.68
6	245	2.74	7.51	4.98	24.84	−73.26	5367.34

(d) Cubic model
y_1: 89.19; y_2: 1186.42; y_3: 8333.36;

(e) The sums represent the sum of the squares of the errors.

75. False. See Example 5, page 263.

77. True. See page 262.

79. $4x - 3y - 9 = 0 \Rightarrow y = \frac{1}{3}(4x - 9)$

Line: slope $\frac{4}{3}$. Matches (b)

Intercepts: $(0, -3)$. $\left(\frac{9}{4}, 0\right)$

81. $y = 25 - 2.25x$

Slope -2.25, y-intercept 25. Matches (f)

83. $y - 3 = 0$ horizontal line. Matches (d).

Intercept $(0, 3)$

85.
$$\begin{array}{r|rrrr} -4 & 4 & 4 & -39 & 36 \\ & & -16 & 48 & -36 \\ \hline & 4 & -12 & 9 & 0 \end{array}$$

$$\frac{4x^3 + 4x^2 - 39x + 36}{x + 4} = 4x^2 - 12x + 9, x \neq -4$$

87.
$$\begin{array}{r|rrrr} 4 & 2 & -8 & 3 & -9 \\ & & 8 & 0 & 12 \\ \hline & 2 & 0 & 3 & 3 \end{array}$$

$$\frac{2x^3 - 8x^2 + 3x - 9}{x - 4}$$

$$= 2x^2 + 3 + \frac{3}{x - 4}$$

89. $f(x) = 2^{x-1} + 5$

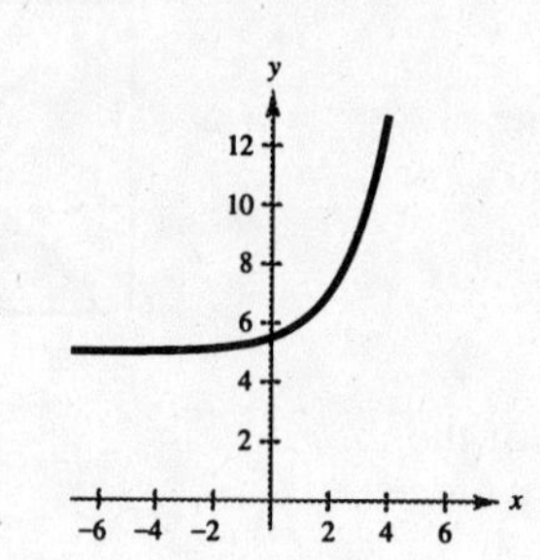

91. $f(x) = 3^x - 4$

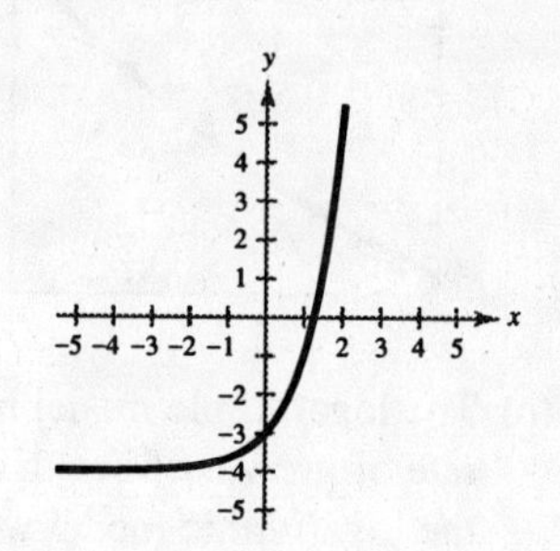

Review Exercises for Chapter 3

Solutions to Odd-Numbered Exercises

1. $(1.45)^{2\pi} \approx 10.325$

3. $(1.59)^{-2\sqrt{3}} \approx 0.201$

5. $f(x) = 4^x$

Intercept: $(0, 1)$

Horizontal asymptote: x-axis

Increasing on: $(-\infty, \infty)$

Matches graph (e).

7. $f(x) = -4^x$

Intercept: $(0, -1)$

Horizontal asymptote: x-axis

Decreasing on: $(-\infty, \infty)$

Matches graph (b).

9. $f(x) = 4^{-x} - 1$

Intercept: $(0, 0)$

Decreasing on: $(-\infty, \infty)$

Matches graph (d).

11. $f(x) = 6^x$

Intercept: $(0, 1)$

Increasing horizontal asymptote: x-axis

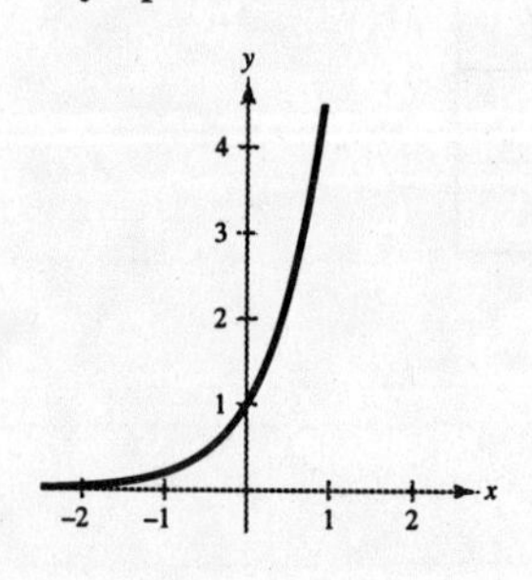

13. $g(x) = 6^{-x} = \left(\frac{1}{6}\right)^x$

Intercept: $(0, 1)$

Decreasing horizontal asymptote: x-axis

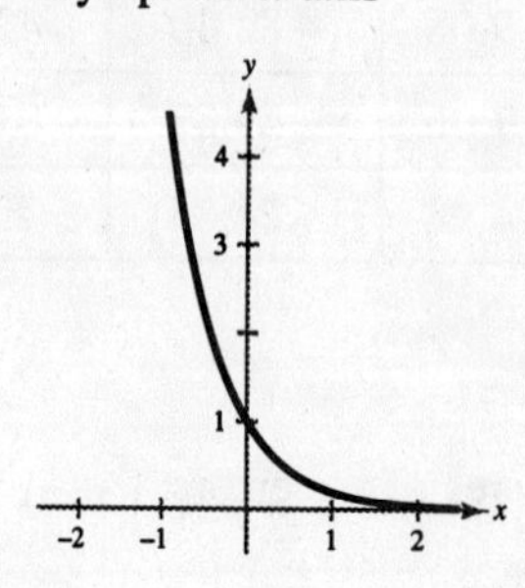

15. $h(x) = e^{-x/2}$

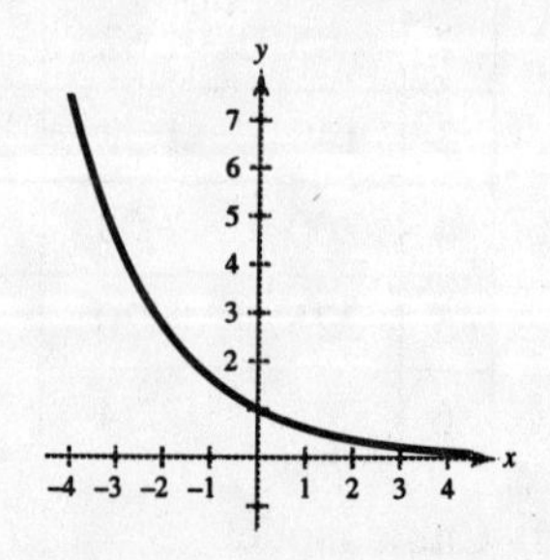

x	-2	-1	0	1	2
y	2.72	1.65	1	0.61	0.37

17. $h(x) = 2 - e^{-x/2}$

Horizontal asymptote: $y = 2$

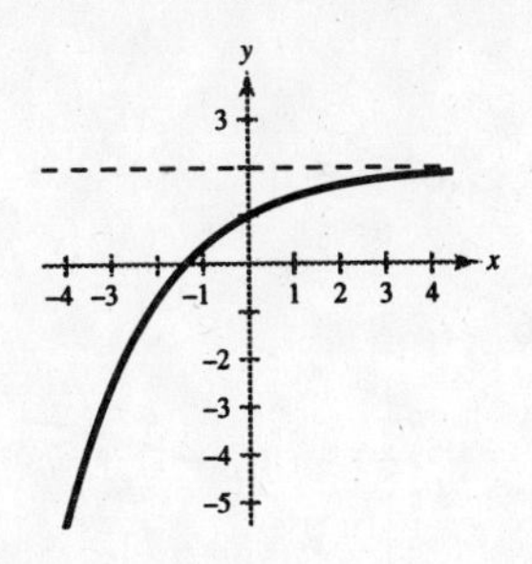

19. $s(t) = 4e^{-2/t}, t > 0$

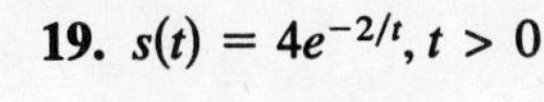

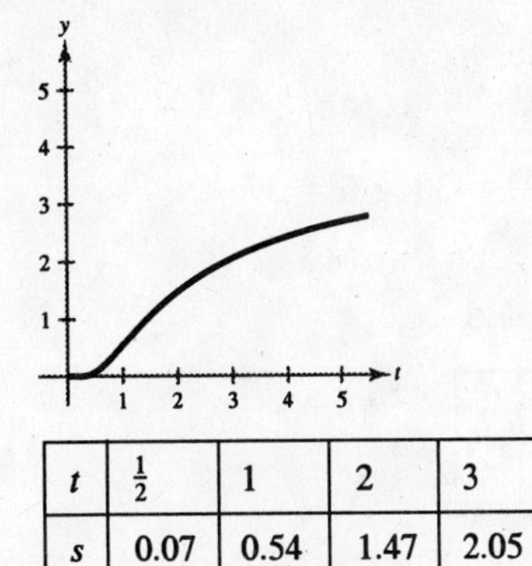

t	$\frac{1}{2}$	1	2	3	4
s	0.07	0.54	1.47	2.05	2.43

21. $g(t) = 8 - 0.5e^{-t/4}$

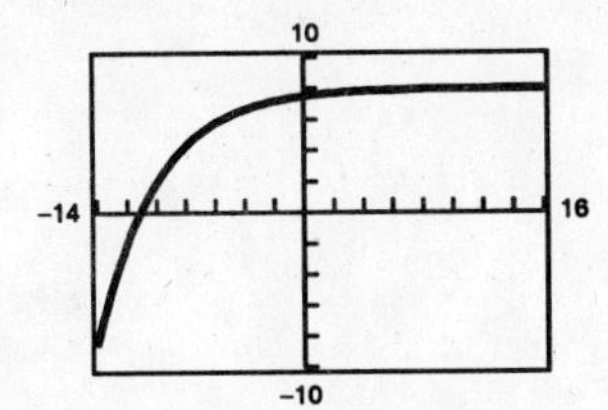

Horizontal asymptote: $y = 8$

23. $g(x) = 200e^{4/x}$

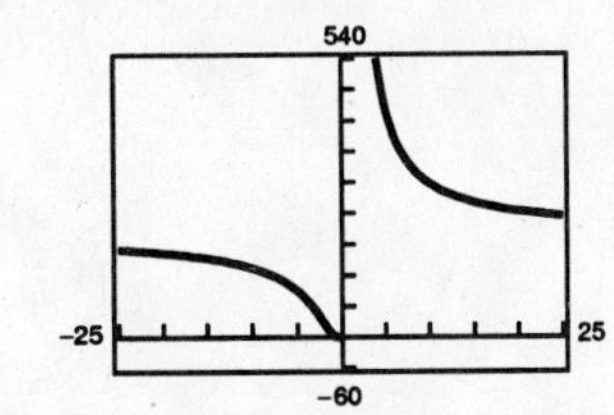

Vertical asymptote: $x = 0$

Horizontal asymptote: $y = 200$

25. $f(x) = \dfrac{10}{1 + 2^{-0.005x}}$

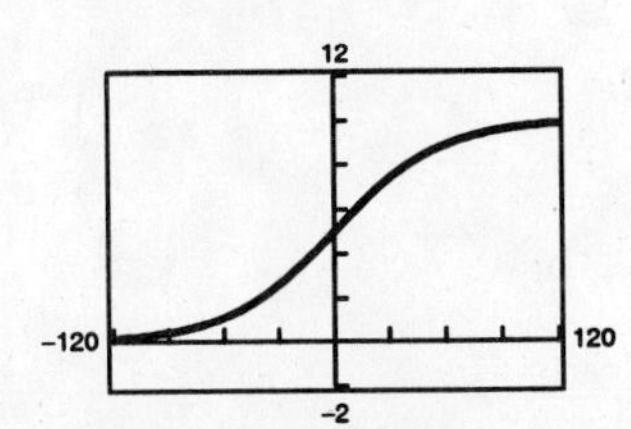

Horizontal asymptotes:
$y = 0,\ y = 10$

27. $e^{-8/3} \approx 0.069$

29. $6e^{4.12} \approx 369.355$

31. $200{,}000 = Pe^{0.08t}$

$$P = \frac{200{,}000}{e^{0.08t}}$$

t	1	10	20	30	40	50
P	\$184,623.27	\$89,865.79	\$40,379.30	\$18,143.59	\$8,152.44	\$3,663.13

33. $V(t) = 26{,}000\left(\dfrac{3}{4}\right)^t$

(a) 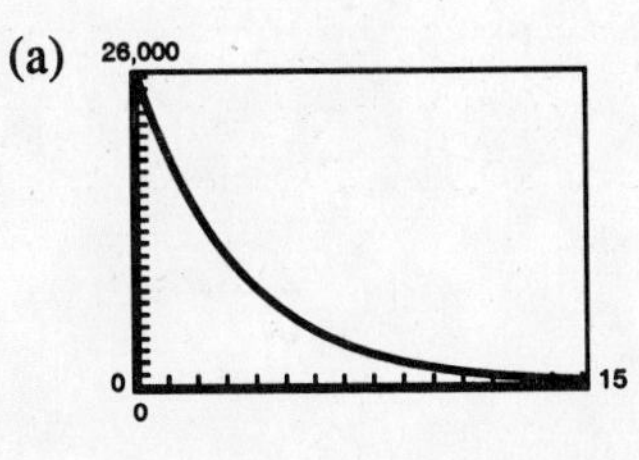

(b) For $t = 2$, $V(2) = \$14{,}625$

(c) The car depreciates most rapidly at the beginning, which is realistic.

35. $4^3 = 64$

$\log_4 64 = 3$

37. $25^{3/2} = 125$

$\log_{25}125 = \dfrac{3}{2}$

39. $\log_8 8^{-0.34} = -0.34 \log_8 8 = -0.34$

41. $\log_6 216 = \log_6 6^3 = 3 \log_6 6 = 3$

43. $\log_{36}\left(\dfrac{1}{6}\right) = \log_{36}1 - \log_{36}6 = -\log_{36}(36)^{1/2} = -\dfrac{1}{2}$

45. $g(x) = -\log_2 x + 5$

x	$\frac{1}{2}$	1	2	4
y	6	5	4	3

47. $f(x) = \log_2(x - 1) + 6$

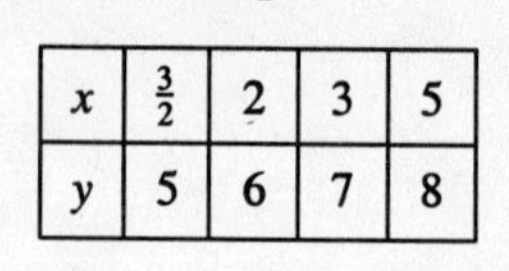

x	$\frac{3}{2}$	2	3	5
y	5	6	7	8

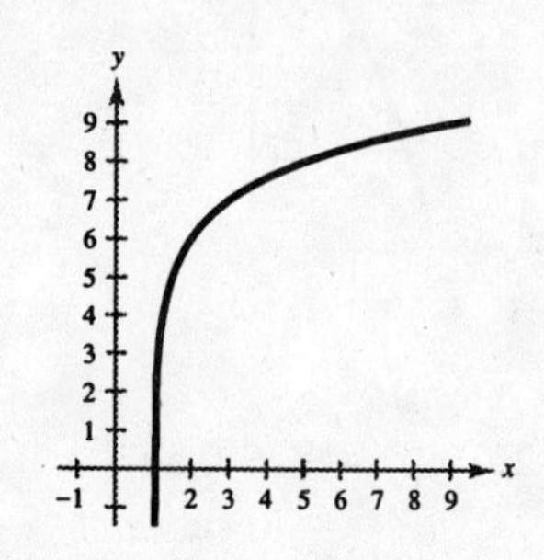

49. $f(x) = \ln x + 3$

Domain: $(0, \infty)$

Vertical asymptote: $x = 0$

x	1	2	3	$\frac{1}{2}$	$\frac{1}{4}$
$f(x)$	3	3.69	4.10	2.31	1.61

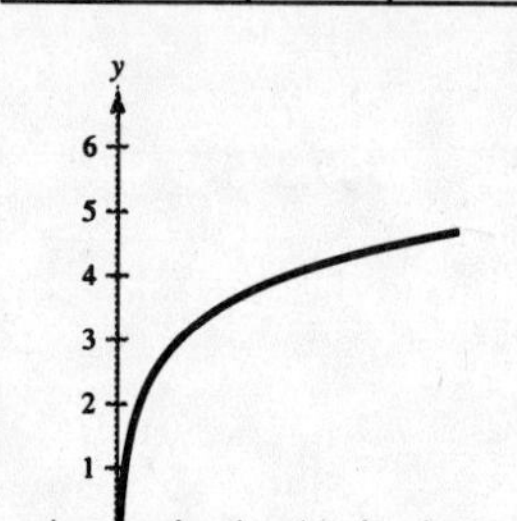

51. $h(x) = \frac{1}{2}\ln x$

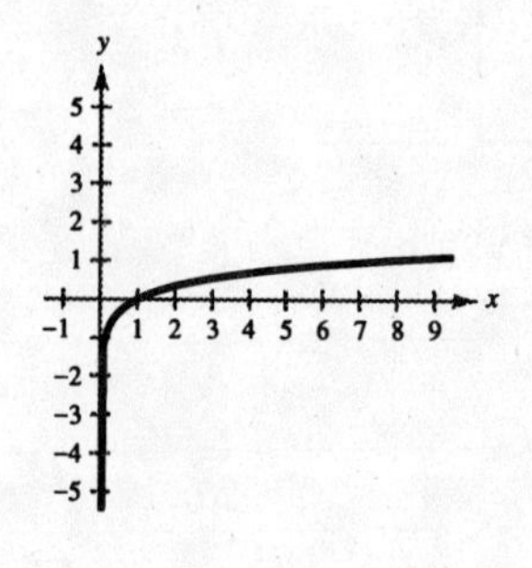

53. $y = \log_{10}(x^2 + 1)$

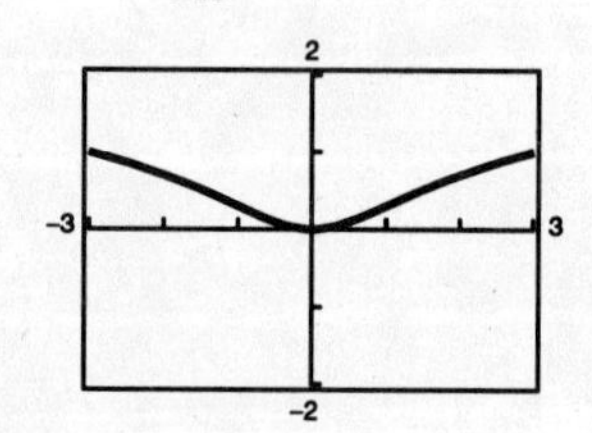

Domain: all x

55. $y = \sqrt{x}\ln(x + 1)$

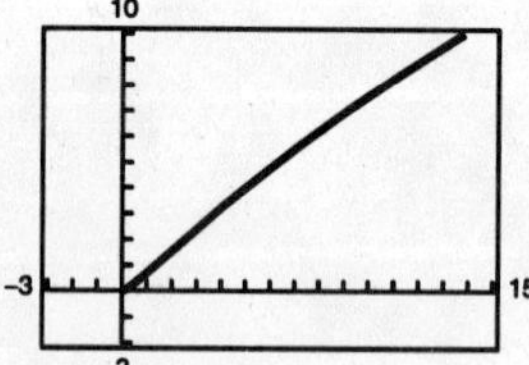

Domain: $x \geq 0$

57. $\ln e^7 = 7$

59. $6 \ln e^{-3} = 6(-3) \ln e = -18$

61. $t = 50 \log_{10} \dfrac{18{,}000}{18{,}000 - h}$

(a) $0 \le h < 18{,}000$

(b)

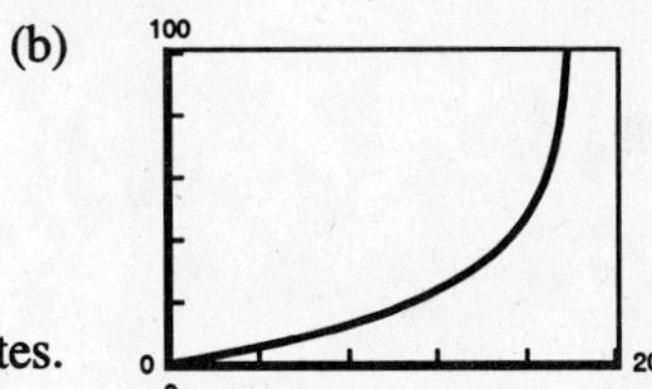

Vertical asymptote: $h = 18{,}000$

(c) The plane climbs at a slower rate as it approaches its absolute ceiling.

(d) If $h = 4000$, $t = 50 \log_{10} \dfrac{18{,}000}{18{,}000 - 4000} \approx 5.46$ minutes.

63. $\log_4 9 = \dfrac{\log_{10} 9}{\log_{10} 4} \approx 1.585$

$\log_4 9 = \dfrac{\ln 9}{\ln 4} \approx 1.585$

65. $\log_{12} 200 = \dfrac{\log_{10} 200}{\log_{10} 12} \approx 2.132$

$\log_{12} 200 = \dfrac{\ln 200}{\ln 12} \approx 2.132$

67. $\ln 20 = \ln[4 \cdot 5] = \ln 4 + \ln 5$

69. $\ln\left(\frac{5}{64}\right) = \ln 5 - \ln 64 = \ln 5 - \ln 4^3 = \ln 5 - 3 \ln 4$

71. $\log_b 25 = \log_b 5^2 = 2 \log_b 5 \approx 2(0.8271) = 1.6542$

73. $\log_b \sqrt{3} = \log_b 3^{1/2} = \frac{1}{2} \log_b 3 \approx \frac{1}{2}(0.5646) = 0.2823$

75. $\log_5 5x^2 = \log_5 5 + \log_5 x^2$

$= 1 + 2 \log_5 x$

77. $\log_{10} \dfrac{5\sqrt{y}}{x^2} = \log_{10} 5\sqrt{y} - \log_{10} x^2$

$= \log_{10} 5 + \log_{10}\sqrt{y} - \log_{10} x^2$

$= \log_{10} 5 + \dfrac{1}{2} \log_{10} y - 2 \log_{10} x$

79. $\ln[(x^2 + 1)(x - 1)] = \ln(x^2 + 1) + \ln(x - 1)$

81. $\log_2 5 + \log_2 x = \log_2 5x$

83. $\dfrac{1}{2} \ln|2x - 1| - 2 \ln|x + 1| = \ln\sqrt{|2x - 1|} - \ln|x + 1|^2$

$= \ln \dfrac{\sqrt{|2x - 1|}}{(x + 1)^2}$

85. $\ln 3 + \dfrac{1}{3} \ln(4 - x^2) - \ln x = \ln\left[\dfrac{3(4 - x^2)^{1/3}}{x}\right] = \ln\left[\dfrac{3\sqrt[3]{4 - x^2}}{x}\right]$

87. (a)

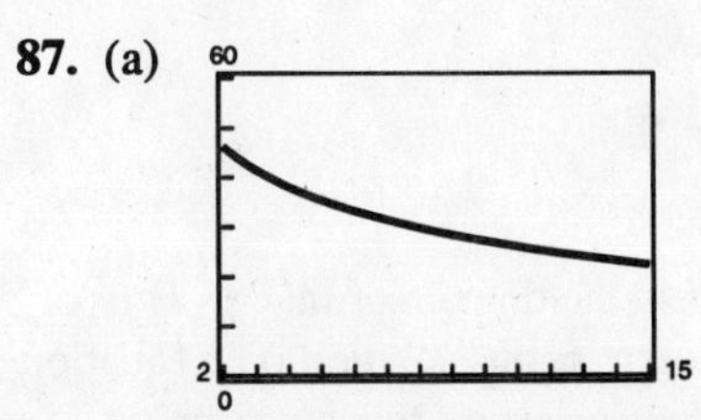

(b)

h	4	6	8	10	12	14
s	38	33.2	29.8	27.2	25	23.2

(c) As the depth increases, the number of miles of roads cleared decreases.

89. $8^x = 512 = 8^3 \Rightarrow x = 3$

91. $6^x = \frac{1}{216} = \frac{1}{6^3} = 6^{-3} \Rightarrow x = -3$

93. $\log_7 x = 4 \Rightarrow x = 7^4 = 2401$

95. $e^x = 12$

$x = \ln 12 \approx 2.485$

97. $3e^{-5x} = 132$

$e^{-5x} = 44$

$-5x = \ln 44$

$x = -\frac{\ln 44}{5} \approx -0.757$

99. $e^x + 13 = 35$

$e^x = 22$

$x = \ln 22 \approx 3.091$

101. $-4(5^x) = -68$

$5^x = 17$

$x \ln 5 = \ln 17$

$x = \frac{\ln 17}{\ln 5} \approx 1.760$

103. $e^{2x} - 7e^x + 10 = 0$

$(e^x - 5)(e^x - 2) = 0$

$e^x = 5 \Rightarrow x = \ln 5 \approx 1.609$

$e^x = 2 \Rightarrow x = \ln 2 \approx 0.693$

105. $\ln 3x = 8.2$

$3x = e^{8.2}$

$x = \frac{e^{8.2}}{3} \approx 1213.650$

107. $2 \ln 4x = 15$

$\ln 4x = \frac{15}{2}$

$4x = e^{15/2}$

$x = \frac{1}{4}e^{15/2} \approx 452.011$

109. $\ln x - \ln 3 = 2$

$\ln \frac{x}{3} = 2$

$\frac{x}{3} = e^2$

$x = 3e^2 \approx 22.167$

111. $\ln\sqrt{x+1} = 2$

$\frac{1}{2}\ln(x + 1) = 2$

$\ln(x + 1) = 4$

$x + 1 = e^4$

$x = e^4 - 1 \approx 53.598$

113. $\log(x - 1) = \log(x - 2) - \log(x + 2)$

$\log(x - 1) = \log\left(\frac{x - 2}{x + 2}\right)$

$x - 1 = \frac{x - 2}{x + 2}$

$(x - 1)(x + 2) = x - 2$

$x^2 + x - 2 = 2 - 2$

$x^2 = 0$

$x = 0$

Since $x = 0$ is not in the domain of $\ln(x - 1)$ or of $\ln(x - 2)$, it is an extraneous solution. The equation has no solution. You can verify this by graphing each side of the equation and observing that the two curves do not intersect.

115. $\log_{10}(1 - x) = -1$

$10^{-1} = 1 - x$

$x = 1 - 10^{-1} = 0.9$

117. $3(7550) = 7550e^{0.0725t}$

$$3 = e^{0.0725t}$$

$$\ln 3 = 0.0725$$

$$t = \frac{\ln 3}{0.0725} \approx 15.2 \text{ years}$$

119. $y = 3e^{-2x/3}$ Decreasing exponential. Matches (e).

121. $y = \ln(x + 3)$ logarithmic function shifted to left. Matches (f).

123. $y = 2e^{-(x+4)^2/3}$ Gaussian model. Matches (a).

125. $17{,}000 = 12{,}620e^{0.0118t}$

$$\frac{17000}{12620} = e^{0.0118t}$$

$$\ln\left(\frac{1700}{1262}\right) = 0.0118t$$

$$t = \frac{1}{0.0118}\ln\left(\frac{1700}{1262}\right) \approx 25.2 \text{ years, or } 2025$$

127. (a) $20{,}000 = 10{,}000e^{r(12)}$

$$2 = e^{12r}$$

$$\ln 2 = 12r$$

$$r = \frac{\ln 2}{12} \approx 0.0577 \text{ or } 5.78\%$$

(b) $10{,}000e^{0.0577(1)} \approx \$10{,}593.97$

(c) $\dfrac{10{,}593.97}{10{,}000} = 1.059 \text{ or } 5.9\%$

129. $N = \dfrac{157}{1 + 5.4e^{-0.12t}}$

(a) When $N = 50$:

$$50 = \frac{157}{1 + 5.4e^{-0.12t}}$$

$$1 + 5.4e^{-0.12t} = \frac{157}{50}$$

$$5.4e^{-0.12t} = \frac{107}{50}$$

$$e^{-0.12t} = \frac{107}{270}$$

$$-0.12t = \ln\frac{107}{270}$$

$$t = \frac{\ln(107/270)}{-0.12} \approx 7.7 \text{ weeks}$$

(b) When $N = 75$:

$$75 = \frac{157}{1 + 5.4e^{-0.12t}}$$

$$1 + 5.4e^{-0.12t} = \frac{157}{75}$$

$$5.4e^{-0.12t} = \frac{82}{75}$$

$$e^{-0.12t} = \frac{82}{405}$$

$$-0.12^t = \ln\frac{82}{405}$$

$$t = \frac{\ln(82/405)}{-0.12} \approx 13.3 \text{ weeks}$$

131. $y = ae^{bx}$

$$2 = ae^{b(0)} \implies a = 2$$

$$3 = 2e^{b(4)}$$

$$1.5 = e^{4b}$$

$$\ln 1.5 = 4b \implies b \approx 0.1014$$

Thus, $y \approx 2e^{0.1014x}$.

133. $y = ae^{bx}$

$$\tfrac{1}{2} = ae^{b(0)} \implies a = \tfrac{1}{2}$$

$$5 = \tfrac{1}{2}e^{b(5)}$$

$$10 = e^{5b}$$

$$\ln 10 = 5b \implies b \approx 0.4605$$

Thus, $y = \frac{1}{2}e^{0.4605x}$.

135. $y = 234.6839(0.8746)^x$

$= 234.684e^{-0.134x}$

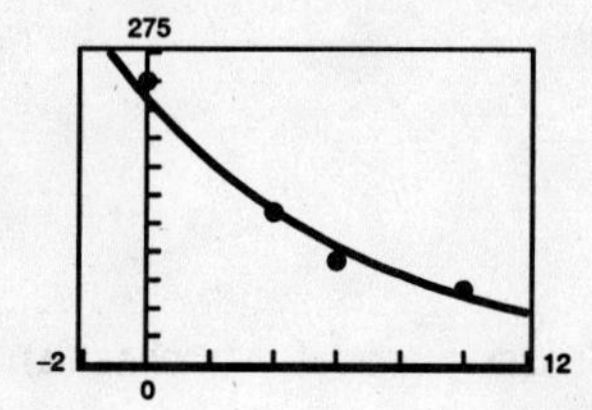

137. True; by the inverse properties, $\log_b b^{2x} = 2x$.

139. False; $\ln x + \ln y = \ln(xy) \neq \ln(x + y)$

141. True. $\log_{10}\left(\frac{10}{x}\right) = \log_{10}10 - \log_{10}x = 1 - \log_{10}x.$

Chapter 3 Practice Test

1. Solve for x: $x^{3/5} = 8$

2. Solve for x: $3^{x-1} = \frac{1}{81}$

3. Graph $f(x) = 2^{-x}$ by hand.

4. Graph $g(x) = e^x + 1$ by hand.

5. If \$5000 is invested at 9% interest, find the amount after three years if the interest is compounded

   (a) monthly (b) quarterly (c) continuously.

6. Write the equation in logarithmic form: $7^{-2} = \frac{1}{49}$

7. Solve for x: $x - 4 = \log_2 \frac{1}{64}$

8. Given $\log_b 2 = 0.3562$ and $\log_b 5 = 0.8271$, evaluate $\log_b \sqrt[4]{8/25}$.

9. Write $5 \ln x - \frac{1}{2} \ln y + 6 \ln z$ as a single logarithm.

10. Using your calculator and the change of base formula, evaluate $\log_9 28$.

11. Use your calculator to solve for N: $\log_{10} N = 0.6646$

12. Graph $y = \log_4 x$ by hand.

13. Determine the domain of $f(x) = \log_3(x^2 - 9)$.

14. Graph $y = \ln(x - 2)$ by hand.

15. True or false: $\dfrac{\ln x}{\ln y} = \ln(x - y)$

16. Solve for x: $5^x = 41$

17. Solve for x: $x - x^2 = \log_5 \frac{1}{25}$

18. Solve for x: $\log_2 x + \log_2(x - 3) = 2$

19. Solve for x: $\dfrac{e^x + e^{-x}}{3} = 4$

20. Six thousand dollars is deposited into a fund at an annual percentage rate of 13%. Find the time required for the investment to double if the interest is compounded continuously.

21. Use a graphing utility to find the points of intersection of the graphs of $y = \ln(3x)$ and $y = e^x - 4$.

22. Use a graphing utility to find the power model $y = ax^b$ for the data $(1, 1), (2, 5), (3, 8)$, and $(4, 17)$.

CHAPTER 4
Trigonometric Functions

CHAPTER 4
Trigonometric Functions

Section 4.1 Radian and Degree Measure

Solutions to Odd-Numbered Exercises

1.

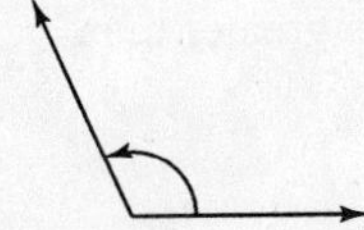

The angle shown is approximately 2 radians.

3.

The angle shown is approximately -3 radians.

5. (a) Since $0 < \frac{\pi}{5} < \frac{\pi}{2}$, $\frac{\pi}{5}$ lies in Quadrant I.

(b) Since $\pi < \frac{7\pi}{5} < \frac{3\pi}{2}$, $\frac{7\pi}{5}$ lies in Quadrant III.

7. (a) Since $-\frac{\pi}{2} < -\frac{\pi}{12} < 0$, $-\frac{\pi}{12}$ lies in Quadrant IV.

(b) Since $-\frac{3\pi}{2} < -\frac{11\pi}{9} < -\pi$, $-\frac{11\pi}{9}$ lies in Quadrant II.

9. (a) Since $\pi < 3.5 < \frac{3\pi}{2}$, 3.5 lies in Quadrant III.

(b) Since $\frac{\pi}{2} < 2.25 < \pi$, 2.25 lies in Quadrant II.

11. (a)

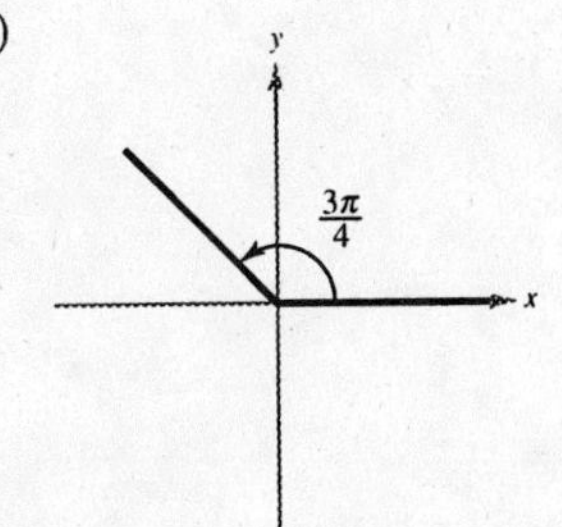

(b)

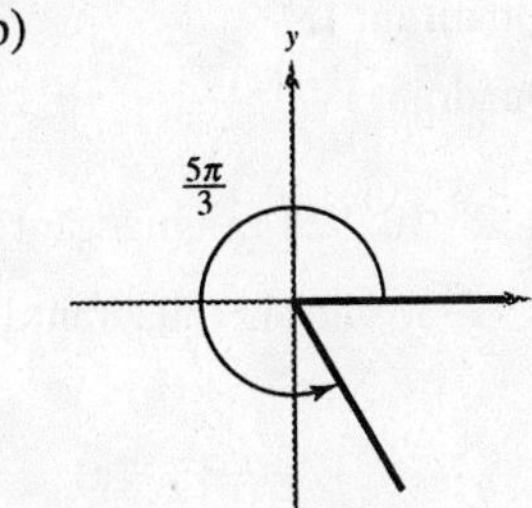

13. (a)

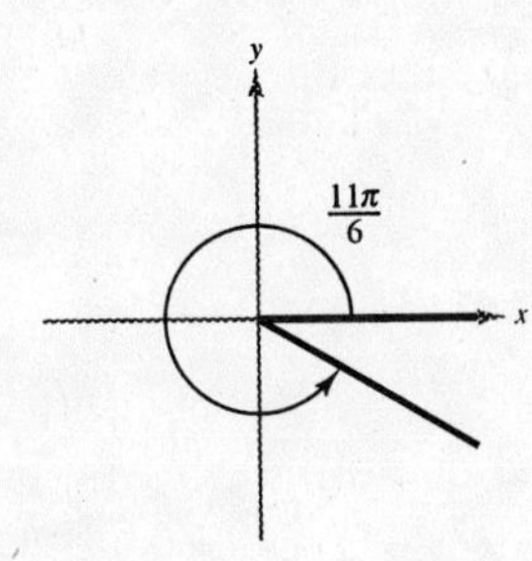

(b)

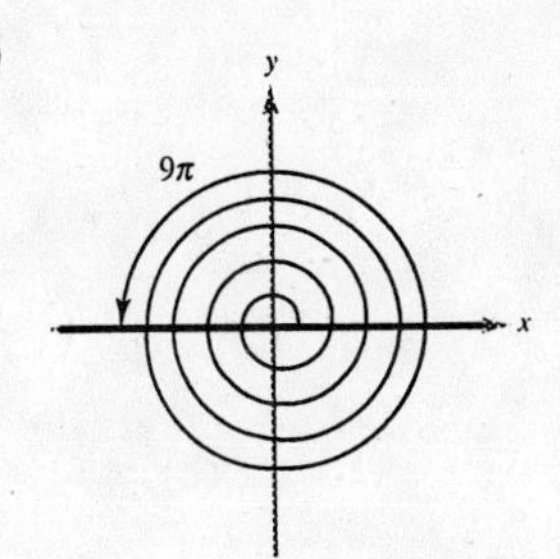

15. (a) Coterminal angles for $\frac{\pi}{12}$

$$\frac{\pi}{12} + 2\pi = \frac{25\pi}{12}$$

$$\frac{\pi}{12} - 2\pi = -\frac{23\pi}{12}$$

(b) Coterminal angles for $\frac{2\pi}{3}$

$$\frac{2\pi}{3} + 2\pi = \frac{8\pi}{3}$$

$$\frac{2\pi}{3} - 2\pi = -\frac{4\pi}{3}$$

17. (a) Coterminal angles for $-\dfrac{11\pi}{4}$

$$-\frac{11\pi}{4} + 4\pi = \frac{5\pi}{4}$$

$$-\frac{11\pi}{4} + 2\pi = -\frac{3\pi}{4}$$

(b) Coterminal angles for $-\dfrac{2\pi}{15}$

$$-\frac{2\pi}{15} + 2\pi = \frac{28\pi}{15}$$

$$-\frac{2\pi}{15} - 2\pi = -\frac{32\pi}{15}$$

19. (a) Complement: $\dfrac{\pi}{2} - \dfrac{\pi}{3} = \dfrac{\pi}{6}$

Supplement: $\pi - \dfrac{\pi}{3} = \dfrac{2\pi}{3}$

(b) Complement: Not possible; $\dfrac{3\pi}{4}$ is greater than $\dfrac{\pi}{2}$.

Supplement: $\pi - \dfrac{3\pi}{4} = \dfrac{\pi}{4}$

21. (a) Complement: $\dfrac{\pi}{2} - 1 \approx 0.57$

Supplement: $\pi - 1 \approx 2.14$

(b) Complement: none $\left(2 > \dfrac{\pi}{2}\right)$

Supplement: $\pi - 2 \approx 1.14$

23.

The angle shown is approximately 210°.

25.

The angle shown is approximately $-45°$.

27. (a) Since $90° < 150° < 180°$, 150° lies in Quadrant II.

(b) Since $270° < 282° < 360°$, 282° lies in Quadrant IV.

29. (a) Since $-180° < -132°\ 50' < -90°$, $-132°\ 50'$ lies in Quadrant III.

(b) Since $-360° < -336°\ 30' < -270°$, $-336°\ 30'$ lies in Quadrant I.

31. (a)

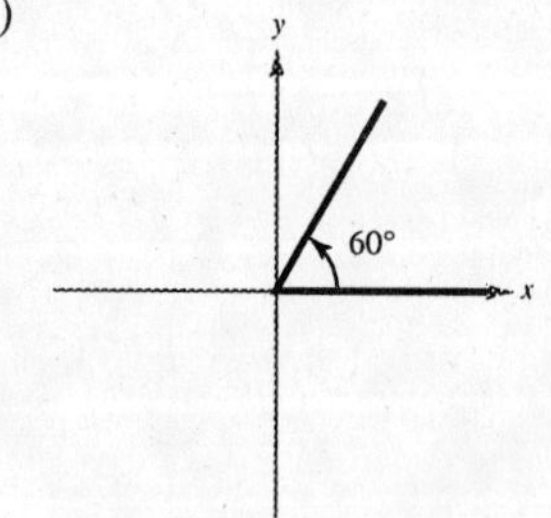

(b)

33. (a)

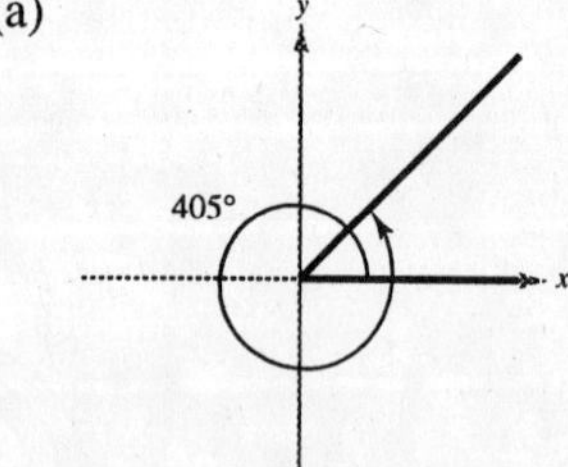

(b)

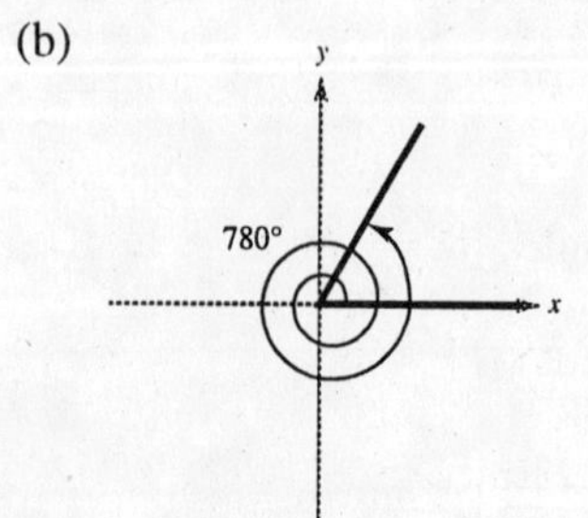

35. (a) Coterminal angles for $52°$

$52° + 360° = 412°$

$52° - 360° = -308°$

(b) Coterminal angles for $-36°$

$-36° + 360° = 324°$

$-36° - 360° = -396°$

37. (a) Coterminal angles for $300°$

$300° + 360° = 660°$

$300° - 360° = -60°$

(b) Coterminal angles for $230°$

$230° + 360° = 590°$

$230° - 360° = -130°$

39. (a) Complement of $24°$: $90° - 24° = 66°$

Supplement of $24°$: $180° - 24° = 156°$

(b) Complement of $126°$: Not possible because $126° > 90°$

Supplement of $126°$: $180° - 126° = 54°$

41. (a) Complement: $90° - 79° = 11°$

Supplement: $180° - 79° = 101°$

(b) Complement: does not exist

Supplement: $180° - 150° = 30°$

43. (a) $30° = 30\left(\frac{\pi}{180}\right) = \frac{\pi}{6}$

(b) $150° = 150\left(\frac{\pi}{180}\right) = \frac{5\pi}{6}$

45. (a) $-20° = -20\left(\frac{\pi}{180}\right) = -\frac{\pi}{9}$

(b) $-240° = -240\left(\frac{\pi}{180}\right) = -\frac{4\pi}{3}$

47. $115° = 115\left(\frac{\pi}{180}\right) \approx 2.007$ radians

49. $-216.35° = -216.35\left(\frac{\pi}{180}\right) \approx -3.776$ radians

51. $642° = 642\left(\frac{\pi}{180}\right) \approx 11.205$ radians

53. $-0.78° = -0.78\left(\frac{\pi}{180}\right) \approx -0.014$ radians

55. (a) $\frac{3\pi}{2} = \frac{3\pi}{2}\left(\frac{180}{\pi}\right)^{\circ} = 270°$

(b) $-\frac{7\pi}{6} = -\frac{7\pi}{6}\left(\frac{180}{\pi}\right)^{\circ} = -210°$

57. (a) $\frac{7\pi}{3} = \frac{7\pi}{3}\left(\frac{180°}{\pi}\right) = 420°$

(b) $-\frac{13\pi}{60} = -\frac{13\pi}{60}\left(\frac{180°}{\pi}\right) = -39°$

59. $\frac{\pi}{7} = \frac{\pi}{7}\left(\frac{180}{\pi}\right) \approx 25.714°$

61. $\frac{25\pi}{8} = \frac{25\pi}{8}\left(\frac{180}{\pi}\right) = 562.5°$

63. $-4.2\pi = -4.2\pi\left(\frac{180}{\pi}\right) = -756°$

65. $-2 = -2\left(\frac{180}{\pi}\right) \approx -114.592°$

67. (a) $64° \, 45' = 64° + \left(\frac{45}{60}\right)^{\circ} = 64.75°$

(b) $-124° \, 30' = -124° - \left(\frac{30}{60}\right)^{\circ} = -124.5°$

69. (a) $85° \, 18' 30'' = 85° + \left(\frac{18}{60}\right)^{\circ} + \left(\frac{30}{3600}\right)^{\circ} \approx 85.308°$

(b) $-408° \, 16' 25'' = -408° - \left(\frac{16}{60}\right)^{\circ} - \left(\frac{25}{3600}\right)^{\circ} \approx -408.274°$

71. (a) $280.6° = 280° + 0.6(60)' = 280° \, 36'$

(b) $-115.8° = -115° - 0.8(60)' = -115° \, 48'$

73. (a) $4.5 = 4.5\left(\frac{180}{\pi}\right)° \approx 257° \, 49' \, 51.628''$

(b) $-3.58 = -3.58\left(\frac{180}{\pi}\right)° \approx -205° \, 7' \, 8.006''$

75. $s = r\theta$

$6 = 5\theta$

$\theta = \frac{6}{5}$ radians

77. $s = r\theta$

$32 = 7\theta$

$\theta = \frac{32}{7} = 4\frac{4}{7}$ radians

79. $s = r\theta$

$8 = 15\theta$

$\theta = \frac{8}{15}$ radians

81. $s = r\theta$

$35 = 14.5\theta$

$\theta = \frac{70}{29} \approx 2.414$ radians

83. $s = r\theta$, θ in radians

$s = 14(180)\left(\frac{\pi}{180}\right) = 14\pi \approx 43.982$ inches

85. $s = r\theta$

$s = 6\left(\frac{2\pi}{3}\right) = 4\pi \approx 12.57$ meters

87. $\theta = 42° \, 7' \, 15'' - 25° \, 46' \, 37'' = 16° \, 20' \, 38'' \approx 0.2853$ radian

$s = r\theta = 4000(0.2853) \approx 1141.02$ miles

89. $\theta = \frac{s}{r} = \frac{600}{6378} \approx 0.094$ radian $\approx 5.39°$

91. $\theta = \frac{s}{r} = \frac{2.5}{6} = \frac{25}{60} = \frac{5}{12}$ radian $\approx 23.87°$

93. (a) single axel: $1\frac{1}{2}$ revolutions $= 360° + 180° = 540°$

$= 2\pi + \pi = 3\pi$ radians

(b) double axel: $2\frac{1}{2}$ revolutions $= 720° + 180° = 900°$

$= 4\pi + \pi = 5\pi$ radians

(c) triple axel: $3\frac{1}{2}$ revolutions $= 1260°$

$= 7\pi$ radians

95. (a) 40 miles per hour $= 40\frac{(5280)}{60} = 3520$ feet per minute

Circumference of tire is $C = 2.5\pi$ feet

Number of revolutions per minute is $r = \frac{3520}{2.5\pi} = \frac{1408}{\pi} \approx 448.2$ revolutions per minute

(b) The angular speed is $\frac{\theta}{t}$:

$\theta = \frac{3520}{2.5\pi}(2\pi) = 2816$ radians

Angular speed $= \frac{2816 \text{ radians}}{1 \text{ minute}} = 2816$ radians/minute

97. speed $=$ (360 revolutions/minute)(2π(1.68) inches/revolution)

$= 1209.6\pi$ inches/minute

$= 20.16\pi$ inches/second

99. False, 1 radian $= \left(\dfrac{180}{\pi}\right)^\circ \approx 57.3^\circ$, so one radian is much larger than one degree.

101. True: $\dfrac{2\pi}{3} + \dfrac{\pi}{4} + \dfrac{\pi}{12} = \dfrac{8\pi + 3\pi + \pi}{12} = \pi = 180^\circ$

103. Two angles in standard position are coterminal angles if they have the same initial and terminal sides. For example, 30° and 390° are coterminal.

105. $A = \dfrac{1}{2}r^2\theta = \dfrac{1}{2}(10)^2 \cdot \dfrac{\pi}{3} = \dfrac{50}{3}\pi\ m^2$

107. $A = \frac{1}{2}r^2\theta,\ s = r\theta$

(a) $\theta = 0.8 \implies A = \frac{1}{2}r^2(0.8) = 0.4r^2$ Domain: $r > 0$

$s = r\theta = r(0.8)$ Domain: $r > 0$

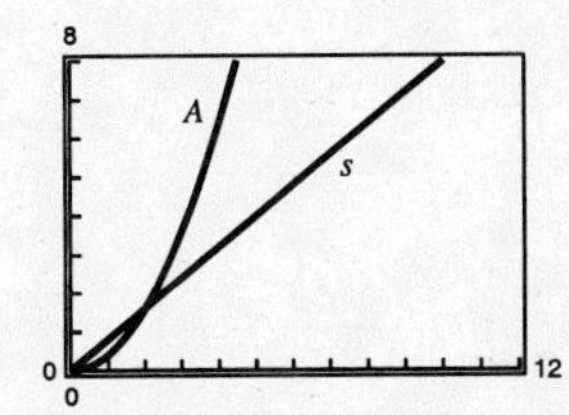

The area function changes more rapidly for $r > 1$ because it is quadratic and the arc length function is linear.

(b) $r = 10 \implies A = \frac{1}{2}(10^2)\theta = 50\theta$ Domain: $0 < \theta < 2\pi$

$s = r\theta = 10\theta$ Domain: $0 < \theta < 2\pi$

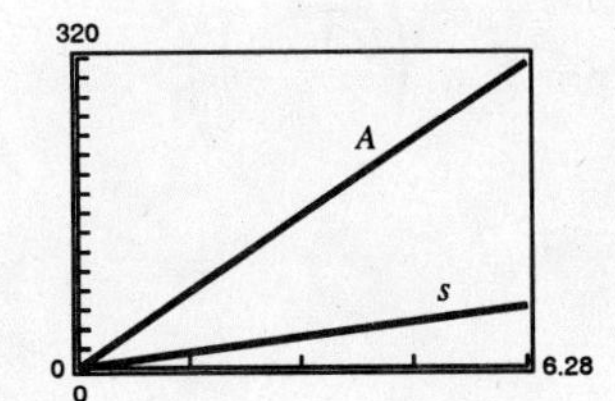

109.

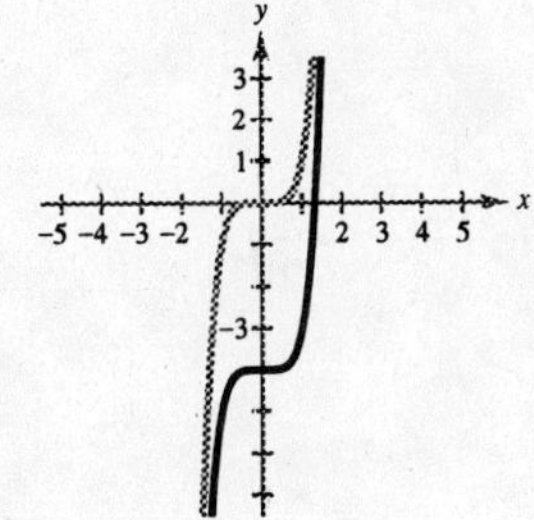

111.

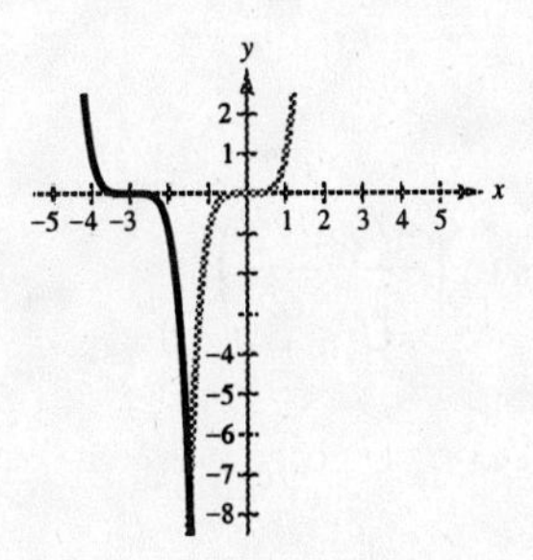

113.

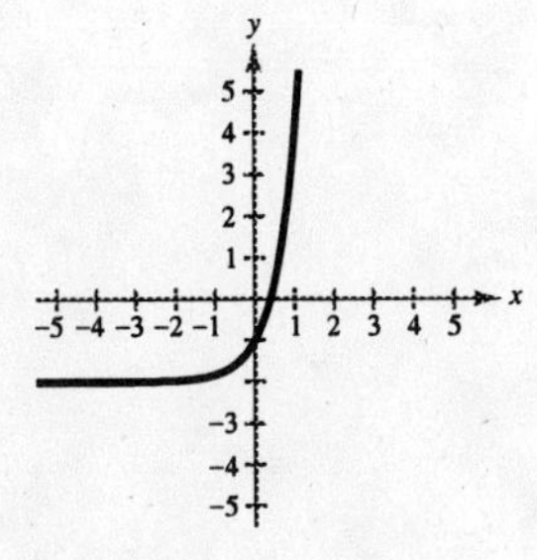

115.

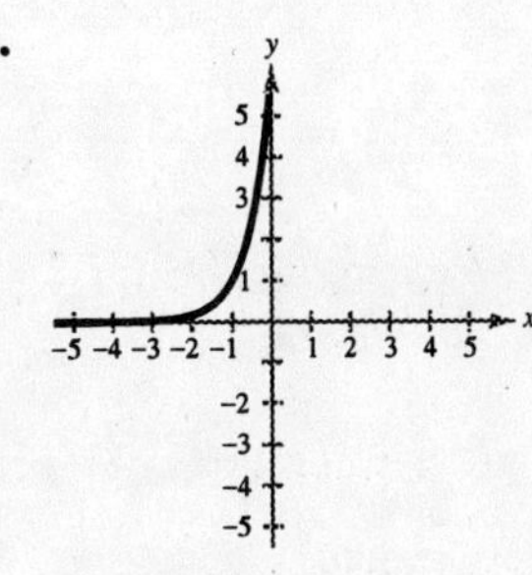

Section 4.2 Trigonometric Functions: The Unit Circle

Solutions to Odd-Numbered Exercises

1. $\sin t = y = \frac{15}{17}$

$\cos t = x = -\frac{8}{17}$

$\tan t = \frac{y}{x} = -\frac{15}{8}$

$\cot t = \frac{x}{y} = -\frac{8}{15}$

$\sec t = \frac{1}{x} = -\frac{17}{8}$

$\csc t = \frac{1}{y} = \frac{17}{15}$

3. $\sin t = y = -\frac{5}{13}$

$\cos t = x = \frac{12}{13}$

$\tan t = \frac{y}{x} = -\frac{5}{12}$

$\cot t = \frac{x}{y} = -\frac{12}{5}$

$\sec t = \frac{1}{x} = \frac{13}{12}$

$\csc t = \frac{1}{y} = -\frac{13}{5}$

5. $t = \frac{\pi}{4}$ corresponds to $\left(\frac{\sqrt{2}}{2}, \frac{\sqrt{2}}{2}\right)$.

7. $t = \frac{7\pi}{6}$ corresponds to $\left(-\frac{\sqrt{3}}{2}, -\frac{1}{2}\right)$.

9. $t = \frac{4\pi}{3}$ corresponds to $\left(-\frac{1}{2}, -\frac{\sqrt{3}}{2}\right)$.

11. $t = \frac{3\pi}{2}$ corresponds to $(0, -1)$.

13. $t = \frac{\pi}{4}$ corresponds to $\left(\frac{\sqrt{2}}{2}, \frac{\sqrt{2}}{2}\right)$.

$\sin t = y = \frac{\sqrt{2}}{2}$

$\cos t = x = \frac{\sqrt{2}}{2}$

$\tan t = \frac{y}{x} = 1$

15. $t = -\frac{\pi}{6}$ corresponds to $\left(\frac{\sqrt{3}}{2}, -\frac{1}{2}\right)$.

$\sin t = y = -\frac{1}{2}$

$\cos t = x = \frac{\sqrt{3}}{2}$

$\tan t = \frac{y}{x} = -\frac{\sqrt{3}}{3}$

17. $t = -\frac{7\pi}{4}$ corresponds to $\left(\frac{\sqrt{2}}{2}, \frac{\sqrt{2}}{2}\right)$

$\sin t = y = \frac{\sqrt{2}}{2}$

$\cos t = x = \frac{\sqrt{2}}{2}$

$\tan t = \frac{y}{x} = 1$

19. $t = \frac{11\pi}{6}$ corresponds to $\left(\frac{\sqrt{3}}{2}, -\frac{1}{2}\right)$.

$\sin t = y = -\frac{1}{2}$

$\cos t = x = \frac{\sqrt{3}}{2}$

$\tan t = \frac{y}{x} = -\frac{\sqrt{3}}{3}$

21. $t = -\frac{3\pi}{2}$ corresponds to $(0, 1)$.

$\sin t = y = 1$

$\cos t = x = 0$

$\tan t = \frac{y}{x}$ is undefined.

23. $t = \frac{3\pi}{4}$ corresponds to $\left(-\frac{\sqrt{2}}{2}, \frac{\sqrt{2}}{2}\right)$.

$\sin t = y = \frac{\sqrt{2}}{2}$ $\quad$ $\csc t = \frac{1}{y} = \sqrt{2}$

$\cos t = x = -\frac{\sqrt{2}}{2}$ $\quad$ $\sec t = \frac{1}{x} = -\sqrt{2}$

$\tan t = \frac{y}{x} = -1$ $\quad$ $\cot t = \frac{x}{y} = -1$

25. $t = \frac{\pi}{2}$ corresponds to $(0, 1)$.

$\sin t = y = 1$ $\quad$ $\csc t = \frac{1}{y} = 1$

$\cos t = x = 0$ $\quad$ $\sec t = \frac{1}{x}$ is undefined.

$\tan t = \frac{y}{x}$ is undefined. $\quad$ $\cot t = \frac{x}{y} = 0$

27. $t = -\frac{\pi}{3}$ corresponds to $\left(\frac{1}{2}, -\frac{\sqrt{3}}{2}\right)$

$\sin t = y = -\frac{\sqrt{3}}{2}$, $\cos t = x = \frac{1}{2}$, $\tan t = \frac{y}{x} = -\sqrt{3}$

$\csc t = \frac{1}{y} = -\frac{2}{\sqrt{3}}$, $\sec t = \frac{1}{x} = 2$, $\cot t = \frac{x}{y} = -\frac{1}{\sqrt{3}}$

29. $\sin 5\pi = \sin \pi = 0$

31. $\cos \frac{8\pi}{3} = \cos \frac{2\pi}{3} = -\frac{1}{2}$

33. $\cos(-3\pi) = \cos \pi = -1$

35. $\sin\left(-\frac{9\pi}{4}\right) = \sin\left(-\frac{\pi}{4}\right) = -\frac{\sqrt{2}}{2}$

37. $\sin t = \frac{1}{3}$

(a) $\sin(-t) = -\sin t = -\frac{1}{3}$

(b) $\csc(-t) = -\csc t = -3$

39. $\cos(-t) = -\frac{1}{5}$

(a) $\cos t = \cos(-t) = -\frac{1}{5}$

(b) $\sec(-t) = \frac{1}{\cos(-t)} = -5$

41. $\sin t = \frac{4}{5}$

(a) $\sin(\pi - t) = \sin t = \frac{4}{5}$

(b) $\sin(t + \pi) = -\sin t = -\frac{4}{5}$

43. $\sin \frac{\pi}{4} \approx 0.7071$

45. $\csc 1.3 \approx 1.0378$

47. $\cos(-1.7) \approx -0.1288$

49. $\csc 0.8 = \frac{1}{\sin 0.8} \approx 1.3940$

51. $\sec 22.8 = \frac{1}{\cos 22.8} \approx -1.4486$

53. (a) $\sin 5 \approx -1$

(b) $\cos 2 \approx -0.4$

55. (a) $\sin t = 0.25$

$t \approx 0.25$ or 2.89

(b) $\cos t = -0.25$

$t \approx 1.82$ or 4.46

57. $I = 5e^{-2(0.7)} \sin(0.7) \approx 0.794$

59. $y(t) = \frac{1}{4}\cos 6t$

(a) $y(0) = \frac{1}{4}\cos 0 = 0.2500$ feet

(b) $y\left(\frac{1}{4}\right) = \frac{1}{4}\cos\frac{3}{2} \approx 0.0177$ feet

(c) $y\left(\frac{1}{2}\right) = \frac{1}{4}\cos 3 \approx -0.2475$ feet

61. $\cos 1.5 \approx 0.0707$, $2\cos 0.75 \approx 1.4634$

Thus, $\cos 2t \neq 2\cos t$

63. False. $\sin\left(\dfrac{-4\pi}{3}\right) = \dfrac{\sqrt{3}}{2} > 0$

65. (a) The points have y-axis symmetry.

(b) $\sin t_1 = \sin(\pi - t_1)$ since they have the same y-value.

(c) $-\cos t_1 = \cos(\pi - t_1)$ since the x-values have the opposite signs.

67. $\cos\theta = x = \cos(-\theta)$

$$\sin\theta = \frac{1}{x} = \sec(-\theta)$$

$$\sin\theta = y$$

$$\sin(-\theta) = -y = -\sin\theta$$

$$\sec\theta = \frac{1}{y}$$

$$\sec(-\theta) = -\frac{1}{y} = -\sec\theta$$

$$\tan\theta = \frac{y}{x}$$

$$\tan(-\theta) = \frac{-y}{x} = -\tan\theta$$

$$\cot\theta = \frac{x}{y}$$

$$\cot(-\theta) = \frac{x}{-y} = -\cot\theta$$

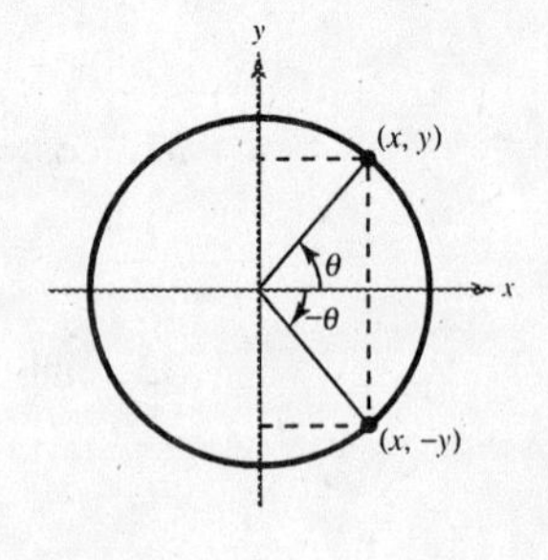

69. $f(t) = \sin t$ and $g(t) = \tan t$

Both f and g are odd functions.

$h(t) = f(t)g(t) = \sin t \tan t$

$h(-t) = \sin(-t)\tan(-t)$

$= (-\sin t)(-\tan t)$

$= \sin t \tan t = h(t)$

The function $h(t) = f(t)g(t)$ is even.

71. $f(x) = \frac{1}{4}x^3 + 1$

$$y = \frac{1}{4}x^3 + 1$$

$$x = \frac{1}{4}y^3 + 1$$

$$x - 1 = \frac{1}{4}y^3$$

$$4(x - 1) = y^3$$

$$y = \sqrt[3]{4(x - 1)}$$

$$f^{-1}(x) = \sqrt[3]{4(x - 1)}$$

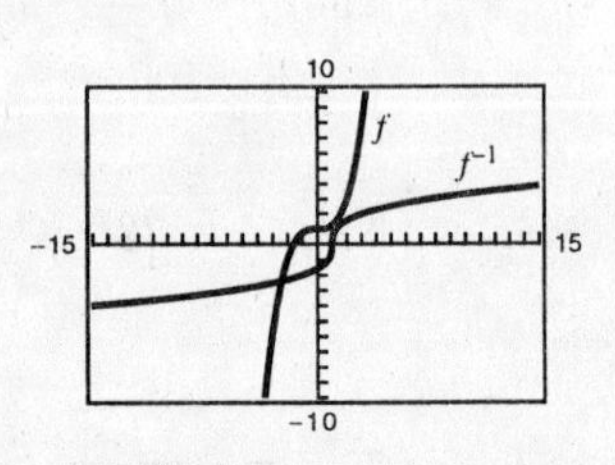

73. $f(x) = \frac{2x}{x+1}, x > -1$

$$y = \frac{2x}{x+1}, x > -1$$
$$x = \frac{2y}{y+1}$$
$$xy + x = 2y$$
$$x = 2y - xy$$
$$x = y(2 - x)$$
$$\frac{x}{2-x} = y, x < 2$$
$$f^{-1}(x) = \frac{x}{2-x}, x < 2$$

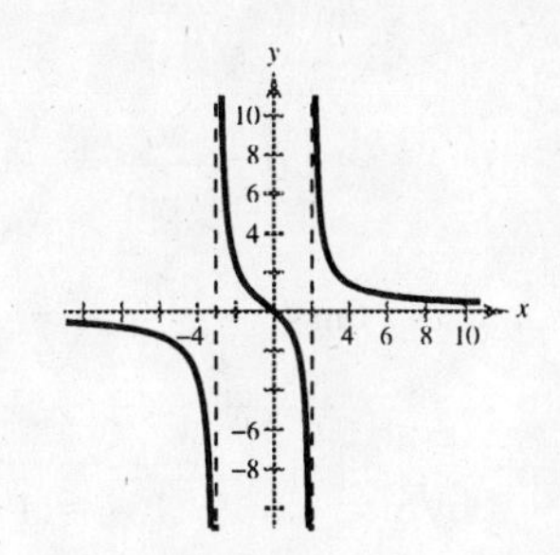

75. $f(x) = \frac{5x}{x^2 - x - 6} = \frac{5x}{(x+3)(x-2)}$

Asymptotes: $x = -3, x = 2, y = 0$

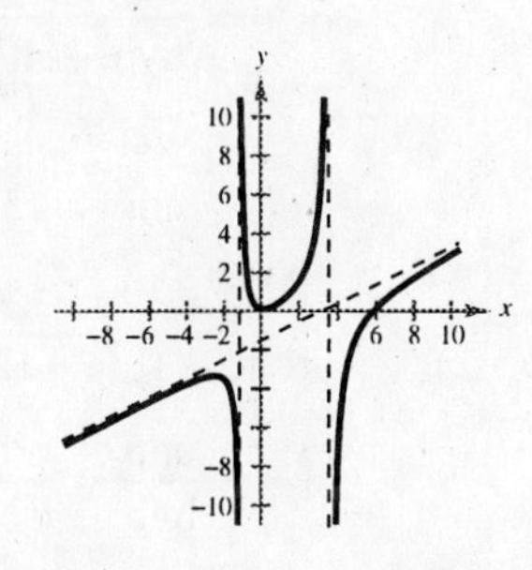

77. $f(x) = \frac{x^3 - 6x^2 + x - 1}{2x^2 - 5x - 8} = \frac{x}{2} - \frac{7}{4} - \frac{15(x+4)}{4(2x^2 - 5x - 8)}$

Slant asymptote: $y = \frac{x}{2} - \frac{7}{4}$

Vertical asymptotes: $x \approx 3.608, x \approx -1.108$

79. $C(10) \approx 69.95(1.045)^{10} \approx \$108.63$

81. (a) $p(0) = \frac{1200}{1 + 3e^0} = \frac{1200}{4} = 300$

(b) $p(5) \approx 570$ (c) $p = 800$ when $t \approx 8.96$ years

Section 4.3 Right Triangle Trigonometry

- You should know the right triangle definition of trigonometric functions.

(a) $\sin\theta = \frac{\text{opp}}{\text{hyp}}$ (b) $\cos\theta = \frac{\text{adj}}{\text{hyp}}$ (c) $\tan\theta = \frac{\text{opp}}{\text{adj}}$

(d) $\csc\theta = \frac{\text{hyp}}{\text{opp}}$ (e) $\sec\theta = \frac{\text{hyp}}{\text{adj}}$ (f) $\cot\theta = \frac{\text{adj}}{\text{opp}}$

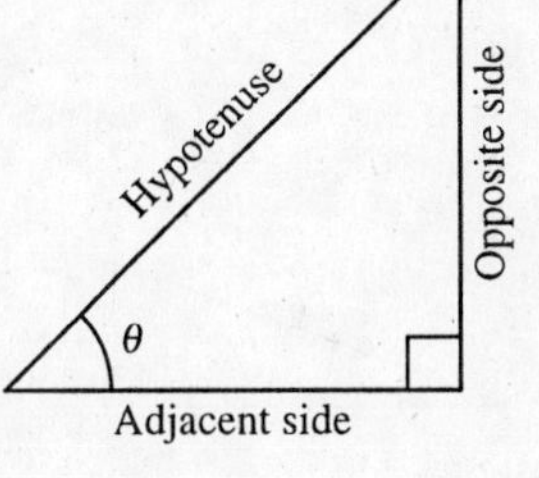

- You should know the following identities.

(a) $\sin\theta = \frac{1}{\csc\theta}$ (b) $\csc\theta = \frac{1}{\sin\theta}$ (c) $\cos\theta = \frac{1}{\sec\theta}$

(d) $\sec\theta = \frac{1}{\cos\theta}$ (e) $\tan\theta = \frac{1}{\cot\theta}$ (f) $\cot\theta = \frac{1}{\tan\theta}$

(g) $\tan\theta = \frac{\sin\theta}{\cos\theta}$ (h) $\cot\theta = \frac{\cos\theta}{\sin\theta}$ (i) $\sin^2\theta + \cos^2\theta = 1$

(j) $1 + \tan^2\theta = \sec^2\theta$ (k) $1 + \cot^2\theta = \csc^2\theta$

- You should know that two acute angles α and β are complementary if $\alpha + \beta = 90°$, and cofunctions of complementary angles are equal.
- You should know the trigonometric function values of 30°, 45°, and 60°, or be able to construct triangles from which you can determine them.

Solutions to Odd-Numbered Exercises

1.

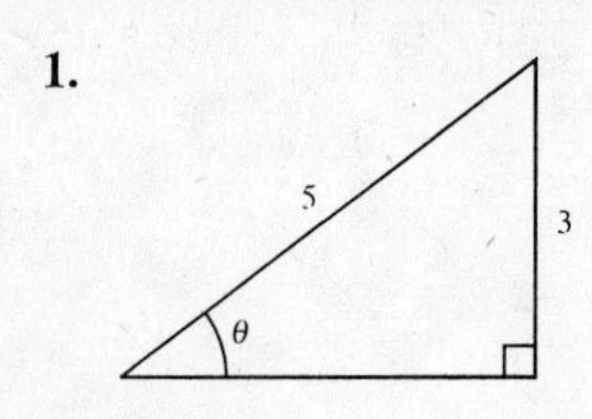

$$\text{adj} = \sqrt{5^2 - 3^2} = \sqrt{16} = 4$$

$$\sin\theta = \frac{\text{opp}}{\text{hyp}} = \frac{3}{5} \qquad \csc\theta = \frac{\text{hyp}}{\text{opp}} = \frac{5}{3}$$

$$\cos\theta = \frac{\text{adj}}{\text{hyp}} = \frac{4}{5} \qquad \sec\theta = \frac{\text{hyp}}{\text{adj}} = \frac{5}{4}$$

$$\tan\theta = \frac{\text{opp}}{\text{adj}} = \frac{3}{4} \qquad \cot\theta = \frac{\text{adj}}{\text{opp}} = \frac{4}{3}$$

3.

$$\text{hyp} = \sqrt{8^2 + 15^2} = 17$$

$$\sin\theta = \frac{\text{opp}}{\text{hyp}} = \frac{8}{17} \qquad \csc\theta = \frac{\text{hyp}}{\text{opp}} = \frac{17}{8}$$

$$\cos\theta = \frac{\text{adj}}{\text{hyp}} = \frac{15}{17} \qquad \sec\theta = \frac{\text{hyp}}{\text{adj}} = \frac{17}{15}$$

$$\tan\theta = \frac{\text{opp}}{\text{adj}} = \frac{8}{15} \qquad \cot\theta = \frac{\text{adj}}{\text{opp}} = \frac{15}{8}$$

5.

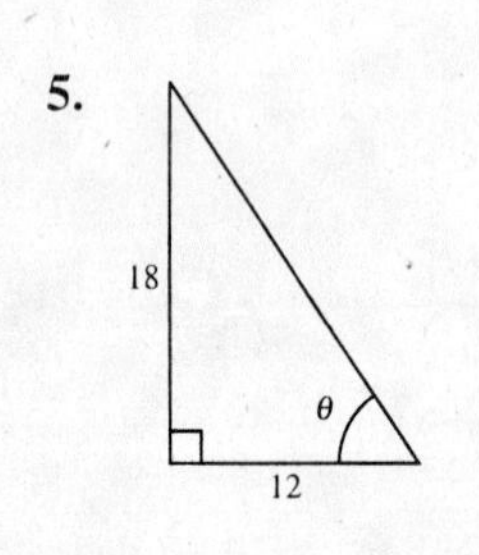

$$\text{hyp} = \sqrt{18^2 + 12^2} = \sqrt{468} = 6\sqrt{13}$$

$$\sin\theta = \frac{\text{opp}}{\text{hyp}} = \frac{18}{6\sqrt{13}} = \frac{3}{\sqrt{13}} = \frac{3\sqrt{13}}{13} \qquad \csc\theta = \frac{\text{hyp}}{\text{opp}} = \frac{\sqrt{13}}{3}$$

$$\cos\theta = \frac{\text{adj}}{\text{hyp}} = \frac{12}{6\sqrt{13}} = \frac{2}{\sqrt{13}} = \frac{2\sqrt{13}}{13} \qquad \sec\theta = \frac{\text{hyp}}{\text{adj}} = \frac{\sqrt{13}}{2}$$

$$\tan\theta = \frac{\text{opp}}{\text{adj}} = \frac{18}{12} = \frac{3}{2} \qquad \cot\theta = \frac{\text{adj}}{\text{opp}} = \frac{2}{3}$$

7.

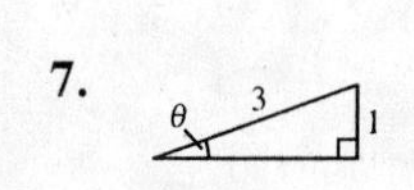

$$\text{adj} = \sqrt{3^2 - 1^2} = \sqrt{8} = 2\sqrt{2}$$

$$\sin\theta = \frac{\text{opp}}{\text{hyp}} = \frac{1}{3} \qquad \csc\theta = \frac{\text{hyp}}{\text{opp}} = 3$$

$$\cos\theta = \frac{\text{adj}}{\text{hyp}} = \frac{2\sqrt{2}}{3} \qquad \sec\theta = \frac{\text{hyp}}{\text{adj}} = \frac{3}{2\sqrt{2}} = \frac{3\sqrt{2}}{4}$$

$$\tan\theta = \frac{\text{opp}}{\text{adj}} = \frac{1}{2\sqrt{2}} = \frac{\sqrt{2}}{4} \qquad \cot\theta = \frac{\text{adj}}{\text{opp}} = 2\sqrt{2}$$

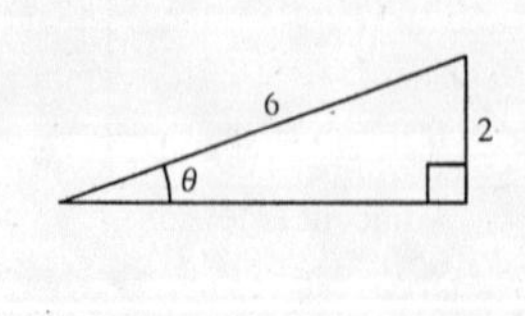

$$\text{adj} = \sqrt{6^2 - 2^2} = \sqrt{32} = 4\sqrt{2}$$

$$\sin\theta = \frac{\text{opp}}{\text{hyp}} = \frac{2}{6} = \frac{1}{3} \qquad \csc\theta = \frac{\text{hyp}}{\text{opp}} = \frac{6}{2} = 3$$

$$\cos\theta = \frac{\text{adj}}{\text{hyp}} = \frac{4\sqrt{2}}{6} = \frac{2\sqrt{2}}{3} \qquad \sec\theta = \frac{\text{hyp}}{\text{adj}} = \frac{6}{4\sqrt{2}} = \frac{3}{2\sqrt{2}} = \frac{3\sqrt{2}}{4}$$

$$\tan\theta = \frac{\text{opp}}{\text{adj}} = \frac{2}{4\sqrt{2}} = \frac{1}{2\sqrt{2}} = \frac{\sqrt{2}}{4} \qquad \cot\theta = \frac{\text{adj}}{\text{opp}} = \frac{4\sqrt{2}}{2} = 2\sqrt{2}$$

The function values are the same since the triangles are similar and the corresponding sides are proportional.

9. 

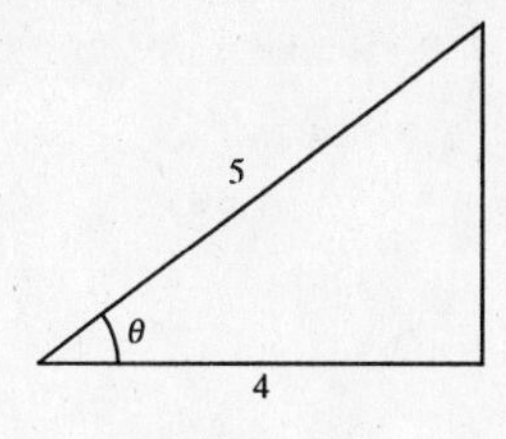

$\text{opp} = \sqrt{10^2 - 8^2} = 6$

$\sin\theta = \dfrac{\text{opp}}{\text{hyp}} = \dfrac{6}{10} = \dfrac{3}{5}$ $\qquad$ $\csc\theta = \dfrac{\text{hyp}}{\text{opp}} = \dfrac{10}{6} = \dfrac{5}{3}$

$\cos\theta = \dfrac{\text{adj}}{\text{hyp}} = \dfrac{8}{10} = \dfrac{4}{5}$ $\qquad$ $\sec\theta = \dfrac{\text{hyp}}{\text{adj}} = \dfrac{10}{8} = \dfrac{5}{4}$

$\tan\theta = \dfrac{\text{opp}}{\text{adj}} = \dfrac{6}{8} = \dfrac{3}{4}$ $\qquad$ $\cot\theta = \dfrac{\text{adj}}{\text{opp}} = \dfrac{8}{6} = \dfrac{4}{3}$

$\text{opp} = \sqrt{2.5^2 - 2^2} = 1.5$

$\sin\theta = \dfrac{\text{opp}}{\text{hyp}} = \dfrac{1.5}{2.5} = \dfrac{3}{5}$ $\qquad$ $\csc\theta = \dfrac{\text{hyp}}{\text{opp}} = \dfrac{2.5}{1.5} = \dfrac{5}{3}$

$\cos\theta = \dfrac{\text{adj}}{\text{hyp}} = \dfrac{2}{2.5} = \dfrac{4}{5}$ $\qquad$ $\sec\theta = \dfrac{\text{hyp}}{\text{adj}} = \dfrac{2.5}{2} = \dfrac{5}{4}$

$\tan\theta = \dfrac{\text{opp}}{\text{adj}} = \dfrac{1.5}{2} = \dfrac{3}{4}$ $\qquad$ $\cot\theta = \dfrac{\text{adj}}{\text{opp}} = \dfrac{2}{1.5} = \dfrac{4}{3}$

The function values are the same since the triangles are similar and the corresponding sides are proportional.

11. Given: $\sin\theta = \dfrac{5}{6} = \dfrac{\text{opp}}{\text{hyp}}$

$5^2 + (\text{adj})^2 = 6^2$

$\text{adj} = \sqrt{11}$

$\cos\theta = \dfrac{\sqrt{11}}{6}$

$\tan\theta = \dfrac{5}{\sqrt{11}} = \dfrac{5\sqrt{11}}{11}$

$\cot\theta = \dfrac{\sqrt{11}}{5}$

$\sec\theta = \dfrac{6}{\sqrt{11}} = \dfrac{6\sqrt{11}}{11}$

$\csc\theta = \dfrac{6}{5}$

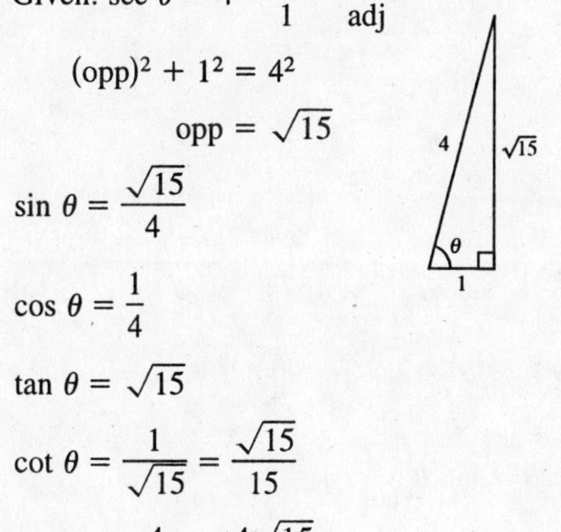

13. Given: $\sec\theta = 4 = \dfrac{4}{1} = \dfrac{\text{hyp}}{\text{adj}}$

$(\text{opp})^2 + 1^2 = 4^2$

$\text{opp} = \sqrt{15}$

$\sin\theta = \dfrac{\sqrt{15}}{4}$

$\cos\theta = \dfrac{1}{4}$

$\tan\theta = \sqrt{15}$

$\cot\theta = \dfrac{1}{\sqrt{15}} = \dfrac{\sqrt{15}}{15}$

$\csc\theta = \dfrac{4}{\sqrt{15}} = \dfrac{4\sqrt{15}}{15}$

15. Given: $\tan\theta = 3 = \dfrac{3}{1} = \dfrac{\text{opp}}{\text{adj}}$

$3^2 + 1^2 = (\text{hyp})^2$

$\text{hyp} = \sqrt{10}$

$\sin\theta = \dfrac{3\sqrt{10}}{10}$

$\cos\theta = \dfrac{\sqrt{10}}{10}$

$\cot\theta = \dfrac{1}{3}$

$\sec\theta = \sqrt{10}$

$\csc\theta = \dfrac{\sqrt{10}}{3}$

17. Given: $\cot\theta = \dfrac{9}{4} = \dfrac{\text{adj}}{\text{hyp}}$

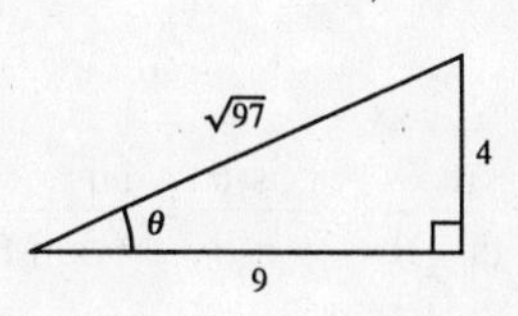

$4^2 + 9^2 = (\text{hyp})^2$

$\text{hyp} = \sqrt{97}$

$\sin\theta = \dfrac{4}{\sqrt{97}} = \dfrac{4\sqrt{97}}{97}$

$\cos\theta = \dfrac{9}{\sqrt{97}} = \dfrac{9\sqrt{97}}{97}$

$\tan\theta = \dfrac{4}{9}$

$\sec\theta = \dfrac{\sqrt{97}}{9}$

$\csc\theta = \dfrac{\sqrt{97}}{4}$

19. $\sin 60° = \dfrac{\sqrt{3}}{2}, \ \cos 60° = \dfrac{1}{2}$

(a) $\tan 60° = \dfrac{\sin 60°}{\cos 60°} = \sqrt{3}$

(b) $\sin 30° = \cos 60° = \dfrac{1}{2}$

(c) $\cos 30° = \sin 60° = \dfrac{\sqrt{3}}{2}$

(d) $\cot 60° = \dfrac{\cos 60°}{\sin 60°} = \dfrac{1}{\sqrt{3}} = \dfrac{\sqrt{3}}{3}$

21. $\csc \theta = 3, \ \sec \theta = \dfrac{3\sqrt{2}}{4}$

(a) $\sin \theta = \dfrac{1}{\csc \theta} = \dfrac{1}{3}$

(b) $\cos \theta = \dfrac{1}{\sec \theta} = \dfrac{2\sqrt{2}}{3}$

(c) $\tan \theta = \dfrac{\sin \theta}{\cos \theta} = \dfrac{1/3}{(2\sqrt{2})/3} = \dfrac{\sqrt{2}}{4}$

(d) $\sec(90° - \theta) = \csc \theta = 3$

23. $\cos \alpha = \dfrac{1}{4}$

(a) $\sec \alpha = \dfrac{1}{\cos \alpha} = 4$

(b) $\sin^2 \alpha + \cos^2 \alpha = 1$

$$\sin^2 \alpha + \left(\frac{1}{4}\right)^2 = 1$$

$$\sin^2 \alpha = \frac{15}{16}$$

$$\sin \alpha = \pm\frac{\sqrt{15}}{4}$$

(c) $\cot \alpha = \dfrac{\cos \alpha}{\sin \alpha} = \pm\dfrac{1/4}{\sqrt{15}/4} = \pm\dfrac{1}{\sqrt{15}} = \pm\dfrac{\sqrt{15}}{15}$

(d) $\sin(90° - \alpha) = \cos \alpha = \dfrac{1}{4}$

25. $\tan \theta \cot \theta = \tan \theta\left(\dfrac{1}{\tan \theta}\right) = 1$

27. $\tan \alpha \cos \alpha = \left(\dfrac{\sin \alpha}{\cos \alpha}\right)\cos \alpha = \sin \alpha$

29.
$$\begin{aligned}(1 + \cos \theta)(1 - \cos \theta) &= 1 - \cos^2 \theta \\ &= (\sin^2 \theta + \cos^2 \theta) - \cos^2 \theta \\ &= \sin^2 \theta\end{aligned}$$

31.
$$\begin{aligned}\frac{\sin \theta}{\cos \theta} + \frac{\cos \theta}{\sin \theta} &= \frac{\sin^2 \theta + \cos^2 \theta}{\sin \theta \cos \theta} \\ &= \frac{1}{\sin \theta \cos \theta} \\ &= \frac{1}{\sin \theta} \cdot \frac{1}{\cos \theta} \\ &= \csc \theta \sec \theta\end{aligned}$$

33. (a) $\cos 60° = \dfrac{1}{2}$

(b) $\tan \dfrac{\pi}{6} = \dfrac{1}{\sqrt{3}} = \dfrac{\sqrt{3}}{3}$

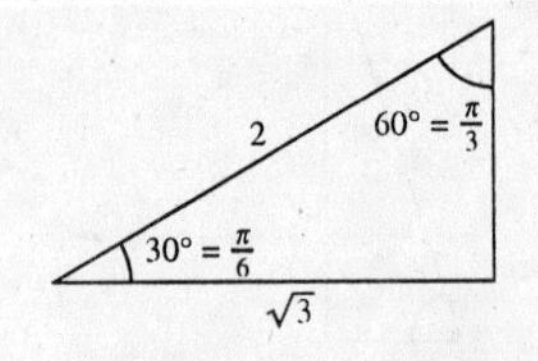

35. (a) $\cot \frac{\pi}{4} = \cot 45° = 1$

(b) $\cos 45° = \frac{1}{\sqrt{2}} = \frac{\sqrt{2}}{2}$

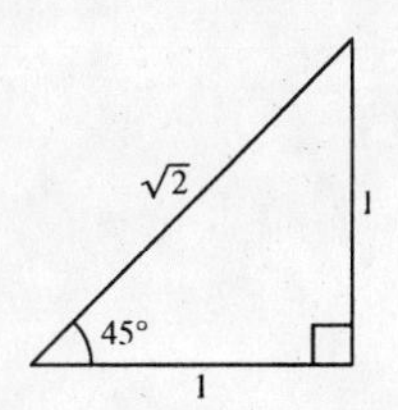

37. (a) $\cos \frac{\pi}{6} = \cos 30° = \frac{\sqrt{3}}{2}$

(b) $\sec 60° = \frac{1}{\cos 60°} = 2$

39. (a) $\sin 25° \approx 0.4226$

(b) $\cos 65° \approx 0.4226$

Note: $\sin 25° = \cos(90° - 25°) = \cos 65°$

41. (a) $\sec 42° \, 12' = \sec 42.2° = \frac{1}{\cos 42.2°} \approx 1.3499$

(b) $\csc 48° \, 7' = \frac{1}{\sin\left(48 + \frac{7}{60}\right)°} \approx 1.3432$

43. Make sure that your calculator is in radian mode.

(a) $\cot \frac{\pi}{16} = \frac{1}{\tan(\pi/16)} \approx 5.0273$

(b) $\tan \frac{\pi}{16} \approx 0.1989$

45. Make sure that your calculator is in radian mode.

(a) $\csc 1 = \frac{1}{\sin 1} \approx 1.1884$

(b) $\tan \frac{1}{2} \approx 0.5463$

47. (a) $\sin \theta = \frac{1}{2} \implies \theta = 30° = \frac{\pi}{6}$

(b) $\csc \theta = 2 \implies \theta = 30° = \frac{\pi}{6}$

49. (a) $\sec \theta = 2 \implies \theta = 60° = \frac{\pi}{3}$

(b) $\cot \theta = 1 \implies \theta = 45° = \frac{\pi}{4}$

51. (a) $\csc \theta = \frac{2\sqrt{3}}{3} \implies \theta = 60° = \frac{\pi}{3}$

(b) $\sin \theta = \frac{\sqrt{2}}{2} \implies \theta = 45° = \frac{\pi}{4}$

53. (a) $\sin \theta = 0.8191 \implies \theta \approx 55° \approx 0.960$ radian

(b) $\cos \theta = 0.0175 \implies \theta \approx 89° \approx 1.553$ radians

55. (a) $\tan \theta = 1.1920 \implies \theta \approx 50° \approx 0.873$ radian

(b) $\tan \theta = 0.4663 \implies \theta \approx 25° \approx 0.436$ radian

57. $\tan 30° = \frac{y}{105}$

$y = 105 \cdot \tan 30° = 105\frac{\sqrt{3}}{3} = 35\sqrt{3} \approx 60.6218$

59. $\cot 60° = \frac{x}{38}$

$\frac{\sqrt{3}}{3} = \frac{x}{38}$

$\frac{38\sqrt{3}}{3} = x$

61. $\sin 50° = \dfrac{y}{15}$

$y = 15 \cdot \sin 50° \approx 11.4907 \approx 11.5$

63. (a)

Not drawn to scale

(b) $\tan \theta = \dfrac{6}{3}$ and $\tan \theta = \dfrac{h}{135}$

Thus, $\dfrac{6}{3} = \dfrac{h}{135}$.

(c) $\dfrac{135 \cdot 6}{3} = h = 270$ feet

65. $\tan \theta = \dfrac{\text{opp}}{\text{adj}}$

$\tan 58° = \dfrac{w}{100}$

$w = 100 \tan 58° \approx 160.0$ feet

67. (a)

(b) $\sin \theta = \dfrac{\text{opp}}{\text{hyp}}$

$\sin \theta = \dfrac{10/3}{20} = \dfrac{1}{6}$

(c) $\sin \theta = \dfrac{1}{6} \Longrightarrow \theta = 9.59°$

69. $\tan \theta = \dfrac{\text{opp}}{\text{adj}}$

$\tan 80° = \dfrac{h}{75}$

$h = 75 \tan 80° \approx 425.3$ meters

$\cos \theta = \dfrac{\text{adj}}{\text{hyp}}$

$\cos 80° = \dfrac{75}{d}$

$d = 75 \dfrac{1}{\cos 80°} \approx 431.9$ meters

71. $\tan 3° = \dfrac{x}{15}$

$x = 15 \tan 3°$

$d = 5 + 2x$

$= 5 + 2(15 \tan 3°)$

≈ 6.57 centimeters

73. $x \approx 2.588,\ y \approx 9.659$

$\sin \theta = \dfrac{y}{10} \approx 0.97$ $\qquad \csc \theta = \dfrac{10}{y} \approx 1.04$

$\cos \theta = \dfrac{x}{10} \approx 0.26$ $\qquad \sec \theta = \dfrac{10}{x} \approx 3.86$

$\tan \theta = \dfrac{y}{x} \approx 3.73$ $\qquad \cot \theta = \dfrac{x}{y} \approx 0.27$

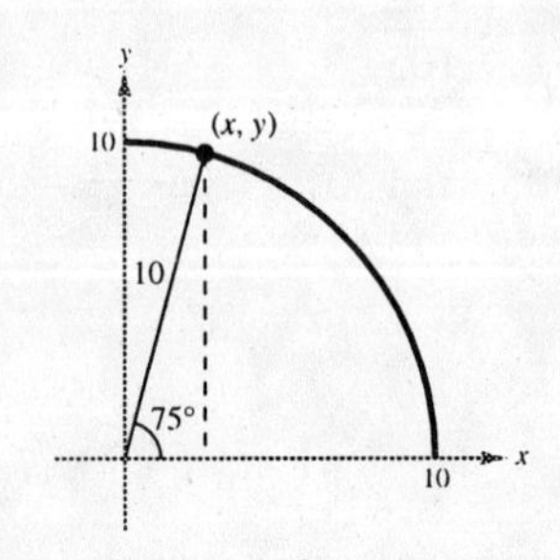

75. $\sec 30° = \csc 60°$

True, because $\sec(90° - \theta) = \csc \theta$.

77. $\cot^2 10° - \csc^2 10° = -1$

True, because
$$1 + \cot^2 \theta = \csc^2 \theta$$
$$\cot^2 \theta = \csc^2 \theta - 1$$
$$\cot^2 \theta - \csc^2 \theta = -1.$$

79.

θ	0°	20°	40°	60°	80°
$\cos \theta$	1	0.9397	0.7660	0.5000	0.1736
$\sin(90° - \theta)$	1	0.9397	0.7660	0.5000	0.1736

It seems that $\cos \theta = \sin(90° - \theta)$ for all θ.

θ and $90° - \theta$ are called complementary angles.

81. $y = -x - 9$

Intercepts: $(0, -9), (-9, 0)$

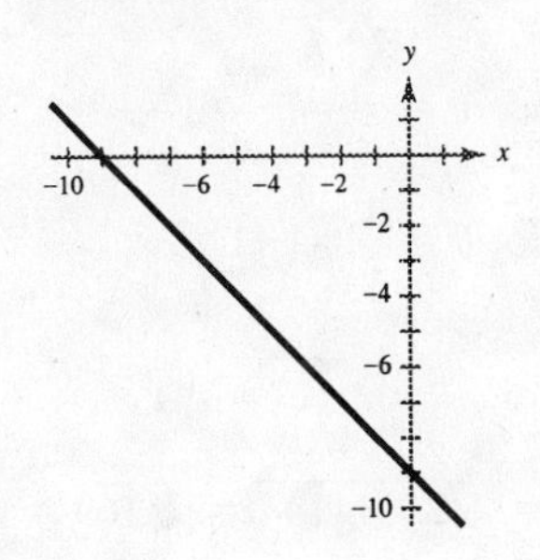

83. $-3x + 8y = 16$

Intercepts: $(0, 2), \left(-\frac{16}{3}, 0\right)$

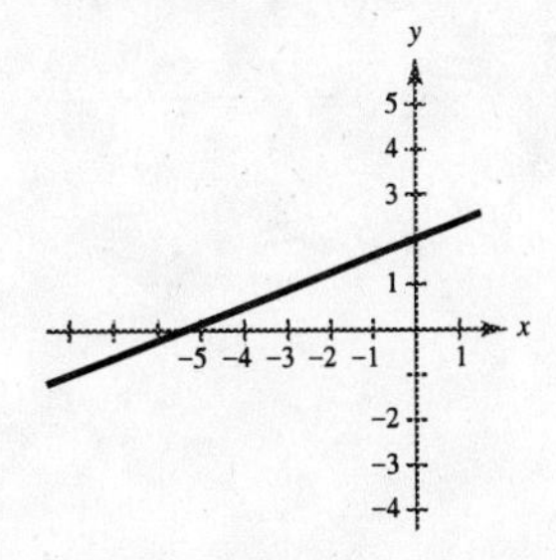

85. 146° lies in quadrant II

87. $-310° 30'$ lies in quadrant I

Section 4.4 Trigonometric Functions of Any Angle

- Know the Definitions of Trigonometric Functions of Any Angle.
  If θ is in standard position, (x, y) a point on the terminal side and $r = \sqrt{x^2 + y^2} \neq 0$, then:

  $\sin \theta = \dfrac{y}{r}$ $\qquad \csc \theta = \dfrac{r}{y},\ y \neq 0$

  $\cos \theta = \dfrac{x}{r}$ $\qquad \sec \theta = \dfrac{r}{x},\ x \neq 0$

  $\tan \theta = \dfrac{y}{x},\ x \neq 0$ $\qquad \cot \theta = \dfrac{x}{y},\ y \neq 0$
- You should know the signs of the trigonometric functions in each quadrant.
- You should know the trigonometric function values of the quadrant angles $0, \dfrac{\pi}{2}, \pi$, and $\dfrac{3\pi}{2}$.
- You should be able to find reference angles.
- You should be able to evaluate trigonometric functions of any angle. (Use reference angles.)
- You should know that the period of sine and cosine is 2π.
- You should know which trigonometric functions are odd and even.

  Even: $\cos x$ and $\sec x$

  Odd: $\sin x$, $\tan x$, $\cot x$, $\csc x$

Solutions to Odd-Numbered Exercises

1. (a) $(x, y) = (4, 3)$

$r = \sqrt{16 + 9} = 5$

$\sin\theta = \frac{y}{r} = \frac{3}{5}$ $\csc\theta = \frac{r}{y} = \frac{5}{3}$

$\cos\theta = \frac{x}{r} = \frac{4}{5}$ $\sec\theta = \frac{r}{x} = \frac{5}{4}$

$\tan\theta = \frac{y}{x} = \frac{3}{4}$ $\cot\theta = \frac{x}{y} = \frac{4}{3}$

(b) $(x, y) = (-8, -15)$

$r = \sqrt{64 + 225} = 17$

$\sin\theta = \frac{y}{r} = -\frac{15}{17}$ $\csc\theta = \frac{r}{y} = -\frac{17}{15}$

$\cos\theta = \frac{x}{r} = -\frac{8}{17}$ $\sec\theta = \frac{r}{x} = -\frac{17}{8}$

$\tan\theta = \frac{y}{x} = \frac{15}{8}$ $\cot\theta = \frac{x}{y} = \frac{8}{15}$

3. (a) $(x, y) = (-\sqrt{3}, -1)$

$r = \sqrt{3 + 1} = 2$

$\sin\theta = \frac{y}{r} = -\frac{1}{2}$ $\csc\theta = \frac{r}{y} = -2$

$\cos\theta = \frac{x}{r} = \frac{-\sqrt{3}}{2}$ $\sec\theta = \frac{r}{x} = \frac{-2\sqrt{3}}{3}$

$\tan\theta = \frac{y}{x} = \frac{\sqrt{3}}{3}$ $\cot\theta = \frac{x}{y} = \sqrt{3}$

(b) $(x, y) = (-2, 2)$

$r = \sqrt{4 + 4} = 2\sqrt{2}$

$\sin\theta = \frac{y}{r} = \frac{\sqrt{2}}{2}$ $\csc\theta = \frac{r}{y} = \sqrt{2}$

$\cos\theta = \frac{x}{r} = -\frac{\sqrt{2}}{2}$ $\sec\theta = \frac{r}{x} = -\sqrt{2}$

$\tan\theta = \frac{y}{x} = -1$ $\cot\theta = \frac{x}{y} = -1$

5. $(x, y) = (7, 24)$

$r = \sqrt{49 + 576} = 25$

$\sin\theta = \frac{y}{r} = \frac{24}{25}$ $\csc\theta = \frac{r}{y} = \frac{25}{24}$

$\cos\theta = \frac{x}{r} = \frac{7}{25}$ $\sec\theta = \frac{r}{x} = \frac{25}{7}$

$\tan\theta = \frac{y}{x} = \frac{24}{7}$ $\cot\theta = \frac{x}{y} = \frac{7}{24}$

7. $(x, y) = (5, -12)$

$r = \sqrt{5^2 + (-12)^2} = \sqrt{25 + 144} = \sqrt{169} = 13$

$\sin\theta = \frac{y}{r} = -\frac{12}{13}$ $\csc\theta = \frac{r}{y} = -\frac{13}{12}$

$\cos\theta = \frac{x}{r} = \frac{5}{13}$ $\sec\theta = \frac{r}{x} = \frac{13}{5}$

$\tan\theta = \frac{y}{x} = -\frac{12}{5}$ $\cot\theta = \frac{x}{y} = -\frac{5}{12}$

9. $(x, y) = (-4, 10)$

$r = \sqrt{16 + 100} = 2\sqrt{29}$

$\sin\theta = \frac{y}{r} = \frac{5\sqrt{29}}{29}$ $\csc\theta = \frac{r}{y} = \frac{\sqrt{29}}{5}$

$\cos\theta = \frac{x}{r} = -\frac{2\sqrt{29}}{29}$ $\sec\theta = \frac{r}{x} = -\frac{\sqrt{29}}{2}$

$\tan\theta = \frac{y}{x} = -\frac{5}{2}$ $\cot\theta = \frac{x}{y} = -\frac{2}{5}$

11. $(x, y) = (-2, 9)$

$r = \sqrt{(-2)^2 + 9^2} = \sqrt{4 + 81} = \sqrt{85}$

$\sin\theta = \frac{y}{r} = \frac{9}{\sqrt{85}} = \frac{9\sqrt{85}}{85}$ $\csc\theta = \frac{\sqrt{85}}{9}$

$\cos\theta = \frac{x}{r} = \frac{-2}{\sqrt{85}} = -\frac{2\sqrt{85}}{85}$ $\sec\theta = \frac{r}{x} = -\frac{\sqrt{85}}{2}$

$\tan\theta = \frac{y}{x} = -\frac{9}{2}$ $\cot\theta = \frac{x}{y} = -\frac{2}{9}$

13. $\sin\theta < 0 \Rightarrow \theta$ lies in Quadrant III or in Quadrant IV.

$\cos\theta < 0 \Rightarrow \theta$ lies in Quadrant II or in Quadrant III.

$\sin\theta < 0$ *and* $\cos\theta < 0 \Rightarrow \theta$ lies in Quadrant III.

15. $\sin\theta > 0 \Rightarrow \theta$ lies in Quadrant I or in Quadrant II.

$\tan\theta < 0 \Rightarrow \theta$ lies in Quadrant II or in Quadrant IV.

$\sin\theta > 0$ *and* $\tan\theta < 0 \Rightarrow \theta$ lies in Quadrant II.

17. $\cot\theta > 0 \Rightarrow \theta$ lies in Quadrant I or Quadrant III

$\cos\theta > 0 \Rightarrow \theta$ lies in Quadrant I or Quadrant IV

$\cot\theta > 0$ and $\cos\theta > 0 \Rightarrow \theta$ lies in Quadrant I

19. $\sin\theta = \frac{y}{r} = \frac{3}{5} \Rightarrow x^2 = 25 - 9 = 16$

θ in Quadrant II $\Rightarrow x = -4$

$\sin\theta = \frac{y}{r} = \frac{3}{5}$ $\qquad$ $\csc\theta = \frac{r}{y} = \frac{5}{3}$

$\cos\theta = \frac{x}{r} = -\frac{4}{5}$ $\qquad$ $\sec\theta = \frac{r}{x} = -\frac{5}{4}$

$\tan\theta = \frac{y}{x} = -\frac{3}{4}$ $\qquad$ $\cot\theta = \frac{x}{y} = -\frac{4}{3}$

21. $\sin\theta < 0 \Rightarrow y < 0$

$\tan\theta = \frac{y}{x} = \frac{-15}{8} \Rightarrow r = 17$

$\sin\theta = \frac{y}{r} = -\frac{15}{17}$ $\qquad$ $\csc\theta = \frac{r}{y} = -\frac{17}{15}$

$\cos\theta = \frac{x}{r} = \frac{8}{17}$ $\qquad$ $\sec\theta = \frac{r}{x} = \frac{17}{8}$

$\tan\theta = \frac{y}{x} = -\frac{15}{8}$ $\qquad$ $\cot\theta = \frac{x}{y} = -\frac{8}{15}$

23. $\sec\theta = \frac{r}{x} = \frac{2}{-1} \Rightarrow y^2 = 4 - 1 = 3$

$\sin\theta \geq 0 \Rightarrow y = \sqrt{3}$

$\sin\theta = \frac{y}{r} = \frac{\sqrt{3}}{2}$ $\qquad$ $\csc\theta = \frac{r}{y} = \frac{2\sqrt{3}}{3}$

$\cos\theta = \frac{x}{r} = -\frac{1}{2}$ $\qquad$ $\sec\theta = \frac{r}{x} = -2$

$\tan\theta = \frac{y}{x} = -\sqrt{3}$ $\qquad$ $\cot\theta = \frac{x}{y} = -\frac{\sqrt{3}}{3}$

25. $\sin\theta = 0 \Rightarrow \theta = n\pi$

$\sec\theta = -1 \Rightarrow \theta = \pi$

$\sin\theta = \frac{y}{r} = \frac{0}{r} = 0$ $\qquad$ $\csc\theta = \frac{r}{y}$ is undefined.

$\cos\theta = \frac{x}{r} = \frac{-r}{r} = -1$ $\qquad$ $\sec\theta = \frac{r}{x} = -1$

$\tan\theta = \frac{y}{x} = \frac{0}{x} = 0$ $\qquad$ $\cot\theta = \frac{x}{y}$ is undefined.

27. To find a point on the terminal side of θ, use any point on the line $y = -x$ that lies in Quadrant II. $(-1, 1)$ is one such point.

$x = -1, y = 1, r = \sqrt{2}$

$\sin\theta = \frac{1}{\sqrt{2}} = \frac{\sqrt{2}}{2}$ $\qquad$ $\csc\theta = \sqrt{2}$

$\cos\theta = -\frac{1}{\sqrt{2}} = -\frac{\sqrt{2}}{2}$ $\qquad$ $\sec\theta = -\sqrt{2}$

$\tan\theta = -1$ $\qquad$ $\cot\theta = -1$

29. To find a point on the terminal side of θ, use any point on the line $y = 2x$ that lies in Quadrant III. $(-1, -2)$ is one such point.

$x = -1, y = -2, r = \sqrt{5}$

$\sin\theta = -\frac{2}{\sqrt{5}} = -\frac{2\sqrt{5}}{5}$ $\qquad$ $\csc\theta = \frac{\sqrt{5}}{-2} = -\frac{\sqrt{5}}{2}$

$\cos\theta = -\frac{1}{\sqrt{5}} = -\frac{\sqrt{5}}{5}$ $\qquad$ $\sec\theta = \frac{\sqrt{5}}{-1} = -\sqrt{5}$

$\tan\theta = \frac{-2}{-1} = 2$ $\qquad$ $\cot\theta = \frac{-1}{-2} = \frac{1}{2}$

31. $(x, y) = (-1, 0)$

$$\sec \pi = \frac{r}{x} = \frac{1}{-1} = -1$$

33. $(x, y) = (0, 1)$

$$\cot \frac{\pi}{2} = \frac{x}{y} = \frac{0}{1} = 0$$

35. $(x, y) = (1, 0)$

$$\sec \theta = \frac{r}{x} = \frac{1}{1} = 1$$

37. $(x, y) = (-1, 0)$

$$\cot \pi = \frac{x}{y} = -\frac{1}{0} \text{ undefined}$$

39. $\theta = 208°$

$\theta' = 208° - 180° = 28°$

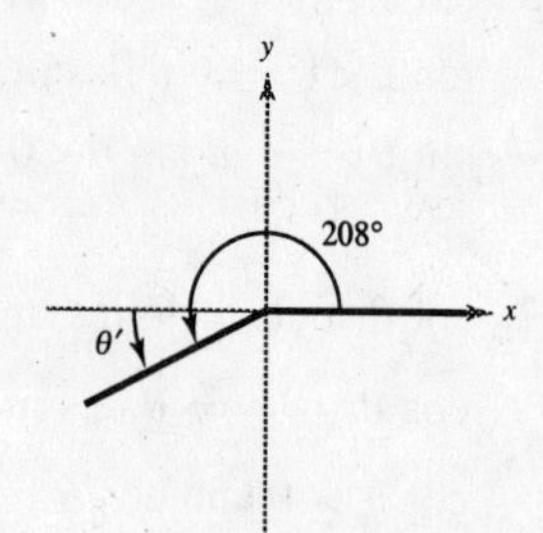

41. $\theta = -245°$

$360° - 245° = 115°$ (coterminal angle)

$\theta' = 180° - 115° = 65°$

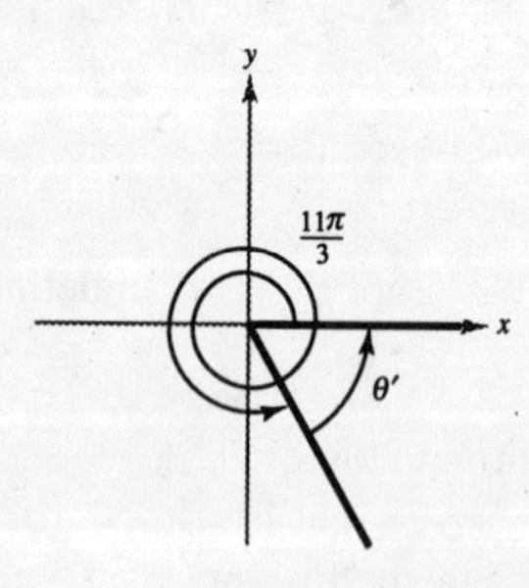

43. $\theta = -292°$

$\theta' = 360° - 292° = 68°$

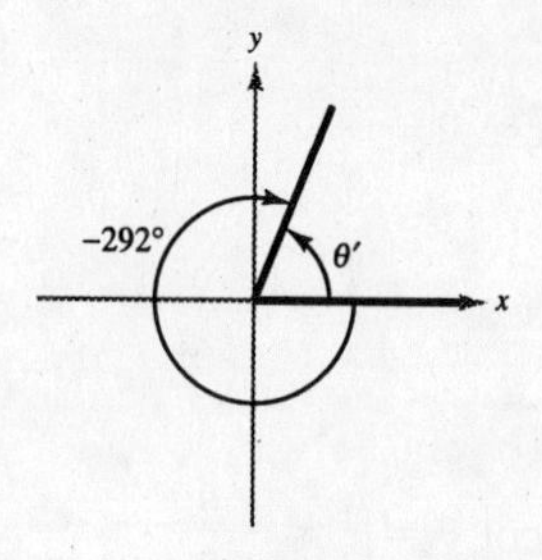

45. $\theta = \frac{11\pi}{3}$ coterminal with $\frac{5\pi}{3}$.

$$\theta' = 2\pi - \frac{5\pi}{3} = \frac{\pi}{3}$$

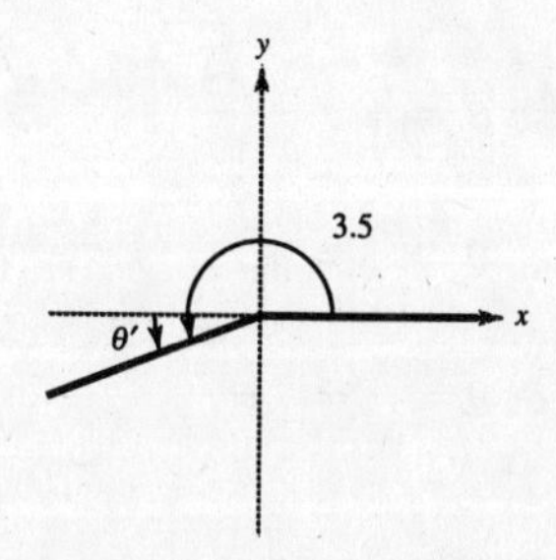

47. $\theta = 3.5$

$\theta' = 3.5 - \pi \approx 0.3584$

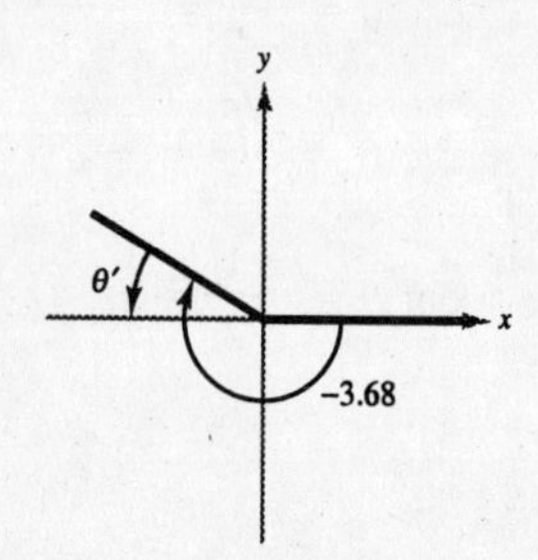

49. $\theta = -3.68$ coterminal with $2\pi - 3.68 \approx 2.6032$

$\theta' = \pi - (2\pi - 3.68) \approx 0.5384$

51. $\theta' = 45°$,

Quadrant III

$$\sin 225° = -\sin 45° = -\frac{\sqrt{2}}{2}$$

$$\cos 225° = -\cos 45° = -\frac{\sqrt{2}}{2}$$

$$\tan 225° = \tan 45° = 1$$

53. $\theta = -750°$ coterminal with $330°$. Quadrant IV

$\theta' = 360° - 330° = 30°$

$\sin(-750°) = -\sin 30° = -\frac{1}{2}$

$\cos(-750°) = \cos 30° = \frac{\sqrt{3}}{2}$

$\tan(-750°) = -\tan 30° = -\frac{\sqrt{3}}{3}$

55. $\theta = -240°$ coterminal with $120°$. Quadrant II

$\theta' = 180° - 120° = 60°$

$\sin(-240°) = \sin 60° = \frac{\sqrt{3}}{2}$

$\cos(-240°) = -\cos 60° = -\frac{1}{2}$

$\tan(-240°) = -\tan 60° = -\sqrt{3}$

57. $\theta = \frac{5\pi}{3}$. Quadrant IV

$\theta' = 2\pi - \frac{5\pi}{3} = \frac{\pi}{3}$

$\sin\left(\frac{5\pi}{3}\right) = -\sin\left(\frac{\pi}{3}\right) = -\frac{\sqrt{3}}{2}$

$\cos\left(\frac{5\pi}{3}\right) = \cos\left(\frac{\pi}{3}\right) = \frac{1}{2}$

$\tan\left(\frac{5\pi}{3}\right) = -\tan\left(\frac{\pi}{3}\right) = -\sqrt{3}$

59. $\theta' = \frac{\pi}{6}$, Quadrant IV

$\sin\left(-\frac{\pi}{6}\right) = -\sin\frac{\pi}{6} = -\frac{1}{2}$

$\cos\left(-\frac{\pi}{6}\right) = \cos\frac{\pi}{6} = \frac{\sqrt{3}}{2}$

$\tan\left(-\frac{\pi}{6}\right) = -\tan\frac{\pi}{6} = -\frac{\sqrt{3}}{3}$

61. $\theta' = \frac{\pi}{4}$, Quadrant II

$\sin\frac{11\pi}{4} = \sin\frac{\pi}{4} = \frac{\sqrt{2}}{2}$

$\cos\frac{11\pi}{4} = -\cos\frac{\pi}{4} = -\frac{\sqrt{2}}{2}$

$\tan\frac{11\pi}{4} = -\tan\frac{\pi}{4} = -1$

63. $\theta = -\frac{7\pi}{6}$. Quadrant II

$\theta' = \frac{\pi}{6}$

$\sin\left(-\frac{7\pi}{6}\right) = \sin\left(\frac{\pi}{6}\right) = \frac{1}{2}$

$\cos\left(-\frac{7\pi}{6}\right) = -\cos\left(\frac{\pi}{6}\right) = -\frac{\sqrt{3}}{2}$

$\tan\left(-\frac{7\pi}{6}\right) = -\tan\left(\frac{\pi}{6}\right) = -\frac{\sqrt{3}}{3}$

65. $\sin 10° \approx 0.1736$

67. $\tan 240° \approx 1.7321$

69. $\cos(-110°) \approx -0.3420$

71. $\sec(-280°) = \frac{1}{\cos(-280°)} \approx 5.7588$

73. $\sin 0.65 \approx 0.6052$

75. $\tan\frac{\pi}{9} \approx 0.3640$

77. $\csc\left(-\frac{8\pi}{9}\right) = \frac{1}{\sin\left(-\frac{8\pi}{9}\right)} \approx -2.9238$

79. (a) $\sin\theta = \frac{1}{2} \Rightarrow$ reference angle is $30°$ or $\frac{\pi}{6}$ and θ is in Quadrant I or Quadrant II.

Values in degrees: $30°, 150°$

Values in radian: $\frac{\pi}{6}, \frac{5\pi}{6}$

(b) $\sin\theta = -\frac{1}{2} \Rightarrow$ reference angle is $30°$ or $\frac{\pi}{6}$ and θ is in Quadrant III or Quadrant IV.

Values in degrees: $210°, 330°$

Values in radians: $\frac{7\pi}{6}, \frac{11\pi}{6}$

81. (a) $\csc\theta = \frac{2\sqrt{3}}{3} \Rightarrow$ reference angle is $60°$ or $\frac{\pi}{3}$ and θ is in Quadrant I or Quadrant II.

Values in degrees: $60°, 120°$

Values in radians: $\frac{\pi}{3}, \frac{2\pi}{3}$

(b) $\cot\theta = -1 \Rightarrow$ reference angle is $45°$ or $\frac{\pi}{4}$ and θ is in Quadrant II or Quadrant IV.

Values in degrees: $135°, 315°$

Values in radians: $\frac{3\pi}{4}, \frac{7\pi}{4}$

83. (a) $\sec\theta = -\frac{2\sqrt{3}}{3} \Rightarrow$ reference angle is $\frac{\pi}{6}$ or $30°$, and θ is in Quadrant II or Quadrant III.

Value in degrees: $150°, 210°$

Value in radians: $\frac{5\pi}{6}, \frac{7\pi}{6}$

(b) $\cos\theta = -\frac{1}{2} \Rightarrow$ reference angle is $\frac{\pi}{3}$ or $60°$, and θ is in Quadrant II or Quadrant III.

Value in degrees: $120°, 240°$

Value in radians: $\frac{2\pi}{3}, \frac{4\pi}{3}$

85. $\sin\theta = 0.8191$

Quadrant I: $\theta = \sin^{-1} 0.8191 \approx 54.99°$

Quadrant II: $\theta = 180° - \sin^{-1} 0.8191 \approx 125.01°$

87. $\tan\theta = 0.6524$

Quadrant I: $\theta = \tan^{-1}(0.6524) \approx 33.1204°$

Quadrant III: $\theta = \tan^{-1}(0.6524) + 180° \approx 213.1204°$

89. $\sec\theta = -1.2241 \Rightarrow \cos\theta = -0.8169$

Quadrant II: $\theta \approx 144.7783°$

Quadrant III: $\theta \approx 215.2217°$

91. $\sin\theta = -0.4793$

Quadrant III: $\theta \approx 208.6397$

Quadrant IV: $\theta \approx 331.3603$

93. $\cos\theta = 0.9848 \implies \theta \approx 10.0026°$

Quadrant I: $\theta = \cos^{-1}(0.9848) \approx 10.0026°$

Quadrant IV: $\theta \approx 360° - 10.0026° = 349.9974°$

95. $\tan\theta = 1.192 \implies \theta \approx 50.0058°$

Quadrant I: $\theta = \tan^{-1}1.192 \approx 50.0058°$

Quadrant III: $\theta \approx 180° + 50.0058° \approx 230.0058°$

97. $\sin\theta = -\frac{3}{5}$

$\sin^2\theta + \cos^2\theta = 1$

$\cos^2\theta = 1 - \sin^2\theta$

$\cos^2\theta = 1 - \left(-\frac{3}{5}\right)^2$

$\cos^2\theta = 1 - \frac{9}{25}$

$\cos^2\theta = \frac{16}{25}$

$\cos\theta > 0$ in Quadrant IV.

$\cos\theta = \frac{4}{5}$

99. $\tan\theta = \dfrac{3}{2}$

$\sec^2\theta = 1 + \tan^2\theta$

$\sec^2\theta = 1 + \left(\dfrac{3}{2}\right)^2$

$\sec^2\theta = 1 + \dfrac{9}{4}$

$\sec^2\theta = \dfrac{13}{4}$

$\sec\theta < 0$ in Quadrant III.

$\sec\theta = -\dfrac{\sqrt{13}}{2}$

101. $\cos\theta = \dfrac{5}{8}$

$\cos\theta = \dfrac{1}{\sec\theta} \implies \sec\theta = \dfrac{1}{\cos\theta}$

$\sec\theta = \dfrac{1}{5/8} = \dfrac{8}{5}$

103. (a) $t = 1$

$T = 45 - 23\cos\left[\dfrac{2\pi}{365}(1 - 32)\right] \approx 25.2°\text{ F}$

(b) $t = 185$

$T = 45 - 23\cos\left[\dfrac{2\pi}{365}(185 - 32)\right] \approx 65.1°\text{ F}$

(c) $t = 291$

$T = 45 - 23\cos\left[\dfrac{2\pi}{365}(291 - 32)\right] \approx 50.8°\text{ F}$

105. $\sin\theta = \dfrac{6}{d} \implies d = \dfrac{6}{\sin\theta}$

(a) $\theta = 30°$

$d = \dfrac{6}{\sin 30°} = \dfrac{6}{(1/2)} = 12$ miles

(b) $\theta = 90°$

$d = \dfrac{6}{\sin 90°} = \dfrac{6}{1} = 6$ miles

(c) $\theta = 120°$

$d = \dfrac{6}{\sin 120°} \approx 6.9$ miles

107. False. $\sin\left(\dfrac{7\pi}{6}\right) \neq \sin\left(\dfrac{11\pi}{6}\right)$

109. (a)

θ	$0°$	$20°$	$40°$	$60°$	$80°$
$\sin\theta$	0	0.3420	0.6428	0.8660	0.9848
$\sin(180° - \theta)$	0	0.3420	0.6428	0.8660	0.9848

(b) It appears that $\sin\theta = \sin(180° - \theta)$ for all θ.

111. Answers will vary.

113. $y = 2^{x-1}$

x	-1	0	1	2	3
y	$\frac{1}{4}$	$\frac{1}{2}$	1	2	4

Intercept: $\left(0, \frac{1}{2}\right)$

Asymptote: $y = 0$

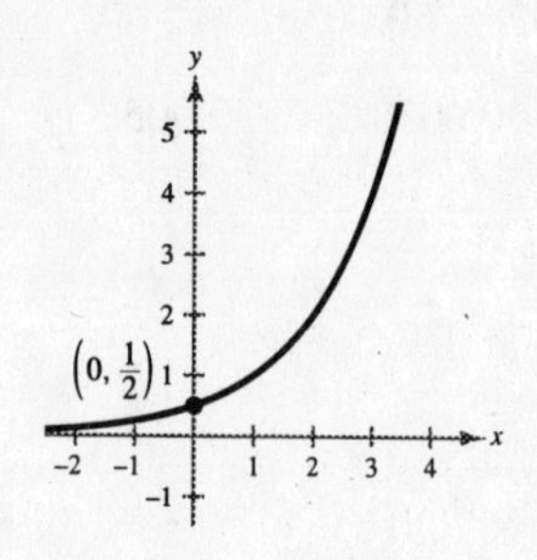

115. $y = \ln(x - 1)$

Domain: $x - 1 > 0 \implies x > 1$

x	1.1	1.5	2	3	4
y	-2.30	-0.69	0	0.69	1.10

Intercept: $(2, 0)$

Asymptote: $x = 1$

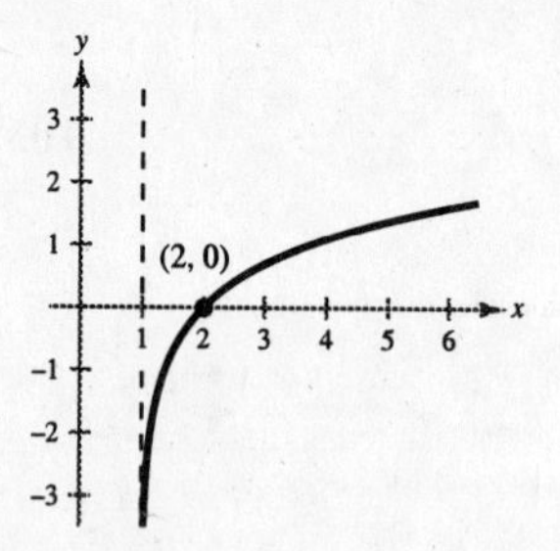

117. $4^{3-x} = 726$

$3 - x = \log_4 726$

$$x = 3 - \log_4 726 = 3 - \frac{\ln 726}{\ln 4} \approx -1.752$$

119. $\ln x = -6$

$$x = e^{-6} \approx 0.002479 \approx 0.002$$

121.

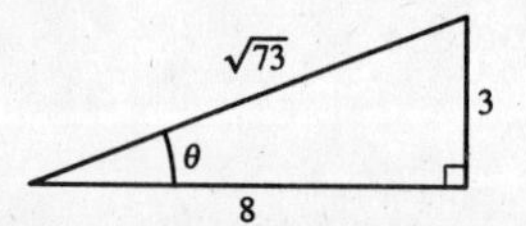

$\cot\theta = \dfrac{\text{adj}}{\text{opp}} = \dfrac{8}{3}$

$\text{hyp} = \sqrt{8^2 + 3^2} = \sqrt{73}$

$\sin\theta = \dfrac{\text{opp}}{\text{hyp}} = \dfrac{3}{\sqrt{73}} = \dfrac{3\sqrt{73}}{73}$

$\cos\theta = \dfrac{\text{adj}}{\text{hyp}} = \dfrac{8}{\sqrt{73}} = \dfrac{8\sqrt{73}}{73}$

$\tan\theta = \dfrac{\text{opp}}{\text{adj}} = \dfrac{3}{8}$

$\sec\theta = \dfrac{\text{hyp}}{\text{adj}} = \dfrac{\sqrt{73}}{8}$

$\csc\theta = \dfrac{\text{hyp}}{\text{opp}} = \dfrac{\sqrt{73}}{3}$

123.

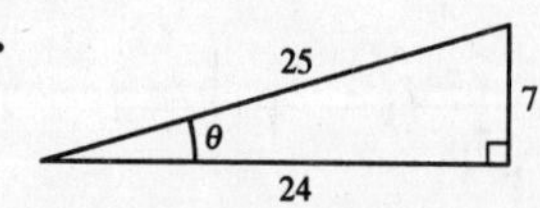

$\tan\theta = \dfrac{7}{24} = \dfrac{\text{opp}}{\text{adj}}$

$\text{hyp} = \sqrt{7^2 + 24^2} = 25$

$\sin\theta = \dfrac{\text{opp}}{\text{hyp}} = \dfrac{7}{25}$

$\cos\theta = \dfrac{\text{adj}}{\text{hyp}} = \dfrac{24}{25}$

$\cot\theta = \dfrac{\text{adj}}{\text{opp}} = \dfrac{24}{7}$

$\sec\theta = \dfrac{\text{hyp}}{\text{adj}} = \dfrac{25}{24}$

$\csc\theta = \dfrac{\text{hyp}}{\text{opp}} = \dfrac{25}{7}$

Section 4.5 Graphs of Sine and Cosine Functions

- You should be able to graph $y = a\sin(bx - c)$ and $y = a\cos(bx - c)$.
- Amplitude: $|a|$
- Period: $\dfrac{2\pi}{|b|}$
- Shift: Solve $bx - c = 0$ and $bx - c = 2\pi$.
- Key increments: $\dfrac{1}{4}$ (period)

Solutions to Odd-Numbered Exercises

1. $y = 3\sin 2x$

Period: $\dfrac{2\pi}{2} = \pi$

Amplitude: $|3| = 3$

Xmin = -2π
Xmax = 2π
Xscl = $\pi/2$
Ymin = -4
Ymax = 4
Yscl = 1

3. $y = \dfrac{5}{2}\cos\dfrac{x}{2}$

Period: $\dfrac{2\pi}{1/2} = 4\pi$

Amplitude $\left|\dfrac{5}{2}\right| = \dfrac{5}{2}$

Xmin = -4π
Xmax = 4π
Xscl = π
Ymin = -3
Ymax = 3
Yscl = 1

5. $y = \dfrac{2}{3}\sin \pi x$

Period: $\dfrac{2\pi}{\pi} = 2$

Amplitude: $\left|\dfrac{2}{3}\right| = \dfrac{2}{3}$

Xmin = $-\pi$
Xmax = π
Xscl = $\pi/2$
Ymin = -1
Ymax = 1
Yscl = .5

7. $y = -2\sin x$

Period: $\dfrac{2\pi}{1} = 2\pi$

Amplitude: $|-2| = 2$

9. $y = 3\sin 6x$

Period: $\dfrac{2\pi}{6} = \dfrac{\pi}{3}$

Amplitude: $|3| = 3$

11. $y = \dfrac{1}{4}\cos\dfrac{2x}{3}$

Period: $\dfrac{2\pi}{2/3} = 3\pi$

Amplitude: $\left|\dfrac{1}{4}\right| = \dfrac{1}{4}$

13. $y = 3\sin 4\pi x$

Period: $\dfrac{2\pi}{4\pi} = \dfrac{1}{2}$

Amplitude: $|3| = 3$

15. $f(x) = \sin x$

$g(x) = \sin(x - \pi)$

The graph of g is a horizontal shift to the right π units of the graph of f (a phase shift).

17. $f(x) = \cos 2x$

$g(x) = -\cos 2x$

The graph of g is a reflection in the x-axis of the graph of f.

19. $f(x) = \cos x$

$g(x) = -5\cos x$

The graph of g is five times the amplitude of f, and reflected in the x-axis.

21. $f(x) = \sin x$

$f(x) = 4 = \sin x$

The graph of g is a vertical shift upward of 4 units of the graph of f.

23. The graph of g has twice the amplitude as the graph of f. The period is the same.

25. The graph of g is a horizontal shift π units to the right of the graph of f.

27. $f(x) = -2 \sin x$

Period: 2π

Amplitude: 2

$g(x) = 4 \sin x$

Period: 2π

Amplitude: 4.

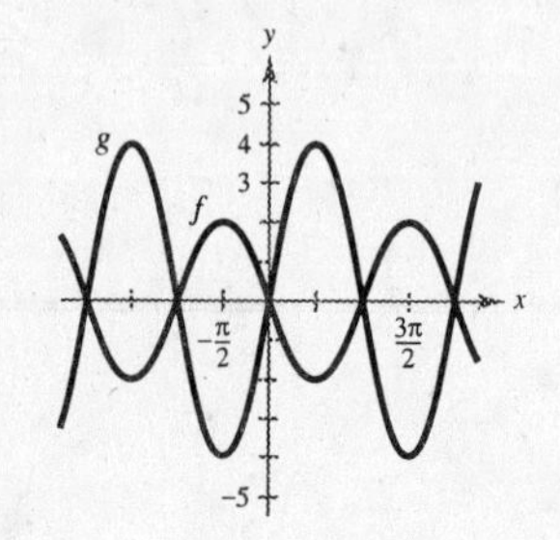

29. $f(x) = \cos x$

Period: 2π

Amplitude: 1

$g(x) = 4 = \cos x$ is a vertical shift of the graph of $f(x)$ four units upward.

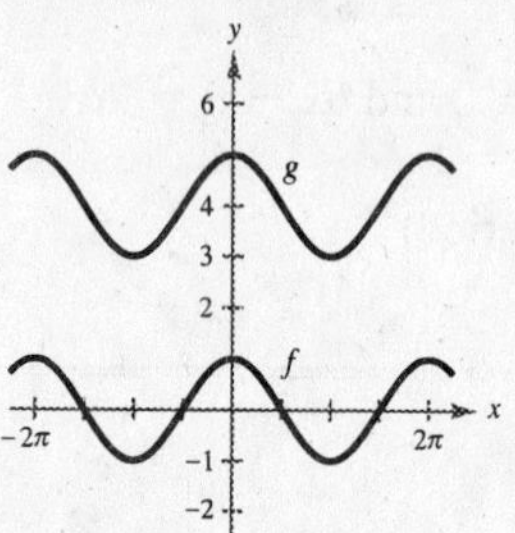

31. $f(x) = -\frac{1}{2} \sin \frac{x}{2}$

Period: 4π

Amplitude: $\frac{1}{2}$

$g(x) = 5 - \frac{1}{2} \sin \frac{x}{2}$ is the graph of $f(x)$ shifted vertically five units upward.

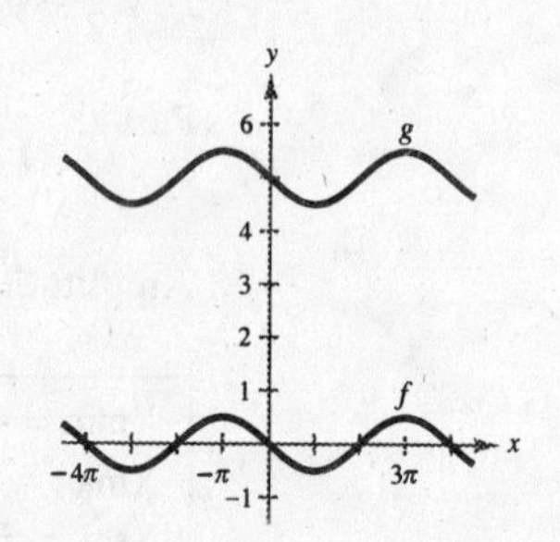

33. $f(x) = 2 \cos x$

Period: 2π

Amplitude: 2

$g(x) = 2 \cos(x = \pi)$ is the graph of $f(x)$ shifted π units to the left.

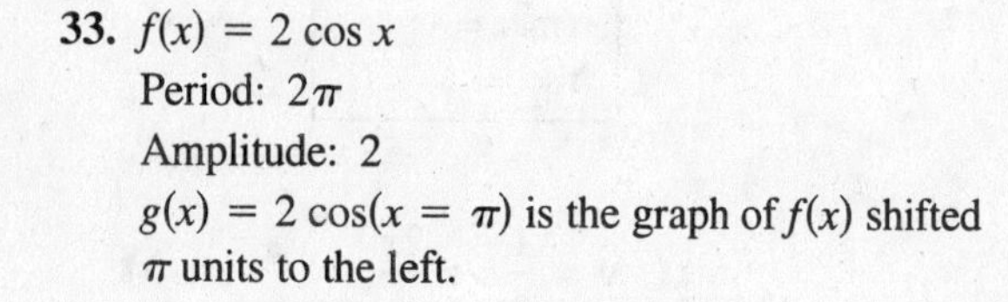

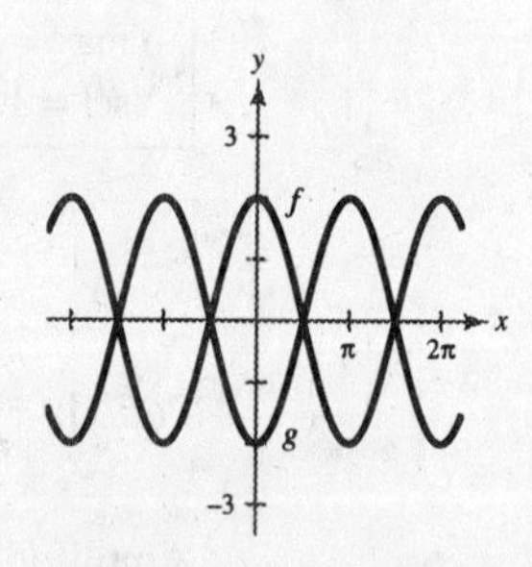

35. Since sine and cosine are cofunctions and x and $x - (\pi/2)$ are complementary, we have

$$\sin x = \cos\left(x - \frac{\pi}{2}\right).$$

Period: 2π

Amplitude: 1

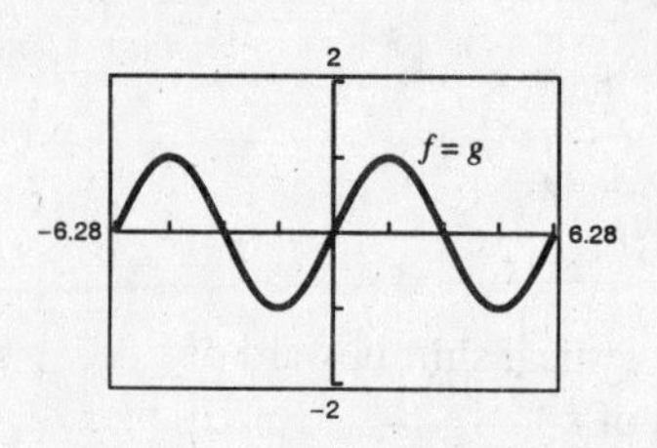

37. $f(x) = \cos x$

$$g(x) = -\sin\left(x - \frac{\pi}{2}\right) = \sin\left(\frac{\pi}{2} - x\right) = \cos x$$

Thus, $f(x) = g(x)$.

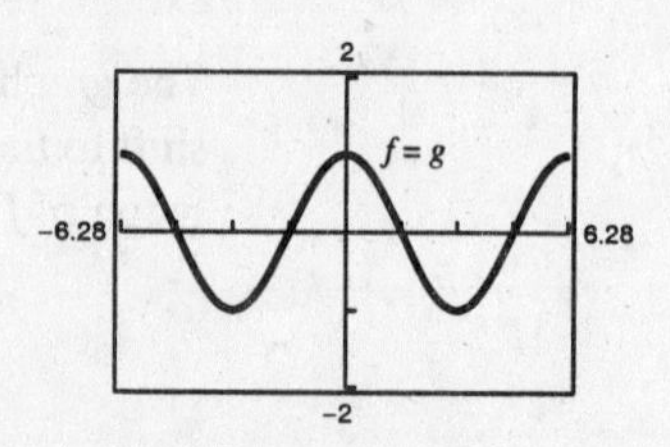

39. $y = -2 \sin 4x.$

Period: $\dfrac{2\pi}{4} = \dfrac{\pi}{2}.$

Amplitude: 2

Key points: $(0, 0), \left(\dfrac{\pi}{8}, -2\right), \left(\dfrac{\pi}{4}, 0\right), \left(\dfrac{3\pi}{8}, 2\right), \left(\dfrac{\pi}{2}, 0\right)$

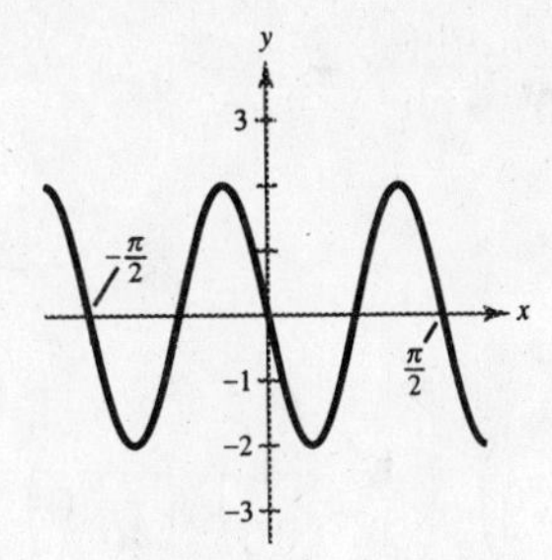

41. $y = \cos 2\pi x$

Period: $\dfrac{2\pi}{2\pi} = 1$

Amplitude: 1

Key points: $(0, 1), \left(\dfrac{1}{4}, 0\right), \left(\dfrac{1}{2}, 0\right), \left(\dfrac{3}{4}, 0\right), (1, 1)$

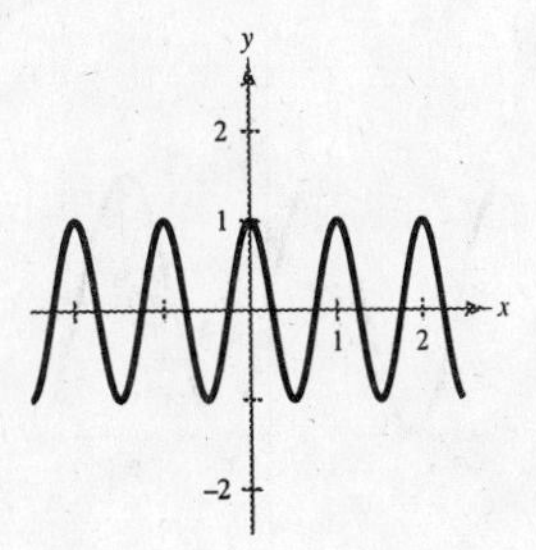

43. $y = -2 \sin \dfrac{2\pi x}{3}$

Period: $\dfrac{2\pi}{\dfrac{2\pi}{3}} = 3$

Amplitude: 2

Key points: $(0, 0), \left(\dfrac{3}{4}, -2\right), \left(\dfrac{3}{2}, 0\right), \left(\dfrac{9}{4}, 2\right), (3, 0)$

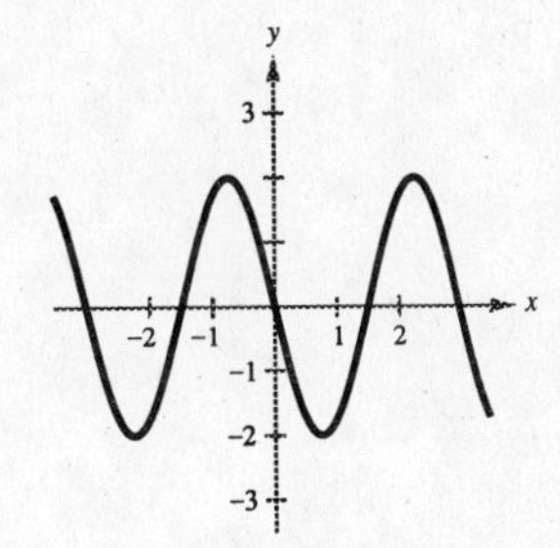

45. $y = \sin\left(x - \dfrac{\pi}{4}\right)$; $a = 1,\ b = 1,\ c = \dfrac{\pi}{4}$

Period: 2π

Amplitude: 1

Shift: Set $x - \dfrac{\pi}{4} = 0$ and $x - \dfrac{\pi}{4} = 2\pi$

$x = \dfrac{\pi}{4}$ $\qquad$ $x = \dfrac{9\pi}{4}$

Key points: $\left(\dfrac{\pi}{4}, 0\right), \left(\dfrac{3\pi}{4}, 1\right), \left(\dfrac{5\pi}{4}, 0\right), \left(\dfrac{7\pi}{4}, -1\right), \left(\dfrac{9\pi}{4}, 0\right)$

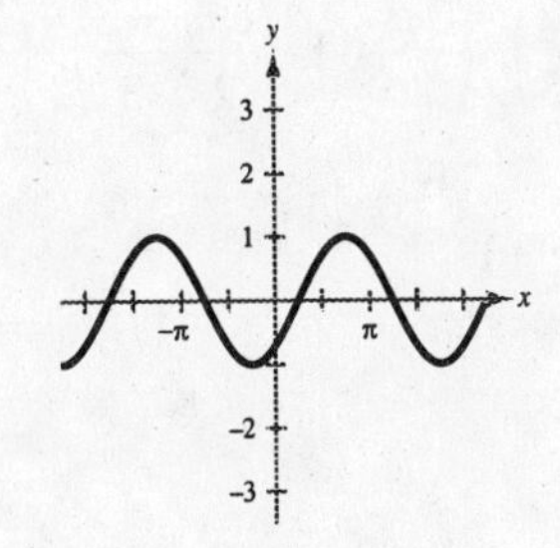

47. $y = 8 \cos (x = \pi)$

Period: 2π

Amplitude: 8

Key points: $(-\pi, 8), \left(-\dfrac{\pi}{2}, 0\right), (0, -8), \left(\dfrac{\pi}{2}, 0\right), (\pi, 8)$

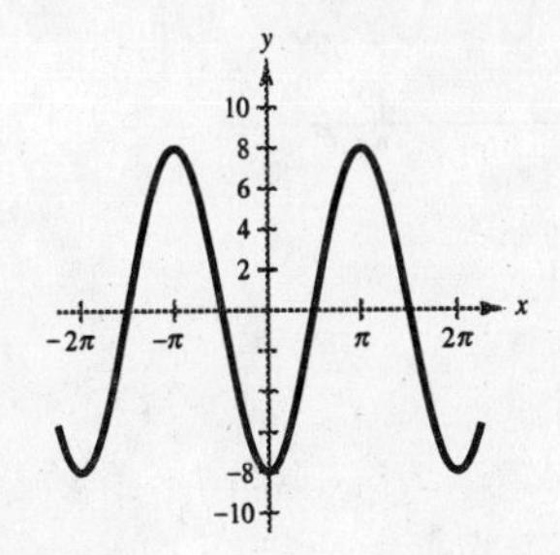

49. $y = \frac{1}{10}\cos(60\pi x)$; $a = \frac{1}{10}$, $b = 60\pi$, $c = 0$

Period: $\frac{2\pi}{60\pi} = \frac{1}{30}$

Amplitude: $\frac{1}{10}$

Key points: $\left(0, \frac{1}{10}\right)$, $\left(\frac{1}{120}, 0\right)$, $\left(\frac{1}{60}, -\frac{1}{10}\right)$, $\left(\frac{1}{40}, 0\right)$, $\left(\frac{1}{30}, \frac{1}{10}\right)$

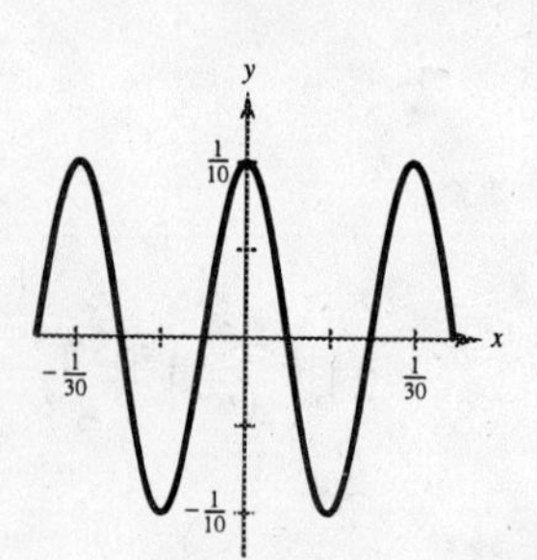

51. $y = 2 - 2\sin\frac{2\pi x}{3}$

Vertical shift 2 units upward of the graph in Exercise #43.

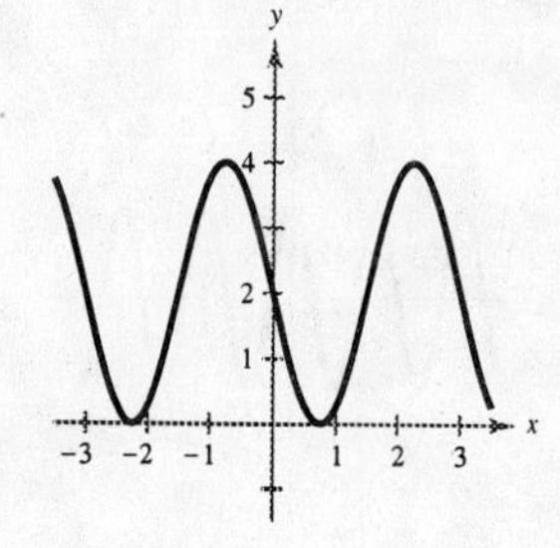

53. $y = \frac{2}{3}\cos\left(\frac{x}{2} - \frac{\pi}{4}\right)$; $a = \frac{2}{3}$, $b = \frac{1}{2}$, $c = \frac{\pi}{4}$

Period: 4π

Amplitude: $\frac{2}{3}$

Shift: Set $\frac{x}{2} - \frac{\pi}{4} = 0$ and $\frac{x}{2} - \frac{\pi}{4} = 2\pi$

$x = \frac{\pi}{2}$ $\qquad$ $x = \frac{9\pi}{2}$

Key points: $\left(\frac{\pi}{2}, \frac{2}{3}\right)$, $\left(\frac{3\pi}{2}, 0\right)$, $\left(\frac{5\pi}{2}, \frac{-2}{3}\right)$, $\left(\frac{7\pi}{2}, 0\right)$, $\left(\frac{9\pi}{2}, \frac{2}{3}\right)$

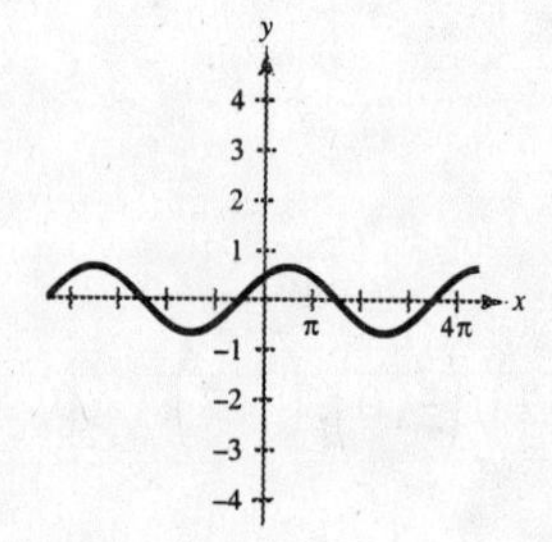

55. $y = -2\sin(4x = \pi)$

Amplitude: 2

period: $\frac{\pi}{2}$

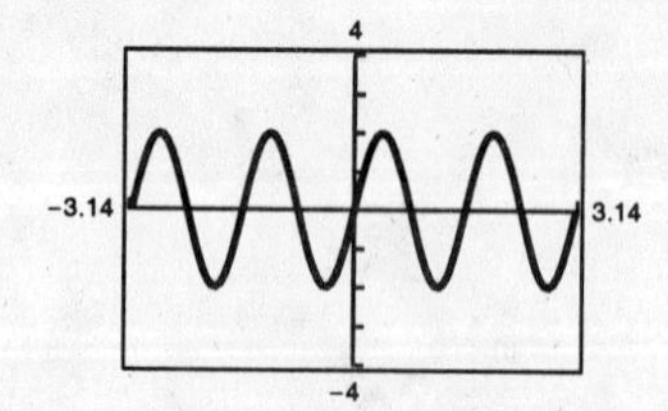

57. $y = \cos\left(2\pi x - \frac{\pi}{2}\right) = 1$

Amplitude: 1

period: 1

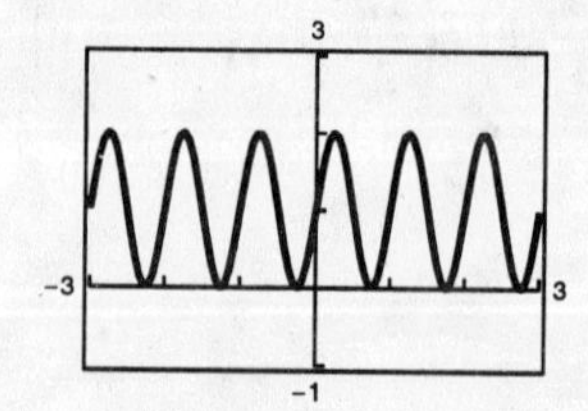

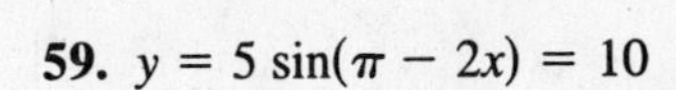

59. $y = 5\sin(\pi - 2x) = 10$

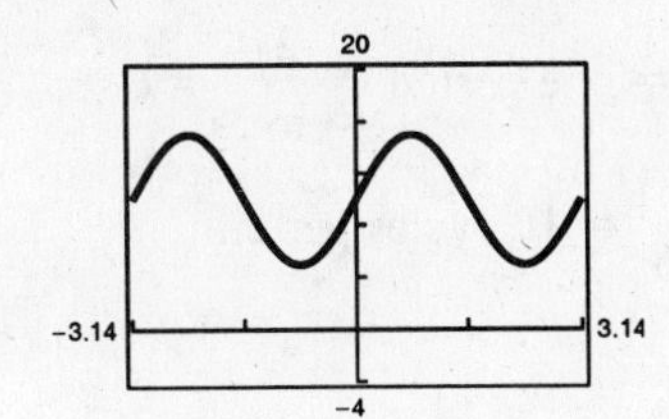

Amplitude: 5

Period: π

61. $y = \dfrac{1}{100}\sin 120\,\pi t$

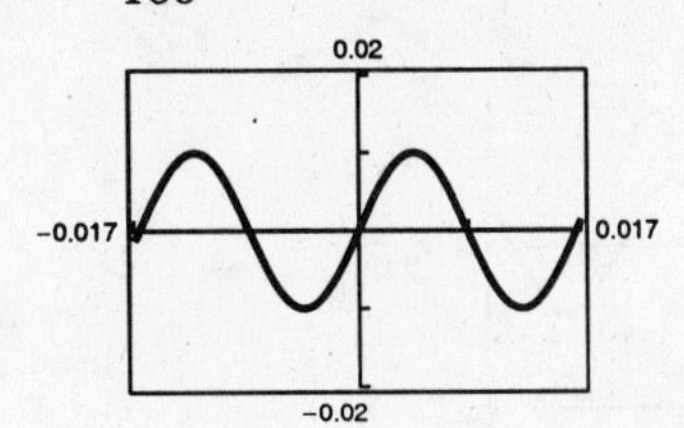

Amplitude: $\frac{1}{100}$

Period: $\frac{1}{60}$

63. $f(x) = a\cos x = d$

Amplitude: $\frac{1}{2}[8 - 0] = 4$

Since $f(x)$ is the graph of $g(x) = 4\cos x$ reflected about the x-axis and shifted vertically 4 units upward, we have $a = -4$ and $d = 4$. Thus, $f(x) = -4\cos x = 4 = 4 - 4\cos x$.

65. $f(x) = a\cos x = d$

Amplitude: $\frac{1}{2}[7 - (-5)] = 6$

Graph of f is the graph of $g(x) = 6\cos x$ reflected about the x-axis and shifted vertically 1 unit upward. Thus $f(x) = -6\cos x = 1$.

67. $y = a\sin(bx - c)$

Amplitude: $|a| = |3|$.

Since the graph is reflected about the x-axis, we have $a = -3$.

Period: $\dfrac{2\pi}{b} = \pi \Rightarrow b = 2$

Phase shift: $c = 0$

Thus, $y = -3\sin 2x$.

69. $y = a\sin(bx = c)$

Amplitude: $a = 1$

Period: $2\pi \Rightarrow b = 1$

Phase shift: $bx = c = 0$ when $x = \dfrac{\pi}{4}$

$$(1)\frac{\pi}{4} = c = 0 \Rightarrow c = -\frac{\pi}{4}$$

Thus, $y = \sin\left(x - \dfrac{\pi}{4}\right)$.

71. $y_1 = \sin x$

$y_2 = -\dfrac{1}{2}$

In the interval $[-2\pi, 2\pi]$, $\sin x = -\dfrac{1}{2}$ when

$$x = -\frac{5\pi}{6}, -\frac{\pi}{6}, \frac{7\pi}{6}, \frac{11\pi}{6}.$$

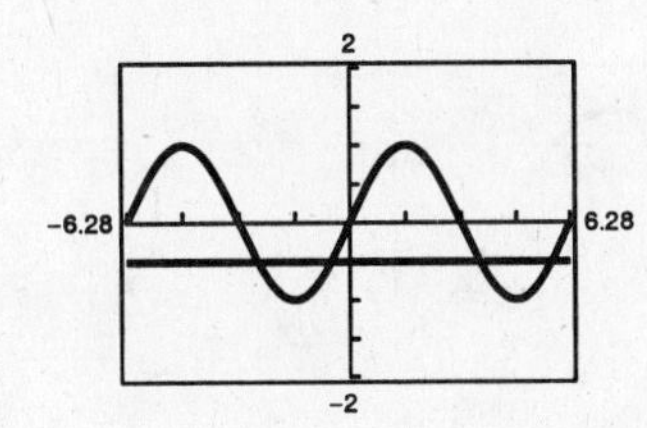

73. $y_1 = \cos x$

$y_2 = \dfrac{\sqrt{2}}{2}$

In the interval $[-2\pi, 2\pi]$, $\cos x = \dfrac{\sqrt{2}}{2}$ when $x = \pm\dfrac{\pi}{4}, \pm\dfrac{7\pi}{4}$.

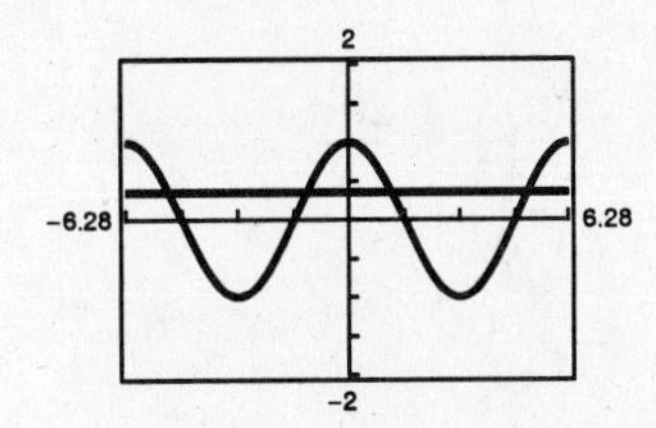

75. $y = 0.85 \sin \dfrac{\pi t}{3}$

(a)

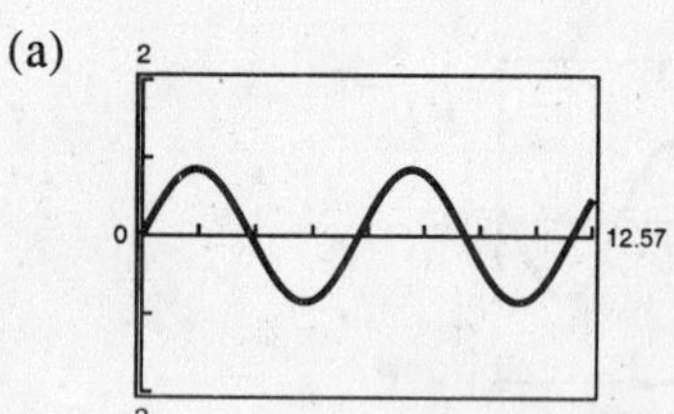

(b) Time for one cycle = one period $= \dfrac{2\pi}{\pi/3} = 6$ sec

(c) Cycles per min $= \dfrac{60}{6} = 10$ cycles per min

77. $S = 22.3 - 3.4 \cos \dfrac{\pi t}{6}, \quad 1 \le t \le 12$

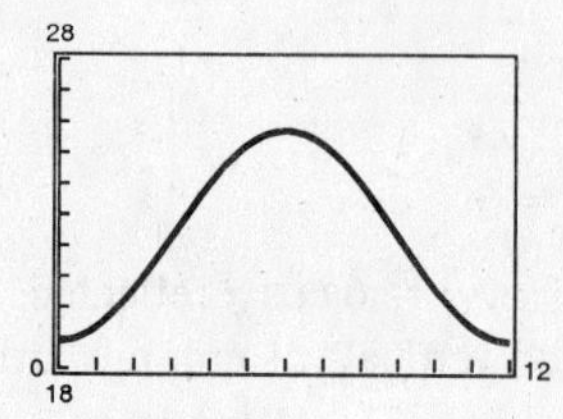

Maximum sales: June

Minimum sales: December

79. $h = 25 \sin \dfrac{\pi}{15}(t - 75) = 30$

(a)

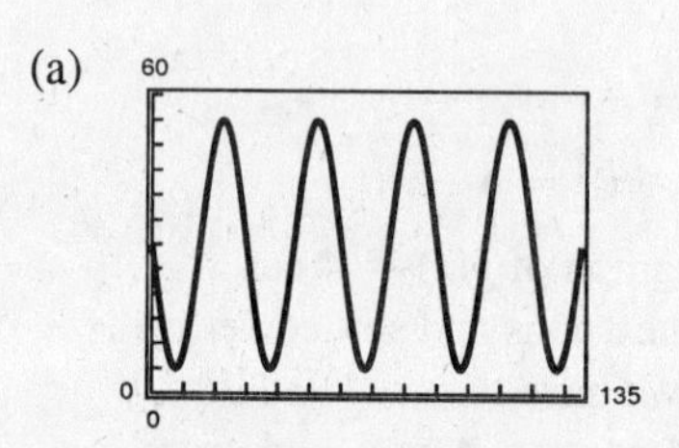

(b) Minimum: $30 - 25 = 5$ feet

Maximum: $30 = 25 = 55$ feet

81. $C = 30.3 = 21.6 \sin\left(\dfrac{2\pi t}{365} = 10.9\right)$

(a) period $= \dfrac{2\pi}{b} = \dfrac{2\pi}{(2\pi/365)} = 365$ days

This is to be expected: 365 days = 1 year

(b) The constant 30.3 gallons is the average daily fuel consumption.

(c)

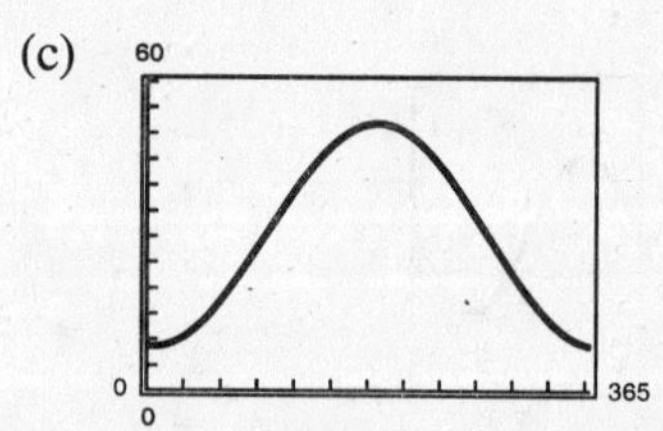

Consumption exceeds 40 gallons/day when $124 \le x \le 252$. (Graph C together with $y = 40$).

83. (a) $p = 2.55 \cos\left(\dfrac{\pi t}{6} - \dfrac{\pi}{12}\right) = 3.45$

(b)

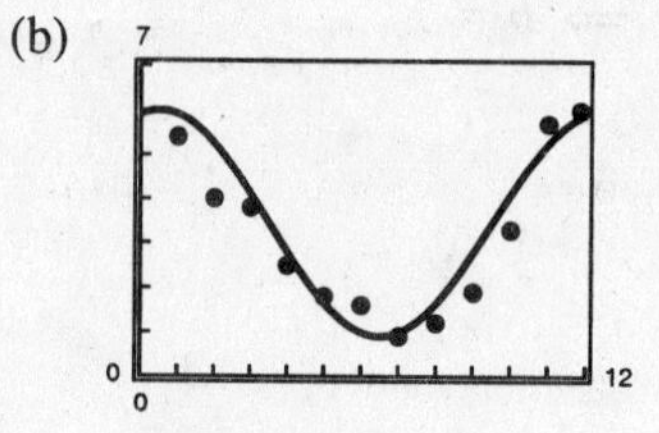

The graph fits the data fairly well.

85. True. The period is $\dfrac{2\pi}{\left(\dfrac{3}{10}\right)} = \dfrac{20\pi}{3}$

87. The amplitude changes from $\dfrac{1}{2}$ to $\dfrac{3}{2}$ to 3. The graph of $y = -3 \sin x$ is a reflection in the x-axis of the graph of $g(x) = 3 \sin x$

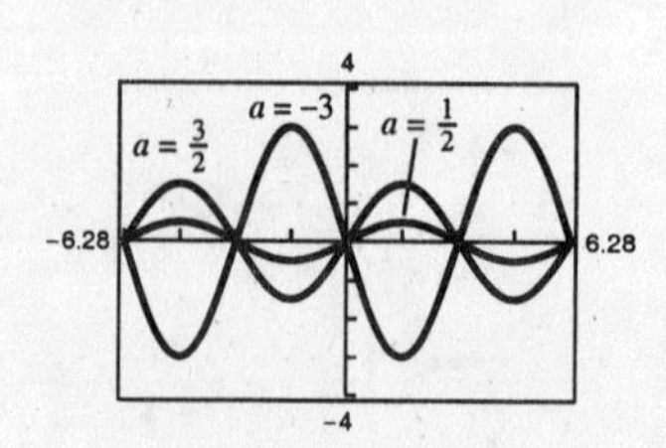

89. The period of the sign function changes from 4π to $\frac{4\pi}{3}$ to $\frac{\pi}{2}$.

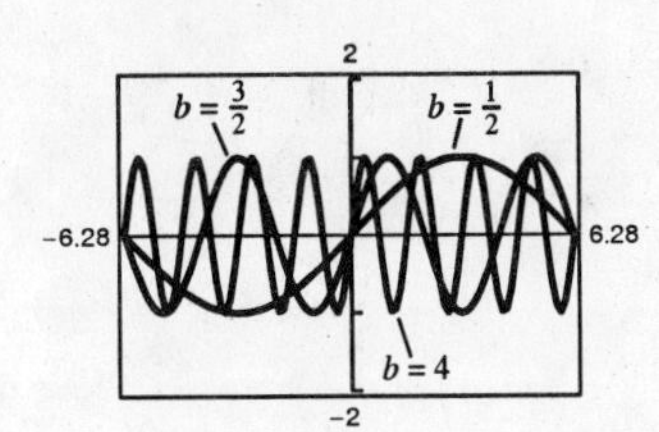

91. (a) $h(x) = \cos^2 x$ is even.

(b) $g(x) = \sin^2 x$ is even.

(c) $h(x) = \sin x \cos x$ is odd.

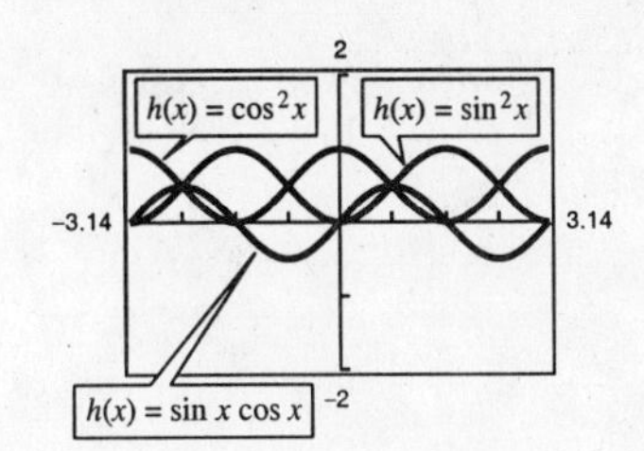

93. $f(x) = 1 - \frac{1}{2}x^2$ is the parabola opening downward. $g(x) = \cos x$ is periodic.

95. $f(x) = \dfrac{5x - 3}{x} = 5 - \dfrac{3}{x}$

Asymptotes: $x = 0$, $y = 5$

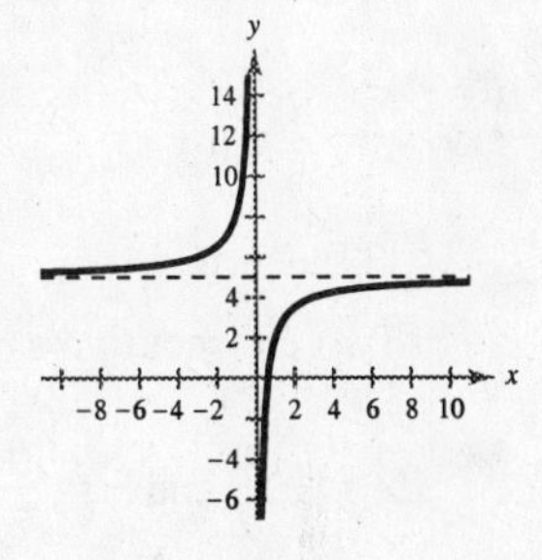

97. $f(x) = \dfrac{x^2 - 7x + 12}{x + 2} = x - 9 + \dfrac{30}{x + 2}$

Asymptotes: $x = -2$, $y = x - 9$

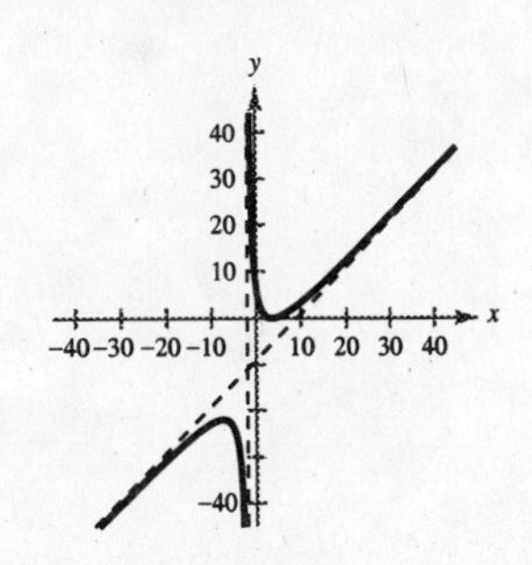

99. $-\frac{\pi}{9} = -\frac{\pi}{9}\left(\frac{180}{\pi}\right) = -20°$

101. $-0.48 = -0.48\left(\frac{180}{\pi}\right) \approx -27.502°$

Section 4.6 Graphs of Other Trigonometric Functions

- You should be able to graph:

  $y = a\tan(bx - c)$ $\quad$ $y = a\cot(bx - c)$

  $y = a\sec(bx - c)$ $\quad$ $y = a\csc(bx - c)$

- When graphing $y = a\sec(bx - c)$ or $y = a\csc(bx - c)$ you should know to first graph $y = a\cos(bx - c)$ or $y = a\sin(bx - c)$ since
  (a) The intercepts of sine and cosine are vertical asymptotes of cosecant and secant.
  (b) The maximums of sine and cosine are local minimums of cosecant and secant.
  (c) The minimums of sine and cosine are local maximums of cosecant and secant.
- You should be able to graph using a damping factor.

Solutions to Odd-Numbered Exercises

1. $y = \sec \dfrac{x}{2}$

Period: $\dfrac{2\pi}{1/2} = 4\pi$

Matches graph (g).

3. $y = \tan 2x$

Period: $\dfrac{\pi}{2}$

Matches graph (f).

5. $y = \cot \dfrac{\pi x}{2}$

Period: $\dfrac{\pi}{\pi/2} = 2$

Matches graph (b).

7. $y = -\csc x$

Period: 2π

Matches graph (e).

9. $y = \dfrac{1}{3} \tan x$

Period: π

Two consecutive asymptotes:

$x = -\dfrac{\pi}{2}$ and $x = \dfrac{\pi}{2}$

x	$-\dfrac{\pi}{4}$	0	$\dfrac{\pi}{4}$
y	$-\dfrac{1}{3}$	0	$\dfrac{1}{3}$

11. $y = -2 \tan 2x$

Period: $\dfrac{\pi}{2}$

Two consecutive asymptotes:

$2x = -\dfrac{\pi}{2} \Rightarrow x = -\dfrac{\pi}{4}$

$2x = \dfrac{\pi}{2} \Rightarrow x = \dfrac{\pi}{4}$

x	$-\dfrac{\pi}{8}$	0	$\dfrac{\pi}{8}$
y	2	0	-2

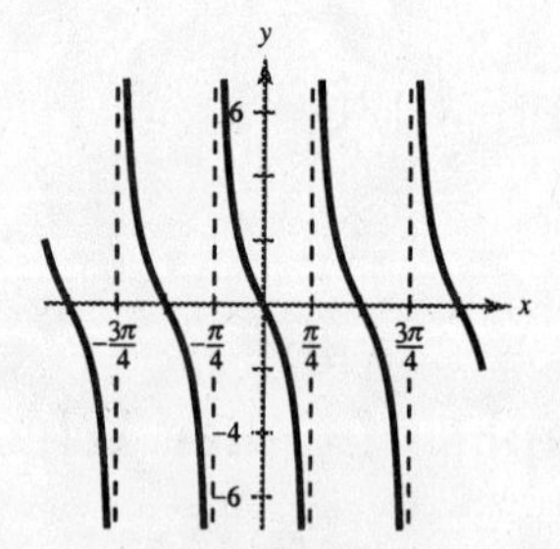

13. $y = -\dfrac{1}{2} \sec x$

Graph $y = -\dfrac{1}{2} \cos x$ first.

Period: 2π

One cycle: 0 to 2π

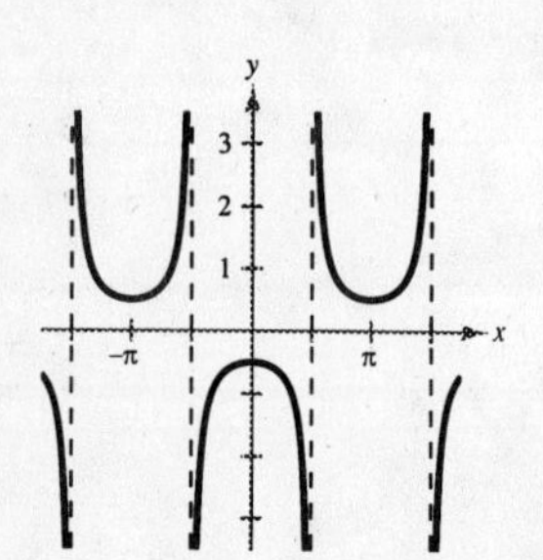

15. $y = -\sec \pi x$

Graph $y = -\cos \pi x$ first.

Period: $\dfrac{2\pi}{\pi} = 2$

One cycle: 0 to 2

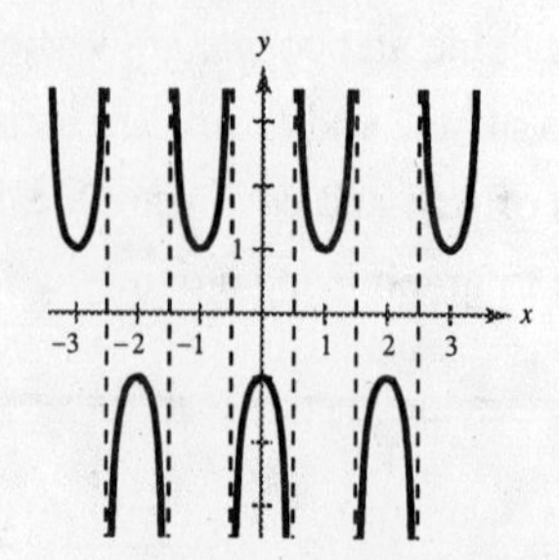

17. $y = \sec \pi x - 3$

Reflect the graph in Exercise #15 about the x-axis and then shift it vertically down three units.

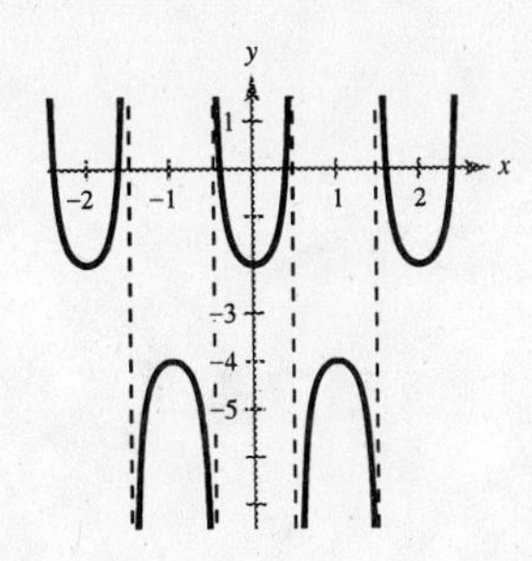

19. $y = \csc \dfrac{x}{2}$

Graph $y = \sin \dfrac{x}{2}$ first.

Period: $\dfrac{2\pi}{1/2} = 4\pi$

One cycle: 0 to 4π

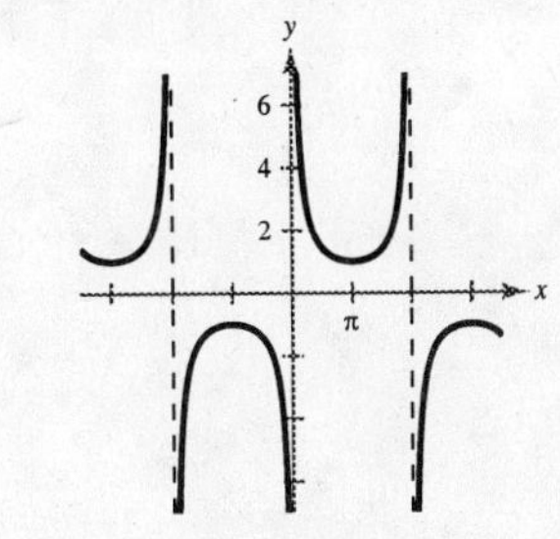

21. $y = \dfrac{1}{2} \cot \dfrac{x}{2}$

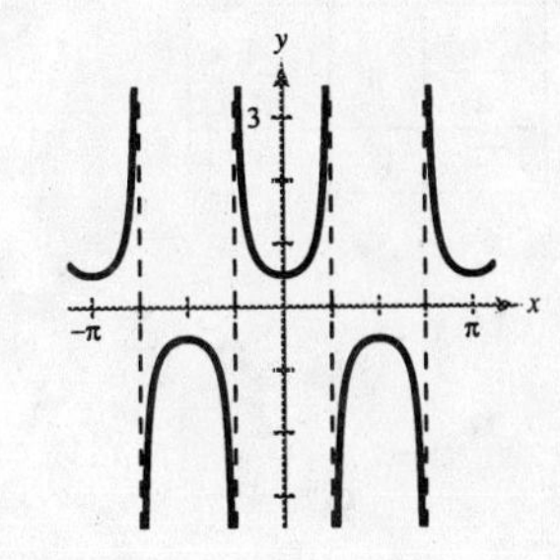

Period: $\dfrac{\pi}{1/2} = 2\pi$

Two consecutive asymptotes:

$\dfrac{x}{2} = 0 \Rightarrow x = 0$

$\dfrac{x}{2} = \pi \Rightarrow x = 2\pi$

x	$\frac{\pi}{2}$	π	$\frac{3\pi}{2}$
y	$\frac{1}{2}$	0	$-\frac{1}{2}$

23. $y = \dfrac{1}{2} \sec 2x$

Graph $y = \dfrac{1}{2} \cos 2x$ first.

Period: $\dfrac{2\pi}{2} = \pi$

One cycle: 0 to π

25. $y = 2 \tan \dfrac{\pi x}{4}$

Period: $\dfrac{\pi}{\pi/4} = 4$

Two consecutive asymptotes:

$\dfrac{\pi x}{4} = -\dfrac{\pi}{2} \Rightarrow x = -2$

$\dfrac{\pi x}{4} = \dfrac{\pi}{2} \Rightarrow x = 2$

x	-1	0	1
y	-2	0	2

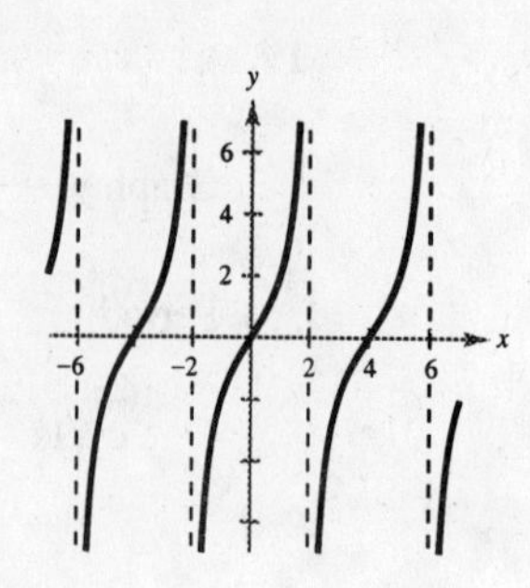

27. $y = \csc(\pi - x)$

Graph $y = \sin(\pi - x)$ first.

Period: 2π

Shift: Set $\pi - x = 0$ and $\pi - x = 2\pi$

$x = \pi \qquad x = -\pi$

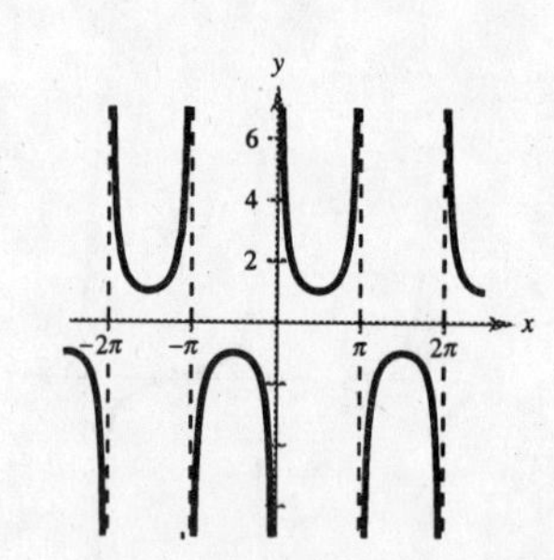

29. $y = 2 \cot\left(x - \dfrac{\pi}{2}\right).$

Period: π

Two consecutive asymptotes: $x - \dfrac{\pi}{2} = 0 \Longrightarrow x = \dfrac{\pi}{2}$

$x - \dfrac{\pi}{2} = \pi \Longrightarrow x = \dfrac{3\pi}{2}$

x	$\dfrac{3\pi}{4}$	π	$\dfrac{5\pi}{4}$
y	2	0	-2

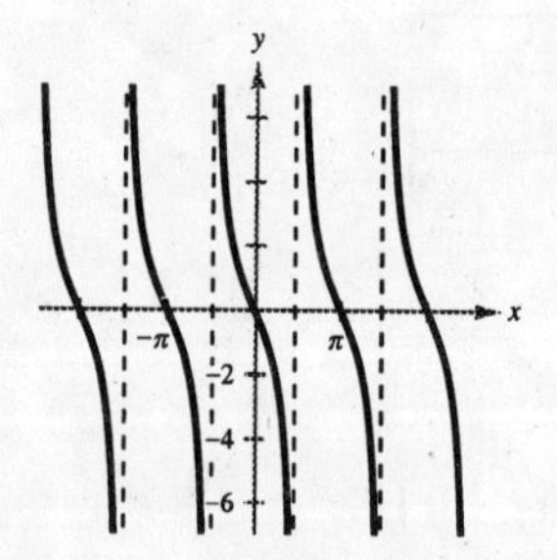

31. $y = \dfrac{1}{3} \tan \dfrac{x}{3}$

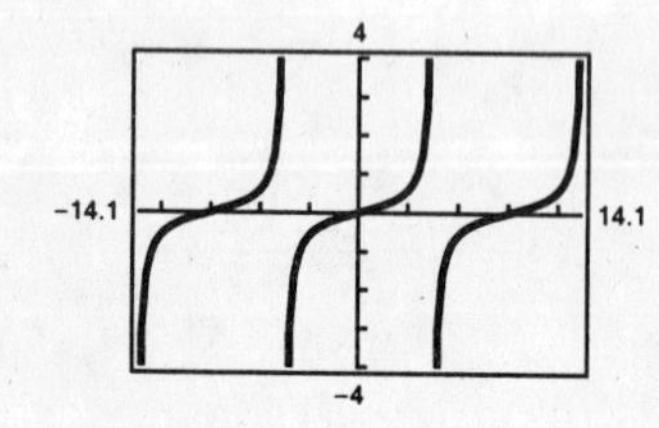

33. $y = -2 \sec 4x$

$= \dfrac{-2}{\cos 4x}$

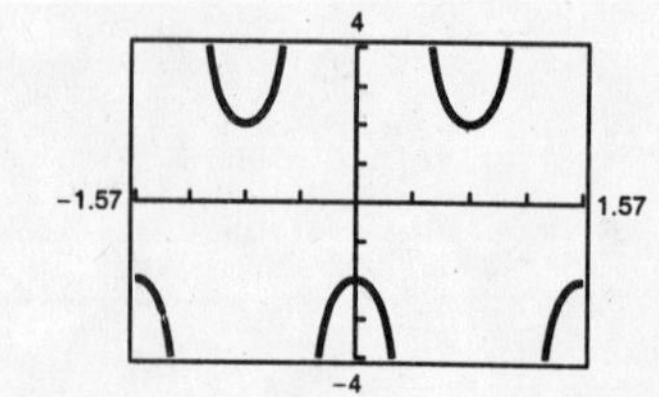

35. $y = \tan\left(x - \dfrac{\pi}{4}\right)$

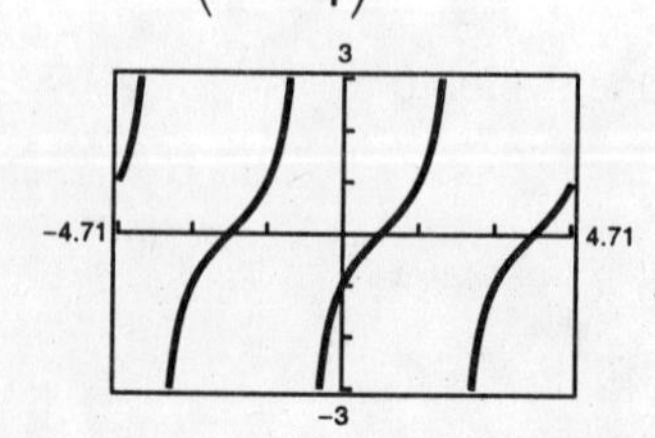

37. $y = \frac{1}{4}\cot\left(x + \frac{\pi}{2}\right)$

$= \dfrac{1}{4\tan\left(x + \frac{\pi}{2}\right)}$

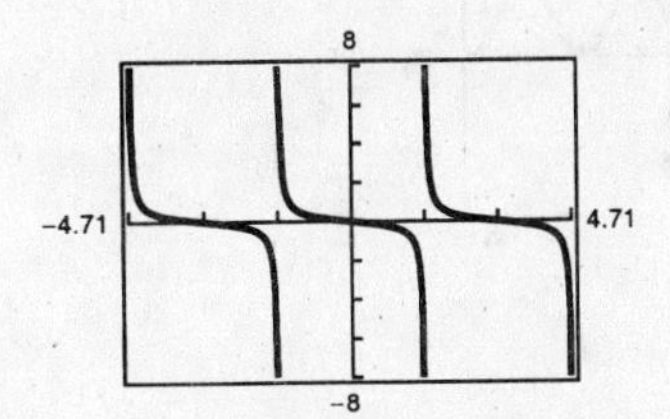

39. $y = \frac{1}{2}\sec(2x - \pi)$

$y = \dfrac{1}{2\cos(2x - \pi)}$

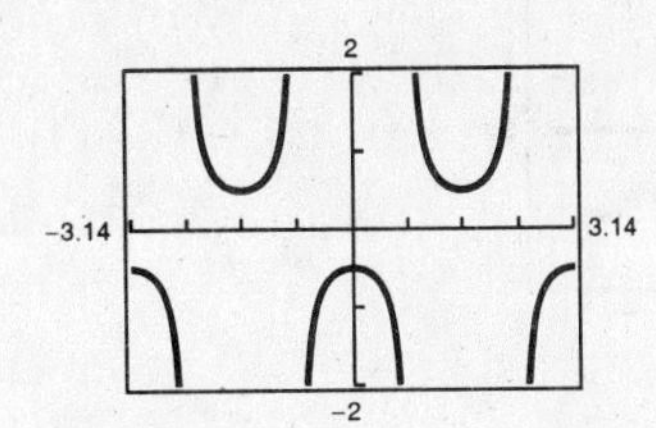

41. $\tan x = 1$

$x = -\frac{7\pi}{4}, -\frac{3\pi}{4}, \frac{\pi}{4}, \frac{5\pi}{4}$

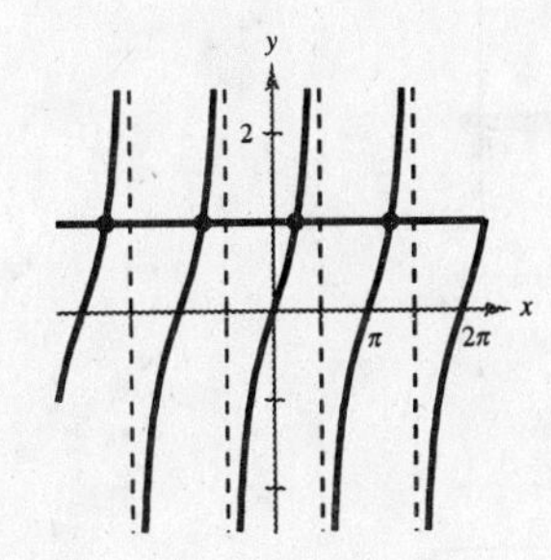

43. $\sec x = -2$

$x = \pm\frac{2\pi}{3}, \pm\frac{4\pi}{3}$

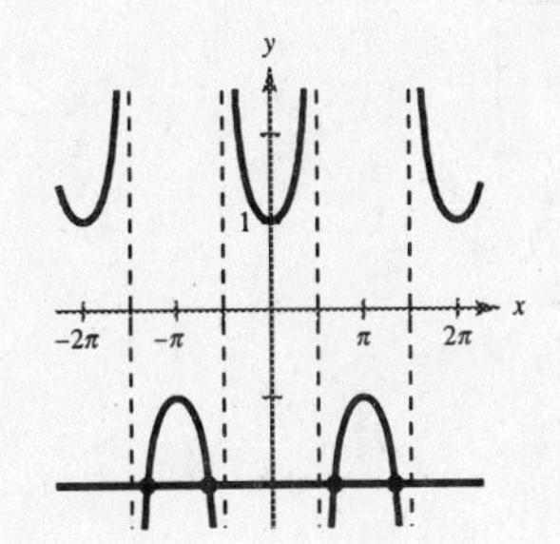

45. The graph of $f(x) = \sec x$ has y-axis symmetry. Thus, the function is even.

47. $y_1 = \sin x \csc x$ and $y_2 = 1$

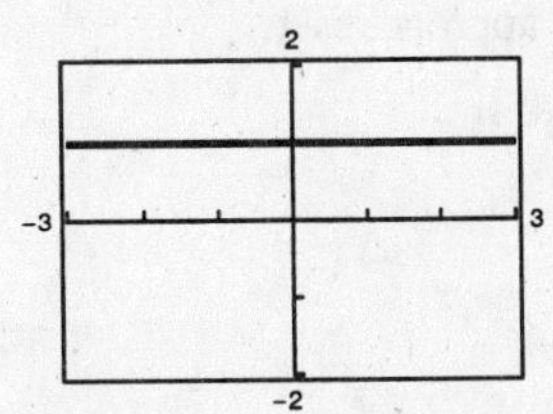

Not equivalent because y_1, is not defined at 0

$\sin x \csc x = \sin x\left(\frac{1}{\sin x}\right) = 1, \sin x \neq 0$

49. $y_1 = \dfrac{\cos x}{\sin x}$ and $y_2 = \cot x = \dfrac{1}{\tan x}$

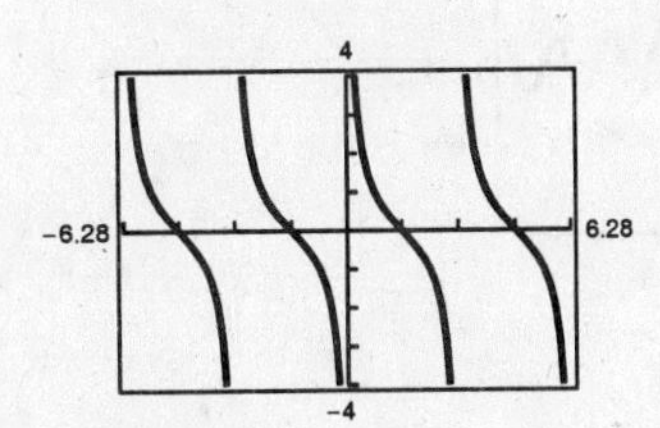

Equivalent

$\cot x = \dfrac{\cos x}{\sin x}$

51. $f(x) = x\cos x$

As $x \to 0, f(x) \to 0$.

Matches graph (d).

53. $g(x) = |x| \sin x$

As $x \to 0, g(x) \to 0$.

Matches graph (b).

55. $f(x) = \sin x + \cos\left(x + \dfrac{\pi}{2}\right), g(x) = 0$

$f(x) = g(x)$

The graph is the line $y = 0$.

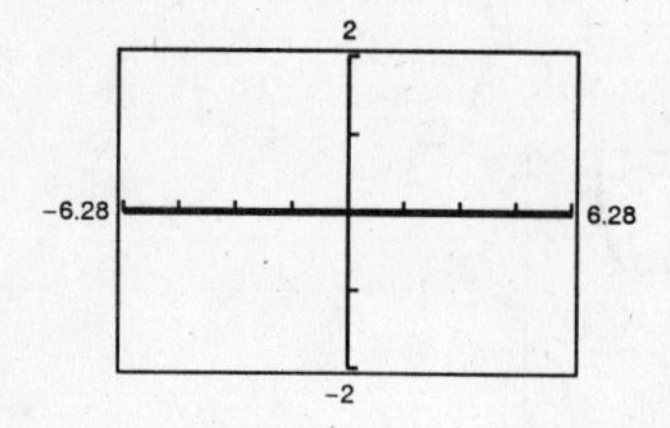

57. $f(x) = \sin^2 x, g(x) = \dfrac{1}{2}(1 - \cos 2x)$

$f(x) = g(x)$

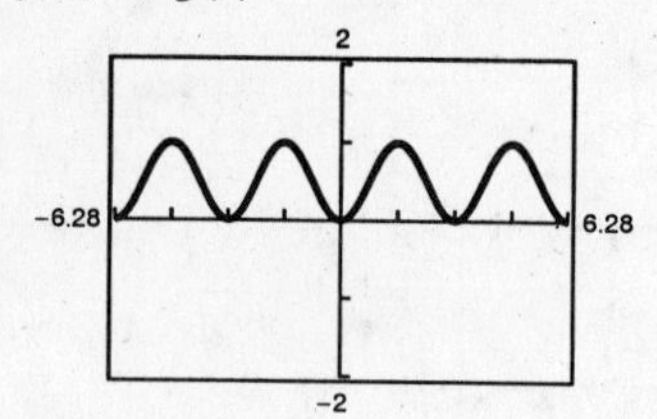

59. $f(x) = e^{-x} \cos x$

Damping factor: e^{-x}

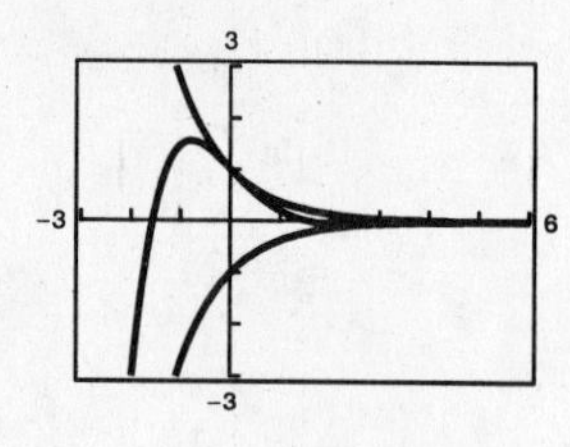

As $x \longrightarrow \infty, f(x) \longrightarrow 0$.

61. $f(x) = 2^{-x/4} \cos \pi x$

$-2^{-x/4} \leq f(x) \leq 2^{-2x/4}$

The damping factor is $y = 2^{-x/4}$.

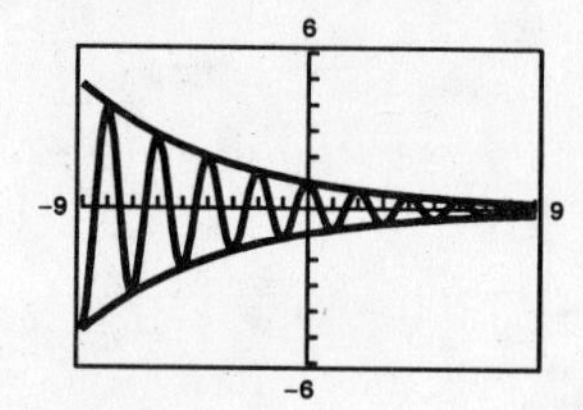

As $x \longrightarrow \infty, f \longrightarrow 0$

63. $y = \dfrac{6}{x} + \cos x$

As $x \longrightarrow 0$, from the right, $y \longrightarrow \infty$.

As $x \longrightarrow 0$. from the left, $y \longrightarrow -\infty$.

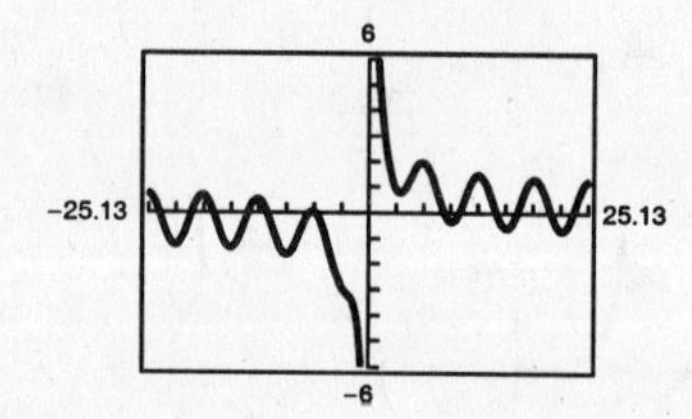

65.

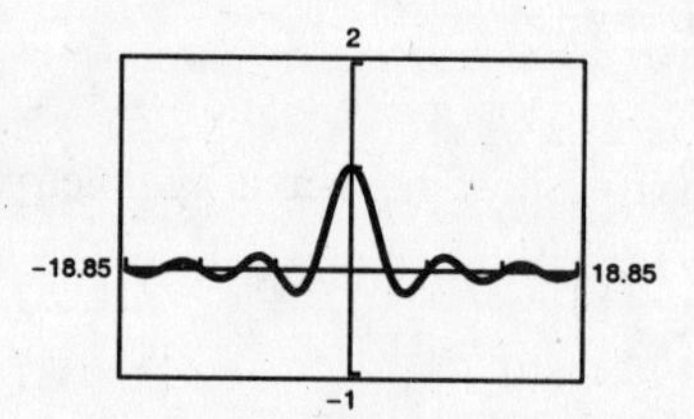

As x tends to 0, $\dfrac{\sin x}{x}$ approaches 1.

67. $f(x) = \dfrac{\tan x}{x}$

As $x \rightarrow 0, f(x) \rightarrow 1$

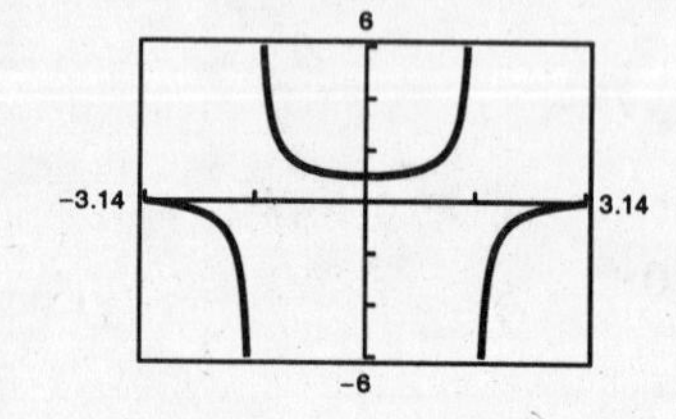

69. $\tan x = \dfrac{5}{d}$

$d = \dfrac{5}{\tan x} = 5 \cot x$

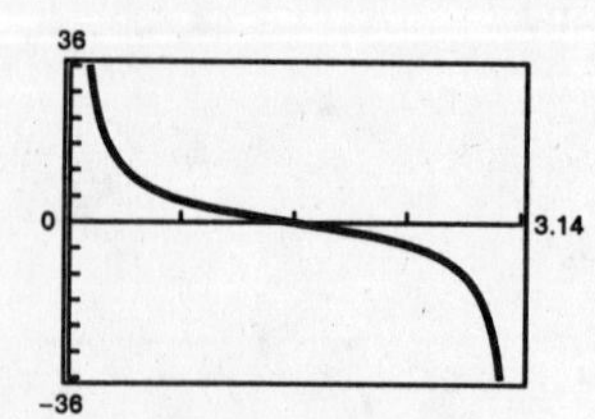

71. As the predator population increases, the number of prey decrease. When the number of prey is small, the number of predators decreases.

73. (a)

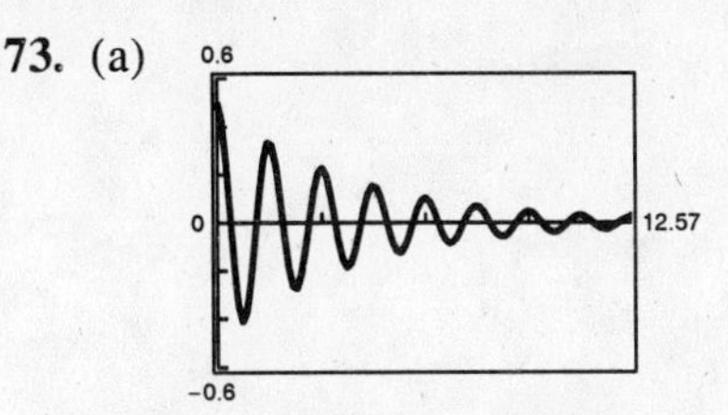

(b) The displacement function is approximately periodic, but damped. It approaches 0 as t increases.

75. $S = 52 + 5t - 28 \cos \dfrac{\pi t}{6}$

(a)

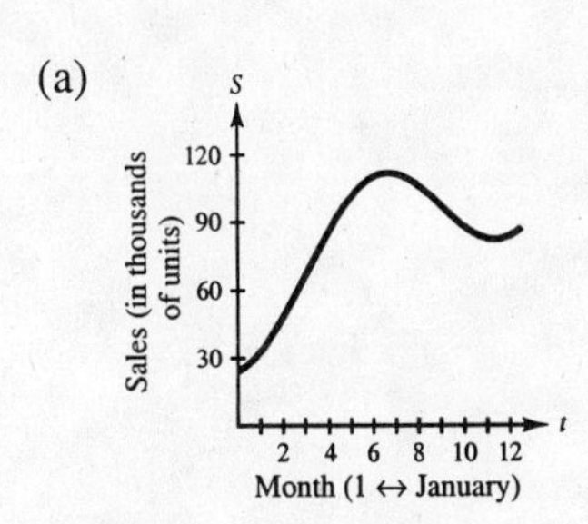

(b) least sales: January
($t = 1, S \approx 32.75$ thousand units)

greatest sales: June
($t \approx 6.66, S \approx 111.64$ thousand units)

77. (a) If a spring of less stiffness is used, then c will be less than 8.2.

(b) If the effect of friction is decreased, then b will be greater than 0.22: $0.22 < b < 1$.

79. True. $-\dfrac{3\pi}{2} + \pi = -\dfrac{\pi}{2}$ and $x = -\dfrac{\pi}{2}$ is a vertical asymptote for the tangent function.

81. True. $2^x \sin x \to 0$ as $x \to -\infty$.

83. For $f(x) = \csc x$, as x approaches π from the left, f approaches ∞. As x approaches π from the right, f approaches $-\infty$.

85. (a) $y_1 = \dfrac{4}{\pi}\left(\sin(\pi x) + \dfrac{1}{3}\sin(3\pi x)\right)$

$y_2 = \dfrac{4}{\pi}\left(\sin(\pi x) + \dfrac{1}{3}\sin(3\pi x) + \dfrac{1}{5}\sin(5\pi x)\right)$

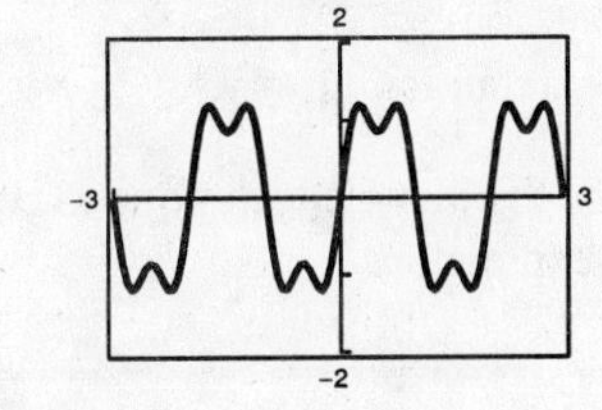

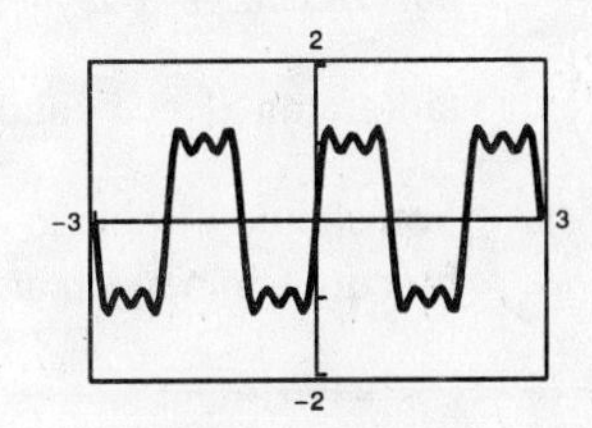

(b) $y_3 = \dfrac{4}{\pi}\left(\sin(\pi x) + \dfrac{1}{3}\sin(3\pi x) + \dfrac{1}{5}\sin(5\pi x) + \dfrac{1}{7}\sin(7\pi x)\right)$

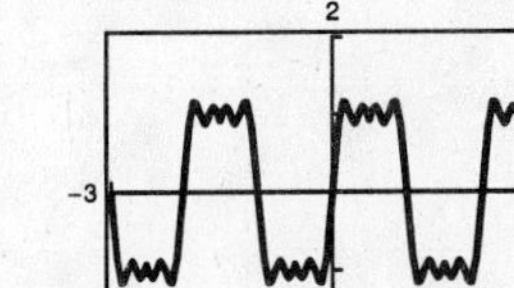

(c) $y_4 = \dfrac{4}{\pi}\left(\sin(\pi x) + \dfrac{1}{3}\sin(3\pi x) + \dfrac{1}{5}\sin(5\pi x) + \dfrac{1}{7}\sin(7\pi x) + \dfrac{1}{9}\sin(9\pi x)\right)$

87. Not one-to-one

89. One-to-one.

$$y = \sqrt[3]{x - 5}$$
$$x = \sqrt[3]{y - 5}$$
$$x^3 = y - 5$$
$$y = x^3 + 5$$
$$f^{-1}(x) = x^3 + 5$$

91. hyp $= \sqrt{9^2 + 14^2} = \sqrt{277}$

$\sin\theta = \dfrac{9}{\sqrt{277}}$ $\quad\cos\theta = \dfrac{14}{\sqrt{277}}$

$\tan\theta = \dfrac{9}{14}$ $\quad\cot\theta = \dfrac{14}{9}$

$\csc\theta = \dfrac{\sqrt{277}}{9}$ $\quad\sec\theta = \dfrac{\sqrt{277}}{14}$

Section 4.7 Inverse Trigonometric Functions

- You should know the definitions, domains, and ranges of $y = \arcsin x$, $y = \arccos x$, and $y = \arctan x$.

Function	Domain	Range
$y = \arcsin x \implies x = \sin y$	$-1 \le x \le 1$	$-\frac{\pi}{2} \le y \le \frac{\pi}{2}$
$y = \arccos x \implies x = \cos y$	$-1 \le x \le 1$	$0 \le y \le \pi$
$y = \arctan x \implies x = \tan y$	$-\infty < x < \infty$	$-\frac{\pi}{2} < x < \frac{\pi}{2}$

- You should know the inverse properties of the inverse trigonometric functions.

  $\sin(\arcsin x) = x,\ -1 \le x \le 1$ and $\arcsin(\sin y) = y,\ -\frac{\pi}{2} \le y \le \frac{\pi}{2}$

  $\cos(\arccos x) = x,\ -1 \le x \le 1$ and $\arccos(\cos y) = y,\ 0 \le y \le \pi$

  $\tan(\arctan x) = x$ and $\arctan(\tan y) = y,\ -\frac{\pi}{2} < y < \frac{\pi}{2}$

- You should be able to use the triangle technique to convert trigonometric functions of inverse trigonometric functions into algebraic expressions.

Solutions to Odd-Numbered Exercises

1. (a)

x	-1.0	-0.8	-0.6	-0.4	-0.2
y	-1.5708	-0.9273	-0.6435	-0.4115	-0.2014

x	0	0.2	0.4	0.6	0.8	1
y	0	0.2014	0.4115	0.6435	0.9273	1.5708

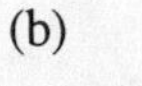
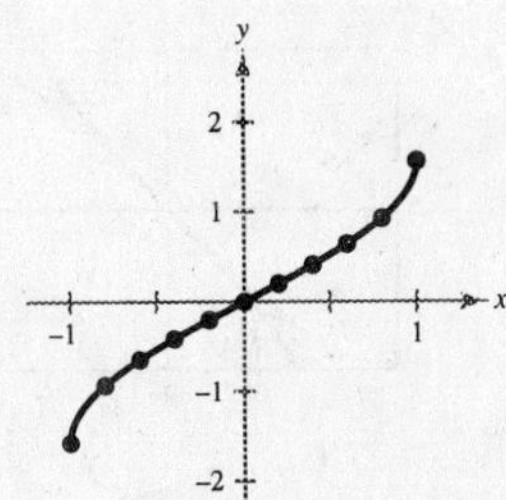

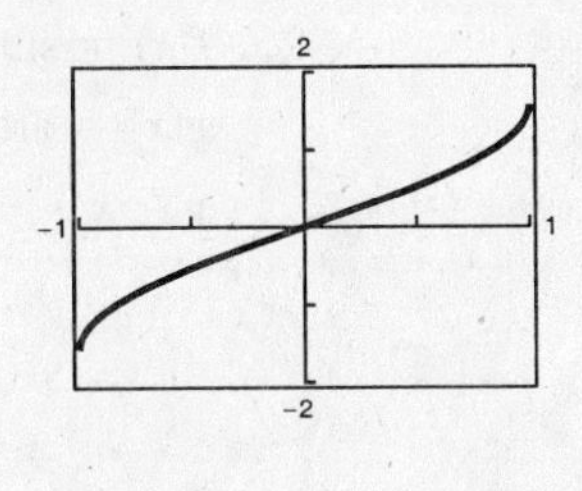

(d) (0, 0), Symmetric to the origin

3. (a)

x	-10	-8	-6	-4	-2
y	-1.4711	-1.4464	-1.4056	-1.3258	-1.1071

x	0	2	4	6	8	10
y	0	1.1071	1.3258	1.4056	1.4464	1.4711

(b)

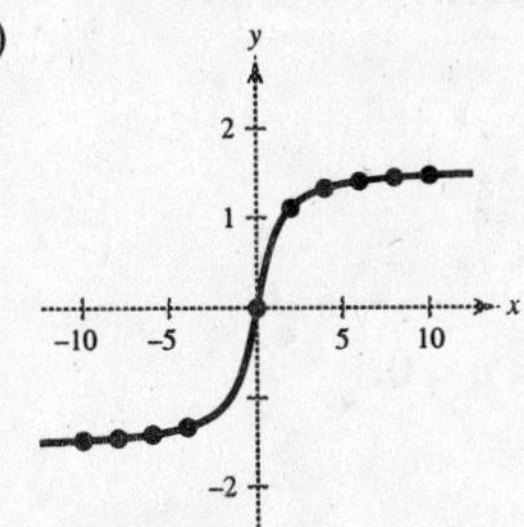

(c)

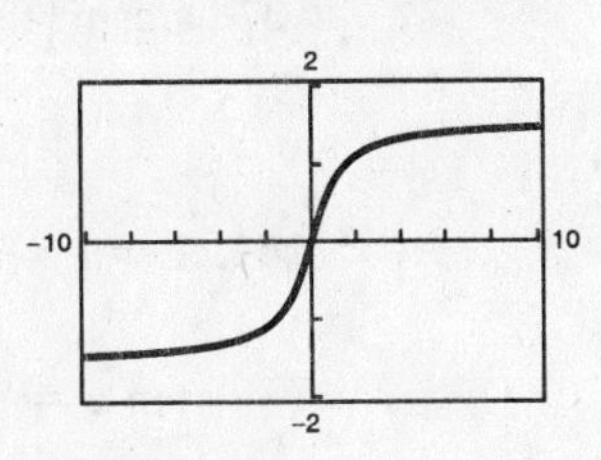

(d) Horizontal asymptotes: $y = \pm\frac{\pi}{2}$

5. $\tan\left(-\frac{\pi}{4}\right) = -1 \Rightarrow \arctan(-1) = -\frac{\pi}{4}$

7. $\arcsin(-1) = -\frac{\pi}{2} \Rightarrow \sin\left(-\frac{\pi}{2}\right) = -1$

9. (a) $y = \arccos\frac{1}{2} \Rightarrow \cos y = \frac{1}{2}$ for $0 \le y \le \pi \Rightarrow y = \frac{\pi}{3} \approx 1.047$

(b) $y = \arccos 0 \Rightarrow \cos y = 0$ for $0 \le y \le \pi \Rightarrow y = \frac{\pi}{2} \approx 1.571$

11. (a) $\arccos\left(-\frac{\sqrt{2}}{2}\right) = \frac{3\pi}{4} \approx 2.356$

(b) $\arcsin\left(-\frac{\sqrt{2}}{2}\right) = -\frac{\pi}{4} \approx -0.785$

13. (a) $y = \arccos\left(-\frac{1}{2}\right) \Rightarrow \cos y = -\frac{1}{2}$ for $0 \le y \le \pi \Rightarrow y = \frac{2\pi}{3} \approx 2.094$

(b) $y = \arcsin\frac{\sqrt{2}}{2} \Rightarrow \sin y = \frac{\sqrt{2}}{2}$ for $-\frac{\pi}{2} \le y \le \frac{\pi}{2} \Rightarrow y = \frac{\pi}{4} \approx 0.785$

15. (a) $\arcsin(-1) = -\frac{\pi}{2} \approx -1.571$

(b) $\arccos(1) = 0$

17. $y = \arccos x \quad (-1, \pi), \left(-\frac{1}{2}, \frac{2\pi}{3}\right), \left(\frac{\sqrt{3}}{2}, \frac{\pi}{6}\right)$

$x = \cos y$

19. (a) $\arcsin(-0.75) \approx -.85$

(b) $\arccos(-0.7) \approx 2.35$

21. (a) $\arcsin 0.41 \approx 0.42$

(b) $\arccos 0.36 \approx 1.20$

23. (a) $\arctan 0.98 \approx 0.78$

(b) $\arctan 4.7 \approx 1.36$

25. $f(x) = \sin x$

$g(x) = \arcsin x$

$y = x$

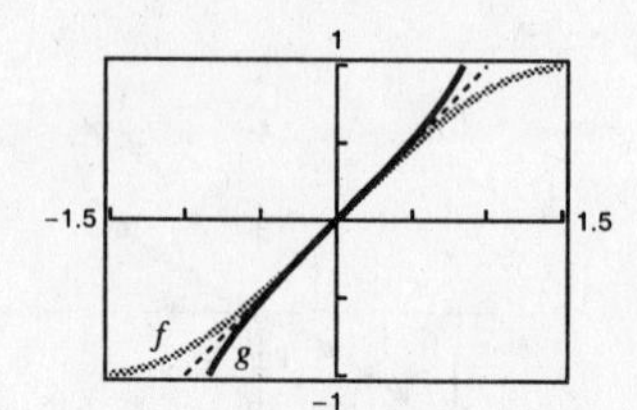

27. $\cos\theta = \frac{4}{x}$

$\theta = \arccos\frac{4}{x}$

29. $\tan\theta = \frac{x+1}{10}$

$\theta = \arctan\left(\frac{x+1}{10}\right)$

31. $\tan(\arctan 35) = 35$

33. $\sin(\arcsin(-0.1)) = -0.1$

35. $\arccos\left(\cos\frac{7\pi}{2}\right) = \arccos(0) = \frac{\pi}{2}$

37. $\arcsin\left(\sin\frac{7\pi}{4}\right) = \arcsin\left(-\frac{\sqrt{2}}{2}\right) = -\frac{\pi}{4}$

39. Let $u = \arcsin\frac{3}{5}$,

$\sin u = \frac{3}{5}, 0 < u < \frac{\pi}{2}$.

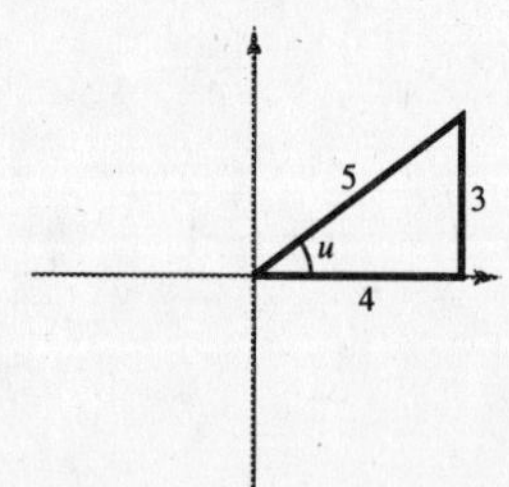

$\sec\left(\arcsin\frac{3}{5}\right) = \sec u = \frac{\text{hyp}}{\text{adj}} = \frac{5}{4}$

41. Let $u = \arctan\left(-\frac{12}{5}\right)$

$\tan u = -\frac{12}{5}, -\frac{\pi}{2} < u < 0$.

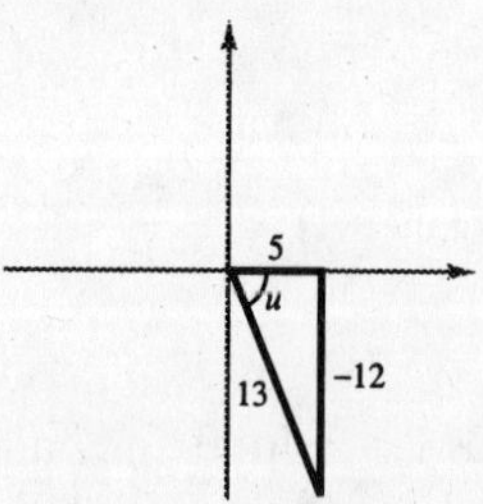

$\csc\left[\arctan\left(-\frac{12}{5}\right)\right] = \csc u = \frac{\text{hyp}}{\text{opp}} = -\frac{13}{12}$

43. Let $u = \arcsin\left(-\frac{3}{8}\right)$,

$\sin u = -\frac{3}{8}, -\frac{\pi}{2} < u < 0.$

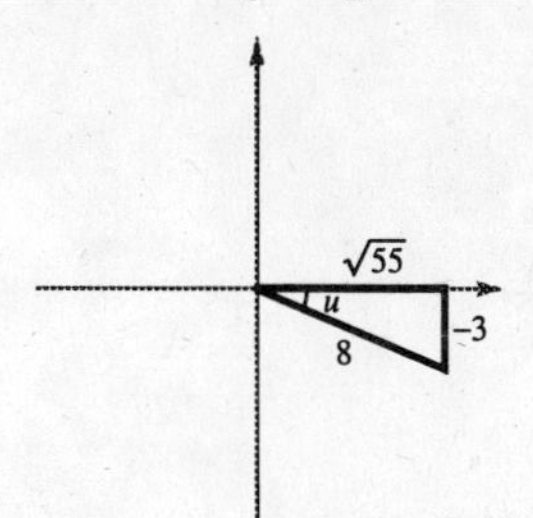

$$\tan\left[\arcsin\left(-\frac{3}{8}\right)\right] = \tan u = -\frac{3}{\sqrt{55}} = -\frac{3\sqrt{55}}{55}$$

45. Let $u = \arctan\frac{6}{11}$,

$\tan u = \frac{6}{11}, 0 < u < \frac{\pi}{2}.$

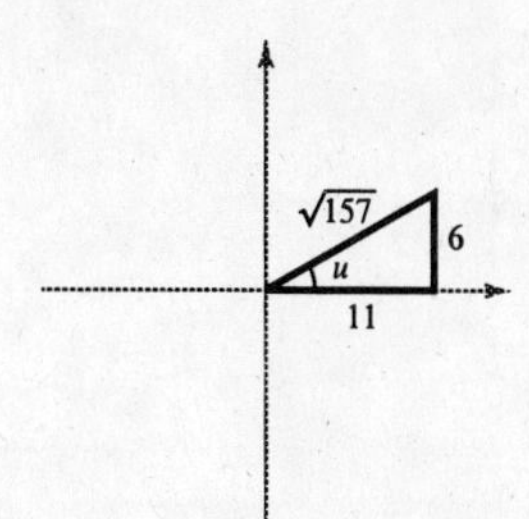

$$\cot\left(\arctan\frac{6}{11}\right) = \cot u = \frac{\text{adj}}{\text{opp}} = \frac{11}{6}$$

47. Let $u = \arctan x$,

$\tan u = x = \frac{x}{1}.$

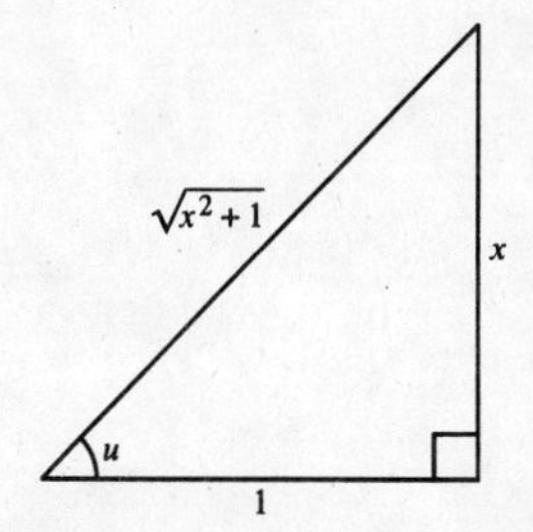

$$\sin(\arctan x) = \sin u = \frac{\text{opp}}{\text{hyp}} = \frac{x}{\sqrt{x^2 + 1}}$$

49. Let $u = \arcsin(x - 1)$,

$\sin u = x - 1 = \frac{x - 1}{1}.$

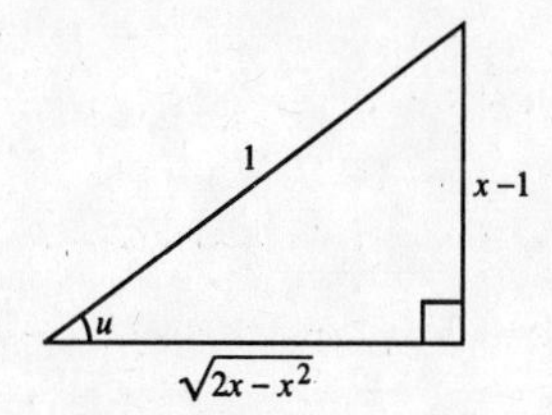

$$\sec[\arcsin(x - 1)] = \sec u = \frac{\text{hyp}}{\text{adj}} = \frac{1}{\sqrt{2x - x^2}}$$

51. Let $u = \arctan\frac{1}{x}$,

$\tan u = \frac{1}{x}.$

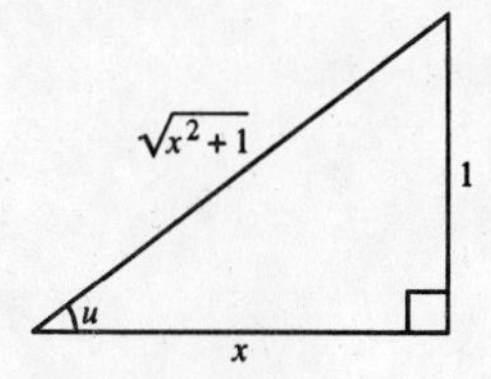

$$\cot\left(\arctan\frac{1}{x}\right) = \cot u = \frac{\text{adj}}{\text{opp}} = x$$

53. Let $u = \arcsin\frac{x - h}{r}$,

$\sin u = \frac{x - h}{r}.$

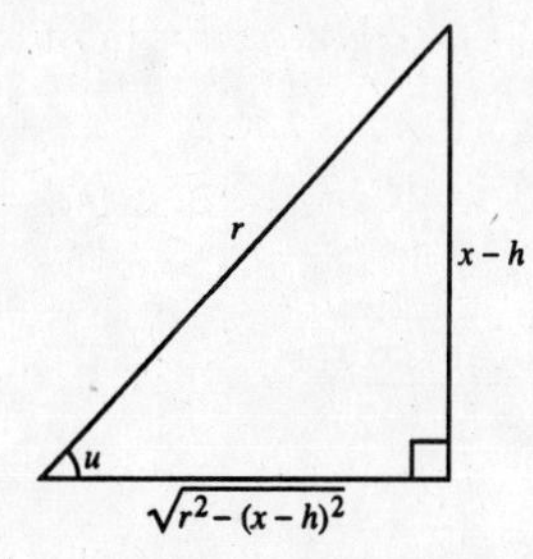

$$\cos\left(\arcsin\frac{x - h}{r}\right) = \cos u = \frac{\sqrt{r^2 - (x - h)^2}}{r}$$

55. $f(x) = \tan\left(\arccos \frac{x}{2}\right)$

$g(x) = \frac{\sqrt{4 - x^2}}{x}$

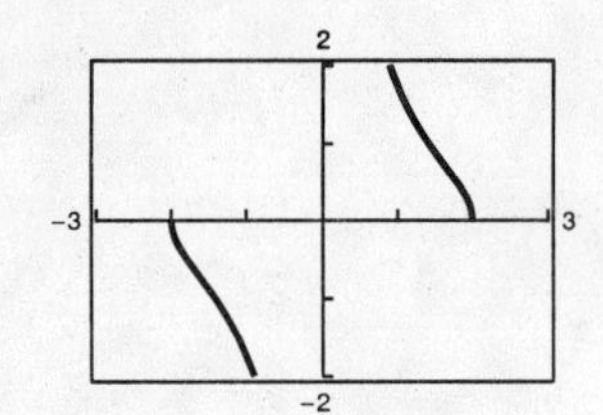

Asymptote: $x = 0$

These are equal because:

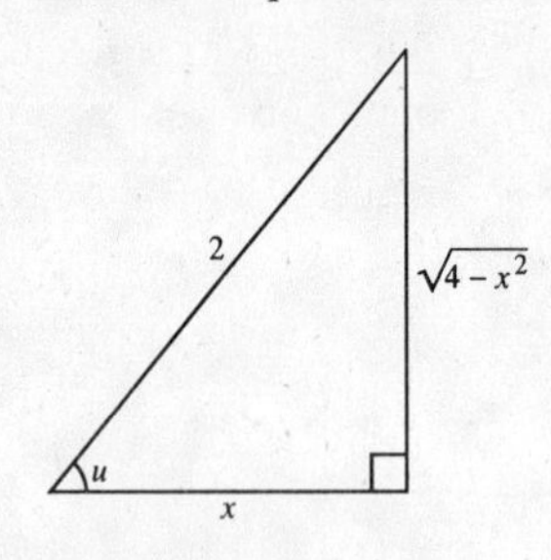

Let $u = \arccos \frac{x}{2}$.

$\tan\left(\arccos \frac{x}{2}\right) = \tan u = \frac{\sqrt{4 - x^2}}{x}$

57. If $\arcsin \frac{\sqrt{36 - x^2}}{6} = u$,

then $\sin u = \frac{\sqrt{36 - x^2}}{6}$.

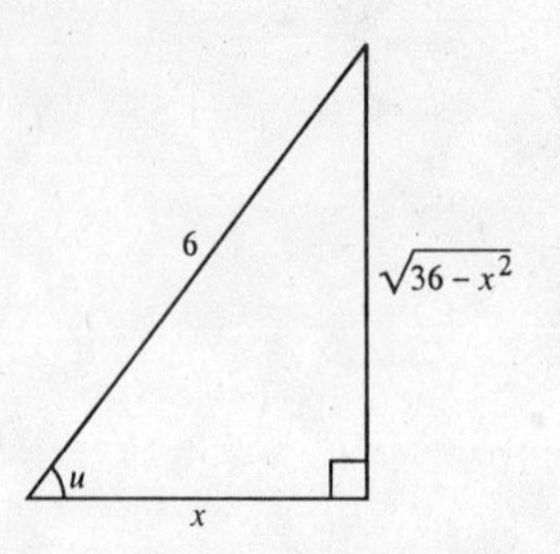

$\arcsin \frac{\sqrt{36 - x^2}}{6} = \arccos \frac{x}{6}$

59. If $\arccos \frac{x - 2}{2} = u$,

then $\cos u = \frac{x - 2}{2}$.

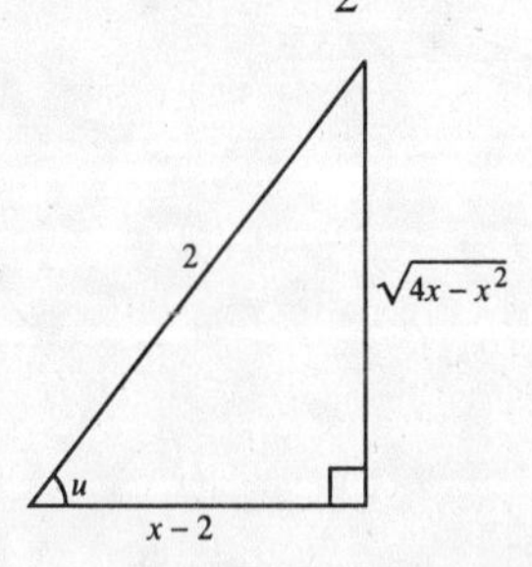

$\arccos \frac{x - 2}{2} = \arctan \frac{\sqrt{4x - x^2}}{x - 2}$

61. $y = \arcsin \frac{x}{2}$

Domain: $-2 \le x \le 2$

Range: $-\frac{\pi}{2} \le y \le \frac{\pi}{2}$

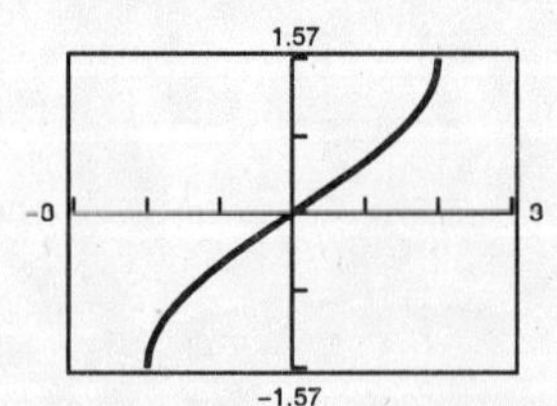

63. $g(t) = \arccos(t + 2)$

Domain: $-3 \le t \le -1$

This is the graph of $y = \arccos t$ shifted two units to the left.

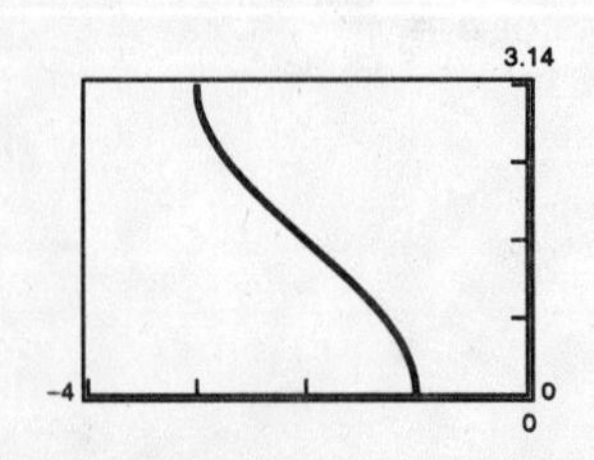

65. $f(x) = \pi + \arctan x$

Domain: $(-\infty, \infty)$

Range: $\left(\frac{\pi}{2}, \frac{3\pi}{2}\right)$

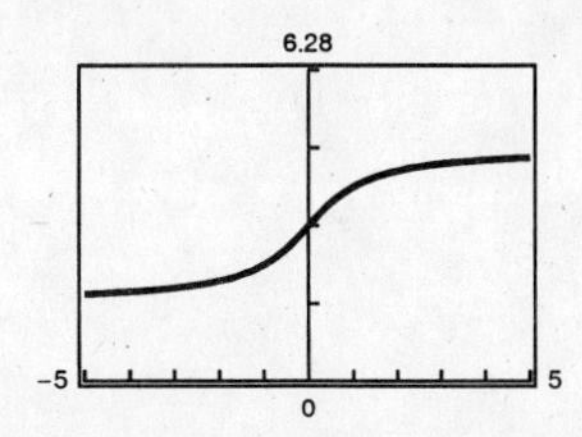

67. $f(x) = \arccos \frac{x}{4}$

Domain: $[-4, 4]$

Range: $[0, \pi]$

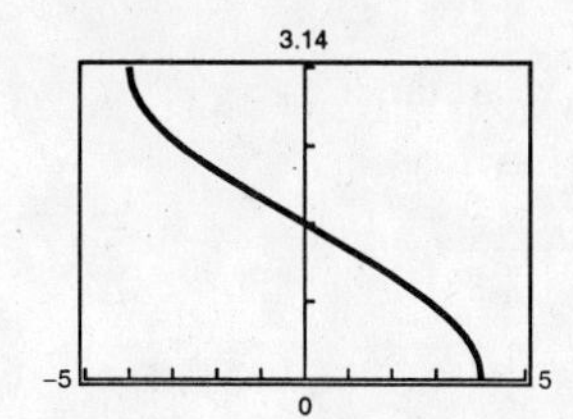

69. $f(t) = 4 \cos \pi t + 3 \sin \pi t$

$= \sqrt{4^2 + 3^2} \sin\left(\pi t + \arctan \frac{4}{3}\right)$

$= 5 \sin\left(\pi t + \arctan \frac{4}{3}\right)$

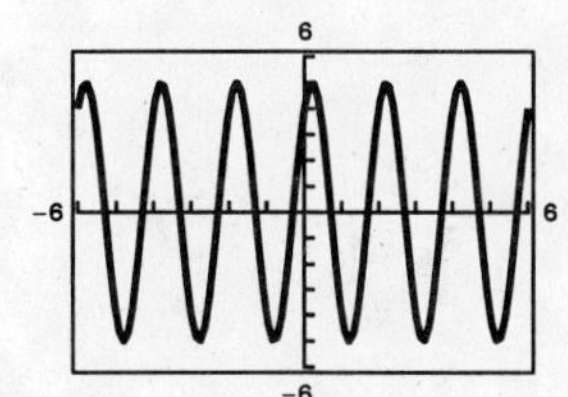

The graph suggests that $A \cos \omega t + B \sin \omega t = \sqrt{A^2 + b^2} \sin\left(\omega t + \arctan \frac{A}{B}\right)$ is true.

71. (a) $\tan \theta = \frac{s}{750}$

$\theta = \arctan\left(\frac{s}{750}\right)$

(b) When $s = 400$, $\theta = \arctan\left(\frac{400}{750}\right) \approx 0.4900 \ (\approx 28.07°)$

When $s = 1600$, $\theta = \arctan\left(\frac{1600}{750}\right) \approx 1.1325 \ (\approx 64.89°)$

73. (a) $\theta = \arctan\left(\frac{20}{41}\right) \approx 26.0°$ (.45 rad)

(b) $\tan(26.0°) = \frac{h}{50}$

$h = 50 \tan(26.0°) \approx 24.4$ feet

75. Area $=$ arctan b $-$ arctan a

(a) $a = 0, b = 1$

Area $= \arctan 1 - \arctan 0 = \dfrac{\pi}{4} - 0 = \dfrac{\pi}{4}$

(c) $a = 0, b = 3$

Area $= \arctan 3 - \arctan 0$

$\approx 1.25 - 0 = 1.25$

$= 1.25$

(b) $a = -1, b = 1$

Area $= \arctan 1 - \arctan(-1)$

$= \dfrac{\pi}{4} - \left(-\dfrac{\pi}{4}\right) = \dfrac{\pi}{2}$

(d) $a = -1, b = 3$

Area $= \arctan 3 - \arctan(-1)$

$\approx 1.25 - \left(-\dfrac{\pi}{4}\right) \approx 2.03$

77. (a) $\tan \theta = \dfrac{x}{20}$

$\theta = \arctan \dfrac{x}{20}$

(b) When $x = 5$,

$\theta = \arctan \dfrac{5}{20} \approx 14.0°$, (0.24 rad).

When $x = 12$, $\theta = \arctan \dfrac{12}{20} \approx 31.0°. = (0.54 \text{ rad.})$.

79. False. $\tan x = \dfrac{\sin x}{\cos x}$.

81. $y = \operatorname{arcsec} x$ if and only if $\sec y = x$ where $x \le -1 \cup x \ge 1$ and $0 \le y < \pi/2$ and $\pi/2 < y \le \pi$. The domain of y, arcsec x is $(-\infty, -1] \cup [1, \infty)$ and the range is $[0, \pi/2) \cup (\pi/2, \pi]$.

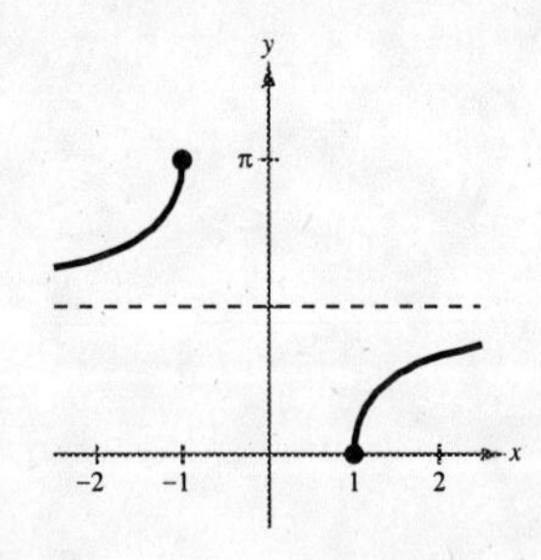

83. (a) $y = \operatorname{arcsec} \sqrt{2} \Longrightarrow \sec y = \sqrt{2}$ and $0 \le y < \dfrac{\pi}{2} \cup \dfrac{\pi}{2} < y \le \pi \Longrightarrow y = \dfrac{\pi}{4}$

(b) $y = \operatorname{arcsec} 1 \Longrightarrow \sec y = 1$ and $0 \le y < \dfrac{\pi}{2} \cup \dfrac{\pi}{2} < y \le \pi \Longrightarrow y = 0$

(c) $y = \operatorname{arccot}\left(-\sqrt{3}\right) \Longrightarrow \cot y = -\sqrt{3}$ and $0 < y < \pi \Longrightarrow y = \dfrac{5\pi}{6}$

(d) $y = \operatorname{arccsc} 2 \Longrightarrow \csc y = 2$ and $-\dfrac{\pi}{2} \le y < 0 \cup 0 < y \le \dfrac{\pi}{2} \Longrightarrow y = \dfrac{\pi}{6}$

85.

$$y = \arctan(-x)$$

$$\tan y = -x,\ -\frac{\pi}{2} < y < \frac{\pi}{2}$$

$$-\tan y = x$$

$$\tan(-y) = x,\ -\frac{\pi}{2} < -y < \frac{\pi}{2}$$

$$\arctan(\tan(-y)) = \arctan x$$

$$-y = \arctan x$$

$$y = -\arctan x$$

87. $y_2 = \dfrac{\pi}{2} - y_1$

$$\arctan x + \arctan\frac{1}{x} = y_1 + y_2$$

$$= y_1 + \left(\frac{\pi}{2} - y_1\right) = \frac{\pi}{2}$$

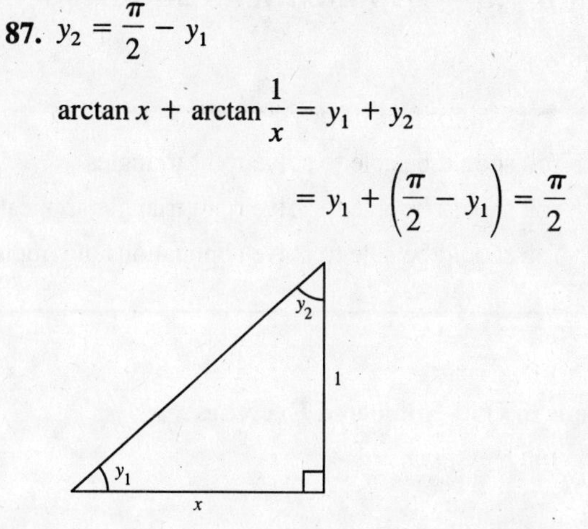

89. $\arcsin x = \arcsin\dfrac{x}{1} = \arctan\dfrac{x}{\sqrt{1 - x^2}}$

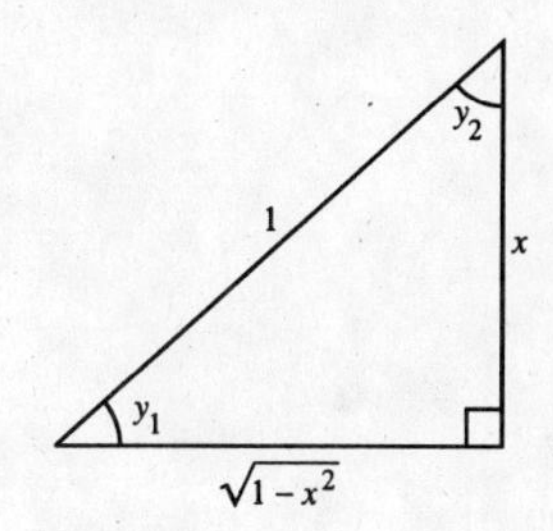

91. $-585°$ coterminal with $720° - 585° = 135°$.

Quadrant II

Reference angle $\theta' = 45°$

$$\sin(-585°) = \frac{\sqrt{2}}{2}$$

$$\cos(-585°) = -\frac{\sqrt{2}}{2}$$

$$\tan(-585°) = -1$$

93. $-\dfrac{19\pi}{4}$ coterminal with $6\pi - \dfrac{19\pi}{4} = \dfrac{5\pi}{4}$.

Quadrant III

Reference angle $\theta' = \dfrac{\pi}{4}$

$$\sin\left(-\frac{19\pi}{4}\right) = -\frac{\sqrt{2}}{2}$$

$$\cos\left(-\frac{19\pi}{4}\right) = -\frac{\sqrt{2}}{2}$$

$$\tan\left(-\frac{19\pi}{4}\right) = 1$$

95. $y = 2 - \cos(x + \pi)$

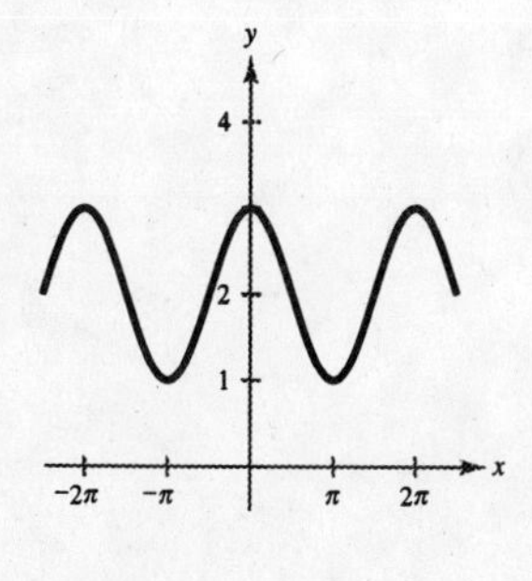

97. $y = 4\cot\left(\dfrac{1}{2}x\right) = \dfrac{4}{\tan\left(\dfrac{1}{2}x\right)}$.

Asymptotes: $x = 0, x = 2\pi, x = -2\pi, \ldots$

Section 4.8 Applications and Models

- You should be able to solve right triangles.
- You should be able to solve right triangle applications.
- You should be able to solve applications of simple harmonic motion: $d = a \sin wt$ or $d = a \cos wt$

Solutions to Odd-Numbered Exercises

1. Given: $A = 25°, b = 10$

$$\tan A = \frac{a}{b} \Rightarrow a = b \tan A = 10 \tan 25° \approx 4.66$$

$$\cos A = \frac{b}{c} \Rightarrow c = \frac{b}{\cos A} = \frac{10}{\cos(25°)} \approx 11.03$$

$B = 90° - 25° = 65°$

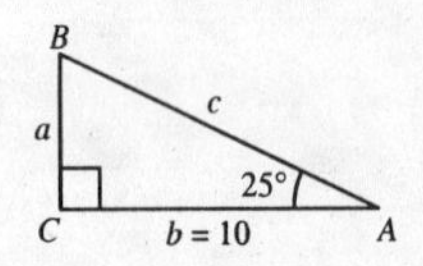

3. Given: $B = 71°,\ b = 24$

$$\tan B = \frac{b}{a} \Rightarrow a = \frac{b}{\tan B} = \frac{24}{\tan 71°} \approx 8.26$$

$$\sin B = \frac{b}{c} \Rightarrow c = \frac{b}{\sin B} = \frac{24}{\sin 71°} \approx 25.38$$

$A = 90° - 71° = 19°$

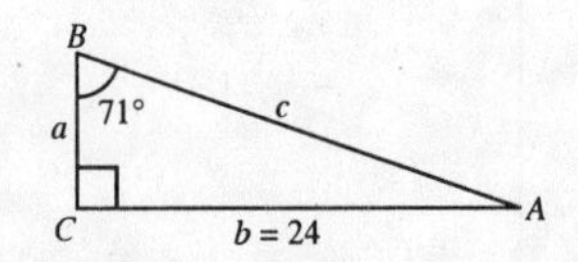

5. Given: $a = 6, b = 16$

$$c^2 = a^2 + b^2 \Rightarrow c = \sqrt{292} \approx 17.09$$

$$\tan A = \frac{a}{b} = \frac{6}{16} \Rightarrow A = \arctan\left(\frac{3}{8}\right) \approx 20.56°$$

$B = 90° - 20.56° = 69.44°$

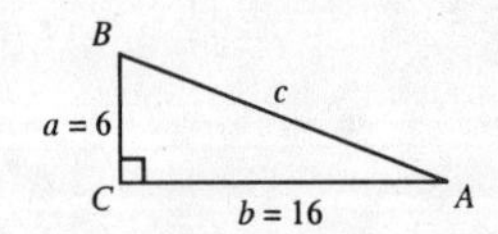

7. Given: $b = 18, c = 42$

$$a = \sqrt{c^2 - b^2} = \sqrt{1440} \approx 37.95$$

$$\cos A = \frac{18}{42} \Rightarrow A = \arccos\left(\frac{3}{7}\right) \approx 64.62°$$

$B = 90° - 64.62° = 25.38°$

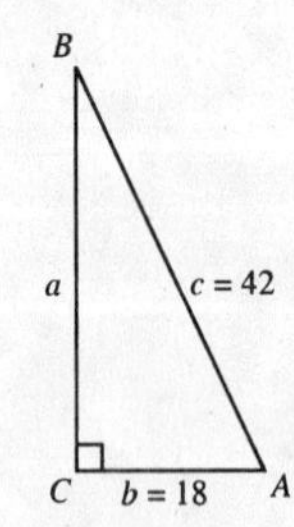

9. $A = 12° \, 15', \; c = 430.5$

$B = 90° - 12° \, 15' = 77° \, 45'$

$\sin 12° \, 15' = \dfrac{a}{430.5}$

$a = 430.5 \sin 12° \, 15' \approx 91.34$

$\cos 12° \, 15' = \dfrac{b}{430.5}$

$b = 430.5 \cos 12° \, 15' \approx 420.70$

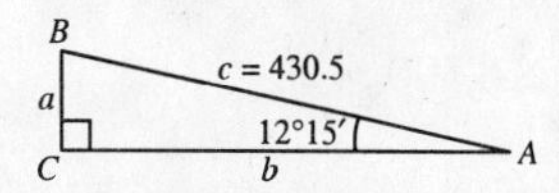

11. $\tan \theta = \dfrac{h}{\frac{1}{2}b}$

$h = \dfrac{1}{2}b \tan \theta$

$h = \dfrac{1}{2}(4) \tan 52° \approx 2.56$ in.

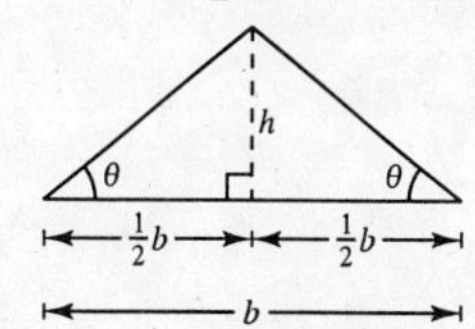

13. $\tan \theta = \dfrac{h}{\frac{1}{2}b} \Rightarrow h = \dfrac{1}{2}b \tan \theta = \dfrac{1}{2}(14.2) \tan (41.6°) \approx 6.30$ feet

15. (a) $\tan \theta = \dfrac{60}{L}$

$L = \dfrac{60}{\tan \theta}$

$= 60 \cot \theta$

(b)

θ	10°	20°	30°	40°	50°
L	340	165	104	72	50

(c) No, the shadow lengths do not increase in equal increments. The cotangent function is not linear.

17. (a) $\sin \theta = \dfrac{h}{20}$

$h = 20 \sin \theta$

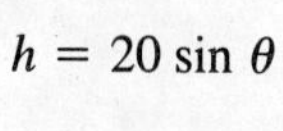

(b)

θ	60°	65°	70°	75°	80°
h	17.3	18.1	18.8	19.3	19.7

19. (a)

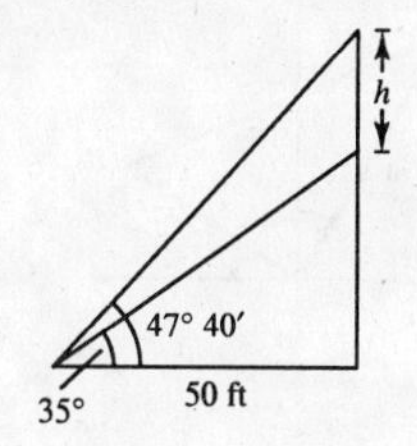

(b) Let the height of the church $= x$ and the height of the church and steeple $= y$. Then:

$\tan 35° = \dfrac{x}{50}$ and $\tan 47° \, 40' = \dfrac{y}{50}$

$x = 50 \tan 35°$ and $y = 50 \tan 47° \, 40'$

$h = y - x = 50(\tan 47° \, 40' - \tan 35°)$

(c) $h \approx 19.9$ feet

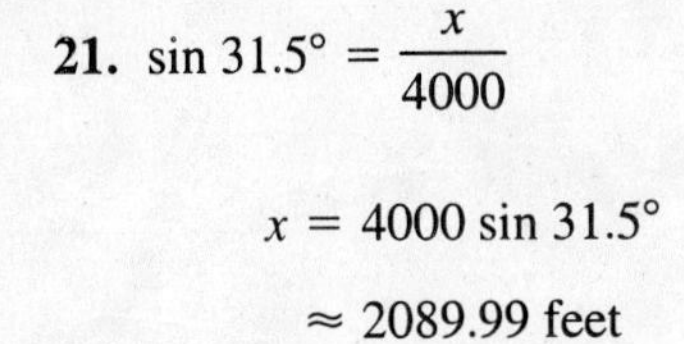

21. $\sin 31.5° = \dfrac{x}{4000}$

$x = 4000 \sin 31.5°$

≈ 2089.99 feet

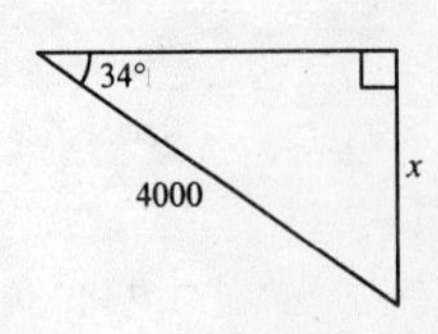

23. $\tan \theta = \dfrac{75}{95}$

$\theta = \arctan\left(\dfrac{15}{19}\right) \approx 38.29°$

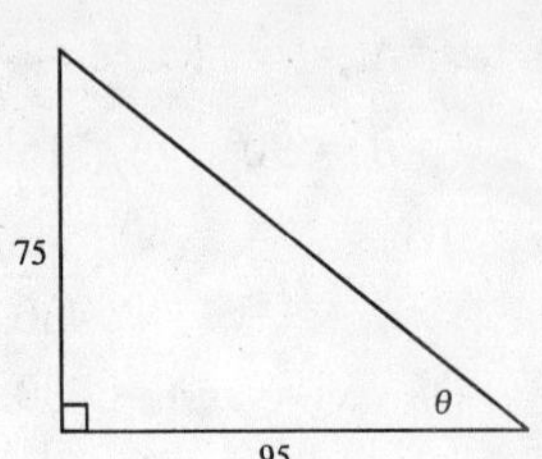

25. $\sin \theta = \dfrac{4000}{4150}$

$\theta = \arcsin\left(\dfrac{4000}{4150}\right)$

$\theta \approx 74.5°$

$\alpha = 90° - 74.5° = 15.5°$

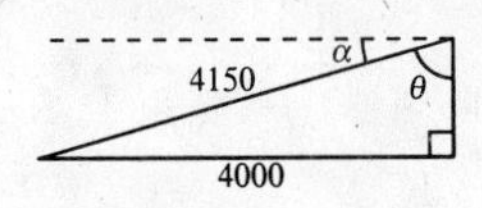

27. Since the airplane speed is

$$\left(275\,\frac{\text{ft}}{\text{sec}}\right)\left(60\,\frac{\text{sec}}{\text{min}}\right) = 16{,}500\,\frac{\text{ft}}{\text{min}},$$

after one minute its distance travelled is 16,500 feet.

$\sin 18° = \dfrac{a}{16{,}500}$

$a = 16{,}500 \sin 18°$

≈ 5099 ft

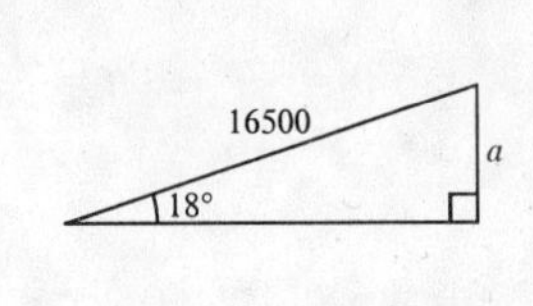

29. $\sin 9.5° = \dfrac{x}{4}$

$x = 4 \sin 9.5° \approx 0.66$ mile

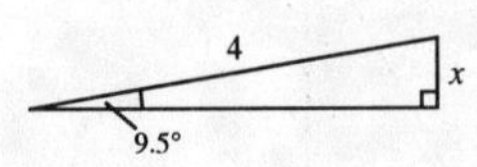

31. The plane has traveled $550(1.5) = 825$ miles

$\sin 32° = \dfrac{a}{825} \Longrightarrow a \approx 437.2$ miles north

$\cos 32° = \dfrac{b}{825} \Longrightarrow b \approx 699.6$ miles east

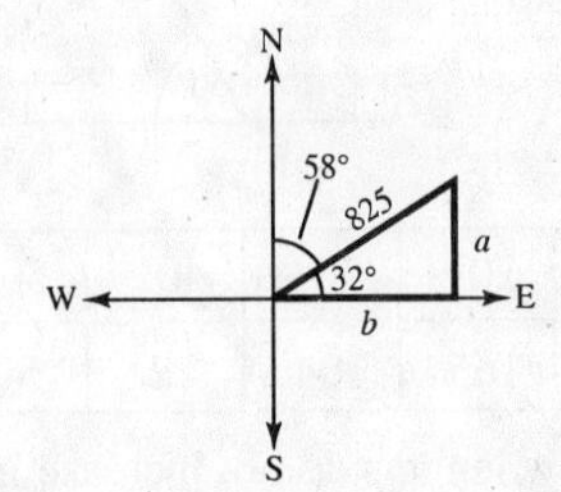

33. $\theta = 32°$, $\phi = 68°$. Note: ABC form a right triangle.

(a) $\alpha = 90° - 32° = 58°$

Bearing from A to C: N 58° E

(b) $\beta = \theta = 32°$

$\gamma = 90° - \phi = 22°$

$C = \beta + \gamma = 54°$

$\tan C = \dfrac{d}{50} \Longrightarrow \tan 54° = \dfrac{d}{50} \Longrightarrow d \approx 68.82$ yd

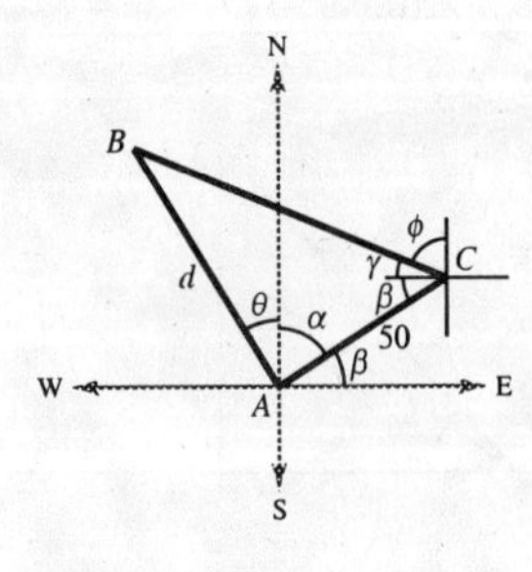

35. $\tan \theta = \dfrac{50}{35} \Longrightarrow \theta \approx 55.0°$

Bearing: N 55.0° W

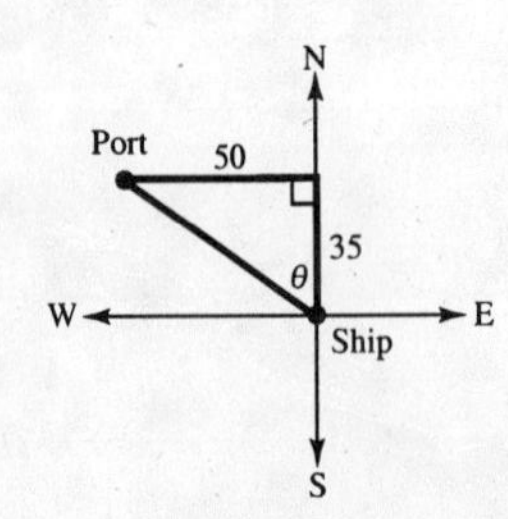

37. $\tan 6.5^\circ = \dfrac{350}{d} \Rightarrow d \approx 3071.91$ ft

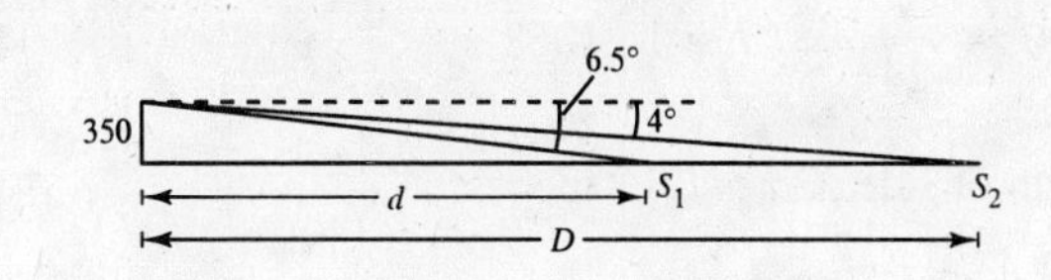

$\tan 4^\circ = \dfrac{350}{D} \Rightarrow D \approx 5005.23$ ft

Distance between ships: $D - d \approx 1933.3$ ft

39. $\tan 57^\circ = \dfrac{a}{x} \Rightarrow x = a \cot 57^\circ$

$\tan 16^\circ = \dfrac{a}{x + (55/6)}$

$\tan 16^\circ = \dfrac{a}{a \cot 57^\circ + (55/6)}$

$\cot 16^\circ = \dfrac{a \cot 57^\circ + (55/6)}{a}$

$a \cot 16^\circ - a \cot 57^\circ = \dfrac{55}{6} \Rightarrow a \approx 3.23$ miles

$\approx 17{,}054$ ft

41. L_1: $3x - 2y = 5 \Rightarrow y = \dfrac{3}{2}x - \dfrac{5}{2} \Rightarrow m_1 = \dfrac{3}{2}$

L_2: $x + y = 1 \Rightarrow y = -x + 1 \Rightarrow m_2 = -1$

$\tan \alpha = \left|\dfrac{-1 - (3/2)}{1 + (-1)(3/2)}\right| = \left|\dfrac{-5/2}{-1/2}\right| = 5$

$\alpha = \arctan 5 \approx 78.7^\circ$

43. The diagonal of the base has a length of $\sqrt{a^2 + a^2} = \sqrt{2}a$.

Now, we have:

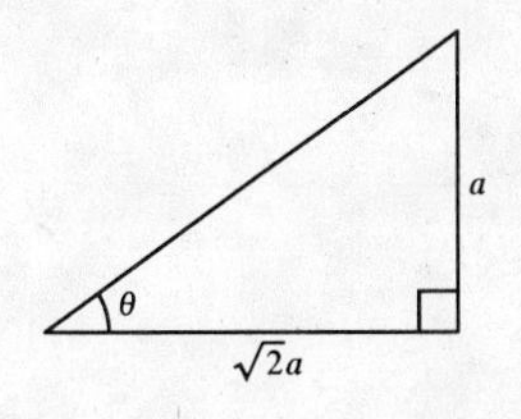

$\tan \theta = \dfrac{a}{\sqrt{2}a} = \dfrac{1}{\sqrt{2}}$

$\theta = \arctan \dfrac{1}{\sqrt{2}}$

$\theta \approx 35.3^\circ$

45. $\cos 30^\circ = \dfrac{b}{r}$

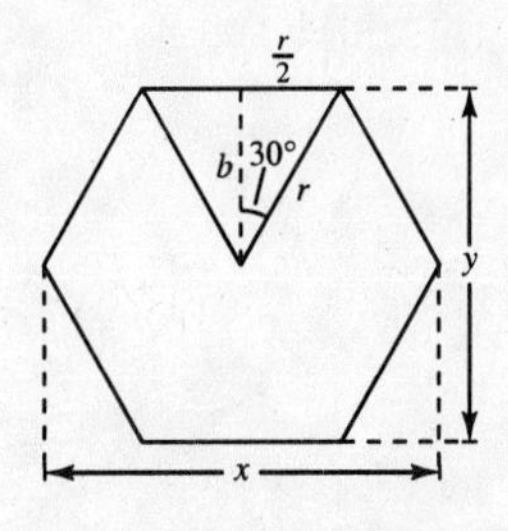

$b = \cos 30^\circ r$

$b = \dfrac{\sqrt{3}r}{2}$

$y = 2b = 2\left(\dfrac{\sqrt{3}r}{2}\right) = \sqrt{3}r$

47. $\sin 36° = \dfrac{d}{24} \Rightarrow d \approx 14.1068$

Length of side: $2d \approx 28.2$ inches

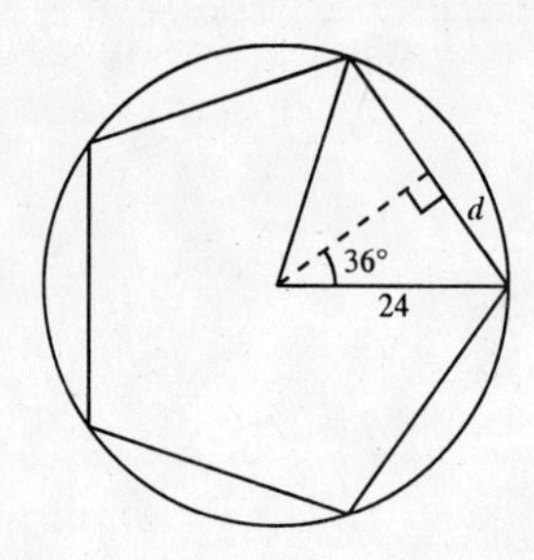

49. $\tan 35° = \dfrac{b}{10}$

$b = 10 \tan 35° \approx 7$

$\cos 35 = \dfrac{10}{a}$

$a = \dfrac{10}{\cos 35°} \approx 12.2$

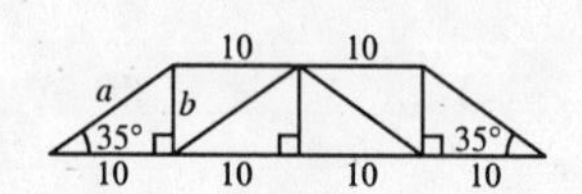

51. $d = 4 \cos 8\pi t$

(a) Maximum displacement = amplitude = 4

(b) Frequency $= \dfrac{\omega}{2\pi} = \dfrac{8\pi}{2\pi}$

$= 4$ cycles per unit of time

(c) $8\pi t = \dfrac{\pi}{2} \Rightarrow t = \dfrac{1}{16}$

53. $d = \dfrac{1}{16} \sin 140\pi t$

(a) Maximum displacement = amplitude $= \dfrac{1}{16}$

(b) Frequency $= \dfrac{\omega}{2\pi} = \dfrac{140\pi}{2\pi}$

$= 70$ cycles per unit of time

(c) $140\pi t = \pi \Rightarrow t = \dfrac{1}{140}$

55. $d = 0$ when $t = 0$, $a = 8$, period $= 2$

Use $d = a \sin wt$ since $d = 0$ when $t = 0$

$\dfrac{2\pi}{w} = 2 \Rightarrow w = \pi$

Thus, $d = 8 \sin \pi t$

57. $d = 3$ when $t = 0$, $a = 3$, Period $= 1.8$

Use $d = a \cos wt$ since $d = 3$ when $t = 0$

$\dfrac{2\pi}{w} = 1.8 \Rightarrow w = \dfrac{10}{9}\pi$

Thus, $d = 3 \cos\left(\dfrac{10}{9}\pi t\right)$

59. $d = a \sin \omega t$

Period $= \dfrac{2\pi}{\omega} = \dfrac{1}{\text{frequency}}$

$\dfrac{2\pi}{\omega} = \dfrac{1}{264}$

$\omega = 2\pi(264) = 528\pi$

61. $y = \dfrac{1}{4} \cos 16t,\ t > 0$

(a)

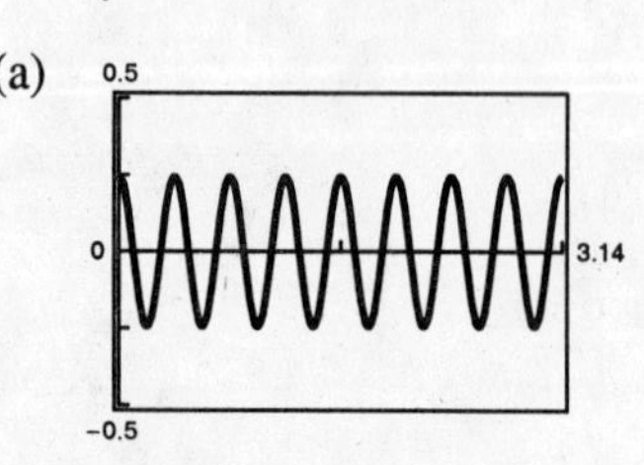

(b) Period: $\dfrac{2\pi}{16} = \dfrac{\pi}{8}$ seconds

(c) $\dfrac{1}{4} \cos 16t = 0$ when $16t = \dfrac{\pi}{2} \Rightarrow t = \dfrac{\pi}{32}$ seconds.

63. $S(t) + 18.09 + 1.41 \sin\left(\dfrac{\pi t}{6} + 4.60\right)$

(a)

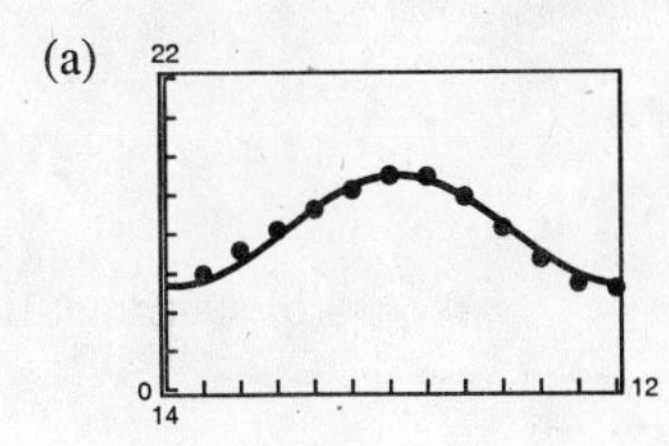

(b) The period is 12 months, which corresponds to 1 year.

(c) The amplitude is 1.41. This gives the maximum change in time from the average time (18.09) of sunset.

65. (a)

(b) $a + \dfrac{1}{2}(14.30 - 1.70) + 6.3$

$\dfrac{2\pi}{b} + 12 \Rightarrow b + \dfrac{\pi}{6}$

Shift: $d + 14.3 - 6.3 + 8$

$S + d + a \cos bt$

$S + 8 + 6.3 \cos\left(\dfrac{\pi t}{6}\right)$

The model is a good fit.

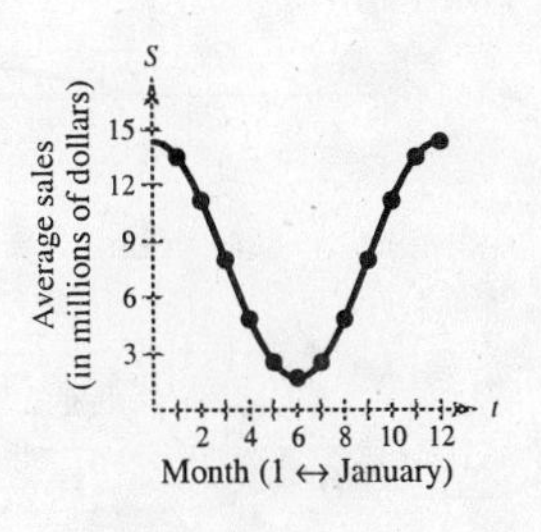

(c) Period: $\dfrac{2\pi}{\pi/6} + 12$

This corresponds to the 12 months in a year. Since the sales of outerwear is seasonal, this is reasonable.

(d) The amplitude represents the maximum displacement from the average sale of 8 million dollars. Sales are greatest in December (cold weather + holidays) and least in June.

67. False. The other acute angle is $90° - 48.1° + 41.9°$. Then $\tan(41.9°) + \dfrac{\text{opp}}{\text{adj}} + \dfrac{a}{22.56} \Rightarrow a + 22.56 \cdot \tan(41.9°)$

69. $\dfrac{x^2}{x+6} - \dfrac{1}{2x+1} + \dfrac{2x^3 + x^2 - x - 6}{(x+6)(2x+1)}$

71. $\dfrac{3x^2 - 13x + 4}{x^2 + 8x + 12} \cdot \dfrac{x^3 + 3x^2 - 18x}{12x^2 - 4x} + \dfrac{(x-4)(3x-1)}{(x+2)(x+6)} \cdot \dfrac{x(x-3)(x+6)}{4x(3x-1)}$

$+ \dfrac{(x-4)(x-3)}{4(x+2)}, x \neq 0, -6, \dfrac{1}{3}$

73. $e^{2x} + 54$

$2x + \ln 54$

$x + \frac{1}{2}\ln 54 \approx 1.994$

75. $\ln(x^2 + 1) + 3.2$

$x^2 + 1 + e^{3.2}$

$x + \pm\sqrt{e^{3.2} - 1} \approx \pm 4.851$

77. arc cos $0.13 \approx 1.44$ or $82.53°$

79. $\arcsin(-0.11) \approx -0.11$ or $-6.32°$

Review Exercises for Chapter 4

Solutions to Odd-Numbered Exercises

1. 40° or 0.7 radians

3. 250° or 4.4 radians

5. (a)

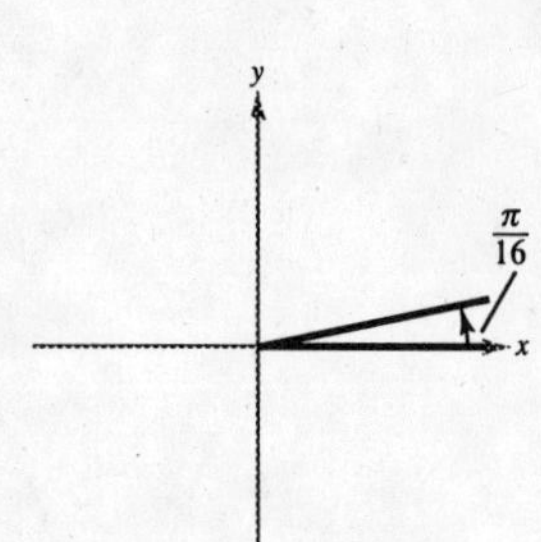

(b) Quadrant I

(c) $\frac{\pi}{16} + 2\pi = \frac{33\pi}{16}$

$\frac{\pi}{16} - 2\pi = -\frac{31\pi}{16}$

7. (a)

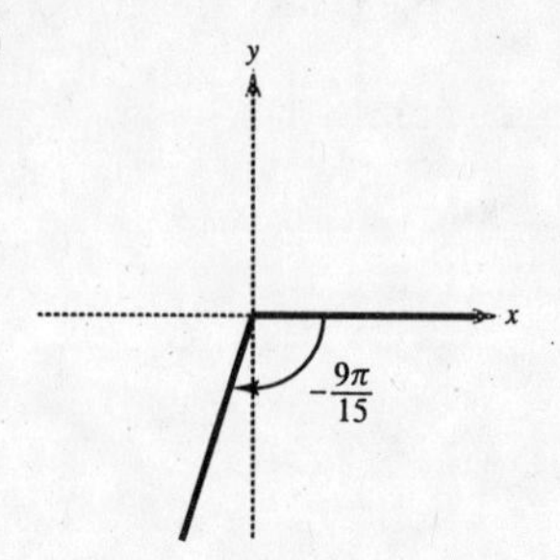

(b) Quadrant III

(c) $-\frac{9\pi}{15} + 2\pi = \frac{21\pi}{15}$

$-\frac{9\pi}{15} - 2\pi = -\frac{39\pi}{15}$

9. Complement: $\frac{\pi}{2} - \frac{\pi}{8} = \frac{3\pi}{8}$

Supplement: $\pi - \frac{\pi}{8} = \frac{7\pi}{8}$

11. Complement: $\frac{\pi}{2} - \frac{3\pi}{10} = \frac{\pi}{5}$

Supplement: $\pi - \frac{3\pi}{10} = \frac{7\pi}{10}$

13. $\frac{5\pi \text{ rad}}{7} = \frac{5\pi \text{ rad}}{7} \cdot \frac{180°}{\pi \text{ rad}} \approx 128.57°$

15. $-3.5 \text{ rad} = -3.5 \text{ rad} \cdot \frac{180°}{\pi \text{ rad}} \approx -200.54°$

17. (a)

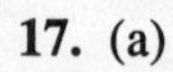

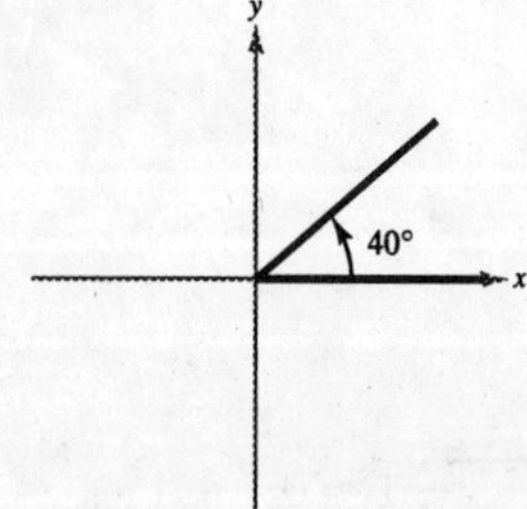

(b) Quadrant I

(c) $40° + 360° = 400°$

$40° - 360° = -320°$

19. (a)

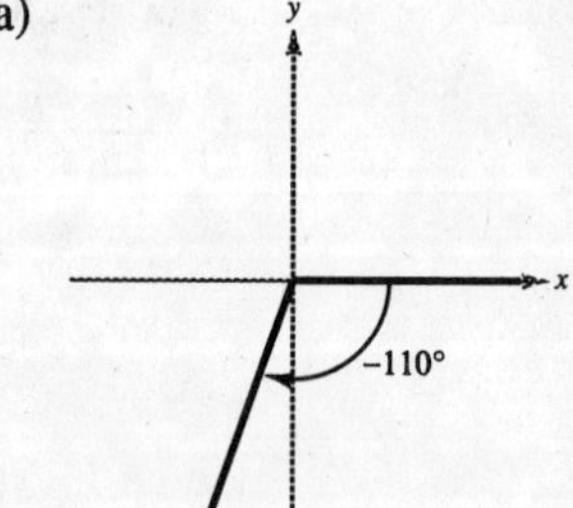

(b) Quadrant III

(c) $-110° + 360° = 250°$

$-110° - 360° = -470°$

21. Complement: $90° - 8° = 82°$

Supplement: $180° - 8° = 172°$

23. Complement: not possible

Supplement: $180° - 171° = 9°$

25. $135°\,16'45'' = \left(135 + \frac{16}{60} + \frac{45}{3600}\right)° \approx 135.28°$

27. $5°\,22'\,53'' = \left(5 + \frac{22}{60} + \frac{53}{3600}\right)° \approx 5.38°$

29. $135.29° = 135° + (0.29)(60)'$

$= 135°\,17'\,24''$

31. $-85.36° = -[85 + 0.36(60')] = -85°21'36''$

33. $480° = 480° \cdot \dfrac{\pi \text{ rad}}{180°} = \dfrac{8\pi}{3} \text{ rad} \approx 8.3776 \text{ rad}$

35. $-33° \, 45' = -33.75° = -33.75° \cdot \dfrac{\pi \text{ rad}}{180°} = -\dfrac{3\pi}{16} \text{ rad} \approx -0.5890 \text{ rad}$

37. $s = r\theta$

$25 = 12\theta$

$\theta = \dfrac{25}{12} \approx 2.083$

39. $s = r\theta$

$s = 20(138°)\dfrac{\pi}{180°}$

$s \approx 48.171 \text{ m}$

41. $s = r\theta$

$600 = 6378\theta$

$\theta \approx 0.094 \approx 5.39°$

43. $\cot t = \cos \dfrac{2\pi}{3} = -\dfrac{1}{2}$

$\sin t = \sin \dfrac{2\pi}{3} = \dfrac{\sqrt{3}}{2}$

$(x, y) = \left(-\dfrac{1}{2}, \dfrac{\sqrt{3}}{2}\right)$

45. $\cos \dfrac{5\pi}{6} = -\dfrac{\sqrt{3}}{2}$

$\sin \dfrac{5\pi}{6} = \dfrac{1}{2}$

$(x, y) = \left(-\dfrac{\sqrt{3}}{2}, \dfrac{1}{2}\right)$

47. $\sin \dfrac{7\pi}{6} = -\dfrac{1}{2}$

$\cos \dfrac{7\pi}{6} = -\dfrac{\sqrt{3}}{2}$

$\tan \dfrac{7\pi}{6} = \dfrac{1}{\sqrt{3}} = \dfrac{\sqrt{3}}{3}$

$\cot \dfrac{7\pi}{6} = \sqrt{3}$

$\sec \dfrac{7\pi}{6} = -\dfrac{2}{\sqrt{3}} = -\dfrac{2\sqrt{3}}{3}$

$\csc \dfrac{7\pi}{6} = -2$

49. $\sin\left(-\dfrac{2\pi}{3}\right) = \sin\left(\dfrac{4\pi}{3}\right) = -\dfrac{\sqrt{3}}{2}$

$\cos\left(-\dfrac{2\pi}{3}\right) = -\dfrac{1}{2}$

$\tan\left(-\dfrac{2\pi}{3}\right) = \sqrt{3}$

$\cot\left(-\dfrac{2\pi}{3}\right) = \dfrac{1}{\sqrt{3}} = \dfrac{\sqrt{3}}{3}$

$\sec\left(-\dfrac{2\pi}{3}\right) = -2$

$\csc\left(-\dfrac{2\pi}{3}\right) = -\dfrac{2}{\sqrt{3}} = -\dfrac{2\sqrt{3}}{3}$

51. $\sin\left(\dfrac{11\pi}{4}\right) = \sin\left(\dfrac{3\pi}{4}\right) = \dfrac{\sqrt{2}}{2}$

53. $\sin\left(-\dfrac{17\pi}{6}\right) = \sin\left(\dfrac{7\pi}{6}\right) = -\dfrac{1}{2}$

55. $\cot 2.3 = \dfrac{1}{\tan 2.3} \approx -0.89$

57. $\cos \dfrac{5\pi}{3} = \dfrac{1}{2}$

59. The hypotenuse is $\sqrt{12^2 + 10^2} = \sqrt{244} = 2\sqrt{61}$.

$\sin \theta = \dfrac{\text{opp}}{\text{hyp}} = \dfrac{10}{2\sqrt{61}} = \dfrac{5}{\sqrt{61}} = \dfrac{5\sqrt{61}}{61}$

$\cos \theta = \dfrac{\text{adj}}{\text{hyp}} = \dfrac{12}{2\sqrt{61}} = \dfrac{6}{\sqrt{61}} = \dfrac{6\sqrt{61}}{61}$

$\csc \theta = \dfrac{1}{\sin \theta} = \dfrac{\sqrt{61}}{5}$

$\sec \theta = \dfrac{1}{\cos \theta} = \dfrac{\sqrt{61}}{6}$

$\tan \theta = \dfrac{\text{opp}}{\text{adj}} = \dfrac{10}{12} = \dfrac{5}{6}$

$\cot \theta = \dfrac{1}{\tan \theta} = \dfrac{6}{5}$

61. The oppposite side is $\sqrt{9^2 - 4^2} = \sqrt{81 - 16} = \sqrt{65}$

$\sin\theta = \dfrac{\text{opp}}{\text{hyp}} = \dfrac{\sqrt{65}}{9}$ $\quad\cos\theta = \dfrac{\text{adj}}{\text{hyp}} = \dfrac{4}{9}$

$\csc\theta = \dfrac{1}{\sin\theta} = \dfrac{9}{\sqrt{65}} = \dfrac{9\sqrt{65}}{65}$ $\quad\sec\theta = \dfrac{1}{\cos\theta} = \dfrac{9}{4}$

$\tan\theta = \dfrac{\text{opp}}{\text{adj}} = \dfrac{\sqrt{65}}{4}$ $\quad\cot\theta = \dfrac{1}{\tan\theta} = \dfrac{4}{\sqrt{65}} = \dfrac{4\sqrt{65}}{65}$

63. $\csc\theta\tan\theta = \dfrac{1}{\sin\theta}\cdot\dfrac{\sin\theta}{\cos\theta} = \dfrac{1}{\cos\theta} = \sec\theta$

65. (a) $\cos 84° \approx 0.1045$ (b) $\sin 6° \approx 0.1045$

67. (a) $\cos\dfrac{\pi}{4} \approx 0.7071$ (c) $\sec\dfrac{\pi}{4} \approx 1.4142$

69. $\sin 50° = \dfrac{h}{12}$

$h = 12\sin 50°$

≈ 9.2 m

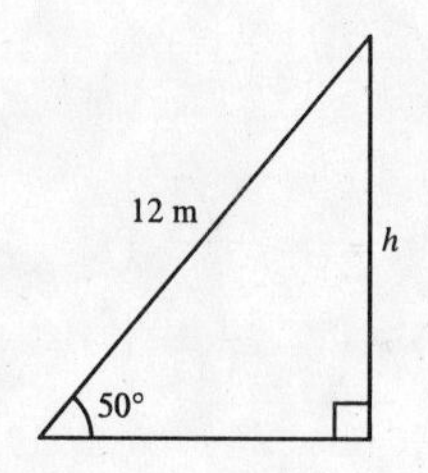

71. $x = 12, y = 16$

$r = \sqrt{144 + 256} = \sqrt{400} = 20$

$\sin\theta = \dfrac{y}{r} = \dfrac{4}{5}$ $\quad\csc\theta = \dfrac{r}{y} = \dfrac{5}{4}$

$\cos\theta = \dfrac{x}{r} = \dfrac{3}{5}$ $\quad\sec\theta = \dfrac{r}{x} = \dfrac{5}{3}$

$\tan\theta = \dfrac{y}{x} = \dfrac{4}{3}$ $\quad\cot\theta = \dfrac{x}{y} = \dfrac{3}{4}$

73. $x = -7,\ y = 2,\ r = \sqrt{49 + 4} = \sqrt{53}$

$\sin\theta = \dfrac{y}{r} = \dfrac{2}{\sqrt{53}} = \dfrac{2\sqrt{53}}{53}$ $\quad\csc\theta = \dfrac{\sqrt{53}}{2}$

$\cos\theta = \dfrac{x}{r} = -\dfrac{7}{\sqrt{53}} = -\dfrac{7\sqrt{53}}{53}$ $\quad\sec\theta = -\dfrac{\sqrt{53}}{7}$

$\tan\theta = \dfrac{y}{x} = -\dfrac{2}{7}$ $\quad\cot\theta = -\dfrac{7}{2}$

75. $x = \dfrac{2}{3}, y = \dfrac{5}{2}$

$r = \sqrt{\left(\dfrac{2}{3}\right)^2 + \left(\dfrac{5}{2}\right)^2} = \dfrac{\sqrt{241}}{6}$

$\sin\theta = \dfrac{y}{r} = \dfrac{(5/2)}{(\sqrt{241}/6)} = \dfrac{15}{\sqrt{241}} = \dfrac{15\sqrt{241}}{241}$

$\cos\theta = \dfrac{x}{r} = \dfrac{(2/3)}{(\sqrt{241}/6)} = \dfrac{4}{\sqrt{241}} = \dfrac{4\sqrt{241}}{241}$

$\tan\theta = \dfrac{y}{x} = \dfrac{(5/2)}{(2/3)} = \dfrac{15}{4}$

$\csc\theta = \dfrac{r}{y} = \dfrac{(\sqrt{241}/6)}{(5/2)} = \dfrac{2\sqrt{241}}{30} = \dfrac{\sqrt{241}}{15}$

$\sec\theta = \dfrac{r}{x} = \dfrac{(\sqrt{241}/6)}{(2/3)} = \dfrac{\sqrt{241}}{4}$

$\cot\theta = \dfrac{x}{y} = \dfrac{(2/3)}{(5/2)} = \dfrac{4}{15}$

77. $\sec\theta = \dfrac{6}{5}$, $\tan\theta < 0 \implies \theta$ is in Quadrant IV.

$r = 6, x = 5, y = -\sqrt{36 - 25} = -\sqrt{11}$

$\sin\theta = \dfrac{y}{r} = -\dfrac{\sqrt{11}}{6}$ $\quad\csc\theta = -\dfrac{6\sqrt{11}}{11}$

$\cos\theta = \dfrac{x}{r} = \dfrac{5}{6}$ $\quad\sec\theta = \dfrac{6}{5}$

$\tan\theta = \dfrac{y}{x} = -\dfrac{\sqrt{11}}{5}$ $\quad\cot\theta = -\dfrac{5\sqrt{11}}{11}$

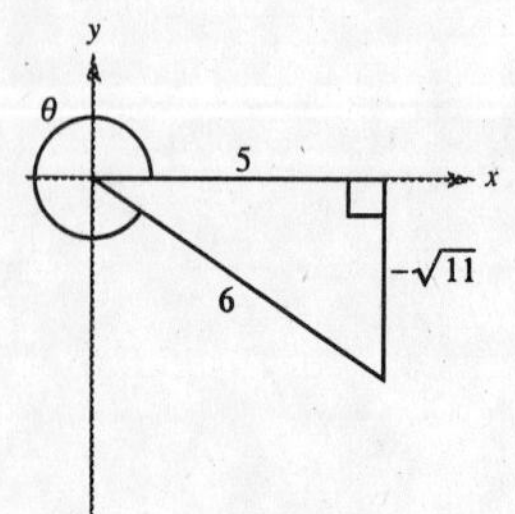

79. $\sin\theta = \frac{3}{8}$, $\cos\theta < 0 \implies \theta$ is in Quadrant II.

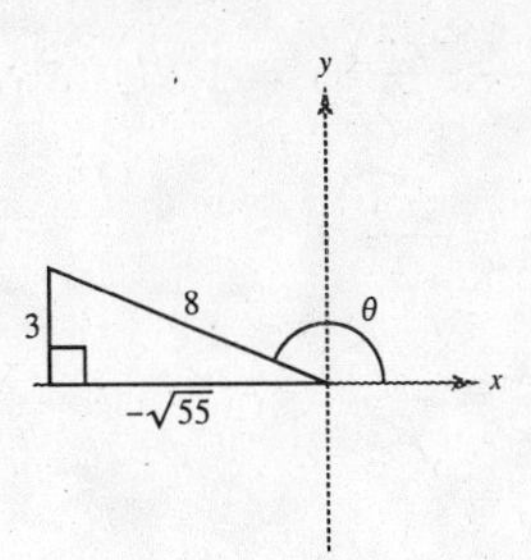

$y = 3, r = 8, x = -\sqrt{55}$

$\sin\theta = \frac{y}{r} = \frac{3}{8}$ $\qquad \csc\theta = \frac{8}{3}$

$\cos\theta = \frac{x}{r} = -\frac{\sqrt{55}}{8}$ $\qquad \sec\theta = -\frac{8}{\sqrt{55}} = -\frac{8\sqrt{55}}{55}$

$\tan\theta = \frac{y}{x} = -\frac{3}{\sqrt{55}} = -\frac{3\sqrt{55}}{55}$ $\qquad \cot\theta = -\frac{\sqrt{55}}{3}$

81. $\tan\frac{\pi}{3} = \sqrt{3}$

83. $\sin -\frac{5\pi}{3} = \sin\frac{\pi}{3} = \frac{\sqrt{3}}{2}$

85. $\cos 495° = -\cos 45° = -\frac{\sqrt{2}}{2}$

87. $\tan 33° \approx 0.65$

89. $\sec\frac{12\pi}{5} = \frac{1}{\cos(12\pi/5)} \approx 3.24$

91. Reference angle: $264° - 180° = 84°$

93. Coterminal angle: $2\pi - \frac{6\pi}{5} = \frac{4\pi}{5}$

Reference angle: $\pi - \frac{4\pi}{5} = \frac{\pi}{5}$

95. 240° is in Quadrant III with reference angle 60°.

$\sin 240° = -\sin 60° = -\frac{\sqrt{3}}{2}$

$\cos 240° = -\cos 60° = -\frac{1}{2}$

$\tan 240° = \frac{-\frac{\sqrt{3}}{2}}{-\frac{1}{2}} = \sqrt{3}$

97. $-210°$ is coterminal with 150° in Quadrant II with reference angle 30°.

$\sin(-210°) = \sin(30°) = \frac{1}{2}$

$\cos(-210°) = -\cos(30°) = -\frac{\sqrt{3}}{2}$

$\tan(-210°) = \frac{\frac{1}{2}}{-\frac{\sqrt{3}}{2}} = -\frac{1}{\sqrt{3}} = -\frac{\sqrt{3}}{3}$

99. $-\frac{9\pi}{4}$ is coterminal with $\frac{7\pi}{4}$ in Quadrant IV with reference angle $\frac{\pi}{4}$.

$\sin\left(-\frac{9\pi}{4}\right) = -\sin\left(\frac{\pi}{4}\right) = -\frac{\sqrt{2}}{2}$

$\cos\left(-\frac{9\pi}{4}\right) = \cos\left(\frac{\pi}{4}\right) = \frac{\sqrt{2}}{2}$

$\tan\left(-\frac{9\pi}{4}\right) = \frac{-\frac{\sqrt{2}}{2}}{\frac{\sqrt{2}}{2}} = -1$

101. $-\frac{\pi}{2}$ is coterminal with $\frac{3\pi}{2}$.

$\sin\left(-\frac{\pi}{2}\right) = -1$

$\cos\left(-\frac{\pi}{2}\right) = 0$

$\tan\left(-\frac{\pi}{2}\right)$ is undefined

103. $(x, y) = \left(-\frac{1}{2}, \frac{\sqrt{3}}{2}\right), r = 1$

$$\sin\frac{2\pi}{3} = \frac{y}{r} = \frac{\sqrt{3}}{2}$$

$$\cos\frac{2\pi}{3} = \frac{x}{r} = -\frac{1}{2}$$

$$\tan\frac{2\pi}{3} = \frac{y}{x} = -\sqrt{3}$$

105. $(x, y) = \left(-\frac{\sqrt{3}}{2}, -\frac{1}{2}\right), r = 1$

$$\sin\frac{7\pi}{6} = \frac{y}{r} = -\frac{1}{2}$$

$$\cos\frac{7\pi}{6} = \frac{x}{r} = -\frac{\sqrt{3}}{2}$$

$$\tan\frac{7\pi}{6} = \frac{y}{x} = \frac{1}{\sqrt{3}} = \frac{\sqrt{3}}{3}$$

107. $f(x) = 3 \sin x$, Amplitude: 3

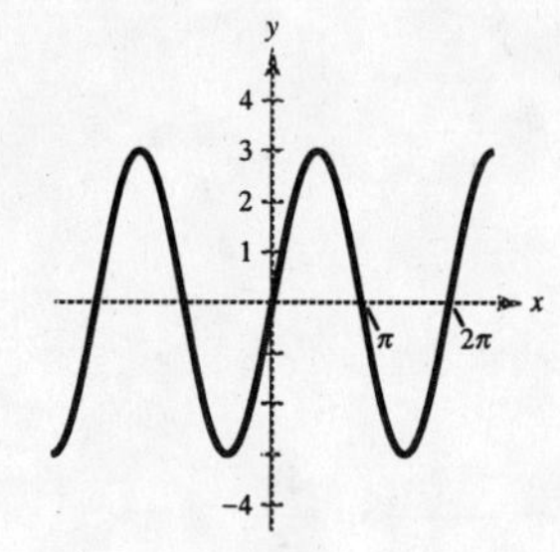

109. $f(x) = \frac{7}{2} \sin x$, Amplitude: $\frac{7}{2}$

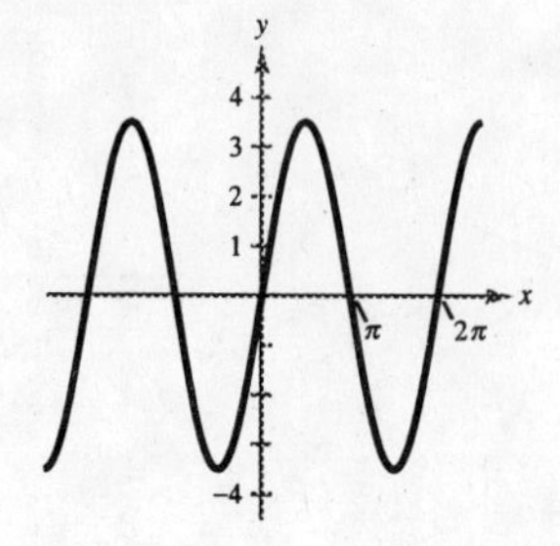

111. Period: $\frac{2\pi}{\pi}, = 2$ amplitude: 5

113. Period: $\frac{2\pi}{\pi} = \pi$, amplitude: 3.4

115. $y = 3 \cos 2\pi x$

Amplitude: 3

Period: $\frac{2\pi}{2\pi} = 1$

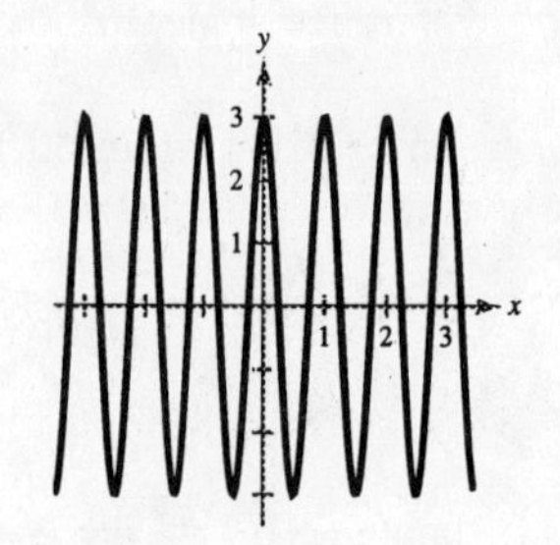

117. $f(x) = 5 \sin\frac{2x}{5}$

Amplitude: 5

Period: $\frac{2\pi}{2/5} = 5\pi$

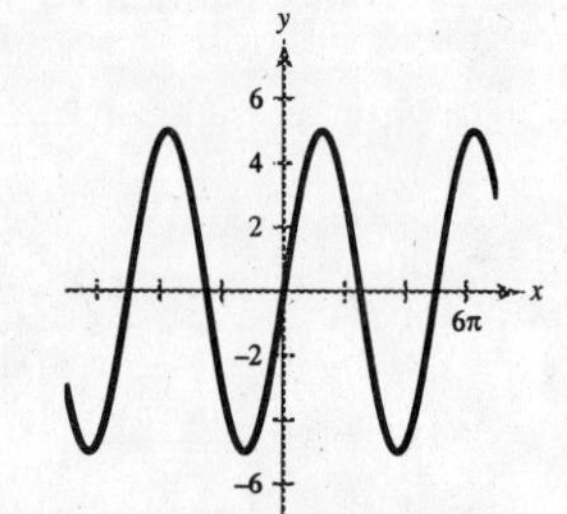

119. $f(x) = -\frac{5}{2}\cos\left(\frac{x}{4}\right)$

Amplitude: $\frac{5}{2}$

Period: $\frac{2\pi}{\frac{1}{4}} = 8\pi$

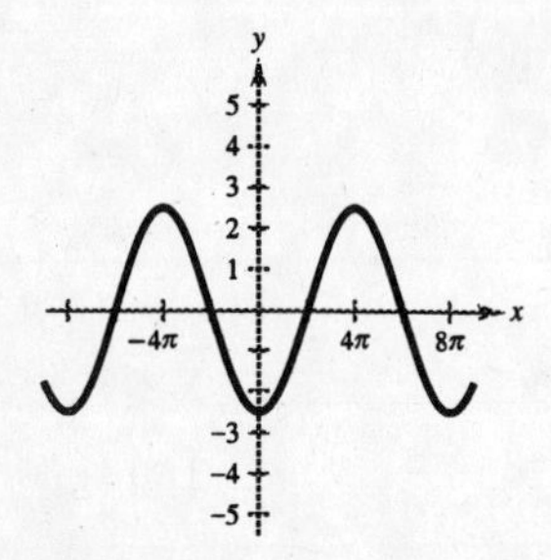

121. $f(x) = \frac{5}{2}\sin(x - \pi)$

Amplitude: $\frac{5}{2}$

Period: 2π

Shift:

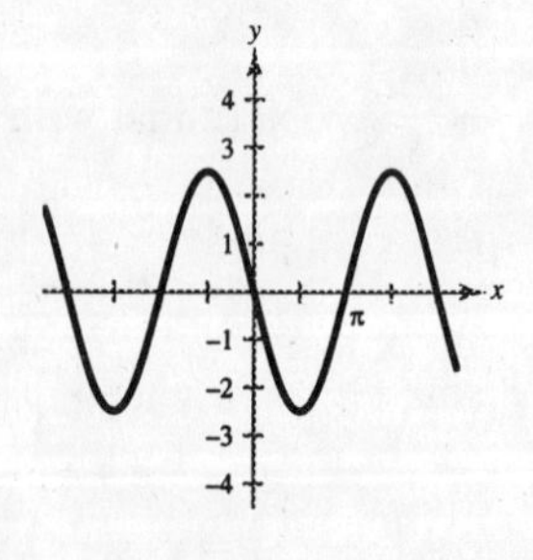

$$x - \pi = 0 \text{ and } x - \pi = 2\pi$$

$$x = \pi \qquad\qquad x = 3\pi$$

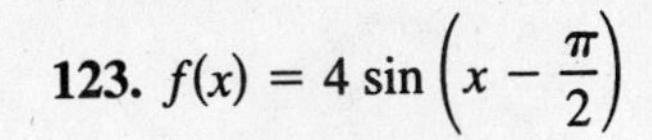

123. $f(x) = 4\sin\left(x - \frac{\pi}{2}\right)$

Amplitude: 4

Period: 2π

Shift: $\frac{\pi}{2}$ to right

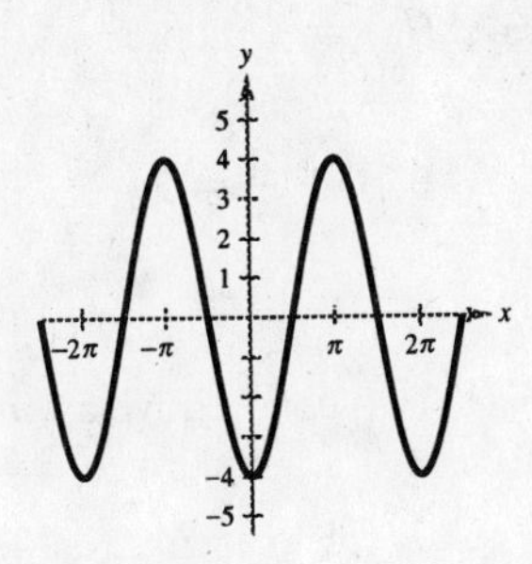

125. $f(x) = -3\cos\left(\frac{x}{2} - \frac{\pi}{2}\right)$

Amplitude: 3

Period: $\dfrac{2\pi}{\frac{1}{2}} = 4\pi$

Shift: $\frac{x}{2} - \frac{\pi}{4} = 0 \Rightarrow x = \frac{\pi}{2}$

$\frac{x}{2} - \frac{\pi}{4} = 2\pi \Rightarrow x = \frac{5\pi}{2}$

127. $f(x) = -2\cos\left(x - \frac{\pi}{4}\right)$

129. $f(x) = -4\cos\left(2x - \frac{\pi}{2}\right)$

131. $S = 48.4 - 6.1\cos\frac{\pi t}{6}$

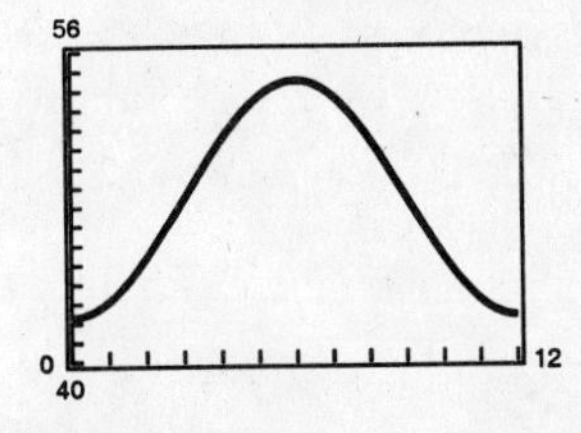

Maximum sales: $t = 6$ (June)

Minimum sales: $t = 12$ (December)

133. $f(x) = \frac{1}{2}\tan\frac{x}{2}$

135. $f(x) = -\tan \dfrac{\pi x}{4}$

Period $= \dfrac{\pi}{(\pi/4)} = 4$

Asymptotes: $x = -2, x = 2$

Reflected in x-axis

x	-1	0	1
y	1	0	-1

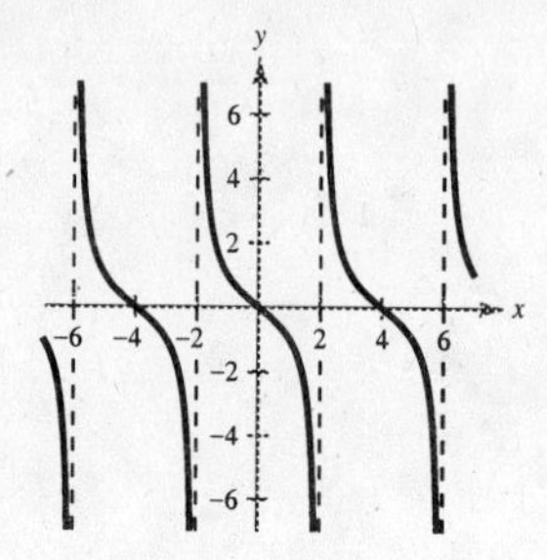

137. $f(x) = -\dfrac{1}{4} \tan \dfrac{\pi x}{2}$

Period: $\dfrac{\pi}{\frac{\pi}{2}} = 2$

Two consecutive asymptotes:

$$\frac{\pi x}{2} = -\frac{\pi}{2} \Longrightarrow x = -1$$

$$\frac{\pi x}{2} = \frac{\pi}{2} \Longrightarrow x = 1$$

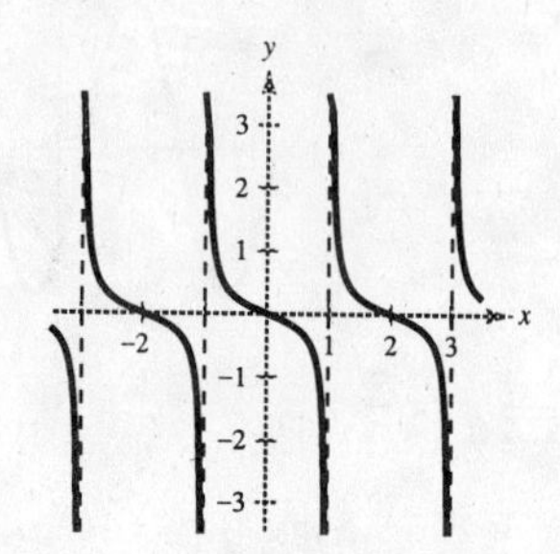

139. $f(x) = \dfrac{1}{4} \tan\left(x - \dfrac{\pi}{2}\right)$

Period: π

Two consecutive asymptotes:

$$x - \frac{\pi}{2} = -\frac{\pi}{2} \Longrightarrow x = 0$$

$$x - \frac{\pi}{2} = \frac{\pi}{2} \Longrightarrow x = \pi$$

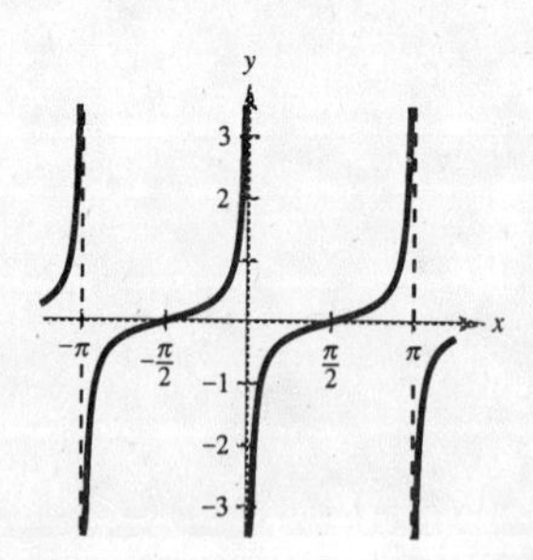

141. $f(x) = 3 \cot \dfrac{x}{2}$

Period: $\dfrac{\pi}{\frac{1}{2}} = 2\pi$

Two consecutive asymptotes:

$$\frac{x}{2} = 0 \Longrightarrow x = 0$$

$$\frac{x}{2} = \pi \Longrightarrow x = 2\pi$$

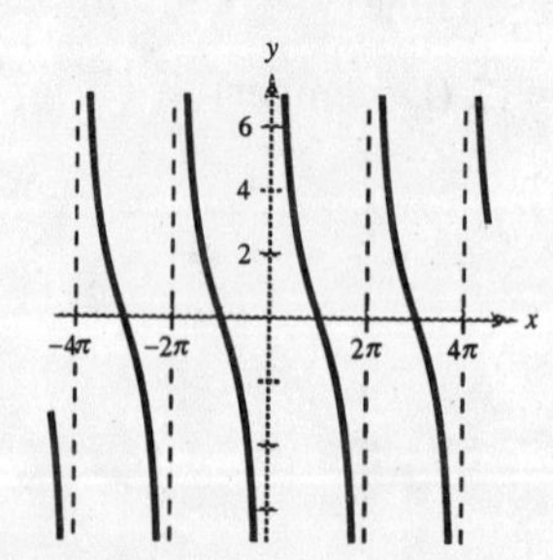

143. $f(x) = \frac{1}{2}\cot\left(x - \frac{\pi}{2}\right)$

Period: π

Two consecutive asymptotes:

$$x - \frac{\pi}{2} = 0 \Rightarrow x = \frac{\pi}{2}$$

$$x - \frac{\pi}{2} = \pi \Rightarrow x = \frac{3\pi}{2}$$

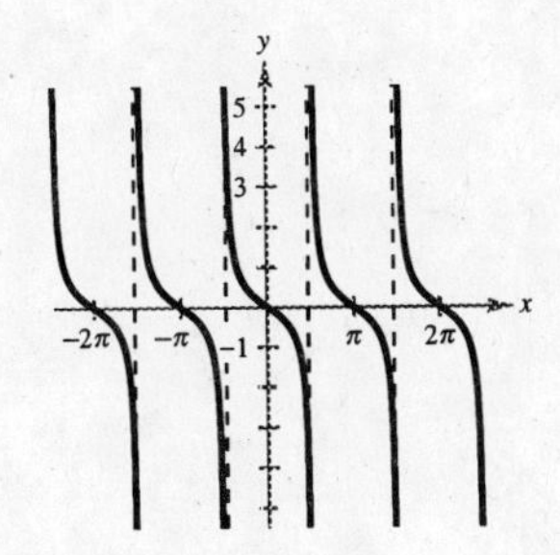

145. $f(x) = \frac{1}{4}\sec x$

Period: 2π

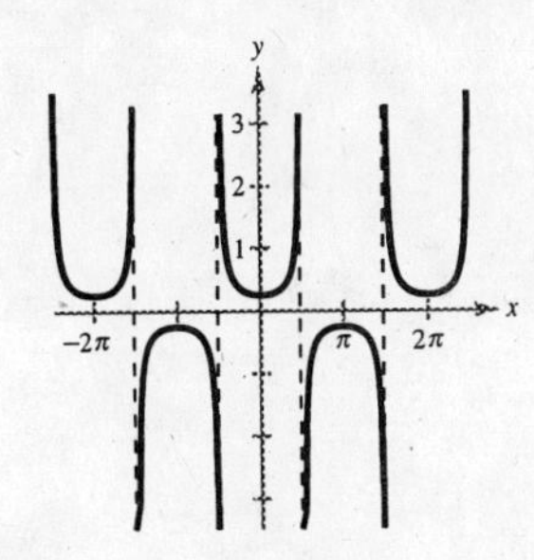

147. $f(x) = \frac{1}{4}\csc 2x$

Period: π

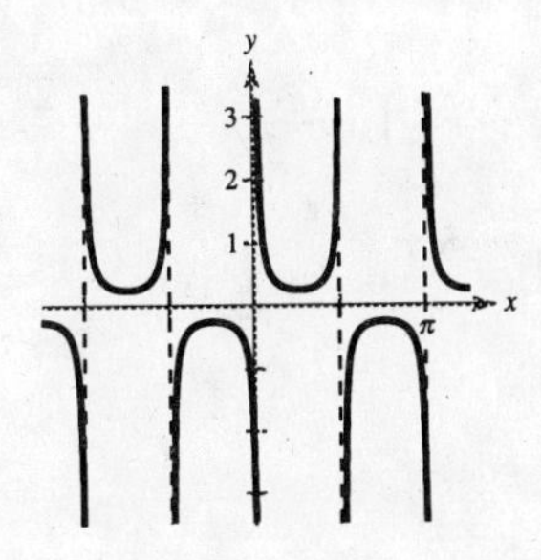

149. $f(x) = \sec\left(x - \frac{\pi}{4}\right)$

Secant function shifted $\frac{\pi}{4}$ to right

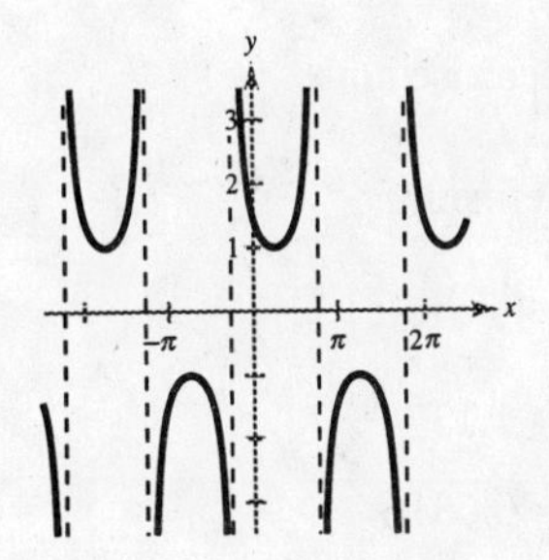

151. $f(x) = 2\sec(x - \pi)$

$y = 2\sec x$ shifted π to the right

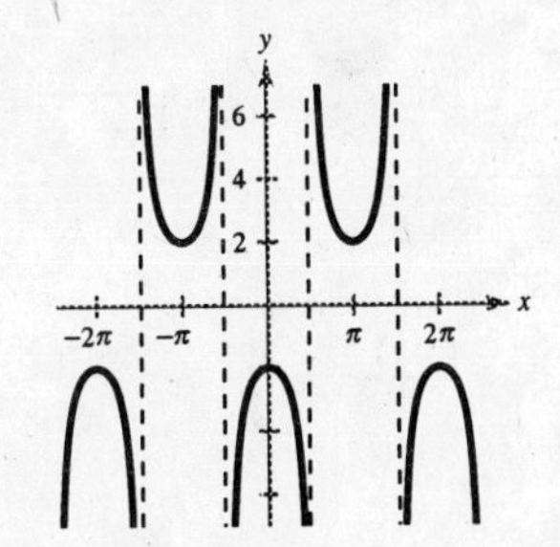

153. $f(x) = \csc\left(3x - \frac{\pi}{2}\right)$

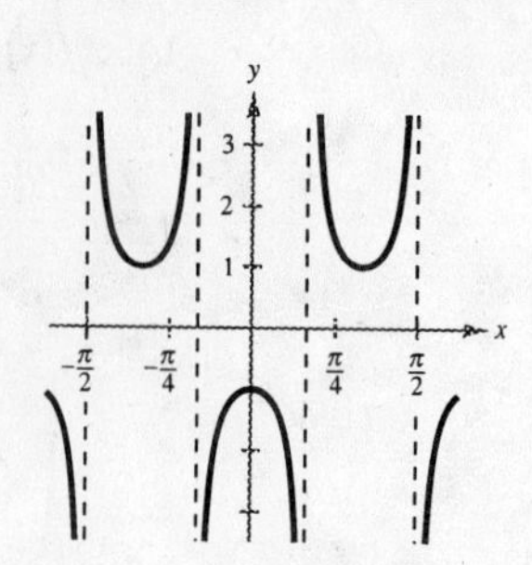

Graph: $y = \sin\left(3x - \frac{\pi}{2}\right)$ first.

Period: $\frac{2\pi}{3}$

Shift: $3x - \frac{\pi}{2} = 0$ and $3x - \frac{\pi}{2} = 2\pi$

$x = \frac{\pi}{6}$ $\qquad x = \frac{5\pi}{6}$

155. $f(x) = e^x \sin 2x$ Damping factor: $y = e^x$

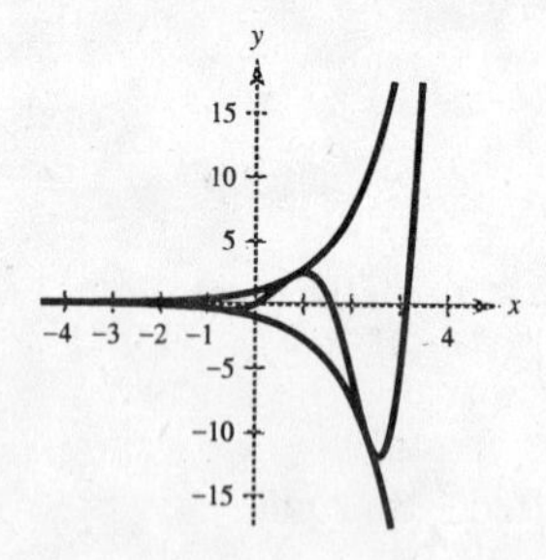

157. $f(x) = x \sin \pi x$ Dumping factor: $y = x$

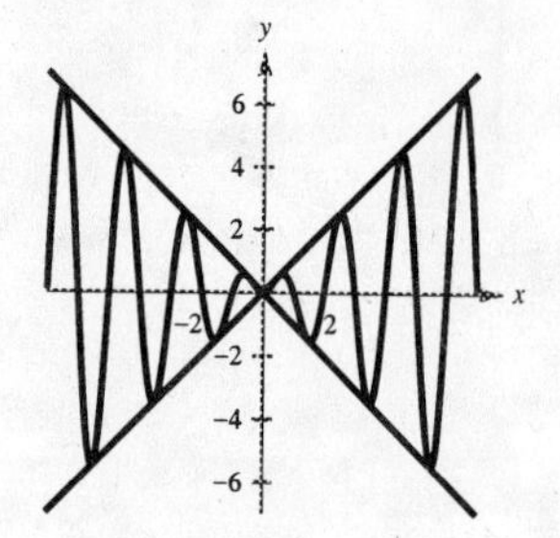

159. (a) $\arcsin 1 = \frac{\pi}{2}$ because $\sin \frac{\pi}{2} = 1$

(b) arcsin 4 is undefined

161. (a) $\arccos\left(\frac{\sqrt{2}}{2}\right) = \frac{\pi}{4}$ because $\cos \frac{\pi}{4} = \frac{\sqrt{2}}{2}$

(b) $\arccos\left(-\frac{\sqrt{3}}{2}\right) = \frac{5\pi}{6}$ because $\cos \frac{5\pi}{6} = -\frac{\sqrt{3}}{2}$

163. (a) $\arccos(0.42) \approx 1.1374$

(b) $\arcsin(0.63) \approx 0.6816$

165. (a) $\arctan(-12) \approx -1.4877$

(b) $\arctan(21) \approx 1.5232$

167. $\sin\theta = \frac{x+3}{16} \Longrightarrow \theta = \arcsin\left(\frac{x+3}{16}\right)$

169. Let $y = \arcsin(x - 1)$. Then,

$\sin y = (x - 1) = \frac{x-1}{1}$ and

$\sec y = \frac{1}{\sqrt{-x^2 + 2x}} = \frac{\sqrt{-x^2 + 2x}}{-x^2 + 2x}.$

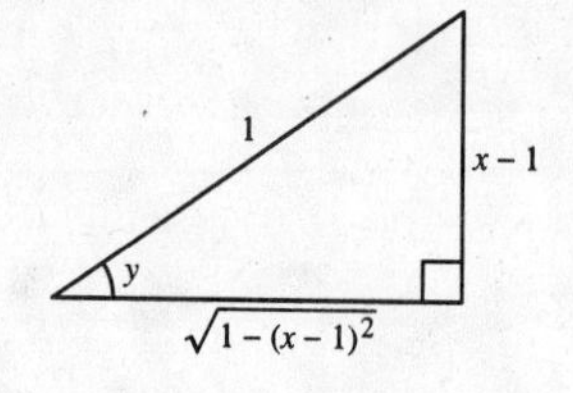

171. Let $y = \arccos \frac{x^2}{4 - x^2}$ Then $\cos y = \frac{x^2}{4 - x^2}$ and

$$\sin y = \frac{\sqrt{(4 - x^2)^2 - (x^2)^2}}{4 - x^2}$$

$$= \frac{\sqrt{16 - 8x^2}}{4 - x^2}$$

$$= \frac{2\sqrt{4 - 2x^2}}{4 - x^2}.$$

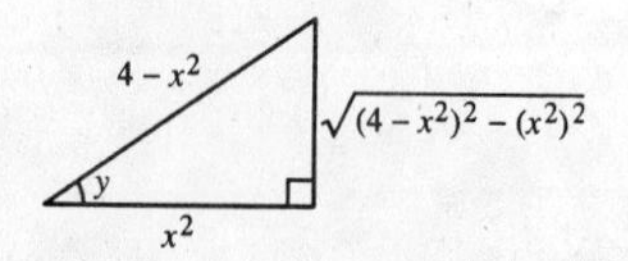

173. $\tan 1° \, 10' = \dfrac{a}{3.5}$

$a = 3.5 \tan 1° \, 10'$

≈ 0.071 km

175. $r = 490(1.8) = 882$ miles

$882 \cos(46°) \approx 612.7$ miles north

$882 \sin(46°) \approx 634.5$ miles west

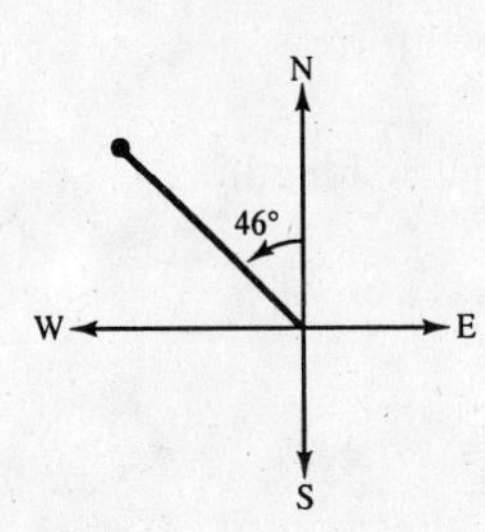

177. Use cosine model with amplitude 3 feet.

Period $=$ 15 seconds.

$y = 3 \cos\left(\dfrac{2\pi}{15} t\right)$

179. False

181. False. $\dfrac{3\pi}{4}$ is not in the range of the arctangent function.

183. If the radius increases, the speed of the tip increases.

Chapter 4 Practice Test

1 Express 350° in radian measure.

2. Express $(5\pi)/9$ in degree measure.

3. Convert 135° 14′ 12″ to decimal form.

4. Convert −22.569° to D° M′ S″ form.

5. If $\cos\theta = \frac{2}{3}$, use the trigonometric identities to find $\tan\theta$.

6. Find θ given $\sin\theta = 0.9063$.

7. Solve for x in the figure below.

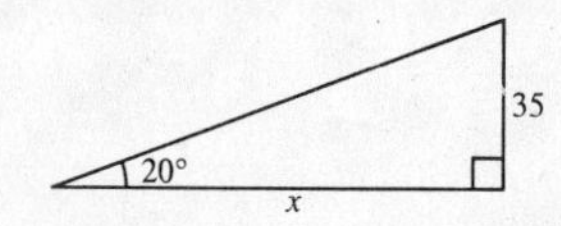

8. Find the magnitude of the reference angle for $\theta = (6\pi)/5$.

9. Evaluate csc 3.92.

10. Find $\sec\theta$ given that θ lies in Quadrant III and $\tan\theta = 6$.

11. Graph $y = 3\sin\frac{\pi}{2}$.

12. Graph $y = -2\cos(x - \pi)$.

13. Graph $y = \tan 2x$.

14. Graph $y = -\csc\left(x + \frac{\pi}{4}\right)$.

15. Graph $y = 2x + \sin x$, using a graphing calculator.

16. Graph $y = 3x\cos x$, using a graphing calculator.

17. Evaluate arcsin 1.

18. Evaluate $\arctan(-3)$.

19. Evaluate $\sin\left(\arccos\frac{4}{\sqrt{35}}\right)$.

20. Write an algebraic expression for $\cos\left(\arcsin\frac{x}{4}\right)$.

For Exercises 21–23, solve the right triangle.

21. $A = 40°,\ c = 12$

22. $B = 6.84°,\ a = 21.3$

23. $a = 5,\ b = 9$

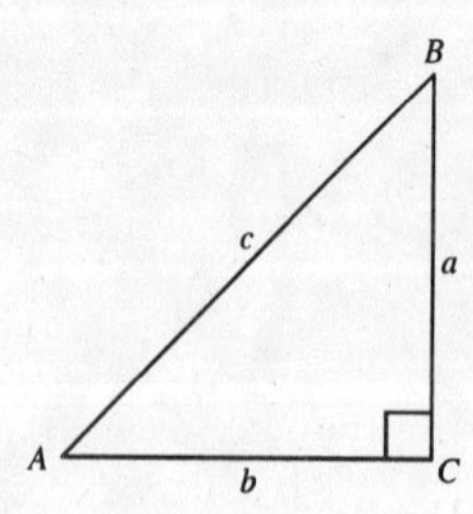

24. A 20-foot ladder leans against the side of a barn. Find the height of the top of the ladder if the angle of elevation of the ladder is 67°.

25. An observer in a lighthouse 250 feet above sea level spots a ship off the shore. If the angle of depression to the ship is 5°, how far out is the ship?

CHAPTER 5
Analytic Trigonometry

CHAPTER 5
Analytic Trigonometry

Section 5.1 Using Fundamental Identities

- You should know the fundamental trigonometric identities.
  (a) Reciprocal Identities

$$\sin u = \frac{1}{\csc u} \qquad \csc u = \frac{1}{\sin u}$$

$$\cos u = \frac{1}{\sec u} \qquad \sec u = \frac{1}{\cos u}$$

$$\tan u = \frac{1}{\cot u} = \frac{\sin u}{\cos u} \qquad \cot u = \frac{1}{\tan u} = \frac{\cos u}{\sin u}$$

  (b) Pythagorean Identities

$$\sin^2 u + \cos^2 u = 1$$

$$1 + \tan^2 u = \sec^2 u$$

$$1 + \cot^2 u = \csc^2 u$$

  (c) Cofunction Identities

$$\sin\left(\frac{\pi}{2} - u\right) = \cos u \qquad \cos\left(\frac{\pi}{2} - u\right) = \sin u$$

$$\tan\left(\frac{\pi}{2} - u\right) = \cot u \qquad \cot\left(\frac{\pi}{2} - u\right) = \tan u$$

$$\sec\left(\frac{\pi}{2} - u\right) = \csc u \qquad \csc\left(\frac{\pi}{2} - u\right) = \sec u$$

  (d) Negative Angle Identities

$$\sin(-x) = -\sin x \qquad \csc(-x) = -\csc x$$

$$\cos(-x) = \cos x \qquad \sec(-x) = \sec x$$

$$\tan(-x) = -\tan x \qquad \cot(-x) = -\cot x$$

- You should be able to use these fundamental identities to find function values.
- You should be able to convert trigonometric expressions to equivalent forms by using the fundamental identities.
- You should be able to check your answers with a graphing utility.

Solutions to Odd-Numbered Exercises

1. $\sin x = \frac{\sqrt{3}}{2}, \cos x = \frac{1}{2} \Rightarrow x$ is in Quadrant I

$$\tan x = \frac{\sin x}{\cos x} = \frac{\frac{\sqrt{3}}{2}}{\frac{1}{2}} = \sqrt{3}$$

$$\cot x = \frac{1}{\tan x} = \frac{1}{\sqrt{3}} = \frac{\sqrt{3}}{3}$$

$$\sec x = \frac{1}{\cos x} = 2$$

$$\csc x = \frac{1}{\sin x} = \frac{2}{\sqrt{3}} = \frac{2\sqrt{3}}{3}$$

3. $\sec \theta = \sqrt{2}, \sin\theta = -\frac{\sqrt{2}}{2} \Rightarrow \theta$ is in Quadrant IV.

$$\cos \theta = \frac{1}{\sec \theta} = \frac{1}{\sqrt{2}} = \frac{\sqrt{2}}{2}$$

$$\tan \theta = \frac{\sin \theta}{\cos \theta} = \frac{-\sqrt{2}/2}{\sqrt{2}/2} = -1$$

$$\cot \theta = \frac{1}{\tan \theta} = -1$$

$$\csc \theta = -\sqrt{2}$$

5. $\tan x = \frac{7}{24}, \sec x = \frac{-25}{24} \Rightarrow x$ is in Quadrant III

$$\cot x = \frac{24}{7}$$

$$\cos x = -\frac{24}{25}$$

$$\sin x = -\sqrt{1 - \cos^2 x} = -\frac{7}{25}$$

$$\csc x = \frac{1}{\sin x} = -\frac{25}{7}$$

7. $\sec \phi = -1, \sin \phi = 0 \Rightarrow \phi = \pi$

$\cos \phi = -1$

$\tan \phi = 0$

$\cot \phi$ is undefined.

$\csc \phi$ is undefined.

9. $\sin(-x) = -\sin x = -\frac{2}{3} \Rightarrow \sin x = \frac{2}{3}$

$\sin x = \frac{2}{3}, \tan x = -\frac{2\sqrt{5}}{5} \Rightarrow x$ is in Quadrant II.

$$\cos x = -\sqrt{1 - \sin^2 x} = -\sqrt{1 - \frac{4}{9}} = -\frac{\sqrt{5}}{3}$$

$$\cot x = \frac{1}{\tan x} = -\frac{\sqrt{5}}{2}$$

$$\sec x = \frac{1}{\cos x} = -\frac{3\sqrt{5}}{5}$$

$$\csc x = \frac{1}{\sin x} = \frac{3}{2}$$

11. $\tan \theta = 4, \sin \theta < 0 \Rightarrow \theta$ is in Quadrant III

$$\sec \theta = -\sqrt{\tan^2 \theta + 1} = -\sqrt{17}$$

$$\cos \theta = -\frac{1}{\sqrt{17}} = -\frac{\sqrt{17}}{17}$$

$$\cot \theta = \frac{1}{4}$$

$$\sin \theta = -\sqrt{1 - \cos^2 \theta} = -\sqrt{1 - \frac{1}{17}} = -\frac{4}{\sqrt{17}}$$

$$= \frac{-4\sqrt{17}}{17}$$

$$\csc \theta = -\frac{\sqrt{17}}{4}$$

13. $\sin \theta = -1, \cot \theta = 0 \Rightarrow \theta = \frac{3\pi}{2}$

$\cos \theta = \sqrt{1 - \sin^2 \theta} = 0$

$\sec \theta$ is undefined.

$\tan \theta$ is undefined.

$\csc \theta = -1$

15. By looking at the basic graphs of $\sin x$ and $\csc x$, we see that as $x \longrightarrow \frac{\pi^-}{2}$, $\sin x \longrightarrow 1$ and $\csc x \longrightarrow 1$.

17. By looking at the basic graphs of $\tan x$ and $\cot x$, we see that as $x \longrightarrow \frac{\pi}{2}^-$, $\tan x \longrightarrow \infty$ and $\cot x \longrightarrow 0$.

19. $\csc x \sin x = \frac{1}{\sin x} \sin x = 1$. Matches (d)

21. $\tan^2 x - \sec^2 x = \tan^2 x - (\tan^2 x + 1) = -1$

The expression is matched with (a).

23. $\frac{\sin(-x)}{\cos(-x)} = \frac{-\sin x}{\cos x} = -\tan x$

The expression is matched with (e).

25. $\cos x \csc x = \frac{\cos x}{\sin x} = \cot x$. Matches (b)

27. $\sec^4 x - \tan^4 x = (\sec^2 x + \tan^2 x)(\sec^2 x - \tan^2 x)$

$= (\sec^2 x + \tan^2 x)(1) = \sec^2 x + \tan^2 x$

The expression is matched with (f).

29. $\frac{\sec^2 x - 1}{\sin^2 x} = \frac{\tan^2 x}{\sin^2 x} = \frac{\sin^2 x}{\cos^2 x} \cdot \frac{1}{\sin^2 x} = \sec^2 x$

The expression is matched with (e).

31. $\cot x \sin x = \frac{\cos x}{\sin x} \sin x = \cos x$

33. $\sin \phi(\csc \phi - \sin \phi) = \sin \phi \csc \phi - \sin^2 \phi$

$= \sin \phi \cdot \frac{1}{\sin \phi} - \sin^2 \phi$

$= 1 - \sin^2 \phi$

$= \cos^2 \phi$

35. $\frac{\cot x}{\csc x} = \frac{\cos x/\sin x}{1/\sin x}$

$= \frac{\cos x}{\sin x} \cdot \frac{\sin x}{1} = \cos x$

37. $\sec \alpha \frac{\sin \alpha}{\tan \alpha} = \frac{1}{\cos \alpha}(\sin \alpha) \cot \alpha$

$= \frac{1}{\cos \alpha}(\sin \alpha)\left(\frac{\cos \alpha}{\sin \alpha}\right) = 1$

39. $\frac{\sin(-x)}{\cos x} = -\frac{\sin x}{\cos x} = -\tan x$

41. $\sin\left(\frac{\pi}{2} - x\right)\csc x = \cos x \cdot \frac{1}{\sin x} = \cot x$

43. $\frac{\cos^2 y}{1 - \sin y} = \frac{1 - \sin^2 y}{1 - \sin y}$

$= \frac{(1 + \sin y)(1 - \sin y)}{1 - \sin y}$

$= 1 + \sin y$

45. $\tan \phi \csc \phi = \frac{\sin \phi}{\cos \phi} \cdot \frac{1}{\sin \phi} = \frac{1}{\cos \phi} = \sec \phi$

47. $\frac{\csc \theta}{\sec \theta} + \frac{\cos \theta}{\sin \theta} = \frac{\cos \theta}{\sin \theta} + \frac{\cos \theta}{\sin \theta} = \cot \theta + \cot \theta$

$= 2 \cot \theta$

49. $1 - \frac{\sin^2 \theta}{1 - \cos \theta} = \frac{1 - \cos \theta - \sin^2 \theta}{1 - \cos \theta} = \frac{\cos^2 \theta - \cos \theta}{1 - \cos \theta}$

$= \frac{\cos \theta(\cos \theta - 1)}{1 - \cos \theta} = -\cos \theta$

51. $\dfrac{\cot(-\theta)}{\csc\theta} = \dfrac{\cos(-\theta)}{\sin(-\theta)}\sin\theta = \dfrac{\cos\theta}{-\sin\theta}\sin\theta = -\cos\theta$

53.
$$\begin{aligned}\sin\theta + \cos\theta\cot\theta &= \sin\theta + \cos\theta\frac{\cos\theta}{\sin\theta}\\ &= \frac{\sin^2\theta + \cos^2\theta}{\sin\theta}\\ &= \frac{1}{\sin\theta}\\ &= \csc\theta\end{aligned}$$

55.
$$\begin{aligned}\frac{\cos\theta}{1-\sin\theta} &= \frac{\cos\theta}{1-\sin\theta}\cdot\frac{1+\sin\theta}{1+\sin\theta}\\ &= \frac{\cos\theta(1+\sin\theta)}{1-\sin^2\theta}\\ &= \frac{\cos\theta(1+\sin\theta)}{\cos^2\theta}\\ &= \frac{1+\sin\theta}{\cos\theta}\\ &= \sec\theta + \tan\theta\end{aligned}$$

57. $\dfrac{\sin\theta}{\csc\theta} + \dfrac{\cos\theta}{\sec\theta} = \sin^2\theta + \cos^2\theta = 1$

59.
$$\begin{aligned}\frac{1+\cos\theta}{\sin\theta} + \frac{\sin\theta}{1+\cos\theta} &= \frac{1 + 2\cos\theta + \cos^2\theta + \sin^2\theta}{\sin\theta(1+\cos\theta)}\\ &= \frac{2 + 2\cos\theta}{\sin\theta(1+\cos\theta)}\\ &= \frac{2(1+\cos\theta)}{\sin\theta(1+\cos\theta)}\\ &= \frac{2}{\sin\theta} = 2\csc\theta\end{aligned}$$

61. $\ln|\csc\theta| = \ln\left|\dfrac{1}{\sin\theta}\right| = \ln|\sin\theta|^{-1} = -\ln|\sin\theta|$

63. $\cot^2 x - \cot^2 x\cos^2 x = \cot^2 x(1-\cos^2 x) = \dfrac{\cos^2 x}{\sin^2 x}\sin^2 x = \cos^2 x$

65.
$$\begin{aligned}\sin^2 x\sec^2 x - \sin^2 x &= \sin^2 x(\sec^2 x - 1)\\ &= \sin^2 x\tan^2 x\end{aligned}$$

67.
$$\begin{aligned}\tan^4 x + 2\tan^2 x + 1 &= (\tan^2 x + 1)^2\\ &= (\sec^2 x)^2\\ &= \sec^4 x\end{aligned}$$

69.
$$\begin{aligned}\sin^4 x - \cos^4 x &= (\sin^2 x + \cos^2 x)(\sin^2 x - \cos^2 x)\\ &= (1)(\sin^2 x - \cos^2 x)\\ &= \sin^2 x - \cos^2 x\end{aligned}$$

71.
$$\begin{aligned}(\sin x + \cos x)^2 &= \sin^2 x + 2\sin x\cos x + \cos^2 x\\ &= (\sin^2 x + \cos^2 x) + 2\sin x\cos x\\ &= 1 + 2\sin x\cos x\end{aligned}$$

73. $(\sec x + 1)(\sec x - 1) = \sec^2 x - 1 = \tan^2 x$

75. $\dfrac{1}{1+\cos x}+\dfrac{1}{1-\cos x}=\dfrac{1-\cos x+1+\cos x}{(1+\cos x)(1-\cos x)}$

$=\dfrac{2}{1-\cos^2 x}$

$=\dfrac{2}{\sin^2 x}$

$=2\csc^2 x$

77. $\dfrac{\cos x}{1+\sin x}+\dfrac{1+\sin x}{\cos x}=\dfrac{\cos^2 x+(1+\sin x)^2}{\cos x(1+\sin x)}$

$=\dfrac{2+2\sin x}{\cos x(1+\sin x)}$

$=\dfrac{2(1+\sin x)}{\cos x(1+\sin x)}$

$=\dfrac{2}{\cos x}$

$=2\sec x$

79. $\dfrac{\sin^2 y}{1-\cos y}=\dfrac{1-\cos^2 y}{1-\cos y}$

$=\dfrac{(1+\cos y)(1-\cos y)}{1-\cos y}$

$=1+\cos y$

81. $\dfrac{3}{\sec x-\tan x}\cdot\dfrac{\sec x+\tan x}{\sec x+\tan x}=\dfrac{3(\sec x+\tan x)}{\sec^2 x-\tan^2 x}$

$=\dfrac{3(\sec x+\tan x)}{1}$

$=3(\sec x+\tan x)$

83. $y_1=\cos\left(\dfrac{\pi}{2}-x\right),\ y_2=\sin x$

x	0.2	0.4	0.6	0.8	1.0	1.2	1.4
y_1	0.1987	0.3894	0.5646	0.7174	0.8415	0.9320	0.9854
y_2	0.1987	0.3894	0.5646	0.7174	0.8415	0.9320	0.9854

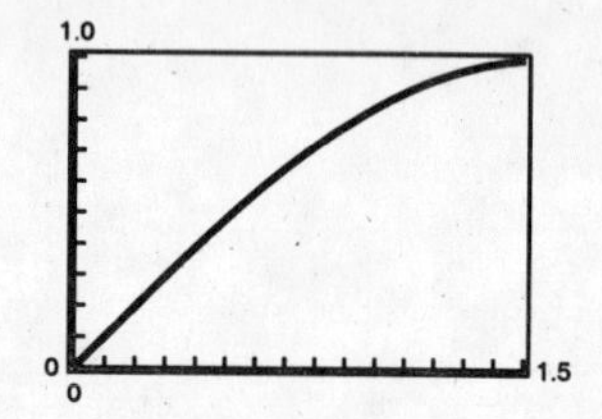

Conjecture: $y_1=y_2$

85. $y_1=\dfrac{\cos x}{1-\sin x},\ y_2=\dfrac{1+\sin x}{\cos x}$

x	0.2	0.4	0.6	0.8	1.0	1.2	1.4
y_1	1.2230	1.5085	1.8958	2.4650	3.4082	5.3319	11.6814
y_2	1.2230	1.5085	1.8958	2.4650	3.4082	5.3319	11.6814

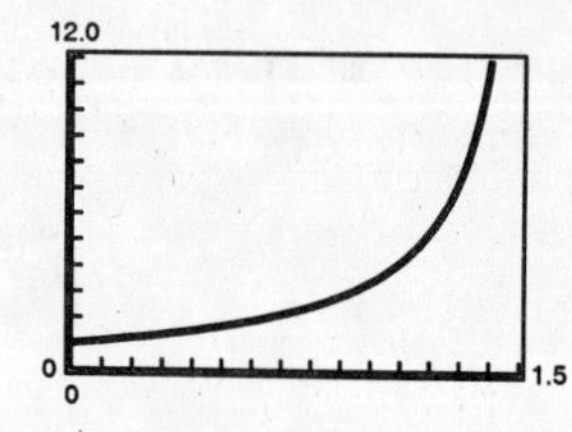

Conjecture: $y_1=y_2$

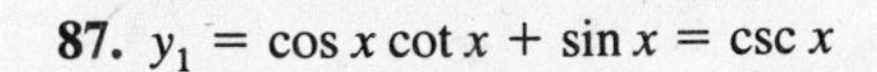

87. $y_1 = \cos x \cot x + \sin x = \csc x$

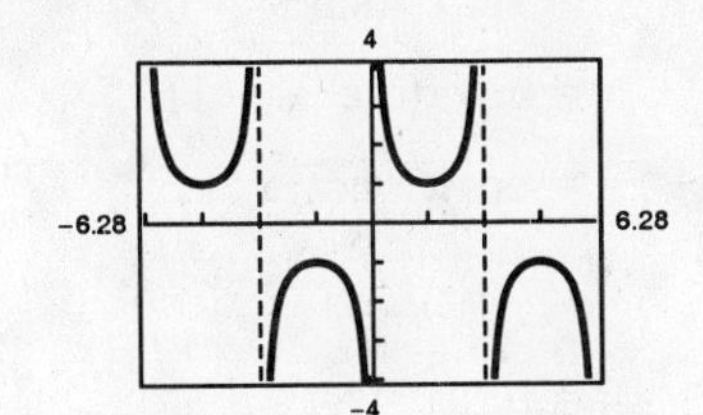

89. $y_1 = \sec x - \dfrac{\cos x}{1 + \sin x} = \tan x$

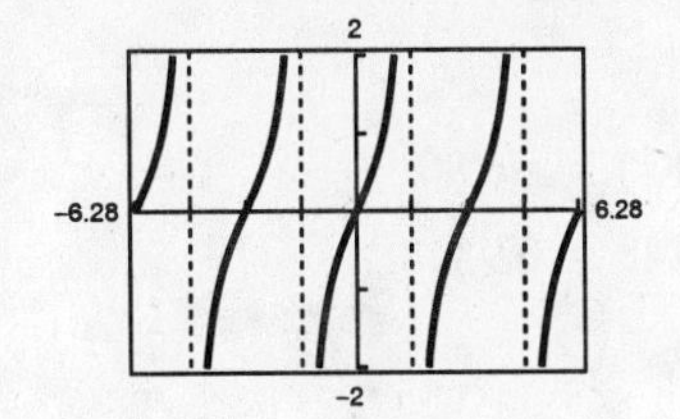

91.
$$\begin{aligned}\sqrt{25 - x^2} &= \sqrt{25 - (5\sin\theta)^2},\ x = 5\sin\theta\\ &= \sqrt{25 - 25\sin^2\theta}\\ &= \sqrt{25(1 - \sin^2\theta)}\\ &= \sqrt{25\cos^2\theta}\\ &= 5\cos\theta\end{aligned}$$

93.
$$\begin{aligned}\sqrt{x^2 - 9} &= \sqrt{(3\sec\theta)^2 - 9},\ x = 3\sec\theta\\ &= \sqrt{9\sec^2\theta - 9}\\ &= \sqrt{9(\sec^2\theta - 1)}\\ &= \sqrt{9\tan^2\theta}\\ &= 3\tan\theta\end{aligned}$$

95.
$$\begin{aligned}\sqrt{x^2 + 25} &= \sqrt{(5\tan\theta)^2 + 25},\ x = 5\tan\theta\\ &= \sqrt{25\tan^2\theta + 25}\\ &= \sqrt{25(\tan^2\theta + 1)}\\ &= \sqrt{25\sec^2\theta}\\ &= 5\sec\theta\end{aligned}$$

97. $\sin\theta = \sqrt{1 - \cos^2\theta}$

Let $y_1 = \sin x$ and $y_2 = \sqrt{1 - \cos^2 x},\ 0 \le x < 2\pi$.

$y_1 = y_2$ for $0 \le x \le \pi$, so we have

$\sin\theta = \sqrt{1 - \cos^2\theta}$ for $0 \le \theta \le \pi$.

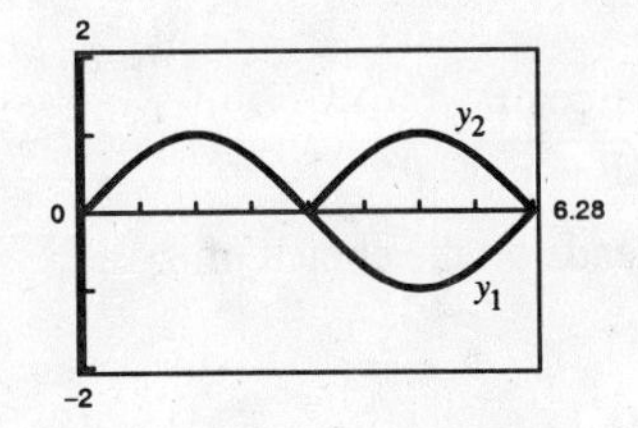

99. $\sec\theta = \sqrt{1 + \tan^2\theta}$

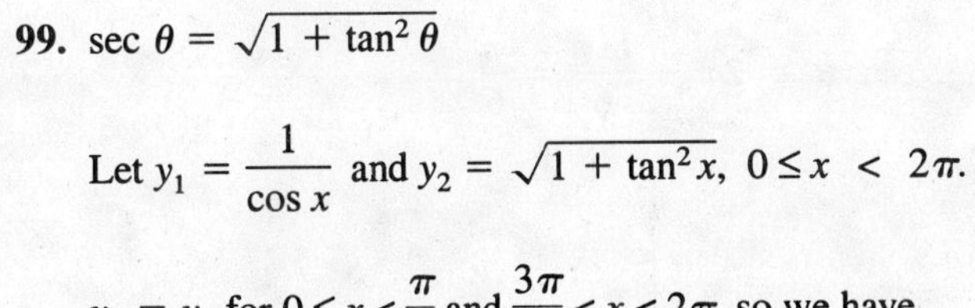

Let $y_1 = \dfrac{1}{\cos x}$ and $y_2 = \sqrt{1 + \tan^2 x},\ 0 \le x < 2\pi$.

$y_1 = y_2$ for $0 \le x < \dfrac{\pi}{2}$ and $\dfrac{3\pi}{2} < x < 2\pi$, so we have

$\sin\theta = \sqrt{1 + \tan^2\theta}$ for $0 \le \theta < \dfrac{\pi}{2}$ and $\dfrac{3\pi}{2} < \theta < 2\pi$.

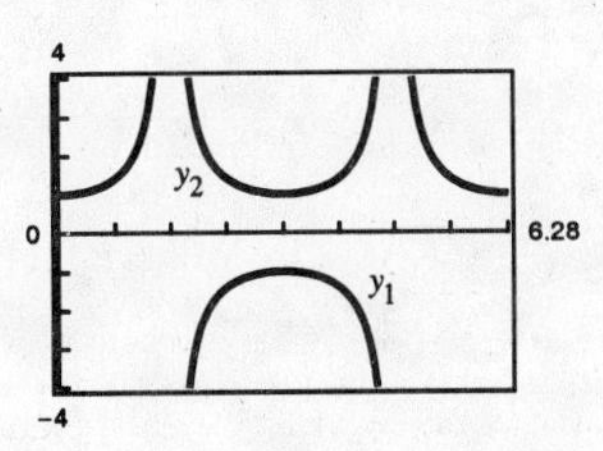

101. $\ln|\cos\theta| - \ln|\sin\theta| = \ln\dfrac{|\cos\theta|}{|\sin\theta|} = \ln|\cot\theta|$

103. $\ln(1 + \sin x) - \ln|\sec x| = \ln\left|\dfrac{1 + \sin x}{\sec x}\right| = \ln|\cos x + \cos x \cdot \sin x|$

105. (a) $\csc^2 132^\circ - \cot^2 132^\circ \approx 1.8107 - 0.8107 = 1$

(b) $\csc^2\dfrac{2\pi}{7} - \cot^2\dfrac{2\pi}{7} \approx 1.6360 - 0.6360 = 1$

107. $\cos\left(\frac{\pi}{2} - \theta\right) = \sin\theta$

(a) $\theta = 80°$

$\cos(90° - 80°) = \sin 80°$

$0.9848 = 0.9848$

(b) $\theta = 0.8$

$\cos\left(\frac{\pi}{2} - 0.8\right) = \sin 0.8$

$0.7174 = 0.7174$

109. $\csc x \cot x - \cos x = \frac{1}{\sin x} \cdot \frac{\cos x}{\sin x} - \cos x$

$= \cos x(\csc^2 x - 1)$

$= \cos x \cdot \cot^2 x$

111. False. $\frac{1}{5\cos\theta} = \frac{1}{5}\sec\theta$

113. False. $\sin\theta\csc\phi \neq 1$ unless $\theta = \phi$

115. $\cos\theta$

$\sin\theta = \pm\sqrt{1 - \cos^2\theta}$

$\tan\theta = \frac{\sin\theta}{\cos\theta} = \pm\frac{\sqrt{1-\cos^2\theta}}{\cos\theta}$

$\csc\theta = \frac{1}{\sin\theta} = \pm\frac{1}{\sqrt{1-\cos^2\theta}}$

$\sec\theta = \frac{1}{\cos\theta}$

$\cot\theta = \frac{1}{\tan\theta} = \pm\frac{\cos\theta}{\sqrt{1-\cos^2\theta}}$

The sign + or − depends on the choice of θ.

117. $\theta = 341°$

$\theta' = 360° - 341° = 19°$

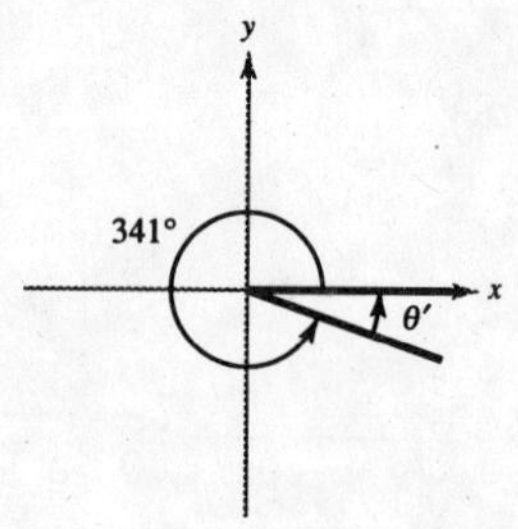

119. $\theta = -212°$ is coterminal with $148°$

$\theta' = 180° - 148° = 32°$

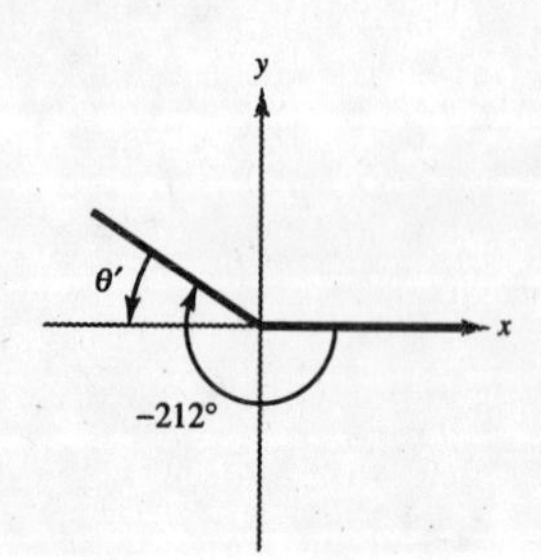

121. $\theta = \frac{35\pi}{6}$ is coterminal with $\frac{11\pi}{6}$

$\theta' = 2\pi - \frac{11\pi}{6} = \frac{\pi}{6}$

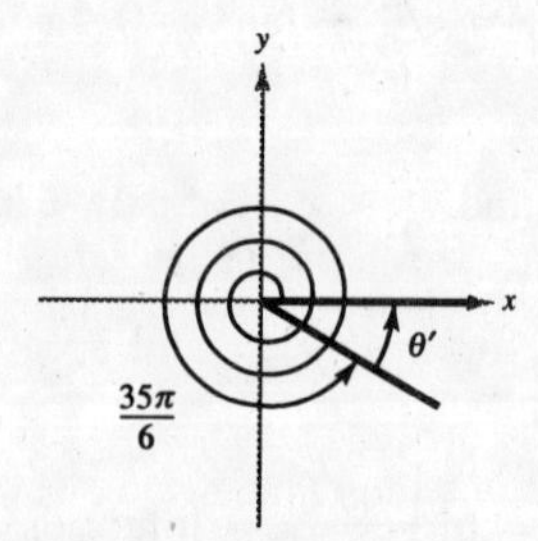

123. $f(x) = -2\tan\frac{\pi x}{2}$

Period: $\frac{\pi}{\frac{\pi}{2}} = 2$

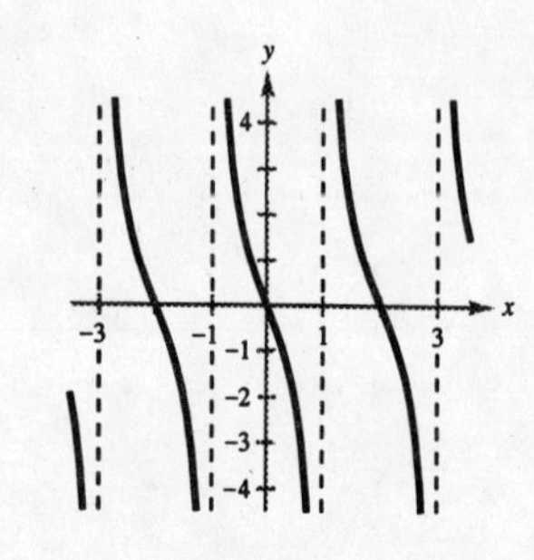

125. $f(x) = \frac{3}{2}\cos(x - \pi) + 3$

Amplitude: $\frac{3}{2}$

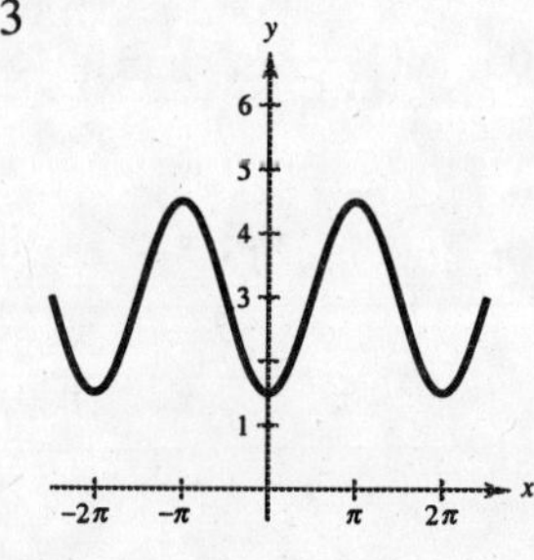

127. $\sin A = \dfrac{a}{c} \Rightarrow a = c \cdot \sin A = 20 \sin 28° \approx 9.39$

$B = 90° - A° = 62°$

$\cos A = \dfrac{b}{c} \Rightarrow b = c \cdot \cos A \approx 17.66$

129. $a = \sqrt{c^2 - b^2} = \sqrt{12.54^2 - 6.2^2} \approx 10.90$

$\sin B = \dfrac{b}{c} = \dfrac{6.2}{12.54} \Rightarrow B \approx 29.63°$

$A = 90° - 29.63° = 60.37°$

Section 5.2 Verifying Trigonometric Identities

- You should know the difference between an expression, a conditional equation, and an identity.
- You should be able to solve trigonometric identities, using the following techniques.
  - (a) Work with *one* side at a time. Do not "cross" the equal sign.
  - (b) Use algebraic techniques such as combining fractions, factoring expressions, rationalizing denominators, and squaring binomials.
  - (c) Use the fundamental identities.
  - (d) Convert all the terms into sines and cosines.

Solutions to Odd-Numbered Exercises

1. $\sin t \csc t = \sin t\left(\dfrac{1}{\sin t}\right) = 1$

3. $\dfrac{\csc^2 x}{\cot x} = \dfrac{1}{\sin^2 x} \cdot \dfrac{\sin x}{\cos x} = \dfrac{1}{\sin x \cdot \cos x}$

$= \csc x \cdot \sec x$

5. $\cos^2 \beta - \sin^2 \beta = (1 - \sin^2 \beta) - \sin^2 \beta$

$= 1 - 2 \sin^2 \beta$

7. $\tan^2 \theta + 6 = (\sec^2 \theta - 1) + 6$

$= \sec^2 \theta + 5$

9. $\cos x + \sin x \tan x = \cos x + \sin x \cdot \dfrac{\sin x}{\cos x}$

$= \dfrac{\cos^2 x + \sin^2 x}{\cos x}$

$= \dfrac{1}{\cos x}$

$= \sec x$

11.

x	0.2	0.4	0.6	0.8	1.0	1.2	1.4
y_1	4.835	2.1785	1.2064	0.6767	0.3469	0.1409	0.0293
y_2	4.835	2.1785	1.2064	0.6767	0.3469	0.1409	0.0293

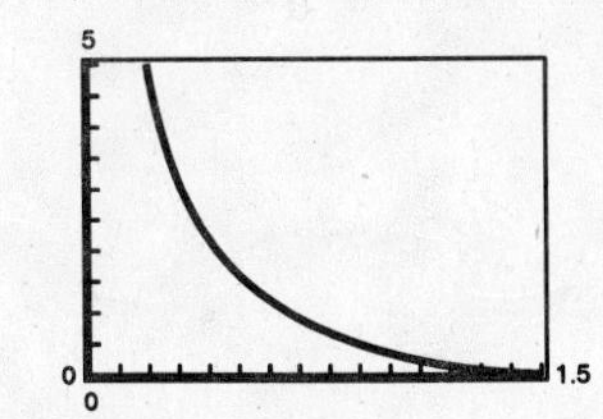

$\dfrac{1}{\sec x \tan x} = \cos x \cdot \dfrac{\cos x}{\sin x}$

$= \dfrac{\cos^2 x}{\sin x}$

$= \dfrac{1 - \sin^2 x}{\sin x}$

$= \dfrac{1}{\sin x} - \sin x$

$= \csc x - \sin x$

13.

x	0.2	0.4	0.6	0.8	1.0	1.2	1.4
y_1	4.835	2.1785	1.2064	0.6767	0.3469	0.1409	0.0293
y_2	4.835	2.1785	1.2064	0.6767	0.3469	0.1409	0.0293

$$\begin{aligned}\csc x - \sin x &= \frac{1}{\sin x} - \sin x \\ &= \frac{1 - \sin^2 x}{\sin x} \\ &= \frac{\cos^2 x}{\sin x} \\ &= \cos x \cdot \frac{\cos x}{\sin x} \\ &= \cos x \cdot \cot x\end{aligned}$$

15.

x	0.2	0.4	0.6	0.8	1.0	1.2	1.4
y_1	5.0335	2.5679	1.7710	1.3940	1.1884	1.0729	1.0148
y_2	5.0335	2.5679	1.7710	1.3940	1.1884	1.0729	1.0148

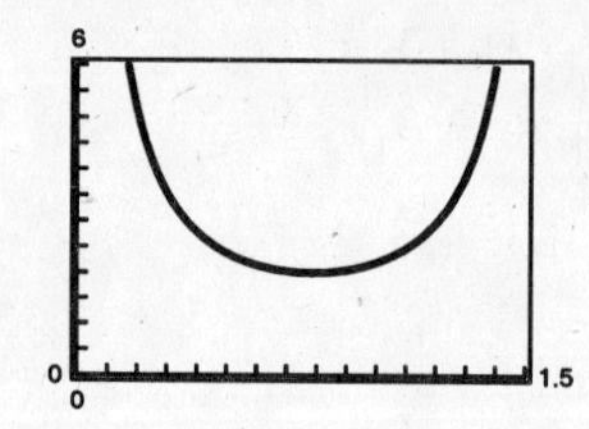

$$\begin{aligned}\sin x + \cos x \cot x &= \sin x + \cos x \frac{\cos x}{\sin x} \\ &= \frac{\sin^2 x + \cos^2 x}{\sin x} \\ &= \frac{1}{\sin x} \\ &= \csc x\end{aligned}$$

17.

x	0.2	0.4	0.6	0.8	1.0	1.2	1.4
y_1	5.1359	2.7880	2.1458	2.0009	2.1995	2.9609	5.9704
y_2	5.1359	2.7880	2.1458	2.0009	2.1995	2.9609	5.9704

$$\begin{aligned}\frac{1}{\tan x} + \frac{1}{\cot x} &= \frac{\cot x + \tan x}{\tan x \cdot \cot x} \\ &= \cot x + \tan x\end{aligned}$$

19. The error is in line 1: $\cot(-x) \neq \cot x$.

21. Missing step: $(\sec^2 x - 1)^2 = (\tan^2 x)^2 = \tan^4 x$

23. $\sin^{1/2} x \cos x - \sin^{5/2} x \cos x = \sin^{1/2} x \cos x(1 - \sin^2 x) = \sin^{1/2} x \cos x \cdot \cos^2 x = \cos^3 x\sqrt{\sin x}$

25.
$$\begin{aligned}\tan\left(\frac{\pi}{2} - x\right)\sec x &= \cot x \cdot \sec x \\ &= \frac{\cos x}{\sin x} \cdot \frac{1}{\cos x} \\ &= \frac{1}{\sin x} \\ &= \csc x\end{aligned}$$

27. $\dfrac{\sec(-x)}{\csc(-x)} = \dfrac{\dfrac{1}{\cos(-x)}}{\dfrac{1}{\sin(-x)}} = \dfrac{\sin(-x)}{\cos(-x)} = -\dfrac{\sin x}{\cos x} = -\tan x$

29. $$\frac{\cos(-\theta)}{1+\sin(-\theta)} = \frac{\cos\theta}{1-\sin\theta}\cdot\frac{1+\sin\theta}{1+\sin\theta}$$
$$= \frac{\cos\theta(1+\sin\theta)}{1-\sin^2\theta}$$
$$= \frac{\cos\theta(1+\sin\theta)}{\cos^2\theta}$$
$$= \frac{1+\sin\theta}{\cos\theta}$$
$$= \frac{1}{\cos\theta} + \frac{\sin\theta}{\cos\theta}$$
$$= \sec\theta + \tan\theta$$

31. $$\frac{\sin x\cos y + \cos x\sin y}{\cos x\cos y - \sin x\sin y} = \frac{\dfrac{\sin x\cos y}{\cos x\cos y} + \dfrac{\cos x\sin y}{\cos x\cos y}}{\dfrac{\cos x\cos y}{\cos x\cos y} - \dfrac{\sin x\sin y}{\cos x\cos y}} = \frac{\tan x + \tan y}{1 - \tan x\tan y}$$

33. $$\frac{\tan x + \cot y}{\tan x\cot y} = \frac{\dfrac{1}{\cot x} + \dfrac{1}{\tan y}}{\dfrac{1}{\cot x}\cdot\dfrac{1}{\tan y}} = \frac{\dfrac{\tan y + \cot x}{\cot x\cdot\tan y}}{\dfrac{1}{\cot x\cdot\tan y}} = \tan y + \cot x$$

35. $$\sqrt{\frac{1+\sin\theta}{1-\sin\theta}} = \sqrt{\frac{1+\sin\theta}{1-\sin\theta}\cdot\frac{1+\sin\theta}{1+\sin\theta}}$$
$$= \sqrt{\frac{(1+\sin\theta)^2}{1-\sin^2\theta}}$$
$$= \sqrt{\frac{(1+\sin\theta)^2}{\cos^2\theta}}$$
$$= \frac{1+\sin\theta}{|\cos\theta|}$$

Note: Check your answer with a graphing utility. What happens if you leave off the absolute value?

37. $\cos^2 x + \cos^2\left(\frac{\pi}{2} - x\right) = \cos^2 x + \sin^2 x = 1$

39. $\sec x \cdot \sin\left(\frac{\pi}{2} - x\right) = \sec x \cdot \cos x = 1$

41. $$2\sec^2 x - 2\sec^2 x\sin^2 x - \sin^2 x - \cos^2 x = 2\sec^2 x(1-\sin^2 x) - (\sin^2 x + \cos^2 x)$$
$$= 2\sec^2 x(\cos^2 x) - 1$$
$$= 2\cdot\frac{1}{\cos^2 x}\cdot\cos^2 x - 1$$
$$= 2 - 1$$
$$= 1$$

43. $$2 + \cos^2 x - 3\cos^4 x = (1-\cos^2 x)(2 + 3\cos^2 x)$$
$$= \sin^2 x(2 + 3\cos^2 x)$$

45. $\csc^4 x - 2\csc^2 x + 1 = (\csc^2 x - 1)^2$

$= (\cot^2 x)^2 = \cot^4 x$

47. $\sec^4 \theta - \tan^4 \theta = (\sec^2 \theta + \tan^2 \theta)(\sec^2 \theta - \tan^2 \theta)$

$= (1 + \tan^2 \theta + \tan^2 \theta)(1)$

$= 1 + 2\tan^2 \theta$

49. $\dfrac{\sin \beta}{1 - \cos \beta} \cdot \dfrac{1 + \cos \beta}{1 + \cos \beta} = \dfrac{\sin \beta(1 + \cos \beta)}{1 - \cos^2 \beta}$

$= \dfrac{\sin \beta(1 + \cos \beta)}{\sin^2 \beta} = \dfrac{1 + \cos \beta}{\sin \beta}$

51. $\dfrac{\tan^3 \alpha - 1}{\tan \alpha - 1} = \dfrac{(\tan \alpha - 1)(\tan^2 \alpha + \tan \alpha + 1)}{\tan \alpha - 1} = \tan^2 \alpha + \tan \alpha + 1$

53. It appears that $y_1 = 1$. Analytically,

$\dfrac{1}{\cot x + 1} + \dfrac{1}{\tan x + 1} = \dfrac{\tan x + 1 + \cot x + 1}{(\cot x + 1)(\tan x + 1)}$

$= \dfrac{\tan x + \cot x + 2}{\cot x \tan x + \cot x + \tan x + 1}$

$= \dfrac{\tan x + \cot x + 2}{\tan x + \cot x + 2}$

$= 1.$

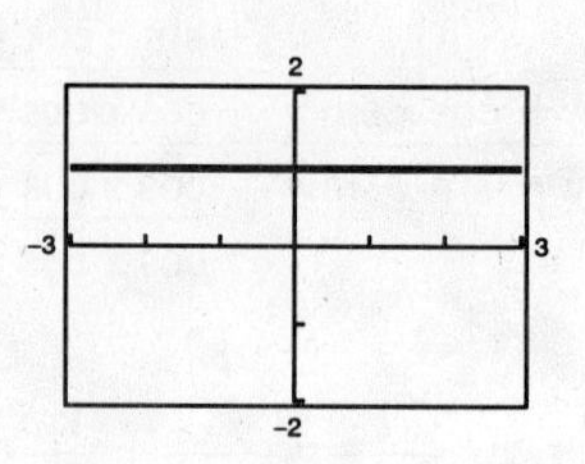

55. It appears that $y_1 = \sin x$. Analytically,

$\dfrac{1}{\sin x} - \dfrac{\cos^2 x}{\sin x} = \dfrac{1 - \cos^2 x}{\sin x} = \dfrac{\sin^2 x}{\sin x} = \sin x.$

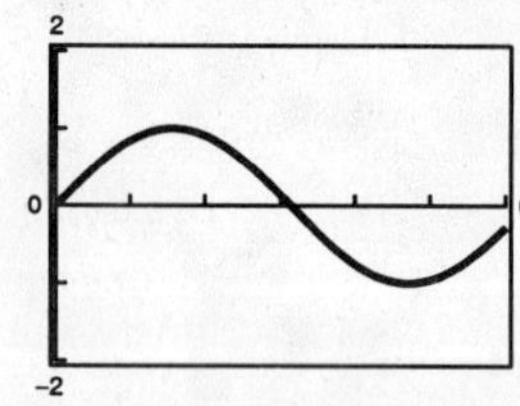

57. $\ln|\cot \theta| = \ln\left|\dfrac{\cos \theta}{\sin \theta}\right|$

$= \ln\dfrac{|\cos \theta|}{|\sin \theta|}$

$= \ln|\cos \theta| - \ln|\sin \theta|$

59. $-\ln(1 + \cos \theta) = \ln(1 + \cos \theta)^{-1}$

$= \left[\ln\dfrac{1}{1 + \cos \theta} \cdot \dfrac{1 - \cos \theta}{1 - \cos \theta}\right]$

$= \ln\dfrac{1 - \cos \theta}{1 - \cos^2 \theta}$

$= \ln\dfrac{1 - \cos \theta}{\sin^2 \theta}$

$= \ln(1 - \cos \theta) - \ln \sin^2 \theta$

$= \ln(1 - \cos \theta) - 2\ln|\sin \theta|$

61. $\sin^2 25 + \sin^2 65 = \sin^2 25 + \cos^2 25° = 1$

63. $\cos^2 20° + \cos^2 52° + \cos^2 38° + \cos^2 70° = \cos^2 20° + \cos^2 52^2 + \sin^2(90° - 38°) + \sin^2(90° - 70°)$

$= \cos^2 20° + \cos^2 52^2 + \sin^2 52° + \sin^2 20°$

$= (\cos^2 20° + \sin^2 20°) + (\cos^2 52° + \sin^2 52°)$

$= 1 + 1$

$= 2$

65. $\tan^5 x = \tan^3 x \cdot \tan^2 x$

$= \tan^3 x(\sec^2 x - 1)$

$= \tan^3 x \sec^2 x - \tan^3 x$

67. $(\sin^2 x - \sin^4 x)\cos x = \sin^2 x(1 - \sin^2 x)\cos x$

$= \sin^2 x \cdot \cos^2 x \cdot \cos x$

$= \cos^3 x \sin^2 x$

69. $\mu W \cos\theta = W \sin\theta$

$$\mu = \frac{W\sin\theta}{W\cos\theta} = \frac{\sin\theta}{\cos\theta} = \tan\theta, W \neq 0$$

71. $\cos x - \csc x \cdot \cot x = \cos x - \dfrac{1}{\sin x} \cdot \dfrac{\cos x}{\sin x}$

$= \cos x\left[1 - \dfrac{1}{\sin^2 x}\right]$

$= \cos x(1 - \csc^2 x)$

$= \cos x(-\cot^2 x)$

$= -\cos x \cdot \cot^2 x$

73. True. $f(x) = \cos x$ and $g(x) = \sec x$ are even

75. False. For example, $\sin(1^2) \neq \sin^2(1)$

77. $\sin\theta = \sqrt{1 - \cos^2\theta}$.

True identity is $\sin\theta = \pm\sqrt{1 - \cos^2\theta}$.

For example, $\sin\theta \neq \sqrt{1 - \cos^2\theta}$ for $\theta = \dfrac{3\pi}{2}$:

$\sin\left(\dfrac{3\pi}{2}\right) = -1 \neq \sqrt{1 - 0} = 1$

79. $\sqrt{\sin^2 x + \cos^2 x} \neq \sin x + \cos x$

The left side is 1 for any x, but the right side is not necessarily 1. For example, the equation is not true for $x = \pi/4$.

81. $\sin\left[\dfrac{(12n + 1)\pi}{6}\right] = \sin\left[\dfrac{1}{6}(12n\pi + \pi)\right]$

$= \sin\left(2n\pi + \dfrac{\pi}{6}\right)$

$= \sin\dfrac{\pi}{6} = \dfrac{1}{2}$

Thus, $\sin\left[\dfrac{(12n + 1)\pi}{6}\right] = \dfrac{1}{2}$ for all integers n.

83. $(x - i)(x + i)(x - 4i)(x + 4i) = (x^2 + 1)(x^2 + 16)$

$= x^4 + 17x^2 + 16$

85. $x^2(x - 2)(x - (1 - i))(x - (1 + i)) = (x^3 - 2x^2)((x - 1)^2 + 1)$

$= (x^3 - 2x^2)(x^2 - 2x + 2)$

$= x^5 - 4x^4 + 6x^3 - 4x^2$

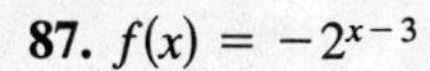

87. $f(x) = -2^{x-3}$

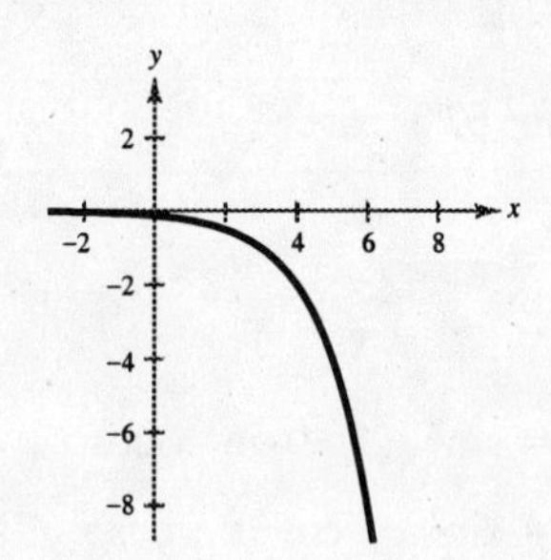

89. $f(x) = 5^{-x} - 2$

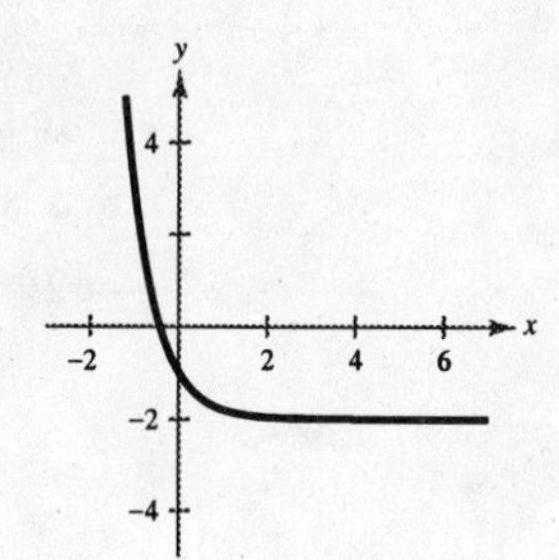

91. $s = r\theta$

$$\theta = \frac{s}{r} = \frac{26}{11} \approx 2.3636 \text{ radians}$$

93. Quadrant III

95. Quadrant III

Section 5.3 Solving Trigonometric Equations

- You should be able to identify and solve trigonometric equations.
- A trigonometric equation is a conditional equation. It is true for a specific set of values.
- To solve trigonometric equations, use algebraic techniques such as collecting like terms, taking square roots, factoring, squaring, converting to quadratic form, using formulas, and using inverse functions. Study the examples in this section.
- Use your graphing utility to calculate solutions and verify results.

Solutions to Odd-Numbered Exercises

1. $2\cos x - 1 = 0$

(a) $2\cos\frac{\pi}{3} - 1 = 2\left(\frac{1}{2}\right) - 1 = 0$

(b) $2\cos\frac{5\pi}{3} - 1 = 2\left(\frac{1}{2}\right) - 1 = 0$

3. $3\tan^2 2x - 1 = 0$

(a) $$3\left[\tan\left(\frac{2\pi}{12}\right)\right]^2 - 1 = 3\tan^2\frac{\pi}{6} - 1$$
$$= 3\left(\frac{1}{\sqrt{3}}\right)^2 - 1$$
$$= 0$$

(b) $$3\left[\tan\left(\frac{10\pi}{12}\right)\right]^2 - 1 = 3\tan^2\frac{5\pi}{6} - 1$$
$$= 3\left(-\frac{1}{\sqrt{3}}\right)^2 - 1$$
$$= 0$$

5. $2\cos^2 x + 3\cos x + 1 = 0$

(a) $x = \frac{4\pi}{3}$: $2\cos^2\left(\frac{4\pi}{3}\right) + 3\cos\left(\frac{4\pi}{3}\right) + 1 = 2\left(\frac{1}{2}\right)^2 - 3\left(\frac{1}{2}\right) + 1 = 0$

(b) $x = \pi$: $2\cos^2 \pi + 3\cos \pi + 1 = 2(-1)^2 - 3 + 1 = 0$

7. $y = \sin\frac{\pi x}{2} + 1$

From the graph in the textbook we see that the curve has x-intercepts at $x = -1$ and at $x = 3$.

9. $y = \tan^2\left(\frac{\pi x}{6}\right) - 3$

From the graph in the textbook we see that the curve has x-intercepts at $x = \pm 2$.

11. $2\cos x + 1 = 0$

$$2\cos x = -1$$
$$\cos x = -\frac{1}{2}$$
$$x = \frac{2\pi}{3}$$
$$\text{or } x = \frac{4\pi}{3}$$

13. $\sqrt{3}\sec x - 2 = 0$

$$\sqrt{3}\sec x = 2$$
$$\sec x = \frac{2}{\sqrt{3}}$$
$$\cos x = \frac{\sqrt{3}}{2}$$
$$x = \frac{\pi}{6} \text{ or } x = \frac{11\pi}{6}$$

15. $3\csc^2 x - 4 = 0$

$$\csc^2 x = \frac{4}{3}$$
$$\csc x = \pm\frac{2}{\sqrt{3}}$$
$$\sin x = \pm\frac{\sqrt{3}}{2}$$
$$x = \frac{\pi}{3}, \frac{2\pi}{3}, \frac{4\pi}{3}, \frac{5\pi}{3}$$

17. $2\sin^2 2x = 1$

$$\sin 2x = \pm\frac{1}{\sqrt{2}} = \pm\frac{\sqrt{2}}{2}$$
$$2x = \frac{\pi}{4},\ 2x = \frac{3\pi}{4},\ 2x = \frac{5\pi}{4},\ 2x = \frac{7\pi}{4},$$
$$2x = \frac{9\pi}{4},\ 2x = \frac{11\pi}{4},\ 2x = \frac{13\pi}{4},\ 2x = \frac{15\pi}{4}$$

Thus, $x = \frac{\pi}{8}, \frac{3\pi}{8}, \frac{5\pi}{8}, \frac{7\pi}{8}, \frac{9\pi}{8}, \frac{11\pi}{8}, \frac{13\pi}{8}, \frac{15\pi}{8}$ (8 solutions)

19. $4\cos^2 x - 3 = 0$

$$\cos^2 x = \frac{3}{4}$$
$$\cos x = \pm\frac{\sqrt{3}}{2}$$
$$x = \frac{\pi}{6}, \frac{5\pi}{6}, \frac{7\pi}{6}, \frac{11\pi}{6}$$

21.

$$\sin^2 x = 3\cos^2 x$$
$$\sin^2 x - 3(1 - \sin^2 x) = 0$$
$$4\sin^2 x = 3$$
$$\sin x = \pm\frac{\sqrt{3}}{2}$$
$$x = \frac{\pi}{3}, \frac{2\pi}{3}, \frac{4\pi}{3}, \frac{5\pi}{3}$$

23. $(3\tan^2 x - 1)(\tan^2 x - 3) = 0$

$3\tan^2 x - 1 = 0$ or $\tan^2 x - 3 = 0$

$\tan x = \pm\frac{1}{\sqrt{3}}$ $\qquad$ $\tan x = \pm\sqrt{3}$

$x = \frac{\pi}{6}, \frac{7\pi}{6}$ $\qquad$ $x = \frac{\pi}{3}, \frac{4\pi}{3}$

$\text{or } x = \frac{5\pi}{6}, \frac{11\pi}{6}$ $\qquad$ $\text{or } x = \frac{2\pi}{3}, \frac{5\pi}{3}$

25. $\cos^3 x = \cos x$

$\cos^3 x - \cos x = 0$

$\cos x(\cos^2 x - 1) = 0$

$\cos x = 0$ or $\cos^2 x - 1 = 0$

$x = \frac{\pi}{2}, \frac{3\pi}{2}$ $\qquad$ $\cos x = \pm 1$

$x = 0, \pi$

27. $3\tan^3 x - \tan x = 0$

$\tan x(3\tan^2 x - 1) = 0$

$\tan x = 0$ or $3\tan^2 x - 1 = 0$

$x = 0, \pi$ $\qquad$ $\tan x = \pm\frac{\sqrt{3}}{3}$

$x = \frac{\pi}{6}, \frac{5\pi}{6}, \frac{7\pi}{6}, \frac{11\pi}{6}$

29. $\sec^2 x - \sec x - 2 = 0$

$(\sec x - 2)(\sec x + 1) = 0$

$\sec x - 2 = 0$ or $\sec x + 1 = 0$

$\sec x = 2$ $\qquad$ $\sec x = -1$

$x = \frac{\pi}{3}, \frac{5\pi}{3}$ $\qquad$ $x = \pi$

31. $2\sin x + \csc x = 0$

$2\sin x + \frac{1}{\sin x} = 0$

$2\sin^2 x + 1 = 0$

Since $2\sin^2 x + 1 > 0$, there are no solutions.

33. $\csc x + \cot x = 1$

$\frac{1}{\sin x} + \frac{\cos x}{\sin x} = 1$

$1 + \cos x = \sin x$

$(1 + \cos x)^2 = \sin^2 x$

$1 + 2\cos x + \cos^2 x = 1 - \cos^2 x$

$2\cos^2 x + 2\cos x = 0$

$2\cos x(\cos x + 1) = 0$

$\cos x = 0$ or $\cos x = -1$

$x = \frac{\pi}{2}, \frac{3\pi}{2}$ $\qquad$ $x = \pi$

($3\pi/2$ is extraneous.) $\qquad$ (π is extraneous.)

$x = \pi/2$ is the only solution.

35. $\cos\left(\frac{x}{2}\right) = \frac{\sqrt{2}}{2}$

$\frac{x}{2} = \frac{\pi}{4} + 2n\pi$

$x = \frac{\pi}{2} + 4n\pi$

$x = \frac{\pi}{2}$

37. $\frac{1 + \cos x}{1 - \cos x} = 0$

$1 + \cos x = 0$

$\cos x = -1$

$x = \pi$

39.
$$\begin{aligned} 2\sec^2 x + \tan^2 x - 3 &= 0 \\ 2(\tan^2 x + 1) + \tan^2 x - 3 &= 0 \\ 3\tan^2 x - 1 &= 0 \\ \tan x &= \pm\frac{\sqrt{3}}{3} \\ x &= \frac{\pi}{6}, \frac{5\pi}{6}, \frac{7\pi}{6}, \frac{11\pi}{6} \end{aligned}$$

41.
$$\begin{aligned} \sec^2 x + \tan x &= 3 \\ (1 + \tan^2 x) + \tan x &= 3 \\ \tan^2 x + \tan x - 2 &= 0 \\ (\tan x + 2)(\tan x - 1) &= 0 \end{aligned}$$

$\tan x = -2$ or $\tan x = 1$

$x \approx 2.0344, 5.1760$ or $x = \dfrac{\pi}{4}, \dfrac{5\pi}{4}$

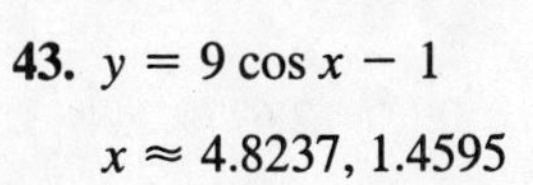

43. $y = 9\cos x - 1$

$x \approx 4.8237, 1.4595$

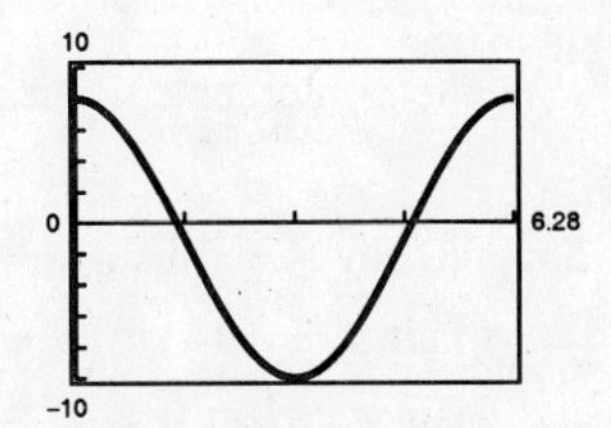

45. $y = 4\sin^2 x - 2\cos x - 1$

$x \approx 0.8614, 5.4218$

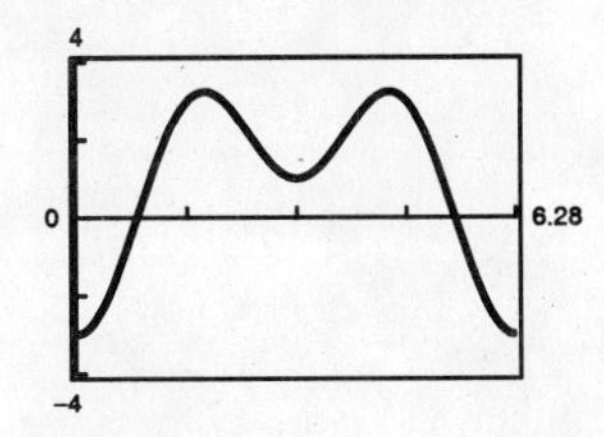

47. $4\sin^3 x + 2\sin^2 x - 2\sin x - 1 = 0$

Graph $y = 4\sin^3 x + 2\sin^2 x - 2\sin x - 1$.

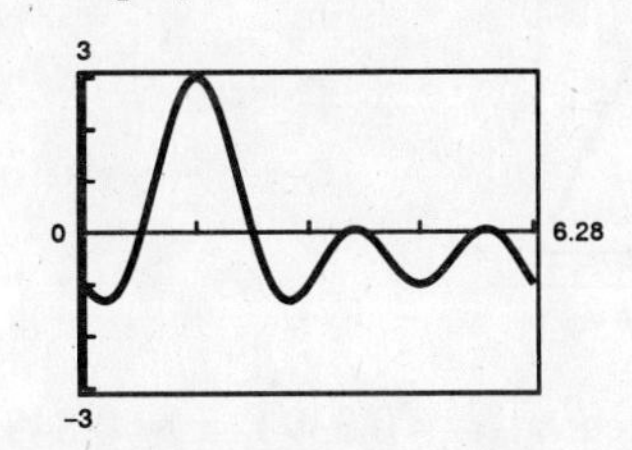

By altering the y-range to Ymin $= -.5$ and Ymax $= .5$, you see that there are 6 solutions: 0.7854, 2.3562, 3.6652, 3.9270, 5.4978, 5.7596.

49. $\dfrac{\cos x \cot x}{1 - \sin x} = 3$

Graph $y = \dfrac{\cos x}{(1 - \sin x)\tan x} - 3$.

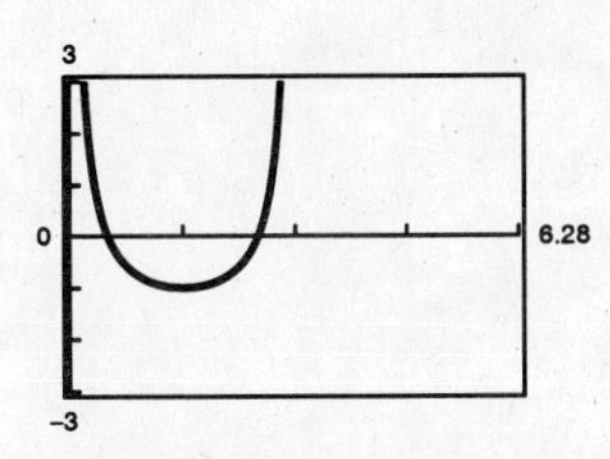

The solutions are approximately $x \approx 0.5236$, $x \approx 2.6180$

51. $x\cos x - 1 = 0$

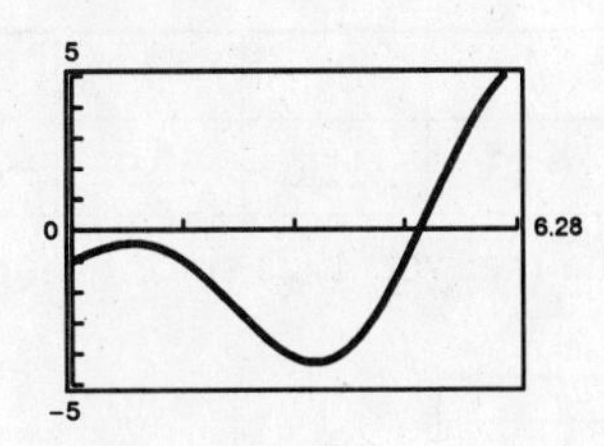

$x \approx 4.9172$

53. $\sec^2 x + 0.5 \tan x - 1 = 0$

Graph $y_1 = \dfrac{1}{(\cos x)^2} + 0.5 \tan x - 1.$

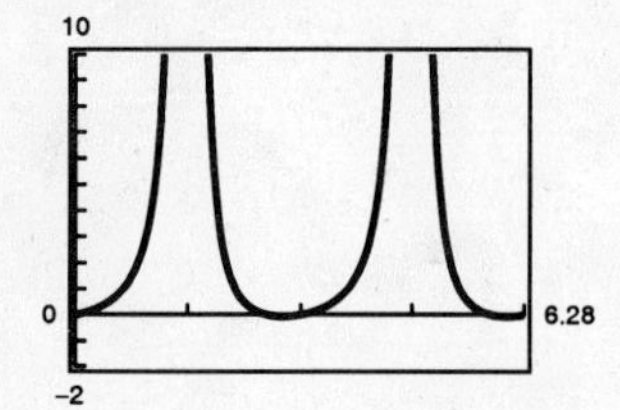

The x-intercepts occur at $x = 0, x \approx 2.6779, x = 3.1416$ and $x \approx 5.8195.$

55. $12 \sin^2 x - 13 \sin x + 3 = 0$

$(3 \sin x - 1)(4 \sin x - 3) = 0$

$3 \sin x - 1 = 0$ or $4 \sin x - 3 = 0$

$\sin x = \frac{1}{3}$ $\qquad$ $\sin x = \frac{3}{4}$

$x = 0.3398, \; 2.8018$ $\qquad$ $x = 0.8481, \; 2.2935$

Graph $y_1 = 12 \sin^2 x - 13 \sin x + 3.$

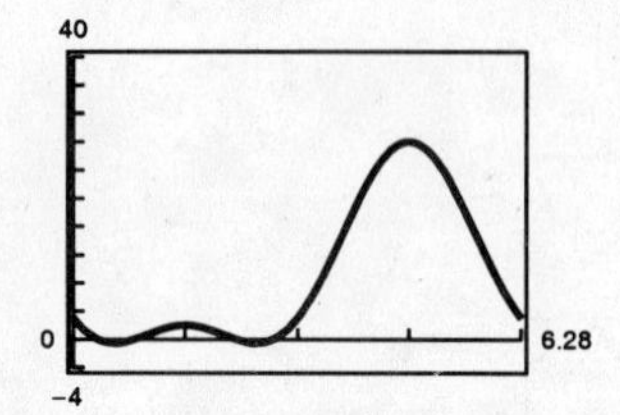

The x-intercepts occur at $x \approx 0.3398, \; x \approx 0.8481, \; x \approx 2.2935,$ and $x \approx 2.8018.$

57. $y = 3 \tan^2 x + 5 \tan x - 4, \left(-\dfrac{\pi}{2}, \dfrac{\pi}{2}\right)$

$x \approx -1.154, 0.535$

59. $y = 4 \cos^2 x - 2 \sin x + 1, \left[-\dfrac{\pi}{2}, \dfrac{\pi}{2}\right]$

$x \approx 1.110$

61. (a)

x	0	1	2	3	4	5	6
$f(x)$	Undef.	0.83	−1.36	−2.93	−4.46	−6.34	−13.02

The zero is in the interval (1, 2) since f changes signs in the interval.

(b)

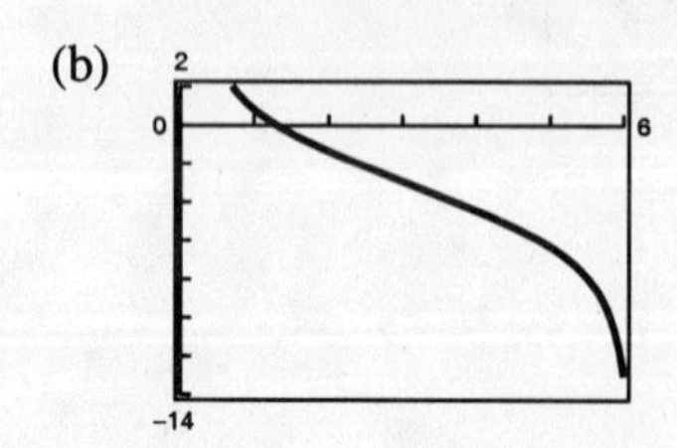

The interval is the same as part (a).

(c) 1.3065

63. (a)

x	0	1	2	3	4	5	6
$f(x)$	-1	1.39	1.65	-0.70	-1.94	-2.00	-1.48

The zeros are in the intervals (0, 1) and (2, 3) since f changes signs in these intervals.

(b)

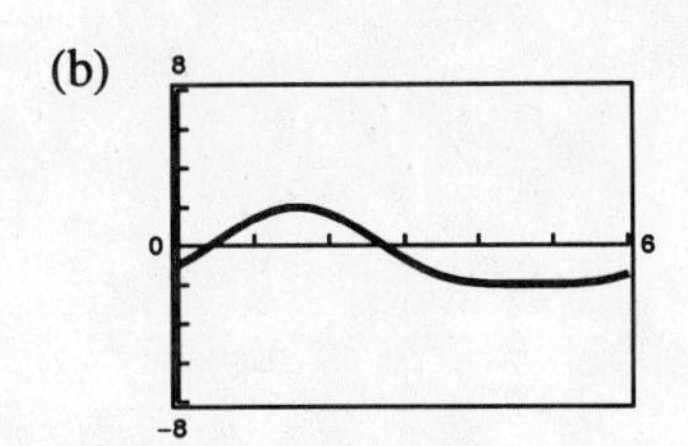

The intervals are the same as part (a).

(c) 0.4271, 2.7145

65. (a) $f(x) = \sin x + \cos x$

Maximum: $\left(\frac{\pi}{4}, \sqrt{2}\right)$

Minimum: $\left(\frac{5\pi}{4}, -\sqrt{2}\right)$

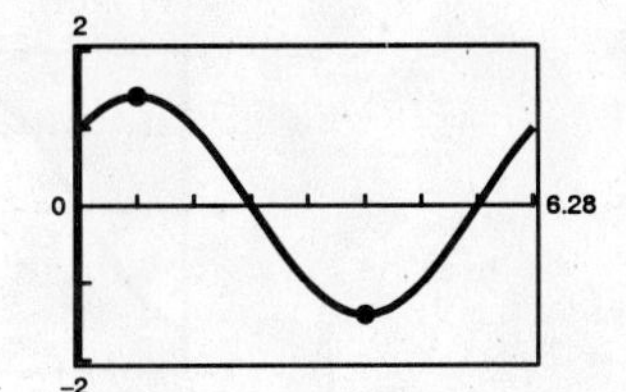

(b) $\cos x - \sin x = 0$

$$\cos x = \sin x$$

$$1 = \frac{\sin x}{\cos x}$$

$$\tan x = 1$$

$$x = \frac{\pi}{4}, \frac{5\pi}{4}$$

$$f\left(\frac{\pi}{4}\right) = \sin\frac{\pi}{4} + \cos\frac{\pi}{4} = \frac{\sqrt{2}}{2} + \frac{\sqrt{2}}{2} = \sqrt{2}$$

$$f\left(\frac{5\pi}{4}\right) = \sin\frac{5\pi}{4} + \cos\frac{5\pi}{4} = -\sin\frac{\pi}{4} + \left(-\cos\frac{\pi}{4}\right) = -\frac{\sqrt{2}}{2} - \frac{\sqrt{2}}{2} = -\sqrt{2}$$

Therefore, the maximum point in the interval $[0, 2\pi)$ is $(\pi/4, \sqrt{2})$ and the minimum point is $(5\pi/4, -\sqrt{2})$.

67. (a) $f(x) = x \sin 2x$

Maximum: (3.989, 3.958)

Minimum: (5.543, -5.520)

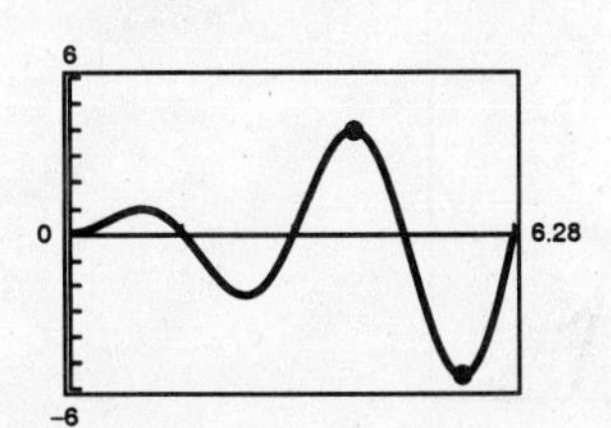

(b) $2x \cos 2x + \sin 2x = 0$

$y = 2x \cos 2x + \sin 2x$

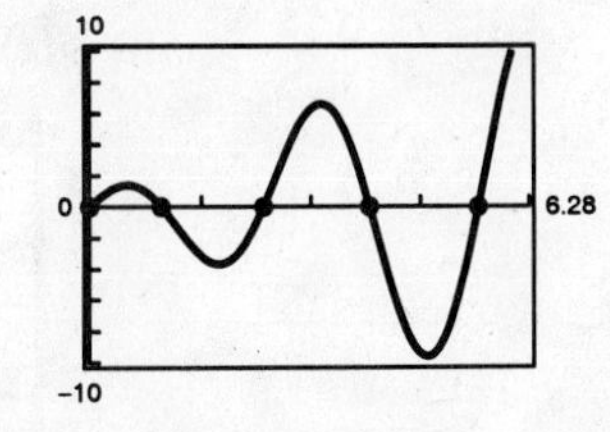

5 solutions: 0, 1.014, 2.457, 3.989, 5.543.

The fourth and fifth correspond to the maximum and minimum found in part (a).

69. $f(x) = \tan \frac{\pi x}{4}$. $\tan 0 = 0$, but 0 is not positive. By graphing $y = \tan \frac{\pi x}{4} - x$, you see that the smallest positive fixed point is $x = 1$.

71. $f(x) = \cos \frac{1}{x}$

(a) The domain of $f(x)$ is all real numbers except 0.

(b) The graph has y-axis symmetry and a horizontal asymptote at $y = 1$.

(c) As $x \to 0$, $f(x)$ oscillates between -1 and 1.

(d) There are an infinite number of solutions in the interval $[-1, 1]$.

(e) The greatest solution appears to occur at $x \approx 0.6366$.

73.

$$y = \frac{1}{12}(\cos 8t - 3 \sin 8t)$$

$$\frac{1}{12}(\cos 8t - 3 \sin 8t) = 0$$

$$\cos 8t = 3 \sin 8t$$

$$\frac{1}{3} = \tan 8t$$

$$8t = 0.32175 + n\pi$$

$$t = 0.04 + \frac{n\pi}{8}$$

In the interval $0 \le t \le 1$, $t = 0.04, 0.43$, and 0.83.

75. $D = 31 \sin\left(\frac{2\pi}{365}t - 1.4\right)$

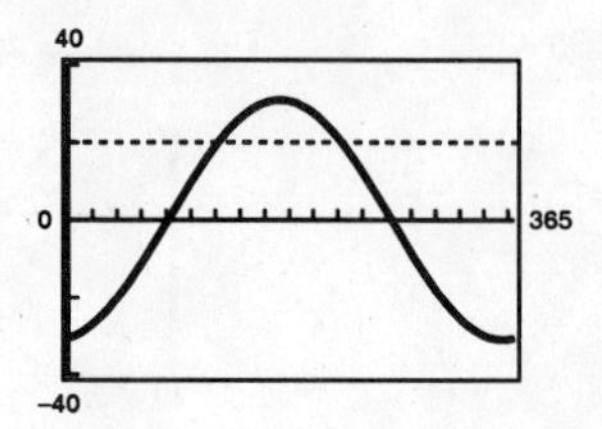

$D > 20°$ for $122 \le t \le 223$

77.

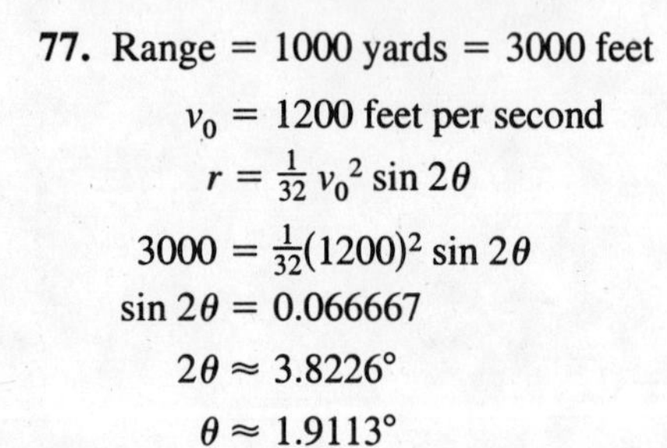

$$\text{Range} = 1000 \text{ yards} = 3000 \text{ feet}$$

$$v_0 = 1200 \text{ feet per second}$$

$$r = \tfrac{1}{32} v_0^2 \sin 2\theta$$

$$3000 = \tfrac{1}{32}(1200)^2 \sin 2\theta$$

$$\sin 2\theta = 0.066667$$

$$2\theta \approx 3.8226°$$

$$\theta \approx 1.9113°$$

79. $y_1 = 1.56e^{-0.22t} \cos 4.9t$ intersects $y_2 = -1$ at $t \approx 1.96$

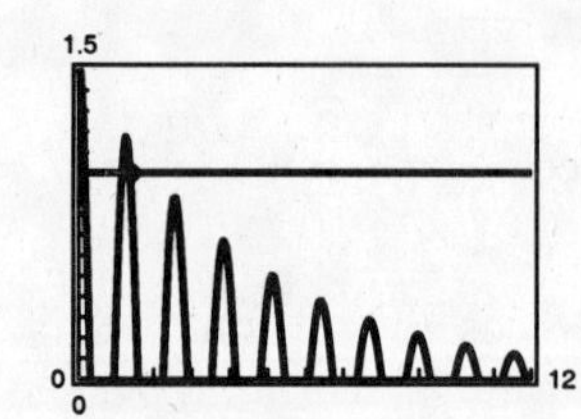

The displacement does not exceed one inch from equilibrium after $t = 1.96$ seconds.

81. $f(x) = 3 \sin(0.6x - 2)$

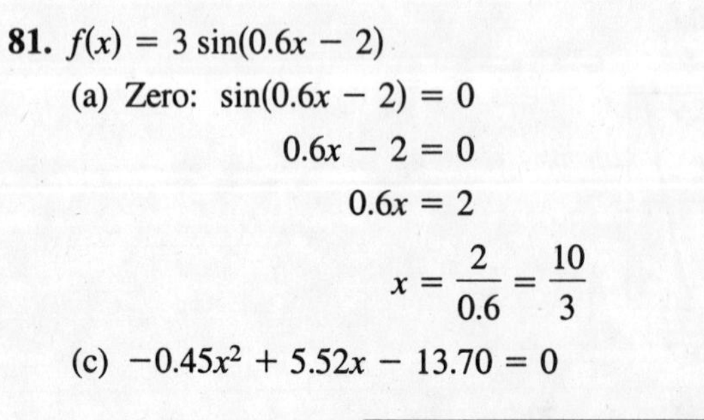

(a) Zero: $\sin(0.6x - 2) = 0$

$$0.6x - 2 = 0$$

$$0.6x = 2$$

$$x = \frac{2}{0.6} = \frac{10}{3}$$

(b) $g(x) = -0.45x^2 + 5.52x - 13.70$

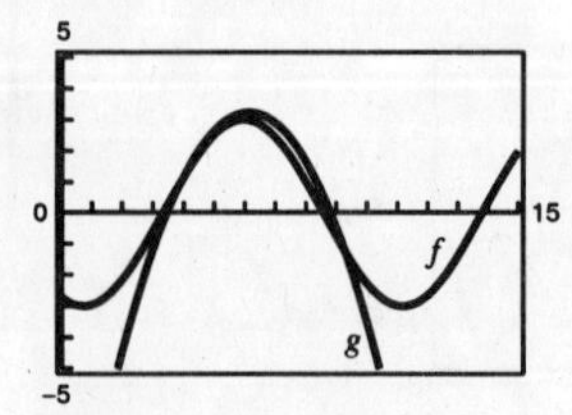

(c) $-0.45x^2 + 5.52x - 13.70 = 0$

$$x = \frac{-5.52 \pm \sqrt{(5.52)^2 - 4(-0.45)(-13.70)}}{2(-0.45)}$$

$x \approx 3.46, 8.81$

The zero of g on $[0, 6]$ is 3.46. The zero is close to the zero $\frac{10}{3} \approx 3.33$ of f.

For $3.5 \le x \le 6$ the approximation appears to be good.

83. False. There might not be periodicity, as in the equation $\sin(x^2) = 0$

85. $124° = 124°\left(\dfrac{\pi}{180°}\right)$

≈ 2.164 radians

87. $-0.41° = -0.41\left(\dfrac{\pi}{180°}\right)$

≈ -0.007 radians

89. $\tan 30° = \dfrac{14}{x} \Rightarrow x = \dfrac{14}{\tan 30°} = \dfrac{14}{\frac{\sqrt{3}}{3}} \approx 24.249$

91. $\sin 40° = \dfrac{16}{x} \Rightarrow x = \dfrac{16}{\sin 40°} \approx 24.892$

93. $f(x) = \dfrac{1}{4}\sin\left(x - \dfrac{\pi}{2}\right)$

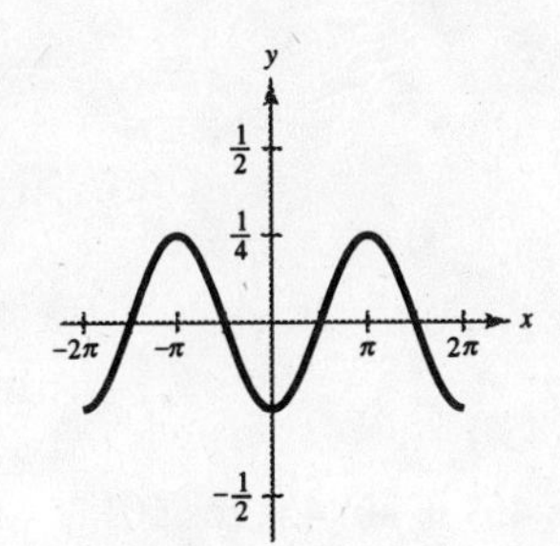

95. $f(x) = \dfrac{1}{2}\cot\left(x - \dfrac{\pi}{4}\right)$

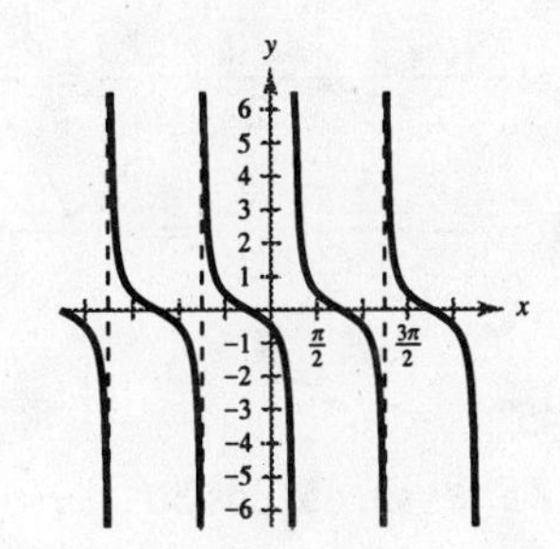

Section 5.4 Sum and Difference Formulas

- You should memorize the sum and difference formulas.

  $\sin(u \pm v) = \sin u \cos v \pm \cos u \sin v$

  $\cos(u \pm v) = \cos u \cos v \mp \sin u \sin v$

  $\tan(u \pm v) = \dfrac{\tan u \pm \tan v}{1 \mp \tan u \tan v}$
- You should be able to use these formulas to find the values of the trigonometric functions of angles whose sums or differences are special angles.
- You should be able to use these formulas to solve trigonometric equations.

Solutions to Odd-Numbered Exercises

1. (a) $\cos\left(\dfrac{\pi}{6} + \dfrac{\pi}{3}\right) = \cos\dfrac{\pi}{6}\cos\dfrac{\pi}{3} - \sin\dfrac{\pi}{6}\sin\dfrac{\pi}{3}$

$= \dfrac{\sqrt{3}}{2} \cdot \dfrac{1}{2} - \dfrac{1}{2} \cdot \dfrac{\sqrt{3}}{2} = 0$

(b) $\cos\dfrac{\pi}{6} + \cos\dfrac{\pi}{3} = \dfrac{\sqrt{3}}{2} + \dfrac{1}{2} = \dfrac{\sqrt{3}+1}{2}$

3. (a) $\sin\left(\frac{7\pi}{6} - \frac{\pi}{3}\right) = \sin\frac{5\pi}{6} = \sin\frac{\pi}{6} = \frac{1}{2}$ (b) $\sin\frac{7\pi}{6} - \sin\frac{\pi}{3} = -\frac{1}{2} - \frac{\sqrt{3}}{2} = \frac{-1-\sqrt{3}}{2}$

5. (a) $\cos(0° + 135°) = \cos 0° \cos 135° - \sin 0° \sin 135°$

$$= \cos 135° = -\frac{\sqrt{2}}{2}$$

(b) $\cos 0° + \cos 135° = 1 - \frac{\sqrt{2}}{2}$

7. (a) $\sin(315° - 60°) = \sin 315 \cos 60° - \cos 315° \sin 60°$

$$= -\frac{\sqrt{2}}{2}\frac{1}{2} - \frac{\sqrt{2}}{2} \cdot \frac{\sqrt{3}}{2} = \frac{-\sqrt{2}-\sqrt{6}}{4}$$

(b) $\sin 315° - \sin 60° = -\frac{\sqrt{2}}{2} - \frac{\sqrt{3}}{2} = -\frac{\sqrt{2}-\sqrt{3}}{2}$

9. $\sin 75° = \sin(30° + 45°)$

$$= \sin 30° \cos 45° + \sin 45° \cos 30°$$

$$= \frac{1}{2} \cdot \frac{\sqrt{2}}{2} + \frac{\sqrt{2}}{2} \cdot \frac{\sqrt{3}}{2}$$

$$= \frac{\sqrt{2}}{4}(1 + \sqrt{3})$$

$\cos 75° = \cos(30° + 45°)$

$$= \cos 30° \cos 45° - \sin 30° \sin 45°$$

$$= \frac{\sqrt{3}}{2} \cdot \frac{\sqrt{2}}{2} - \frac{1}{2} \cdot \frac{\sqrt{2}}{2}$$

$$= \frac{\sqrt{2}}{4}(\sqrt{3} - 1)$$

$\tan 75° = \tan(30° + 45°)$

$$= \frac{\tan 30° + \tan 45°}{1 - \tan 30° \tan 45°}$$

$$= \frac{(\sqrt{3}/3) + 1}{1 - (\sqrt{3}/3)} = \frac{\sqrt{3}+3}{3-\sqrt{3}} \cdot \frac{3+\sqrt{3}}{3+\sqrt{3}}$$

$$= \frac{6\sqrt{3}+12}{6} = \sqrt{3} + 2$$

11. $\sin 105° = \sin(60° + 45°)$

$$= \sin 60° \cos 45° + \sin 45° \cos 60°$$

$$= \frac{\sqrt{3}}{2} \cdot \frac{\sqrt{2}}{2} + \frac{\sqrt{2}}{2} \cdot \frac{1}{2}$$

$$= \frac{\sqrt{2}}{4}(\sqrt{3} + 1)$$

$\cos 105° = \cos(60° + 45°)$

$$= \cos 60° \cos 45° - \sin 60° \sin 45°$$

$$= \frac{1}{2} \cdot \frac{\sqrt{2}}{2} - \frac{\sqrt{3}}{2} \cdot \frac{\sqrt{2}}{2}$$

$$= \frac{\sqrt{2}}{4}(1 - \sqrt{3})$$

$\tan 105° = \tan(60° + 45°)$

$$= \frac{\tan 60° + \tan 45°}{1 - \tan 60° \tan 45°}$$

$$= \frac{\sqrt{3}+1}{1-\sqrt{3}} = \frac{\sqrt{3}+1}{1-\sqrt{3}} \cdot \frac{1+\sqrt{3}}{1+\sqrt{3}}$$

$$= \frac{4+2\sqrt{3}}{-2} = -2 - \sqrt{3}$$

13. $\sin 195° = \sin(225° - 30°)$

$= \sin 225° \cos 30° - \sin 30° \cos 225°$

$= -\sin 45° \cos 30° + \sin 30° \cos 45°$

$= -\frac{\sqrt{2}}{2} \cdot \frac{\sqrt{3}}{2} + \frac{\sqrt{2}}{2} \cdot \frac{1}{2}$

$= \frac{\sqrt{2}}{4}(1 - \sqrt{3})$

$\cos 195° = \cos(225° - 30°)$

$= \cos 225° \cos 30° + \sin 225° \sin 30°$

$= -\cos 45° \cos 30° - \sin 45° \sin 30°$

$= \frac{\sqrt{2}}{2} \cdot \frac{\sqrt{2}}{2} - \frac{\sqrt{2}}{2} \cdot \frac{1}{2}$

$= -\frac{\sqrt{2}}{4}(\sqrt{3} + 1)$

$\tan 195° = \tan(225° - 30°)$

$= \frac{\tan 225° - \tan 30°}{1 + \tan 225° \tan 30°}$

$= \frac{\tan 45° - \tan 30°}{1 + \tan 45° \tan 30°}$

$= \frac{1 - (\sqrt{3}/3)}{1 + (\sqrt{3}/3)} = \frac{3 - \sqrt{3}}{3 + \sqrt{3}} \cdot \frac{3 - \sqrt{3}}{3 - \sqrt{3}}$

$= \frac{12 - 6\sqrt{3}}{6} = 2 - \sqrt{3}$

15. $\sin \frac{11\pi}{12} = \sin\left(\frac{3\pi}{4} + \frac{\pi}{6}\right)$

$= \sin \frac{3\pi}{4} \cos \frac{\pi}{6} + \sin \frac{\pi}{6} \cos \frac{3\pi}{4}$

$= \frac{\sqrt{2}}{2} \cdot \frac{\sqrt{3}}{2} + \frac{1}{2}\left(-\frac{\sqrt{2}}{2}\right)$

$= \frac{\sqrt{2}}{4}(\sqrt{3} - 1)$

$\cos \frac{11\pi}{12} = \cos\left(\frac{3\pi}{4} + \frac{\pi}{6}\right)$

$= \cos \frac{3\pi}{4} \cos \frac{\pi}{6} - \sin \frac{3\pi}{4} \sin \frac{\pi}{6}$

$= -\frac{\sqrt{2}}{2} \cdot \frac{\sqrt{3}}{2} - \frac{\sqrt{2}}{2} \cdot \frac{1}{2}$

$= -\frac{\sqrt{2}}{4}(\sqrt{3} + 1)$

$\tan \frac{11\pi}{4} = \tan\left(\frac{3\pi}{4} + \frac{\pi}{6}\right)$

$= \frac{\tan(3\pi/4) + \tan(\pi/6)}{1 - \tan(3\pi/4)\tan(\pi/6)}$

$= \frac{-1 + (\sqrt{3}/3)}{1 - (-1)(\sqrt{3}/3)}$

$= \frac{-3 + \sqrt{3}}{3 + \sqrt{3}} \cdot \frac{3 - \sqrt{3}}{3 - \sqrt{3}}$

$= \frac{-12 + 6\sqrt{3}}{6} = -2 + \sqrt{3}$

17. $-\frac{\pi}{12} = \frac{\pi}{6} - \frac{\pi}{4}$

$\sin\left(-\frac{\pi}{12}\right) = \sin\left(\frac{\pi}{6} - \frac{\pi}{4}\right) = \sin \frac{\pi}{6} \cos \frac{\pi}{4} - \sin \frac{\pi}{4} \cos \frac{\pi}{6}$

$= \frac{1}{2} \cdot \frac{\sqrt{2}}{2} - \frac{\sqrt{2}}{2} \cdot \frac{\sqrt{3}}{2} = \frac{\sqrt{2}}{4}(1 - \sqrt{3})$

$\cos\left(-\frac{\pi}{12}\right) = \cos\left(\frac{\pi}{6} - \frac{\pi}{4}\right) = \cos \frac{\pi}{6} \cos \frac{\pi}{4} + \sin \frac{\pi}{6} \sin \frac{\pi}{4}$

$= \frac{\sqrt{3}}{2} \cdot \frac{\sqrt{2}}{2} + \frac{1}{2} \cdot \frac{\sqrt{2}}{2} = \frac{\sqrt{2}}{4}(\sqrt{3} + 1)$

$\tan\left(-\frac{\pi}{12}\right) = \tan\left(\frac{\pi}{6} - \frac{\pi}{4}\right) = \frac{\tan(\pi/6) - \tan(\pi/4)}{1 + \tan(\pi/6)\tan(\pi/4)}$

$= \frac{(\sqrt{3}/3) - 1}{1 + (\sqrt{3}/3)} = \frac{\sqrt{3} - 3}{\sqrt{3} + 3} \cdot \frac{\sqrt{3} - 3}{\sqrt{3} - 3} = \frac{12 - 6\sqrt{3}}{-6} = -2 + \sqrt{3}$

19. $\cos 40° \cos 15° - \sin 40° \sin 15° = \cos(40° + 15°) = \cos 55°$

21. $\sin 340° \cos 50° - \cos 340° \sin 50° = \sin(340° - 50°) = \sin 290°$

23. $\dfrac{\tan 325^\circ - \tan 86^\circ}{1 + \tan 325^\circ \tan 86^\circ} = \tan(325^\circ - 86^\circ) = \tan 239^\circ$

25. $\sin 3 \cos 1.2 - \cos 3 \sin 1.2 = \sin(3 - 1.2) = \sin 1.8$

27. $\cos\dfrac{\pi}{7}\cos\dfrac{\pi}{5} - \sin\dfrac{\pi}{7}\sin\dfrac{\pi}{5} = \cos\left(\dfrac{\pi}{7} + \dfrac{\pi}{5}\right)$

$$= \cos\frac{12\pi}{35}$$

29.

x	0.2	0.4	0.6	0.8	1.0	1.2	1.4
y_1	0.9801	0.9211	0.8253	0.6967	0.5403	0.3624	0.1700
y_2	0.9801	0.9211	0.8253	0.6967	0.5403	0.3624	0.1700

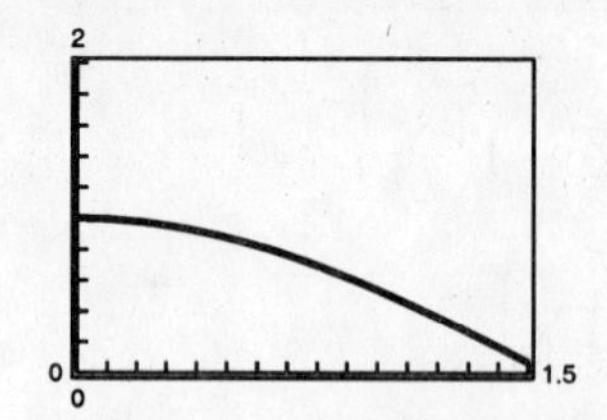

$$\begin{aligned} y_1 &= \sin\left(\frac{\pi}{2} + x\right) \\ &= \sin\frac{\pi}{2}\cos x + \sin x \cdot \cos\frac{\pi}{2} \\ &= \cos x \\ &= y_2 \end{aligned}$$

31.

x	0.2	0.4	0.6	0.8	1.0	1.2	1.4
y_1	0.6621	0.7978	0.9017	0.9696	0.9989	0.9883	0.9384
y_2	0.6621	0.7978	0.9017	0.9696	0.9989	0.9883	0.9384

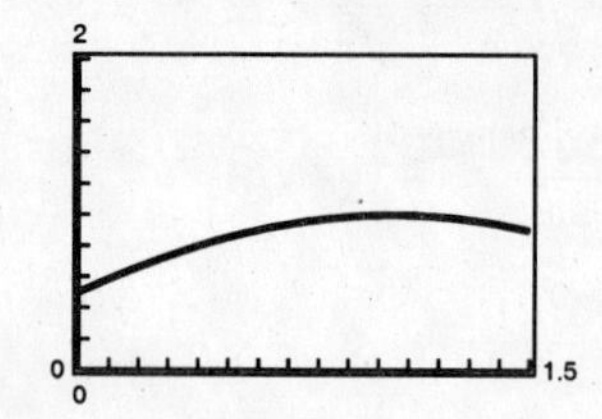

$$\begin{aligned} y_1 &= \sin\left(\frac{\pi}{6} + x\right) \\ &= \sin\frac{\pi}{6}\cos x + \sin x \cdot \cos\frac{\pi}{6} \\ &= \frac{1}{2}\cos x + \frac{\sqrt{3}}{2}\sin x \\ &= \frac{1}{2}(\cos x + \sqrt{3}\sin x) \\ &= y_2 \end{aligned}$$

33.

x	0.2	0.4	0.6	0.8	1.0	1.2	1.4
y_1	0.9605	0.8484	0.6812	0.4854	0.2919	0.1313	0.0289
y_2	0.9605	0.8484	0.6812	0.4854	0.2919	0.1313	0.0289

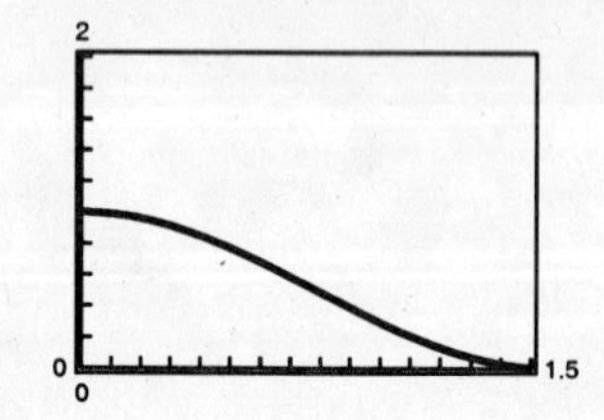

$$\begin{aligned} y_1 &= \cos(x + \pi)\cos(x - \pi) \\ &= (\cos x \cdot \cos \pi - \sin x \cdot \sin \pi) \\ &\quad [\cos x \cos \pi + \sin x \sin \pi] \\ &= [-\cos x][-\cos x] \\ &= \cos^2 x \\ &= y_2 \end{aligned}$$

For Exercises 35–37, we have:

$\sin u = \frac{5}{13}$, u is in Quadrant I $\Rightarrow$ $\cos u = \frac{12}{13}$

$\cos v = -\frac{3}{5}$, v is in Quadrant II $\Rightarrow$ $\sin v = \frac{4}{5}$

35. $\sin(u + v) = \sin u \cos v + \cos u \sin v$

$$= \left(\tfrac{5}{13}\right)\left(-\tfrac{3}{5}\right) + \left(\tfrac{12}{13}\right)\left(\tfrac{4}{5}\right)$$

$$= \tfrac{33}{65}$$

37. $\cos(u + v) = \cos u \cos v - \sin u \sin v$

$$= \left(\tfrac{12}{13}\right)\left(-\tfrac{3}{5}\right) - \left(\tfrac{5}{13}\right)\left(\tfrac{4}{5}\right)$$

$$= \tfrac{-56}{65}$$

For Exercises 39–41, we have:

$\sin u = \frac{7}{25}$, u is in Quadrant II $\Rightarrow$ $\cos u = \frac{-24}{25}$

$\cos v = \frac{4}{5}$, v is in Quadrant IV $\Rightarrow$ $\sin v = -\frac{3}{5}$

39. $\cos(u + v) = \cos u \cdot \cos v - \sin u \cdot \sin v$

$$= \left(-\frac{24}{25}\right)\left(\frac{4}{5}\right) - \left(\frac{7}{25}\right)\left(-\frac{3}{5}\right)$$

$$= \frac{-96 + 21}{125} = \frac{-75}{125} = \frac{-3}{5}$$

41. $\sin(v - u) = \sin v \cdot \cos u - \sin u \cdot \cos v$

$$= \left(-\frac{3}{5}\right)\left(-\frac{24}{25}\right) - \left(\frac{7}{25}\right)\left(\frac{4}{5}\right)$$

$$= \frac{72 - 28}{125} = \frac{44}{125}$$

43. $\cos(\pi - \theta) + \sin\left(\frac{\pi}{2} + \theta\right) = \cos \pi \cos \theta + \sin \pi \sin \theta + \sin \frac{\pi}{2} \cos \theta + \sin \theta \cos \frac{\pi}{2}$

$$= (-1)(\cos \theta) + (0)(\sin \theta) + (1)(\cos \theta) + (\sin \theta)(0) = -\cos \theta + \cos \theta = 0$$

45. $\tan(x + \pi) - \tan(\pi - x) = \frac{\tan x + \tan \pi}{1 - \tan x \cdot \tan \pi} - \frac{\tan \pi - \tan x}{1 + \tan \pi \tan x}$

$$= \frac{\tan x}{1} - \left(-\frac{\tan x}{1}\right)$$

$$= 2 \tan x$$

47. $\sin(x + y) + \sin(x - y) = \sin x \cos y + \sin y \cos x + \sin x \cos y - \sin y \cos x = 2 \sin x \cos y$

49. $\cos(x + y)\cos(x - y) = [\cos x \cos y - \sin x \sin y][\cos x \cos y + \sin x \sin y]$

$$= \cos^2 x \cos^2 y - \sin^2 x \sin^2 y$$

$$= \cos^2 x(1 - \sin^2 y) - \sin^2 x \sin^2 y$$

$$= \cos^2 x - \sin^2 y(\cos^2 x + \sin^2 x)$$

$$= \cos^2 x - \sin^2 y$$

51. $\sin(\arcsin x + \arccos x) = \sin(\arcsin x)\cos(\arccos x) + \sin(\arccos x)\cos(\arcsin x)$

$$= x \cdot x + \sqrt{1-x^2} \cdot \sqrt{1-x^2}$$
$$= x^2 + 1 - x^2$$
$$= 1$$

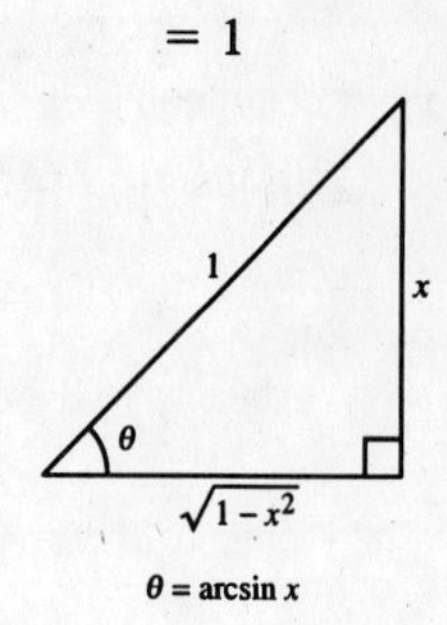

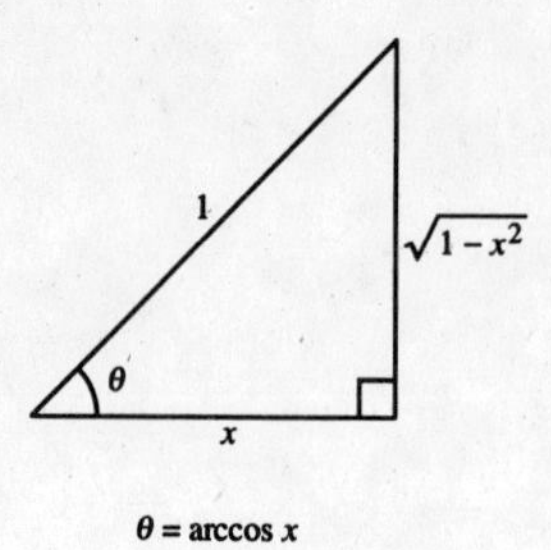

53. Let:

$u = \arctan 2x$ and $v = \arccos x$

$\tan u = 2x$ $\qquad \cos v = x$

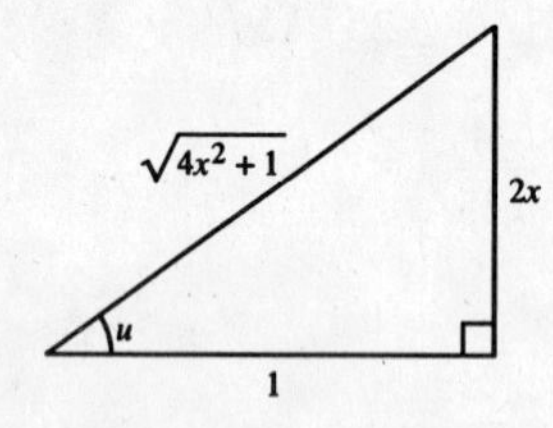

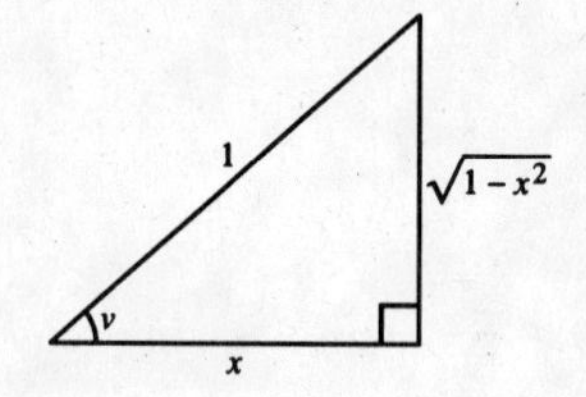

$$\sin(\arctan 2x - \arccos x) = \sin(u - v)$$
$$= \sin u \cos v - \cos u \sin v$$
$$= \frac{2x}{\sqrt{4x^2+1}}(x) - \frac{1}{\sqrt{4x^2+1}}\left(\sqrt{1-x^2}\right)$$
$$= \frac{2x^2 - \sqrt{1-x^2}}{\sqrt{4x^2+1}}$$

55.
$$\sin\left(x + \frac{\pi}{3}\right) + \sin\left(x - \frac{\pi}{3}\right) = 1$$
$$\sin x \cos\frac{\pi}{3} + \cos x \sin\frac{\pi}{3} + \sin x \cos\frac{\pi}{3} - \cos x \sin\frac{\pi}{3} = 1$$
$$2 \sin x(0.5) = 1$$
$$\sin x = 1$$
$$x = \frac{\pi}{2}$$

57.

$$\cos\left(x + \frac{\pi}{4}\right) - \cos\left(x - \frac{\pi}{4}\right) = 1$$

$$\cos x \cos \frac{\pi}{4} - \sin x \sin \frac{\pi}{4} - \left(\cos x \cos \frac{\pi}{4} + \sin x \sin \frac{\pi}{4}\right) = 1$$

$$-2 \sin x\left(\frac{\sqrt{2}}{2}\right) = 1$$

$$-\sqrt{2} \sin x = 1$$

$$\sin x = -\frac{1}{\sqrt{2}}$$

$$\sin x = -\frac{\sqrt{2}}{2}$$

$$x = \frac{5\pi}{4}, \frac{7\pi}{4}$$

59.

$$\tan(x + \pi) + 2 \sin(x + \pi) = 0$$

$$\frac{\tan x + \tan \pi}{1 - \tan x \tan \pi} + 2(\sin x \cos \pi + \cos x \sin \pi) = 0$$

$$\frac{\tan x + 0}{1 - \tan x(0)} + 2[\sin x(-1) + \cos x(0)] = 0$$

$$\frac{\tan x}{1} - 2 \sin x = 0$$

$$\frac{\sin x}{\cos x} = 2 \sin x$$

$$\sin x = 2 \sin x \cos x$$

$$\sin x(1 - 2 \cos x) = 0$$

$$\sin x = 0 \quad \text{or} \quad \cos x = \frac{1}{2}$$

$$x = 0, \pi \qquad x = \frac{\pi}{3}, \frac{5\pi}{3}$$

61. Graph $y_1 = \cos\left(x + \frac{\pi}{4}\right) + \cos\left(x - \frac{\pi}{4}\right)$ and $y_2 = 1$.

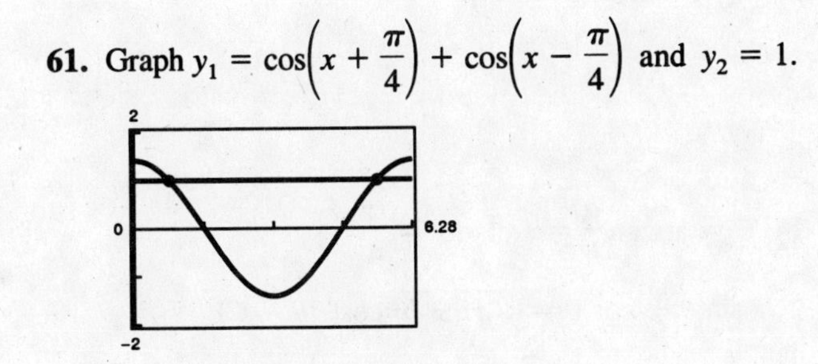

The points of intersection occur at $x \approx 0.7854$ and $x \approx 5.4978$.

63. $\tan(x + \pi) - \cos\left(x + \dfrac{\pi}{2}\right) = 0$

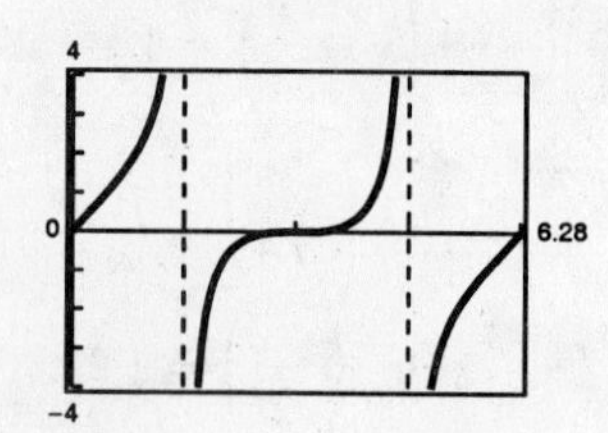

Answers: $(0, 0), (3.14, 0) \implies x = 0, \pi$

65. $y_1 + y_2 = A\cos 2\pi\left(\dfrac{t}{T} - \dfrac{x}{\lambda}\right) + A\cos 2\pi\left(\dfrac{t}{T} + \dfrac{x}{\lambda}\right)$

$$= A\left[\cos\left(\frac{2\pi t}{T}\right)\cos\left(\frac{2\pi x}{\lambda}\right) + \sin\left(\frac{2\pi t}{T}\right)\sin\left(\frac{2\pi x}{\lambda}\right)\right] + A\left[\cos\left(\frac{2\pi t}{T}\right)\cos\left(\frac{2\pi x}{\lambda}\right) - \sin\left(\frac{2\pi t}{T}\right)\sin\left(\frac{2\pi x}{\lambda}\right)\right]$$

$$= 2A\cos\left(\frac{2\pi}{T}\right)\cos\left(\frac{\pi}{\lambda}\right)$$

67. False. See page 404.

69. False. $\sin 75° = \sin(30° + \sin 45°)$

$$= \sin 30°\cos 45° + \cos 30° 45°$$

$$= \frac{1}{2}\frac{\sqrt{2}}{2} + \frac{\sqrt{3}}{2}\frac{\sqrt{2}}{2}$$

$$= \frac{\sqrt{2} + \sqrt{6}}{4}$$

71. $\cos(n\pi + \theta) = \cos n\pi \cos\theta - \sin n\pi \sin\theta$

$= (-1)^n(\cos\theta) - (0)(\sin\theta)$

$= (-1)^n(\cos\theta)$, where n is an integer.

73. $C = \arctan\dfrac{b}{a} \implies \tan C = \dfrac{b}{a} \implies \sin C = \dfrac{b}{\sqrt{a^2 + b^2}}, \cos C = \dfrac{a}{\sqrt{a^2 + b^2}}$

$$\sqrt{a^2 + b^2}\sin(B\theta + C) = \sqrt{a^2 + b^2}\left(\sin B\theta \cdot \frac{a}{\sqrt{a^2 + b^2}} + \frac{b}{\sqrt{a^2 + b^2}} \cdot \cos B\theta\right) = a\sin B\theta + b\cos B\theta$$

75. $\sin\theta + \cos\theta$

$a = 1,\ b = 1,\ B = 1$

(a) $C = \arctan\dfrac{b}{a} = \arctan 1 = \dfrac{\pi}{4}$

$\sin\theta + \cos\theta = \sqrt{a^2 + b^2}\sin(B\theta + C)$

$= \sqrt{2}\sin\left(\theta + \dfrac{\pi}{4}\right)$

(b) $C = \arctan\dfrac{a}{b} = \arctan 1 = \dfrac{\pi}{4}$

$\sin\theta + \cos\theta = \sqrt{a^2 + b^2}\cos(B\theta - C)$

$= \sqrt{2}\cos\left(\theta - \dfrac{\pi}{4}\right)$

77. $12\sin 3\theta + 5\cos 3\theta$

$a = 12,\ b = 5,\ B = 3$

(a) $C = \arctan\dfrac{b}{a} = \arctan\dfrac{5}{12} \approx 0.3948$

$12\sin 3\theta + 5\cos 3\theta = \sqrt{a^2 + b^2}\sin(B\theta + C)$

$\approx 13\sin(3\theta + 0.3948)$

(b) $C = \arctan\dfrac{a}{b} = \arctan\dfrac{12}{5} \approx 1.1760$

$12\sin 3\theta + 5\cos 3\theta = \sqrt{a^2 + b^2}\cos(B\theta - C)$

$\approx 13\cos(3\theta - 1.1760)$

79. $C = \arctan \dfrac{b}{a} = \dfrac{\pi}{2} \implies a = 0$

$\sqrt{a^2 + b^2} = 2 \implies b = 2$

$B = 1$

$2 \sin\left(\theta + \dfrac{\pi}{2}\right) = (0)(\sin \theta) + (2)(\cos \theta) = 2 \cos \theta$

81. From the figure, it appears that $u + v = w$. Assume that u, v, and w are all in Quadrant I. From the figure:

$\tan u = \dfrac{s}{3s} = \dfrac{1}{3}$

$\tan v = \dfrac{s}{2s} = \dfrac{1}{2}$

$\tan w = \dfrac{s}{s} = 1$

$\tan(u + v) = \dfrac{\tan u + \tan v}{1 - \tan u \tan v} = \dfrac{1/3 + 1/2}{1 - (1/3)(1/2)} = \dfrac{5/6}{1 - (1/6)} = 1 = \tan w.$

Thus, $\tan(u + v) = \tan w$. Because u, v, and w are all in Quadrant I, we have

$$\arctan[\tan(u + v)] = \arctan[\tan w]$$
$$u + v = w.$$

83. $\tan(\pi + \theta) = \dfrac{\tan \pi + \tan \theta}{1 - \tan \pi \tan \theta}$

$= \dfrac{0 + \tan \theta}{1 - (0) \tan \theta}$

$= \tan \theta$

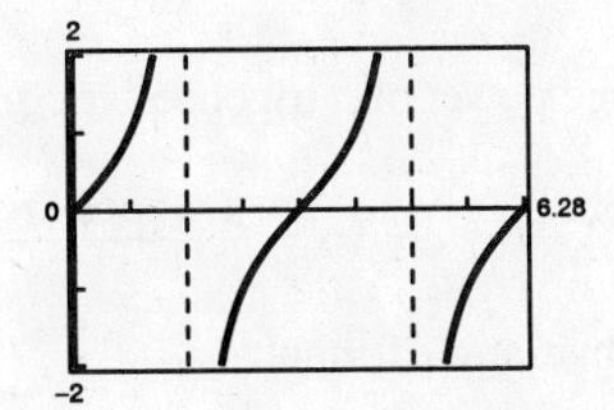

85. $f(h) = \dfrac{\cos\left(\dfrac{\pi}{6} + h\right) - \cos \dfrac{\pi}{6}}{h}$

$g(h) = \cos \dfrac{\pi}{6}\left(\dfrac{\cos h - 1}{h}\right) - \sin \dfrac{\pi}{6}\left(\dfrac{\sin h}{h}\right)$

(a) The domains are both $(-\infty, 0)$, $(0, \infty)$.

(b)

h	0.01	0.02	0.05	0.1	0.2	0.5
$f(h)$	−0.5043	−0.5086	−0.5214	−0.5424	−0.5830	−0.6915
$g(h)$	−0.5043	−0.5086	−0.5214	−0.5424	−0.5830	−0.6915

(c)

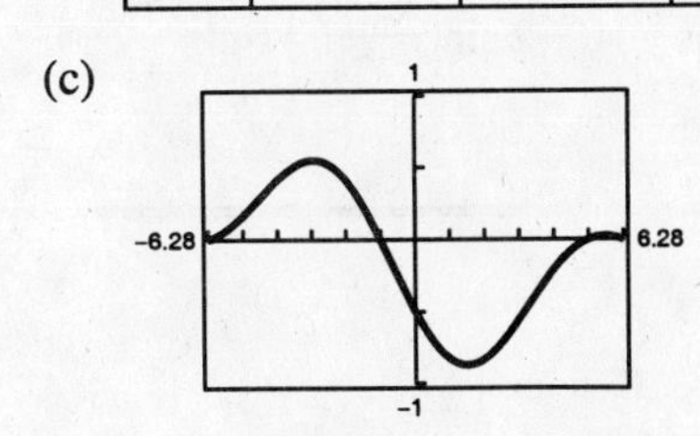

(d) The values tend to $y = -\dfrac{1}{2}$.

87. $x = 0$: $y = -\frac{1}{2}(0 - 10) + 14 = 5 + 14 = 19$. y-intercept: $(0, 19)$

$y = 0$: $0 = -\frac{1}{2}(x - 10) + 14 = -\frac{1}{2}x + 19 \Rightarrow x = 38$. x-intercept: $(38, 0)$

89. $x = 0$: $|2(0) - 9| - 5 = 9 - 5 = 4$. y-intercept: $(0, 4)$

$y = 0$: $|2x - 9| = 5 \Rightarrow x = 7, 2$. x-intercepts: $(2, 0), (7, 0)$

91. $\arccos\left(\frac{\sqrt{3}}{2}\right) = \frac{\pi}{6}$ because $\cos \frac{\pi}{6} = \frac{\sqrt{3}}{2}$

93. $\arcsin 1 = \frac{\pi}{2}$ because $\sin \frac{\pi}{2} = 1$.

Section 5.5 Multiple-Angle and Product-Sum Formulas

- You should know the following double-angle formulas.
  - (a) $\sin 2u = 2 \sin u \cos u$
  - (b) $\cos 2u = \cos^2 u - \sin^2 u$
    $= 2 \cos^2 u - 1$
    $= 1 - 2 \sin^2 u$
  - (c) $\tan 2u = \dfrac{2 \tan u}{1 - \tan^2 u}$
- You should be able to reduce the power of a trigonometric function.
  - (a) $\sin^2 u = \dfrac{1 - \cos 2u}{2}$ (b) $\cos^2 u = \dfrac{1 + \cos 2u}{2}$ (c) $\tan^2 u = \dfrac{1 - \cos 2u}{1 + \cos 2u}$
- You should be able to use the half-angle formulas.
  - (a) $\sin \dfrac{u}{2} = \pm\sqrt{\dfrac{1 - \cos u}{2}}$ (b) $\cos \dfrac{u}{2} = \pm\sqrt{\dfrac{1 + \cos u}{2}}$ (c) $\tan \dfrac{u}{2} = \dfrac{1 - \cos u}{\sin u} = \dfrac{\sin u}{1 + \cos u}$
- You should be able to use the product-sum formulas.
  - (a) $\sin u \sin v = \frac{1}{2}[\cos(u - v) - \cos(u + v)]$ (b) $\cos u \cos v = \frac{1}{2}[\cos(u - v) + \cos(u + v)]$
  - (c) $\sin u \cos v = \frac{1}{2}[\sin(u + v) + \sin(u - v)]$ (d) $\cos u \sin v = \frac{1}{2}[\sin(u + v) - \sin(u - v)]$
- You should be able to use the sum-product formulas.
  - (a) $\sin x + \sin y = 2 \sin\left(\frac{x + y}{2}\right) \cos\left(\frac{x - y}{2}\right)$ (b) $\sin x - \sin y = 2 \cos\left(\frac{x + y}{2}\right) \sin\left(\frac{x - y}{2}\right)$
  - (c) $\cos x + \cos y = 2 \cos\left(\frac{x + y}{2}\right) \cos\left(\frac{x - y}{2}\right)$ (d) $\cos x - \cos y = -2 \sin\left(\frac{x + y}{2}\right) \sin\left(\frac{x - y}{2}\right)$

Solutions to Odd-Numbered Exercises

Figure for Exercises 1–7

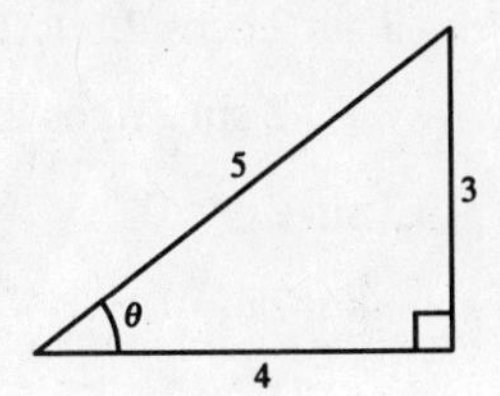

$\sin \theta = \frac{3}{5}$

$\cos \theta = \frac{4}{5}$

$\tan \theta = \frac{3}{4}$

1. $\sin \theta = \frac{3}{5}$

3. $$\begin{aligned} \cos 2\theta &= 2\cos^2 \theta - 1 \\ &= 2\left(\tfrac{4}{5}\right)^2 - 1 \\ &= \tfrac{32}{25} - \tfrac{25}{25} \\ &= \tfrac{7}{25} \end{aligned}$$

5. $$\begin{aligned} \tan 2\theta &= \frac{2\tan\theta}{1 - \tan^2\theta} \\ &= \frac{2(3/4)}{1 - (3/4)^2} \\ &= \frac{3/2}{1 - (9/16)} \\ &= \frac{3}{2} \cdot \frac{16}{7} \\ &= \frac{24}{7} \end{aligned}$$

7. $$\begin{aligned} \csc 2\theta &= \frac{1}{\sin 2\theta} \\ &= \frac{1}{2\sin\theta\cos\theta} \\ &= \frac{1}{2(3/5)(4/5)} \\ &= \frac{25}{24} \end{aligned}$$

9. Solutions: 0, 1.047, 3.142, 5.236

$$\sin 2x - \sin x = 0$$

$$2\sin x\cos x - \sin x = 0$$

$$\sin x(2\cos x - 1) = 0$$

$\sin x = 0$ or $2\cos x - 1 = 0$

$x = 0, \pi$ $\qquad \cos x = \dfrac{1}{2}$

$\qquad x = \dfrac{\pi}{3}, \dfrac{5\pi}{3}$

$$x = 0, \frac{\pi}{3}, \pi, \frac{5\pi}{3}$$

11. Solutions: 0.1263, 1.4445, 3.2679, 4.5860

13. Solutions: 1.047, 3.142, 5.236

$$\cos 2x = -\cos x$$
$$2\cos^2 x - 1 = -\cos x$$
$$2\cos^2 x + \cos x - 1 = 0$$
$$(2\cos x - 1)(\cos x + 1) = 0$$
$$2\cos x = 1 \quad \text{or} \quad \cos x = -1$$
$$\cos x = \frac{1}{2} \qquad x = \pi$$
$$x = \frac{\pi}{3}, \frac{5\pi}{3}$$

15. Solutions: 0, 1.571, 3.142, 4.712

$$\sin 4x = -2\sin 2x$$
$$\sin 4x + 2\sin 2x = 0$$
$$2\sin 2x\cos 2x + 2\sin 2x = 0$$
$$2\sin 2x(\cos 2x + 1) = 0$$
$$2\sin 2x = 0 \quad \text{or} \quad \cos 2x + 1 = 0$$
$$\sin 2x = 0 \qquad \cos 2x = -1$$
$$2x = n\pi \qquad 2x = \pi + 2n\pi$$
$$x = \frac{n}{2}\pi \qquad x = \frac{\pi}{2} + n\pi$$
$$x = 0, \frac{\pi}{2}, \pi, \frac{3\pi}{2} \qquad x = \frac{\pi}{2}, \frac{3\pi}{2}$$

17. $8\sin x\cos x = 4(2\sin x\cos x) = 4\sin 2x$

19. $5 - 10\sin^2 x = 5(1 - 2\sin^2 x) = 5\cos 2x$

21. $\sin u = \frac{3}{5},\ 0 < u < \frac{\pi}{2} \implies \cos u = \frac{4}{5}$

$$\sin 2u = 2\sin u\cos u = 2 \cdot \frac{3}{5} \cdot \frac{4}{5} = \frac{24}{25}$$
$$\cos 2u = \cos^2 u - \sin^2 u = \frac{16}{25} - \frac{9}{25} = \frac{7}{25}$$
$$\tan 2u = \frac{2\tan u}{1 - \tan^2 u} = \frac{2(3/4)}{1 - (9/16)} = \frac{24}{7}$$

23. $\tan u = \frac{1}{2},\ \pi < u < \frac{3\pi}{2} \implies \sin u = -\frac{1}{\sqrt{5}}$ and

$$\cos u = -\frac{2}{\sqrt{5}}$$
$$\sin 2u = 2\sin u\cos u = 2\left(-\frac{1}{\sqrt{5}}\right)\left(-\frac{2}{\sqrt{5}}\right) = \frac{4}{5}$$
$$\cos 2u = \cos^2 u - \sin^2 u = \left(-\frac{2}{\sqrt{5}}\right)^2 - \left(-\frac{1}{\sqrt{5}}\right)^2 = \frac{3}{5}$$
$$\tan 2u = \frac{2\tan u}{1 - \tan^2 u} = \frac{2(1/2)}{1 - (1/4)} = \frac{4}{3}$$

25. $\cos^4 x = (\cos^2 x)(\cos^2 x) = \left(\frac{1 + \cos 2x}{2}\right)\left(\frac{1 + \cos 2x}{2}\right) = \frac{1 + 2\cos 2x + \cos^2 2x}{4}$

$$= \frac{1 + 2\cos 2x + (1 + \cos 4x)/2}{4}$$
$$= \frac{2 + 4\cos 2x + 1 + \cos 4x}{8}$$
$$= \frac{3 + 4\cos 2x + \cos 4x}{8}$$
$$= \frac{1}{8}(3 + 4\cos 2x + \cos 4x)$$

27. $(\sin^2 x)(\cos^2 x) = \left(\frac{1 - \cos 2x}{2}\right)\left(\frac{1 + \cos 2x}{2}\right)$

$= \frac{1 - \cos^2 2x}{4}$

$= \frac{1}{4}\left(1 - \frac{1 + \cos 4x}{2}\right)$

$= \frac{1}{8}(2 - 1 - \cos 4x)$

$= \frac{1}{8}(1 - \cos 4x)$

29. $\sin^2 x \cos^4 x = \sin^2 x \cos^2 x \cos^2 x = \left(\frac{1 - \cos 2x}{2}\right)\left(\frac{1 + \cos 2x}{2}\right)\left(\frac{1 + \cos 2x}{2}\right)$

$= \frac{1}{8}(1 - \cos 2x)(1 + \cos 2x)(1 + \cos 2x)$

$= \frac{1}{8}(1 - \cos^2 2x)(1 + \cos 2x)$

$= \frac{1}{8}(1 + \cos 2x - \cos^2 2x - \cos^3 2x)$

$= \frac{1}{8}\left[1 + \cos 2x - \left(\frac{1 + \cos 4x}{2}\right) - \cos 2x\left(\frac{1 + \cos 4x}{2}\right)\right]$

$= \frac{1}{16}[2 + 2\cos 2x - 1 - \cos 4x - \cos 2x - \cos 2x \cos 4x]$

$= \frac{1}{16}\left[1 + \cos 2x - \cos 4x - \left(\frac{1}{2}\cos 2x + \frac{1}{2}\cos 6x\right)\right]$

$= \frac{1}{32}(2 + 2\cos 2x - 2\cos 4x - \cos 2x - \cos 6x)$

$= \frac{1}{32}(2 + \cos 2x - 2\cos 4x - \cos 6x)$

Figure for Exercises 31 – 35

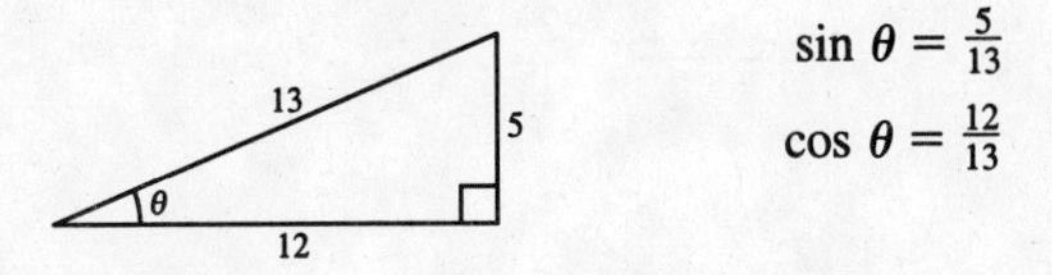

$\sin\theta = \frac{5}{13}$

$\cos\theta = \frac{12}{13}$

31. $\cos\frac{\theta}{2} = \sqrt{\frac{1 + \cos\theta}{2}} = \sqrt{\frac{1 + (12/13)}{2}} = \sqrt{\frac{25}{26}} = \frac{5}{\sqrt{26}} = \frac{5\sqrt{26}}{26}$

33. $\tan\frac{\theta}{2} = \frac{\sin\theta}{1 + \cos\theta} = \frac{5/13}{1 + (12/13)} = \frac{5}{25} = \frac{1}{5}$

35. $\csc\frac{\theta}{2} = \frac{1}{\sin(\theta/2)} = \frac{1}{\sqrt{(1 - \cos\theta)/2}} = \frac{1}{\sqrt{\left(1 - \frac{12}{13}\right)/2}} = \frac{1}{\sqrt{1/26}} = \sqrt{26}$

37. $2\sin\frac{\theta}{2}\cos\frac{\theta}{2} = \sin\theta = \frac{5}{13}$

39. $\sin 15° = \sin\left(\frac{1}{2}\cdot 30°\right) = \sqrt{\frac{1-\cos 30°}{2}} = \sqrt{\frac{1-\sqrt{3}/2}{2}} = \frac{1}{2}\sqrt{2-\sqrt{3}}$

$\cos 15° = \cos\left(\frac{1}{2}\cdot 30°\right) = \sqrt{\frac{1+\cos 30°}{2}} = \sqrt{\frac{1+\sqrt{3}/2}{2}} = \frac{1}{2}\sqrt{2+\sqrt{3}}$

$\tan 15° \tan\left(\frac{1}{2}\cdot 30°\right) = \frac{\sin 30°}{1+\cos 30°} = \frac{\frac{1}{2}}{1+\frac{\sqrt{3}}{2}} = \frac{1}{2+\sqrt{3}} = 2-\sqrt{3}$

41. $\sin 112° 30' = \sin\left(\frac{1}{2}\cdot 225°\right) = \sqrt{\frac{1-\cos 225°}{2}} = \sqrt{\frac{1+(\sqrt{2}/2)}{2}} = \frac{1}{2}\sqrt{2+\sqrt{2}}$

$\cos 112° 30' = \cos\left(\frac{1}{2}\cdot 225°\right) = -\sqrt{\frac{1+\cos 225°}{2}} = -\sqrt{\frac{1-(\sqrt{2}/2)}{2}} = -\frac{1}{2}\sqrt{2-\sqrt{2}}$

$\tan 112° 30' = \tan\left(\frac{1}{2}\cdot 225°\right) = \frac{\sin 225°}{1+\cos 225°} = \frac{-\sqrt{2}/2}{1-(\sqrt{2}/2)} = -1-\sqrt{2}$

43. $\sin\frac{\pi}{8} = \sin\left[\frac{1}{2}\left(\frac{\pi}{4}\right)\right] = \sqrt{\frac{1-\cos(\pi/4)}{2}} = \frac{1}{2}\sqrt{2-\sqrt{2}}$

$\cos\frac{\pi}{8} = \cos\left[\frac{1}{2}\left(\frac{\pi}{4}\right)\right] = \sqrt{\frac{1+\cos(\pi/4)}{2}} = \frac{1}{2}\sqrt{2+\sqrt{2}}$

$\tan\frac{\pi}{8} = \tan\left[\frac{1}{2}\left(\frac{\pi}{4}\right)\right] = \frac{\sin(\pi/4)}{1+\cos(\pi/4)} = \frac{\sqrt{2}/2}{1+(\sqrt{2}/2)} = \sqrt{2}-1$

45. $\sin\frac{3\pi}{8} = \sin\left(\frac{1}{2}\cdot\frac{3\pi}{4}\right) = \sqrt{\frac{1-\cos(3\pi/4)}{2}} = \sqrt{\frac{1+\sqrt{2}/2}{2}} = \frac{1}{2}\sqrt{2+\sqrt{2}}$

$\cos\frac{3\pi}{8} = \cos\left(\frac{1}{2}\cdot\frac{3\pi}{4}\right) = \sqrt{\frac{1+\cos 3\pi/4}{2}} = \sqrt{\frac{1-\sqrt{2}/2}{2}} = \frac{1}{2}\sqrt{2-\sqrt{2}}$

$\tan\frac{3\pi}{8} = \tan\left(\frac{1}{2}\cdot\frac{3\pi}{4}\right) = \frac{\sin\frac{3\pi}{4}}{1+\cos\frac{3\pi}{4}} = \frac{\frac{\sqrt{2}}{2}}{1-\frac{\sqrt{2}}{2}} = \frac{\sqrt{2}}{2-\sqrt{2}} = \sqrt{2}+1$

47. $\sin u = \frac{5}{13},\ \frac{\pi}{2} < u < \pi \implies \cos u = -\frac{12}{13}$

$\sin\left(\frac{u}{2}\right) = \sqrt{\frac{1-\cos u}{2}} = \sqrt{\frac{1+(12/13)}{2}} = \frac{5\sqrt{26}}{26}$

$\cos\left(\frac{u}{2}\right) = \sqrt{\frac{1+\cos u}{2}} = \sqrt{\frac{1-(12/13)}{2}} = \frac{\sqrt{26}}{26}$

$\tan\left(\frac{u}{2}\right) = \frac{\sin u}{1+\cos u} = \frac{5/13}{1-(12/13)} = \frac{5}{1} = 5$

49. $\tan u = -\frac{8}{5}, \frac{3\pi}{2} < u < 2\pi$. Quadrant IV

$\sin u = -\frac{8}{\sqrt{89}}, \cos u = \frac{5}{\sqrt{89}}$

$$\sin\left(\frac{u}{2}\right) = \sqrt{\frac{1-\cos u}{2}} = \sqrt{\frac{1-(5/\sqrt{89})}{2}}$$

$$= \sqrt{\frac{\sqrt{89}-5}{2\sqrt{89}}} = \sqrt{\frac{89-5\sqrt{89}}{178}}$$

$$\cos\left(\frac{u}{2}\right) = -\sqrt{\frac{1+\cos u}{2}} = -\sqrt{\frac{1+(5/\sqrt{89})}{2}} = -\sqrt{\frac{\sqrt{89}+5}{2\sqrt{89}}} = -\sqrt{\frac{89+5\sqrt{89}}{178}}$$

$$\tan\left(\frac{u}{2}\right) = \frac{1-\cos u}{\sin u} = \frac{1-\frac{5}{\sqrt{89}}}{\frac{-8}{\sqrt{89}}} = \frac{5-\sqrt{89}}{8}$$

y
x
u
−8
5
√89

51. $$\sqrt{\frac{1-\cos 6x}{2}} = |\sin 3x|$$

53. $$-\sqrt{\frac{1-\cos 8x}{1+\cos 8x}} = -\frac{\sqrt{(1-\cos 8x)/2}}{\sqrt{(1+\cos 8x)/2}}$$

$$= -\frac{|\sin 4x|}{|\cos 4x|}$$

$$= -|\tan 4x|$$

55. $$\sin\frac{x}{2} - \cos x = 0$$

$$\pm\sqrt{\frac{1-\cos x}{2}} = \cos x$$

$$\frac{1-\cos x}{2} = \cos^2 x$$

$$0 = 2\cos^2 x + \cos x - 1$$

$$= (2\cos x - 1)(\cos x + 1)$$

$\cos x = \frac{1}{2}$ or $\cos x = -1$

$x = \frac{\pi}{3}, \frac{5\pi}{3}$ $\qquad x = \pi$

By checking these values in the original equations, we see that $x = \pi/3$ and $x = 5\pi/3$ are the only solutions. $x = \pi$ is extraneous.

57. $$\cos\frac{x}{2} - \sin x = 0$$

$$\pm\sqrt{\frac{1+\cos x}{2}} = \sin x$$

$$\frac{1+\cos x}{2} = \sin^2 x$$

$$1 + \cos x = 2\sin^2 x$$

$$1 + \cos x = 2 - 2\cos^2 x$$

$$2\cos^2 x + \cos x - 1 = 0$$

$$(2\cos x - 1)(\cos x + 1) = 0$$

$2\cos x - 1 = 0$ or $\cos x + 1 = 0$

$\cos x = \frac{1}{2}$ $\qquad \cos x = -1$

$x = \frac{\pi}{3}, \frac{5\pi}{3}$ $\qquad x = \pi$

$x = \frac{\pi}{3}, \pi, \frac{5\pi}{3}$

$\pi/3$, π, and $5\pi/3$ are all solutions to the equation.

59. $$6\sin\frac{\pi}{3}\cos\frac{\pi}{3} = 6\cdot\frac{1}{2}\left[\sin\left(\frac{\pi}{3}+\frac{\pi}{3}\right) + \sin\left(\frac{\pi}{3}-\frac{\pi}{3}\right)\right]$$

$$= 3\left[\sin\frac{2\pi}{3} + \sin 0\right] = 3\sin\frac{2\pi}{3}$$

61. $\sin 5\theta \cos 3\theta = \frac{1}{2}[\sin(5\theta + 3\theta) + \sin(5\theta - 3\theta) = \frac{1}{2}(\sin 8\theta + \sin 2\theta)$

63. $5\cos(-5\beta)\cos 3\beta = 5 \cdot \frac{1}{2}[\cos(-5\beta - 3\beta) + \cos(-5\beta + 3\beta)] = \frac{5}{2}[\cos(-8\beta) + \cos(-2\beta)]$

$= \frac{5}{2}(\cos 8\beta + \cos 2\beta)$

65. $\sin 60° + \sin 30° = 2\sin\left(\frac{60° + 30°}{2}\right)\cos\left(\frac{60° - 30°}{2}\right) = 2\sin 45° \cos 15°$

67. $\sin 5\theta - \sin\theta = 2\cos\left(\frac{5\theta + \theta}{2}\right)\sin\left(\frac{5\theta - \theta}{2}\right) = 2\cos 3\theta \cdot \sin 2\theta$

69. $\sin(\alpha + \beta) - \sin(\alpha - \beta) = 2\cos\left(\frac{\alpha + \beta + \alpha - \beta}{2}\right)\sin\left(\frac{\alpha + \beta - \alpha + \beta}{2}\right) = 2\cos\alpha\sin\beta$

71. $\cos\left(\theta + \frac{\pi}{2}\right) - \cos\left(\theta - \frac{\pi}{2}\right) = -2\sin\left(\frac{\theta + (\pi/2) + \theta - (\pi/2)}{2}\right)\sin\left(\frac{\theta + (\pi/2) - \theta + (\pi/2)}{2}\right)$

$= -2\sin\theta\sin\frac{\pi}{2} = -2\sin\theta$

73. $\sin 6x + \sin 2x = 0$

$2\sin\left(\frac{6x + 2x}{2}\right)\cos\left(\frac{6x - 2x}{2}\right) = 0$

$\sin 4x \cos 2x = 0$

$\sin 4x = 0$ or $\cos 2x = 0$

$4x = n\pi \qquad 2x = \frac{\pi}{2} + n\pi$

$x = \frac{n\pi}{4} \qquad x = \frac{\pi}{4} + \frac{n\pi}{2}$

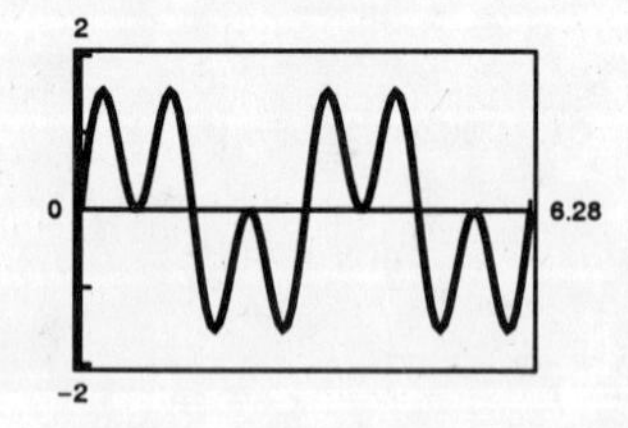

In the interval we have

$x = 0, \frac{\pi}{4}, \frac{\pi}{2}, \frac{3\pi}{4}, \pi, \frac{5\pi}{4}, \frac{3\pi}{2}, \frac{7\pi}{4}.$

75. $\frac{\cos 2x}{\sin 3x - \sin x} - 1 = 0$

$\frac{\cos 2x}{\sin 3x - \sin x} = 1$

$\frac{\cos 2x}{2\cos 2x \sin x} = 1$

$2\sin x = 1$

$\sin x = \frac{1}{2}$

$x = \frac{\pi}{6}, \frac{5\pi}{6}$

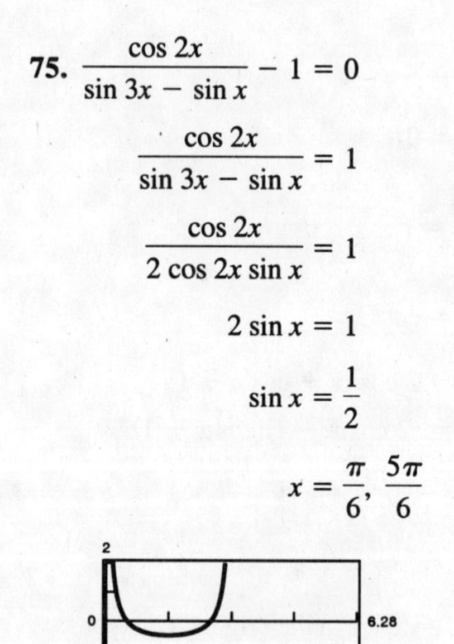

77. $\sin^2\alpha = \left(\frac{5}{13}\right)^2 = \frac{25}{169}$

$\sin^2\alpha = 1 - \cos^2\alpha = 1 - \left(\frac{12}{13}\right)^2 = 1 - \frac{144}{169} = \frac{25}{169}$

79. $\sin \alpha \cos \beta = \left(\frac{5}{13}\right)\left(\frac{4}{5}\right) = \frac{4}{13}$

$\sin \alpha \cos \beta = \cos\left(\frac{\pi}{2} - \alpha\right) \sin\left(\frac{\pi}{2} - \beta\right) = \left(\frac{5}{13}\right)\left(\frac{4}{5}\right) = \frac{4}{13}$

81. $\csc 2\theta = \frac{1}{\sin 2\theta}$

$= \frac{1}{2 \sin \theta \cos \theta}$

$= \frac{1}{\sin \theta} \cdot \frac{1}{2 \cos \theta}$

$= \frac{\csc \theta}{2 \cos \theta}$

83. $\cos^2 2\alpha - \sin^2 2\alpha = \cos[2(2\alpha)]$

$= \cos 4\alpha$

85. $(\sin x + \cos x)^2 = \sin^2 x + 2 \sin x \cos x + \cos^2 x$

$= (\sin^2 x + \cos^2 x) + 2 \sin x \cos x$

$= 1 + \sin 2x$

87. $\sec \frac{u}{2} = \pm \frac{1}{\cos (u/2)}$

$= \pm \sqrt{\frac{2}{1 + \cos u}}$

$= \pm \sqrt{\frac{2 \sin u}{\sin u(1 + \cos u)}}$

$= \pm \sqrt{\frac{2 \sin u}{\sin u + \sin u \cos u}}$

$= \pm \sqrt{\frac{(2 \sin u)/(\cos u)}{(\sin u)/(\cos u) + (\sin u \cos u)/(\cos u)}}$

$= \pm \sqrt{\frac{2 \tan u}{\tan u + \sin u}}$

89. $\cos 3\beta = \cos(2\beta + \beta)$

$= \cos 2\beta \cos \beta - \sin 2\beta \sin \beta$

$= (\cos^2 \beta - \sin^2 \beta) \cos \beta - 2 \sin \beta \cos \beta \sin \beta$

$= \cos^3 \beta - \sin^2 \beta \cos \beta - 2 \sin^2 \beta \cos \beta$

$= \cos^3 \beta - 3 \sin^2 \beta \cos \beta$

91. $\frac{\cos 4x - \cos 2x}{2 \sin 3x} = \frac{-2 \sin\left(\frac{4x + 2x}{2}\right) \sin\left(\frac{4x - 2x}{2}\right)}{2 \sin 3x}$

$= \frac{-2 \sin 3x \sin x}{2 \sin 3x}$

$= -\sin x$

93. $\sin^2 x = \frac{1 - \cos 2x}{2} = \frac{1}{2} - \frac{\cos 2x}{2}$

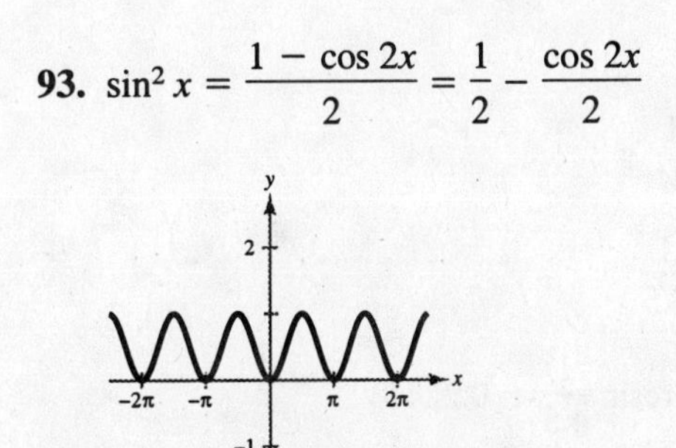

95. (a) $y = 4\sin\dfrac{x}{2} + \cos x$

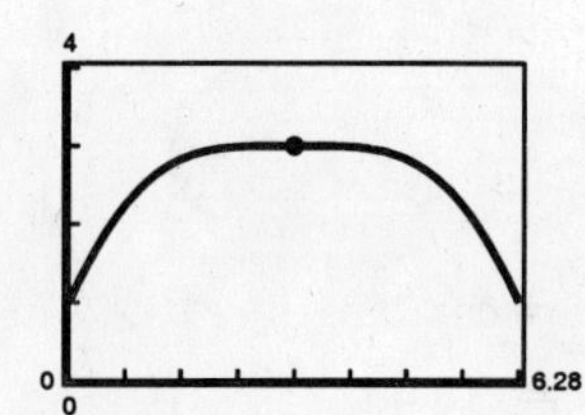

Maximum: $(\pi, 3)$

(b)
$$2\cos\frac{x}{2} - \sin x = 0$$
$$2\left(\pm\sqrt{\frac{1+\cos x}{2}}\right) = \sin x$$
$$4\left(\frac{1+\cos x}{2}\right) = \sin^2 x$$
$$2(1+\cos x) = 1 - \cos^2 x$$
$$\cos^2 x + 2\cos x + 1 = 0$$
$$(\cos x + 1)^2 = 0$$
$$\cos x = -1$$
$$x = \pi$$

97. (a) $y = \cos\dfrac{x}{2} + \sin 2x$

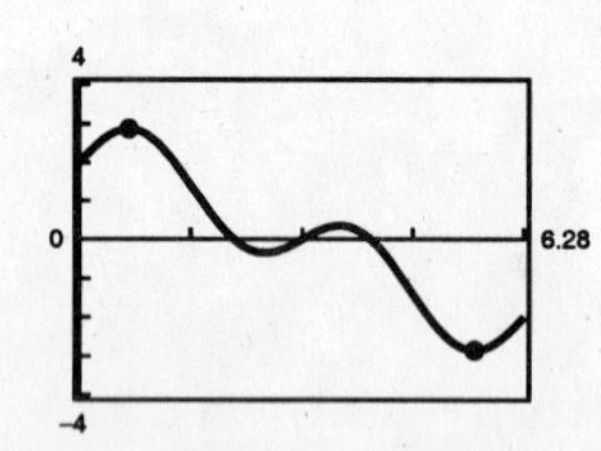

Maximum: $(0.699, 2.864)$

Minimum: $(5.584, -2.864)$

(b) $2\cos 2x - \sin\dfrac{x}{2} = 0$ has 4 zeros on $[0, 2\pi)$. Two of the zeros are $x = 0.699$ and $x = 5.584$. (The other two are 2.608, 3.675)

99. $\sin(2\arcsin x) = 2\sin(\arcsin x)\cos(\arcsin x) = 2x\sqrt{1-x^2}$

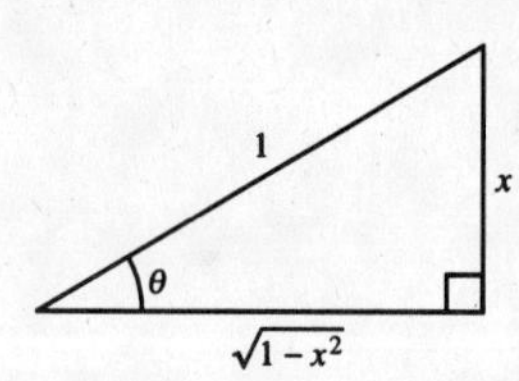

101. $\cos(2\arcsin x) = 1 - 2\sin^2(\arcsin x)$
$$= 1 - 2x^2$$

103.
$$r = \tfrac{1}{32}v_0 \sin 2\theta$$
$$= \tfrac{1}{32}v_0(2\sin\theta\cos\theta)$$
$$= \tfrac{1}{16}v_0\sin\theta\cos\theta$$

105.
$$\sin\frac{\theta}{2} = \frac{1}{M}$$
$$\sin\frac{\theta}{2} = \frac{1}{4.5}$$
$$\frac{\theta}{2} = \arcsin\frac{1}{4.5} \approx 0.2241$$
$$\theta \approx 0.4482 \text{ or } 25.68^\circ$$

107. False. If $x = \pi$, $\sin\dfrac{x}{2} = \sin\dfrac{\pi}{2} = 1$, whereas
$$-\sqrt{\frac{1-\cos\pi}{2}} = -1$$

109. $f(x) = 2\sin x\left[2\cos^2\left(\frac{x}{2}\right) - 1\right]$

(a)

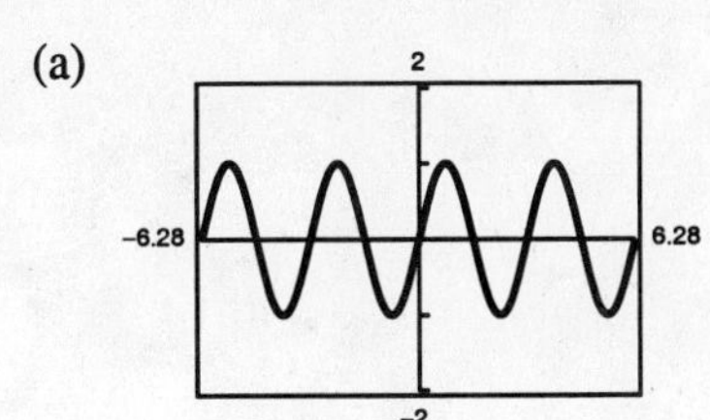

(b) The graph appears to be that of $y = \sin 2x$.

(c) $2\sin x\left[2\cos^2\left(\frac{x}{2}\right) - 1\right] = 2\sin x\left[2\frac{1+\cos x}{2} - 1\right]$

$= 2\sin x\,[\cos x]$

$= \sin 2x$

111. (a) Complement: $\frac{\pi}{2} - \frac{\pi}{18} = \frac{8\pi}{18} = \frac{4\pi}{9}$

Supplement: $\pi - \frac{\pi}{18} = \frac{17\pi}{18}$

(b) Complement: $\frac{\pi}{2} - \frac{9\pi}{20} = \frac{\pi}{20}$

Supplement: $\pi - \frac{9\pi}{20} = \frac{11\pi}{20}$

113. $f(x) = \frac{5}{2}\sin\frac{x}{2}$

Period: 4π

Amplitude: $\frac{5}{2}$

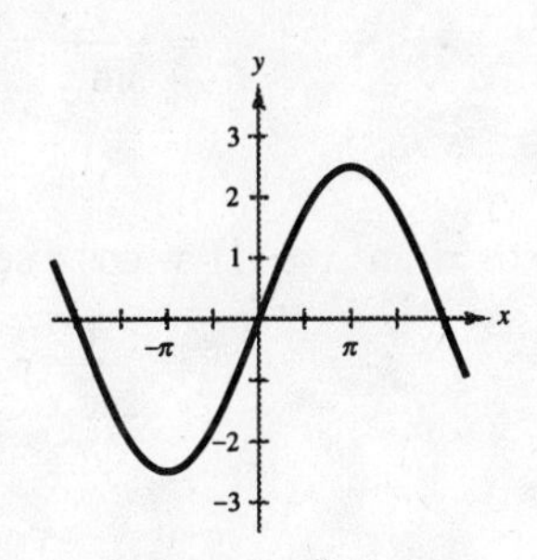

115. $f(x) = \frac{1}{4}\sec\frac{\pi x}{4}$

Period: $\frac{2\pi}{\frac{\pi}{2}} = 4$

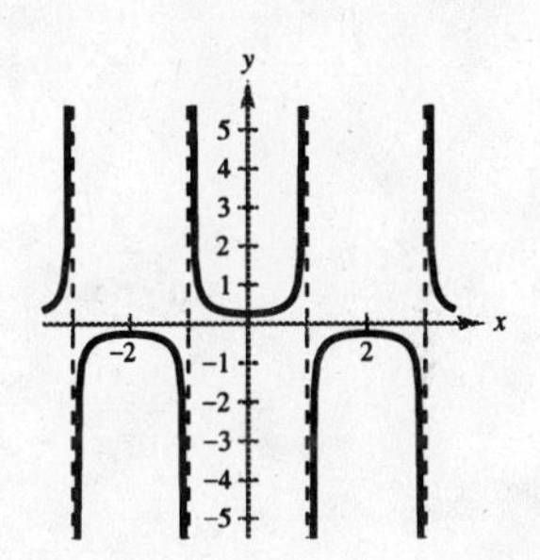

Review Exercises for Chapter 5

Solutions to Odd-Numbered Exercises

1. $\frac{1}{\cos x} = \sec x$

3. $\frac{1}{\sec x} = \cos x$

5. $\frac{\cos x}{\sin x} = \cot x$

7. $\csc\left(\frac{\pi}{2} - x\right) = \sec x$

9. $\sec(-x) = \sec x$

11. $\sin x = \frac{3}{5}$, $\cos x = \frac{4}{5}$, Quadrant I

$\tan x = \frac{\sin x}{\cos x} = \frac{\frac{3}{5}}{\frac{4}{5}} = \frac{3}{4}$

$\cot x = \frac{4}{3}$

$\csc x = \frac{5}{3}$

$\sec x = \frac{5}{4}$

13. $\sin\left(\frac{\pi}{2} - x\right) = \cos x = \frac{1}{\sqrt{2}} = \frac{\sqrt{2}}{2}$,

$\sin x = -\frac{1}{\sqrt{2}} = -\frac{\sqrt{2}}{2}$. Quadrant IV

$\tan x = -1$

$\cot x = -1$

$\sec x = \sqrt{2}$

$\csc x = -\sqrt{2}$

15. $\dfrac{1}{\cot^2 x + 1} = \dfrac{1}{\csc^2 x} = \sin^2 x$

17. $\dfrac{\sin^2 \alpha - \cos^2 \alpha}{\sin^2 \alpha - \sin \alpha \cos \alpha} = \dfrac{(\sin \alpha + \cos \alpha)(\sin \alpha - \cos \alpha)}{\sin \alpha(\sin \alpha - \cos \alpha)}$

$= \dfrac{\sin \alpha + \cos \alpha}{\sin \alpha}$

$= 1 + \cot \alpha$

19. $\tan^2 \theta (\csc^2 \theta - 1) = \tan^2 \theta(\cot^2 \theta)$

$= \tan^2 \theta\left(\dfrac{1}{\tan^2 \theta}\right)$

$= 1$

21. $\tan\left(\dfrac{\pi}{2} - x\right)\sec x = \cot x \cdot \sec x$

$= \dfrac{\cos x}{\sin x} \cdot \dfrac{1}{\cos x}$

$= \dfrac{1}{\sin x} = \csc x$

23. $\sin^{-1/2} \cos x = \dfrac{\cos x}{\sin^{1/2} x}$

$= \dfrac{\cos x}{\sqrt{\sin x}} \cdot \dfrac{\sqrt{\sin x}}{\sqrt{\sin x}}$

$= \dfrac{\cos x}{\sin x}\sqrt{\sin x}$

$= \cot x\sqrt{\sin x}$

25. $\cos x(\tan^2 x + 1) = \cos x \sec^2 x$

$= \dfrac{1}{\sec x}\sec^2 x$

$= \sec x$

27. $\sin^3 \theta + \sin \theta \cos^2 \theta = \sin \theta(\sin^2 \theta + \cos^2 \theta)$

$= \sin \theta$

29. $\sin^5 x \cos^2 x = \sin^4 x \cos^2 x \sin x$

$= (1 - \cos^2 x)^2 \cos^2 x \sin x$

$= (1 - 2\cos^2 x + \cos^4 x) \cos^2 x \sin x$

$= (\cos^2 x - 2\cos^4 x + \cos^6 x) \sin x$

31. $\sqrt{\dfrac{1 - \sin \theta}{1 + \sin \theta}} = \sqrt{\dfrac{1 - \sin \theta}{1 + \sin \theta} \cdot \dfrac{1 - \sin \theta}{1 - \sin \theta}}$

$= \sqrt{\dfrac{(1 - \sin \theta)^2}{1 - \sin^2 \theta}} = \sqrt{\dfrac{(1 - \sin \theta)^2}{\cos^2 \theta}} = \dfrac{|1 - \sin \theta|}{|\cos \theta|} = \dfrac{1 - \sin \theta}{|\cos \theta|}$

Note: We can drop the absolute value on $1 - \sin \theta$ since it is always nonnegative.

33. $\dfrac{\csc(-x)}{\sec(-x)} = -\dfrac{\csc x}{\sec x}$

$= -\dfrac{\cos x}{\sin x} = -\cot x$

35. $\sin^2 x + \sin^2\left(\dfrac{\pi}{2} - x\right) = \sin^2 x + \cos^2 x = 1$

37. $2\cos^2 4x - 1 = 0$

(a) $x = \frac{\pi}{16}$: $2\cos^2\left(\frac{4\pi}{16}\right) - 1 = 2\left(\frac{\sqrt{2}}{2}\right)^2 - 1 = 1 - 1 = 0$

(b) $x = \frac{3\pi}{16}$: $2\cos^2\left(4 \cdot \frac{3\pi}{16}\right) - 1 = 2\left(\frac{\sqrt{2}}{2}\right)^2 - 1 = 1 - 1 = 0$

39. $2\sin x - 1 = 0$

$$\sin x = \frac{1}{2}$$

$$x = \frac{\pi}{6}, \frac{5\pi}{6}$$

41. $\sin x = \sqrt{3} - \sin x$

$$2\sin x = \sqrt{3}$$

$$\sin x = \frac{\sqrt{3}}{2}$$

$$x = \frac{\pi}{3}, \frac{2\pi}{3}$$

43. $3\sqrt{3}\tan x = 3$

$$\tan x = \frac{1}{\sqrt{3}}$$

$$x = \frac{\pi}{6}, \frac{7\pi}{6}$$

45. $3\csc^2 x = 4$

$$\csc^2 x = \frac{4}{3}$$

$$\sin^2 x = \frac{3}{4}$$

$$\sin x = \pm\frac{\sqrt{3}}{2}$$

$$x = \frac{\pi}{3}, \frac{2\pi}{3}, \frac{4\pi}{3}, \frac{5\pi}{3}$$

47. $4\cos^2 x - 3 = 0$

$$\cos^2 x = \frac{3}{4}$$

$$\cos x = \pm\frac{\sqrt{3}}{2}$$

$$x = \frac{\pi}{6}, \frac{5\pi}{6}, \frac{7\pi}{6}, \frac{11\pi}{6}$$

49. $\sin x - \tan x = 0$

$$\sin x - \frac{\sin x}{\cos x} = 0$$

$$\sin x \cos x - \sin x = 0$$

$$\sin x(\cos x - 1) = 0$$

$\sin x = 0$ or $\cos x - 1 = 0$

$x = 0, \pi$ $\qquad$ $\cos x = 1$

$x = 0$

51. $2\cos^2 x - \cos x - 1 = 0$

$$(2\cos x + 1)(\cos x - 1) = 0$$

$2\cos x + 1 = 0$ or $\cos x - 1 = 0$

$\cos x = -\frac{1}{2}$ $\qquad$ $\cos x = 1$

$x = \frac{2\pi}{3}, \frac{4\pi}{3}$ $\qquad$ $x = 0$

53. $\cos^2 x + \sin x = 1$

$$1 - \sin^2 x + \sin x = 1$$

$$\sin x(\sin x - 1) = 0$$

$\sin x = 0$ or $\sin x = 1$

$x = 0, \pi$ $\qquad$ $x = \frac{\pi}{2}$

55. $2\sin 2x = \sqrt{2}$

$$\sin 2x = \frac{\sqrt{2}}{2}$$

$$2x = \frac{\pi}{4}, \frac{3\pi}{4}, \frac{9\pi}{4}, \frac{11\pi}{4}$$

$$x = \frac{\pi}{8}, \frac{3\pi}{8}, \frac{9\pi}{8}, \frac{11\pi}{8}$$

57. $\cos 4x(\cos x - 1) = 0$

$\cos 4x = 0$ or $\cos x - 1 = 0$

$4x = \frac{\pi}{2}, \frac{3\pi}{2}, \frac{5\pi}{2}, \frac{7\pi}{2}, \frac{9\pi}{2}, \frac{11\pi}{2}, \frac{13\pi}{2}, \frac{15\pi}{2},$

or $\cos x = 1$

$x = \frac{\pi}{8}, \frac{3\pi}{8}, \frac{5\pi}{8}, \frac{7\pi}{8}, \frac{9\pi}{8}, \frac{11\pi}{8}, \frac{13\pi}{8}, \frac{15\pi}{8}, 0$

59

$$\cos 4x - 7\cos 2x = 8$$
$$2\cos^2 2x - 1 - 7\cos 2x = 8$$
$$2\cos^2 2x - 7\cos 2x - 9 = 0$$
$$(2\cos 2x - 9)(\cos 2x + 1) = 0$$

$2\cos 2x - 9 = 0$ or $\cos 2x + 1 = 0$

$\cos 2x = \frac{9}{2}$ $\quad$ $\cos 2x = -1$

No solution $\quad$ $2x = \pi + 2n\pi$

$x = \frac{\pi}{2} + n\pi$

$x = \frac{\pi}{2}, \frac{3\pi}{2}$

61. $\sin^2 x - 2\sin x = 0$

$\sin x(\sin x - 2) = 0$

$\sin x = 0$ or $\sin x = 2$ (impossible)

$x = 0, \pi$

63. $\tan^2\theta + \tan\theta - 12 = 0$

$(\tan\theta + 4)(\tan\theta - 3) = 0$

$\tan\theta = -4$ or $\tan\theta = 3$

$\theta = 1.8158, 4.9574, 1.2490, 4.3906$

65. $\sin 285° = \sin(225° + 60°) = \sin 225° \cos 60° + \cos 225° \sin 60°$

$$= -\frac{\sqrt{2}}{2}\cdot\frac{1}{2} - \frac{\sqrt{2}}{2}\cdot\frac{\sqrt{3}}{2} = -\frac{\sqrt{2} - \sqrt{6}}{4}$$

$\cos 285° = \cos(225° + 60°) = \cos 225° \cos 60° - \sin 225° \sin 60°$

$$= -\frac{\sqrt{2}}{2}\frac{1}{2} + \frac{\sqrt{2}}{2}\frac{\sqrt{3}}{2} = \frac{\sqrt{6} - \sqrt{2}}{4}$$

$$\tan 285° = \frac{\sin 285°}{\cos 285°} = -\frac{\sqrt{2} - \sqrt{6}}{\sqrt{6} - \sqrt{2}} = \frac{\sqrt{6} + \sqrt{2}}{\sqrt{2} - \sqrt{6}} = -2 - \sqrt{3}$$

67. $\sin\frac{5\pi}{12} = \sin\left(\frac{2\pi}{3} - \frac{\pi}{4}\right) = \sin\frac{2\pi}{3}\cos\frac{\pi}{4} - \cos\frac{2\pi}{3}\sin\frac{\pi}{4}$

$$= \frac{\sqrt{3}}{2}\cdot\frac{\sqrt{2}}{2} + \frac{1}{2}\frac{\sqrt{2}}{2} = \frac{\sqrt{6} + \sqrt{2}}{4}$$

$\cos\frac{5\pi}{12} = \cos\left(\frac{2\pi}{3} - \frac{\pi}{4}\right) = \cos\frac{2\pi}{3}\cos\frac{\pi}{4} + \sin\frac{2\pi}{3}\sin\frac{\pi}{4}$

$$= -\frac{1}{2}\frac{\sqrt{2}}{2} + \frac{\sqrt{3}}{2}\frac{\sqrt{2}}{2} = \frac{\sqrt{6} - \sqrt{2}}{4}$$

$$\tan\frac{5\pi}{12} = \frac{\sin\frac{5\pi}{12}}{\cos\frac{5\pi}{12}} = \frac{\sqrt{6} + \sqrt{2}}{\sqrt{6} - \sqrt{2}} = 2 + \sqrt{3}$$

69. $\sin 140° \cos 50° + \cos 140° \sin 50° = \sin(140° + 50°) = \sin 190°$

71. $\frac{\tan 25° + \tan 10°}{1 - \tan 25° \tan 10°} = \tan(25° + 10°) = \tan 35°$

For Exercises 73–77:

$$\sin u = \frac{3}{4},\ u \text{ in Quadrant II} \implies \cos u = -\frac{\sqrt{7}}{4}$$

$$\cos v = -\frac{5}{13},\ v \text{ in Quadrant II} \implies \sin v = \frac{12}{13}$$

73. $\sin(u + v) = \sin u \cos v + \cos u \sin v$

$$= \left(\frac{3}{4}\right)\left(-\frac{5}{13}\right) + \left(-\frac{\sqrt{7}}{4}\right)\left(\frac{12}{13}\right)$$
$$= -\frac{15}{52} - \frac{12\sqrt{7}}{52}$$
$$= \frac{-3(5 + 4\sqrt{7})}{52}$$

75. $\cos(u - v) = \cos u \cos v + \sin u \sin v$

$$= \left(-\frac{\sqrt{7}}{4}\right)\left(-\frac{5}{13}\right) + \left(\frac{3}{4}\right)\left(\frac{12}{13}\right)$$
$$= \frac{5\sqrt{7} + 36}{52}$$

77. $\cos(u + v) = \cos u \cos v - \sin u \sin u$

$$= \left(-\frac{\sqrt{7}}{4}\right)\left(-\frac{5}{13}\right) - \left(\frac{3}{4}\right)\left(\frac{12}{13}\right)$$
$$= \frac{5\sqrt{7} - 36}{52}$$

79. $\cos\left(x + \frac{\pi}{2}\right) = \cos x \cos \frac{\pi}{2} - \sin x \sin \frac{\pi}{2}$

$$= (\cos x)(0) - (\sin x)(1)$$
$$= -\sin x$$

81. $\cot\left(\frac{\pi}{2} - x\right) = \frac{\cos[(\pi/2) - x]}{\sin[(\pi/2) - x]}$

$$= \frac{\cos(\pi/2) \cos x + \sin(\pi/2) \sin x}{\sin(\pi/2) \cos x - \sin x \cos(\pi/2)}$$
$$= \frac{\sin x}{\cos x}$$
$$= \tan x$$

83. $\cos 3x = \cos(2x + x)$

$$= \cos 2x \cos x - \sin 2x \sin x$$
$$= (\cos^2 x - \sin^2 x) \cos x - 2 \sin x \cos x \sin x$$
$$= \cos^3 x - 3 \sin^2 x \cos x$$
$$= \cos^3 x - 3 \cos x(1 - \cos^2 x)$$
$$= \cos^3 x - 3 \cos x + 3 \cos^3 x$$
$$= 4 \cos^3 x - 3 \cos x$$

85. $\sin\left(x + \frac{\pi}{2}\right) - \sin\left(x - \frac{\pi}{2}\right) = \sqrt{3}$

$$\cos x + \cos x = \sqrt{3}$$
$$\cos x = \frac{\sqrt{3}}{2}$$
$$x = \frac{\pi}{6}, \frac{11\pi}{6}$$

87. $6 \sin x \cos x = 3[2 \sin x \cos x] = 3 \sin 2x$

89. $1 - 4 \sin^2 x \cos^2 x = 1 - (2 \sin x \cos x)^2$

$$= 1 - \sin^2 2x = \cos^2 2x$$

91. $\frac{\sin 2x}{\cos^2 x - \sin^2 x} = \frac{\sin 2x}{\cos 2x} = \tan 2x$

93. $\sin u = -\frac{5}{7},\ \pi < u < \frac{3\pi}{2}$. Quadrant III

$$\cos^2 u = 1 - \left(-\frac{5}{7}\right)^2 = \frac{24}{49} \Rightarrow \cos u = -\frac{2\sqrt{6}}{7}$$

$$\sin 2u = 2\sin u \cos u = 2\left(-\frac{5}{7}\right)\left(-\frac{2\sqrt{6}}{7}\right) = \frac{20\sqrt{6}}{49}$$

$$\cos 2u = 1 - 2\sin^2 u = 1 - 2\left(-\frac{5}{7}\right)^2 = 1 - \frac{50}{49} = -\frac{1}{49}$$

$$\tan 2u = \frac{\sin 2u}{\cos 2u} = \frac{20\sqrt{6}}{-1} = -20\sqrt{6}$$

95. $r = \frac{1}{32}v_0^2 \sin 2\theta$

$$100 = \frac{1}{32}(80)^2 \sin 2\theta$$

$$\sin 2\theta = 0.5$$

$$2\theta = 30^\circ \quad \text{or} \quad 2\theta = 180^\circ - 30^\circ = 150^\circ$$

$$\theta = 15^\circ \qquad \theta = 75^\circ$$

97. $$\cos^4 x \sin^4 x = \left(\frac{1+\cos 2x}{2}\right)^2\left(\frac{1-\cos 2x}{2}\right)^2$$

$$= \frac{(1 + 2\cos 2x + \cos^2 2x)(1 - 2\cos 2x + \cos^2 2x)}{8}$$

$$= \frac{\left(1 + 2\cos 2x + \frac{1+\cos 4x}{2}\right)\left(1 - 2\cos 2x + \frac{1+\cos 4x}{2}\right)}{8}$$

$$= \frac{(3 + 4\cos 2x + \cos 4x)(3 - 4\cos 2x + \cos 4x)}{16}$$

$$= \frac{1}{128}(\cos 8x - 4\cos 4x + 3)$$

99. $$\sin^4 2x = \frac{1 - \cos^2 4x}{2} = \frac{1}{8}(\cos 8x - 4\cos 4x + 3)$$

101. $67^\circ 30' = 67.5^\circ = \frac{1}{2}(135^\circ)$. Quadrant I

$$\sin(67^\circ 30') = \sin\left(\frac{1}{2}135^\circ\right) = \sqrt{\frac{1 - \cos 135^\circ}{2}} = \sqrt{\frac{1 + \sqrt{2}/2}{2}} = \frac{\sqrt{2+\sqrt{2}}}{2}$$

$$\cos(67^\circ 30') = \cos\left(\frac{1}{2}135^\circ\right) = \sqrt{\frac{1 + \cos 135^\circ}{2}} = \sqrt{\frac{1 - \sqrt{2}/2}{2}} = \frac{\sqrt{2-\sqrt{2}}}{2}$$

$$\tan(67^\circ 30') = \tan\left(\frac{1}{2}135^\circ\right) = \frac{\sin 135^\circ}{1 + \cos 135^\circ} = \frac{\frac{\sqrt{2}}{2}}{1 - \frac{\sqrt{2}}{2}} = \frac{\sqrt{2}}{2 - \sqrt{2}} = \sqrt{2} + 1$$

103. $\frac{11\pi}{12} = \frac{1}{2}\left(\frac{11\pi}{6}\right)$. Quadrant II.

$$\sin\left(\frac{11\pi}{12}\right) = \sin\left(\frac{1}{2}\frac{11\pi}{6}\right) = \sqrt{\frac{1 - \cos 11\pi/6}{2}} = \sqrt{\frac{1 - \sqrt{3}/2}{2}} = \frac{\sqrt{2-\sqrt{3}}}{2}$$

$$\cos\left(\frac{11\pi}{12}\right) = \sin\left(\frac{1}{2}\frac{11\pi}{6}\right) = -\sqrt{\frac{1 + \cos 11\pi/6}{2}} = -\sqrt{\frac{1 + \sqrt{3}/2}{2}} = -\frac{\sqrt{2+\sqrt{3}}}{2}$$

$$\tan\frac{11\pi}{6} = \tan\left(\frac{1}{2}\cdot\frac{11\pi}{6}\right) = \frac{\sin\frac{11\pi}{6}}{1 + \cos\frac{11\pi}{6}} = \frac{-\frac{1}{2}}{1 + \frac{\sqrt{3}}{2}} = \frac{-1}{2 + \sqrt{3}} = -2 + \sqrt{3}$$

105. $\dfrac{\sin 6x}{1+\cos 6x} = \tan\left(\dfrac{6x}{2}\right) = \tan 3x$

107. $\dfrac{\sec x - 1}{\tan x} = \dfrac{(1/\cos x) - 1}{\sin x/\cos x} = \dfrac{1-\cos x}{\sin x} = \tan\dfrac{x}{2}$

109. $\cos 3\theta + \cos 2\theta = 2\cos\left(\dfrac{3\theta+2\theta}{2}\right)\cos\left(\dfrac{3\theta-2\theta}{2}\right)$

$= 2\cos\dfrac{5\theta}{2}\cos\dfrac{\theta}{2}$

111. $\cos\dfrac{3\pi}{4} - \cos\dfrac{\pi}{4} = -2\sin\left(\dfrac{\dfrac{3\pi}{4}+\dfrac{\pi}{4}}{2}\right)\sin\left(\dfrac{\dfrac{3\pi}{4}-\dfrac{\pi}{4}}{2}\right)$

$= -2\sin\left(\dfrac{\pi}{2}\right)\sin\left(\dfrac{\pi}{4}\right) = -\sqrt{2}$

113. $\sin 3\alpha \sin 2\alpha = \frac{1}{2}[\cos(3\alpha - 2\alpha) - \cos(3\alpha + 2\alpha)$

$= \frac{1}{2}(\cos\alpha - \cos 5\alpha)$

115. $6\sin\dfrac{\pi}{4}\cos\dfrac{\pi}{4} = 3\left[2\sin\dfrac{\pi}{4}\cos\dfrac{\pi}{4}\right] = 3\sin\dfrac{\pi}{2} = 3$

117. $y = 1.5\sin 8t - 0.5\cos 8t$

(a) $a = \dfrac{3}{2},\ b = -\dfrac{1}{2},\ B = 8,\ C = \arctan\left(-\dfrac{1/2}{3/2}\right)$

$y = \sqrt{(3/2)^2 + (1/2)^2}\,\sin\left(8t + \arctan -\dfrac{1}{3}\right)$

$y = \dfrac{1}{2}\sqrt{10}\,\sin\left(8t - \arctan\dfrac{1}{3}\right)$

(b)

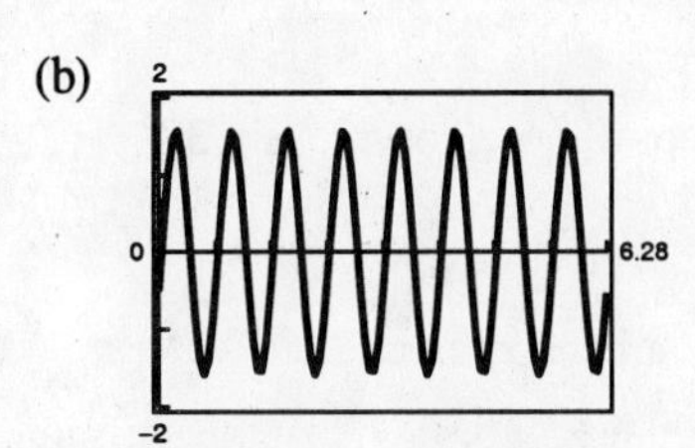

(c) The amplitude is $\dfrac{\sqrt{10}}{2}$.

(d) Frequency $= \dfrac{1}{\text{period}} = \dfrac{4}{\pi}$

119. $\sin(x+y) = \sin x + \sin y$. False.

$\sin(x+y) = \sin x\cos y + \cos x\sin y$

121. $4\sin 45^\circ \cos 15^\circ = 1 + \sqrt{3}$. True.

$4\sin 45^\circ\cos 15^\circ = 4\left(\dfrac{1}{2}[\sin(45^\circ + 15^\circ) + \sin(45^\circ - 15^\circ)]\right)$

$= 2[\sin 60^\circ + \sin 30^\circ]$

$= 2\left[\dfrac{\sqrt{3}}{2} + \dfrac{1}{2}\right]$

$= 2\left(\dfrac{\sqrt{3}+1}{2}\right)$

$= 1 + \sqrt{3}$

123. No. $\cos\theta = \pm\sqrt{1 - \sin^2\theta}$

125. $y_1 = \dfrac{\cos 3x}{\cos x}$

$y_2 = (2\sin x)^2$

From the graphs, $y_3 = -y_2 + 1 = y_1$

Chapter 5 Practice Test

1 Find the value of the other five trigonometric functions, given $\tan x = \frac{4}{11}$, $\sec x < 0$.

2. Simplify $\frac{\sec^2 x + \csc^2 x}{\csc^2 x(1 + \tan^2 x)}$.

3. Rewrite as a single logarithm and simplify $\ln|\tan \theta| - \ln|\cot \theta|$.

4. True or false:

$$\cos\left(\frac{\pi}{2} - x\right) = \frac{1}{\csc x}$$

5. Factor and simplify: $\sin^4 x + (\sin^2 x) \cos^2 x$

6. Multiply and simplify: $(\csc x + 1)(\csc x - 1)$

7. Rationalize the denominator and simplify:

$$\frac{\cos^2 x}{1 - \sin x}$$

8. Verify:

$$\frac{1 + \cos \theta}{\sin \theta} + \frac{\sin \theta}{1 + \cos \theta} = 2 \csc \theta$$

9. Verify:

$$\tan^4 x + 2 \tan^2 x + 1 = \sec^4 x$$

10. Use the sum or difference formulas to determine:

(a) $\sin 105°$ (b) $\tan 15°$

11. Simplify: $(\sin 42°) \cos 38° - (\cos 42°) \sin 38°$

12. Verify $\tan\left(\theta + \frac{\pi}{4}\right) = \frac{1 + \tan \theta}{1 - \tan \theta}$.

13. Write $\sin(\arcsin x - \arccos x)$ as an algebraic expression in x.

14. Use the double-angle formulas to determine:

(a) $\cos 120°$ (b) $\tan 300°$

15. Use the half-angle formulas to determine:

(a) $\sin 22.5°$ (b) $\tan \frac{\pi}{12}$

16. Given $\sin = 4/5$, θ lies in Quadrant II, find $\cos \theta/2$.

17. Use the power-reducing identities to write $(\sin^2 x) \cos^2 x$ in terms of the first power of cosine.

18. Rewrite as a sum: $6(\sin 5\theta) \cos 2\theta$.

19. Rewrite as a product: $\sin(x + \pi) + \sin(x - \pi)$.

20. Verify $\frac{\sin 9x + \sin 5x}{\cos 9x - \cos 5x} = -\cot 2x$.

21. Verify:

$$(\cos u) \sin v = \tfrac{1}{2}[\sin(u + v) - \sin(u - v)].$$

22. Find all solutions in the interval $[0, 2\pi)$:

$$4 \sin^2 x = 1$$

23. Find all solutions in the interval $[0, 2\pi)$:

$$\tan^2 \theta + \left(\sqrt{3} - 1\right) \tan \theta - \sqrt{3} = 0$$

24. Find all solutions in the interval $[0, 2\pi)$:

$$\sin 2x = \cos x$$

25. Use the Quadratic Formula to find all solutions in the interval $[0, 2\pi)$:

$$\tan^2 x - 6 \tan x + 4 = 0$$

CHAPTER 6
Additional Topics in Trigonometry

CHAPTER 6
Additional Topics in Trigonometry

Section 6.1 Law of Sines

- If ABC is any oblique triangle with sides a, b, and c, then the Law of Sines says
$$\frac{a}{\sin A} = \frac{b}{\sin B} = \frac{c}{\sin C}.$$
- You should be able to use the Law of Sines to solve an oblique triangle for the remaining three parts, given:
  (a) Two angles and any side (AAS or ASA)
  (b) Two sides and an angle opposite one of them (SSA)
    1. If A is acute and $h = b \sin A$:
       (a) $a < h$, no triangle is possible.
       (b) $a = h$ or $a > b$, one triangle is possible.
       (c) $h < a < b$, two triangles are possible.
    2. If A is obtuse and $h = b \sin A$:
       (a) $a \le b$, no triangle is possible.
       (b) $a > b$, one triangle is possible.
- The area of any triangle equals one-half the product of the lengths of two sides times the sine of their included angle.
$$A = \tfrac{1}{2}ab \sin C = \tfrac{1}{2}ac \sin B = \tfrac{1}{2}bc \sin A$$

Solutions to Odd-Numbered Exercises

1. Given: $A = 30°, B = 45°, a = 12$

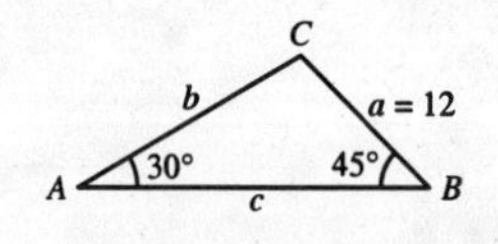

$C = 180° - 30° - 45° = 105°$

$$b = \frac{a}{\sin A}(\sin B) = \frac{12}{\sin 30°}(\sin 45°) = 12\sqrt{2} \approx 16.97$$

$$c = \frac{a}{\sin A}(\sin C) = \frac{12}{\sin 30°}(\sin 105°) \approx 23.18$$

3. Given: $A = 10°, B = 60°, a = 4.5$

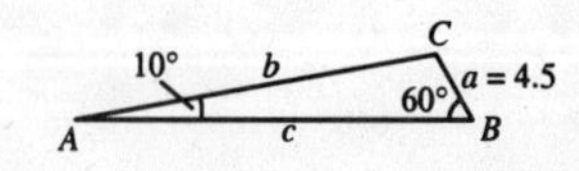

$C = 180° - 10° - 60° = 110°$

$$b = \frac{a}{\sin A}(\sin B) = \frac{4.5}{\sin 10°}(\sin 60°) \approx 22.44$$

$$c = \frac{a}{\sin A}(\sin C) = \frac{4.5}{\sin 10°}(\sin 110°) \approx 24.35$$

5. Given: $A = 36°, a = 10, b = 4$

$$\sin B = \frac{b \sin A}{a} = \frac{4 \sin 36°}{10} \approx 0.2351 \implies B \approx 13.60°$$

$$C = 180° - A - B = 180° - 36° - 13.60° = 130.40°$$

$$c = \frac{a}{\sin A}(\sin C) = \frac{10}{\sin 36°}(\sin 130.40°) \approx 12.96$$

7. Given: $A = 150°,\ C = 20°,\ a = 200$

$$B = 180° - A - C = 180° - 150° - 20° = 10°$$

$$b = \frac{a}{\sin A}(\sin B) = \frac{200}{\sin 150°}(\sin 10°) \approx 69.46$$

$$c = \frac{a}{\sin A}(\sin C) = \frac{200}{\sin 150°}(\sin 20°) \approx 136.81$$

9. Given: $A = 83°\,20',\ C = 54.6°,\ c = 18.1$

$$B = 180° - A - C = 180° - 83°\,20' - 54°\,36' = 42°\,4'$$

$$a = \frac{c}{\sin C}(\sin A) = \frac{18.1}{\sin 54.6°}(\sin 83°\,20') \approx 22.05$$

$$b = \frac{c}{\sin C}(\sin B) = \frac{18.1}{\sin 54.6°}(\sin 42°\,4') \approx 14.88$$

11. Given: $B = 15°\,30',\ a = 4.5,\ b = 6.8$

$$\sin A = \frac{a \sin B}{b} = \frac{4.5 \sin 15°\,30'}{6.8} \approx 0.17685 \implies A \approx 10°\,11'$$

$$C = 180° - A - B \approx 180° - 10°\,11' - 15.5° = 154°\,19'$$

$$c = \frac{b}{\sin B}(\sin C) = \frac{6.8}{\sin 15°\,30'}(\sin 154°\,19') \approx 11.03$$

13. Given: $A = 110°\,15',\ a = 48,\ b = 16$

$$\sin B = \frac{b \sin A}{a} = \frac{16 \sin 110°\,15'}{48} \approx 0.31273 \implies B \approx 18°\,13'$$

$$C = 180° - A - B \approx 180° - 110°\,15' - 18°\,13' = 51°\,32'$$

$$c = \frac{a}{\sin A}(\sin C) = \frac{48}{\sin 110°\,15'}(\sin 51°\,32') \approx 40.05$$

15. Given: $a = 4.5,\ b = 12.8,\ A = 58°$

$$h = 12.8 \sin 58° \approx 10.86$$

Since $a < h$, no triangle is formed.

17. Given: $A = 58°, a = 11.4, b = 12.8$

$$\sin B = \frac{b \sin A}{a} = \frac{12.8 \sin 58°}{11.4} \approx 0.9522 \Rightarrow B \approx 72.21° \text{ or } 107.79°$$

Case 1

$B \approx 72.21°$

$C = 180° - 58° - 72.21° = 49.79°$

$$c = \frac{a}{\sin A}(\sin C) = \frac{11.4}{\sin 58°}(\sin 49.79°)$$

≈ 10.27

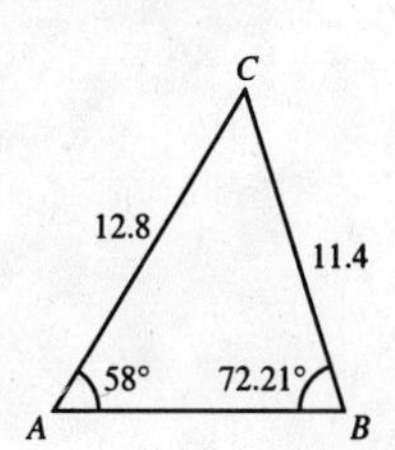

Case 2

$B \approx 107.79$

$C = 180° - 58° - 107.79° = 14.21°$

$$c = \frac{a}{\sin A}(\sin C) = \frac{11.4}{\sin 58°}(\sin 14.21°)$$

≈ 3.30

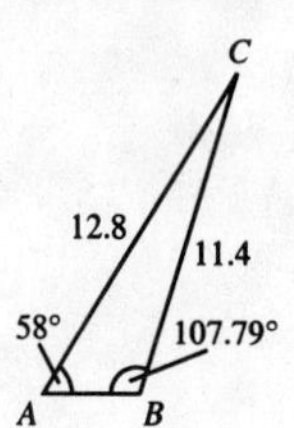

19. Given: $A = 110°, a = 125, b = 100$

$$\sin B = \frac{b \sin A}{a} = \frac{100 \sin 110°}{125} \approx 0.75175 \Rightarrow B \approx 48.74°$$

$C = 180° - A - B = 21.26°$

$$c = \frac{a}{\sin A}(\sin C) = \frac{125 \sin 21.26°}{\sin 110°} \approx 48.23$$

21. Given: $A = 36°,\ a = 5$

(a) One solution if $b \le 5$ or $b = \dfrac{5}{\sin 36°}$.

(b) Two solutions if $5 < b < \dfrac{5}{\sin 36°}$.

(c) No solution if $b > \dfrac{5}{\sin 36°}$.

23. Area $= \frac{1}{2}ab \sin C$

$= \frac{1}{2}(6)(10) \sin(110°)$

≈ 28.2 square units

25. Area $= \frac{1}{2}bc \sin A$

$= \frac{1}{2}(67)(85) \sin(38° 45')$

≈ 1782.3 square units

27. Area $= \frac{1}{2}ac \sin B$

$= \frac{1}{2}(92)(30) \sin 130°$

≈ 1057.1 square units

29. Angle $\measuredangle CAB = 70°$

Angle $B = 20° + 14° = 34°$

(b) $\dfrac{16}{\sin 70°} = \dfrac{h}{\sin 34°}$ (c) $h = \dfrac{16 \sin 34°}{\sin 70°} \approx 9.52$ meters

(a) 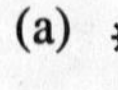

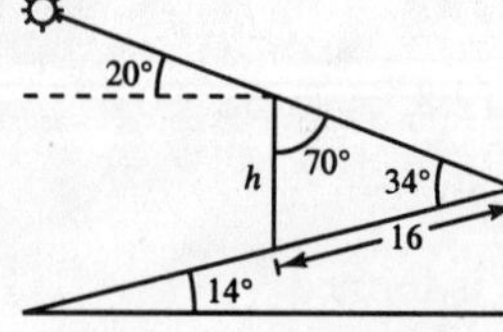

31. $\sin A = \dfrac{a \sin B}{b} = \dfrac{500 \sin 46°}{840} \approx 0.4282 \Rightarrow A \approx 25°$

The bearing from C to A is S 65° W.

33. (a)

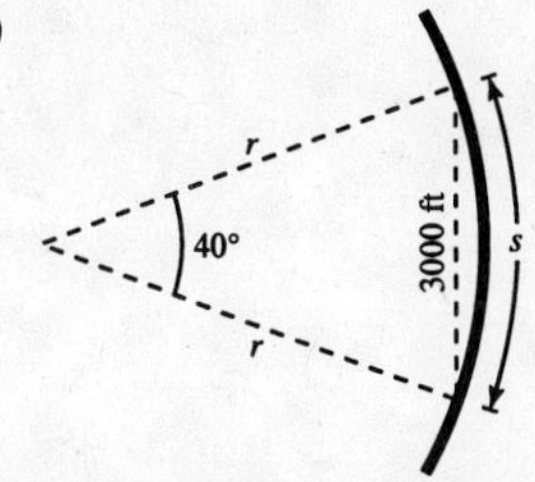

(b) $r = \dfrac{3000 \sin[1/2(180° - 40°)]}{\sin 40°} \approx 4385.71$ feet

(c) $s \approx 40°\left(\dfrac{\pi}{180°}\right)4385.71 \approx 3061.80$ feet

35. $A = 65° - 28° = 37°$

$c = 30$

$B = 180° - 16.5° - 65° = 98.5°$

$C = 180° - 37° - 98.5° = 44.5°$

$a = \dfrac{c}{\sin C}(\sin A) = \dfrac{30}{\sin 44.5°}(\sin 37°) \approx 25.8$ km to B

$b = \dfrac{c}{\sin C}(\sin B) = \dfrac{30}{\sin 44.5°}(\sin 98.5°) \approx 42.3$ km to A

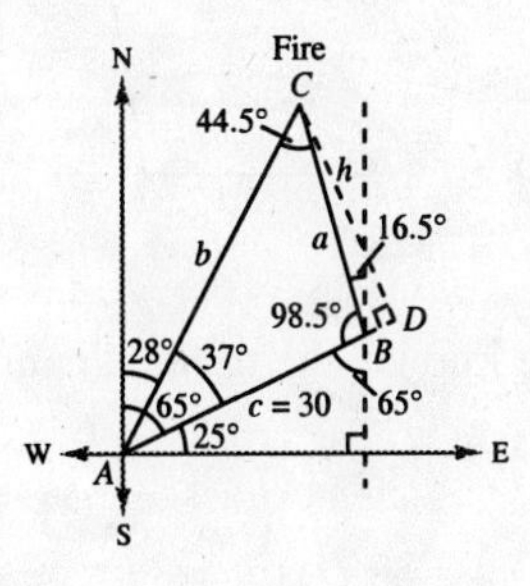

37. $A = 90° - 62° = 28°$,

$B = 90° + 38° = 128°$, $c = 5$

$C = 180° - 128° - 28° = 24°$

$a = \dfrac{c}{\sin C}(\sin A) = \dfrac{5}{\sin 24°}(\sin 28°) \approx 5.77$

$d = a \sin(90° - 38°) \approx 5.77 \sin 52° \approx 4.55$ miles

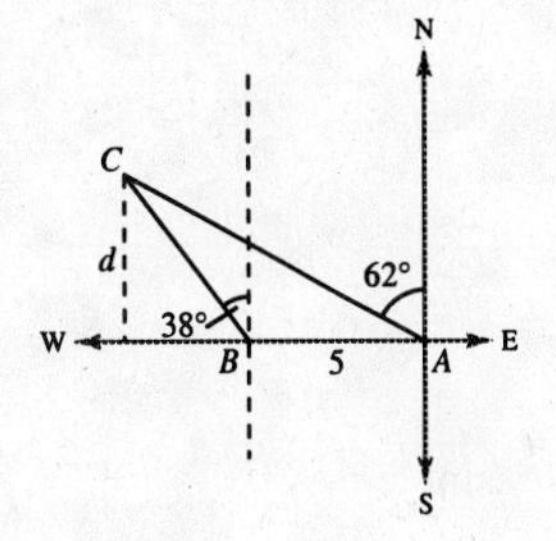

39. (a) $\dfrac{6}{\sin \theta} = \dfrac{1.5}{\sin C}$, $\sin \theta \neq 0$

$\sin C = \dfrac{1.5 \sin \theta}{6} \Rightarrow C = \arcsin \dfrac{1.5 \sin \theta}{6}$

$B = 180° - \theta - \arcsin \dfrac{1.5 \sin \theta}{6}$

$\dfrac{7.5 - d}{\sin B} = \dfrac{6}{\sin \theta}$

$d = 7.5 - \dfrac{6 \sin\left(180° - \theta - \arcsin \dfrac{1.5 \sin \theta}{6}\right)}{\sin \theta}$

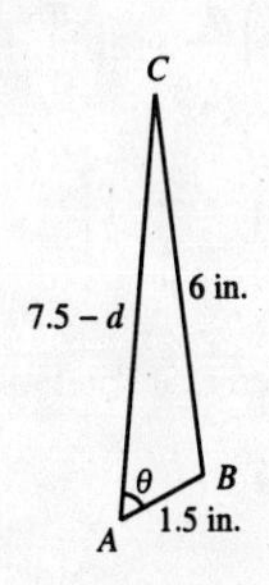

—CONTINUED—

39. —CONTINUED—

For $\theta = 0°, C = 0°, B = 180° \Rightarrow 7.5 - d = 1.5 + 6 \Rightarrow d = 0.$

θ	0°	45°	90°	135°	180°
d	0	0.5338	1.6905	2.6552	3

For $\theta = 180°, C = 0°, B = 0° \Rightarrow 7.5 - d = 6 - 1.5 \Rightarrow d = 3.$

(b) $\theta = 5°$

$$d = 7.5 - \frac{6 \sin\left(180° - 5° - \arcsin\dfrac{1.5 \sin 5°}{6}\right)}{\sin 5°} \approx 0.0071 \text{ inch}$$

41. $\alpha = 180 - /\phi + 180 - \theta) = \theta - \phi$

$$\frac{d}{\sin \phi} = \frac{2}{\sin \alpha}$$

$$d = \frac{2 \sin \phi}{\sin/\phi - \theta)}$$

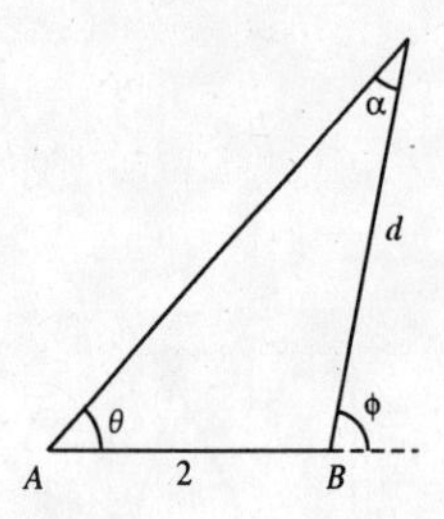

43. False. If the 3 angles are known, the triangle cannot be solved.

45. False. See page 428.

47. $\tan \theta = \dfrac{\sin \theta}{\cos \theta} = -\dfrac{12}{5}$ $\qquad \sec \theta = \dfrac{13}{5}$

$\cot \theta = -\dfrac{5}{12}$ $\qquad \csc \theta = -\dfrac{13}{12}$

49. $\tan \theta = -\dfrac{1}{11}; \cos \theta = -\dfrac{11}{\sqrt{122}} = -\dfrac{11\sqrt{122}}{122}$

$\sin \theta = \left(-\dfrac{1}{11}\right)\left(-\dfrac{11}{\sqrt{122}}\right) = \dfrac{1}{\sqrt{122}} = \dfrac{\sqrt{122}}{122};$

$\csc \theta = \sqrt{122}$

51. $\sec^2 x(\csc^2 x - 1) = \sec^2 x(\cot^2 x)$

$$= \frac{1}{\sin^2 x} = \csc^2 x$$

53. $6 \sin 8\,\theta \cos 3\theta = 6\left(\tfrac{1}{2}\right)[\sin(8\theta + 3\theta) + \sin(8\theta - 3\theta)]$

$$= 3[\sin 11\theta + \sin 5\theta]$$

55. $3 \cos \dfrac{\pi}{6} \sin \dfrac{5\pi}{3} = 3\left(\dfrac{1}{2}\right)\left[\sin\left(\dfrac{\pi}{6} + \dfrac{5\pi}{3}\right) - \sin\left(\dfrac{\pi}{6} - \dfrac{5\pi}{3}\right)\right]$

$$= \frac{3}{2}\left[\sin\left(\frac{11\pi}{6}\right) - \sin\left(-\frac{3\pi}{2}\right)\right]$$

$$= \frac{3}{2}\left[-\frac{1}{2} - 1\right] = -\frac{9}{4}$$

Section 6.2 Law of Cosines

- If ABC is any oblique triangle with sides a, b, and c, then the Law of Cosines says:

  (a) $a^2 = b^2 + c^2 - 2bc \cos A$ or $\cos A = \dfrac{b^2 + c^2 - a^2}{2bc}$

  (b) $b^2 = a^2 + c^2 - 2ac \cos B$ or $\cos B = \dfrac{a^2 + c^2 - b^2}{2ac}$

  (c) $c^2 = a^2 + b^2 - 2ab \cos C$ or $\cos C = \dfrac{a^2 + b^2 - c^2}{2ab}$

- You should be able to use the Law of Cosines to solve an oblique triangle for the remaining three parts, given:

  (a) Three sides (SSS)

  (b) Two sides and their included angle (SAS)

- Given any triangle with sides of length a, b, and c, then the area of the triangle is

  $\text{Area} = \sqrt{s(s-a)(s-b)(s-c)}$, where $s = \dfrac{a+b+c}{2}$. (Heron's Formula)

Solutions to Odd-Numbered Exercises

1. Given: $a = 6,\ b = 8,\ c = 12$

$$\cos A = \frac{b^2 + c^2 - a^2}{2bc} = \frac{64 + 144 - 36}{2(8)(12)} \approx 0.8958 \implies A \approx 26.4°$$

$$\sin B = \frac{b \sin A}{a} \approx \frac{8 \sin 26.4°}{6} \approx 0.5928 \implies B \approx 36.3°$$

$$C \approx 180° - 26.4° - 36.3° = 117.3°$$

3. Given: $A = 50°, b = 15, c = 30$

$$a^2 = b^2 + c^2 - 2bc \cos A = 225 + 900 - 2(15)(30) \cos 50° \approx 546.49 \implies a \approx 23.4$$

$$\cos B = \frac{a^2 + c^2 - b^2}{2ac} \approx 0.8708 \implies B \approx 29.5°$$

$$C = 180° - A - B = 180° - 50° - 29.5° = 100.5°$$

5. Given: $a = 9,\ b = 12,\ c = 15$

$$\cos C = \frac{a^2 + b^2 - c^2}{2ab} = \frac{81 + 144 - 225}{2(9)(12)} = 0 \implies C = 90°$$

$$\sin A = \frac{9}{15} = \frac{3}{5} \implies A \approx 36.9°$$

$$B \approx 180° - 90° - 36.9° = 53.1°$$

7. Given: $a = 75.4, b = 48, c = 48$

$$\cos A = \frac{b^2 + c^2 - a^2}{2bc} = \frac{48^2 + 48^2 - 75.4^2}{2(48)(48)} = -0.2338 \Rightarrow A \approx 103.5°$$

$$\sin B = \frac{b \sin A}{a} = \frac{48 \sin(103.5°)}{75.4} = 0.6190 \Rightarrow B \approx 38.2$$

$C = B \approx 38.2$ (Because of roundoff error, $A + B + C \neq 360°$)

9. Given: $B = 8° \, 15' = 8.25°, a = 26, c = 18$

$$b^2 = a^2 + c^2 - 2ac \cos B = 26^2 + 18^2 - 2(26)(18) \cos(8.25) \approx 73.6863 \Rightarrow b \approx 8.6$$

$$\sin C = \frac{c \sin B}{b} = \frac{18 \sin(8.25)}{8.6} \approx 0.3 \Rightarrow C \approx 17.5°$$

$$A = 180° - B - C = 180° - 8.25° - 17.5° = 154.25$$

11. $d^2 = 4^2 + 8^2 - 2(4)(8) \cos 30° \approx 24.57 \Rightarrow d \approx 4.96$

$2\phi = 360° - 2\theta \Rightarrow \phi = 150°$

$c^2 = 4^2 + 8^2 - 2(4)(8) \cos 150° \approx 135.43$

$c \approx 11.64$

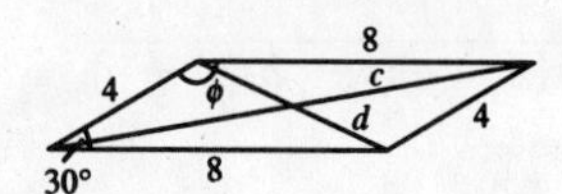

13. $\cos \phi = \dfrac{10^2 + 14^2 - 20^2}{2(10)(14)}$

$\phi \approx 111.8$

$2\theta \approx 360° - 2(111.80°)$

$\theta = 68.2°$

$d^2 = 10^2 + 14^2 - 2(10)(14) \cos 68.2°$

$d \approx 13.86$

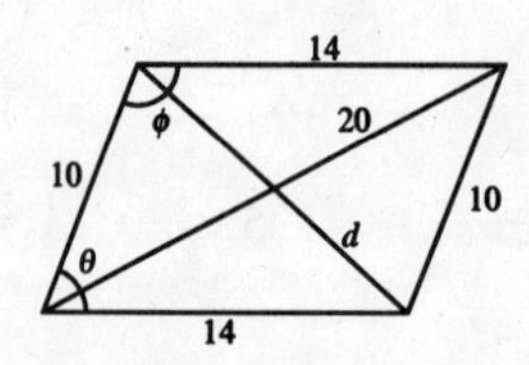

15. $\cos \alpha = \dfrac{(9)^2 + (10)^2 - (6)^2}{2(9)(10)}$

$\alpha = 36.3°$

$\cos \beta = \dfrac{6^2 + 10^2 - 9^2}{2(6)(10)}$

$\beta \approx 62.7°$

$z = 180° - \alpha - \beta \approx 80.9$

$\mu = 180° - z \approx 99.1$

$b^2 = 9^2 + 6^2 - 2(9)(6)(\cos 99.0°)$

$b \approx 11.58$

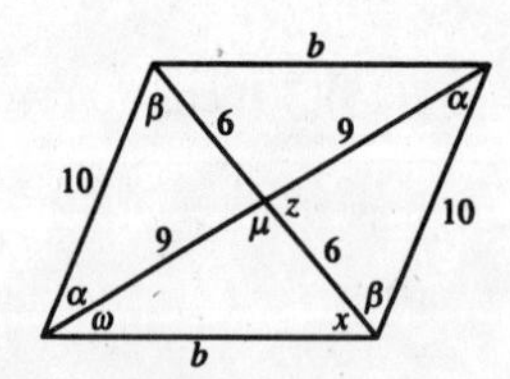

$\cos \omega = \dfrac{9^2 + 11.58^2 - 6^2}{2(9)(11.58)}$

$\omega \approx 30.8°$

$\theta = \alpha + \omega \approx 67.1°$

$\cos x = \dfrac{6^2 + 11.58^2 - 9^2}{2(6)(11.58)}$

$x \approx 50.1$

$\phi = \beta + x \approx 112.8°$

17. Given: $a = 5, b = 9, c = 10$

$$s = \frac{a+b+c}{2} = 12$$

$$\text{Area} = \sqrt{s(s-a)(s-b)(s-c)} = \sqrt{12(7)(3)(2)} \approx 22.45 \text{ square units}$$

19. Given: $a = 3.5, b = 10.2, c = 9$

$$s = \frac{a+b+c}{2} = 11.35$$

$$\text{Area} = \sqrt{s(s-a)(s-b)(s-c)} = \sqrt{11.35(7.85)(1.15)(2.35)} \approx 15.52 \text{ square units}$$

21. $a = 20,\ b = 20,\ c = 10 \implies s = \dfrac{20+20+10}{2} = 25$

$$\text{Area} = \sqrt{25(5)(5)(15)} \approx 96.82$$

23.

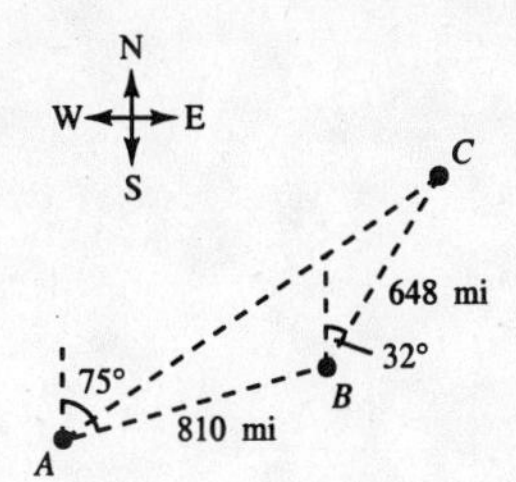

$$B = 105° + 32° = 137°$$

$$b^2 = a^2 + c^2 - 2ac \cdot \cos B$$

$$= 648^2 + 810^2 - 2(648)(810)\cos(137°)$$

$$= 1{,}843{,}749.862$$

$$b = 1357.8 \text{ miles}$$

From the Law of Sines, $\dfrac{a}{\sin A} = \dfrac{b}{\sin B} \implies \sin A = \dfrac{a}{b}\sin B = \dfrac{648}{1357.8}\sin(137°) \approx 0.32548$

$$\implies A \approx 19° \implies \text{Bearing S } 56° \text{ W}$$

25. Angle at $B = 180° - 80° = 100°$

$b^2 = 240^2 + 380^2 - 2(240)(380)\cos 100° \approx 233{,}673.4 \implies b \approx 483.4$ meters

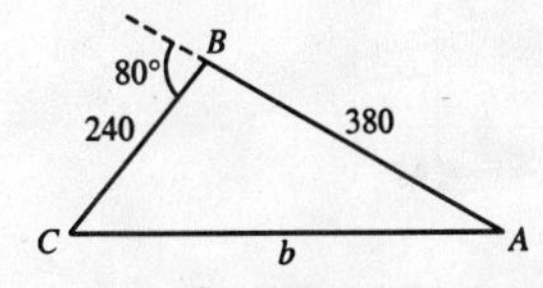

27. $C = 180° - 53° - 67° = 60°$

$c^2 = a^2 + b^2 - 2ab\cos C = 36^2 + 48^2 - 2(36)(48)(0.5) = 1872$

$c \approx 43.3$ mi

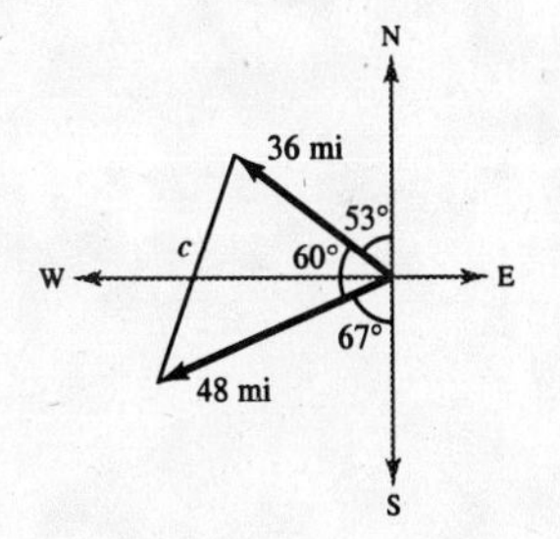

29. (a) $\cos\theta = \dfrac{273^2 + 178^2 - 235^2}{2(273)(178)}$

$\theta \approx 58.4^\circ$

Bearing: N 58.4° W

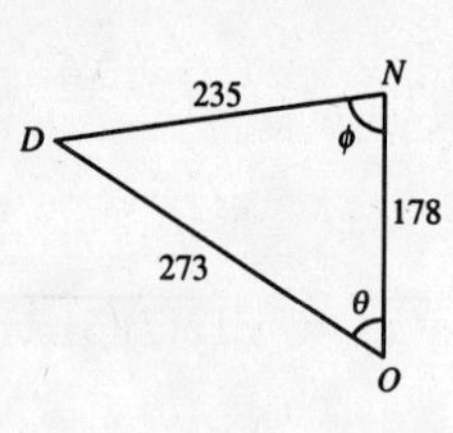

(b) $\cos\phi = \dfrac{235^2 + 178^2 - 273^2}{2(235)(178)}$

$\phi \approx 81.5^\circ$

Bearing: S 81.5° W

31. $d^2 = 60.5^2 + 90^2 - 2(60.5)(90)\cos 45^\circ \approx 4059.9 \implies d \approx 63.7$ ft

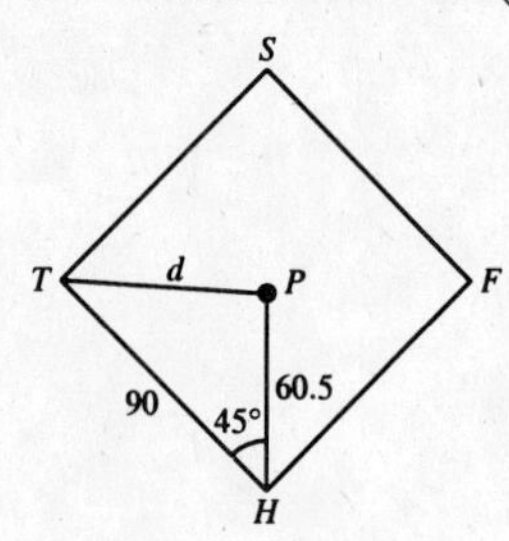

33. $\overline{RS} = \sqrt{8^2 + 10^2} = \sqrt{164} = 2\sqrt{41} \approx 12.8$ ft

$\overline{PQ} = \frac{1}{2}\sqrt{16^2 + 10^2} = \frac{1}{2}\sqrt{356} = \sqrt{89} \approx 9.4$ ft

$\tan P = \frac{10}{16}$

$P = \arctan\frac{5}{8} \approx 32.0^\circ$

$\overline{QS} = \sqrt{8^2 + 9.4^2 - 2(8)(9.4)\cos 32^\circ} \approx \sqrt{24.81} \approx 5.0$ ft

35. (a) $7^2 = 1.5^2 + x^2 - 2(1.5)(x)\cos\theta$

$49 = 2.25 + x^2 - 3x\cos\theta$

(b) $x^2 - 3x\cos\theta = 46.75$

$$x^2 - 3x\cos\theta + \left(\frac{3\cos\theta}{2}\right)^2 = 46.75 + \left(\frac{3\cos\theta}{2}\right)^2$$

$$\left[x - \frac{3\cos\theta}{2}\right]^2 = \frac{187}{4} - \frac{9\cos^2\theta}{4}$$

$$x - \frac{3\cos\theta}{2} = \pm\sqrt{\frac{187 + 9\cos^2\theta}{4}}$$

Choosing the positive values of x, we have
$x = \frac{1}{2}\left(3\cos\theta + \sqrt{9\cos^2\theta + 187}\right)$.

(c)

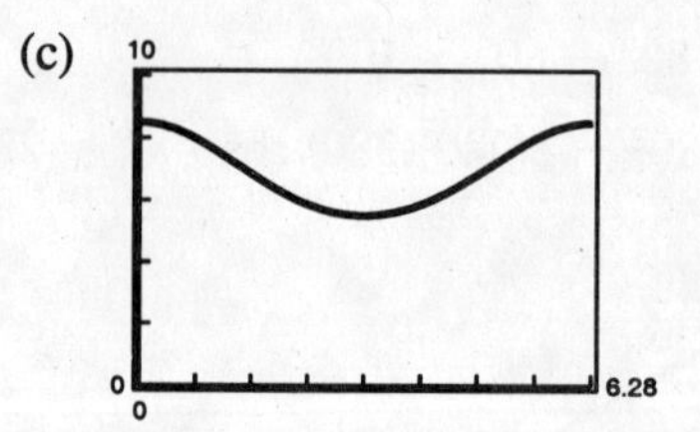

(d) $x = \dfrac{1}{2}\left(3\cos\pi + \sqrt{9\cos^2\pi + 187}\right)$

$= 5.5$

≈ 6 inches

37. $A = 180^\circ - 40^\circ - 20^\circ = 120^\circ$

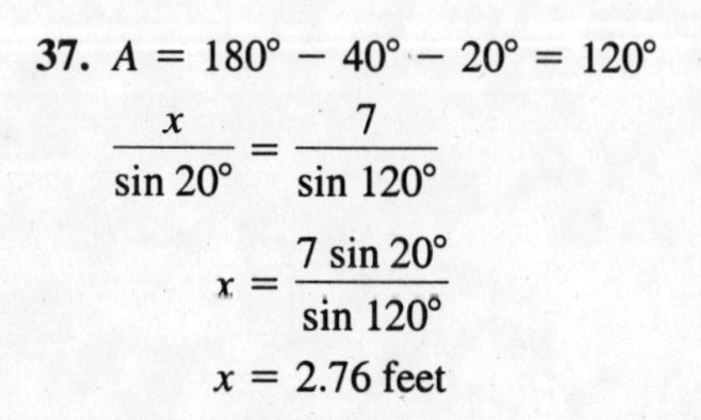

$$\frac{x}{\sin 20^\circ} = \frac{7}{\sin 120^\circ}$$

$$x = \frac{7\sin 20^\circ}{\sin 120^\circ}$$

$x = 2.76$ feet

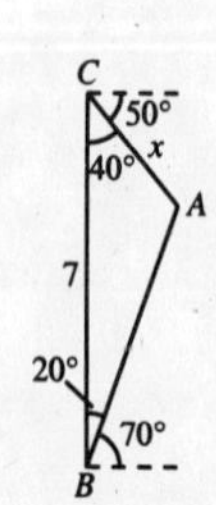

39. False. This is not a triangle! $5 + 10 < 16$

41. (a) Working with $\triangle OBC$, we have $\cos \alpha = \dfrac{a/2}{R}$.

This implies that $2R = a/\cos \alpha$. Since we know that

$$\frac{a}{\sin A} = \frac{b}{\sin B} = \frac{c}{\sin C},$$

we can complete the proof by showing that $\cos \alpha = \sin A$. The solution of the system

$$A + B + C = 180^\circ$$
$$\alpha - C + A = \beta$$
$$\alpha + \beta = B$$

is $\alpha = 90^\circ - A$. Therefore:

$$2R = \frac{a}{\cos \alpha} = \frac{a}{\cos(90^\circ - A)} = \frac{a}{\sin A}.$$

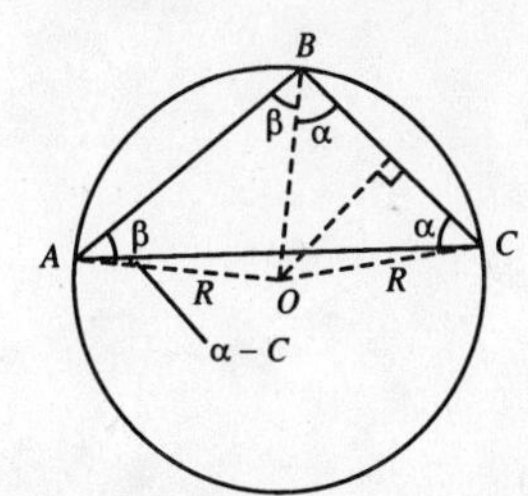

(b) By Heron's Formula, the area of the triangle is

$$\text{Area} = \sqrt{s(s-a)(s-b)(s-c)}.$$

We can also find the area by dividing the area into six triangles and using the fact that the area is 1/2 the base times the height. Using the figure as given, we have

$$\text{Area} = \frac{1}{2}xr + \frac{1}{2}xr + \frac{1}{2}yr + \frac{1}{2}yr + \frac{1}{2}zr + \frac{1}{2}zr$$
$$= r(x + y + z)$$
$$= rs.$$

Therefore: $rs = \sqrt{s(s-a)(s-b)(s-c)} \Rightarrow$

$$r = \sqrt{\frac{(s-a)(s-b)(s-c)}{s}}.$$

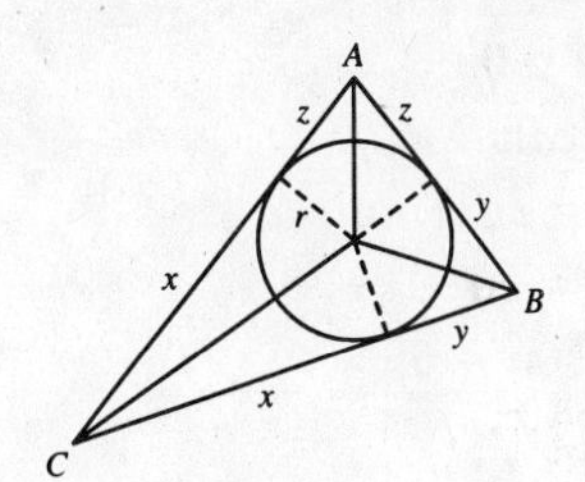

43. Given: $a = 200$ ft, $b = 250$ ft, $c = 325$ ft

$$s = \frac{200 + 250 + 325}{2} \approx 387.5$$

Radius of the inscribed circle: $r = \sqrt{\dfrac{(s-a)(s-b)(s-c)}{s}} = \sqrt{\dfrac{(187.5)(137.5)(62.5)}{387.5}} \approx 64.5$ ft

Circumference of an inscribed circle: $C = 2\pi r \approx 2\pi(64.5) \approx 405$ ft

45. $3 \sec x + 4 = 10$

$$\sec x = 2$$
$$\cos x = \frac{1}{2}$$
$$x = \frac{\pi}{3}, \frac{5\pi}{3}$$

47. $\dfrac{3}{\cos x \sin x} = 2\sqrt{3} \csc x = 2\sqrt{3}\left(\dfrac{1}{\sin x}\right)$

$$\frac{3}{\cos x} = 2\sqrt{3}$$
$$\cos x = \frac{3}{2\sqrt{3}} = \frac{\sqrt{3}}{2}$$
$$x = \frac{\pi}{6}, \frac{11\pi}{6}$$

49. $\cos \dfrac{5\pi}{6} - \cos \dfrac{\pi}{3} = -2 \sin\left(\dfrac{\frac{5\pi}{6} + \frac{\pi}{3}}{2}\right) \sin\left(\dfrac{\frac{5\pi}{6} - \frac{\pi}{3}}{2}\right)$

$$= -2 \sin\left(\frac{7\pi}{12}\right) \sin\left(\frac{\pi}{4}\right)$$

Section 6.3 Vectors in the Plane

- A vector **v** is the collection of all directed line segments that are equivalent to a given directed line segment $\overrightarrow{PQ}$.
- You should be able to *geometrically* perform the operations of vector addition and scalar multiplication.
- The component form of the vector with initial point $P = (p_1, p_2)$ and terminal point $Q = (q_1, q_2)$ is
  $$\overrightarrow{PQ} = \langle q_1 - p_1, q_2 - p_2 \rangle = \langle v_1, v_2 \rangle = \mathbf{v}.$$
- The magnitude of $\mathbf{v} = \langle v_1, v_2 \rangle$ is given by $\|\mathbf{v}\| = \sqrt{v_1^2 + v_2^2}$.
- You should be able to perform the operations of scalar multiplication and vector addition in component form.
- You should know the following properties of vector addition and scalar multiplication.
  (a) $\mathbf{u} + \mathbf{v} = \mathbf{v} + \mathbf{u}$
  (b) $(\mathbf{u} + \mathbf{v}) + \mathbf{w} = \mathbf{u} + (\mathbf{v} + \mathbf{w})$
  (c) $\mathbf{u} + \mathbf{0} = \mathbf{u}$
  (d) $\mathbf{u} + (-\mathbf{u}) = \mathbf{0}$
  (e) $c(d\mathbf{u}) = (cd)\mathbf{u}$
  (f) $(c + d)\mathbf{u} = c\mathbf{u} + d\mathbf{u}$
  (g) $c(\mathbf{u} + \mathbf{v}) = c\mathbf{u} + c\mathbf{v}$
  (h) $1(\mathbf{u}) = \mathbf{u}, 0\mathbf{u} = \mathbf{0}$
  (i) $\|c\mathbf{v}\| = |c|\,\|\mathbf{v}\|$
- A unit vector in the direction of **v** is given by $\mathbf{u} = \dfrac{\mathbf{v}}{\|\mathbf{v}\|}$.
- The standard unit vectors are $\mathbf{i} = \langle 1, 0 \rangle$ and $\mathbf{j} = \langle 0, 1 \rangle$. $\mathbf{v} = \langle v_1, v_2 \rangle$ can be written as $\mathbf{v} = v_1\mathbf{i} + v_2\mathbf{j}$.
- A vector **v** with magnitude $\|\mathbf{v}\|$ and direction θ can be written as $\mathbf{v} = a\mathbf{i} + b\mathbf{j} = \mathbf{v}(\cos\theta)\mathbf{i} + \mathbf{v}(\sin\theta)\mathbf{j}$ where $\tan\theta = b/a$.

Solutions to Odd-Numbered Exercises

1. $\mathbf{v} = \langle 6 - 2, 5 - 4 \rangle = \langle 4, 1 \rangle = \mathbf{v}$

3. Initial point: $(0, 0)$
Terminal point: $(4, 3)$
$$\mathbf{v} = \langle 4 - 0, 3 - 0 \rangle = \langle 4, 3 \rangle$$
$$\|\mathbf{v}\| = \sqrt{4^2 + 3^2} = 5$$

5. Initial point: $(2, 2)$
Terminal point: $(-1, 4)$
$$\mathbf{v} = \langle -1 - 2, 4 - 2 \rangle = \langle -3, 2 \rangle$$
$$\|\mathbf{v}\| = \sqrt{(-3)^2 + 2^2} = \sqrt{13}$$

7. Initial point: $(3, -2)$
Terminal point: $(3, 3)$
$$\mathbf{v} = \langle 3 - 3, 3 - (-2) \rangle = \langle 0, 5 \rangle$$
$$\|\mathbf{v}\| = 5$$

9. Initial point: $\left(\frac{5}{2}, 1\right)$

Terminal point: $\left(-2, -\frac{3}{2}\right)$
$$\mathbf{v} = \left\langle -2 - \frac{5}{2}, -\frac{3}{2} - 1 \right\rangle = \left\langle -\frac{9}{2}, -\frac{5}{2} \right\rangle$$
$$\|\mathbf{v}\| = \sqrt{\left(-\frac{9}{2}\right)^2 + \left(-\frac{5}{2}\right)^2} = \sqrt{\frac{81 + 25}{4}} = \frac{1}{2}\sqrt{106}$$

11. Initial point: $(-3, -5)$

Terminal point: $(5, 1)$

$\mathbf{v} = \langle 5 - (-3), 1 - (-5) \rangle = \langle 8, 6 \rangle$

$\|\mathbf{v}\| = \sqrt{8^2 + 6^2} = \sqrt{100} = 10$

13. Initial point: $(-4.2, 5)$

Terminal point: $(3.7, -12.9)$

$\mathbf{v} = \langle 3.7, -(-4.2), -12.9 - 5 \rangle = \langle 7.9, -17.9 \rangle$

$\|\mathbf{v}\| = \sqrt{7.9^2 + (-17.9)^2} \approx 19.6$

15.

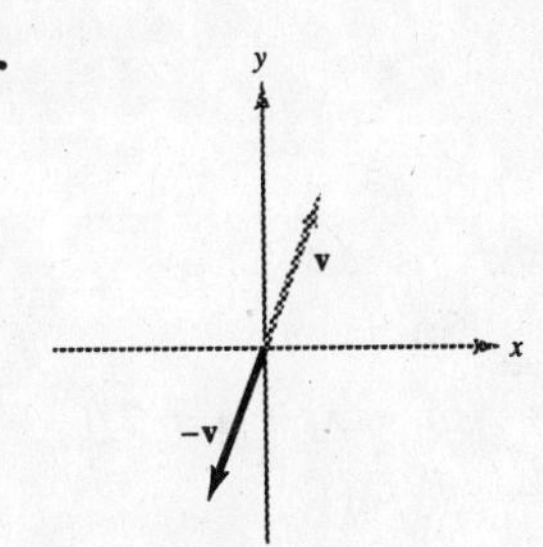

17.

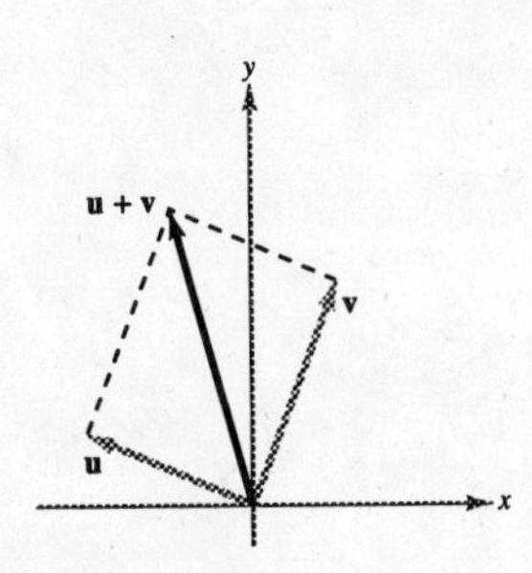

19. $\mathbf{u} + 2\mathbf{v}$

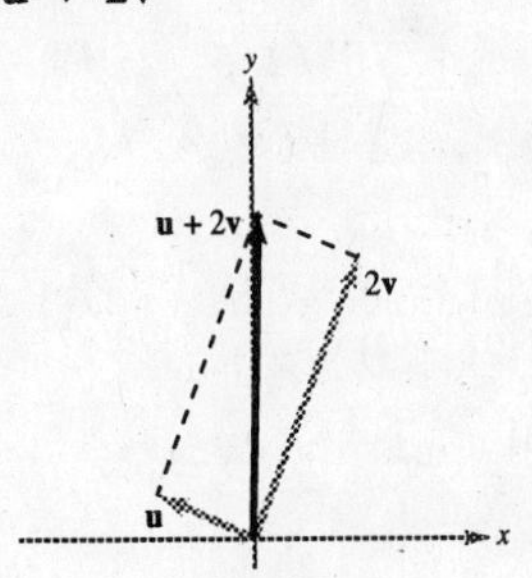

21. $\mathbf{u} = \langle 4, 2 \rangle, \mathbf{v} = \langle 7, 1 \rangle$

(a) $\mathbf{u} + \mathbf{v} = \langle 11, 3 \rangle$

(b) $\mathbf{u} - \mathbf{v} = \langle -3, 1 \rangle$

(c) $2\mathbf{u} - 3\mathbf{v} = \langle 8, 4 \rangle - \langle 21, 3 \rangle = \langle -13, 1 \rangle$

(d) $\mathbf{v} + 4\mathbf{u} = \langle 7, 1 \rangle + \langle 16, 8 \rangle = \langle 23, 9 \rangle$

23. $\mathbf{u} = \langle -5, -2 \rangle, \mathbf{v} = \langle 1, -3 \rangle$

(a) $\mathbf{u} + \mathbf{v} = \langle -4, -5 \rangle$

(b) $\mathbf{u} - \mathbf{v} = \langle -6, 1 \rangle$

(c) $2\mathbf{u} - 3\mathbf{v} = \langle -10, -4 \rangle - \langle 3, -9 \rangle = \langle -13, 5 \rangle$

(d) $\mathbf{v} + 4\mathbf{u} = \langle 1, -3 \rangle + \langle -20, -8 \rangle$
$= \langle -19, -11 \rangle$

25. $\mathbf{u} = \mathbf{i} + \mathbf{j}, \mathbf{v} = 2\mathbf{i} - 3\mathbf{j}$

(a) $\mathbf{u} + \mathbf{v} = 3\mathbf{i} - 2\mathbf{j}$

(b) $\mathbf{u} - \mathbf{v} = -\mathbf{i} + 4\mathbf{j}$

(c) $2\mathbf{u} - 3\mathbf{v} = (2\mathbf{i} + 2\mathbf{j}) - (6\mathbf{i} - 9\mathbf{j})$
$= -4\mathbf{i} + 11\mathbf{j}$

(d) $\mathbf{v} + 4\mathbf{u} = (2\mathbf{i} - 3\mathbf{j}) + (4\mathbf{i} + 4\mathbf{j})$
$= 6\mathbf{i} + \mathbf{j}$

27. $\|\langle 6, 0 \rangle\| = 6$

unit vector $= \frac{1}{6}\langle 6, 0 \rangle = \langle 1, 0 \rangle$

29. $\|\mathbf{v}\| = \|\langle -4, 4 \rangle\| = \sqrt{16 + 16} = 4\sqrt{2}$

unit vector $= \dfrac{1}{4\sqrt{2}}\langle -4, 4 \rangle = \left\langle -\dfrac{1}{\sqrt{2}}, \dfrac{1}{\sqrt{2}} \right\rangle = \left\langle -\dfrac{\sqrt{2}}{2}, \dfrac{\sqrt{2}}{2} \right\rangle$

31. $\|\mathbf{v}\| = \sqrt{(-24)^2 + 7^2} = 25$

unit vector $= \dfrac{1}{25}\langle -24, -7 \rangle = \left\langle -\dfrac{24}{25}, -\dfrac{7}{25} \right\rangle$

33. $\mathbf{u} = \dfrac{1}{\|\mathbf{v}\|}\mathbf{v}$

$= \dfrac{1}{\sqrt{16 + 9}}(4\mathbf{i} - 3\mathbf{j}) = \dfrac{1}{5}(4\mathbf{i} - 3\mathbf{j})$

$= \dfrac{4}{5}\mathbf{i} - \dfrac{3}{5}\mathbf{j}$

35. $\mathbf{u} = \dfrac{1}{2}(2\mathbf{j}) = \mathbf{j}$

37. $5\left(\frac{1}{\|\mathbf{v}\|}\mathbf{v}\right) = 5\left(\frac{1}{\sqrt{3^2+3^2}}\langle 3, 3\rangle\right)$

$= 5\left(\frac{1}{3\sqrt{2}}\langle 3, 3\rangle\right)$

$= \left\langle \frac{5}{\sqrt{2}}, \frac{5}{\sqrt{2}} \right\rangle$

$= \left\langle \frac{5\sqrt{2}}{2}, \frac{5\sqrt{2}}{2} \right\rangle$

39. $7\left(\frac{1}{\|\mathbf{v}\|}\mathbf{v}\right) = 7\left(\frac{1}{\sqrt{3^2+4^2}}\langle 3, 4\rangle\right)$

$= \frac{7}{5}\langle 3, 4\rangle$

$= \left\langle \frac{21}{5}, \frac{28}{5} \right\rangle$

41. $8\left(\frac{1}{\|\mathbf{v}\|}\mathbf{v}\right) = 8\left(\frac{1}{2}\langle -2, 0\rangle\right)$

$= 4\langle -2, 0\rangle$

$= \langle -8, 0\rangle = -8\mathbf{i}$

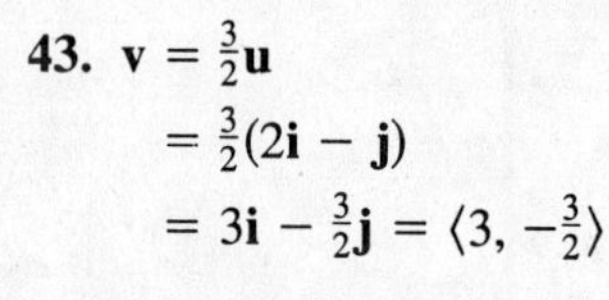

43. $\mathbf{v} = \frac{3}{2}\mathbf{u}$

$= \frac{3}{2}(2\mathbf{i} - \mathbf{j})$

$= 3\mathbf{i} - \frac{3}{2}\mathbf{j} = \langle 3, -\frac{3}{2}\rangle$

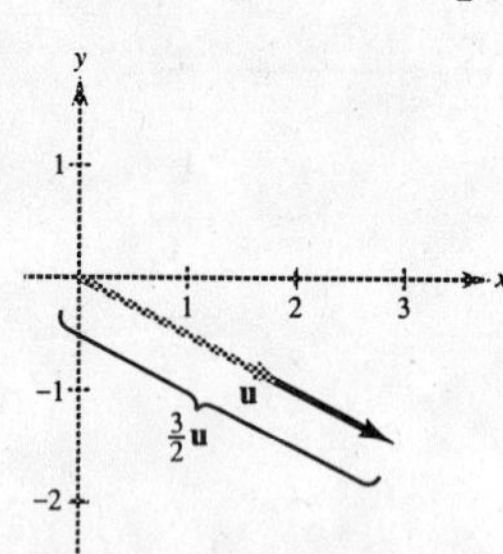

45. $\mathbf{v} = \mathbf{u} + 2\mathbf{w}$

$= (2\mathbf{i} - \mathbf{j}) + 2(\mathbf{i} + 2\mathbf{j})$

$= 4\mathbf{i} + 3\mathbf{j} = \langle 4, 3\rangle$

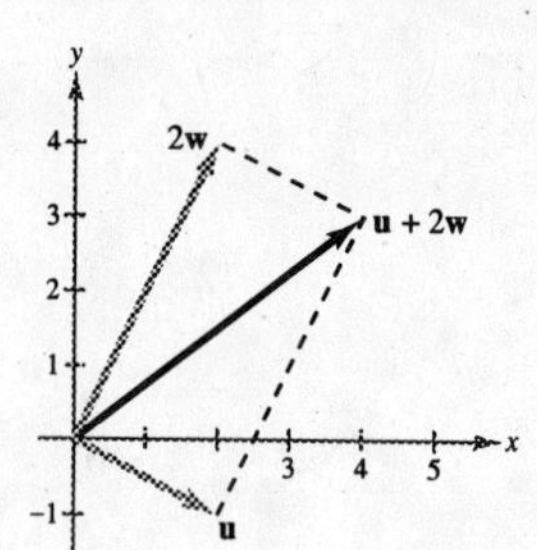

47. $\mathbf{v} = (3\mathbf{u} + \mathbf{w})$

$= (6\mathbf{i} - 3\mathbf{j} + \mathbf{i} + 2\mathbf{j})$

$= \langle 7, -1\rangle$

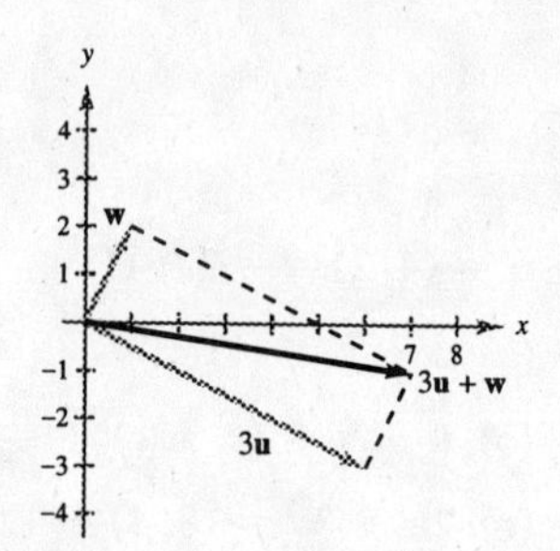

49. $\mathbf{v} = 5(\cos 30°\mathbf{i} + \sin 30\mathbf{j})$

$\|\mathbf{v}\| = 5, \ \theta = 30°$

51. $\mathbf{v} = 6\mathbf{i} - 6\mathbf{j}$

$\|\mathbf{v}\| = \sqrt{6^2 + (-6)^2} = \sqrt{72} = 6\sqrt{2}$

$\tan\theta = \frac{-6}{6} = -1$

Since $\mathbf{v}$ lies in Quadrant IV, $\theta = 315°$.

53. $\mathbf{v} = -2\mathbf{i} + 5\mathbf{j}$

$\|\mathbf{v}\| = \sqrt{(-2)^2 + 5^2} = \sqrt{29}$

$\tan\theta = -\frac{5}{2}$

Since $\mathbf{v}$ lies in Quadrant II,
$\theta \approx 111.8°$.

55. $\mathbf{v} = \langle 3\cos 0°, 3\sin 0°\rangle$

$= \langle 3, 0\rangle$

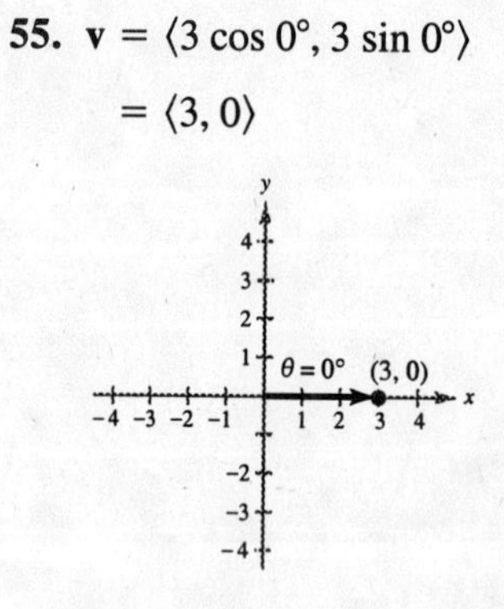

57. $\mathbf{v} = \langle 3\sqrt{2}\cos 150°,$
$3\sqrt{2}\sin 150°\rangle$

$= \left\langle -\frac{3\sqrt{6}}{2}, \frac{3\sqrt{2}}{2} \right\rangle$

59. $\mathbf{v} = 2\left(\frac{1}{\sqrt{3^2+1^2}}\right)(\mathbf{i} + 3\mathbf{j})$

$= \frac{2}{\sqrt{10}}(\mathbf{i} + 3\mathbf{j})$

$= \frac{\sqrt{10}}{5}\mathbf{i} + \frac{3\sqrt{10}}{5}\mathbf{j} = \left\langle \frac{\sqrt{10}}{5}, \frac{3\sqrt{10}}{5} \right\rangle$

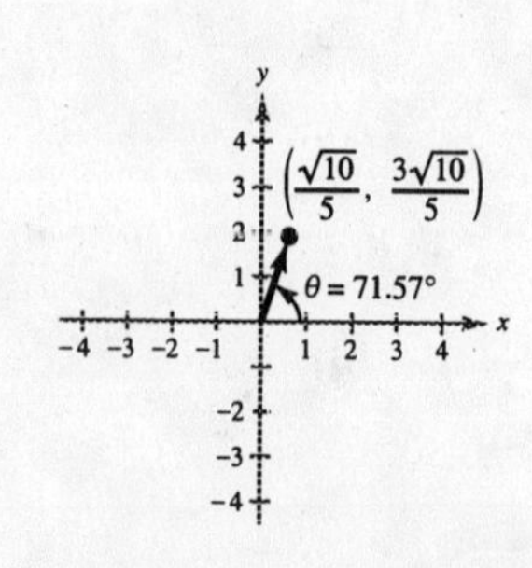

61. $\mathbf{u} = \langle 5\cos 60°, 5\sin 60° \rangle = \left\langle \frac{5}{2}, \frac{5\sqrt{3}}{2} \right\rangle$

$\mathbf{v} = \langle 5\cos 90°, 5\sin 90° \rangle = \langle 0, 5 \rangle$

$\mathbf{u} + \mathbf{v} = \left\langle \frac{5}{2}, \frac{5\sqrt{3}}{2} \right\rangle + \langle 0, 5 \rangle = \left\langle \frac{5}{2}, 5 + \frac{5}{2}\sqrt{3} \right\rangle$

63. $\mathbf{u} = \langle 20\cos 45°, 20\sin 45° \rangle = \langle 10\sqrt{2}, 10\sqrt{2} \rangle$

$\mathbf{v} = \langle 50\cos 150°, 50\sin 150° \rangle = \langle -25\sqrt{3}, 25 \rangle$

$\mathbf{u} + \mathbf{v} = \langle 10\sqrt{2} - 25\sqrt{3}, 10\sqrt{2} + 25 \rangle$

65.

$$\mathbf{v} = \mathbf{i} + \mathbf{j}$$
$$\mathbf{w} = 2(\mathbf{i} - \mathbf{j})$$
$$\mathbf{u} = \mathbf{v} - \mathbf{w} = -\mathbf{i} + 3\mathbf{j}$$
$$\|\mathbf{v}\| = \sqrt{2}$$
$$\|\mathbf{w}\| = 2\sqrt{2}$$
$$\|\mathbf{v} - \mathbf{w}\| = \sqrt{10}$$
$$\cos\alpha = \frac{\|\mathbf{v}\|^2 + \|\mathbf{w}\|^2 - \|\mathbf{v} - \mathbf{w}\|^2}{2\|\mathbf{v}\|\,\|\mathbf{w}\|} = \frac{2 + 8 - 10}{2\sqrt{2} \cdot 2\sqrt{2}} = 0$$
$$\alpha = 90°$$

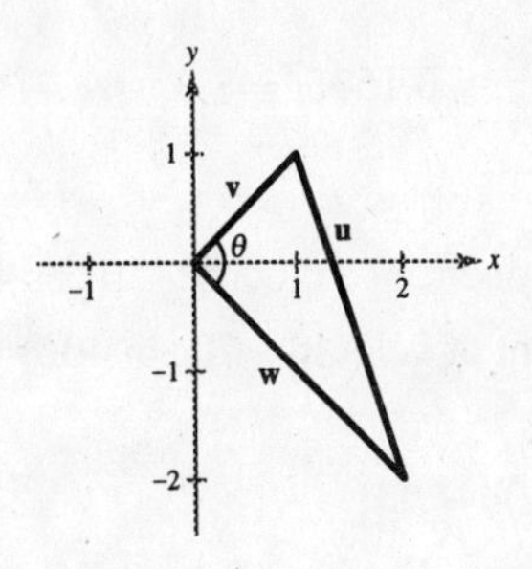

67.

$$\mathbf{v} = \mathbf{i} + \mathbf{j}$$
$$\mathbf{w} = 3\mathbf{i} - \mathbf{j}$$
$$\mathbf{u} = \mathbf{v} - \mathbf{w} = -2\mathbf{i} + 2\mathbf{j}$$
$$\cos\alpha = \frac{\|\mathbf{v}\|^2 + \|\mathbf{w}\|^2 - \|\mathbf{v} - \mathbf{w}\|^2}{2\|\mathbf{v}\|\,\|\mathbf{w}\|} = \frac{2 + 10 - 8}{2\sqrt{2}\sqrt{10}} \approx 0.4472$$
$$\alpha = 63.4°$$

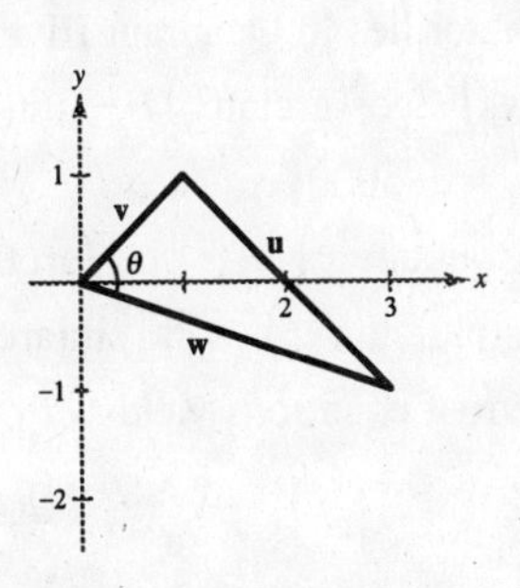

69.

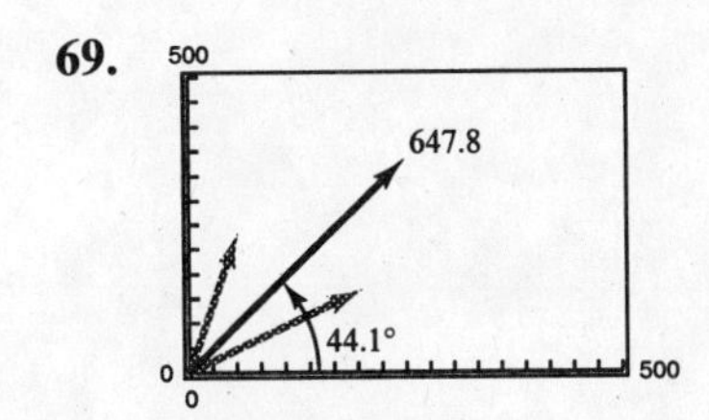

$$\mathbf{u} = 400\cos 25°\mathbf{i} + 400\sin 25°\mathbf{j}$$
$$\mathbf{v} = 300\cos 70°\mathbf{i} + 300\sin 70°\mathbf{j}$$
$$\mathbf{u} + \mathbf{v} \approx 465.13\mathbf{i} + 450.96\mathbf{j}$$
$$\|\mathbf{u} + \mathbf{v}\| \approx \sqrt{(465.13)^2 + (450.96)^2} \approx 647.8$$
$$\alpha = \arctan\left(\frac{450.96}{465.13}\right) \approx 44.1°$$

71. Force One: $\mathbf{u} = 45\mathbf{i}$

Force Two: $\mathbf{v} = 60\cos\theta\mathbf{i} + 60\sin\theta\mathbf{j}$

Resultant Force: $\mathbf{u} + \mathbf{v} = (45 + 60\cos\theta)\mathbf{i} + 60\sin\theta\mathbf{j}$

$$\|\mathbf{u} + \mathbf{v}\| = \sqrt{(45 + 60\cos\theta)^2 + (60\sin\theta)^2} = 90$$
$$2025 + 5400\cos\theta + 3600 = 8100$$
$$5400\cos\theta = 2475$$
$$\cos\theta = \frac{2475}{5400} \approx 0.4583$$
$$\theta \approx 62.7°$$

73.

$$\mathbf{u} = (2000 \cos 30°)\,\mathbf{i} + (2000 \sin 30°\mathbf{j})$$
$$\approx 1732.05\mathbf{i} + 1000\mathbf{j}$$
$$\mathbf{v} = (900 \cos(-45°))\mathbf{i} + (900 \sin(-45°)\mathbf{j})$$
$$\approx 636.4\mathbf{i} + -636.4\mathbf{j}$$
$$\mathbf{u} + \mathbf{v} \approx 2368.4\mathbf{i} + 363.6\mathbf{j}$$
$$\|\mathbf{u} + \mathbf{v}\| \approx \sqrt{(2368.4)^2 + (363.6)^2} \approx 2396.19$$
$$\tan\theta = \frac{363.6}{2368.4} \approx 0.1535 \implies \theta \approx 8.7°$$

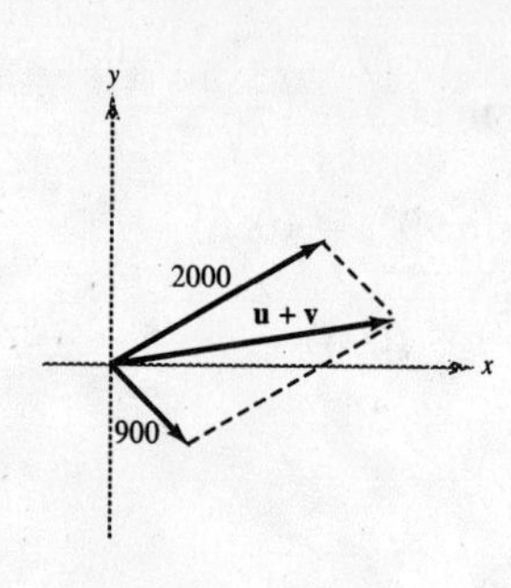

75. Horizontal component of velocity: $70 \cos 40° \approx 53.62$ ft/sec

Vertical component of velocity: $70 \sin 40° \approx 45.0$ ft/sec

77. Rope $\overrightarrow{AC}$: $\mathbf{u} = 10\mathbf{i} - 24\mathbf{j}$

The vector lies in Quadrant IV and its reference angle is $\arctan\left(\frac{12}{5}\right)$.

$$\mathbf{u} = \|\mathbf{u}\|\left[\cos\left(\arctan\tfrac{12}{5}\right)\mathbf{i} - \sin\left(\arctan\tfrac{12}{5}\right)\mathbf{j}\right]$$

Rope $\overrightarrow{BC}$: $\mathbf{v} = -20\mathbf{i} - 24\mathbf{j}$

The vector lies in Quadrant III and its reference angle is $\arctan\left(\frac{6}{5}\right)$.

$$\mathbf{v} = \|\mathbf{v}\|\left[-\cos\left(\arctan\tfrac{6}{5}\right)\mathbf{i} - \sin\left(\arctan\tfrac{6}{5}\right)\mathbf{j}\right]$$

Resultant: $\mathbf{u} + \mathbf{v} = -5000\mathbf{j}$

$$\|\mathbf{u}\| \cos\left(\arctan\tfrac{12}{5}\right) - \|\mathbf{v}\| \cos\left(\arctan\tfrac{6}{5}\right) = 0$$
$$-\|\mathbf{u}\| \sin\left(\arctan\tfrac{12}{5}\right) - \|\mathbf{v}\| \sin\left(\arctan\tfrac{6}{5}\right) = -5000$$

Solving this system of equations yields: $T_{AC} = \|\mathbf{u}\| \approx 3611.1$ pounds

$T_{BC} = \|\mathbf{v}\| \approx 2169.5$ pounds

79. (a) Tow line 1: $\mathbf{u} = \|\mathbf{u}\|\,(\cos 20°\mathbf{i} + \sin 20°\mathbf{j})$

Tow line 2: $\mathbf{v} = \|\mathbf{u}\|\,(\cos(-20°)\mathbf{i} + \sin(-20°)\mathbf{j})$

Resultant: $\mathbf{u} + \mathbf{v} = 6000\mathbf{i} = [\|\mathbf{u}\| \cos 20° + \|\mathbf{u}\| \cos(-20°)]\mathbf{i}$

$$\implies 6000 = 2\,\|\mathbf{u}\| \cos 20°$$
$$\implies \|\mathbf{u}\| \approx 3192.5 \text{ lb}$$

(b) $\mathbf{u} + \mathbf{v} = 6000\mathbf{i} = 2\,\|\mathbf{u}\| \cos\theta \implies T = \|\mathbf{u}\| = 3000 \sec\theta$. Domain: $0° \le \theta <$ [illegible]

(c)

θ	10°	20°	30°	40°	50°	60°
T	3046.3	3192.5	3464.1	2916.2	4667.2	6000.0

(d)

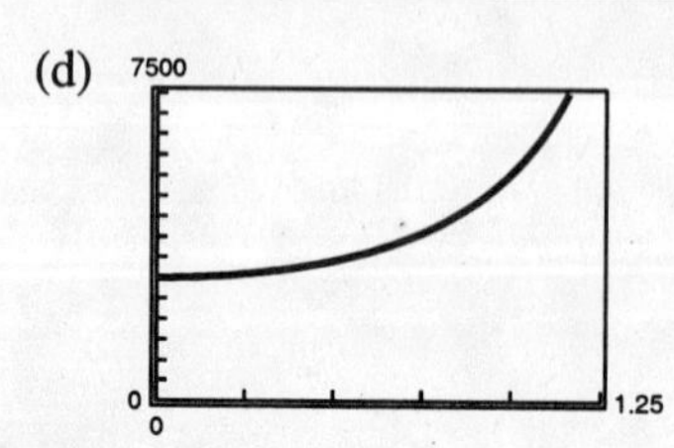

(e) The tension increases because the component in the direction of the motion of the barge decreases.

81. Airspeed: $\mathbf{v} = 860(\cos 302°\mathbf{i} + \sin 302°\mathbf{j})$

Groundspeed: $\mathbf{u} = 800(\cos 310°\mathbf{i} + \sin 310°\mathbf{j})$

$\mathbf{w} + \mathbf{v} = \mathbf{u}$

$\mathbf{w} = \mathbf{u} - \mathbf{v} = 800(\cos 310°\mathbf{i} + \sin 310°\mathbf{j}) - 860(\cos 302°\mathbf{i} + \sin 302°\mathbf{j}) \approx 58.50\mathbf{i} + 116.49\mathbf{j}$

$\|\mathbf{w}\| = \sqrt{58.50^2 + 116.49^2} \approx 130.35$ km/hr

$$\theta = \arctan\left(\frac{116.49}{58.50}\right) \approx 63.3°$$

Direction: N 26.7° E

83. (a) $\mathbf{u} = 220\mathbf{i}, \quad \mathbf{v} = 150 \cos 30°\mathbf{i} + 150 \sin 30°\mathbf{j}$

$$\mathbf{u} + \mathbf{v} = \left(220 + 75\sqrt{3}\right)\mathbf{i} + 75\mathbf{j}$$

$$\|\mathbf{u} + \mathbf{v}\| = \sqrt{\left(220 + 75\sqrt{3}\right)^2 + 75^2} \approx 357.85 \text{ newtons}$$

$$\tan\theta = \frac{75}{220 + 75\sqrt{3}} \implies \theta \approx 12.1°$$

(b) $\mathbf{u} + \mathbf{v} = 220\mathbf{i} + (150 \cos \theta\mathbf{i} + 150 \sin \theta\mathbf{j})$

$$M = \|\mathbf{u} + \mathbf{v}\| = \sqrt{(220^2 + 150^2(\cos^2\theta + \sin^2\theta) + 2(220)(150)\cos\theta}$$

$$= \sqrt{70{,}900 + 66{,}000 \cos\theta}$$

$$= 10\sqrt{709 + 660 \cos\theta}$$

$$\alpha = \arctan\left(\frac{15 \sin\theta}{22 + 15 \cos\theta}\right)$$

(c)

θ	0°	30°	60°	90°	120°	150°	180°
M	370.0	357.9	322.3	266.3	194.7	117.2	70.0
α	0°	12.1°	23.8°	34.3°	41.9°	39.8°	0°

(d)

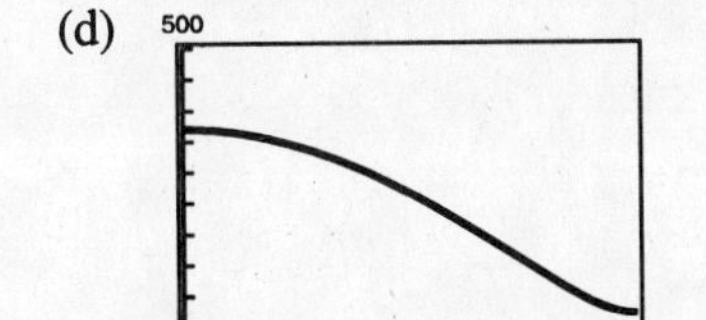

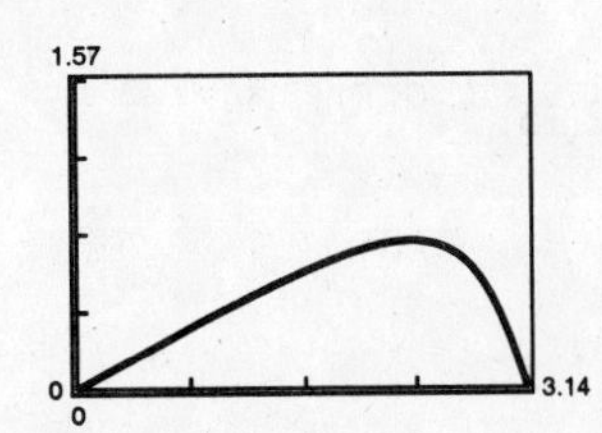

(e) For increasing θ the two vectors tend to work against each other resulting in a decrease in the magnitude of the resultant.

85. True. See page 444

87. True. In fact, $a = b = 0$.

89. (a) The angle between them is 0°.

(b) The angle between them is 180°.

(c) No. At most it can be equal to the sum when the angle between them is 0°.

91. Let $\mathbf{v} = (\cos\theta)\mathbf{i} + (\sin\theta)\mathbf{j}$

$\|\mathbf{v}\| = \sqrt{\cos^2\theta + \sin^2\theta} = \sqrt{1} = 1$

Therefore, $\mathbf{v}$ is a unit vector for any value of θ.

93. $\mathbf{u} = \langle 5 - 1, 2 - 6 \rangle = \langle 4, -4 \rangle$

$\mathbf{v} = \langle 9 - 4, 4 - 5 \rangle = \langle 5, -1 \rangle$

$\mathbf{u} - \mathbf{v} = \langle -1, -3 \rangle$

$\mathbf{v} - \mathbf{u} = \langle 1, 3 \rangle$

95. $\sin\theta = \dfrac{x}{7} \implies \sqrt{49 - x^2} = 7\cos\theta$

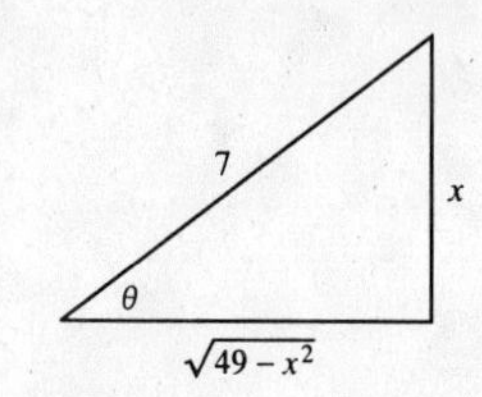

97. $\cot\theta = \dfrac{x}{10} \implies \sqrt{x^2 + 100} = 10 \cdot \csc\theta$

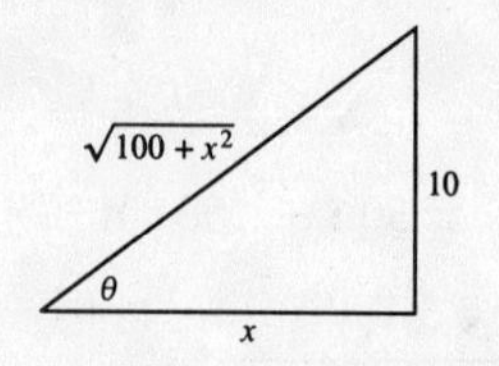

99. Given: $A = 32°, a = 11, B = 61°$

$C = 180° - 32° - 61° = 87°$

$b = \dfrac{a}{\sin A}(\sin B) = \dfrac{11}{\sin 32°}(\sin 61°) \approx 18.16$

$c = \dfrac{a}{\sin A}(\sin C) = \dfrac{11}{\sin 32°}(\sin 87°) \approx 20.73$

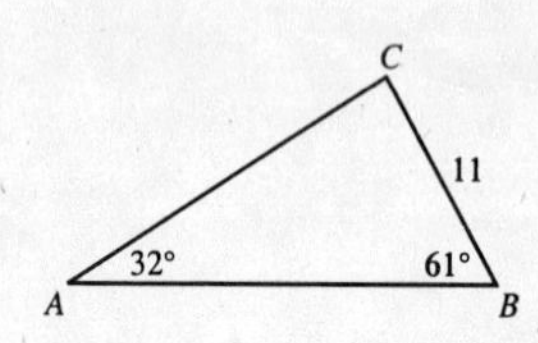

101. Given: $a = 12, b = 15, c = 24$

$\cos A = \dfrac{b^2 + c^2 - a^2}{2bc} = \dfrac{15^2 + 24^2 - 12^2}{2(15)(24)} = 0.9125 \implies A \approx 24.1°$

$\sin B = \dfrac{b \sin A}{a} \approx 0.5104 \implies B \approx 30.7°$

$C = 180° - 24.1° - 30.7° = 125.2°$

Section 6.4 Vectors and Dot Products

- Know the definition of the dot product of $\mathbf{u} = \langle u_1, u_2 \rangle$ and $\mathbf{v} = \langle v_1, v_2 \rangle$.

  $\mathbf{u} \cdot \mathbf{v} = u_1 v_1 + u_2 v_2$
- Know the following properties of the dot product:
  1. $\mathbf{u} \cdot \mathbf{v} = \mathbf{v} \cdot \mathbf{u}$
  2. $\mathbf{0} \cdot \mathbf{v} = 0$
  3. $\mathbf{u} \cdot (\mathbf{v} + \mathbf{w}) = \mathbf{u} \cdot \mathbf{v} + \mathbf{u} \cdot \mathbf{w}$
  4. $\mathbf{v} \cdot \mathbf{v} = \|\mathbf{v}\|^2$
  5. $c(\mathbf{u} \cdot \mathbf{v}) = c\mathbf{u} \cdot \mathbf{v} = \mathbf{u} \cdot c\mathbf{v}$
- If θ is the angle between two nonzero vectors $\mathbf{u}$ and $\mathbf{v}$, then

  $$\cos\theta = \frac{\mathbf{u} \cdot \mathbf{v}}{\|\mathbf{u}\|\,\|\mathbf{v}\|}.$$
- The vectors $\mathbf{u}$ and $\mathbf{v}$ are orthogonal if $\mathbf{u} \cdot \mathbf{v} = 0$.
- Know the definition of vector components. $\mathbf{u} = \mathbf{w}_1 + \mathbf{w}_2$ where $\mathbf{w}_1$ and $\mathbf{w}_2$ are orthogonal, and $\mathbf{w}_1$ is parallel to $\mathbf{v}$. $\mathbf{w}_1$ is called the projection of $\mathbf{u}$ onto $\mathbf{v}$ and is denoted by

  $$\mathbf{w}_1 = \text{proj}_{\mathbf{v}}\mathbf{u} = \left(\frac{\mathbf{u} \cdot \mathbf{v}}{\|\mathbf{v}\|^2}\right)\mathbf{v}.$$

  Then we have $\mathbf{w}_2 = \mathbf{u} - \mathbf{w}_1$.
- Know the definition of work.
  1. Projection form: $W = \|\text{proj}_{\overrightarrow{PQ}}\mathbf{F}\|\,\|\overrightarrow{PQ}\|$
  2. Dot product form: $W = \mathbf{F} \cdot \overrightarrow{PQ}$

Solutions to Odd-Numbered Exercises

1. $\mathbf{u} = \langle 3, 6 \rangle, \mathbf{v} = \langle 2, -4 \rangle$

$\mathbf{u} \cdot \mathbf{v} = 3 \cdot 2 + 6(-4) = -18$

3. $\mathbf{u} = 4\mathbf{i} - 7\mathbf{j}, \mathbf{v} = \mathbf{i} - \mathbf{j}$

$\mathbf{u} \cdot \mathbf{v} = 4(1) + (-7)(-1) = 11$

5. $\mathbf{u} = \langle 2, 2 \rangle$

$\mathbf{u} \cdot \mathbf{u} = 2(2) + 2(2) = 8$

The result is a scalar.

7. $\mathbf{u} = \langle 2, 2 \rangle, \mathbf{v} = \langle -3, 4 \rangle, w = \langle 1, -4 \rangle$

$\mathbf{u} \cdot \mathbf{v} = 2(-3) + 2(4) = 2$

$(\mathbf{u} \cdot \mathbf{v})\mathbf{w} = 2\langle 1, -4 \rangle = \langle 2, -8 \rangle$, vector

9. $\mathbf{u} = \langle 2, 2 \rangle, \mathbf{v} = \langle -3, 4 \rangle$

$\mathbf{u} \cdot 2\mathbf{v} = 2\mathbf{u} \cdot \mathbf{v} = 2(2) = 4$, scalar

11. $\mathbf{u} = \langle -5, 12 \rangle$

$\|\mathbf{u}\| = \sqrt{\mathbf{u} \cdot \mathbf{u}} = \sqrt{(-5)^2 + 12^2} = 13$

13. $\mathbf{u} = 20\mathbf{i} + 25\mathbf{j}$

$\|\mathbf{u}\| = \sqrt{(20)^2 + (25)^2} = \sqrt{1025} = 5\sqrt{41}$

15. $\mathbf{u} = 6\mathbf{j}$

$\|\mathbf{u}\| = \sqrt{\mathbf{u} \cdot \mathbf{u}} = \sqrt{6(6)} = 6$

17. $\mathbf{u} = \langle -1, 0 \rangle, \mathbf{v} = \langle 0, 2 \rangle$

$$\cos\theta = \frac{\mathbf{u} \cdot \mathbf{v}}{\|\mathbf{u}\|\,\|\mathbf{v}\|} = \frac{0}{(1)(2)} = 0 \implies \theta = 90^\circ$$

19. $\mathbf{u} = 3\mathbf{i} + 4\mathbf{j}, \mathbf{v} = -2\mathbf{i} + 3\mathbf{j}$

$$\cos\theta = \frac{\mathbf{u}\cdot\mathbf{v}}{\|\mathbf{u}\|\,\|\mathbf{v}\|} = \frac{-6+12}{(5)(\sqrt{13})} = \frac{6}{5\sqrt{13}}$$

$$\theta = \arccos\left(\frac{6}{5\sqrt{13}}\right) \approx 70.56°$$

21. $\mathbf{u} = 2\mathbf{i}, \mathbf{v} = -3\mathbf{j}$

$$\cos\theta = \frac{\mathbf{u}\cdot\mathbf{v}}{\|\mathbf{u}\|\,\|\mathbf{v}\|} = \frac{0}{(2)(3)} = 0 \Rightarrow \theta = 90°$$

23.

$$\mathbf{u} = \left(\cos\frac{\pi}{3}\right)\mathbf{i} + \left(\sin\frac{\pi}{3}\right)\mathbf{j} = \frac{1}{2}\mathbf{i} + \frac{\sqrt{3}}{2}\mathbf{j}$$

$$\mathbf{v} = \left(\cos\frac{3\pi}{4}\right)\mathbf{i} + \left(\sin\frac{3\pi}{4}\right)\mathbf{j} = -\frac{\sqrt{2}}{2}\mathbf{i} + \frac{\sqrt{2}}{2}\mathbf{j}$$

$$\|\mathbf{u}\| = \|\mathbf{v}\| = 1$$

$$\cos\theta = \frac{\mathbf{u}\cdot\mathbf{v}}{\|\mathbf{u}\|\,\|\mathbf{v}\|} = \mathbf{u}\cdot\mathbf{v} = \left(\frac{1}{2}\right)\left(-\frac{\sqrt{2}}{2}\right) + \left(\frac{\sqrt{3}}{2}\right)\left(\frac{\sqrt{2}}{2}\right) = \frac{-\sqrt{2}+\sqrt{6}}{4}$$

$$\theta = \arccos\left(\frac{-\sqrt{2}+\sqrt{6}}{4}\right) = 75° = \frac{5\pi}{12}$$

25. $\mathbf{u} = 3\mathbf{i} + 4\mathbf{j},\ \mathbf{v} = -7\mathbf{i} + 5\mathbf{j}$

$$\cos\theta = \frac{\mathbf{u}\cdot\mathbf{v}}{\|\mathbf{u}\|\,\|\mathbf{v}\|} = -\frac{1}{(5)(\sqrt{74})} \Rightarrow \theta \approx 91.33°$$

27. $\mathbf{u} = 5\mathbf{i} + 5\mathbf{j},\ \mathbf{v} = -8\mathbf{i} + 8\mathbf{j}$

$$\cos\theta = \frac{\mathbf{u}\cdot\mathbf{v}}{\|\mathbf{u}\|\,\|\mathbf{v}\|} = 0 \Rightarrow \theta = 90°$$

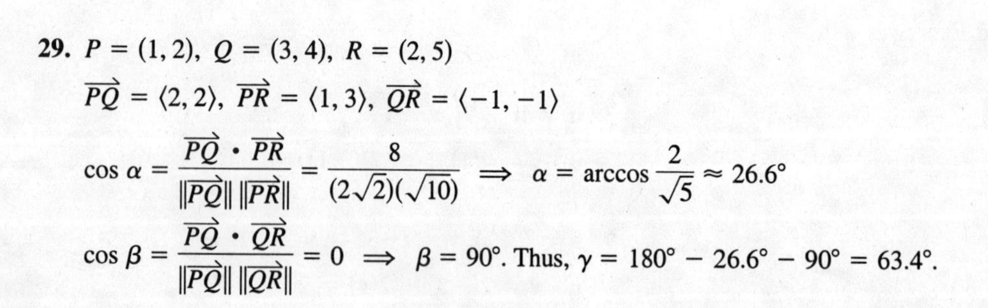

29. $P = (1, 2),\ Q = (3, 4),\ R = (2, 5)$

$\overrightarrow{PQ} = \langle 2, 2\rangle,\ \overrightarrow{PR} = \langle 1, 3\rangle,\ \overrightarrow{QR} = \langle -1, -1\rangle$

$$\cos\alpha = \frac{\overrightarrow{PQ}\cdot\overrightarrow{PR}}{\|\overrightarrow{PQ}\|\,\|\overrightarrow{PR}\|} = \frac{8}{(2\sqrt{2})(\sqrt{10})} \Rightarrow \alpha = \arccos\frac{2}{\sqrt{5}} \approx 26.6°$$

$$\cos\beta = \frac{\overrightarrow{PQ}\cdot\overrightarrow{QR}}{\|\overrightarrow{PQ}\|\,\|\overrightarrow{QR}\|} = 0 \Rightarrow \beta = 90°.\ \text{Thus, } \gamma = 180° - 26.6° - 90° = 63.4°.$$

31.

$$\mathbf{u}\cdot\mathbf{v} = \|\mathbf{u}\|\,\|\mathbf{v}\|\cos\theta$$

$$= (4)(10)\cos\frac{2\pi}{3}$$

$$= 40\left(-\frac{1}{2}\right)$$

$$= -20$$

33. $\mathbf{u} = \langle -12, 30\rangle,\ \mathbf{v} = \left\langle \frac{1}{2}, -\frac{5}{4}\right\rangle$

$\mathbf{u} = -24\mathbf{v} \Rightarrow$ **u and v are parallel.**

35. $\mathbf{u} = \frac{1}{4}(3\mathbf{i} - \mathbf{j}),\ \mathbf{v} = 5\mathbf{i} + 6\mathbf{j}$

$\mathbf{u} \neq k\mathbf{v} \implies$ Not parallel

$\mathbf{u} \cdot \mathbf{v} \neq 0 \implies$ Not orthogonal

Neither

37. $\mathbf{u} = 2\mathbf{i} - 2\mathbf{j},\ \mathbf{v} = -\mathbf{i} - \mathbf{j}$

$\mathbf{u} \cdot \mathbf{v} = 0 \implies$ **u** and **v** are orthogonal.

39. $\mathbf{u} = \langle 3, 4\rangle,\ \mathbf{v} = \langle 8, 2\rangle$

$$\mathbf{w}_1 = \text{proj}_{\mathbf{v}}\mathbf{u} = \left(\frac{\mathbf{u} \cdot \mathbf{v}}{\|\mathbf{v}\|^2}\right)\mathbf{v} = \left(\frac{32}{68}\right)\mathbf{v} = \frac{8}{17}\langle 8, 2\rangle = \frac{16}{17}\langle 4, 1\rangle$$

$$\mathbf{w}_2 = \mathbf{u} - \mathbf{w}_1 = \langle 3, 4\rangle - \frac{16}{17}\langle 4, 1\rangle = \frac{13}{17}\langle -1, 4\rangle$$

41. $\mathbf{u} = \langle 0, 3\rangle,\ \mathbf{v} = \langle 2, 15\rangle$

$$\mathbf{w}_1 = \text{proj}_{\mathbf{v}}\mathbf{u} = \left(\frac{\mathbf{u} \cdot \mathbf{v}}{\|\mathbf{v}\|^2}\right)\mathbf{v} = \frac{45}{229}\langle 2, 15\rangle$$

$$\mathbf{w}_2 = \mathbf{u} - \mathbf{w}_1 = \langle 0, 3\rangle - \frac{45}{229}\langle 2, 15\rangle = \left\langle -\frac{90}{229}, \frac{12}{229}\right\rangle = \frac{6}{229}\langle -15, 2\rangle$$

43. $\text{proj}_{\mathbf{v}}\mathbf{u} = \mathbf{u}$ since they are parallel.

$$\text{proj}_{\mathbf{v}}\mathbf{u} = \frac{\mathbf{u} \cdot \mathbf{v}}{\|\mathbf{v}\|^2}\mathbf{v} = \frac{18 + 18}{36 + 16}\mathbf{v} = \frac{26}{52}\langle 6, 4\rangle = \langle 3, 2\rangle = \mathbf{u}$$

45. $\text{proj}_{\mathbf{v}}\mathbf{u} = \mathbf{0}$ since they are perpendicular.

$$\text{proj}_{\mathbf{v}}\mathbf{u} = \frac{\mathbf{u} \cdot \mathbf{v}}{\|\mathbf{v}\|^2}\mathbf{v} = \mathbf{0}, \text{ since } \mathbf{u} \cdot \mathbf{v} = 0.$$

47. $\mathbf{u} = \langle 4, 7\rangle$

For **v** to be orthogonal to **u**, $\mathbf{u} \cdot \mathbf{v}$ must be equal 0.
Two possibilities: $\langle 7, -4\rangle$ and $\langle -7, 4\rangle$

49. $\mathbf{u} = \frac{1}{2}\mathbf{i} - \frac{3}{4}\mathbf{j}$

For **v** to be orthogonal to **u**, $\mathbf{u} \cdot \mathbf{v}$ must equal 0.
Two possibilities: $\left\langle \frac{3}{4}, \frac{1}{2}\right\rangle$ and $\left\langle -\frac{3}{4}, -\frac{1}{2}\right\rangle$

51. $W = \|\text{proj}_{\overrightarrow{PQ}}\mathbf{v}\|\,\|\overrightarrow{PQ}\|$ where $\overrightarrow{PQ} = \langle 4, 7\rangle$ and $\mathbf{v} = \langle 1, 4\rangle$.

$$\text{proj}_{\overrightarrow{PQ}}\mathbf{v} = \left(\frac{\mathbf{v} \cdot \overrightarrow{PQ}}{\|\overrightarrow{PQ}\|^2}\right)\overrightarrow{PQ} = \left(\frac{32}{65}\right)\langle 4, 7\rangle$$

$$W = \|\text{proj}_{\overrightarrow{PQ}}\mathbf{v}\|\,\|\overrightarrow{PQ}\| = \left(\frac{32\sqrt{65}}{65}\right)(\sqrt{65}) = 32$$

53. $\mathbf{u} = \langle 1245, 2600\rangle,\ \mathbf{v} = \langle 12.20, 8.50\rangle$

$\mathbf{u} \cdot \mathbf{v} = 1245(12.20) + 2600(8.50) = \$37{,}289$

This gives the total revenue that can be earned by selling all of the units.

55. (a) $\mathbf{F} = -36{,}000\mathbf{j}$ Gravitational force

$\mathbf{v} = (\cos 10°)\mathbf{i} + (\sin 10°)\mathbf{j}$

$$\mathbf{w}_1 = \text{proj}_{\mathbf{v}}\mathbf{F} = \left(\frac{\mathbf{F} \cdot \mathbf{v}}{\|\mathbf{v}\|^2}\right)\mathbf{v} = (\mathbf{F} \cdot \mathbf{v})\mathbf{v} \approx -6251.3\mathbf{v}$$

The magnitude of this force is 6251.3, therefore a force of 6251.3 pounds is needed to keep the truck from rolling down the hill.

(b) $\mathbf{w}_2 = \mathbf{F} - \mathbf{w}_1 = -36{,}000\mathbf{j} + 6251.3\,(\cos 10°\mathbf{i} + \sin 10°\mathbf{j})$

$= [(6251.3 \cos 10°)\mathbf{i} + (6251.3 \sin 10° - 36{,}000)\mathbf{j}]$

$\|\mathbf{w}_2\| \approx 35{,}453.1$ pounds

57. $W = (245)(3) = 735$ Newton-meters

59. $W = (\cos 30°)(45)(20) \approx 779.4$ foot-pounds

61. $W = (\cos 25°)(20)(40) \approx 725.05$ ft/pounds

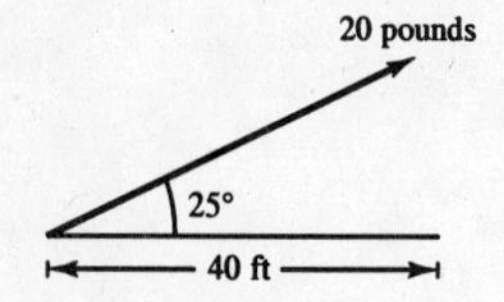

63. True. $\mathbf{u} \cdot \mathbf{v} = 0$

65. $\mathbf{u} \cdot \mathbf{v} = 0 \Rightarrow$ they are orthogonal(unit vectors)

67. (a) $\text{proj}_{\mathbf{v}}\, \mathbf{u} = \mathbf{u} \Rightarrow \mathbf{u}$ and $\mathbf{v}$ are parallel

(b) $\text{proj}_{\mathbf{v}}\, \mathbf{u} = 0 \Rightarrow \mathbf{u}$ and $\mathbf{v}$ are orthogonal

69. Let $\mathbf{u}$ and $\mathbf{v}$ be two sides of the rhombus $\|\mathbf{u}\| = \|\mathbf{v}\|$. The diagonals are $\mathbf{u} + \mathbf{v}$ and $\mathbf{u} - \mathbf{v}$.

$$\begin{aligned}(\mathbf{u} + \mathbf{v}) \cdot (\mathbf{u} - \mathbf{v}) &= \mathbf{u} \cdot \mathbf{u} + \mathbf{v} \cdot \mathbf{u} - \mathbf{u} \cdot \mathbf{v} - \mathbf{v} \cdot \mathbf{v} \\ &= \|\mathbf{u}\|^2 - \|\mathbf{v}\|^2 \\ &= 0\end{aligned}$$

Hence, the diagonals are perpendicular.

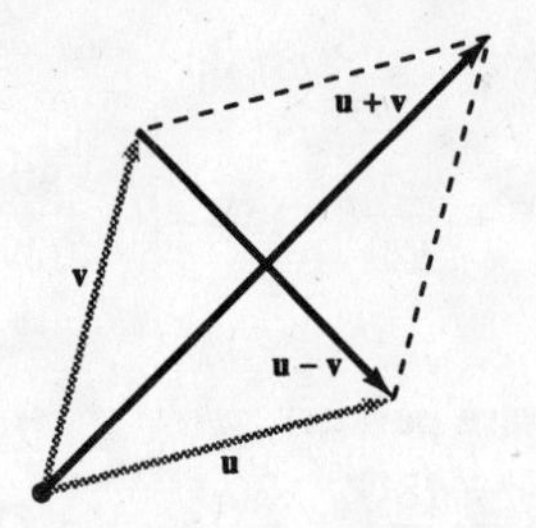

71. (a) $\mathbf{0} \cdot \mathbf{v} = \langle 0, 0\rangle \cdot \langle v_1, v_2\rangle = 0v_1 + 0v_2 = 0$

(b)
$$\begin{aligned}\mathbf{u} \cdot (\mathbf{v} + \mathbf{w}) &= \langle u_1, u_2\rangle \cdot \langle v_1 + w_1, v_2 + w_2\rangle \\ &= u_1(v_1 + w_1) + u_2(v_2 + w_2) \\ &= u_1v_1 + u_1w_1 + u_2v_2 + u_2w_2 \\ &= \langle u_1, u_2\rangle \cdot \langle v_1, v_2\rangle + \langle u_1, u_2\rangle \cdot \langle w_1, w_2\rangle \\ &= \mathbf{u} \cdot \mathbf{v} + \mathbf{u} \cdot \mathbf{w}\end{aligned}$$

73.
$$\begin{aligned}2\cos(x + \pi) + 2\cos(x - \pi) &= 0 \\ -2\cos x - 2\cos x &= 0 \\ -4\cos x &= 0 \\ \cos x &= 0 \\ x &= \frac{\pi}{2}, \frac{3\pi}{2}\end{aligned}$$

75.
$$\sin\left(x - \frac{\pi}{3}\right) - \sin\left(x + \frac{\pi}{3}\right) = \frac{3}{2}$$

$$\begin{aligned}\left[\sin x \cos\frac{\pi}{3} - \cos x \sin\frac{\pi}{3}\right] - \left[\sin x \cdot \cos\frac{\pi}{3} + \cos x \sin\frac{\pi}{3}\right] &= \frac{3}{2} \\ (-2\cos x)\left(\frac{\sqrt{3}}{2}\right) &= \frac{3}{2} \\ \cos x &= -\frac{\sqrt{3}}{2} \\ x &= \frac{5\pi}{6}, \frac{7\pi}{6}\end{aligned}$$

77. $s = \dfrac{a+b+c}{2} = \dfrac{8+15+16}{2} = 19.5$

Area $= \sqrt{s(s-a)(s-b)(s-c)} = \sqrt{19.5(19.5-8)(19.5-15)(19.5-16)} \approx 59.43$ sq. units

79. (a) $\mathbf{u} + \mathbf{v} = \langle 3, 0\rangle + \langle 4, -1\rangle = \langle 7, -1\rangle$

(b) $2\mathbf{v} - \mathbf{u} = 2\langle 4, -1\rangle - \langle 3, 0\rangle = \langle 5, -2\rangle$

(c) $3\mathbf{u} - 5\mathbf{v} = 3\langle 3, 0\rangle - 5\langle 4, -1\rangle = \langle -11, 5\rangle$

81. (a) $\mathbf{u} + \mathbf{v} = \langle -2, -2\rangle + \langle -4, 5\rangle = \langle -6, 3\rangle$

(b) $2\mathbf{v} - \mathbf{u} = 2\langle -4, 5\rangle - \langle -2, -2\rangle = \langle -6, 12\rangle$

(c) $3\mathbf{u} - 5\mathbf{v} = 3\langle -2, -2\rangle - 5\langle -4, 5\rangle = \langle 14, -31\rangle$

83. The car will cost $(1.04)(23{,}500) = \$24{,}440$ in one month, \$940 over the present price. This is more than the \$725 interest penalty. Buy now.

85. Let x be the number of people presently in the group. Each share is $\dfrac{250{,}000}{x}$.

Also, $\dfrac{250{,}000}{x} - 6250 = \dfrac{250{,}000}{x+2}$

Solving this equation, $x = 8$.

Section 6.5 Trigonometric Form of a Complex Number

- You should be able to graphically represent complex numbers.
- The absolute value of the complex numbers $z = a + bi$ is $|z| = \sqrt{a^2 + b^2}$.
- The trigonometric form of the complex number $z = a + bi$ is $z = r(\cos\theta + i\sin\theta)$ where
  - (a) $a = r\cos\theta$
  - (b) $b = r\sin\theta$
  - (c) $r = \sqrt{a^2 + b^2}$; r is called the modulus of z.
  - (d) $\tan\theta = b/a$; θ is called the argument of z.
- Given $z_1 = r_1(\cos\theta_1 + i\sin\theta_1)$ and $z_2 = r_2(\cos\theta_2 + i\sin\theta_2)$:
  - (a) $z_1 z_2 = r_1 r_2[\cos(\theta_1 + \theta_2) + i\sin(\theta_1 + \theta_2)]$
  - (b) $\dfrac{z_1}{z_2} = \dfrac{r_1}{r_2}[\cos(\theta_1 - \theta_2) + i\sin(\theta_1 - \theta_2)]$, $z_2 \neq 0$
- You should know DeMoivre's Theorem: If $z = r(\cos\theta + i\sin\theta)$, then for any positive integer n, $z^n = r^n(\cos n\theta + i\sin n\theta)$.
- You should know that for any positive integer n, $z = r(\cos\theta + i\sin\theta)$ has n distinct nth roots given by

  $$\sqrt[n]{r}\left[\cos\left(\frac{\theta + 2\pi k}{n}\right) + i\sin\left(\frac{\theta + 2\pi k}{n}\right)\right]$$

  where $k = 0, 1, 2, \ldots, n - 1$.

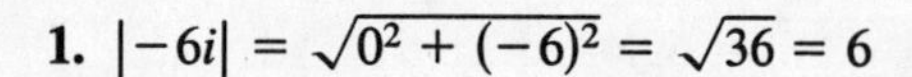

Solutions to Odd-Numbered Exercises

1. $|-6i| = \sqrt{0^2 + (-6)^2} = \sqrt{36} = 6$

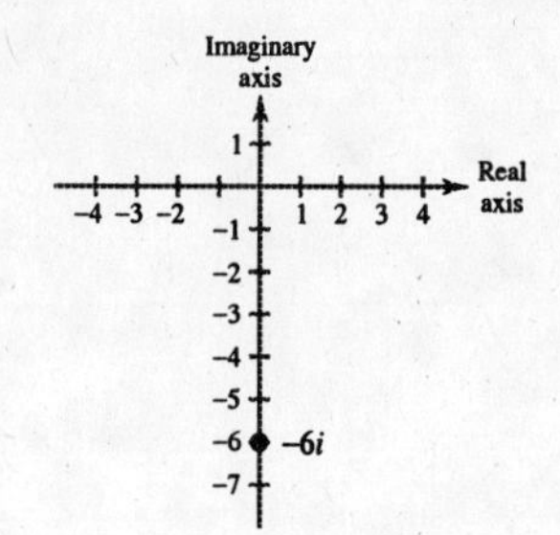

3. $|-4| = \sqrt{(-4)^2 + 0^2} = \sqrt{16} = 4$

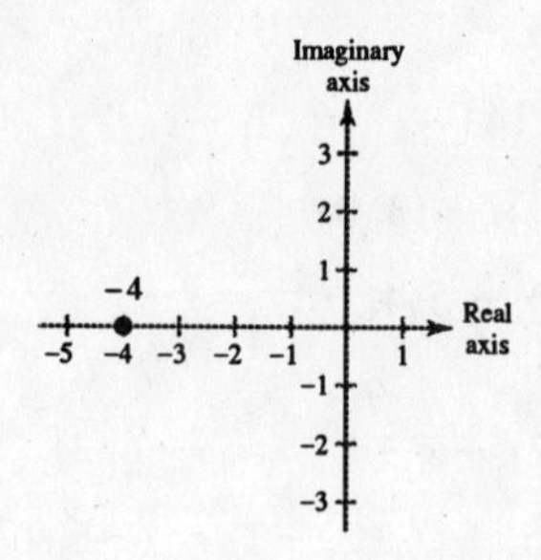

5. $|-4 + 4i| = \sqrt{(-4)^2 + (4)^2}$

$= \sqrt{32} = 4\sqrt{2}$

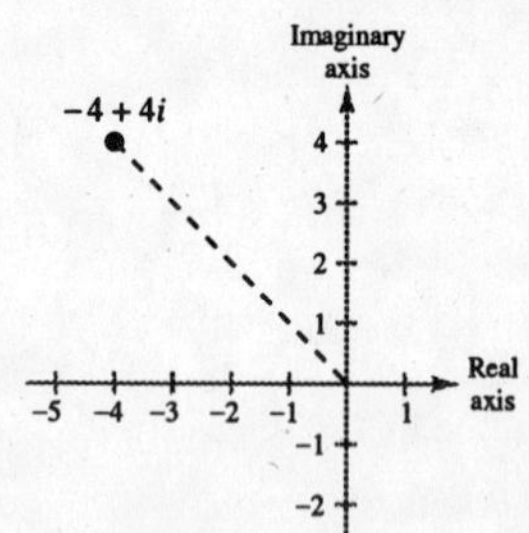

7. $|9 - 7i| = \sqrt{9^2 + (-7)^2} = \sqrt{130}$

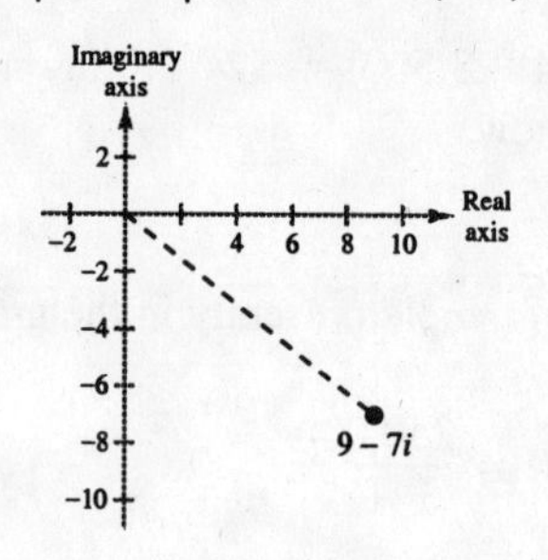

9. $z = 3i$

$r = \sqrt{0^2 + 3^2} = \sqrt{9} = 3$

$\tan\theta = \dfrac{3}{0}$, undefined $\Rightarrow \theta = \dfrac{\pi}{2}$

$z = 3\left(\cos\dfrac{\pi}{2} + i\sin\dfrac{\pi}{2}\right)$

11. $z = -2 - 2i$

$r = \sqrt{(-2)^2 + (-2)^2} = \sqrt{8} = 2\sqrt{2}$

$\tan\theta = \dfrac{-2}{-2} = 1$, θ is in Quadrant III.

$\theta = \dfrac{5\pi}{4}$

$z = 2\sqrt{2}\left(\cos\dfrac{5\pi}{4} + i\sin\dfrac{5\pi}{4}\right)$

13. $z = 5 - 5i$

$r = \sqrt{5^2 + (-5)^2} = \sqrt{50} = 5\sqrt{2}$

$\tan\theta = -\dfrac{5}{5} = -1 \Rightarrow \theta = \dfrac{7\pi}{4}$

$z = 5\sqrt{2}\left(\cos\dfrac{7\pi}{4} + i\sin\dfrac{7\pi}{4}\right)$

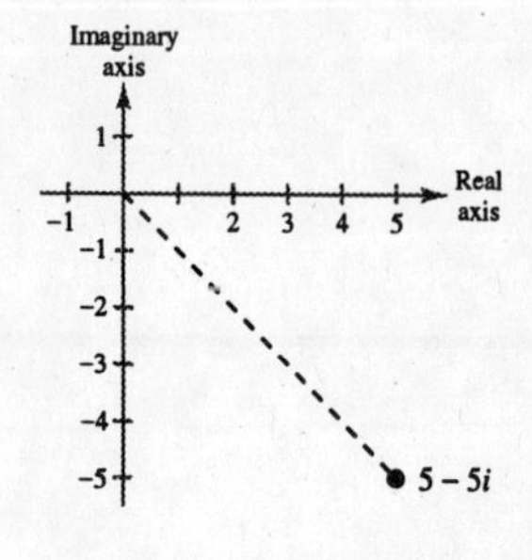

15. $z = \sqrt{3} + i$

$r = \sqrt{\left(\sqrt{3}\right)^2 + 1^2} = \sqrt{4} = 2$

$\tan\theta = \dfrac{1}{\sqrt{3}} = \dfrac{\sqrt{3}}{3} \Rightarrow \theta = \dfrac{\pi}{6}$

$z = 2\left(\cos\dfrac{\pi}{6} + i\sin\dfrac{\pi}{6}\right)$

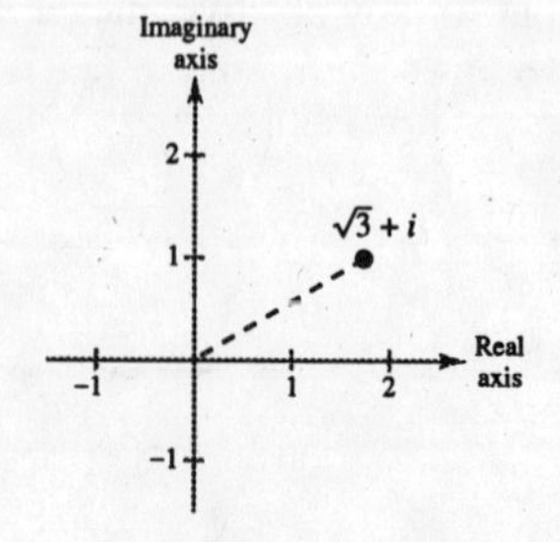

17. 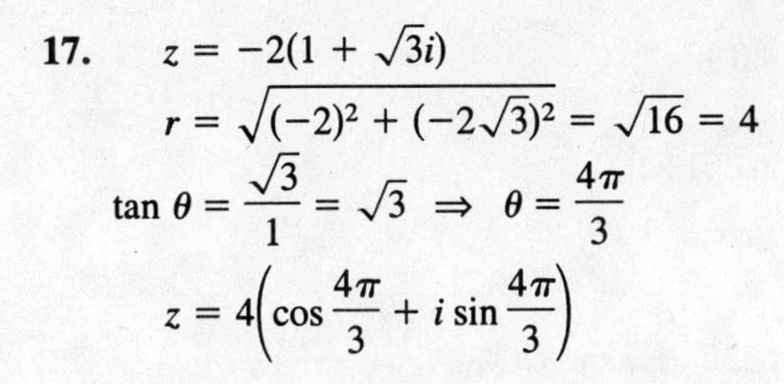

$z = -2(1 + \sqrt{3}i)$

$r = \sqrt{(-2)^2 + (-2\sqrt{3})^2} = \sqrt{16} = 4$

$\tan\theta = \dfrac{\sqrt{3}}{1} = \sqrt{3} \Rightarrow \theta = \dfrac{4\pi}{3}$

$z = 4\left(\cos\dfrac{4\pi}{3} + i\sin\dfrac{4\pi}{3}\right)$

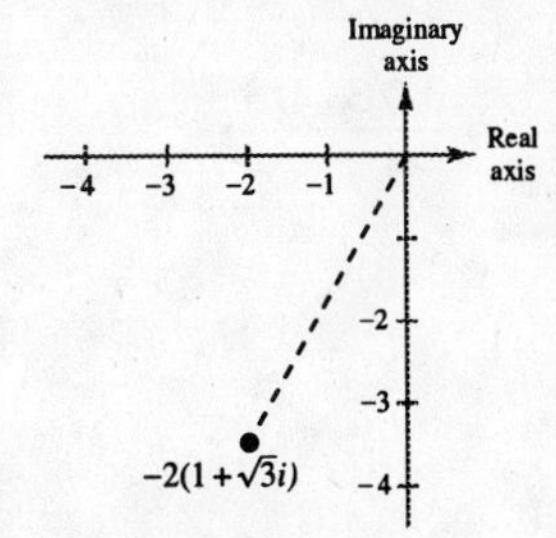

19. $z = 8i$

$r = \sqrt{0^2 + 8^2} = \sqrt{64} = 8$

$\tan\theta = \dfrac{8}{0}$ undefined $\Rightarrow \theta = \dfrac{\pi}{2}$

$z = 8\left(\cos\dfrac{\pi}{2} + i\sin\dfrac{\pi}{2}\right)$

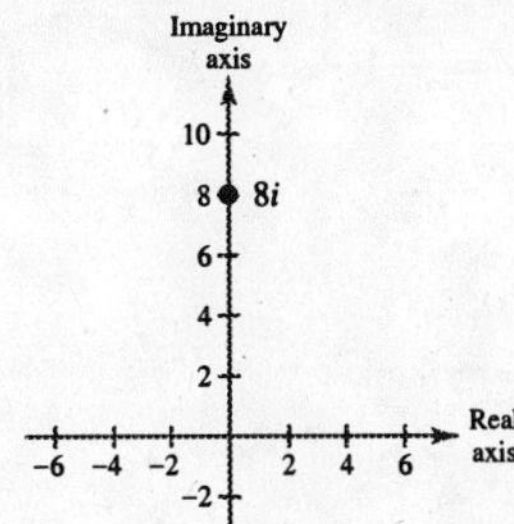

21. $z = -7 + 4i$

$r = \sqrt{(-7)^2 + (4)^2} = \sqrt{65}$

$\tan\theta = \dfrac{4}{-7} \Rightarrow \theta = 2.62$

$z \approx \sqrt{65}(\cos 2.62 + i\sin 2.62)(150.26°)$

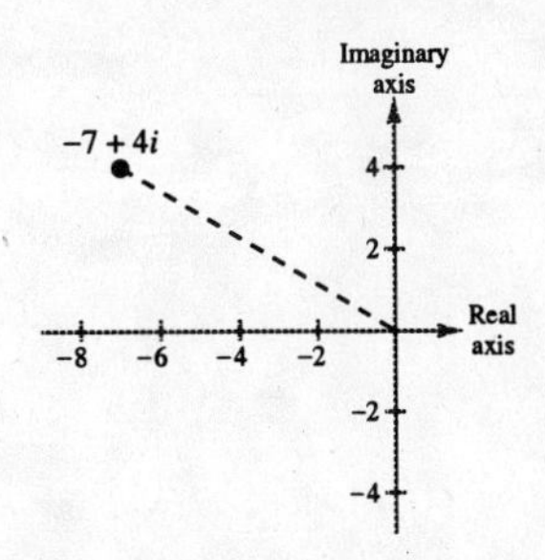

23. $z = 7 + 0i$

$r = \sqrt{(7)^2 + (0)^2} = \sqrt{49} = 7$

$\tan\theta = \dfrac{0}{7} = 0 \Rightarrow \theta = 0$

$z = 7(\cos 0 + i\sin 0)$

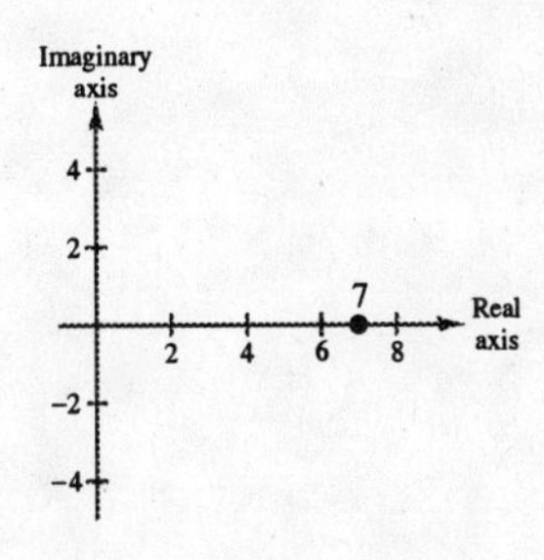

25. $z = 1 + 10i$

$r = \sqrt{1^2 + 10^2} = \sqrt{101}$

$\tan\theta = \dfrac{10}{1} = 10 \Rightarrow \theta \approx 1.47(84.29°)$

$z = \sqrt{101}(\cos 1.47 + i\sin 1.47)$

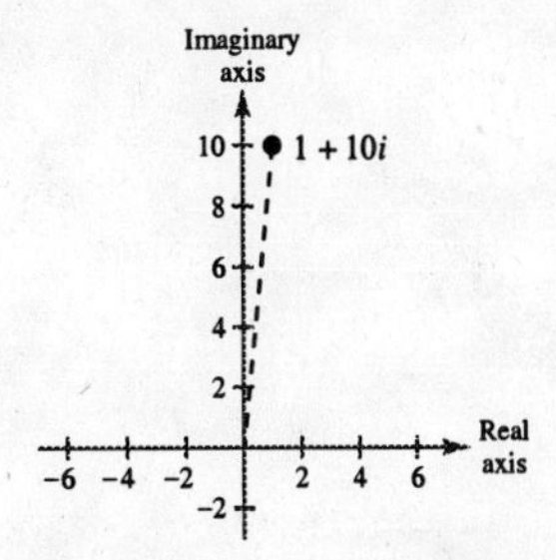

27. 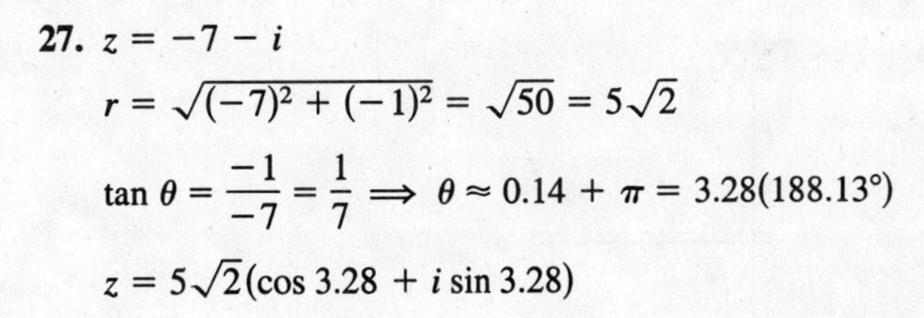

$z = -7 - i$

$r = \sqrt{(-7)^2 + (-1)^2} = \sqrt{50} = 5\sqrt{2}$

$\tan\theta = \dfrac{-1}{-7} = \dfrac{1}{7} \Rightarrow \theta \approx 0.14 + \pi = 3.28(188.13°)$

$z = 5\sqrt{2}(\cos 3.28 + i\sin 3.28)$

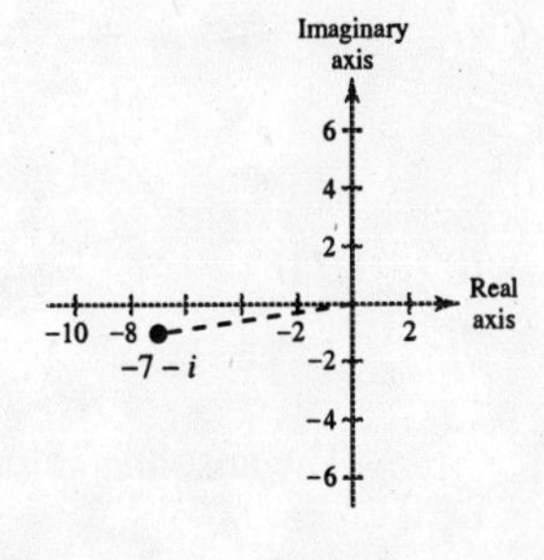

29. $5 + 2i \approx 5.39(\cos 0.38 + i \sin 0.38) \approx 5.39(\cos 21.80° + i \sin 21.80°)$

31. $3\sqrt{2} - 7i \approx 8.19(\cos 5.26 + i \sin 5.26) \approx 8.19(\cos 301.22° + i \sin 301.22°)$

33. $2(\cos 120° + i \sin 120°) = 2\left(-\frac{1}{2} + \frac{\sqrt{3}}{2}i\right)$

$= -1 + \sqrt{3}i$

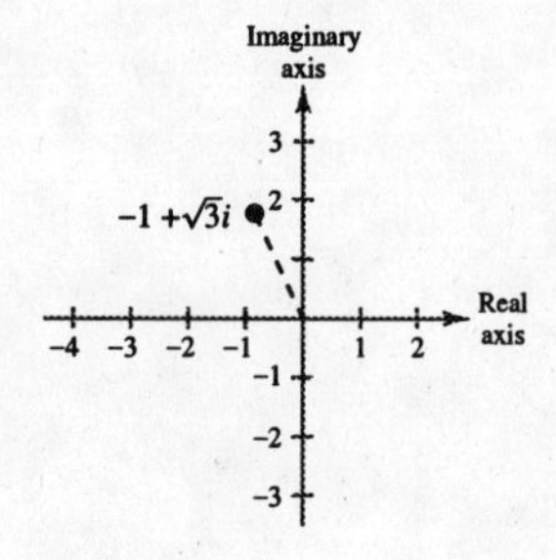

35. $\frac{3}{2}(\cos 330° + i \sin 330°) = \frac{3}{2}\left(\frac{\sqrt{3}}{2} - \frac{1}{2}i\right)$

$= \frac{3\sqrt{3}}{4} - \frac{3}{4}i$

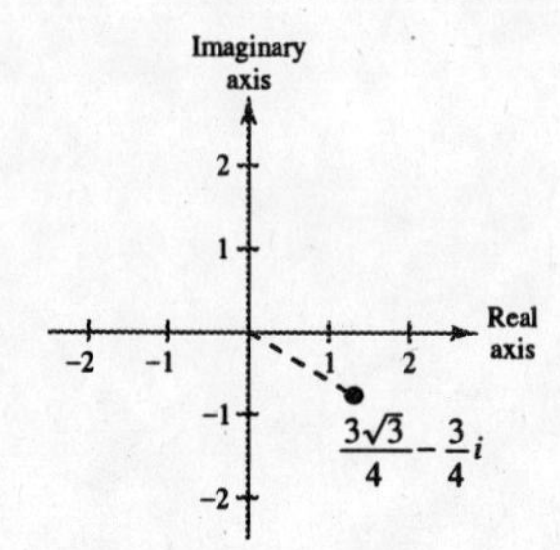

37. $3.75\left(\cos \frac{3\pi}{4} + i \sin \frac{3\pi}{4}\right) = -\frac{15\sqrt{2}}{8} + \frac{15\sqrt{2}}{8}i$

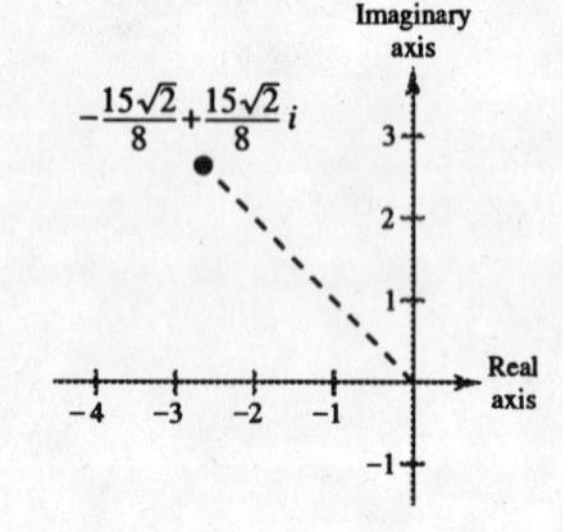

39. $4\left(\cos \frac{3\pi}{2} + i \sin \frac{3\pi}{2}\right) = 4(0 - i) = -4i$

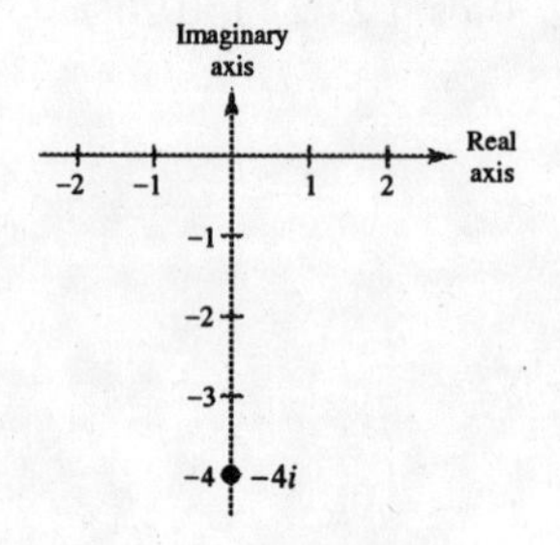

41. $3[\cos(18° 45') + i \sin(18° 45')] \approx 2.8408 + 0.9643i$

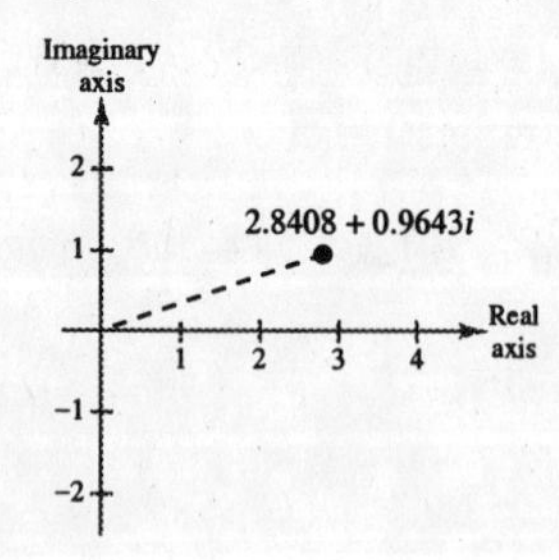

43. $5\left(\cos \frac{\pi}{9} + i \sin \frac{\pi}{9}\right) \approx 4.70 + 1.71i$

45. $9(\cos 58° + i \sin 58°) \approx 4.77 + 7.63i$

47. $z^3 = \frac{\sqrt{2}}{2}(-1 + i)$, $z^2 = i$, $z = \frac{\sqrt{2}}{2}(1 + i)$, $z^4 = -1$

The absolute value of each power is 1.

49. $\left[3\left(\cos\frac{\pi}{3} + i\sin\frac{\pi}{3}\right)\right]\left[4\left(\cos\frac{\pi}{6} + i\sin\frac{\pi}{6}\right)\right] = (3)(4)\left[\cos\left(\frac{\pi}{3} + \frac{\pi}{6}\right) + i\sin\left(\frac{\pi}{6} + \frac{\pi}{3}\right)\right]$

$= 12\left(\cos\frac{\pi}{2} + i\sin\frac{\pi}{2}\right)$

51. $\left[\frac{5}{3}(\cos 140° + i\sin 140°)\right]\left[\frac{2}{3}(\cos 60° + i\sin 60°)\right] = \left(\frac{5}{3}\right)\left(\frac{2}{3}\right)[\cos(140° + 60°) + i\sin(140° + 60°)]$

$= \frac{10}{9}(\cos 200° + i\sin 200°)$

53. $\left[\frac{11}{20}(\cos 290° + i\sin 290°)\right]\left[\frac{2}{5}(\cos 200° + i\sin 200°)\right] = \left(\frac{11}{20}\right)\left(\frac{2}{5}\right)[\cos(290° + 200°) + i\sin(290° + 200°)]$

$= \frac{11}{50}(\cos 490° + i\sin 490°)$

$= \frac{11}{50}(\cos 130° + i\sin 130°)$

55. $\dfrac{\cos 50° + i\sin 50°}{\cos 20° + i\sin 20°} = \cos(50° - 20°) + i\sin(50° - 20°)$

$= \cos 30° + i\sin 30°$

57. $\dfrac{2(\cos 120° + i\sin 120°)}{4(\cos 40° + i\sin 40°)} = \frac{1}{2}[\cos(120° - 40°) + i\sin(120° - 40°)]$

$= \frac{1}{2}(\cos 80° + i\sin 80°)$

59. $\dfrac{18(\cos 54° + i\sin 54°)}{3(\cos 102° + i\sin 102°)} = 6(\cos(54° - 102°) + i\sin(54° - 102°))$

$= 6(\cos(-48°) + i\sin(-48°))$

61. (a) $2 + 2i = 2\sqrt{2}(\cos 45° + i\sin 45°)$

$1 - i = \sqrt{2}[\cos(-45°) + i\sin(-45°)]$

(b) $(2 + 2i)(1 - i) = [2\sqrt{2}(\cos 45° + i\sin 45°)][\sqrt{2}(\cos(-45°) + i\sin(-45°))] = 4(\cos 0° + i\sin 0°) = 4.$

(c) $(2 + 2i)(1 - i) = 2 - 2i + 2i - 2i^2 = 2 + 2 = 4$

63. (a) $-2i = 2[\cos(-90°) + i\sin(-90°)]$

$1 + i = \sqrt{2}(\cos 45° + i\sin 45°)$

(b) $-2i(1 + i) = 2[\cos(-90°) + i\sin(-90°)][\sqrt{2}(\cos 45° + i\sin 45°)]$

$= 2\sqrt{2}[\cos(-45°) + i\sin(-45°)]$

$= 2\sqrt{2}\left[\frac{1}{\sqrt{2}} - \frac{1}{\sqrt{2}}i\right] = 2 - 2i$

(c) $-2i(1 + i) = -2i - 2i^2 = -2i + 2 = 2 - 2i$

65. (a) $5 = 5(\cos 0° + i\sin 0°)$

$2 + 3i \approx \sqrt{13}(\cos 56.31° + i\sin 56.31°)$

(b) $\dfrac{5}{2 + 3i} \approx \dfrac{5(\cos 0° + i\sin 0°)}{\sqrt{13}(\cos 56.31° + i\sin 56.31°)} = \dfrac{5\sqrt{13}}{13}(\cos -56.31° + i\sin -56.31) \approx 0.7692 - 1.154i$

(c) $\dfrac{5}{2 + 3i} = \dfrac{5}{2 + 3i} \cdot \dfrac{2 - 3i}{2 - 3i} = \dfrac{10 - 15i}{13} = \dfrac{10}{13} - \dfrac{15}{13}i \approx 0.7692 - 1.154i$

67. Let $z = x + iy$ such that:

$|z| = 2 \Rightarrow 2 = \sqrt{x^2 + y^2}$

$\Rightarrow 4 = x^2 + y^2$: circle with radius of 2

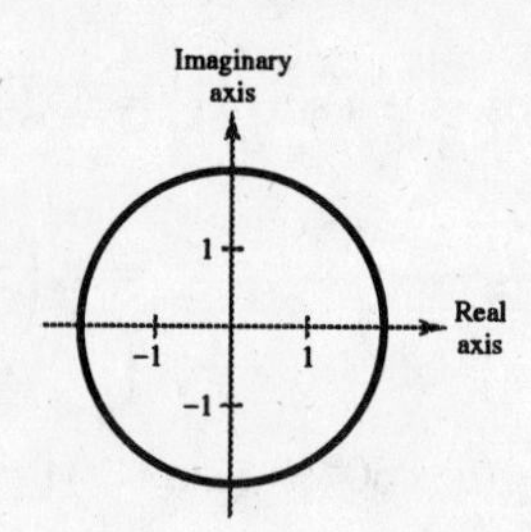

69. $\theta = \dfrac{\pi}{6}$

Let $z = x + iy$ such that $\tan\dfrac{\pi}{6} = \dfrac{y}{x} \Rightarrow \dfrac{y}{x} = \dfrac{1}{\sqrt{3}} \Rightarrow y = \dfrac{1}{\sqrt{3}}x$; line

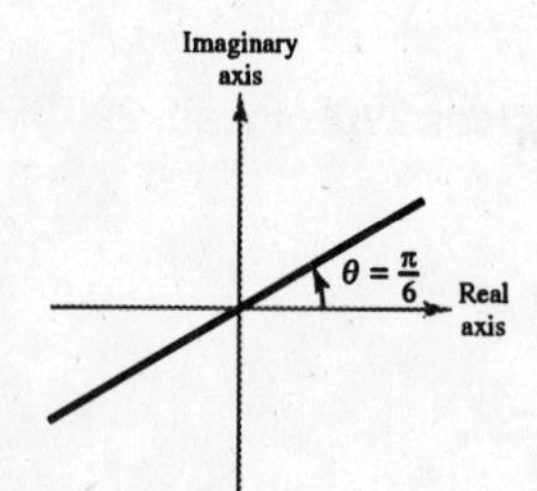

71.
$$\begin{aligned}(1 + i)^3 &= \left[\sqrt{2}\left(\cos\frac{\pi}{4} + i\sin\frac{\pi}{4}\right)\right]^3\\ &= (\sqrt{2})^3\left(\cos\frac{3\pi}{4} + i\sin\frac{3\pi}{4}\right)\\ &= 2\sqrt{2}\left(-\frac{\sqrt{2}}{2} + \frac{\sqrt{2}}{2}i\right)\\ &= -2 + 2i\end{aligned}$$

73.
$$\begin{aligned}(-1 + i)^{10} &= \left[\sqrt{2}\left(\cos\frac{3\pi}{4} + i\sin\frac{3\pi}{4}\right)\right]^{10}\\ &= (\sqrt{2})^{10}\left(\cos\frac{30\pi}{4} + i\sin\frac{30\pi}{4}\right)\\ &= 32\left[\cos\left(\frac{3\pi}{2} + 6\pi\right) + i\sin\left(\frac{3\pi}{2} + 6\pi\right)\right]\\ &= 32\left(\cos\frac{3\pi}{2} + i\sin\frac{3\pi}{2}\right)\\ &= 32[0 + i(-1)]\\ &= -32i\end{aligned}$$

75.
$$\begin{aligned}2(\sqrt{3} + i)^5 &= 2\left[2\left(\cos\frac{\pi}{6} + i\sin\frac{\pi}{6}\right)\right]^5\\ &= 2\left[2^5\left(\cos\frac{5\pi}{6} + i\sin\frac{5\pi}{6}\right)\right]\\ &= 64\left(-\frac{\sqrt{3}}{2} + \frac{1}{2}i\right)\\ &= -32\sqrt{3} + 32i\end{aligned}$$

77. $[5(\cos 20^\circ + i\sin 20^\circ)]^3 = 5^3(\cos 60^\circ + i\sin 60^\circ) = \dfrac{125}{2} + \dfrac{125\sqrt{3}}{2}i$

79.
$$\begin{aligned}\left(\cos\frac{5\pi}{4} + i\sin\frac{5\pi}{4}\right)^{10} &= \cos\frac{25\pi}{2} + i\sin\frac{25\pi}{2}\\ &= \cos\left(12\pi + \frac{\pi}{2}\right) + i\sin\left(12\pi + \frac{\pi}{2}\right) = \cos\frac{\pi}{2} + i\sin\frac{\pi}{2} = i\end{aligned}$$

81.
$$\begin{aligned}[4(\cos 2.8 + i\sin 2.8)]^5 &= 4^5(\cos 14 + i\sin 14)\\ &\approx 140.02 + 1014.38i\end{aligned}$$

83. $(3 - 2i)^5 = -597 - 122i$

85. $[3(\cos 15° + i \sin 15°)]^4 = 81(\cos 60° + i \sin 60°)$

$= \dfrac{81}{2} + \dfrac{81\sqrt{3}}{2}i$

87. $\left[-\dfrac{1}{2}(1 + \sqrt{3}i)\right]^6 = \left[\cos\dfrac{4\pi}{3} + i \sin\dfrac{4\pi}{3}\right]^6$

$= \cos 8\pi + i \sin 8\pi$

$= 1$

89. (a) In trigonometric form we have: $2(\cos 30° + i \sin 30°)$

$2(\cos 150° + i \sin 150°)$

$2(\cos 270° + i \sin 270°)$

(b) There are three roots evenly spaced around a circle of radius 2. Therefore, they represent the cube roots of some number of modulus 8. Cubing them shows that they are all cube roots of $8i$.

(c) $[2(\cos 30° + i \sin 30°)]^3 = 8i$

$[2(\cos 150° + i \sin 150°)]^3 = 8i$

$[2(\cos 270° + i \sin 270°)]^3 = 8i$

91. (a) Square roots of $5(\cos 120° + i \sin 120°)$:

$\sqrt{5}\left[\cos\left(\dfrac{120° + 360°k}{2}\right) + i \sin\left(\dfrac{120° + 360°k}{2}\right)\right],\ k = 0, 1$

$\sqrt{5}(\cos 60° + i \sin 60°)$

$\sqrt{5}(\cos 240° + i \sin 240°)$

(b)

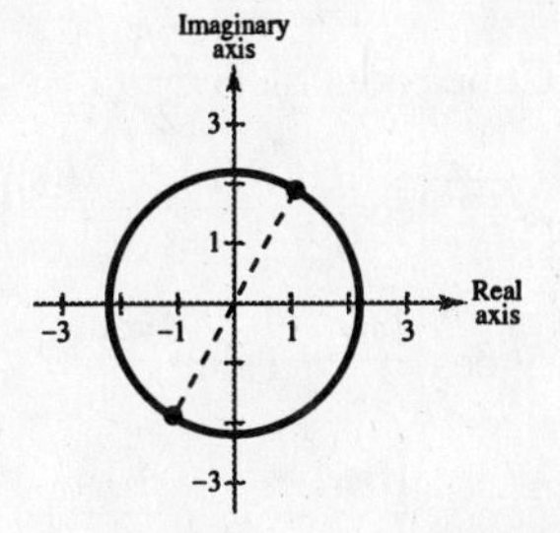

(c) $\dfrac{\sqrt{5}}{2} + \dfrac{\sqrt{15}}{2}i,\ -\dfrac{\sqrt{5}}{2} - \dfrac{\sqrt{15}}{2}i$

93. (a) Fourth roots of $16\left(\cos\dfrac{4\pi}{3} + i \sin\dfrac{4\pi}{3}\right)$:

$\sqrt[4]{16}\left[\cos\left(\dfrac{(4\pi/3) + 2k\pi}{4}\right) + i \sin\left(\dfrac{(4\pi/3) + 2k\pi}{4}\right)\right],\ k = 0, 1, 2, 3$

$2\left(\cos\dfrac{\pi}{3} + i \sin\dfrac{\pi}{3}\right)$

$2\left(\cos\dfrac{5\pi}{6} + i \sin\dfrac{5\pi}{6}\right)$

$2\left(\cos\dfrac{4\pi}{3} + i \sin\dfrac{4\pi}{3}\right)$

$2\left(\cos\dfrac{11\pi}{6} + i \sin\dfrac{11\pi}{6}\right)$

(b)

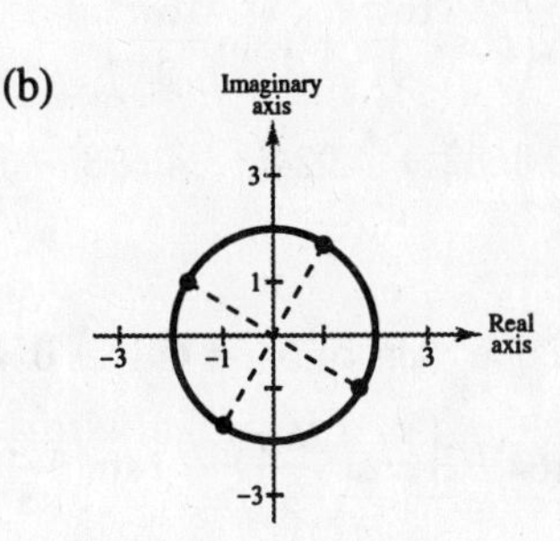

(c) $1 + \sqrt{3}i,\ -\sqrt{3} + i,\ -1 - \sqrt{3}i,\ \sqrt{3} - i$

95. (a) Cube roots of $-27i = 27\left(\cos\frac{3\pi}{2} + i\sin\frac{3\pi}{2}\right)$

$$(27)^{1/3}\left[\cos\left(\frac{\frac{3\pi}{2} + 2k\pi}{3}\right) + i\sin\left(\frac{\frac{3\pi}{2} + 2k\pi}{3}\right)\right],\ k = 0, 1, 2$$

$$3\left(\cos\frac{\pi}{2} + i\sin\frac{\pi}{2}\right)$$

$$3\left(\cos\frac{7\pi}{6} + i\sin\frac{7\pi}{6}\right)$$

$$3\left(\cos\frac{11\pi}{6} + i\sin\frac{11\pi}{6}\right)$$

(c) $3i,\ -\frac{3\sqrt{3}}{2} - \frac{3}{2}i,\ \frac{3\sqrt{3}}{2} - \frac{3}{2}i$

(b)

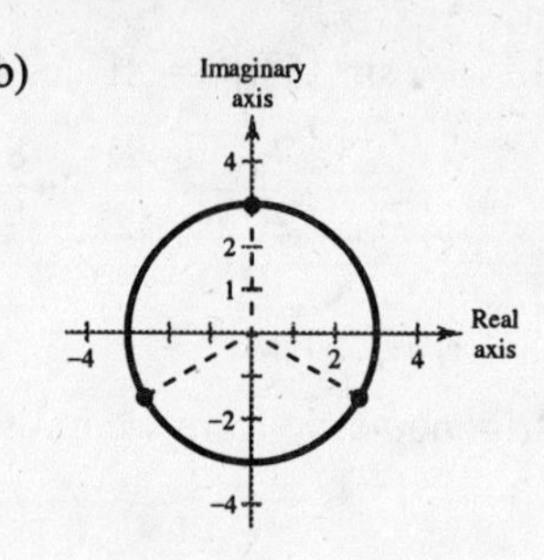

97. (a) Cube roots of $-\frac{125}{2}(1 + \sqrt{3}i) = 125\left(\cos\frac{4\pi}{3} + i\sin\frac{4\pi}{3}\right)$:

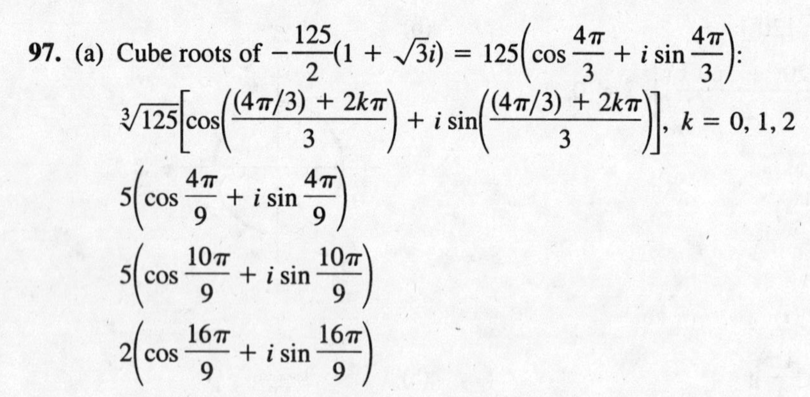

$$\sqrt[3]{125}\left[\cos\left(\frac{(4\pi/3) + 2k\pi}{3}\right) + i\sin\left(\frac{(4\pi/3) + 2k\pi}{3}\right)\right],\ k = 0, 1, 2$$

$$5\left(\cos\frac{4\pi}{9} + i\sin\frac{4\pi}{9}\right)$$

$$5\left(\cos\frac{10\pi}{9} + i\sin\frac{10\pi}{9}\right)$$

$$2\left(\cos\frac{16\pi}{9} + i\sin\frac{16\pi}{9}\right)$$

(c) $0.8682 + 4.924i,\ -4.698 - 1.710i,\ 3.830 - 3.214i$

(b)

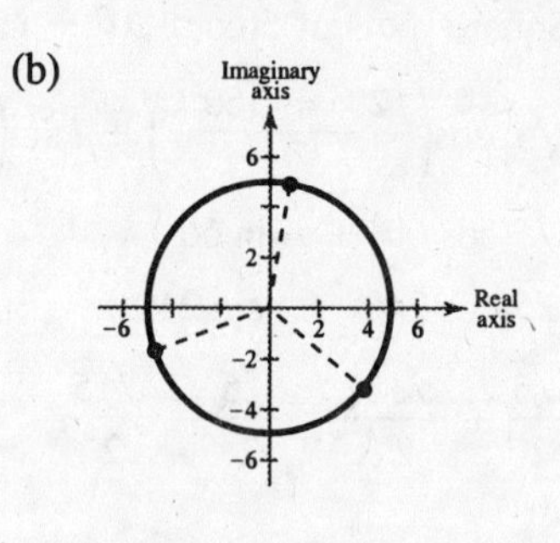

99. (a) Cube roots of $64 = 64(\cos 0 + i\sin 0)$

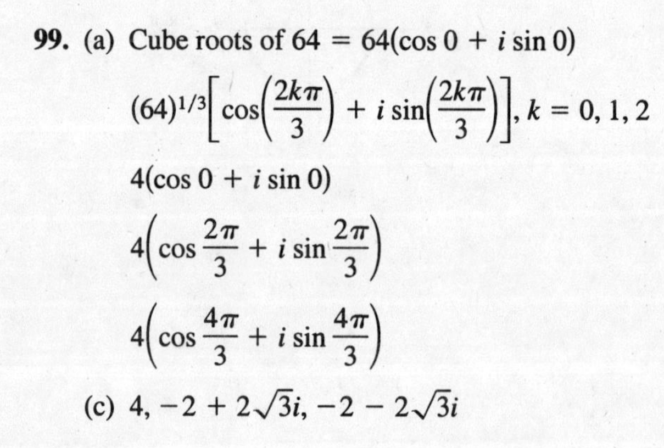

$$(64)^{1/3}\left[\cos\left(\frac{2k\pi}{3}\right) + i\sin\left(\frac{2k\pi}{3}\right)\right],\ k = 0, 1, 2$$

$$4(\cos 0 + i\sin 0)$$

$$4\left(\cos\frac{2\pi}{3} + i\sin\frac{2\pi}{3}\right)$$

$$4\left(\cos\frac{4\pi}{3} + i\sin\frac{4\pi}{3}\right)$$

(c) $4,\ -2 + 2\sqrt{3}i,\ -2 - 2\sqrt{3}i$

(b)

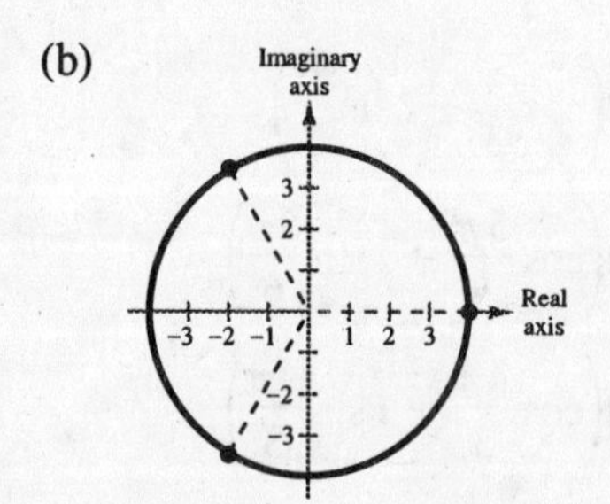

101. (a) Fifth roots of $1 = \cos 0 + i \sin 0$:

$\cos \frac{2k\pi}{5} + i \sin \frac{2k\pi}{5}, k = 0, 1, 2, 3, 4$

$\cos 0 + i \sin 0$

$\cos \frac{2\pi}{5} + i \sin \frac{2\pi}{5}$

$\cos \frac{4\pi}{5} + i \sin \frac{4\pi}{5}$

$\cos \frac{6\pi}{5} + i \sin \frac{6\pi}{5}$

$\cos \frac{8\pi}{5} + i \sin \frac{8\pi}{5}$

(b)

(c) $1, 0.3090 + 0.9511i, -0.8090 + 0.5878i, -0.8090 - 0.5878i, 0.3090 - 0.9511i$

103. (a) Cube roots of $-125 = 125(\cos 180° + i \sin 180°)$ are:

$5(\cos 60° + i \sin 60°)$

$5(\cos 180° + i \sin 180°)$

$5(\cos 300° + i \sin 300°)$

(b)

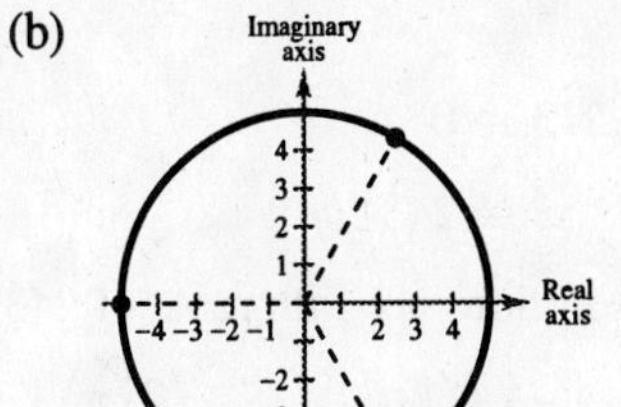

(c) $\frac{5}{2} + \frac{5\sqrt{3}}{2}i, -5, \frac{5}{2} - \frac{5\sqrt{3}}{2}i$

105. (a) Fifth roots of $128(-1 + i) = 128\sqrt{2}(\cos 135° + i \sin 135°)$ are:

$2\sqrt[5]{4\sqrt{2}}\,(\cos 27° + i \sin 27°)$

$2\sqrt[5]{4\sqrt{2}}\,(\cos 99° + i \sin 99°)$

$2\sqrt[5]{4\sqrt{2}}\,(\cos 171° + i \sin 171°)$

$2\sqrt[5]{4\sqrt{2}}\,(\cos 243° + i \sin 243°)$

$2\sqrt[5]{4\sqrt{2}}\,(\cos 315° + i \sin 315°)$

(b)

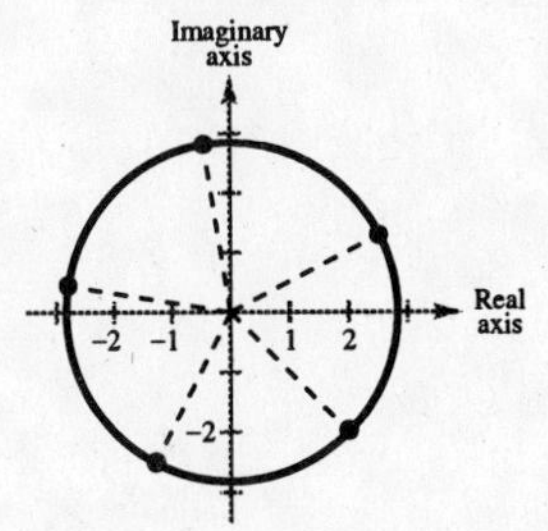

(c) $2.52 + 1.28i, -0.44 + 2.79i, -2.79 + 0.44i,$
$-1.28 - 252i, 2 - 2i$

107. $x^4 - i = 0$

$x^4 = i$

The solutions are the fourth roots of $i = \cos\frac{\pi}{2} + i\sin\frac{\pi}{2}$:

$$\sqrt[4]{1}\left[\cos\left(\frac{(\pi/2) + 2k\pi}{4}\right) + i\sin\left(\frac{(\pi/2) + 2k\pi}{4}\right)\right],\ k = 0, 1, 2, 3$$

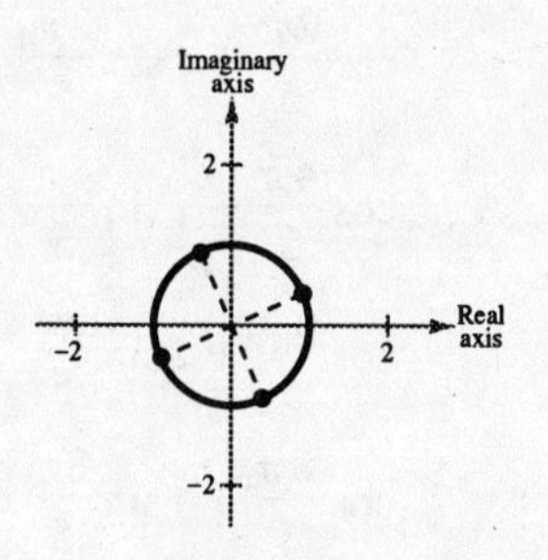

$\cos\frac{\pi}{8} + i\sin\frac{\pi}{8}$

$\cos\frac{5\pi}{8} + i\sin\frac{5\pi}{8}$

$\cos\frac{9\pi}{8} + i\sin\frac{9\pi}{8}$

$\cos\frac{13\pi}{8} + i\sin\frac{13\pi}{8}$

109. $x^5 - 243 = 0$

$x^5 = 243$

The solutions are fifth roots of $243 = 243(\cos 0 + i\sin 0)$

$$\sqrt[5]{243}\left[\cos\left(\frac{2k\pi}{5}\right) + i\sin\left(\frac{2k\pi}{5}\right)\right],\ k = 0, 1, 2, 3, 4$$

$3(\cos 0 + i\sin 0) = 3$

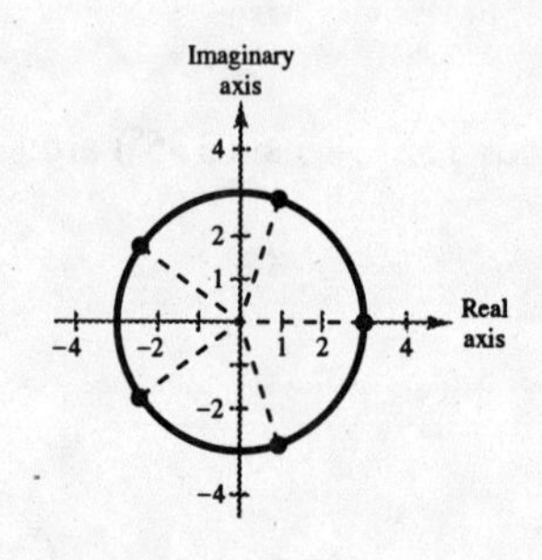

$3\left(\cos\frac{2\pi}{5} + i\sin\frac{2\pi}{5}\right)$

$3\left(\cos\frac{4\pi}{5} + i\sin\frac{4\pi}{5}\right)$

$3\left(\cos\frac{6\pi}{5} + i\sin\frac{6\pi}{5}\right)$

$3\left(\cos\frac{8\pi}{5} + i\sin\frac{8\pi}{5}\right)$

111. $x^3 + 64i = 0$

$x^3 = -64i$

The solutions are the third roots of $-64i$:

$$\sqrt[3]{64}\left[\cos\left(\frac{(3\pi/2) + 2k\pi}{3}\right) + i\sin\left(\frac{(3\pi/2) + 2k\pi}{3}\right)\right],\ k = 0, 1, 2$$

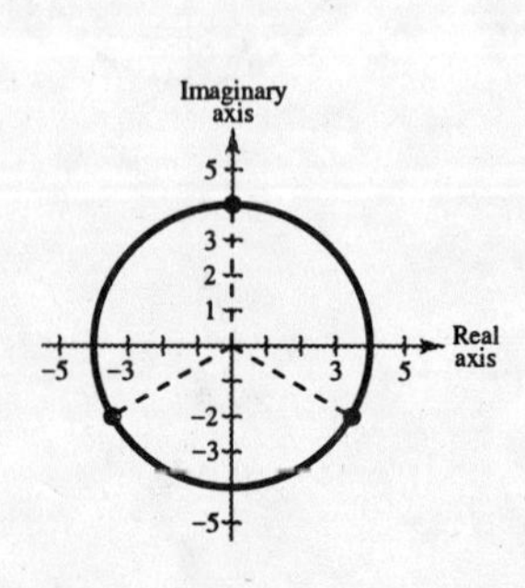

$4\left(\cos\frac{\pi}{2} + i\sin\frac{\pi}{2}\right) = 4i$

$4\left(\cos\frac{7\pi}{6} + i\sin\frac{7\pi}{6}\right) = -2\sqrt{3} - 2i$

$4\left(\cos\frac{11\pi}{6} + i\sin\frac{11\pi}{6}\right) = 2\sqrt{3} - 2i$

113. $x^3 - (1 - i) = 0$

$x^3 = 1 - i = \sqrt{2}(\cos 315° + i \sin 315°)$

The solutions are the cube roots of $1 - i$:

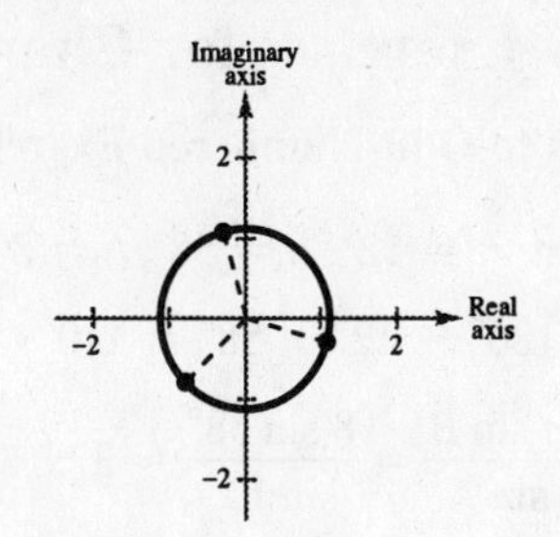

$$\sqrt[3]{\sqrt{2}}\left[\cos\left(\frac{315° + 360°k}{3}\right) + i \sin\left(\frac{315° + 360°k}{3}\right)\right], \; k = 0, 1, 2$$

$\sqrt[6]{2}(\cos 105° + i \sin 105°)$

$\sqrt[6]{2}(\cos 225° + i \sin 225°)$

$\sqrt[6]{2}(\cos 345° + i \sin 345°)$

115. True. $\left[\frac{1}{2}(1 - \sqrt{3}i)\right]^9 = \left[\frac{1}{2} - \frac{\sqrt{3}}{2}i\right]^9 = -1$

117. $$\frac{z_1}{z_2} = \frac{r_1(\cos \theta_1 + i \sin \theta_1)}{r_2(\cos \theta_2 + i \sin \theta_2)} \cdot \frac{\cos \theta_2 - i \sin \theta_2}{\cos \theta_2 - i \sin \theta_2}$$

$$= \frac{r_1}{r_2(\cos^2 \theta_2 + \sin^2 \theta_2)}[\cos \theta_1 \cos \theta_2 + \sin \theta_1 \sin \theta_2 + i(\sin \theta_1 \cos \theta_2 - \sin \theta_2 \cos \theta_1)]$$

$$= \frac{r_1}{r_2}[\cos(\theta_1 - \theta_2) + i \sin(\theta_1 - \theta_2)]$$

119. (a) $z\bar{z} = [r(\cos \theta + i \sin \theta)][r(\cos(-\theta) + i \sin(-\theta)]$

$= r^2[\cos(\theta - \theta) + i \sin(\theta - \theta)]$

$= r^2[\cos 0 + i \sin 0]$

$= r^2$

(b) $$\frac{z}{\bar{z}} = \frac{r(\cos \theta + i \sin \theta)}{r[\cos(-\theta) + i \sin(-\theta)]}$$

$= \frac{r}{r}[\cos(\theta - (-\theta)) + i \sin(\theta - (-\theta))]$

$= \cos 2\theta + i \sin 2\theta$

121. $\sin 28.1° = \frac{h}{18} \Rightarrow h = 18 \sin 28.1° \approx 8.48$

123. unit vector $= \frac{\langle 10, 0 \rangle}{10} = \langle 1, 0 \rangle$

125. unit vector $= \frac{12\mathbf{i} - 5\mathbf{j}}{13} = \frac{12}{13}\mathbf{i} - \frac{5}{13}\mathbf{j}$

127. $\mathbf{v} \cdot \mathbf{v} = \langle 5, -4 \rangle \cdot \langle 5, -4 \rangle = 25 + 16 = 41$

129. $3\mathbf{u} \cdot \mathbf{v} = \langle -3, 9 \rangle \cdot \langle 5, -4 \rangle = -15 - 36 = -51$

131. $\mathbf{u} \cdot \mathbf{v} = \langle -7, 35 \rangle \cdot \langle 3, -15 \rangle = -21 - 525 = -546 \Rightarrow$ not orthogonal

$\frac{-7}{3}\langle 3, -15 \rangle = \langle 7, 35 \rangle \Rightarrow$ parallel

Review Exercises for Chapter 6

Solutions to Odd-Numbered Exercises

1. Given: $A = 12°, B = 58°, a = 8$

$C = 180° - 12° - 58° = 110°$

$$b = \frac{a \sin B}{\sin A} = \frac{8 \sin 58°}{\sin 12°} \approx 32.6$$

$$c = \frac{a \sin C}{\sin A} = \frac{8 \sin 110°}{\sin 12°} \approx 36.2$$

3. Given: $A = 75°,\ a = 2.5,\ b = 16.5$

$$\sin B = \frac{b \sin A}{a} = \frac{16.5 \sin 75°}{2.5} \approx \frac{16.5(0.9659)}{2.5} \approx 5.375 \implies \text{no triangle formed}$$

No solution

5. Given: $B = 115°, a = 9, b = 14.5$

$$\sin A = \frac{a \sin B}{b} = \frac{9 \sin 115°}{14.5} \approx 0.5625 \implies A \approx 34.2°$$

$C \approx 180° - 115° - 34.2° = 30.8°$

$$c = \frac{b}{\sin B}(\sin C) = \frac{14.5}{\sin 115°}(\sin 30.8°) \approx 8.2$$

7. Given: $A = 15°,\ a = 5,\ b = 10$

$$\sin B = \frac{b \sin A}{a} = \frac{10 \sin 15°}{5} \approx \frac{10(0.2588)}{5} \approx 0.5176 \implies B \approx 31.2° \text{ or } 148.8°$$

Case 1: $B \approx 31.2°$

$C \approx 180° - 15° - 31.2° = 133.8°$

$$c = \frac{a \sin C}{\sin A} \approx 13.9$$

Case 2: $B \approx 148.8°$

$C \approx 180° - 15° - 148.8° = 16.2°$

$$c = \frac{a \sin C}{\sin A} \approx 5.39$$

9. Given: $B = 25°,\ a = 6.2,\ b = 4$

$$\sin A = \frac{a \sin B}{b} \approx 0.6551 \implies A \approx 40.9° \text{ or } 139.1°$$

Case 1: $A \approx 40.9°$

$C \approx 180° - 25° - 40.9° = 114.1°$

$c \approx 8.6$

Case 2: $A \approx 139.1°$

$C \approx 180° - 25° - 139.1° = 15.9°$

$c \approx 2.6$

11. $A = 27°,\ b = 5,\ c = 8$

$$\text{Area} = \frac{1}{2}bc \sin A = \frac{1}{2}(5)(8)(0.4540) = 9.08 \text{ sq. units}$$

13. $\text{Area} = \frac{1}{2}ab \sin C$

$= \frac{1}{2}(29)(18) \sin 122°$

≈ 221.34 square units

15. $\alpha = 180° - 31° = 149°$

$\phi = 180° - 149° - 17° = 14°$

$$x = \frac{50 \sin 17°}{\sin \phi} = \frac{50 \sin 17°}{\sin 14°} \approx 60.43$$

$h = x \sin 31°$

$\approx 50.43(0.5150) \approx 31.1$ meters

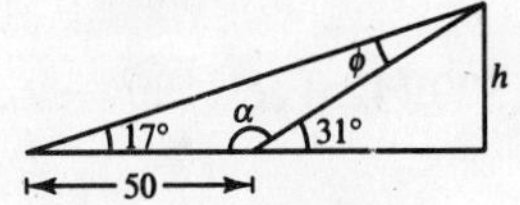

17. $\sin 28° = \dfrac{h}{75}$

$h = 75 \sin 28° \approx 35.21$ feet

$\cos 28° = \dfrac{x}{75}$

$x = 75 \cos 28° \approx 66.2211$ feet

$\tan 45° = \dfrac{H}{x}$

$H = x \tan 45° \approx 66.22$ feet

Height of tree: $H - h \approx 31$ feet

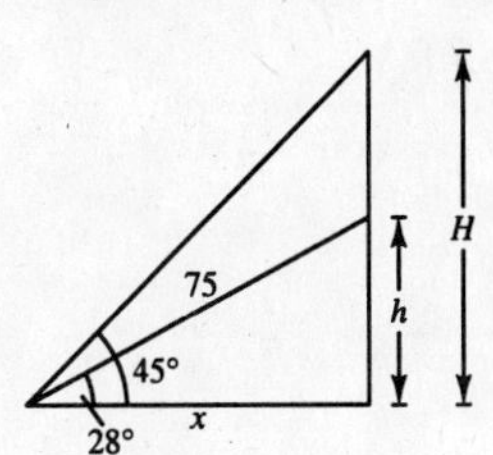

19. Given: $a = 5,\ b = 8,\ c = 10$

$$\cos C = \frac{a^2 + b^2 - c^2}{2ab} = \frac{25 + 64 - 100}{80} \approx -0.1375 \implies C \approx 97.9°$$

$$\sin A = \frac{a \sin C}{c} \approx \frac{5(0.9905)}{10} \approx 0.4953 \implies A \approx 29.7°$$

$$\sin B = \frac{b \sin C}{c} \approx \frac{8(0.9905)}{10} \approx 0.7924 \implies B \approx 52.4°$$

21. Given: $a = 6, b = 9, C = 45°$

$c^2 = a^2 + b^2 - 2ab \cos 45° \approx 36 + 81 - 2(6)(9)(0.7071) \approx 40.63 \implies c \approx 6.374$

$$\cos B = \frac{a^2 + c^2 - b^2}{2ac} \approx \frac{36 + 40.63 - 81}{2(6)(6.374)} \approx -0.0571 \implies B \approx 93.3°$$

$A \approx 180° - 45° - 93.3° = 41.7°$

23. Given: $B = 110°,\ a = 4,\ c = 4$

$b^2 = a^2 + c^2 - 2ac \cos B \approx 16 + 16 - 2(4)(4)(-0.3420) \approx 42.94 \implies b \approx 6.6$

$$\sin A = \frac{a \sin B}{b} = \frac{4 \sin 110°}{6.6} \approx \frac{4(0.9397)}{6.6} \approx 0.5736 \implies A \approx 35°$$

$c = a \implies C = A \approx 35°$

25. Given: $a = 42, b = 25, c = 58$

$$\cos C = \frac{a^2 + b^2 - c^2}{2ab} = \frac{42^2 + 25^2 - 58^2}{2(42)(25)} \approx -0.4643 \implies C \approx 117.7°$$

$$\cos B = \frac{a^2 + c^2 - b^2}{2ac} = \frac{42^2 + 58^2 - 25^2}{2(42)(58)} \approx 0.9243 \implies B \approx 22.4°$$

$A = 180° - 117.7° - 22.4° = 39.9°$

27. Given: $B = 150°,\ a = 10,\ c = 20$

$$b^2 = a^2 + c^2 - 2ac\cos B \approx 100 + 400 - 400(-0.8660) \approx 846.4 \implies b \approx 29.1$$

$$\sin C = \frac{c\sin B}{b} \approx \frac{20(0.5)}{29.09} \approx 0.3437 \implies C \approx 20.1°$$

$$\sin A = \frac{a\sin B}{b} \approx \frac{10(0.5)}{29.09} \approx 0.1719 \implies A \approx 9.9°$$

29. $a^2 = 5^2 + 8^2 - 2(5)(8)\cos 152°$

$\approx 159.6 \implies a \approx 12.63$ ft

$b^2 = 5^2 + 8^2 - 2(5)(8)\cos 28°$

$\approx 18.36 \implies b \approx 4.285$ ft

a
5
8
8
5
b
28°

31. $b^2 = a^2 + c^2 - 2ac\cos B$

$= 300^2 + 425^2 - 2(300)(425)\cos(180° - 65°)$

≈ 378392.66

$b \approx 615.1$ meters

33. $a = 4,\ b = 5,\ c = 7$

$$s = \frac{a+b+c}{2} = \frac{4+5+7}{2} = 8$$

$$\text{Area} = \sqrt{s(s-a)(s-b)(s-c)}$$

$$= \sqrt{8(4)(3)(1)} \approx 9.798 \text{ square units}$$

35. $a = 64.8, b = 49.2, c = 24.1$

$$s = \frac{a+b+c}{2} = \frac{64.8 + 49.2 + 24.1}{2} = 69.05$$

$$\text{Area} = \sqrt{s(s-a)(s-b)(s-c)} = \sqrt{69.05(4.25)(19.85)(44.95)}$$

$$\approx 511.7 \text{ square units}$$

37. $\mathbf{u} = \langle 4 - (-2), 6 - 1\rangle = \langle 6, 5\rangle$

$\mathbf{v} = \langle 6 - 0, 3 - (-2)\rangle = \langle 6, 5\rangle$

$\mathbf{u} = \mathbf{v}$

39. Initial point: $(-5, 4)$

Terminal point: $(2, -1)$

$\mathbf{v} = \langle 2 - (-5),\ -1, -4\rangle = \langle 7, -5\rangle$

41. Initial point: $(0, 10)$

Terminal point: $(7, 3)$

$\mathbf{v} = \langle 7 - 0, 3 - 10\rangle = \langle 7, -7\rangle$

43. $\langle 8\cos 120°, 8\sin 120°\rangle = (-4, 4\sqrt{3})$

45. $-\mathbf{w} = -\langle 4, 5\rangle = \langle -4, -5\rangle$

47. $\mathbf{w} + 2\mathbf{u} = \langle 4, 5\rangle + 2\langle -1, -3\rangle = \langle 2, -1\rangle$

49. $3\mathbf{v} + 3\mathbf{w} = 3\langle -3, 6\rangle + 3\langle 4, 5\rangle = \langle 3, 33\rangle$

51. $\mathbf{u} = 6\mathbf{i} - 5\mathbf{j}$

$$\frac{1}{\|\mathbf{u}\|}\mathbf{u} = \frac{1}{\sqrt{6^2+5^2}}(6\mathbf{i} - 5\mathbf{j}) = \frac{6}{\sqrt{61}}\mathbf{i} - \frac{5}{\sqrt{61}}\mathbf{j}$$
$$= \left\langle \frac{6}{\sqrt{61}}, -\frac{5}{\sqrt{61}} \right\rangle$$

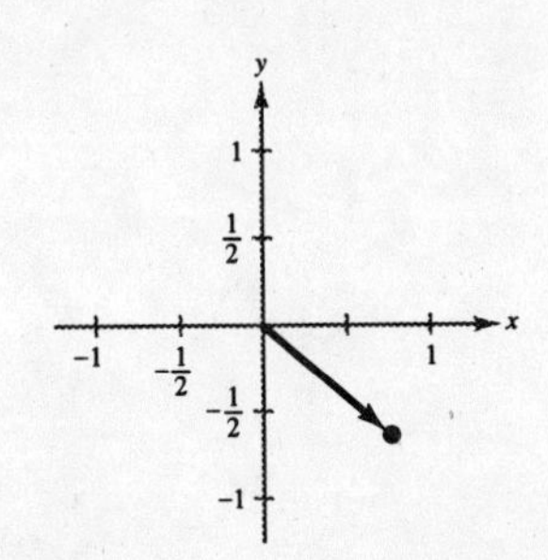

53. $3\mathbf{v} = 3(10\mathbf{i} + 3\mathbf{j}) = 30\mathbf{i} + 9\mathbf{j} = \langle 30, 9 \rangle$

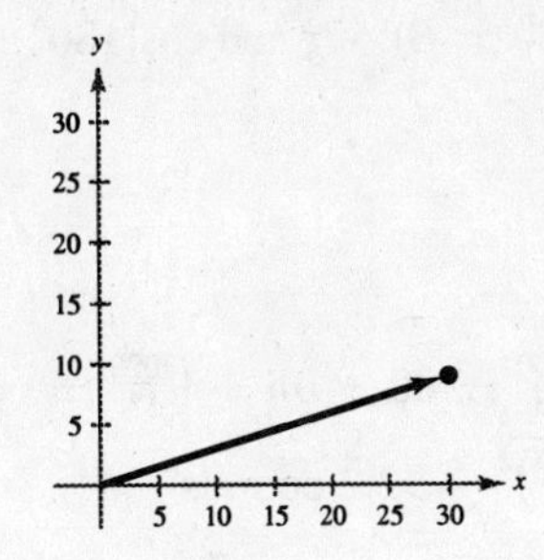

55. $\mathbf{u} = 6\mathbf{i} - 5\mathbf{j}$, $\mathbf{v} = 10\mathbf{i} + 3\mathbf{j}$

$$4\mathbf{u} - 5\mathbf{v} = (24\mathbf{i} - 20\mathbf{j}) - (50\mathbf{i} + 15\mathbf{j}) = -26\mathbf{i} - 35\mathbf{j}$$
$$= \langle -26, -35 \rangle$$

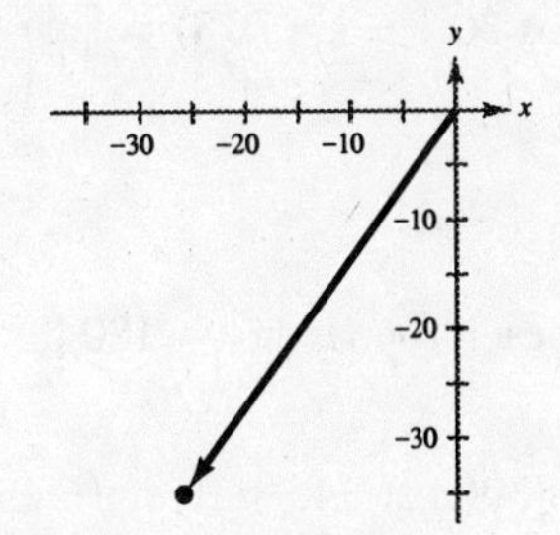

57. $\|\mathbf{u}\| = 6$. unit vector: $\frac{1}{6}\langle 0, -6 \rangle = \langle 0, -1 \rangle$

59. $\|\mathbf{v}\| = \sqrt{5^2 + (-2)^2} = \sqrt{29}$.

unit vector: $\frac{1}{\sqrt{29}}\langle 5, -2 \rangle = \left\langle \frac{5}{\sqrt{29}}, \frac{-2}{\sqrt{29}} \right\rangle$

61. $P(7, -4), Q(-3, 2)$

$$\overrightarrow{PQ} = \langle -3 - 7, 2 - (-4) \rangle = \langle -10, 6 \rangle$$
$$\|\overrightarrow{PQ}\| = \sqrt{(-10)^2 + (6)^2} = \sqrt{136} = 2\sqrt{34}$$
$$\frac{\overrightarrow{PQ}}{\|\overrightarrow{PQ}\|} = \frac{1}{2\sqrt{34}}\langle -10, 6 \rangle = \frac{1}{\sqrt{34}}\langle -5, 3 \rangle$$

63. $P(0, 3), Q(5, -8)$

$$\overrightarrow{PQ} = \langle 5 - 0, -8 - 3 \rangle = \langle 5, -11 \rangle$$
$$\|\overrightarrow{PQ}\| = \sqrt{5^2 + (-11)^2} = \sqrt{146}$$
$$\frac{\overrightarrow{PQ}}{\|\overrightarrow{PQ}\|} = \frac{1}{\sqrt{146}}\langle 5, -11 \rangle$$

65. $\mathbf{u} = \langle 1 - (-8), -5 - 3 \rangle = \langle 9, -8 \rangle = 9\mathbf{i} - 8\mathbf{j}$

67. $\mathbf{v} = -10\mathbf{i} + 10\mathbf{j}$

$$\|\mathbf{v}\| = \sqrt{(-10)^2 + (10)^2} = \sqrt{200} = 10\sqrt{2}$$
$$\tan\theta = \frac{10}{-10} = -1 \implies \theta = 135° \text{ since}$$

$\mathbf{v}$ is in Quadrant II.

$$\mathbf{v} = 10\sqrt{2}(\cos 135°\,\mathbf{i} + \sin 135°\mathbf{j})$$

69.

$$\mathbf{u} = 15[(\cos 20°)\mathbf{i} + (\sin 20°)\mathbf{j}]$$
$$\mathbf{v} = 20[(\cos 63°)\mathbf{i} + (\sin 63°)\mathbf{j}]$$
$$\mathbf{u} + \mathbf{v} \approx 23.1752\mathbf{i} + 22.9504\mathbf{j}$$
$$\|\mathbf{u} + \mathbf{v}\| \approx 32.62$$
$$\tan\theta = \frac{22.9504}{23.1752} \implies \theta \approx 44.72°$$

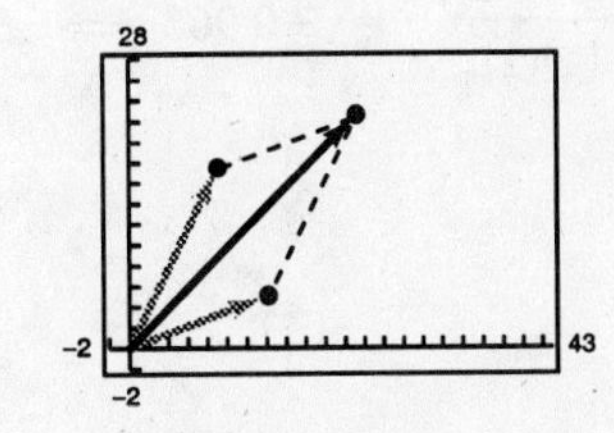

71. $\tan \alpha = \frac{12}{5} \implies \sin \alpha = \frac{12}{13}$ and $\cos \alpha = \frac{5}{13}$

$\tan \beta = \frac{3}{4} \implies \sin(180 - \beta) = \frac{3}{5}$ and $\cos(180° - \beta) = -\frac{4}{5}$

$\mathbf{u} = 250\left(\frac{5}{13}\mathbf{i} + \frac{12}{13}\mathbf{j}\right)$

$\mathbf{v} = 100\left(-\frac{4}{5}\mathbf{i} + \frac{3}{5}\mathbf{j}\right)$

$\mathbf{w} = 200(0\mathbf{i} - \mathbf{j})$

$\mathbf{r} = \mathbf{u} + \mathbf{v} + \mathbf{w} = \left(\frac{1250}{13} - 80 + 0\right)\mathbf{i} + \left(\frac{3000}{13} + 60 - 200\right)\mathbf{j} = \frac{210}{13}\mathbf{i} + \frac{1180}{13}\mathbf{j}$

$\|\mathbf{r}\| = \sqrt{\left(\frac{210}{13}\right)^2 + \left(\frac{1180}{13}\right)^2} \approx 92.2$ pounds

$\tan \theta = \frac{1180}{210} \implies \theta \approx 79.9°$

73. Rope One: $\mathbf{u} = \|\mathbf{u}\|(\cos 30°\mathbf{i} - \sin 30°\mathbf{j}) = \|\mathbf{u}\|\left(\frac{\sqrt{3}}{2}\mathbf{i} - \frac{1}{2}\mathbf{j}\right)$

Rope Two: $\mathbf{v} = \|\mathbf{u}\|(-\cos 30°\mathbf{i} - \sin 30°\mathbf{j}) = \|\mathbf{u}\|\left(-\frac{\sqrt{3}}{2}\mathbf{i} - \frac{1}{2}\mathbf{j}\right)$

Resultant: $\mathbf{u} + \mathbf{v} = -\|\mathbf{u}\|\mathbf{j} = -180\mathbf{j}$

$\|\mathbf{u}\| = 180$

Therefore, the tension on each rope is $\|\mathbf{u}\| = 180$ lb.

75. Airplane velocity: $\mathbf{u} = 430(\cos(-45°)\mathbf{i} + \sin(-45°)\mathbf{j})$

Wind velocity: $\mathbf{w} = 35(\cos 60°\mathbf{i} + \sin 60°\mathbf{j})$

$$\begin{aligned}\mathbf{u} + \mathbf{w} &= (430\cos(-45) + 35\cos 60)\mathbf{i} + (430\sin(-45) + 35\sin 60)\\ &= 321.56\mathbf{i} - 273.75\mathbf{j}\end{aligned}$$

$\|\mathbf{u} + \mathbf{w}\| = 422.3$ mph

$\theta = \arctan\left(\frac{-273.75}{321.56}\right) = -40.4°$

Direction: S 49.6° E

77. $\mathbf{u} \cdot \mathbf{v} = \langle 6, -1\rangle \cdot \langle 2, 5\rangle = 6(2) + (-1)(5) = 7$

79. $$\begin{aligned}(\mathbf{u} \cdot \mathbf{v})\mathbf{u} &= (-12 + 3)\langle 6, -3\rangle = -9\langle 6, -3\rangle\\ &= \langle -54, 27\rangle\end{aligned}$$

81. $4\mathbf{u} \cdot \mathbf{v} = 4(-9) = -36$

83. $\mathbf{u} = \cos\frac{7\pi}{4}\mathbf{i} + \sin\frac{7\pi}{4}\mathbf{j} = \left\langle \frac{1}{\sqrt{2}}, -\frac{1}{\sqrt{2}}\right\rangle$

$\mathbf{v} = \cos\frac{5\pi}{6}\mathbf{i} + \sin\frac{5\pi}{6}\mathbf{j} = \left\langle -\frac{\sqrt{3}}{2}, \frac{1}{2}\right\rangle$

$$\cos\theta = \frac{\mathbf{u} \cdot \mathbf{v}}{\|\mathbf{u}\|\,\|\mathbf{v}\|} = \frac{\frac{-\sqrt{3}}{2\sqrt{2}} - \frac{1}{2\sqrt{2}}}{(1)(1)} \approx -0.966 \implies \theta \approx 165° \text{ or } \frac{11\pi}{12}$$

85. $\mathbf{u} = \langle 2\sqrt{2}, -4\rangle, \mathbf{v} = \langle -\sqrt{2}, 1\rangle$

$$\cos\theta = \frac{\mathbf{u}\cdot\mathbf{v}}{\|\mathbf{u}\|\,\|\mathbf{v}\|} = \frac{-8}{(\sqrt{24})(\sqrt{3})} \Rightarrow \theta \approx 160.5°$$

87.

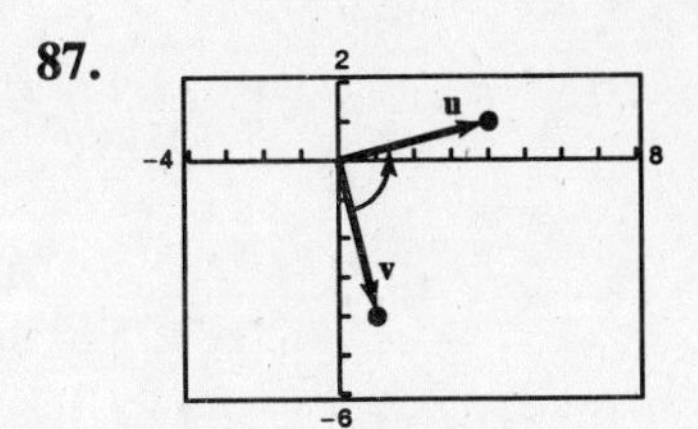

$$\cos\theta = \frac{\mathbf{u}\cdot\mathbf{v}}{\|\mathbf{u}\|\,\|\mathbf{v}\|} = 0 \Rightarrow \theta = 90°$$

89.

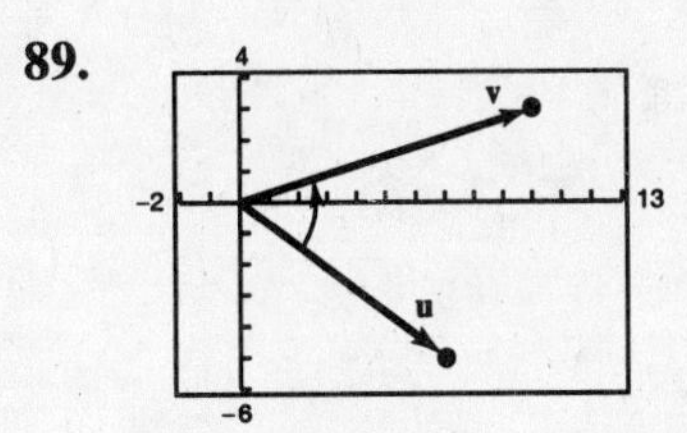

$$\cos\theta = \frac{\mathbf{u}\cdot\mathbf{v}}{\|\mathbf{u}\|\,\|\mathbf{v}\|} = \frac{70-15}{\sqrt{74}\sqrt{109}} \approx 0.612 \Rightarrow \theta \approx 52.2°$$

91. $\mathbf{u} = \langle 39, -12\rangle, \mathbf{v} = \langle -26, 8\rangle$

$\mathbf{u}\cdot\mathbf{v} = 39(-26) + (-12)(8) = -1110 \neq 0 \Rightarrow$ **u** and **v** are not orthogonal.

$\mathbf{v} = -\frac{2}{3}\mathbf{u} \Rightarrow$ **u** and **v** are parallel.

93. $\mathbf{u} = \langle 8, 5\rangle, \mathbf{v} = \langle -2, 4\rangle$

$\mathbf{u}\cdot\mathbf{v} = 8(-2) + 5(4) = 4 \neq 0$

u and **v** are not orthogonal.

$\mathbf{u} \neq k\mathbf{v} \Rightarrow$ **u** and **v** are not parallel.

Neither

95. $\mathbf{u} = \langle -4, 3\rangle,\ \mathbf{v} = \langle -8, -2\rangle$

(a) $\text{proj}_{\mathbf{v}}\mathbf{u} = \left(\frac{\mathbf{u}\cdot\mathbf{v}}{\|\mathbf{v}\|^2}\right)\mathbf{v} = \left(\frac{26}{68}\right)\langle -8, -2\rangle$

$= -\frac{13}{17}\langle 4, 1\rangle$

(b) $\mathbf{u} - \text{proj}_{\mathbf{v}}\,\mathbf{u} = \left\langle -\frac{16}{17}, \frac{64}{17}\right\rangle \cdot \mathbf{u} = \left\langle -\frac{52}{17}, -\frac{13}{17}\right\rangle + \left\langle -\frac{16}{17}, \frac{64}{17}\right\rangle$

97. $\mathbf{u} = \langle 2, 7\rangle, \mathbf{v} = \langle 1, -1\rangle$

$\text{proj}_{\mathbf{v}}\,\mathbf{u} = \left(\frac{\mathbf{u}\cdot\mathbf{v}}{\|\mathbf{v}\|^2}\right)\mathbf{v} = \frac{-5}{2}\langle 1, -1\rangle = \left\langle -\frac{5}{2}, \frac{5}{2}\right\rangle$

$\mathbf{u} - \text{proj}_{\mathbf{v}}\,\mathbf{u} = \langle 2, 7\rangle - \left\langle -\frac{5}{2}, \frac{5}{2}\right\rangle = \left\langle \frac{9}{2}, \frac{9}{2}\right\rangle$

$\mathbf{u} = \left\langle -\frac{5}{2}, \frac{5}{2}\right\rangle + \left\langle \frac{9}{2}, \frac{9}{2}\right\rangle$

99. 48 inches = 4 feet

Work = 18,000(4) = 72,000 ft/lb

101. $|-i| = 1$

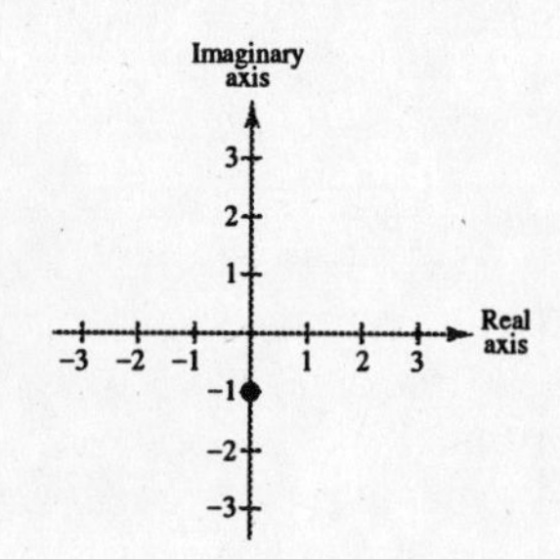

103. $|7 - 5i| = \sqrt{7^2 + (-5)^2} = \sqrt{74}$

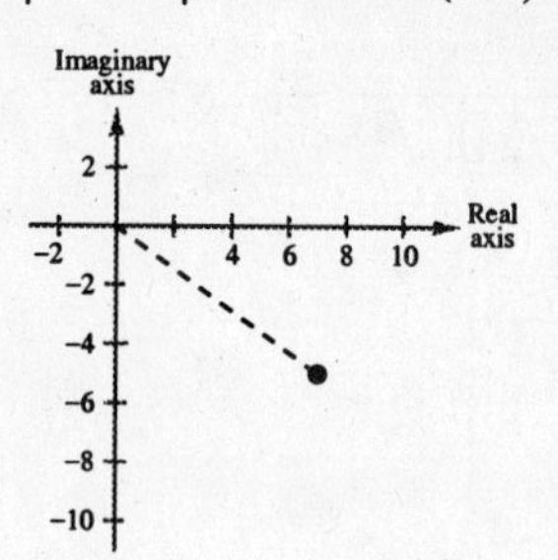

105. $z = 5 - 2i,\ r = \sqrt{25 + 4} = \sqrt{29},\ \theta = \arctan\left(-\frac{2}{5}\right) \approx 338.2°$

$z = \sqrt{29}(\cos 338.2° + i \sin 338.2°)$

107. $5 - 5i$

$$r = \sqrt{5^2 + (-5)^2} = \sqrt{50} = 5\sqrt{2}$$

$\tan\theta = \dfrac{-5}{5} - -1 \Rightarrow \theta \approx 315°$ since the complex number is in Quadrant IV.

$5 - 5i = 5\sqrt{2}(\cos 315° + i \sin 315°)$

109. $5 + 12i$

$$r = \sqrt{5^2 + 12^2} = \sqrt{169} = 13$$

$\tan\theta = \frac{12}{5} \Rightarrow \theta \approx 67.38°$ since the number is in Quadrant I.

$5 + 12i \approx 13(\cos 67.38° + i \sin 67.38°)$

111. (a) $z_1 = 2\sqrt{3} - 2i = 4(\cos 330° + i \sin 330°)$

$z_2 = -10i = 10(\cos 270° + i \sin 270°)$

(b) $z_1 z_2 = [4(\cos 330° + i \sin 330°)][10(\cos 270° + i \sin 270°)]$

$= 40(\cos 600° + i \sin 600°)$

$= 40(\cos 240° + i \sin 240°)$

$\approx -20.00 - 34.64i$

$$\frac{z_1}{z_2} = \frac{4(\cos 330° + i \sin 330°)}{10(\cos 270° + i \sin 270°)}$$

$$= \frac{2}{5}(\cos 60° + i \sin 60°)$$

113. $\left[\frac{5}{2}\left(\cos\frac{\pi}{2} + i \sin\frac{\pi}{2}\right)\right]\left[4\left(\cos\frac{\pi}{2} + i \sin\frac{\pi}{4}\right)\right] = 10\left[\cos\frac{3\pi}{4} + i \sin\frac{3\pi}{4}\right] = 10\left[-\frac{\sqrt{2}}{2} + \frac{\sqrt{2}}{2}i\right] = -5\sqrt{2} + 5\sqrt{2}i$

115. $\dfrac{20(\cos 320° + i \sin 320°)}{5(\cos 80° + i \sin 80°)} = 4[\cos 240° + i \sin 240°]$

117. $\left[5\left(\cos\frac{\pi}{12} + i \sin\frac{\pi}{12}\right)\right]^4 = 5^4\left(\frac{4\pi}{12} + i \sin\frac{4\pi}{12}\right)$

$$= 625\left(\cos\frac{\pi}{3} + \text{i} \sin\frac{\pi}{3}\right)$$

$$= 625\left(\frac{1}{2} + \frac{\sqrt{3}}{2}i\right)$$

$$= \frac{625}{2} + \frac{625\sqrt{3}}{2}i$$

119. $(2 + 3i)^6 = [\sqrt{13}(\cos 56.3° + i \sin 56.3°)]^6$

$= 13^3(\cos 337.9° + i \sin 337.9°)$

$\approx 13^3(0.9263 - 0.3769i)$

$\approx 2035 - 828i$

121. Sixth roots of $-729i = 729\left(\cos \dfrac{3\pi}{2} + i \sin \dfrac{3\pi}{2}\right)$:

$$\sqrt[6]{729}\left(\cos \frac{(3\pi/2) + 2k\pi}{6} + i \sin \frac{(3\pi/2) + 2k\pi}{6}\right), k = 1, 2, 3, 4, 5$$

$$3\left(\cos \frac{\pi}{4} + i \sin \frac{\pi}{4}\right)$$

$$3\left(\cos \frac{7\pi}{12} + i \sin \frac{7\pi}{12}\right)$$

$$3\left(\cos \frac{11\pi}{12} + i \sin \frac{11\pi}{12}\right)$$

$$3\left(\cos \frac{5\pi}{4} + i \sin \frac{5\pi}{4}\right)$$

$$3\left(\cos \frac{19\pi}{12} + i \sin \frac{19\pi}{12}\right)$$

$$3\left(\cos \frac{23\pi}{12} + i \sin \frac{23\pi}{12}\right)$$

123. $x^4 + 256 = 0$

$x^4 = -256 = 256(\cos \pi + i \sin \pi)$

$$\sqrt[4]{-256} = 4\left[\cos\left(\frac{\pi + 2\pi k}{4}\right) + i \sin\left(\frac{\pi + 2\pi k}{4}\right)\right], k = 0, 1, 2, 3$$

$$4\left(\cos \frac{\pi}{4} + i \sin \frac{\pi}{4}\right) = \frac{4\sqrt{2}}{2} + \frac{4\sqrt{2}}{2}i = 2\sqrt{2} + 2\sqrt{2}\,i$$

$$4\left(\cos \frac{3\pi}{4} + i \sin \frac{3\pi}{4}\right) = -\frac{4\sqrt{2}}{2} + \frac{4\sqrt{2}}{2}i = -2\sqrt{2} + 2\sqrt{2}\,i$$

$$4\left(\cos \frac{5\pi}{4} + i \sin \frac{5\pi}{4}\right) = -\frac{4\sqrt{2}}{2} - \frac{4\sqrt{2}}{2}i = -2\sqrt{2} - 2\sqrt{2}\,i$$

$$4\left(\cos \frac{7\pi}{4} + i \sin \frac{7\pi}{4}\right) = \frac{4\sqrt{2}}{2} - \frac{4\sqrt{2}}{2}i = 2\sqrt{2} - 2\sqrt{2}\,i$$

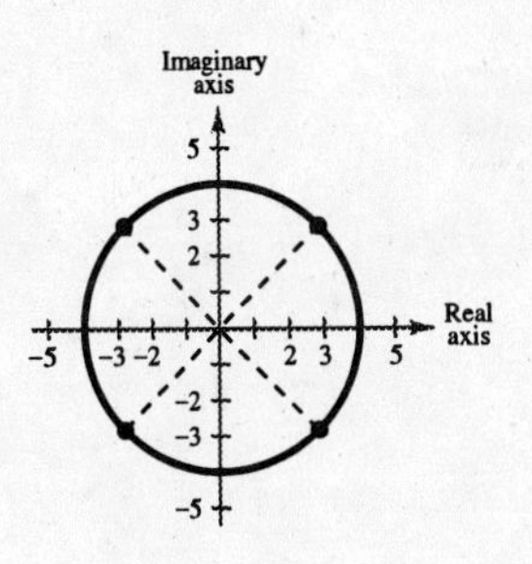

125. $x^3 + 8i = 0$

$$x^3 = -8i$$

$$-8i = 8\left(\cos\frac{3\pi}{2} + i\sin\frac{3\pi}{2}\right)$$

$$\sqrt[3]{-8i} = \sqrt[3]{8}\left[\cos\frac{(3\pi/2) + 2\pi k}{3} + i\sin\frac{(3\pi/2) + 2\pi k}{3}\right],\ k = 0, 1, 2$$

$$2\left(\cos\frac{\pi}{2} + i\sin\frac{\pi}{2}\right) = 2i$$

$$2\left(\cos\frac{7\pi}{6} + i\sin\frac{7\pi}{6}\right) = -\sqrt{3} - i$$

$$2\left(\cos\frac{11\pi}{6} + i\sin\frac{11\pi}{6}\right) = \sqrt{3} - i$$

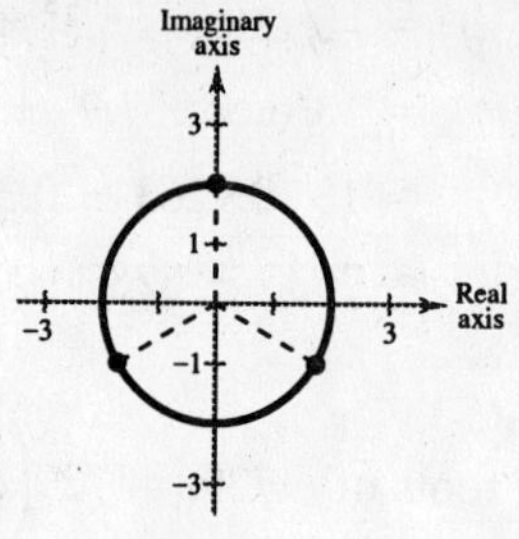

Imaginary axis
3
1
-3
-1
3
Real axis
-3

127. True

129. Length and direction characterize vectors in plane.

131. (a) The three other 4th roots are not shown.

(b) The modulus of each is 2, and the arguments are 120°, 210°, and 300°.

Chapter 6 Practice Test

For Exercises 1 and 2, use the Law of Sines to find the remaining sides and angles of the triangle.

1. $A = 40°,\ B = 12°,\ b = 100$

2. $C = 150°,\ a = 5,\ c = 20$

3. Find the area of the triangle: $a = 3,\ b = 6,\ C = 130°$.

4. Determine the number of solutions to the triangle: $1 = 10,\ b = 35,\ A = 22.5°$.

For Exercises 5 and 6, use the Law of Cosines to find the remaining sides and angles off the triangle.

5. $a = 49,\ b = 53,\ c = 38$

6. $C = 29°,\ a = 100,\ b = 300$

7. Use Heron's Formula to find the area of the triangle: $a = 4.1,\ b = 6.8,\ c = 5.5$.

8. A ship travels 40 miles due east, then adjusts its course 12° southward. After traveling 70 miles in that direction, how far is the ship from its point of departure?

9. $\mathbf{w}$ is $4\mathbf{u} - 7\mathbf{v}$ where $\mathbf{u} = 3\mathbf{i} + \mathbf{j}$ and $\mathbf{v} = -\mathbf{i} + 2\mathbf{j}$. Find $\mathbf{w}$.

10. Find a unit vector in the direction of $\mathbf{v} = 5\mathbf{i} - 3\mathbf{j}$.

11. Find the dot product and the angle between $\mathbf{u} = 6\mathbf{i} + 5\mathbf{j}$ and $\mathbf{v} = 2\mathbf{i} - 3\mathbf{j}$.

12. $\mathbf{v}$ is a vector of magnitude 4 making an angle of 30° with the positive x-axis. Find $\mathbf{v}$ in component form.

13. Find the projection of $\mathbf{u}$ onto $\mathbf{v}$ given $\mathbf{u} = \langle 3, -1\rangle$ and $\mathbf{v} = \langle -2, 4\rangle$.

14. Give the trigonometric form of $z = 5 - 5i$.

15. Give the standard form of $z = 6(\cos 225° + i \sin 225°)$

16. Multiply $[7 \cos 23° + i \sin 23°)][4(\cos 7° + i \sin 7°)]$.

17. Divide $\dfrac{9\left(\cos \frac{5\pi}{4} + i \sin \frac{5\pi}{4}\right)}{3(\cos \pi + i \sin \pi)}$.

18. Find $(2 + 2i)^8$.

19. Find the cube roots of $8\left(\cos \frac{\pi}{3} + i \sin \frac{\pi}{3}\right)$.

20. Find all the solutions to $x^4 + i = 0$.

CHAPTER 7
Systems of Equations and Inequalities

CHAPTER 7
Systems of Equations and Inequalities

Section 7.1 Solving Systems of Equations

- You should be able to solve systems of equations by the method of substitution.
  1. Solve one of the equations for one of the variables.
  2. Substitute this expression into the other equation and solve.
  3. Back-substitute into the first equation to find the value of the other variable.
  4. Check your answer in each of the original equations.
- You should be able to find solutions graphically. (See Example 5 in textbook.)

Solutions to Odd-Numbered Exercises

1. (a) $4(0) - (-3) \stackrel{?}{=} 1$

$6(0) + (-3) \stackrel{?}{=} -6$

$3 \neq 1$

$-3 \neq -6$

No, $(0, -3)$ is not a solution.

(b) $4(-1) - (-5) \stackrel{?}{=} 1$

$6(-1) + (-5) \stackrel{?}{=} -6$

$1 = 1$

$-11 \neq -6$

No, $(-1, -5)$ is not a solution.

(c) $4\left(-\frac{3}{2}\right) - (3) \stackrel{?}{=} 1$

$6\left(-\frac{3}{2}\right) + (3) \stackrel{?}{=} -6$

$-9 \neq 1$

$-6 = -6$

No, $\left(-\frac{3}{2}, 3\right)$ is not a solution.

(d) $4\left(-\frac{1}{2}\right) - (-3) \stackrel{?}{=} 1$

$6\left(-\frac{1}{2}\right) + (-3) \stackrel{?}{=} -6$

$1 = 1$

$-6 = -6$

Yes, $\left(-\frac{1}{2}, -3\right)$ is a solution.

3. (a) $0 \stackrel{?}{=} -2e^{-2}$

$3(-2) - 0 \stackrel{?}{=} 2$

$0 \neq -2e^{-2}$

$-6 \neq 2$

No, $(-2, 0)$ is not a solution.

(b) $-2 \stackrel{?}{=} -2e^{0}$

$3(0) - (-2) \stackrel{?}{=} 2$

$-2 = -2$

$2 = 2$

Yes, $(0, -2)$ is a solution.

(c) $-3 \stackrel{?}{=} -2e^{0}$

$3(0) - (-3) \stackrel{?}{=} 2$

$-3 \neq -2$

$3 \neq 2$

No, $(0, -3)$ is not a solution.

(d) $-5 \stackrel{?}{=} -2e^{-1}$

$3(-1) - (-5) \stackrel{?}{=} 2$

$-5 \neq -2e^{-1}$

$2 = 2$

No, $(-1, -5)$ is not a solution.

5. $2x + y = 6$ Equation 1

$-x + y = 0$ Equation 2

Solve for y in Equation 1: $y = 6 - 2x$

Substitute for y in Equation 2: $-x + (6 - 2x) = 0$

Solve for x: $-3x + 6 = 0 \Rightarrow x = 2$

Back-substitute $x = 2$: $y = 6 - 2(2) = 2$

Answer: $(2, 2)$

7. $x - y = -4$ Equation 1

$x^2 - y = -2$ Equation 2

Solve for y in Equation 1: $y = x + 4$

Substitute for y in Equation 2: $x^2 - (x + 4) = -2$

Solve for x: $x^2 - x - 2 = 0 \Rightarrow (x + 1)(x - 2) = 0 \Rightarrow x = -1, 2$

Back-substitute $x = -1$: $y = -1 + 4 = 3$

Back-substitute $x = 2$: $y = 2 + 4 = 6$

Answers: $(-1, 3), (2, 6)$

9. $3x + y = 2$ Equation 1

$x^3 - 2 + y = 0$ Equation 2

Solve for y in Equation 1: $y = 2 - 3x$

Substitute for y in Equation 2: $x^3 - 2 + (2 - 3x) = 0$

Solve for x: $x^3 - 3x = 0 \Rightarrow x(x^2 - 3) = 0 \Rightarrow x = 0, \pm\sqrt{3}$

Back-substitute: $x = 0: y = 2$

$x = \sqrt{3}: y = 2 - 3\sqrt{3}$

$x = -\sqrt{3}: y = 2 + 3\sqrt{3}$

Solutions: $(0, 2), \left(\sqrt{3}, 2 - 3\sqrt{3}\right), \left(-\sqrt{3}, 2 + 3\sqrt{3}\right)$

11. $x^2 + y = 0$ Equation 1

$x^2 - 4x - y = 0$ Equation 2

Solve for y in Equation 1: $y = -x^2$

Substitute for y in Equation 2: $x^2 - 4x - (-x^2) = 0$

Solve for x: $2x^2 - 4x = 0 \Rightarrow 2x(x - 2) = 0 \Rightarrow x = 0, 2$

Back-substitute $x = 0$: $y = -0^2 = 0$

Back-substitute $x = 2$: $y = -2^2 = -4$

Answers: $(0, 0), (2, -4)$

13. $-\frac{7}{2}x - y = -18$ Equation 1

$8x^2 - 2y^3 = 0$ Equation 2

Solve for x in Equation 1: $-\frac{7}{2}x = y - 18 \implies x = -\frac{2}{7}y + \frac{36}{7}$

Substitute for x in Equation 2: $8\left(-\frac{2}{7}y + \frac{36}{7}\right)^2 - 2y^3 = 0$

Solve for y: $-2y^3 + 8\left(\frac{4}{49}y^2 - \frac{144}{49}y + \frac{36^2}{49}\right) = 0$

$$49y^3 - 16y^2 + 576y - 5184 = 0$$

$$(y - 4)(49y^2 + 180y + 1296) = 0$$

Hence, $y = 4$ and $x = -\frac{2}{7}(4) + \frac{36}{7} = 4$.

Solution: $(4, 4)$

15. $x - y = 0$ Equation 1

$5x - 3y = 10$ Equation 2

Substitute for y in Equation 2: $5x - 3x = 10$

Solve for x: $2x = 10 \implies x = 5$

Back-substitute in Equation 1: $y = x = 5$

Answer: $(5, 5)$

17. $2x - y + 2 = 0$ Equation 1

$4x + y - 5 = 0$ Equation 2

Solve for y in Equation 1: $y = 2x + 2$

Substitute for y in Equation 2: $4x + (2x + 2) - 5 = 0$

Solve for x: $4x + (2x + 2) - 5 = 0 \implies 6x - 3 = 0 \implies x = \frac{1}{2}$

Back-substitute $x = \frac{1}{2}$: $y = 2x + 2 = 2\left(\frac{1}{2}\right) + 2 = 3$

Answer: $\left(\frac{1}{2}, 3\right)$

19. $1.5x + 0.8y = 2.3 \implies 15x + 8y = 23$

$0.3x - 0.2y = 0.1 \implies 3x - 2y = 1$

Solve for y in Equation 2: $-2y = 1 - 3x$

$$y = \frac{3x - 1}{2}$$

Substitute for y in Equation 1: $15x + 8\left(\frac{3x - 1}{2}\right) = 23$

$$15x + 12x - 4 = 23$$

$$27x = 27$$

$$x = 1$$

Then, $y = \dfrac{3x - 1}{2} = \dfrac{3(1) - 1}{2} = 1$. Solution: $(1, 1)$

21. $\frac{1}{5}x + \frac{1}{2}y = 8$ Equation 1

$x + y = 20$ Equation 2

Solve for x in Equation 2: $x = 20 - y$

Substitute for x in Equation 1: $\frac{1}{5}(20 - y) + \frac{1}{2}y = 8$

Solve for y: $4 + \frac{3}{10}y = 8 \implies y = \frac{40}{3}$

Back-substitute $y = \frac{40}{3}$: $x = 20 - y$

$$= 20 - \frac{40}{3} = \frac{20}{3}$$

Answer: $\left(\frac{20}{3}, \frac{40}{3}\right)$

23. $8x + 4y = 7$ Equation 1

$2x + y = 0$ Equation 2

Solve for y in Equation 2: $y = -2x$

Substitute for y in Equation 1: $8x + 4(-2x) = 7$

Solve for x: $8x - 8x = 7$

$0 \neq 7$ Inconsistent

No solution

25. $-\frac{5}{3}x + y = 5$ Equation 1

$-5x + 3y = 6$ Equation 2

Solve for y in Equation 1: $y = 5 + \frac{5}{3}x$

Substitute for y in Equation 2: $-5x + 3\left(5 + \frac{5}{3}x\right) = 6$

Solve for x: $-5x + 15 + 5x = 6$

$15 \neq 6$ Inconsistent

No solution

27. $x - y = 0$ Equation 1

$2x + y = 0$ Equation 2

Solve for y in Equation 1: $y = x$

Substitute for y in Equation 2: $2x + x = 0$

Solve for x: $3x = 0 \implies x = 0$

Back-substitute $x = 0$: $y = x = 0$

Answer: $(0, 0)$

29. $-x + 2y = 2$

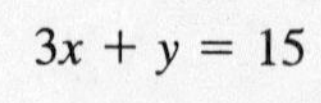

$3x + y = 15$

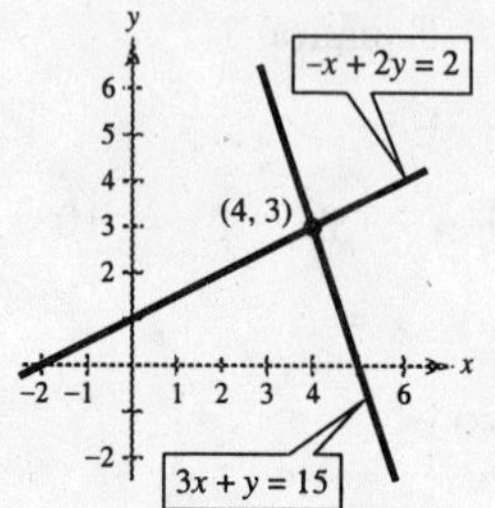

Point of intersection: $(4, 3)$

31. $x - 3y = -2$

$5x + 3y = 17$

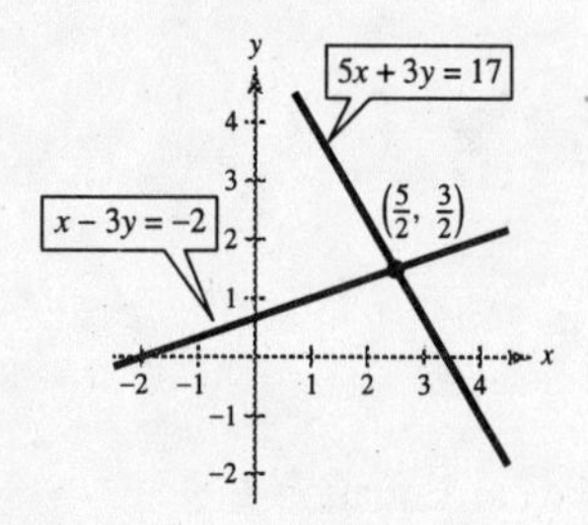

Point of intersection: $\left(\frac{5}{2}, \frac{3}{2}\right)$

33. $x + y = 4$

$x^2 + y^2 - 4x = 0$

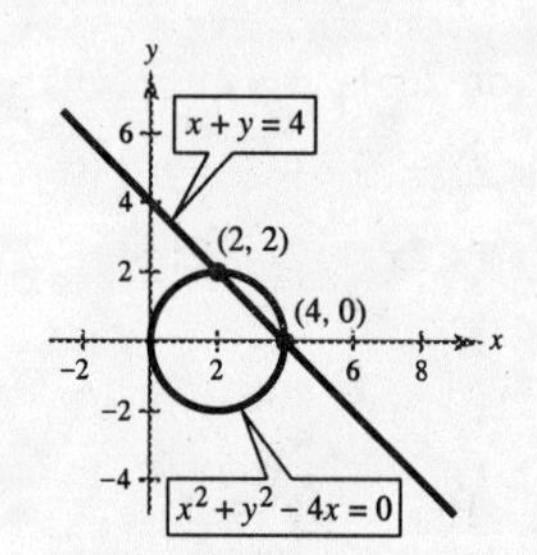

Points of intersection: $(2, 2), (4, 0)$

35. $x - y + 3 = 0$

$x^2 - 4x + 7 = y$

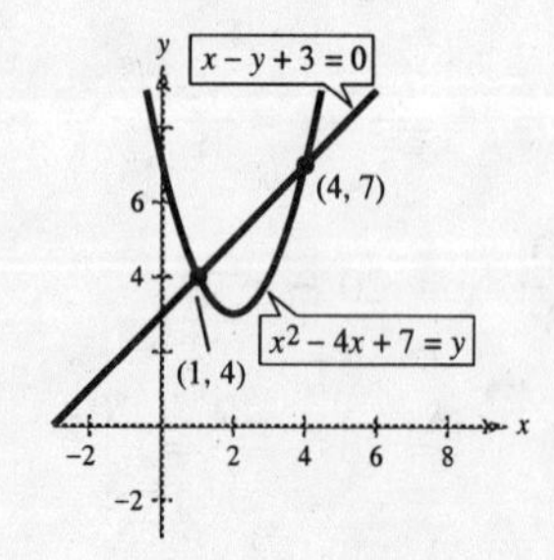

Points of intersection: $(1, 4), (4, 7)$

37. $7x + 8y = 24 \implies y_1 = -\frac{7}{8}x + 3$

$x - 8y = 8 \implies y_2 = \frac{1}{8}x - 1$

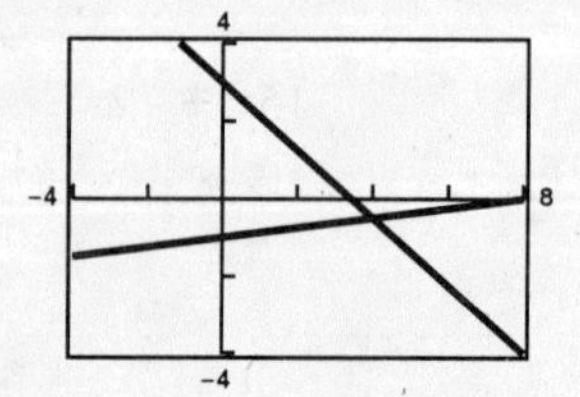

Point of intersection: $\left(4, -\frac{1}{2}\right)$

39. $2x - y + 3 = 0 \Rightarrow y_1 = 2x + 3$

$x^2 + y^2 - 4x = 0 \Rightarrow y_2 = \sqrt{4x - x^2},\ y_3 = -\sqrt{4x - x^2}$

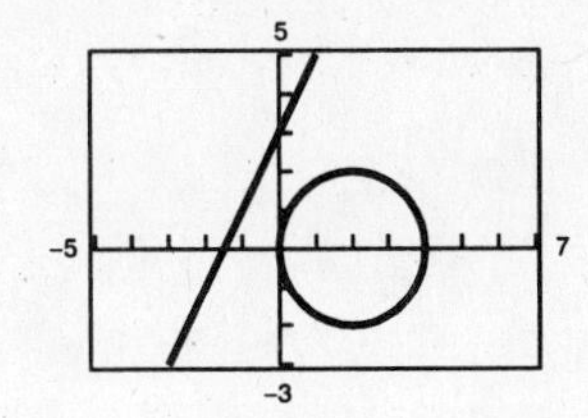

No points of intersection

41. $x^2 + y^2 = 8 \Rightarrow y_1 = \sqrt{8 - x^2}$ and $y_2 = -\sqrt{8 - x^2}$

$y = x^2 \Rightarrow y_3 = x^2$

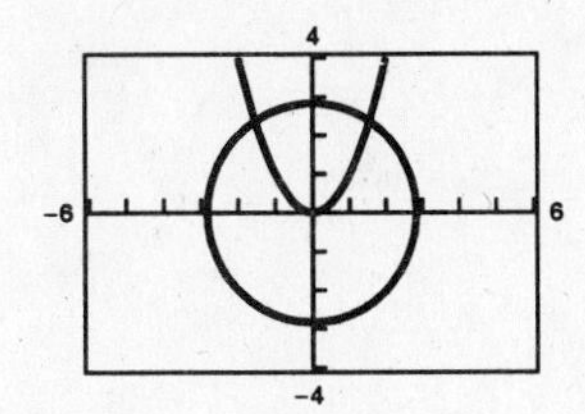

Points of intersection: $\left(\pm\sqrt{\dfrac{-1 + \sqrt{33}}{2}}, \dfrac{-1 + \sqrt{33}}{2}\right) \approx (\pm 1.54, 2.37)$

43. $y = e^x$

$x - y + 1 = 0 \Rightarrow y = x + 1$

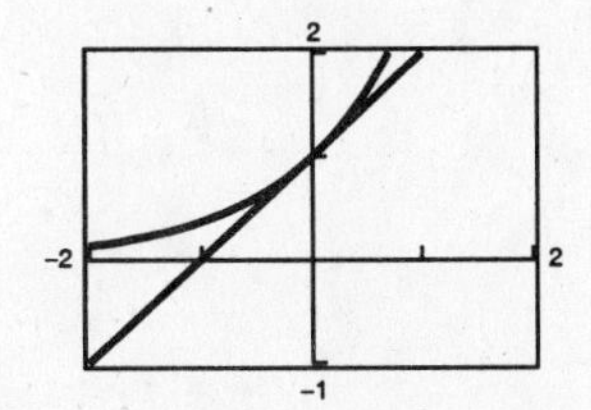

Point of intersection: (0, 1)

45. $x + 2y = 8 \Rightarrow y_1 = 4 - \dfrac{1}{2}x$

$y = \log_2 x \Rightarrow y_2 = \dfrac{\ln x}{\ln 2}$

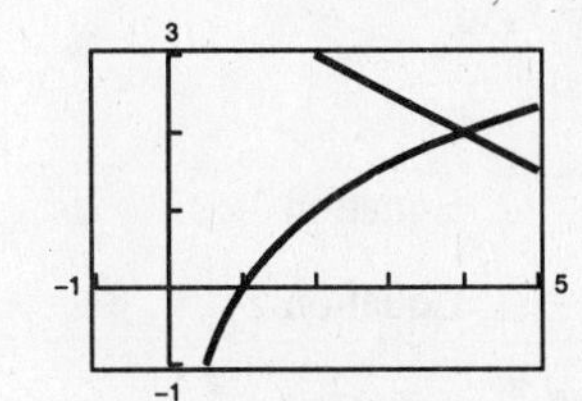

Point of intersection: (4, 2)

47. $y = \sqrt{x}$

$y = x$

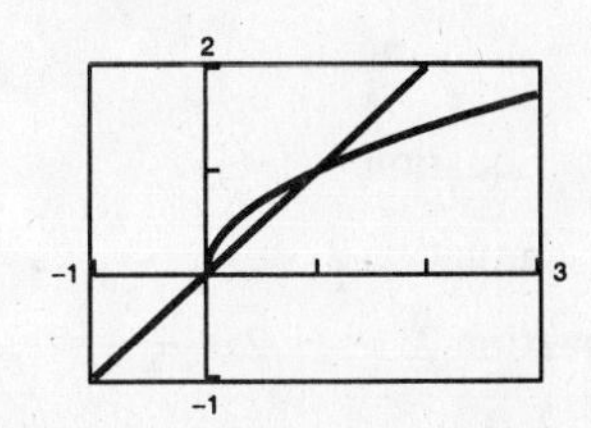

Points of intersection: (0, 0), (1, 1)

49. $x^2 + y^2 = 169 \Rightarrow y_1 = \sqrt{169 - x^2}$ and $y_2 = -\sqrt{169 - x^2}$

$x^2 - 8y = 104 \Rightarrow y_3 = \frac{1}{8}x^2 - 13$

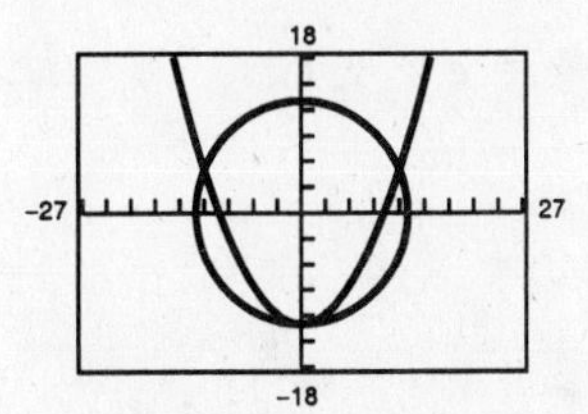

Points of intersection: $(0, -13), (\pm 12, 5)$

51. $y = 2x$ Equation 1

$y = x^2 + 1$ Equation 2

Substitute for y in Equation 2: $2x = x^2 + 1$

Solve for x: $x^2 - 2x + 1 = (x - 1)^2 = 0 \Rightarrow x = 1$

Back-substitute $x = 1$ in Equation 1: $y = 2x = 2$

Answer: $(1, 2)$

53. $3x - 7y + 6 = 0$ Equation 1

$x^2 - y^2 = 4$ Equation 2

Solve for y in Equation 1: $y = \dfrac{3x + 6}{7}$

Solve for y in Equation 2: $x^2 - \left(\dfrac{3x + 6}{7}\right)^2 = 4$

Solve for x: $x^2 - \left(\dfrac{9x^2 + 36x + 36}{49}\right) = 4$

$$49x^2 - (9x^2 + 36x + 36) = 196$$

$$40x^2 - 36x - 232 = 0$$

$$10x^2 - 9x - 58 = 0 \Rightarrow x = \frac{9 \pm \sqrt{81 + 40(58)}}{20} \Rightarrow x = \frac{29}{10}, -2$$

Back-substitute $x = \dfrac{29}{10}$: $y = \dfrac{3x + 6}{7} = \dfrac{3(29/10) + 6}{7} = \dfrac{21}{10}$

Back-substitute $x = -2$: $y = \dfrac{3x + 6}{7} = 0$

Answers: $\left(\dfrac{29}{10}, \dfrac{21}{10}\right), (-2, 0)$

55. $x - 2y = 4$ Equation 1

$x^2 - y = 0$ Equation 2

Solve for y in Equation 2: $y = x^2$

Substitute for y in Equation 1: $x - 2x^2 = 4$

Solve for x: $0 = 2x^2 - x + 4$

No real solutions, the discriminant in the Quadratic Formula is negative.

Inconsistent, No solution

57. $y - e^{-x} = 1 \Rightarrow y = e^{-x} + 1$

$y - \ln x = 3 \Rightarrow y = \ln x + 3$

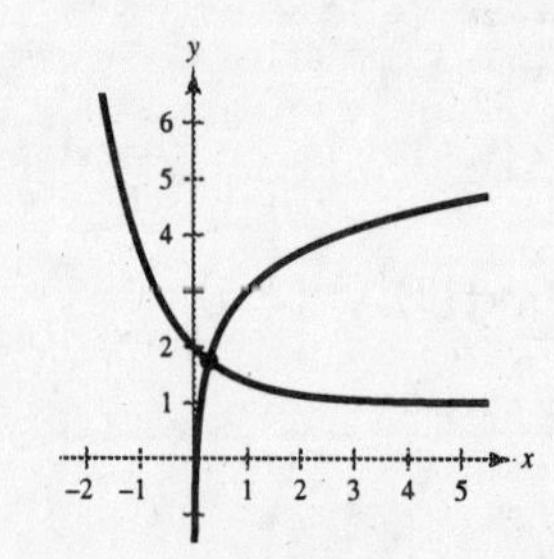

Point of intersection: Approximately $(0.287, 1.751)$

59. $y = x^3 - 2x^2 + 1$ Equation 1

$y = 1 - x^2$ Equation 2

Substitute for y in Equation 2: $x^3 - 2x^2 + 1 = 1 - x^2$

Solve for x: $x^3 - x^2 = 0$

$$x^2(x - 1) = 0 \Rightarrow x = 0, 1$$

Back-substitute: $x = 0 \Rightarrow y = 1$

$x = 1 \Rightarrow y = 0$

Solutions: $(0, 1), (1, 0)$

61. $xy - 1 = 0$ Equation 1

$2x - 4y + 7 = 0$ Equation 2

Solve for y in Equation 1: $y = \dfrac{1}{x}$

Substitute for y in Equation 2: $2x - 4\left(\dfrac{1}{x}\right) + 7 = 0$

Solve for x: $2x^2 - 4 + 7x = 0 \implies (2x - 1)(x + 4) = 0 \implies x = \dfrac{1}{2}, -4$

Back-substitute $x = \dfrac{1}{2}$: $y = \dfrac{1}{1/2} = 2$

Back-substitute $x = -4$: $y = \dfrac{1}{-4} = -\dfrac{1}{4}$

Answers: $\left(\dfrac{1}{2}, 2\right), \left(-4, -\dfrac{1}{4}\right)$

63. $C = 8650x + 250{,}000,\ R = 9950x$

$R = C$

$9950x = 8650x + 250{,}000$

$1300x = 250{,}000$

$x \approx 192$ units

$R \approx \$1{,}910{,}400$

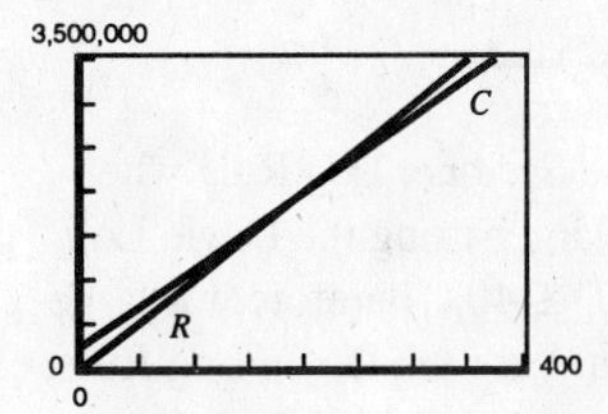

65. $C = 5.5\sqrt{x} + 10{,}000,\ R = 3.29x$

$$R = C$$

$$3.29x = 5.5\sqrt{x} + 10{,}000$$

$$3.29x - 10{,}000 = 5.5\sqrt{x}$$

$$10.8241x^2 - 65{,}800x + 100{,}000{,}000 = 30.25x$$

$$10.8241x^2 - 65{,}830.25x + 100{,}000{,}000 = 0$$

$$x \approx 3133 \text{ units}$$

In order for the revenue to break even with the cost, 3133 units must be sold, $R \approx \$10{,}308$.

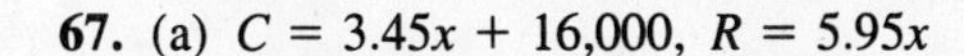

67. (a) $C = 3.45x + 16{,}000,\ R = 5.95x$

(b)

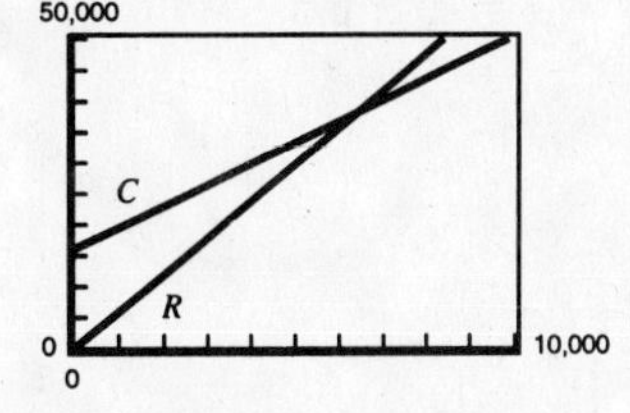

$x \approx 6400$ units

(c) $R = C$

$5.95x = 3.45x + 16{,}000$

$2.50x = 16{,}000$

$x \approx 6400$ units

69. $0.06x = 0.03x + 250$

$0.03x = 250$

$x \approx \$8333.33$

To make the straight commission offer better, you would have to sell more than \$8333.33 per week.

71. (a) $x + y = 20{,}000$

$0.065x + 0.085y = 1600$

(b) As x increases, y decreases and the amount of interest decreases.

(c) The curves intersect at $x = 5000$. Thus, \$5000 should be invested at 6.5%.

73. $V = (D - 4)^2, \ 5 \le D \le 40$

$V = 0.79D^2 - 2D - 4, \ 5 \le D \le 40$

(a)

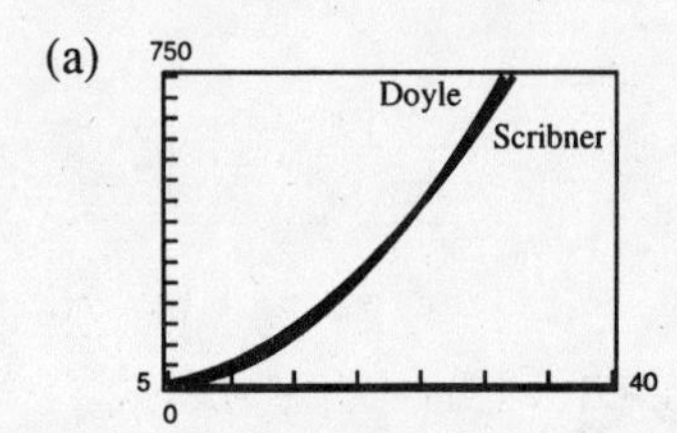

(b) The two graphs intersect at $D = 24.72$. Algebraically:

$$(D - 4)^2 = 0.79D^2 - 2D - 4$$
$$D^2 - 8D + 16 = 0.79D^2 - 2D - 4$$
$$0.21D^2 - 6D + 20 = 0$$
$$D \approx 24.72, \ 3.9$$

Since $5 \le D \le 40$, the scales agree when $D \approx 24.72$ inches.

(c) V is larger using the Scribner Log Rule when $5 \le D \le 24.7$. V is larger using the Doyle Log Rule when $24.7 \le D \le 40$. Therefore, for large diameters, you would use the Doyle Log Rule.

75. $2l + 2w = 280 \Rightarrow l + w = 140$

$w = l - 20 \Rightarrow l + (l - 20) = 140$

$2l = 160$

$l = 80$

$w = l - 20 = 80 - 20 = 60$

Dimensions: 60×80 centimeters

77. $2l + 2w = 210 \Rightarrow \quad l + w = 105$

$l = \frac{3}{2}w \Rightarrow \frac{3}{2}w + w = 105$

$\frac{5}{2}w = 105$

$w = 42$

$l = \frac{3}{2}(42) = 63$

Dimensions: 42×63 feet

79. $A = \frac{1}{2}bh$

$1 = \frac{1}{2}a^2$

$a^2 = 2$

$a = \sqrt{2}$

The dimensions are $b = h = \sqrt{2}$ and hypotenuse $= 2$.

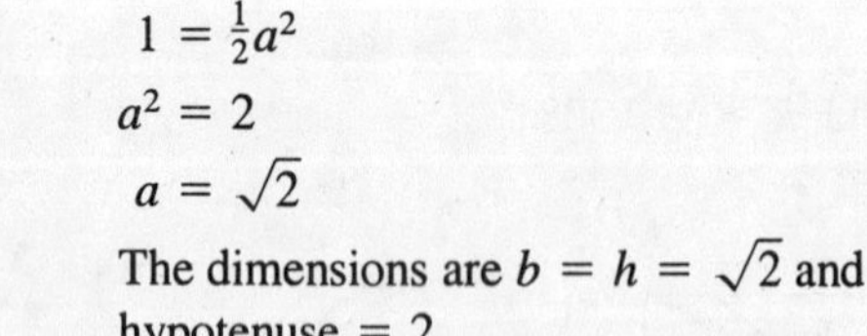

81. (a) Linear model: $E = 0.342t + 10.723$

Quadratic model:

$E = 0.01625t^2 + 0.196t + 11.00$

(b)

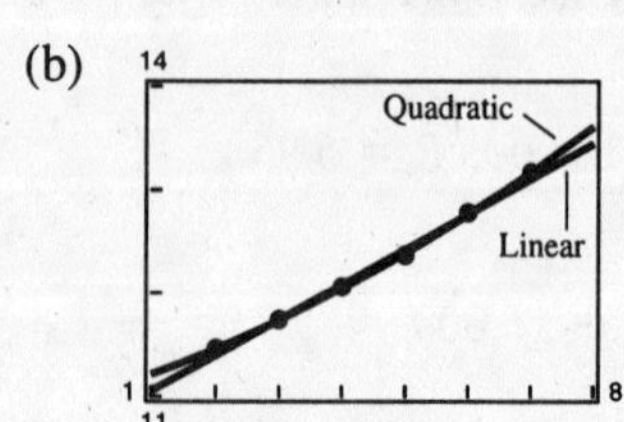

(c) The models intersect at (6.2, 12.8) and (2.8, 11.7).

(d) For 2000, $t = 10$.

Linear model: $E \approx \$14.14$

Quadratic model: $E \approx \$14.59$

The linear model seems more accurate.

83. False. There could be four points of intersection. For example, $x^2 + y^2 = 4$ and $y = x^2 - 3$.

85. The advantage of the method of substitution over the graphical method is that substitution gives an exact answer.

87. (a)

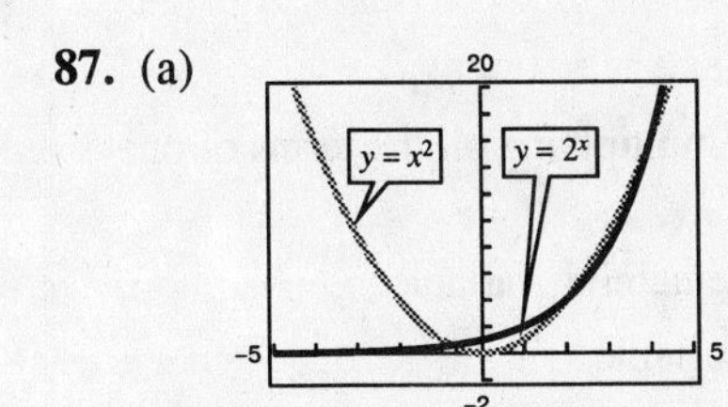

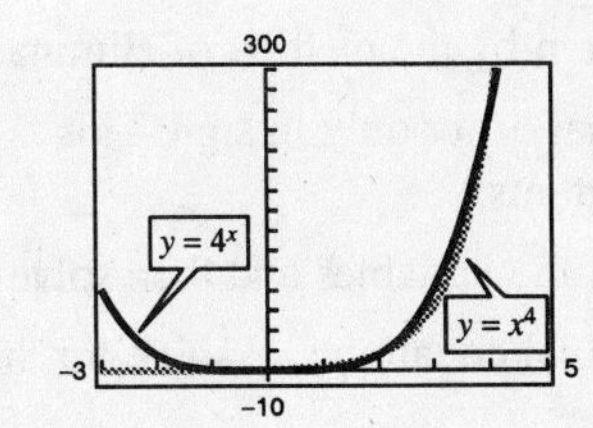

(b) Based on the graphs in part (a) it appears that for $b > 1$, there are three points of intersection for the graphs of $y = b^x$ and $y = x^b$ when b is an even number.

89. $(3.5, 4), (10, 6)$

$$m = \frac{6-4}{10-3.5} = \frac{2}{6.5}$$

$$y - 6 = \frac{2}{6.5}(x - 10)$$

$$6.5y - 39 = 2x - 20$$

$$4x - 13y + 38 = 0$$

91. $(4, -2), (4, 5)$

$$x = 4$$

93. $\left(-\frac{7}{3}, 8\right), \left(\frac{5}{2}, \frac{1}{2}\right)$

$$m = \frac{8 - (1/2)}{-(7/3) - (5/2)} = \frac{15/2}{-29/6} = -\frac{45}{29}$$

$$y - \frac{1}{2} = -\frac{45}{29}\left(x - \frac{5}{2}\right)$$

$$29y - \frac{29}{2} = -45x + \frac{225}{2}$$

$$45x + 29y - 127 = 0$$

95. Domain: all $x \neq -\frac{2}{3}$

Vertical asymptote: $x = -\frac{2}{3}$

Horizontal asymptote: $y = \frac{2}{3}$

97. $f(x) = 3 - \dfrac{2}{x^2} = \dfrac{3x^2 - 2}{x^2}$

Domain: all $x \neq 0$

Vertical asymptote: $x = 0$

Horizontal asymptote: $y = 3$

99.

$$8(10^{2x}) = 28$$

$$10^{2x} = \frac{28}{8} = \frac{7}{2}$$

$$2x \cdot \ln 10 = \ln\left(\frac{7}{2}\right)$$

$$x = \frac{\ln\left(\frac{7}{2}\right)}{2\ln(10)} \approx 0.272$$

101. $\log_{10}(x + 3) - \log_{10} x = \log_{10}(x + 8)$

$$\log_{10}\frac{x+3}{x} = \log_{10}(x + 8)$$

$$\frac{x+3}{x} = x + 8$$

$$x + 3 = x^2 + 8x$$

$$x^2 + 7x - 3 = 0$$

$$x = \frac{-7 \pm \sqrt{49 - 4(-3)}}{2} = -\frac{7}{2} \pm \frac{\sqrt{61}}{2}$$

Solution: $x = -\frac{7}{2} + \frac{\sqrt{61}}{2} \approx 0.405$ $\left(-\frac{7}{2} - \frac{\sqrt{61}}{2} \text{ is extraneous.}\right)$

Section 7.2 Systems of Linear Equations in Two Variables

- You should be able to solve a linear system by the method of elimination.
  1. Obtain coefficients for either x or y that differ only in sign. This is done by multiplying all the terms of one or both equations by appropriate constants.
  2. Add the equations to eliminate one of the variables and then solve for the remaining variable.
  3. Use back-substitution into either original equation and solve for the other variable.
  4. Check your answer.
- You should know that for a system of two linear equations, one of the following is true.
  (a) There are infinitely many solutions; the lines are identical. The system is consistent.
  (b) There is no solution; the lines are parallel. The system is inconsistent.
  (c) There is one solution; the lines intersect at one point. The system is consistent.

Solutions to Odd-Numbered Exercises

1. $2x = y = 5$ Equation 1

$x - y = 1$ Equation 2

Add to eliminate y: $3x = 6 \Rightarrow x = 2$

Substitute $x = 2$ in Equation 2: $2 - y = 1 \Rightarrow y = 1$

Answer: $(2, 1)$

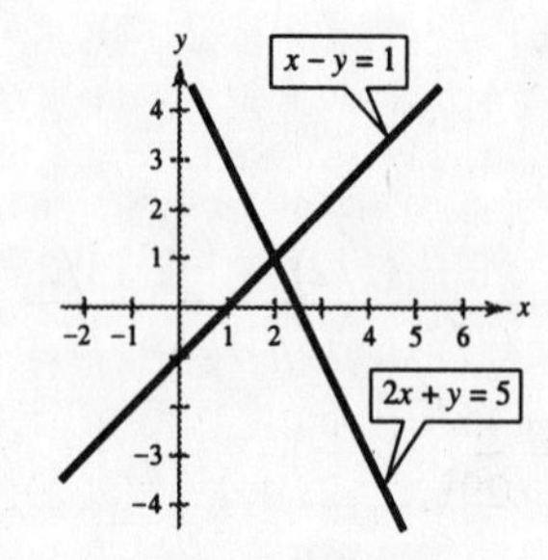

3. $x = \ y = 0$ Equation 1

$3x = 2y = 1$ Equation 2

Multiply Equation 1 by -2: $-2x - 2y = 0$

Add this to Equation 2 to eliminate y: $x = 1$

Substitute $x = 1$ in Equation 1: $1 = y = 0 \Rightarrow y = -1$

Answer: $(1, -1)$

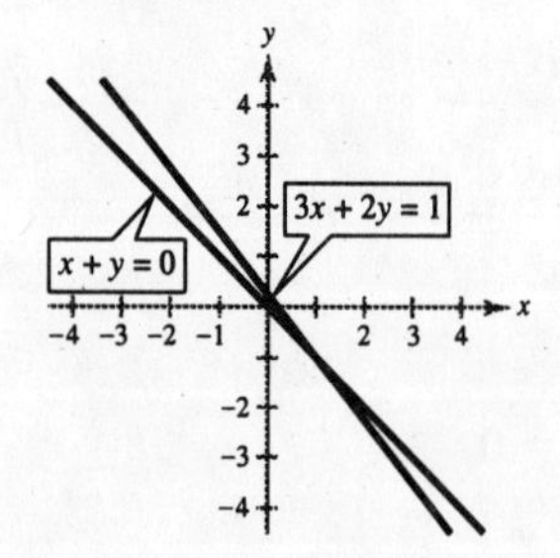

5.

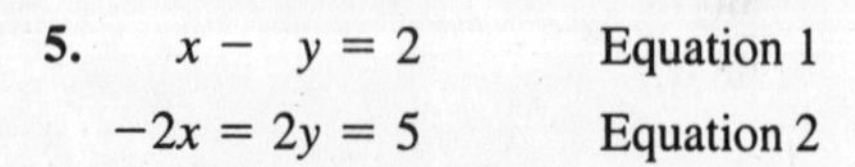

$x - \ y = 2$ Equation 1

$-2x = 2y = 5$ Equation 2

Multiply Equation 1 by 2: $2x - 2y = 4$

Add this to Equation 2: $0 = 9$

There are no solutions.

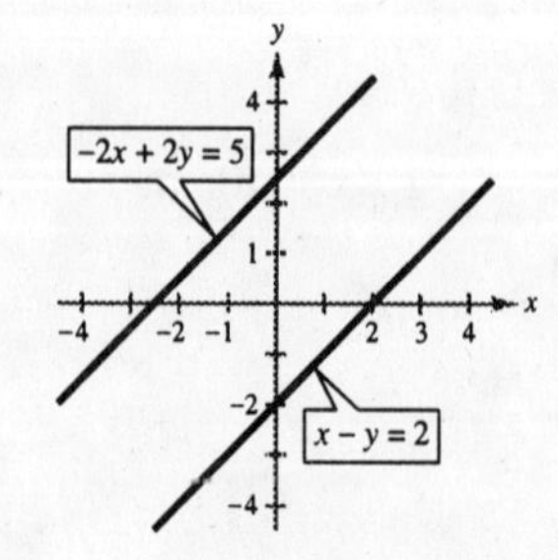

7. $3x - 2y = \quad 5$ Equation 1

$-6x = 4y = -10$ Equation 2

Multiply Equation 1 by 2 and add to Equation 2: $0 = 0$

The equations are dependent. There are infinitely many solutions. The solutions consist of all (x, y) satisfying $3x - 2y = 5$.

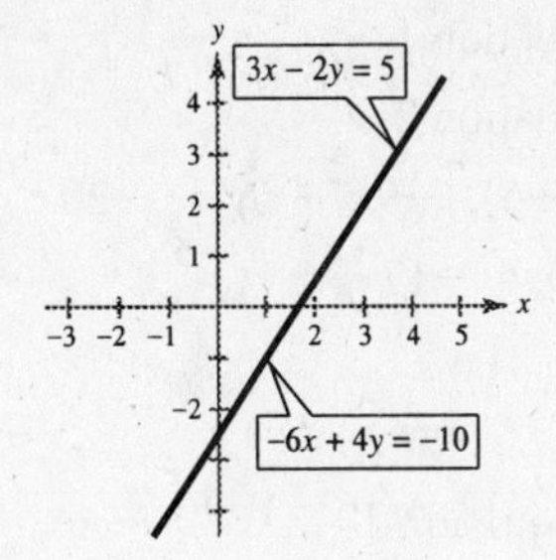

9. $9x = 3y = 1$ Equation 1

$3x - 6y = 5$ Equation 2

Multiply Equation 2 by (-3): $\quad 9x = \ 3y = \quad 1$

$$-9x = 18y = -15$$

Add to eliminate x: $21y = -14 \implies y = -\frac{2}{3}$

Substitute $y = -\frac{2}{3}$ in Equation 1: $9x = 3\left(-\frac{2}{3}\right) = 1$

$$x = \frac{1}{3}$$

Answer: $\left(\frac{1}{3}, -\frac{2}{3}\right)$

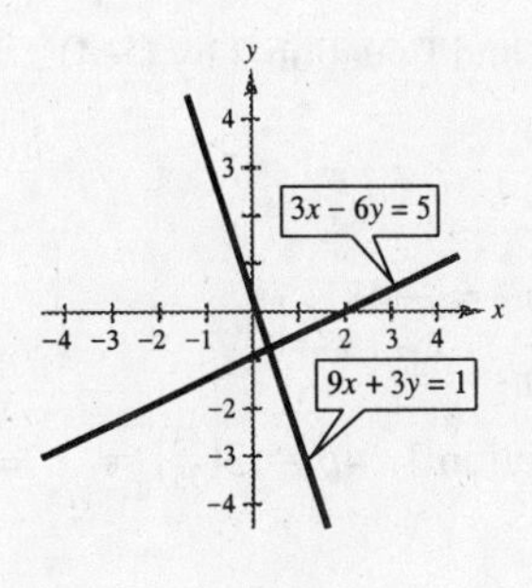

11. $x = 2y = 4$ Equation 1

$x - 2y = 1$ Equation 2

Add to eliminate y:

$2x = 5$

$x = \frac{5}{2}$

Substitute $x = \frac{5}{2}$ in Equation 1:

$\frac{5}{2} = 2y = 4 \implies y = \frac{3}{4}$

Answer: $\left(\frac{5}{2}, \frac{3}{4}\right)$

13. $2x = 3y = 18$ Equation 1

$5x - \ y = 11$ Equation 2

Multiply Equation 2 by 3: $15x - 3y = 33$

Add this to Equation 1 to eliminate y:

$17x = 51 \implies x = 3$

Substitute $x = 3$ in Equation 1:

$6 = 3y = 18 \implies y = 4$

Answer: $(3, 4)$

15. $3x = 2y = 10$ Equation 1

$2x = 5y = \ 3$ Equation 2

Multiply Equation 1 by 2 and Equation 2 by (-3):

$6x = \ 4y = 20$

$-6x - 15y = -9$

Add to eliminate x: $-11y = 11 \implies y = -1$

Substitute $y = -1$ in Equation 1:

$3x - 2 = 10 \implies x = 4$

Answer: $(4, -1)$

17. $5u = 6v = 24$ Equation 1

$3u = 5v = 18$ Equation 2

Multiply Equation 1 by 3 and Equation 2 by (-5): $\quad 15u = 18v = \quad 72$

$$-15u - 25v = -90$$

Add to eliminate u: $-7v = -18 \implies v = \frac{18}{7}$

Substitute $v = \frac{18}{7}$ in Equation 2: $3u = 5\left(\frac{18}{7}\right) = 18 \implies u = \frac{12}{7}$

Answer: $\left(\frac{12}{7}, \frac{18}{7}\right)$

19. $1.8x = 1.2y = 4$ Equation 1

$9x = 6y = 3$ Equation 2

Multiply Equation 1 by (-5): $-9x - 6y = -20$

Add this to Equation 2: $0 = -17$

Inconsistent; no solution

21. $4b = 3m = 3$ Equation 1

$3b = 11m = 13$ Equation 2

Multiply Equation 1 by 3 and Equation 2 by (-4):

$$12b = 9m = 9$$

$$-12b - 44m = -52$$

Add to eliminate b: $-35m = -43$

$$m = \tfrac{43}{35}$$

Substitute $m = \frac{43}{35}$ in Equation 1: $4b = 3\left(\frac{43}{35}\right) = 3 \Rightarrow b = -\frac{6}{35}$

Answer: $\left(-\frac{6}{35}, \frac{43}{35}\right)$

23. $\dfrac{x}{4} = \dfrac{y}{6} = 1$

$\dfrac{1}{2}x - \dfrac{1}{2}y = \dfrac{1}{6}$

Multiply both equations by 12 to clear fractions.

$3x = 2y = 12$ Equation 1

$6x - 6y = 2$ Equation 2

Multiply equation 1 by 3: $9x = 6y = 36$

Add this to equation 2: $15x = 38 \Rightarrow x = \dfrac{38}{15}$

Substitute $x = \dfrac{38}{15}$ into equation 1:

$$3\left(\frac{38}{15}\right) = 2y = 12 \Rightarrow 2y = 12 - \frac{38}{5} = \frac{22}{5} \Rightarrow y = \frac{11}{5}$$

Answer: $\left(\dfrac{38}{15}, \dfrac{11}{5}\right)$

25. $\dfrac{3}{4}x = y = \dfrac{1}{8}$ Equation 1

$\dfrac{9}{4}x = 3y = \dfrac{3}{8}$ Equation 2

Multiply equation 1 by -3: $-\dfrac{9}{4}x - 3y = -\dfrac{3}{8}$

Add this to equation 2: $0 = 0$

There are an infinite number of solutions.

The solutions consist of all (x, y) satisfying

$\dfrac{3}{4}x = y = \dfrac{1}{8}$, or $6x = 8y = 1$.

27. $\dfrac{x = 3}{4} = \dfrac{y - 1}{3} = 1$ Equation 1

$2x - y = 12$ Equation 2

Multiply Equation 1 by 12 and Equation 2 by 4

$3x = 4y = 7$

$8x - 4y = 48$

Add to eliminate y: $11x = 55 \Rightarrow x = 5$

Substitute $x = 5$ into Equation 2:

$2(5) - y = 12 \Rightarrow y = -2$

Answer: $(5, -2)$

29. $2.5x - 3y = 1.5$ Equation 1

$10x - 12y = 6$ Equation 2

Multiply Equation 1 by (-4):

$-10x = 12y = -6$

Add this to Equation 2 to eliminate x:

$0 = 0$

The solution set consists of all points lying on the line

$10x - 12y = 6.$

Let $x = a$, then $y = \frac{5}{6}a - \frac{1}{2}$.

Answer: $\left(a, \frac{5}{6}a - \frac{1}{2}\right)$, where a is any real number.

31. $0.02x - 0.05y = -0.19$ Equation 1

$0.03x = 0.04y = 0.52$ Equation 2

Multiply Equation 1 by 4 and Equation 2 by 5:

$0.08x - 0.2y = -0.76$

$0.15x = 0.2y = 2.6$

Add these to eliminate y: $0.23x = 1.84 \Longrightarrow x = 8$

Substitute $x = 8$ in Equation 1:

$0.02(8) - 0.05y = -0.19 \Longrightarrow y = 7$

Answer: $(8, 7)$

33. $2x - 5y = 0 \implies y = \frac{2}{5}x$

$x - y = 3 \implies y = x - 3$

The system is consistent. There is one solution, $(5, 2)$.

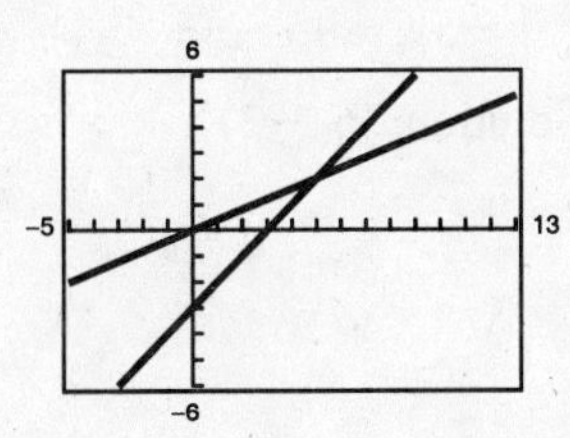

35. $\frac{3}{5}x - y = 3 \implies y = \frac{3}{5}x - 3$

$-3x = 5y = 9 \implies y = \frac{1}{5}(3x = 9) = \frac{3}{5}x = \frac{9}{5}$

The lines are parallel. The system is inconsistent.

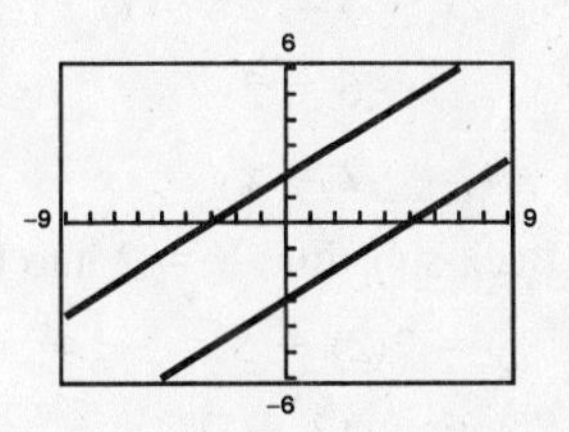

37. $x = 7y = 2 \implies y = \frac{1}{7}(2 - x)$

$4x - y = 9 \implies y = 4x - 9$

The system is consistent. There is one solution, $(2.241, -0.034)$ $= \left(\frac{65}{29}, -\frac{1}{29}\right)$.

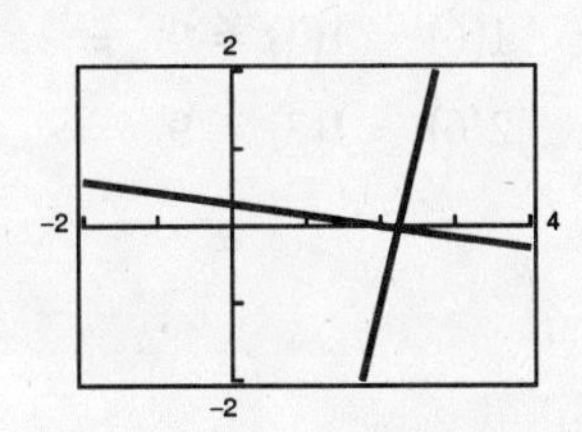

39. $-x = 7y = 3 \implies y = \frac{1}{7}(x = 3)$

$-\frac{1}{7}x = y = 5 \implies y = \frac{1}{7}x = 5$

The lines are parallel. The system is inconsistent.

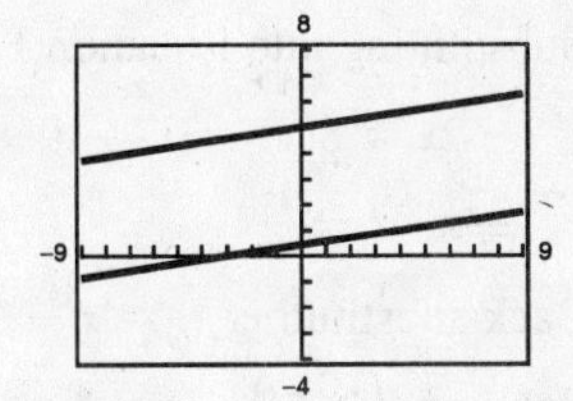

41. $8x = 9y = 42$

$6x - y = 16$

Solution: $(3, 2)$

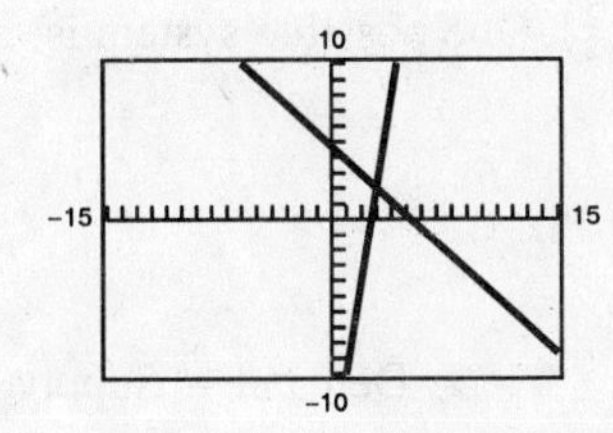

43. $\frac{3}{2}x - \frac{1}{5}y = 8$

$-2x = 3y = 3$

Solution: $(6, 5)$

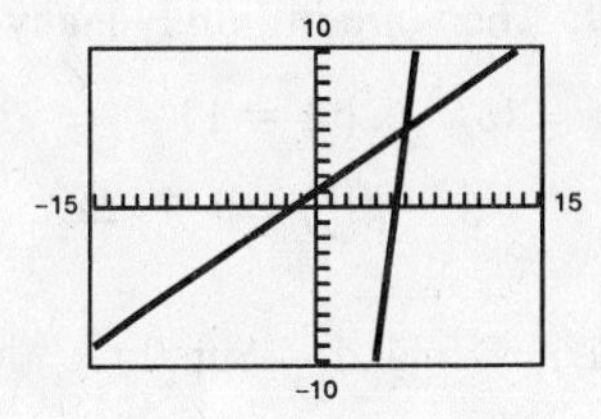

45. $0.5x = 2.2y = 9$

$6x = 0.4y = -22$

Solution: $(-4, 5)$

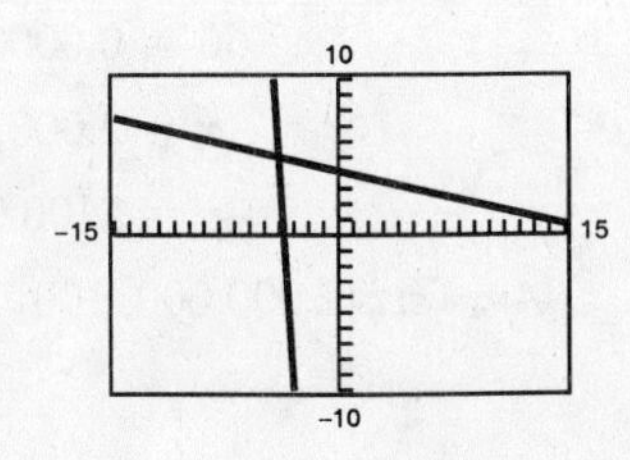

47. $7x - 2y = \quad 24 \implies y = \frac{1}{2}(7x - 24)$

$5x = 6y = -20 \implies y = \dfrac{-20 - 5x}{6}$

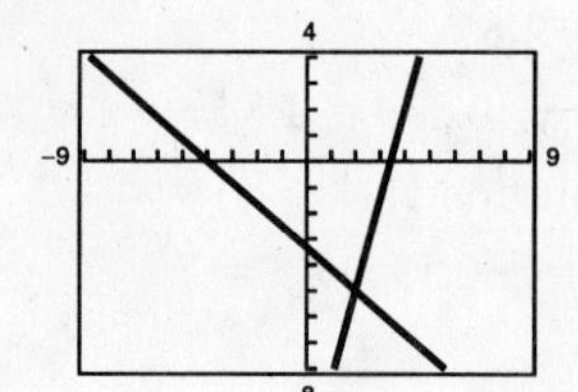

Solution: $(2, -5)$

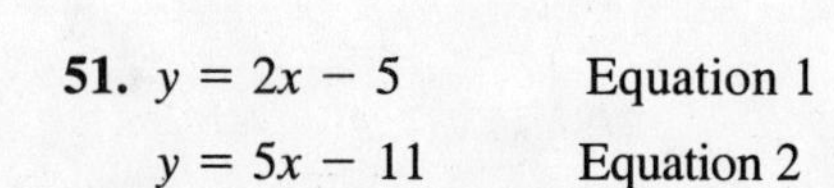

49. $3x - 5y = 7$ Equation 1

$2x = \quad y = 9$ Equation 2

Multiply Equation 2 by 5:

$10x = 5y = 45$

Add this to Equation 1:

$13x = 52 \implies x = 4$

Back-substitute $x = 4$ into Equation 2:

$2(4) = y = 9 \implies y = 1$

Solution: $(4, 1)$

51. $y = 2x - 5$ Equation 1

$y = 5x - 11$ Equation 2

Since both equations are solved for y, set them equal to one another and solve for x.

$2x - 5 = 5x - 11$

$6 = 3x$

$2 = x$

Back-substitute $x = 2$ into Equation 1:

$y = 2(2) - 5 = -1$

Solution: $(2, -1)$

53. $x - 5y = 21$

$6x = 5y = 21$

Adding the equations, $7x = 42 \implies x = 6.$

Back-substituting, $x - 5y = 6 - 5y = 21 \implies$

$-5y = 15 \implies y = -3$

Solution: $(6, -3)$

55. $-2x = 8y = 19$ Equation 1

$y = x - 3$ Equation 2

Substituting into Equation 1,

$-2x = 8(x - 3) = 19 \implies 6x = 43$

$\implies x = \frac{43}{6}$

Back-substituting, $y = x - 3 = \frac{43}{6} - 3 = \frac{25}{6}$

Solution: $\left(\frac{43}{6}, \frac{25}{6}\right)$

57. There are infinitely many systems that have the solution $(6, 3)$. One possible system is

$1(6) = 1(3) = 9 \implies x = y = 9$

$2(6) - 1(3) = 9 \implies 2x - y = 9$

59. There are infinitely many systems that have the solution $\left(3, \frac{5}{2}\right)$. One possible system is:

$2(3) = 2\left(\frac{5}{2}\right) = 11 \implies 2x = 2y = 11$

$3 - 4\left(\frac{5}{2}\right) = -7 \implies x - 4y = -7$

61. Demand = Supply

$50 - 0.5x = 0.125x$

$50 = 0.625x$

$x = 80$ units

$p = \$10$

Answer: $(80, 10)$

63. Demand = Supply

$140 - 0.00002x = 80 = 0.00001x$

$60 = 0.00003x$

$x = 2{,}000{,}000$ units

$p = \$100.00$

Answer: $(2{,}000{,}000, 100)$

65. Let $x =$ the ground speed and $y =$ the wind speed.

$$\begin{array}{llrl} 3.6(x - y) = 1800 & \text{Equation 1} & x - y = & 500 \\ 6(x = y) = 1800 & \text{Equation 2} & x = y = & 600 \\ \hline & & 2x \quad = & 1100 \\ & & x \quad = & 550 \\ & & 550 = y = & 600 \\ & & y = & 50 \end{array}$$

Answer: $x = 550$ mph, $y = 50$ mph

67. Let $x =$ the number of liters at 20%, $y =$ the number of liters at 50%.

(a) $x = \quad y = \quad 10$

$0.2x = 0.5y = 0.3(10)$

(b)

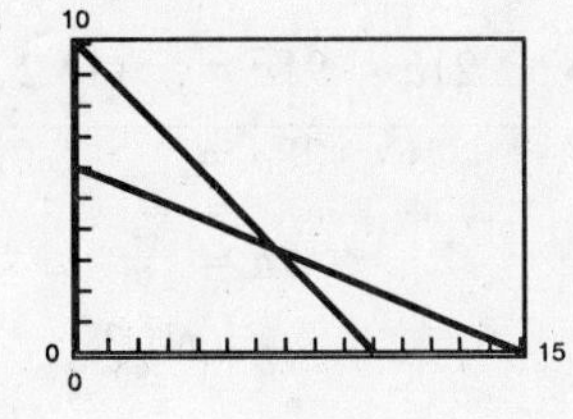

As x increases, y decreases.

(c)

$$\begin{array}{llr} (-2 & \text{Equation 1)} & -2x - 2y = -20 \\ (10 & \text{Equation 2)} & 2x = 5y = \quad 30 \\ \hline & & 3y = \quad 10 \end{array}$$

$$y = \tfrac{10}{3}$$

$$x = \tfrac{10}{3} = 10$$

$$x = \tfrac{20}{3}$$

Answer: $x = 6\frac{2}{3}$ liters at 20%, $y = 3\frac{1}{3}$ liters at 50%

69. Let $x =$ amount invested at 7.5%

Let $y =$ amount invested at 9%

$$\begin{array}{rl} x = \quad y = 12{,}000 & \text{Equation 1} \\ 0.075x = 0.09y = \quad 990 & \text{Equation 2} \end{array}$$

From Equation 1, $y = 12{,}000 - x$. Substituting into Equation 2,

$$0.075x = 0.09(12{,}000 - x) = 990$$

$$-0.015x = -90$$

$$x = 6000$$

Then, $y = 12{,}000 - 6000 = 6000$.

Answer: $x = \$6000$ at 7.5%

$y = \$6000$ at 9%

71. Let $x =$ number of adult tickets sold, $y =$ number of child tickets sold.

$$\begin{array}{llrr} x = \ y = \quad 500 & \text{Equation 1} & -4x - 4y = & -2000.00 \\ 7.5x = 4y = \$3312.50 & \text{Equation 2} & 7.5x = 4y = & 3312.50 \\ \hline & & 3.5x \quad = & 1312.50 \\ & & x \quad = & 375 \\ & & 375 = \quad y = & 500 \\ & & y = & 125 \end{array}$$

Answer: $x = 375$ adult tickets, $y = 125$ child tickets

73. Let $x =$ distance one person drives,

$y =$ distance other person drives.

$x = y = 300$ Equation 1

$y = 3x$ Equation 2

$x = 3x = 300$ Use substitution.

$4x = 300$

$x = 75$

$y = 3x = 225$

Answer: 75 km and 225 km

75. $5b = 10a = 20.2 \Rightarrow -10b - 20a = -40.4$

$10b = 30a = 50.1 \Rightarrow 10b = 30a = 50.1$

$10a = 9.7$

$a = 0.97$

$b = 2.10$

Least squares regression line:

$y = 0.97x = 2.10$

77. $7b = 21a = 35.1 \Rightarrow -21b - 63a = -105.3$

$21b = 91a = 114.2 \Rightarrow 21b = 91a = 114.2$

$28a = 8.9$

$a = \frac{89}{280}$

$b = \frac{1137}{280}$

Least squares regression line: $y = \frac{1}{280}(89x = 1137) \approx 0.318x = 4.061$

79. (a) $4b = 7a = 174 \Rightarrow 28b = 49a = 1218$

$7b = 13.5a = 322 \Rightarrow -28b - 54a = -1288$

Adding, $-5a = -70 \Rightarrow a = 14,\ b = 19.$

Thus, $y = 14x = 19$

(b) Using a graphing utility, you obtain $y = 14x = 19$.

(c)

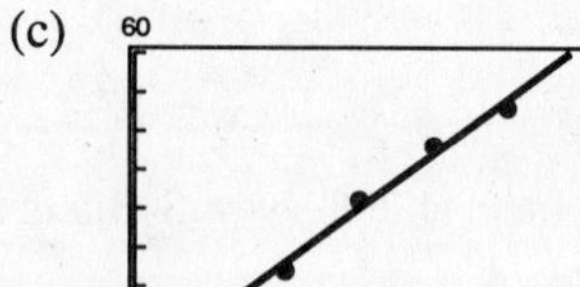

(d) If $x = 1.6$, (160 pounds/acre), $y = 14(1.6) = 19 = 41.4$ bushels per acre.

81. True. A consistent linear system has either one solution or an infinite number of solutions.

83. True. The lines are distinct and parallel.

85. $21x - 20y = 0$ Equation 1

$13x - 12y = 120$ Equation 2

Multiply Equation 2 by $\left(-\frac{5}{3}\right)$: $-\frac{65}{3}x = 20y = -200$

Add this to Equation 1 to eliminate y: $-\frac{2}{3}x = -200 \Rightarrow x = 300$

Substitute $x = 300$ in Equation 1: $21(300) - 20y = 0 \Rightarrow y = 315$

Solution: (300, 315)

The lines are not parallel. The scale on the axes must be changed to see the point of intersection.

87. (a) $x = y = 10$ Equation 1

$x = y = 20$ Equation 2

Subtract Equation 2 from Equation 1:

$0 = -10$

System is inconsistent $\Rightarrow$ no solution

(b) $x = y = 3$ Equation 1

$2x = 2y = 6$ Equation 2

Multiply Equation 1 by (-2): $-2x - 2y = -6$

Add this to Equation 2: $0 = 0$ (dependent)

The system has an infinite number of solutions.

89. $15x = 3y = 6$ Equation 1

$-10x = ky = 9$ Equation 2

Multiply Equation 1 by $\frac{2}{3}$: $10x = 2y = 4$

Add this to Equation 2: $ky = 2y = 13$

$(k = 2)y = 13$

The system is inconsistent if $k = -2$.

91.

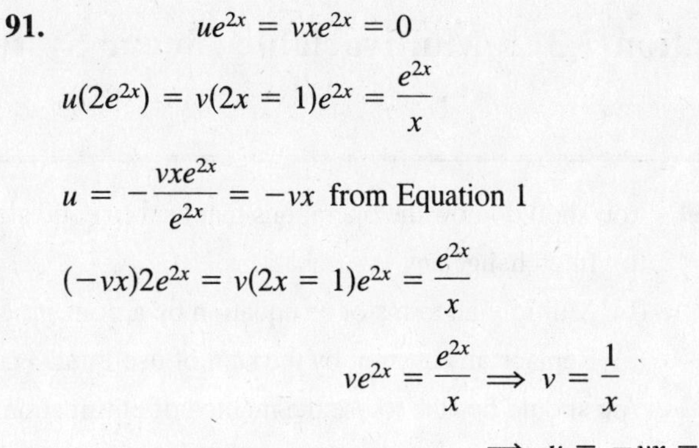

$$ue^{2x} = vxe^{2x} = 0$$

$$u(2e^{2x}) = v(2x = 1)e^{2x} = \frac{e^{2x}}{x}$$

$$u = -\frac{vxe^{2x}}{e^{2x}} = -vx \text{ from Equation 1}$$

$$(-vx)2e^{2x} = v(2x = 1)e^{2x} = \frac{e^{2x}}{x}$$

$$ve^{2x} = \frac{e^{2x}}{x} \Rightarrow v = \frac{1}{x}$$

$$\Rightarrow u = -vx = -1$$

93.

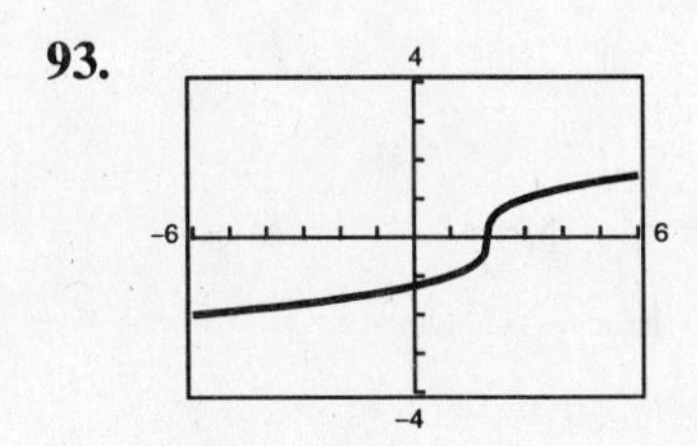

Domain: all x

Range: all y

95.

Domain: all $x \neq \pm 3$

Range: all $y \neq 0$

97. $2(x - 3) > -5x = 1$

$7x > 7$

$x > 1$

99. $-6 \leq 3x - 10 < 6$

$4 \leq 3x < 16$

$\frac{4}{3} \leq x < \frac{16}{3}$

101. $|x = 10| \geq -3$ is true for all x, since $|x = 10| \geq 0$.

103. $3x^2 = 12x > 0$

$3x(x = 4) > 0$

Critical numbers: $0, -4$. Checking the three intervals, you obtain $x < -4$ and $x > 0$.

105. $\ln x - 5\ln(x = 3) = \ln x - \ln(x = 3)^5$

$$= \ln \frac{x}{(x = 3)^5}$$

107. $\frac{1}{4}\log_6 3 = \frac{1}{4}\log_6 x = \log_6 3^{1/4} = \log_6 x^{1/4}$

$$= \log_6(3x)^{1/4}$$

109. $-12x = 9y = 51$

$-x - 7y = -19$

From the second equation, $x = -7y = 19$.

Then $-12(-7y = 19) = 9y = 51$

$84y - 228 = 9y = 51$

$93y = 279$

$y = 3$ $\quad x = -7(3) = 19 = -2$ $\quad$ *Answer:* $(-2, 3)$

Section 7.3 Multivariable Linear Systems

- You should know the operations that lead to equivalent systems of linear equations:
  (a) Interchange any two equations.
  (b) Multiply all terms of an equation by a nonzero constant.
  (c) Replace an equation by the sum of itself and a constant multiple of any other equation in the system.
- You should be able to use the method of elimination.

Solutions to Odd-Numbered Exercises

1. (a) $3(2) - 5 + 0 \stackrel{?}{=} 1$ Yes

$2(2) - 3(0) \stackrel{?}{=} -14$ No

$5(5) + 2(0) \stackrel{?}{=} 8$ No

No, $(2, 5, 0)$ is not a solution.

(b) $3(-2) - 0 + 4 \stackrel{?}{=} 1$ No

$2(-2) - 3(4) \stackrel{?}{=} -14$ No

$5(0) + 2(4) \stackrel{?}{=} 8$ Yes

No, $(-2, 0, 4)$ is not a solution.

(c) $3(0) - (-1) + 3 \stackrel{?}{=} 1$ No

$2(0) - 3(3) \stackrel{?}{=} -14$ No

$5(-1) + 2(3) \stackrel{?}{=} 8$ No

No, $(0, -1, 3)$ is not a solution.

(d) $3(-1) - 0 + 4 \stackrel{?}{=} 1$ Yes

$2(-1) - 3(4) \stackrel{?}{=} -14$ Yes

$5(0) + 2(4) \stackrel{?}{=} 8$ Yes

Yes, $(-1, 0, 4)$ is a solution.

3. (a) $4(0) + 1 - 1 \stackrel{?}{=} 0$ Yes

$-8(0) - 6(1) + 1 \stackrel{?}{=} -\frac{7}{4}$ No

$3(0) - 1 \stackrel{?}{=} -\frac{9}{4}$ No

No, $(0, 1, 1)$ is not a solution.

(b) $4\left(-\frac{3}{2}\right) + \frac{5}{4} - \left(-\frac{5}{4}\right) \stackrel{?}{=} 0$ No

$-8\left(-\frac{3}{2}\right) - 6\left(\frac{5}{4}\right) + \left(-\frac{5}{4}\right) \stackrel{?}{=} -\frac{7}{4}$ No

$3\left(-\frac{3}{2}\right) - \left(\frac{5}{4}\right) \stackrel{?}{=} -\frac{9}{4}$ No

No, $\left(-\frac{3}{2}, \frac{5}{4}, -\frac{5}{4}\right)$ is not a solution.

(c) $4\left(-\frac{1}{2}\right) + \frac{3}{4} - \left(-\frac{5}{4}\right) \stackrel{?}{=} 0$ Yes

$-8\left(-\frac{1}{2}\right) - 6\left(\frac{3}{4}\right) - \frac{5}{4} \stackrel{?}{=} -\frac{7}{4}$ Yes

$3\left(-\frac{1}{2}\right) - \frac{3}{4} \stackrel{?}{=} -\frac{9}{4}$ Yes

Yes, $\left(-\frac{1}{2}, \frac{3}{4}, -\frac{5}{4}\right)$ is a solution.

(d) $4\left(-\frac{1}{2}\right) + 2 - 0 \stackrel{?}{=} 0$ Yes

$-8\left(-\frac{1}{2}\right) - 6(2) + 0 \stackrel{?}{=} -\frac{7}{2}$ No

$3\left(-\frac{1}{2}\right) - 2 \stackrel{?}{=} -\frac{9}{4}$ No

No, $\left(-\frac{1}{2}, 2, 0\right)$ is not a solution.

5. $2x - y + 5z = 24$ Equation 1

$y + 2z = 4$ Equation 2

$z = 6$ Equation 3

Back-substitute $z = 6$ into Equation 2

$y + 2(6) = 4$

$y = -8$

Back-substitute $y = -8$ and $z = 6$ into Equation 1

$2x - (-8) + 5(6) = 24$

$2x = -14$

$x = -7$

Answer: $(-7, -8, 6)$

7. $2x + y - 3z = 10$ Equation 1

$y = -2$ Equation 2

$y - z = 4$ Equation 3

Back-substitute $y = -2$ into Equation 3

$-2 - z = 4$

$z = -6$

Back-substitute $y = -2$ and $z = -6$ into Equation 1,

$2x - 2 - 3(-6) = 10$

$2x = -6$

$x = -3$

Answer: $(-3, -2, -6)$

9. $4x - 2y + z = 8$ Equation 1

$2z = 4$ Equation 2

$-y + z = 4$ Equation 3

From Equation 2 we have $z = 2$. Back-substitute $z = 2$ into Equation 3.

$$-y + 2 = 4$$
$$y = -2$$

Back-substitute $y = -2$ and $z = 2$ into Equation 1.

$$4x - 2(-2) + 2 = 8$$
$$4x + 6 = 8$$
$$x = \tfrac{1}{2}$$

Answer: $\left(\frac{1}{2}, -2, 2\right)$

11. $x - 2y + 3z = 5$ Equation 1

$-x + 3y - 5z = 4$ Equation 2

$2x \quad - 3z = 0$ Equation 3

Add Equation 1 to Equation 2.

$y - 2z = 9$ New Equation 2

This is the first step in putting the system in row-echelon form.

13. $x + y + z = 6$ Equation 1

$2x - y + z = -1$ Equation 2

$3x \quad - z = -7$ Equation 3

$x + y + z = 6$

$-3y - z = -13$ -2 Eq. 1 + Eq. 2

$-3y - 4z = -25$ -3 Eq. 1 + Eq. 3

$x + y + z = 6$

$-3y - z = -13$

$-3z = -12$ $-$Eq. 2 + Eq. 3

$-3z = -12 \Rightarrow z = 4$

$-3y - 4 = -13 \Rightarrow y = 3$

$x + 3 + 4 = 6 \Rightarrow x = -1$

Answer: $(-1, 3, 4)$

15. $2x \quad + 2z = 6$ Equation 1

$5x + 3y \quad = 11$ Equation 2

$3y - 4z = 1$ Equation 3

$x \quad + z = 3$ $\frac{1}{2}$ Eq. 1

$5x + 3y \quad = 11$

$3y - 4z = 1$

$x \quad + z = 3$

$3y - 5z = -4$ -5 Eq. 1 + Eq. 2

$3y - 4z = 1$

$x \quad + z = 3$

$3y - 5z = -4$

$z = 5$ $-$Eq. 2 + Eq. 3

$3y - 5(5) = -4 \Rightarrow y = 7$

$x + 5 = 3 \Rightarrow x = -2$

Answer: $(-2, 7, 5)$

17.

$$\begin{aligned} 6y + 4z &= -18 && \text{Equation 1} \\ 3x + 3y &= 9 && \text{Equation 2} \\ 2x - 3z &= 12 && \text{Equation 3} \end{aligned}$$

$$\begin{aligned} 3x + 3y &= 9 && \text{Interchange equations} \\ 6y + 4z &= -18 && \text{1 and 2} \\ 2x - 3z &= 12 \end{aligned}$$

$$\begin{aligned} x + y &= 3 && \tfrac{1}{3}\text{ (New Eq. 1)} \\ 6y + 4z &= -18 \\ 2x - 3z &= 12 \end{aligned}$$

$$\begin{aligned} x + y &= 3 \\ 6y + 4z &= -18 \\ -2y - 3z &= 6 \end{aligned}$$

$$\begin{aligned} x + y &= 3 \\ -2y - 3z &= 6 && \text{Interchange} \\ 6y + 4z &= -18 && \text{the equations} \end{aligned}$$

$$\begin{aligned} x + y &= 3 \\ -2y - 3z &= 6 \\ -5z &= 0 \end{aligned}$$

$$\begin{aligned} -5z = 0 &\implies z = 0 \\ -2y - 3(0) = 6 &\implies y = -3 \\ x + (-3) = 3 &\implies x = 6 \end{aligned}$$

Answer: $(6, -3, 0)$

19.

$$\begin{aligned} x + y - 2z &= 3 && \text{Interchange} \\ 3x - 2y + 4z &= 1 && \text{the equations.} \\ 2x - 3y + 6z &= 8 \end{aligned}$$

$$\begin{aligned} x + y - 2z &= 3 \\ -5y + 10z &= -8 && -3\text{Eq.1} + \text{Eq.2} \\ -5y + 10z &= 2 && -2\text{Eq.1} + \text{Eq.3} \end{aligned}$$

$$\begin{aligned} x + y - 2z &= 3 \\ -5y + 10z &= -8 \\ 0 &= 10 && \rightarrow\leftarrow -\text{Eq.2} + \text{Eq.3} \end{aligned}$$

No solution, inconsistent

21.

$$\begin{aligned} 3x + 3y + 5z &= 1 \\ 3x + 5y + 9z &= 0 \\ 5x + 9y + 17z &= 0 \end{aligned}$$

$$\begin{aligned} 6x + 6y + 10z &= 2 && 2\text{ Eq.1} \\ 3x + 5y + 9z &= 0 \\ 5x + 9y + 17z &= 0 \end{aligned}$$

$$\begin{aligned} x - 3y - 7z &= 2 && -\text{Eq.3} + \text{Eq.1} \\ 3x + 5y + 9z &= 0 \\ 5x + 9y + 17z &= 0 \end{aligned}$$

$$\begin{aligned} x - 3y - 7z &= 2 \\ 14y + 30z &= -6 && -3\text{Eq.1} + \text{Eq.2} \\ 24y + 52z &= -10 && -5\text{Eq.1} + \text{Eq.3} \end{aligned}$$

$$\begin{aligned} x - 3y - 7z &= 2 \\ 84y + 180z &= -36 && 6\text{Eq.2} \\ 84y + 182z &= -35 && 3.5\text{Eq.3} \end{aligned}$$

$$\begin{aligned} x - 3y - 7z &= 2 \\ 84y + 180z &= -36 \\ 2z &= 1 && -\text{Eq.2} + \text{Eq.3} \end{aligned}$$

$$\begin{aligned} 2z = 1 &\implies x = \tfrac{1}{2} \\ 84y + 180\left(\tfrac{1}{2}\right) = -36 &\implies y = -\tfrac{3}{2} \\ x - 3\left(-\tfrac{3}{2}\right) - 7\left(\tfrac{1}{2}\right) = 2 &\implies x = 1 \end{aligned}$$

Answer: $\left(1, -\frac{3}{2}, \frac{1}{2}\right)$

23. $x + 2y - 7z = -4$

$2x + y + z = 13$

$3x + 9y - 36z = -33$

$x + 2y - 7z = -4$

$-3y + 15z = 21$ $\quad -2\text{Eq.}1 + \text{Eq.}2$

$3y - 15z = -21$ $\quad -3\text{Eq.}1 + \text{Eq.}3$

$x + 2y - 1z = -4$

$-3y + 15z = 21$

$0 = 0$ $\quad \text{Eq.}2 + \text{Eq.}3$

$x + 2y - 7z = -4$

$y - 5z = -7$ $\quad \frac{1}{3}\text{Eq.}2$

$x + 3z = 10$ $\quad -2\text{Eq.}2 + \text{Eq.}1$

$y - 5z = -7$

Let $z = a$, then:

$y = 5a - 7$

$x = -3a + 10$

Answer: $(-3a + 10, 5a - 7, a)$

25. $3x - 3y + 6z = 6$

$x + 2y - z = 5$

$5x - 8y + 13z = 7$

$x - y + 2z = 2$ $\quad \frac{1}{3}\text{Eq.}1$

$3y - 3z = 3$ $\quad -\text{Eq.}1 + \text{Eq.}2$

$-3y + 3z = -3$ $\quad -5\text{Eq.}1 + \text{Eq.}3$

$x - y + 2z = 2$

$y - z = 1$ $\quad \frac{1}{3}\text{Eq.}2$

$0 = 0$ $\quad \text{Eq.}2 + \text{Eq.}3$

$x + z = 3$ $\quad \text{Eq.}2 + \text{Eq.}1$

$y - z = 1$

Let $z = a$, then:

$y = a + 1$

$x = -a + 3$

Answer: $(-a + 3, a + 1, a)$

27. $x - 2y + 3z = 4$ Equation 1

$3x - y + 2z = 0$ Equation 2

$x + 3y - 4z = -2$ Equation 3

$x - 2y + 3z = 4$

$5y - 7z = -12$ $\quad -3\text{ Eq. }1 + \text{Eq. }2$

$5y - 7z = -6$ $\quad -1\text{ Eq. }1 + \text{Eq. }3$

$x - 2y + 3z = 4$

$5y - 7z = -12$

$0 = 6$ $\quad -\text{Eq. }2 + \text{Eq. }3$

No solution. Inconsistent.

29. $x + 4z = 1$

$x + y + 10z = 10$

$2x - y + 2z = -5$

$x + 4z = 1$

$y + 6z = 9$ $\quad -\text{Eq.}1 + \text{Eq.}2$

$-y - 6z = -7$ $\quad -2\text{Eq.}1 + \text{Eq.}3$

$x + 4z = 1$

$y + 6z = 9$

$0 = 2$ $\quad \rightarrow \leftarrow \text{Eq.}2 + \text{Eq.}3$

No solution, inconsistent.

31. $2x + 3y = 0$

$4x + 3y - z = 0$

$8x + 3y + 3z = 0$

$2x + 3y = 0$

$-3y - z = 0$ $\quad -2\text{Eq.}1 + \text{Eq.}2$

$-9y + 3z = 0$ $\quad -4\text{Eq.}1 + \text{Eq.}3$

$2x + 3y = 0$

$-3y - z = 0$

$6z = 0$ $\quad -3\text{Eq.}2 + \text{Eq.}3$

$6z = 0 \Rightarrow z = 0$

$-3y - 0 = 0 \Rightarrow y = 0$

$2x + 3(0) = 0 \Rightarrow x = 0$

Answer: $(0, 0, 0)$

33. $x - 2y + 5z = 2$

$4x \quad - z = 0$

Let $z = a$, then $x = \frac{1}{4}a$.

$\frac{1}{4}a - 2y + 5a = 2$

$a - 8y + 20a = 8$

$-8y = -21a + 8$

$y = \frac{21}{8}a - 1$

Answer: $\left(\frac{1}{4}a, \frac{21}{8}a - 1, a\right)$

To avoid fractions, we could go back and let $z = 8a$, then $4x - 8a = 0 \Rightarrow x = 2a$.

$2a - 2y + 5(8a) = 2$

$-2y + 42a = 2$

$y = 21a - 1$

Answer: $(2a, 21a - 1, 8a)$

35. $2x - 3y + z = -2$

$-4x + 9y = 7$

$2x - 3y + z = -2$

$3y + 2z = 3$ 2Eq.1 + Eq.2

$2x + 3z = 1$ Eq.2 + Eq.1

$3y + 2z = 3$

Let $z = a$, then:

$y = -\frac{2}{3}a + 1$

$x = -\frac{3}{2}a + \frac{1}{2}$

Answer: $\left(-\frac{3}{2}a + \frac{1}{2}, -\frac{2}{3}a + 1, a\right)$

37. $23x + 4y - z = 0$ Interchange

$12x + 5y + z = 0$ the equations.

$x + 6y + 3z = 0$ 2Eq.2 − Eq.1

$-67y - 35z = 0$ −12Eq.1 + Eq.2

To avoid fractions, let $z = 67a$, then:

$-67y - 35(67a) = 0$

$y = -35a$

$x + 6(-35a) + 3(67a) = 0$

$x = 9a$

Answer: $(9a, -35a, 67a)$

39. $x + 3w = 4$

$2y - z - w = 0$

$3y - 2w = 1$

$2x - y + 4z = 5$

$x + 3w = 4$

$2y - z - w = 0$

$3y - 2w = 1$

$-y + 4z - 6w = -3$ −2Eq.1 + Eq.4

$x + 3w = 4$

$y - 4z + 6w = 3$ −Eq.4 and interchange the equations.

$2y - z - w = 0$

$3y - 2w = 1$

$x + 3w = 4$

$y - 4z + 6w = 3$

$7z - 13w = -6$ −Eq.2 + Eq.3

$12z - 20w = -8$ −3Eq.2 + Eq.4

$x + 3w = 4$

$y - 4z + 6w = 3$

$z + 3w = -2$ $-\frac{1}{2}$Eq.4 + Eq.3

$12z - 20w = -8$

$x + 3w = 4$

$y - 4z + 6w = 3$

$z - 3w = -2$

$16w = 16$ −12Eq.3 + Eq.4

$16w = 16 \Rightarrow w = 1$

$z - 3(1) = -2 \Rightarrow z = 1$

$y - 4(1) + 6(1) = 3 \Rightarrow y = 1$

$x + 3(1) = 4 \Rightarrow x = 1$

Answer: $(1, 1, 1, 1)$

41. There are an infinite number of linear systems that has $(4, -1, 2)$ as their solution.
One such system is as follows:

$$3(4) + (-1) - (2) = 9 \Longrightarrow 3x + y - z = 9$$
$$(4) + 2(-1) - (2) = 0 \Longrightarrow x + 2y - z = 0$$
$$-(4) + (-1) + 3(2) = 1 \Longrightarrow -x + y + 3z = 1$$

43. There are infinite numbers of linear systems that have $\left(3, -\frac{1}{2}, \frac{7}{4}\right)$ as their solution.
One such system is as follows:

$$1(3) + 2\left(-\tfrac{1}{2}\right) + 4\left(\tfrac{7}{4}\right) = 9 \Rightarrow x + 2y + 4z = 9$$
$$4\left(-\tfrac{1}{2}\right) + 8\left(\tfrac{7}{4}\right) = 12 \Rightarrow 4y + 8z = 12$$
$$4\left(\tfrac{7}{4}\right) = 7 \Rightarrow 4z = 7$$

45. Plane: $2x + 3y + 4z = 12$

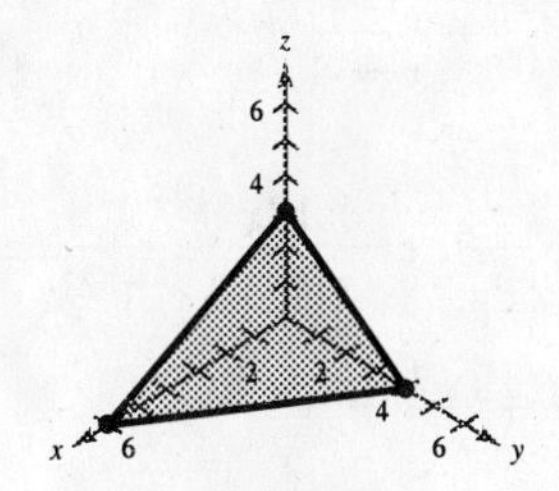

Four points are:
$(6, 0, 0),\ (0, 4, 0),\ (0, 0, 3),\ (4, 0, 1)$

47. Plane: $2x + y + z = 4$

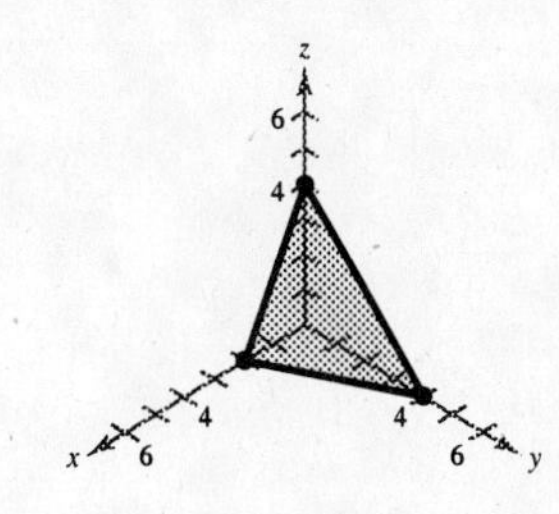

Four points are:
$(2, 0, 0),\ (0, 4, 0),\ (0, 0, 4),\ (0, 2, 2)$

49. $$\frac{7}{x^2 - 14x} = \frac{7}{x(x - 14)} = \frac{A}{x} + \frac{B}{x - 14}$$

51. $$\frac{12}{x^3 - 10x^2} = \frac{12}{x^2(x - 10)} = \frac{A}{x} + \frac{B}{x^2} + \frac{C}{x - 10}$$

53. $$\frac{4x^2 + 3}{(x - 5)^3} = \frac{A}{(x - 5)} + \frac{B}{(x - 5)^2} + \frac{C}{(x - 5)^3}$$

55. $$\frac{1}{x^2 - 1} = \frac{A}{x + 1} + \frac{B}{x - 1}$$
$$1 = A(x - 1) + B(x + 1)$$

Let $x = -1$: $1 = -2A \Rightarrow A = -\dfrac{1}{2}$

Let $x = 1$: $1 = 2B \Rightarrow B = \dfrac{1}{2}$

$$\frac{1}{x^2 - 1} = \frac{1/2}{x - 1} - \frac{1/2}{x + 1} = \frac{1}{2}\left[\frac{1}{x - 1} - \frac{1}{x + 1}\right]$$

57. $$\frac{1}{x^2 + x} = \frac{A}{x} + \frac{B}{x + 1}$$
$$1 = A(x + 1) + Bx$$

Let $x = 0$: $1 = A$

Let $x = -1$: $1 = -B \Rightarrow B = -1$

$$\frac{1}{x^2 + x} = \frac{1}{x} - \frac{1}{x + 1}$$

59. $$\frac{1}{2x^2 + x} = \frac{A}{2x + 1} + \frac{B}{x}$$
$$1 = Ax + B(2x + 1)$$

Let $x = -\dfrac{1}{2}$: $1 = -\dfrac{1}{2}A \Rightarrow A = -2$

Let $x = 0$: $1 = B$

$$\frac{1}{2x^2 + x} = \frac{1}{x} - \frac{2}{2x + 1}$$

61. $$\frac{3}{x^2 + x - 2} = \frac{A}{x - 1} + \frac{B}{x + 2}$$
$$3 = A(x + 2) + B(x - 1)$$

Let $x = 1$: $3 = 3A \Rightarrow A = 1$

Let $x = -2$: $3 = -3B \Rightarrow B = -1$

$$\frac{3}{x^2 + x - 2} = \frac{1}{x - 1} - \frac{1}{x + 2}$$

63. $\dfrac{x^2 + 12x + 12}{x^3 - 4x} = \dfrac{A}{x} + \dfrac{B}{x + 2} + \dfrac{C}{x - 2}$

$x^2 + 12x + 12 = A(x + 2)(x - 2) + Bx(x - 2) + Cx(x + 2)$

Let $x = 0$: $12 = -4A \implies A = -3$

Let $x = -2$: $-8 = 8B \implies B = -1$

Let $x = 2$: $40 = 8C \implies C = 5$

$$\frac{x^2 + 12x + 12}{x^3 - 4x} = -\frac{3}{x} - \frac{1}{x + 2} + \frac{5}{x - 2}$$

65. $\dfrac{4x^2 + 2x - 1}{x^2(x + 1)} = \dfrac{A}{x} + \dfrac{B}{x^2} + \dfrac{C}{x + 1}$

$4x^2 + 2x - 1 = Ax(x + 1) + B(x + 1) + Cx^2$

Let $x = 0$: $-1 = B$

Let $x = -1$: $1 = C$

Let $x = 1$: $5 = 2A + 2B + C$

$5 = 2A - 2 + 1$

$6 = 2A$

$3 = A$

$$\frac{4x^2 + 2x - 1}{x^2(x + 1)} = \frac{3}{x} - \frac{1}{x^2} + \frac{1}{x + 1}$$

67. $\dfrac{3x}{(x - 3)^2} = \dfrac{A}{x - 3} + \dfrac{B}{(x - 3)^2}$

$3x = A(x - 3) + B$

Let $x = 3$: $9 = B$

Let $x = 0$: $0 = -3A + B$

$0 = -3A + 9$

$3 = A$

$$\frac{3x}{(x - 3)^2} = \frac{3}{x - 3} + \frac{9}{(x - 3)^2}$$

69. $\dfrac{2x^3 - x^2 + x + 5}{x^2 + 3x + 2} = 2x - 7 + \dfrac{18x + 19}{(x + 1)(x + 2)}$

$\dfrac{18x + 19}{(x + 1)(x + 2)} = \dfrac{A}{x + 1} + \dfrac{B}{x + 2}$

$18x + 19 = A(x + 2) + B(x + 1)$

Let $x = -2$: $-17 = -B \implies B = 17$

Let $x = -1$: $1 = A$

$$\frac{2x^3 - x^2 + x + 5}{x^2 + 3x + 2} = 2x - 7 + \frac{1}{x + 1} + \frac{17}{x + 2}$$

71. $\dfrac{x^4}{(x - 1)^3} = \dfrac{x^4}{x^3 - 3x^2 + 3x - 1} = x + 3 + \dfrac{6x^2 - 8x + 3}{(x - 1)^3}$

$\dfrac{6x^2 - 8x + 3}{(x - 1)^3} = \dfrac{A}{x - 1} + \dfrac{B}{(x - 1)^2} + \dfrac{C}{(x - 1)^3}$

$6x^2 - 8x + 3 = A(x - 1)^2 + B(x - 1) + C$

Let $x = 1$: $1 = C$

$6x^2 - 8x + 3 = Ax^2 - 2Ax + A + Bx - B + 1$

$6x^2 - 8x + 3 = Ax^2 + (-2A + B)x + (A - B + 1)$

Equating coefficients of like powers:

$6 = A$, $-8 = -2A + B$ and $3 = A - B + 1$

$-8 = -12 + B \qquad 3 = 6 - B + 1$

$4 = B \qquad 4 = B$

$$\frac{x^4}{(x - 1)^3} = x + 3 + \frac{6}{x - 1} + \frac{4}{(x - 1)^2} + \frac{1}{(x - 1)^3}$$

73. $\dfrac{5 - x}{2x^2 + x - 1} = \dfrac{A}{2x - 1} + \dfrac{B}{x + 1}$

$-x + 5 = A(x + 1) + B(2x - 1)$

Let $x = \dfrac{1}{2}$: $\dfrac{9}{2} = \dfrac{3}{2}A \implies A = 3$

Let $x = -1$: $6 = -3B \implies B = -2$

$$\frac{5 - x}{2x^2 + x - 1} = \frac{3}{2x - 1} - \frac{2}{x + 1}$$

75. $\dfrac{x-1}{x^3+x^2} = \dfrac{A}{x} + \dfrac{B}{x^2} + \dfrac{C}{x+1}$

$x - 1 = Ax(x+1) + B(x+1) + Cx^2$

Let $x = -1$: $-2 = C$

Let $x = 0$: $-1 = B$

Let $x = 1$: $0 = 2A + 2B + C$

$0 = 2A - 2 - 2$

$2 = A$

$\dfrac{x-1}{x^3+x^2} = \dfrac{2}{x} - \dfrac{1}{x^2} - \dfrac{2}{x+1}$

77. $\dfrac{2x^3 - 4x^2 - 15x + 5}{x^2 - 2x - 8} = 2x + \dfrac{x+5}{(x+2)(x-4)}$

$\dfrac{x+5}{(x+2)(x-4)} = \dfrac{A}{x+2} + \dfrac{B}{x-4}$

$x + 5 = A(x-4) + B(x+2)$

Let $x = -2$: $3 = -6A \implies A = -\dfrac{1}{2}$

Let $x = 4$: $9 = 6B \implies B = \dfrac{3}{2}$

$\dfrac{2x^3 - 4x^2 - 15x + 5}{x^2 - 2x - 8} = 2x + \dfrac{1}{2}\left[\dfrac{3}{x-4} - \dfrac{1}{x+2}\right]$

79. $\dfrac{x-12}{x(x-4)} = \dfrac{A}{x} + \dfrac{B}{x-4}$

$x - 12 = A(x-4) + Bx$

Let $x = 0$: $-12 = -4A \implies A = 3$

Let $x = 4$: $-8 = 4B \implies B = -2$

$\dfrac{x-12}{x(x-4)} = \dfrac{3}{x} - \dfrac{2}{x-4}$

$y = \dfrac{x-12}{x(x-4)}$

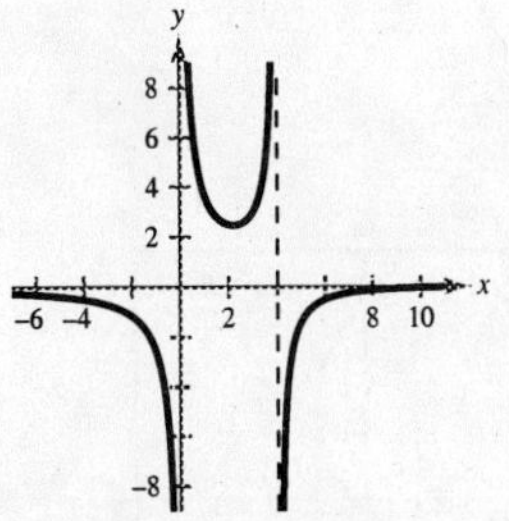

Vertical asymptotes: $x = 0$ and $x = 4$

$y = \dfrac{3}{x}, y = -\dfrac{2}{x-4}$

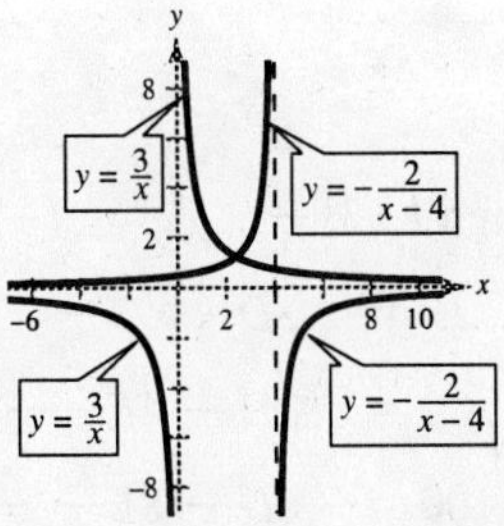

Vertical asymptotes: $x = 0$ and $x = 4$

The combination of the vertical asymptotes of the terms of the decompositions are the same as the vertical asymptotes of the rational function.

81. $s = \frac{1}{2}at^2 + v_0t + s_0$

$(1, 128), (2, 80), (3, 0)$

$128 = \frac{1}{2}a + v_0 + s_0 \implies a + 2v_0 + 2s_0 = 256$

$80 = 2a + 2v_0 + s_0 \implies 2a + 2v_0 + s_0 = 80$

$0 = \frac{9}{2}a + 3v_0 + s_0 \implies 9a + 6v_0 + 2s_0 = 0$

Solving this system yields $a = -32, v_0 = 0, s_0 = 144$.

Thus, $s = \frac{1}{2}(-32)t^2 + (0)t + 144$

$= -16t^2 + 144.$

83. $s = \frac{1}{2}at^2 + v_0t + s_0$

$(1, 452), (2, 260), (3, 116)$

$452 = \frac{1}{2}a + v_0 + s_0 \implies a + 2v_0 + 2s_0 = 904$

$260 = 2a + 2v_0 + s_0 \implies 2a + 2v_0 + s_0 = 260$

$116 = \frac{9}{2}a + 3v_0 + s_0 \implies 9a + 6v_0 + 2s_0 = 232$

Solving this system yields
$a = 48, v_0 = -264, s_0 = 692.$

Thus, $s = \frac{1}{2}(48)t^2 + (-264)t + 692$

$= 24t^2 - 264t + 692.$

85. $y = ax^2 + bx + c$ passing through $(0, 0)$, $(2, -2)$, $(4, 0)$

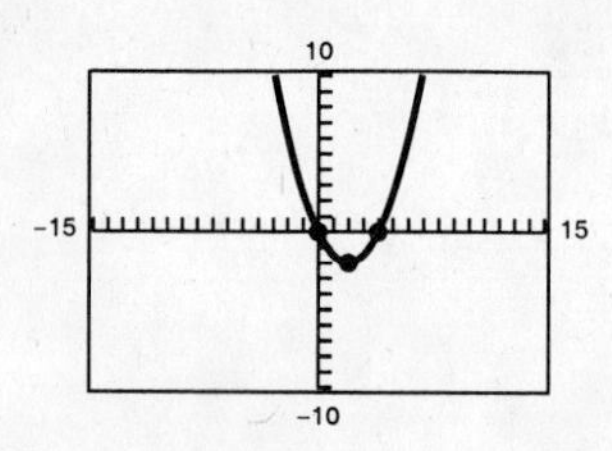

$(0, 0)$: $0 = c$

$(2, -2)$: $-2 = 4a + 2b + c \Rightarrow -1 = 2a + b$

$(4, 0)$: $0 = 16a + 4b + c \Rightarrow 0 = 4a + b$

Answer: $a = \frac{1}{2}, b = -2, c = 0$

The equation of the parabola is $y = \frac{1}{2}x^2 - 2x$.

87. $y = ax^2 + bx + c$ passing through $(2, 0)$, $(3, -1)$, $(4, 0)$

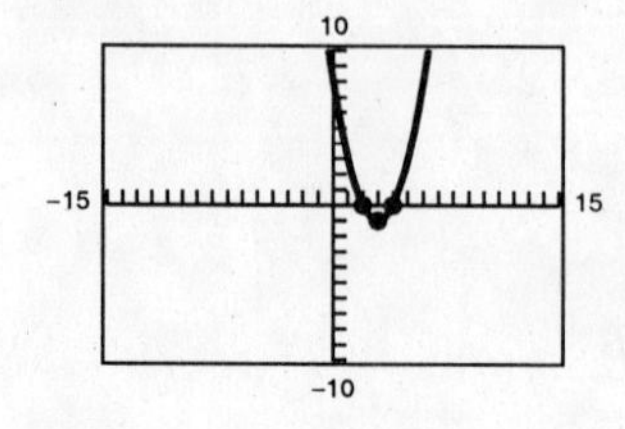

$(2, 0)$: $0 = 4a + 2b + c$

$(3, -1)$: $-1 = 9a + 3b + c \Rightarrow -1 = 5a + b$

$(4, 0)$: $0 = 16a + 4b + c \Rightarrow 0 = 12a + 2b$

Answer: $a = 1, b = -6, c = 8$

The equation of the parabola is $y = x^2 - 6x + 8$.

89. $x^2 + y^2 + Dx + Ey + F = 0$ passing through $(0, 0)$, $(2, 2)$, $(4, 0)$

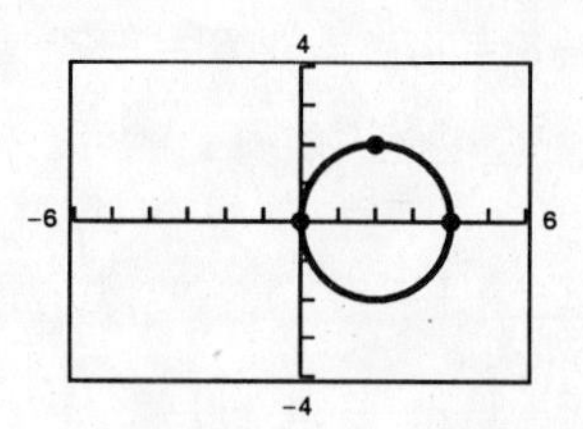

$(0, 0)$: $F = 0$

$(2, 2)$: $8 + 2D + 2E + F = 0 \Rightarrow D + E = -4$

$(4, 0)$: $16 + 4D + F = 0 \Rightarrow D = -4$ and $E = 0$

The equation of the circle is $x^2 + y^2 - 4x = 0$.

To graph, let $y_1 = \sqrt{4x - x^2}$ and $y_2 = -\sqrt{4x - x^2}$.

91. $x^2 + y^2 + Dx + Ey + F = 0$ passing through $(-3, -1)$, $(2, 4)$, $(-6, 8)$

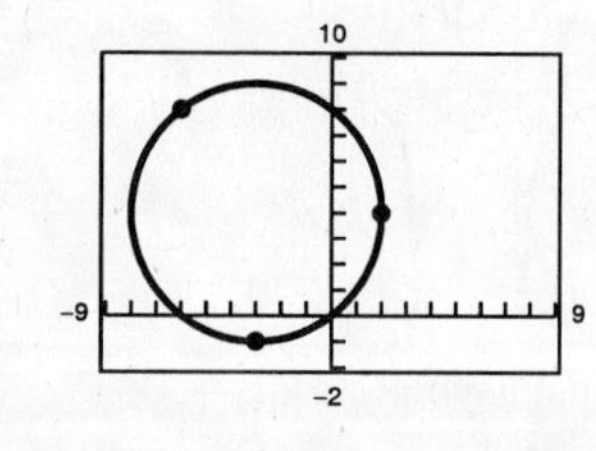

$(-3, -1)$: $10 - 3D - E + F = 0 \Rightarrow 10 = 3D + E - F$

$(2, 4)$: $20 + 2D + 4E + F = 0 \Rightarrow 20 = -2D - 4E - F$

$(-6, 8)$: $100 - 6D + 8E + F = 0 \Rightarrow 100 = 6D - 8E - F$

Answer: $D = 6, E = -8, F = 0$

The equation of the circle is $x^2 + y^2 + 6x - 8y = 0$. To graph, complete the squares first, then solve for y.

$$(x^2 + 6x + 9) + (y^2 - 8y + 16) = 0 + 9 + 16$$

$$(x + 3)^2 + (y - 4)^2 = 25$$

$$(y - 4)^2 = 25 - (x + 3)^2$$

$$y - 4 = \pm\sqrt{25 - (x + 3)^2}$$

$$y = 4 \pm \sqrt{25 - (x + 3)^2}$$

Let $y_1 = 4 + \sqrt{25 - (x + 3)^2}$ and $y_2 = 4 - \sqrt{25 - (x + 3)^2}$.

93. Let x = amount at 5%.
Let y = amount at 6%.
Let z = amount at 7%.

$$\begin{aligned} x + y + z &= 16{,}000 \\ 0.05x + 0.06y + 0.07z &= 990 \\ x + 3000 &= z \\ y + 2000 &= z \end{aligned}$$

$$\begin{aligned} (z - 3000) + (z - 2000) + z &= 16{,}000 \\ 3z &= 21{,}000 \\ z &= 7000 \end{aligned}$$

$x = 4000$, $y = 5000$

Check: $0.05(4000) + 0.06(5000) + 0.07(7000) = 990$

Answer: $x = \$4000$ at 5%,
$y = \$5000$ at 6%,
$z = \$7000$ at 7%

95. Let x = amount at 8%.
Let y = amount at 9%.
Let z = amount at 10%.

$$\begin{aligned} x + y + z &= 775{,}000 \\ 0.08x + 0.09y + 0.10z &= 67{,}000 \\ x &= 4z \end{aligned}$$

$$\begin{aligned} y + 5z &= 775{,}000 \\ 0.09y + 0.42z &= 67{,}000 \\ z &\approx 91{,}666.67 \end{aligned}$$

$y = 775{,}000 - 5z = 316{,}666.67$
$x = 4z = 366{,}666.67$

Answer: $x = \$366{,}666.67$ at 8%
$y = \$316{,}666.67$ at 9%
$z = \$91{,}666.67$ at 10%

97. Let C = amount in certificates of deposit.
Let M = amount in municipal bonds.
Let B = amount in blue-chip stocks.
Let G = amount in growth or speculative stocks.

$$\begin{aligned} C + M + B + G &= 500{,}000 \\ 0.10C + 0.08M + 0.12B + 0.13G &= 0.10(500{,}000) \\ B + G &= \tfrac{1}{4}(500{,}000) \end{aligned}$$

This system has infinitely many solutions.

Let $G = s$, then $B = 125{,}000 - s$
$M = 125{,}000 + \frac{1}{2}s$
$C = 250{,}000 - \frac{1}{2}s$.

Answer:
$\left(250{,}000 - \frac{1}{2}s,\ 125{,}000 + \frac{1}{2}s,\ 125{,}000 - s,\ s\right)$,
where $0 \le s \le 125{,}000$.

One possible solution is to let $s = 50{,}000$.
Certificates of deposit: \$225,000
Municipal bonds: \$150,000
Blue-chip stocks: \$75,000
Growth or speculative stocks: \$50,000

99. Let x = gallons of spray X.
Let y = gallons of spray Y.
Let z = gallons of spray Z.

Chemical A: $\frac{1}{5}x + \frac{1}{2}z = 12$
Chemical B: $\frac{2}{5}x + \frac{1}{2}z = 16$ $\Rightarrow x = 20, z = 16$
Chemical C: $\frac{2}{5}x + y = 26$ $\Rightarrow y = 18$

Answer: 20 liters of spray X
18 liters of spray Y
16 liters of spray Z

101.

	Product	
Truck	A	B
Large	6	3
Medium	4	4
Small	0	3

Possible solutions:
(1) 4 medium trucks
(2) 2 large trucks, 1 medium truck, 2 small trucks
(3) 3 large trucks, 1 medium truck, 1 small truck
(4) 3 large trucks, 3 small trucks

103.

$$\begin{aligned} t_1 - 2t_2 \quad &= 0 \\ t_1 \quad - 2a &= 128 \Rightarrow \quad 2t_2 - 2a = 128 \\ t_2 + a &= 32 \Rightarrow -2t_2 - 2a = -64 \end{aligned}$$

$$-4a = 64$$

$$a = -16$$

$$t_2 = 48$$

$$t_1 = 96$$

Answer: $t_1 = 96$ lb, $t_2 = 48$ lb, $a = -16$ ft/sec^2

105. Least squares regression parabola through $(-4, 5), (-2, 6), (2, 6), (4, 2)$

$$\begin{aligned} 4c \quad + 40a &= 19 \\ 40b \quad &= -12 \\ 40c \quad + 544a &= 160 \end{aligned}$$

Solving this system yields

$a = -\frac{5}{24}$, $b = -\frac{3}{10}$, and $c = \frac{41}{6}$.

Thus, $y = -\frac{5}{24}x^2 - \frac{3}{10}x + \frac{41}{6}$.

107. Least squares regression parabola through $(0, 0), (2, 2), (3, 6), (4, 12)$

$$\begin{aligned} 4c + 9b + 29a &= 20 \\ 9c + 29b + 99a &= 70 \\ 29c + 99b + 353a &= 254 \end{aligned}$$

Solving this system yields

$a = 1, b = -1$, and $c = 0$. Thus, $y = x^2 - x$.

109. (a) Least squares regression parabola through $(20, 25), (30, 55), (40, 105), (50, 188), (60, 300)$

Using a graphing utility,

$y = 0.14x^2 - 4.43x + 58.40$.

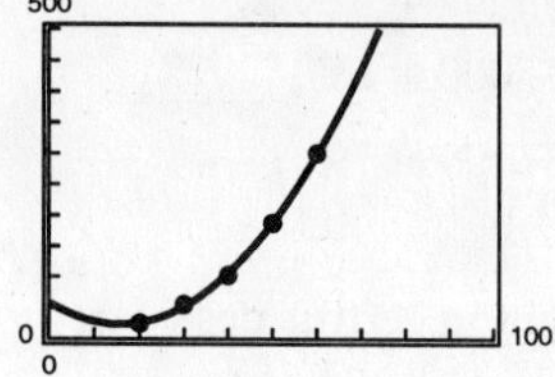

(b) When $x = 70$, $y \approx 434.3$ feet.

111. (a) $\dfrac{2000(4 - 3x)}{(11 - 7x)(7 - 4x)} = \dfrac{A}{11 - 7x} + \dfrac{B}{7 - 4x}, \quad 0 \le x \le 1$

$$2000(4 - 3x) = A(7 - 4x) + B(11 - 7x)$$

Let $x = \dfrac{11}{7}$: $-\dfrac{10{,}000}{7} = \dfrac{5}{7}A \Rightarrow A = -2000$

Let $x = \dfrac{7}{4}$: $-2500 = -\dfrac{5}{4}B \Rightarrow B = 2000$

$$\frac{2000(4 - 3x)}{(11 - 7x)(7 - 4x)} = \frac{-2000}{11 - 7x} + \frac{2000}{7 - 4x} = \frac{2000}{7 - 4x} - \frac{2000}{11 - 7x}$$

(b) $y_1 = \dfrac{2000}{7 - 4x}$

$y_2 = \dfrac{2000}{11 - 7x}$

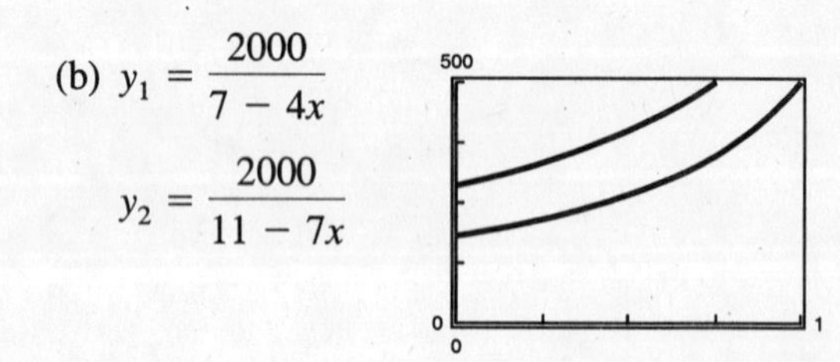

113. False. The coefficient of y in the second equation is not 1.

115. False. The correct form is

$$\frac{A}{x + 10} + \frac{B}{x - 10} + \frac{C}{(x - 10)^2}.$$

117. $\dfrac{1}{a^2 - x^2} = \dfrac{A}{a+x} + \dfrac{B}{a-x}$, a is a constant

$$1 = A(a - x) + B(a + x)$$

Let $x = -a$: $1 = 2aA \implies A = \dfrac{1}{2a}$

Let $x = a$: $1 = 2aB \implies B = \dfrac{1}{2a}$

$$\frac{1}{a^2 - x^2} = \frac{1}{2a}\left[\frac{1}{a+x} + \frac{1}{a-x}\right]$$

119. $\dfrac{1}{y(a-y)} = \dfrac{A}{y} + \dfrac{B}{a-y}$

$$1 = A(a - y) + By$$

Let $y = 0$: $1 = aA \implies A = \dfrac{1}{a}$

Let $y = a$: $1 = aB \implies B = \dfrac{1}{a}$

$$\frac{1}{y(a-y)} = \frac{1}{a}\left(\frac{1}{y} + \frac{1}{a-y}\right)$$

121. No, they are not equivalent. The constant in the second equation should be -11 and the coefficient of z in the third equation should be 2.

123. $\left.\begin{aligned} y + \lambda &= 0 \\ x + \lambda &= 0 \end{aligned}\right\} \implies x = y = -\lambda$

$x + y - 10 = 0 \implies 2x - 10 = 0$

$x = 5$

$y = 5$

$\lambda = -5$

125. $2x - 2x\lambda = 0 \implies x = x\lambda$

$-2y + \lambda = 0 \implies 2y = \lambda$

$y - x^2 = 0 \implies y = x^2$

From the first equation, $x = 0$ or $\lambda = 1$.

If $x = 0$, then $y = 0^2 = 0$ and $\lambda = 0$.

If $x \neq 0$, then $\lambda = 1 \implies y = \frac{1}{2}$ and $x = \pm\sqrt{\frac{1}{2}}$.

Thus, the solutions are:

(1) $x = y = \lambda = 0$

(2) $x = \dfrac{\sqrt{2}}{2}$, $y = \dfrac{1}{2}$, $\lambda = 1$

(3) $x = -\dfrac{\sqrt{2}}{2}$, $y = \dfrac{1}{2}$, $\lambda = 1$

127.

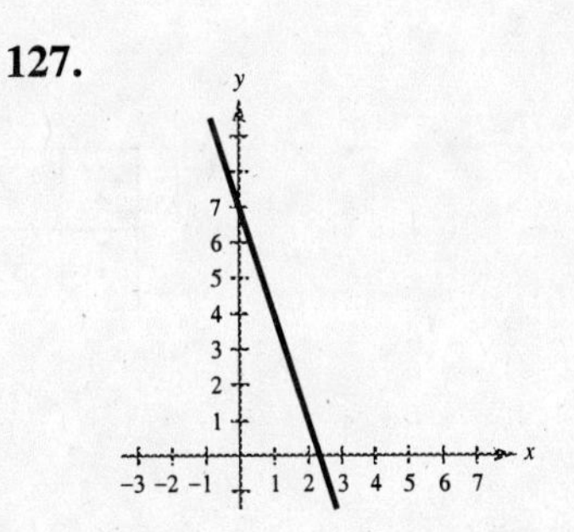

129.

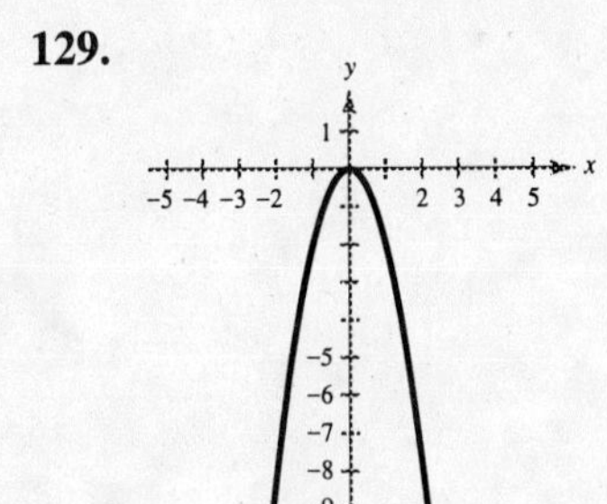

131.

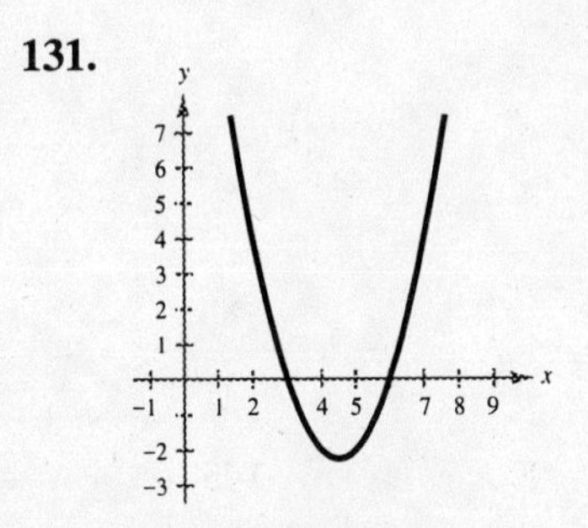

133.

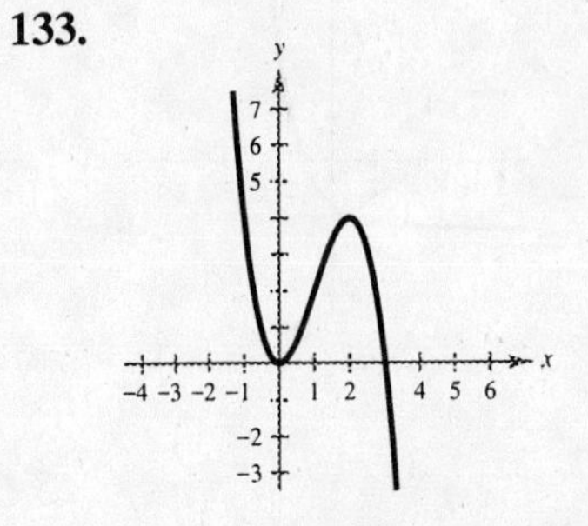

135. $f(x) = x^3 + x^2 - 12x = x(x^2 + x - 12) = x(x + 4)(x - 3) \implies x = 0, -4, 3$

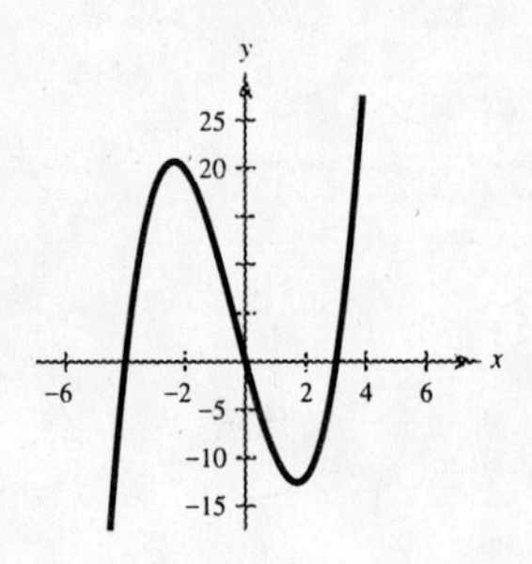

137. $f(x) = 2x^3 + 5x^2 - 21x - 36 = (2x + 3)(x + 4)(x - 3) \implies x = -\frac{3}{2}, -4, 3$

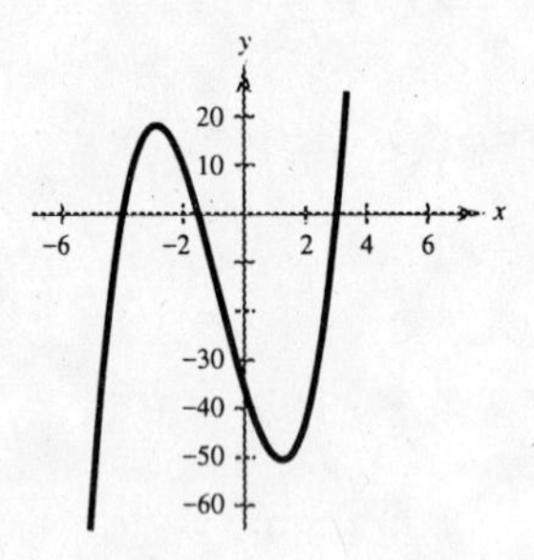

139. $y = 4^{-x-4} - 5$

x	-7	-6	-5	-4	-3	-1	0
y	59	11	-1	-4	-4.75	-4.98	-4.996

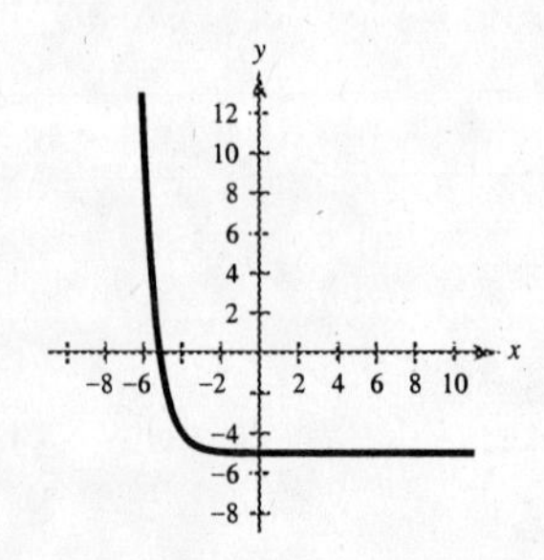

141. $y = 2.9^{0.8x} - 3$

x	-3	-2	-1	0	1	2	3
y	-2.9	-2.8	-2.6	-2	$-.66$	2.5	9.9

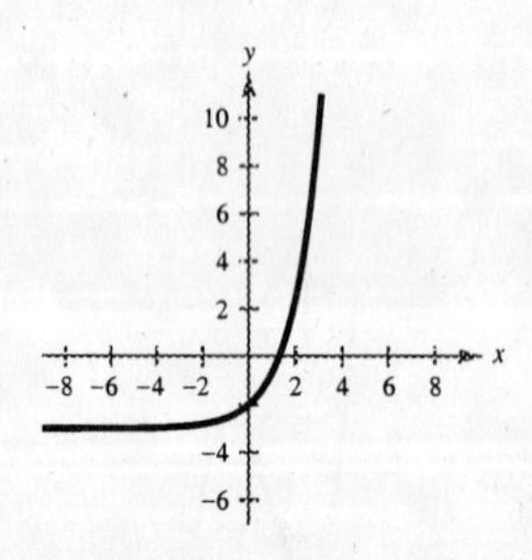

143. $3x + 3y = 7$

$3x + 5y = 3$

Multiplying the first equation by (-1) and adding gives $2y = -4 \implies y = -2$ and $3x + 3(-2) = 7 \implies x = \frac{13}{3}$.

Answer: $\left(\frac{13}{3}, -2\right)$

145. $2x + y = 120$

$x + 2y = 120$

Multiplying the first equation by (-2) and adding yields $-3x = -120 \implies x = 40$. Then $2(40) + y = 120 \implies y = 40$.

Answer: $(40, 40)$

Section 7.4 Systems of Inequalities

- You should be able to sketch the graph of an inequality in two variables:
  (a) Replace the inequality with an equal sign and graph the equation. Use a dashed line for < or >, a solid line for ≤ or ≥.
  (b) Test a point in each region formed by the graph. If the point satisfies the inequality, shade the whole region.

Solutions to Odd-Numbered Exercises

1. $x < 2$

Vertical boundary

Matches graph (g).

3. $2x + 3y \geq 6$

$y \geq -\frac{2}{3}x + 2$

Line with negative slope

Matches (a).

5. $x^2 + y^2 < 9$

Circular boundary

Matches (e).

7. $xy > 1$ or $y > \dfrac{1}{x}$

Matches (f).

9. $x \geq 2$

Using a solid line, graph the vertical line $x = 2$ and shade to the right of this line.

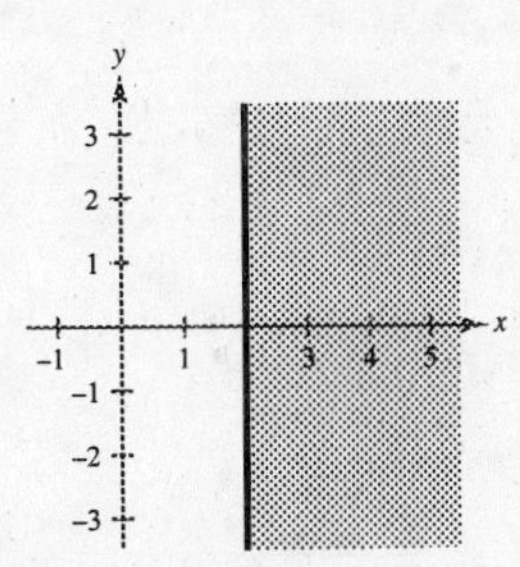

11. $y \geq -1$

Using a solid line, graph the horizontal line $y = -1$ and shade above this line.

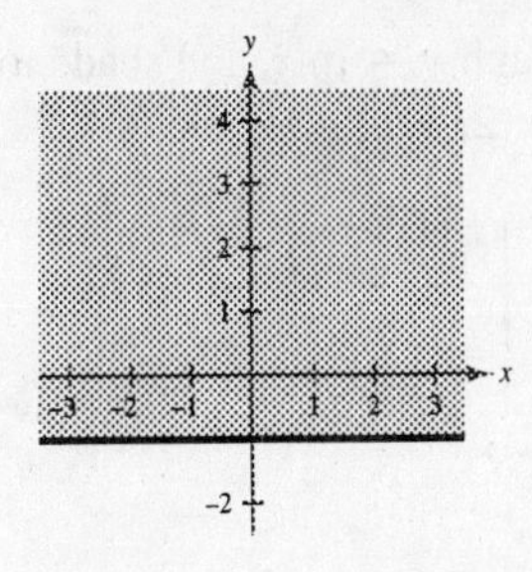

13. $y < 2 - x$

Using a dashed line, graph $y = 2 - x$, and then shade below the line. (Use (0, 0) as a test point.)

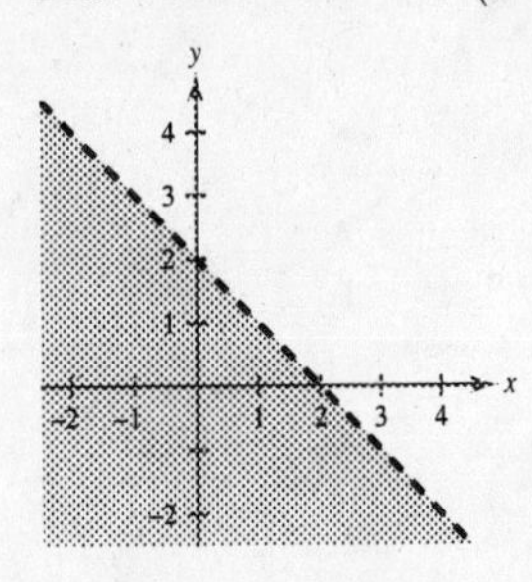

15. $2y - x \geq 4$

Using a solid line, graph $2y - x = 4$, and then shade above the line. (Use $(0, 0)$ as a test point.)

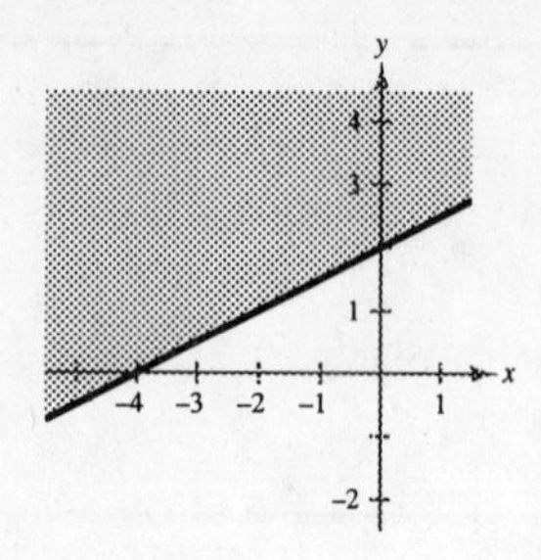

17. $y^2 - x < 0$

$y^2 < x$

Using a dashed line, graph the parabola $y^2 = x$, and then shade inside. (Use $(1, 0)$ as a test point.)

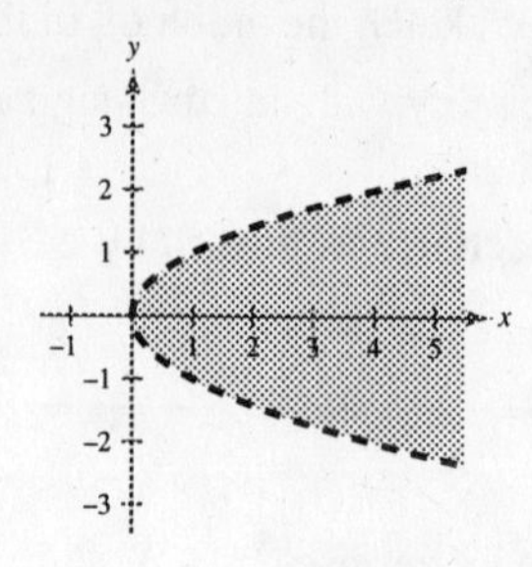

19. $(x + 1)^2 + y^2 < 9$

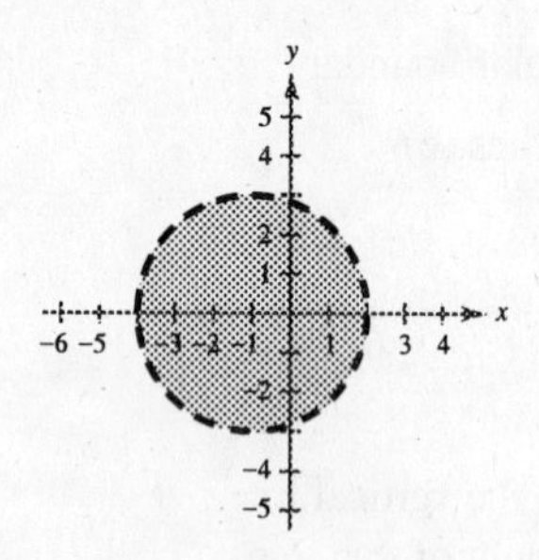

21. $y \geq \frac{2}{3}x - 1$

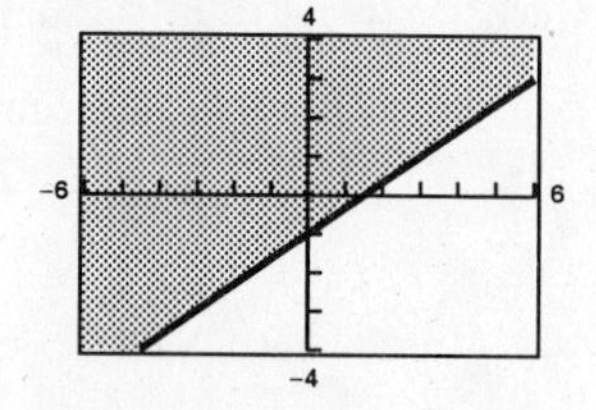

23. $y < -3.8x + 1.1$

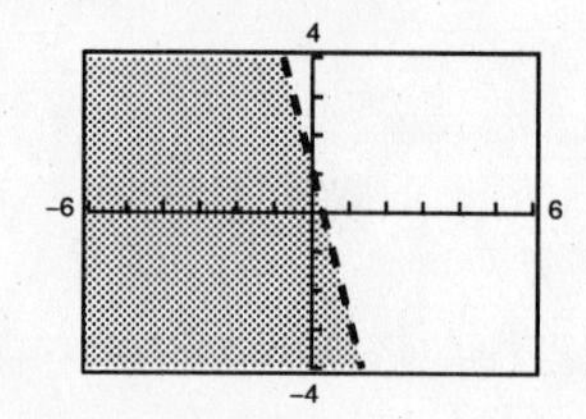

25. $x^2 + 5y - 10 \leq 0$

$y \leq 2 - \frac{x^2}{5}$

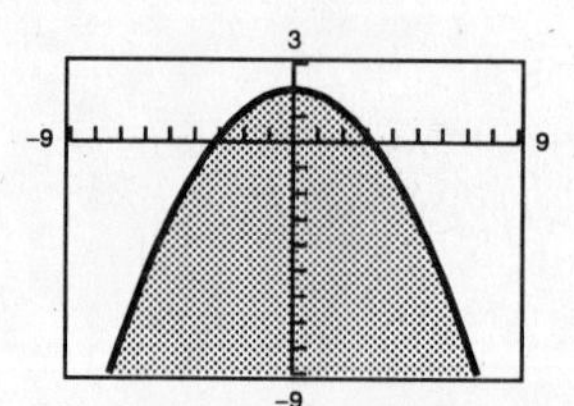

27. $y \leq \frac{1}{1 + x^2}$

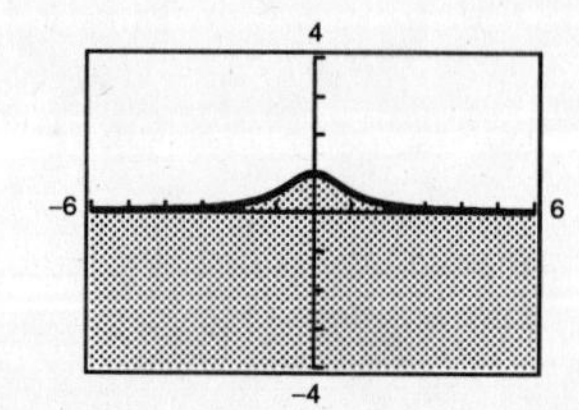

29. $y < \ln x$

Using a dashed line, graph $y = \ln x$, and shade to the right of the curve. (Use $(2, 0)$ as a test point.)

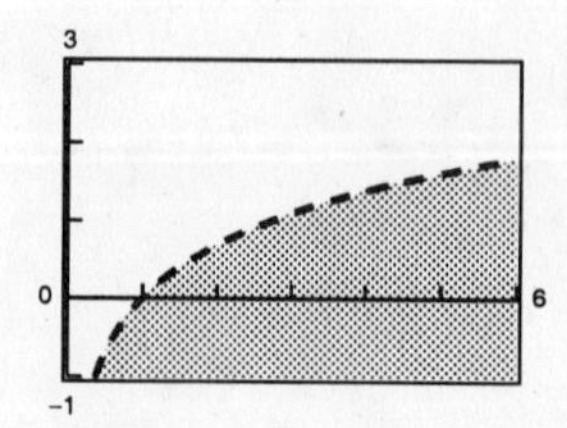

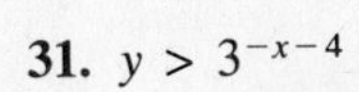

31. $y > 3^{-x-4}$

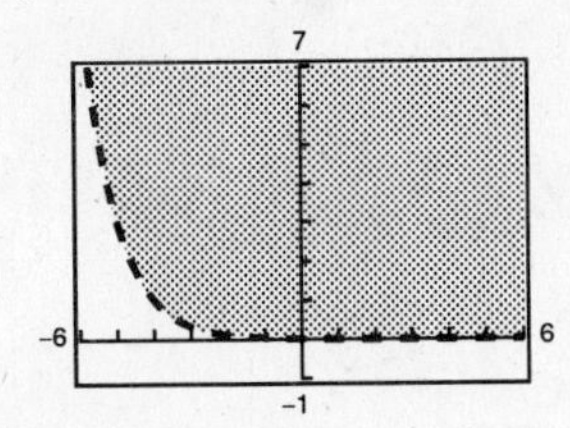

33. The line through (0, 2) and (3, 0) is $y = -\frac{2}{3}x + 2$. For the shaded region above the line, we have:

$$y \geq -\frac{2}{3}x + 2$$

$$3y \geq -2x + 6$$

$$2x + 3y \geq 6$$

$$\frac{x}{3} + \frac{y}{2} \geq 1$$

35. The circle shown is $x^2 + y^2 = 9$. For the shaded region inside the circle, we have $x^2 + y^2 \leq 9$.

37. (a) (0, 2) is a solution: $-2(0) + 5(2) \geq 3$

$2 < 4$

$-4(0) + 2(2) < 7$

(b) $(-6, 4)$ is not a solution: $4 \not< 4$

(c) $(-8, -2)$ is not a solution:
$-4(-8) + 2(-2) \not< 7$

(d) $(-3, 2)$ is not a solution: $-4(-3) + 2(2) \not< 7$

39.

$$x + y \leq 1$$
$$-x + y \leq 1$$
$$y \geq 0$$

First, find the points of intersection of each pair of equations.

Vertex A	*Vertex B*	*Vertex C*
$x + y = 1$	$x + y = 1$	$-x + y = 1$
$-x + y = 1$	$y = 0$	$y = 0$
(0, 1)	(1, 0)	$(-1, 0)$

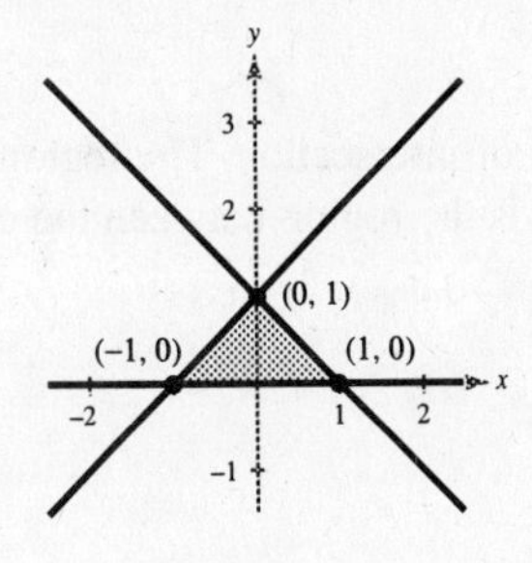

41.

$$-3x + 2y < 6$$
$$x - 4y > -2$$
$$2x + y < 3$$

First, find the points of intersection of each pair of equations.

Vertex A	*Vertex B*	*Vertex C*
$-3x + 2y = 6$	$-3x + 2y = 6$	$x - 4y = -2$
$x - 4y = -2$	$2x + y = 3$	$2x + y = 3$
$(-2, 0)$	(0, 3)	$\left(\frac{10}{9}, \frac{7}{9}\right)$

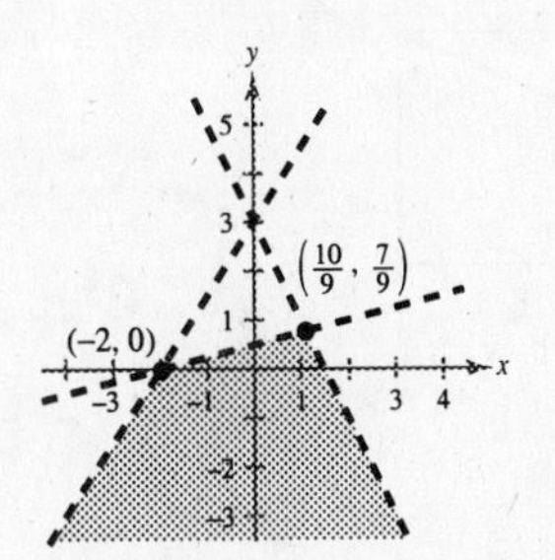

43.

$$2x + y > 2$$
$$6x + 3y < 2$$

The lines are parallel. There are no points of intersection. There is no region in common to both inequalities.

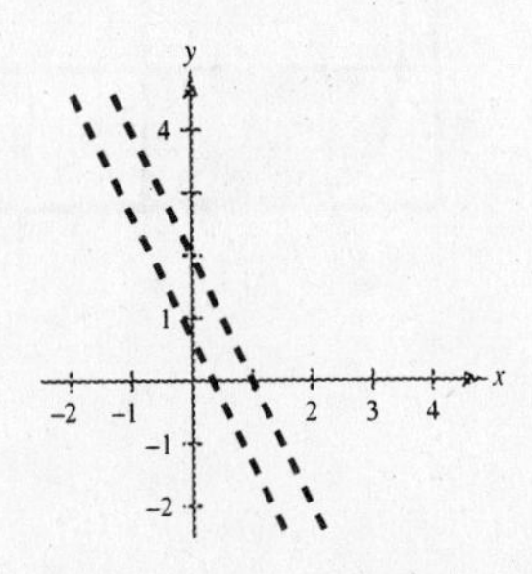

45. $-x^2 + y \ge 5 \implies y \ge x^2 + 5$

$\frac{1}{4}x + y < 3 \implies y < 3 - \frac{1}{4}x$

The curves given by $y = x^2 + 5$ and $y = 3 - \frac{1}{4}x$ do not intersect.

$$x^2 + 5 = 3 - \frac{1}{4}x$$

$$x^2 + \frac{1}{4}x + 2 = 0 \qquad \text{no real solutions}$$

There is no region common to both inequalities.

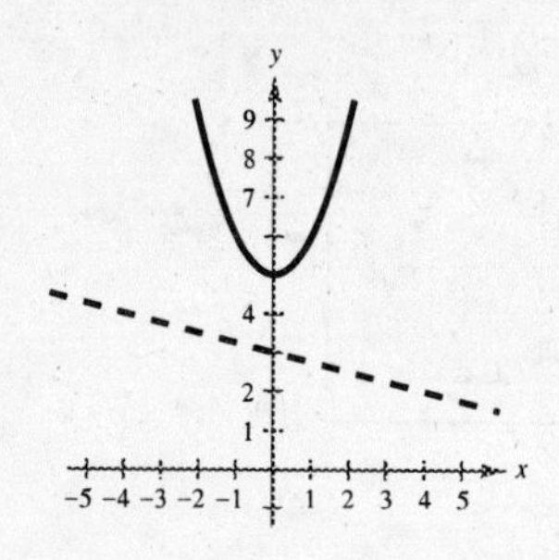

47. $x > y^2$

$x < y + 2$

Points of intersection:

$$y^2 = y + 2$$

$$y^2 - y - 2 = 0$$

$$(y + 1)(y - 2) = 0$$

$$y = -1, 2$$

$(1, -1), (4, 2)$

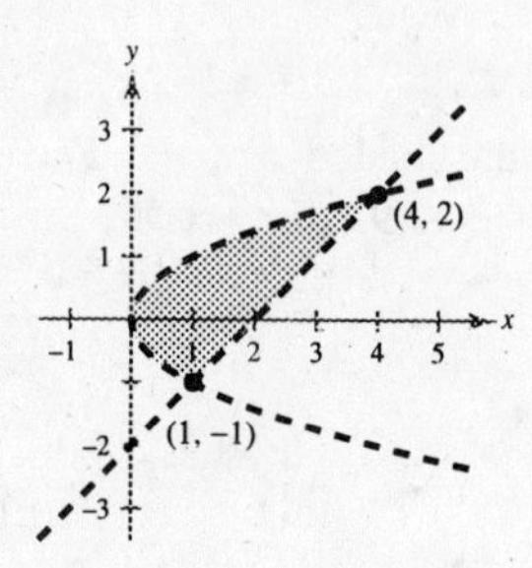

49. $x^2 + y^2 \le 9$

$x^2 + y^2 \ge 1$

There are no points of intersection. The region in common to both inequalities is the region between the circles.

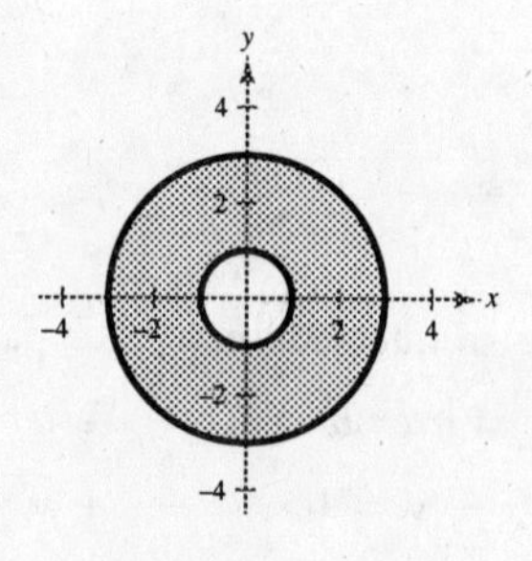

51. $y \le \sqrt{3x} + 1$

$y \ge x^2 + 1$

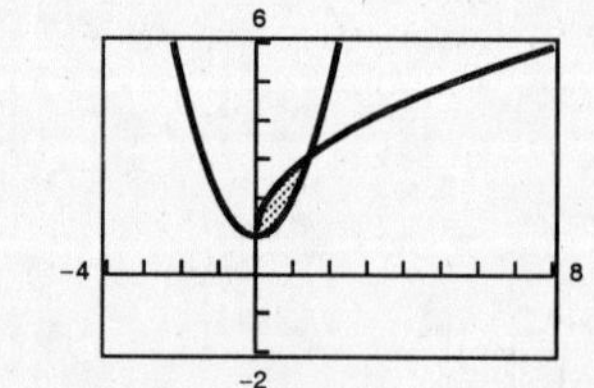

53. $y < x^3 - 2x + 1$

$y > -2x$

$x \le 1$

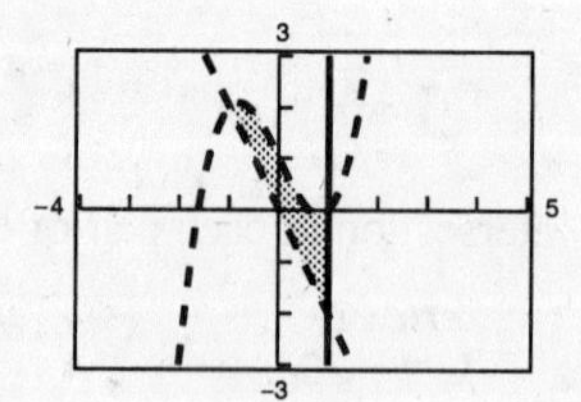

55. $x^2y \ge 1$

$0 < x \le 4$

$y \le 4$

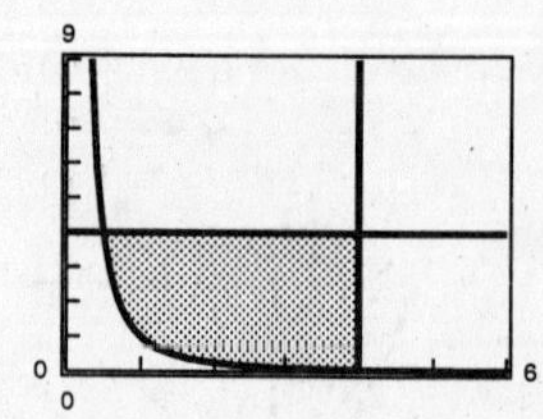

57. $y \le -x + 4 \implies \frac{x}{4} + \frac{y}{4} \le 1$

$x \ge 0 \qquad x \ge 0$

$y \ge 0 \qquad y \ge 0$

59. $(0, 4), (4, 0)$ Line: $y \le 4 - x$

$(0, 2), (8, 0)$ Line: $y \le -\frac{1}{4}x + 2$

$x \ge 0,\ y \ge 0$

61. Rectangular region with vertices at $(2, 1), (5, 1), (5, 7)$, and $(2, 7)$

$x \ge 2$

$x \le 5$

$y \ge 1$

$y \le 7$

Thus, $2 \le x \le 5,\ 1 \le y \le 7$.

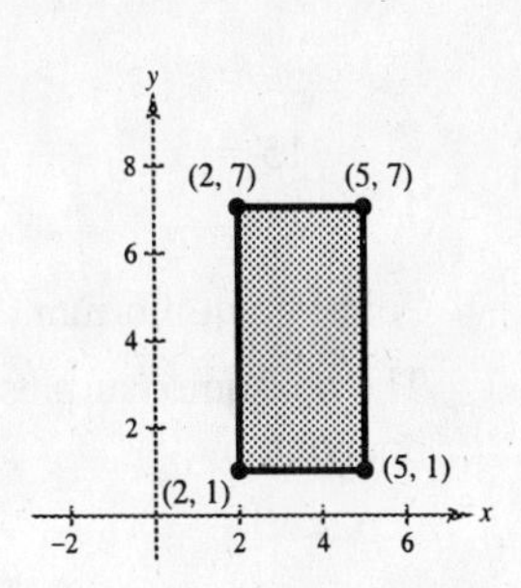

63. Triangle with vertices at $(0, 0), (5, 0), (2, 3)$

$(0, 0), (5, 0)$

Line: $y \ge 0$

$(0, 0), (2, 3)$

Line: $y \le \frac{3}{2}x$

$(2, 3), (5, 0)$

Line: $y \le -x + 5$

65. Account constraints:

$x \ge 5000$

$y \ge 5000$

$2x \le y$

$x + y \le 20{,}000$

25,000

0 25,000

0

67. Assembly center constraint: $x + \frac{3}{2}y \le 12$

Finishing center constraint: $\frac{4}{3}x + \frac{3}{2}y \le 15$

Point of intersection: $(9, 2)$

Physical constraints: $x \ge 0$ and $y \ge 0$

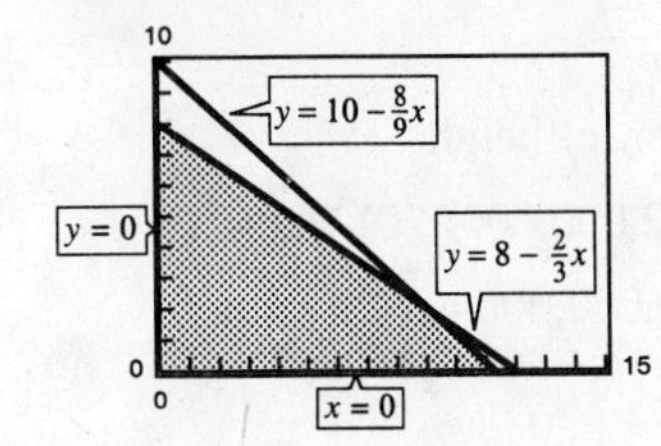

69. x = number of ounces of food X

y = number of ounces of food Y

Calcium: $20x + 10y \ge 280$

Iron: $15x + 10y \ge 160$

Vitamin B: $10x + 20y \ge 180$

$x \ge 0$

$y \ge 0$

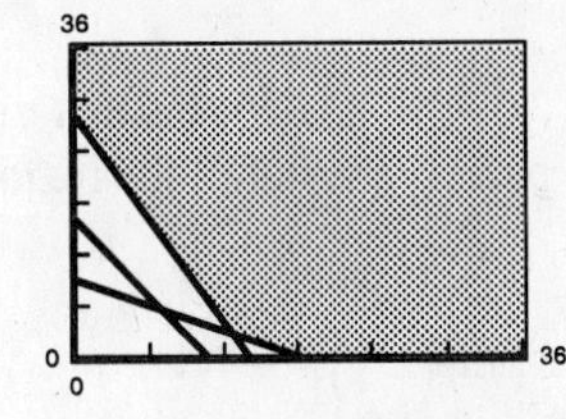

71. x = radius of smaller circle

y = radius of larger circle

(a) Constraints on circles: $\pi y^2 - \pi x^2 \ge 10$

$x > 0$

$y > x$

(b)

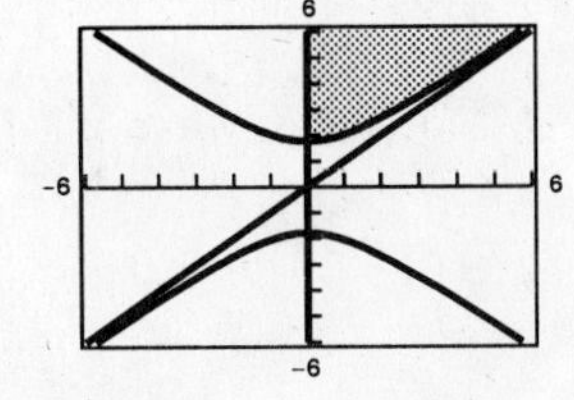

(c) The line is an asymptote to the boundary. The larger the circles, the closer the radii can be and the constraint still be satisfied.

73. Demand = Supply

$60 - x = 10 + \frac{7}{3}x$

$50 = \frac{10}{3}x$

$15 = x$

$45 = p$

Point of equilibrium: $(15, 45)$

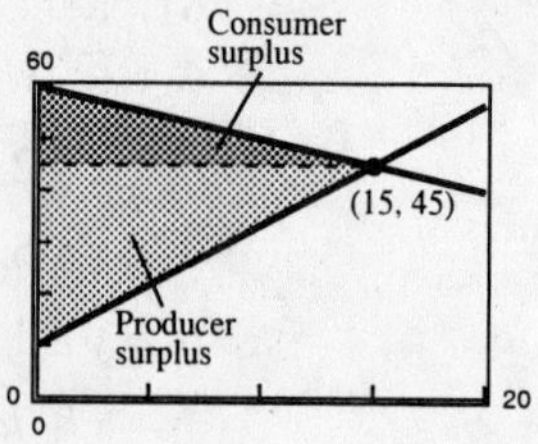

The consumer surplus is the area of the triangle bounded by

$p \le 60 - x$

$p \ge 45$

$x \ge 0.$

Consumer surplus $= \frac{1}{2}(\text{base})(\text{height})$

$= \frac{1}{2}(15)(15)$

$= \frac{225}{2}$

$= 112.5$

The producer surplus is the area of the triangle bounded by

$p \ge 10 + \frac{7}{3}x$

$p \le 45$

$x \ge 0.$

Producer surplus $= \frac{1}{2}(\text{base})(\text{height})$

$= \frac{1}{2}(15)(35)$

$= \frac{525}{2}$

$= 262.5$

75. Demand = Supply

$140 - 0.00002x = 80 + 0.00001x$

$60 = 0.00003x$

$2,000,000 = x$

$100 = p$

Point of equilibrium: $(2,000,000, 100)$

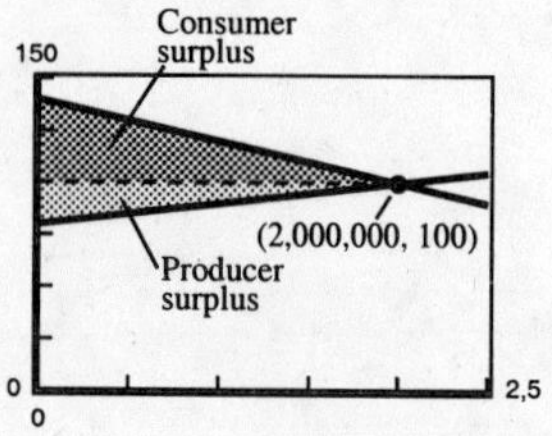

The consumer surplus is the area of the triangle bounded by

$p \le 140 - 0.00002x$

$p \ge 100$

$x \ge 0.$

Consumer surplus $= \frac{1}{2}(\text{base})(\text{height})$

$= \frac{1}{2}(2,000,000)(40)$

$= 40,000,000$ or \$40 million

The producer surplus is the area of the triangle bounded by

$p \ge 80 + 0.00001x$

$p \le 100$

$x \ge 0.$

Producer surplus $= \frac{1}{2}(\text{base})(\text{height})$

$= \frac{1}{2}(2,000,000)(20)$

$= 20,000,000$ or \$20 million

77. False. The inequality $3x + y^2 \ge 2$ means that the area to the left of the parabola is not in the region.

79. The solution set of a system of inequalities is usually a region in the xy-plane.

81. $y - 0 = \frac{-1 - 0}{3 - (-8)}[x - (-8)]$

$y = -\frac{1}{11}(x + 8)$

$11y + x = -8$

83. $y + 5.2 = \frac{0.8 + 5.2}{-2.6 - 3.4}(x - 3.4)$

$y + 5.2 = -x + 3.4$

$y + x = -1.8$

Section 7.5 Linear Programming

- To solve a linear programming problem:
  1. Sketch the solution set for the system of constraints.
  2. Find the vertices of the region.
  3. Test the objective function at each of the vertices.

Solutions to Odd-Numbered Exercises

1. $z = 3x + 5y$

At $(0, 6)$: $z = 3(0) + 5(6) = 30$

At $(0, 0)$: $z = 3(0) + 5(0) = 0$

At $(6, 0)$: $z = 3(6) + 5(0) = 18$

The minimum value is 0 at $(0, 0)$.
The maximum value is 30 at $(0, 6)$.

3. $z = 10x + 7y$

At $(0, 6)$: $z = 10(0) + 7(6) = 42$

At $(0, 0)$: $z = 10(0) + 7(0) = 0$

At $(6, 0)$: $z = 10(6) + 7(0) = 60$

The minimum value is 0 at $(0, 0)$.
The maximum value is 60 at $(6, 0)$.

5. $z = 5x + 2y$

At $(0, 5)$: $z = 5(0) + 2(5) = 10$

At $(4, 0)$: $z = 5(4) + 2(0) = 20$

At $(3, 4)$: $z = 5(3) + 2(4) = 23$

At $(0, 0)$: $z = 5(0) + 2(0) = 0$

The minimum value is 0 at $(0, 0)$.
The maximum value is 23 at $(3, 4)$.

7. $z = 5x + 0.5y$

At $(0, 5)$: $z = 5(0) + \frac{5}{2} = \frac{5}{2}$

At $(4, 0)$: $z = 5(4) + \frac{0}{2} = 20$

At $(3, 4)$: $z = 5(3) + \frac{4}{2} = 17$

At $(0, 0)$: $z = 5(0) + \frac{0}{2} = 0$

The minimum value is 0 at $(0, 0)$.
The maximum value is 20 at $(4, 0)$.

9. $z = 10x + 7y$

At $(0, 45)$: $z = 10(0) + 7(45) = 315$

At $(30, 45)$: $z = 10(30) + 7(45) = 615$

At $(60, 20)$: $z = 10(60) + 7(20) = 740$

At $(60, 0)$: $z = 10(60) + 7(0) = 600$

At $(0, 0)$: $z = 10(0) + 7(0) = 0$

The minimum value is 0 at $(0, 0)$.

The maximum value is 740 at $(60, 20)$.

11. $z = 25x + 30y$

At $(0, 45)$: $z = 25(0) + 30(45) = 1350$

At $(30, 45)$: $z = 25(30) + 30(45) = 2100$

At $(60, 20)$: $z = 25(60) + 30(20) = 2100$

At $(60, 0)$: $z = 25(60) + 30(0) = 1500$

At $(0, 0)$: $z = 25(0) + 30(0) = 0$

The minimum value is 0 at $(0, 0)$.
The maximum value is 2100 at any point along the line segment connecting $(30, 45)$ and $(60, 20)$.

13. $z = 6x + 10y$

At $(0, 2)$: $z = 6(0) + 10(2) = 20$

At $(5, 0)$: $z = 6(5) + 10(0) = 30$

At $(0, 0)$: $z = 6(0) + 10(0) = 0$

The minimum value is 0 at $(0, 0)$.
The maximum value is 30 at $(5, 0)$.

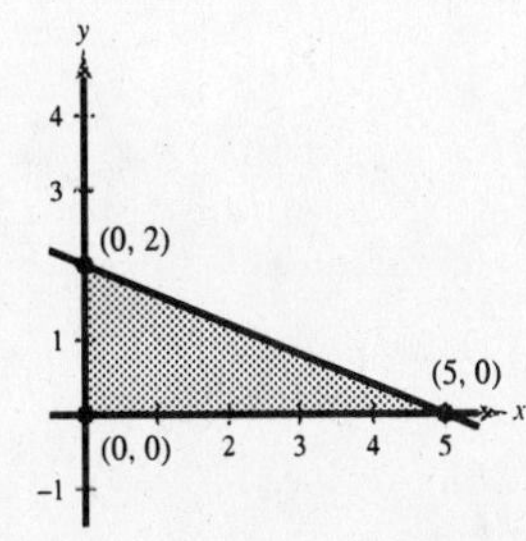

15. $z = 9z + 24y$

At $(0, 2)$: $z = 9(0) + 24(2) = 48$

At $(5, 0)$: $z = 9(5) + 24(0) = 45$

At $(0, 0)$: $z = 9(0) + 24(0) = 0$

The minimum value is 0 at $(0, 0)$.

The maximum value is 48 at $(0, 2)$.

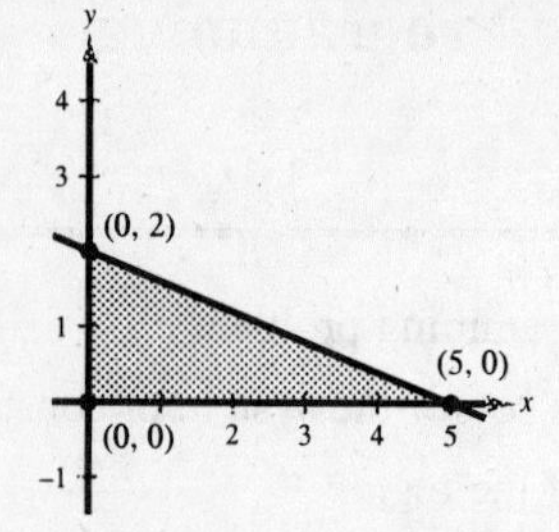

17. $z = 4x + 3y$

At $(3, 0)$: $z = 4(3) + 3(0) = 12$

At $(5, 3)$: $z = 4(5) + 3(3) = 29$

At $(0, 4)$: $z = 4(0) + 3(4) = 12$

At $(0, 2)$: $z = 4(0) + 3(2) = 6$

The minimum value is 6 at $(0, 2)$.

The maximum value is 29 at $(5, 3)$.

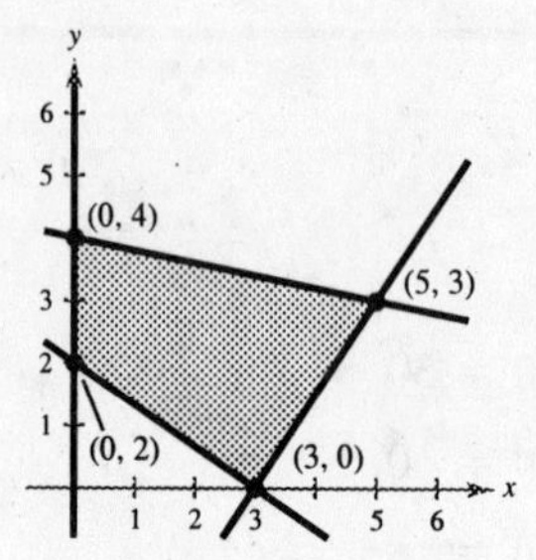

19. $z = 3x + 7y$ (same region as Exercise 17)

At $(3, 0)$: $z = 3(3) + 7(0) = 9$

At $(5, 3)$: $z = 3(5) + 7(3) = 36$

At $(0, 4)$: $z = 3(0) + 7(4) = 28$

At $(0, 2)$: $z = 3(0) + 7(2) = 14$

The minimum value is 9 at $(3, 0)$.

The maximum value is 36 at $(5, 3)$.

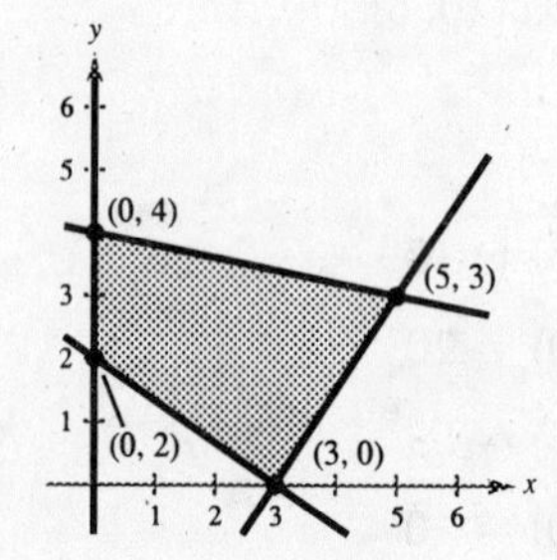

21. $z = 4x + y$

At $(36, 0)$: $z = 4(36) + 0 = 144$

At $(40, 0)$: $z = 4(40) + 0 = 160$

At $(24, 8)$: $z = 4(24) + 8 = 104$

The minimum value is 104 at $(24, 8)$.

The maximum value is 160 at $(40, 0)$.

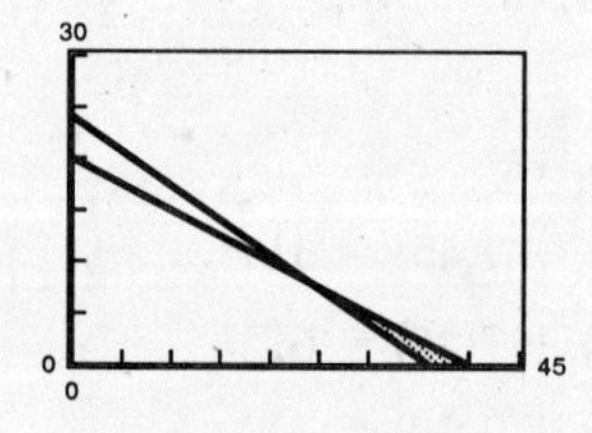

23. $z = x + 4y$

At $(36, 0)$: $z = 36 + 4(0) = 36$

At $(40, 0)$: $z = 40 + 4(0) = 40$

At $(24, 8)$: $z = 24 + 4(8) = 56$

The minimum value is 36 at $(36, 0)$.

The maximum value is 56 at $(24, 8)$.

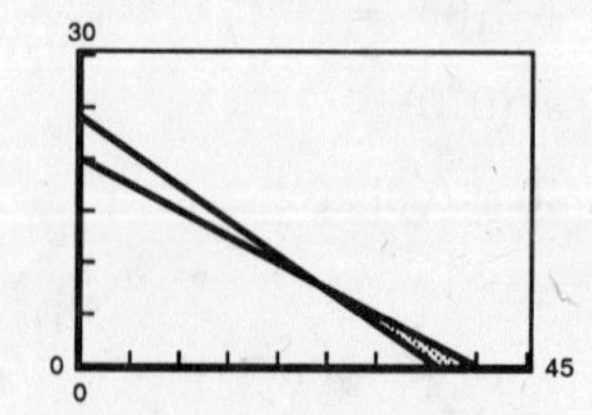

25. $z = 2x + 3y$

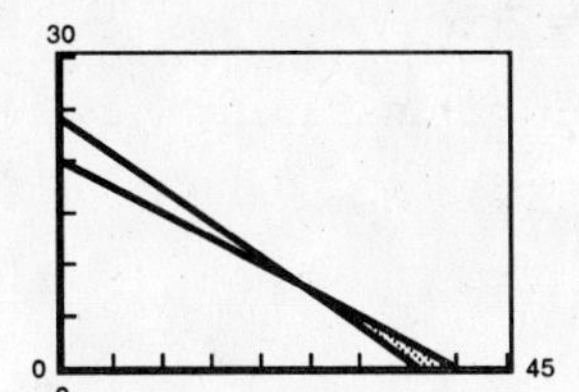

At $(36, 0)$: $z = 2(36) + 3(0) = 72$

At $(40, 0)$: $z = 2(40) + 3(0) = 80$

At $(24, 8)$: $z = 2(24) + 3(8) = 72$

Minimum at any point on the line segment joining (36, 0) and (24, 8): 72.

Maximum at (40, 0): 80.

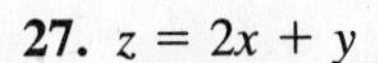

27. $z = 2x + y$

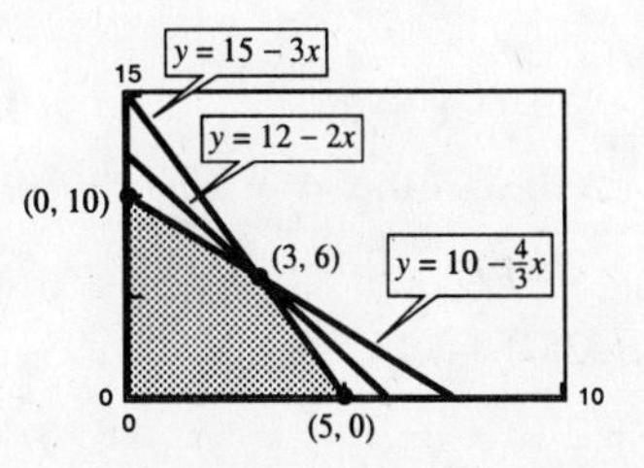

At $(0, 10)$: $z = 2(0) + (10) = 10$

At $(3, 6)$: $z = 2(3) + (6) = 12$

At $(5, 0)$: $z = 2(5) + (0) = 10$

At $(0, 0)$: $z = 2(0) + (0) = 0$

The maximum value is 12 at (3, 6).

29. $z = x + y$

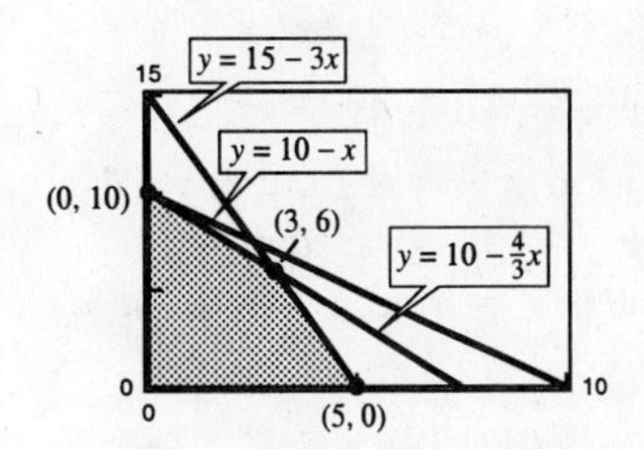

At $(0, 10)$: $z = (0) + (10) = 10$

At $(3, 6)$: $z = (3) + (6) = 9$

At $(5, 0)$: $z = (5) + (0) = 5$

At $(0, 0)$: $z = (0) + (0) = 0$

The maximum value is 10 at (0, 10).

31. $z = x + 5y$

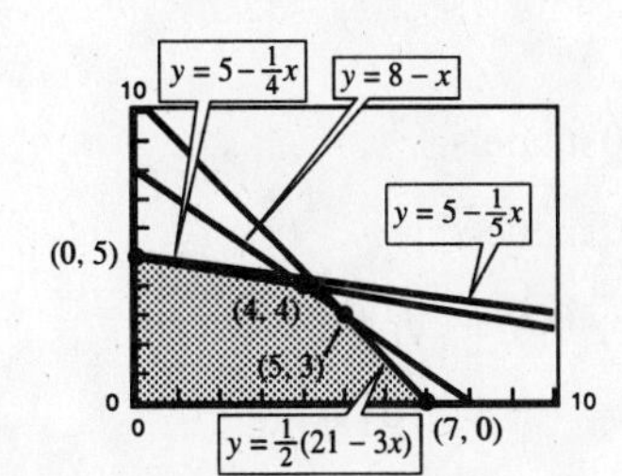

At $(0, 5)$: $z = 0 + 5(5) = 25$

At $(4, 4)$: $z = 4 + 5(4) = 24$

At $(5, 3)$: $z = 5 + 5(3) = 20$

At $(7, 0)$: $z = 7 + 5(0) = 7$

The maximum value is 25 at (0, 5).

33. $z = 4x + 5y$

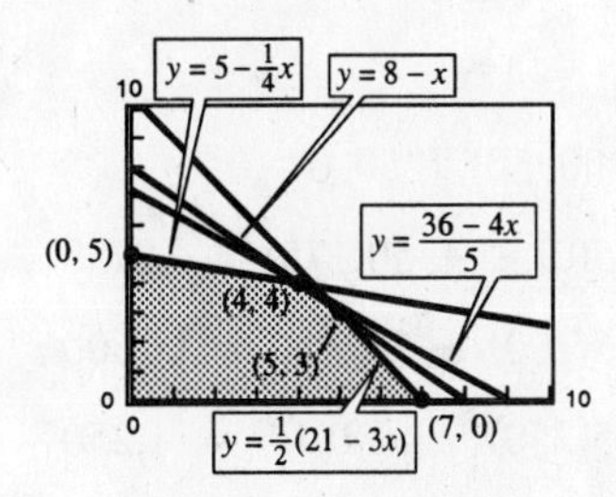

At $(0, 5)$: $z = 4(0) + 5(5) = 25$

At $(4, 4)$: $z = 4(4) + 5(4) = 36$

At $(5, 3)$: $z = 4(5) + 5(3) = 35$

At $(7, 0)$: $z = 4(7) + 5(0) = 28$

The maximum value is 36 at (4, 4).

35. Objective function: $z = 2.5x + y$

Constraints: $x \geq 0$, $y \geq 0$, $3x + 5y \leq 15$, $5x + 2y \leq 10$

At $(0, 0)$: $z = 0$

At $(2, 0)$: $z = 5$

At $\left(\frac{20}{19}, \frac{45}{19}\right)$: $z = \frac{95}{19} = 5$

At $(0, 3)$: $z = 3$

z is the maximum at any point on the line $5x + 2y = 10$ between the points $(2, 0)$ and $\left(\frac{20}{19}, \frac{45}{19}\right)$.

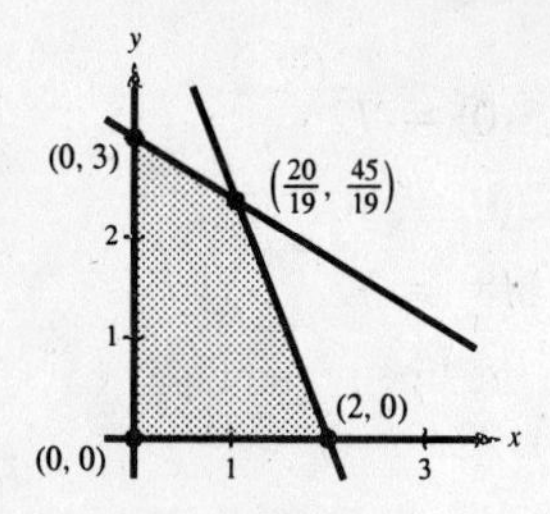

37. Objective function: $z = -x + 2y$

Constraints: $x \geq 0$, $y \geq 0$, $x \leq 10$, $x + y \leq 7$

At $(0, 0)$: $z = -0 + 2(0) = 0$

At $(0, 7)$: $z = -0 + 2(7) = 14$

At $(7, 0)$: $z = -7 + 2(0) = -7$

The constraint $x \leq 10$ is extraneous.

The maximum value of 14 occurs at $(0, 7)$.

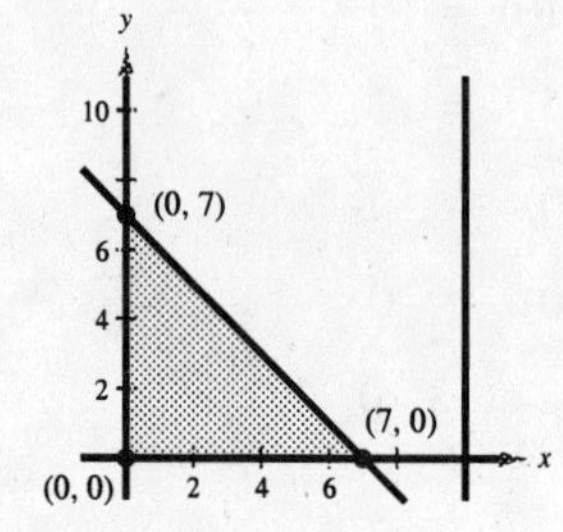

39. Objective function: $z = 3x + 4y$

Constraints: $x \geq 0$, $y \geq 0$, $x + y \leq 1$, $2x + y \leq 4$

The constraint $2x + y \leq 4$ is extraneous.

The maximum value of $z = 4$ occurs at $(0, 1)$.

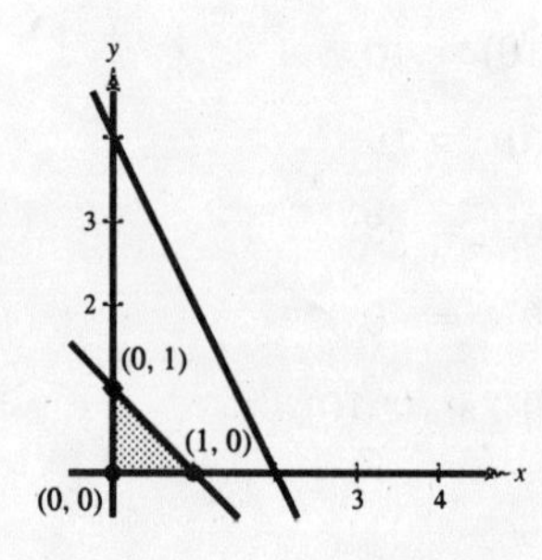

41. x = number of \$250 models

y = number of \$400 models

Constraints: $250x + 400y \leq 70{,}000$

$x + y \leq 250$

$x \geq 0$

$y \geq 0$

Objective function: $P = 45x + 50y$

Vertices: $(0, 175)$, $(200, 50)$, $(250, 0)$, $(0, 0)$

At $(0, 175)$: $P = 45(0) + 50(175) = 8750$

At $(200, 50)$: $P = 45(200) + 50(50) = 11{,}500$

At $(250, 0)$: $P = 45(250) + 50(0) = 11{,}250$

At $(0, 0)$: $P = 45(0) + 50(0) = 0$

To maximize the profit, the merchant should stock 200 units of the model costing \$250 and 50 units of the model costing \$400. Then the maximum profit would be \$11,500.

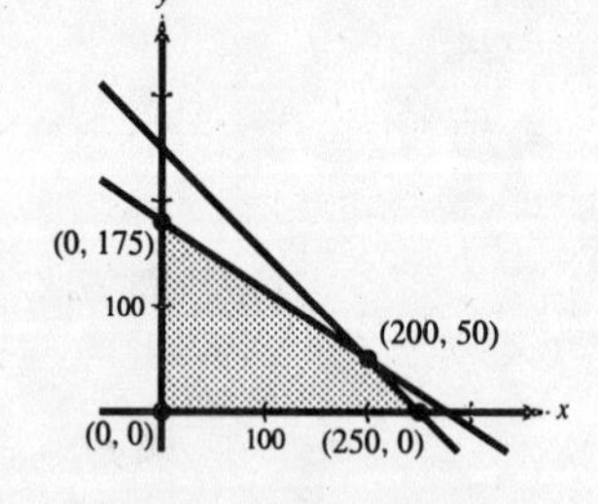

43. x = fraction of type A

y = fraction of type B

Constraints: $80x + 92y \leq 90$

$x + y \leq 1$

$x \geq 0$

$y \geq 0$

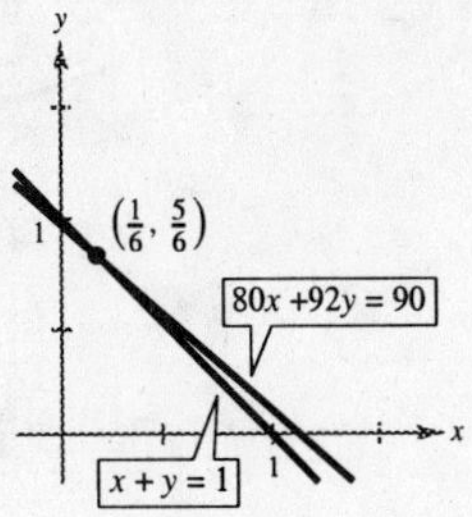

Objective function: $C = 1.13x + 1.28y$

Vertices: $\left(\frac{1}{6}, \frac{5}{6}\right)$

At $\left(\frac{1}{6}, \frac{5}{6}\right)$: $C = (1.13)\left(\frac{1}{6}\right) + (1.28)\left(\frac{5}{6}\right) = 1.255$

The minimum cost is \$1.26 and occurs with a mixture that is $\frac{1}{6}$ A and $\frac{5}{6}$ B.

45. Objective function: $R = 1000x + 300y$

At (0, 0): $R = 1000(0) + 300(0) = 0$

At (0, 40): $R = 1000(0) + 300(40) = 12{,}000$

At (8, 8): $R = 1000(8) + 300(8) = 10{,}400$

At (9, 0): $R = 1000(9) + 300(0) = 9000$

The revenue will be maximum (\$12,000) if the firm does 0 audits and 40 tax returns.

47. x = fraction of Model A

y = fraction of Model B

Constraints: $2.5x + 3y \leq 4000$

$2x + y \leq 2500$

$0.75x + 1.25y \leq 1500$

$x \geq 0$

$y \geq 0$

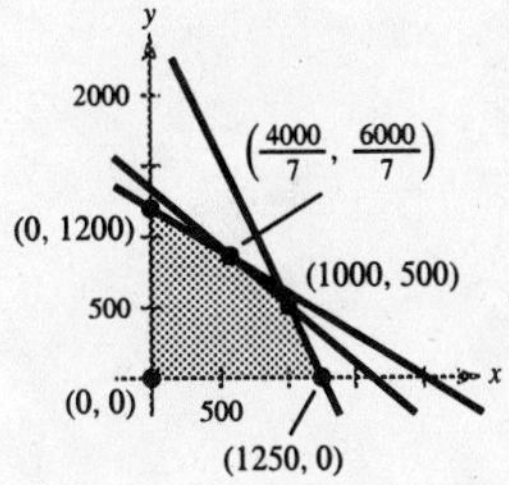

Objective function: $P = 50x + 52y$

Vertices: (0, 0), (0, 1200), $\left(\frac{4000}{7}, \frac{6000}{7}\right)$, (1000, 500), (1250, 0)

At (0, 0): $P = (50)(0) + 52(0) = 0$

At (0, 1200): $P = 50(0) + 52(1200) = 62{,}400$

At $\left(\frac{4000}{7}, \frac{6000}{7}\right)$: $P = 50\left(\frac{4000}{7}\right) + 52\left(\frac{6000}{7}\right) \approx 73{,}142.86$

At (1000, 500): $P = 50(1000) + 52(500) = 76{,}000$

At (1250, 0): $P = 50(1250) + 52(0) = 62{,}500$

The maximum profit (\$76,000) occurs when 1000 units of Model A and 500 units of Model B are produced.

49. True. The line joining (4, 7) and (8, 3) is $x + y = 11$. Both points (4.5, 6.5) and (7.8, 3.2) lie on this line.

51. There are an infinite number of objective functions that would have a maximum at (0, 4). One such objective function is $z = x + 5y$.

53. There are an infinite number of objective functions that would have a maximum at (5, 0). One such objective function is $z = 4x + y$.

55. Constraints: $x \geq 0, y \geq 0, x + 3y \leq 15, 4x + y \leq 16$

Vertex	Value of $z = 3x + ty$
$(0, 0)$	$z = 0$
$(0, 5)$	$z = 5t$
$(3, 4)$	$z = 9 + 4t$
$(4, 0)$	$z = 12$

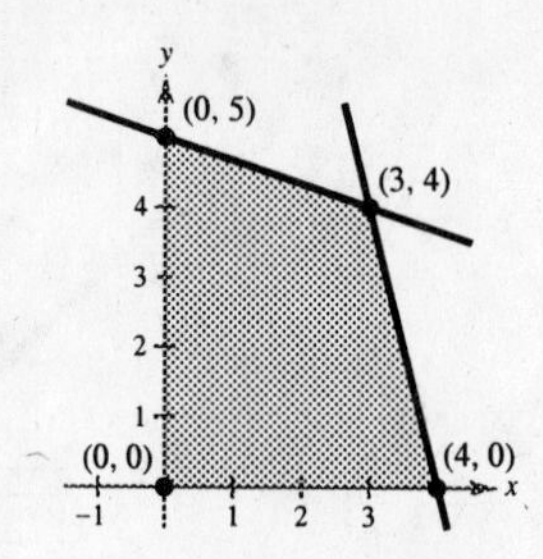

(a) For the maximum value to be at $(0, 5)$, $z = 5t$ must be greater than $z = 9 + 4t$ and $z = 12$.

$5t > 9 + 4t$ and $5t > 12$

$t > 9$ $\qquad t > \frac{12}{5}$

Thus, $t > 9$.

(b) For the maximum value to be at $(3, 4)$, $z = 9 + 4t$ must be greater than $z = 5t$ and $z = 12$.

$9 + 4t > 5t$ and $9 + 4t > 12$

$9 > t$ $\qquad 4t > 3$

$\qquad t > \frac{3}{4}$

Thus, $\frac{3}{4} < t < 9$.

57. $e^{2x} + 2e^x - 15 = 0$ Quadratic in e^x

$(e^x + 5)(e^x - 3) = 0$

$e^x + 5 = 0 \implies e^x = -5$ Impossible

$e^x - 3 = 0 \implies e^x = 3 \implies x = \ln 3 \approx 1.099$

59. $8(62 - e^{x/4}) = 192$

$62 - e^{x/4} = 24$

$e^{x/4} = 38$

$\frac{x}{4} = \ln 38$

$x = 4 \ln 38 \approx 14.550$

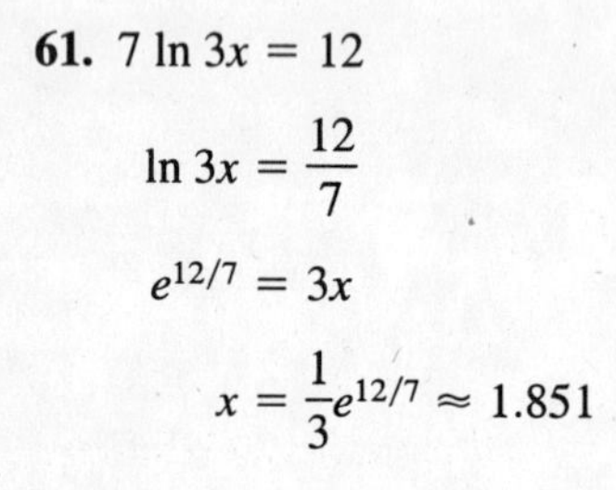

61. $7 \ln 3x = 12$

$\ln 3x = \frac{12}{7}$

$e^{12/7} = 3x$

$x = \frac{1}{3}e^{12/7} \approx 1.851$

63. $-3x + 5y = -9$ Equation 1

$x - 4y = 10$ Equation 2

3 times Equation 2 added to Equation 1 produces

$-7y = 21 \implies y = -3.$

Then $x = 4y + 10 = 4(-3) + 10 = -2$

Answer: $(-2, -3)$

65. $4x + 5y = 6$ Equation 1

$5x - 11y = -27$ Equation 2

5 times Equation 1 and 4 times Equation 2 produces

$20x + 25y = 30$

$20x - 44y = -108$

Subtracting, $69y = 138 \implies y = 2$

Then $4x + 5(2) = 6 \implies x = -1$

Answer: $(-1, 2)$

67. $2x + y \leq 7$

$y < 4$

$x \geq -1$

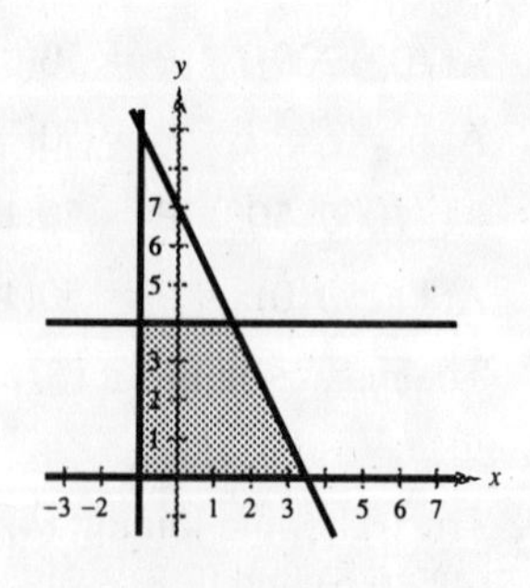

69. $x^2 + y \le 1$

$y - 3x \ge -3$

$y \ge -3$

$y = 1 - x^2 = -3$

$x^2 = 4$

$x = \pm 2$

$(2, -3), (-2, -3)$

$y = 3x - 3 = -3$

$x = 0$

$(0, -3)$

$y = 1 - x^2 = 3x - 3$

$x^2 + 3x - 4 = 0$

$(x + 4)(x - 1) = 0$

$(-4, -15), (1, 0)$

Review Exercises for Chapter 7

Solutions to Odd-Numbered Exercises

1. $x + y = 2 \Rightarrow \qquad y = 2 - x$

$x - y = 0 \Rightarrow x - (2 - x) = 0$

$2x - 2 = 0$

$x = 1$

$y = 2 - 1 = 1$

Solution: $(1, 1)$

3. $x^2 - y^2 = 9$

$x - y = 1 \Rightarrow \quad x = y + 1$

$(y + 1)^2 - y^2 = 9$

$2y + 1 = 9$

$y = 4$

$x = 5$

Solution: $(5, 4)$

5. $y = 2x^2$

$y = x^4 - 2x^2 \Rightarrow 2x^2 = x^4 - 2x^2$

$0 = x^4 - 4x^2$

$0 = x^2(x^2 - 4)$

$0 = x^2(x + 2)(x - 2)$

$x = 0, x = -2, x = 2$

$y = 0, y = 8, y = 8$

Solutions: $(0, 0), (-2, 8), (2, 8)$

7. $5x + 6y = 7 \Rightarrow y = \frac{1}{6}(7 - 5x)$

$-x - 4y = 0 \Rightarrow y = -\frac{x}{4}$

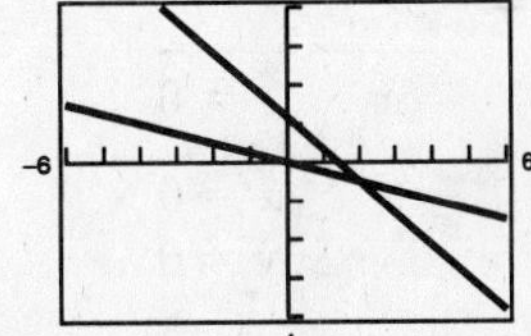

Point of Intersection:

$\left(2, -\frac{1}{2}\right)$

9. $y^2 - 2y + x = 0 \Rightarrow (y - 1)^2 = 1 - x \Rightarrow y = 1 \pm \sqrt{1 - x}$

$x + y = 0 \Rightarrow \qquad y = -x$

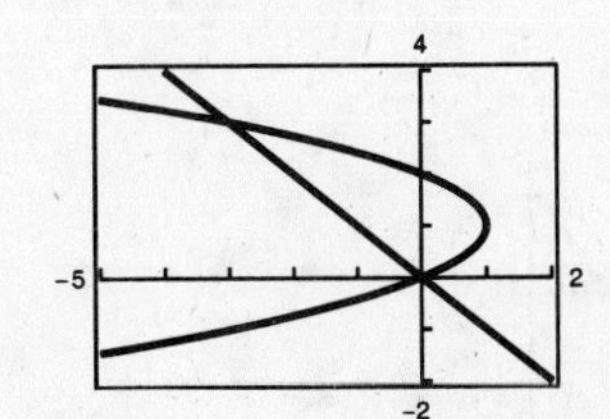

Points of intersection: $(0, 0)$ and $(-3, 3)$

11. $y = 2(6 - x)$

$y = 2^{x-2}$

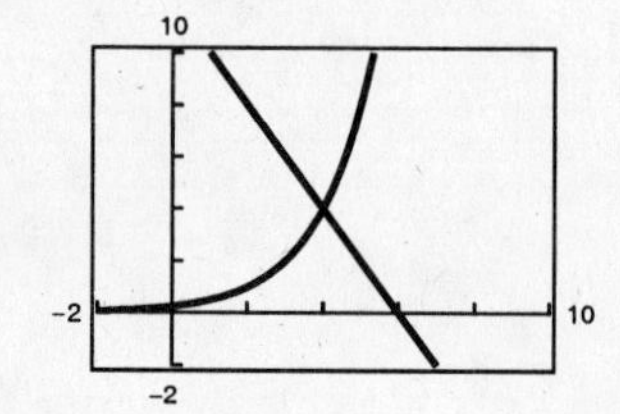

Point of intersection: $(4, 4)$

13. Revenue $= 4.95x$

Cost $= 2.85x + 10{,}000$

Break even when Revenue $=$ Cost

$$4.95x = 2.85x + 10{,}000$$
$$2.10x = 10{,}000$$
$$x \approx 4762 \text{ units}$$

15.
$$2l + 2w = 480$$
$$l = 1.50w$$
$$2(1.50w) + 2w = 480$$
$$5w = 480$$
$$w = 96$$
$$l = 144$$

The dimensions are 96×144 meters.

17. $2x - y = 2 \Rightarrow 16x - 8y = 16$

$6x + 8y = 39 \Rightarrow 6x + 8y = 39$

$22x = 55$

$x = \frac{55}{22} = \frac{5}{2}$

$y = 3$

Solution: $\left(\frac{5}{2}, 3\right)$

19. $0.2x + 0.3y = 0.14 \Rightarrow 20x + 30y = 14 \Rightarrow 20x + 30y = 14$

$0.4x + 0.5y = 0.20 \Rightarrow 4x + 5y = 2 \Rightarrow -20x - 25y = -10$

$5y = 4$

$y = \frac{4}{5}$

$x = -\frac{1}{2}$

Solution: $\left(-\frac{1}{2}, \frac{4}{5}\right)$ or $(-0.5, 0.8)$

21. $3x - 2y = 0 \Rightarrow 3x - 2y = 0$

$3x + 2(y + 5) = 10 \Rightarrow 3x + 2y = 0$

$6x = 0$

$x = 0$

$y = 0$

Solution: $(0, 0)$

23. $1.25x - 2y = 3.5 \Rightarrow 5x - 8y = 14$

$5x - 8y = 14 \Rightarrow -5x + 8y = -14$

$0 = 0$

Infinite number of solutions

Let $y = a$, then $5x - 8a = 14 \Rightarrow x = \frac{14}{5} + \frac{8}{5}a$.

Solution: $\left(\frac{14}{5} + \frac{8}{5}a, a\right)$

25. $3x + 2y = 0 \Rightarrow y = -\frac{3}{2}x$

$x - y = 4 \Rightarrow y = x - 4$

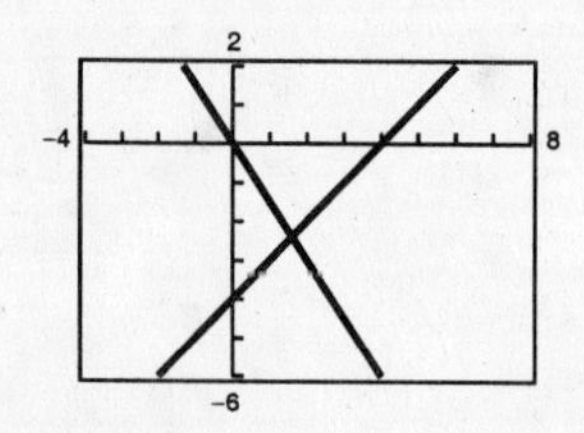

Consistent.

Answer: $(1.6, -2.4)$

27. $\frac{1}{4}x - \frac{1}{5}y = 2 \Rightarrow y = \frac{5}{4}x - 10$

$-5x + 4y = 8 \Rightarrow y = \frac{1}{4}(8 + 5x) = \frac{5}{4}x + 2$

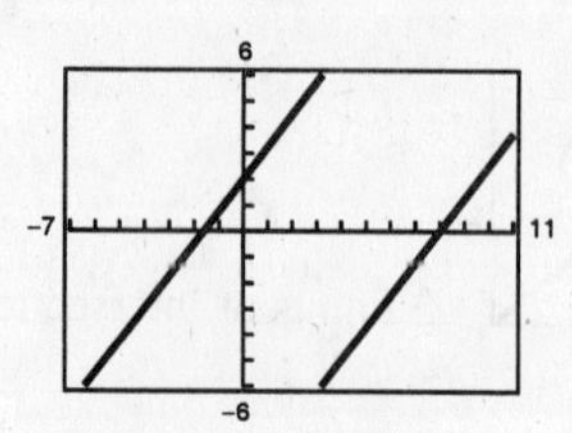

Inconsistent. Lines are parallel.

29. $8x - 2y = 11 \Rightarrow y = \frac{1}{2}(8x - 11)$

$-4x + y = -5.5 \Rightarrow y = 4x - 5.5$

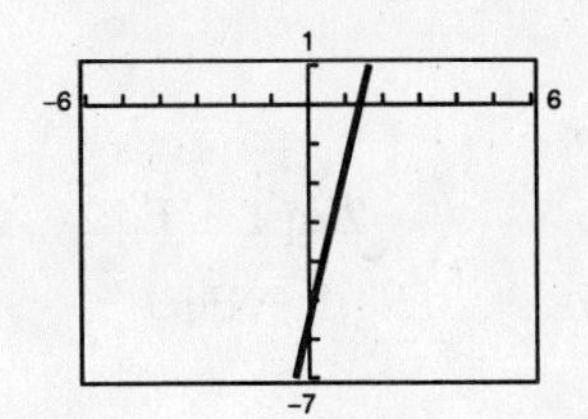

System is consistent. Lines coincide.

Solution: all points on line $y = 4x - 5.5$

31. Let x = speed of the slower plane.

Let y = speed of the faster plane.

Then, distance of first plane + distance of second plane = 275 miles.

(rate of first plane)(time) + (rate of second plane)(time) = 275 miles

$$x\left(\tfrac{40}{60}\right) + y\left(\tfrac{40}{60}\right) = 275$$

$$y = x + 25$$

$$\tfrac{2}{3}x + \tfrac{2}{3}(x + 25) = 275$$

$$4x + 50 = 825$$

$$4x = 775$$

$$x = 193.75 \text{ mph}$$

$$y = x + 25 = 218.75 \text{ mph}$$

33. Demand = Supply

$$37 - 0.0002x = 22 + 0.00001x$$

$$15 = 0.00021x$$

$$x = \frac{500{,}000}{7}$$

Point of equilibrium: $p = \dfrac{159}{7}$

$$\left(\frac{500{,}000}{7}, \frac{159}{7}\right)$$

35. $x - 4y + 3z = 3$ Equation 1

$-y + z = -1$ Equation 2

$z = -5$ Equation 3

Substitute $z = -5$ into Equation 2: $-y + (-5) = -1 \Rightarrow -y = 4 \Rightarrow y = -4$

Substitute $z = -5$ and $y = -4$ into Equation 1:

$$x - 4(-4) + 3(-5) = 3$$

$$x = -16 + 15 + 3$$

$$x = 2$$

Answer: $(2, -4, -5)$

37.
$$\begin{aligned} x + 3y - z &= 13 \\ 2x - 5z &= 23 \\ 4x - y - 2z &= 14 \end{aligned}$$

$$\begin{aligned} x + 3y - z &= 13 \\ -6y - 3z &= -3 \\ -13y + 2z &= -38 \end{aligned}$$

$$\begin{aligned} x + 3y - z &= 13 \\ -6y - 3z &= -3 \\ \tfrac{17}{2}z &= -\tfrac{63}{2} \end{aligned}$$

$\frac{17}{2}z = -\frac{63}{2} \Rightarrow z = -\frac{63}{17}$

$-6y - 3\left(-\frac{63}{17}\right) = -3 \Rightarrow y = \frac{40}{17}$

$x + 3\left(\frac{40}{17}\right) - \left(-\frac{63}{17}\right) = 13 \Rightarrow x = \frac{38}{17}$

Solution: $\left(\frac{38}{17}, \frac{40}{17}, -\frac{63}{17}\right)$

39.
$$\begin{aligned} x - 2y + z &= -6 \\ 2x - 3y &= -7 \\ -x + 3y - 3z &= 11 \end{aligned}$$

$$\begin{aligned} x - 2y + z &= -6 \\ y - 2z &= 5 && -2\text{Eq.1} + \text{Eq.2} \\ y - 2z &= 5 && \text{Eq.1} + \text{Eq.3} \end{aligned}$$

$$\begin{aligned} x - 2y + z &= -6 \\ y - 2z &= 5 \\ 0 &= 0 && -\text{Eq.2} + \text{Eq.3} \end{aligned}$$

Let $z = a$, then $y = 2a + 5$.

$x - 2(2a + 5) + a = -6$

$x - 3a - 10 = -6$

$x = 3a + 4$

Solution: $(3a + 4, 2a + 5, a)$ where a is any real number.

41. $y = ax^2 + bx + c$ through $(0, -5)$, $(1, -2)$, and $(2, 5)$

$(0, -5)$: $-5 = c$

$(1, -2)$: $-2 = a + b + c \Rightarrow a + b = 3$

$(2, 5)$: $5 = 4a + 2b + c \Rightarrow 2a + b = 5$

$$\begin{aligned} 2a + b &= 5 \\ -a - b &= -3 \\ \hline a &= 2 \\ b &= 1 \end{aligned}$$

The equation of the parabola is $y = 2x^2 + x - 5$.

43. $5x - 12y + 7z = 16$ Equation 1

$3x - 7y + 4z = 9$ Equation 2

3 times Eq. 1 and (-5) times Eq. 2:

$15x - 36y + 21z = 48$

$-15x + 35y - 20z = -45$

Adding, $-y + z = 3 \Rightarrow y = z - 3$.

$5x - 12(z - 3) + 7z = 16$

$5x - 5z + 36 = 16$

$5x = 5z - 20$

$x = z - 4$

Infinite number of solutions of form

$(z - 4, z - 3, z)$

where z is any real number.

45. Plane: $2x - 4y + z = 8$

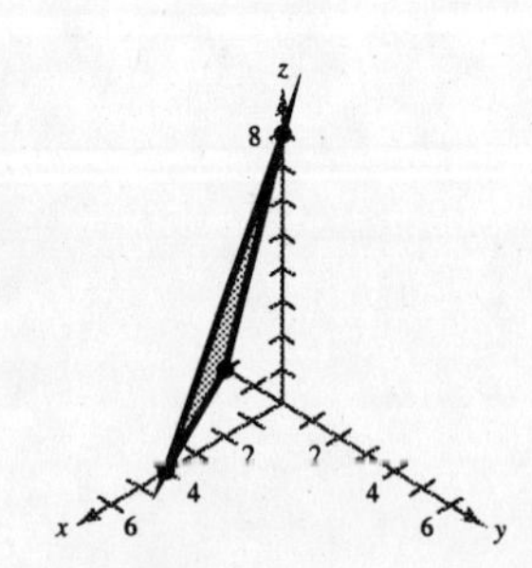

4 points on plane: $(4, 0, 0)$, $(0, -2, 0)$, $(0, 0, 8)$, $(1, 0, 6)$

47. $\dfrac{4 - x}{x^2 + 6x + 8} = \dfrac{A}{x + 2} + \dfrac{B}{x + 4}$

$4 - x = A(x + 4) + B(x + 2)$

Let $x = -2$: $6 = 2A \Rightarrow A = 3$

Let $x = -4$: $8 = -2B \Rightarrow B = -4$

$\dfrac{4 - x}{x^2 + 6x + 8} = \dfrac{3}{x + 2} - \dfrac{4}{x + 4}$

49. $\dfrac{x^2}{x^2 + 2x - 15} = 1 - \dfrac{2x - 15}{x^2 + 2x - 15} = 1 + \dfrac{A}{x + 5} + \dfrac{B}{x - 3}$

$-2x + 15 = A(x - 3) + B(x + 5)$

Let $x = -5$: $25 = -8A \implies A = -\dfrac{25}{8}$

Let $x = 3$: $9 = 8B \implies B = \dfrac{9}{8}$

$\dfrac{x^2}{x^2 + 2x - 15} = 1 + \dfrac{9}{8(x - 3)} - \dfrac{25}{8(x + 5)}$

51. $\dfrac{x^2 + 2x}{x^3 - x^2 + x - 1} = \dfrac{A}{x - 1} + \dfrac{Bx + C}{x^2 + 1}$

$x^2 + 2x = A(x^2 + 1) + (Bx + C)(x - 1)$

Let $x = 1$: $3 = 2A \implies A = \dfrac{3}{2}$

Let $x = 0$: $0 = A - C \implies C = \dfrac{3}{2}$

Let $x = 2$: $8 = 5A + 2B + C$

$$8 = \left(\frac{15}{2}\right) + 2B + \left(\frac{3}{2}\right) \implies B = -\frac{1}{2}.$$

$\dfrac{x^2 + 2x}{x^3 - x^2 + x - 1} = \dfrac{3/2}{x - 1} + \dfrac{-(1/2)x + 3/2}{x^2 + 1} = \dfrac{1}{2}\left(\dfrac{3}{x - 1} - \dfrac{x - 3}{x^2 + 1}\right)$

53. Let x = gallons of spray X

Let y = gallons of spray Y

Let z = gallons of spray Z

Chemical A: $\frac{1}{5}x + \quad \frac{1}{3}z = 6$ Eq. 1

Chemical B: $\frac{2}{5}x + \quad \frac{1}{3}z = 8$ Eq. 2

Chemical C: $\frac{2}{5}x + y + \frac{1}{3}z = 13$ Eq. 3

Subtracting Eq. 2 − Eq. 1 gives $\frac{1}{5}x = 2 \implies x = 10$.

Then $z = 12$ and $y = 5$.

Answer: 10 gallons of spray X
5 gallons of spray Y
12 gallons of spray Z

55. $x \le 6$

57. $y \le 5 - \frac{1}{2}x$

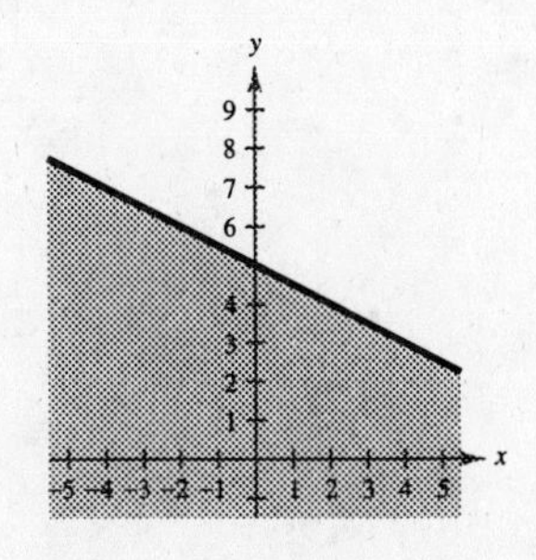

59. $y - 4x^2 > -1$

$y > 4x^2 - 1$

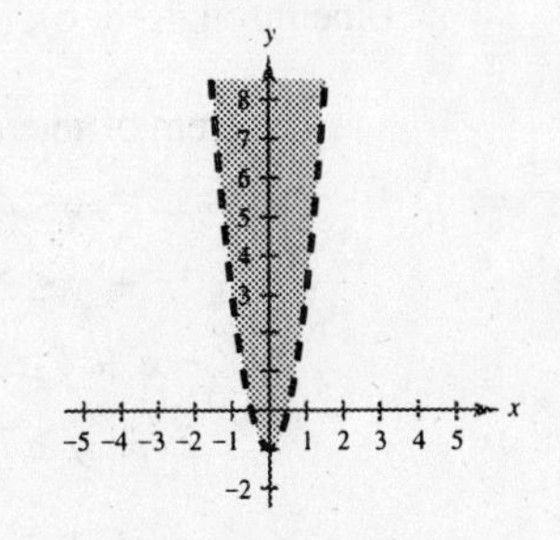

61. Inside circle and above line: matches (d)

63. Outside circle and above line: matches (c)

65. $x + 2y \le 160$

$3x + y \le 180$

$x \ge 0$

$y \ge 0$

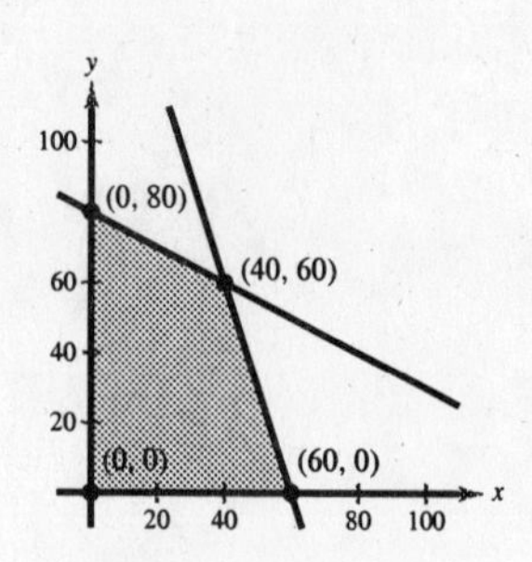

Vertex A	Vertex B	Vertex C	Vertex D	Vertex E	Vertex F
$x + 2y = 160$	$x + 2y = 160$	$3x + y = 180$	$x = 0$	$x + 2y = 160$	$3x + y = 180$
$3x + y = 180$	$x = 0$	$y = 0$	$y = 0$	$y = 0$	$x = 0$
$(40, 60)$	$(0, 80)$	$(60, 0)$	$(0, 0)$	$(160, 0)$	$(0, 180)$
				Outside the region	Outside the region

67. $y < x + 1$

$y > x^2 - 1$

Vertices:

$x + 1 = x^2 - 1$

$0 = x^2 - x - 2 = (x + 1)(x - 2)$

$x = -1$ or $x = 2$

$y = 0$ $\quad y = 3$

$(-1, 0)$ $\quad (2, 3)$

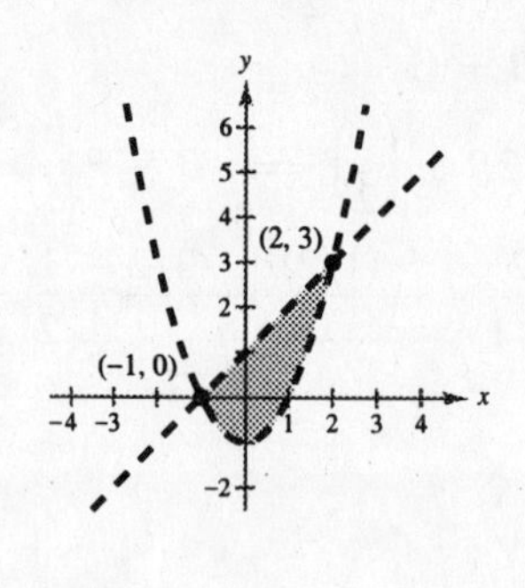

69. $2x - 3y \ge 0$

$2x - y \le 8$

$y \ge 0$

Vertex A	Vertex B	Vertex C
$2x - 3y = 0$	$2x - 3y = 0$	$2x - y = 8$
$2x - y = 8$	$y = 0$	$y = 0$
$(6, 4)$	$(0, 0)$	$(4, 0)$

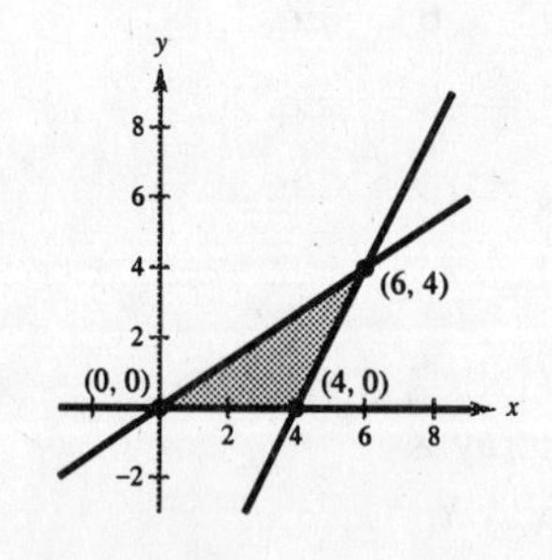

71

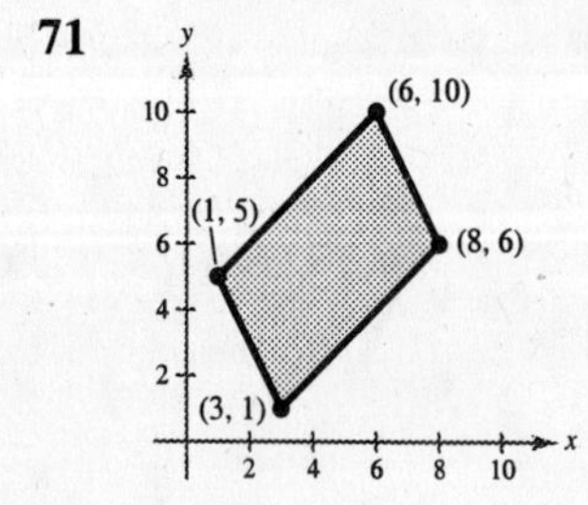

Line through (1, 5), (3, 1): $2x + y = 7$

Line through (1, 5), (6, 10): $-x + y = 4$

Line through (6, 10), (8, 6): $2x + y = 22$

Line through (8, 6), (3, 1): $-x + y = -2$

System of inequalities:

$-x + y \le 4$

$2x + y \le 22$

$-x + y \ge -2$

$2x + y \ge 7$

73. Let $x =$ the number of bushels for Harrisburg, and $y =$ the number of bushels for Philadelphia.

$x \geq 400$

$y \geq 600$

$x + y \leq 1500$

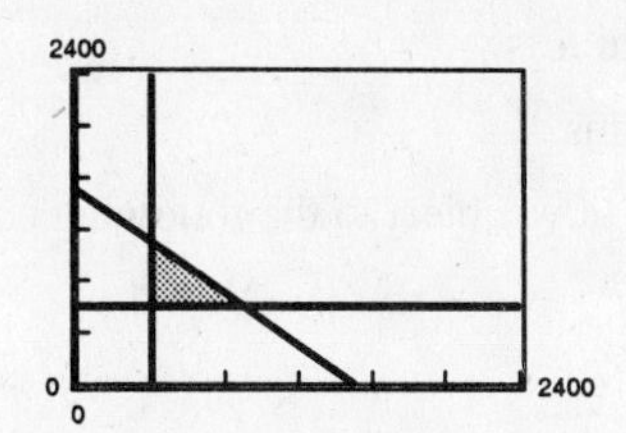

75. Demand = Supply

$160 - 0.0001x = 70 + 0.0002x$

$90 = 0.0003x$

$x = 300{,}000$ units

$p = \$130$

Point of equilibrium: (300,000, 130)

Consumer surplus: $\frac{1}{2}(300{,}000)(30) = \$4{,}500{,}000$

Producer surplus: $\frac{1}{2}(300{,}000)(60) = \$9{,}000{,}000$

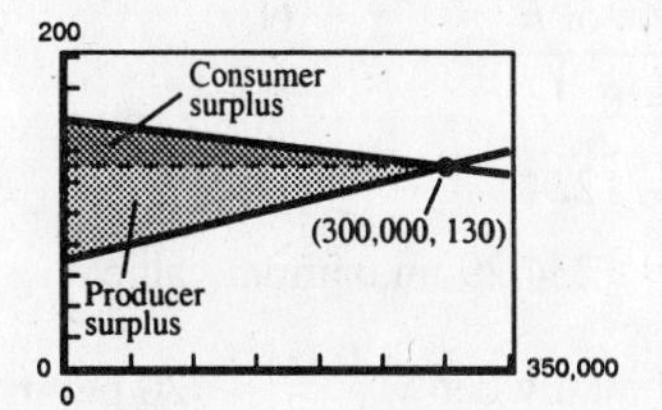

77. Maximize $z = 3x + 4y$ subject to the following constraints.

$x \geq 0$

$y \geq 0$

$2x + 5y \leq 50$

$4x + y \leq 28$

Vertex	Value of $z = 3x + 4y$
(0, 0)	$z = 0$
(0, 10)	$x = 40$
(5, 8)	$z = 47$, maximum value
(7, 0)	$z = 21$

79. Minimize $z = 1.75x + 2.25y$ subject to the following constraints.

$2x + y \geq 25$

$3x + 2y \geq 45$

$x \geq 0$

$y \geq 0$

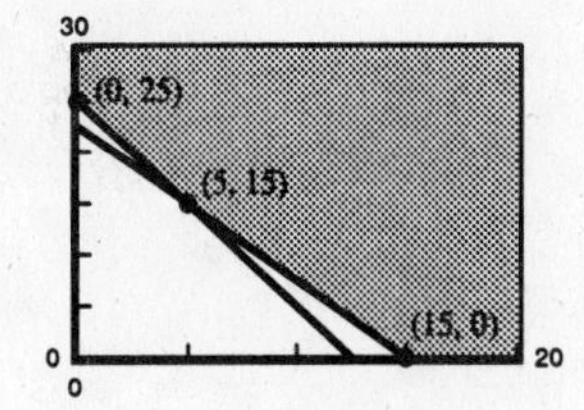

Vertex	Value of $z = 1.75x + 2.25y$
(0, 25)	$z = 56.25$
(5, 15)	$z = 42.5$
(15, 0)	$z = 26.25$, minimum value

81. Let x = number of haircuts.

Let y = number of perms.

Maximize $R = 17x + 60y$ subject to the following constraints.

$$x \geq 0$$

$$y \geq 0$$

$$\left(\frac{20}{60}\right)x + \left(\frac{70}{60}\right)y \leq 24 \implies 2x + 7y \leq 144$$

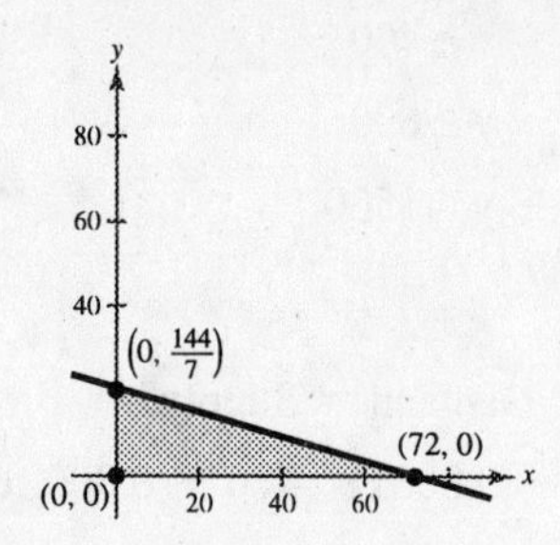

Vertex	Value of $R = 17x + 60y$
$(0, 0)$	$R = 0$
$(72, 0)$	$R = 1224$
$\left(0, \frac{144}{7}\right)$	$R \approx 1234.29$, maximum value

The revenue is maximum when $y = \frac{144}{7} \approx 20$ perms. (Round down since the student cannot work more than 24 hours. Note: Because we rounded down, the student would have enough time left to do 2 haircuts.)

Maximum Revenue = $17(2) + 60(20) = \$1234$

83. False. There is no solution to the system.

Chapter 7 Practice Test

For Exercises 1–3, solve the given system by the method of substitution.

1. $x + y = 1$
$3x - y = 15$

2. $x - 3y = -3$
$x^2 + 6y = 5$

3. $x + y + z = 6$
$2x - y + 3z = 0$
$5x + 2y - z = -3$

4. Find the two numbers whose sum is 110 and product is 2800.

5. Find the dimensions of a rectangle if its perimeter is 170 feet and its area is 2800 square feet.

For Exercises 6–7, solve the linear system by elimination.

6. $2x + 15y = 4$
$x - 3y = 23$

7. $x + y = 2$
$38x - 19y = 7$

8. Use a graphing utility to graph the two equations. Use the graph to approximate the solution of the system. Verify your answer analytically.

$0.4x + 0.5y = 0.112$
$0.3x - 0.7y = -0.131$

9. Herbert invests $17,000 in two funds that pay 11% and 13% simple interest, respectively. If he receives $2080 in yearly interest, how much is invested in each fund?

10. Find the least squares regression line for the points $(4, 3), (1, 1), (-1, -2),$ and $(-2, -1)$.

For Exercises 11–13, solve the system of equations.

11. $x + y = -2$
$2x - y + z = 11$
$4y - 3z = -20$

12. $4x - y + 5z = 4$
$2x + y - z = 0$
$2x + 4y + 8z = 0$

13. $3x + 2y - z = 5$
$6x - y + 5z = 2$

14. Find the equation of the parabola $y = ax^2 + bx + c$ passing through the points $(0, -1), (1, 4)$ and $(2, 13)$.

15. Find the position equation $s = \frac{1}{2}at^2 + v_0t + s_0$ given that $s = 12$ feet after 1 second, $s = 5$ feet after 2 seconds, and $s = 4$ after 3 seconds.

16. Graph $x^2 + y^2 \geq 9$.

17. Graph the solution of the system.

$x + y \leq 6$
$x \geq 2$
$y \geq 0$

18. Derive a set of inequalities to describe the triangle with vertices $(0, 0), (0, 7),$ and $(2, 3)$.

19. Find the maximum value of the objective function, $z = 30z + 26y$, subject to the following constraints.

$x \geq 0$
$y \geq 0$
$2x + 3y \leq 21$
$5x + 3y \leq 30$

20. Graph the system of inequalities.

$x^2 + y^2 \leq 4$
$(x - 2)^2 + y^2 \geq 4$

CHAPTER 8
Matrices and Determinants

CHAPTER 8
Matrices and Determinants

Section 8.1 Matrices and Systems of Equations

- You should be able to use elementary row operations to produce a row-echelon form (or reduced row-echelon form) of a matrix.
  1. Interchange two rows
  2. Multiply a row by a nonzero constant
  3. Add a multiple of one row to another row
- You should be able to use either Gaussian elimination with back-substitution or Gauss-Jordan elimination to solve a system of linear equations.

Solutions to Odd-Numbered Exercises

1. Since the matrix has three rows and two columns, its order is 3×2.

3. Since the matrix has three rows and one column, its order is 3×1.

5. Since the matrix has two rows and two columns, its order is 2×2.

7.
$$\begin{aligned} 4x - 5y &= 33 \\ -x + 5y &= -27 \end{aligned}$$

$$\left[\begin{array}{rr|r} 4 & -5 & 33 \\ -1 & 5 & -27 \end{array}\right]$$

9.
$$\begin{aligned} x + 10y - 2z &= 2 \\ 5x - 3y + 4z &= 0 \\ 2x + y \quad\quad &= 6 \end{aligned}$$

$$\left[\begin{array}{rrr|r} 1 & 10 & -2 & 2 \\ 5 & -3 & 4 & 0 \\ 2 & 1 & 0 & 6 \end{array}\right]$$

11.
$$\left[\begin{array}{rr|r} 1 & 2 & 7 \\ 2 & -3 & 4 \end{array}\right]$$

$$\begin{aligned} x + 2y &= 7 \\ 2x - 3y &= 4 \end{aligned}$$

13.
$$\left[\begin{array}{rrrr|r} 9 & 12 & 3 & 0 & 0 \\ -2 & 18 & 5 & 2 & 10 \\ 1 & 7 & -8 & 0 & -4 \end{array}\right]$$

$$\begin{aligned} 9x + 12y + 3z \quad\quad &= 0 \\ -2x + 18y + 5z + 2w &= 10 \\ x + 7y - 8z \quad\quad &= -4 \end{aligned}$$

15.
$$\begin{bmatrix} 1 & 0 & 0 & 0 \\ 0 & 1 & 1 & 5 \\ 0 & 0 & 0 & 0 \end{bmatrix}$$

This matrix is in reduced row-echelon form.

17.
$$\begin{bmatrix} 2 & 0 & 4 & 0 \\ 0 & -1 & 3 & 6 \\ 0 & 0 & 1 & 5 \end{bmatrix}$$

The first nonzero entries in rows one and two are not one. The matrix is not in row-echelon form.

19.
$$\begin{bmatrix} 1 & 4 & 3 \\ 2 & 10 & 5 \end{bmatrix}$$

$$-2R_1 + R_2 \rightarrow \begin{bmatrix} 1 & 4 & 3 \\ 0 & \boxed{2} & -1 \end{bmatrix}$$

21. $\begin{bmatrix} 1 & 1 & 4 & -1 \\ 3 & 8 & 10 & 3 \\ -2 & 1 & 12 & 6 \end{bmatrix}$ $\begin{matrix} \\ -3R_1 + R_2 \rightarrow \\ 2R_1 + R_3 \rightarrow \end{matrix}\begin{bmatrix} 1 & 1 & 4 & -1 \\ 0 & 5 & \boxed{-2} & \boxed{6} \\ 0 & 3 & \boxed{20} & \boxed{4} \end{bmatrix}$ $\begin{matrix} \\ \frac{1}{5}R_2 \rightarrow \\ \\ \end{matrix}\begin{bmatrix} 1 & 1 & 4 & -1 \\ 0 & 1 & -\frac{2}{5} & \frac{6}{5} \\ 0 & 3 & 20 & 4 \end{bmatrix}$

23. Add 5 times Row 2 to Row 1.

25. Interchange Rows 1 and 2.

27. $\begin{bmatrix} 1 & 2 & 3 \\ 2 & -1 & -4 \\ 3 & 1 & -1 \end{bmatrix}$

(a) $\begin{bmatrix} 1 & 2 & 3 \\ 0 & -5 & -10 \\ 3 & 1 & -1 \end{bmatrix}$

(b) $\begin{bmatrix} 1 & 2 & 3 \\ 0 & -5 & -10 \\ 0 & -5 & -10 \end{bmatrix}$

(c) $\begin{bmatrix} 1 & 2 & 3 \\ 0 & -5 & -10 \\ 0 & 0 & 0 \end{bmatrix}$

(d) $\begin{bmatrix} 1 & 2 & 3 \\ 0 & 1 & 2 \\ 0 & 0 & 0 \end{bmatrix}$

(e) $\begin{bmatrix} 1 & 0 & -1 \\ 0 & 1 & 2 \\ 0 & 0 & 0 \end{bmatrix}$ This matrix is in reduced row-echelon form.

29. (See Exercise 27.) (Answer is series of screens.)

(a)
```
*row+(-2,[A],1,2)
      [[1  2    3]
       [0 -5  -10]
       [3  1   -1]]
```

(b)
```
*row+(-3,[B],1,3)
      [[1  2    3]
       [0 -5  -10]
       [0 -5  -10]]
```

(c)
```
*row+(-1,[C],2,3)
      [[1  2    3]
       [0 -5  -10]
       [0  0    0]]
```

(d)
```
*row(-1/5,[D],2)
        [[1 2 3]
         [0 1 2]
         [0 0 0]]
```

(e)
```
*row+(-2,[E],2,1)
        [[1 0 -1]
         [0 1  2]
         [0 0  0]]
```

31. $\begin{bmatrix} 1 & 1 & 0 & 5 \\ -2 & -1 & 2 & -10 \\ 3 & 6 & 7 & 14 \end{bmatrix}$

$\begin{matrix} \\ 2R_1 + R_2 \rightarrow \\ -3R_1 + R_3 \rightarrow \end{matrix}\begin{bmatrix} 1 & 1 & 0 & 5 \\ 0 & 1 & 2 & 0 \\ 0 & 3 & 7 & -1 \end{bmatrix}$

$\begin{matrix} \\ \\ -3R_2 + R_3 \rightarrow \end{matrix}\begin{bmatrix} 1 & 1 & 0 & 5 \\ 0 & 1 & 2 & 0 \\ 0 & 0 & 1 & -1 \end{bmatrix}$

33. $\begin{bmatrix} 1 & -1 & -1 & 1 \\ 5 & -4 & 1 & 8 \\ -6 & 8 & 18 & 0 \end{bmatrix}$

$\begin{matrix} \\ -5R_1 + R_2 \rightarrow \\ 6R_1 + R_3 \rightarrow \end{matrix}\begin{bmatrix} 1 & -1 & -1 & 1 \\ 0 & 1 & 6 & 3 \\ 0 & 2 & 12 & 6 \end{bmatrix}$

$\begin{matrix} \\ \\ -2R_2 + R_3 \rightarrow \end{matrix}\begin{bmatrix} 1 & -1 & -1 & 1 \\ 0 & 1 & 6 & 3 \\ 0 & 0 & 0 & 0 \end{bmatrix}$

35. $\begin{bmatrix} 3 & 3 & 3 \\ -1 & 0 & -4 \\ 2 & 4 & -2 \end{bmatrix}$

$\frac{1}{3}R_1 \rightarrow \begin{bmatrix} 1 & 1 & 1 \\ -1 & 0 & -4 \\ 2 & 4 & -2 \end{bmatrix}$

$\begin{matrix} \\ R_2 + R_1 \rightarrow \\ -2R_1 + R_3 \rightarrow \end{matrix} \begin{bmatrix} 1 & 1 & 1 \\ 0 & 1 & -3 \\ 0 & 2 & -4 \end{bmatrix}$

$\begin{matrix} -R_2 + R_1 \rightarrow \\ \\ -2R_2 + R_3 \rightarrow \end{matrix} \begin{bmatrix} 1 & 0 & 4 \\ 0 & 1 & -3 \\ 0 & 0 & 2 \end{bmatrix}$

$\begin{matrix} \\ \\ \frac{1}{2}R_3 \rightarrow \end{matrix} \begin{bmatrix} 1 & 0 & 4 \\ 0 & 1 & -3 \\ 0 & 0 & 1 \end{bmatrix}$

$\begin{matrix} -4R_3 + R_1 \rightarrow \\ 3R_3 + R_2 \rightarrow \\ \\ \end{matrix} \begin{bmatrix} 1 & 0 & 0 \\ 0 & 1 & 0 \\ 0 & 0 & 1 \end{bmatrix}$

37. $\begin{bmatrix} -3 & 5 & 1 & 12 \\ 1 & -1 & 1 & 4 \end{bmatrix}$

$\begin{matrix} R_1 \rightarrow \\ R_2 \rightarrow \end{matrix} \begin{bmatrix} 1 & -1 & 1 & 4 \\ -3 & 5 & 1 & 12 \end{bmatrix}$

$\begin{matrix} \\ 3R_1 + R_2 \rightarrow \end{matrix} \begin{bmatrix} 1 & -1 & 1 & 4 \\ 0 & 2 & 4 & 24 \end{bmatrix}$

$\begin{matrix} \\ \frac{1}{2}R_2 \end{matrix} \begin{bmatrix} 1 & -1 & 1 & 4 \\ 0 & 1 & 2 & 12 \end{bmatrix}$

$\begin{matrix} R_2 + R_1 \rightarrow \\ \\ \end{matrix} \begin{bmatrix} 1 & 0 & 3 & 16 \\ 0 & 1 & 2 & 12 \end{bmatrix}$

39. $x - 2y = 4$

$y = -3$

$x - 2(-3) = 4$

$x = -2$

Answer: $(-2, -3)$

41. $x - y + 2z = 4$

$y - z = 2$

$z = -2$

$y - (-2) = 2$

$y = 0$

$x - 0 + 2(-2) = 4$

$x = 8$

Answer: $(8, 0, -2)$

43. $\left[\begin{array}{cc|c} 1 & 0 & 7 \\ 0 & 1 & -5 \end{array}\right]$

$x = 7$

$y = -5$

Answer: $(7, -5)$

45. $\left[\begin{array}{ccc|c} 1 & 0 & 0 & -4 \\ 0 & 1 & 0 & -8 \\ 0 & 0 & 1 & 2 \end{array}\right]$

$x = -4$

$y = -8$

$z = 2$

Answer: $(-4, -8, 2)$

47. $x + 2y = 7$

$2x + y = 8$

$\left[\begin{array}{cc|c} 1 & 2 & 7 \\ 2 & 1 & 8 \end{array}\right] \begin{matrix} \\ -2R_1 + R_2 \rightarrow \end{matrix} \left[\begin{array}{cc|c} 1 & 2 & 7 \\ 0 & -3 & -6 \end{array}\right] \begin{matrix} \\ -\frac{1}{3}R_2 \rightarrow \end{matrix} \left[\begin{array}{cc|c} 1 & 2 & 7 \\ 0 & 1 & 2 \end{array}\right]$

$y = 2$

$x + 2(2) = 7 \implies x = 3$

Answer: $(3, 2)$

49. $-3x + 5y = -28$

$3x + 4y = 10$

$4x - 8y = 40$

$$\begin{bmatrix} -3 & 5 & \vdots & -28 \\ 3 & 4 & \vdots & 10 \\ 4 & -8 & \vdots & 40 \end{bmatrix}$$

$$\begin{matrix} R_3 + R_1 \rightarrow \\ \\ \\ \end{matrix} \begin{bmatrix} 1 & -3 & \vdots & 12 \\ 3 & 4 & \vdots & 10 \\ 4 & -8 & \vdots & 40 \end{bmatrix}$$

$$\begin{matrix} \\ -3R_1 + R_2 \rightarrow \\ -4R_1 + R_3 \rightarrow \end{matrix} \begin{bmatrix} 1 & -3 & \vdots & 12 \\ 0 & 13 & \vdots & -26 \\ 0 & 4 & \vdots & -8 \end{bmatrix}$$

$$\begin{matrix} \\ \frac{1}{13}R_2 \rightarrow \\ -4R_2 + R_3 \rightarrow \end{matrix} \begin{bmatrix} 1 & -3 & \vdots & 12 \\ 0 & 1 & \vdots & -2 \\ 0 & 0 & \vdots & 0 \end{bmatrix}$$

$y = -2$

$x = 3(-2) + 12 = 6$

Answer: $(6, -2)$

51. $8x - 4y = 13$

$5x + 2y = 7$

$$\begin{bmatrix} 8 & -4 & 13 \\ 5 & 2 & 7 \end{bmatrix} \qquad \begin{matrix} 3R_1 \rightarrow \\ 5R_2 \rightarrow \end{matrix} \begin{bmatrix} 24 & -12 & 39 \\ 25 & 10 & 35 \end{bmatrix}$$

$$\begin{matrix} -R_2 + R_1 \rightarrow \\ \frac{1}{5}R_2 \rightarrow \end{matrix} \begin{bmatrix} -1 & -22 & 4 \\ 5 & 2 & 7 \end{bmatrix}$$

$$\begin{matrix} \\ 5R_1 + R_2 \rightarrow \end{matrix} \begin{bmatrix} -1 & -22 & 4 \\ 0 & -108 & 27 \end{bmatrix}$$

$$\begin{matrix} -1R_1 \rightarrow \\ -\frac{1}{108}R_2 \rightarrow \end{matrix} \begin{bmatrix} 1 & 22 & -4 \\ 0 & 1 & -\frac{1}{4} \end{bmatrix}$$

$y = -\frac{1}{4}$

$x + 22\left(-\frac{1}{4}\right) = -4 \quad \Rightarrow \quad x = \frac{3}{2}$

Answer: $\left(\frac{3}{2}, -\frac{1}{4}\right)$

53. $-x + 2y = 1.5$

$2x - 4y = 3.0$

$$\begin{bmatrix} -1 & 2 & \vdots & 1.5 \\ 2 & -4 & \vdots & 3.0 \end{bmatrix}$$

$$\begin{matrix} \\ 2R_1 + R_2 \rightarrow \end{matrix} \begin{bmatrix} -1 & 2 & \vdots & 1.5 \\ 0 & 0 & \vdots & 6.0 \end{bmatrix}$$

The system is inconsistent and there is no solution.

55. $x \quad - 3z = -2$

$3x + y - 2z = 5$

$2x + 2y + z = 4$

$$\begin{bmatrix} 1 & 0 & -3 & \vdots & -2 \\ 3 & 1 & -2 & \vdots & 5 \\ 2 & 2 & 1 & \vdots & 4 \end{bmatrix}$$

$$\begin{matrix} \\ -3R_1 + R_2 \rightarrow \\ -2R_1 + R_3 \rightarrow \end{matrix} \begin{bmatrix} 1 & 0 & -3 & \vdots & -2 \\ 0 & 1 & 7 & \vdots & 11 \\ 0 & 2 & 7 & \vdots & 8 \end{bmatrix}$$

$$\begin{matrix} \\ \\ -2R_2 + R_3 \rightarrow \end{matrix} \begin{bmatrix} 1 & 0 & -3 & \vdots & -2 \\ 0 & 1 & 7 & \vdots & 11 \\ 0 & 0 & -7 & \vdots & -14 \end{bmatrix}$$

$$\begin{matrix} \\ \\ -\frac{1}{7}R_3 \rightarrow \end{matrix} \begin{bmatrix} 1 & 0 & -3 & \vdots & -2 \\ 0 & 1 & 7 & \vdots & 11 \\ 0 & 0 & 1 & \vdots & 2 \end{bmatrix}$$

$z = 2$

$y + 7(2) = 11 \Rightarrow y = -3$

$x - 3(2) = -2 \Rightarrow x = 4$

Answer: $(4, -3, 2)$

57. $x + y - 5z = 3$

$x \quad - 2z = 1$

$2x - y - z = 0$

$$\begin{bmatrix} 1 & 1 & -5 & \vdots & 3 \\ 1 & 0 & -2 & \vdots & 1 \\ 2 & -1 & -1 & \vdots & 0 \end{bmatrix}$$

$$\begin{matrix} \\ -R_1 + R_2 \rightarrow \\ -2R_1 + R_3 \rightarrow \end{matrix} \begin{bmatrix} 1 & 1 & -5 & \vdots & 3 \\ 0 & -1 & 3 & \vdots & -2 \\ 0 & -3 & 9 & \vdots & -6 \end{bmatrix}$$

$$\begin{matrix} \\ \\ -3R_2 + R_3 \rightarrow \end{matrix} \begin{bmatrix} 1 & 1 & -5 & \vdots & 3 \\ 0 & -1 & 3 & \vdots & -2 \\ 0 & 0 & 0 & \vdots & 0 \end{bmatrix}$$

$$\begin{matrix} R_2 + R_1 \rightarrow \\ -R_2 \rightarrow \\ \\ \end{matrix} \begin{bmatrix} 1 & 0 & -2 & \vdots & 1 \\ 0 & 1 & -3 & \vdots & 2 \\ 0 & 0 & 0 & \vdots & 0 \end{bmatrix}$$

Let $z = a$, any real number

$y - 3a = 2 \Rightarrow y = 3a + 2$

$x - 2a = 1 \Rightarrow x = 2a + 1$

Answer: $(2a + 1, 3a + 2, a)$

59. $x + 2y = 0$

$-x - y = 0$

$$\begin{bmatrix} 1 & 2 & \vdots & 0 \\ -1 & -1 & \vdots & 0 \end{bmatrix}$$

$$\begin{matrix} \\ R_1 + R_2 \rightarrow \end{matrix} \begin{bmatrix} 1 & 2 & \vdots & 0 \\ 0 & 1 & \vdots & 0 \end{bmatrix}$$

$y = 0$

$x + 2(0) = 0 \Rightarrow$

$x = 0$

Answer: $(0, 0)$

61. $3x + 3y + 12z = 6$

$x + y + 4z = 2$

$2x + 5y + 20z = 10$

$-x + 2y + 8z = 4$

$$\begin{bmatrix} 3 & 3 & 12 & \vdots & 6 \\ 1 & 1 & 4 & \vdots & 2 \\ 2 & 5 & 20 & \vdots & 10 \\ -1 & 2 & 8 & \vdots & 4 \end{bmatrix} \Rightarrow \begin{bmatrix} 1 & 0 & 0 & \vdots & 0 \\ 0 & 0 & 0 & \vdots & 0 \\ 0 & 1 & 4 & \vdots & 2 \\ 0 & 0 & 0 & \vdots & 0 \end{bmatrix}$$

Let $z = a$, any real number

$y = -4a + 2$

$x = 0$

Answer: $(0, -4a + 2, a)$

63. $$\begin{bmatrix} 2 & 1 & -1 & 2 & \vdots & -6 \\ 3 & 4 & 0 & 1 & \vdots & 1 \\ 1 & 5 & 2 & 6 & \vdots & -3 \\ 5 & 2 & -1 & -1 & \vdots & 3 \end{bmatrix} \text{ row reduces to } \begin{bmatrix} 1 & 0 & 0 & 0 & \vdots & 1 \\ 0 & 1 & 0 & 0 & \vdots & 0 \\ 0 & 0 & 1 & 0 & \vdots & 4 \\ 0 & 0 & 0 & 1 & \vdots & -2 \end{bmatrix}$$

Answer: $(1, 0, 4, -2)$

65. $x + y + z = 0$

$2x + 3y + z = 0$

$3x + 5y + z = 0$

$$\begin{bmatrix} 1 & 1 & 1 & \vdots & 0 \\ 2 & 3 & 1 & \vdots & 0 \\ 3 & 5 & 1 & \vdots & 0 \end{bmatrix} \Rightarrow \begin{bmatrix} 1 & 0 & 2 & \vdots & 0 \\ 0 & 1 & -1 & \vdots & 0 \\ 0 & 0 & 0 & \vdots & 0 \end{bmatrix}$$

Let $z = a$, any real number

$y = a$

$x = -2a$

Answer: $(-2a, a, a)$

67. Yes, the systems yield the same solutions.

(a) $z = -3$; $y = 5(-3) + 16 = 1$;
$x = 2(1) - (-3) - 6 = -1$

Answer: $(-1, 1, -3)$

(b) $z = -3$, $y = -3(-3) - 8 = 1$,
$x = -1 + 2(-3) + 6 = -1$

Answer: $(-1, 1, -3)$

69. No, solutions are different.

(a) $z = 8$, $y = 7(8) - 54 = 2$,
$x = 4(2) - 5(8) + 27 = -5$

Answer: $(-5, 2, 8)$

(b) $z = 8$, $y = -5(8) + 42 = 2$,
$x = 6(2) - 8 + 15 = 19$

Answer: $(19, 2, 8)$

71. $f(x) = ax^2 + bx + c$

$f(1) = a + b + c = 8$

$f(2) = 4a + 2b + c = 13$

$f(3) = 9a + 3b + c = 20$

$$\begin{bmatrix} 1 & 1 & 1 & \vdots & 8 \\ 4 & 2 & 1 & \vdots & 13 \\ 9 & 3 & 1 & \vdots & 20 \end{bmatrix}$$

$$\begin{array}{r} \\ -4R_1 + R_2 \rightarrow \\ -9R_1 + R_3 \rightarrow \end{array} \begin{bmatrix} 1 & 1 & 1 & \vdots & 8 \\ 0 & -2 & -3 & \vdots & -19 \\ 0 & -6 & -8 & \vdots & -52 \end{bmatrix}$$

$$\begin{array}{r} \\ -\frac{1}{2}R_2 \rightarrow \\ -3R_2 + R_3 \rightarrow \end{array} \begin{bmatrix} 1 & 1 & 1 & \vdots & 8 \\ 0 & 1 & \frac{3}{2} & \vdots & \frac{19}{2} \\ 0 & 0 & 1 & \vdots & 5 \end{bmatrix}$$

$c = 5$

$b + \frac{3}{2}(5) = \frac{19}{2} \Rightarrow b = 2(a) 670 = c$

$a + 2 + 5 = 8 \Rightarrow a = 1$

Answer: $y = x^2 + 2x + 5$

73. $f(x) = ax^3 + bx^2 + cx + d$

$f(-2) = -8a + 4b - 2c + d = 2$

$f(-1) = -a + b - c + d = -\frac{1}{4}$

$f(1) = a + b + c + d = -\frac{7}{4}$

$f(2) = 8a + 4b + 2c + d = 2$

Solving this system, you obtain

$a = \frac{1}{4}$, $b = 1$, $c = -1$ and $d = -2$.

Thus, $y = \frac{1}{4}x^3 + x^2 - x - 2$.

75. $f(x) = ax^4 + bx^3 + cx^2 + dx + e$

$f(-2) = 16a - 8b + 4c - 2d + e = 0$

$f(-1) = a - b + c - d + e = 3$

$f(0) = e = 0$

$f(1) = a + b + c + d + e = 3$

$f(2) = 16a + 8b + 4c + 2d + e = 0$

Solving this system, you obtain

$a = -1$, $b = 0$, $c = 4$, $d = e = 0$.

Thus, $y = -x^4 + 4x^2$.

77. $x =$ amount at 8%

$y =$ amount at 9%

$z =$ amount at 12%

$$\begin{aligned} x + y + z &= 1{,}500{,}000 \\ 0.08x + 0.09y + 0.12z &= 133{,}000 \\ x - 4z &= 0 \end{aligned}$$

$$\begin{bmatrix} 1 & 1 & 1 & \vdots & 1{,}500{,}000 \\ 0.08 & 0.09 & 0.12 & \vdots & 133{,}000 \\ 1 & 0 & -4 & \vdots & 0 \end{bmatrix}$$

$$\begin{matrix} \\ -0.08R_1 + R_2 \rightarrow \\ -R_1 + R_3 \rightarrow \end{matrix} \begin{bmatrix} 1 & 1 & 1 & \vdots & 1{,}500{,}000 \\ 0 & 0.01 & 0.04 & \vdots & 13{,}000 \\ 0 & -1 & -5 & \vdots & -1{,}500{,}000 \end{bmatrix}$$

$$\begin{matrix} \\ 100R_2 \\ R_2 + R_3 \end{matrix} \begin{bmatrix} 1 & 1 & 1 & \vdots & 1{,}500{,}000 \\ 0 & 1 & 4 & \vdots & 1{,}300{,}000 \\ 0 & 0 & -1 & \vdots & -200{,}000 \end{bmatrix}$$

$$-z = -200{,}000 \implies z = 200{,}000$$

$$y + 4(200{,}000) = 1{,}300{,}000 \implies y = 500{,}000$$

$$x + (500{,}000) + (200{,}000) = 1{,}500{,}000 \implies x = 800{,}000$$

Answer: \$800,000 at 8%, \$500,000 at 9%, \$200,000 at 12%

79.

$$\begin{aligned} I_1 - I_2 + I_3 &= 0 \\ 2I_1 + 2I_2 &= 7 \\ 2I_2 + 4I_3 &= 8 \end{aligned}$$

$$\begin{bmatrix} 1 & -1 & 1 & \vdots & 0 \\ 2 & 2 & 0 & \vdots & 7 \\ 0 & 2 & 4 & \vdots & 8 \end{bmatrix}$$

$$\begin{matrix} \\ -2R_1 + R_2 \rightarrow \\ \\ \end{matrix} \begin{bmatrix} 1 & -1 & 1 & \vdots & 0 \\ 0 & 4 & -2 & \vdots & 7 \\ 0 & 2 & 4 & \vdots & 8 \end{bmatrix}$$

$$\begin{matrix} \\ R_3 \rightarrow \\ R_2 \rightarrow \end{matrix} \begin{bmatrix} 1 & -1 & 1 & \vdots & 0 \\ 0 & 2 & 4 & \vdots & 8 \\ 0 & 4 & -2 & \vdots & 7 \end{bmatrix}$$

$$\begin{matrix} \\ \frac{1}{2}R_2 \rightarrow \\ \\ \end{matrix} \begin{bmatrix} 1 & -1 & 1 & \vdots & 0 \\ 0 & 1 & 2 & \vdots & 4 \\ 0 & 4 & -2 & \vdots & 7 \end{bmatrix}$$

$$\begin{matrix} \\ \\ -4R_2 + R_3 \rightarrow \end{matrix} \begin{bmatrix} 1 & -1 & 1 & \vdots & 0 \\ 0 & 1 & 2 & \vdots & 4 \\ 0 & 0 & -10 & \vdots & -9 \end{bmatrix}$$

$$\begin{matrix} \\ \\ -\frac{1}{10}R_3 \rightarrow \end{matrix} \begin{bmatrix} 1 & -1 & 1 & \vdots & 0 \\ 0 & 1 & 2 & \vdots & 4 \\ 0 & 0 & 1 & \vdots & \frac{9}{10} \end{bmatrix}$$

$I_3 = \frac{9}{10}$ amperes

$I_2 + 2\left(\frac{9}{10}\right) = 4 \implies I_2 = \frac{11}{5}$ amperes

$I_1 - \frac{11}{5} + \frac{9}{10} = 0 \implies I_1 = \frac{13}{10}$ amperes

81. (a) The points are $(5, 421)$, $(6, 595)$, $(7, 512)$.

If $y = at^2 + bt + c$, then

$$\begin{aligned} 25a + 5b + c &= 421 \\ 36a + 6b + c &= 595 \\ 49a + 7b + c &= 512 \end{aligned}$$

Solving this system, you obtain

$a = -128.5$

$b = 1587.5$

$c = -4304.0$

Thus, $y = -128.5t^2 + 1587.5t - 4304$

(b)

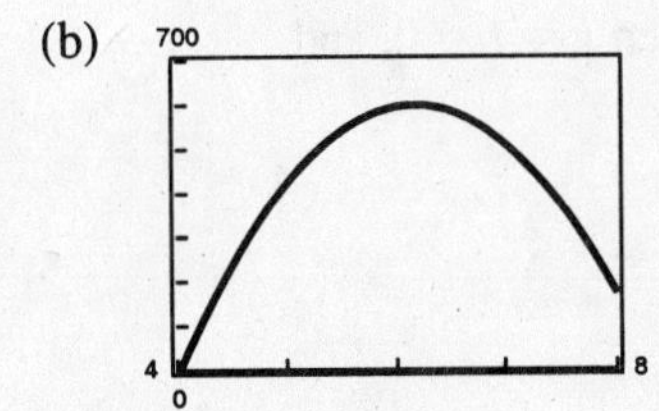

(c) For 1998, $t = 8$ and $y \approx 172$ million dollars.

83. (a) $x_1 + x_3 = 600$

$x_1 = x_2 + x_4 \Rightarrow x_1 - x_2 - x_4 = 0$

$x_2 + x_5 = 500$

$x_3 + x_6 = 600$

$x_4 + x_7 = x_6 \Rightarrow x_4 - x_6 + x_7 = 0$

$x_5 + x_7 = 500$

$$\begin{bmatrix} 1 & 0 & 1 & 0 & 0 & 0 & 0 & \vdots & 600 \\ 1 & -1 & 0 & -1 & 0 & 0 & 0 & \vdots & 0 \\ 0 & 1 & 0 & 0 & 1 & 0 & 0 & \vdots & 500 \\ 0 & 0 & 1 & 0 & 0 & 1 & 0 & \vdots & 600 \\ 0 & 0 & 0 & 1 & 0 & -1 & 1 & \vdots & 0 \\ 0 & 0 & 0 & 0 & 1 & 0 & 1 & \vdots & 500 \end{bmatrix}$$

$$\begin{matrix} \\ -R_1 + R_2 \rightarrow \\ R_2 + R_3 \rightarrow \\ R_3 + R_4 \rightarrow \\ R_4 + R_5 \rightarrow \\ -R_5 + R_6 \rightarrow \end{matrix} \begin{bmatrix} 1 & 0 & 1 & 0 & 0 & 0 & 0 & \vdots & 600 \\ 0 & -1 & -1 & -1 & 0 & 0 & 0 & \vdots & -600 \\ 0 & 0 & -1 & -1 & 1 & 0 & 0 & \vdots & -100 \\ 0 & 0 & 0 & -1 & 1 & 1 & 0 & \vdots & 500 \\ 0 & 0 & 0 & 0 & 1 & 0 & 1 & \vdots & 500 \\ 0 & 0 & 0 & 0 & 0 & 0 & 0 & \vdots & 0 \end{bmatrix}$$

$$\begin{matrix} \\ -R_3 + R_2 \rightarrow \\ -R_4 + R_3 \rightarrow \\ -R_4 \rightarrow \\ \\ \\ \end{matrix} \begin{bmatrix} 1 & 0 & 1 & 0 & 0 & 0 & 0 & \vdots & 600 \\ 0 & -1 & 0 & 0 & -1 & 0 & 0 & \vdots & -500 \\ 0 & 0 & -1 & 0 & 0 & -1 & 0 & \vdots & -600 \\ 0 & 0 & 0 & 1 & -1 & -1 & 0 & \vdots & -500 \\ 0 & 0 & 0 & 0 & 1 & 0 & 1 & \vdots & 500 \\ 0 & 0 & 0 & 0 & 0 & 0 & 0 & \vdots & 0 \end{bmatrix}$$

Let $x_7 = t$ and $x_6 = s$, then:

$x_5 = 500 - t$

$x_4 = -500 + s + (500 - t) = s - t$

$x_3 = 600 - s$

$x_2 = 500 - (500 - t) = t$

$x_1 = 600 - (600 - s) = s$

(b) If $x_6 = x_7 = 0$, then $s = t = 0$, and

$x_1 = 0$

$x_2 = 0$

$x_3 = 600$

$x_4 = 0$

$x_5 = 500$

$x_6 = x_7 = 0$

(c) If $x_5 = 1000$ and $x_6 = 0$, then $s = 0$ and $t = -500$.

Thus, $x_1 = 0$

$x_2 = -500$

$x_3 = 600$

$x_4 = 500$

$x_5 = 1000$

$x_6 = 0$

$x_7 = -500$

85. (a) $200 + x_2 = x_1 \Rightarrow -x_1 - x_2 = 200$

$x_2 + 100 = x_4 \Rightarrow -x_2 + x_4 = 100$

$x_4 + 200 = x_3 \Rightarrow x_3 - x_4 = 200$

$x_1 + 100 = x_3 \Rightarrow x_1 - x_3 = -100$

Solving this system, you obtain:

$x_1 = 100 + t$

$x_2 = -100 + t$

$x_3 = 200 + t$

$x_4 = t$

(b) If $x_4 = 0$, then $t = 0$ and

$x_1 = 100, x_2 = -100, x_3 = 200.$

(c) If $x_4 = 100$, then $t = 100$ and $x_1 = 200, x_2 = 0,$

$x_3 = 300.$

87. False. The row of zeros is not at the bottom.

89. $x + 3z = -2$ Equation 1

$y + 4z = 1$ Equation 2

(Equation 1) + (Equation 2) $\rightarrow$ new Equation 1

(Equation 1) + 2(Equation 2) $\rightarrow$ new Equation 2

2(Equation 1) + (Equation 2) $\rightarrow$ new Equation 3

$x + y + 7z = -1$

$x + 2y + 11z = 0$

$2x + y + 10z = -3$

91. (a) In the row-echelon form of an augmented matrix that corresponds to an inconsistent system of linear equations, there exists a row consisting of all zeros except for the entry in the last column.

(b) In the row-echelon form of an augmented matrix that corresponds to a system with an infinite number of solutions, there are fewer rows with nonzero entries than there are variables. Nor does the last row consist of all zeros except for the entry in the last column.

93. $f(x) = \dfrac{4x}{5x^2 + 2}$ Horizontal asymptote: $y = 0$

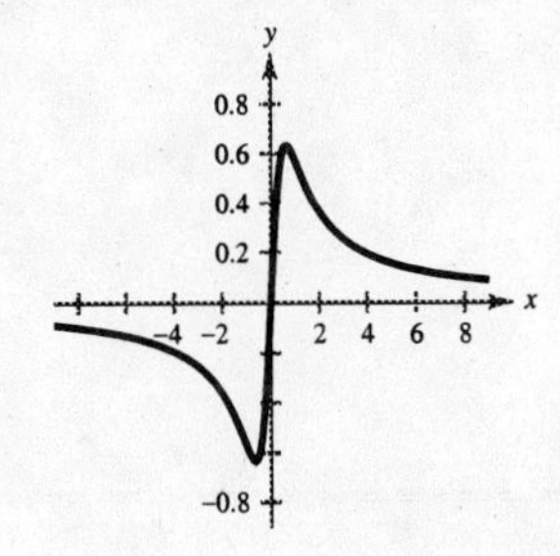

95. $f(x) = \dfrac{x^2 - 36}{x + 1}$ Vertical asymptote: $x = -1$, Slant asymptote: $y = x - 1$

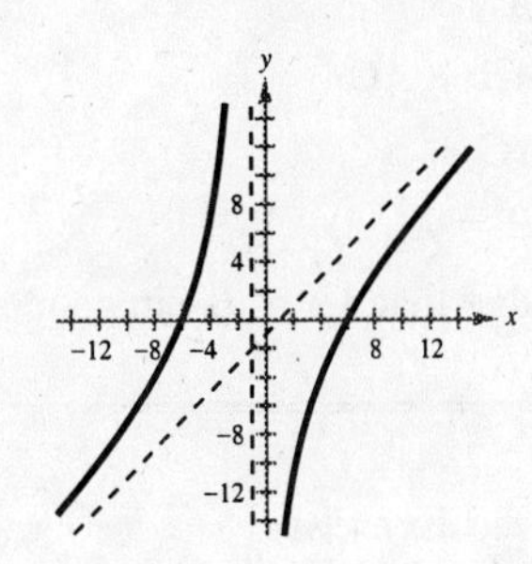

97. $g(x) = 3^{-x+2}$

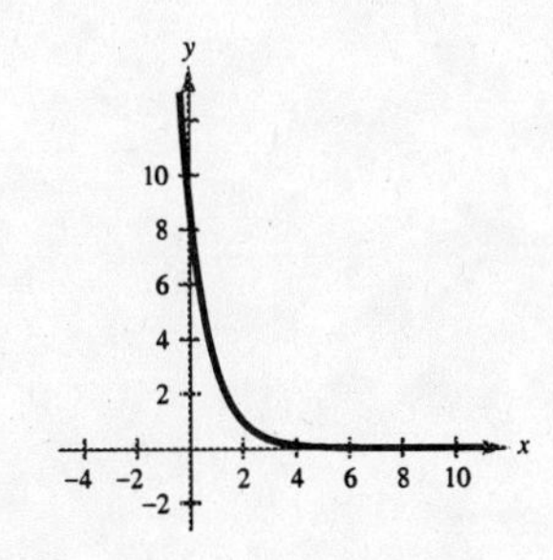

99. $f(x) = 3 + \ln x$

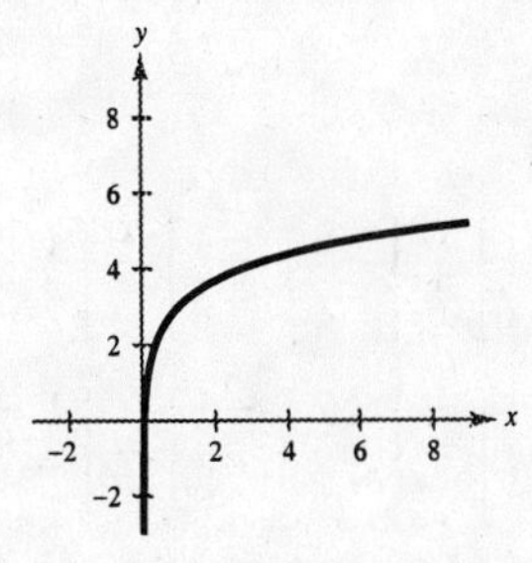

101. $3x - 8y = 47 \Rightarrow 15x - 40y = 235$

$-5x + 9y = -48 \Rightarrow -15x + 27y = -144$

Adding, $-13y = 91 \Rightarrow y = -7$. Then $3x = 8(-7) + 47 = -9 \Rightarrow x = -3$

Answer: $(-3, -7)$

103. $6x - y + 4z = -11$

$5x + 4y - 2z = 0$

$-8x - y + 5z = 18$

Using a graphing utility, you obtain $x = -2$, $y = 3$, $z = 1$.

Section 8.2 Operations with Matrices

- $A = B$ if and only if they have the same order and $a_{ij} = b_{ij}$.
- You should be able to perform the operations of matrix addition, scalar multiplication, and matrix multiplication.
- Some properties of matrix addition and scalar multiplication are:
  (a) $A + B = B + A$
  (b) $A + (B + C) = (A + B) + C$
  (c) $(cd)A = c(dA)$
  (d) $1A = A$
  (e) $c(A + B) = cA + cB$
  (f) $(c + d)A = cA + dA$
- Some properties of matrix multiplication are:
  (a) $A(BC) = (AB)C$
  (b) $A(B + C) = AB + AC$
  (c) $(A + B)C = AC + BC$
  (d) $c(AB) = (cA)B = A(cB)$
- You should remember that $AB \neq BA$ in general.

Solutions to Odd-Numbered Exercises

1. $x = -4,\ y = 22$

3. $2x + 7 = 5 \implies x = -1$

$3y = 12 \implies y = 4$

$3z - 14 = 4 \implies z = 6$

5. (a) $A + B = \begin{bmatrix} 1 & -1 \\ 2 & -1 \end{bmatrix} + \begin{bmatrix} 2 & -1 \\ -1 & 8 \end{bmatrix} = \begin{bmatrix} 1+2 & -1-1 \\ 2-1 & -1+8 \end{bmatrix} = \begin{bmatrix} 3 & -2 \\ 1 & 7 \end{bmatrix}$

(b) $A - B = \begin{bmatrix} 1 & -1 \\ 2 & -1 \end{bmatrix} - \begin{bmatrix} 2 & -1 \\ -1 & 8 \end{bmatrix} = \begin{bmatrix} 1-2 & -1+1 \\ 2+1 & -1-8 \end{bmatrix} = \begin{bmatrix} -1 & 0 \\ 3 & -9 \end{bmatrix}$

(c) $3A = 3\begin{bmatrix} 1 & -1 \\ 2 & -1 \end{bmatrix} = \begin{bmatrix} 3(1) & 3(-1) \\ 3(2) & 3(-1) \end{bmatrix} = \begin{bmatrix} 3 & -3 \\ 6 & -3 \end{bmatrix}$

(d) $3A - 2B = \begin{bmatrix} 3 & -3 \\ 6 & -3 \end{bmatrix} - 2\begin{bmatrix} 2 & -1 \\ -1 & 8 \end{bmatrix} = \begin{bmatrix} 3 & -3 \\ 6 & -3 \end{bmatrix} + \begin{bmatrix} -4 & 2 \\ 2 & -16 \end{bmatrix} = \begin{bmatrix} -1 & -1 \\ 8 & -19 \end{bmatrix}$

7. $A = \begin{bmatrix} 8 & -1 \\ 2 & 3 \\ -4 & 5 \end{bmatrix},\ B = \begin{bmatrix} 1 & 6 \\ -1 & -5 \\ 1 & 10 \end{bmatrix}$

(a) $A + B = \begin{bmatrix} 9 & 5 \\ 1 & -2 \\ -3 & 15 \end{bmatrix}$ (b) $A - B = \begin{bmatrix} 7 & -7 \\ 3 & 8 \\ -5 & -5 \end{bmatrix}$ (c) $3A = \begin{bmatrix} 24 & -3 \\ 6 & 9 \\ -12 & 15 \end{bmatrix}$

(d) $3A - 2B = \begin{bmatrix} 24 & -3 \\ 6 & 9 \\ -12 & 15 \end{bmatrix} - \begin{bmatrix} 2 & 12 \\ -2 & -10 \\ 2 & 20 \end{bmatrix} = \begin{bmatrix} 22 & -15 \\ 8 & 19 \\ -14 & -5 \end{bmatrix}$

9. $A = \begin{bmatrix} 2 & 2 & -1 & 0 & 1 \\ 1 & 1 & -2 & 0 & -1 \end{bmatrix}$, $B = \begin{bmatrix} 1 & 1 & -1 & 1 & 0 \\ -3 & 4 & 9 & -6 & -7 \end{bmatrix}$

(a) $A + B = \begin{bmatrix} 3 & 3 & -2 & 1 & 1 \\ -2 & 5 & 7 & -6 & -8 \end{bmatrix}$

(b) $A - B = \begin{bmatrix} 1 & 1 & 0 & -1 & 1 \\ 4 & -3 & -11 & 6 & 6 \end{bmatrix}$

(c) $3A = \begin{bmatrix} 6 & 6 & -3 & 0 & 3 \\ 3 & 3 & -6 & 0 & -3 \end{bmatrix}$

(d) $3A - 2B = \begin{bmatrix} 6 & 6 & -3 & 0 & 3 \\ 3 & 3 & -6 & 0 & -3 \end{bmatrix} - \begin{bmatrix} 2 & 2 & -2 & 2 & 0 \\ -6 & 8 & 18 & -12 & -14 \end{bmatrix} = \begin{bmatrix} 4 & 4 & -1 & -2 & 3 \\ 9 & -5 & -24 & 12 & 11 \end{bmatrix}$

11. $A = \begin{bmatrix} 6 & 0 & 3 \\ -1 & -4 & 0 \end{bmatrix}$, $B = \begin{bmatrix} 8 & -1 \\ 4 & -3 \end{bmatrix}$

(a) $A + B$ not possible (b) $A - B$ not possible (c) $3A = \begin{bmatrix} 18 & 0 & 9 \\ -3 & -12 & 0 \end{bmatrix}$ (d) $3A - 2B$ not possible

13. $\begin{bmatrix} -4 & 0 \\ 5 & -6 \end{bmatrix} + \begin{bmatrix} 9 & 1 \\ -2 & -2 \end{bmatrix} + \begin{bmatrix} -12 & -10 \\ 14 & 6 \end{bmatrix} = \begin{bmatrix} 5 & 1 \\ 3 & -8 \end{bmatrix} + \begin{bmatrix} -12 & -10 \\ 14 & 6 \end{bmatrix} = \begin{bmatrix} -7 & -9 \\ 17 & -2 \end{bmatrix}$

15. $4\left(\begin{bmatrix} -4 & 0 & 1 \\ 0 & 2 & 3 \end{bmatrix} - \begin{bmatrix} 2 & 1 & -2 \\ 3 & -6 & 0 \end{bmatrix}\right) = 4\begin{bmatrix} -6 & -1 & 3 \\ -3 & 8 & 3 \end{bmatrix} = \begin{bmatrix} -24 & -4 & 12 \\ -12 & 32 & 12 \end{bmatrix}$

17. $-3\left(\begin{bmatrix} 0 & -3 \\ 7 & 2 \end{bmatrix} + \begin{bmatrix} -6 & 3 \\ 8 & 1 \end{bmatrix}\right) - \begin{bmatrix} 4 & -4 \\ 7 & -9 \end{bmatrix} = -3\begin{bmatrix} -6 & 0 \\ 15 & 3 \end{bmatrix} - \begin{bmatrix} 4 & -4 \\ 7 & -9 \end{bmatrix} = \begin{bmatrix} 18 & 0 \\ -45 & -9 \end{bmatrix} - \begin{bmatrix} 4 & -4 \\ 7 & -9 \end{bmatrix}$

$= \begin{bmatrix} 14 & 4 \\ -52 & 0 \end{bmatrix}$

19. $\frac{3}{7}\begin{bmatrix} 2 & 5 \\ -1 & -4 \end{bmatrix} = \begin{bmatrix} 0.857 & 2.143 \\ -0.429 & -1.714 \end{bmatrix}$

21. $\begin{bmatrix} -1.581 & -3.739 \\ -4.252 & -13.249 \\ 9.713 & -0.362 \end{bmatrix}$

23. $X = 3\begin{bmatrix} -2 & -1 \\ 1 & 0 \\ 3 & 4 \end{bmatrix} - 2\begin{bmatrix} 0 & 3 \\ 2 & 0 \\ -4 & -1 \end{bmatrix} = \begin{bmatrix} -6 & -3 \\ 3 & 0 \\ 9 & -12 \end{bmatrix} - \begin{bmatrix} 0 & 6 \\ 4 & 0 \\ -8 & -2 \end{bmatrix} = \begin{bmatrix} -6 & -9 \\ -1 & 0 \\ 17 & -10 \end{bmatrix}$

25. $X = -\frac{3}{2}A + \frac{1}{2}B = -\frac{3}{2}\begin{bmatrix} -2 & -1 \\ 1 & 0 \\ 3 & -4 \end{bmatrix} + \frac{1}{2}\begin{bmatrix} 0 & 3 \\ 2 & 0 \\ -4 & -1 \end{bmatrix} = \begin{bmatrix} 3 & 3 \\ -\frac{1}{2} & 0 \\ -\frac{13}{2} & \frac{11}{2} \end{bmatrix}$

27. (a) $AB = \begin{bmatrix} 1 & 2 \\ 5 & 2 \end{bmatrix}\begin{bmatrix} 2 & -1 \\ -1 & 8 \end{bmatrix} = \begin{bmatrix} 2-2 & -1+16 \\ 10-2 & -5+16 \end{bmatrix} = \begin{bmatrix} 0 & 15 \\ 8 & 11 \end{bmatrix}$

(b) $BA = \begin{bmatrix} 2 & -1 \\ -1 & 8 \end{bmatrix}\begin{bmatrix} 1 & 2 \\ 5 & 2 \end{bmatrix} = \begin{bmatrix} 2-5 & 4-2 \\ -1+40 & -2+16 \end{bmatrix} = \begin{bmatrix} -3 & 2 \\ 39 & 14 \end{bmatrix}$

(c) $A^2 = \begin{bmatrix} 1 & 2 \\ 5 & 2 \end{bmatrix}\begin{bmatrix} 1 & 2 \\ 5 & 2 \end{bmatrix} = \begin{bmatrix} 1+10 & 2+4 \\ 5+10 & 10+4 \end{bmatrix} = \begin{bmatrix} 11 & 6 \\ 15 & 14 \end{bmatrix}$

29. (a) $AB = \begin{bmatrix} 3 & -1 \\ 1 & 3 \end{bmatrix}\begin{bmatrix} 1 & -3 \\ 3 & 1 \end{bmatrix} = \begin{bmatrix} 3-3 & -9-1 \\ 1+9 & -3+3 \end{bmatrix} = \begin{bmatrix} 0 & -10 \\ 10 & 0 \end{bmatrix}$

(b) $BA = \begin{bmatrix} 1 & -3 \\ 3 & 1 \end{bmatrix}\begin{bmatrix} 3 & -1 \\ 1 & 3 \end{bmatrix} = \begin{bmatrix} 3-3 & -1-9 \\ 9+1 & -3+3 \end{bmatrix} = \begin{bmatrix} 0 & -10 \\ 10 & 0 \end{bmatrix}$

(c) $A^2 = \begin{bmatrix} 3 & -1 \\ 1 & 3 \end{bmatrix}\begin{bmatrix} 3 & -1 \\ 1 & 3 \end{bmatrix} = \begin{bmatrix} 9-1 & -3-3 \\ 3=3 & -1+9 \end{bmatrix} = \begin{bmatrix} 8 & -6 \\ 6 & 8 \end{bmatrix}$

31. (a) $AB = \begin{bmatrix} 1 & -1 & 7 \\ 2 & -1 & 8 \\ 3 & 1 & -1 \end{bmatrix}\begin{bmatrix} 1 & 1 & 2 \\ 2 & 1 & 1 \\ 1 & -3 & 2 \end{bmatrix} = \begin{bmatrix} 1-2+7 & 1-1-21 & 2-1+14 \\ 2-2+8 & 2-1-24 & 4-1+16 \\ 3+2-1 & 3+1+3 & 6+1-2 \end{bmatrix} = \begin{bmatrix} 6 & -21 & 15 \\ 8 & -23 & 19 \\ 4 & 7 & 5 \end{bmatrix}$

(b) $BA = \begin{bmatrix} 1 & 1 & 2 \\ 2 & 1 & 1 \\ 1 & -3 & 2 \end{bmatrix}\begin{bmatrix} 1 & -1 & 7 \\ 2 & -1 & 8 \\ 3 & 1 & -1 \end{bmatrix} = \begin{bmatrix} 1+2+6 & -1-1+2 & 7+8-2 \\ 2+2+3 & -2-1+1 & 14+8-1 \\ 1-6+6 & -1+3+2 & 7-24-2 \end{bmatrix} = \begin{bmatrix} 9 & 0 & 13 \\ 7 & -2 & 21 \\ 1 & 4 & -19 \end{bmatrix}$

(c) $A^2 = \begin{bmatrix} 1 & -1 & 7 \\ 2 & -1 & 8 \\ 3 & 1 & -1 \end{bmatrix}\begin{bmatrix} 1 & -1 & 7 \\ 2 & -1 & 8 \\ 3 & 1 & -1 \end{bmatrix} = \begin{bmatrix} 1-2+21 & -1+1+7 & 7-8-7 \\ 2-2+24 & -2+1+8 & 14-8-8 \\ 3+2-3 & -3-1-1 & 21+8+1 \end{bmatrix} = \begin{bmatrix} 20 & 7 & -8 \\ 24 & 7 & -2 \\ 2 & -5 & 30 \end{bmatrix}$

33. A is 3×2 and B is $3 \times 3 \implies AB$ is not defined.

35. $AB = \begin{bmatrix} -1 & 6 \\ -4 & 5 \\ 0 & 3 \end{bmatrix}\begin{bmatrix} 2 & 3 \\ 0 & 9 \end{bmatrix} = \begin{bmatrix} -2 & 51 \\ -8 & 33 \\ 0 & 27 \end{bmatrix}$

37. A is 3×3, B is $3 \times 3 \implies AB$ is 3×3.

$$AB = \begin{bmatrix} 5 & 0 & 0 \\ 0 & -8 & 0 \\ 0 & 0 & 7 \end{bmatrix}\begin{bmatrix} \frac{1}{5} & 0 & 0 \\ 0 & -\frac{1}{8} & 0 \\ 0 & 0 & \frac{1}{2} \end{bmatrix} = \begin{bmatrix} 1 & 0 & 0 \\ 0 & 1 & 0 \\ 0 & 0 & \frac{7}{2} \end{bmatrix}$$

39. A is 2×1 and B is $1 \times 4 \implies AB$ is 2×4.

$$AB = \begin{bmatrix} 10 \\ 11 \end{bmatrix}\begin{bmatrix} 4 & -2 & -1 & 8 \end{bmatrix} = \begin{bmatrix} 40 & -20 & -10 & 80 \\ 44 & -22 & -11 & 88 \end{bmatrix}$$

41. $AB = \begin{bmatrix} 70 & -17 & 73 \\ 32 & 11 & 6 \\ 16 & -38 & 70 \end{bmatrix}$

43. $\begin{bmatrix} -3 & 8 & -6 & 8 \\ -12 & 15 & 9 & 6 \\ 5 & -1 & 1 & 5 \end{bmatrix}\begin{bmatrix} 3 & 1 & 6 \\ 24 & 15 & 14 \\ 16 & 10 & 21 \\ 8 & -4 & 10 \end{bmatrix} = \begin{bmatrix} 151 & 25 & 48 \\ 516 & 279 & 387 \\ 47 & -20 & 87 \end{bmatrix}$

45. A is 2×4 and B is $2 \times 4 \implies AB$ is not defined.

47. $\left(\begin{bmatrix} 3 & 1 \\ 0 & -2 \end{bmatrix}\begin{bmatrix} 1 & 0 \\ -2 & 2 \end{bmatrix}\right)\begin{bmatrix} 1 & 0 \\ 2 & 4 \end{bmatrix} = \begin{bmatrix} 1 & 2 \\ 4 & -4 \end{bmatrix}\begin{bmatrix} 1 & 0 \\ 2 & 4 \end{bmatrix} = \begin{bmatrix} 5 & 8 \\ -4 & -16 \end{bmatrix}$

49. $\begin{bmatrix} 0 & 2 & -2 \\ 4 & 1 & 2 \end{bmatrix}\left(\begin{bmatrix} 4 & 0 \\ 0 & -1 \\ -1 & 2 \end{bmatrix} + \begin{bmatrix} -2 & 3 \\ -3 & 5 \\ 0 & -3 \end{bmatrix}\right) = \begin{bmatrix} 0 & 2 & -2 \\ 4 & 1 & 2 \end{bmatrix}\begin{bmatrix} 2 & 3 \\ -3 & 4 \\ -1 & -1 \end{bmatrix} = \begin{bmatrix} -4 & 10 \\ 3 & 14 \end{bmatrix}$

51. $\begin{bmatrix} 1 & 2 & \vdots & 4 \\ 3 & 2 & \vdots & 0 \end{bmatrix}$

(a) $\begin{bmatrix} 1 & 2 \\ 3 & 2 \end{bmatrix}\begin{bmatrix} 2 \\ 1 \end{bmatrix} = \begin{bmatrix} 4 \\ 8 \end{bmatrix} \Rightarrow \begin{bmatrix} 2 \\ 1 \end{bmatrix}$ is not a solution.

(b) $\begin{bmatrix} 1 & 2 \\ 3 & 2 \end{bmatrix}\begin{bmatrix} -2 \\ 3 \end{bmatrix} = \begin{bmatrix} 4 \\ 0 \end{bmatrix} \Rightarrow \begin{bmatrix} -2 \\ 3 \end{bmatrix}$ is a solution.

(c) $\begin{bmatrix} 1 & 2 \\ 3 & 2 \end{bmatrix}\begin{bmatrix} -4 \\ 4 \end{bmatrix} = \begin{bmatrix} 4 \\ -4 \end{bmatrix} \Rightarrow \begin{bmatrix} -4 \\ 4 \end{bmatrix}$ is not a solution.

(d) $\begin{bmatrix} 1 & 2 \\ 3 & 2 \end{bmatrix}\begin{bmatrix} 2 \\ -3 \end{bmatrix} = \begin{bmatrix} -4 \\ 0 \end{bmatrix} \Rightarrow \begin{bmatrix} 2 \\ -3 \end{bmatrix}$ is not a solution.

53. $\begin{bmatrix} -2 & -3 & \vdots & -6 \\ 4 & 2 & \vdots & 20 \end{bmatrix}$

(a) $\begin{bmatrix} -2 & -3 \\ 4 & 2 \end{bmatrix}\begin{bmatrix} 3 \\ 0 \end{bmatrix} = \begin{bmatrix} -6 \\ 12 \end{bmatrix} \Rightarrow \begin{bmatrix} 3 \\ 0 \end{bmatrix}$ is not a solution.

(b) $\begin{bmatrix} -2 & -3 \\ 4 & 2 \end{bmatrix}\begin{bmatrix} 6 \\ -2 \end{bmatrix} = \begin{bmatrix} -6 \\ 20 \end{bmatrix} \Rightarrow \begin{bmatrix} 6 \\ -2 \end{bmatrix}$ is a solution.

(c) $\begin{bmatrix} -2 & -3 \\ 4 & 2 \end{bmatrix}\begin{bmatrix} -6 \\ 6 \end{bmatrix} = \begin{bmatrix} -6 \\ -12 \end{bmatrix} \Rightarrow \begin{bmatrix} -6 \\ 6 \end{bmatrix}$ is not a solution.

(d) $\begin{bmatrix} -2 & -3 \\ 4 & 2 \end{bmatrix}\begin{bmatrix} 4 \\ 2 \end{bmatrix} = \begin{bmatrix} -14 \\ 20 \end{bmatrix} \Rightarrow \begin{bmatrix} 4 \\ 2 \end{bmatrix}$ is not a solution.

55. $\begin{bmatrix} 1 & -3 & 4 & \vdots & 3 \\ 2 & -5 & 0 & \vdots & 19 \\ -2 & 4 & -5 & \vdots & -6 \end{bmatrix}$ Let $A = \begin{bmatrix} 1 & -3 & 4 \\ 2 & -5 & 0 \\ -2 & 4 & -5 \end{bmatrix}$.

(a) $A\begin{bmatrix} 3 \\ 2 \\ -2 \end{bmatrix} = \begin{bmatrix} -11 \\ -4 \\ 12 \end{bmatrix} \Rightarrow \begin{bmatrix} 3 \\ 2 \\ -2 \end{bmatrix}$ is not a solution.

(b) $A\begin{bmatrix} 2 \\ -3 \\ -2 \end{bmatrix} = \begin{bmatrix} 3 \\ 19 \\ -6 \end{bmatrix} \Rightarrow \begin{bmatrix} 2 \\ -3 \\ -2 \end{bmatrix}$ is a solution.

(c) $A\begin{bmatrix} 1 \\ 3 \\ -2 \end{bmatrix} = \begin{bmatrix} -16 \\ -13 \\ 20 \end{bmatrix} \Rightarrow \begin{bmatrix} 1 \\ 3 \\ -2 \end{bmatrix}$ is not a solution.

(d) $A\begin{bmatrix} 3 \\ -3 \\ 2 \end{bmatrix} = \begin{bmatrix} 20 \\ 21 \\ -28 \end{bmatrix} \Rightarrow \begin{bmatrix} 3 \\ -3 \\ 2 \end{bmatrix}$ is not a solution.

57. (a) $A = \begin{bmatrix} -1 & 1 \\ -2 & 1 \end{bmatrix}$, $X = \begin{bmatrix} x_1 \\ x_2 \end{bmatrix}$, $B = \begin{bmatrix} 4 \\ 0 \end{bmatrix}$

(b) By Gauss-Jordan elimination on

$$\begin{bmatrix} -1 & 1 & \vdots & 4 \\ -2 & 1 & \vdots & 0 \end{bmatrix} \begin{matrix} -R_1 \rightarrow \\ 2R_1 + R_2 \rightarrow \end{matrix} \begin{bmatrix} 1 & -1 & \vdots & -4 \\ 0 & -1 & \vdots & -8 \end{bmatrix} \begin{matrix} R_2 + R_1 \rightarrow \\ -R_2 \rightarrow \end{matrix} \begin{bmatrix} 1 & 0 & \vdots & 4 \\ 0 & 1 & \vdots & 8 \end{bmatrix},$$

we have $x_1 = 4$ and $x_2 = 8$. Thus, $X = \begin{bmatrix} 4 \\ 8 \end{bmatrix}$.

59. (a) $A = \begin{bmatrix} -2 & -3 \\ 6 & 1 \end{bmatrix}, X = \begin{bmatrix} x_1 \\ x_2 \end{bmatrix}, B = \begin{bmatrix} -4 \\ -36 \end{bmatrix}$

(b) $\left[\begin{array}{cc:c} -2 & -3 & -4 \\ 6 & 1 & -36 \end{array}\right]$

$3R_1 + R_2 \left[\begin{array}{cc:c} -2 & -3 & -4 \\ 0 & -8 & -48 \end{array}\right]$

$\left(-\frac{1}{8}\right)R_2 \left[\begin{array}{cc:c} -2 & -3 & -4 \\ 0 & 1 & 6 \end{array}\right]$

$3R_2 + R_1 \left[\begin{array}{cc:c} -2 & 0 & 14 \\ 0 & 1 & 6 \end{array}\right]$

$-\frac{1}{2}R_1 \left[\begin{array}{cc:c} 1 & 0 & -7 \\ 0 & 1 & 6 \end{array}\right]$

$x_1 = -7, x_2 = 6$ Thus, $X = \begin{bmatrix} -7 \\ 6 \end{bmatrix}$.

61. (a) $A = \begin{bmatrix} 1 & -2 & 3 \\ -1 & 3 & -1 \\ 2 & -5 & 5 \end{bmatrix}, X = \begin{bmatrix} x_1 \\ x_2 \\ x_3 \end{bmatrix}, B = \begin{bmatrix} 9 \\ -6 \\ 17 \end{bmatrix}$

(b) $\left[\begin{array}{ccc:c} 1 & -2 & 3 & 9 \\ -1 & 3 & -1 & -6 \\ 2 & -5 & 5 & 17 \end{array}\right]$

$\begin{array}{r} \\ R_1 + R_2 \rightarrow \\ -2R_1 + R_3 \rightarrow \end{array} \left[\begin{array}{ccc:c} 1 & -2 & 3 & 9 \\ 0 & 1 & 2 & 3 \\ 0 & -1 & -1 & -1 \end{array}\right]$

$\begin{array}{r} 2R_2 + R_1 \rightarrow \\ \\ R_2 + R_3 \rightarrow \end{array} \left[\begin{array}{ccc:c} 1 & 0 & 7 & 15 \\ 0 & 1 & 2 & 3 \\ 0 & 0 & 1 & 2 \end{array}\right]$

$\begin{array}{r} -7R_3 + R_1 \rightarrow \\ -2R_3 + R_2 \rightarrow \\ \end{array} \left[\begin{array}{ccc:c} 1 & 0 & 0 & 1 \\ 0 & 1 & 0 & -1 \\ 0 & 0 & 1 & 2 \end{array}\right]$

$x_1 = 1, x_2 = -1, x_3 = 2$. Thus, $X = \begin{bmatrix} 1 \\ -1 \\ 2 \end{bmatrix}$.

63. (a) $A = \begin{bmatrix} 1 & -5 & 2 \\ -3 & 1 & -1 \\ 0 & -2 & 5 \end{bmatrix}, X = \begin{bmatrix} x_1 \\ x_2 \\ x_3 \end{bmatrix}, B = \begin{bmatrix} -20 \\ 8 \\ -16 \end{bmatrix}$

(b) $\left[\begin{array}{ccc:c} 1 & -5 & 2 & -20 \\ -3 & 1 & -1 & 8 \\ 0 & -2 & 5 & -16 \end{array}\right]$ $3R_1 + R_2 \left[\begin{array}{ccc:c} 1 & -5 & 2 & -20 \\ 0 & -14 & 5 & -52 \\ 0 & -2 & 5 & -16 \end{array}\right]$

$R_2 \leftrightarrow R_3 \left[\begin{array}{ccc:c} 1 & -5 & 2 & -20 \\ 0 & -2 & 5 & -16 \\ 0 & -14 & 5 & -52 \end{array}\right]$

$-7R_2 + R_3 \left[\begin{array}{ccc:c} 1 & -5 & 2 & -20 \\ 0 & -2 & 5 & -16 \\ 0 & 0 & -30 & 60 \end{array}\right]$

$-\frac{1}{30}R_3 \left[\begin{array}{ccc:c} 1 & -5 & 2 & -20 \\ 0 & -2 & 5 & -16 \\ 0 & 0 & 1 & -2 \end{array}\right]$

$\begin{array}{r} -2R_3 + R_1 \\ -5R_3 + R_2 \\ \end{array} \left[\begin{array}{ccc:c} 1 & -5 & 0 & -16 \\ 0 & -2 & 0 & -6 \\ 0 & 0 & 1 & -2 \end{array}\right]$

$\left(-\frac{1}{2}\right)R_2 \left[\begin{array}{ccc:c} 1 & -5 & 0 & -16 \\ 0 & 1 & 0 & 3 \\ 0 & 0 & 1 & -2 \end{array}\right]$

$5R_2 + R_1 \left[\begin{array}{ccc:c} 1 & 0 & 0 & -1 \\ 0 & 1 & 0 & 3 \\ 0 & 0 & 1 & -2 \end{array}\right]$

$x_1 = -1, x_2 = 3, x_3 = -2$. Thus, $X = \begin{bmatrix} -1 \\ 3 \\ -2 \end{bmatrix}$.

65. $A = \begin{bmatrix} 2 & 0 \\ 4 & 5 \end{bmatrix}$

$$f(A) = A^2 - 5A + 2 = \begin{bmatrix} 2 & 0 \\ 4 & 5 \end{bmatrix}\begin{bmatrix} 2 & 0 \\ 4 & 5 \end{bmatrix} - 5\begin{bmatrix} 2 & 4 \\ 0 & 5 \end{bmatrix} + 2\begin{bmatrix} 1 & 0 \\ 0 & 1 \end{bmatrix} = \begin{bmatrix} -4 & 0 \\ 8 & 2 \end{bmatrix}$$

67. $A = \begin{bmatrix} 3 & 1 & 4 \\ 0 & 2 & 6 \\ 0 & 0 & 5 \end{bmatrix}$

$$f(A) = \begin{bmatrix} 3 & 1 & 4 \\ 0 & 2 & 6 \\ 0 & 0 & 5 \end{bmatrix}^3 - 10\begin{bmatrix} 3 & 1 & 4 \\ 0 & 2 & 6 \\ 0 & 0 & 5 \end{bmatrix}^2 + 31\begin{bmatrix} 3 & 1 & 4 \\ 0 & 2 & 6 \\ 0 & 0 & 5 \end{bmatrix} - 30\begin{bmatrix} 1 & 0 & 0 \\ 0 & 1 & 0 \\ 0 & 0 & 1 \end{bmatrix} = \begin{bmatrix} 0 & 0 & 0 \\ 0 & 0 & 0 \\ 0 & 0 & 0 \end{bmatrix}$$

For 69–77, A is of order 2×3, B is of order 2×3, C is of order 3×2 and D is of order 2×2.

69. $A + 2C$ is not possible. A and C are not of the same order.

71. AB is not possible. The number of columns of A does not equal the number of rows of B.

73. $BC - D$ is possible. The resulting order is 2×2.

75. (CA) is 3×3 so $(CA)D$ is not possible.

77. $D(A - 3B)$ is possible. The resulting order is 2×3.

79. $1.20\begin{bmatrix} 60 & 40 & 20 \\ 30 & 90 & 60 \end{bmatrix} = \begin{bmatrix} 72 & 48 & 24 \\ 36 & 108 & 72 \end{bmatrix}$

81. $BA = [3.75 \quad 7.00]\begin{bmatrix} 100 & 75 & 75 \\ 125 & 150 & 100 \end{bmatrix} = [\$1250.00 \quad \$1331.25 \quad \$981.25]$

The entries in the last matrix represent the profit for both crops at each of the three outlets.

83. $ST = \begin{bmatrix} 3 & 2 & 2 & 3 & 0 \\ 0 & 2 & 3 & 4 & 3 \\ 4 & 2 & 1 & 3 & 2 \end{bmatrix}\begin{bmatrix} 840 & 1100 \\ 1200 & 1350 \\ 1450 & 1650 \\ 2650 & 3000 \\ 3050 & 3200 \end{bmatrix} = \begin{bmatrix} \$15,770 & \$18,300 \\ \$26,500 & \$29,250 \\ \$21,260 & \$24,150 \end{bmatrix}$

The entries represent the wholesale and retail prices of the inventory at each outlet.

85. $P^2 = \begin{bmatrix} 0.6 & 0.1 & 0.1 \\ 0.2 & 0.7 & 0.1 \\ 0.2 & 0.2 & 0.8 \end{bmatrix}\begin{bmatrix} 0.6 & 0.1 & 0.1 \\ 0.2 & 0.7 & 0.1 \\ 0.2 & 0.2 & 0.8 \end{bmatrix} = \begin{bmatrix} 0.40 & 0.15 & 0.15 \\ 0.28 & 0.53 & 0.17 \\ 0.32 & 0.32 & 0.68 \end{bmatrix}$

This product represents the changes in party affiliation after *two* elections.

87. True

89. False. The product is $\begin{bmatrix} 2 & 2 \\ -3 & -3 \\ 7 & 7 \end{bmatrix}$.

91. $A = \begin{bmatrix} 3 & 3 \\ 4 & 4 \end{bmatrix}, B = \begin{bmatrix} 1 & -1 \\ -1 & 1 \end{bmatrix}$

$AB = \begin{bmatrix} 3 & 3 \\ 4 & 4 \end{bmatrix}\begin{bmatrix} 1 & -1 \\ -1 & 1 \end{bmatrix} = \begin{bmatrix} 0 & 0 \\ 0 & 0 \end{bmatrix}$

$AB = O$ but $A \neq O$ and $B \neq O$.

93. $A = \begin{bmatrix} 0 & -i \\ i & 0 \end{bmatrix}$

$A^2 = \begin{bmatrix} 0 & -i \\ i & 0 \end{bmatrix}\begin{bmatrix} 0 & -i \\ i & 0 \end{bmatrix} = \begin{bmatrix} 1 & 0 \\ 0 & 1 \end{bmatrix} = I,$

the identity matrix.

95. (a) $A = \begin{bmatrix} 0 & 2 \\ 0 & 0 \end{bmatrix}, B = \begin{bmatrix} 0 & 2 & 3 \\ 0 & 0 & 4 \\ 0 & 0 & 0 \end{bmatrix}$

(b) A^2 and B^3 are both zero matrices.

(c) If A is 4×4, then A^4 will be the zero matrix.

(d) If A is $n \times n$, then A^n is the zero matrix.

97. $8x^2 - 10x - 3 = 0$

$(4x + 1)(2x - 3) = 0$

$x = -\frac{1}{4}, \frac{3}{2}$

99. $3x^3 + 22x^2 - 45x = 0$

$x(3x^2 + 22x - 45) = 0$

$x(x + 9)(3x - 5) = 0$

$x = 0, -9, \frac{5}{3}$

101. $2x^3 - 5x^2 - 12x + 30 = 0$

$x^2(2x - 5) - 6(2x - 5) = 0$

$(x^2 - 6)(2x - 5) = 0$

$x = \pm\sqrt{6}, \frac{5}{2}$

103. $\log_2\left(\frac{12}{x}\right) = \log_2 12 - \log_2 x = \log_2(2^2 \cdot 3) - \log_2 x$

$= \log_2 2^2 + \log_2 3 - \log_2 x = 2 + \log_2 3 - \log_2 x$

105. $\ln\left(\frac{x^2 - 9}{x^4}\right) = \ln[(x - 3)(x + 3)] - \ln x^4$

$= \ln(x - 3) + \ln(x + 3) - 4 \ln x$

107. $\ln x - 3[\ln(x + 6) + \ln(x - 6)] = \ln x - 3[\ln(x + 6)(x - 6)]$

$= \ln x - \ln(x^2 - 36)^3$

$= \ln\left[\frac{x}{(x^2 - 36)^3}\right]$

109. $\frac{3}{2} \ln 7t^4 - \frac{3}{5} \ln t^5 = \ln(7t^4)^{3/2} - \ln(t^5)^{3/5}$

$= \ln(7^{3/5}t^6) - \ln t^3$

$= \ln\left[\frac{7^{3/2}t^6}{t^3}\right] = \ln(7^{3/2}t^3)$

111. $\begin{bmatrix} 3 & -8 & 1 & \vdots & -8 \\ -2 & -4 & -3 & \vdots & -11 \\ 4 & 9 & -5 & \vdots & -10 \end{bmatrix}$ row reduces to $\begin{bmatrix} 1 & 0 & 0 & \vdots & -1 \\ 0 & 1 & 0 & \vdots & 1 \\ 0 & 0 & 1 & \vdots & 3 \end{bmatrix}$

Answer: $x = -1, y = 1, z = 3$

Section 8.3 The Inverse of a Square Matrix

- You should be able to find the inverse, if it exists, of a square matrix.
  (a) Write the $n \times 2n$ matrix that consists of the given matrix A on the left and the $n \times n$ identity matrix I on the right to obtain $[A \;\vdots\; I]$. Note that we separate the matrices A and I by a dotted line. We call this process **adjoining** the matrices A and I.
  (b) If possible, row reduce A to I using elementary row operations on the *entire* matrix $[A \;\vdots\; I]$. The result will be the matrix $[I \;\vdots\; A^{-1}]$. If this is not possible, then A is not invertible.
  (c) Check your work by multiplying to see that $AA^{-1} = I = A^{-1}A$.
- You should be able to use inverse matrices to solve systems of equation.
- You should be able to find inverses using a graphing utility.

Solutions to Odd-Numbered Exercises

1. $AB = \begin{bmatrix} 2 & 1 \\ 5 & 3 \end{bmatrix}\begin{bmatrix} 3 & -1 \\ -5 & 2 \end{bmatrix} = \begin{bmatrix} 2(3) + 1(-5) & 2(-1) + 1(2) \\ 5(3) + 3(-5) & 5(-1) + 3(2) \end{bmatrix} = \begin{bmatrix} 1 & 0 \\ 0 & 1 \end{bmatrix}$

$BA = \begin{bmatrix} 3 & -1 \\ -5 & 2 \end{bmatrix}\begin{bmatrix} 2 & 1 \\ 5 & 3 \end{bmatrix} = \begin{bmatrix} 3(2) + (-1)(5) & 3(1) + (-1)(3) \\ -5(2) + 2(5) & -5(1) + 2(3) \end{bmatrix} = \begin{bmatrix} 1 & 0 \\ 0 & 1 \end{bmatrix}$

3. $AB = \begin{bmatrix} 1 & 2 \\ 3 & 4 \end{bmatrix}\begin{bmatrix} -2 & 1 \\ \frac{3}{2} & -\frac{1}{2} \end{bmatrix} = \begin{bmatrix} -2 + 3 & 1 - 1 \\ -6 + 6 & 3 - 2 \end{bmatrix} = \begin{bmatrix} 1 & 0 \\ 0 & 1 \end{bmatrix}$

$BA = \begin{bmatrix} -2 & 1 \\ \frac{3}{2} & -\frac{1}{2} \end{bmatrix}\begin{bmatrix} 1 & 2 \\ 3 & 4 \end{bmatrix} = \begin{bmatrix} -2 + 3 & -4 + 4 \\ \frac{3}{2} - \frac{3}{2} & 3 - 2 \end{bmatrix} = \begin{bmatrix} 1 & 0 \\ 0 & 1 \end{bmatrix}$

5. $AB = \begin{bmatrix} 2 & -17 & 11 \\ -1 & 11 & -7 \\ 0 & 3 & -2 \end{bmatrix}\begin{bmatrix} 1 & 1 & 2 \\ 2 & 4 & -3 \\ 3 & 6 & -5 \end{bmatrix}$

$= \begin{bmatrix} 2 - 34 + 33 & 2 - 68 + 66 & 4 + 51 - 55 \\ -1 + 22 - 21 & -1 + 44 - 42 & -2 - 33 + 35 \\ 6 - 6 & 12 - 12 & -9 + 10 \end{bmatrix} = \begin{bmatrix} 1 & 0 & 0 \\ 0 & 1 & 0 \\ 0 & 0 & 1 \end{bmatrix}$

$BA = \begin{bmatrix} 1 & 1 & 2 \\ 2 & 4 & -3 \\ 3 & 6 & -5 \end{bmatrix}\begin{bmatrix} 2 & -17 & 11 \\ -1 & 11 & -7 \\ 0 & 3 & -2 \end{bmatrix} = \begin{bmatrix} 2 - 1 & -17 + 11 + 6 & 11 - 7 - 4 \\ 4 - 4 & -34 + 44 - 9 & 22 - 28 + 6 \\ 6 - 6 & -51 + 66 - 15 & 33 - 42 + 10 \end{bmatrix} = \begin{bmatrix} 1 & 0 & 0 \\ 0 & 1 & 0 \\ 0 & 0 & 1 \end{bmatrix}$

7. $AB = \frac{1}{3}\begin{bmatrix} -2 & 2 & 3 \\ 1 & -1 & 0 \\ 0 & 1 & 4 \end{bmatrix}\begin{bmatrix} -4 & -5 & 3 \\ -4 & -8 & 3 \\ 1 & 2 & 0 \end{bmatrix} = \frac{1}{3}\begin{bmatrix} -8 + 8 + 3 & 10 - 16 + 6 & -6 + 6 \\ -4 + 4 & -5 + 8 & 3 - 3 \\ -4 + 4 & -8 + 8 & 3 \end{bmatrix}$

$= \frac{1}{3}\begin{bmatrix} 3 & 0 & 0 \\ 0 & 3 & 0 \\ 0 & 0 & 3 \end{bmatrix} = \begin{bmatrix} 1 & 0 & 0 \\ 0 & 1 & 0 \\ 0 & 0 & 1 \end{bmatrix}$

$BA = \frac{1}{3}\begin{bmatrix} -4 & -5 & 3 \\ -4 & -8 & 3 \\ 1 & 2 & 0 \end{bmatrix}\begin{bmatrix} -2 & 2 & 3 \\ 1 & -1 & 0 \\ 0 & 1 & 4 \end{bmatrix} = \frac{1}{3}\begin{bmatrix} 8 - 5 & -8 + 5 + 3 & -12 + 12 \\ 8 - 8 & -8 + 8 + 3 & -12 + 12 \\ -2 + 2 & 2 - 2 & 3 \end{bmatrix} = \begin{bmatrix} 1 & 0 & 0 \\ 0 & 1 & 0 \\ 0 & 0 & 1 \end{bmatrix}$

9. $AB = \begin{bmatrix} -1 & -4 \\ 1 & 2 \end{bmatrix}\begin{bmatrix} 1 & 2 \\ -\frac{1}{2} & -\frac{1}{2} \end{bmatrix} = \begin{bmatrix} 1 & 0 \\ 0 & 1 \end{bmatrix}; BA = \begin{bmatrix} 1 & 0 \\ 0 & 1 \end{bmatrix}$

11. $AB = \begin{bmatrix} 1.6 & 2 \\ -3.5 & -4.5 \end{bmatrix}\begin{bmatrix} 22.5 & 10 \\ -17.5 & -8 \end{bmatrix} = \begin{bmatrix} 1 & 0 \\ 0 & 1 \end{bmatrix}; BA = \begin{bmatrix} 1 & 0 \\ 0 & 1 \end{bmatrix}$

13. $[A \ \vdots \ I] = \left[\begin{array}{cc:cc} 2 & 0 & 1 & 0 \\ 0 & 3 & 0 & 1 \end{array}\right]$

$\begin{array}{l} \frac{1}{2}R_1 \rightarrow \\ \frac{1}{3}R_2 \rightarrow \end{array} \left[\begin{array}{cc:cc} 1 & 0 & \frac{1}{2} & 0 \\ 0 & 1 & 0 & \frac{1}{3} \end{array}\right] = [I \ \vdots \ A^{-1}]$

$A^{-1} = \begin{bmatrix} \frac{1}{2} & 0 \\ 0 & \frac{1}{3} \end{bmatrix} = \frac{1}{6}\begin{bmatrix} 3 & 0 \\ 0 & 2 \end{bmatrix}$

15. $[A \ \vdots \ I] = \left[\begin{array}{cc:cc} 1 & -2 & 1 & 0 \\ 2 & -3 & 0 & 1 \end{array}\right]$

$-2R_1 + R_2 \rightarrow \left[\begin{array}{cc:cc} 1 & -2 & 1 & 0 \\ 0 & 1 & -2 & 1 \end{array}\right]$

$2R_2 + R_1 \rightarrow \left[\begin{array}{cc:cc} 1 & 0 & -3 & 2 \\ 0 & 1 & -2 & 1 \end{array}\right] = [I \ \vdots \ A^{-1}]$

$A^{-1} = \begin{bmatrix} -3 & 2 \\ -2 & 1 \end{bmatrix}$

17. $[A \ \vdots \ I] = \left[\begin{array}{cc:cc} -1 & 1 & 1 & 0 \\ -2 & 1 & 0 & 1 \end{array}\right]$

$-2R_1 + R_2 \rightarrow \left[\begin{array}{cc:cc} -1 & 1 & 1 & 0 \\ 0 & -1 & -2 & 1 \end{array}\right]$

$R_2 + R_1 \rightarrow \left[\begin{array}{cc:cc} -1 & 0 & -1 & 1 \\ 0 & -1 & -2 & 1 \end{array}\right]$

$\begin{array}{l} -R_1 \rightarrow \\ -R_2 \rightarrow \end{array} \left[\begin{array}{cc:cc} 1 & 0 & 1 & -1 \\ 0 & 1 & 2 & -1 \end{array}\right] = [I \ \vdots \ A^{-1}]$

$A^{-1} = \begin{bmatrix} 1 & -1 \\ 2 & -1 \end{bmatrix}$

19. $A = \begin{bmatrix} 2 & 7 & 1 \\ -3 & -9 & 2 \end{bmatrix}$

A has no inverse because it is not square.

21.
$\left[\begin{array}{ccc:ccc} 1 & 1 & 1 & 1 & 0 & 0 \\ 3 & 5 & 4 & 0 & 1 & 0 \\ 3 & 6 & 5 & 0 & 0 & 1 \end{array}\right]$

$\begin{array}{r} \\ -3R_1 + R_2 \rightarrow \\ -3R_1 + R_3 \rightarrow \end{array} \left[\begin{array}{ccc:ccc} 1 & 1 & 1 & 1 & 0 & 0 \\ 0 & 2 & 1 & -3 & 1 & 0 \\ 0 & 3 & 2 & -3 & 0 & 1 \end{array}\right]$

$\begin{array}{r} -R_2 + R_1 \rightarrow \\ \frac{1}{2}R_2 \rightarrow \\ -3R_2 + R_3 \rightarrow \end{array} \left[\begin{array}{ccc:ccc} 1 & 0 & \frac{1}{2} & \frac{5}{2} & -\frac{1}{2} & 0 \\ 0 & 1 & \frac{1}{2} & -\frac{3}{2} & \frac{1}{2} & 0 \\ 0 & 0 & \frac{1}{2} & \frac{3}{2} & -\frac{3}{2} & 1 \end{array}\right]$

$\begin{array}{r} -R_3 + R_1 \rightarrow \\ -R_3 + R_2 \rightarrow \\ 2R_3 \rightarrow \end{array} \left[\begin{array}{ccc:ccc} 1 & 0 & 0 & 1 & 1 & -1 \\ 0 & 1 & 0 & -3 & 2 & -1 \\ 0 & 0 & 1 & 3 & -3 & 2 \end{array}\right]$

$A^{-1} = \begin{bmatrix} 1 & 1 & -1 \\ -3 & 2 & -1 \\ 3 & -3 & 2 \end{bmatrix}$

23. $[A \ \vdots \ I] = \left[\begin{array}{ccc:ccc} 1 & 0 & 0 & 1 & 0 & 0 \\ 3 & 4 & 0 & 0 & 1 & 0 \\ 2 & 5 & 5 & 0 & 0 & 1 \end{array}\right]$

$\begin{array}{r} \\ -3R_1 + R_2 \rightarrow \\ -2R_1 + R_3 \rightarrow \end{array} \left[\begin{array}{ccc:ccc} 1 & 0 & 0 & 1 & 0 & 0 \\ 0 & 4 & 0 & -3 & 1 & 0 \\ 0 & 5 & 5 & -2 & 0 & 1 \end{array}\right]$

$\begin{array}{r} \\ \\ -\frac{5}{4}R_2 + R_3 \rightarrow \end{array} \left[\begin{array}{ccc:ccc} 1 & 0 & 0 & 1 & 0 & 0 \\ 0 & 4 & 0 & -3 & 1 & 0 \\ 0 & 0 & 5 & \frac{7}{4} & -\frac{5}{4} & 1 \end{array}\right]$

$\begin{array}{r} \\ \frac{1}{4}R_2 \rightarrow \\ \frac{1}{5}R_3 \rightarrow \end{array} \left[\begin{array}{ccc:ccc} 1 & 0 & 0 & 1 & 0 & 0 \\ 0 & 1 & 0 & -\frac{3}{4} & \frac{1}{4} & 0 \\ 0 & 0 & 1 & \frac{7}{20} & -\frac{1}{4} & \frac{1}{5} \end{array}\right]$

$= [I \ \vdots \ A^{-1}]$

$A^{-1} = \frac{1}{20}\begin{bmatrix} 20 & 0 & 0 \\ -15 & 5 & 0 \\ 7 & -5 & 4 \end{bmatrix} = \begin{bmatrix} 1 & 0 & 0 \\ -0.75 & 0.25 & 0 \\ 0.35 & -0.25 & 0.2 \end{bmatrix}$

25. $[A \ \vdots \ I] = \left[\begin{array}{cccc:cccc} -8 & 0 & 0 & 0 & 1 & 0 & 0 & 0 \\ 0 & 1 & 0 & 0 & 0 & 1 & 0 & 0 \\ 0 & 0 & 4 & 0 & 0 & 0 & 1 & 0 \\ 0 & 0 & 0 & -5 & 0 & 0 & 0 & 1 \end{array}\right]$

$$\begin{array}{l} -\frac{1}{8}R_1 \rightarrow \\ \\ \frac{1}{4}R_3 \rightarrow \\ -\frac{1}{5}R_4 \rightarrow \end{array} \left[\begin{array}{cccc:cccc} 1 & 0 & 0 & 0 & -\frac{1}{8} & 0 & 0 & 0 \\ 0 & 1 & 0 & 0 & 0 & 1 & 0 & 0 \\ 0 & 0 & 1 & 0 & 0 & 0 & \frac{1}{4} & 0 \\ 0 & 0 & 0 & 1 & 0 & 0 & 0 & -\frac{1}{5} \end{array}\right]$$

$$= [I \ \vdots \ A^{-1}]$$

$$A^{-1} = \begin{bmatrix} -\frac{1}{8} & 0 & 0 & 0 \\ 0 & 1 & 0 & 0 \\ 0 & 0 & \frac{1}{4} & 0 \\ 0 & 0 & 0 & -\frac{1}{5} \end{bmatrix}$$

27. $A = \begin{bmatrix} 1 & 2 & -1 \\ 3 & 7 & -10 \\ -5 & -7 & -15 \end{bmatrix}$

$$A^{-1} = \begin{bmatrix} -175 & 37 & -13 \\ 95 & -20 & 7 \\ 14 & -3 & 1 \end{bmatrix}$$

29. $A = \begin{bmatrix} 1 & 1 & 2 \\ 3 & 1 & 0 \\ -2 & 0 & 3 \end{bmatrix}$

$$A^{-1} = \frac{1}{2}\begin{bmatrix} -3 & 3 & 2 \\ 9 & -7 & -6 \\ -2 & 2 & 2 \end{bmatrix}$$

31. $A = \begin{bmatrix} -\frac{1}{2} & \frac{3}{4} & \frac{1}{4} \\ 1 & 0 & -\frac{3}{2} \\ 0 & -1 & \frac{1}{2} \end{bmatrix}$

$$A^{-1} = \begin{bmatrix} -12 & -5 & -9 \\ -4 & -2 & -4 \\ -8 & -4 & -6 \end{bmatrix}$$

33. $A = \begin{bmatrix} 0.1 & 0.2 & 0.3 \\ -0.3 & 0.2 & 0.2 \\ 0.5 & 0.4 & 0.4 \end{bmatrix}$

$$A^{-1} = \frac{5}{11}\begin{bmatrix} 0 & -4 & 2 \\ -22 & 11 & 11 \\ 22 & -6 & -8 \end{bmatrix}$$

35. $A = \begin{bmatrix} 1 & 0 & 3 & 0 \\ 0 & 2 & 0 & 4 \\ 1 & 0 & 3 & 0 \\ 0 & 2 & 0 & 4 \end{bmatrix}$

A^{-1} does not exist.

37. $A = \begin{bmatrix} -1 & 0 & 1 & 0 \\ 0 & 2 & 0 & -1 \\ 2 & 0 & -1 & 0 \\ 0 & -1 & 0 & 1 \end{bmatrix}$

$$A^{-1} = \begin{bmatrix} 1 & 0 & 1 & 0 \\ 0 & 1 & 0 & 1 \\ 2 & 0 & 1 & 0 \\ 0 & 1 & 0 & 2 \end{bmatrix}$$

39. $\begin{bmatrix} -4 & -6 \\ 2 & 3 \end{bmatrix}^{-1}$ does not exist because

$ad - bc = (-4)(3) - (-6)(2) = 0.$

41. $\begin{bmatrix} \frac{7}{2} & -\frac{3}{4} \\ \frac{1}{5} & \frac{4}{5} \end{bmatrix} = \dfrac{1}{\left(\frac{7}{2}\right)\left(\frac{4}{5}\right) - \left(-\frac{3}{4}\right)\left(\frac{1}{5}\right)}\begin{bmatrix} \frac{4}{5} & \frac{3}{4} \\ -\frac{1}{5} & \frac{7}{2} \end{bmatrix}$

$$= \frac{20}{59}\begin{bmatrix} \frac{4}{5} & \frac{3}{4} \\ -\frac{1}{5} & \frac{7}{2} \end{bmatrix} = \frac{1}{59}\begin{bmatrix} 16 & 15 \\ -4 & 70 \end{bmatrix}$$

43. $\begin{bmatrix} x \\ y \end{bmatrix} = \begin{bmatrix} -3 & 2 \\ -2 & 1 \end{bmatrix}\begin{bmatrix} 5 \\ 10 \end{bmatrix} = \begin{bmatrix} 5 \\ 0 \end{bmatrix}$

Answer: $(5, 0)$

45. $\begin{bmatrix} x \\ y \end{bmatrix} = \begin{bmatrix} -3 & 2 \\ -2 & 1 \end{bmatrix} = \begin{bmatrix} 4 \\ 2 \end{bmatrix} = \begin{bmatrix} -8 \\ -6 \end{bmatrix}$

Answer: $(-8, -6)$

47. $\begin{bmatrix} x \\ y \\ z \end{bmatrix} = \begin{bmatrix} 1 & 1 & -1 \\ -3 & 2 & -1 \\ 3 & -3 & 2 \end{bmatrix}\begin{bmatrix} 0 \\ 5 \\ 2 \end{bmatrix} = \begin{bmatrix} 3 \\ 8 \\ -11 \end{bmatrix}$

Answer: $(3, 8, -11)$

49. $\begin{bmatrix} x_1 \\ x_2 \\ x_3 \\ x_4 \end{bmatrix} = \begin{bmatrix} -24 & 7 & 1 & -2 \\ -10 & 3 & 0 & -1 \\ -29 & 7 & 3 & -2 \\ 12 & -3 & -1 & 1 \end{bmatrix}\begin{bmatrix} 0 \\ 1 \\ -1 \\ 2 \end{bmatrix} = \begin{bmatrix} 2 \\ 1 \\ 0 \\ 0 \end{bmatrix}$

Answer: $(2, 1, 0, 0)$

51. $A = \begin{bmatrix} 3 & 4 \\ 5 & 3 \end{bmatrix}$

$$A^{-1} = \frac{1}{9 - 20}\begin{bmatrix} 3 & -4 \\ -5 & 3 \end{bmatrix}$$

$$\begin{bmatrix} x \\ y \end{bmatrix} = -\frac{1}{11}\begin{bmatrix} 3 & -4 \\ -5 & 3 \end{bmatrix}\begin{bmatrix} -2 \\ 4 \end{bmatrix} = -\frac{1}{11}\begin{bmatrix} -22 \\ 22 \end{bmatrix} = \begin{bmatrix} 2 \\ -2 \end{bmatrix}$$

Answer: $(2, -2)$

53. $A = \begin{bmatrix} -0.4 & 0.8 \\ 2 & -4 \end{bmatrix}$

$$A^{-1} = \frac{1}{1.6 - 1.6}\begin{bmatrix} -4 & -0.8 \\ -2 & -0.4 \end{bmatrix} \Rightarrow A^{-1} \text{ does not exist}$$

[The system actually has no solution.]

55. $A = \begin{bmatrix} -\frac{1}{4} & \frac{3}{8} \\ \frac{3}{2} & \frac{3}{4} \end{bmatrix}$

$$A^{-1} = \begin{bmatrix} -1 & \frac{1}{2} \\ 2 & \frac{1}{3} \end{bmatrix}$$

$$\begin{bmatrix} x \\ y \end{bmatrix} = A^{-1}b = \begin{bmatrix} -1 & \frac{1}{2} \\ 2 & \frac{1}{3} \end{bmatrix}\begin{bmatrix} -2 \\ -12 \end{bmatrix} = \begin{bmatrix} -4 \\ -8 \end{bmatrix}$$

Answer: $(-4, -8)$

57. $A = \begin{bmatrix} 4 & -1 & 1 \\ 2 & 2 & 3 \\ 5 & -2 & 6 \end{bmatrix}$

$$A^{-1} = \frac{1}{55}\begin{bmatrix} 18 & 4 & -5 \\ 3 & 19 & -10 \\ -14 & 3 & 10 \end{bmatrix}$$

$$\begin{bmatrix} x \\ y \\ z \end{bmatrix} = \frac{1}{55}\begin{bmatrix} 18 & 4 & -5 \\ 3 & 19 & -10 \\ -14 & 3 & 10 \end{bmatrix}\begin{bmatrix} -5 \\ 10 \\ 1 \end{bmatrix} = \frac{1}{55}\begin{bmatrix} -55 \\ 165 \\ 110 \end{bmatrix} = \begin{bmatrix} -1 \\ 3 \\ 2 \end{bmatrix}$$

Answer: $(-1, 3, 2)$

59. $A = \begin{bmatrix} 5 & -3 & 2 \\ 2 & 2 & -3 \\ -1 & 7 & -8 \end{bmatrix}$

A^{-1} does not exist. [The system actually has an infinite number of solutions of the form

$x = 0.3125t + 0.8125$

$y = 1.1875t + 0.6875$

$z = t$

where t is any real number.]

61. $A = \begin{bmatrix} 3 & -2 & 1 \\ -4 & 1 & -3 \\ 1 & -5 & 1 \end{bmatrix}$

$$A^{-1} = \begin{bmatrix} 0.56 & 0.12 & -0.2 \\ -0.04 & -0.08 & -0.2 \\ -0.76 & -0.52 & 0.2 \end{bmatrix}$$

$$\begin{bmatrix} x \\ y \\ z \end{bmatrix} = A^{-1}\begin{bmatrix} -29 \\ 37 \\ -24 \end{bmatrix} = \begin{bmatrix} -7 \\ 3 \\ -2 \end{bmatrix}$$

63. $\begin{bmatrix} 7 & -3 & 0 & 2 & \vdots & 41 \\ -2 & 1 & 0 & -1 & \vdots & -13 \\ 4 & 0 & 1 & -2 & \vdots & 12 \\ -1 & 1 & 0 & -1 & \vdots & -8 \end{bmatrix}$ row reduces to $\begin{bmatrix} 1 & 0 & 0 & 0 & \vdots & 5 \\ 0 & 1 & 0 & 0 & \vdots & 0 \\ 0 & 0 & 1 & 0 & \vdots & -2 \\ 0 & 0 & 0 & 1 & \vdots & 3 \end{bmatrix}$

Solution: $(5, 0, -2, 3)$

For 65 and 67 use $A = \begin{bmatrix} 1 & 1 & 1 \\ 0.065 & 0.07 & 0.09 \\ 0 & 2 & -1 \end{bmatrix}$. Using the methods of this section, we have $A^{-1} = \frac{1}{11}\begin{bmatrix} 50 & -600 & -4 \\ -13 & 200 & 5 \\ -26 & 400 & -1 \end{bmatrix}$.

65. $X = A^{-1}B = \frac{1}{11}\begin{bmatrix} 50 & -600 & -4 \\ -13 & 200 & 5 \\ -26 & 400 & -1 \end{bmatrix}\begin{bmatrix} 25{,}000 \\ 1900 \\ 0 \end{bmatrix} = \begin{bmatrix} 10{,}000 \\ 5000 \\ 10{,}000 \end{bmatrix}$

Answer: \$10,000 in AAA bonds, \$5000 in A bonds, \$10,000 in B bonds

67. $X = A^{-1}B = \frac{1}{11}\begin{bmatrix} 50 & -600 & -4 \\ -13 & 200 & 5 \\ -26 & 400 & -1 \end{bmatrix}\begin{bmatrix} 12{,}000 \\ 835 \\ 0 \end{bmatrix} = \begin{bmatrix} 9000 \\ 1000 \\ 2000 \end{bmatrix}$

Answer: \$9000 in AAA bonds, \$1000 in A bonds, \$2000 in B bonds

69. $A = \begin{bmatrix} 2 & 0 & 4 \\ 0 & 1 & 4 \\ 1 & 1 & -1 \end{bmatrix} \qquad A^{-1} = \frac{1}{14}\begin{bmatrix} 5 & -4 & 4 \\ -4 & 6 & 8 \\ 1 & 2 & -2 \end{bmatrix}$

$$\begin{bmatrix} I_1 \\ I_2 \\ I_3 \end{bmatrix} = \frac{1}{14}\begin{bmatrix} 5 & -4 & 4 \\ -4 & 6 & 8 \\ 1 & 2 & -2 \end{bmatrix}\begin{bmatrix} 14 \\ 28 \\ 0 \end{bmatrix} = \begin{bmatrix} -3 \\ 8 \\ 5 \end{bmatrix}$$

Answer: $I_1 = -3$ amps, $I_2 = 8$ amps, $I_3 = 5$ amps

71. True. $AA^{-1} = A^{-1}A = I$

73. True

75. $AA^{-1} = \begin{bmatrix} a & b \\ c & d \end{bmatrix}\left(\frac{1}{ad-bc}\right)\begin{bmatrix} d & -b \\ -c & a \end{bmatrix} = \frac{1}{ad-bc}\begin{bmatrix} a & b \\ c & d \end{bmatrix}\begin{bmatrix} d & -b \\ -c & a \end{bmatrix}$

$$= \frac{1}{ad-bc}\begin{bmatrix} ad-bc & 0 \\ 0 & ad-bc \end{bmatrix} = \begin{bmatrix} 1 & 0 \\ 0 & 1 \end{bmatrix}$$

$$A^{-1}A = \frac{1}{ad-bc}\begin{bmatrix} d & -b \\ -c & a \end{bmatrix}\begin{bmatrix} a & b \\ c & d \end{bmatrix} = \frac{1}{ad-bc}\begin{bmatrix} ad-bc & 0 \\ 0 & ad-bc \end{bmatrix} = \begin{bmatrix} 1 & 0 \\ 0 & 1 \end{bmatrix}$$

(a) $\begin{bmatrix} 5 & -2 \\ 2 & 3 \end{bmatrix}^{-1} = \frac{1}{19}\begin{bmatrix} 3 & 2 \\ -2 & 5 \end{bmatrix}$

(b) $\begin{bmatrix} 7 & 12 \\ -8 & -5 \end{bmatrix} = \frac{1}{61}\begin{bmatrix} -5 & -12 \\ 8 & 7 \end{bmatrix}$

77. $x^3 + 5x^2 - 6x - 30 = 0$

$$x^2(x + 5) - 6(x + 5) = 0$$

$$(x^2 - 6)(x + 5) = 0$$

$$x = \pm\sqrt{6}, -5$$

79. $x^4 - x^3 - 10x^2 - 2x - 24 = 0$

$$(x - 4)(x + 3)(x^2 + 2) = 0$$

$$x = 4, -3$$

81. $f(x) = -3^{x-5} + 3$

83. $f(x) = 4 + e^{x-4}$

x	-2	0	5	6	7
y	2.9995	2.996	2	0	-6

x	-2	0	3	4	5
y	4.002	4.018	4.368	5	6.718

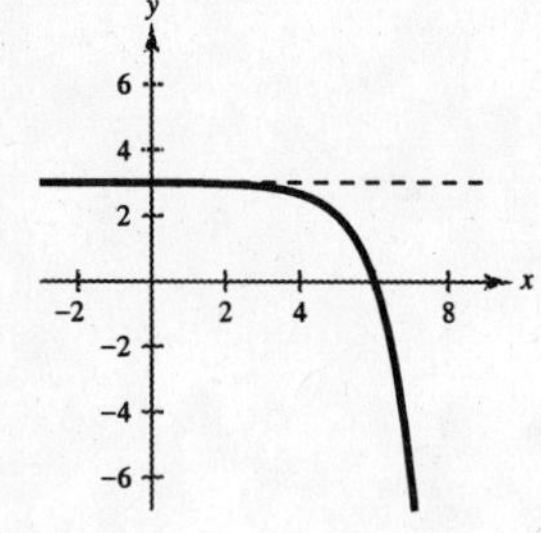

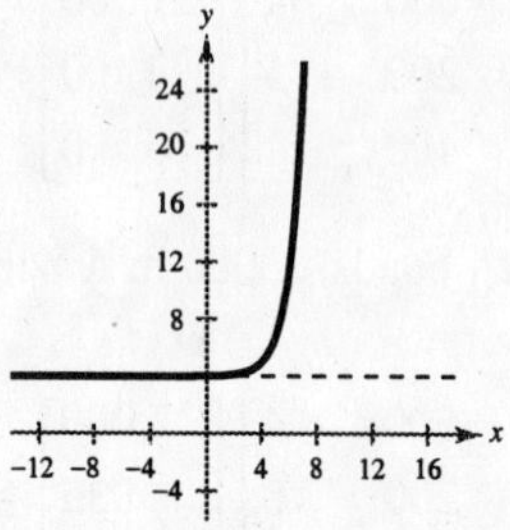

85. $-3\begin{bmatrix} -4 & 6 \\ 2 & -8 \\ 1 & 12 \end{bmatrix} = \begin{bmatrix} 12 & -18 \\ -6 & 24 \\ -3 & -36 \end{bmatrix}$

87. $\begin{bmatrix} 2 & -7 \\ -3 & -1 \end{bmatrix} - 5\begin{bmatrix} -1 & 4 \\ 8 & -5 \end{bmatrix} = \begin{bmatrix} 7 & -27 \\ -43 & 24 \end{bmatrix}$

89. $\begin{bmatrix} -2 & 0 \\ -1 & 4 \end{bmatrix}\begin{bmatrix} -3 & 1 \\ 3 & 2 \end{bmatrix} = \begin{bmatrix} 6 & -2 \\ 15 & 7 \end{bmatrix}$

Section 8.4 The Determinant of a Square Matrix

- You should be able to determine the determinant of a matrix of order 2×2 by using the products of the diagonals.
- You should be able to use expansion by cofactors to find the determinant of a matrix of order 3 or greater.
- The determinant of a triangular matrix equals the product of the entries on the main diagonal.
- You should be able to calculate determinants using a graphing utility.

Solutions to Odd-Numbered Exercises

1. $|7| = 7$

3. $\begin{vmatrix} 1 & 4 \\ 7 & 3 \end{vmatrix} = (1)(3) - (4)(7) = 3 - 28 = -25$

5. $\begin{vmatrix} 6 & 2 \\ -5 & 3 \end{vmatrix} = 6(3) - (2)(-5) = 18 + 10 = 28$

7. $\begin{vmatrix} -7 & 6 \\ \frac{1}{2} & 3 \end{vmatrix} = -7(3) - 6(\frac{1}{2}) = -21 - 3 = -24$

9. $\begin{vmatrix} 2 & -1 & 0 \\ 4 & 2 & 1 \\ 4 & 2 & 1 \end{vmatrix} = 2\begin{vmatrix} 2 & 1 \\ 2 & 1 \end{vmatrix} - 4\begin{vmatrix} -1 & 0 \\ 2 & 1 \end{vmatrix} + 4\begin{vmatrix} -1 & 0 \\ 2 & 1 \end{vmatrix} = 2(0) - 4(-1) + 4(-1) = 0$

11. $\begin{vmatrix} -1 & 2 & -5 \\ 0 & 3 & 4 \\ 0 & 0 & 3 \end{vmatrix} = (-1)(3)(3) = -9$ (Upper Triangular)

13. $\begin{vmatrix} 0.3 & 0.2 & 0.2 \\ 0.2 & 0.2 & 0.2 \\ -0.4 & 0.4 & 0.3 \end{vmatrix} = -0.002$

15. $\begin{bmatrix} 3 & 4 \\ 2 & -5 \end{bmatrix}$

(a) $M_{11} = -5$

$M_{12} = 2$

$M_{21} = 4$

$M_{22} = 3$

(b) $C_{11} = M_{11} = -5$

$C_{12} = -M_{12} = -2$

$C_{21} = -M_{21} = -4$

$C_{22} = M_{22} = 3$

17. $\begin{bmatrix} 3 & -2 & 8 \\ 3 & 2 & -6 \\ -1 & 3 & 6 \end{bmatrix}$

(a) $M_{11} = \begin{vmatrix} 2 & -6 \\ 3 & 4 \end{vmatrix} = 12 + 18 = 30$

$M_{12} = \begin{vmatrix} 3 & -6 \\ -1 & 6 \end{vmatrix} = 18 - 6 = 12$

$M_{13} = \begin{vmatrix} 3 & 2 \\ -1 & 3 \end{vmatrix} = 9 + 2 = 11$

$M_{21} = \begin{vmatrix} -2 & 8 \\ 3 & 6 \end{vmatrix} = -12 - 24 = -36$

$M_{22} = \begin{vmatrix} 3 & 8 \\ -1 & 6 \end{vmatrix} = 18 + 8 = 26$

$M_{23} = \begin{vmatrix} 3 & -2 \\ -1 & 3 \end{vmatrix} = 9 - 2 = 7$

$M_{31} = \begin{vmatrix} -2 & 8 \\ 2 & -6 \end{vmatrix} = 12 - 16 = -4$

$M_{32} = \begin{vmatrix} 3 & 8 \\ 3 & -6 \end{vmatrix} = -18 - 24 = -42$

$M_{33} = \begin{vmatrix} 3 & -2 \\ 3 & 2 \end{vmatrix} = 6 + 6 = 12$

(b) $C_{11} = (-1)^2 M_{11} = 30$

$C_{12} = (-1)^3 M_{12} = -12$

$C_{13} = (-1)^4 M_{13} = 11$

$C_{21} = (-1)^3 M_{21} = 36$

$C_{22} = (-1)^4 M_{22} = 26$

$C_{23} = (-1)^5 M_{23} = -7$

$C_{31} = (-1)^4 M_{31} = -4$

$C_{32} = (-1)^5 M_{32} = 42$

$C_{33} = (-1)^6 M_{33} = 12$

19. (a) $\begin{vmatrix} -3 & 2 & 1 \\ 4 & 5 & 6 \\ 2 & -3 & 1 \end{vmatrix} = -3\begin{vmatrix} 5 & 6 \\ -3 & 1 \end{vmatrix} - 2\begin{vmatrix} 4 & 6 \\ 2 & 1 \end{vmatrix} + \begin{vmatrix} 4 & 5 \\ 2 & -3 \end{vmatrix} = -3(23) - 2(-8) - 22 = -75$

(b) $\begin{vmatrix} -3 & 2 & 1 \\ 4 & 5 & 6 \\ 2 & -3 & 1 \end{vmatrix} = -2\begin{vmatrix} 4 & 6 \\ 2 & 1 \end{vmatrix} + 5\begin{vmatrix} -3 & 1 \\ 2 & 1 \end{vmatrix} + 3\begin{vmatrix} -3 & 1 \\ 4 & 6 \end{vmatrix} = -2(-8) + 5(-5) + 3(-22) = -75$

21. (a) $\begin{vmatrix} 6 & 0 & -3 & 5 \\ 4 & 13 & 6 & -8 \\ -1 & 0 & 7 & 4 \\ 8 & 6 & 0 & 2 \end{vmatrix} = -4\begin{vmatrix} 0 & -3 & 5 \\ 0 & 7 & 4 \\ 6 & 0 & 2 \end{vmatrix} + 13\begin{vmatrix} 6 & -3 & 5 \\ -1 & 7 & 4 \\ 8 & 0 & 2 \end{vmatrix} - 6\begin{vmatrix} 6 & 0 & 5 \\ -1 & 0 & 4 \\ 8 & 6 & 2 \end{vmatrix} - 8\begin{vmatrix} 6 & 0 & -3 \\ -1 & 0 & 7 \\ 8 & 6 & 0 \end{vmatrix}$

$= -4(-282) + 13(-298) - 6(-174) - 8(-234) = 170$

(b) $\begin{vmatrix} 6 & 0 & -3 & 5 \\ 4 & 13 & 6 & -8 \\ -1 & 0 & 7 & 4 \\ 8 & 6 & 0 & 2 \end{vmatrix} = 0\begin{vmatrix} 4 & 6 & -8 \\ -1 & 7 & 4 \\ 8 & 0 & 2 \end{vmatrix} + 13\begin{vmatrix} 6 & -3 & 5 \\ -1 & 7 & 4 \\ 8 & 0 & 2 \end{vmatrix} + 0\begin{vmatrix} 6 & -3 & 5 \\ 4 & 6 & -8 \\ 8 & 0 & 2 \end{vmatrix} + 6\begin{vmatrix} 6 & -3 & 5 \\ 4 & 6 & -8 \\ -1 & 7 & 4 \end{vmatrix}$

$= 0 + 13(-298) + 0 + 6(674) = 170$

23. Expand by Column 3.

$\begin{vmatrix} 1 & 4 & -2 \\ 3 & 2 & 0 \\ -1 & 4 & 3 \end{vmatrix} = -2\begin{vmatrix} 3 & 2 \\ -1 & 4 \end{vmatrix} + 3\begin{vmatrix} 1 & 4 \\ 3 & 2 \end{vmatrix} = -2(14) + 3(-10) = -58$

25. $\begin{vmatrix} 2 & 4 & 6 \\ 0 & 3 & 1 \\ 0 & 0 & -5 \end{vmatrix} = (2)(3)(-5) = -30$ (Upper Triangular)

27. Expand by Column 3.

$\begin{vmatrix} 2 & 6 & 6 & 2 \\ 2 & 7 & 3 & 6 \\ 1 & 5 & 0 & 1 \\ 3 & 7 & 0 & 7 \end{vmatrix} = 6\begin{vmatrix} 2 & 7 & 6 \\ 1 & 5 & 1 \\ 3 & 7 & 7 \end{vmatrix} - 3\begin{vmatrix} 2 & 6 & 2 \\ 1 & 5 & 1 \\ 3 & 7 & 7 \end{vmatrix} = 6(-20) - 3(16) = -168$

29. Expand by Column 2.

$\begin{vmatrix} 3 & 2 & 4 & -1 & 5 \\ -2 & 0 & 1 & 3 & 2 \\ 1 & 0 & 0 & 4 & 0 \\ 6 & 0 & 2 & -1 & 0 \\ 3 & 0 & 5 & 1 & 0 \end{vmatrix} = -2\begin{vmatrix} -2 & 1 & 3 & 2 \\ 1 & 0 & 4 & 0 \\ 6 & 2 & -1 & 0 \\ 3 & 5 & 1 & 0 \end{vmatrix} = (-2)(-2)\begin{vmatrix} 1 & 0 & 4 \\ 6 & 2 & -1 \\ 3 & 5 & 1 \end{vmatrix} = 4(103) = 412$

31. $\begin{vmatrix} 4 & 0 & 0 & 0 \\ 6 & -5 & 0 & 0 \\ 1 & 3 & 1 & 0 \\ 1 & -2 & 7 & 3 \end{vmatrix} = (4)(-5)(2)(3) = -60$ (Lower Triangular)

33. $\det(A) = (8)(-3)(-1)(9)(1) = 216$ (Upper Triangular)

35. $\begin{vmatrix} 1 & -1 & 8 & 4 \\ 2 & 6 & 0 & -4 \\ 2 & 0 & 2 & 6 \\ 0 & 2 & 8 & 0 \end{vmatrix} = -336$

37. $\begin{vmatrix} 3 & -2 & 4 & 3 & 1 \\ -1 & 0 & 2 & 1 & 0 \\ 5 & -1 & 0 & 3 & 2 \\ 4 & 7 & -8 & 0 & 0 \\ 1 & 2 & 3 & 0 & 2 \end{vmatrix} = 410$

39. (a) $\begin{vmatrix} -1 & 0 \\ 0 & 3 \end{vmatrix} = -3$

(b) $\begin{vmatrix} 2 & 0 \\ 0 & -1 \end{vmatrix} = -2$

(c) $\begin{bmatrix} -1 & 0 \\ 0 & 3 \end{bmatrix}\begin{bmatrix} 2 & 0 \\ 0 & -1 \end{bmatrix} = \begin{bmatrix} -2 & 0 \\ 0 & -3 \end{bmatrix}$

(d) $\begin{vmatrix} -2 & 0 \\ 0 & -3 \end{vmatrix} = 6$ [Note: $|AB| = |A|\,|B|$]

41. (a) $\begin{vmatrix} -1 & 2 & 1 \\ 1 & 0 & 1 \\ 0 & 1 & 0 \end{vmatrix} = 2$

(b) $\begin{vmatrix} -1 & 0 & 0 \\ 0 & 2 & 0 \\ 0 & 0 & 3 \end{vmatrix} = -6$

(c) $\begin{bmatrix} -1 & 2 & 1 \\ 1 & 0 & 1 \\ 0 & 1 & 0 \end{bmatrix}\begin{bmatrix} -1 & 0 & 0 \\ 0 & 2 & 0 \\ 0 & 0 & 3 \end{bmatrix} = \begin{bmatrix} 1 & 4 & 3 \\ -1 & 0 & 3 \\ 0 & 2 & 0 \end{bmatrix}$

(d) $\begin{vmatrix} 1 & 4 & 3 \\ -1 & 0 & 3 \\ 0 & 2 & 0 \end{vmatrix} = -12$ [Note: $|AB| = |A|\,|B|$]

43. (a) $|A| = -25$

(b) $|B| = -220$

(c) $AB = \begin{bmatrix} -7 & -16 & -1 & -28 \\ -4 & -14 & -11 & 8 \\ 13 & 4 & 4 & -4 \\ -2 & 3 & 2 & 2 \end{bmatrix}$

(d) $|AB| = 5500$ [Note: $|AB| = |A|\,|B|$]

45. $\begin{vmatrix} w & x \\ y & z \end{vmatrix} = wz - xy$

$-\begin{vmatrix} y & z \\ w & x \end{vmatrix} = -(xy - wz) = wz - xy$

Thus, $\begin{vmatrix} w & x \\ y & z \end{vmatrix} = -\begin{vmatrix} y & z \\ w & x \end{vmatrix}$.

47. $\begin{vmatrix} w & x \\ y & z \end{vmatrix} = wz - xy$

$\begin{vmatrix} w & x + cw \\ y & z + cy \end{vmatrix} = w(z + cy) - y(x + cw) = wz - xy$

Thus, $\begin{vmatrix} w & x \\ y & z \end{vmatrix} = \begin{vmatrix} w & x + cw \\ y & z + cy \end{vmatrix}$.

49. $\begin{vmatrix} 1 & x & x^2 \\ 1 & y & y^2 \\ 1 & z & z^2 \end{vmatrix} = \begin{vmatrix} y & y^2 \\ z & z^2 \end{vmatrix} - \begin{vmatrix} x & x^2 \\ z & z^2 \end{vmatrix} + \begin{vmatrix} x & x^2 \\ y & y^2 \end{vmatrix}$

$$= (yz^2 - y^2z) - (xz^2 - x^2z) + (xy^2 - x^2y)$$
$$= yz^2 - xz^2 - y^2z + x^2z + xy(y - x)$$
$$= z^2(y - x) - z(y^2 - x^2) + xy(y - x)$$
$$= z^2(y - x) - z(y - x)(y + x) + xy(y - x)$$
$$= (y - x)[z^2 - z(y + x) + xy]$$
$$= (y - x)[z^2 - zy - zx + xy]$$
$$= (y - x)[z^2 - zx - zy + xy]$$
$$= (y - x)[z(z - x) - y(z - x)]$$
$$= (y - x)(z - x)(z - y)$$

51. $\begin{vmatrix} x - 1 & 2 \\ 3 & x - 2 \end{vmatrix} = 0$

$$(x - 1)(x - 2) - 6 = 0$$
$$x^2 - 3x - 4 = 0$$
$$(x + 1)(x - 4) = 0$$
$$x = -1 \text{ or } x = 4$$

53. $\begin{vmatrix} 4u & -1 \\ -1 & 2v \end{vmatrix} = 8uv - 1$

55. $\begin{vmatrix} e^{2x} & e^{3x} \\ 2e^{2x} & 3e^{3x} \end{vmatrix} = 3e^{5x} - 2e^{5x} = e^{5x}$

57. $\begin{vmatrix} x & \ln x \\ 1 & \frac{1}{x} \end{vmatrix} = 1 - \ln x$

59. True. Expand along the row of zeros.

61. Let $A = \begin{bmatrix} 1 & 3 \\ -2 & 4 \end{bmatrix}$ and $B = \begin{bmatrix} -4 & 0 \\ 3 & 5 \end{bmatrix}$.

$$|A| = \begin{vmatrix} 1 & 3 \\ -2 & 4 \end{vmatrix} = 10, \ |B| = \begin{vmatrix} -4 & 0 \\ 3 & 5 \end{vmatrix} = -20$$

$$A + B = \begin{bmatrix} -3 & 3 \\ 1 & 9 \end{bmatrix}, \ |A + B| = \begin{vmatrix} -3 & 3 \\ 1 & 9 \end{vmatrix} = -30$$

Thus, $|A + B| \neq |A| + |B|$. Your answer may differ, depending on how you choose A and B.

63. A square matrix is a square array of numbers. The determinant of a square matrix is a real number.

65. (a) Columns 2 and 3 are interchanged.

(b) Rows 1 and 3 are interchanged.

67. (a) 5 is factored out of the first row of A.

(b) 4 and 3 are factored out of columns 2 and 3.

69. From Equation 2, $y = 2 - 2x$. Substituting into Equation 1, $5x + 7(2 - 2x) = 23 \implies -9x = 9 \implies x = -1$ and $y = 2 - 2(-1) = 4$

Answer: $(-1, 4)$

71. 3 times Equation 2 added to Equation 1 produces $-27x = 0 \implies x = 0$. Then $-15y = 3 \implies y = -\frac{1}{5}$.

Answer: $\left(0, -\frac{1}{5}\right)$

73.

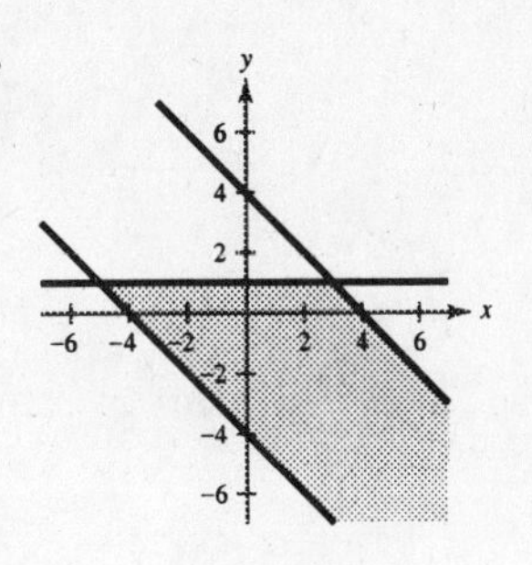

75. $\begin{bmatrix} -5 & -8 \\ 3 & 6 \end{bmatrix}^{-1} = \begin{bmatrix} -1 & -\frac{4}{3} \\ \frac{1}{2} & \frac{5}{6} \end{bmatrix}$

77. $\begin{bmatrix} -6 & 2 & 0 \\ 1 & 3 & -2 \\ -2 & 0 & 1 \end{bmatrix}^{-1} = \begin{bmatrix} -\frac{1}{4} & \frac{1}{6} & \frac{1}{3} \\ -\frac{1}{4} & \frac{1}{2} & 1 \\ -\frac{1}{2} & \frac{1}{3} & \frac{5}{3} \end{bmatrix}$

Section 8.5 Applications of Matrices and Determinants

- You should be able to find the area of a triangle with vertices (x_1, y_1), (x_2, y_2), and (x_3, y_3).

$$\text{Area} = \pm\frac{1}{2}\begin{vmatrix} x_1 & y_1 & 1 \\ x_2 & y_2 & 1 \\ x_3 & y_3 & 1 \end{vmatrix}$$

  The $\pm$ symbol indicates that the appropriate sign should be chosen so that the area is positive.

- You should be able to test to see if three points, (x_1, y_1), (x_2, y_2), and (x_3, y_3), are collinear.

$$\begin{vmatrix} x_1 & y_1 & 1 \\ x_2 & y_2 & 1 \\ x_3 & y_3 & 1 \end{vmatrix} = 0, \text{ if and only if they are collinear.}$$

- You should be able to use Cramer's Rule to solve a system of linear equations.
- Now you should be able to solve a system of linear equations by substitution, elimination, elementary row operations on an augmented matrix, using the inverse matrix, or Cramer's Rule.
- You should be able to encode and decode messages by using an invertible $n \times n$ matrix.

Solutions to Odd-Numbered Exercises

1. Vertices: $(-2, -3)$, $(2, -3)$, $(0, 4)$

$$\frac{1}{2}\begin{vmatrix} -2 & -3 & 1 \\ 2 & -3 & 1 \\ 0 & 4 & 1 \end{vmatrix} = \frac{1}{2}\left(-2\begin{vmatrix} -3 & 1 \\ 4 & 1 \end{vmatrix} - 2\begin{vmatrix} -3 & 1 \\ 4 & 1 \end{vmatrix}\right) = \frac{1}{2}(14 + 14). \text{ Area} = 14 \text{ square units}$$

3. Vertices: (0, 0), (3, 1), (1, 5)

$$\frac{1}{2}\begin{vmatrix} 0 & 0 & 1 \\ 3 & 1 & 1 \\ 1 & 5 & 1 \end{vmatrix} = \frac{1}{2}\begin{vmatrix} 3 & 1 \\ 1 & 5 \end{vmatrix} = 7. \text{ Area} = 7 \text{ square units}$$

5. Vertices: $\left(0, \frac{1}{2}\right), \left(\frac{5}{2}, 0\right), (4, 3)$

$$\frac{1}{2}\begin{vmatrix} 0 & \frac{1}{2} & 1 \\ \frac{5}{2} & 0 & 1 \\ 4 & 3 & 1 \end{vmatrix} = \frac{1}{2}\left(2 + \frac{15}{2} - \frac{5}{4}\right) = \frac{33}{8}. \text{ Area} = \frac{33}{8} \text{ square units}$$

7. Vertices: (4, 5), (6, 1), (7, 9)

$$\frac{1}{2}\begin{vmatrix} 4 & 5 & 1 \\ 6 & 1 & 1 \\ 7 & 9 & 1 \end{vmatrix} = \frac{1}{2}(20) = 10. \text{ Area} = 10 \text{ square units}$$

9. $4 = \pm\frac{1}{2}\begin{vmatrix} -5 & 1 & 1 \\ 0 & 2 & 1 \\ -2 & x & 1 \end{vmatrix}$

$$\pm 8 = -5\begin{vmatrix} 2 & 1 \\ x & 1 \end{vmatrix} - 2\begin{vmatrix} 1 & 1 \\ 2 & 1 \end{vmatrix}$$

$$\pm 8 = -5(2 - x) - 2(-1)$$

$$\pm 8 = 5x - 8$$

$$x = \frac{8 \pm 8}{5}$$

$$x = \frac{16}{5} \quad \text{OR} \quad x = 0$$

11. Points: $(3, -1), (0, -3), (12, 5)$

$$\begin{vmatrix} 3 & -1 & 1 \\ 0 & -3 & 1 \\ 12 & 5 & 1 \end{vmatrix} = \begin{vmatrix} 3 & -1 & 1 \\ 0 & -3 & 1 \\ 0 & 9 & -3 \end{vmatrix} = 3\begin{vmatrix} -3 & 1 \\ 9 & -3 \end{vmatrix} = 0$$

The points are collinear.

13. Points: $\left(2, -\frac{1}{2}\right), (-4, 4), (6, -3)$

$$\begin{vmatrix} 2 & -\frac{1}{2} & 1 \\ -4 & 4 & 1 \\ 6 & -3 & 1 \end{vmatrix} = \begin{vmatrix} -4 & \frac{5}{2} & 0 \\ -10 & 7 & 0 \\ 6 & -3 & 1 \end{vmatrix} = 3\begin{vmatrix} -4 & \frac{5}{2} \\ -10 & 7 \end{vmatrix} = -3 \neq 0$$

The points are not collinear.

15. $\begin{vmatrix} 2 & -5 & 1 \\ 4 & x & 1 \\ 5 & -2 & 1 \end{vmatrix} = 0$

$$2\begin{vmatrix} x & 1 \\ -2 & 1 \end{vmatrix} + 5\begin{vmatrix} 4 & 1 \\ 5 & 1 \end{vmatrix} + \begin{vmatrix} 4 & x \\ 5 & -2 \end{vmatrix} = 0$$

$$2(x + 2) + 5(-1) + (-8 - 5x) = 0$$

$$-3x - 9 = 0$$

$$x = -3$$

17. $3x + 4y = -2$

$5x + 3y = 4$

$$x = \frac{\begin{vmatrix} -2 & 4 \\ 4 & 3 \end{vmatrix}}{\begin{vmatrix} 3 & 4 \\ 5 & 3 \end{vmatrix}} = \frac{-22}{-11} = 2$$

$$y = \frac{\begin{vmatrix} 3 & -2 \\ 5 & 4 \end{vmatrix}}{\begin{vmatrix} 3 & 4 \\ 5 & 3 \end{vmatrix}} = \frac{22}{-11} = -2 \qquad \textit{Answer: } (2, -2)$$

19. $-7x + 11y = -1$

$3x - 9y = 9$

$$x = \frac{\begin{vmatrix} -1 & 11 \\ 9 & -9 \end{vmatrix}}{\begin{vmatrix} -7 & 11 \\ 3 & -9 \end{vmatrix}} = \frac{-90}{30} = -3$$

$$y = \frac{\begin{vmatrix} -7 & -1 \\ 3 & 9 \end{vmatrix}}{\begin{vmatrix} -7 & 11 \\ 3 & -9 \end{vmatrix}} = \frac{-60}{30} = -2$$

Answer: $(-3, -2)$

21. $-0.4x + 0.8y = 1.6$

$0.2x + 0.3y = 2.2$

$$D = \begin{vmatrix} -0.4 & 0.8 \\ 0.2 & 0.3 \end{vmatrix} = -0.28$$

$$x = \frac{\begin{vmatrix} 1.6 & 0.8 \\ 2.2 & 0.3 \end{vmatrix}}{-0.28} = \frac{-1.18}{-0.28} = \frac{32}{7}$$

$$y = \frac{\begin{vmatrix} -0.4 & 1.6 \\ 0.2 & 2.2 \end{vmatrix}}{-0.28} = \frac{-1.20}{-0.28} = \frac{30}{7}$$

Answer: $\left(\frac{32}{7}, \frac{30}{7}\right)$

23. $4x - y + z = -5$

$2x + 2y + 3z = 10$

$5x - 2y + 6z = 1$

$$D = \begin{vmatrix} 4 & -1 & 1 \\ 2 & 2 & 3 \\ 5 & -2 & 6 \end{vmatrix} = 55$$

$$x = \frac{\begin{vmatrix} -5 & -1 & 1 \\ 10 & 2 & 3 \\ 1 & -2 & 6 \end{vmatrix}}{55} = \frac{-55}{55} = -1, \quad y = \frac{\begin{vmatrix} 4 & -5 & 1 \\ 2 & 10 & 3 \\ 5 & 1 & 6 \end{vmatrix}}{55} = \frac{165}{55} = 3, \quad z = \frac{\begin{vmatrix} 4 & -1 & -5 \\ 2 & 2 & 10 \\ 5 & -2 & 1 \end{vmatrix}}{55} = \frac{110}{55} = 2$$

Answer: $(-1, 3, 2)$

25. $3x + 3y + 5z = 1$

$3x + 5y + 9z = 2$

$5x + 9y + 17z = 4$

$$D = \begin{vmatrix} 3 & 3 & 5 \\ 3 & 5 & 9 \\ 5 & 9 & 17 \end{vmatrix} = 4$$

$$x = \frac{\begin{vmatrix} 1 & 3 & 5 \\ 2 & 5 & 9 \\ 4 & 9 & 17 \end{vmatrix}}{4} = 0, \quad y = \frac{\begin{vmatrix} 3 & 1 & 5 \\ 3 & 2 & 9 \\ 5 & 4 & 17 \end{vmatrix}}{4} = -\frac{1}{2}, \quad z = \frac{\begin{vmatrix} 3 & 3 & 1 \\ 3 & 5 & 2 \\ 5 & 9 & 4 \end{vmatrix}}{4} = \frac{1}{2}$$

Answer: $\left(0, -\frac{1}{2}, \frac{1}{2}\right)$

27. Vertices: $\overset{A}{(0, 25)}, \overset{B}{(10, 0)}, \overset{C}{(28, 5)}$

$$\frac{1}{2}\begin{vmatrix} 0 & 25 & 1 \\ 10 & 0 & 1 \\ 28 & 5 & 1 \end{vmatrix} = 250.$$ Area = 250 square miles

29. The uncoded row matrices are the rows of the 7 × 3 matrix on the left.

$$\begin{array}{ccc} T & R & O \\ U & B & L \\ E & & I \\ N & & R \\ I & V & E \\ R & & C \\ I & T & Y \end{array} \begin{bmatrix} 20 & 18 & 15 \\ 21 & 2 & 12 \\ 5 & 0 & 9 \\ 14 & 0 & 18 \\ 9 & 22 & 5 \\ 18 & 0 & 3 \\ 9 & 20 & 25 \end{bmatrix} \begin{bmatrix} 1 & -1 & 0 \\ 1 & 0 & -1 \\ -6 & 2 & 3 \end{bmatrix} = \begin{bmatrix} -52 & 10 & 27 \\ -49 & 3 & 34 \\ -49 & 13 & 27 \\ -94 & 22 & 54 \\ 1 & 1 & -7 \\ 0 & -12 & 9 \\ -121 & 41 & 55 \end{bmatrix}$$

Answer: [−52, 10, 27], [−49, 3, 34], [−49, 13, 27], [−94, 22, 54], [1, 1, −7], [0, −12, 9], [−121, 41, 55]

In Exercises 31 and 33, use the matrix $A = \begin{bmatrix} 1 & 2 & 2 \\ 3 & 7 & 9 \\ -1 & -4 & -7 \end{bmatrix}$.

31. S H O W _ M E _ T H E _ M O N E Y _

[19 8 15] [23 0 13] [5 0 20] [8 5 0] [13 15 14] [5 25 0]

$[19 \;\; 8 \;\; 15]A = [28 \;\; 34 \;\; 5]$

$[23 \;\; 0 \;\; 13]A = [10 \;\; -6 \;\; -45]$

$[5 \;\; 0 \;\; 20]A = [-15 \;\; -70 \;\; -130]$

$[8 \;\; 5 \;\; 0]A = [23 \;\; 51 \;\; 61]$

$[13 \;\; 15 \;\; 14]A = [44 \;\; 75 \;\; 63]$

$[5 \;\; 25 \;\; 0]A = [80 \;\; 185 \;\; 235]$

Cryptogram: 38 74 75 10 −6 −45 −15 −70 −130 23 51 61 44 75 63 80 185 235

33. H A P P Y _ B I R T H D A Y _

[8 1 16] [16 25 0] [2 9 18] [20 8 4] [1 25 0]

$[8 \;\; 1 \;\; 16]A = [5 \;\; -41 \;\; -87]$

$[16 \;\; 25 \;\; 0]A = [91 \;\; 207 \;\; 257]$

$[2 \;\; 9 \;\; 18]A = [11 \;\; -5 \;\; -41]$

$[20 \;\; 8 \;\; 4]A = [40 \;\; 80 \;\; 84]$

$[1 \;\; 25 \;\; 0]A = [76 \;\; 177 \;\; 227]$

Cryptogram: −5 −41 −87 91 207 257 11 −5 −41 40 80 84 76 177 227

35. $A^{-1} = \begin{bmatrix} 1 & 2 \\ 3 & 5 \end{bmatrix}^{-1} = \begin{bmatrix} -5 & 2 \\ 3 & -1 \end{bmatrix}$

$$\begin{bmatrix} 11 & 21 \\ 64 & 112 \\ 25 & 50 \\ 29 & 53 \\ 23 & 46 \\ 40 & 75 \\ 55 & 92 \end{bmatrix} \begin{bmatrix} -5 & 2 \\ 3 & -1 \end{bmatrix} = \begin{bmatrix} 8 & 1 \\ 16 & 16 \\ 25 & 0 \\ 14 & 5 \\ 23 & 0 \\ 25 & 5 \\ 1 & 18 \end{bmatrix} \begin{array}{cc} H & A \\ P & P \\ Y & \\ N & E \\ W & \\ Y & E \\ A & R \end{array}$$

Message: HAPPY NEW YEAR

37. $A^{-1} = \begin{bmatrix} 1 & 2 & 2 \\ 3 & 7 & 9 \\ -1 & -4 & -7 \end{bmatrix}^{-1} = \begin{bmatrix} -13 & 6 & 4 \\ 12 & -5 & -3 \\ -5 & 2 & 1 \end{bmatrix}$

$$\begin{bmatrix} 20 & 17 & -15 \\ -12 & -56 & -104 \\ 1 & -25 & -65 \\ 62 & 143 & 181 \end{bmatrix} \begin{bmatrix} -13 & 6 & 4 \\ 12 & -5 & -3 \\ -5 & 2 & 1 \end{bmatrix} = \begin{bmatrix} 19 & 5 & 14 \\ 4 & 0 & 16 \\ 12 & 1 & 14 \\ 5 & 19 & 0 \end{bmatrix} \begin{matrix} \text{S E N} \\ \text{D \quad P} \\ \text{L A N} \\ \text{E S \quad} \end{matrix}$$

Message: SEND PLANES

39. Let A be the 2×2 matrix needed to decode the message.

$$\begin{bmatrix} -18 & -18 \\ 1 & 16 \end{bmatrix} A = \begin{bmatrix} 0 & 18 \\ 15 & 14 \end{bmatrix} \begin{matrix} \text{\quad R} \\ \text{O N} \end{matrix}$$

$$A = \begin{bmatrix} -18 & -18 \\ 1 & 16 \end{bmatrix}^{-1} \begin{bmatrix} 0 & 18 \\ 15 & 14 \end{bmatrix} = \begin{bmatrix} -\frac{8}{135} & -\frac{1}{15} \\ \frac{1}{270} & \frac{1}{15} \end{bmatrix} \begin{bmatrix} 0 & 18 \\ 15 & 14 \end{bmatrix} = \begin{bmatrix} -1 & -2 \\ 1 & 1 \end{bmatrix}$$

$$\begin{bmatrix} 8 & 21 \\ -15 & -10 \\ -13 & -13 \\ 5 & 10 \\ 5 & 25 \\ 5 & 19 \\ -1 & 6 \\ 20 & 40 \\ -18 & -18 \\ 1 & 16 \end{bmatrix} \begin{bmatrix} -1 & -2 \\ 1 & 1 \end{bmatrix} = \begin{bmatrix} 13 & 5 \\ 5 & 20 \\ 0 & 13 \\ 5 & 0 \\ 20 & 15 \\ 14 & 9 \\ 7 & 8 \\ 20 & 0 \\ 0 & 18 \\ 15 & 14 \end{bmatrix} \begin{matrix} \text{M E} \\ \text{E T} \\ \text{\quad M} \\ \text{E \quad} \\ \text{T O} \\ \text{N I} \\ \text{G H} \\ \text{T \quad} \\ \text{\quad R} \\ \text{O N} \end{matrix}$$

Message: MEET ME TONIGHT RON

41. True. Cramer's Rule requires that the determinant of the coefficient matrix be nonzero.

43. Answers will vary.

45. $y + 6 = \dfrac{10 - (-6)}{-2 - 0}(x - 0) = -8x$

$y = -8x - 6$

47. $y - 12 = \dfrac{12 - 2}{-4 - 4}(x + 4) = -\dfrac{5}{4}(x + 4)$

$4y - 48 = -5x - 20$

$4y + 5x = 28$

49. $f(x) = \dfrac{2x}{x^2 + 3x - 18} = \dfrac{2x}{(x + 6)(x - 3)}$

Vertical asymptotes: $x = -6, 3$

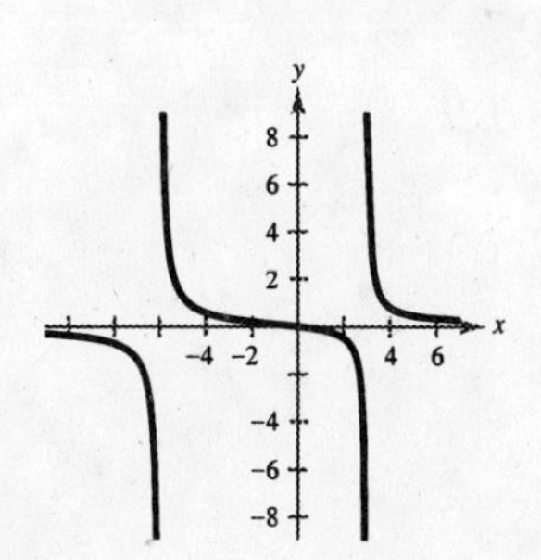

51. $\begin{bmatrix} 5 & -1 & -1 & \vdots & 7 \\ -2 & 3 & 1 & \vdots & -5 \\ 4 & 10 & -5 & \vdots & -37 \end{bmatrix}$

row reduces to

$\begin{bmatrix} 1 & 0 & 0 & \vdots & 2 \\ 0 & 1 & 0 & \vdots & -2 \\ 0 & 0 & 1 & \vdots & 5 \end{bmatrix}$

Answer: $(2, -2, 5)$

Review Exercises for Chapter 8

Solutions to Odd-Numbered Exercises

1. Order 3×1

3. Order 1×1

5. Order 4×2

7. $\begin{bmatrix} 3 & -10 & \vdots & 15 \\ 5 & 4 & \vdots & 22 \end{bmatrix}$

9. $\begin{bmatrix} 8 & -7 & 4 & \vdots & 12 \\ 3 & -5 & 2 & \vdots & 20 \\ 5 & 3 & -3 & \vdots & 26 \end{bmatrix}$

11. $\begin{bmatrix} 5 & 1 & 7 & \vdots & -9 \\ 4 & 2 & 0 & \vdots & 10 \\ 9 & 4 & 2 & \vdots & 3 \end{bmatrix}$

$$\begin{aligned} 5x + y + 7z &= -9 \\ 4x + 2y &= 10 \\ 9x + 4y + 2z &= 3 \end{aligned}$$

13. $\begin{bmatrix} 0 & 1 & 1 \\ 1 & 2 & 3 \\ 2 & 2 & 2 \end{bmatrix}$

$$\begin{array}{r} R_1 + R_2 \rightarrow \\ -R_1 + R_2 \rightarrow \\ -2R_1 + R_3 \rightarrow \end{array} \begin{bmatrix} 1 & 3 & 4 \\ 0 & -1 & -1 \\ 0 & -4 & -6 \end{bmatrix}$$

$$\begin{array}{r} 3R_2 + R_1 \rightarrow \\ -R_2 \rightarrow \\ -4R_2 + R_3 \rightarrow \end{array} \begin{bmatrix} 1 & 0 & 1 \\ 0 & 1 & 1 \\ 0 & 0 & -2 \end{bmatrix}$$

$$\begin{array}{r} -R_3 + R_1 \rightarrow \\ -R_3 + R_2 \rightarrow \\ -\frac{1}{2}R_3 \rightarrow \end{array} \begin{bmatrix} 1 & 0 & 0 \\ 0 & 1 & 0 \\ 0 & 0 & 1 \end{bmatrix}$$

15. $\begin{bmatrix} 3 & -2 & 1 & 0 \\ 4 & -3 & 0 & 1 \end{bmatrix} \Rightarrow \begin{bmatrix} 1 & 0 & 3 & -2 \\ 0 & 1 & 4 & -3 \end{bmatrix}$

17. $\begin{bmatrix} 1 & 3 & 4 \\ 0 & 1 & 1 \\ 2 & 4 & 6 \end{bmatrix} \Rightarrow \begin{bmatrix} 1 & 0 & 1 \\ 0 & 1 & 1 \\ 0 & 0 & 0 \end{bmatrix}$

19. $\begin{bmatrix} 5 & 4 & \vdots & 2 \\ -1 & 1 & \vdots & -22 \end{bmatrix}$

$$\begin{array}{r} 4R_2 + R_1 \rightarrow \\ R_1 + R_2 \rightarrow \end{array} \begin{bmatrix} 1 & 8 & \vdots & -86 \\ 0 & 9 & \vdots & -108 \end{bmatrix}$$

$9y = -108$

$y = -12$

$x = -8(-12) - 86 = 10$

Answer: $(10, -12)$

21. $\begin{bmatrix} 2 & 1 & \vdots & 0.3 \\ 3 & -1 & \vdots & -1.3 \end{bmatrix}$

$$-R_1 + R_2 \rightarrow \begin{bmatrix} 2 & 1 & \vdots & 0.3 \\ 1 & -2 & \vdots & -1.6 \end{bmatrix}$$

$$\begin{bmatrix} 1 & -2 & \vdots & -1.6 \\ 2 & 1 & \vdots & 0.3 \end{bmatrix}$$

$$-2R_1 + R_2 \rightarrow \begin{bmatrix} 1 & -2 & \vdots & -1.6 \\ 0 & 5 & \vdots & 3.5 \end{bmatrix}$$

$5y = 3.5 \Rightarrow y = 0.7$

$x = 2(0.7) - 1.6 = -0.2$

$x = -0.2,\ y = 0.7$

23. $\begin{bmatrix} 2 & 3 & 3 & \vdots & 3 \\ 6 & 6 & 12 & \vdots & 13 \\ 12 & 9 & -1 & \vdots & 2 \end{bmatrix}$

$\begin{matrix} \\ -3R_1 + R_2 \rightarrow \\ -6R_1 + R_3 \rightarrow \end{matrix} \begin{bmatrix} 2 & 3 & 3 & \vdots & 3 \\ 0 & -3 & 3 & \vdots & 4 \\ 0 & -3 & -25 & \vdots & -24 \end{bmatrix}$

$\begin{matrix} R_2 + R_1 \rightarrow \\ \\ -R_2 + R_3 \rightarrow \end{matrix} \begin{bmatrix} 2 & 0 & 6 & \vdots & 7 \\ 0 & -3 & 3 & \vdots & 4 \\ 0 & 0 & -28 & \vdots & -28 \end{bmatrix}$

$\begin{matrix} \frac{1}{2}R_1 \rightarrow \\ -\frac{1}{3}R_2 \rightarrow \\ -\frac{1}{28}R_3 \rightarrow \end{matrix} \begin{bmatrix} 1 & 0 & 3 & \vdots & \frac{7}{2} \\ 0 & 1 & -1 & \vdots & -\frac{4}{3} \\ 0 & 0 & 1 & \vdots & 1 \end{bmatrix}$

$z = 1$

$y - 1 = -\frac{4}{3} \Rightarrow y = -\frac{1}{3}$

$x + 3(1) = \frac{7}{2} \Rightarrow x = \frac{1}{2}$

Answer: $\left(\frac{1}{2}, -\frac{1}{3}, 1\right)$

25. $\begin{bmatrix} 3 & 21 & -29 & \vdots & -1 \\ 2 & 15 & -21 & \vdots & 0 \end{bmatrix}$

$-R_2 + R_1 \rightarrow \begin{bmatrix} 1 & 6 & -8 & \vdots & -1 \\ 2 & 15 & -21 & \vdots & 0 \end{bmatrix}$

$\begin{matrix} \\ -2R_1 + R_2 \rightarrow \end{matrix} \begin{bmatrix} 1 & 6 & -8 & \vdots & -1 \\ 0 & 3 & -5 & \vdots & 2 \end{bmatrix}$

$-2R_2 + R_1 \rightarrow \begin{bmatrix} 1 & 0 & 2 & \vdots & -5 \\ 0 & 3 & -5 & \vdots & 2 \end{bmatrix}$

Let $z = a$, any real number.

$3y - 5a = 2 \Rightarrow \frac{5}{3}a + \frac{2}{3}$

$x + 2a = -5 \Rightarrow -2a - 5$

Answer: $\left(-2a - 5, \frac{5}{3}a + \frac{2}{3}, a\right)$

27. $\begin{bmatrix} -1 & 1 & 2 & \vdots & 1 \\ 2 & 3 & 1 & \vdots & -2 \\ 5 & 4 & 2 & \vdots & 4 \end{bmatrix}$

$\begin{matrix} -R_1 \rightarrow \\ 2R_1 + R_2 \rightarrow \\ 5R_1 + R_3 \rightarrow \end{matrix} \begin{bmatrix} 1 & -1 & -2 & \vdots & -1 \\ 0 & 5 & 5 & \vdots & 0 \\ 0 & 9 & 12 & \vdots & 9 \end{bmatrix}$

$\begin{matrix} R_2 + R_1 \rightarrow \\ \frac{1}{5}R_2 \rightarrow \\ -9R_2 + R_3 \rightarrow \end{matrix} \begin{bmatrix} 1 & 0 & -1 & \vdots & -1 \\ 0 & 1 & 1 & \vdots & 0 \\ 0 & 0 & 3 & \vdots & 9 \end{bmatrix}$

$\begin{matrix} R_3 + R_1 \rightarrow \\ -R_3 + R_2 \rightarrow \\ \frac{1}{3}R_3 \rightarrow \end{matrix} \begin{bmatrix} 1 & 0 & 0 & \vdots & 2 \\ 0 & 1 & 0 & \vdots & -3 \\ 0 & 0 & 1 & \vdots & 3 \end{bmatrix}$

$x = 2, y = -3, z = 3$

Answer: $(2, -3, 3)$

29. $\begin{bmatrix} 2 & -1 & 9 & \vdots & -8 \\ -1 & -3 & 4 & \vdots & -15 \\ 5 & 2 & -1 & \vdots & 17 \end{bmatrix}$

$\begin{bmatrix} 1 & 3 & -4 & \vdots & 15 \\ 0 & -7 & 17 & \vdots & -38 \\ 0 & -13 & 19 & \vdots & -58 \end{bmatrix}$

$\begin{bmatrix} 1 & 3 & -4 & \vdots & 15 \\ 0 & -7 & 17 & \vdots & -38 \\ 0 & 0 & -\frac{88}{7} & \vdots & \frac{88}{7} \end{bmatrix}$

$z = -1$

$-7y = -17(-1) - 38 = -21 \Rightarrow y = 3$

$x = -3(3) + 4(-1) + 15 = 2$

Answers: $(2, 3, -1)$

31. $\begin{bmatrix} 1 & -3 & \vdots & -2 \\ 1 & 1 & \vdots & 2 \end{bmatrix}$ reduces to $\begin{bmatrix} 1 & 0 & \vdots & 1 \\ 0 & 1 & \vdots & 1 \end{bmatrix}$

Answer: $(1, 1)$

33. $\begin{bmatrix} 1 & 2 & -1 & \vdots & 7 \\ 0 & -1 & -1 & \vdots & 4 \\ 4 & 0 & -1 & \vdots & 16 \end{bmatrix}$ reduces to $\begin{bmatrix} 1 & 0 & 0 & \vdots & 3 \\ 0 & 1 & 0 & \vdots & 0 \\ 0 & 0 & 1 & \vdots & -4 \end{bmatrix}$

Answer: $(3, 0, -4)$

35. $\begin{bmatrix} 3 & -1 & 5 & -2 & \vdots & -44 \\ 1 & 6 & 4 & -1 & \vdots & 1 \\ 5 & -1 & 1 & 3 & \vdots & -15 \\ 0 & 4 & -1 & -8 & \vdots & 58 \end{bmatrix}$ reduces to $\begin{bmatrix} 1 & 0 & 0 & 0 & \vdots & 2 \\ 0 & 1 & 0 & 0 & \vdots & 6 \\ 0 & 0 & 1 & 0 & \vdots & -10 \\ 0 & 0 & 0 & 1 & \vdots & -3 \end{bmatrix}$

Answer: $(2, 6, -10, -3)$

37. $x = 12$

$y = -7$

39. $x + 3 = 5x - 1 \Rightarrow x = 1$

$-4y = -44 \Rightarrow y = 11$

$y + 5 = 16 \Rightarrow y = 11$

$6x = 6 \Rightarrow x = 1$

Answer: $x = 1, y = 11$

41. $\begin{bmatrix} 7 & 3 \\ -1 & 5 \end{bmatrix} + \begin{bmatrix} 10 & -20 \\ 14 & -3 \end{bmatrix} = \begin{bmatrix} 17 & -17 \\ 13 & 2 \end{bmatrix}$

43. $\begin{bmatrix} 2 & 1 & 0 \\ 0 & 5 & -4 \end{bmatrix} - 3\begin{bmatrix} 5 & 3 & -6 \\ 0 & -2 & 5 \end{bmatrix} = \begin{bmatrix} 2 & 1 & 0 \\ 0 & 5 & -4 \end{bmatrix} - \begin{bmatrix} 15 & 9 & -18 \\ 0 & -6 & 15 \end{bmatrix}$

$= \begin{bmatrix} -13 & -8 & 18 \\ 0 & 11 & -19 \end{bmatrix}$

45. $-\begin{bmatrix} 8 & -1 & 8 \\ -2 & 4 & 12 \\ 0 & -6 & 0 \end{bmatrix} - 5\begin{bmatrix} -2 & 0 & -4 \\ 3 & -1 & 1 \\ 6 & 12 & -8 \end{bmatrix} = \begin{bmatrix} -8 & 1 & -8 \\ 2 & -4 & -12 \\ 0 & 6 & 0 \end{bmatrix} + \begin{bmatrix} 10 & 0 & 20 \\ -15 & 5 & -5 \\ -30 & -60 & 40 \end{bmatrix}$

$= \begin{bmatrix} 2 & 1 & 12 \\ -13 & 1 & -17 \\ -30 & -54 & 40 \end{bmatrix}$

47. $3\begin{bmatrix} 8 & -2 & 5 \\ 1 & 3 & -1 \end{bmatrix} + 6\begin{bmatrix} 4 & -2 & -3 \\ 2 & 7 & 6 \end{bmatrix} = \begin{bmatrix} 48 & -18 & -3 \\ 15 & 51 & 33 \end{bmatrix}$

49. $X = 3A - 2B = 3\begin{bmatrix} -4 & 0 \\ 1 & -5 \\ -3 & 2 \end{bmatrix} - 2\begin{bmatrix} 1 & 2 \\ -2 & 1 \\ 4 & 4 \end{bmatrix} = \begin{bmatrix} -14 & -4 \\ 7 & -17 \\ -17 & -2 \end{bmatrix}$

51. $X = \frac{1}{3}[B - 2A] = \frac{1}{3}\left(\begin{bmatrix} 1 & 2 \\ -2 & 1 \\ 4 & 4 \end{bmatrix} - 2\begin{bmatrix} -4 & 0 \\ 1 & -5 \\ -3 & 2 \end{bmatrix}\right) = \frac{1}{3}\begin{bmatrix} 9 & 2 \\ -4 & 11 \\ 10 & 0 \end{bmatrix}$

53. $\begin{bmatrix} 1 & 2 \\ 5 & -4 \\ 6 & 0 \end{bmatrix}\begin{bmatrix} 6 & -2 & 8 \\ 4 & 0 & 0 \end{bmatrix} = \begin{bmatrix} 1(6) + 2(4) & 1(-2) + 2(0) & 1(8) + 2(0) \\ 5(6) + (-4)(4) & 5(-2) + (-4)(0) & 5(8) + (-4)(0) \\ 6(6) + (0)(4) & 6(-2) + (0)(0) & 6(8) + (0)(0) \end{bmatrix}$

$= \begin{bmatrix} 14 & -2 & 8 \\ 14 & -10 & 40 \\ 36 & -12 & 48 \end{bmatrix}$

55. $\begin{bmatrix} 1 & 5 & 6 \\ 2 & -4 & 0 \end{bmatrix}\begin{bmatrix} 6 & 4 \\ -2 & 0 \\ 8 & 0 \end{bmatrix} = \begin{bmatrix} 1(6) + 5(-2) + 6(8) & 1(4) + 5(0) + 6(0) \\ 2(6) - 4(-2) + 0(8) & 2(4) - 4(0) + 0(0) \end{bmatrix}$

$= \begin{bmatrix} 44 & 4 \\ 20 & 8 \end{bmatrix}$

57. $\begin{bmatrix} 2 & 1 \\ 6 & 0 \end{bmatrix}\left(\begin{bmatrix} 4 & 2 \\ -3 & 1 \end{bmatrix} + \begin{bmatrix} -2 & 4 \\ 0 & 4 \end{bmatrix}\right) = \begin{bmatrix} 2 & 1 \\ 6 & 0 \end{bmatrix}\begin{bmatrix} 2 & 6 \\ -3 & 5 \end{bmatrix}$

$= \begin{bmatrix} 2(2) + 1(-3) & 2(6) + 1(5) \\ 6(2) + 0 & 6(6) + 0 \end{bmatrix}$

$= \begin{bmatrix} 1 & 17 \\ 12 & 36 \end{bmatrix}$

59. $\begin{bmatrix} 4 & 1 \\ 11 & -7 \\ 12 & 3 \end{bmatrix}\begin{bmatrix} 3 & -5 & 6 \\ 2 & -2 & -2 \end{bmatrix} = \begin{bmatrix} 14 & -22 & 22 \\ 19 & -41 & 80 \\ 42 & -66 & 66 \end{bmatrix}$

61. $\begin{bmatrix} 5 & 4 \\ -1 & 1 \end{bmatrix}\begin{bmatrix} x \\ y \end{bmatrix} = \begin{bmatrix} 2 \\ -22 \end{bmatrix} \Rightarrow \begin{bmatrix} 5x + 4y \\ -x + y \end{bmatrix} = \begin{bmatrix} 2 \\ -22 \end{bmatrix} \Rightarrow \begin{aligned} 5x + 4y &= 2 \\ -x + y &= -22 \end{aligned}$

63. $BA = \begin{bmatrix} 10.25 & 14.50 & 17.75 \end{bmatrix}\begin{bmatrix} 8200 & 7400 \\ 6500 & 9800 \\ 5400 & 4800 \end{bmatrix} = \begin{bmatrix} 274{,}150 & 303{,}150 \end{bmatrix}$

This 1×2 matrix represents the total value of the products in each of the two warehouses.

65. $AB = \begin{bmatrix} -4 & -1 \\ 7 & 2 \end{bmatrix}\begin{bmatrix} -2 & -1 \\ 7 & 4 \end{bmatrix} = \begin{bmatrix} 1 & 0 \\ 0 & 1 \end{bmatrix}$; $BA = I_2$

67. $AB = \begin{bmatrix} 1 & 1 & 0 \\ 1 & 0 & 1 \\ 6 & 2 & 3 \end{bmatrix}\begin{bmatrix} -2 & -3 & 1 \\ 3 & 3 & -1 \\ 2 & 4 & -1 \end{bmatrix} = \begin{bmatrix} 1 & 0 & 0 \\ 0 & 1 & 0 \\ 0 & 0 & 1 \end{bmatrix}$; $BA = I_3$

69. $\left[\begin{array}{cc:cc} -6 & 5 & 1 & 0 \\ -5 & 4 & 0 & 1 \end{array}\right]$ row reduces to $\left[\begin{array}{cc:cc} 1 & 0 & 4 & -5 \\ 0 & 1 & 5 & -6 \end{array}\right]$

$\begin{bmatrix} -6 & 5 \\ -5 & 4 \end{bmatrix}^{-1} = \begin{bmatrix} 4 & -5 \\ 5 & -6 \end{bmatrix}$

71. $\left[\begin{array}{ccc:ccc} -1 & -2 & -2 & 1 & 0 & 0 \\ 3 & 7 & 9 & 0 & 1 & 0 \\ 1 & 4 & 7 & 0 & 0 & 1 \end{array}\right]$ row reduces to $\left[\begin{array}{ccc:ccc} 1 & 0 & 0 & 13 & 6 & -4 \\ 0 & 1 & 0 & -12 & -5 & 3 \\ 0 & 0 & 1 & 5 & 2 & -1 \end{array}\right]$

$\begin{bmatrix} -1 & -2 & -2 \\ 3 & 7 & 9 \\ 1 & 4 & 7 \end{bmatrix}^{-1} = \begin{bmatrix} 13 & 6 & -4 \\ -12 & -5 & 3 \\ 5 & 2 & -1 \end{bmatrix}$

73. $\begin{bmatrix} 2 & 6 \\ 3 & -6 \end{bmatrix}^{-1} = \begin{bmatrix} \frac{1}{5} & \frac{1}{5} \\ \frac{1}{10} & -\frac{1}{15} \end{bmatrix}$

75. $\begin{bmatrix} 2 & 0 & 3 \\ -1 & 1 & 1 \\ 2 & -2 & 1 \end{bmatrix}^{-1} = \begin{bmatrix} \frac{1}{2} & -1 & -\frac{1}{2} \\ \frac{1}{2} & -\frac{2}{3} & -\frac{5}{6} \\ 0 & \frac{2}{3} & \frac{1}{3} \end{bmatrix}$

77. $\begin{bmatrix} -7 & 2 \\ -8 & 2 \end{bmatrix}^{-1} = \dfrac{1}{(-7)(2) - (2)(-8)} \begin{bmatrix} 2 & -2 \\ 8 & -7 \end{bmatrix} = \begin{bmatrix} 1 & -1 \\ 4 & -\frac{7}{2} \end{bmatrix}$

79. $\begin{bmatrix} -6 & -5 \\ 3 & 3 \end{bmatrix}^{-1} = \dfrac{1}{(-6)(3) - (-5)(3)} \begin{bmatrix} 3 & 5 \\ -3 & -6 \end{bmatrix} = \begin{bmatrix} -1 & -\frac{5}{3} \\ 1 & 2 \end{bmatrix}$

81. $\begin{bmatrix} -1 & 20 \\ \frac{3}{10} & -6 \end{bmatrix}^{-1} = \dfrac{1}{(-1)(-6) - (20)\left(\frac{3}{10}\right)} \begin{bmatrix} -6 & -20 \\ -\frac{3}{10} & -1 \end{bmatrix} = \dfrac{1}{0} \begin{bmatrix} -6 & -20 \\ -\frac{3}{10} & -1 \end{bmatrix}$

Inverse does not exist.

83. $\begin{bmatrix} -1 & 4 \\ 2 & -7 \end{bmatrix}^{-1} = \begin{bmatrix} 7 & 4 \\ 2 & 1 \end{bmatrix}$

$\begin{bmatrix} x \\ y \end{bmatrix} = \begin{bmatrix} 7 & 4 \\ 2 & 1 \end{bmatrix} \begin{bmatrix} 8 \\ -5 \end{bmatrix} = \begin{bmatrix} 36 \\ 11 \end{bmatrix}$

Answer: $(36, 11)$

85. $\begin{bmatrix} -3 & 10 \\ 5 & -17 \end{bmatrix}^{-1} = \begin{bmatrix} -17 & -10 \\ -5 & -3 \end{bmatrix}$

$\begin{bmatrix} x \\ y \end{bmatrix} = \begin{bmatrix} -17 & -10 \\ -5 & -3 \end{bmatrix} \begin{bmatrix} 8 \\ -13 \end{bmatrix} = \begin{bmatrix} -6 \\ -1 \end{bmatrix}$

Answer: $(-6, -1)$

87. $\begin{bmatrix} 3 & 2 & -1 \\ 1 & -1 & 2 \\ 5 & 1 & 1 \end{bmatrix}^{-1} = \begin{bmatrix} -1 & -1 & 1 \\ 3 & \frac{8}{3} & -\frac{7}{3} \\ 2 & \frac{7}{3} & -\frac{5}{3} \end{bmatrix}$

$\begin{bmatrix} x \\ y \\ z \end{bmatrix} = \begin{bmatrix} -1 & -1 & 1 \\ 3 & \frac{8}{3} & -\frac{7}{3} \\ 2 & \frac{7}{3} & -\frac{5}{3} \end{bmatrix} \begin{bmatrix} 6 \\ -1 \\ 7 \end{bmatrix} = \begin{bmatrix} 2 \\ -1 \\ -2 \end{bmatrix}$

Answer: $(2, -1, -2)$

89. $\begin{bmatrix} -2 & 1 & 2 \\ -1 & -4 & 1 \\ 0 & -1 & -1 \end{bmatrix}^{-1} = \begin{bmatrix} -\frac{5}{9} & \frac{1}{9} & -1 \\ \frac{1}{9} & -\frac{2}{9} & 0 \\ -\frac{1}{9} & \frac{2}{9} & -1 \end{bmatrix}$

$\begin{bmatrix} x \\ y \\ z \end{bmatrix} = \begin{bmatrix} -\frac{5}{9} & \frac{1}{9} & -1 \\ \frac{1}{9} & -\frac{2}{9} & 0 \\ -\frac{1}{9} & \frac{2}{9} & -1 \end{bmatrix} \begin{bmatrix} -13 \\ -11 \\ 0 \end{bmatrix} = \begin{bmatrix} 6 \\ 1 \\ 1 \end{bmatrix}$

Answer: $(6, 1, -1)$

91. $x + 2y = -1$

$3x + 4y = -5$

$\begin{bmatrix} 1 & 2 \\ 3 & 4 \end{bmatrix}^{-1} = \begin{bmatrix} -2 & 1 \\ \frac{3}{2} & -\frac{1}{2} \end{bmatrix} \Rightarrow \begin{bmatrix} x \\ y \end{bmatrix} = \begin{bmatrix} -2 & 1 \\ \frac{3}{2} & -\frac{1}{2} \end{bmatrix} \begin{bmatrix} -1 \\ -5 \end{bmatrix} = \begin{bmatrix} -3 \\ 1 \end{bmatrix}$

$x = -3, y = 1$

Answer: $(-3, 1)$

93. $-3x - 3y - 4z = -2$

$y + z = -1$

$4x + 3y + 4z = -1$

$\begin{bmatrix} -3 & -3 & -4 \\ 0 & 1 & 1 \\ 4 & 3 & 4 \end{bmatrix}^{-1} = \begin{bmatrix} 1 & 0 & 1 \\ 4 & 4 & 3 \\ 4 & -3 & -3 \end{bmatrix} \Rightarrow \begin{bmatrix} x \\ y \\ z \end{bmatrix} = \begin{bmatrix} 1 & 0 & 1 \\ 4 & 4 & 3 \\ -4 & -3 & -3 \end{bmatrix} \begin{bmatrix} 2 \\ -1 \\ -1 \end{bmatrix} = \begin{bmatrix} 1 \\ 1 \\ -2 \end{bmatrix}$

$x = 1, y = 1, z = -2$

Answer: $(1, 1, -2)$

95. $-x + y + z = 6$
$4x - 3y + z = 20$
$2x - y + 3z = 8$

$\begin{bmatrix} -1 & 1 & 1 \\ 4 & -3 & 1 \\ 2 & -1 & 3 \end{bmatrix}^{-1}$ does not exist.

The system is inconsistent and has no solution.

97. $\begin{vmatrix} 8 & 5 \\ 2 & -4 \end{vmatrix} = 8(-4) - 2(5) = -42$

99. $\begin{vmatrix} 50 & -30 \\ 10 & 5 \end{vmatrix} = 50(5) - (-30)(10) = 550$

101. $A = \begin{bmatrix} 2 & -1 \\ 7 & 4 \end{bmatrix}$

Minors: $M_{11} = 4$ $M_{21} = -1$
$M_{12} = 7$ $M_{22} = 2$

Cofactors: $C_{11} = 4$ $C_{21} = 1$
$C_{12} = -7$ $C_{22} = 2$

103. $A = \begin{bmatrix} 3 & 2 & -1 \\ -2 & 5 & 0 \\ 1 & 8 & 6 \end{bmatrix}$

Minors: $M_{11} = \begin{vmatrix} 5 & 0 \\ 8 & 6 \end{vmatrix} = 30, M_{12} = \begin{vmatrix} -2 & 0 \\ 1 & 6 \end{vmatrix} = -12, M_{13} = \begin{vmatrix} -2 & 5 \\ 1 & 8 \end{vmatrix} = -21$

$M_{21} = \begin{vmatrix} 2 & -1 \\ 8 & 6 \end{vmatrix} = 20, M_{22} = \begin{vmatrix} 3 & -1 \\ 1 & 6 \end{vmatrix} = 19, M_{23} = \begin{vmatrix} 3 & 2 \\ 1 & 8 \end{vmatrix} = 22$

$M_{31} = \begin{vmatrix} 2 & -1 \\ 5 & 0 \end{vmatrix} = 5, M_{32} = \begin{vmatrix} 3 & -1 \\ -2 & 0 \end{vmatrix} = -2, M_{33} = \begin{vmatrix} 3 & 2 \\ -2 & 5 \end{vmatrix} = 19$

Cofactors: $C_{11} = 30, C_{12} = 12, C_{13} = -21$
$C_{21} = -20, C_{22} = 19, C_{23} = -22$
$C_{31} = 5, C_{32} = 2, C_{33} = 19$

105. (a) $0 + 12\begin{vmatrix} 5 & -3 \\ 1 & 3 \end{vmatrix} - 4\begin{vmatrix} 5 & 0 \\ 1 & 6 \end{vmatrix} = 12(18) - 4(30) = 96$

(b) $0 + 12\begin{vmatrix} 5 & -3 \\ 1 & 3 \end{vmatrix} - 6\begin{vmatrix} 5 & -3 \\ 0 & 4 \end{vmatrix} = 12(18) - 6(20) = 96$

107. $\begin{vmatrix} -2 & 4 & 1 \\ -6 & 0 & 2 \\ 5 & 3 & 4 \end{vmatrix} = 6\begin{vmatrix} 4 & 1 \\ 3 & 4 \end{vmatrix} - 2\begin{vmatrix} -2 & 4 \\ 5 & 3 \end{vmatrix} = 6(13) - 2(-26) = 130$

109. $\begin{vmatrix} 1 & 0 & -2 \\ 0 & 1 & 0 \\ -2 & 0 & 1 \end{vmatrix} = 1\begin{vmatrix} 1 & -2 \\ -2 & 1 \end{vmatrix} = 1(1) - (-2)(-2) = 1 - 4 = -3$

111. $\begin{vmatrix} 3 & 0 & -4 & 0 \\ 0 & 8 & 1 & 2 \\ 6 & 1 & 8 & 2 \\ 0 & 3 & -4 & 1 \end{vmatrix} = 3\begin{vmatrix} 8 & 1 & 2 \\ 1 & 8 & 2 \\ 3 & -4 & 1 \end{vmatrix} + (-4)\begin{vmatrix} 0 & 8 & 2 \\ 6 & 1 & 2 \\ 0 & 3 & 1 \end{vmatrix}$ (Expansion along Row 1)

$= 3[3(8 - (-8)) - 1(1 - 6) + 2(-4 - 24)] - 4[0 - 6(8 - 6) + 0]$

$= 3[128 + 5 - 56] - 4[-12]$

$= 279$

113. $\begin{vmatrix} 5 & 3 & 0 & 6 \\ 4 & 6 & 4 & 12 \\ 0 & 2 & -3 & 4 \\ 0 & 1 & -2 & 2 \end{vmatrix} = 5\begin{vmatrix} 6 & 4 & 12 \\ 2 & -3 & 4 \\ 1 & -2 & 2 \end{vmatrix} - 4\begin{vmatrix} 3 & 0 & 6 \\ 2 & -3 & 4 \\ 1 & -2 & 2 \end{vmatrix}$

$= 5\left[6\begin{vmatrix} -3 & 4 \\ -2 & 2 \end{vmatrix} - 4\begin{vmatrix} 2 & 4 \\ 1 & 2 \end{vmatrix} + 12\begin{vmatrix} 2 & -3 \\ 1 & -2 \end{vmatrix}\right] - 4\left[-3\begin{vmatrix} 3 & 6 \\ 1 & 2 \end{vmatrix} + 2\begin{vmatrix} 3 & 6 \\ 2 & 4 \end{vmatrix}\right]$

$= 5[6(2) - 4(0) + 12(-1)] - 4[-3(0) + 2(0)] = 0$

115. $\det(A) = 8(-1)(4)(3) = -96$ (Upper Triangular)

117. $(1, 0),\ (5, 0),\ (5, 8)$

$\frac{1}{2}\begin{vmatrix} 1 & 0 & 1 \\ 5 & 0 & 1 \\ 5 & 8 & 1 \end{vmatrix} = \frac{1}{2}(32) = 16$

Area $= 16$ square units

119. $\frac{1}{2}\begin{vmatrix} \frac{1}{2} & 1 & 1 \\ 2 & -\frac{5}{2} & 1 \\ \frac{3}{2} & 1 & 1 \end{vmatrix} = \frac{1}{2}\left(\frac{7}{2}\right) = \frac{7}{4}$

Area $= \frac{7}{4}$ square units

121. $\begin{vmatrix} -1 & 7 & 1 \\ 3 & -9 & 1 \\ -3 & 15 & 1 \end{vmatrix} = 0$

The points are collinear.

123. $\begin{vmatrix} 9 & -10 & 1 \\ 4 & -1 & 1 \\ 1 & 5 & 1 \end{vmatrix} = -3$

The points are not collinear.

125. $x = \dfrac{\begin{vmatrix} 5 & 2 \\ 1 & 1 \end{vmatrix}}{\begin{vmatrix} 1 & 2 \\ -1 & 1 \end{vmatrix}} = \dfrac{3}{3} = 1$

$y = \dfrac{\begin{vmatrix} 1 & 5 \\ -1 & 1 \end{vmatrix}}{\begin{vmatrix} 1 & 2 \\ -1 & 1 \end{vmatrix}} = \dfrac{6}{3} = 2$

Answer: $(1, 2)$

127. $x = \dfrac{\begin{vmatrix} 6 & -2 \\ -23 & 3 \end{vmatrix}}{\begin{vmatrix} 5 & -2 \\ -11 & 3 \end{vmatrix}} = \dfrac{-28}{-7} = 4 \qquad y = \dfrac{\begin{vmatrix} 5 & 6 \\ -11 & -23 \end{vmatrix}}{\begin{vmatrix} 5 & -2 \\ -11 & 3 \end{vmatrix}} = \dfrac{-49}{-7} = 7$

Answer: $(4, 7)$

129. $x = \dfrac{\begin{vmatrix} -11 & 3 & -5 \\ -3 & -1 & 1 \\ 15 & -4 & 6 \end{vmatrix}}{\begin{vmatrix} -2 & 3 & -5 \\ 4 & -1 & 1 \\ -1 & -4 & 6 \end{vmatrix}} = \dfrac{-14}{14} = -1$

$y = \dfrac{\begin{vmatrix} -2 & -11 & -5 \\ 4 & -3 & 1 \\ -1 & 15 & 6 \end{vmatrix}}{14} = \dfrac{56}{14} = 4$

$z = \dfrac{\begin{vmatrix} -2 & 3 & -11 \\ 4 & -1 & -3 \\ -1 & -4 & 15 \end{vmatrix}}{14} = \dfrac{70}{14} = 5$

Answer: $(-1, 4, 5)$

131. $x = \dfrac{\begin{vmatrix} 5 & 6 \\ 11 & 14 \end{vmatrix}}{\begin{vmatrix} 3 & 6 \\ 6 & 14 \end{vmatrix}} = \dfrac{4}{6} = \dfrac{2}{3}$

$y = \dfrac{\begin{vmatrix} 3 & 5 \\ 6 & 11 \end{vmatrix}}{\begin{vmatrix} 3 & 6 \\ 6 & 14 \end{vmatrix}} = \dfrac{3}{6} = \dfrac{1}{2}$

Answer: $\left(\frac{2}{3}, \frac{1}{2}\right)$

133. Cramer's Rule does not apply because the determinant of the coefficient matrix is zero.

135. x = number of carnations

y = number of roses

$$x + y = 12$$
$$0.75x + 1.50y = 12.00$$

$$\begin{bmatrix} 1 & 1 & \vdots & 12 \\ 0.75 & 1.50 & \vdots & 12 \end{bmatrix} \quad -0.75R_1 + R_2 \begin{bmatrix} 1 & 1 & \vdots & 12 \\ 0 & 0.75 & \vdots & 3 \end{bmatrix}$$

$0.75y = 3 \implies y = 4$

$x + (4) = 12 \implies x = 8$

Answer: 8 carnations, 4 roses

137. $(-1, 2),\ (0, 3),\ (1, 6)$

$f(x) = ax^2 + bx + c$

$$\left.\begin{array}{l} f(-1) = a - b + c = 2 \implies a - b = -1 \\ f(0) = c = 3 \\ f(1) = a + b + c = 6 \implies a + b = 3 \end{array}\right\} a = 1, b = 2$$

Thus, $y = x^2 + 2x + 3$.

139. x = number of units produced

y = number of units sold

$x - y = 0$

$-3.75x + 5.25y = 25{,}000$

$$\begin{bmatrix} 1 & -1 \\ -3.75 & 5.25 \end{bmatrix} \begin{bmatrix} x \\ y \end{bmatrix} = \begin{bmatrix} 0 \\ 25{,}000 \end{bmatrix}$$

$$D = \begin{vmatrix} 1 & -1 \\ -3.75 & 5.25 \end{vmatrix} = 1.5$$

$$y = \dfrac{\begin{vmatrix} 1 & 0 \\ -3.75 & 25{,}000 \end{vmatrix}}{1.5} \approx 16{,}667 \text{ units must be sold.}$$

141. L O O K _ O U T _ B E L O W _

[12 15 15][11 0 15] [21 20 0] [2 5 12] [15 23 0]

$[12 \quad 15 \quad 15]A = [-21 \quad 6 \quad 0]$

$[11 \quad 0 \quad 15]A = [-68 \quad 8 \quad 45]$

$[21 \quad 20 \quad 0]A = [102 \quad -42 \quad -60]$

$[2 \quad 5 \quad 12]A = [-53 \quad 20 \quad 21]$

$[15 \quad 23 \quad 0]A = [99 \quad -30 \quad -69]$

Cryptogram: −21 6 0 −68 8 45 102 −42 −60
−53 20 21 99 −30 −69

143. $\begin{bmatrix} -5 & 11 & -2 \\ 370 & -265 & 225 \\ -57 & 48 & -33 \\ 32 & -15 & 20 \\ 245 & -171 & 147 \end{bmatrix} \begin{bmatrix} -1 & 2 & -3 \\ 2 & 1 & 0 \\ 4 & -2 & 5 \end{bmatrix} = \begin{bmatrix} 19 & 5 & 5 \\ 0 & 25 & 15 \\ 21 & 0 & 6 \\ 18 & 9 & 4 \\ 1 & 25 & 0 \end{bmatrix} \begin{matrix} S & E & E \\ \_ & Y & O \\ U & \_ & F \\ R & I & D \\ A & Y & \_ \end{matrix}$

Message: SEE YOU FRIDAY

145. False. Determinants are defined for square matrices.

147. The row operations on matrices are equivalent to the operations used in the method of elimination. See page 554.

Chapter 8 Practice Test

1. Write the matrix in reduced row-echelon form.

$$\begin{bmatrix} 1 & -2 & 4 \\ 3 & -5 & 9 \end{bmatrix}$$

For Exercises 2–4, use matrices to solve the system of equations.

2. $3x + 5y = 3$
$2x - y = -11$

3. $2x + 3y = -3$
$3x + 2y = 8$
$x + y = 1$

4. $x + 3z = -5$
$2x + y = 0$
$3x + y - z = 3$

5. Multiply $\begin{bmatrix} 1 & 4 & 5 \\ 2 & 0 & -3 \end{bmatrix}\begin{bmatrix} 1 & 6 \\ 0 & -7 \\ -1 & 2 \end{bmatrix}$

6. Given $A = \begin{bmatrix} 9 & 1 \\ -4 & 8 \end{bmatrix}$ and $B = \begin{bmatrix} 6 & -2 \\ 3 & 5 \end{bmatrix}$, find $3A - 5B$.

7. Find $f(A)$:

$f(x) = x^2 - 7x + 8,\ A = \begin{bmatrix} 3 & 0 \\ 7 & 1 \end{bmatrix}$

8. True or false:

$(A + B)(A + 3B) = A^2 + 4AB + 3B^2$ where A and B are matrices.

(Assume that A^2, AB, and B^2 exist.)

For Exercises 9–10, find the inverse of the matrix, if it exists.

9. $\begin{bmatrix} 1 & 2 \\ 3 & 5 \end{bmatrix}$

10. $\begin{bmatrix} 1 & 1 & 1 \\ 3 & 6 & 5 \\ 6 & 10 & 8 \end{bmatrix}$

11. Use an inverse matrix to solve the systems.

(a) $x + 2y = 4$
$3x + 5y = 1$

(b) $x + 2y = 3$
$3x + 5y = -2$

For Exercises 12–13, find the determinant of the matrix.

12. $\begin{bmatrix} 6 & -1 \\ 3 & 4 \end{bmatrix}$

13. $\begin{bmatrix} 1 & 3 & -1 \\ 5 & 9 & 0 \\ 6 & 2 & -5 \end{bmatrix}$

14. Use a graphing utility to find the determinant of the matrix.

$$\begin{bmatrix} 1 & 4 & 2 & 3 \\ 0 & 1 & -2 & 0 \\ 3 & 5 & -1 & 1 \\ 2 & 0 & 6 & 1 \end{bmatrix}$$

15. Evaluate $\begin{vmatrix} 6 & 4 & 3 & 0 & 6 \\ 0 & 5 & 1 & 4 & 8 \\ 0 & 0 & 2 & 7 & 3 \\ 0 & 0 & 0 & 9 & 2 \\ 0 & 0 & 0 & 0 & 1 \end{vmatrix}$.

16. Use a determinant to find the area of the triangle with vertices $(0, 7)$, $(5, 0)$, and $(3, 9)$.

17. Use a determinant to find the equation of the line through $(2, 7)$ and $(-1, 4)$.

For Exercises 18–20, use Cramer's Rule to find the indicated value.

18. Find x.

$$\begin{aligned} 6x - 7y &= 4 \\ 2x + 5y &= 11 \end{aligned}$$

19. Find z.

$$\begin{aligned} 3x \quad + z &= 1 \\ y + 4z &= 3 \\ x - y \quad &= 2 \end{aligned}$$

20. Find y.

$$\begin{aligned} 721.4x - 29.1y &= 33.77 \\ 45.9x + 105.6y &= 19.85 \end{aligned}$$

CHAPTER 9
Sequences, Series, and Probability

CHAPTER 9
Sequences, Series, and Probability

Section 9.1 Sequences and Series

- Given the general nth term in a sequence, you should be able to find, or list, some of the terms.
- You should be able to find an expression for the nth term of a sequence.
- You should be able to use and evaluate factorials.
- You should be able to use sigma notation for a sum.

Solutions to Odd-Numbered Exercises

1. $a_n = 2n + 5$

$a_1 = 2(1) + 5 = 7$

$a_2 = 2(2) + 5 = 9$

$a_3 = 2(3) + 5 = 11$

$a_4 = 2(4) + 5 = 13$

$a_5 = 2(5) + 5 = 15$

3. $a_n = 2^n$

$a_1 = 2^1 = 2$

$a_2 = 2^2 = 4$

$a_3 = 2^3 = 8$

$a_4 = 2^4 = 16$

$a_5 = 2^5 = 32$

5. $a_n = (-2)^n$

$a_1 = (-2)^1 = -2$

$a_3 = (-2)^2 = 4$

$a_3 = (-2)^3 = -8$

$a_4 = (-2)^4 = 16$

$a_5 = (-2)^5 = -32$

7. $a_n = \dfrac{n+1}{n}$

$a_1 = \dfrac{1+1}{1} = 2$

$a_2 = \dfrac{3}{2}$

$a_3 = \dfrac{4}{3}$

$a_4 = \dfrac{5}{4}$

$a_5 = \dfrac{6}{5}$

9. $a_n = \dfrac{6n}{3n^2 - 1}$

$a_1 = \dfrac{6(1)}{3(1)^2 - 1} = 3$

$a_2 = \dfrac{6(2)}{3(2)^2 - 1} = \dfrac{12}{11}$

$a_3 = \dfrac{6(3)}{3(3)^2 - 1} = \dfrac{9}{13}$

$a_4 = \dfrac{6(4)}{3(4)^2 - 1} = \dfrac{24}{47}$

$a_5 = \dfrac{6(5)}{3(5)^2 - 1} = \dfrac{15}{37}$

11. $a_n = \dfrac{1 + (-1)^n}{n}$

$a_1 = 0$

$a_2 = \dfrac{2}{2} = 1$

$a_3 = 0$

$a_4 = \dfrac{2}{4} = \dfrac{1}{2}$

$a_5 = 0$

13. $a_n = 3 - \frac{1}{2^n}$

$a_1 = 3 - \frac{1}{2} = \frac{5}{2}$

$a_2 = 3 - \frac{1}{4} = \frac{11}{4}$

$a_3 = 3 - \frac{1}{8} = \frac{23}{8}$

$a_4 = 3 - \frac{1}{16} = \frac{47}{16}$

$a_5 = 3 - \frac{1}{32} = \frac{95}{32}$

15. $a_n = \frac{1}{n^{3/2}}$

$a_1 = \frac{1}{1} = 1$

$a_2 = \frac{1}{2^{3/2}}$

$a_3 = \frac{1}{3^{3/2}}$

$a_4 = \frac{1}{4^{3/2}} = \frac{1}{8}$

$a_5 = \frac{1}{5^{3/2}}$

17. $a_n = \frac{3^n}{n!}$

$a_1 = \frac{3^1}{1!} = \frac{3}{1} = 3$

$a_2 = \frac{3^2}{2!} = \frac{9}{2}$

$a_3 = \frac{27}{6} = \frac{9}{2}$

$a_4 = \frac{81}{24} = \frac{27}{8}$

$a_5 = \frac{243}{120} = \frac{81}{40}$

19. $a_n = \frac{(-1)^n}{n^2}$

$a_1 = \frac{-1}{1} = -1$

$a_2 = \frac{1}{4}$

$a_3 = \frac{-1}{9}$

$a_4 = \frac{1}{16}$

$a_5 = \frac{-1}{25}$

21. $a_n = (2n - 1)(2n + 1)$

$a_1 = (1)(3) = 3$

$a_2 = (3)(5) = 15$

$a_3 = (5)(7) = 35$

$a_4 = (7)(9) = 63$

$a_5 = (9)(11) = 99$

23. $a_{25} = (-1)^{25}[3(25) - 2] = -73$

25. $a_n = \frac{2^n}{n!}$

$a_{10} = \frac{2^{10}}{10!} = \frac{1024}{3,628,800} = \frac{4}{14,175}$

27. $a_n = \frac{4n}{2n^2 - 3}$

$a_{12} = \frac{4(12)}{2(12)^2 - 3} = \frac{48}{285} = \frac{16}{95}$

29. $a_1 = 28$ and $a_{k+1} = a_k - 4$

$a_1 = 28$

$a_2 = a_1 - 4 = 28 - 4 = 24$

$a_3 = a_2 - 4 = 24 - 4 = 20$

$a_4 = a_3 - 4 = 20 - 4 = 16$

$a_5 = a_4 - 4 = 16 - 4 = 12$

31. $a_1 = 3$ and $a_{k+1} = 2(a_k - 1)$

$a_1 = 3$

$a_2 = 2(a_1 - 1) = 2(3 - 1) = 4$

$a_3 = 2(a_2 - 1) = 2(4 - 1) = 6$

$a_4 = 2(a_3 - 1) = 2(6 - 1) = 10$

$a_5 = 2(a_4 - 1) = 2(10 - 1) = 18$

33. $a_1 = 2, a_2 = 6, a_{k+2} = a_{k+1} + 2a_k$

$a_3 = a_2 + 2a_1 = 6 + 2(2) = 10$

$a_4 = a_3 + 2a_2 = 10 + 2(6) = 22$

$a_5 = a_4 + 2a_3 = 22 + 2(10) = 42$

35. $a_n = \frac{2}{3}n$

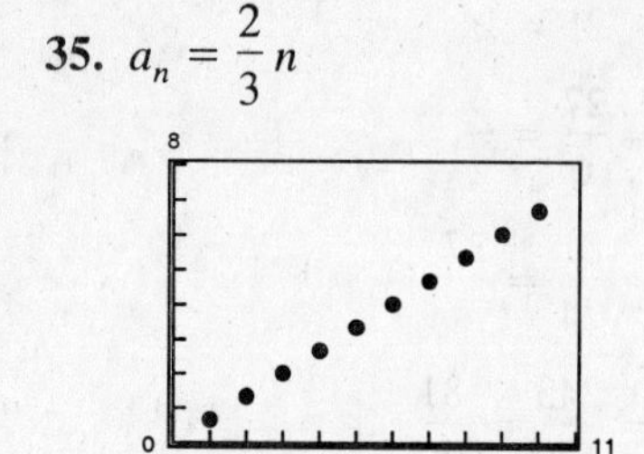

37. $a_n = 16(-0.5)^{n-1}$

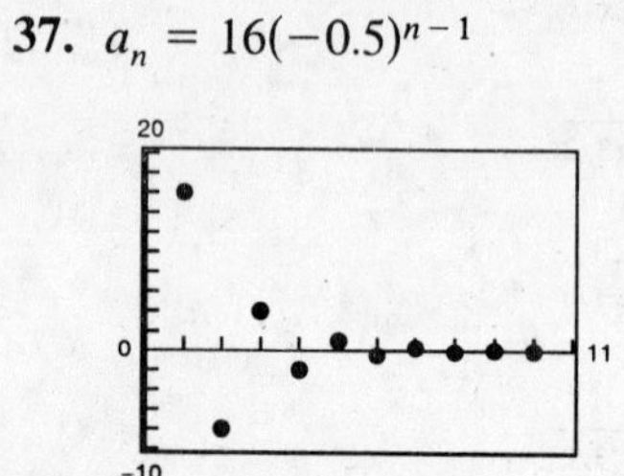

39. $a_n = \frac{2n}{n+1}$

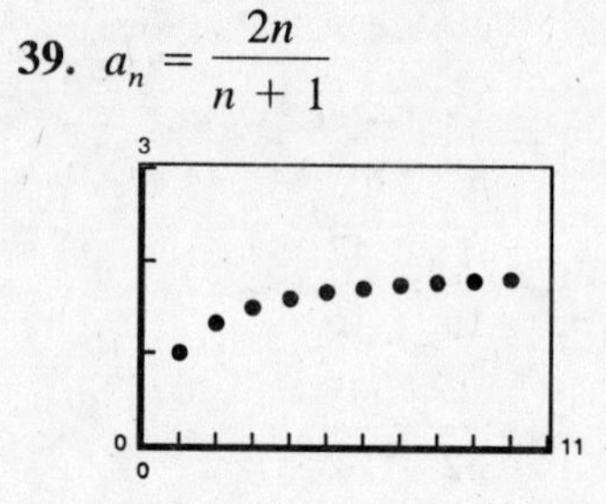

41. $a_n = 2(3n - 1) + 5$

n	1	2	3	4	5	6	7	8	9	10
a_n	9	15	21	27	33	39	45	51	57	63

43. $a_n = \frac{6^n}{n!}$

n	1	2	3	4	5	6	7	8	9	10
a_n	6	18	36	54	64.8	64.8	55.543	41.657	22.771	16.663

45. $a_n = 1 + \frac{n+1}{n}$

n	1	2	3	4	5	6	7	8	9	10
a_n	3	2.5	2.33	2.25	2.2	2.17	2.14	2.13	2.11	2.1

47. $a_n = \frac{8}{n+1}$

$a_n \to 0$ as $n \to \infty$

$a_1 = 4, \ a_{10} = \frac{8}{11}$

Matches graph (c).

49. $a_n = 4(0.5)^{n-1}$

$a_n \to 0$ as $n \to \infty$

$a_1 = 4, \ a_{10} \approx 0.008$

Matches graph (d).

51. $1, 4, 7, 10, 13, \ldots$

$a_n = 1 + (n - 1)3 = 3n - 2$

53. $0, 3, 8, 15, 24, \ldots$

$a_n = n^2 - 1$

55. $\frac{2}{3}, \frac{3}{4}, \frac{4}{5}, \frac{5}{6}, \frac{6}{7}, \ldots$

$a_n = \frac{n+1}{n+2}$

57. $\frac{1}{2}, \frac{-1}{4}, \frac{1}{8}, \frac{-1}{16}, \ldots$

$a_n = \frac{(-1)^{n+1}}{2^n}$

59. $1 + \frac{1}{1}, 1 + \frac{1}{2}, 1 + \frac{1}{3}, 1 + \frac{1}{4}, 1 + \frac{1}{5}, \ldots$

$a_n = 1 + \frac{1}{n}$

61. $1, \frac{1}{2}, \frac{1}{6}, \frac{1}{24}, \frac{1}{120}, \ldots$

$a_n = \frac{1}{n!}$

63. $1, 3, 1, 3, 1, 3, \ldots$

$a_n = 2 + (-1)^n$

65. $a_1 = 6$ and $a_{k+1} = a_k + 2$

$a_1 = 6$

$a_2 = a_1 + 2 = 6 + 2 = 8$

$a_3 = a_2 + 2 = 8 + 2 = 10$

$a_4 = a_3 + 2 = 10 + 2 = 12$

$a_5 = a_4 + 2 = 12 + 2 = 14$

In general, $a_n = 2n + 4$.

67. $a_1 = 81$ and $a_{k+1} = \frac{1}{3}a_k$

$a_1 = 81$

$a_2 = \frac{1}{3}a_1 = \frac{1}{3}(81) = 27$

$a_3 = \frac{1}{3}a_2 = \frac{1}{3}(27) = 9$

$a_4 = \frac{1}{3}a_3 = \frac{1}{3}(9) = 3$

$a_5 = \frac{1}{3}a_4 = \frac{1}{3}(3) = 1$

In general, $a_n = 81\left(\frac{1}{3}\right)^{n-1} = 81(3)\left(\frac{1}{3}\right)^n = \frac{243}{3^n}$.

69. $\frac{3!}{6!} = \frac{3!}{6 \cdot 5 \cdot 4 \cdot 3!} = \frac{1}{6 \cdot 5 \cdot 4} = \frac{1}{120}$

71. $\frac{10!}{8!} = \frac{10 \cdot 9 \cdot 8!}{8!} = 90$

73. $\frac{12!}{4!8!} = \frac{12 \cdot 11 \cdot 10 \cdot 9 \cdot 8!}{4!8!} = \frac{12 \cdot 11 \cdot 10 \cdot 9}{4 \cdot 3 \cdot 2} = 495$

75. $\frac{(n+1)!}{n!} = \frac{(n+1)n!}{n!} = n + 1$

77. $\frac{(2n-1)!}{(2n+1)!} = \frac{(2n-1)!}{(2n+1)(2n)(2n-1)!}$

$= \frac{1}{2n(2n+1)}$

79. $\sum_{i=1}^{5}(2i + 1) = (2 + 1) + (4 + 1) + (6 + 1) + (8 + 1) + (10 + 1) = 35$

81. $\sum_{k=1}^{4} 10 = 10 + 10 + 10 + 10 = 40$

83. $\sum_{i=0}^{4} i^2 = 0^2 + 1^2 + 2^2 + 3^2 + 4^2 = 30$

85. $\sum_{k=0}^{3} \frac{1}{k^2 + 1} = \frac{1}{1} + \frac{1}{1+1} + \frac{1}{1+4} + \frac{1}{9+1} = \frac{9}{5}$

87. $\sum_{i=1}^{4}[(i-1)^2 + (i+1)^3] = [(0)^2 + (2)^3] + [(1)^2 + (3)^3] + [(2)^2 + (4)^3] + [(3)^2 + (5)^3] = 238$

89. $\sum_{i=1}^{4} 2^i = 2^1 + 2^2 + 2^3 + 2^4 = 30$

91. $\sum_{j=1}^{6}(24 - 3j) = 81$

93. $\displaystyle\sum_{k=0}^{4} \frac{(-1)^k}{k+1} = \frac{47}{60}$

95. $\displaystyle\frac{1}{3(1)} + \frac{1}{3(2)} + \frac{1}{3(3)} + \cdots + \frac{1}{3(9)} = \sum_{i=1}^{9} \frac{1}{3i} \approx 0.94299$

97. $\displaystyle\left[2\left(\frac{1}{8}\right) + 3\right] + \left[2\left(\frac{2}{8}\right) + 3\right] + \left[2\left(\frac{3}{8}\right) + 3\right] + \cdots + \left[2\left(\frac{8}{8}\right) + 3\right] = \sum_{i=1}^{8}\left[2\left(\frac{i}{8}\right) + 3\right] = 33$

99. $\displaystyle 3 - 9 + 27 - 81 + 243 - 729 = \sum_{i=1}^{6} (-1)^{i+1}3^i = -546$

101. $\displaystyle\frac{1}{1^2} - \frac{1}{2^2} + \frac{1}{3^2} - \frac{1}{4^2} + \cdots + -\frac{1}{20^2} = \sum_{i=1}^{20} \frac{(-1)^{i+1}}{i^2} \approx 0.82128$

103. $\displaystyle\frac{1}{4} + \frac{3}{8} + \frac{7}{16} + \frac{15}{32} + \frac{31}{64} = \sum_{i=1}^{5} \frac{2^i - 1}{2^{i+1}} = \frac{129}{64} \approx 2.0156$

105. $\displaystyle\sum_{i=1}^{4} 5\left(\tfrac{1}{2}\right)^i = 4.6875 = \tfrac{75}{16}$

107. $\displaystyle\sum_{n=1}^{3} 4\left(-\tfrac{1}{2}\right)^n = -1.5 = -\tfrac{3}{2}$

109. $\displaystyle\sum_{i=1}^{\infty} 6\left(\tfrac{1}{10}\right)^i = 6[0.1 + 0.01 + 0.001 + \cdots]$

$= 6[0.111\ldots]$

$= 0.666\ldots$

$= \tfrac{2}{3}$

111. $\displaystyle\sum_{k=1}^{\infty} \left(\tfrac{1}{10}\right)^k = 0.1 + 0.11 + 0.111 + \cdots$

$= 0.11111$

$= \tfrac{1}{9}$

113. $\displaystyle A_n = 5000\left(1 + \frac{0.08}{4}\right)^n, n = 1, 2, 3, \ldots$

(a) $A_1 = \$5100.00$

$A_2 = \$5202.00$

$A_3 = \$5306.04$

$A_4 = \$5412.16$

$A_5 = \$5520.40$

$A_6 = \$5630.81$

$A_7 = \$5743.43$

$A_8 = \$5858.30$

(b) $A_{40} = \$11{,}040.20$

115. $a_n = 696.39 + 66.44n - 2.37n^2, n = -1, 0, \ldots 6$

n	-1	0	1	2	3	4	5	6
a_n	627.58	696.39	760.46	819.79	874.38	924.23	969.34	1009.71

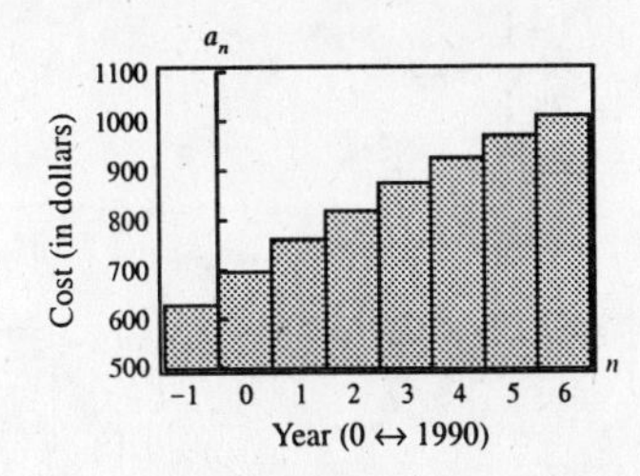

According to the graph, hospital costs are increasing.

117. $\sum_{n=0}^{8} [1215.16 + 608.19n - 114.83n^2 + 11n^3] = \$23{,}661.96$ million

119. True

121. $a_1 = 1, a_2 = 1, a_{k+2} = a_{k+1} + a_k$

$a_1 = 1$ $\qquad b_1 = \frac{1}{1} = 1$

$a_2 = 1$ $\qquad b_2 = \frac{2}{1} = 2$

$a_3 = 1 + 1 = 2$ $\qquad b_3 = \frac{3}{2}$

$a_4 = 2 + 1 = 3$ $\qquad b_4 = \frac{5}{3}$

$a_5 = 3 + 2 = 5$ $\qquad b_5 = \frac{8}{5}$

$a_6 = 5 + 3 = 8$ $\qquad b_6 = \frac{13}{8}$

$a_7 = 8 + 5 = 13$ $\qquad b_7 = \frac{21}{13}$

$a_8 = 13 + 8 = 21$ $\qquad b_8 = \frac{34}{21}$

$a_9 = 21 + 13 = 34$ $\qquad b_9 = \frac{55}{34}$

$a_{10} = 34 + 21 = 55$ $\qquad b_{10} = \frac{89}{55}$

$a_{11} = 55 + 34 = 89$

$a_{12} = 89 + 55 = 144$

123. $a_n = n^2 - n + 11$

$a_1 = 11$

$a_2 = 13$

$a_3 = 17$

$a_4 = 23$

$a_5 = 31$

The terms seem to be prime numbers. However, $a_{11} = 121$ is not prime.

125. $a_n = \dfrac{(-1)^n x^{2n+1}}{2n + 1}$

$a_1 = \dfrac{-x^3}{3}$

$a_2 = \dfrac{x^5}{5}$

$a_3 = -\dfrac{x^7}{7}$

$a_4 = \dfrac{x^9}{9}$

$a_5 = \dfrac{-x^{11}}{11}$

127. $a_n = \dfrac{(-1)x^{2n+1}}{(2n + 1)!}$

$a_1 = \dfrac{-x^3}{3!}$

$a_2 = \dfrac{x^5}{5!}$

$a_3 = -\dfrac{x^7}{7!}$

$a_4 = \dfrac{x^9}{9!}$

$a_5 = \dfrac{-x^{11}}{11!}$

129. $\begin{bmatrix} 2 & 1 & 3 & \vdots & -3 \\ -1 & 5 & 0 & \vdots & 14 \\ -3 & -6 & -7 & \vdots & -7 \end{bmatrix}$

131. (a) $A - B = \begin{bmatrix} 10 & 19 \\ -12 & -5 \end{bmatrix}$

(b) $2B - 3A = \begin{bmatrix} -30 & -45 \\ 28 & 4 \end{bmatrix}$

(c) $AB = \begin{bmatrix} 56 & -43 \\ 48 & 114 \end{bmatrix}$

(d) $BA = \begin{bmatrix} 48 & -72 \\ 36 & 122 \end{bmatrix}$

133. (a) $A - B = \begin{bmatrix} -1 & 0 & 0 \\ 2 & 0 & 4 \\ 1 & -1 & 1 \end{bmatrix}$

(b) $2B - 3A = \begin{bmatrix} 3 & -4 & 0 \\ -9 & -1 & -10 \\ -2 & 3 & -5 \end{bmatrix}$

(c) $AB = \begin{bmatrix} 12 & 0 & -8 \\ 1 & 21 & 2 \\ -6 & -1 & 8 \end{bmatrix}$

(d) $BA = \begin{bmatrix} 20 & 4 & 8 \\ 2 & 15 & -4 \\ 1 & -6 & 6 \end{bmatrix}$

135. $\det\begin{bmatrix} -4 & 11 \\ 13 & 20 \end{bmatrix} = (-4)(20) - 11(13) = -223$

137. $\det(A) = 664$

Section 9.2 Arithmetic Sequences and Partial Sums

- You should be able to recognize an arithmetic sequence, find its common difference, and find its *n*th term.
- You should be able to find the *n*th partial sum of an arithmetic sequence with common difference *d* using the formula

$$S_n = \frac{n}{2}(a_1 + a_n).$$

Solutions to Odd-Numbered Exercises

1. 10, 8, 6, 4, 2, . . .

Arithmetic sequence, $d = -2$

3. $3, \frac{5}{2}, 2, \frac{3}{2}, 1, \ldots$

Arithmetic sequence, $d = -\frac{1}{2}$

5. $-24, -16, -8, 0, 8$

Arithmetic sequence, $d = 8$

7. 3.7, 4.3, 4.9, 5.5, 6.1, . . .

Arithmetic sequence, $d = 0.6$

9. $a_n = 8 + 13n$

21, 34, 47, 60, 73

Arithmetic sequence, $d = 13$

11. $a_n = \dfrac{1}{n+1}$

$\frac{1}{2}, \frac{1}{3}, \frac{1}{4}, \frac{1}{5}, \frac{1}{6}$

Not an arithmetic sequence

13. $a_n = 150 - 7n$

$143, 136, 129, 122, 115$

Arithmetic sequence, $d = -7$

15. $a_n = 3 + \frac{(-1)^n 2}{n}$

$1, 4, \frac{7}{3}, \frac{7}{2}, \frac{13}{5}$

Not an arithmetic sequence

17. $a_1 = 15, a_{k+1} = a_k + 9$

$a_2 = 15 + 9 = 24$

$a_3 = 24 + 9 = 33$

$a_4 = 33 + 9 = 42$

$a_5 = 42 + 9 = 51$

$d = 9, a_n = 6 + 9n$

19. $a_1 = \frac{7}{2}, a_{k+1} = a_k - \frac{1}{4}$

$a_2 = \frac{7}{2} - \frac{1}{4} = \frac{13}{4}$

$a_3 = \frac{13}{4} - \frac{1}{4} = \frac{12}{4} = 3$

$a_4 = \frac{12}{4} - \frac{1}{4} = \frac{11}{4}$

$a_5 = \frac{11}{4} - \frac{1}{4} = \frac{10}{4} = \frac{5}{2}$

$d = -\frac{1}{4}, a_n = \frac{15}{4} - \frac{1}{4}n$

21. $a_1 = 5, d = 6$

$a_1 = 5$

$a_2 = 5 + 6 = 11$

$a_3 = 11 + 6 = 17$

$a_4 = 17 + 6 = 23$

$a_5 = 23 + 6 = 29$

23. $a_1 = -2.6, d = -0.4$

$a_1 = -2.6$

$a_2 = -2.6 + (-0.4) = -3.0$

$a_3 = -3.0 + (-0.4) = -3.4$

$a_4 = -3.4 + (-0.4) = -3.8$

$a_5 = -3.8 + (-0.4) = -4.2$

25. $a_8 = 26, a_{12} = 42$

$26 = a_8 = a_1 + (n - 1)d = a_1 + 7d$

$42 = a_{12} = a_1 + (n - 1)d = a_1 + 11d$

Answer: $d = 4, a_1 = -2$

$a_1 = -2$

$a_2 = -2 + 4 = 2$

$a_3 = 2 + 4 = 6$

$a_4 = 6 + 4 = 10$

$a_5 = 10 + 4 = 14$

27. $a_3 = 19, a_{15} = -1.7$

$a_{15} = a_3 + 12d$

$-1.7 = 19 + 12d \Rightarrow d = -1.725$

$a_3 = a_1 + 2d \Rightarrow 19 = a_1 + 2(-1.725) \Rightarrow a_1 = 22.45$

$a_2 = a_1 - 1.725 = 20.725$

$a_3 = 19$

$a_4 = 19 - 1.725 = 17.275$

$a_5 = 17.275 - 1.725 = 15.55$

29. $a_1 = 5, a_2 = 11 \Rightarrow d = 6$

$a_{10} = a_1 + 9d = 5 + 9(6) = 59$

31. $a_1 = 2, a_2 = -2 \Rightarrow d = -4$

$a_{14} = a_1 + 13d = 2 + 13(-4) = -50$

33. $a_1 = 4.2, a_2 = 6.6 \Rightarrow d = 2.4$

$a_7 = a_1 + 6d = 4.2 + 6(2.4) = 18.6$

35. $a_1 = 1,\ d = 3$

$a_n = a_1 + (n - 1)d = 1 + (n - 1)(3) = 3n - 2$

37. $a_1 = 100,\ d = -8$

$a_n = a_1 + (n - 1)d = 100 + (n - 1)(-8) = 108 - 8n$

39. $4, \frac{3}{2}, -1, -\frac{7}{2}, \ldots$

$d = -\frac{5}{2}$

$a_n = a_1 + (n - 1)d = 4 + (n - 1)\left(-\frac{5}{2}\right) = \frac{13}{2} - \frac{5}{2}n$

41. $a_1 = 5,\ a_4 = 15$

$a_4 = a_1 + 3d \Rightarrow 15 = 5 + 3d \Rightarrow d = \frac{10}{3}$

$a_n = a_1 + (n - 1)d = 5 + (n - 1)\left(\frac{10}{3}\right) = \frac{10}{3}n + \frac{5}{3}$

43. $a_3 = 94,\ a_6 = 85$

$a_6 = a_3 + 3d \Rightarrow 85 = 94 + 3d \Rightarrow d = -3$

$a_1 = a_3 - 2d \Rightarrow a_1 = 94 - 2(-3) = 100$

$a_n = a_1 + (n - 1)d = 100 + (n - 1)(-3) = 103 - 3n$

45. $a_n = -\frac{2}{3}n + 6$

$d = -\frac{2}{3}$ so the sequence is decreasing,

and $a_1 = 5\frac{1}{3}$.

Matches (b).

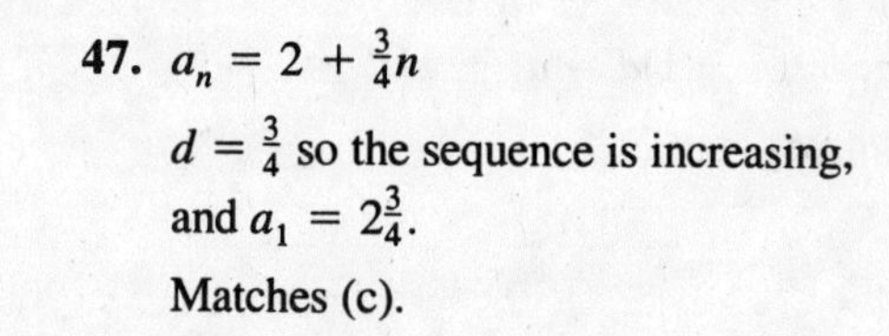

47. $a_n = 2 + \frac{3}{4}n$

$d = \frac{3}{4}$ so the sequence is increasing,

and $a_1 = 2\frac{3}{4}$.

Matches (c).

49. $a_n = 15 - \frac{3}{2}n$

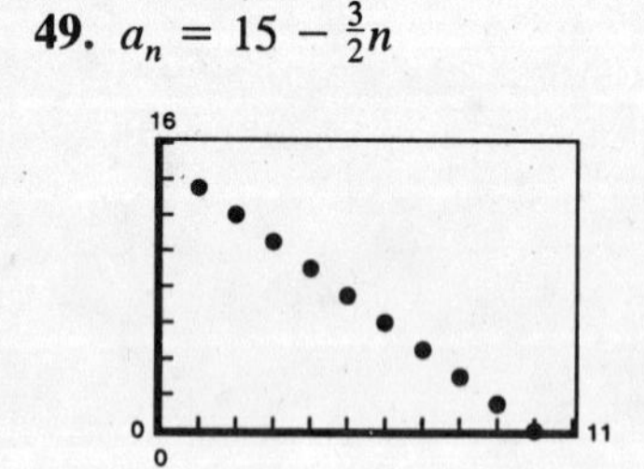

51. $a_n = 0.2n + 3$

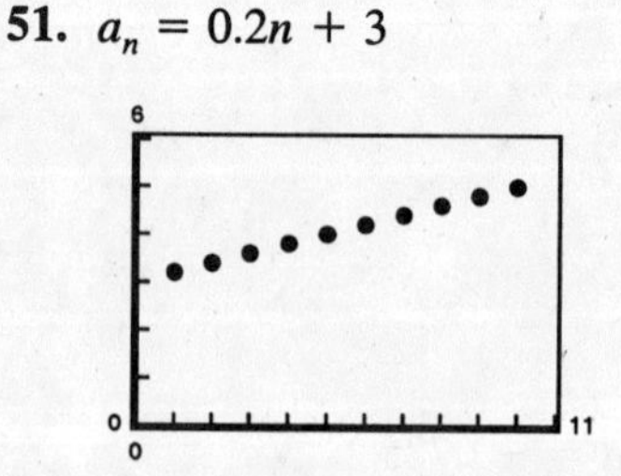

53. $a_n = 4n - 5$

n	1	2	3	4	5	6	7	8	9	10
a_n	-1	3	7	11	15	19	23	27	31	35

55. $a_n = 20 - \frac{3}{4}n$

n	1	2	3	4	5	6	7	8	9	10
a_n	19.25	18.5	17.75	17	16.25	15.5	14.75	14	13.25	12.5

57. $a_n = 1.5 + 0.005n$

n	1	2	3	4	5	6	7	8	9	10
a_n	1.505	1.51	1.515	1.52	1.525	1.53	1.535	1.54	1.545	1.55

59. $a_1 = 8, a_2 = 26 \Rightarrow d = 18$

$a_{10} = a_1 + 9d = 8 + 9(18) = 170$

$S_{10} = \frac{10}{2}(a_1 + a_{10}) = 5(8 + 170) = 890$

61. $a_1 = 0.5, a_2 = 1.3 \Rightarrow d = 0.8$

$a_{10} = a_1 + 9d = 0.5 + 9(0.8) = 7.7$

$S_{10} = \frac{10}{2}(a_1 + a_{10}) = 5(0.5 + 7.7) = 41$

63. $a_1 = 100, a_{25} = 220$

$a_{25} = a_1 + 24d \Rightarrow d = 5$

$S_{25} = \frac{25}{2}(a_1 + a_{25}) = 12.5(100 + 220) = 4000$

65. $a_1 = 1,\ a_{50} = 50,\ n = 50$

$\sum_{n=1}^{50} n = \frac{50}{2}(1 + 50) = 1275$

67. $a_1 = 5,\ a_{100} = 500, n = 100$

$\sum_{n=1}^{100} 5n = \frac{100}{2}(5 + 500) = 25{,}250$

69. $\sum_{n=11}^{30} n - \sum_{n=1}^{10} n = \frac{20}{2}(11 + 30) - \frac{10}{2}(1 + 10) = 355$

71. $a_1 = 4,\ a_{500} = 503, n = 500$

$\sum_{n=1}^{500} (n + 3) = \frac{500}{2}(4 + 503) = 126{,}750$

73. $a_1 = 7, a_{20} = 45,\ n = 20$

$\sum_{n=1}^{20} (2n + 5) = \frac{20}{2}(7 + 45) = 520$

75. $a_0 = 1000,\ a_{50} = 750,\ n = 51$

$\sum_{n=0}^{50} (1000 - 5n) = \frac{51}{2}(1000 + 750) = 44{,}625$

77. $a_1 = \frac{742}{3},\ a_{60} = 90,\ n = 60$

$\sum_{i=1}^{60} \left(250 - \frac{8}{3}i\right) = \frac{60}{2}\left(\frac{742}{3} + 90\right) = 10{,}120$

79. $a_1 = 1,\ a_{100} = 199,\ n = 100$

$\sum_{n=1}^{100} (2n - 1) = \frac{100}{2}(1 + 199) = 10{,}000$

81. (a) $a_1 = 32{,}500,\ d = 1500$

$a_6 = a_1 + 5d = 32{,}500 + 5(1500) = \$40{,}000$

(b) $S_6 = \frac{6}{2}[32{,}500 + 40{,}000] = \$217{,}500$

(c) first year: \$32,500; second year: \$34,000; third year: \$35,500; fourth year: \$37,000; fifth year: \$38,500; sixth year: \$40,000

\$32,500 + \$34,000 + \$35,500 + \$37,000 + \$38,500 + \$40,000 = \$217,500

83. $a_1 = 20,\ d = 4,\ n = 30$

$a_{30} = 20 + 29(4) = 136$

$S_{30} = \frac{30}{2}(20 + 36) = 2340$ seats

85. $a_1 = 14, a_{18} = 31$

$S_{18} = \frac{18}{2}(14 + 31) = 405$ bricks

87. $a_1 = 25,\ a_2 = 25 + 2 = 27$, etc. $\Rightarrow$ $d = 2$ and $n = 15$.

$a_{15} = 2(15) + 23 = 53$

$S_{15} = \frac{15}{2}(25 + 53) = \frac{15}{2} \cdot 78 = 585$ seats

89. $(1 + 2 + \cdots + 12) + (1 + 2 + \cdots + 12) = \frac{12}{2}(1 + 12) \times 2 = 12 \cdot 13 = 156$ times

91. True. Given a_1 and a_2, you know $d = a_2 - a_1$. Thus, $a_n = a_1 + (n - 1)d$.

93. $a_1 = x$

$a_2 = x + 2x = 3x$

$a_3 = 3x + 2x = 5x$

$a_4 = 7x$

$a_5 = 9x$

$a_6 = 11x$

$a_7 = 13x$

$a_8 = 15x$

$a_9 = 17x$

$a_{10} = 19x$

95. (a) $1 + 3 = 4$

$1 + 3 + 5 = 9$

$1 + 3 + 5 + 7 = 16$

$1 + 3 + 5 + 7 + 9 = 25$

$1 + 3 + 5 + 7 + 9 + 11 = 36$

(b) $S_n = n^2$

$S_7 = 1 + 3 + 5 + 7 + 9 + 11 + 13 = 49 = 7^2$

(c) $S_n = \frac{n}{2}[1 + (2n - 1)] = \frac{n}{2}(2n) = n^2$

97. Let $S_n = \frac{n}{2}(a_1 + a_n)$ be the sum of the first n terms of the original sequence. If the first term is increased by 5, then the new sum is

$$S' = \frac{n}{2}(a_1 + 5 + a_n + 5) = \frac{n}{2}(a_1 + a_n + 10)$$

$$= \frac{n}{2}(a_1 + a_n) + \frac{n}{2}(10)$$

$$= \frac{n}{2}(a_1 + a_n) + 5n$$

$$= S_n + 5n$$

99. $\begin{bmatrix} -1 & 4 & 10 & \vdots & 4 \\ 5 & -3 & 1 & \vdots & 31 \\ 8 & 2 & -3 & \vdots & -5 \end{bmatrix}$ row reduces to $\begin{bmatrix} 1 & 0 & 0 & \vdots & 2 \\ 0 & 1 & 0 & \vdots & -6 \\ 0 & 0 & 1 & \vdots & 3 \end{bmatrix}$

Answer: $(2, -6, 3)$

101. $\begin{vmatrix} -1 & 2 & 1 \\ 5 & 1 & 1 \\ 3 & 8 & 1 \end{vmatrix} = 40$

Area $= \frac{1}{2}(40) = 20$ square units

103. $\dfrac{6!8!}{14!} = \dfrac{6!8!}{14 \cdot 13 \cdot 12 \cdot 11 \cdot 10 \cdot 9 \cdot 8!} = \dfrac{6 \cdot 5 \cdot 4 \cdot 3 \cdot 2}{14 \cdot 13 \cdot 12 \cdot 11 \cdot 10 \cdot 9} = \dfrac{1}{3003}$

Section 9.3 Geometric Sequences and Series

- You should be able to identify a geometric sequence, find its common ratio, and find the nth term.
- You should be able to find the nth partial sum of a geometric sequence with common ratio r using the formula.

$$S_n = a_1\left(\frac{1 - r^n}{1 - r}\right)$$

- You should know that if $|r| < 1$, then

$$\sum_{n=1}^{\infty} a_1 r^{n-1} = \frac{a_1}{1 - r}.$$

Solutions to Odd-Numbered Exercises

1. $5, 15, 45, 135, \ldots$

Geometric sequence, $r = 3$

3. $6, 18, 30, 42, \ldots$

Not a geometric sequence

(Note: It is an arithmetic sequence with $d = 12$.)

5. $1, -\frac{1}{2}, \frac{1}{4}, -\frac{1}{8}, \ldots$

Geometric sequence, $r = -\frac{1}{2}$

7. $\frac{1}{2}, \frac{2}{3}, \frac{3}{4}, \frac{4}{5}, \ldots$

Not a geometric sequence

9. $1, \frac{1}{2}, \frac{1}{3}, \frac{1}{4}, \ldots$

Not a geometric sequence

11. $a_1 = 8, r = 3$

$a_2 = 8(3) = 24$

$a_3 = 24(3) = 72$

$a_4 = 72(3) = 216$

$a_5 = 216(3) = 648$

13. $a_1 = 1, r = \frac{1}{2}$

$a_1 = 1$

$a_2 = 1\left(\frac{1}{2}\right) = \frac{1}{2}$

$a_3 = \frac{1}{2}\left(\frac{1}{2}\right) = \frac{1}{4}$

$a_4 = \frac{1}{4}\left(\frac{1}{2}\right) = \frac{1}{8}$

$a_5 = \frac{1}{8}\left(\frac{1}{2}\right) = \frac{1}{16}$

15. $a_1 = 5, r = -\frac{1}{10}$

$a_1 = 5$

$a_2 = 5\left(-\frac{1}{10}\right) = -\frac{1}{2}$

$a_3 = \left(-\frac{1}{2}\right)\left(-\frac{1}{10}\right) = \frac{1}{20}$

$a_4 = \frac{1}{20}\left(-\frac{1}{10}\right) = -\frac{1}{200}$

$a_5 = \left(-\frac{1}{200}\right)\left(-\frac{1}{10}\right) = \frac{1}{2000}$

17. $a_1 = 3.5, r = 5$

$a_2 = 3.5(5) = 17.5$

$a_3 = 17.5(5) = 87.5$

$a_4 = 87.5(5) = 437.5$

$a_5 = 437.5(5) = 2187.5$

19. $a_1 = 1, r = e$

$a_1 = 1$

$a_2 = 1(e) = e$

$a_3 = (e)(e) = e^2$

$a_4 = (e^2)(e) = e^3$

$a_5 = (e^3)(e) = e^4$

21. $a_1 = 64, a_{k+1} = \frac{1}{2}a_k$

$a_1 = 64$

$a_2 = \frac{1}{2}(64) = 32$

$a_3 = \frac{1}{2}(32) = 16$

$a_4 = \frac{1}{2}(16) = 8$

$a_5 = \frac{1}{2}(8) = 4$

$r = \frac{1}{2}, a_n = 64\left(\frac{1}{2}\right)^{n-1} = 128\left(\frac{1}{2}\right)^n$

23. $a_1 = 4,\ a_{k+1} = 3a_k$

$a_1 = 4$

$a_2 = 3(4) = 12$

$a_3 = 3(12) = 36$

$a_4 = 3(36) = 108$

$a_5 = 3(108) = 324$

$r = 3,\ a_n = 4(3)^{n-1} = \frac{4}{3}(3)^n$

25. $a_k = 6,\ a_{k+1} = -\frac{3}{2}a_k$

$a_1 = 6$

$a_2 = -\frac{3}{2}(6) = -9$

$a_3 = -\frac{3}{2}(-9) = \frac{27}{2}$

$a_4 = -\frac{3}{2}\left(\frac{27}{2}\right) = -\frac{81}{4}$

$a_5 = -\frac{3}{2}\left(-\frac{81}{4}\right) = \frac{243}{8}$

$r = -\frac{3}{2},\ a_n = 6\left(-\frac{3}{2}\right)^{n-1}$

27. $a_1 = 4,\ r = \frac{1}{2},\ n = 10$

$a_n = a_1 r^{n-1}$

$a_{10} = 4\left(\frac{1}{2}\right)^9 = \left(\frac{1}{2}\right)^7 = \frac{1}{128}$

29. $a_1 = 6,\ r = -\dfrac{1}{3},\ n = 12$

$a_n = a_1 r^{n-1}$

$a_{12} = 6\left(-\dfrac{1}{3}\right)^{11} = \dfrac{-2}{3^{10}}$

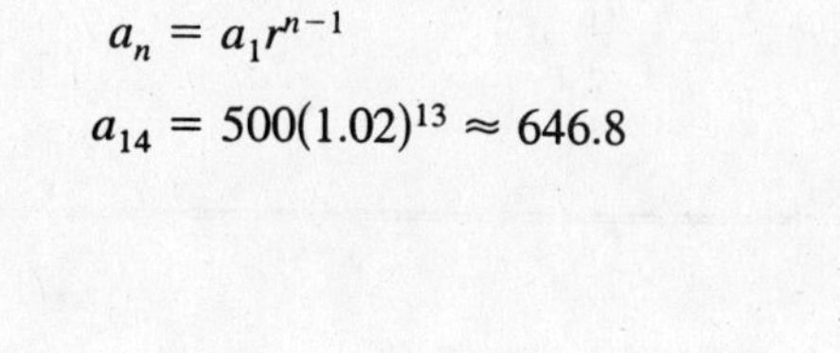

31. $a_1 = 500,\ r = 1.02,\ n = 14$

$a_n = a_1 r^{n-1}$

$a_{14} = 500(1.02)^{13} \approx 646.8$

33. $a_1 = 16,\ a_4 = \dfrac{27}{4},\ n = 3$

$\dfrac{27}{4} = 16r^3 \implies r = \dfrac{3}{4}$

$a_n = a_1 r^{n-1}$

$a_3 = 16\left(\dfrac{3}{4}\right)^2 = 9$

35. $a_2 = a_1 r = -18 \implies a_1 = \dfrac{-18}{r}$

$a_5 = a_1 r^4 = (a_1 r)r^3 = -18r^3 = \dfrac{2}{3} \implies r = -\dfrac{1}{3}$

$a_1 = \dfrac{-18}{r} = \dfrac{-18}{-1/3} = 54$

$a_6 = a_1 r^5 = 54\left(\dfrac{-1}{3}\right)^5 = \dfrac{54}{243} = -\dfrac{2}{9}$

37. $r = \frac{21}{7} = 3.$

$a_9 = a_1 r^{9-1} = 7(3)^8 = 45{,}927$

39. $r = \frac{30}{5} = 6$

$a_{10} = a_1 r^{10-1} = 5(6)^9 = 50{,}388{,}480$

41. $r = \dfrac{\frac{3}{4}}{\frac{3}{16}} = 4$

$a_{12} = a_1 r^{12-1} = \dfrac{3}{16}(4)^{11} = 786{,}432$

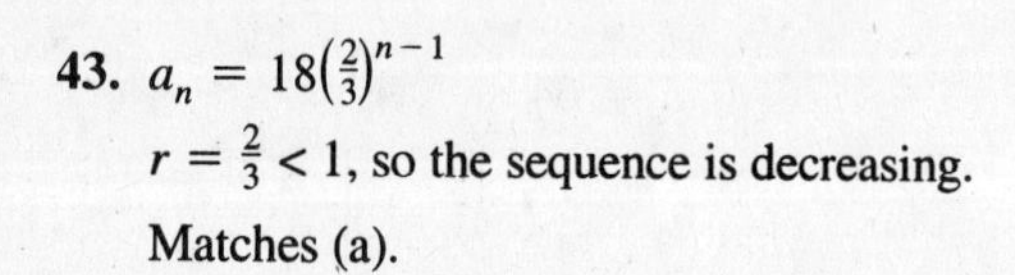

43. $a_n = 18\left(\frac{2}{3}\right)^{n-1}$

$r = \frac{2}{3} < 1$, so the sequence is decreasing.

Matches (a).

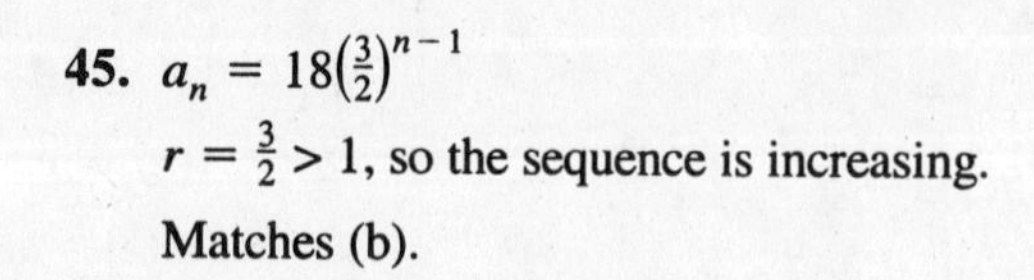

45. $a_n = 18\left(\frac{3}{2}\right)^{n-1}$

$r = \frac{3}{2} > 1$, so the sequence is increasing.

Matches (b).

47. $a_n = 12(-0.75)^{n-1}$

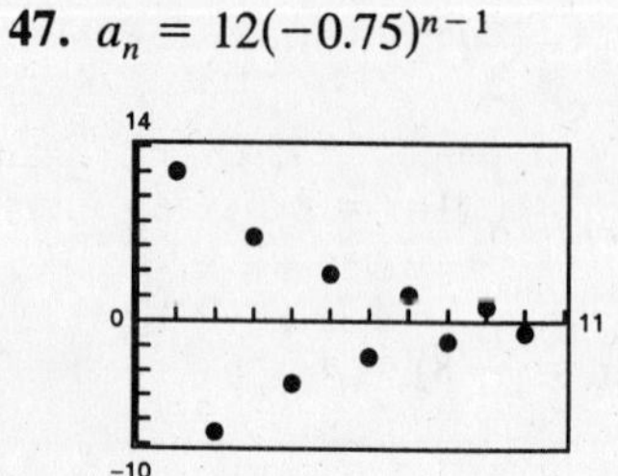

49. $a_n = 2(1.3)^{n-1}$

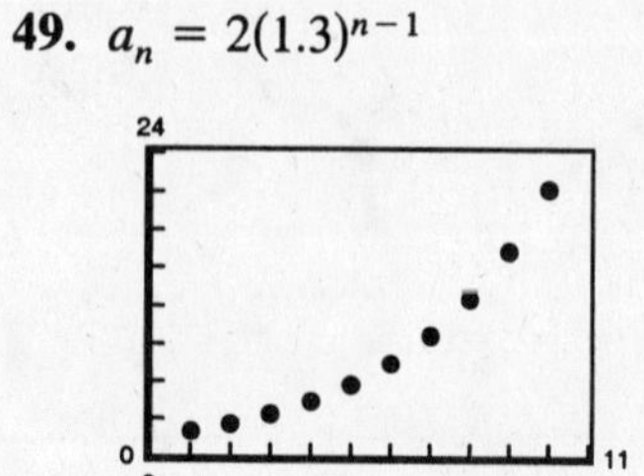

51. $8, -4, 2, -1, \frac{1}{2}$

$S_1 = 8$

$S_2 = 8 + (-4) = 4$

$S_3 = 8 + (-4) + 2 = 6$

$S_4 = 8 + (-4) + 2 + (-1) = 5$

53. $\sum_{n=1}^{\infty} 16\left(-\frac{1}{2}\right)^{n-1}$

n	1	2	3	4	5	6	7	8	9	10
S_n	16	24	28	30	31	31.5	31.75	31.875	31.9375	31.96875

55. $\sum_{n=1}^{9} 2^{n-1} \Rightarrow a_1 = 1,\ r = 2$

$$S_9 = \frac{1(1 - 2^9)}{1 - 2} = 511$$

57. $\sum_{i=1}^{7} 64\left(-\frac{1}{2}\right)^{i-1} \Rightarrow a_1 = 64,\ r = -\frac{1}{2}$

$$S_7 = 64\left[\frac{1 - (-1/2)^7}{1 - (-1/2)}\right] = \frac{128}{3}\left[1 - \left(-\frac{1}{2}\right)^7\right] = 43$$

59. $\sum_{n=0}^{20} 3\left(\frac{3}{2}\right)^n = \sum_{n=1}^{21} 3\left(\frac{3}{2}\right)^{n-1} \Rightarrow a_1 = 3,\ r = \frac{3}{2}$

$$S_{21} = 3\left[\frac{1 - (3/2)^{21}}{1 - (3/2)}\right] = -6\left[1 - \left(\frac{3}{2}\right)^{21}\right] \approx 29{,}921.31$$

61. $\sum_{i=1}^{10} 8\left(-\frac{1}{4}\right)^{i-1} \Rightarrow a_1 = 8,\ r = -\frac{1}{4}$

$$S_{10} = 8\left[\frac{1 - \left(-1/4\right)^{10}}{1 - \left(-1/4\right)}\right] = \frac{32}{5}\left[1 - \left(-\frac{1}{4}\right)^{10}\right] \approx 6.4$$

63. $\sum_{n=0}^{5} 300(1.06)^n = \sum_{n=1}^{6} 300(1.06)^{n-1} \Rightarrow a_1 = 300,\ r = 1.06$

$$S_6 = 300\left[\frac{1 - (1.06)^6}{1 - 1.06}\right] \approx 2092.60$$

65. $5 + 15 + 45 + \cdots + 3645$

$r = 3$ and $3645 = 5(3)^{n-1} \Rightarrow n = 7$

Thus, the sum can be written as $\sum_{n=1}^{7} 5(3)^{n-1}$.

67. $2 - \frac{1}{2} + \frac{1}{8} - \ldots + \frac{1}{2048}$

$r = -\frac{1}{4}$ and $\frac{1}{2048} = 2\left(-\frac{1}{4}\right)^{n-1} \Rightarrow n = 7$

$\sum_{n=1}^{7} 2\left(-\frac{1}{4}\right)^{n-1}$

69. $a_1 = 1,\ r = \frac{1}{2}$

$$\sum_{n=0}^{\infty}\left(\frac{1}{2}\right)^n = \frac{a_1}{1 - r} = \frac{1}{1 - (1/2)} = 2$$

71. $a_1 = 1,\ r = -\frac{1}{2}$

$$\sum_{n=0}^{\infty}\left(-\frac{1}{2}\right)^n = \sum_{n=1}^{\infty}\left(-\frac{1}{2}\right)^{n-1} = \frac{a_1}{1-r} = \frac{1}{1-(-1/2)} = \frac{2}{3}$$

73. $a_1 = 4,\ r = \frac{1}{4}$

$$\sum_{n=0}^{\infty}4\left(\frac{1}{4}\right)^n = \frac{a_1}{1-r} = \frac{4}{1-(1/4)} = \frac{16}{3}$$

75. $\sum_{n=1}^{\infty}2\left(\frac{7}{3}\right)^{n-1}$ does not have a finite sum $\left(\frac{7}{3} > 1\right)$

77. $\sum_{n=0}^{\infty}(0.4)^n.\ a_1 = 1, r = 0.4$

$$\sum_{n=0}^{\infty}(0.4)^n = \frac{a_1}{1-r} = \frac{1}{1-0.4} = \frac{1}{0.6} = \frac{10}{6} = \frac{5}{3}$$

79. $a = -3, r = 0.9$

$$\sum_{n=0}^{\infty} -3(0.9)^n = \frac{a_1}{1-r} = \frac{-3}{1-0.9} = \frac{-3}{0.1} = -30$$

81. $8 + 6 + \frac{9}{2} + \frac{27}{8} + \cdots = \sum_{n=0}^{\infty}8\left(\frac{3}{4}\right)^n = \frac{8}{1-3/4} = 32$

83. $3 - 1 + \frac{1}{3} - \frac{1}{9} + \ldots = \sum_{n=0}^{\infty}3\left(-\frac{1}{3}\right)^n = \frac{a_1}{1-r} = \frac{3}{1-\left(-\frac{1}{3}\right)} = 3\left(\frac{3}{4}\right) = \frac{9}{4}$

85. $0.\overline{36} = \sum_{n=0}^{\infty}0.36(0.01)^n = \frac{0.36}{1-0.01} = \frac{0.36}{0.99} = \frac{36}{99} = \frac{4}{11}$

87. $0.3\overline{18} = 0.3 + \sum_{n=0}^{\infty}0.018(0.01)^n = \frac{3}{10} + \frac{0.018}{1-0.01}$

$$= \frac{3}{10} + \frac{0.018}{0.99} = \frac{3}{10} + \frac{18}{990} = \frac{3}{10} + \frac{2}{110}$$

$$= \frac{35}{110} = \frac{7}{22}$$

89. $A = P\left(1 + \frac{r}{n}\right)^{nt} = 1000\left(1 + \frac{0.08}{n}\right)^{n(10)}$

(a) $n = 1,\quad A = 1000(1 + 0.08)^{10} \approx \$2158.92$

(b) $n = 2,\quad A = 1000\left(1 + \frac{0.08}{2}\right)^{2(10)} \approx \$2191.12$

(c) $n = 4,\quad A = 1000\left(1 + \frac{0.08}{4}\right)^{4(10)} \approx \$2208.04$

(d) $n = 12,\quad A = 1000\left(1 + \frac{0.08}{12}\right)^{12(10)} \approx \$2219.64$

(e) $n = 365,\quad A = 1000\left(1 + \frac{0.08}{365}\right)^{365(10)} \approx \$2225.35$

91. $V_5 = 155{,}000(0.70)^5 = \$26{,}050.85$

93. $A = \sum_{n=1}^{60} 100\left(1 + \frac{0.06}{12}\right)^n = 100\left(1 + \frac{0.06}{12}\right) \cdot \frac{\left[1 - \left(1 + \frac{0.06}{12}\right)^{60}\right]}{\left[1 - \left(1 + \frac{0.06}{12}\right)\right]} \approx \$7011.89$

95. Let $N = 12t$ be the total number of deposits.

$$A = P\left(1 + \frac{r}{12}\right) + P\left(1 + \frac{r}{12}\right)^2 + \cdots + P\left(1 + \frac{r}{12}\right)^N$$

$$= \left(1 + \frac{r}{12}\right)\left[P + P\left(r + \frac{r}{12}\right) + \cdots + P\left(1 + \frac{r}{12}\right)^{N-1}\right]$$

$$= P\left(1 + \frac{r}{12}\right)\sum_{n=1}^{N}\left(1 + \frac{r}{12}\right)^{n-1}$$

$$= P\left(1 + \frac{r}{12}\right)\frac{1 - \left(1 + \frac{r}{12}\right)^N}{1 - \left(1 + \frac{r}{12}\right)}$$

$$= P\left(1 + \frac{r}{12}\right)\left(-\frac{12}{r}\right)\left[1 - \left(1 + \frac{r}{12}\right)^N\right]$$

$$= P\left(\frac{12}{r} + 1\right)\left[-1 + \left(1 + \frac{r}{12}\right)^N\right]$$

$$= P\left[\left(1 + \frac{r}{12}\right)^N - 1\right]\left(1 + \frac{12}{r}\right)$$

$$= P\left[\left(1 + \frac{r}{12}\right)^{12t} - 1\right]\left(1 + \frac{12}{r}\right)$$

97. $P = \$50,\ r = 7\%,\ t = 20$ years

(a) Compounded monthly: $A = 50\left[\left(1 + \frac{0.07}{12}\right)^{12(20)} - 1\right]\left(1 + \frac{12}{0.07}\right) \approx \$26{,}198.27$

(b) Compounded continuously: $A = \frac{50e^{0.07/12}(e^{0.07(20)} - 1)}{e^{0.07/12} - 1} \approx \$26{,}263.88$

99. $P = \$100,\ r = 10\%,\ t = 40$ years

(a) Compounded monthly: $A = 100\left[\left(1 + \frac{0.10}{12}\right)^{12(40)} - 1\right]\left(1 + \frac{12}{0.10}\right) \approx \$637{,}678.02$

(b) Compounded continuously: $A = \frac{100e^{0.10/12}(e^{(0.10)(40)} - 1)}{e^{0.10/12} - 1} \approx \$645{,}861.43$

101. $P = W\sum_{n=1}^{12t}\left[\left(1+\frac{r}{12}\right)^{-1}\right]^n$

$$= W\left(1+\frac{r}{12}\right)^{-1}\left[\frac{1-\left(1+\frac{r}{12}\right)^{-12t}}{1-\left(1-\frac{r}{12}\right)^{-1}}\right]$$

$$= W\left(\frac{1}{1+\frac{r}{12}}\right)\frac{\left[1-\left(1+\frac{r}{12}\right)^{-12t}\right]}{1-\frac{1}{\left(1+\frac{r}{12}\right)}}$$

$$= W\frac{\left[1-\left(1+\frac{r}{12}\right)^{-12t}\right]}{\left(1+\frac{r}{12}\right)-1}$$

$$= W\left(\frac{12}{r}\right)\left[1-\left(1+\frac{r}{12}\right)^{-12t}\right]$$

103. $\sum_{n=0}^{5}\frac{16^2}{4}\left(\frac{1}{2}\right)^n \approx 126$

Total area of shaded region is approximately 126 square inches.

105. $\sum_{n=0}^{6} 3.978e^{0.11n}$, $a_1 = 3.978$, $r = e^{0.11}$

$$\text{Sum} = a_1\left(\frac{1-r^n}{1-r}\right) = 3.978\,\frac{1-e^{0.11(7)}}{1-e^{0.11}} \approx 39.68 \text{ billion}$$

107. $S_n = \sum_{i=1}^{n} 0.01(2)^{i-1}$

$S_{29} = \$5,368,709.11$

$S_{30} = \$10,737,418.23$

$S_{31} = \$21,474,836.47$

109. False. See definition page 638.

111. True. The sequence is $a, a, a, \ldots$ which is arithmetic ($d = 0$).

113. $a_1 = 8$

$$a_2 = 8\left(\frac{2x}{3}\right) = \frac{16}{3}x$$

$$a_3 = \frac{16}{3}x\left(\frac{2x}{3}\right) = \frac{32x^2}{9} = \frac{2^5x^2}{3^2}$$

$$a_4 = \frac{2^5x^2}{3^2}\left(\frac{2x}{3}\right) = \frac{2^6x^3}{3^3}$$

$$a_5 = \frac{2^7x^4}{3^4}$$

115. $a_1 = \frac{1}{2}$

$a_2 = \frac{1}{2}(7x) = \frac{7x}{2}$

$a_3 = \frac{7x}{2}(7x) = \frac{7^2x^2}{2}$

$a_4 = \frac{7^3x^3}{2}$

$a_5 = \frac{7^4x^4}{2}$

117. $a_1 = 6, r = 3e^x, n = 8$

$a_n = a_1r^{n-1}$

$a_8 = 6(3e^x)^7 = 13{,}122e^{7x}$

119. $a_1 = 4, r = \frac{4x}{3}, n = 6$

$a_n = a_1r^{n-1}$

$a_6 = 4\left(\frac{4x}{3}\right)^5 = \frac{4096}{243}x^5$

121. $f(x) = 2\left[\frac{1 - 0.8^x}{1 - 0.8}\right]$

$\sum_{n=0}^{\infty} 2\left(\frac{4}{5}\right)^n = \frac{2}{1 - \frac{4}{5}} = 10$

The horizontal asymptote of $f(x)$ is $y = 10$. This corresponds to the sum of the series.

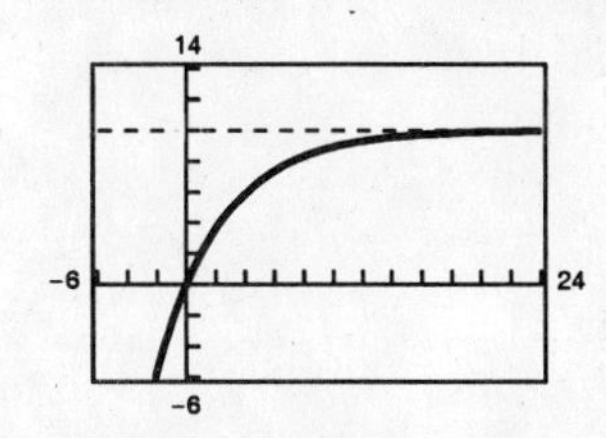

123. To use the first two terms of a geometric series to find the nth term, first divide the second term by the first term to obtain the constant ratio. The nth term is the first term multiplied by the common ratio raised to the $(n - 1)$ power.

$r = \frac{a_2}{a_1}, a_n = a_1r^{n-1}$

125. $\text{Time} = \frac{\text{Distance}}{\text{Speed}} = \frac{200}{50} + \frac{200}{42} = 200\left[\frac{92}{2100}\right]$ hours

$\text{Speed} = \frac{\text{Distance}}{\text{Time}} = \frac{400}{200\left[\frac{92}{2100}\right]} = \frac{2(2100)}{92} \approx 45.65$ mph

127. Your friend mows at the rate of $\frac{1}{4}$ lawns/hour, and your rate is $\frac{1}{6}$ lawns/hour. Together, the time would be

$\frac{1}{\frac{1}{4} + \frac{1}{6}} = \frac{1}{\frac{10}{24}} = \frac{24}{10} = 2.4$ hours.

129. $-4\begin{bmatrix} 7 & 2 \\ -6 & 0 \end{bmatrix} + \begin{bmatrix} -5 & 5 \\ 3 & 1 \end{bmatrix} = \begin{bmatrix} -33 & -3 \\ 27 & 1 \end{bmatrix}$

131. $\begin{bmatrix} 2 & 6 & -8 \\ 12 & 4 & -20 \\ 4 & 2 & 10 \end{bmatrix}$

133. $\sum_{i=0}^{6} 4i^2 = 4[0 + 1 + 2^2 + 3^2 + 4^2 + 5^2 + 6^2] = 364$

135. $\sum_{k=0}^{4} \frac{2}{k^2 + 2} = \frac{2}{2} + \frac{2}{3} + \frac{2}{6} + \frac{2}{11} + \frac{2}{18} = \frac{227}{99} \approx 2.293$

Section 9.4 Mathematical Induction

- You should be sure that you understand the principle of mathematical induction. If P_n is a statement involving the positive integer n, where P_1 is true and the truth of P_k implies the truth of P_{k+1}, then P_n is true for all positive integers n.
- You should be able to verify (by induction) the formulas for the sums of powers of integers and be able to use these formulas.
- You should be able to work with finite differences.

Solutions to Odd-Numbered Exercises

1. $P_k = \dfrac{5}{k(k+1)}$

$$P_{k+1} = \frac{5}{(k+1)[(k+1)+1]} = \frac{5}{(k+1)(k+2)}$$

3. $P_k = \dfrac{k^2(k+3)^2}{6}$

$$P_{k+1} = \frac{(k+1)^2[(k+1)+3]^2}{6} = \frac{(k+1)^2(k+4)^2}{6}$$

5. $P_k = 1 + 6 + 11 + \cdots + [5(k-1) - 4] + [5k - 4]$

$$P_{k+1} = 1 + 6 + 11 + \cdots + [5k - 4] + [5(k+1) - 4]$$
$$= 1 + 6 + 11 + \cdots + [5k - 4] + [5k + 1]$$

7. 1. When $n = 1$, $S_1 = 2 = 1(1 + 1)$.

2. Assume that

$$S_k = 2 + 4 + 6 + 8 + \cdots + 2k = k(k+1).$$

Then,

$$S_{k+1} = 2 + 4 + 6 + 8 + \cdots + 2k + 2(k+1)$$
$$= S_k + 2(k+1) = k(k+1) + 2(k+1) = (k+1)(k+2).$$

We conclude by mathematical induction that the formula is valid for all positive integer values of n.

9. 1. When $n = 1$, $S_1 = 3 = \frac{1}{2}(5(1) + 1)$

2. Assume that $S_k = 3 + 8 + 13 + \cdots + (5k - 2) = \frac{k}{2}(5k + 1)$

Then: $S_{k+1} = 3 + 8 + 13 + \cdots + (5k - 2) + [5(k+1) - 2]$

$$= S_k + [5k + 3] = \frac{k}{2}(5k + 1) + 5k + 3$$
$$= \frac{1}{2}[5k^2 + 11k + 6] = \frac{1}{2}(k+1)(5k+6)$$
$$= \frac{1}{2}(k+1)(5(k+1) + 1)$$

We conclude by mathematical induction that the formula is valid for all positive integers n.

11. 1. When $n = 1$, $S_1 = 1 = 2^1 - 1$.

2. Assume that

$$S_k = 1 + 2 + 2^2 + 2^3 + \cdots + 2^{k-1} = 2^k - 1.$$

Then,

$$S_{k+1} = 1 + 2 + 2^2 + 2^3 + \cdots + 2^{k-1} + 2^k$$
$$= S_k + 2^k = 2^k - 1 + 2^k = 2(2^k) - 1 = 2^{k+1} - 1.$$

Therefore, by mathematical induction, the formula is valid for all positive integer values of n.

13. 1. When $n = 1$, $S_1 = 1 = \dfrac{1(1 + 1)}{2}$.

2. Assume that

$$S_k = 1 + 2 + 3 + 4 + \cdots + k = \frac{k(k + 1)}{2}.$$

Then,

$$S_{k+1} = 1 + 2 + 3 + 4 + \cdots + k + (k + 1)$$
$$= S_k + (k + 1) = \frac{k(k + 1)}{2} + \frac{2(k + 1)}{2} = \frac{(k + 1)(k + 2)}{2}.$$

Therefore, we conclude that this formula holds for all positive integer values of n.

15. 1. When $n = 1$, $S_1 = 1^2 = \dfrac{1(2(1) - 1)(2(1) + 1)}{3}$

2. Assume that $S = 1^2 + 3^2 + \cdots + (2k - 1)^2 = \dfrac{k(2k - 1)(2k + 1)}{3}$

Then, $S_{k+1} = 1^2 + 3^2 + \cdots + (2k - 1)^2 + (2k + 1)^2$

$$= S_k + (2k + 1)^2 = \frac{k(2k - 1)(2k + 1)}{3} + (2k + 1)^2$$

$$= (2k + 1)\left[\frac{k(2k - 1)}{3} + (2k + 1)\right] = \frac{2k + 1}{3}[2k^2 - k + 6k + 3]$$

$$= \frac{2k + 1}{3}(2k + 3)(k + 1) = \frac{(k + 1)(2(k + 1) - 1)(2(k + 1) + 1)}{3}$$

Therefore, we conclude that the formula is valid for all positive integers n.

17. 1. When $n = 1$,

$$S_1 = 1^4 = \frac{1(1+1)(2 \cdot 1 + 1)(3 \cdot 1^2 + 3 \cdot 1 - 1)}{30}.$$

2. Assume that $S_k = \sum_{i=1}^{k} i^4 = \frac{k(k+1)(2k+1)(3k^2+3k-1)}{30}$.

Then, $S_{k+1} = S_k + (k+1)^4$

$$= \frac{k(k+1)(2k+1)(3k^2+3k-1)}{30} + (k+1)^4 = \frac{k(k+1)(2k+1)(3k^2+3k-1) + 30(k+1)^4}{30}$$

$$= \frac{(k+1)[k(2k+1)(3k^2+3k-1) + 30(k+1)^3]}{30} = \frac{(k+1)(6k^4+39k^3+91k^2+89k+30)}{30}$$

$$= \frac{(k+1)(k+2)(2k+3)(3k^2+9k+5)}{30} = \frac{(k+1)(k+2)(2(k+1)+1)(3(k+1)^2+3(k+1)-1)}{30}.$$

Therefore, we conclude that this formula holds for all positive integer values of n.

19. 1. When $n = 1$, $S_1 = 2 = \frac{1(2)(3)}{3}$.

2. Assume that

$$S_k = 1(2) + 2(3) + 3(4) + \cdots + k(k+1) = \frac{k(k+1)(k+2)}{3}.$$

Then,

$$S_{k+1} = 1(2) + 2(3) + 3(4) + \cdots + k(k+1) + (k+1)(k+2)$$

$$= S_k + (k+1)(k+2) = \frac{k(k+1)(k+2)}{3} + \frac{3(k+1)(k+2)}{3}$$

$$= \frac{(k+1)(k+2)(k+3)}{3}.$$

Thus, this formula is valid for all positive integer values of n.

21. 1. When $n = 4$, $4! = 24$ and $2^4 = 16$, thus $4! > 2^4$.

2. Assume $k! > 2^k$, $k > 4$. Then, $(k+1)! = k!(k+1) > 2^k(2)$ since $k + 1 > 2$. Thus, $(k+1)! > 2^{k+1}$.

Therefore, by mathematical induction, the formula is valid for all integers n such that $n \geq 4$.

23. 1. When $n = 2, \frac{1}{\sqrt{1}} + \frac{1}{\sqrt{2}} \approx 1.707$ and $\sqrt{2} \approx 1.414$, thus $\frac{1}{\sqrt{1}} + \frac{1}{\sqrt{2}} > \sqrt{2}$.

2. Assume

$$\frac{1}{\sqrt{1}} + \frac{1}{\sqrt{2}} + \frac{1}{\sqrt{3}} + \cdots + \frac{1}{\sqrt{k}} > \sqrt{k}, k > 2.$$

Then,

$$\frac{1}{\sqrt{1}} + \frac{1}{\sqrt{2}} + \frac{1}{\sqrt{3}} + \cdots + \frac{1}{\sqrt{k}} + \frac{1}{\sqrt{k+1}} > \sqrt{k} + \frac{1}{\sqrt{k+1}}.$$

Now we need to show that

$$\sqrt{k} + \frac{1}{\sqrt{k+1}} > \sqrt{k+1}, \ k > 2.$$

This is true because

$$\sqrt{k(k+1)} > k$$
$$\sqrt{k(k+1)} + 1 > k + 1$$
$$\frac{\sqrt{k(k+1)} + 1}{\sqrt{k+1}} > \frac{k+1}{\sqrt{k+1}}$$
$$\sqrt{k} + \frac{1}{\sqrt{k+1}} > \sqrt{k+1}.$$

Therefore,

$$\frac{1}{\sqrt{1}} + \frac{1}{\sqrt{2}} + \frac{1}{\sqrt{3}} + \cdots + \frac{1}{\sqrt{k}} + \frac{1}{\sqrt{k+1}} > \sqrt{k+1}.$$

Therefore, by mathematical induction, the formula is valid for all integers n such that $n \geq 2$.

25. 1. When $n = 1, 1 + a \geq a$ since $1 > 0$.

2. Assume $(1 + a)^k \geq ka$

Then $(1 + a)^{k+1} = (1 + a)^k(1 + a)$

$\geq ka(1 + a)$

$= ka + ka^2$

$\geq ka + k \quad$ (because $a > 1$)

$= (k + 1)a$

Therefore, by mathematical induction, the inequality is valid for all integers $n \geq 1$.

27. 1. When $n = 1, (ab)^1 = a^1b^1 = ab$.

2. Assume that $(ab)^k = a^kb^k$.

Then, $(ab)^{k+1} = (ab)^k(ab)$

$= a^kb^kab$

$= a^{k+1}b^{k+1}$.

Thus, $(ab)^n = a^nb^n$.

29. 1. When $n = 1, (x_1)^{-1} = x_1^{-1}$.

2. Assume that

$$(x_1x_2x_3 \cdots x_k)^{-1} = x_1^{-1}x_2^{-1}x_3^{-1} \cdots x_k^{-1}.$$

Then,

$$(x_1x_2x_3 \cdots x_kx_{k+1})^{-1} = [(x_1x_2x_3 \cdots x_k)x_{k+1}]^{-1}$$
$$= (x_1x_2x_3 \ldots x_k)^{-1}x_{k+1}^{-1}$$
$$= x_1^{-1}x_2^{-1}x_3^{-1} \cdots x_k^{-1}x_{k+1}^{-1}.$$

Thus, the formula is valid.

31. 1. When $n = 1$, $x(y_1) = xy_1$.

2. Assume that

$$x(y_1 + y_2 + \cdots + y_k) = xy_1 + xy_2 + \cdots + xy_k.$$

Then,

$$\begin{aligned} xy_1 + xy_2 + \cdots + xy_k + xy_{k+1} &= x(y_1 + y_2 + \cdots + y_k) + xy_{k+1} \\ &= x[(y_1 + y_2 + \cdots + y_k) + y_{k+1}] \\ &= x(y_1 + y_2 + \cdots + y_k + y_{k+1}). \end{aligned}$$

Hence, the formula holds.

33. 1. When $n = 1$, $[1^3 + 3(1)^2 + 2(1)] = 6$ and 3 is a factor.

2. Assume that 3 is a factor of $(k^3 + 3k^2 + 2k)$. Then,

$$\begin{aligned} [(k+1)^3 + 3(k+1)^2 + 2(k+1)] &= k^3 + 3k^2 + 3k + 1 + 3k^2 + 6k + 3 + 2k + 2 \\ &= (k^3 + 3k^2 + 2k) + (3k^2 + 9k + 6) \\ &= (k^3 + 3k^2 + 2k) + 3(k^2 + 3k + 2). \end{aligned}$$

Since 3 is a factor of $(k^3 + 3k^2 + 2k)$ by our assumption, and 3 it is a factor of $3(k^2 + 3k + 2)$ then 3 is a factor of the whole sum.

Thus, 3 is a factor of $(n^3 + 3n^2 + 2n)$ for every positive integer n.

35. 1. When $n = 2$, $[9^2 - 8(2) - 1] = 64$ and 64 is a factor.

2. Assume that 64 is a factor of $(9^k - 8k - 1)$. Then $9^{k+1} - 8(k+1) - 1 = 9^k \cdot 9 - 8k - 9$ $= 9[9^k - 1 - 8k] + 64k$ Since 64 is a factor of $(9^k - 8k - 1)$ is a factor of $9(9^k - 8k - 1) + 64k$.

Therefore, by mathematical induction, the statement is true for all integers $n \geq 2$.

37. $a_0 = 10, \; a_n = 4a_{n-1}$

$a_0 = 10$

$a_1 = 4(10) = 40$

$a_2 = 4(40) = 160$

$a_3 = 4(160) = 640$

$a_4 = 4(640) = 2560$

39. $a_0 = 0, \; a_1 = 2, \; a_n = a_{n-1} + 2a_{n-2}$

$a_0 = 0$

$a_1 = 2$

$a_2 = 2 + 2(0) = 2$

$a_3 = 2 + 2(2) = 6$

$a_4 = 6 + 2(2) = 10$

41. $a_1 = 2, \; a_n = n - a_{n-1}$

$a_1 = 2$

$a_2 = n - a_1 = 2 - 2 = 0$

$a_3 = n - a_2 = 3 - 0 = 3$

$a_4 = n - a_3 = 4 - 3 = 1$

$a_5 = n - a_5 = 5 - 1 = 4$

a_n: 2 0 3 1 4

First differences: −2 3 −2 3

Second differences: 5 −5 5

Since neither the first differences nor the second differences are equal, the sequence does not have a linear or quadratic model.

43. $a_2 = -3, \; a_n = -2a_{n-1}$

$a_2 = -3$

$a_3 = -2a_2 = -2(-3) = 6$

$a_4 = -2a_3 = -2(6) = -12$

$a_5 = -2a_4 = -2(-12) = 24$

$a_6 = -2a_5 = -2(24) = -48$

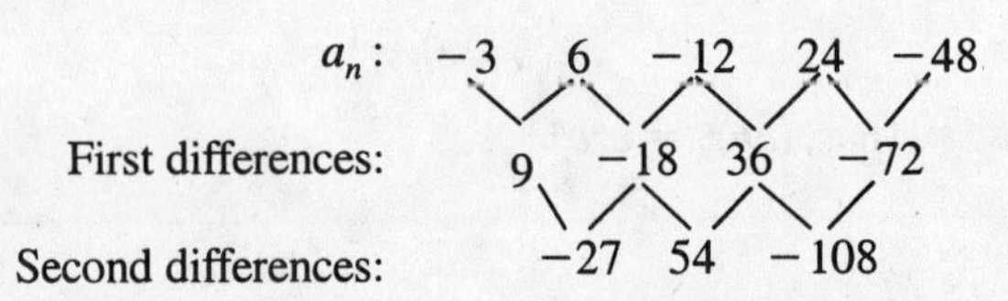

Since neither the first nor the second differences are equal, the sequence does not have a linear or quadratic model.

45. $a_0 = 2, \ a_n = (a_{n-1})^2$

$a_0 = 2$

$a_1 = a_0^2 = 2^2 = 4$

$a_2 = a_1^2 = 4^2 = 16$

$a_3 = a_2^2 = 16^2 = 256$

$a_4 = a_3^2 = 256^2 = 65{,}536$

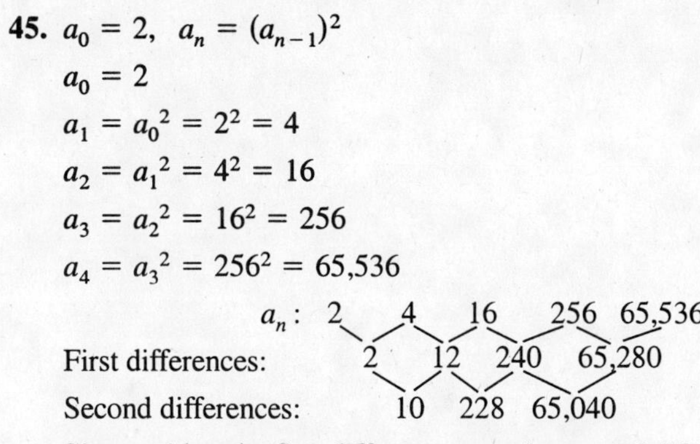

a_n: 2 4 16 256 65,536

First differences: 2 12 240 65,280

Second differences: 10 228 65,040

Since neither the first differences nor the second differences are equal, the sequence does not have a linear or quadratic model.

47. $a_1 = 0, \ a_n = a_{n-1} + 2n$

$a_1 = 0$

$a_2 = a_1 + 2(2) = 0 + 4 = 4$

$a_3 = a_2 + 2(3) = 4 + 6 = 10$

$a_4 = a_3 + 2(4) = 10 + 8 = 18$

$a_5 = a_4 + 2(5) = 18 + 10 = 28$

a_n: 0 4 10 18 28

First differences: 4 6 8 10

Second differences: 2 2 2

Since the second differences are equal, the sequence has a quadratic model.

49. $a_0 = 0, \ a_n = a_{n-1} - 1$

$a_0 = 0$

$a_1 = a_0 - 1 = 0 - 1 = -1$

$a_2 = a_1 - 1 = -1 - 1 = -2$

$a_3 = a_2 - 1 = -2 - 1 = -3$

$a_4 = a_3 - 1 = -3 - 1 = -4$

a_n: 0 −1 −2 −3 −4

First differences: −1 −1 −1 −1

Second differences: 0 0 0

Since the first differences are equal, the sequence has a linear model.

51. $a_0 = 7, \ a_1 = 6, \ a_3 = 10$

Let $a_n = an^2 + bn + c$. Thus,

$a_0 = a(0)^2 + b(0) + c = 7 \implies c = 7$

$a_1 = a(1)^2 + b(1) + c = 6 \implies a + b + c = 6$

$a + b = -1$

$a_3 = {}_a(3)^2 + b(3) + c = 10 \implies 9a + 3b + c = 10$

$9a + 3b = 3$

$3a + b = 1$

By elimination:

$$\begin{aligned} -a - b &= 1 \\ 3a + b &= 1 \\ \hline 2a &= 2 \\ a = 1 &\implies b = -2 \end{aligned}$$

Thus, $a_n = n^2 - 2n + 7$.

53. $a_0 + 3,\ a_2 + 0,\ a_6 + 36$

Let $a_n + an^2 + bn + c$. Thus,

$$\begin{aligned}
a_0 + a(0)^2 + b(0) + c + 3 &\Rightarrow \quad c + 3\\
a_2 + a(2)^2 + b(2) + c + 0 &\Rightarrow 4a + 2b + c + 0\\
& 4a + 2b \quad + -3\\
a_6 + a(6)^2 + b(6) + c + 36 &\Rightarrow 36a + 6b + c + 36\\
& 36a + 6b \quad + 33\\
& 12a + 2b \quad + 11
\end{aligned}$$

By elimination:
$$\begin{aligned}
-4a - 2b &+ 3\\
12a + 2b &+ 11\\
8a \quad &+ 14\\
a + \tfrac{7}{4} &\Rightarrow b + -5
\end{aligned}$$

Thus, $a_n + \frac{7}{4}n^2 - 5n + 3$.

55. False. See page 653.

57. See the domino illustration and Figure 9.11.

59. $\left[\begin{array}{rr:r} 1 & -1 & 2\\ -4 & 5 & -3 \end{array}\right]$ row reduces to $\left[\begin{array}{rr:r} 1 & 0 & 7\\ 0 & 1 & 5 \end{array}\right]$

Answers: $(7, 5)$

61.
$$\begin{aligned}
y &+ x^2\\
-3x + 2y + 2 \quad \Rightarrow \quad -3x &+ 2x^2 + 2\\
2x^2 - 3x - 2 &+ 0\\
(2x + 1)(x - 2) &+ 0\\
x + -\tfrac{1}{2} \ \text{ or } \ x &+ 2
\end{aligned}$$

$x + -\frac{1}{2} \Rightarrow y + \frac{1}{4},\ x + 2 \Rightarrow y + 4$

Points of intersection: $\left(-\frac{1}{2}, \frac{1}{4}\right), (2, 4)$

63.
$$\begin{aligned}
x - \ y \qquad &+ -1\\
x + 2y - 2z &+ \ 3\\
3x - \ y + 2z &+ \ 3
\end{aligned}$$

Using an augmented matrix, we have

$$\left[\begin{array}{rrr:r} 1 & -1 & 0 & -1\\ 1 & 2 & -2 & 3\\ 3 & -1 & 2 & 3 \end{array}\right]$$

$$\begin{array}{r} \\ -R_1 + R_2 \rightarrow \\ -3R_1 + R_3 \rightarrow \end{array}\left[\begin{array}{rrr:r} 1 & -1 & 0 & -1\\ 0 & 3 & -2 & 4\\ 0 & 2 & 2 & 6 \end{array}\right]$$

$$\begin{array}{r} \\ -R_3 + R_2 \rightarrow \\ \frac{1}{2}R_3 \rightarrow \end{array}\left[\begin{array}{rrr:r} 1 & -1 & 0 & -1\\ 0 & 1 & -4 & -2\\ 0 & 1 & 1 & 3 \end{array}\right]$$

$$\begin{array}{r} R_2 + R_1 \rightarrow \\ \\ -R_2 + R_3 \rightarrow \end{array}\left[\begin{array}{rrr:r} 1 & 0 & -4 & -3\\ 0 & 1 & -4 & -2\\ 0 & 0 & 5 & 5 \end{array}\right]$$

$$\begin{array}{r} 4R_3 + R_1 \rightarrow \\ 4R_3 + R_2 \rightarrow \\ \frac{1}{5}R_3 \rightarrow \end{array}\left[\begin{array}{rrr:r} 1 & 0 & 0 & 1\\ 0 & 1 & 0 & 2\\ 0 & 0 & 1 & 1 \end{array}\right]$$

Thus, $x + 1, y + 2, z + 1$.

Answer: $(1, 2, 1)$

65. $\begin{bmatrix} -3 & 1 & 5 & : & 25 \\ 1 & -2 & 3 & : & 7 \\ 2 & 3 & -1 & : & 0 \end{bmatrix}$ row reduces to $\begin{bmatrix} 1 & 0 & 0 & : & -1 \\ 0 & 1 & 0 & : & 2 \\ 0 & 0 & 1 & : & 4 \end{bmatrix}$

Answer: $(-1, 2, 4)$

67. $\begin{vmatrix} 7 & 6 \\ -4 & 2 \end{vmatrix} = 14 - (-24) = 38$

69. $(2x^2 - 1)^2 = 4x^4 - 4x^2 + 1$

71. $(5 - 4x)^3 = -64x^3 + 240x^2 - 300x + 125$

Section 9.5 The Binomial Theorem

- You should be able to use the Binomial Theorem

$$(x + y)^n = x^n + nx^{n-1}y + \frac{n(n-1)}{2!}x^{n-2}y^2 + \cdots + {}_nC_r x^{n-r}y^r + \cdots + y^n$$

where ${}_nC_r = \dfrac{n!}{(n-r)!r!}$, to expand$(x + y)^n$.

- You should be able to use Pascal's Triangle.

Solutions to Odd-Numbered Exercises

1. ${}_7C_5 = \dfrac{7!}{2!5!} = \dfrac{7 \cdot 6 \cdot 5!}{2 \cdot 5!} = \dfrac{42}{2} = 21$

3. $\binom{12}{0} = {}_{12}C_0 = \dfrac{12!}{0!12!} = 1$

5. ${}_{20}C_{15} = \dfrac{20!}{15!5!} = \dfrac{20 \cdot 19 \cdot 18 \cdot 17 \cdot 16}{5 \cdot 4 \cdot 3 \cdot 2 \cdot 1} = 15{,}504$

7. ${}_{14}C_1 = \dfrac{14!}{13!1!} = \dfrac{14!13!}{13!} = 14$

9. $\binom{100}{98} = {}_{100}C_{98} = \dfrac{100!}{98!2!} = \dfrac{100 \cdot 99}{2 \cdot 1} = 4950$

11. ${}_{100}C_2 = \dfrac{100!}{2!98!} = \dfrac{100 \cdot 99}{2 \cdot 1} = 4950$

13. ${}_{32}C_{28} = 35{,}960$

15. ${}_{22}C_9 = 497{,}420$

17. ${}_{41}C_{36} = 749{,}398$

19.

1
1 1
1 2 1
1 3 3 1
1 4 6 4 1
1 5 10 10 5 1
1 6 15 20 15 6 1
1 7 (21) 35 35 21 7 1

${}_7C_2 = 21$, the 3rd entry in the 7th row.

21.

1
1 1
1 2 1
1 3 3 1
1 4 6 4 1
1 5 10 10 5 1
1 6 15 20 15 6 1
1 7 21 35 35 21 7 1
1 8 28 56 70 (56) 28 8 1

${}_8C_5 = 56$, the 6th entry in the 8th row.

23. $(x+1)^4 = {}_4C_0x^4 + {}_4C_1x^3(1) + {}_4C_2x^2(1)^2 + {}_4C_3x(1)^3 + {}_4C_4(1)^4$
$= x^4 + 4x^3 + 6x^2 + 4x + 1$

25. $(a+3)^3 = {}_3C_0a^3 + {}_3C_1a^2(3) + {}_3C_2a(3)^2 + {}_3C_3(3)^3$
$= a^3 + 3a^2(3) + 3a(3)^2 + (3)^3$
$= a^3 + 9a^2 + 27a + 27$

27. $(y-2)^4 = {}_4C_0y^4 - {}_4C_1y^3(2) + {}_4C_2y^2(2)^2 - {}_4C_3y(2)^3 + {}_4C_4(2)^4$
$= y^4 - 4y^3(2) + 6y^2(4) - 4y(8) + 16$
$= y^4 - 8y^3 + 24y^2 - 32y + 16$

29. $(x+y)^5 = {}_5C_0x^5 + {}_5C_1x^4y + {}_5C_2x^3y^2 + {}_5C_3x^2y^3 + {}_5C_4xy^4 + {}_5C_5y^5$
$= x^5 + 5x^4y + 10x^3y^2 + 10x^2y^3 + 5xy^4 + y^5$

31. $(r+2s)^6 = {}_6C_0r^6 + {}_6C_1r^5(2s) + {}_6C_2r^4(2s)^2 + {}_6C_3r^3(2s)^3 + {}_6C_4r^2(2s)^4$
$+ {}_6C_5r(2s)^5 + {}_6C_6(2s)^6$
$= r^6 + 12r^5s + 60r^4s^2 + 160r^3s^3 + 240r^2s^4 + 192rs^5 + 64s^6$

33. $(x-y)^5 = {}_5C_0x^5 - {}_5C_1x^4y + {}_5C_2x^3y^2 - {}_5C_3x^2y^3 - {}_5C_4xy^4 - {}_5C_5y^5$
$= x^5 - 5x^4y + 10x^3y^2 - 10x^2y^3 + 5xy^4 - y^5$

35. $(1-4x)^3 = {}_3C_01^3 - {}_3C_11^2(4x) + {}_3C_21(4x)^2 - {}_3C_3(4x)^3$
$= 1 - 3(4x) + 3(4x)^2 - (4x)^3$
$= 1 - 12x + 48x^2 - 64x^3$

37. $(x^2+5)^4 = {}_4C_0(x^2)^4 + {}_4C_1(x^2)^3(5) + {}_4C_2(x^2)^2(5)^2 + {}_4C_3(x^2)(5)^3 + {}_4C_4(5)^4$
$= x^8 + 4x^6(5) + 6x^4(25) + 4x^2(125) + 625$
$= x^8 + 20x^6 + 150x^4 + 500x^2 + 625$

39. $\left(\frac{1}{x}+y\right)^5 = {}_5C_0\left(\frac{1}{x}\right)^5 + {}_5C_1\left(\frac{1}{x}\right)^4y + {}_5C_2\left(\frac{1}{x}\right)^3y^2 + {}_5C_3\left(\frac{1}{x}\right)^2y^3 + {}_5C_4\left(\frac{1}{x}\right)y^4 + {}_5C_5y^5$
$= \frac{1}{x^5} + \frac{5y}{x^4} + \frac{10y^2}{x^3} + \frac{10y^3}{x^2} + \frac{5y^4}{x} + y^5$

41. $2(x-3)^4 + 5(x-3)^2 = 2[x^4 - 4(x^3)(3) + 6(x^2)(3^2) - 4(x)(3^3) + 3^4] + 5[x^2 - 2(x)(3) + 3^2]$
$= 2(x^4 - 12x^3 + 54x^2 - 108x + 81) + 5(x^2 - 6x + 9)$
$= 2x^4 - 24x^3 + 113x^2 - 246x + 207$

43. $-3(x-2)^3 - 4(x+1)^6$

$= [-3x^3 + 18x^2 - 36x + 24] - [4x^6 + 24x^5 + 60x^4 + 80x^3 + 60x^2 + 24x + 4]$

$= -4x^6 - 24x^5 - 60x^4 - 83x^3 - 42x^2 - 60x + 20$

45. 5th Row of Pascal's Triangle: 1 5 10 10 5 1

$(3t - s)^5 = 1(3t)^5 + 5(3t)^4(-s) + 10(3t)^3(-s)^2 + 10(3t)^2(-s)^3 + 5(3t)(-s)^4 + 1(-5)^5$

$= 243t^5 - 405t^4s + 270t^3s^2 - 90t^2s^3 + 15ts^4 - s^5$

47. 4th Row of Pascal's Triangle: 1 4 6 4 1

$(3 - 2z)^4 = 3^4 - 4(3)^3(2z) + 6(3)^2(2z)^2 - 4(3)(2z)^3 + (2z)^4$

$= 81 - 216z + 216z^2 - 96z^3 + 16z^4$

49. The term involving x^4 in the expansion of $(x+3)^{12}$ is ${}_{12}C_8x^4(3)^8 = 495x^4(3)^8 = 3{,}247{,}695x^4$

51. The term involving x^8y^2 in the expansion of $(x - 2y)^{10}$ is

${}_{10}C_2x^8(-2y)^2 = \dfrac{10!}{2!8!} \cdot 4x^8y^2 = 180x^8y^2$. The coefficient is 180.

53. The term involving x^6y^3 in $(3x - 2y)^9$ is ${}_9C_3(3x)^6(-2y)^3 = 84(3)^6(-2)^3x^6y^3 = -489{,}888$

55. The coefficient of $x^8y^6 = (x^2)^4y^6$ in the expansion of $(x^2 + y)^{10}$ is ${}_{10}C_6 = 210$.

57. $(\sqrt{x} + 5)^4 = (\sqrt{x})^4 + 4(\sqrt{x})^3(5) + 6(\sqrt{x})^2(5)^2 + 4(\sqrt{x})(5^3) + 5^4$

$= x^2 + 20x\sqrt{x} + 150x + 500\sqrt{x} + 625$

$= x^2 + 20x^{3/2} + 150x + 500x^{1/2} + 625$

59. $(x^{2/3} - y^{1/3})^3 = (x^{2/3})^3 - 3(x^{2/3})^2(y^{1/3}) + 3(x^{2/3})(y^{1/3})^2 - (y^{1/3})^3$

$= x^2 - 3x^{4/3}y^{1/3} + 3x^{2/3}y^{2/3} - y$

61. $\dfrac{f(x+h) - f(x)}{h} = \dfrac{(x+h)^3 - x^3}{h}$

$= \dfrac{x^3 + 3x^2h + 3xh^2 + h^3 - x^3}{h}$

$= \dfrac{h(3x^2 + 3xh + h^2)}{h}$

$= 3x^2 + 3xh + h^2,\ h \neq 0$

63. $\dfrac{f(x+h)-f(x)}{h} = \dfrac{\sqrt{x+h}-\sqrt{x}}{h}$

$= \dfrac{\sqrt{x+h}-\sqrt{x}}{h} \cdot \dfrac{\sqrt{x+h}+\sqrt{x}}{\sqrt{x+h}+\sqrt{x}}$

$= \dfrac{(x+h)-x}{h\left[\sqrt{x+h}+\sqrt{x}\right]}$

$= \dfrac{1}{\sqrt{x+h}+\sqrt{x}}, h \neq 0$

65. $(1+i)^4 = {}_4C_0 1^4 + {}_4C_1(1)^3 i + {}_4C_2(1)^2 i^2 + {}_4C_3 1 \cdot i^3 + {}_4C_4 i^4$

$= 1 + 4i - 6 - 4i + 1$

$= -4$

67. $(2-3i)^6 = {}_6C_0 2^6 - {}_6C_1 2^5(3i) + {}_6C_2 2^4(3i)^2 - {}_6C_3 2^3(3i)^3 + {}_6C_4 2^2(3i)^4 - {}_6C_5 2(3i)^5 + {}_6C_6(3i)^6$

$= 64 - 576i - 2160 + 4320i + 4860 - 2916i - 729$

$= 2035 + 828i$

69. $\left(-\dfrac{1}{2} + \dfrac{\sqrt{3}}{2}i\right)^3 = \dfrac{1}{8}(-1+\sqrt{3}i)^3$

$= \dfrac{1}{8}\left[(-1)^3 + 3(-1)^2(\sqrt{3}i) + 3(-1)(\sqrt{3}i)^2 + (\sqrt{3}i)^3\right]$

$= \dfrac{1}{8}\left[-1 + 3\sqrt{3}i + 9 - 3\sqrt{3}i\right]$

$= 1$

71. $(1.02)^8 = (1+0.02)^8 = 1 + 8(0.02) + 28(0.02)^2 + 56(0.02)^3 + 70(0.02)^4 + 56(0.02)^5$

$+ 28(0.02)^6 + 8(0.02)^7 + (0.02)^8$

$= 1 + 0.16 + 0.0112 + 0.000448 + \cdots \approx 1.172$

73. $(2.99)^{12} = (3-0.01)^{12}$

$= 3^{12} - 12(3)^{11}(0.01) + 66(3)^{10}(0.01)^2 - 220(3)^9(0.01)^3 + 495(3)^8(0.01)^4$

$- 792(3)^7(0.01)^5 + 924(3)^6(0.01)^6 - 792(3)^5(0.01)^7 + 495(3)^4(0.01)^8$

$- 220(3)^3(0.01)^9 + 66(3)^2(0.01)^{10} - 12(3)(0.01)^{11} + (0.01)^{12}$

$\approx 510{,}568.785$

75. $f(x) = x^3 - 4x$

$g(x) = f(x + 6)$

$= (x + 6)^3 - 4(x + 6)$

$= x^3 + 3x^2(6) + 3x(6)^2 + 6^3 - 4x - 24$

$= x^3 + 18x^2 + 104x + 192$

The graph of g is the same as the graph of f shifted 6 units to the left.

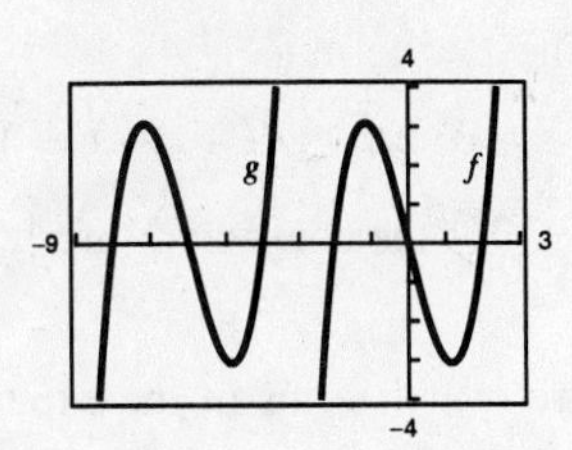

77. $f(x) = -x^2 + 3x + 2$

$g(x) = f(x - 2)$

$= -(x - 2)^2 + 3(x - 2) + 2$

$= -x^2 + 4x - 4 + 3x - 6 + 2$

$= -x^2 + 7x - 8$

The graph of g is the same as the graph of f shifted 2 units to the right

79. (a) ${}_{12}C_5 = 792$

(b) $({}_6C_5)^2 = 36$

(c) ${}_{11}C_5 + {}_{11}C_4 = 792$

(d) ${}_6C_5 + {}_6C_5 = 12$

(a) and (c) are equal.

81. $f(x) = (1 - x)^3$

$g(x) = 1 - 3x$

$h(x) = 1 - 3x + 3x^2$

$p(x) = 1 - 3x + 3x^2 - x^3$

Since $p(x)$ is the expansion of $f(x)$, they have the same graph.

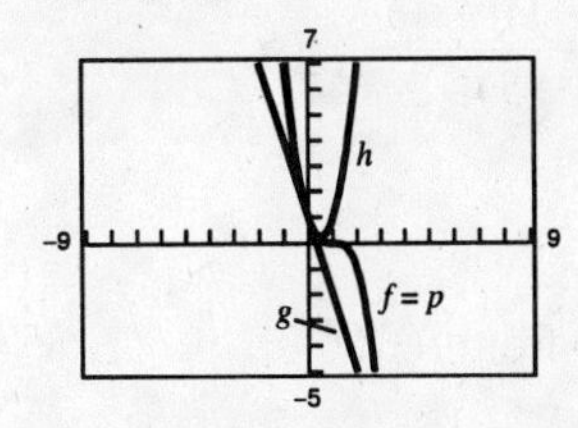

83. ${}_7C_4\left(\frac{1}{2}\right)^4\left(\frac{1}{2}\right)^3 = 35\left(\frac{1}{16}\right)\left(\frac{1}{8}\right) \approx 0.273$

85. ${}_8C_4\left(\frac{1}{3}\right)^4\left(\frac{2}{3}\right)^4 = 70\left(\frac{1}{81}\right)\left(\frac{16}{81}\right) \approx 0.171$

87. (a) $g(t) = f(t + 10)$

$= 0.0348(t + 10)^2 + 5.1083(t + 10) + 41.0250$

$= 0.0348(t^2 + 20t + 100) + 5.1083t + 51.083 + 41.0250$

$= 0.0348t^2 + 5.8043t + 95.588$

(b)

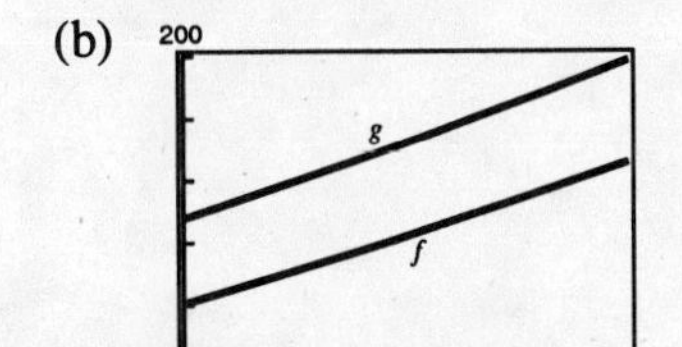

89. False. The x^4y^8 term is

$${}_{12}C_4x^4(-2y)^8 = 495x^4(-2)^8y^8 = 126{,}720x^4y^8$$

[Note 7920 is the coefficient of x^8y^4]

91. Answers will vary. See page 658.

93. There are $n + 1$ terms in the expansion of $(x + y)^n$.

95. $${}_nC_{n-r} = \frac{n!}{[n-(n-r)]!(n-r)!} = \frac{n!}{r!(n-r)!} = \frac{n!}{(n-r)!r!} = {}_nC_r$$

97. $${}_nC_r + {}_nC_{r-1} = \frac{n!}{(n-r)!r!} + \frac{n!}{(n-r+1)!(r-1)!}$$

$$= \frac{n!(n-r+1)}{(n-r)!r!(n-r+1)} + \frac{n!}{(n-r+1)!(r-1)!}\frac{r}{r}$$

$$= \frac{n!(n-r+1)}{(n-r+1)!r!} + \frac{n!r}{(n-r+1)!r!}$$

$$= \frac{n!(n-r+1+r)}{(n-r+1)!r!}$$

$$= \frac{n!(n+1)}{(n-r+1)!r!}$$

$$= \frac{(n+1)!}{(n+1-r)!r!} = {}_{n+1}C_r$$

99. $g(x) = f(x) + 8$

$g(x)$ is shifted 8 units up from $f(x)$.

101. $g(x) = f(-x)$

$g(x)$ is the reflection of $f(x)$ in the y-axis.

103. $$A + B = \begin{bmatrix} -2 & -1 & -3 \\ 8 & 0 & 4 \\ -1 & -2 & 7 \end{bmatrix}$$

105. $$-3A - 5B = \begin{bmatrix} 6 & 11 & 15 \\ -30 & -2 & -16 \\ 13 & 10 & -33 \end{bmatrix}$$

107. $$AB = \begin{bmatrix} 9 & 11 & 12 \\ -13 & -25 & -5 \\ -5 & -18 & -6 \end{bmatrix}$$

109. $$\begin{bmatrix} -6 & 5 \\ -5 & 4 \end{bmatrix}^{-1} = \frac{1}{-24+25}\begin{bmatrix} 4 & -5 \\ 5 & -6 \end{bmatrix} = \begin{bmatrix} 4 & -5 \\ 5 & -6 \end{bmatrix}$$

Section 9.6 Counting Principles

- You should know The Fundamental Counting Principle.
- ${}_nP_r = \dfrac{n!}{(n-r)!}$ is the number of permutations of n elements taken r at a time.
- Given a set of n objects that has n_1 of one kind, n_2 of a second kind, and so on, the number of distinguishable permutations is

$$\frac{n!}{n_1!n_2!\cdots n_k!}.$$

- ${}_nC_r = \dfrac{n!}{(n-r)!r!}$ is the number of combinations of n elements taken r at a time.

Solutions to Odd-Numbered Exercises

1. Odd integers: 1, 3, 5, 7, 9, 11

6 ways

3. Prime integers: 2, 3, 5, 7, 11

5 ways

5. Divisible by 4: 4, 8, 12

3 ways

7. Sum is 10: $1 + 9, 2 + 8, 3 + 7, 4 + 6, 5 + 5, 6 + 4, 7 + 3, 8 + 2, 9 + 1$

9 ways

9. Amplifiers: 4 choices

Compact disc players: 6 choices

Speakers: 8 choices

Total: $4 \cdot 6 \cdot 8 = 192$ ways

11. Chemist: 3 choices

Statistician: 6 choices

Total: $3 \cdot 6 = 18$ ways

13. $10! = 3{,}628{,}800$ ways

15. 1st Position: 2 choices

2nd Position: 3 choices

3rd Position: 2 choices

4th Position: 1 choice

Total: $2 \cdot 3 \cdot 2 \cdot 1 = 12$ ways

Label the four people A, B, C, and D and suppose that A and B are willing to take the first position. The twelve combinations are as follows.

ABCD	BACD
ABDC	BADC
ACBD	BCAD
ACDB	BCDA
ADBC	BDAC
ADCB	BDCA

17. $26 \cdot 26 \cdot 10 \cdot 10 \cdot 10 \cdot 10 = 6{,}760{,}000$

19. (a) $9 \cdot 10 \cdot 10 = 900$

(b) $9 \cdot 9 \cdot 8 = 648$

(c) $9 \cdot 10 \cdot 2 = 180$

(d) $10 \cdot 10 \cdot 10 - 400 = 600$

21. $40^3 = 64{,}000$

23. (a) $6 \cdot 5 \cdot 4 \cdot 3 \cdot 2 \cdot 1 = 720$

(b) $6 \cdot 1 \cdot 4 \cdot 1 \cdot 2 \cdot 1 = 48$

25. ${}_nP_r = \dfrac{n!}{(n-r)!}$

So, ${}_4P_4 = \dfrac{4!}{0!} = 4! = 24.$

27. ${}_8P_3 = \dfrac{8!}{5!} = 8 \cdot 7 \cdot 6 = 336$

29. ${}_5P_4 = \dfrac{5!}{1!} = 120$

31. $14 \cdot {}_nP_3 = {}_{n+2}P_4$ Note $n \geq 3$ for this to be defined.

$$14\left[\frac{n!}{(n-3)!}\right] = \frac{(n+2)!}{(n-2)!}$$

$14n(n-1)(n-2) = (n+2)(n+1)n(n-1)$ (We can divide here by $n(n-1)$ since $n \neq 0, n \neq 1$.)

$$14n - 28 = n^2 + 3n + 2$$

$$0 = n^2 - 11n + 30$$

$$0 = (n-5)(n-6)$$

$$n = 5 \quad \text{or} \quad n = 6$$

33. ${}_{20}P_6 = 27{,}907{,}200$

35. ${}_{120}P_4 = 197{,}149{,}680$

37. ${}_{20}C_4 = 4845$

39. $5! = 120$ ways

41. (a) $2^4 = 16$ characters

(b) $2 + 2^2 + 2^3 = 14$ characters

43. $\dfrac{7!}{2!1!3!1!} = \dfrac{7!}{2!3!} = 420$

45. $\dfrac{7!}{2!1!1!1!1!1!} = \dfrac{7!}{2!} = 7 \cdot 6 \cdot 5 \cdot 4 \cdot 3 = 2520$

47.

ABCD	BACD	CABD	DABC
ABDC	BADC	CADB	DACB
ACBD	BCAD	CBAD	DBAC
ACDB	BCDA	CBDA	DBCA
ADBC	BDAC	CDAB	DCAB
ADCB	BDCA	CDBA	DCBA

49. ${}_6C_2 = 15$

The 15 ways are listed below.

AB, AC, AD, AE, AF,
BC, BD, BE, BF, CD,
CE, CF, DE, DF, EF

51. ${}_{20}C_4 = 4845$ groups

53. ${}_{40}C_6 = 3{,}838{,}380$ ways

55. ${}_{100}C_5 = 75{,}287{,}520$ subsets

57. ${}_9C_2 = 36$ lines

59. (a) ${}_{12}C_4 = 495$ ways

(b) $({}_5C_2)({}_7C_2) = (10)(21) = 210$ ways

61. (a) ${}_8C_4 = \frac{8!}{4!4!} = 70$ ways

(b) There are $2^4 = 16$ ways that a group of four can be formed without any couples in the group. Therefore, if at least one couple is to be in the group, there are $70 - 16 = 54$ ways that could occur.

(c) $2 \cdot 2 \cdot 2 \cdot 2 = 16$ ways

63. ${}_5C_2 - 5 = 10 - 5 = 5$ diagonals

65. ${}_8C_2 - 8 = 28 - 8 = 20$ diagonals

67. False. This is an example of a combination.

69. False. for example, ${}_1P_1 = {}_1C_1 = 1$

71. The symbol ${}_nP_r$ means the number of ways to choose and order r elements out of a set of n elements.

73. (b) ${}_{10}P_6$ is larger than ${}_{10}C_6$ because the permutations count different orderings as distinct.

75. $${}_nC_n = \frac{n!}{(n-n)!n!} = \frac{n!}{0!n!} = \frac{n!}{n!0!} = \frac{n!}{(n-0)!0!} = {}_nC_0$$

77. $${}_nC_r = \frac{n!}{(n-r)!r!}$$
$$= \frac{1}{r!}\left[\frac{n!}{(n-r)^1}\right]$$
$$= \frac{{}_nP_r}{r!}$$

79. $$\frac{4}{t} + \frac{3}{2t} = 1$$
$$\frac{8+3}{2t} = 1$$
$$11 = 2t$$
$$t = \frac{11}{2} = 5.5$$

81. $$e^{x/3} = 16$$
$$\frac{x}{3} = \ln 16$$
$$x = 3 \ln 16 \approx 8.32$$

83. $$x = \frac{\begin{vmatrix} 35 & 1 \\ 10 & 2 \end{vmatrix}}{\begin{vmatrix} 8 & 1 \\ 6 & 2 \end{vmatrix}} = \frac{60}{10} = 6$$
$$y = \frac{\begin{vmatrix} 8 & 35 \\ 6 & 10 \end{vmatrix}}{\begin{vmatrix} 8 & 1 \\ 6 & 2 \end{vmatrix}} = \frac{-130}{10} = -13$$

Answer: $(6, -13)$

85. $$x = \frac{\begin{vmatrix} -74 & -11 \\ 8 & -4 \end{vmatrix}}{\begin{vmatrix} 10 & -11 \\ -8 & -4 \end{vmatrix}} = \frac{384}{-128} = -3$$
$$y = \frac{\begin{vmatrix} 10 & -74 \\ -8 & 8 \end{vmatrix}}{\begin{vmatrix} 10 & -11 \\ -8 & -4 \end{vmatrix}} = \frac{-512}{-128} = 4$$

Answer: $(-3, 4)$

87. $(x - 1)^6 = x^6 - 6x^5 + 15x^4 - 20x^3 + 15x^2 - 6x + 1$

89. $(3x - y)^4 = 81x^4 - 108x^3y + 54x^2y^2 - 12xy^3 + y^4$

Section 9.7 Probability

You should know the following basic principles of probability.

- If an event E has $n(E)$ equally likely outcomes and its sample space has $n(S)$ equally likely outcomes, then the probability of event E is

$$P(E) = \frac{n(E)}{n(S)}, \text{ where } 0 \le P(E) \le 1.$$

- If A and B are mutually exclusive events, then $P(A \cup B) = P(A) + P(B)$.
  If A and B are not mutually exclusive events, then $P(A \cup B) = P(A) + P(B) - P(A \cap B)$.
- If A and B are independent events, then the probability that both A and B will occur is $P(A)P(B)$.
- The probability of the complement of an event A is $P(A') = 1 - P(A)$.

Solutions to Odd-Numbered Exercises

1. $\{(H, 1), (H, 2), (H, 3), (H, 4), (H, 5), (H, 6), (T, 1), (T, 2), (T, 3), (T, 4), (T, 5), (T, 6)\}$

3. $\{ABC, ACB, BAC, BCA, CAB, CBA\}$

5. $\{(A, B), (A, C), (A, D), (A, E), (B, C), (B, D), (B, E), (C, D), (C, E), (D, E)\}$

7. $E = \{HTT, THT, TTH\}$

$$P(E) = \frac{n(E)}{n(S)} = \frac{3}{8}$$

9. $E = \{HHH, HHT, HTH, HTT, THH, THT, TTH\}$

$$P(E) = \frac{n(E)}{n(S)} = \frac{7}{8}$$

11. $E = \{K, K, K, K, Q, Q, Q, Q, J, J, J, J\}$

$$P(E) = \frac{n(E)}{n(S)} = \frac{12}{52} = \frac{3}{13}$$

13. $E = \{K, K, Q, Q, J, J\}$

$$P(E) = \frac{n(E)}{n(S)} = \frac{6}{52} = \frac{3}{26}$$

15. $E = \{(1, 4), (2, 3), (3, 2), (4, 1)\}$

$$P(E) = \frac{n(E)}{n(S)} = \frac{4}{36} = \frac{1}{9}$$

17. not $E = \{(6, 6)\}$

$n(E) = n(S) - n(\text{not } E) = 36 - 1 = 35$

$$P(E) = \frac{n(E)}{n(S)} = \frac{35}{36}$$

19. sum is 3 or 5: $E = \{(1, 2), (2, 1), (1, 4), (2, 3), (3, 2), (4, 1)\}$

$$P(E) = \frac{n(E)}{n(S)} = \frac{6}{36} = \frac{1}{6}$$

21. $P(E) = \dfrac{{}_3C_2}{{}_6C_2} = \dfrac{3}{15} = \dfrac{1}{5}$

23. $P(E) = \dfrac{{}_4C_2}{{}_6C_2} = \dfrac{6}{15} = \dfrac{2}{5}$

25. $P(E') = 1 - P(E) = 1 - 0.7 = 0.3$

27. $P(E') = 1 - P(E) = 1 - \frac{1}{3} = \frac{2}{3}$

29. $P(E) = 1 - P(E') = 1 - p = 1 - 0.15 = 0.85$

31. $P(E) = 1 - P(E') = 1 - \frac{13}{20} = \frac{7}{20}$

33. (a) $0.06(1.3) = 0.078$ or 78,000

(b) $\dfrac{0.30}{1.00} = 0.3$

(c) $\dfrac{0.21 + 0.16}{1.00} = 0.37$

(d) $\dfrac{0.07 + 0.03}{1.00} = 0.10$

35. (a) $\frac{290}{500} = 0.58$

(b) $\frac{478}{500} = 0.956$

(c) $\frac{2}{500} = 0.004$

37. (a) $\dfrac{672}{1254}$

(b) $\dfrac{582}{1254}$

(c) $\dfrac{672 - 124}{1254} = \dfrac{548}{1254}$

39. $p + p + 2p = 1$

$p = 0.25$

Taylor: $0.50 = \dfrac{1}{2}$

Moore: $0.25 = \dfrac{1}{4}$

Perez: $0.25 = \dfrac{1}{4}$

41. (a) $\dfrac{{}_{15}C_{10}}{{}_{20}C_{10}} = \dfrac{3003}{184{,}756} = \dfrac{21}{1292} \approx 0.016$

(b) $\dfrac{{}_{15}C_8 \cdot {}_5C_2}{{}_{20}C_{10}} = \dfrac{64{,}350}{184{,}756} = \dfrac{225}{646} \approx 0.348$

(c) $\dfrac{{}_{15}C_9 \cdot {}_5C_1}{{}_{20}C_{10}} + \dfrac{{}_{15}C_{10}}{{}_{20}C_{10}} = \dfrac{25{,}025 + 3003}{184{,}756} = \dfrac{28{,}028}{184{,}756} = \dfrac{49}{323} \approx 0.152$

43. Total ways to insert letters: $4! = 24$ ways

4 correct: 1 way

3 correct: not possible

2 correct: 6 ways

1 correct: 8 ways

0 correct: 9 ways

(a) $\dfrac{8}{24} = \dfrac{1}{3}$

(b) $\dfrac{8 + 6 + 1}{24} = \dfrac{15}{24} = \dfrac{5}{8}$

45. (a) $\dfrac{1}{{}_5P_5} = \dfrac{1}{120}$

(b) $\dfrac{1}{{}_4P_4} = \dfrac{1}{24}$

47. (a) $\dfrac{{}_{74}C_8}{{}_{84}C_8} = 0.3457$

(b) $\dfrac{{}_{20}C_8}{{}_{84}C_8} = 2.89 \times 10^{-6}$

49. (a) $\dfrac{{}_9C_4}{{}_{12}C_4} = \dfrac{126}{495} = \dfrac{14}{55}$ (4 good units)

(b) $\dfrac{({}_9C_2)({}_3C_2)}{{}_{12}C_4} = \dfrac{108}{495} = \dfrac{12}{55}$ (2 good units)

(c) $\dfrac{({}_9C_3)({}_3C_1)}{{}_{12}C_4} = \dfrac{252}{495} = \dfrac{28}{55}$ (3 good units)

At least 2 good units: $\dfrac{12}{55} + \dfrac{28}{55} + \dfrac{14}{55} = \dfrac{54}{55}$

51. (a) $P(EE) = \frac{15}{30} \cdot \frac{15}{30} = \frac{1}{4}$

(b) $P(EO) + P(OE) = \frac{1}{4} + \frac{1}{4} = \frac{1}{2}$

(c) $P(N_1 < 10, N_2 < 10) = \frac{9}{30} \cdot \frac{9}{30} = \frac{9}{100}$

(d) $P(N_1N_1) = \frac{30}{30} \cdot \frac{1}{30} = \frac{1}{30}$

53. (a) $P(SS) = (0.985)^2 \approx 0.9702$

(b) $P(S) = 1 - P(FF) = 1 - (0.015)^2 \approx 0.9998$

(c) $P(FF) = (0.015)^2 \approx 0.0002$

55. (a) $\left(\frac{1}{5}\right)^6 = \frac{1}{15{,}625}$

(b) $\left(\frac{4}{5}\right)^6 = \frac{4096}{15{,}625} = 0.26144$

(c) $1 - 0.262144 = 0.737856 = \frac{11{,}529}{15{,}625}$

57. $(0.32)^2 = 0.1024$

59. $1 - \frac{(45)^2}{(60)^2} = 1 - \left(\frac{45}{60}\right)^2 = 1 - \left(\frac{3}{4}\right)^2 = 1 - \frac{9}{16} = \frac{7}{16}$

61. True. $P(E) + P(E') = 1$

63. (a) As you consider successive people with distinct birthdays, the probabilities must decrease to take into account the birth dates already used. Since the birth dates of people are independent events, multiply the respective probabilities of distinct birthdays.

(b) $\frac{365}{365} \cdot \frac{364}{365} \cdot \frac{363}{365} \cdot \frac{362}{365}$

(c) $P_1 = \frac{365}{365} = 1$

$$P_2 = \frac{365}{365} \cdot \frac{364}{365} = \frac{364}{365} P_1 = \frac{365 - (2 - 1)}{365} P_1$$

$$P_3 = \frac{365}{365} \cdot \frac{364}{365} \cdot \frac{363}{365} = \frac{363}{365} P_2 = \frac{365 - (3 - 1)}{365} P_2$$

$$P_n = \frac{365}{365} \cdot \frac{364}{365} \cdot \frac{363}{365} \cdot \ldots \cdot \frac{365 - (n - 1)}{365} = \frac{365 - (n - 1)}{365} P_{n-1}$$

(d) Q_n is the probability that the birthdays are *not* distinct which is equivalent to at least 2 people having the same birthday.

(e)

n	10	15	20	23	30	40	50
P_n	0.88	0.75	0.59	0.49	0.29	0.11	0.03
Q_n	0.12	0.25	0.41	0.51	0.71	0.89	0.97

(f) 23, See the chart above.

65. $\dfrac{2}{x-5} = 4$

$2 = 4(x-5) = 4x - 20$

$4x = 22$

$x = \dfrac{11}{2}$

67. $\dfrac{3}{x-2} + \dfrac{x}{x+2} = 1$

$3(x+2) + x(x-2) = (x-2)(x+2)$

$3x + 6 + x^2 - 2x = x^2 - 4$

$x = -10$

69. $\ln x = 8 \Rightarrow x = e^8 \approx 2980.96$

71. $4 \ln 6x = 16$

$\ln 6x = 4$

$e^4 = 6x$

$x = \frac{1}{6}e^4 \approx 9.10$

73. $y \geq -3$

$x \geq -1$

$x + y \leq 8$

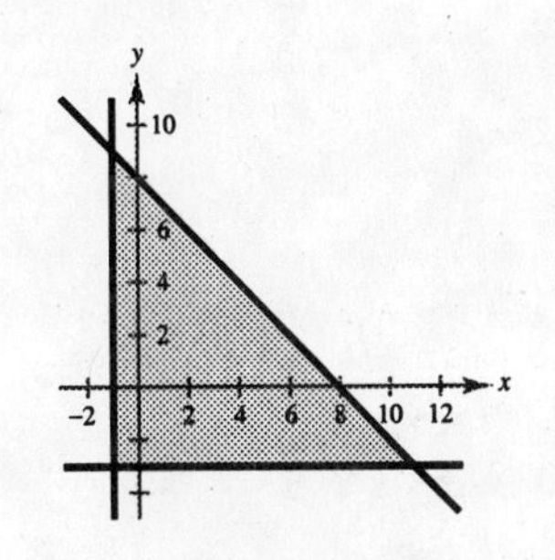

75.

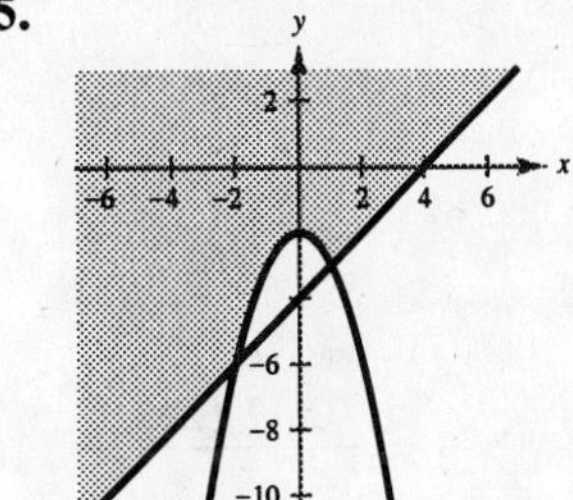

77. ${}_6C_2 = \dfrac{6!}{4!2!} = \dfrac{6 \cdot 5 \cdot 4!}{4!2} = 15$

79. ${}_{11}C_8 = \dfrac{11!}{8!3!} = \dfrac{11 \cdot 10 \cdot 9 \cdot 8!}{8!6} = 165$

Review Exercises for Chapter 9

Solutions to Odd-Numbered Exercises

1. $a_n = 2 + \dfrac{6}{n}$

$a_1 = 2 + \dfrac{6}{1} = 8$

$a_2 = 2 + \dfrac{6}{2} = 5$

$a_3 = 2 + \dfrac{6}{3} = 4$

$a_4 = 2 + \dfrac{6}{4} = \dfrac{7}{2}$

$a_5 = 2 + \dfrac{6}{5} = \dfrac{16}{5}$

3. $a_n = \dfrac{5n}{2n-1}$

$a_1 = \dfrac{5(1)}{2(1)-1} = 5$

$a_2 = \dfrac{5(2)}{2(2)-1} = \dfrac{10}{3}$

$a_3 = \dfrac{5(3)}{2(3)-1} = 3$

$a_4 = \dfrac{5(4)}{2(4)-1} = \dfrac{20}{7}$

$a_5 = \dfrac{5(5)}{2(5)-1} = \dfrac{25}{9}$

5. $a_n = \dfrac{72}{n!}$

$a_1 = \dfrac{72}{1!} = 72$

$a_2 = \dfrac{72}{2!} = 36$

$a_3 = \dfrac{72}{3!} = 12$

$a_4 = \dfrac{72}{4!} = 3$

$a_5 = \dfrac{72}{5!} = \dfrac{3}{5}$

7. $a_n = \frac{3}{2}n$

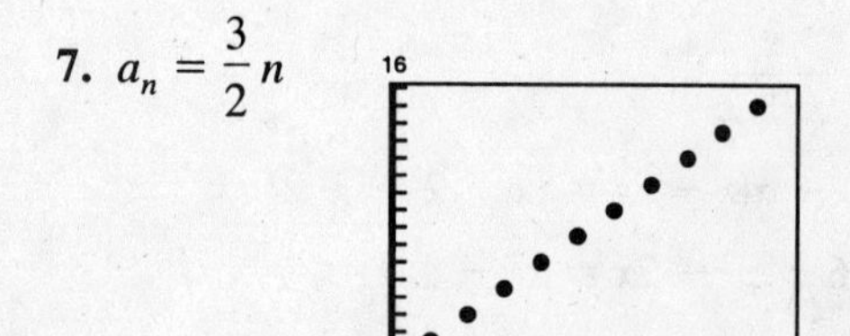

9. $a_n = 4(0.4)^{n-1}$

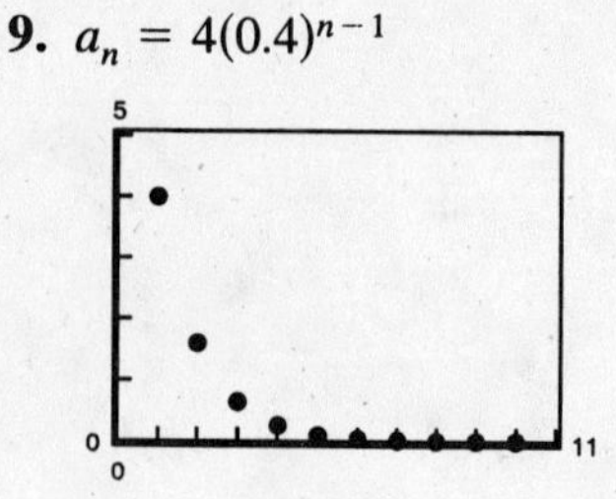

11. $\frac{18!}{20!} = \frac{18!}{20 \cdot 19 \cdot 18!} = \frac{1}{20 \cdot 19} = \frac{1}{380}$

13. $\frac{3!5!}{6!} = \frac{6(5!)}{6 \cdot 5!} = 1$

15. $\sum_{i=1}^{6} 5 = 6(5) = 30$

17. $\sum_{j=1}^{4} \frac{6}{j^2} = \frac{6}{1^2} + \frac{6}{2^2} + \frac{6}{3^2} + \frac{6}{4^2} + = 6 + \frac{3}{2} + \frac{2}{3} + \frac{3}{8} = \frac{205}{24}$

19. $\sum_{k=1}^{10} 2k^3 = 2(1)^3 + 2(2)^3 + 2(3)^3 + \cdots + 2(10)^3 = 6050$

21. $\sum_{n=0}^{10} (n^2 + 3) = \sum_{n=0}^{10} n^2 + \sum_{n=0}^{10} 3 = \frac{10(11)(21)}{6} + 11(3) = 418$

23. $\frac{1}{2(1)} + \frac{1}{2(2)} + \frac{1}{2(3)} + \cdots + \frac{1}{2(20)} = \sum_{k=1}^{20} \frac{1}{2k}$

25. $\frac{1}{2} + \frac{2}{3} + \frac{3}{4} + \cdots + \frac{9}{10} = \sum_{k=1}^{9} \frac{k}{k+1}$

27. (a) $\sum_{k=1}^{4} \frac{5}{10^k} = \frac{5}{10} + \frac{5}{100} + \frac{5}{1000} + \frac{5}{10{,}000} = .5 + .05 + .005 + .0005 = .5555$

(b) $\sum_{k=1}^{\infty} \frac{5}{10^k} = \frac{5}{10} \sum_{k=0}^{\infty} \frac{1}{10^k} = \frac{5}{10} \frac{1}{1 - \frac{1}{10}} = \frac{5}{10} \cdot \frac{10}{9} = \frac{5}{9}$

29. (a) $\sum_{k=1}^{4} 2\left(\frac{1}{100}\right)^k = 2\left(\frac{1}{100}\right) + 2\left(\frac{1}{100}\right)^2 + 2\left(\frac{1}{100}\right)^3 + 2\left(\frac{1}{100}\right)^4$

$= 0.02 + 0.0002 + 0.000002 + 0.00000002$

$= 0.02020202$

(b) $\sum_{k=1}^{\infty} 2\left(\frac{1}{100}\right)^k = \frac{2}{100} \sum_{k=0}^{\infty} \left(\frac{1}{100}\right)^k$

$= \frac{2}{100} \cdot \frac{1}{1 - \left(\frac{1}{100}\right)}$

$= \frac{2}{100} \cdot \frac{100}{99} = \frac{2}{99}$

31. (a) $a_1 = 2500\left(1 + \frac{0.08}{4}\right)^1 = 2500(1.02) = 2550$

$a_2 = 2500(1.02)^2 = 2601$

$a_3 = 2500(1.02)^3 = 2653.02$

$a_4 = 2500(1.02)^4 = 2706.08$

$a_5 = 2500(1.02)^5 = 2760.20$

$a_6 = 2500(1.02)^6 = 2815.41$

$a_7 = 2500(1.02)^7 = 2871.71$

$a_8 = 2500(1.02)^8 = 2929.15$

(b) $a_{40} = 2500(1.02)^{40} = 5520.10$

33. Yes. $d = -2 - 5 = -7$

35. Yes. $d = 1 - \frac{1}{2} = \frac{1}{2}$

37. $a_1 = 3, d = 4$

$a_1 = 3$

$a_2 = 3 + 4 = 7$

$a_3 = 7 + 4 = 11$

$a_4 = 11 + 4 = 15$

$a_5 = 15 + 4 = 19$

39. $a_4 = 10 \quad a_{10} = 28$

$a_{10} = a_4 + 6d$

$28 = 10 + 6d$

$18 = 6d$

$3 = d$

$a_1 = a_4 - 3d$

$a_1 = 10 - 3(3)$

$a_1 = 1$

$a_2 = 1 + 3 = 4$

$a_3 = 4 + 3 = 7$

$a_4 = 7 + 3 = 10$

$a_5 = 10 + 3 = 13$

41. $a_1 = 35, a_{k+1} = a_k - 3$

$a_1 = 35$

$a_2 = a_1 - 3 = 35 - 3 = 32$

$a_3 = a_2 - 3 = 32 - 3 = 29$

$a_4 = a_3 - 3 = 29 - 3 = 26$

$a_5 = a_4 - 3 = 26 - 3 = 23$

$a_n = 35 + (n - 1)(-3) = 38 - 3n$

43. $a_1 = 9, a_{k+1} = a_k + 7$

$a_1 = 9$

$a_2 = a_1 + 7 = 9 + 7 = 16$

$a_3 = a_2 + 7 = 16 + 7 = 23$

$a_4 = a_3 + 7 = 23 + 7 = 30$

$a_5 = a_4 + 7 = 30 + 7 = 37$

$a_n = 9 + (n - 1)(7) = 2 + 7n$

45. $a_n = 100 + (n - 1)(-3) = 103 - 3n$

$$\sum_{n=1}^{20}(103 - 3n) = \sum_{n=1}^{20} 103 - 3\sum_{n=1}^{20} n = 20(103) - 3\left[\frac{(20)(21)}{2}\right] = 1430$$

47. $\sum_{j=1}^{10}(2j - 3) = 2\sum_{j=1}^{10} j - \sum_{j=1}^{10} 3 = 2\left[\frac{10(11)}{2}\right] - 10(3) = 80$

49. $\sum_{k=1}^{11}\left(\frac{2}{3}k + 4\right) = \frac{2}{3}\sum_{k=1}^{11} k + \sum_{k=1}^{11} 4 = \frac{2}{3} \cdot \frac{(11)(12)}{2} + 11(4) = 88$

51. $\sum_{k=1}^{100} 5k = 5\left[\frac{(100)(101)}{2}\right] = 25{,}250$

53. (a) $34{,}000 + 4(2250) = \$43{,}000$

(b) $\sum_{k=1}^{5}[34{,}000 + (k - 1)(2250)] = \sum_{k=1}^{5}(31{,}750 + 2250k) = \$192{,}500$

55. $a_1 = 4,\ r = -\frac{1}{4}$

$a_1 = 4$

$a_2 = 4\left(-\frac{1}{4}\right) = -1$

$a_3 = -1\left(-\frac{1}{4}\right) = \frac{1}{4}$

$a_4 = \frac{1}{4}\left(-\frac{1}{4}\right) = -\frac{1}{16}$

$a_5 = -\frac{1}{16}\left(-\frac{1}{4}\right) = \frac{1}{64}$

57. $a_1 = 9,\ a_3 = 4$

$a_3 = a_1 r^2$

$4 = 9r^2$

$\frac{4}{9} = r^2 \implies r = \pm\frac{2}{3}$

$a_1 = 9$

$a_2 = 9\left(\frac{2}{3}\right) = 6$

$a_3 = 6\left(\frac{2}{3}\right) = 4$

$a_4 = 4\left(\frac{2}{3}\right) = \frac{8}{3}$

$a_5 = \frac{8}{3}\left(\frac{2}{3}\right) = \frac{16}{9}$

OR

$a_1 = 9$

$a_2 = 9\left(-\frac{2}{3}\right) = -6$

$a_3 = -6\left(-\frac{2}{3}\right) = 4$

$a_4 = 4\left(-\frac{2}{3}\right) = -\frac{8}{3}$

$a_5 = -\frac{8}{3}\left(-\frac{2}{3}\right) = \frac{16}{9}$

59. $a_1 = 120,\ a_{k+1} = \frac{1}{3}a_k$

$a_1 = 120$

$a_2 = \frac{1}{3}(120) = 40$

$a_3 = \frac{1}{3}(40) = \frac{40}{3}$

$a_4 = \frac{1}{3}\left(\frac{40}{3}\right) = \frac{40}{9}$

$a_5 = \frac{1}{3}\left(\frac{40}{9}\right) = \frac{40}{27}$

$a_n = 120\left(\frac{1}{3}\right)^{n-1}$

61. $a_1 = 25,\ a_k + 1 = -\frac{3}{5}a_k$

$a_1 = 25$

$a_2 = -\frac{3}{5}(25) = -15$

$a_3 = -\frac{3}{5}(-15) = 9$

$a_4 = -\frac{3}{5}(9) = -\frac{27}{5}$

$a_5 = -\frac{3}{5}\left(-\frac{27}{5}\right) = \frac{81}{25}$

$a_n = 25\left(-\frac{3}{5}\right)^{n-1}$

63. $a_2 = a_1 r$

$-8 = 16r$

$-\frac{1}{2} = r$

$a_n = 16\left(-\frac{1}{2}\right)^{n-1}$

$\sum_{n=1}^{20} 16\left(-\frac{1}{2}\right)^{n-1} = 16\left[\frac{1 - (-1/2)^{20}}{1 - (-1/2)}\right] \approx 10.67$

65. $a_1 = 100, r = 1.05$

$a_n = 100(1.05)^{n-1}$

$$\sum_{n=1}^{20} 100(1.05)^{n-1} = 100\left[\frac{1-(1.05)^{20}}{1-1.05}\right] \approx 3306.60$$

67. $\displaystyle\sum_{i=1}^{7} 2^{i-1} = \frac{1-2^7}{1-2} = 127$

69. $\displaystyle\sum_{n=1}^{7} (-4)^{n-1} = \sum_{n=0}^{6} (-4)^n = \frac{1-(-4)^7}{1-(-4)} = 3277$

71. $\displaystyle\sum_{n=0}^{4} 250(1.02)^n = 250\left(\frac{1-1.02^5}{1-1.02}\right) = 1301.01004$

73. $\displaystyle\sum_{i=1}^{10} 10\left(\frac{3}{5}\right)^{i-1} \approx 24.849$

75. $\displaystyle\sum_{i=1}^{25} 100(1.06)^{i-1} \approx 5486.4512$

77. $\displaystyle\sum_{i=1}^{\infty} \left(\frac{7}{8}\right)^{i-1} = \frac{1}{1-7/8} = 8$

79. $\displaystyle\sum_{k=1}^{\infty} 4\left(\frac{2}{3}\right)^{k-1} = \frac{4}{1-2/3} = 12$

81. (a) $a_t = 120{,}000(0.7)^t$

(b) $a_5 = 120{,}000(0.7)^5 = \$20{,}168.40$

83. $\displaystyle A = \sum_{i=1}^{48} 75\left(1+\frac{0.04}{12}\right)^i = \$3909.96$

85. 1. When $n = 1, 2 = \frac{1}{2}(5(1) - 1)$

2. Assume that $S_k = 2 + 7 + \ldots + (5k - 3) = \frac{k}{2}(5k - 1)$

Then, $S_{k+1} = 2 + 7 + \ldots + (5k - 3) + [5(k + 1) - 3]$

$$= S_k + 5k + 2$$

$$= \frac{k}{2}(5k - 1) + 5k + 2$$

$$= \frac{1}{2}[5k^2 + 9k + 4]$$

$$= \frac{1}{2}[(5k + 4)(k + 1)]$$

$$= \frac{k+1}{2}(5(k + 1) - 1)$$

Therefore, by mathematical induction, the formula is true for all positive integers n.

87. 1. When $n = 1, a = a\left(\frac{1-r}{1-r}\right)$.

2. Assume that

$$S_k = \sum_{i=0}^{k-1} ar^i = \frac{a(1-r^k)}{1-r}.$$

Then,

$$S_{k+1} = \sum_{i=0}^{k} ar^i = \sum_{i=0}^{k-1} ar^i + ar^k = \frac{a(1-r^k)}{1-r} + ar^k$$

$$= \frac{a(1 - r^k + r^k - r^{k+1})}{1-r} = \frac{a(1-r^{k+1})}{1-r}.$$

Therefore, by mathematical induction, the formula is valid for all positive integer values of n.

89. $\sum_{n=1}^{30} n = \frac{30(31)}{2} = 465$

91. $\sum_{n=1}^{7} n^4 = \frac{7(8)(15)(3(7^2) + 3(7) - 1)}{30} = \frac{840(167)}{30} = 4676$

93. $a_1 = f(1) = 5$

$a_2 = a_1 + 5 = 5 + 5 = 10$

$a_3 = a_2 + 5 = 15$

$a_4 = a_3 + 5 = 20$

$a_5 = a_4 + 5 = 25$

n:	1		2		3		4		5
a_n:	5		10		15		20		25
First differences:		5		5		5		5	
Second difference:			0		0		0		

Linear model: $a_n = 5n$

95. $a_1 = f(1) = 16$

$a_2 = a_1 - 1 = 16 - 1 = 15$

$a_3 = a_2 - 1 = 15 - 1 = 14$

$a_4 = 14 - 1 = 13$

$a_5 = 13 - 1 = 12$

n:	1		2		3		4		5
a_n:	16		15		14		13		12
First differences:		-1		-1		-1		-1	
Second difference:			0		0		0		

Linear model: $a_n = 17 - n$

97. ${}_{10}C_8 = 45$

99. $\binom{9}{4} = {}_9C_4 = 126$

101. 4th number in 7th row is ${}_6C_3 = 20$

103. 5th number in 9th row is $\binom{8}{4} = {}_8C_4 = 70$

105. $(a - 3b)^5 = a^5 - 5a^4(3b) + 10a^3(3b)^2 - 10a^2(3b)^3 + 5a(3b)^4 - (3b)^5$

$= a^5 - 15a^4b + 90a^3b^2 - 270a^2b^3 + 405ab^4 - 243b^5$

107. $\left(\frac{x}{2} + y\right)^4 = \left(\frac{x}{2}\right)^4 + 4\left(\frac{x}{2}\right)^3 y + 6\left(\frac{x}{2}\right)^2 y^2 + 4\left(\frac{x}{2}\right)y^3 + y^4$

$= \frac{x^4}{16} + \frac{x^3y}{2} + \frac{3x^2y^2}{2} + 2xy^3 + y^4$

109. $(5 + 2i)^4 = (5)^4 + 4(5)^3(2i) + 6(5)^2(2i)^2 + 4(5)(2i)^3 + (2i)^4$

$= 625 + 1000i + 600i^2 + 160i^3 + 16i^4$

$= 625 + 1000i - 600 - 160i + 16 = 41 + 840i$

111. $E = \{(1, 11), (2, 10), (3, 9), (4, 8), (5, 7), (7, 5), (8, 4), (9, 3), (10, 2), (11, 1)\}$

$n(E) = 10$

113. $(4)(6)(2) = 48$ schedules

115. $\dfrac{8!}{2!2!2!1!1!} = \dfrac{8!}{8} = 7! = 5040$ permutations

117. $10! = 3{,}628{,}800$ ways

119. ${}_{20}C_{15} = 15{,}504$ ways

121. $\frac{10}{10} \cdot \frac{1}{9} = \frac{1}{9}$

123. Chance of rolling a 3 with one die is $\frac{1}{6}$. With two dice $E = \{(1, 5), (2, 4), (3, 3), (4, 2), (5, 1)\}$ and $P(E) = \frac{5}{36}$. The probability of rolling a 3 with one die is higher.

125. (a) $\frac{208}{500} = 0.416$

(b) $\frac{400}{500} = 0.8$

(c) $\frac{37}{500} = 0.074$

127. $(0.8)^3 = 0.512$

129. $1 - P(HHHHH) = 1 - \left(\frac{1}{2}\right)^5 = \dfrac{31}{32}$

131. True.

133. They differ by a minus sign.

(a) $-1, \frac{1}{2}, -\frac{1}{3}, \ldots$ (odd-numbered terms are negative)

(b) $1, -\frac{1}{2}, \frac{1}{3}, \ldots$(even-numbered terms are negative)

135. (a) arithmetic-linear model

(b) geometric-exponential model

137. Answers will vary. See page 620. To define a sequence recursively, you need to be given one or more of the first few terms. All other terms are defined using previous terms.

139. Decreasing sequence. Matches (d)

141. Increasing sequence. Matches (b)

143.

n	100	500	1000	5000
a_n	2.7048	2.7156	2.7169	2.7180

n	10,000	15,000	20,000	25,000
a_n	2.7182	2.7182	2.7182	2.7182

limit of a_n as $n \to \infty$ is $e \approx 2.7182$

145. In the closed interval $[0, 1)$.

147. This means that meterological records indicate that over an extended period of time with similar weather condidions, it will rain 60% of the time.

Chapter 9 Practice Test

1. Write out the first five terms of the sequence $a_n = \dfrac{2n}{(n+2)!}$.

2. Write an expression for the nth term of the sequence $\left\{\frac{4}{3}, \frac{5}{9}, \frac{6}{27}, \frac{7}{81}, \frac{8}{243}, \ldots\right\}$.

3. Find the sum $\sum_{i=1}^{6}(2i - 1)$.

4. Write out the first five terms of the arithmetic sequence where $a_1 = 23$ and $d = -2$.

5. Find a_n for the arithmetic sequence with $a_1 = 12$, $d = 3$, and $n = 50$.

6. Find the sum of the first 200 positive integers.

7. Write out the first five terms of the geometric sequence with $a_1 = 7$ and $r = 2$.

8. Evaluate $\sum_{n=0}^{9} 6\left(\frac{2}{3}\right)^n$.

9. Evaluate $\sum_{n=0}^{\infty}(0.03)^n$.

10. Use mathematical induction to prove that $1 + 2 + 3 + 4 + \cdots + n = \dfrac{n(n+1)}{2}$.

11. Use mathematical induction to prove that $n! > 2^n$, $n \geq 4$.

12. Evaluate ${}_{13}C_4$. Verify with a graphing utility.

13. Expand $(x + 3)^5$.

14. Find the term involving x^7 in $(x - 2)^{12}$.

15. Evaluate ${}_{30}P_4$.

16. How many ways can six people sit at a table with six chairs?

17. Twelve cars run in a race. How many different ways can they come in first, second, and third place? (Assume that there are no ties.)

18. Two six-sided dice are tossed. Find the probability that the total of the two dice is less than 5.

19. Two cards are selected at random form a deck of 52 playing cards without replacement. Find the probability that the first card is a King and the second card is a black ten.

20. A manufacturer has determined that for every 1000 units it produces, 3 will be faulty. What is the probability that an order of 50 units will have one or more faulty units?

CHAPTER 10
Topics in Analytic Geometry

CHAPTER 10
Topics in Analytic Geometry

Section 10.1 Introduction to Conics: Parabolas

- A **parabola** is the set of all points (x, y) that are equidistant from a fixed line **(directrix)** and a fixed point **(focus)** not on the line.
- The standard equation of a parabola with vertex (h, k) and:
  (a) Vertical axis $x = h$ and directrix $y = k - p$ is:
  $(x - h)^2 = 4p(y - k), p \neq 0$
  (b) Horizontal axis $y = k$ and directrix $x = h - p$ is:
  $(y - k)^2 = 4p(x - h), p \neq 0$
- The tangent line to a parabola at a point P makes **equal angles** with:
  (a) the line through P and the focus
  (b) the axis of the parabola

Solutions to Odd-Numbered Exercises

1. $y^2 = -4x$
Vertex: $(0, 0)$
Opens to the left since p is negative.
Matches graph (e).

3. $x^2 = -8y$
Vertex: $(0, 0)$
Opens downward since p is negative.
Matches graph (d).

5. $(y - 1)^2 = 4(x - 3)$
Vertex: $(3, 1)$
Opens to the right since p is positive.
Matches graph (a).

7. $y = \frac{1}{2}x^2$
$x^2 = 2y$
$x^2 = 4\left(\frac{1}{2}\right)y \Rightarrow h = 0, k = 0, p = \frac{1}{2}$
Vertex: $(0, 0)$
Focus: $\left(0, \frac{1}{2}\right)$
Directrix: $y = -\frac{1}{2}$

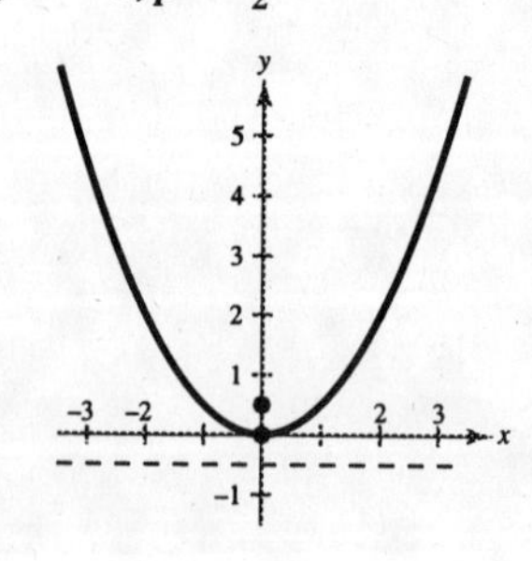

9. $y^2 = -6x$
$y^2 = 4\left(-\frac{3}{2}\right)x \Rightarrow h = 0, k = 0, p = -\frac{3}{2}$
Vertex: $(0, 0)$
Focus: $\left(-\frac{3}{2}, 0\right)$
Directrix: $x = \frac{3}{2}$

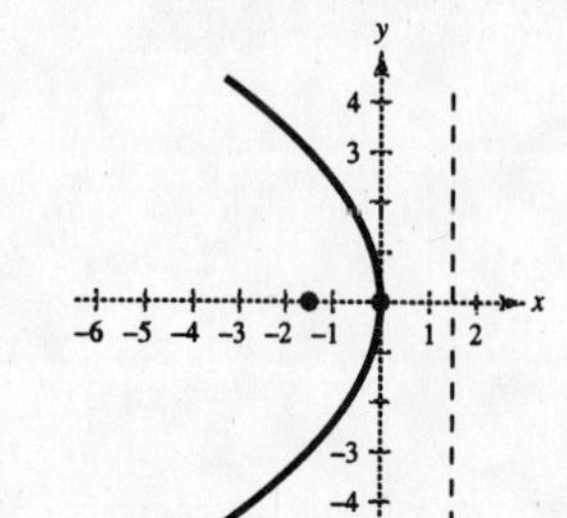

11. $x^2 + 6y = 0$
$x^2 = 4\left(-\frac{3}{2}\right)y \Rightarrow h = 0, k = 0, p = -\frac{3}{2}$
Vertex: $(0, 0)$
Focus: $\left(0, -\frac{3}{2}\right)$
Directrix: $y = \frac{3}{2}$

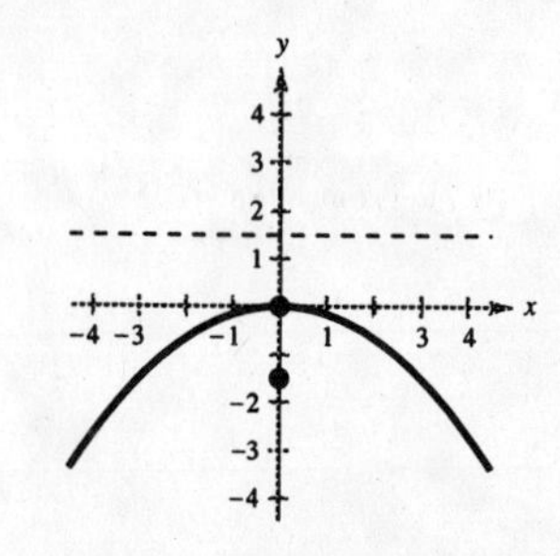

13. $(x - 1)^2 + 8(y + 2) = 0$

$$(x - 1)^2 = 4(-2)(y + 2)$$

$h = 1, k = -2, p = -2$

Vertex: $(1, -2)$

Focus: $(1, -4)$

Directrix: $y = 0$

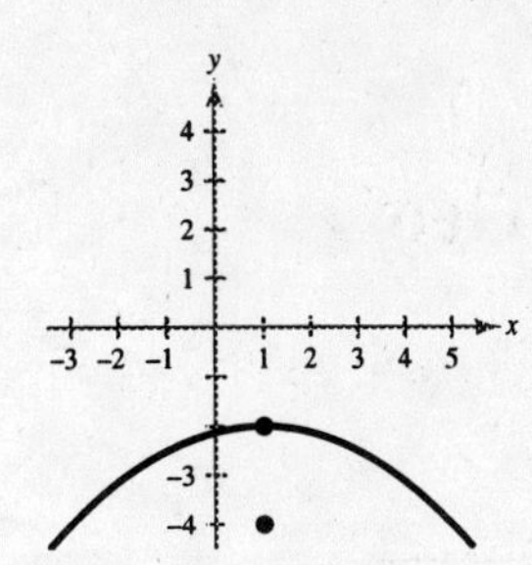

15. $\left(x + \frac{3}{2}\right)^2 = 4(y - 2) \implies h = -\frac{3}{2}, k = 2, p = 1$

Vertex: $\left(-\frac{3}{2}, 2\right)$

Focus: $\left(-\frac{3}{2}, 2 + 1\right) = \left(-\frac{3}{2}, 3\right)$

Directrix: $y = 1$

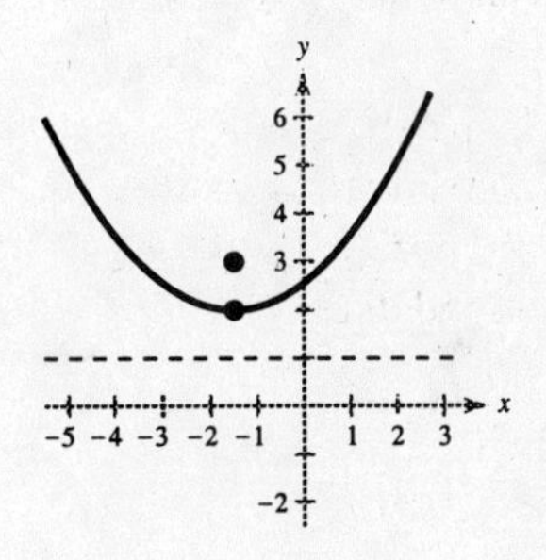

17. $y = \frac{1}{4}(x^2 - 2x + 5)$

$4y - 4 = (x - 1)^2$

$(x - 1)^2 = 4(1)(y - 1)$

$h = 1, k = 1, p = 1$

Vertex: $(1, 1)$

Focus: $(1, 2)$

Directrix: $y = 0$

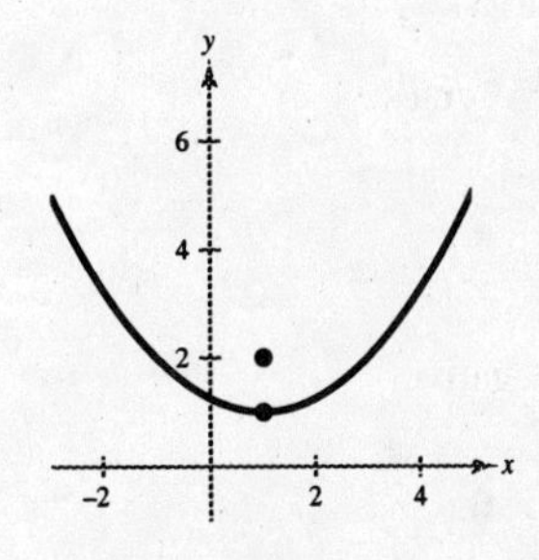

19. $y^2 + 6y + 8x + 25 = 0$

$$y^2 + 6y + 9 = -8x - 25 + 9$$

$$(y + 3)^2 = 4(-2)(x + 2)$$

$h = -2, k = -3, p = -2$

Vertex: $(-2, -3)$

Focus: $(-4, -3)$

Directrix: $x = 0$

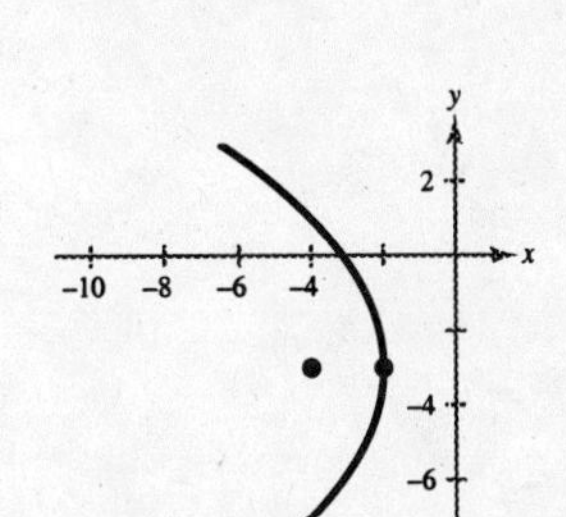

21. $x^2 + 4x + 6y - 2 = 0$

$$x^2 + 4x + 4 = -6y + 2 + 4 = -6y + 6$$

$$(x + 2)^2 = -6(y - 1)$$

$$(x + 2)^2 = 4\left(-\tfrac{3}{2}\right)(y - 1)$$

Vertex: $(-2, 1)$

Focus: $\left(-2, 1 - \frac{3}{2}\right) = \left(-2, -\frac{1}{2}\right)$

Directrix: $y = \frac{5}{2}$

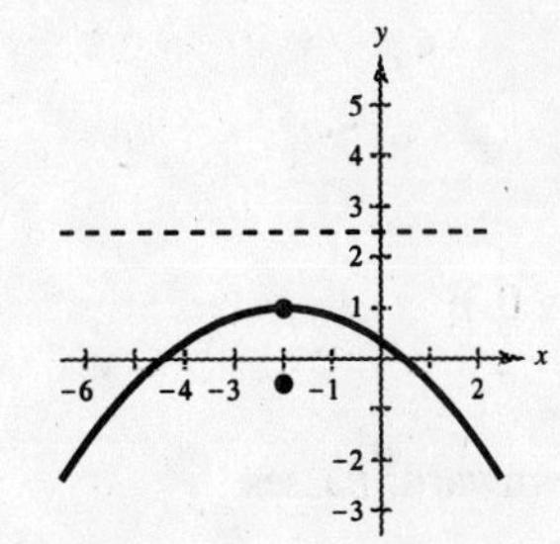

23. $y^2 + x + y = 0$

$y^2 + y + \frac{1}{4} = -x + \frac{1}{4}$

$\left(y + \frac{1}{2}\right)^2 = 4\left(-\frac{1}{4}\right)\left(x - \frac{1}{4}\right)$

$h = \frac{1}{4}, k = -\frac{1}{2}, p = -\frac{1}{4}$

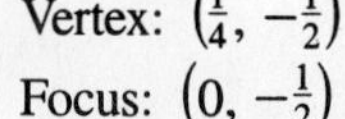

Vertex: $\left(\frac{1}{4}, -\frac{1}{2}\right)$

Focus: $\left(0, -\frac{1}{2}\right)$

Directrix: $x = \frac{1}{2}$

To use a graphing calculator, enter:

$y_1 = -\frac{1}{2} + \sqrt{\frac{1}{4} - x}$

$y_2 = -\frac{1}{2} - \sqrt{\frac{1}{4} - x}$

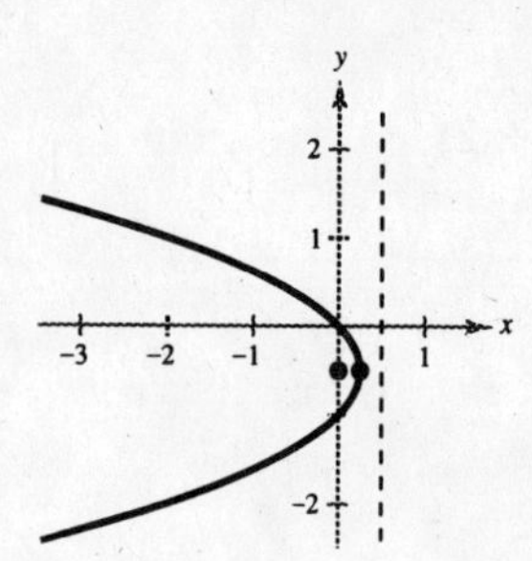

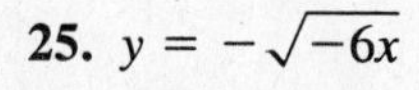

25. $y = -\sqrt{-6x}$

27. Vertex: $(0, 0) \Rightarrow h = 0, k = 0$

Graph opens upward.

$x^2 = 4py$

Point on graph: $(3, 6)$

$3^2 = 4p(6)$

$9 = 24p$

$\frac{3}{8} = p$

Thus, $x^2 = 4\left(\frac{3}{8}\right)y \Rightarrow y = \frac{2}{3}x^2$.

29. Vertex: $(0, 0) \Rightarrow h = 0, k = 0$

Focus: $\left(0, -\frac{3}{2}\right) \Rightarrow p = -\frac{3}{2}$

$(x - h)^2 = 4p(y - k)$

$x^2 = 4\left(-\frac{3}{2}\right)y$

$x^2 = -6y$

31. Vertex: $(0, 0) \Rightarrow h = 0, k = 0$

Focus: $(-2, 0) \Rightarrow p = -2$

$(y - k)^2 = 4p(x - h)$

$y^2 = 4(-2)x$

$y^2 = -8x$

33. Vertex: $(0, 0) \Rightarrow h = 0, k = 0$

Directrix: $y = -1 \Rightarrow p = 1$

$(x - h)^2 = 4p(y - k)$

$(x - 0)^2 = 4(1)(y - 0)$

$x^2 = 4y$ or $y = \frac{1}{4}x^2$

35. Vertex: $(0, 0) \Rightarrow h = 0, k = 0$

Directrix: $x = 2 \Rightarrow p = -2$

$y^2 = 4px$

$y^2 = -8x$

37. Vertex: $(0, 0) \Rightarrow h = 0, k = 0$

Horizontal axis and passes through the point $(4, 6)$

$(y - k)^2 = 4p(x - h)$

$(y - 0)^2 = 4p(x - 0)$

$y^2 = 4px$

$6^2 = 4p(4)$

$36 = 16p \Rightarrow p = \frac{9}{4}$

$y^2 = 4\left(\frac{9}{4}\right)x$

$y^2 = 9x$

39. Vertex: $(3, 1)$ and opens downward. Passes through $(2, 0)$ and $(4, 0)$.

$y = -(x - 2)(x - 4)$

$= -x^2 + 6x - 8$

$= -(x - 3)^2 + 1$

$(x - 3)^2 = -(y - 1)$

41. Vertex: $(-2, 0)$ and opens to the right. Passes through $(0, 2)$.

$(y - 0)^2 = 4p(x + 2)$

$2^2 = 4p(0 + 2)$

$\frac{1}{2} = p$

$y^2 = 4\left(\frac{1}{2}\right)(x + 2)$

$y^2 = 2(x + 2)$

43. Vertex: $(5, 2)$

Focus: $(3, 2)$

Horizontal axis: $p = 3 - 5 = -2$

$(y - 2)^2 = 4(-2)(x - 5)$

$(y - 2)^2 = -8(x - 5)$

45. Vertex: $(0, 4)$

Directrix: $y = 2$

Vertical axis

$p = 4 - 2 = 2$

$(x - 0)^2 = 4(2)(y - 4)$

$x^2 = 8(y - 4)$

47. Focus: $(2, 2)$

Directrix: $x = -2$

Horizontal axis

Vertex: $(0, 2)$

$p = 2 - 0 = 2$

$(y - 2)^2 = 4(2)(x - 0)$

$(y - 2)^2 = 8x$

49. $y^2 - 8x = 0 \Rightarrow y = \pm\sqrt{8x}$

$x - y + 2 = 0 \Rightarrow y = x + 2$

The point of tangency is $(2, 4)$.

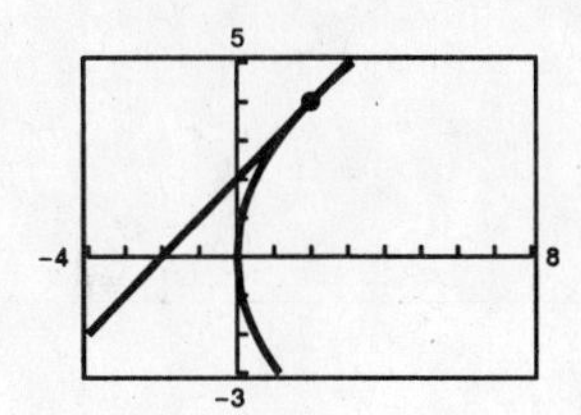

51. $x^2 = 2y$, $(4, 8)$, $p = \frac{1}{2}$, focus: $\left(0, \frac{1}{2}\right)$

Following Example 4, we find the y-intercept $(0, 6)$:

$$d_1 = \frac{1}{2} - b.$$

$$d_2 = \sqrt{(4 - 0)^2 + \left(8 - \frac{1}{2}\right)^2} = \frac{17}{2}$$

$$d_1 = d_2 \quad\Rightarrow\quad \frac{1}{2} - b = \frac{17}{2} \quad\Rightarrow\quad b = -8$$

$$m = \frac{8 - (-8)}{4 - 0} = 4$$

$y = 4x - 8$ Tangent line

Let $y = 0 \;\Rightarrow\; x = 2 \;\Rightarrow\; x$-intercept $(2, 0)$

53. $y = -2x^2 \;\Rightarrow\; x^2 = -\frac{1}{2}y = 4\left(-\frac{1}{8}\right)y \;\Rightarrow\; p = -\frac{1}{8}$, focus: $\left(0, -\frac{1}{8}\right)$

Following Example 4, we find the y-intercept $(0, b)$:

$$d_1 = \frac{1}{8} + b$$

$$d_2 = \sqrt{(-1 - 0)^2 + \left(-2 + \frac{1}{8}\right)^2} = \frac{17}{8}$$

$$d_1 = d_2 \quad\Rightarrow\quad \frac{1}{8} + b = \frac{17}{8} \quad\Rightarrow\quad b = 2$$

$$m = \frac{-2 - 2}{-1 - 0} = 4$$

$y = 4x + 2$

Let $y = 0 \;\Rightarrow\; x = -\frac{1}{2} \;\Rightarrow\; x$-intercept $\left(-\frac{1}{2}, 0\right)$

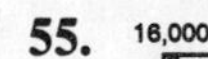

55.

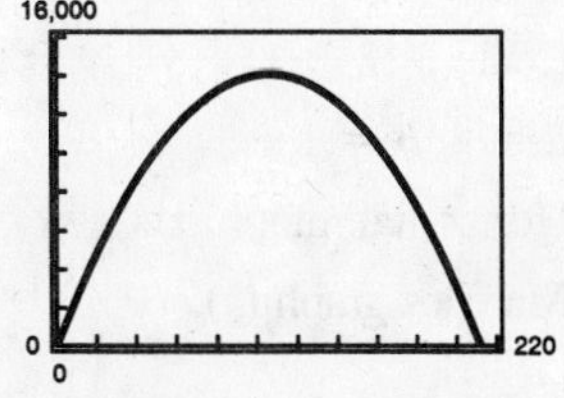

$R = 265x - \frac{5}{4}x^2$ is a maximum (14045) when $x = 106$ units.

57. Vertex: $(0, 0) \Rightarrow h = 0, k = 0$

Focus: $(0, 3.5) \Rightarrow p = 3.5$

$(x - h)^2 = 4p(y - k)$

$(x - 0)^2 = 4(3.5)(y - 0)$

$x^2 = 14y$ or $y = \frac{1}{14}x^2$

59. (a) $x^2 = 4py$ passes through point $\left(16, -\frac{2}{5}\right)$

$256 = 4p\left(-\frac{2}{5}\right) \Rightarrow p = -160$

$x^2 = 4(-160)y$

$x^2 = -640y$ or $y = \frac{-1}{640}x^2$

(b) $-0.1 = \frac{-1}{640}x^2 \quad \Rightarrow \quad x = 8$ feet

61. (a) Escape velocity: $17{,}500\sqrt{2}$

(b) $x^2 = 4p(y - 4100)$ and $p = -4100$.

$x^2 = -16{,}400(y - 4100)$

63. (a) $y = \dfrac{-16}{v^2}x^2 + s$

$= \dfrac{-16}{32^2}x^2 + 75 = -\dfrac{1}{64}x^2 + 75$

(b) $y = 0 = -\dfrac{1}{64}x^2 + 75 \Rightarrow x^2 = (75)(64) \Rightarrow x \approx 69.28$ feet

65. False. It is not possible for a parabola to intersect its directrix. If the graph crossed the directrix there would exist points nearer the directrix than the focus.

67. $\pm 4, \pm 2, \pm 1$

69. $\pm 16, \pm 8, \pm 4, \pm 2, \pm 1, \pm \frac{1}{2}$

Section 10.2 Ellipses

- An **ellipse** is the set of all points (x, y) the sum of whose distances from two distinct fixed points **(foci)** is constant.
- The standard equation of an ellipse with center (h, k) and major and minor axes of lengths $2a$ and $2b$ is:
  (a) $\dfrac{(x-h)^2}{a^2} + \dfrac{(y-k)^2}{b^2} = 1$ if the major axis is horizontal.
  (b) $\dfrac{(x-h)^2}{b^2} + \dfrac{(y-k)^2}{a^2} = 1$ if the major axis is vertical.
- $c^2 = a^2 - b^2$ where c is the distance from the center to a focus.
- The eccentricity of an ellipse is $e = \dfrac{c}{a}$

Solutions to Odd-Numbered Exercises

1. $\dfrac{x^2}{4} + \dfrac{y^2}{9} = 1$

Center: $(0, 0)$

$a = 3, b = 2$

Vertical major axis

Matches graph (b).

3. $\dfrac{x^2}{4} + \dfrac{y^2}{25} = 1$

Center: $(0, 0)$

$a = 5, b = 2$

Vertical major axis

Matches graph (d).

5. $\dfrac{(x-2)^2}{16} + (y+1)^2 = 1$

Center: $(2, -1)$

$a = 4, b = 1$

Horizontal major axis

Matches graph (a).

7. $\dfrac{x^2}{25} + \dfrac{y^2}{16} = 1$

Center: $(0, 0)$

$a = 5, b = 4, c = 3$

Foci: $(\pm 3, 0)$

Vertices: $(\pm 5, 0)$

$e = \dfrac{3}{5}$

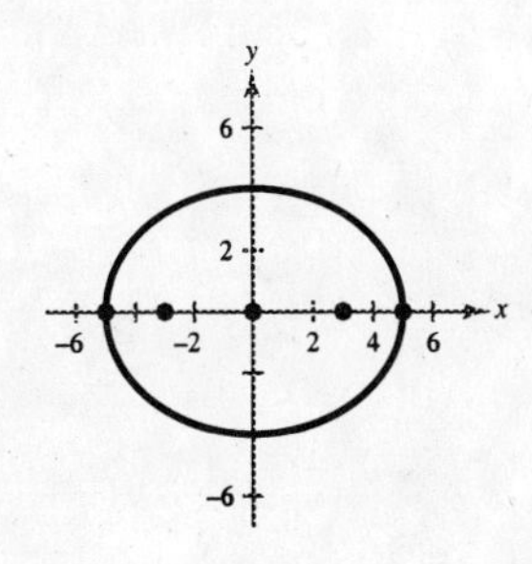

9. $\dfrac{x^2}{5} + \dfrac{y^2}{9} = 1$

Center: $(0, 0)$

$a = 3, b = \sqrt{5}, c = 2$

Foci: $(0, \pm 2)$

Vertices: $(0, \pm 3)$

$e = \dfrac{2}{3}$

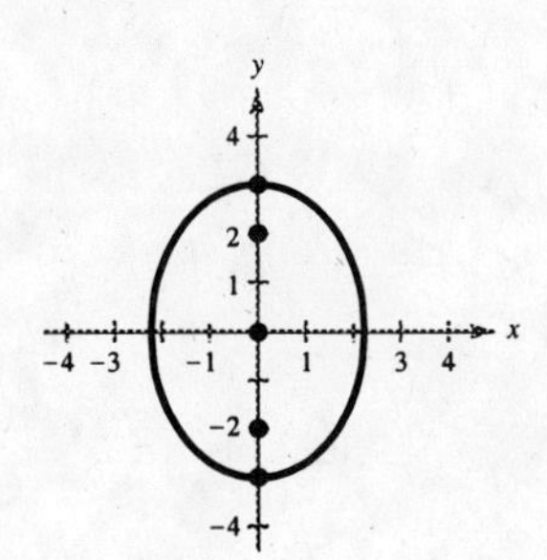

11. $\dfrac{(x + 3)^2}{16} + \dfrac{(y - 5)^2}{25} = 1$

Center: $(-3, 5)$

$a = 5, b = 4, c = 3$

Foci: $(-3, 5 \pm 3) = (-3, 8), (-3, 2)$

Vertices: $(-3, 5 \pm 5) = (-3, 10), (-3, 0)$

$e = \dfrac{3}{5}$

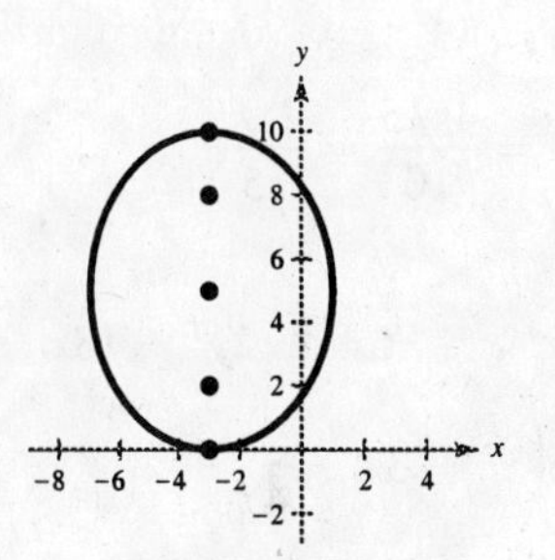

13. $\dfrac{(x + 5)^2}{\frac{9}{4}} + (y - 1)^2 = 1$

Center: $(-5, 1)$

$a = \dfrac{3}{2}, b = 1, c = \sqrt{\dfrac{9}{4} - 1} = \dfrac{\sqrt{5}}{2}$

Foci: $\left(-5 + \dfrac{\sqrt{5}}{2}, 1\right), \left(-5 - \dfrac{\sqrt{5}}{2}, 1\right)$

Vertices: $\left(-5 + \dfrac{3}{2}, 1\right) = \left(-\dfrac{7}{2}, 1\right)$

$\left(-5 - \dfrac{3}{2}, 1\right) = \left(-\dfrac{13}{2}, 1\right)$

$e = \dfrac{\sqrt{5}/2}{3/2} = \sqrt{5}/3$

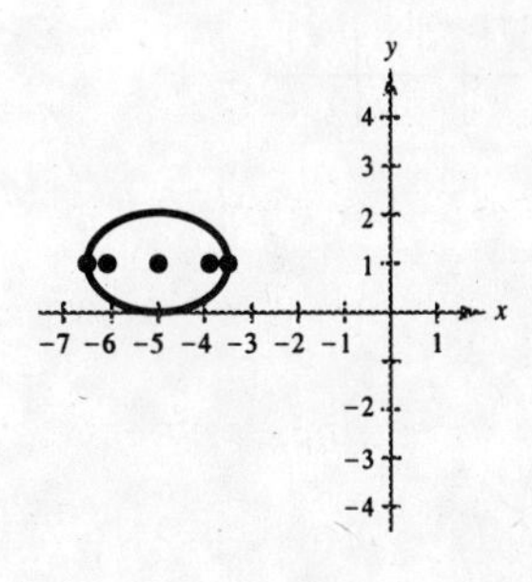

15. $9x^2 + 4y^2 + 36x - 24y + 36 = 0$

$9(x^2 + 4x + 4) + 4(y^2 - 6y + 9) = -36 + 36 + 36$

$\dfrac{(x + 2)^2}{4} + \dfrac{(y - 3)^2}{9} = 1$

$a = 3, b = 2, c = \sqrt{5}$

Center: $(-2, 3)$

Foci: $\left(-2, 3 \pm \sqrt{5}\right)$

Vertices: $(-2, 6), (-2, 0)$

$e = \dfrac{\sqrt{5}}{3}$

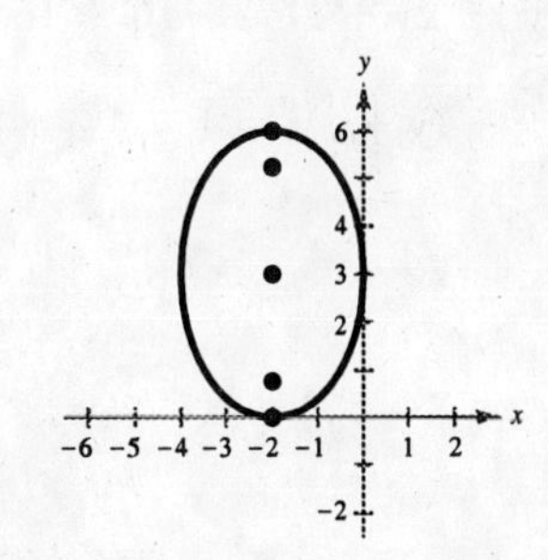

17.

$$x^2 + 5y^2 - 8x - 30y - 39 = 0$$
$$(x^2 - 8x + 16) + 5(y^2 - 6y + 9) = 39 + 16 + 45$$
$$(x - 4)^2 + 5(y - 3)^2 = 100$$
$$\frac{(x - 4)^2}{100} + \frac{(y - 3)^2}{20} = 1$$

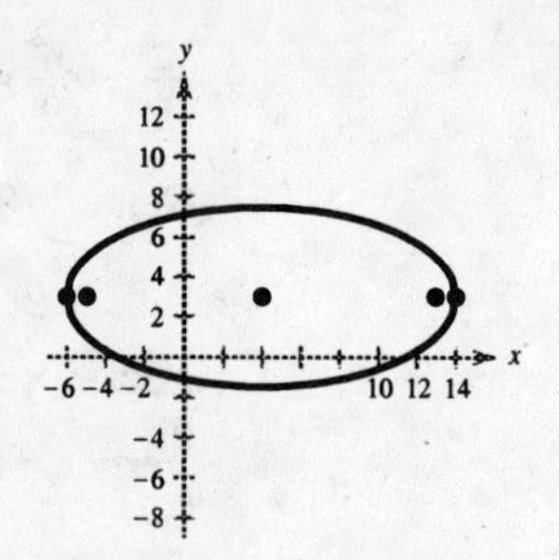

Center: $(4, 3)$

$a = 10, b = 2\sqrt{5}, c = \sqrt{80} = 4\sqrt{5}$

Foci: $\left(4 + 4\sqrt{5}, 3\right), \left(4 - 4\sqrt{5}, 3\right)$

Vertices: $(4 + 10, 3) = (14, 3)$

$(4 - 10, 3) = (-6, 3)$

$$e = \frac{4\sqrt{5}}{10} = \frac{2\sqrt{5}}{5}$$

19.

$$6x^2 + 2y^2 + 18x - 10y + 2 = 0$$
$$6\left(x^2 + 3x + \frac{9}{4}\right) + 2\left(y^2 - 5y + \frac{25}{4}\right) = -2 + \frac{27}{2} + \frac{25}{2}$$
$$6\left(x + \frac{3}{2}\right)^2 + 2\left(y - \frac{5}{2}\right)^2 = 24$$
$$\frac{\left(x + \frac{3}{2}\right)^2}{4} + \frac{\left(y - \frac{5}{2}\right)^2}{12} = 1$$

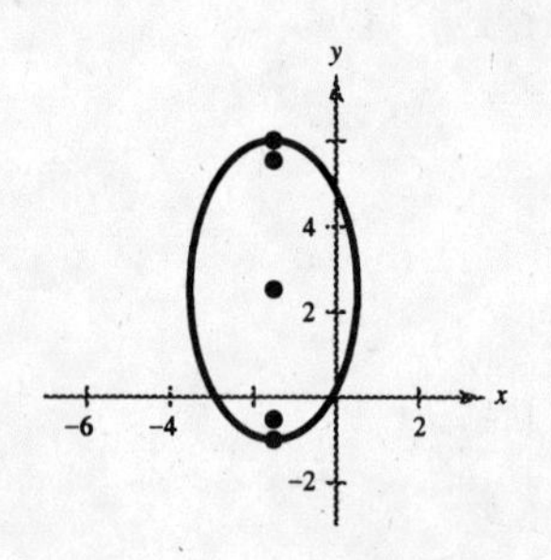

Center: $\left(-\frac{3}{2}, \frac{5}{2}\right)$

$a = 2\sqrt{3}, b = 2, c = 2\sqrt{2}$

Foci: $\left(-\frac{3}{2}, \frac{5}{2} \pm 2\sqrt{2}\right)$

Vertices: $\left(-\frac{3}{2}, \frac{5}{2} \pm 2\sqrt{3}\right)$

$$e = \frac{\sqrt{2}}{\sqrt{3}} = \frac{\sqrt{6}}{3}$$

21.

$$16x^2 + 25y^2 - 32x + 50y + 16 = 0$$
$$16(x^2 - 2x + 1) + 25(y^2 + 2y + 1) = -16 + 16 + 25$$
$$\frac{(x - 1)^2}{25/16} + (y + 1)^2 = 1$$

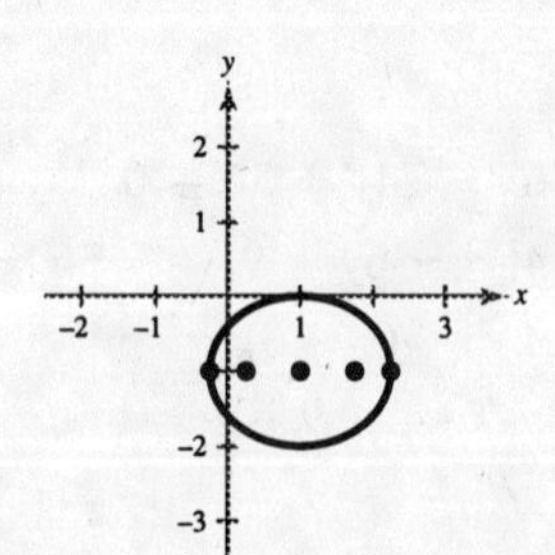

$a = \frac{5}{4}, b = 1, c = \frac{3}{4}$

Center: $(1, -1)$

Foci: $\left(\frac{7}{4}, -1\right), \left(\frac{1}{4}, -1\right)$

Vertices: $\left(\frac{9}{4}, -1\right), \left(-\frac{1}{4}, -1\right)$

$$e = \frac{3}{5}$$

23. $5x^2 + 3y^2 = 15$

$$\frac{x^2}{3} + \frac{y^2}{5} = 1$$

Center: $(0, 0)$

$a = \sqrt{5}, b = \sqrt{3}, c = \sqrt{2}$

Foci: $(0, \pm\sqrt{2})$

Vertices: $(0, \pm\sqrt{5})$

To graph, solve for y.

$$y^2 = \frac{15 - 5x^2}{3}$$

$$y_1 = \sqrt{\frac{15 - 5x^2}{3}}$$

$$y_2 = -\sqrt{\frac{15 - 5x^2}{3}}$$

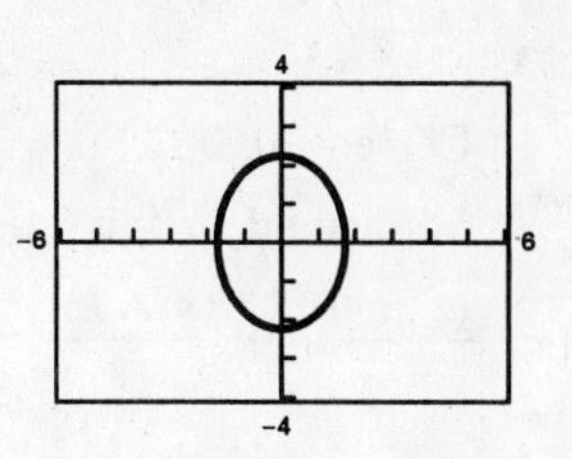

25. $12x^2 + 20y^2 - 12x + 40y - 37 = 0$

$$12\left(x^2 - x + \frac{1}{4}\right) + 20(y^2 + 2y + 1) = 37 + 3 + 20$$

$$\frac{[x - (1/2)]^2}{5} + \frac{(y + 1)^2}{3} = 1$$

$a = \sqrt{5}, b = \sqrt{3}, c = \sqrt{2}$

Center: $\left(\frac{1}{2}, -1\right)$

Foci: $\left(\frac{1}{2} \pm \sqrt{2}, -1\right)$

Vertices: $\left(\frac{1}{2} \pm \sqrt{5}, -1\right)$

$$e = \frac{\sqrt{10}}{5}$$

To graph, solve for y.

$$(y + 1)^2 = 3\left[1 - \frac{(x - 0.5)^2}{5}\right]$$

$$y_1 = -1 + \sqrt{3\left[1 - \frac{(x - 0.5)^2}{5}\right]}$$

$$y_2 = -1 - \sqrt{3\left[1 - \frac{(x - 0.5)^2}{5}\right]}$$

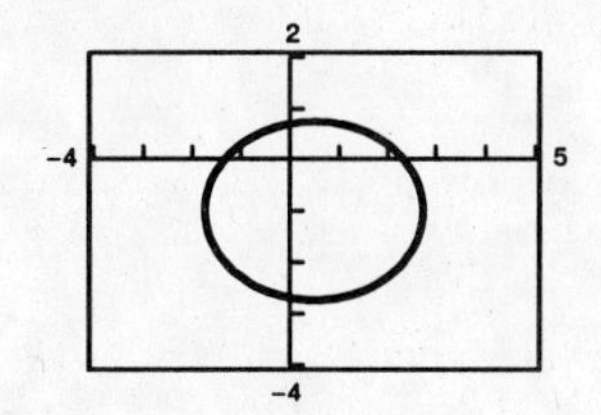

27. Center: $(0, 0)$, $a = 4$, $b = 2$

$$\frac{x^2}{4} + \frac{y^2}{16} = 1$$

29. Vertices: $(\pm 6, 0)$, Foci: $(\pm 2, 0)$

$a = 6, c = 2, b = \sqrt{36 - 4} = \sqrt{32} = 4\sqrt{2}$

Horizontal major axis, center: $(0, 0)$

$$\frac{x^2}{36} + \frac{y^2}{32} = 1$$

31. Foci: $(\pm 5, 0) \Rightarrow c = 5$

Center: $(0, 0)$

Horizontal major axis

Major axis of length $12 \Rightarrow 2a = 12$

$$a = 6$$

$6^2 - b^2 = 5^2 \Rightarrow b^2 = 11$

$$\frac{(x - h)^2}{a^2} + \frac{(y - k)^2}{b^2} = 1$$

$$\frac{x^2}{36} + \frac{y^2}{11} = 1$$

33. Vertices: $(0, \pm 5) \Rightarrow a = 5$

Center: $(0, 0)$

Vertical major axis

$$\frac{(x-h)^2}{b^2} + \frac{(y-k)^2}{a^2} = 1$$

$$\frac{x^2}{b^2} + \frac{y^2}{25} = 1$$

Point: $(4, 2)$

$$\frac{4^2}{b^2} + \frac{2^2}{25} = 1$$

$$\frac{16}{b^2} = 1 - \frac{4}{25} = \frac{21}{25}$$

$$400 = 21b^2$$

$$\frac{400}{21} = b^2$$

$$\frac{x^2}{400/21} + \frac{y^2}{25} = 1$$

$$\frac{21x^2}{400} + \frac{y^2}{25} = 1$$

35. Center: $(2, 3)$

$a = 3, \quad b = 1$

Vertical major axis

$$\frac{(x-h)^2}{b^2} + \frac{(y-k)^2}{a^2} = 1$$

$$\frac{(x-2)^2}{1} + \frac{(y-3)^2}{9} = 1$$

37. Center: $(-2, 3)$, $a = 4$, $b = 3$

$$\frac{(x+2)^2}{16} + \frac{(y-3)^2}{9} = 1$$

39. Center: $(2, 4)$, $a = 2$, $b = \frac{2}{2} = 1$

$$\frac{(x-2)^2}{4} + \frac{(y-4)^2}{1} = 1$$

41. Foci: $(0, 0), (0, 8) \Rightarrow c = 4$

Major axis of length $16 \Rightarrow a = 8$

$b^2 = a^2 - c^2 = 64 - 16 = 48$

Center: $(0, 4) = (h, k)$

$$\frac{(x-h)^2}{b^2} + \frac{(y-k)^2}{a^2} = 1$$

$$\frac{x^2}{48} + \frac{(y-4)^2}{64} = 1$$

43. Vertices: $(3, 1), (3, 9) \Rightarrow a = 4$

Center: $(3, 5)$

Minor axis of length $6 \Rightarrow b = 3$

Vertical major axis

$$\frac{(x-h)^2}{b^2} + \frac{(y-k)^2}{a^2} = 1$$

$$\frac{(x-3)^2}{9} + \frac{(y-5)^2}{16} = 1$$

45. Center: $(0, 4)$

Vertices: $(-4, 4), (4, 4) \Rightarrow a = 4$

$a = 2c \Rightarrow 4 = 2c \Rightarrow c = 2$

$2^2 = 4^2 - b^2 \Rightarrow b^2 = 12$

Horizontal major axis

$$\frac{(x-h)^2}{a^2} + \frac{(y-k)^2}{b^2} = 1$$

$$\frac{x^2}{16} + \frac{(y-4)^2}{12} = 1$$

47. Vertices: $(\pm 5, 0) \Rightarrow a = 5$

Eccentricity: $\frac{3}{5} \Rightarrow c = \frac{3}{5}a = 3$

$b^2 = a^2 - c^2 = 25 - 9 = 16$

Center: $(0, 0) = (h, k)$

$$\frac{(x-h)^2}{a^2} + \frac{(y-k)^2}{b^2} = 1$$

$$\frac{x^2}{25} + \frac{y^2}{16} = 1$$

49. Vertices: $(\pm 3, 0) \Rightarrow a = 3$

Half of minor axis length: $2 \Rightarrow b = 2$

$c^2 = a^2 - b^2 = 9 - 4 = 5 \Rightarrow c = \sqrt{5}$

Place the tacks $\sqrt{5}$ feet from the center: $(\pm\sqrt{5}, 0)$

Length of string: $2a = 2(3) = 6$ feet

51. Area of ellipse = 2 (area of circle)

$$\pi ab = 2\pi r^2$$

$$\pi a(10) = 2\pi(10)^2$$

$$\pi a(10) = 200$$

$$a = 20$$

Length of major axis: $2a = 2(20) = 40$ units

53. Center: $(0, 0) \implies h = 0, k = 0$

$2a = 0.34 + 4.08 = 4.42$

$a = 2.21$

$c = 2.21 - 0.34 = 1.87$

$b^2 = a^2 - c^2 = 4.8841 - 3.4969 = 1.3872$

$$\frac{x^2}{a^2} + \frac{y^2}{b^2} = 1$$

$$\frac{x^2}{4.88} + \frac{y^2}{1.39} = 1$$

55. For $\frac{x^2}{a^2} + \frac{y^2}{b^2} = 1$, we have $c^2 = a^2 - b^2$.

When $x = c$:

$$\frac{c^2}{a^2} + \frac{y^2}{b^2} = 1 \implies y^2 = b^2\left(1 - \frac{a^2 - b^2}{a^2}\right) \implies y^2 = \frac{b^4}{a^2} \implies 2y = \frac{2b^2}{a}.$$

57. $\frac{x^2}{9} + \frac{y^2}{16} = 1$

$a = 4, b = 3, c = \sqrt{7}$

Points on the ellipse: $(\pm 3, 0), (0, \pm 4)$

Length of latus recta: $\frac{2b^2}{a} = \frac{2(3)^2}{4} = \frac{9}{2}$

Additional points; $\left(\pm\frac{9}{4}, -\sqrt{7}\right), \left(\pm\frac{9}{4}, \sqrt{7}\right)$

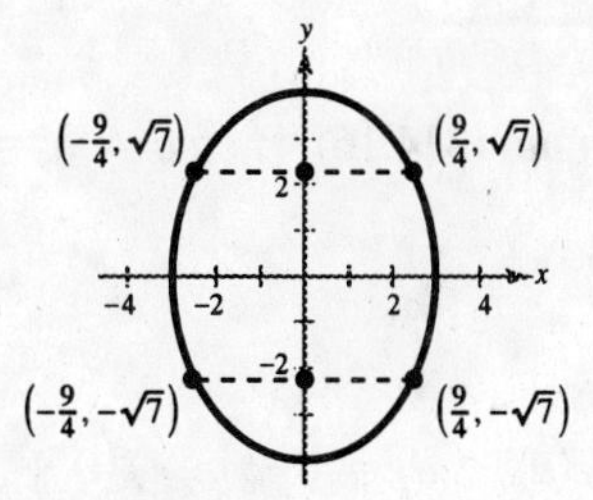

59. $5x^2 + 3y^2 = 15$

$\frac{x^3}{3} + \frac{y^2}{5} = 1$

$a = \sqrt{5}, b = \sqrt{3}, c = \sqrt{2}$

Points on the ellipse: $(\pm\sqrt{3}, 0), (0, \pm\sqrt{5})$

Length of latus recta: $\frac{2b^2}{a} = \frac{2 \cdot 3}{\sqrt{5}} = \frac{6\sqrt{5}}{5}$

Additional points; $\left(\pm\frac{3\sqrt{5}}{5}, -\sqrt{2}\right), \left(\pm\frac{3\sqrt{5}}{5}, \sqrt{2}\right)$

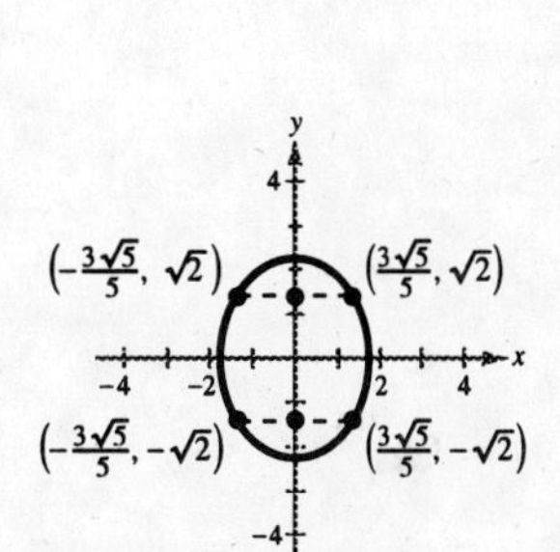

61. True. If $e \approx 1$ then the ellipse is elongated, not circular.

63. False. $c < a$ always

65. $\frac{x^2}{a^2} + \frac{y^2}{b^2} = 1$

(a) $a + b = 20 \Rightarrow b = 20 - a$

$A = \pi ab = \pi a(20 - a)$

(b) $264 = \pi a(20 - a)$

$0 = -\pi a^2 + 20\pi a - 264$

$0 = \pi a^2 - 20\pi a + 264$

$a = 14$ or $a = 6$. The equation of an ellipse with an area of 264 is $\frac{x^2}{196} + \frac{y^2}{36} = 1$.

(c)

a	8	9	10	11	12	13
A	301.6	311.0	314.2	311.0	301.6	285.9

The area is maximum when $a = 10$ and the ellipse is a circle.

(d)

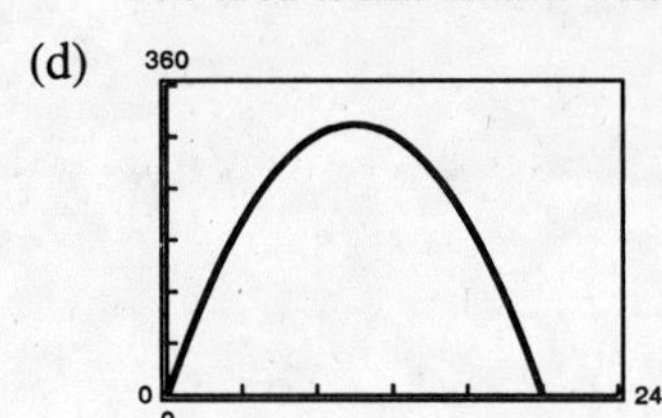

The area is maximum (314.16) when $a = b = 10$ and the ellipse is a circle.

67. Geometric: $r = \frac{1}{2}$

69. Arithmetic: $d = 1$

71. $a_n = dn + c = dn + (a_1 - d) = \left(-\frac{1}{4}\right)n + \left(0 + \frac{1}{4}\right) = \frac{1}{4} - \frac{1}{4}n$

73. $a_n = dn + c$

$a_8 = d8 + c = 72$

$a_3 = d3 + c = 27$

Subtracting, $5d = 45, d = 9$.

$a_1 = a_3 - 2d = 27 - 18 = 9$

Thus, $a_n = 9n + (9 - 9) = 9n$

75. $\sum_{n=0}^{6} (-3)^n = 547$

77. $\sum_{n=0}^{10} 5\left(\frac{4}{3}\right)^n \approx 340.155$

Section 10.3 Hyperbolas

- A **hyperbola** is the set of all points (x, y) the difference of whose distances from two distinct fixed points **(foci)** is constant.
- The standard equation of a hyperbola with center (h, k) and transverse and conjugate axes of lengths $2a$ and $2b$ is:

  (a) $\frac{(x - h)^2}{a^2} - \frac{(y - k)^2}{b^2} = 1$ if the traverse axis is horizontal.

  (b) $\frac{(y - k)^2}{a^2} - \frac{(x - h)^2}{b^2} = 1$ if the traverse axis is vertical.
- $c^2 = a^2 + b^2$ where c is the distance from the center to a focus.

—CONTINUED—

—CONTINUED—

- The asymptotes of a hyperbola are:

  (a) $y = k \pm \frac{b}{a}(x - h)$ if the transverse axis is horizontal.

  (b) $y = k \pm \frac{a}{b}(x - h)$ the transverse axis is vertical.

- The eccentricity of a hyperbola is $e = \frac{c}{a}$.

- To classify a nondegenerate conic from its general equation $Ax^2 + Cy^2 + Dx + Ey + F = 0$:
  (a) If $A = C$ $(A \neq 0, C \neq 0)$, then it is a circle.
  (b) If $AC = 0$ ($A = 0$ or $C = 0$, but not both), then it is a parabola.
  (c) If $AC > 0$, then it is an ellipse.
  (d) If $AC < 0$, then it is a hyperbola.

Solutions to Odd-Numbered Exercises

1. Center: $(0, 0)$

$a = 3, b = 5, c = \sqrt{34}$

Vertical transverse axis

Matches graph (b).

3. Center: $(1, 0)$

$a = 4, b = 2$

Horizontal transverse axis

Matches graph (a).

5. $x^2 - y^2 = 1$

$a = 1, b = 1, c = \sqrt{2}$

Center: $(0, 0)$

Vertices: $(\pm 1, 0)$

Foci: $(\pm\sqrt{2}, 0)$

Asymptotes: $y = \pm x$

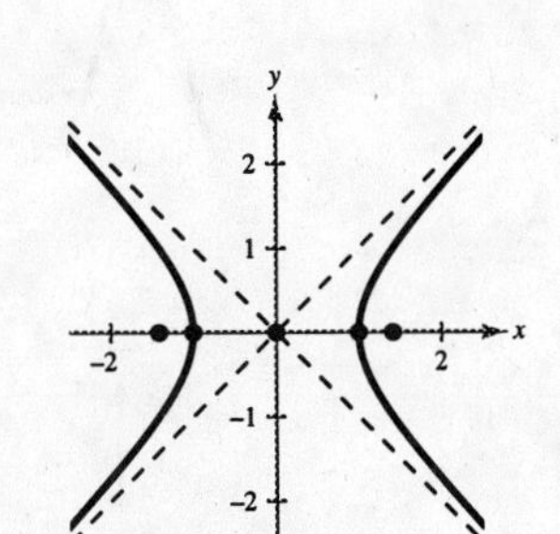

7. $\frac{y^2}{1} - \frac{x^2}{4} = 1$

$a = 1, b = 2, c = \sqrt{5}$

Center: $(0, 0)$

Vertices: $(0, \pm 1)$

Foci: $(0, \pm\sqrt{5})$

Asymptotes: $y = \pm\frac{1}{2}x$

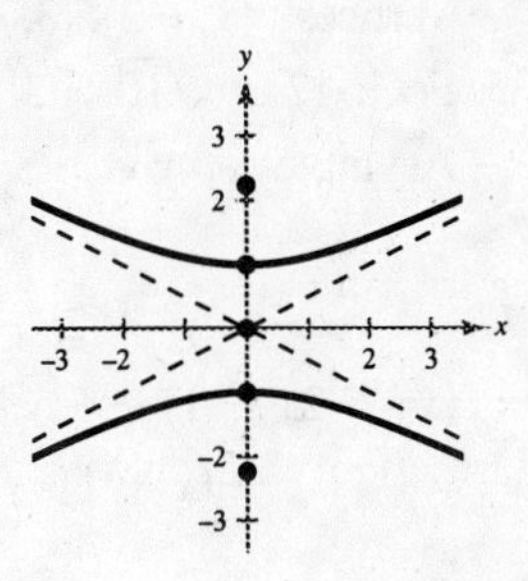

9. $\frac{y^2}{25} - \frac{x^2}{81} = 1$

$a = 5, b = 9, c = \sqrt{a^2 + b^2} = \sqrt{106}$

Center: $(0, 0)$

Vertices: $(0, \pm 5)$

Foci: $(0, \pm\sqrt{106})$

Asymptotes: $y = \pm\frac{a}{b}x = \pm\frac{5}{9}x$

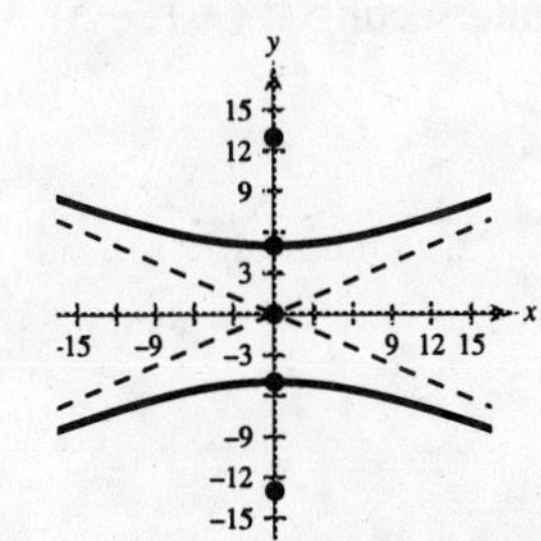

11. $\dfrac{(x-1)^2}{4} - \dfrac{(y+2)^2}{1} = 1$

$a = 2, b = 1, c = \sqrt{5}$

Center: $(1, -2)$

Vertices: $(-1, -2), (3, -2)$

Foci: $\left(1 \pm \sqrt{5}, -2\right)$

Asymptotes: $y = -2 \pm \dfrac{1}{2}(x - 1)$

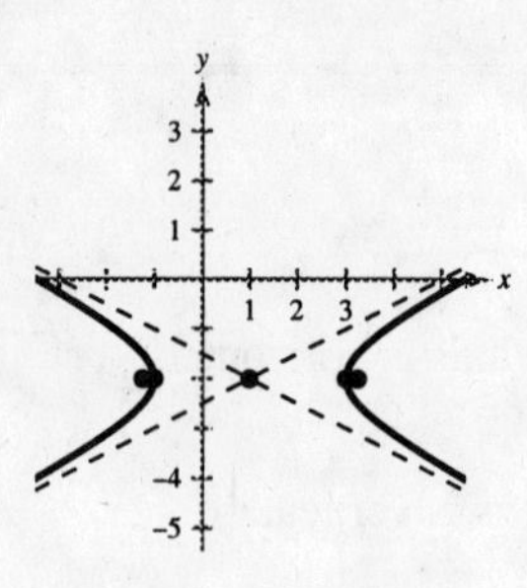

13. $(y + 6)^2 - (x - 2)^2 = 1$

$a = 1, b = 1, c = \sqrt{2}$

Center: $(2, -6)$

Vertices: $(2, -5), (2, -7)$

Foci: $\left(2, -6 \pm \sqrt{2}\right)$

Asymptotes: $y = -6 \pm (x - 2)$

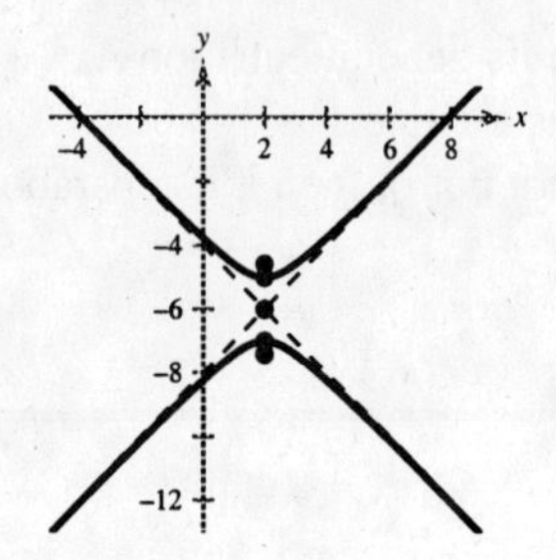

15.

$$9x^2 - y^2 - 36x - 6y + 18 = 0$$

$$9\left(x^2 - 4x + 4\right) - \left(y^2 + 6y + 9\right) = -18 + 36 - 9$$

$$\frac{(x-2)^2}{1} - \frac{(y+3)^2}{9} = 1$$

$a = 1, b = 3, c = \sqrt{10}$

Center: $(2, -3)$

Vertices: $(1, -3), (3, -3)$

Foci: $\left(2 \pm \sqrt{10}, -3\right)$

Asymptotes: $y = -3 \pm 3(x - 2)$

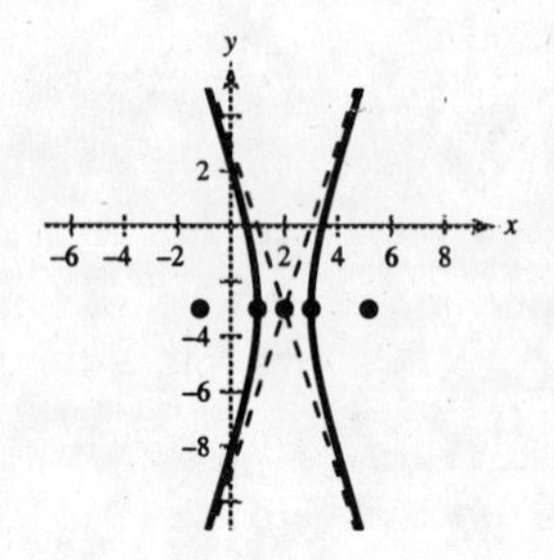

17.

$$x^2 - 9y^2 + 2x - 54y - 80 = 0$$

$$\left(x^2 + 2x + 1\right) - 9\left(y^2 + 6y + 9\right) = 80 + 1 - 81$$

$$(x+1)^2 - 9(y+3)^2 = 0$$

$$y + 3 = \pm\frac{1}{3}(x + 1)$$

Degenerate hyperbola is two lines intersecting at $(-1, -3)$.

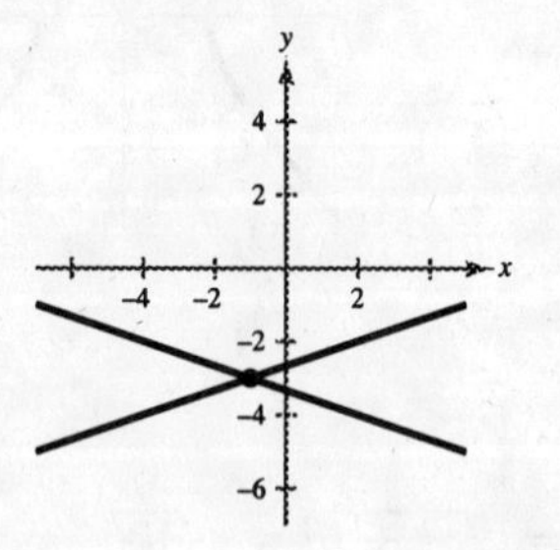

19. $2x^2 - 3y^2 = 6$

$$\frac{x^2}{3} - \frac{y^2}{2} = 1$$

$a = \sqrt{3}, b = \sqrt{2}, c = \sqrt{5}$

Center: $(0, 0)$

Vertices: $\left(\pm\sqrt{3}, 0\right)$

Foci: $\left(\pm\sqrt{5}, 0\right)$

Asymptotes: $y = \pm\sqrt{\frac{2}{3}}x$

To use a graphing calculator, solve first for y.

$$y^2 = \frac{2x^2 - 6}{3}$$

$$\left.\begin{aligned} y_1 &= \sqrt{\frac{2x^2 - 6}{3}} \\ y_2 &= -\sqrt{\frac{2x^2 - 6}{3}} \end{aligned}\right\} \text{Hyperbola}$$

$$\left.\begin{aligned} y_3 &= \sqrt{\frac{2}{3}}x \\ y_4 &= -\sqrt{\frac{2}{3}}x \end{aligned}\right\} \text{Asymptotes}$$

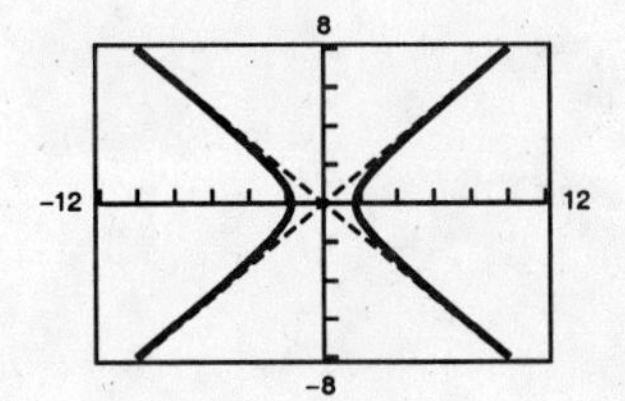

21. $9y^2 - x^2 + 2x + 54y + 62 = 0$

$$9\left(y^2 + 6y + 9\right) - \left(x^2 - 2x + 1\right) = -62 - 1 + 81$$

$$\frac{(y + 3)^2}{2} - \frac{(x - 1)^2}{18} = 1$$

$a = \sqrt{2}, b = 3\sqrt{2}, c = 2\sqrt{5}$

Center: $(1, -3)$

Vertices: $\left(1, -3 \pm \sqrt{2}\right)$

Foci: $\left(1, -3 \pm 2\sqrt{5}\right)$

Asymptotes: $y = -3 \pm \frac{1}{3}(x - 1)$

To use a graphing calculator, solve for y first.

$$9(y + 3)^2 = 18 + (x - 1)^2$$

$$y = -3 \pm \sqrt{\frac{18 + (x - 1)^2}{9}}$$

$$\left.\begin{aligned} y_1 &= -3 + \frac{1}{3}\sqrt{18 + (x - 1)^2} \\ y_2 &= -3 - \frac{1}{3}\sqrt{18 + (x - 1)^2} \end{aligned}\right\} \text{Hyperbola}$$

$$\left.\begin{aligned} y_3 &= -3 + \frac{1}{3}(x - 1) \\ y_4 &= -3 - \frac{1}{3}(x - 1) \end{aligned}\right\} \text{Asymptotes}$$

23. Vertices: $(0, \pm 2) \Rightarrow a = 2$

Foci: $(0, \pm 4) \Rightarrow c = 4$

$b^2 = c^2 - a^2 = 16 - 4 = 12$

Center: $(0, 0) = (h, k)$

$$\frac{(y - k)^2}{a^2} - \frac{(x - h)^2}{b^2} = 1$$

$$\frac{y^2}{4} - \frac{x^2}{12} = 1$$

25. Vertices: $(\pm 1, 0) \Rightarrow a = 1$

Asymptotes: $y = \pm 5x \Rightarrow \frac{b}{a} = 5 \Rightarrow b = 5$

Center: $(0, 0)$

$$\frac{x^2}{1} - \frac{y^2}{25} = 1$$

27. Foci: $(0, \pm 8) \Rightarrow c = 8$

Asymptotes: $y = \pm 4x \Rightarrow \frac{a}{b} = 4 \Rightarrow a = 4b$

Center: $(0, 0) = (h, k)$

$c^2 = a^2 + b^2 \Rightarrow 64 = 16b^2 + b^2$

$$\frac{64}{17} = b^2 \Rightarrow a^2 = \frac{1024}{17}$$

$$\frac{(y-k)^2}{a^2} - \frac{(x-h)^2}{b^2} = 1$$

$$\frac{y^2}{1024/17} - \frac{x^2}{64/17} = 1$$

$$\frac{17y^2}{1024} - \frac{17x^2}{65} = 1$$

29. Vertices: $(2, 0), (6, 0) \Rightarrow a = 2$

Foci: $(0, 0), (8, 0) \Rightarrow c = 4$

$b^2 = c^2 - a^2 = 16 - 4 = 12$

Center: $(4, 0) = (h, k)$

$$\frac{(x-h)^2}{a^2} - \frac{(y-k)^2}{b^2} = 1$$

$$\frac{(x-4)^2}{4} - \frac{y^2}{12} = 1$$

31. Vertices: $(4, 1), (4, 9) \Rightarrow a = 4$

Foci: $(4, 0), (4, 10) \Rightarrow c = 5$

$b^2 = c^2 - a^2 = 25 - 16 = 9$

Center: $(4, 5) = (h, k)$

$$\frac{(y-k)^2}{a^2} - \frac{(x-h)^2}{b^2} = 1$$

$$\frac{(y-5)^2}{16} - \frac{(x-4)^2}{9} = 1$$

33. Vertices: $(2, 3), (2, -3) \Rightarrow a = 3$

Solution point: $(0, 5)$

Center: $(2, 0) = (h, k)$

$$\frac{(y-k)^2}{a^2} - \frac{(x-h)^2}{b^2} = 1$$

$$\frac{y^2}{9} - \frac{(x-2)^2}{b^2} = 1 \Rightarrow b^2 = \frac{9(x-2)^2}{y^2 - 9} = \frac{9(-2)^2}{25 - 9} = \frac{36}{16} = \frac{9}{4}$$

$$\frac{y^2}{9} - \frac{(x-2)^2}{9/4} = 1$$

35. Vertices: $(0, 4), (0, 0)$

Center: $(0, 2)$, $a = 2$

$$\frac{(y-2)^2}{4} - \frac{x^2}{b^2} = 1$$

Passes through $(\sqrt{5}, 5)$: $\frac{(5-2)^2}{4} - \frac{5}{b^2} = 1$

$$\frac{5}{b^2} = \frac{9}{4} - 1 = \frac{5}{4} \Rightarrow b = 2$$

$$\frac{(y-2)^2}{4} - \frac{x^2}{4} = 1$$

37. Vertices: $(1, 2), (3, 2) \Rightarrow a = 1$ Center: $(2, 2)$

Asymptotes: $y = x, y = 4 - x$

$$\frac{b}{a} = 1 \Rightarrow b = 1$$

$$\frac{(x-2)^2}{1} - \frac{(y-2)^2}{1} = 1$$

39. Vertices: $(0, 2), (6, 2) \Rightarrow a = 3$

Asymptotes: $y = \frac{2}{3}x,\ y = 4 - \frac{2}{3}x$

$\frac{b}{a} = \frac{2}{3} \Rightarrow b = 2$

Center: $(3, 2) = (h, k)$

$$\frac{(x-h)^2}{a^2} - \frac{(y-k)^2}{b^2} = 1$$

$$\frac{(x-3)^2}{9} - \frac{(y-2)^2}{4} = 1$$

41. The explosion occurred on the vertical line through $(3300, 1100)$ and $(3300, 0)$.

$d_2 - d_1 = 4(1100) = 4400$

Hence $2a = 4400$

$a = 2200$

$c = 3300$

$b^2 = c^2 - a^2$

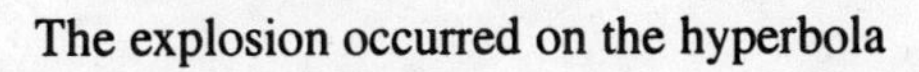
The explosion occurred on the hyperbola

$$\frac{x^2}{a^2} - \frac{y^2}{b^2} = 1.$$

Letting $x = 3300$, $y^2 = b^2\left(\frac{x^2}{a^2} - 1\right)$

$$= (3300^2 - 2200^2)\left(\frac{3300^2}{2200^2} - 1\right)$$

$\Rightarrow\ y = -2750$

$(3300, -2750)$

43. Center: $(0, 0) = (h, k)$
Focus: $(24, 0) \Rightarrow c = 24$
Solution point: $(24, 24)$
$24^2 = a^2 + b^2 \Rightarrow b^2 = 24^2 - a^2$

$$\frac{(x-h)^2}{a^2} - \frac{(y-k)^2}{b^2} = 1$$

$$\frac{x^2}{a^2} - \frac{y^2}{24^2 - a^2} = 1 \Rightarrow \frac{24^2}{a^2} - \frac{24^2}{24^2 - a^2} = 1$$

Solving yields $a^2 = \frac{(3-\sqrt{5})24^2}{2} \approx 220.0124$ and $b^2 \approx 355.9876$. Thus, we have $\frac{x^2}{220.0124} - \frac{y^2}{355.9876} = 1.$

The right vertex is at $(a, 0) \approx (14.83, 0)$.

45. $x^2 + 4y^2 - 6x + 16y + 21 = 0$

$A = 1, C = 4$

$AC = 1(4) = 4 > 0 \Rightarrow$ Ellipse

47. $y^2 - 4y - 4x = 0$

$A = 0, C = 1$

$AC = 0(1) = 0 \Rightarrow$ Parabola

49. $4y^2 - 2x^2 - 4y - 8x - 15 = 0$

$A = -2, C = 4$

$AC = (-2)(4) = -8 < 0 \implies$ Hyperbola

51. $4x^2 + 4y^2 - 16y + 15 = 0$

$A = 4, C = 4$

$A = C \implies$ Circle

53. False. $b \neq 0$ because it is in the denominator.

55. Answers will vary. See Example 3.

57. $\left(3x - \frac{1}{2}\right)(x + 4) = 3x^2 + 12x - \frac{1}{2}x - 2$

$= 3x^2 + \frac{23}{2}x - 2$

59. $[(x + y) + 3]^2 = (x + y)^2 + 6(x + y) + 9$

$= x^2 + 2xy + y^2 + 6x + 6y + 9$

61. $x^2 + 14x + 49 = (x + 7)^2$

63. $6x^3 - 11x^2 - 10x = x(6x^2 - 11x - 10)$

$= x(3x + 2)(2x - 5)$

65. $4 - x + 4x^2 - x^3 = (4 - x) + x^2(4 - x)$

$= (4 - x)(x^2 + 1)$

$= (4 - x)(x + i)(x - i)$

Section 10.4 Rotation and Systems of Quadratic Equations

- The general second-degree equation $Ax^2 + Bxy + Cy^2 + Dx + Ey + F = 0$ can be rewritten as $A'(x')^2 + C'(y')^2 + D'x' + E'y' + F' = 0$ by rotating the coordinate axes through the angle θ, where $\cot 2\theta = (A - C)/B$.
- $x = x' \cos\theta - y' \sin\theta$
  $y = x' \sin\theta + y' \cos\theta$
- The graph of the nondegenerate equation $Ax^2 + Bxy + Cy^2 + Dx + Ey + F = 0$ is:
  (a) An ellipse or circle if $B^2 - 4AC < 0$.
  (b) A parabola if $B^2 - 4AC = 0$.
  (c) A hyperbola if $B^2 - 4AC > 0$.

Solutions to Odd-Numbered Exercises

1. $\theta = 90°$; Point: $(0, 4)$

$x = x' \cos\theta - y' \sin\theta$

$0 = x' \cos 90° - y' \sin 90°$

$0 = y'$

$y = x' \sin\theta + y' \cos\theta$

$4 = x' \sin 90° + y' \cos 90°$

$4 = x'$

Thus, $(x', y') = (4, 0)$.

3. $\theta = 30°$; Point: $(1, 6)$

$$\begin{aligned} x &= x' \cos\theta - y' \sin\theta \\ y &= x' \sin\theta + y' \cos\theta \end{aligned} \implies \begin{cases} 1 = x' \cos 30° - y' \sin 30° \\ 6 = x' \sin 30° + y' \cos 30° \end{cases}$$

Solving this system yields $(x', y') = \left(\frac{6 + \sqrt{3}}{2}, \frac{6\sqrt{3} - 1}{2}\right)$.

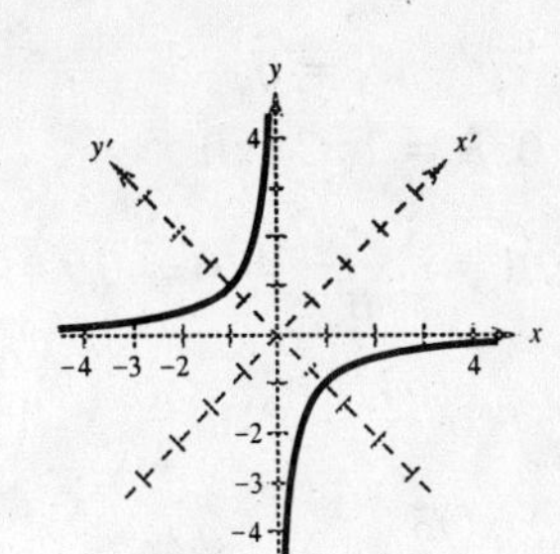

5. $xy + 1 = 0$

$A = 0, B = 1, C = 0$

$$\cot 2\theta = \frac{A - C}{B} = 0 \Rightarrow 2\theta = \frac{\pi}{2} \Rightarrow \theta = \frac{\pi}{4}$$

$$x = x'\cos\frac{\pi}{4} - y'\sin\frac{\pi}{4} = x'\left(\frac{\sqrt{2}}{2}\right) - y'\left(\frac{\sqrt{2}}{2}\right) = \frac{x' - y'}{\sqrt{2}}$$

$$y = x'\sin\frac{\pi}{4} + y'\cos\frac{\pi}{4} = x'\left(\frac{\sqrt{2}}{2}\right) + y'\left(\frac{\sqrt{2}}{2}\right) = \frac{x' + y'}{\sqrt{2}}$$

$$xy + 1 = 0$$

$$\left(\frac{x' - y'}{\sqrt{2}}\right)\left(\frac{x' + y'}{\sqrt{2}}\right) + 1 = 0$$

$$\frac{(y')^2}{2} - \frac{(x')^2}{2} = 1 \text{ Hyperbola}$$

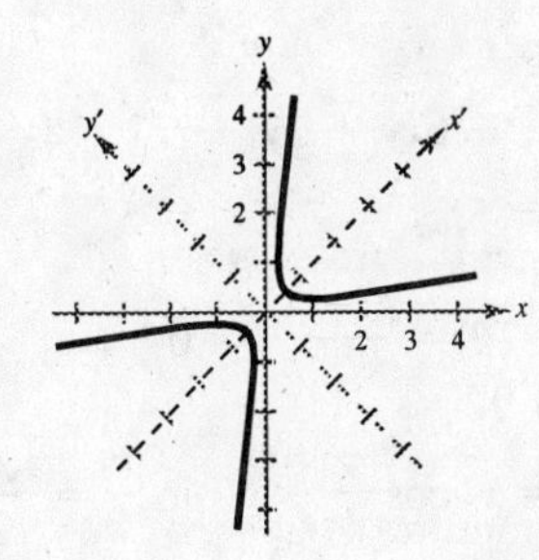

7. $x^2 - 8xy + y^2 + 1 = 0$

$A = 1, B = -8, C = 1$

$$\cot 2\theta = \frac{A - C}{B} = 0 \Rightarrow 2\theta = \frac{\pi}{2} \Rightarrow \theta = \frac{\pi}{4}$$

$$x = x'\cos\frac{\pi}{4} - y'\sin\frac{\pi}{4} = x'\left(\frac{\sqrt{2}}{2}\right) - y'\left(\frac{\sqrt{2}}{2}\right) = \frac{\sqrt{2}}{2}(x' - y')$$

$$y = x'\sin\frac{\pi}{4} + y'\cos\frac{\pi}{4} = x'\left(\frac{\sqrt{2}}{2}\right) + y'\left(\frac{\sqrt{2}}{2}\right) = \frac{\sqrt{2}}{2}(x' + y')$$

$$x^2 - 8xy + y^2 + 1 = 0$$

$$\left[\frac{\sqrt{2}}{2}(x' - y')\right]^2 - 8\left[\frac{\sqrt{2}}{2}(x' - y')\frac{\sqrt{2}}{2}(x' + y')\right] + \left[\frac{\sqrt{2}}{2}(x' + y')\right]^2 + 1 = 0$$

$$\frac{1}{2}(x')^2 + x'y' + \frac{1}{2}(y')^2 - 4[(x')^2 - (y')^2] + \frac{1}{2}(x')^2 + x'y' + \frac{1}{2}(y')^2 + 1 = 0$$

$$-3(x')^2 + 5(y')^2 = -1$$

$$\frac{(x')^2}{1/3} - \frac{(y')^2}{1/5} = 1 \text{ Hyperbola}$$

9. $xy - 2y - 4x = 0$

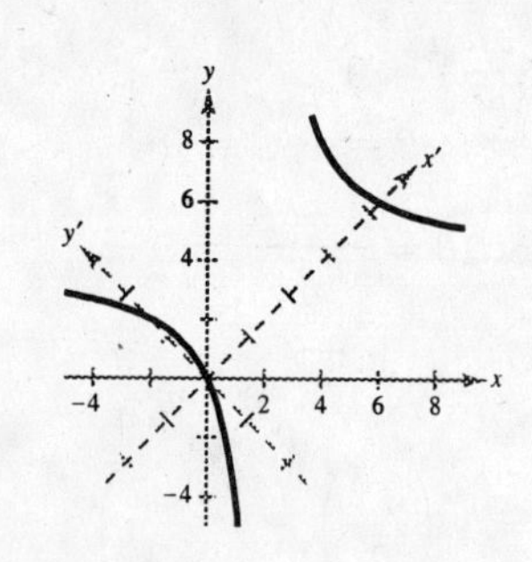

$A = 0, B = 1, C = 0$

$$\cot 2\theta = \frac{A - C}{B} = 0 \Rightarrow 2\theta = \frac{\pi}{2} \Rightarrow \theta = \frac{\pi}{4}$$

$$x = x'\cos\frac{\pi}{4} - y'\sin\frac{\pi}{4} \qquad y = x'\sin\frac{\pi}{4} + y'\cos\frac{\pi}{4}$$

$$= x'\left(\frac{\sqrt{2}}{2}\right) - y'\left(\frac{\sqrt{2}}{2}\right) \qquad = x'\left(\frac{\sqrt{2}}{2}\right) + y'\left(\frac{\sqrt{2}}{2}\right)$$

$$= \frac{x' - y'}{\sqrt{2}} \qquad = \frac{x' + y'}{\sqrt{2}}$$

$$xy - 2y - 4x = 0$$

$$\left(\frac{x' - y'}{\sqrt{2}}\right)\left(\frac{x' + y'}{\sqrt{2}}\right) - 2\left(\frac{x' + y'}{\sqrt{2}}\right) - 4\left(\frac{x' - y'}{\sqrt{2}}\right) = 0$$

$$\frac{(x')^2}{2} - \frac{(y')^2}{2} - \sqrt{2}x' - \sqrt{2}y' - 2\sqrt{2}x' + 2\sqrt{2}y' = 0$$

$$\left[(x')^2 - 6\sqrt{2}x' + (3\sqrt{2})^2\right] - \left[(y')^2 - 2\sqrt{2}y' + (\sqrt{2})^2\right] = 0 + (3\sqrt{2})^2 - (\sqrt{2})^2$$

$$(x' - 3\sqrt{2})^2 - (y' - \sqrt{2})^2 = 16$$

$$\frac{(x' - 3\sqrt{2})^2}{16} - \frac{(y' - \sqrt{2})^2}{16} = 1 \text{ Hyperbola}$$

11. $5x^2 - 6xy + 5y^2 - 12 = 0$

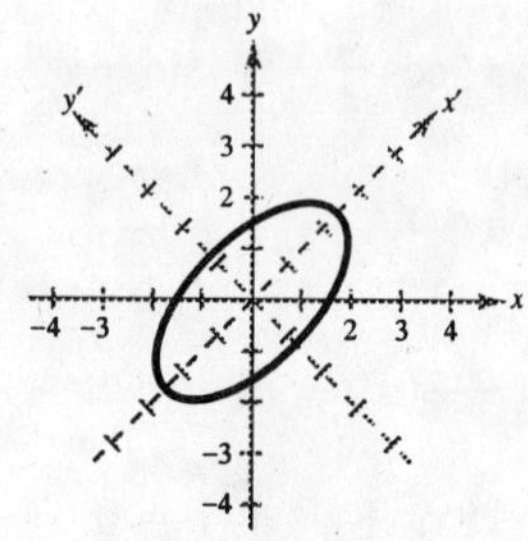

$A = 5, B = -6, C = 5$

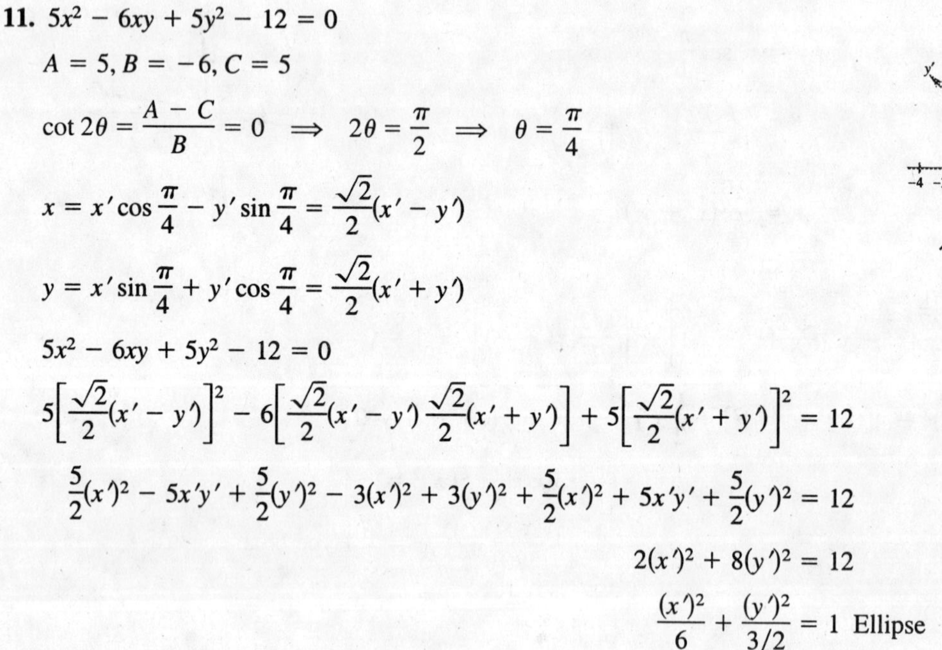

$$\cot 2\theta = \frac{A - C}{B} = 0 \quad \Rightarrow \quad 2\theta = \frac{\pi}{2} \quad \Rightarrow \quad \theta = \frac{\pi}{4}$$

$$x = x'\cos\frac{\pi}{4} - y'\sin\frac{\pi}{4} = \frac{\sqrt{2}}{2}(x' - y')$$

$$y = x'\sin\frac{\pi}{4} + y'\cos\frac{\pi}{4} = \frac{\sqrt{2}}{2}(x' + y')$$

$$5x^2 - 6xy + 5y^2 - 12 = 0$$

$$5\left[\frac{\sqrt{2}}{2}(x' - y')\right]^2 - 6\left[\frac{\sqrt{2}}{2}(x' - y')\frac{\sqrt{2}}{2}(x' + y')\right] + 5\left[\frac{\sqrt{2}}{2}(x' + y')\right]^2 = 12$$

$$\frac{5}{2}(x')^2 - 5x'y' + \frac{5}{2}(y')^2 - 3(x')^2 + 3(y')^2 + \frac{5}{2}(x')^2 + 5x'y' + \frac{5}{2}(y')^2 = 12$$

$$2(x')^2 + 8(y')^2 = 12$$

$$\frac{(x')^2}{6} + \frac{(y')^2}{3/2} = 1 \text{ Ellipse}$$

13. $3x^2 - 2\sqrt{3}xy + y^2 + 2x + 2\sqrt{3}y = 0$

$A = 3, B = -2\sqrt{3}, C = 1$

$\cot 2\theta = \dfrac{A - C}{B} = -\dfrac{1}{\sqrt{3}} \Rightarrow \theta = 60°$

$x = x'\cos 60° - y'\sin 60°$

$= x'\left(\dfrac{1}{2}\right) - y'\left(\dfrac{\sqrt{3}}{2}\right) = \dfrac{x' - \sqrt{3}y'}{2}$

$y = x'\sin\theta + y'\cos\theta = \dfrac{\sqrt{3}x' + y'}{2}$

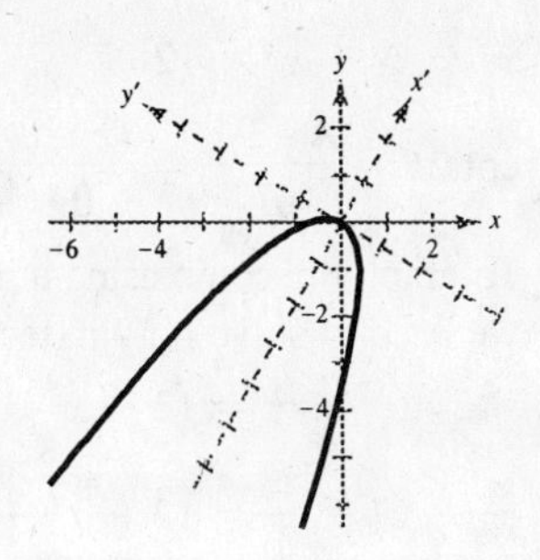

$$3x^2 - 2\sqrt{3}xy + y^2 + 2x + 2\sqrt{3}y = 0$$

$$3\left(\frac{x' - \sqrt{3}y'}{2}\right)^2 - 2\sqrt{3}\left(\frac{x' - \sqrt{3}y'}{2}\right)\left(\frac{\sqrt{3}x' + y'}{2}\right) + \left(\frac{\sqrt{3}x' + y'}{2}\right)^2 + 2\left(\frac{x' - \sqrt{3}y'}{2}\right) + 2\sqrt{3}\left(\frac{\sqrt{3}x' + y'}{2}\right) = 0$$

$$\frac{3(x')^2}{4} - \frac{6\sqrt{3}x'y'}{4} + \frac{9(y')^2}{4} - \frac{6(x')^2}{4} + \frac{4\sqrt{3}x'y'}{4} + \frac{6(y')^2}{4} + \frac{3(x')^2}{4} + \frac{2\sqrt{3}x'y'}{4} + \frac{(y')^2}{4}$$

$$+ x' - \sqrt{3}y' + 3x' + \sqrt{3}y' = 0$$

$$4(y')^2 + 4x' = 0$$

$$x' = -(y')^2$$

Parabola

15. $9x^2 + 24xy + 16y^2 + 90x - 130y = 0$

$A = 9, B = 24, C = 16$

$\cot 2\theta = \dfrac{A - C}{B} = -\dfrac{7}{24} \Rightarrow \theta \approx 53.13°$

$\cos 2\theta = -\dfrac{7}{25}$

$\sin\theta = \sqrt{\dfrac{1 - \cos 2\theta}{2}} = \sqrt{\dfrac{1 - (-7/25)}{2}} = \dfrac{4}{5}$

$\cos\theta = \sqrt{\dfrac{1 + \cos 2\theta}{2}} = \sqrt{\dfrac{1 + (-7/25)}{2}} = \dfrac{3}{5}$

$x = x'\cos\theta - y'\sin\theta$

$= x'\left(\dfrac{3}{5}\right) - y'\left(\dfrac{4}{5}\right) = \dfrac{3x' - 4y'}{5}$

$y = x'\sin\theta + y'\cos\theta$

$= x'\left(\dfrac{4}{5}\right) + y'\left(\dfrac{3}{5}\right)$

$= \dfrac{4x' + 3y'}{5}$

$$9x^2 + 24xy + 16y^2 + 90x - 130y = 0$$

$$9\left(\frac{3x' - 4y'}{5}\right)^2 + 24\left(\frac{3x' - 4y'}{5}\right)\left(\frac{4x' + 3y'}{5}\right) + 16\left(\frac{4x' + 3y'}{5}\right)^2 + 90\left(\frac{3x' - 4y'}{5}\right) - 130\left(\frac{4x' + 3y'}{5}\right) = 0$$

$$\frac{81(x')^2}{25} - \frac{216x'y'}{25} + \frac{144(y')^2}{25} + \frac{288(x')^2}{25} - \frac{168x'y'}{25} - \frac{288(y')^2}{25} + \frac{256(x')^2}{25} + \frac{384x'y'}{25}$$

$$+ \frac{144(y')^2}{25} + 54x' - 72y' - 104x' - 78y' = 0$$

$$25(x')^2 - 50x' - 150y' = 0$$

$$(x')^2 - 2x' + 1 = 6y' + 1$$

$$y' = \frac{(x')^2}{6} - \frac{x'}{3}$$

Parabola

17. $x^2 + xy + y^2 = 12$

$$\cot 2\theta = \frac{A - C}{B} = \frac{1 - 1}{0} = 0 \Rightarrow \theta = \frac{\pi}{4} \text{ or } 45°$$

To graph the conic using a graphing calculator, we need to solve for y in terms of x.

$$y^2 + xy = 12 - x^2$$

$$y^2 + xy + \frac{x^2}{4} = 12 - x^2 + \frac{x^2}{4}$$

$$\left(y + \frac{x}{2}\right)^2 = \frac{48 - 3x^2}{4}$$

$$y = -\frac{x}{2} \pm \frac{\sqrt{48 - 3x^2}}{2}$$

$$\text{Enter } y_1 = \frac{-x + \sqrt{48 - 3x^2}}{2}$$

$$\text{and } y_2 = \frac{-x - \sqrt{48 - 3x^2}}{2}.$$

19. $17x^2 + 32xy - 7y^2 = 75$

$$\cot 2\theta = \frac{A - C}{B} = \frac{17 + 7}{32} = \frac{24}{32} = \frac{3}{4} \Rightarrow \theta \approx 26.57°$$

Solve for y in terms of x by completing the square.

$$-7y^2 + 32xy = -17x^2 + 75$$

$$y^2 - \frac{32}{7}xy = \frac{17}{7}x^2 - \frac{75}{7}$$

$$y^2 - \frac{32}{7}xy + \frac{256}{49}x^2 = \frac{119}{49}x^2 - \frac{525}{49} + \frac{256}{49}x^2$$

$$\left(y - \frac{16}{7}x\right)^2 = \frac{375x^2 - 525}{49}$$

$$y = \frac{16}{7}x \pm \sqrt{\frac{375x^2 - 525}{49}}$$

$$y = \frac{16x \pm 5\sqrt{15x^2 - 21}}{7}$$

$$\text{Use } y_1 = \frac{16x + 5\sqrt{15x^2 - 21}}{7}$$

$$\text{and } y_2 = \frac{16x - 5\sqrt{15x^2 - 21}}{7}.$$

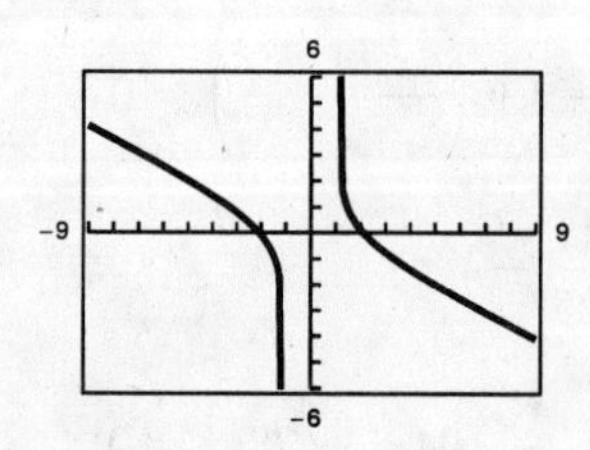

21. $32x^2 + 50xy + 7y^2 = 52$

$$\cot 2\theta = \frac{A - C}{B} = \frac{32 - 7}{50} = \frac{1}{2} \Rightarrow \theta \approx 31.72°$$

Solve for y in terms of x by completing the square.

$$7y^2 + 50xy = 52 - 32x^2$$

$$y^2 + \frac{50}{7}xy = \frac{52 - 32x^2}{7}$$

$$y^2 + \frac{50}{7}xy + \frac{625}{49}x^2 = \frac{52 - 32x^2}{7} + \frac{625x^2}{49}$$

$$\left(y + \frac{25}{7}x\right)^2 = \frac{364 + 401x^2}{49}$$

$$y = -\frac{25x}{7} \pm \frac{\sqrt{364 + 401x^2}}{7}$$

$$\text{Enter } y_1 = \frac{-25x + \sqrt{364 + 401x^2}}{7}$$

$$\text{and } y_2 = \frac{-25x - \sqrt{364 + 401x^2}}{7}.$$

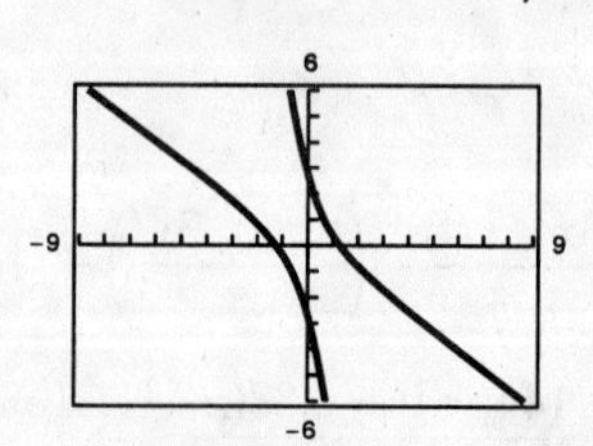

23. $xy + 4 = 0$

$B^2 - 4AC = 1 \Rightarrow$ The graph is a hyperbola.

$$\cot 2\theta = \frac{A - C}{B} = 0 \Rightarrow \theta = 45°$$

Matches graph (e).

25. $-2x^2 + 3xy + 2y^2 + 3 = 0$

$B^2 - 4AC = (3)^2 - 4(-2)(2) = 25 \Rightarrow$ The graph is a hyperbola.

$$\cot 2\theta = \frac{A - C}{B} = -\frac{4}{3} \Rightarrow \theta \approx -18.43°$$

Matches graph (f).

27. $3x^2 + 2xy + y^2 - 10 = 0$

$B^2 - 4AC = (2)^2 - 4(3)(1) = -8 \Rightarrow$ The graph is an ellipse or circle.

$$\cot 2\theta = \frac{A - C}{B} = 1 \Rightarrow \theta = 22.5°$$

Matches graph (d).

29. $16x^2 - 24xy + 9y^2 - 30x - 40y = 0$

(a) $B^2 - 4AC = (-24)^2 - 4(16)(9) = 0 \Rightarrow$ Parabola

(b) $9y^2 + (-24x - 40)y + (16x^2 - 30x) = 0$

$$y = \frac{(24x + 40) \pm \sqrt{(-24x - 40)^2 - 4(9)(16x^2 - 30x)}}{18}$$

$$= \frac{(24x + 40) \pm \sqrt{3000x + 1600}}{18}$$

(c)

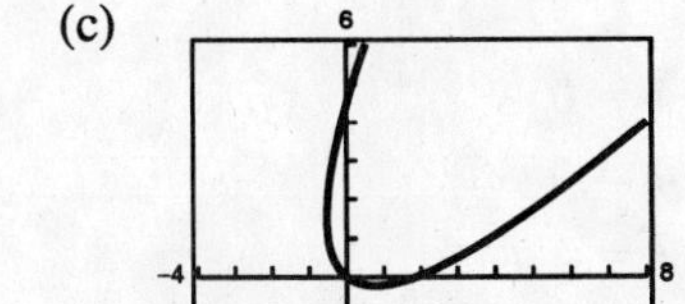

31. $15x^2 - 8xy + 7y^2 - 45 = 0$

(a) $B^2 - 4AC = (-8)^2 - 4(15)(7) = -356$ Ellipse or circle

(b) $7y^2 - 8xy + (15x^2 - 45) = 0$

$$y = \frac{8x \pm \sqrt{(-8x)^2 - 4(7)(15x^2 - 45)}}{14} = \frac{8x \pm \sqrt{1260 - 356x^2}}{14}$$

(c)

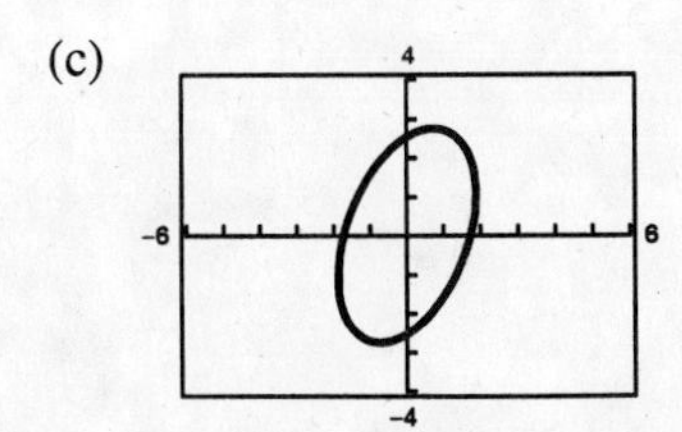

33. $x^2 - 6xy - 5y^2 + 4x - 22 = 0$

(a) $B^2 - 4AC = (-6)^2 - 4(1)(-5) = 56 \implies$ Hyperbola

(b) $-5y^2 - 6xy + (x^2 + 4x - 22) = 0$

$$y = \frac{6x \pm \sqrt{(-6x)^2 - 4(-5)(x^2 + 4x - 22)}}{-10}$$

$$= \frac{6x \pm \sqrt{56x^2 + 80x - 440}}{-10}$$

(c)

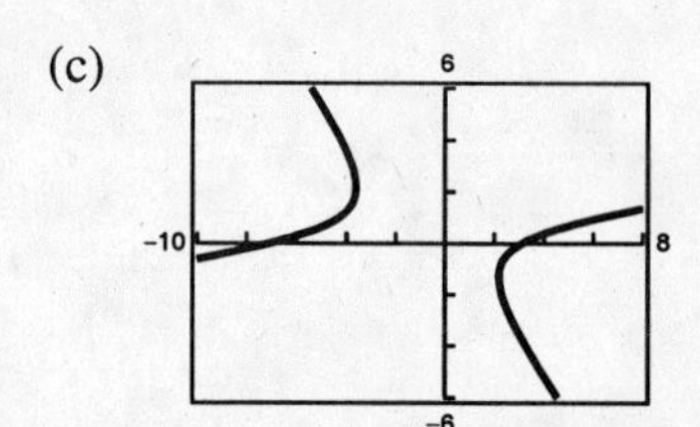

35. $x^2 + 4xy + 4y^2 - 5x - y - 3 = 0$

(a) $B^2 - 4AC = 4^2 - 4(1)(4) = 0 \implies$ Parabola

(b) $4y^2 + (4x - 1)y + (x^2 - 5x - 3) = 0$

$$y = \frac{(1 - 4x) \pm \sqrt{(4x - 1)^2 - 4(4)(x^2 - 5x - 3)}}{8} = \frac{1 - 4x \pm \sqrt{72x + 49}}{8}$$

(c)

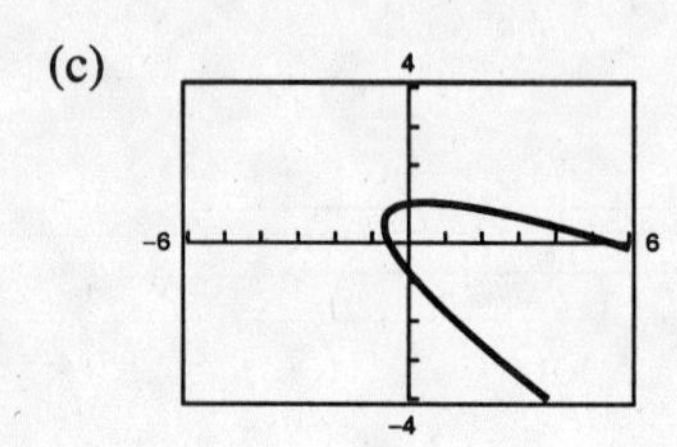

37. $y^2 - 16x^2 = 0$

$y^2 = 16x^2$

$y = \pm 4x$

Two intersecting lines

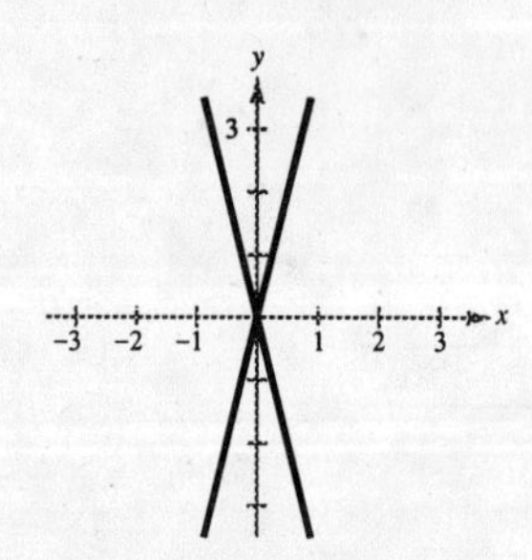

39. $x^2 + 2xy + y^2 - 4 = 0$

$(x + y)^2 - 4 = 0$

$(x + y)^2 = 4$

$x + y = \pm 2$

$y = -x \pm 2$

Two parallel lines

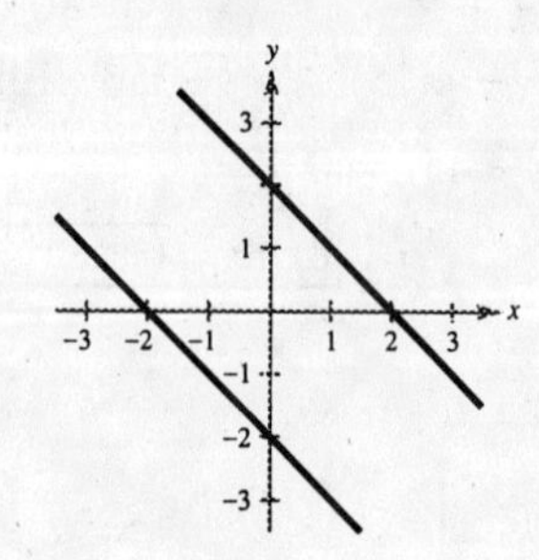

41.

$$-x^2 + y^2 + 4x - 6y + 4 = 0 \Rightarrow (y - 3)^2 - (x - 2)^2 = 1$$
$$x^2 + y^2 - 4x - 6y + 12 = 0 \Rightarrow (x - 2)^2 + (y - 3)^2 = 1$$
$$2y^2 - 12y + 16 = 0$$
$$2(y - 2)(y - 4) = 0$$
$$y = 2 \text{ or } y = 4$$

For $y = 2$: $x^2 + 2^2 - 4x - 6(2) + 12 = 0$

$$x^2 - 4x + 4 = 0$$
$$(x - 2)^2 = 0$$
$$x = 2$$

For $y = 4$: $x^2 + 4^2 - 4x - 6(4) + 12 = 0$

$$x^2 - 4x + 4 = 0$$
$$(x - 2)^2 = 0$$
$$x = 2$$

The points of intersection are $(2, 2)$ and $(2, 4)$.

43.

$$-4x^2 - y^2 - 16x + 24y - 16 = 0$$
$$4x^2 + y^2 + 40x - 24y + 208 = 0$$
$$24x + 192 = 0$$
$$24x = -192$$
$$x = -8$$

For $x = -8$,

$$-4(64) - y^2 - 16(-8) + 24y - 16 = 0$$
$$-y^2 + 24y - 144 = 0$$
$$y^2 - 24y + 144 = 0$$
$$(y - 12)^2 = 0 \Rightarrow y = 12$$

Solution: $(-8, 12)$

45.

$$x^2 - y^2 - 12x + 16y - 64 = 0$$
$$x^2 + y^2 - 12x - 16y + 64 = 0$$
$$2x^2 - 24x = 0$$
$$x^2 - 12x = 0$$
$$x(x - 12) = 0 \Rightarrow x = 0, 12$$

For $x = 0$, $-y^2 + 16y - 64 = 0$

$$y^2 - 16y + 64 = 0$$
$$(y - 8)^2 = 0 \Rightarrow y = 8$$

For $x = 12$, $144 - y^2 - 12(12) + 16y - 64 = 0$

$$-y^2 + 16y - 64 = 0 \Rightarrow y = 8$$

Solutions: $(0, 8)$, $(12, 8)$

47. $$\begin{aligned} -16x^2 - y^2 + 24y - 80 &= 0 \\ 16x^2 + 25y^2 - 400 &= 0 \\ 24y^2 + 24y - 480 &= 0 \\ 24(y+5)(y-4) &= 0 \\ y = -5 \text{ or } y &= 4 \end{aligned}$$

When $y = -5$: $16x^2 + 25(-5)^2 - 400 = 0$

$$16x^2 = -225$$

No real solution

When $y = 4$: $16x^2 + 25(4)^2 - 400 = 0$

$$16x^2 = 0$$

$$x = 0$$

The point of intersection is $(0, 4)$.
In standard form the equations are:

$$\frac{x^2}{4} + \frac{(y-12)^2}{64} = 1$$

$$\frac{x^2}{25} + \frac{y^2}{16} = 1$$

49. $x^2 + y^2 = 4$

$3x - y^2 = 0$

Adding $\quad x^2 + 3x - 4 = 0$

$$(x+4)(x-1) = 0 \implies x = 1, -4$$

For $x = 1, y = \pm\sqrt{3}$

$x = -4$ is impossible

Solutions: $(1, \sqrt{3}), (1, -\sqrt{3})$

51. $$\begin{aligned} x^2 + 2y^2 - 4x + 6y - 5 &= 0 \\ -x + y - 4 = 0 &\implies y = x + 4 \\ x^2 + 2(x+4)^2 - 4x + 6(x+4) - 5 &= 0 \\ x^2 + 2x^2 + 16x + 32 - 4x + 6x + 24 - 5 &= 0 \\ 3x^2 + 18x + 51 &= 0 \end{aligned}$$

No real solutions

53. $$\begin{aligned} xy + x - 2y + 3 = 0 &\Rightarrow y = \frac{-x-3}{x-2} \\ x^2 + 4y^2 - 9 &= 0 \\ x^2 + 4\left(\frac{-x-3}{x-2}\right)^2 &= 9 \\ x^2(x-2)^2 + 4(-x-3)^2 &= 9(x-2)^2 \\ x^2(x^2 - 4x + 4) + 4(x^2 + 6x + 9) &= 9(x^2 - 4x + 4) \\ x^4 - 4x^3 + 4x^2 + 4x^2 + 24x + 36 &= 9x^2 - 36x + 36 \\ x^4 - 4x^3 - x^2 + 60x &= 0 \\ x(x+3)(x^2 - 7x + 20) &= 0 \\ x = 0 \text{ or } x &= -3 \end{aligned}$$

Note: $x^2 - 7x + 20 = 0$ has no real solution.

When $x = 0$: $\quad y = \dfrac{-0-3}{0-2} = \dfrac{3}{2}$

When $x = -3$: $y = \dfrac{-(-3)-3}{-3-2} = 0$

The points of intersection are $\left(0, \frac{3}{2}\right), (-3, 0)$.

55. True. $B^2 - 4AC = 1 - 4k$

If $k < \frac{1}{4}$, then $B^2 - 4AC < 0$.

57. $(x')^2 + (y')^2 = (x\cos\theta + y\sin\theta)^2 + (y\cos\theta - x\sin\theta)^2$

$$= x^2\cos^2\theta + 2xy\cos\theta\sin\theta + y^2\sin^2\theta + y^2\cos^2\theta - 2xy\cos\theta\sin\theta + x^2\sin^2\theta$$

$$= x^2(\cos^2\theta + \sin^2\theta) + y^2(\sin^2\theta + \cos^2\theta) = x^2 + y^2 = r^2$$

59. $g(x) = \dfrac{2}{2 - x}$

Asymptotes: $x = 2$, $y = 0$

Intercepts: $(0, 1)$

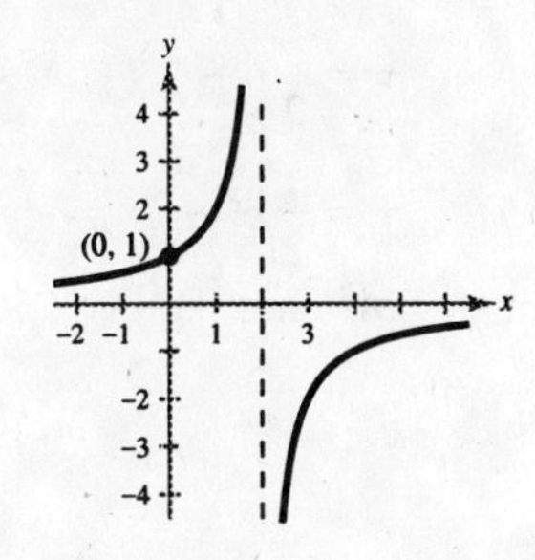

61. $h(t) = \dfrac{t^2}{2 - t} = -t - 2 + \dfrac{4}{2 - t}$

Slant asymptote: $y = -t - 2$

Vertical asymptote: $t = 2$

Intercept: $(0, 0)$

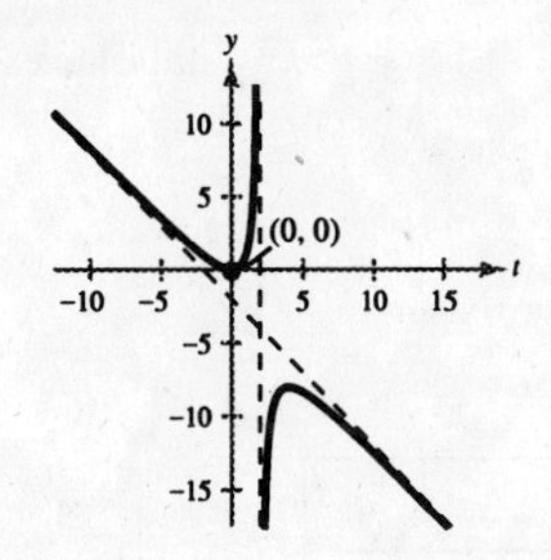

63. (a) $AB = \begin{bmatrix} 1 & -3 \\ 2 & 5 \end{bmatrix}\begin{bmatrix} 0 & 6 \\ 5 & -1 \end{bmatrix} = \begin{bmatrix} -15 & 9 \\ 25 & 7 \end{bmatrix}$

(b) $BA = \begin{bmatrix} 0 & 6 \\ 5 & -1 \end{bmatrix}\begin{bmatrix} 1 & -3 \\ 2 & 5 \end{bmatrix} = \begin{bmatrix} 12 & 30 \\ 3 & -20 \end{bmatrix}$

(c) $A^2 = \begin{bmatrix} 1 & -3 \\ 2 & 5 \end{bmatrix}\begin{bmatrix} 1 & -3 \\ 2 & 5 \end{bmatrix} = \begin{bmatrix} -5 & -18 \\ 12 & 19 \end{bmatrix}$

65. (a) $AB = \begin{bmatrix} 4 & -2 & 5 \end{bmatrix}\begin{bmatrix} 3 \\ -4 \\ 5 \end{bmatrix} = [12 + 8 + 25] = [45]$

(b) $BA = \begin{bmatrix} 3 \\ -4 \\ 5 \end{bmatrix}\begin{bmatrix} 4 & -2 & 5 \end{bmatrix} = \begin{bmatrix} 12 & -6 & 15 \\ -16 & 8 & -20 \\ 20 & -10 & 25 \end{bmatrix}$

(c) A^2 does not exist.

67. $(x + 8)^7 \quad a = 688{,}128$

69. $(x - 4y)^{10} \quad a = 53{,}760$

Section 10.5 Parametric Equations

- If f and g are continuous functions of t on an interval I, then the set of ordered pairs $(f(t), g(t))$ is a *plane curve C*. The equations $x = f(t)$ and $y = g(t)$ are *parametric equations* for C and t is the *parameter.*
- You should be able to graph plane curves with your graphing utility.
- To eliminate the parameter:

  Solve for t in one equation and substitute into the second equation.
- You should be able to find the parametric equations for a graph.

Solutions to Odd-Numbered Exercises

1. $x = t$

$y = t + 2$

$y = x + 2$ line

Matches (c).

3. $x = \sqrt{t}$

$y = t$

$y = x^2$ parabola, $x \geq 0$

Matches (b).

5. $x = \dfrac{1}{t} \Rightarrow t = \dfrac{1}{x}$

$y = t + 2$

$y = \dfrac{1}{x} + 2$

Matches (a).

7. $x = \ln t \Leftrightarrow t = e^x$

$y = \dfrac{1}{2}t - 2$

$y = \dfrac{1}{2}e^x - 2$

Matches (f).

9. $x = \sqrt{t}, y = 2 - t$

(a)

t	0	1	2	3	4
x	0	1	$\sqrt{2}$	$\sqrt{3}$	2
y	2	1	0	-1	-2

(b) Graph by hand

Note: $x \geq 0$

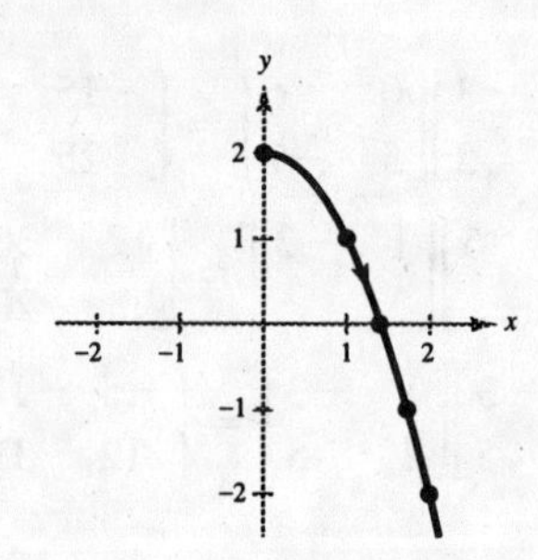

(c)

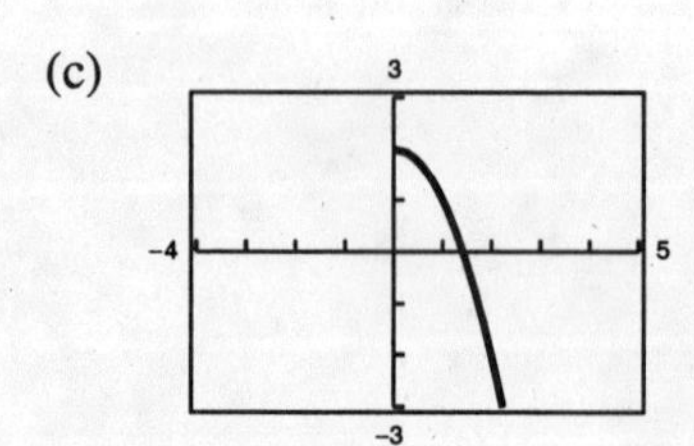

(d) $y = 2 - t = 2 - x^2$ parabola

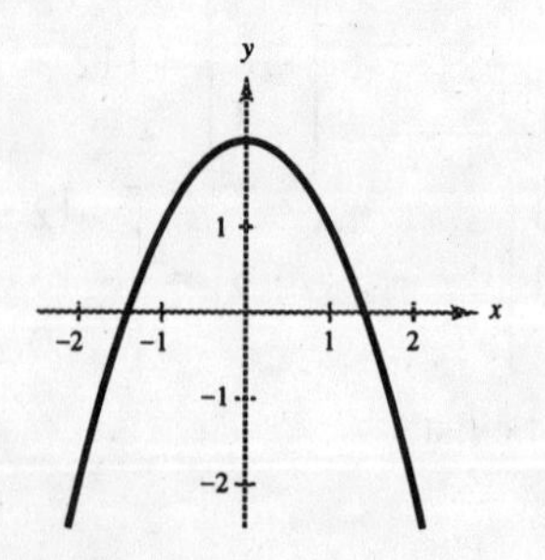

In part (c), $x \geq 0$.

11. $x = t, y = -4t$

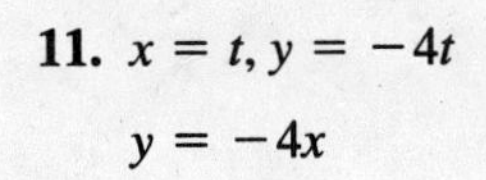

$y = -4x$

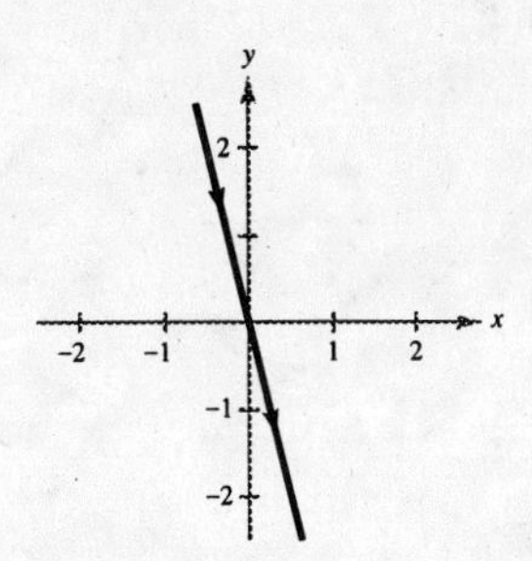

13. $x = 3t + 1, y = 2t - 1$

$y = 2\left(\frac{x-1}{3}\right) - 1 = \frac{2}{3}x - \frac{5}{3}$ or $2x - 3y - 5 = 0$

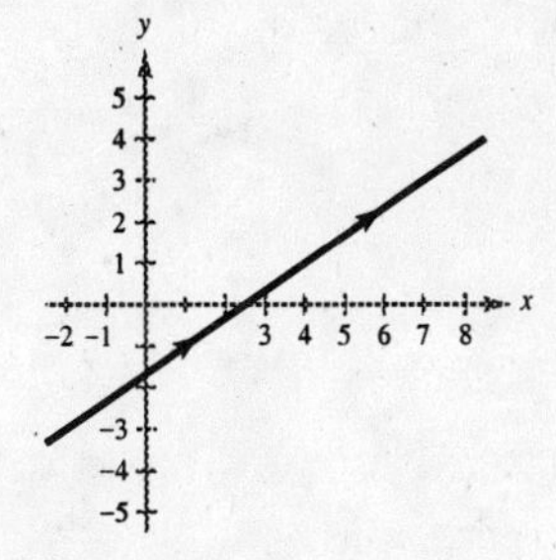

15. $x = \frac{1}{4}t,\ y = t^2$

$y = (4x)^2$

$y = 16x^2$

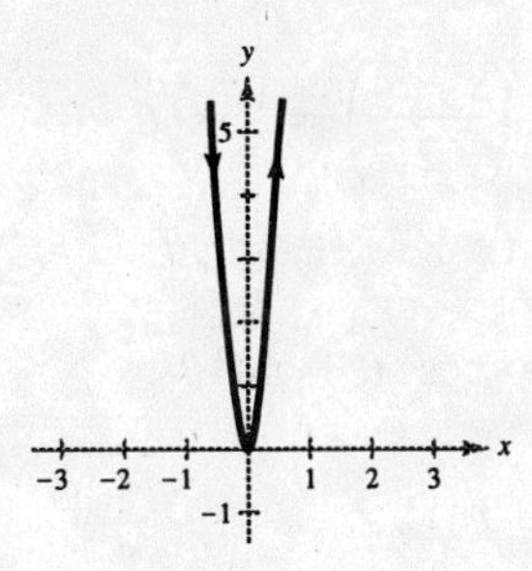

17. $x = t + 5, y = t^2$

$y = (x - 5)^2$

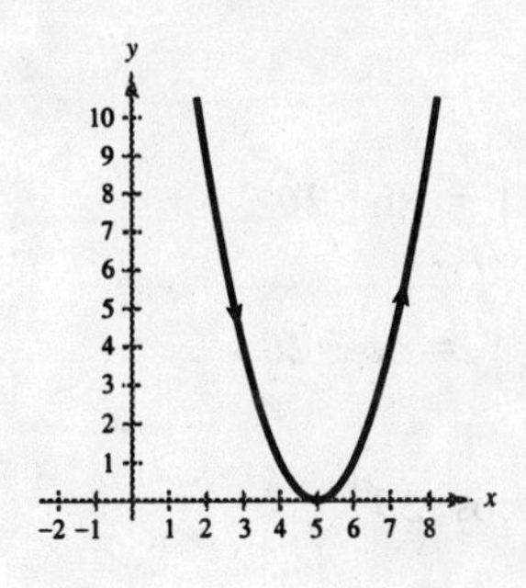

19. $x = 2t$

$y = |t - 2|$

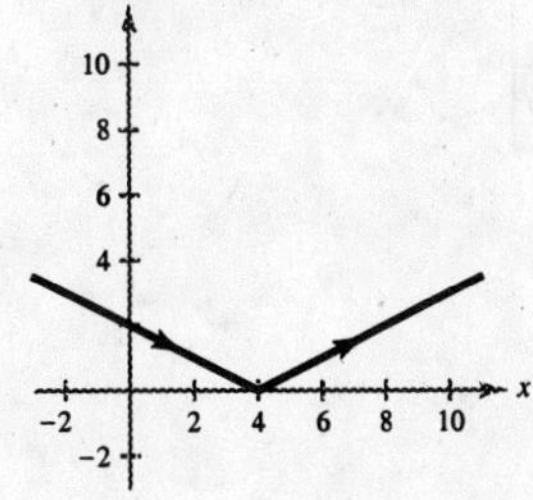

$$t = \frac{x}{2} \implies y = |t - 2|$$

$$= \left|\frac{x}{2} - 2\right|$$

$$= \frac{1}{2}|x - 4|$$

21. $x = 3\cos\theta \implies \left(\frac{x}{3}\right)^2 = \cos^2\theta$

$y = 3\sin\theta \implies \left(\frac{y}{3}\right)^2 = \sin^2\theta$

$\left(\frac{x}{3}\right)^2 + \left(\frac{y}{3}\right)^2 = 1 \implies x^2 + y^2 = 9$

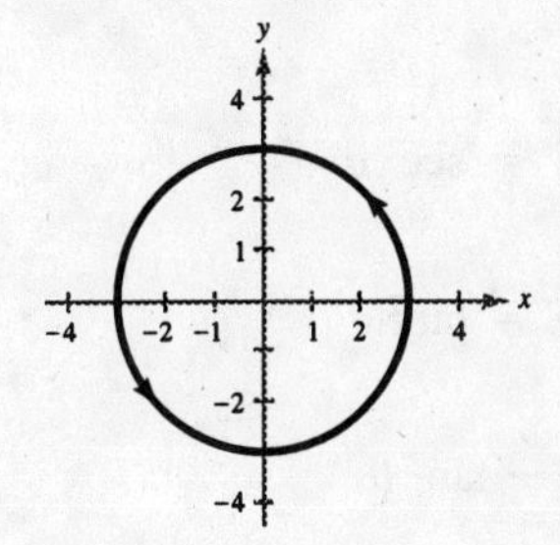

23. $x = e^{-t} \implies \dfrac{1}{x} = e^t$

$y = e^{3t} \implies y = (e^t)^3$

$y = \left(\dfrac{1}{x}\right)^3$

$y = \dfrac{1}{x^3},\ x > 0,\ y > 0$

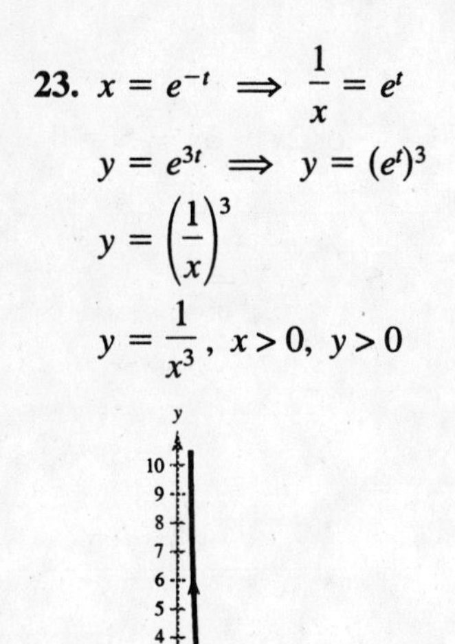

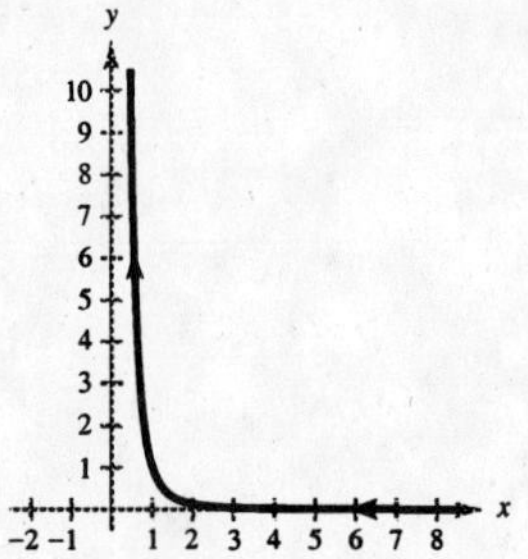

25. $x = t^3 \implies x^{1/3} = t$

$y = 3 \ln t \implies y = \ln t^3$

$y = \ln(x^{1/3})^3$

$y = \ln x$

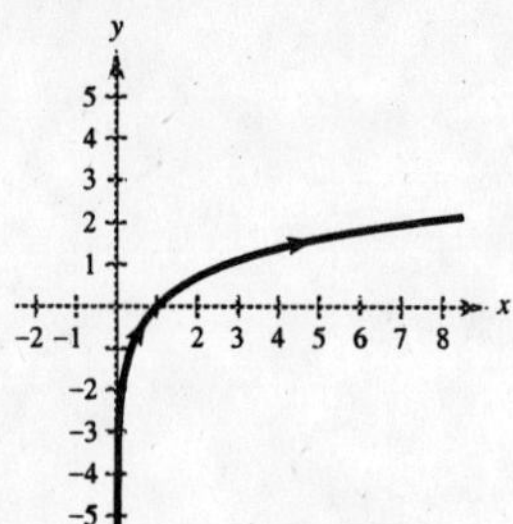

27. $x = 4 \sin 2\theta \implies \left(\dfrac{x}{4}\right)^2 = \sin^2 2\theta$

$y = 2 \cos 2\theta \implies \left(\dfrac{y}{2}\right)^2 = \cos^2 2\theta$

$\dfrac{x^2}{16} + \dfrac{y^2}{4} = \sin^2 2\theta + \cos^2 2\theta$

$\dfrac{x^2}{16} + \dfrac{y^2}{4} = 1$

Ellipse

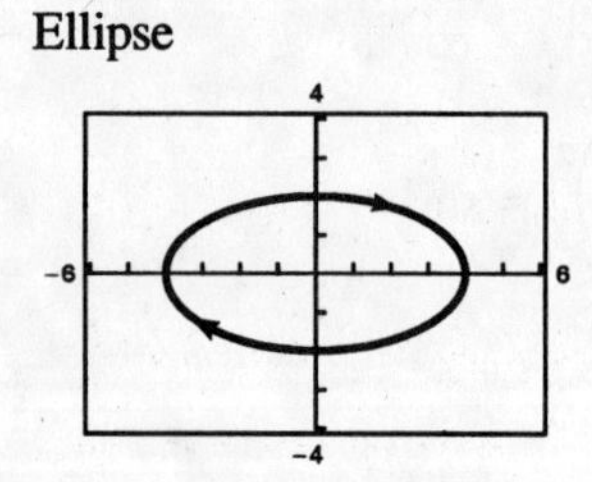

29. $x = 4 + 2 \cos \theta \implies \left(\dfrac{x-4}{2}\right)^2 = \cos^2 \theta$

$y = -1 + \sin \theta \qquad (y+1)^2 = \sin^2 \theta$

$\left(\dfrac{x-4}{2}\right)^2 + (y+1)^2 = \cos^2 \theta + \sin^2 \theta$

$\dfrac{(x-4)^2}{4} + (y+1)^2 = 1$

Ellipse

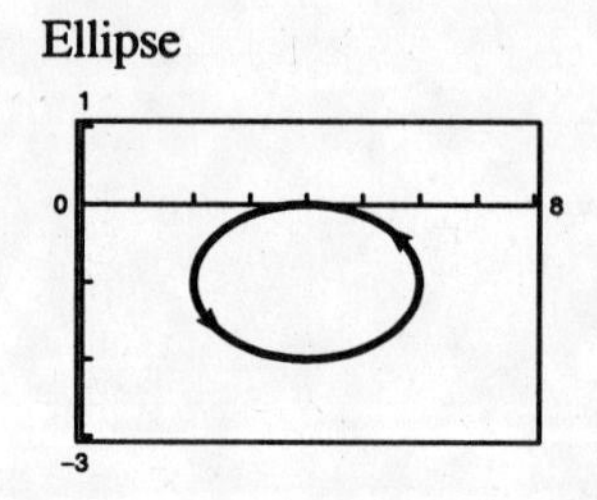

31. $x = 4 \sec \theta \implies \left(\dfrac{x}{4}\right)^2 = \sec^2 \theta$

$y = 3 \tan \theta \implies \left(\dfrac{y}{3}\right)^2 = \tan^2 \theta$

$\left(\dfrac{x}{4}\right)^2 - \left(\dfrac{y}{3}\right)^2 = \sec^2 \theta - \tan^2 \theta$

$\dfrac{x^2}{16} - \dfrac{y^2}{9} = 1$

Hyperbola

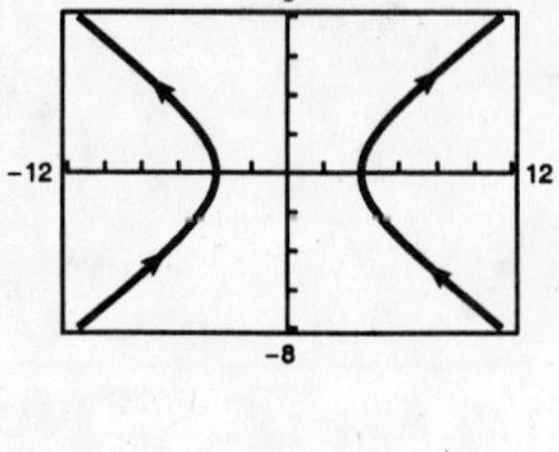

33. $x = \dfrac{t}{2}$

$y = \ln(t^2 + 1)$

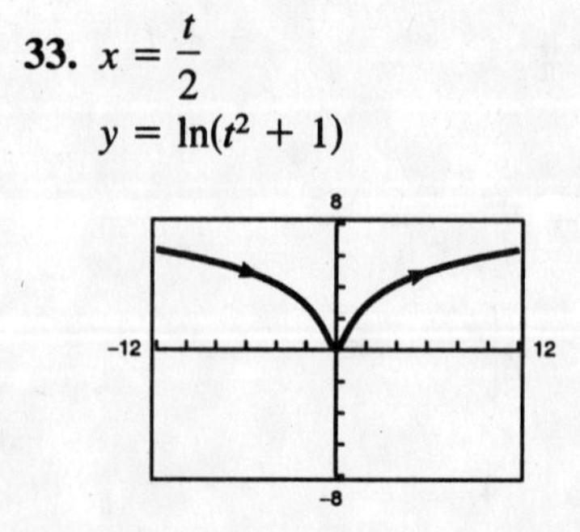

35. By eliminating the parameters in (a)–(d), we get $y = 2x + 1$. They differ from each other in restricted domain and in orientation.

(a) Domain: $-\infty < x < \infty$

Orientation: Left to right

(b) Domain: $-1 \le x \le 1$

Orientation: Depends on θ

(c) Domain: $0 < x < \infty$

Orientation: Right to left

(d) Domain: $0 < x < \infty$

Orientation: Left to right

37. $t = \dfrac{(x - x_1)}{(x_2 - x_1)}$

$y = y_1 + \left(\dfrac{x - x_1}{x_2 - x_1}\right)(y_2 - y_1)$

$\Rightarrow y - y_1 = \left(\dfrac{y_2 - y_1}{x_2 - x_1}\right)(x - x_1)$

39. $x = h + a \cos \theta$

$y = k + b \sin \theta$

$\dfrac{x - h}{a} = \cos \theta, \ \dfrac{y - k}{b} = \sin \theta$

$\dfrac{(x - h)^2}{a^2} + \dfrac{(y - k)^2}{b^2} = 1$

41. $x = x_1 + t(x_2 - x_1) = 0 + t(5 - 0) = 5t$

$y = y_1 + t(y_2 - y_1) = 0 + t(-2 - 0) = -2t$

(Solution not unique.)

43. From Exercise 38:

$x = 2 + 4 \cos \theta$

$y = 1 + 4 \sin \theta$

Solution not unique

45. From Exercise 39:

$a = 5, c = 4$, and hence, $b = 3$.

$x = 5 \cos \theta$

$y = 3 \sin \theta$

Center: $(0, 0)$

Solution not unique

47. $y = 3x - 2$

Examples:

$x = t \qquad x = 2t$

$y = 3t - 2 \qquad y = 6t - 2$

49. $x = 2(\theta - \sin \theta)$

$y = 2(1 - \cos \theta)$

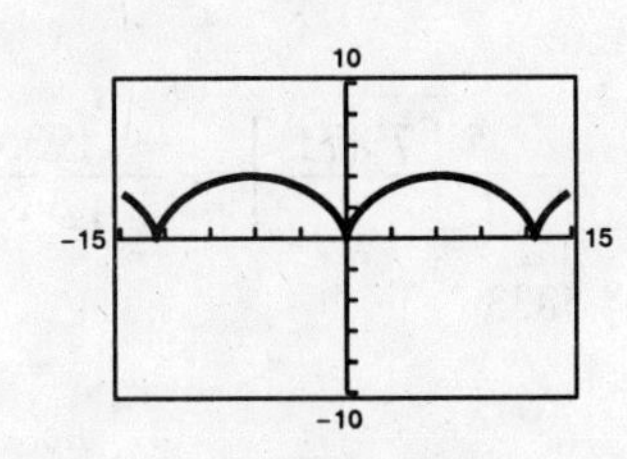

51. $x = 3 \cos^3 \theta, \ y = 3 \sin^3 \theta$

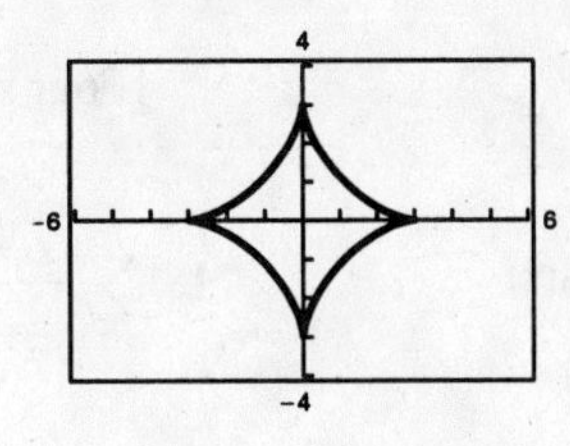

53. $x = 2 \cot \theta, \ y = 2 \sin^2 \theta$

4

-6 6

-4

55. Matches graph (b).

57. Matches graph (d).

59. $x = (v_0 \cos\theta)t,\ y = h + (v_0 \sin\theta)t - 16t^2$

(a) $100 \text{ miles/hour} = \dfrac{100 \text{ mi/hr} \cdot 5280 \text{ ft/mi}}{3600 \text{ sec/hr}}$

$= 146.67 \text{ ft/sec}$

$x = (146.67 \cos\theta)t$

$y = 3 + (146.67 \sin\theta)t - 16t^2$

(b) $\theta = 15°$

$x = (146.67 \cos 15°)t = 141.7t$

$y = 3 + (146.67 \sin 15°)t - 16t^2$

$= 3 + 38.0t - 16t^2$

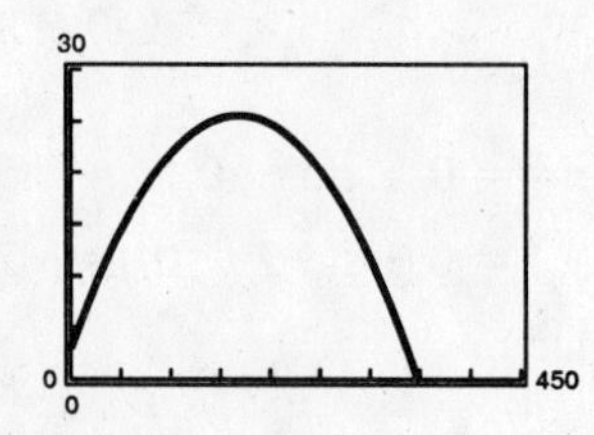

It is not a home run because $y < 10$ when $x = 400$.

(c) $\theta = 23°$

$x = (146.67 \cos 23°)t = 135.0t$

$y = 3 + (146.67 \sin 23°)t - 16t^2$

$= 3 + 57.3t - 16t^2$

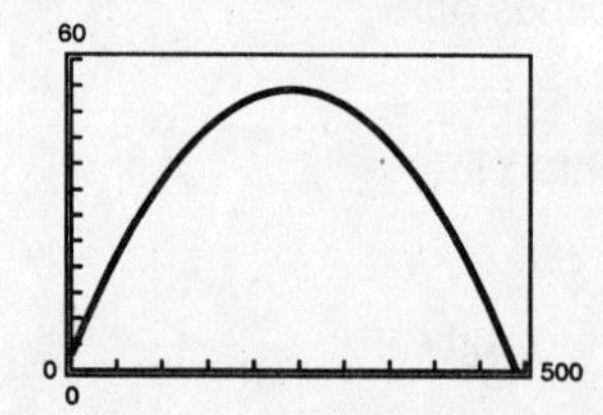

Yes, it is a home run because $y > 10$ when $x = 400$.

(d) $\theta = 19.4°$ is the minimum angle.

61. True

$x = t$ first set

$y = t^2 + 1 = x^2 + 1$

$x = 3t$ second set

$y = 9t^2 + 1 = (3t)^2 + 1 = x^2 + 1$

63. The graph is the same, but the orientation is reversed.

65. $x^2 - 6x + 4 = 0$

$$x = \frac{6 \pm \sqrt{36 - 16}}{2} = 3 \pm \sqrt{5}$$

67. $x^4 - 18x^2 + 18 = 0$

$$x^2 = \frac{18 \pm \sqrt{18^2 - 4(18)}}{2} = 9 \pm \sqrt{63}$$

$$x = \pm\sqrt{9 + \sqrt{63}}, \pm\sqrt{9 - \sqrt{63}}$$

[Four solutions: $x \approx \pm 4.1155, \pm 1.0309$]

69. $$\sum_{n=1}^{200} (n - 8) = \frac{200(201)}{2} - 8(200) = 18{,}500$$

71. $$\sum_{n=1}^{70} \frac{7 - 5n}{12} = \frac{7}{12}(70) - \frac{5}{12}\left[\frac{70(71)}{2}\right] = \frac{-11935}{12}$$

$$\approx -994.5833$$

73. $$\sum_{n=0}^{10} 10\left(\frac{2}{3}\right)^n = 10\frac{1 - \left(\frac{2}{3}\right)^{11}}{1 - \frac{2}{3}} = 30\left[1 - \left(\frac{2}{3}\right)^{11}\right] \approx 29.6532$$

Section 10.6 Polar Coordinates

- In polar coordinates you do not have unique representation of points. The point (r, θ) can be represented by $(r, \theta \pm 2n\pi)$ or by $(-r, \theta \pm (2n + 1)\pi)$ where n is any integer. The pole is represented by $(0, \theta)$ where θ is any angle.
- To convert from polar coordinates to rectangular coordinates, use the following relationships.
  $x = r\cos\theta$
  $y = r\sin\theta$
- To convert from rectangular coordinates to polar coordinates, use the following relationships.
  $r = \pm\sqrt{x^2 + y^2}$
  $\tan\theta = y/x$
  If θ is in the same quadrant as the point (x, y), then r is positive. If θ is in the opposite quadrant as the point (x, y), then r is negative.
- You should be able to convert rectangular equations to polar form and vice versa.

Solutions to Odd-Numbered Exercises

1. Polar coordinates: $\left(4, \dfrac{\pi}{2}\right)$

$x = 4\cos\left(\dfrac{\pi}{2}\right) = 0$

$y = 4\sin\left(\dfrac{\pi}{2}\right) = 4$

Rectangular coordinates: $(0, 4)$

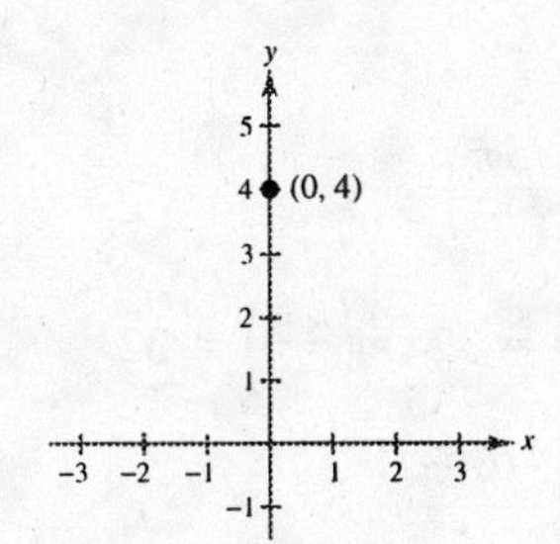

3. Polar coordinates: $\left(-1, \dfrac{5\pi}{4}\right)$

$x = -1\cos\left(\dfrac{5\pi}{4}\right) = \dfrac{\sqrt{2}}{2}$

$y = -1\sin\left(\dfrac{5\pi}{4}\right) = \dfrac{\sqrt{2}}{2}$

Rectangular coordinates: $\left(\dfrac{\sqrt{2}}{2}, \dfrac{\sqrt{2}}{2}\right)$

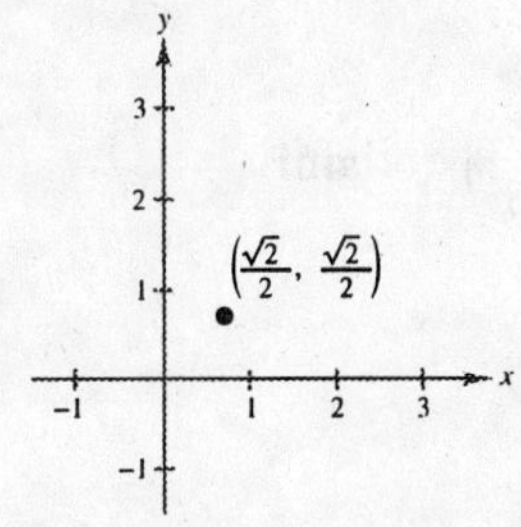

5.

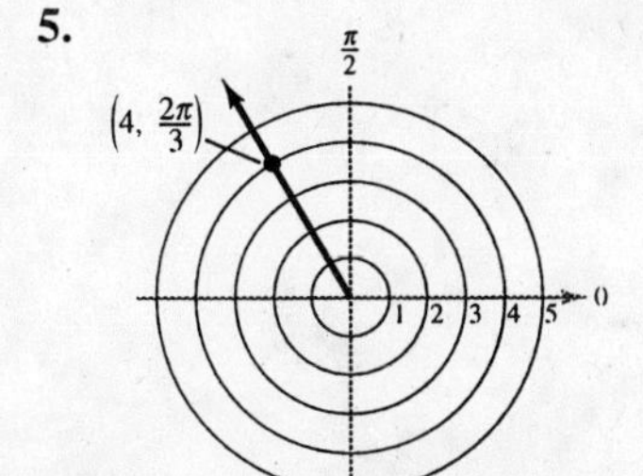

Three additional representations:

$\left(-4, \dfrac{5\pi}{3}\right), \left(4, -\dfrac{4\pi}{3}\right), \left(-4, -\dfrac{\pi}{3}\right)$

7.

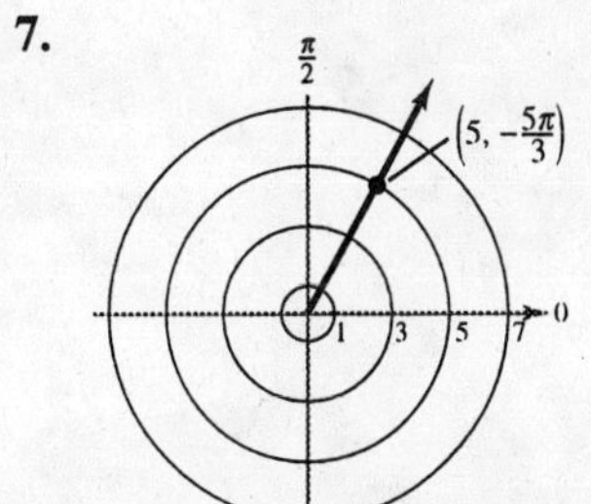

Three additional representations:

$\left(5, \dfrac{\pi}{3}\right), \left(-5, \dfrac{4\pi}{3}\right), \left(-5, -\dfrac{2\pi}{3}\right)$

9.

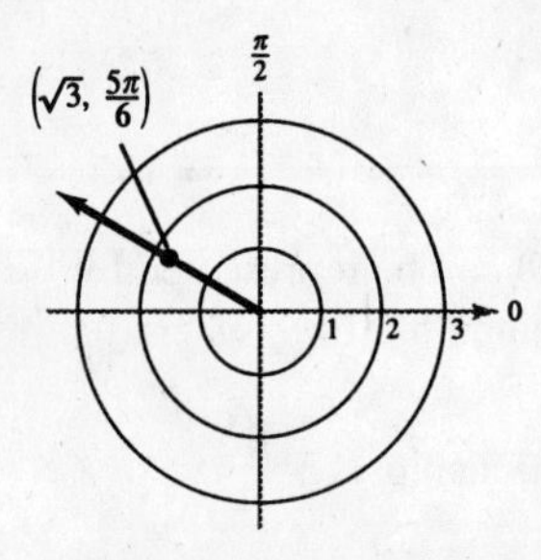

Three additional representations:

$$\left(\sqrt{3}, -\frac{7\pi}{6}\right), \left(-\sqrt{3}, -\frac{\pi}{6}\right), \left(-\sqrt{3}, \frac{11\pi}{6}\right)$$

11.

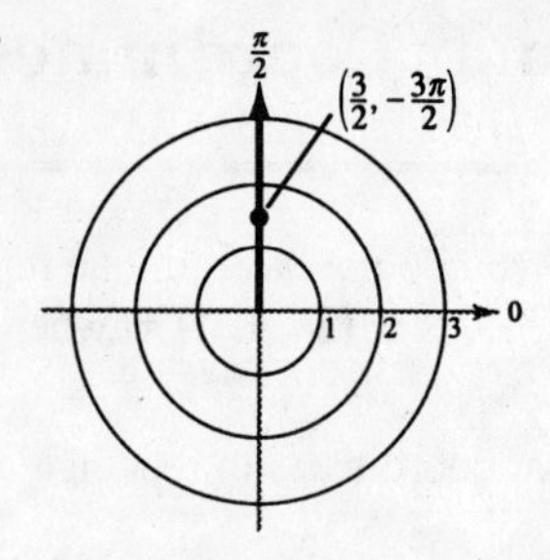

Three additional representations:

$$\left(\frac{3}{2}, \frac{\pi}{2}\right), \left(-\frac{3}{2}, \frac{3\pi}{2}\right), \left(-\frac{3}{2}, -\frac{\pi}{2}\right)$$

13. Polar coordinates: $\left(4, -\frac{\pi}{3}\right)$

$x = 4\cos\left(-\frac{\pi}{3}\right) = 2$

$y = 4\sin\left(-\frac{\pi}{3}\right) = -2\sqrt{3}$

Rectangular coordinates: $(2, -2\sqrt{3})$

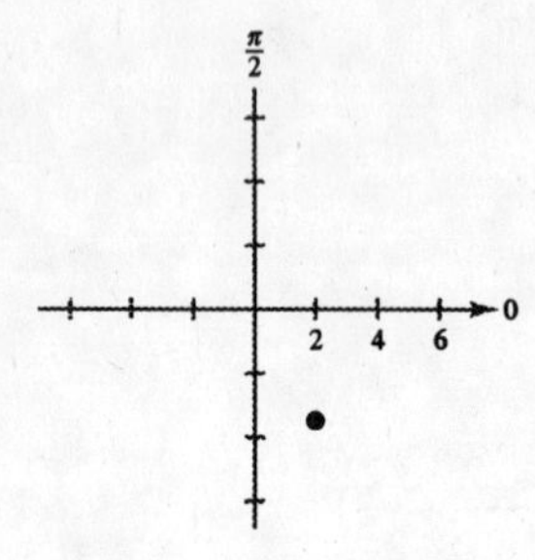

15. Polar coordinates: $\left(-1, \frac{-3\pi}{4}\right)$

$x = -1\cos\left(\frac{-3\pi}{4}\right) = \frac{\sqrt{2}}{2}$

$y = -1\sin\left(\frac{-3\pi}{4}\right) = \frac{\sqrt{2}}{2}$

Rectangular coordinates: $\left(\frac{\sqrt{2}}{2}, \frac{\sqrt{2}}{2}\right)$

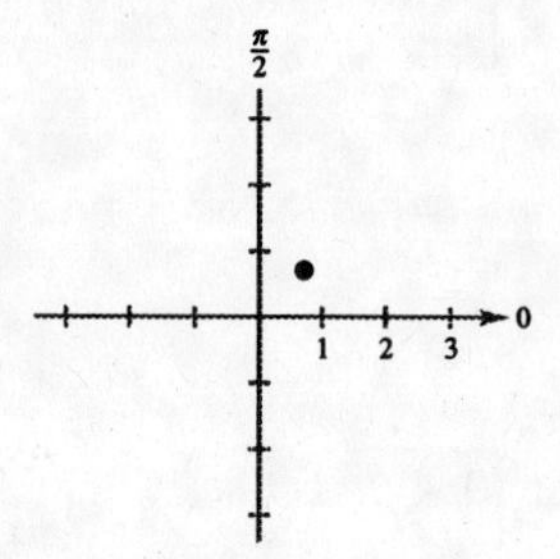

17. Polar coordinates: $\left(0, -\frac{7\pi}{6}\right)$ (origin!)

$x = 0\cos\left(-\frac{7\pi}{6}\right) = 0$

$y = 0\sin\left(-\frac{7\pi}{6}\right) = 0$

Rectangular coordinates: (0, 0)

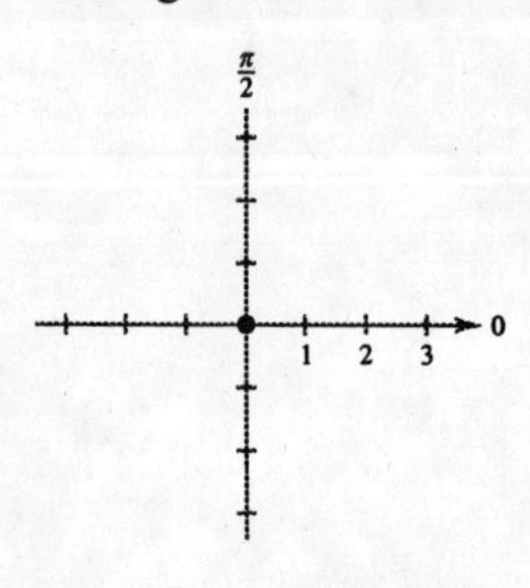

19. Polar coordinates: $\left(32, \frac{5\pi}{2}\right)$

$x = 32\cos\left(\frac{5\pi}{2}\right) = 0, y = 32\sin\left(\frac{5\pi}{2}\right) = 32$

Rectangular coordinates: (0, 32)

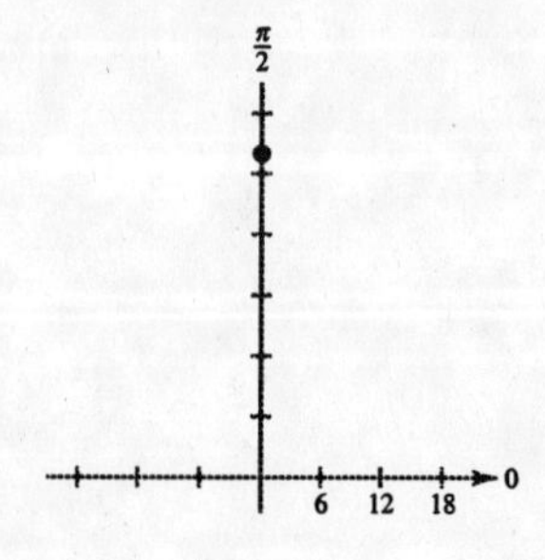

21. Polar coordinates: $(\sqrt{2}, 2.36)$

$x = \sqrt{2}\cos(2.36) \approx -1.004$

$y = \sqrt{2}\sin(2.36) \approx 0.996$

Rectangular coordinates: $(-1.004, 0.996)$

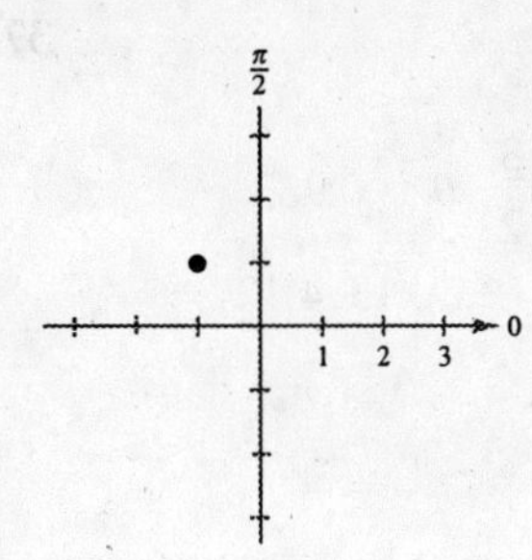

23. $(r, \theta) = \left(2, \frac{3\pi}{4}\right) \implies (x, y) = (-1.414, 1.414) = \left(-\sqrt{2}, \sqrt{2}\right)$

25. $(r, \theta) = (-4.5, 1.3) \implies (x, y) = (-1.204, -4.336)$

27. Rectangular coordinates: $(-7, 0)$

$r = 7,\ \tan\theta = 0,\ \theta = 0$

Polar coordinates: $(7, \pi),\ (-7, 0)$

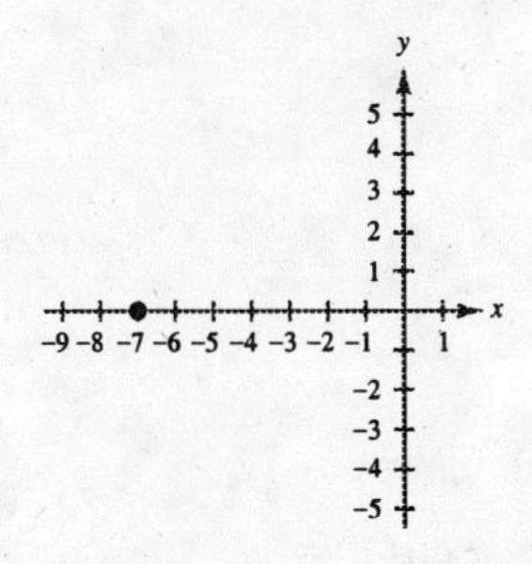

29. Rectangular coordinates: $(1, 1)$

$r = \sqrt{2},\ \tan\theta = 1,\ \theta = \frac{\pi}{4}$

Polar coordinates: $\left(\sqrt{2}, \frac{\pi}{4}\right), \left(-\sqrt{2}, \frac{5\pi}{4}\right)$

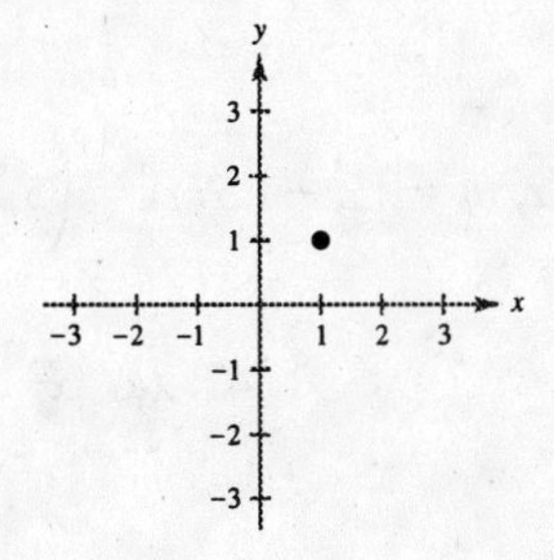

31. Rectangular coordinates: $(-3, 4)$

$r = \sqrt{9 + 16} = 5,\ \tan\theta = -\frac{4}{3},\ \theta \approx 2.214$

Polar coordinates: $(5, 2.214),\ (-5, 5.356)$

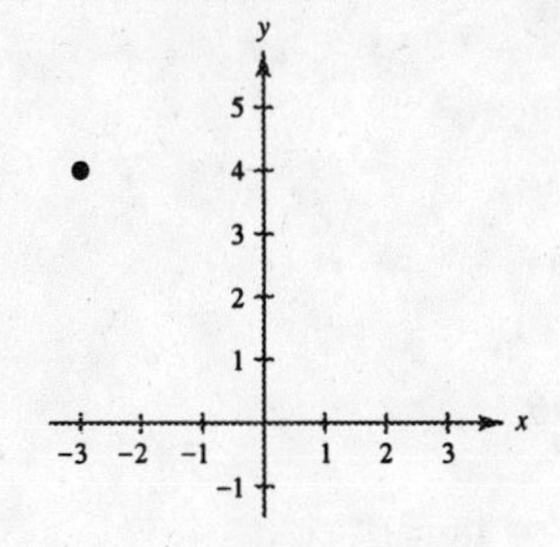

33. Rectangular coordinates: $\left(-\sqrt{3}, -\sqrt{3}\right)$

$r = \sqrt{3 + 3} = \sqrt{6},\ \tan\theta = 1,\ \theta = \frac{\pi}{4}$

Polar coordinates: $\left(\sqrt{6}, \frac{5\pi}{4}\right), \left(-\sqrt{6}, \frac{\pi}{4}\right)$

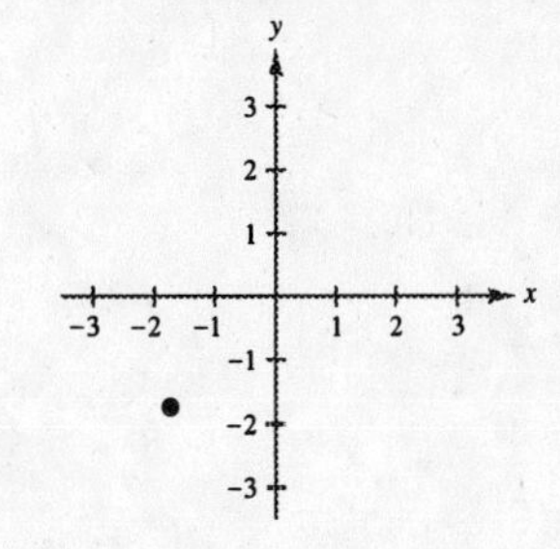

35. Rectangular coordinates: $(4, 6)$

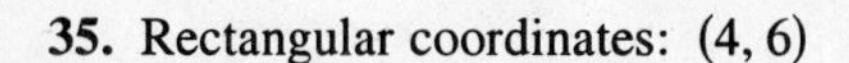

$r = \sqrt{16 + 36} = 2\sqrt{13}$, $\tan \theta = \dfrac{3}{2}$, $\theta \approx 0.983$

Polar coordinates: $(2\sqrt{13}, 0.983)$, $(-2\sqrt{13}, 4.124)$

37. $(x, y) = (3, -2) \implies r = \sqrt{3^2 + (-2)^2} = \sqrt{13}$

$$\theta = \arctan\left(-\tfrac{2}{3}\right) \approx -0.588$$

$(r, \theta) \approx (\sqrt{13}, -0.588)$

39. $(x, y) = (\sqrt{3}, 2) \implies r = \sqrt{3 + 2^2} = \sqrt{7}$

$$\theta = \arctan\left(\frac{2}{\sqrt{3}}\right) \approx 0.857$$

$(r, \theta) \approx (\sqrt{7}, 0.857)$

41. $(x, y) = \left(\dfrac{5}{2}, \dfrac{4}{3}\right) \implies r = \sqrt{\left(\dfrac{5}{2}\right)^2 + \left(\dfrac{4}{3}\right)^2} = \dfrac{17}{6}$

$$\theta = \arctan\left(\frac{4/3}{5/2}\right) \approx 0.490$$

$(r, \theta) \approx \left(\dfrac{17}{6}, 0.490\right)$

43. $(x, y) = (0, -5) \implies (r, \theta) = (5, -1.571) = \left(5, -\dfrac{\pi}{2}\right)$

45. (a) $x^2 + y^2 = 49$

$$r^2 = 49$$
$$r = 7$$

(b) $x^2 + y^2 = a^2 \quad (a > 0)$

$$r = a$$

47. (a) $x^2 + y^2 - 2ax = 0$

$$r^2 - 2a\cos\theta = 0$$
$$r(r - 2a\cos\theta) = 0$$
$$r = 2a\cos\theta$$

(b) $x^2 + y^2 - 2ay = 0$

$$r^2 - 2a\sin\theta = 0$$
$$r(r - 2a\sin\theta) = 0$$
$$r = 2a\sin\theta$$

49. (a) $x = 12$

$$r\cos\theta = 12$$
$$r = 12\sec\theta$$

(b) $x = a$

$$r\cos\theta = a$$
$$r = a\sec\theta$$

51. (a) $xy = 4$

$$(r\cos\theta)(r\sin\theta) = 4$$
$$r^2\cos\theta\sin\theta = 4$$
$$r^2(2\cos\theta\sin\theta) = 8$$
$$r^2\sin 2\theta = 8$$
$$r^2 = 8\csc 2\theta$$

(b) $2xy = 1$

$$2(r\cos\theta)(r\sin\theta) = 1$$
$$r^2(2\cos\theta\sin\theta) = 1$$
$$r^2\sin 2\theta = 1$$
$$r^2 = \csc 2\theta$$

53. (a) $y^2 = x^3$

$$(r\sin\theta)^2 = (r\cos\theta)^3$$
$$\sin^2\theta = r\cos^3\theta$$
$$r = \frac{\sin^2\theta}{\cos^3\theta} = \tan^2\theta\sec\theta$$

(b) $x^2 = y^3$

$$(r\cos\theta)^2 = (r\sin\theta)^3$$
$$\cos^2\theta = r\sin^3\theta$$
$$r = \frac{\cos^2\theta}{\sin^3\theta} = \cot^2\theta\csc\theta$$

55.
$$r = 4\sin\theta$$
$$r^2 = 4r\sin\theta$$
$$x^2 + y^2 = 4y$$
$$x^2 + y^2 - 4y = 0$$

57.
$$\theta = \frac{\pi}{6}$$
$$\tan\theta = \frac{\sqrt{3}}{3}$$
$$\frac{y}{x} = \frac{\sqrt{3}}{3}$$
$$y = \frac{\sqrt{3}}{3}x$$
$$\sqrt{3}x - 3y = 0$$

59.
$$r = 4$$
$$r^2 = 16$$
$$x^2 + y^2 = 16$$

61. $r = -3\csc\theta$

$r\sin\theta = -3$

$y = -3$

63.
$$r^2 = \cos\theta$$
$$r^3 = r\cos\theta$$
$$(x^2 + y^2)^{3/2} = x$$
$$x^2 + y^2 = x^{2/3}$$
$$(x^2 + y^2)^3 = x^2$$

65.
$$r = 2\sin 3\theta$$
$$r = 2(3\sin\theta - 4\sin^3\theta)$$
$$r^4 = 6r^3\sin\theta - 8r^3\sin^3\theta$$
$$(x^2 + y^2)^2 = 6(x^2 + y^2)y - 8y^3$$
$$(x^2 + y^2)^2 = 6x^2y - 2y^3$$

67.
$$r = \frac{1}{1 - \cos\theta}$$
$$r - r\cos\theta = 1$$
$$\sqrt{x^2 + y^2} - x = 1$$
$$x^2 + y^2 = 1 + 2x + x^2$$
$$y^2 = 2x + 1$$

69.
$$r = \frac{6}{2 - 3\sin\theta}$$
$$r(2 - 3\sin\theta) = 6$$
$$2r = 6 + 3r\sin\theta$$
$$2\left(\pm\sqrt{x^2 + y^2}\right) = 6 + 3y$$
$$4(x^2 + y^2) = (6 + 3y)^2$$
$$4x^2 + 4y^2 = 36 + 36y + 9y^2$$
$$4x^2 - 5y^2 - 36y - 36 = 0$$

71.
$$r = 3$$
$$r^2 = 9$$
$$x^2 + y^2 = 9$$

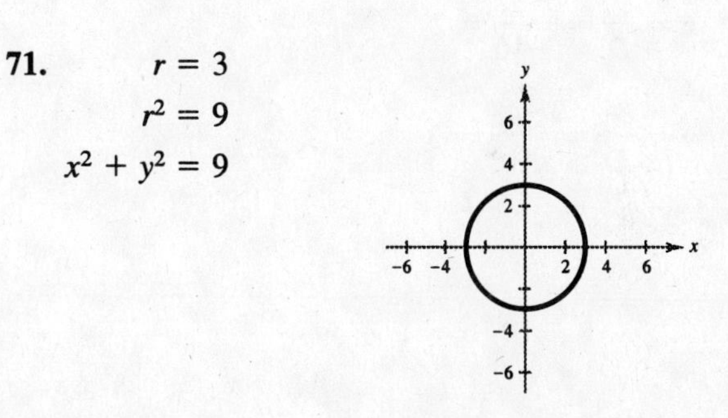

73.
$$\theta = \frac{\pi}{4}$$
$$\tan\theta = \tan\frac{\pi}{4}$$
$$\frac{y}{x} = 1$$
$$y = x$$
$$x - y = 0$$

75.
$$r = 3\sec\theta$$
$$r\cos\theta = 3$$
$$x = 3$$
$$x - 3 = 0$$

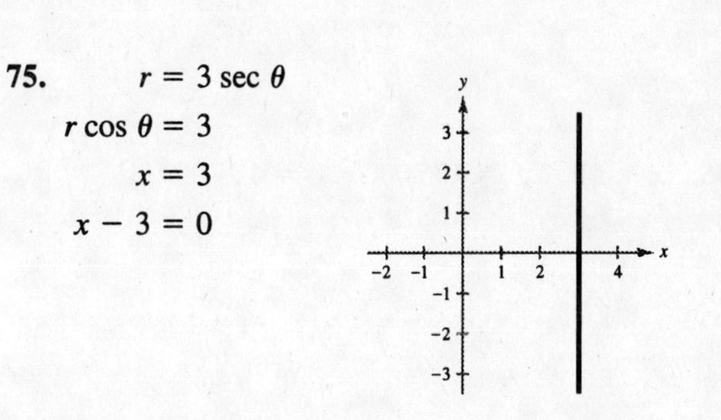

77. True, the distances from the origin are the same.

79. (a) $(r_1, \theta_1) = (x_1, y_1)$ where $x_1 = r_1 \cos \theta_1$ and $y_1 = r_1 \sin \theta_1$.

$(r_2, \theta_2) = (x_2, y_2)$ where $x_2 = r_2 \cos \theta_2$ and $y_2 = r_2 \sin \theta_2$.

Then $x_1^2 + y_1^2 = r_1^2 \cos^2 \theta_1 + r_1^2 \sin^2 \theta_1 = r_1^2$ and $x_2^2 + y_2^2 = r_2^2$. Thus,

$$\begin{aligned} d &= \sqrt{(x_1 - x_2)^2 + (y_1 - y_2)^2} \\ &= \sqrt{x_1^2 - 2x_1x_2 + x_2^2 + y_1^2 - 2y_1y_2 + y_2^2} \\ &= \sqrt{(x_1^2 + y_1^2) + (x_2^2 + y_2^2) - 2(x_1x_2 + y_1y_2)} \\ &= \sqrt{r_1^2 + r_2^2 - 2(r_1r_2 \cos \theta_1 \cos \theta_2 + r_1r_2 \sin \theta_1 \sin \theta_2)} \\ &= \sqrt{r_1^2 + r_2^2 - 2r_1r_2 \cos(\theta_1 - \theta_2)}. \end{aligned}$$

(b) If $\theta_1 = \theta_2$, the points are on the same line through the origin. In this case,

$$d = \sqrt{r_1^2 + r_2^2 - 2r_1r_2 \cos(0)} = \sqrt{(r_1 - r_2)^2} = |r_1 - r_2|.$$

(c) If $\theta_1 - \theta_2 = 90°$, $d = \sqrt{r_1^2 + r_2^2}$, the Pythagorean Theorem.

(d) For instance, $\left(3, \frac{\pi}{6}\right), \left(4, \frac{\pi}{3}\right)$ gives $d \approx 2.053$ and $\left(-3, \frac{7\pi}{6}\right), \left(-4, \frac{4\pi}{3}\right)$ gives $d \approx 2.053$. (same!)

81. $D = \begin{vmatrix} 5 & -7 \\ -3 & 1 \end{vmatrix} = 5 - 21 = -16$

$D_x = \begin{vmatrix} -11 & -7 \\ -3 & 1 \end{vmatrix} = -11 - 21 = -32$

$D_y = \begin{vmatrix} 5 & -11 \\ -3 & -3 \end{vmatrix} = -15 - 33 = -48$

$x = \dfrac{D_x}{D} = \dfrac{-32}{-16} = 2$

$y = \dfrac{D_y}{D} = \dfrac{-48}{-16} = 3$

83. $D = \begin{vmatrix} 3 & -2 & 1 \\ 2 & 1 & -3 \\ 1 & -3 & 9 \end{vmatrix} = 35$

$D_x = \begin{vmatrix} 0 & -2 & 1 \\ 0 & 1 & -3 \\ 8 & -3 & 9 \end{vmatrix} = 40$

$D_y = \begin{vmatrix} 3 & 0 & 1 \\ 2 & 0 & -3 \\ 1 & 8 & 9 \end{vmatrix} = 88$

$D_z = \begin{vmatrix} 3 & -2 & 0 \\ 2 & 1 & 0 \\ 1 & -3 & 8 \end{vmatrix} = 56$

$x = \dfrac{D_x}{D} = \dfrac{40}{35} = \dfrac{8}{7}$

$y = \dfrac{D_y}{D} = \dfrac{88}{35}$

$z = \dfrac{D_z}{D} = \dfrac{56}{35} = \dfrac{8}{5}$

85. $(x + 5)^8$. $ax^3 = 175{,}000x^3$. $a = 175{,}000$

87. $(2x - y)^{12}$. $ax^7y^5 = -101{,}376x^7y^5$. $a = -101{,}376$

Section 10.7 Graphs of Polar Equations

- When graphing polar equations:
  1. Test for symmetry
     (a) $\theta = \pi/2$: Replace (r, θ) by $(r, \pi - \theta)$ or $(-r, -\theta)$.
     (b) Polar axis: Replace (r, θ) by $(r, -\theta)$ or $(-r, \pi - \theta)$.
     (c) Pole: Replace (r, θ) by $(r, \pi + \theta)$ or $(-r, \theta)$.
     (d) $r = f(\sin\theta)$ is symmetric with respect to the line $\theta = \pi/2$.
     (e) $r = f(\cos\theta)$ is symmetric with respect to the polar axis.
  2. Find the θ values for which $|r|$ is maximum.
  3. Find the θ values for which $r = 0$.
  4. Know the different types of polar graphs.
     (a) Limaçons
     $r = a \pm b\cos\theta$
     $r = a \pm b\sin\theta$
     (b) Rose Curves, $n \geq 2$
     $r = a\cos n\theta$
     $r = a\sin n\theta$
     (c) Circles
     $r = a\cos\theta$
     $r = a\sin\theta$
     $r = a$
     (d) Lemniscates
     $r^2 = a^2\cos 2\theta$
     $r^2 = a^2\sin 2\theta$
- You should be able to graph polar equations of the form $r = f(\theta)$ with your graphing utility. If your utility does not have a polar mode, use
  $x = f(t)\cos t$
  $y = f(t)\sin t$
  in parametric mode.

Solutions to Odd-Numbered Exercises

1. $r = 3\cos 2\theta$ is a rose curve.

3. $r = 3\cos\theta$ is a circle.

5. $r = 6\sin 2\theta$ is a rose curve.

7. $r = 10 + 4\cos\theta$

$\theta = \frac{\pi}{2}$: $-r = 10 + 4\cos(-\theta)$

$-r = 10 + 4\cos\theta$ Not an equivalent equation

$r = 10 + 4\cos(\pi - \theta)$

$r = 10 + 4(\cos\pi\cos\theta + \sin\pi\sin\theta)$

$r = 10 - 4\cos\theta$ Not an equivalent equation

Polar axis: $r = 10 + 4\cos(-\theta)$

$r = 10 + 4\cos\theta$ Equivalent equation

Pole: $-r = 10 + 4\cos\theta$ Not an equivalent equation

$r = 10 + 4\cos(\pi + \theta)$

$r = 10 + 4(\cos\pi\cos\theta - \sin\pi\sin\theta)$

$r = 10 - 4\cos\theta$ Not an equivalent equation

Answer: Symmetric with respect to polar axis.

9. $r = \dfrac{6}{1 + \sin\theta}$

$\theta = \dfrac{\pi}{2}$: $r = \dfrac{6}{1 + \sin(\pi - \theta)}$

$r = \dfrac{6}{1 + \sin\pi\cos\theta - \cos\pi\sin\theta}$

$r = \dfrac{6}{1 + \sin\theta}$

Equivalent equation

Polar axis: $r = \dfrac{6}{1 + \sin(-\theta)}$

$r = \dfrac{6}{1 - \sin\theta}$

Not an equivalent equation

$-r = \dfrac{6}{1 + \sin(\pi - \theta)}$

$-r = \dfrac{6}{1 + \sin\theta}$

Not an equivalent equation

The pole: $-r = \dfrac{6}{1 + \sin\theta}$

Not an equivalent equation

$r = \dfrac{6}{1 + \sin(\pi + \theta)}$

$r = \dfrac{6}{1 - \sin\theta}$

Not an equivalent equation

Answer: Symmetric with respect to $\theta = \dfrac{\pi}{2}$.

11. $r = 6\sin\theta$

$\theta = \dfrac{\pi}{2}$: $-r = 6\sin(-\theta)$

$r = 6\sin\theta$

Equivalent equation

Polar axis: $r = 6\sin(-\theta)$

$r = -6\sin\theta$

Not an equivalent equation

$-r = 6\sin(\pi - \theta)$

$-r = 6(\sin\pi\cos\theta - \cos\pi\sin\theta)$

$-r = 6\sin\theta$

Not an equivalent equation

Pole: $-r = 6\sin\theta$

Not an equivalent equation

$r = 6\sin(\pi + \theta)$

$r = -6\sin\theta$

Not an equivalent equation

Answer: Symmetric with respect to $\theta = \dfrac{\pi}{2}$.

13. $r = 4\sec\theta\csc\theta$

$\theta = \dfrac{\pi}{2}$: $-r = 4\sec(-\theta)\csc(-\theta)$

$-r = -4\sec\theta\csc\theta$

$r = 4\sec\theta\csc\theta$

Equivalent equation

Polar axis: $-r = 4\sec(\pi - \theta)\csc(\pi - \theta)$

$-r = 4(-\sec\theta)\csc\theta$

$r = 4\sec\theta\csc\theta$

Equivalent equation

Pole: $r = 4\sec(\pi + \theta)\csc(\pi + \theta)$

$r = 4(-\sec\theta)(-\csc\theta)$

$r = 4\sec\theta\csc\theta$

Equivalent equation

Answer: Symmetric with respect to $\theta = \pi/2$, pole axis, and pole

15. $r^2 = 25 \sin 2\theta$

$\theta = \frac{\pi}{2}$: $\quad (-r)^2 = 25 \sin(2(-\theta))$

$r^2 = -25 \sin 2\theta$ Not an equivalent equation

$r^2 = 25 \sin(2(\pi - \theta))$

$r^2 = 25 \sin(2\pi - 2\theta)$

$r^2 = 25(\sin 2\pi \cos 2\theta - \cos 2\pi \sin 2\theta)$

$r^2 = -25 \sin 2\theta$ Not an equivalent equation

Polar axis: $r^2 = 25 \sin(2(-\theta))$

$r^2 = -25 \sin 2\theta$ Not an equivalent equation

$(-r)^2 = 25 \sin(2(\pi - \theta))$

$r^2 = -25 \sin 2\theta$ Not an equivalent equation

Pole: $\quad (-r)^2 = 25 \sin(2\theta)$

$r^2 = 25 \sin 2\theta$ Equivalent equation

Answer: Symmetric with respect to pole.

17. $|r| = |10(1 - \sin \theta)|$

$= 10|1 - \sin \theta| \leq 10(2) = 20$

$|1 - \sin \theta| = 2$

$1 - \sin \theta = 2$ or $1 - \sin \theta = -2$

$\sin \theta = -1 \qquad \sin \theta = 3$

$\theta = \frac{3\pi}{2} \qquad$ Not possible

Maximum: $|r| = 20$ when $\theta = \frac{3\pi}{2}$.

$r = 0$ when $1 - \sin \theta = 0$

$\sin \theta = 1$

$\theta = \frac{\pi}{2}$.

19. $|r| = |4 \cos 3\theta| = 4\,|\cos 3\theta| \leq 4$

$|\cos 3\theta| = 1$

$\cos 3\theta = \pm 1$

$\theta = 0, \frac{\pi}{3}, \frac{2\pi}{3}, \pi$

Maximum: $|r| = 4$ when $\theta = 0, \frac{\pi}{3}, \frac{2\pi}{3}, \pi$

$r = 0$ when $\cos 3\theta = 0$

$\theta = \frac{\pi}{6}, \frac{\pi}{2}, \frac{5\pi}{6}$.

21. Circle: $r = 5$

23. $\theta = \frac{\pi}{6}$ is a line.

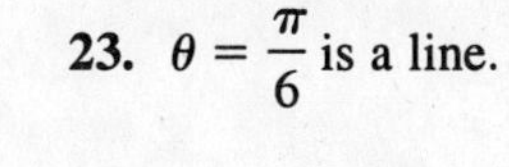

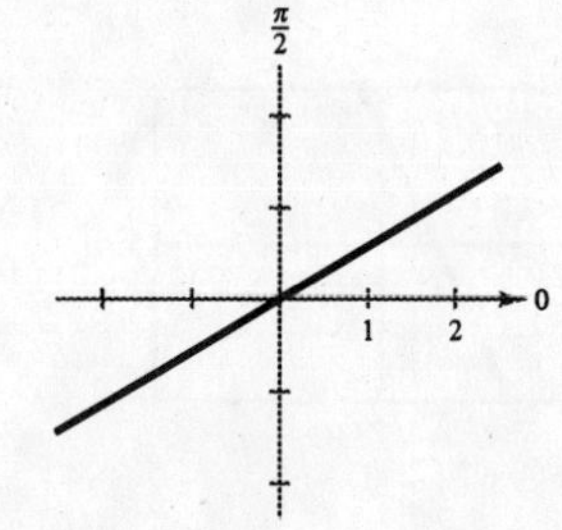

25. $r = 3 \sin \theta$

Symmetric with respect to $\theta = \frac{\pi}{2}$

Circle with radius of $\frac{3}{2}$

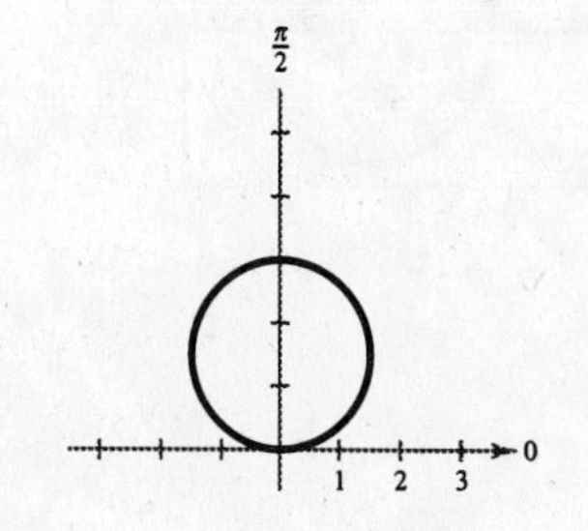

27. $r = 3(1 - \cos\theta)$

Cardioid

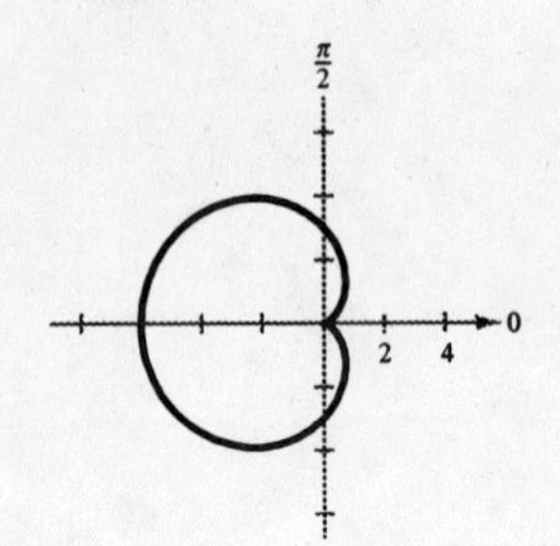

29. $r = 3 - 4\cos\theta$

Limaçon

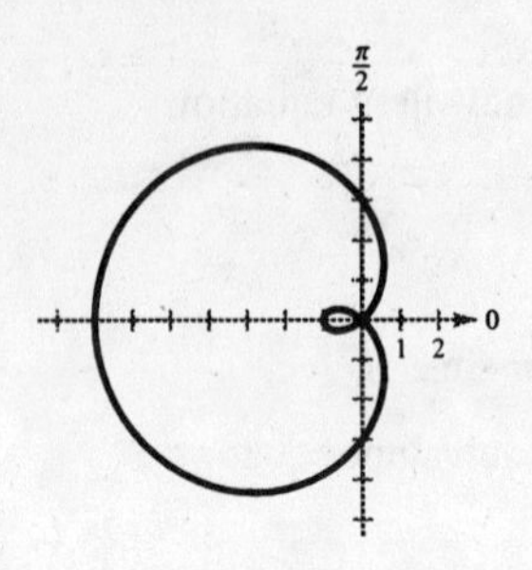

31. $r = 6 + \sin\theta$

Convex limaçon

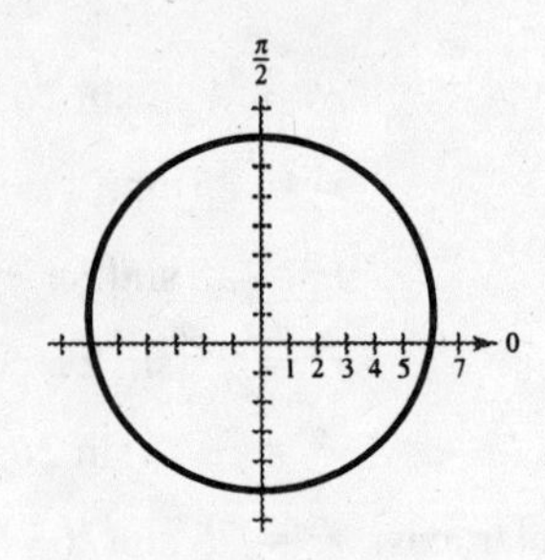

33. $r = 5\cos 3\theta$

Rose curve

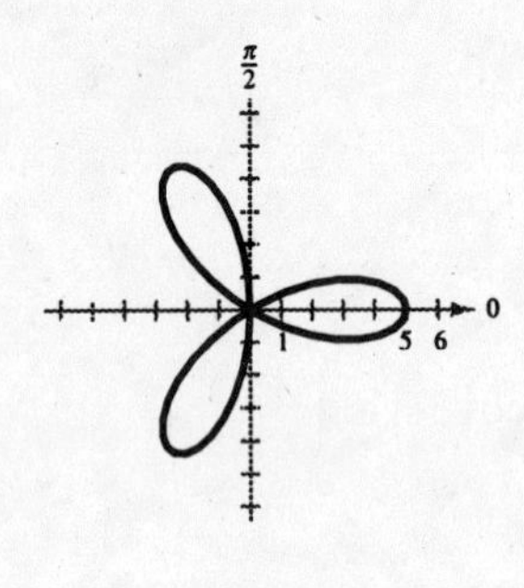

35. $r = 7\sin 2\theta$

Rose curve

4 petals

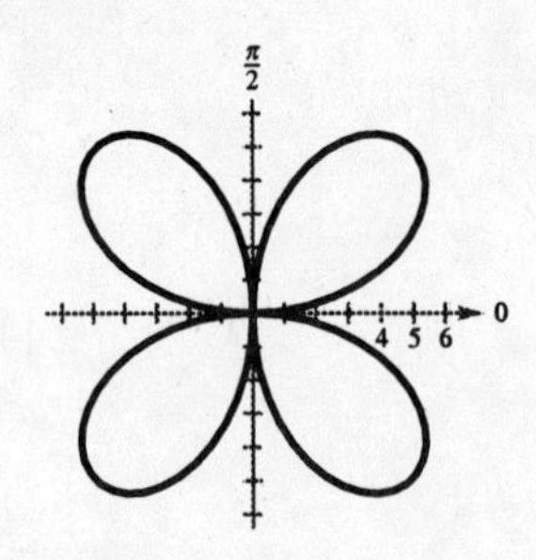

37. $r = \dfrac{\theta}{2}$

Symmetric with respect to $\theta = \dfrac{\pi}{2}$

Spiral

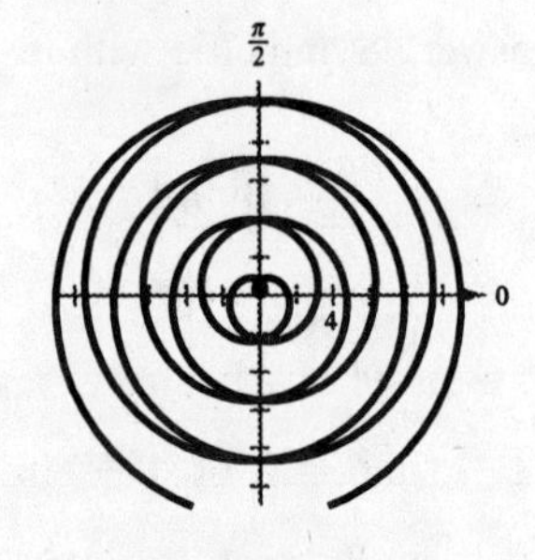

39. $-10\pi \le \theta \le 10\pi$

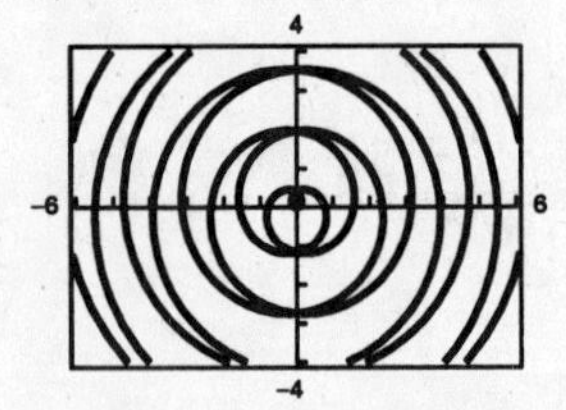

41. $0 \le \theta \le \pi$

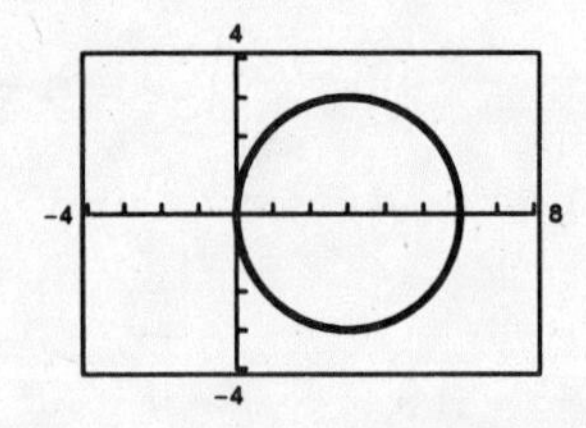

43. $0 \le \theta \le 2\pi$

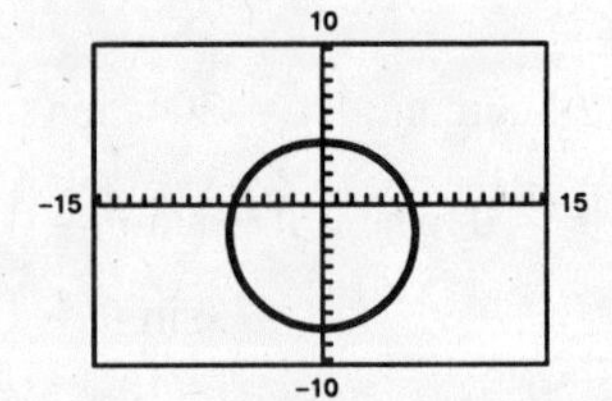

45. $0 \le \theta \le 2\pi$

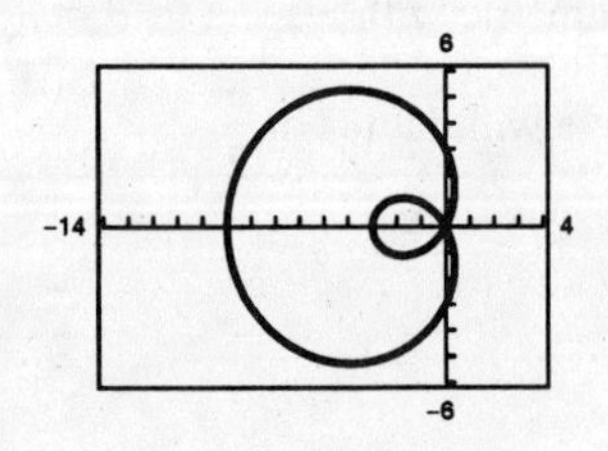

47. $0 \le \theta \le \dfrac{\pi}{2}$

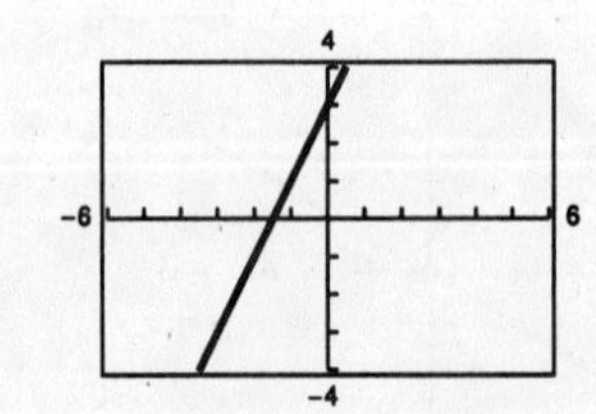

49. $-2\pi \le \theta \le 2\pi$

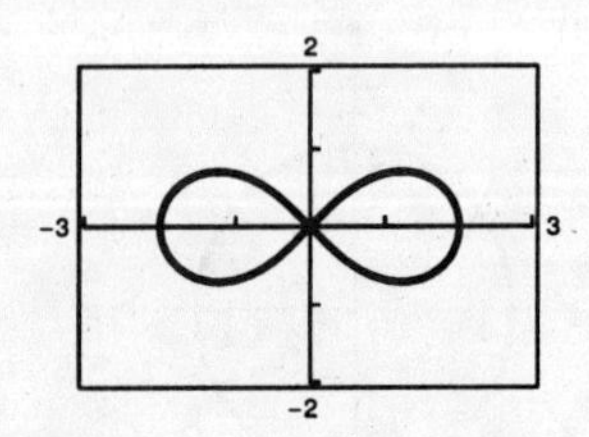

51. $0 \leq \theta \leq \pi$

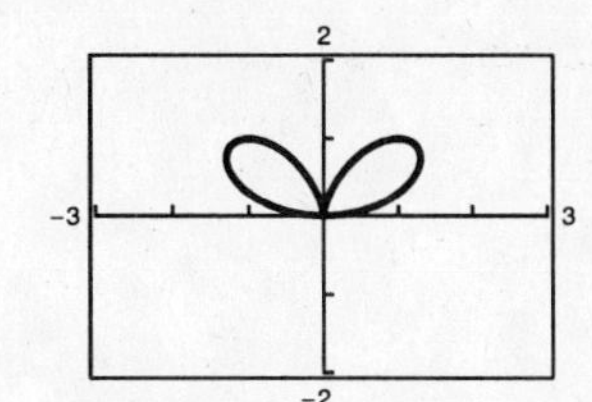

53. $0 \leq \theta < 2\pi$

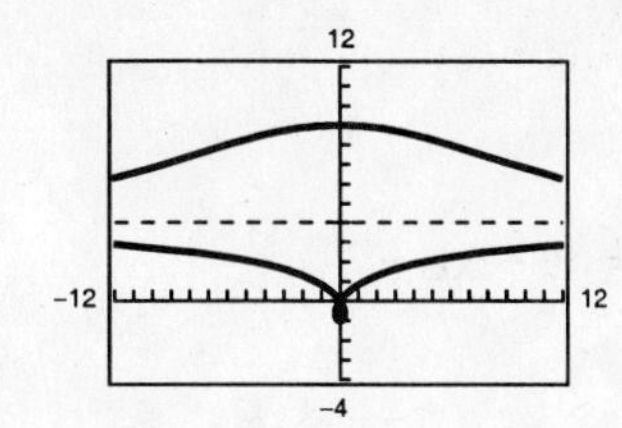

55. $r = 3 - 2\cos\theta,\ 0 \leq \theta < 2\pi$

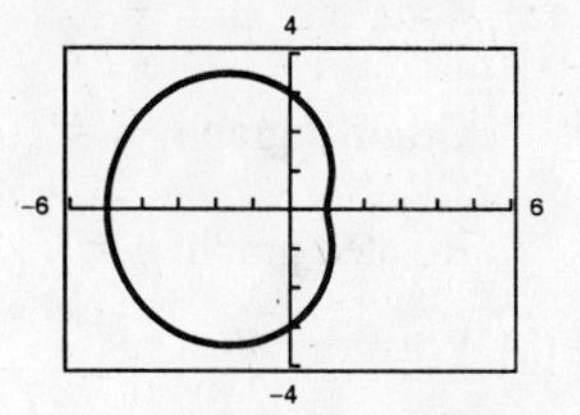

57. $r = 2 + \sin\theta,\ 0 \leq \theta < 2\pi$

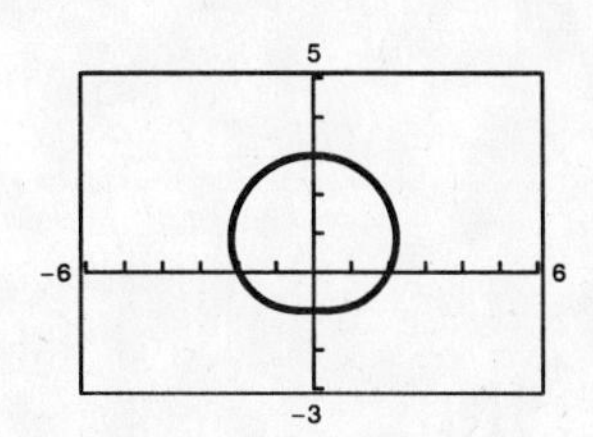

59. $r = 2\cos\left(\dfrac{3\theta}{2}\right),\ 0 \leq \theta < 4\pi$

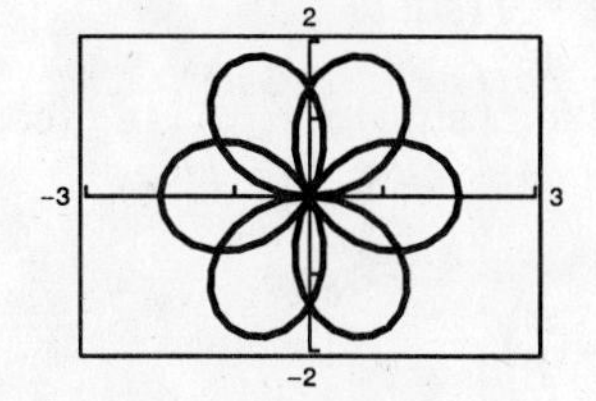

61. $r^2 = 4\sin 2\theta,\ 0 \leq \theta < \dfrac{\pi}{2}$

$\left(\text{Use } r_1 = \sqrt{4\sin 2\theta} \text{ and } r_2 = -\sqrt{4\sin 2\theta}.\right)$

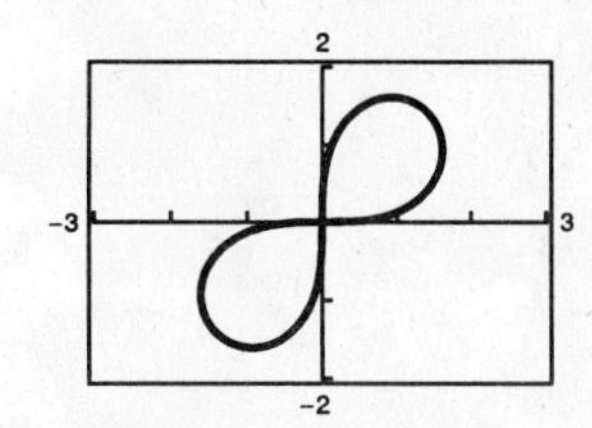

63. $r = 2 - \sec\theta$

$x = -1$ is an asymptote.

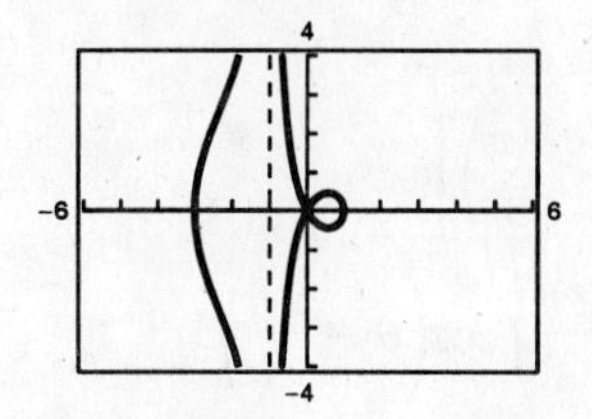

65. $r = \dfrac{2}{\theta}$

$y = 2$ is an asymptote.

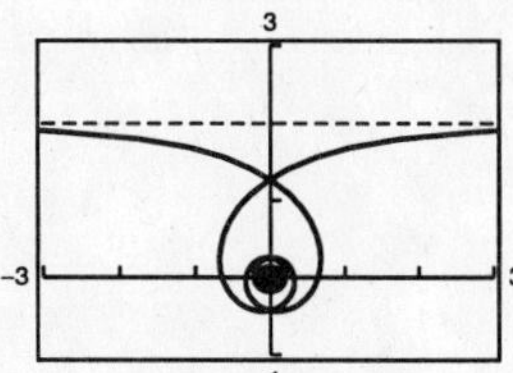

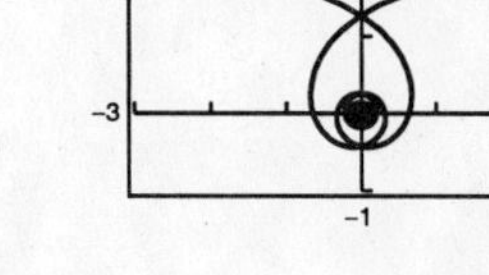

67. False. If $\theta = \dfrac{11\pi}{6}$, $r = 4$

69. False. It has 5 petals

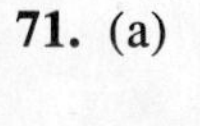

71. (a)

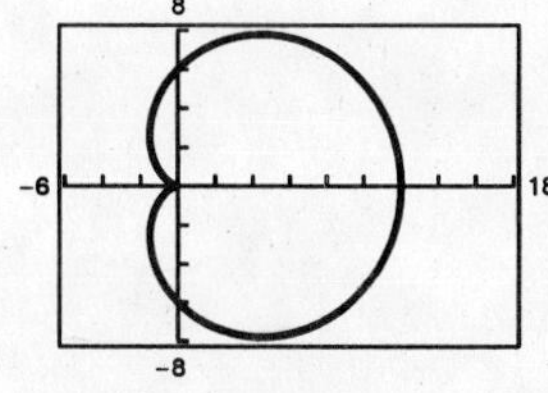

(b)

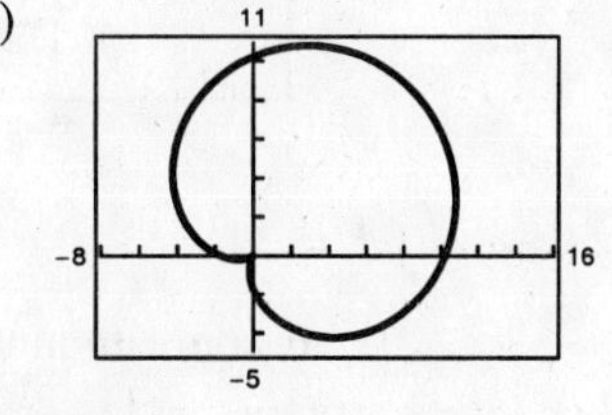

(c)

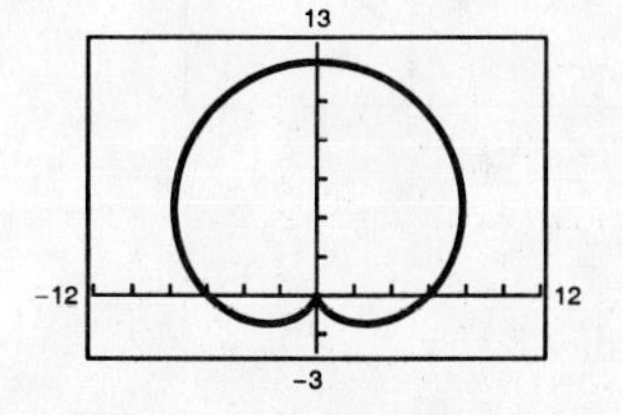

The angle ϕ rotates the graph around the pole. In part (c), $r = 6\left[1 + \cos\left(\theta - \dfrac{\pi}{2}\right)\right] = 6[1 + \sin\theta]$.

73. Use the result of Exercise 72.

(a) Rotation: $\phi = \dfrac{\pi}{2}$

Original graph: $r = f(\sin\theta)$

Rotated graph: $r = f\left(\sin\left(\theta - \dfrac{\pi}{2}\right)\right) = f(-\cos\theta)$

(b) Rotation: $\phi = \pi$

Original graph: $r = f(\sin\theta)$

Rotated graph: $r = f(\sin(\theta - \pi)) = f(-\sin\theta)$

(c) Rotation: $\phi = \pi$

Original graph: $r = f(\sin\theta)$

Rotated graph: $r = f\left(\sin\left(\theta - \dfrac{3\pi}{2}\right)\right) = f(\cos\theta)$

75. (a) $r = 2\sin\left[2\left(\theta - \dfrac{\pi}{6}\right)\right]$

$= 4\sin\left(\theta - \dfrac{\pi}{6}\right)\cos\left(\theta - \dfrac{\pi}{6}\right)$

$= 2\sin\left(2\theta - \dfrac{\pi}{3}\right)$

$= \sin 2\theta - \sqrt{3}\cos 2\theta$

(b) $r = 2\sin\left[2\left(\theta - \dfrac{\pi}{2}\right)\right]$

$= 2\sin(2\theta - \pi)$

$= -2\sin 2\theta$

$= -4\sin\theta\cos\theta$

(c) $r = 2\sin\left[2\left(\theta - \dfrac{2\pi}{3}\right)\right]$

$= 4\sin\left(\theta - \dfrac{2\pi}{3}\right)\cos\left(\theta - \dfrac{2\pi}{3}\right)$

$= 2\sin\left(2\theta - \dfrac{4\pi}{3}\right)$

$= \sqrt{3}\cos 2\theta - \sin 2\theta$

(d) $r = 2\sin[2(\theta - \pi)]$

$= 2\sin(2\theta - 2\pi)$

$= 2\sin 2\theta$

$= 4\sin\theta\cos\theta$

77. $r = 2 + k\cos\theta$

$k = 0$

Circle

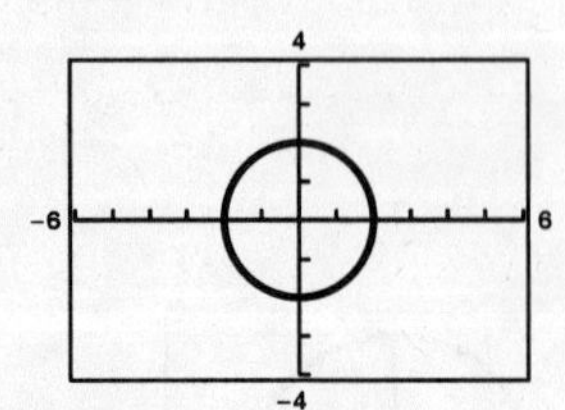

$k = 1$

Convex limaçon

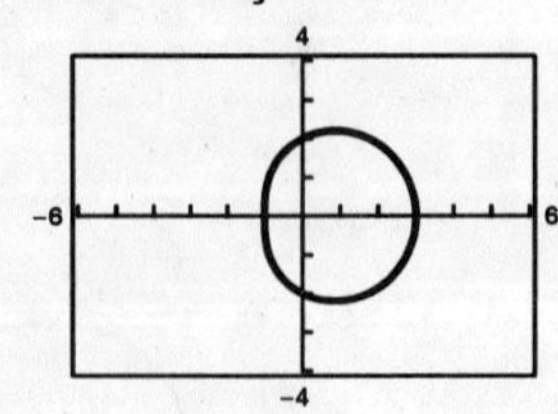

$k = 2$

Cardioid

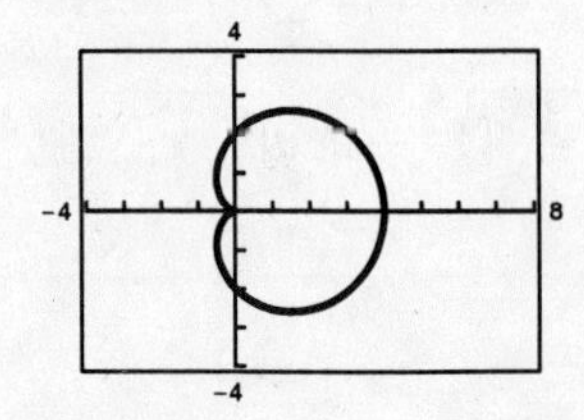

$k = 3$

Limaçon with inner loop

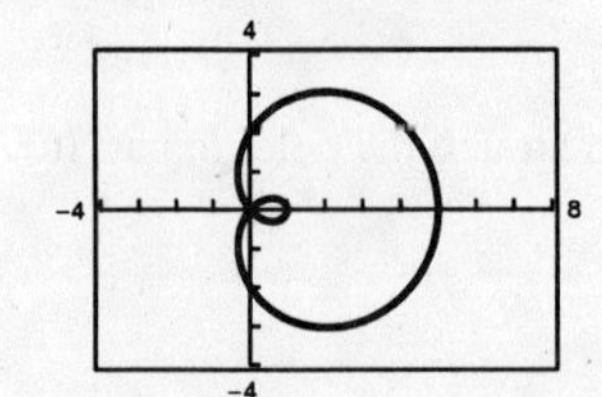

79. $a_1 = 2, a_8 = 23$

$a_8 = a_1 + (8 - 1)d$

$23 = 2 + 7d \Rightarrow d = 3$

$a_1 = 2, a_2 = 5, a_3 = 8, a_4 = 11, a_5 = 14$

$a_n = 2 + (n - 1)3 = 3n - 1$

81. $a_1 = 150, a_{k+1} = a_k - 18$

$a_2 = 150 - 18 = 132$

$a_3 = 132 - 18 = 114$

$a_4 = 114 - 18 = 96$

$a_5 = 96 - 18 = 78$

$d = -18, a_n = 150 + (n - 1)(-18) = 168 - 18n$

83. $\sum_{n=1}^{20} 4n = 4\frac{20(21)}{2} = 840$

85. $\sum_{n=1}^{120} (n + 5) = \frac{120(121)}{2} + 5(120)$

$= 7860$

87. $\sum_{n=0}^{\infty} \left(\frac{1}{8}\right)^n = \frac{1}{1 - \frac{1}{8}} = \frac{8}{7}$

Section 10.8 Polar Equations of Conics

- The graph of a polar equation of the form

  $$r = \frac{ep}{1 \pm e\cos\theta} \quad \text{or} \quad r = \frac{ep}{1 \pm e\sin\theta}$$

  is a conic, where $e > 0$ is the eccentricity and $|p|$ is the distance between the focus (pole) and the directrix.

  (a) If $e < 1$, the graph is an ellipse.

  (b) If $e = 1$, the graph is a parabola.

  (c) If $e > 1$, the graph is a hyperbola.

- Guidelines for finding polar equations of conics:

  (a) Horizontal directrix above the pole: $r = \frac{ep}{1 + e\sin\theta}$

  (b) Horizontal directrix below the pole: $r = \frac{ep}{1 - e\sin\theta}$

  (c) Vertical directrix to the right of the pole: $r = \frac{ep}{1 + e\cos\theta}$

  (d) Vertical directrix to the left of the pole: $r = \frac{ep}{1 - e\cos\theta}$

Solutions to Odd-Numbered Exercises

1.

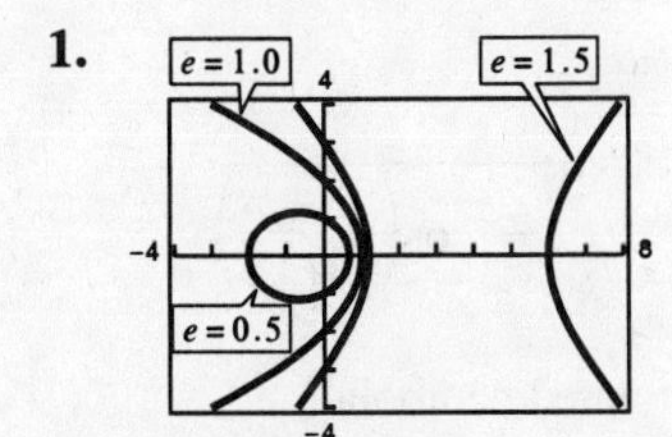

3.

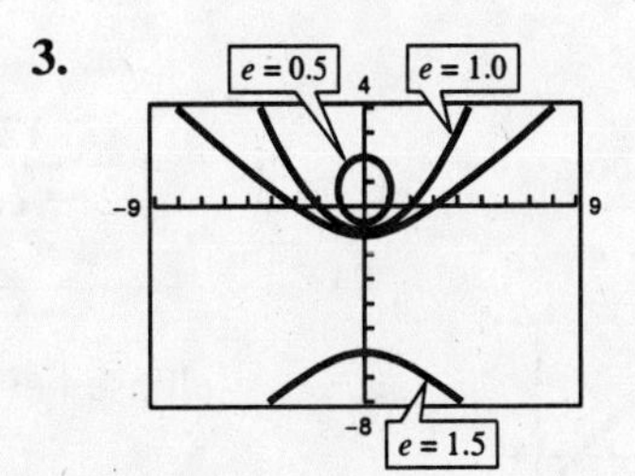

5. Matches (b).
(Parabola $e = 1$)

7. Matches (d).
(Hyperbola $e = 2$)

9. $r = \dfrac{6}{1 - \cos\theta}$

$e = 1$ so the graph is a parabola.

Vertex: $(r, \theta) = (3, \pi)$

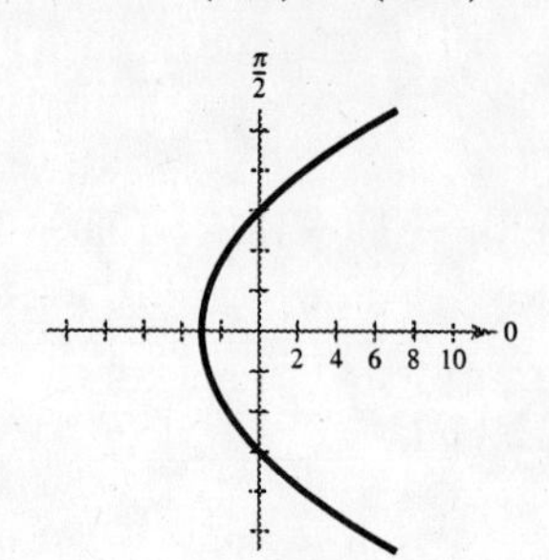

11. $r = \dfrac{4}{4 - \cos\theta} = \dfrac{1}{1 - \frac{1}{4}\cos\theta}$

$e = \dfrac{1}{4}, p = 4$

Ellipse

Vertices:

$(r, \theta) = \left(\dfrac{4}{3}, 0\right), \left(\dfrac{4}{5}, \pi\right)$

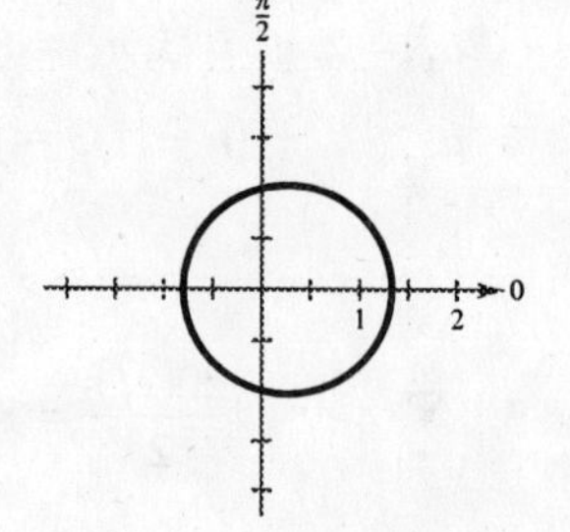

13. $r = \dfrac{8}{4 + 3\sin\theta} = \dfrac{2}{1 + \frac{3}{4}\sin\theta}$

$e = \dfrac{3}{4}$ ellipse

Vertices: $(r, \theta) = \left(\dfrac{8}{7}, \dfrac{\pi}{2}\right), \left(8, \dfrac{3\pi}{2}\right)$

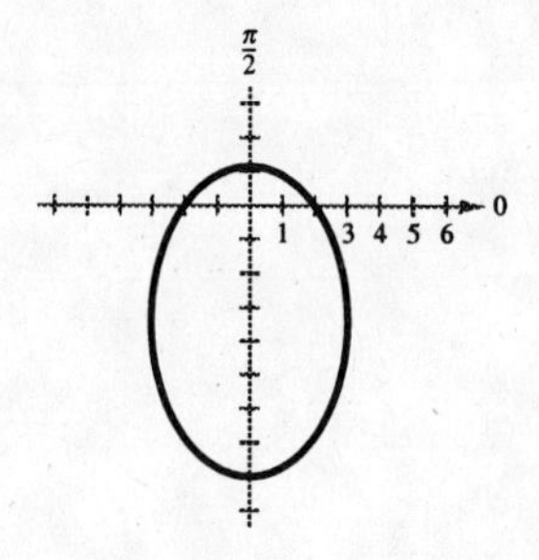

15. $r = \dfrac{4}{2 + 3\sin\theta} = \dfrac{2}{1 + \frac{3}{2}\sin\theta}$

$e = \dfrac{3}{2}$ hyperbola

Vertices: $(r, \theta) = \left(\dfrac{4}{5}, \dfrac{\pi}{2}\right), \left(-4, \dfrac{3\pi}{2}\right)$

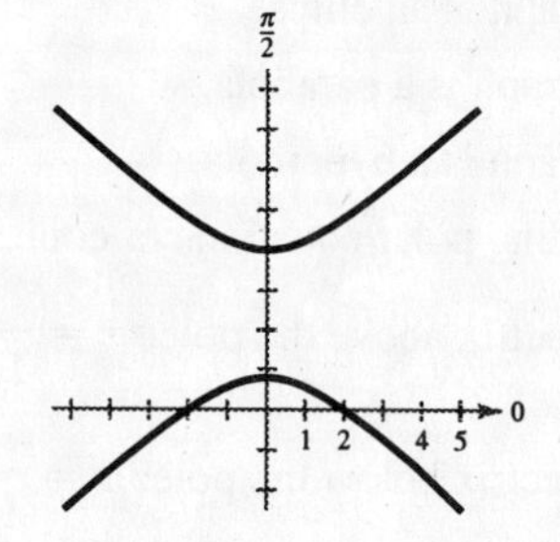

17. $r = \dfrac{3}{4 - 8\cos\theta} = \dfrac{\frac{3}{4}}{1 - 2\cos\theta}$

$e = 2$ hyperbola

Vertices: $(r, \theta) = \left(-\dfrac{3}{4}, 0\right), \left(\dfrac{1}{4}, \pi\right)$

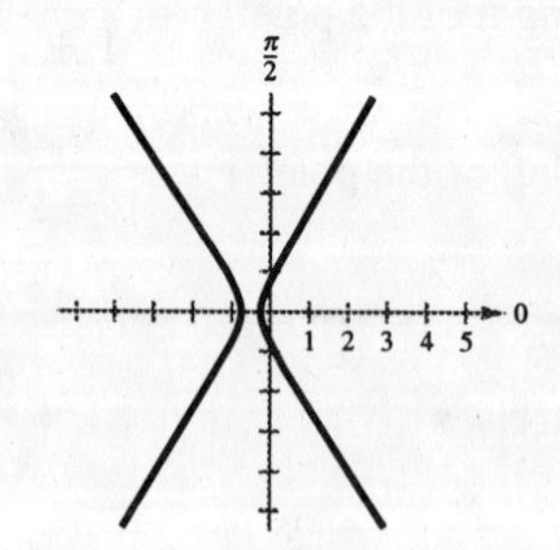

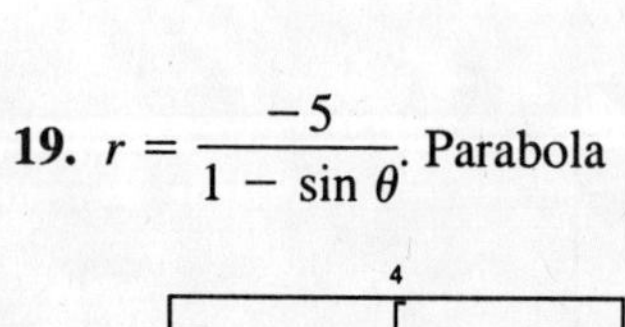

19. $r = \dfrac{-5}{1 - \sin\theta}$. Parabola

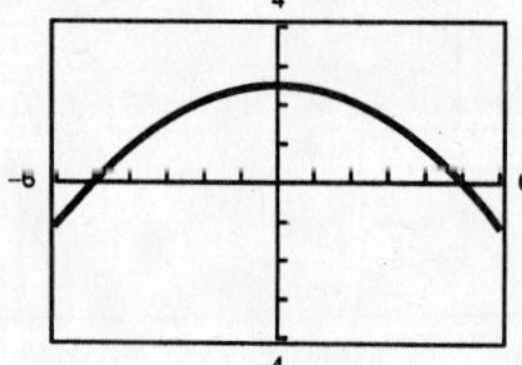

21. $r = \dfrac{12}{2 - \cos\theta} = \dfrac{6}{1 - \frac{1}{2}\cos\theta}$

ellipse $\left(e = \dfrac{1}{2}\right)$

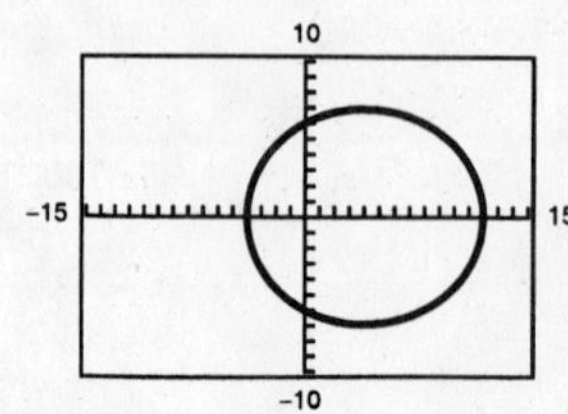

23. $r = \dfrac{6}{1 - \cos\left(\theta - \frac{\pi}{4}\right)}$

rotated parabola

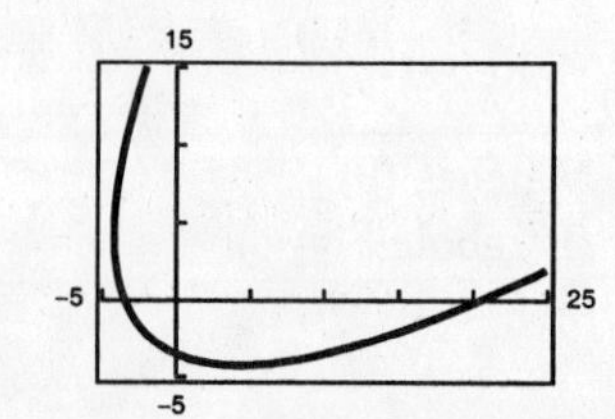

25. $r = \dfrac{8}{4 + 3\sin\left(\theta + \dfrac{\pi}{6}\right)}$ rotated ellipse

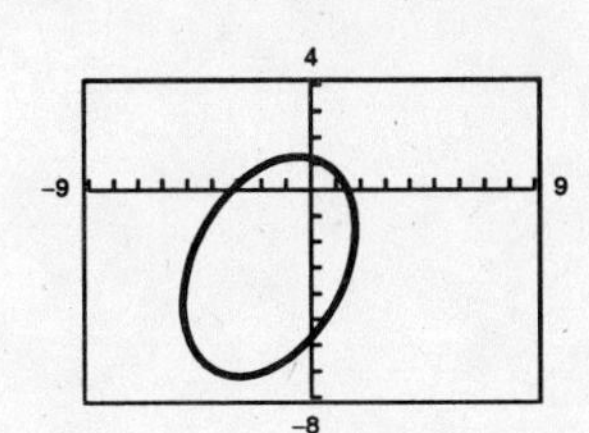

27. $e = 1, x = -1, p = 1$

Vertical directrix to the left of the pole

$$r = \frac{1(1)}{1 - 1\cos\theta} = \frac{1}{1 - \cos\theta}$$

29. $e = \frac{1}{2}, y = 1, p = 1$

Horizontal directrix above the pole

$$r = \frac{(1/2)(1)}{1 + (1/2)\sin\theta} = \frac{1}{2 + \sin\theta}$$

31. $e = 2, x = 1, p = 1$

Vertical directrix to the right of the pole

$$r = \frac{2(1)}{1 + 2\cos\theta} = \frac{2}{1 + 2\cos\theta}$$

33. Vertex: $\left(1, -\dfrac{\pi}{2}\right) \Rightarrow e = 1, p = 2$

Horizontal directrix below the pole

$$r = \frac{1(2)}{1 - 1\sin\theta} = \frac{2}{1 - \sin\theta}$$

35. Center: $(3, \pi)$; $c = 3, a = 5, e = \dfrac{3}{5}$

Vertical directrix to the right of the pole

$$r = \frac{(3/5)p}{1 + (3/5)\cos\theta} = \frac{3p}{5 + 3\cos\theta}$$

$$2 = \frac{3p}{5 + 3\cos 0}$$

$$p = \frac{16}{3}$$

$$r = \frac{3(16/3)}{5 + 3\cos\theta} = \frac{16}{5 + 3\cos\theta}$$

37. Center: $\left(5, \dfrac{3\pi}{2}\right)$; $c = 5, a = 4, e = \dfrac{5}{4}$

Horizontal directrix below the pole

$$r = \frac{(5/4)p}{1 - (5/4)\sin\theta} = \frac{5p}{4 - 5\sin\theta}$$

$$1 = \frac{5p}{4 - 5\sin(3\pi/2)}$$

$$p = \frac{9}{5}$$

$$r = \frac{5(9/5)}{4 - 5\sin\theta} = \frac{9}{4 - 5\sin\theta}$$

39. When $\theta = 0$, $r = c + a = ea + a = a(1 + e)$.

Therefore,

$$a(1 + e) = \frac{ep}{1 - e\cos 0}$$

$$a(1 + e)(1 - e) = ep$$

$$a(1 - e^2) = ep.$$

Thus, $r = \dfrac{ep}{1 - e\cos\theta} = \dfrac{(1 - e^2)a}{1 - e\cos\theta}$.

41. $r = \dfrac{[1 - (0.0167)^2](92.957 \times 10^6)}{1 - 0.0167\cos\theta} \approx \dfrac{9.2931 \times 10^7}{1 - 0.0167\cos\theta}$

Perihelion distance: $r = 92.957 \times 10^6(1 - 0.0167) \approx 9.1405 \times 10^7$

Aphelion distance: $r = 92.957 \times 10^6(1 + 0.0167) \approx 9.4509 \times 10^7$

43. Radius of earth $\approx$ 4000 miles. Choose $r = \dfrac{ep}{1 - e\cos\theta}$.

Vertices: $(126{,}800,\ 0)$ and $(4119,\ \pi)$

$$a = \frac{126{,}800 + 4119}{2} = 65{,}459.5$$

$$c = 65{,}459.5 - 4119 = 61{,}340.5$$

$$e = \frac{c}{a} = \frac{61{,}340.5}{65{,}459.5} \approx 0.937$$

$$2a = \frac{ep}{1 - e\cos 0} + \frac{ep}{1 - e\cos(\pi)} = \frac{ep}{1-e} + \frac{ep}{1+e} = \frac{2ep}{1-e^2}$$

Thus, $p = \dfrac{a(1-e^2)}{e} \approx 8525.2$

Thus, $r = \dfrac{ep}{1 - e\cos\theta} \approx \dfrac{7988.1}{1 - 0.937\cos\theta}$

When $\theta = 60°$, $r \approx 15{,}029$ and the distance from the surface of the earth to the satellite is $15{,}029 - 4000 = 11029$ miles.

45. $r = \dfrac{4}{-3 - 3\sin\theta} = \dfrac{-\frac{4}{3}}{1 + \sin\theta}$

False. The directrix is below the pole.

47.

$$\frac{x^2}{a^2} + \frac{y^2}{b^2} = 1$$

$$\frac{r^2\cos^2\theta}{a^2} + \frac{r^2\sin^2\theta}{b^2} = 1$$

$$\frac{r^2\cos^2\theta}{a^2} + \frac{r^2(1-\cos^2\theta)}{b^2} = 1$$

$$r^2b^2\cos^2\theta + r^2a^2 - r^2a^2\cos^2\theta = a^2b^2$$

$$r^2(b^2 - a^2)\cos^2\theta + r^2a^2 = a^2b^2$$

For an ellipse, $b^2 - a^2 = -c^2$. Hence,

$$-r^2c^2\cos^2\theta + r^2a^2 = a^2b^2$$

$$-r^2\left(\frac{c}{a}\right)^2\cos^2\theta + r^2 = b^2,\ e = \frac{c}{a}$$

$$-r^2e^2\cos^2\theta + r^2 = b^2$$

$$r^2(1 - e^2\cos^2\theta) = b^2$$

$$r^2 = \frac{b^2}{1 - e^2\cos^2\theta}.$$

49. $\dfrac{x^2}{169} + \dfrac{y^2}{144} = 1$

$a = 13, b = 12, c = 5, e = \dfrac{5}{13}$

$$r^2 = \frac{144}{1 - (25/169)\cos^2\theta} = \frac{24{,}336}{169 - 25\cos^2\theta}$$

51. $\dfrac{6!}{3!2!} = \dfrac{6 \cdot 5 \cdot 4 \cdot 3!}{3!2!}$

$= 60$

53. $\dfrac{9!}{2!2!} = 90{,}720$

55. ${}_{12}C_9 = 220$

57. ${}_{10}P_3 = 720$

Review Exercises for Chapter 10

Solutions to Odd-Numbered Exercises

1. Hyperbola

3. Vertex: $(4, 2) = (h, k)$

Focus: $(4, 0) \Rightarrow p = -2$

$(x - h)^2 = 4p(y - k)$

$(x - 4)^2 = -8(y - 2)$

5. Vertex: $(0, 2) = (h, k)$

Directrix: $x = -3 \Rightarrow p = 3$

$(y - k^2) = 4p(x - h)$

$(y - 2)^2 = 12x$

7. $x^2 = -2y = 4\left(-\frac{1}{2}\right)y, p = -\frac{1}{2}$

Focus: $\left(0, -\frac{1}{2}\right)$

$d_1 = \frac{1}{2} + b$

$d_2 = \sqrt{(2 - 0)^2 + \left(-2 + \frac{1}{2}\right)^2} = \frac{5}{2}$

$d_1 = d_2 \Rightarrow \frac{1}{2} + b = \frac{5}{2} \Rightarrow b = 2$

Slope of tangent line: $\frac{b + 2}{0 - 2} = \frac{4}{-2} = -2$

Equation: $y + 2 = -2(x - 2)$

$y = -2x + 2$

x-intercept: $(1, 0)$

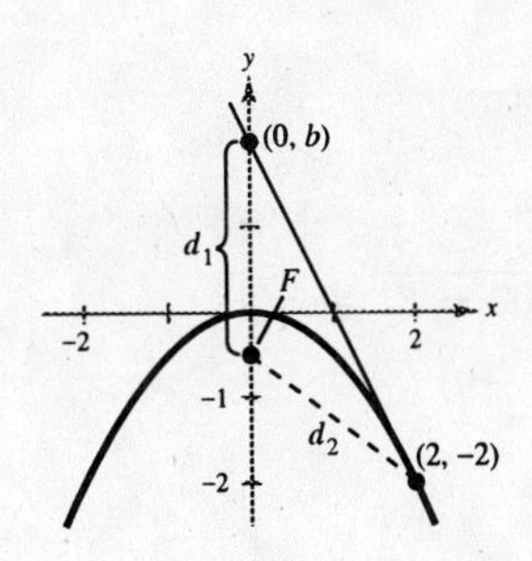

9. Vertex: $(0, 12)$, opens downward

$x^2 = 4p(y - 12)$

$(4, 10)$ on parabola: $16 = 4p(10 - 12) = -8p \Rightarrow p = -2$

$x^2 = -8(y - 12)$

When $y = 0$, $x^2 = 96 \Rightarrow x = 4\sqrt{6}$

Width at ground level: $8\sqrt{6} \approx 19.6$ meters

11. Vertices: $(-3, 0), (7, 0) \Rightarrow a = 5, (h, k) = (2, 0)$

Foci: $(0, 0), (4, 0) \Rightarrow c = 2$

$b^2 = a^2 - c^2 = 25 - 4 = 21$

$\frac{(x - h)^2}{a^2} + \frac{(y - k)^2}{b^2} = 1$

$\frac{(x - 2)^2}{25} + \frac{y^2}{21} = 1$

13. Vertices: $(0, 1), (4, 1) \Rightarrow a = 2, (h, k) = (2, 1)$

Endpoints of minor axis: $(2, 0), (2, 2) \Rightarrow b = 1$

$\frac{(x - h)^2}{a^2} + \frac{(y - k)^2}{b^2} = 1$

$\frac{(x - 2)^2}{4} + (y - 1)^2 = 1$

15. $a = 5, b = 4 \Rightarrow c = \sqrt{25 - 16} = 3.$

Place the foci three feet on either side of the center.

17. $16(x^2 - 2x + 1) + 9(y^2 + 8y + 16) = -16 + 16 + 144$

$$16(x - 1)^2 + 9(y + 4)^2 = 144$$

$$\frac{(x - 1)^2}{9} + \frac{(y + 4)^2}{16} = 1$$

Center: $(1, -4)$

$a = 4, b = 3, c = \sqrt{16 - 9} = \sqrt{7}$

Vertices: $(1, 0), (1, -8)$

Foci: $\left(1, -4 \pm \sqrt{7}\right)$

$e = \dfrac{c}{a} = \dfrac{\sqrt{7}}{4}$

19. Center: $(-2, 1)$

$a = 10, b = 9, c = \sqrt{100 - 81} = \sqrt{19}$

Vertices; $(-2, 11), (-2, -9)$

Foci: $\left(-2, 1 \pm \sqrt{19}\right)$

$e = \dfrac{\sqrt{19}}{10}$

21. Center: $(-2, 3), a = 8, c = 10,$

$b = \sqrt{100 - 64} = 6$

$$\frac{(x + 2)^2}{64} - \frac{(y - 3)^2}{36} = 1$$

23. Foci: $(0, 0), (8, 0) \Rightarrow c = 4, (h, k) = (4, 0)$

Asymptotes: $y = \pm 2(x - 4) \Rightarrow \dfrac{b}{a} = 2, b = 2a$

$b^2 = c^2 - a^2 \Rightarrow 4a^2 = 16 - a^2 \Rightarrow a^2 = \dfrac{16}{5}, b^2 = \dfrac{64}{5}$

$$\frac{(x - h)^2}{a^2} - \frac{(y - k)^2}{b^2} = 1$$

$$\frac{(x - 4)^2}{16/5} - \frac{y^2}{64/5} = 1$$

$$\frac{5(x - 4)^2}{16} - \frac{5y^2}{64} = 1$$

25. $9(x^2 - 2x + 1) - 16(y^2 + 2y + 1) = 151 + 9 - 16$

$$9(x - 1)^2 - 16(y + 1)^2 = 144$$

$$\frac{(x - 1)^2}{16} - \frac{(y + 1)^2}{9} = 1$$

Center: $(1, -1)$ $\quad a = 4, b = 3, c = 5$

Vertices: $(5, -1), (-3, -1)$

Foci: $(6, -1), (-4, -1)$

Asymptotes: $y = -1 \pm \dfrac{3}{4}(x - 1)$

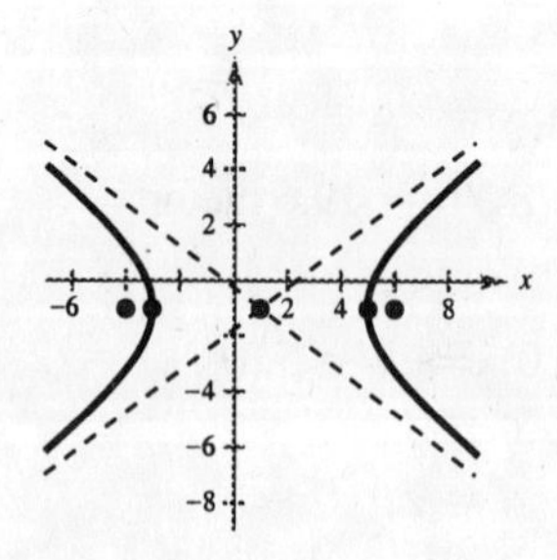

27. $\dfrac{(x - 3)^2}{16} - \dfrac{(y + 5)^2}{4} = 1$

$a = 4, b = 2, c = 2\sqrt{5}$

Center: $(3, -5)$

Vertices: $(7, -5), (-1, -5)$

Foci: $\left(3 \pm 2\sqrt{5}, -5\right)$

Asymptotes: $y = -5 + \dfrac{1}{2}(x - 3)$

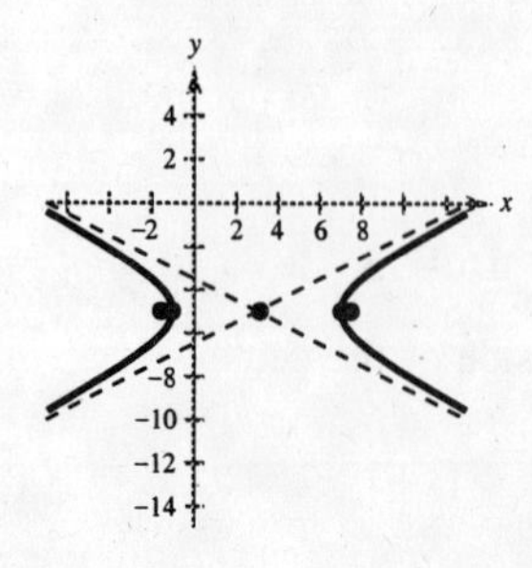

29. $d_2 - d_1 = 186{,}000(0.0005)$

$2a = 93$

$a = 46.5$

$c = 100$

$b = \sqrt{c^2 - a^2}$

$\dfrac{x^2}{a^2} - \dfrac{y^2}{b^2} = 1$

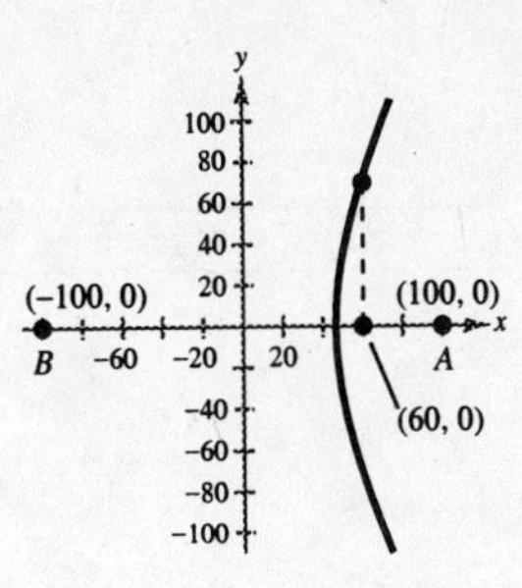

$x = 60 \implies y^2 = b^2\left(\dfrac{x^2}{a^2} - 1\right)$

$= (100^2 - 46.5^2)\left(\dfrac{60^2}{46.5^2} - 1\right)$

≈ 5211.57

$\implies y \approx 72.2$

72.2 miles north

31. $3x^2 + 2y^2 - 12x + 12y + 29 = 0$

$3(x^2 - 4x + 4) + 2(y^2 + 6y + 9) = -29 + 12 + 18$

$3(x - 2)^2 + 2(y + 3)^2 = 1$

Ellipse

33. $xy - 4 = 0$

$A = 0, B = 1, C = 0$

$\cot 2\theta = \dfrac{A - C}{B} = 0 \implies \theta = \dfrac{\pi}{4}$

$x = \dfrac{\sqrt{2}}{2}(x' - y'), y = \dfrac{\sqrt{2}}{2}(x' + y')$

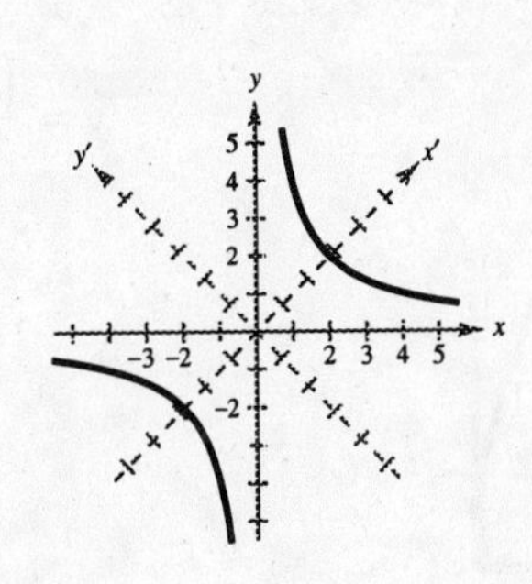

$xy = 4$

$\dfrac{\sqrt{2}}{2}(x' - y')\dfrac{\sqrt{2}}{2}(x' + y') = 4$

$\dfrac{1}{2}(x')^2 - \dfrac{1}{2}(y')^2 = 4$

$\dfrac{(x')^2}{8} - \dfrac{(y')^2}{8} = 1$ Hyperbola

35. $5x^2 - 2xy + 5y^2 - 12 = 0$

$A = 5, B = -2, C = 5$

$\cot 2\theta = 0 \implies \theta = \dfrac{\pi}{4}$

$x = \dfrac{\sqrt{2}}{2}(x' - y'), y = \dfrac{\sqrt{2}}{2}(x' + y')$

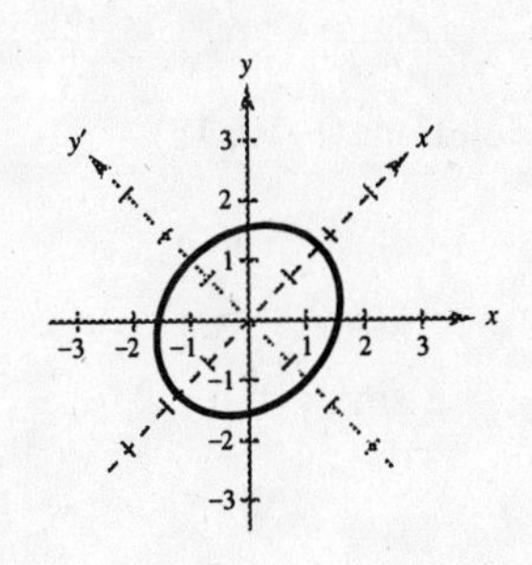

$5x^2 - 2xy + 5y^2 = 12$

$5\left[\dfrac{\sqrt{2}}{2}(x' - y')\right]^2 - 2\left[\dfrac{\sqrt{2}}{2}(x' - y')\right]\left[\dfrac{\sqrt{2}}{2}(x' + y')\right] + 5\left[\dfrac{\sqrt{2}}{2}(x' + y')\right]^2 = 12$

$5\left[\dfrac{1}{2}(x')^2 - x'y' + \dfrac{1}{2}(y')^2\right] - (x')^2 + (y')^2 + 5\left[\dfrac{1}{2}(x')^2 + x'y' + \dfrac{1}{2}(y')^2\right] = 12$

$4(x')^2 + 6(y')^2 = 12$

$\dfrac{(x')^2}{3} + \dfrac{(y')^2}{2} = 1$ Ellipse

37. (a) $B^2 - 4AC = (-8)^2 - 4(16)(1) = 0$ Parabola

(b) $y^2 + (5 - 8x)y + (16x^2 - 10x) = 0$

$$y = \frac{(8x - 5) \pm \sqrt{(5 - 8x)^2 - 4(16x^2 - 10x)}}{2}$$

(c)

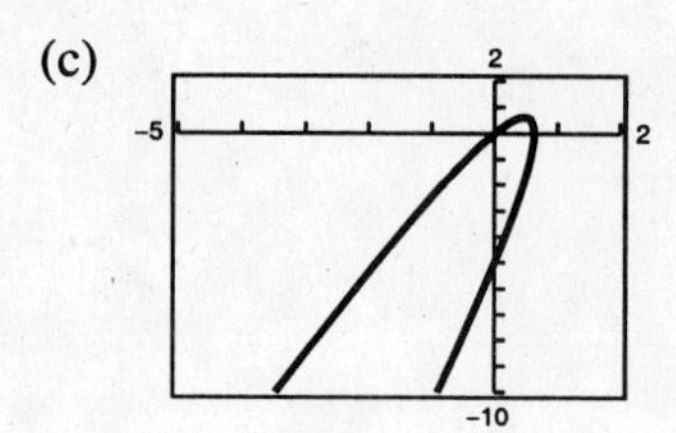

39. (a) $B^2 - 4AC = (2)^2 - 4(1)(1) = 0$ Parabola

(b) $y^2 + \left(2x - 2\sqrt{2}\right)y + \left(x^2 + 2\sqrt{2}x + 2\right) = 0$

$$y = \frac{\left(2\sqrt{2} - 2x\right) \pm \sqrt{(2x - 2\sqrt{2})^2 - 4\left(x^2 + 2\sqrt{2}x + 2\right)}}{2}$$

(c)

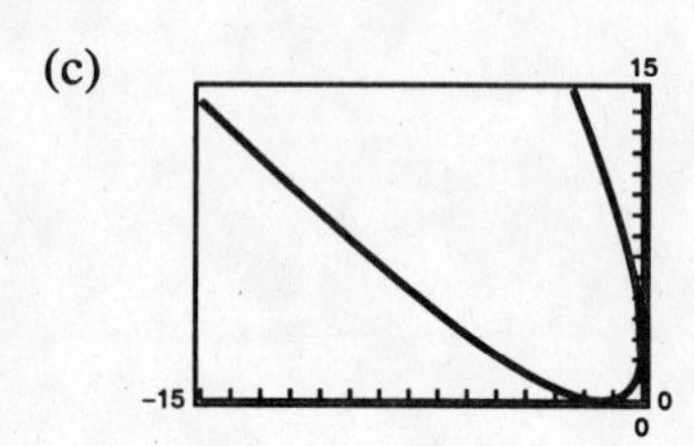

41. Adding the equations, $24x + 240 = 0 \implies x = -10$.

Then,

$$4(100) + y^2 - 560 - 24y + 304 = 0$$
$$y^2 - 24y - 144 = 0$$
$$(y - 12)^2 = 0 \implies y = 12$$

Solution: $(-10, 12)$

43. $x = 3\cos 0 = 3$

$y = 2\sin^2 0 = 0$

45. $x = 3\cos\dfrac{\pi}{6} = \dfrac{3\sqrt{3}}{2}$

$y = 2\sin^2\dfrac{\pi}{6} = \dfrac{1}{2}$

47. $x = 1 + 4t, y = 2 - 3t$

$t = \dfrac{x - 1}{4}$

$y = 2 - 3\left(\dfrac{x - 1}{4}\right)$

$3x + 4y = 11$

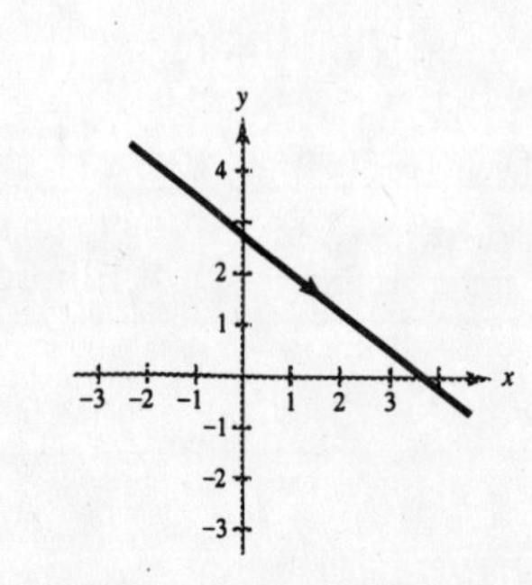

49. $x = \dfrac{1}{t}, y = t^2$

$t = \dfrac{1}{x}$

$y = \dfrac{1}{x^2}$

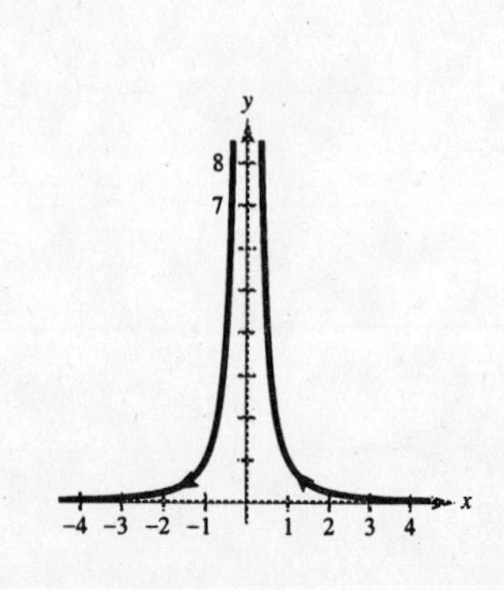

51. $x = 6\cos\theta,\ y = 6\sin\theta$

$\cos\theta = \dfrac{x}{6},\ \sin\theta = \dfrac{y}{6}$

$\dfrac{x^2}{36} + \dfrac{y^2}{36} = 1$

$x^2 + y^2 = 36$

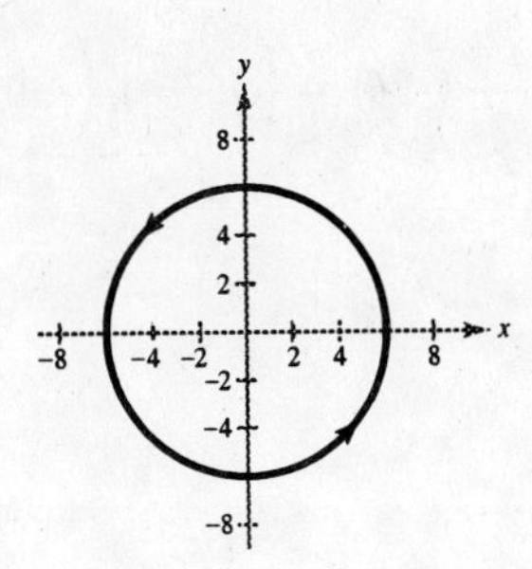

53. $x = \sec\theta,\ y = \tan\theta$

$\tan^2\theta + 1 = \sec^2\theta$

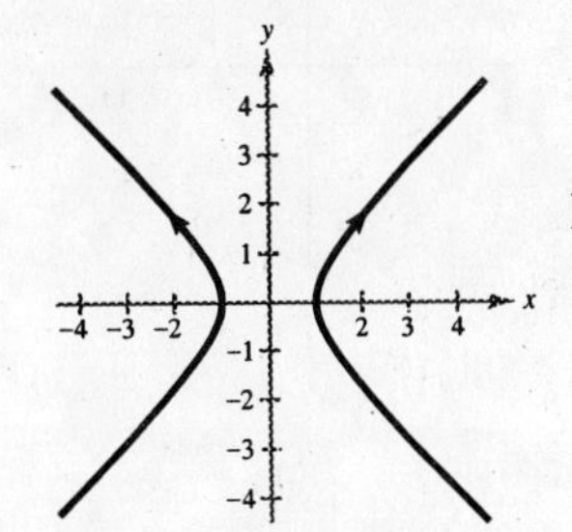

$y^2 + 1 = x^2$

$x^2 - y^2 = 1$

Hyperbola

55. $x = -3 + 4\cos\theta$

$y = 4 + 3\sin\theta$

Note that

$$\frac{(x+3)^2}{16} + \frac{(y-4)^2}{9} = \cos^2\theta + \sin^2\theta = 1$$

57.

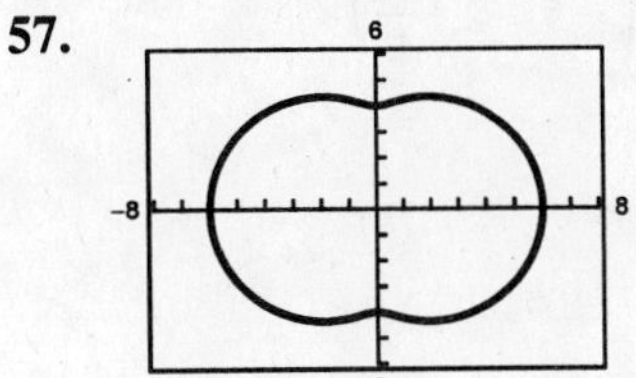

59. $\left(1, -\dfrac{3\pi}{4}\right)$

$\left(-1, \dfrac{\pi}{4}\right)$

$\left(-1, -\dfrac{7\pi}{4}\right)$

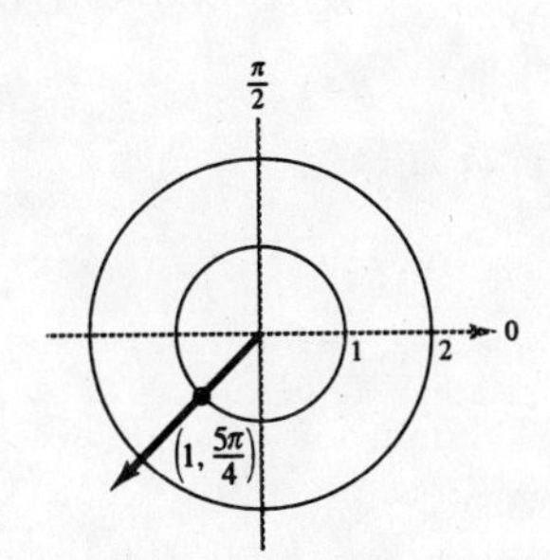

61. $\left(-\dfrac{5}{2}, -\dfrac{11\pi}{6}\right)$

$\left(\dfrac{5}{2}, \dfrac{7\pi}{6}\right)$

$\left(\dfrac{5}{2}, -\dfrac{5\pi}{6}\right)$

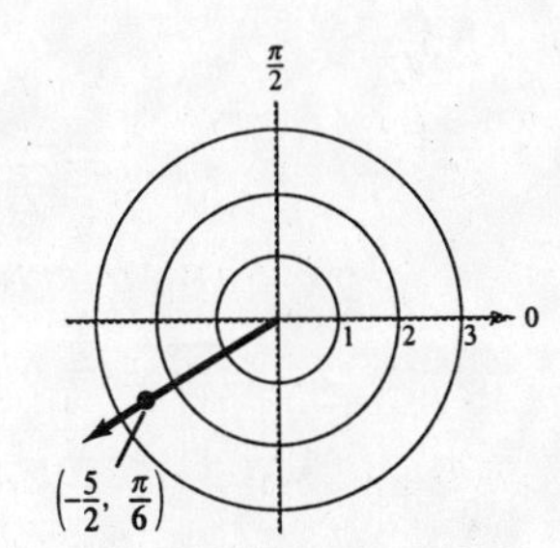

63. $\left(\sqrt{5}, -\dfrac{2\pi}{3}\right)$

$\left(-\sqrt{5}, \dfrac{\pi}{3}\right)$

$\left(-\sqrt{5}, -\dfrac{5\pi}{3}\right)$

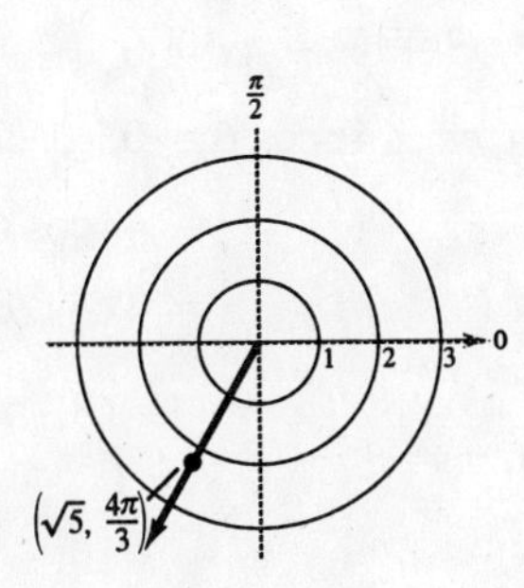

65. $(r, \theta) = \left(5, -\dfrac{7\pi}{6}\right)$

$$(x, y) = \left[5\cos\left(-\frac{7\pi}{6}\right), 5\sin\left(-\frac{7\pi}{6}\right)\right]$$

$$= \left(-\frac{5\sqrt{3}}{2}, \frac{5}{2}\right)$$

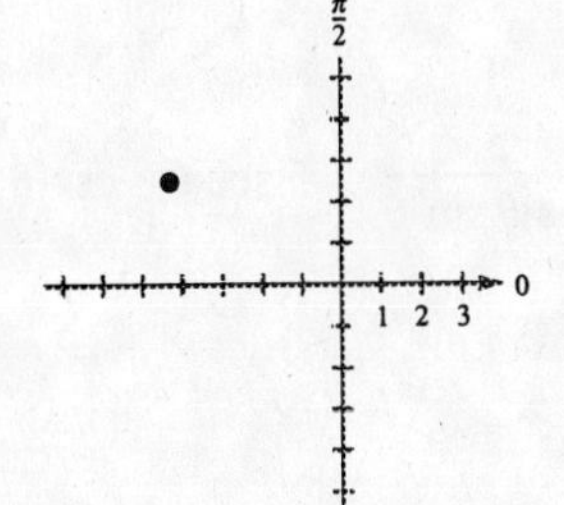

67. $(r, \theta) = \left(12, -\frac{\pi}{2}\right)$

$(x, y) = (0, -12)$

69. $(x, y) = (0, -9)$

$(r, \theta) = \left(9, \frac{3\pi}{2}\right), \left(-9, \frac{\pi}{2}\right)$

71. $(x, y) = (5, -5)$

$(r, \theta) = \left(5\sqrt{2}, \frac{7\pi}{4}\right), \left(-5\sqrt{2}, \frac{3\pi}{4}\right)$

73.
$$r = 5$$
$$\sqrt{x^2 + y^2} = 5$$
$$x^2 + y^2 = 25$$

75.
$$r = 3\cos\theta$$
$$r^2 = 3r\cos\theta$$
$$x^2 + y^2 = 3x$$

77.
$$r^2 = \cos 2\theta$$
$$r^2 = 1 - 2\sin^2\theta$$
$$r^4 = r^2 - 2r^2\sin^2\theta$$
$$(x^2 + y^2)^2 = x^2 + y^2 - 2y^2$$
$$(x^2 + y^2)^2 - x^2 + y^2 = 0$$

79.
$$r = \frac{2}{2 - \sin\theta}$$
$$2r - r\sin\theta = 2$$
$$2\sqrt{x^2 + y^2} = y + 2$$
$$4(x^2 + y^2) = y^2 + 4y + 4$$
$$4x^2 + 3y^2 - 4y - 4 = 0$$

81.
$$x^2 + y^2 = 9$$
$$r^2 = 9$$
$$r = 3$$

83.
$$y = 6$$
$$r\sin\theta = 6$$
$$r = 6\csc\theta$$

85.
$$x^2 + y^2 - 4x = 0$$
$$r^2 - 4r\cos\theta = 0$$
$$r^2 = 4r\cos\theta$$
$$r = 4\cos\theta$$

87.
$$xy = 5$$
$$(r\cos\theta)(r\sin\theta) = 5$$
$$r^2 = \frac{5}{\cos\theta\sin\theta} = 5\sec\theta \cdot \csc\theta$$

89. $r = 5$

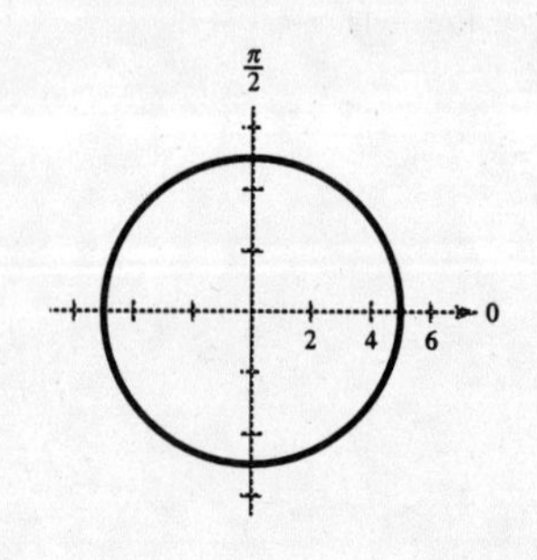

91. $\theta = \dfrac{\pi}{2}$ (y − axis)

93. $r = 5\cos\theta$

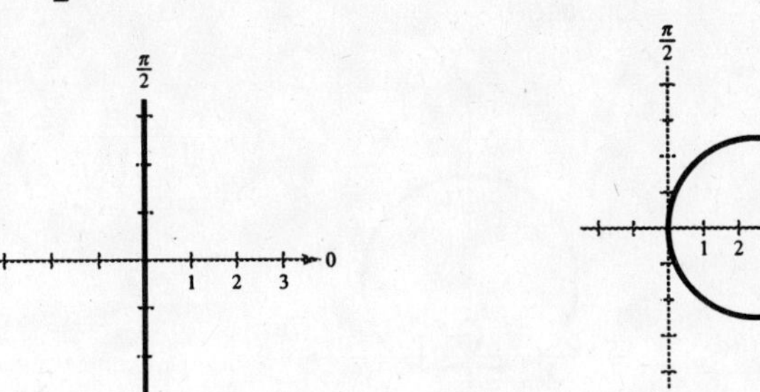

95. $r = -2(1 + \cos\theta)$

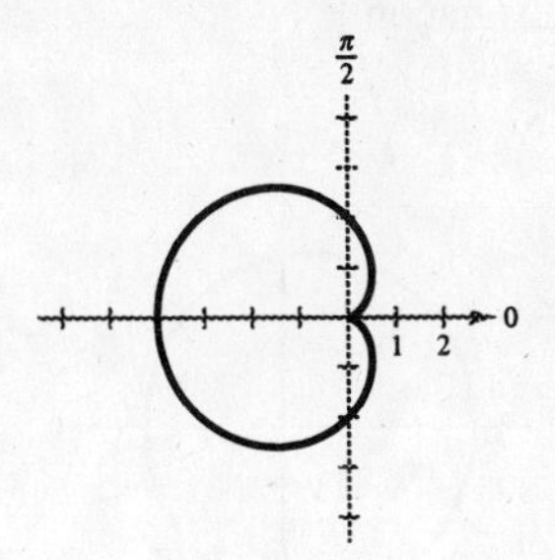

97. $r^2 = 4\sin^2 2\theta \Rightarrow r = \pm 2\sin 2\theta$

Symmetric with respect to $\theta = \pi/2$, polar axis, and pole

Rose curve ($n = 2$) with 4 petals

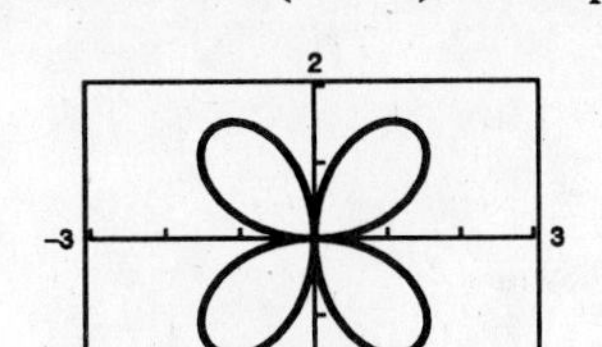

99. $r = \dfrac{3}{\cos(\theta - \pi/4)}$

The graph is a line.

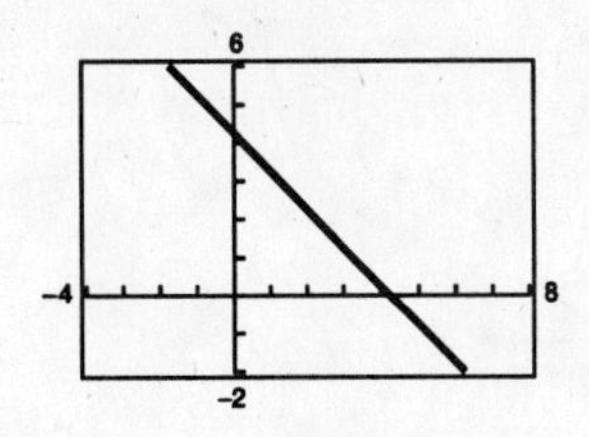

101. $r = 5 + 4\cos\theta$

Symmetry: polar axis

Maximum r-value: $r = 9$ for $\theta = 0, 2\pi$

Zeros of r: none

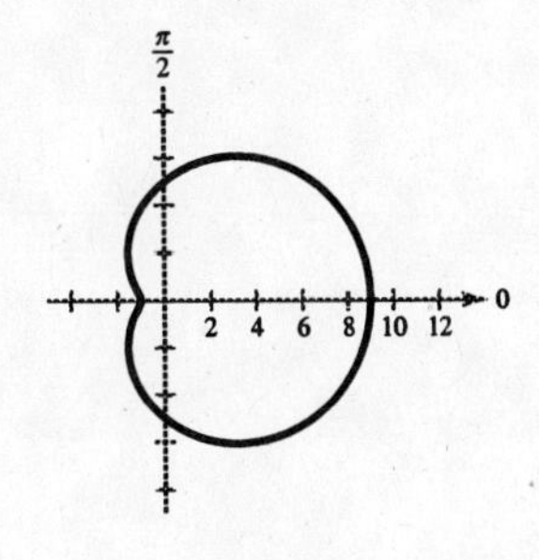

103. $r = -3\cos 2\theta$

Symmetry: polar axis, $\theta = \dfrac{\pi}{2}$, pole

Maximum r-value: 3 for $\theta = 0, \dfrac{\pi}{2}, \pi, \dfrac{3\pi}{2}$

Zeros of r: $\theta = \dfrac{\pi}{4}, \dfrac{3\pi}{4}, \dfrac{5\pi}{4}, \dfrac{7\pi}{4}$

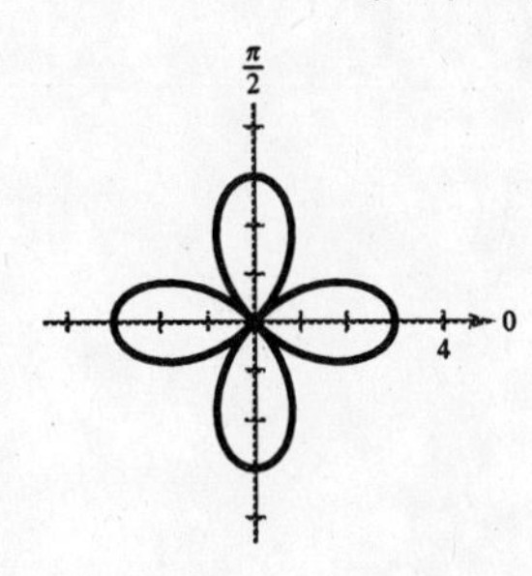

105. $r^2 = \cos 2\theta$

Symmetry: polar axis, $\theta = \dfrac{\pi}{2}$, pole

Maximum r-value: 1 for $\theta = 0, \pi, 2\pi$

Zeros of r: $\theta = \dfrac{\pi}{4}, \dfrac{3\pi}{4}, \dfrac{7\pi}{4}$

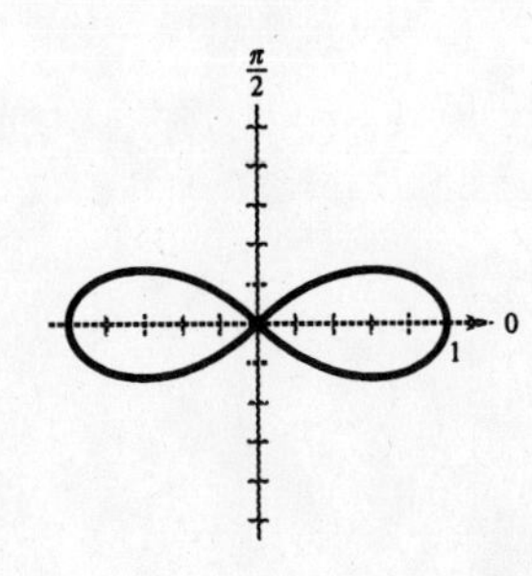

107. $r = 6 - \cos\theta$

Limaçon

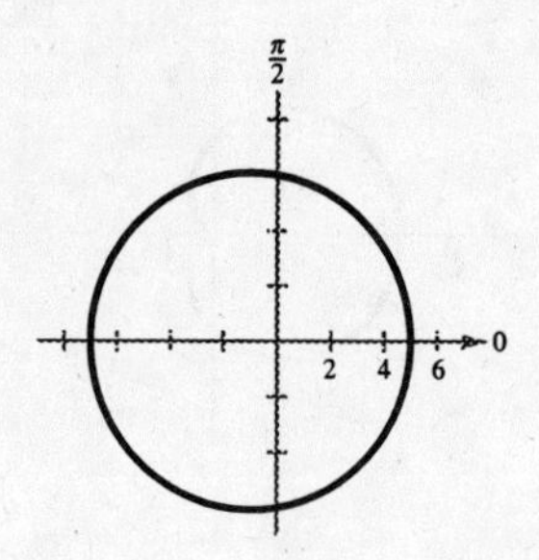

109. $r = 10 + 15\sin\theta$

Limaçon

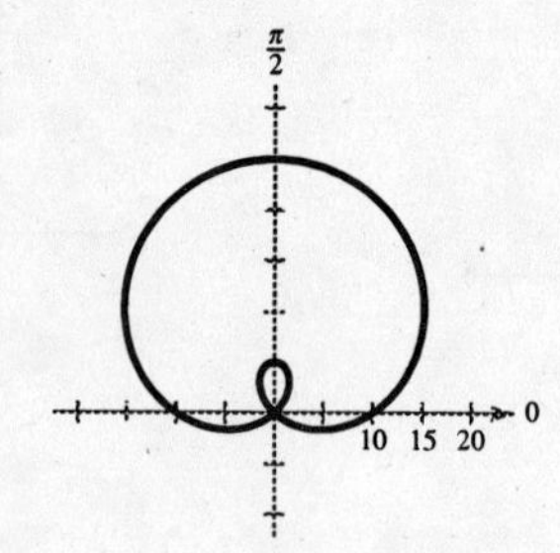

111. $r = 4\sin 5\theta$

5-leaved rose

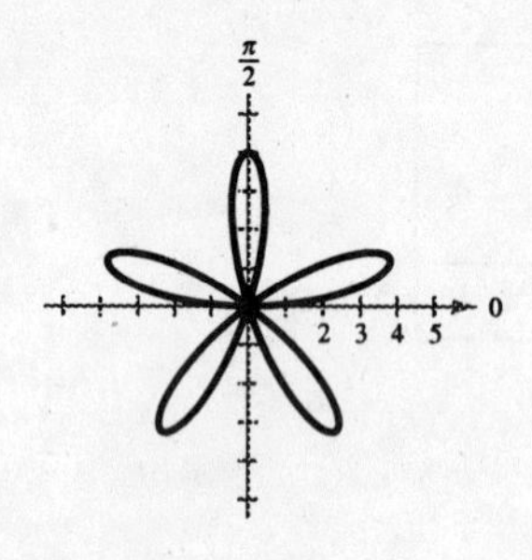

113. $r^2 = 8\cos 2\theta$

Lemniscate

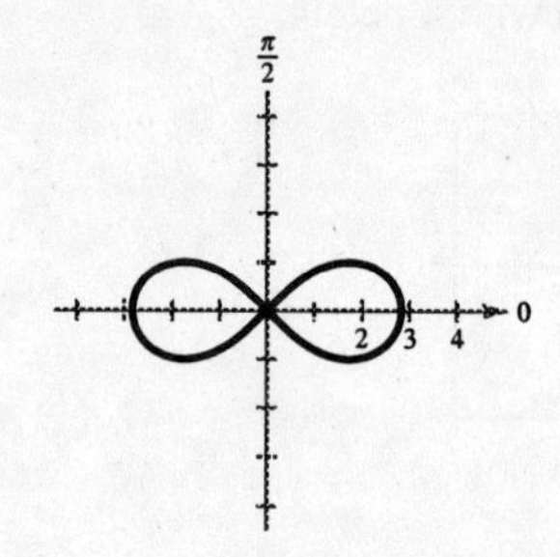

115. $r = \dfrac{2}{1 - \sin\theta}, e = 1$

Parabola symmetric with $\theta = \pi/2$ and the vertex at $(1, 3\pi/2)$

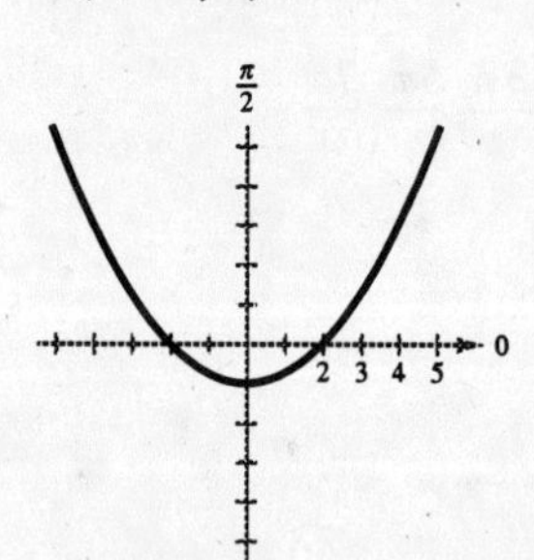

117. $r = \dfrac{4}{5 - 3\cos\theta} = \dfrac{\frac{4}{5}}{1 - \frac{3}{5}\cos\theta}, e = \dfrac{3}{5}$

Ellipse

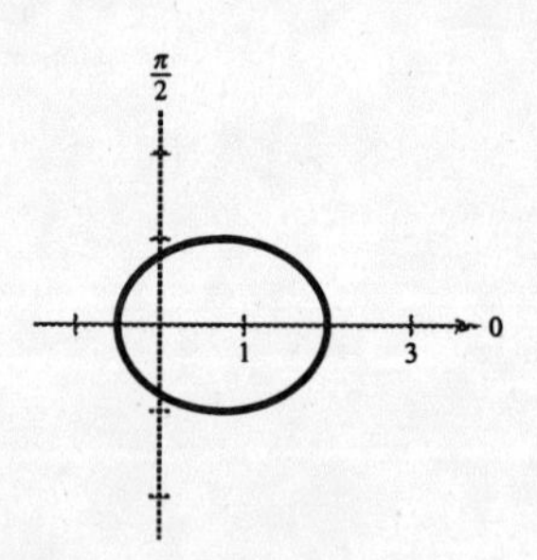

119. $r = \dfrac{5}{6 + 2\sin\theta} = \dfrac{\frac{5}{6}}{1 + \frac{1}{3}\sin\theta}, e = \dfrac{1}{3}$

Ellipse

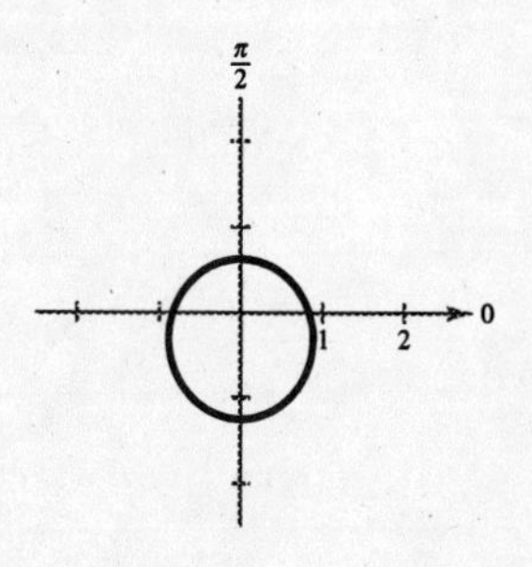

121. Center: $(8, \pi/2)$

Solution point: $(0, 0) \Rightarrow$ Radius $= 8 \Rightarrow a = 16$

$r = a\sin\theta$

$r = 16\sin\theta$

123. Ellipse: $r = \dfrac{ep}{1 - e\cos\theta}$; Vertices: $(5, 0), (1, \pi) \Rightarrow a = 3$; One focus: $(0, 0) \Rightarrow c = 2$

$$e = \frac{c}{a} = \frac{2}{3}, p = \frac{5}{2}$$

$$r = \frac{(2/3)(5/2)}{1 - (2/3)\cos\theta} = \frac{5/3}{1 - (2/3)\cos\theta} = \frac{5}{3 - 2\cos\theta}$$

125. Using a vertical axis, let $r = \dfrac{ep}{1 + e\sin\theta}$.

$$2a = \frac{0.092p}{1 + 0.092} + \frac{0.092p}{1 - 0.092} \approx 0.1856p = 3.05$$

Hence, $p \approx 16.4358$ and $ep \approx 1.5121$

$$r = \frac{1.5121}{1 + 0.092\sin\theta}$$

perihelion (closest) distance: $\theta = \dfrac{\pi}{2} \Rightarrow r = 1.385$ a.u.

aphelion (greatest) distance: $\theta = \dfrac{3\pi}{2} \Rightarrow 1.665$ a.u.

127. False. There exists an infinite number of representations.

129. (a) Vertical translation

(b) Horizontal translation

(c) Reflection in y-axis

(d) Parabola opens more slowly

131. 5. The ellipse becomes more circular and approaches a circle of radius 5.

Chapter 10 Practice Test

1. Find the vertex, focus and directrix of the parabola $x^2 - 6x - 4y + 1 = 0$.

2. Find an equation of the parabola with its vertex at $(2, -5)$ and focus at $(2, -6)$.

3. Find the center, foci, vertices, and eccentricity of the ellipse $x^2 + 4y^2 - 2x + 32y + 61 = 0$.

4. Find an equation of the ellipse with vertices $(0, \pm 6)$ and eccentricity $e = \frac{1}{2}$.

5. Find the center, vertices, foci, and asymptotes of the hyperbola $16y^2 - x^2 - 6x - 128y + 231 = 0$.

6. Find an equation of the hyperbola with vertices at $(\pm 3, 2)$ and foci at $(\pm 5, 2)$.

7. Rotate the axes to eliminate the xy-term. Sketch the graph of the resulting equation, showing both sets of axes.

$5x^2 + 2xy + 5y^2 - 10 = 0$

8. Use the discriminant to determine whether the graph of the equation is a parabola, ellipse, or hyperbola.

(a) $6x^2 - 2xy + y^2 = 0$

(b) $x^2 + 4xy + 4y^2 - x - y + 17 = 0$

For Exercises 9 and 10, eliminate the parameter and write the corresponding rectangular equation.

9. $x = 3 - 2\sin\theta, y = 1 + 5\cos\theta$

10. $x = e^{2t}, y = e^{4t}$

11. Convert the polar point $(\sqrt{2}, (3\pi)/4)$ to rectangular coordinates.

12. Convert the rectangular point $(\sqrt{3}, -1)$ to polar coordinates.

13. Convert the rectangular equation $4x - 3y = 12$ to polar form.

14. Convert the polar equation $r = 5\cos\theta$ to rectangular form.

15. Sketch the graph of $r = 1 - \cos\theta$.

16. Sketch the graph of $r = 5\sin 2\theta$.

17. Sketch the graph of $r = \dfrac{3}{6 - \cos\theta}$.

18. Find a polar equation of the parabola with its vertex at $(6, \pi/2)$ and focus at $(0, 0)$.

CHAPTER 11
Analytic Geometry in Three Dimensions

CHAPTER 11
Analytic Geometry in Three Dimensions

Section 11.1 The Three-Dimensional Coordinate System

- You should be able to plot points in the three-dimensional coordinate system.
- The distance between the points (x_1, y_1, z_1) and (x_2, y_2, z_2) is
  $d = \sqrt{(x_2 - x_1)^2 + (y_2 - y_1)^2 + (z_2 - z_1)^2}.$
- The midpoint of the line segment joining the points (x_1, y_1, z_1) and (x_2, y_2, z_2) is
  $\left(\frac{x_1 + x_2}{2}, \frac{y_1 + y_2}{2}, \frac{z_1 + z_2}{2}\right).$
- The equation of the sphere with center (h, k, j) and radius r is
  $(x - h)^2 + (y - k)^2 + (z - j)^2 = r^2.$
- You should be able to find the trace of a surface in space.

Solutions to Odd-Numbered Exercises

1. $A(-1, 4, 3), B(1, 3, -2), C(-3, 0, -2)$

3.

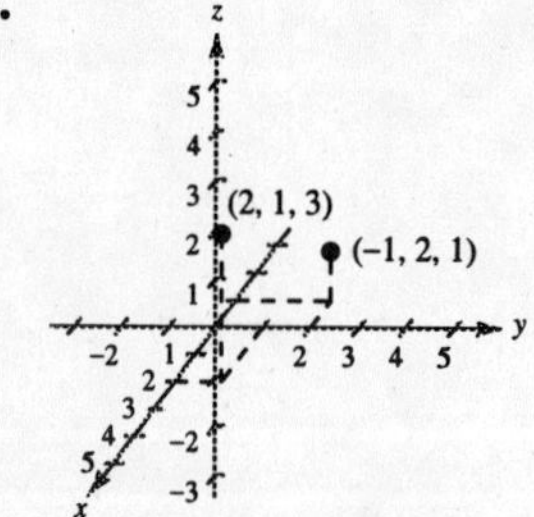

5.

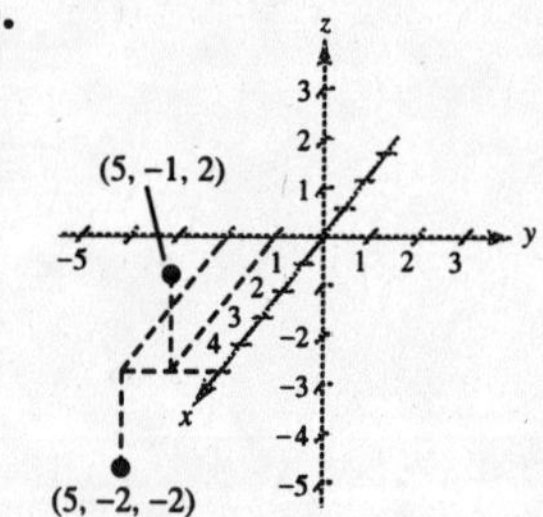

7. $x = -3, y = 3, z = 4$: $(-3, 3, 4)$

9. $y = z = 0, x = 12$,: $(12, 0, 0)$

11. Octant IV

13. Octants I, II, III, IV (above the xy-plane)

15. Octants II, IV, VI, VIII

17.
$$\begin{aligned} d &= \sqrt{(7 - 3)^2 + (4 - 2)^2 + (8 - 5)^2} \\ &= \sqrt{4^2 + 2^2 + 3^2} \\ &= \sqrt{16 + 4 + 9} \\ &= \sqrt{29} \approx 5.385 \end{aligned}$$

19.
$$\begin{aligned} d &= \sqrt{[5 - (-1)]^2 + (-6 - 4)^2 + [2 - (-2)]^2} \\ &= \sqrt{6^2 + (-10)^2 + 4^2} \\ &= \sqrt{36 + 100 + 16} \\ &= \sqrt{152} \\ &= 2\sqrt{38} \approx 12.329 \end{aligned}$$

21. $d = \sqrt{(1-0)^2 + [0-(-3)]^2 + (-10-0)^2}$
$= \sqrt{1 + 9 + 100}$
$= \sqrt{110} \approx 7.416$

23. $d_1 = \sqrt{(-2-0)^2 + (5-0)^2 + (2-2)^2} = \sqrt{4+25} = \sqrt{29}$
$d_2 = \sqrt{(0-0)^2 + (4-0)^2 + (0-2)^2} = \sqrt{16+4} = \sqrt{20}$
$d_3 = \sqrt{(0+2)^2 + (4-5)^2 + (0-2)^2} = \sqrt{4+1+4} = \sqrt{9} = 3$
$d_1^2 = d_2^2 + d_3^2 = 29$

25. $d_1 = \sqrt{(5-1)^2 + (-1+3)^2 + (2+2)^2} = \sqrt{16+4+16} = \sqrt{36} = 6$
$d_2 = \sqrt{(5+1)^2 + (-1-1)^2 + (2-2)^2} = \sqrt{36+4} = \sqrt{40}$
$d_3 = \sqrt{(-1-1)^2 + (1+3)^2 + (2+2)^2} = \sqrt{4+16+16} = \sqrt{36} = 6$
$d_1 = d_3$ Isosceles triangle

27. Midpoint: $\left(\frac{3-3}{2}, \frac{-6+4}{2}, \frac{10+4}{2}\right) = (0, -1, 7)$

29. Midpoint: $\left(\frac{6-4}{2}, \frac{-2+2}{2}, \frac{5+6}{2}\right) = \left(1, 0, \frac{11}{2}\right)$

31. Midpoint: $\left(\frac{-2+7}{2}, \frac{8-4}{2}, \frac{10+2}{2}\right) = \left(\frac{5}{2}, 2, 6\right)$

33. $(x-3)^2 + (y-2)^2 + (z-4)^2 = 16$

35. $(x-0)^2 + (y-4)^2 + (z-3)^2 = 3^2$
$x^2 + (y-4)^2 + (z-3)^2 = 9$

37. Radius $= \frac{\text{diameter}}{2} = 5$: $(x+3)^2 + (y-7)^2 + (z-5)^2 = 5^2 = 25$

39. Center: $\left(\frac{3+0}{2}, \frac{0+0}{2}, \frac{0+6}{2}\right) = \left(\frac{3}{2}, 0, 3\right)$

Radius: $\sqrt{\left(3-\frac{3}{2}\right)^2 + (0-0)^2 + (0-3)^2} = \sqrt{\frac{9}{4} + 9} = \sqrt{\frac{45}{4}}$

Sphere: $\left(x-\frac{3}{2}\right)^2 + (y-0)^2 + (z-3)^2 = \frac{45}{4}$

41. $(x^2 - 4x + 4) + (y^2 + 2y + 1) + (z^2 - 6z + 9) = -10 + 4 + 1 + 9$
$(x-2)^2 + (y+1)^2 + (z-3)^2 = 4$

Center: $(2, -1, 3)$
Radius: 2

43. $(x^2 + 4x + 4) + y^2 + (z^2 - 8z + 16) = -19 + 4 + 16$
$(x+2)^2 + y^2 + (z-4)^2 = 1$

Center: $(-2, 0, 4)$
Radius: 1

45.

$$x^2 + y^2 + z^2 - 2x - \tfrac{2}{3}y - 8z = -\tfrac{73}{9}$$

$$(x^2 - 2x + 1) + \left(y^2 - \tfrac{2}{3}y + \tfrac{1}{9}\right) + (z^2 - 8z + 16) = -\tfrac{73}{9} + 1 + \tfrac{1}{9} + 16$$

$$(x - 1)^2 + \left(y - \tfrac{1}{3}\right)^2 + (z - 4)^2 = 9$$

Center: $\left(1, \tfrac{1}{3}, 4\right)$

Radius: 3

47.

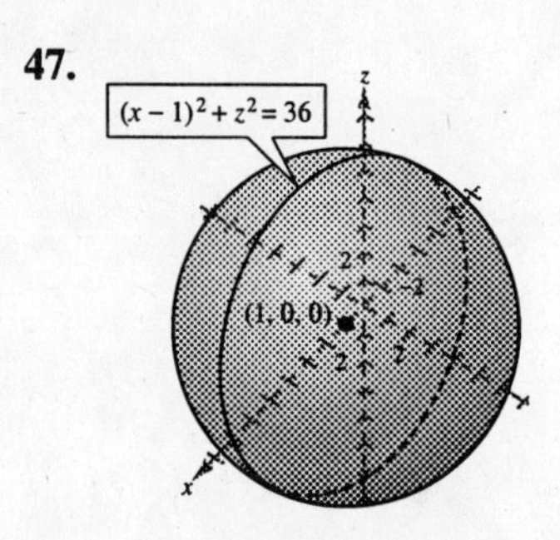

49.

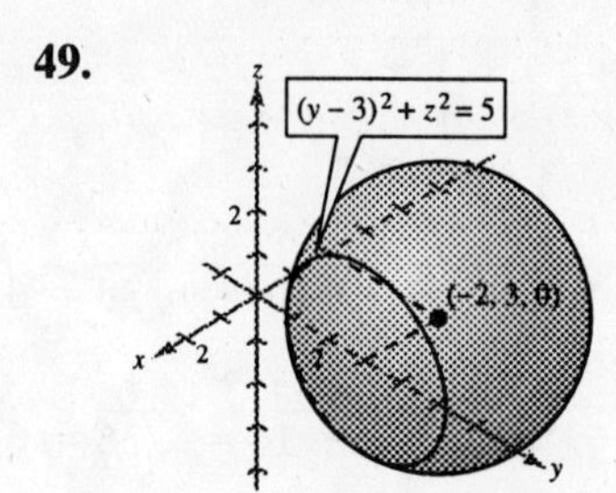

51.

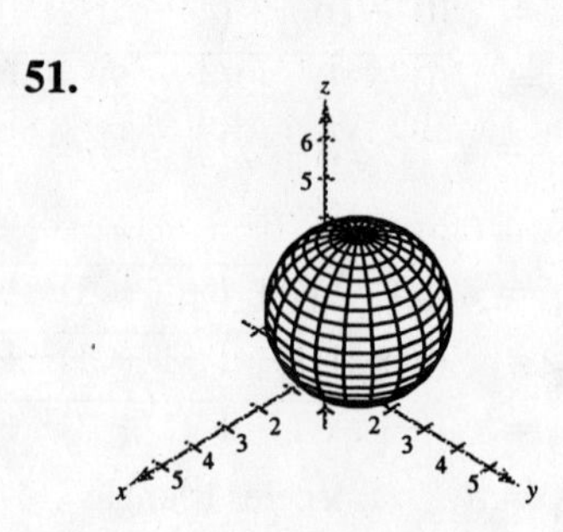

53. The length of each side is 3. Thus $(x, y, z) = (3, 3, 3)$.

55. False. x is the directed distance from the yz-plane to P.

57. In the xy-plane, the z-coordinate is 0.
In the xz-plane, the y-coordinate is 0.
In the yz-plane, the x-coordinate is 0.

59. The trace will be a line in the xy-plane (unless the plane is the xy-plane).

61. $x_2 = 2x_m - x_1 = 2(5) - 3 = 7$

$y_2 = 2y_m - y_1 = 2(8) - 0 = 16$

$z_2 = 2z_m - z_1 = 2(7) - 2 = 12$

$(7, 16, 12)$

63. $(x - h)^2 = 4p(y - k) \qquad p = -5, (h, k) = (-2, 5)$

$(x + 2)^2 = 4(-5)(y - 5)$

$(x + 2)^2 = -20(y - 5)$

65. Center: $(0, 3)$ Vertical major axis length 9 $\implies a = \dfrac{9}{2}$

$c = 3 \implies b^2 = a^2 - c^2 = \dfrac{81}{4} - 9 = \dfrac{45}{4}$

$$\frac{(x - 0)^2}{(45/4)} + \frac{(y - 3)^2}{(81/4)} = 1$$

67. Center: $(3, 5)$ Vertical transverse axis

$a = 4, c = 5, b^2 = c^2 - a^2 = 25 - 16 = 9$

$$\frac{(y - 5)^2}{16} - \frac{(x - 3)^2}{9} = 1$$

Section 11.2 Vectors in Space

- Vectors in space $\mathbf{v} = \langle v_1, v_2, v_3 \rangle$ have many of the same properties as vectors in the plane.
- The dot product of two vectors $\mathbf{u} = \langle u_1, u_2, u_3 \rangle$ and $\mathbf{v} = \langle v_1, v_2, v_3 \rangle$ in space is $\mathbf{u} \cdot \mathbf{v} = u_1v_1 + u_2v_2 + u_3v_3$.
- Two nonzero vectors $\mathbf{u}$ and $\mathbf{v}$ are said to be parallel if there is some scalar c such that $\mathbf{u} = c\mathbf{v}$.
- You should be able to use vectors to solve real life problems.

Solutions to Odd-Numbered Exercises

1. $\mathbf{v} = \langle 0 - 2, 3 - 0, 2 - 1 \rangle = \langle -2, 3, 1 \rangle$

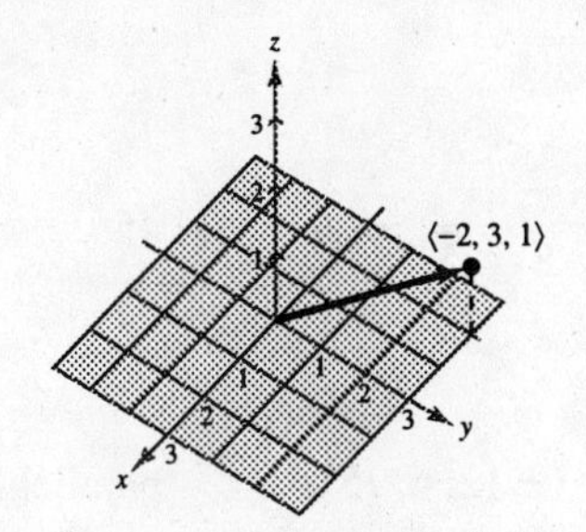

3. (a) $\mathbf{v} = \langle 3 - (-1), 2 - (-2), 5 - 1 \rangle = \langle 4, 4, 4 \rangle$

(b) $\|\mathbf{v}\| = \sqrt{4^2 + 4^2 + 4^2} = \sqrt{48} = 4\sqrt{3}$

(c) Unit vector: $\dfrac{1}{4\sqrt{3}} \langle 4, 4, 4 \rangle = \left\langle \dfrac{\sqrt{3}}{3}, \dfrac{\sqrt{3}}{3}, \dfrac{\sqrt{3}}{3} \right\rangle$

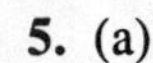

5. (a)

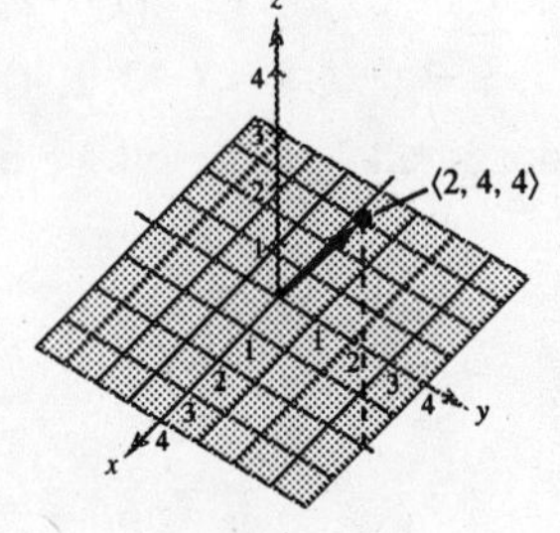

(b)

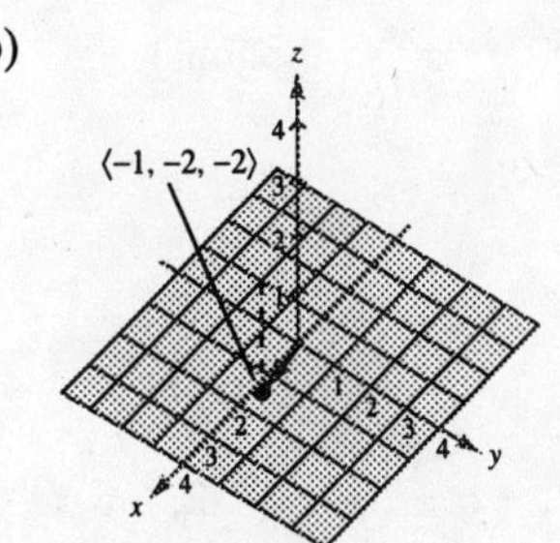

(c)

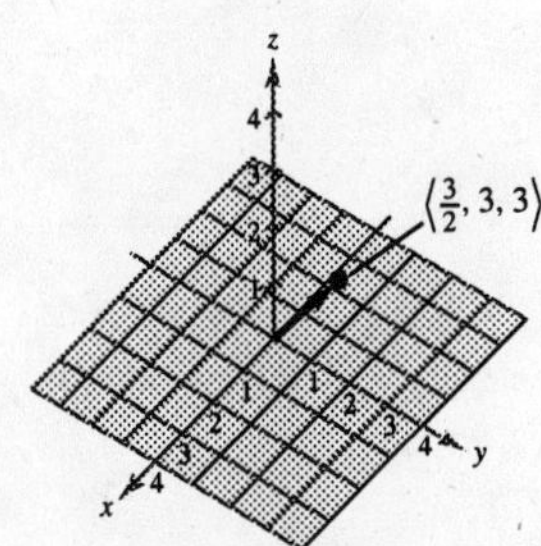

(d)

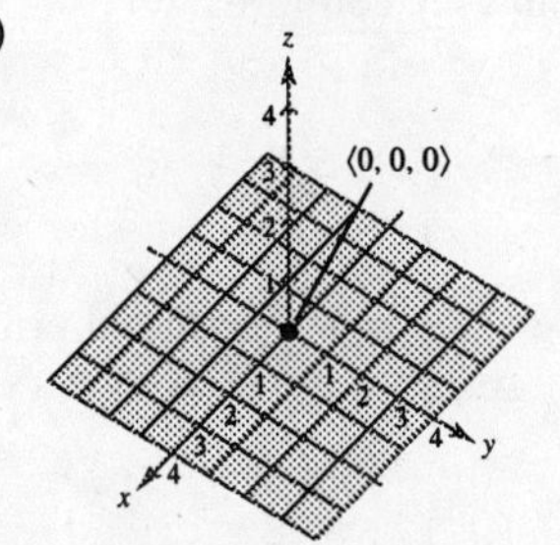

7. $\mathbf{z} = \mathbf{u} - 2\mathbf{v} = \langle -1, 3, 2 \rangle - 2\langle 1, -2, -2 \rangle = \langle -3, 7, 6 \rangle$

9. $2\mathbf{z} - 4\mathbf{u} = \mathbf{w} \Rightarrow \mathbf{z} = \frac{1}{2}(4\mathbf{u} + \mathbf{w}) = \frac{1}{2}(4\langle -1, 3, 2 \rangle + \langle 5, 0, -5 \rangle)$

$= \langle \frac{1}{2}, 6, \frac{3}{2} \rangle$

11. $\|\mathbf{v}\| = \sqrt{4^2 + 1^2 + 4^2} = \sqrt{33}$

13. $\|\mathbf{v}\| = \sqrt{4^2 + (-3)^2 + (-7)^2} = \sqrt{16 + 9 + 49}$

$= \sqrt{74}$

15. $\mathbf{v} = \langle 1-1, 0-(-3), -1-4\rangle = \langle 0, 3, -5\rangle$

$\|\mathbf{v}\| = \sqrt{0 + 3^2 + (-5)^2} = \sqrt{34}$

17. (a) $\dfrac{\mathbf{u}}{\|\mathbf{u}\|} = \dfrac{\langle 8, 3, -1\rangle}{\sqrt{74}} = \dfrac{1}{\sqrt{74}}(8\mathbf{i} + 3\mathbf{j} - \mathbf{k})$

(b) $-\dfrac{1}{\sqrt{74}}(8\mathbf{i} + 3\mathbf{j} - \mathbf{k})$

19. $6\mathbf{u} - 4\mathbf{v} = 6\langle -1, 3, 4\rangle - 4\langle 5, 4.5, -6\rangle$

$= \langle -6, 18, 24\rangle + \langle -20, -18, 24\rangle$

$= \langle -26, 0, 48\rangle$

21. $\mathbf{u} + \mathbf{v} = \langle -1, 3, 4\rangle + \langle 5, 4.5, -6\rangle = \langle 4, 7.5, -2\rangle$

$\|\mathbf{u} + \mathbf{v}\| = \sqrt{4^2 + 7.5^2 + (-2)^2} = \frac{1}{2}\sqrt{305} \approx 34.93$

23. $\|\mathbf{v}\| = \sqrt{5^2 + 4.5^2 + (-6)^2} = \sqrt{\dfrac{325}{4}} = \dfrac{5}{2}\sqrt{13}$

$\dfrac{\mathbf{v}}{\|\mathbf{v}\|} = \dfrac{\langle 5, 4.5, -6\rangle}{\frac{5}{2}\sqrt{13}} = \left\langle \dfrac{2}{\sqrt{13}}, \dfrac{9}{5\sqrt{13}}, \dfrac{-12}{5\sqrt{13}}\right\rangle$

$= \left\langle \dfrac{2\sqrt{13}}{13}, \dfrac{9\sqrt{13}}{65}, \dfrac{-12\sqrt{13}}{65}\right\rangle$

25. $\mathbf{u} \cdot \mathbf{v} = \langle 4, 4, -1\rangle \cdot \langle 2, -5, -8\rangle$

$= 8 - 20 + 8 = -4$

27. $\mathbf{u} \cdot \mathbf{v} = 2(9) + 5(-3) + (-3)(1) = 0$

29. $\cos\theta = \dfrac{\mathbf{u} \cdot \mathbf{v}}{\|\mathbf{u}\|\,\|\mathbf{v}\|} = \dfrac{-8}{\sqrt{8}\sqrt{25}} \Rightarrow \theta \approx 124.45°$

31. $\cos\theta = \dfrac{\mathbf{u} \cdot \mathbf{v}}{\|\mathbf{u}\|\,\|\mathbf{v}\|} = \dfrac{-120}{\sqrt{1700}\sqrt{73}} \Rightarrow \theta \approx 109.92°$

33. $-\dfrac{3}{2}\langle 8, -4, -10\rangle = \langle -12, 6, 15\rangle \Rightarrow$ parallel

35. $\mathbf{u} \cdot \mathbf{v} = 3 - 5 + 2 = 0 \Rightarrow$ orthogonal

37. $\mathbf{v} = \langle 7-5, 3-4, -1-1\rangle = \langle 2, -1, -2\rangle$

$\mathbf{u} = \langle 4-7, 5-3, 3-(-1)\rangle = \langle -3, 2, 4\rangle$

Since **u** and **v** are not parallel, the points are not collinear.

39. $\mathbf{v} = \langle -1-1, 2-3, 5-2\rangle = \langle -2, -1, 3\rangle$

$\mathbf{u} = \langle 3-(-1), 4-2, -1-5\rangle = \langle 4, 2, -6\rangle$

Since $\mathbf{u} = -2\mathbf{v}$, the points are collinear.

41. $\mathbf{v} = \langle 2, -4, 7\rangle = \langle q_1 - 1, q_2 - 5, q_3 - 0\rangle \Rightarrow$

$$\left.\begin{array}{l} 2 = q_1 - 1 \\ -4 = q_2 - 5 \\ 7 = q_3 \end{array}\right\} \Rightarrow \left.\begin{array}{l} q_1 = 3 \\ q_2 = 1 \\ q_3 = 7 \end{array}\right\} \Rightarrow \text{Terminal point is } \langle 3, 1, 7\rangle.$$

43. $\mathbf{v} = \left\langle 4, \frac{3}{2}, -\frac{1}{4}\right\rangle = \left\langle q_1 - 2, q_2 - 1, q_3 + \frac{3}{2}\right\rangle$

$4 = q_1 - 2 \Rightarrow q_1 = 6$

$\frac{3}{2} = q_2 - 1 \Rightarrow q_2 = \frac{5}{2}$

$-\frac{1}{4} = q_3 + \frac{3}{2} \Rightarrow q_3 = -\frac{7}{4}$

Terminal point: $\left(6, \frac{5}{2}, -\frac{7}{4}\right)$

45. $c\mathbf{u} = c\mathbf{i} + 2c\mathbf{j} + 3c\mathbf{k}$

$\|c\mathbf{u}\| = \sqrt{c^2 + 4c^2 + 9c^2} = |c|\sqrt{14} = 3 \Rightarrow c = \pm\dfrac{3}{\sqrt{14}} = \pm\dfrac{3\sqrt{14}}{14}$

47. $\mathbf{v} = \langle q_1, q_2, q_3 \rangle$. Since $\mathbf{v}$ lies in the yz-plane, $q_1 = 0$. Since $\mathbf{v}$ makes an angle of 45°, $q_2 = q_3$. Finally, $\|\mathbf{v}\| = 4$ implies that $q_2^2 + q_3^2 = 16$. Thus, $q_2 = q_3 = 2\sqrt{2}$ and $\mathbf{v} = \langle 0, 2\sqrt{2}, 2\sqrt{2} \rangle$.

49. 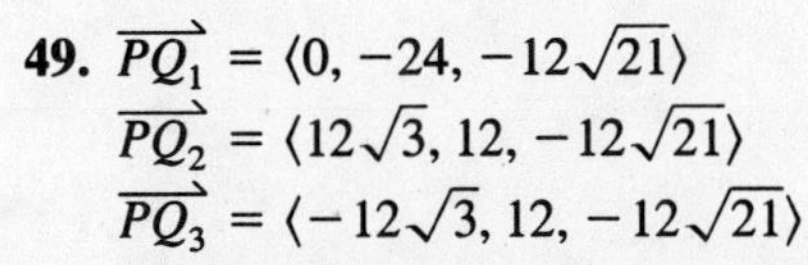

$$\overrightarrow{PQ_1} = \langle 0, -24, -12\sqrt{21} \rangle$$
$$\overrightarrow{PQ_2} = \langle 12\sqrt{3}, 12, -12\sqrt{21} \rangle$$
$$\overrightarrow{PQ_3} = \langle -12\sqrt{3}, 12, -12\sqrt{21} \rangle$$

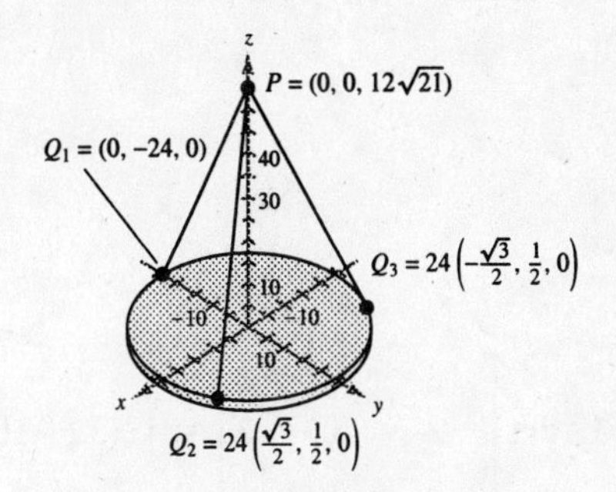

Let $\mathbf{F}_1$, $\mathbf{F}_2$, and $\mathbf{F}_3$ be the tension on each wire. Since $\|\mathbf{F}_1\| = \|\mathbf{F}_2\| = \|\mathbf{F}_3\|$, there exists a constant c such that

$$\mathbf{F}_1 = C\langle 0, -24, -12\sqrt{21} \rangle$$
$$\mathbf{F}_2 = C\langle 12\sqrt{3}, 12, -12\sqrt{21} \rangle$$
$$\mathbf{F}_3 = C\langle -12\sqrt{3}, 12, -12\sqrt{21} \rangle$$

The total force is $-30\mathbf{k} = \mathbf{F}_1 + \mathbf{F}_2 + \mathbf{F}_3 \Rightarrow$ the vertical ($\mathbf{k}$) component satisfies $-10 = -12\sqrt{21}\,c \Rightarrow c = \dfrac{5}{6\sqrt{21}}$. Hence,

$$\mathbf{F}_1 = \left\langle 0, \frac{-20}{\sqrt{21}}, -10 \right\rangle$$
$$\mathbf{F}_2 = \left\langle \frac{10}{\sqrt{7}}, \frac{10}{\sqrt{21}}, -10 \right\rangle$$
$$\mathbf{F} = \left\langle \frac{-10}{\sqrt{7}}, \frac{10}{\sqrt{21}}, -10 \right\rangle$$

$\|\mathbf{F}_1\| = \|\mathbf{F}_2\| = \|\mathbf{F}_3\| \approx 10.91$ pounds

$$Q_1 = (0, -24, 0)$$
$$Q_2 = (20.8, 12, 0)$$
$$Q_3 = (-20.8, 12, 0)$$
$$P = (0, 0, 55)$$

51. True. $\cos\theta = 0 \quad \Rightarrow \quad \theta = 90°$

53. (a)

(b) $\mathbf{w} = a\mathbf{u} + b\mathbf{v} = a\langle 1, 1, 0 \rangle + b\langle 0, 1, 1 \rangle$

$$\mathbf{0} = \langle a, a + b, b \rangle \Rightarrow a = b = 0$$

(c) $\mathbf{w} = \langle 1, 2, 1 \rangle = a\langle 1, 1, 0 \rangle + b\langle 0, 1, 1 \rangle$

$$1 = a$$
$$2 = a + b$$
$$1 = b$$

Hence $a = b = 1$

(d) $\mathbf{w} = \langle 1, 2, 3 \rangle = a\langle 1, 1, 0 \rangle + b\langle 0, 1, 1 \rangle$

$$1 = a$$
$$2 = a + b$$
$$3 = b$$

Impossible

55. If $\mathbf{u} \cdot \mathbf{v} < 0$, then $\cos\theta < 0$ and the angle between $\mathbf{u}$ and $\mathbf{v}$ is obtuse, $180° \le \theta < 90°$.

57. $x = 2t - 1$

$y = -t + 3$

$$y = -\left(\frac{x+1}{2}\right) + 3 = -\frac{x}{2} + \frac{5}{2}$$

Line: $2y + x = 5$

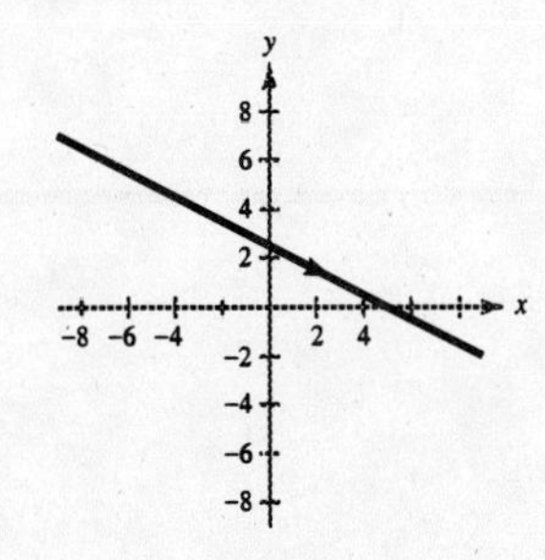

59. $x = t - 1$

$y = 2t^2$

$y = 2(x + 1)^2$ parabola

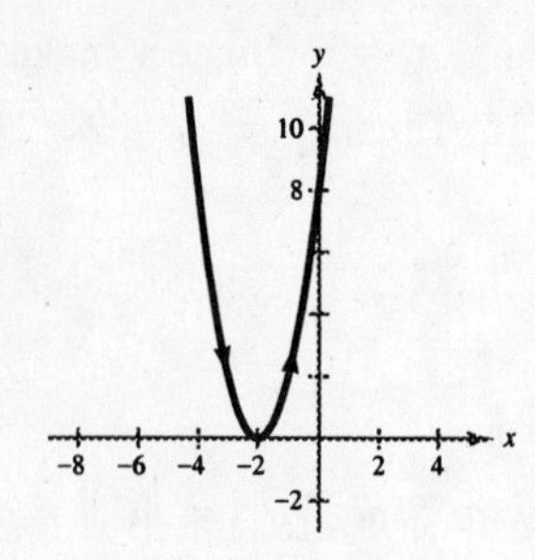

61. $\det\begin{bmatrix} 12 & 4 & -1 \\ -2 & 3 & 2 \\ 5 & 8 & 1 \end{bmatrix} = 12(3 - 16) - 4(-2 - 10) - 1(-16 - 15) = -77$

Section 11.3 The Cross Product of Two Vectors

- The cross product of two vectors $\mathbf{u} = u_1\mathbf{i} + u_2\mathbf{j} + u_3\mathbf{k}$ and $\mathbf{v} = v_1\mathbf{i} + v_2\mathbf{j} + v_3\mathbf{k}$ is given by

$$\mathbf{u} \times \mathbf{v} = (u_2v_3 - u_3v_2)\mathbf{i} - (u_1v_3 - u_3v_1)\mathbf{j} + (u_1v_2 - u_2v_1)\mathbf{k}$$

$$= \begin{vmatrix} \mathbf{i} & \mathbf{j} & \mathbf{k} \\ u_1 & u_2 & u_3 \\ v_1 & v_2 & v_3 \end{vmatrix}.$$

- The cross product satisfies the following algebraic properties.
  - (a) $\mathbf{u} \times \mathbf{v} = -(\mathbf{v} \times \mathbf{u})$
  - (b) $\mathbf{u} \times (\mathbf{v} + \mathbf{w}) = (\mathbf{u} \times \mathbf{v}) + (\mathbf{u} \times \mathbf{w})$
  - (c) $c(\mathbf{u} \times \mathbf{v}) = (c\mathbf{u}) \times \mathbf{v} = \mathbf{u} \times (c\mathbf{v})$
  - (d) $\mathbf{u} \times \mathbf{0} = \mathbf{0} \times \mathbf{u} = \mathbf{0}$
  - (e) $\mathbf{u} \times \mathbf{u} = \mathbf{0}$
  - (f) $\mathbf{u} \cdot (\mathbf{v} \times \mathbf{w}) = (\mathbf{u} \times \mathbf{v}) \cdot \mathbf{w}$
- The following geometric properties of the cross product are valid, where θ is the angle between the vectors **u** and **v**:
  - (a) $\mathbf{u} \times \mathbf{v}$ is orthogonal to both **u** and **v**.
  - (b) $\|\mathbf{u} \times \mathbf{v}\| = \|\mathbf{u}\|\,\|\mathbf{v}\| \sin\theta$
  - (c) $\mathbf{u} \times \mathbf{v} = \mathbf{0}$ if and only if **u** and **v** are scalar multiples.
  - (d) $\|\mathbf{u} \times \mathbf{v}\|$ is the area of the parallelogram having **u** and **v** as sides.
- The absolute value of the triple scalar product is the volume of the parallelepiped having **u**, **v**, and **w** as sides.

$$\mathbf{u} \cdot (\mathbf{v} \times \mathbf{w}) = \begin{vmatrix} u_1 & u_2 & u_3 \\ v_1 & v_2 & v_3 \\ w_1 & w_2 & w_3 \end{vmatrix}$$

Solutions to Odd-Numbered Exercises

1. $\mathbf{j}\times\mathbf{i} = \begin{vmatrix} \mathbf{i} & \mathbf{j} & \mathbf{k} \\ 0 & 1 & 0 \\ 1 & 0 & 0 \end{vmatrix} = -\mathbf{k}$

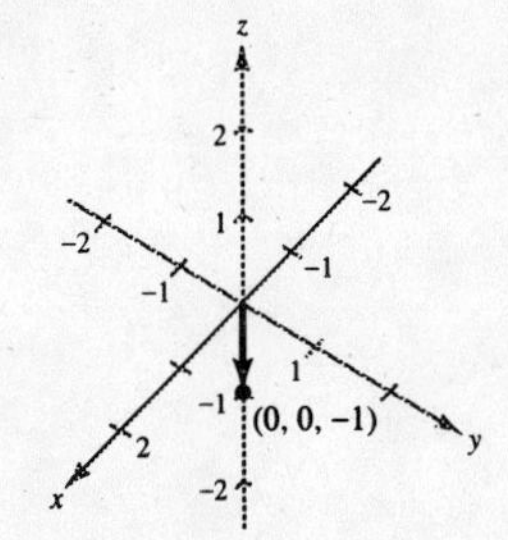

3. $\mathbf{i}\times\mathbf{k} = \begin{vmatrix} \mathbf{i} & \mathbf{j} & \mathbf{k} \\ 1 & 0 & 0 \\ 0 & 0 & 1 \end{vmatrix} = -\mathbf{j}$

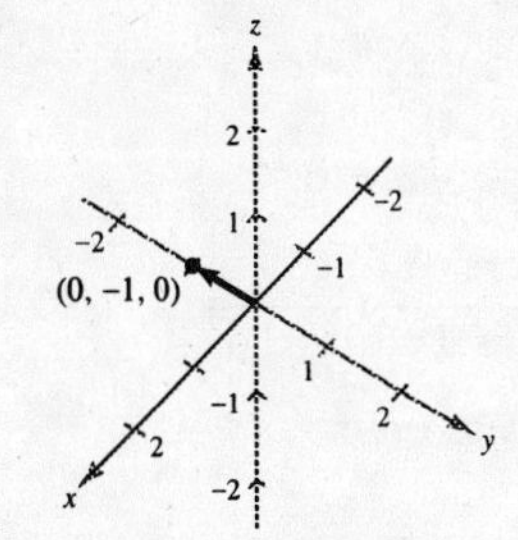

5. $\mathbf{u}\times\mathbf{v} = \begin{vmatrix} \mathbf{i} & \mathbf{j} & \mathbf{k} \\ 1 & -4 & 0 \\ 2 & 6 & 0 \end{vmatrix} = 14\mathbf{k} = \langle 0, 0, 14\rangle$

7. $\mathbf{u}\times\mathbf{v} = \begin{vmatrix} \mathbf{i} & \mathbf{j} & \mathbf{k} \\ 7 & -5 & 2 \\ -1 & 4 & -1 \end{vmatrix} = \langle -3, 5, 23\rangle$

9. $\mathbf{u}\times\mathbf{v} = \begin{vmatrix} \mathbf{i} & \mathbf{j} & \mathbf{k} \\ 6 & 2 & 1 \\ 1 & 3 & -2 \end{vmatrix} = \langle -7, 13, 16\rangle$

$= -7\mathbf{i} + 13\mathbf{j} + 16\mathbf{k}$

11. $\mathbf{u}\times\mathbf{v} = \begin{vmatrix} \mathbf{i} & \mathbf{j} & \mathbf{k} \\ 0 & 0 & 6 \\ -1 & 3 & 1 \end{vmatrix} = \langle -18, -6, 0\rangle$

$= -18\mathbf{i} - 6\mathbf{j}$

13. $\mathbf{u}\times\mathbf{v} = \begin{vmatrix} \mathbf{i} & \mathbf{j} & \mathbf{k} \\ 2 & 4 & 3 \\ 0 & -2 & 1 \end{vmatrix} = \langle 10, -2, -4\rangle$

15. $\mathbf{u}\times\mathbf{v} = \begin{vmatrix} \mathbf{i} & \mathbf{j} & \mathbf{k} \\ 6 & -5 & 1 \\ \frac{1}{3} & -\frac{1}{3} & \frac{2}{3} \end{vmatrix} = \langle -3, -\frac{11}{3}, -\frac{1}{3}\rangle$

$= -3\mathbf{i} - \frac{11}{3}\mathbf{j} - \frac{1}{3}\mathbf{k}$

17. $\mathbf{u}\times\mathbf{v} = \begin{vmatrix} \mathbf{i} & \mathbf{j} & \mathbf{k} \\ -1 & 0 & 1 \\ 0 & 1 & -2 \end{vmatrix} = \langle -1, -2, -1\rangle$

$= -\mathbf{i} - 2\mathbf{j} - \mathbf{k}$

19. $\mathbf{u}\times\mathbf{v} = \begin{vmatrix} \mathbf{i} & \mathbf{j} & \mathbf{k} \\ 3 & 1 & 0 \\ 0 & 1 & 1 \end{vmatrix} = \mathbf{i} - 3\mathbf{j} + 3\mathbf{k}$

$\|\mathbf{u}\times\mathbf{v}\| = \sqrt{19}$

Unit vector $= \dfrac{\mathbf{u}\times\mathbf{v}}{\|\mathbf{u}\times\mathbf{v}\|} = \dfrac{1}{\sqrt{19}}(\mathbf{i} - 3\mathbf{j} + 3\mathbf{k})$

21. $\mathbf{u}\times\mathbf{v} = \begin{vmatrix} \mathbf{i} & \mathbf{j} & \mathbf{k} \\ -2 & 1 & 3 \\ 1 & 4 & 6 \end{vmatrix} = -6\mathbf{i} + 15\mathbf{j} - 9\mathbf{k}$

$\|\mathbf{u}\times\mathbf{v}\| = \sqrt{342}$

Unit vector $= \dfrac{\mathbf{u}\times\mathbf{v}}{\|\mathbf{u}\times\mathbf{v}\|} = \dfrac{1}{\sqrt{342}}(-6\mathbf{i} + 15\mathbf{j} - 9\mathbf{k})$

$= \dfrac{\sqrt{38}}{38}(-2\mathbf{i} + 5\mathbf{j} - 3\mathbf{k})$

23. $\mathbf{u}\times\mathbf{v} = \begin{vmatrix} \mathbf{i} & \mathbf{j} & \mathbf{k} \\ 1 & 1 & -1 \\ 1 & 1 & 1 \end{vmatrix} = 2\mathbf{i} - 2\mathbf{j}$

$\|\mathbf{u}\times\mathbf{v}\| = 2\sqrt{2}$

Unit vector $= \dfrac{\mathbf{u}\times\mathbf{v}}{\|\mathbf{u}\times\mathbf{v}\|} = \dfrac{1}{2\sqrt{2}}(2\mathbf{i} - 2\mathbf{j})$

$= \dfrac{1}{\sqrt{2}}\mathbf{i} - \dfrac{1}{\sqrt{2}}\mathbf{j}$

25. $\mathbf{u} \times \mathbf{v} = \begin{vmatrix} \mathbf{i} & \mathbf{j} & \mathbf{k} \\ 0 & 0 & 1 \\ 1 & 0 & 1 \end{vmatrix} = \mathbf{j}$

Area $= \|\mathbf{u} \times \mathbf{v}\| = \|\mathbf{j}\| = 1$ square units

27. $\mathbf{u} \times \mathbf{v} = \begin{vmatrix} \mathbf{i} & \mathbf{j} & \mathbf{k} \\ 3 & 4 & 6 \\ 2 & -1 & 5 \end{vmatrix} = 26\mathbf{i} + 3\mathbf{j} + 11\mathbf{k}$

Area $= \|\mathbf{u} \times \mathbf{v}\| = \sqrt{26^2 + 3^2 + (-11)^2}$

$= \sqrt{806}$ square units

29. $\mathbf{u} \times \mathbf{v} = \begin{vmatrix} \mathbf{i} & \mathbf{j} & \mathbf{k} \\ 2 & 2 & -3 \\ 0 & 2 & 3 \end{vmatrix} = \langle 12, -6, 4 \rangle$

Area $= \|\mathbf{u} \times \mathbf{v}\| = \sqrt{12^2 + (-6)^2 + 4^2}$

$= 14$ square units

31. (a) $\overrightarrow{AB} = \langle 3 - 2, 1 - (-1), 2 - 4 \rangle = \langle 1, 2, -2 \rangle$ is parallel to $\overrightarrow{DC} = \langle 0 - (-1), 5 - 3, 6 - 8 \rangle = \langle 1, 2, -2 \rangle$.
$\overrightarrow{AD} = \langle -3, 4, 4 \rangle$ is parallel to $\overrightarrow{BC} = \langle -3, 4, 4 \rangle$.

(b) $\overrightarrow{AB} \times \overrightarrow{AD} = \begin{vmatrix} \mathbf{i} & \mathbf{j} & \mathbf{k} \\ 1 & 2 & -2 \\ -3 & 4 & 4 \end{vmatrix} = \langle 16, 2, 10 \rangle$

Area $= \|\overrightarrow{AB} \times \overrightarrow{AD}\| = \sqrt{16^2 + 2^2 + 10^2} = \sqrt{360} = 6\sqrt{10}$ sq. units

(c) $\overrightarrow{AB} \cdot \overrightarrow{AC} = \langle 1, 2, -2 \rangle \cdot \langle -2, -2, 11 \rangle \neq 0 \implies$ not a rectangle

33. $\mathbf{u} = \langle 4, -2, 6 \rangle, \quad \mathbf{v} = \langle -4, 0, 3 \rangle$

$\mathbf{u} \times \mathbf{v} = \begin{vmatrix} \mathbf{i} & \mathbf{j} & \mathbf{k} \\ 4 & -2 & 6 \\ -4 & 0 & 3 \end{vmatrix} = \langle -6, -36, -8 \rangle$

Area $= \frac{1}{2}\|\mathbf{u} \times \mathbf{v}\| = \frac{1}{2}\sqrt{(-6)^2 + (-36)^2 + (-8)^2}$

$= \frac{1}{2}\sqrt{1396} = \sqrt{349}$ sq. units

35. $\mathbf{u} = \langle -2 - 2, -2 - 3, 0 - (-5) \rangle = \langle -4, -5, 5 \rangle$

$\mathbf{v} = \langle 3 - 2, 0 - 3, 6 - (-5) \rangle = \langle 1, -3, 11 \rangle$

$\mathbf{u} \times \mathbf{v} = \begin{vmatrix} \mathbf{i} & \mathbf{j} & \mathbf{k} \\ -4 & -5 & 5 \\ 1 & -3 & 11 \end{vmatrix} = \langle -40, 49, 17 \rangle$

Area $= \frac{1}{2}\|\mathbf{u} \times \mathbf{v}\| = \frac{1}{2}\sqrt{(-40)^2 + 49^2 + 17^2}$

$= \frac{1}{2}\sqrt{4290}$ sq. units

37. $\mathbf{u} \cdot (\mathbf{v} \times \mathbf{w}) = \begin{vmatrix} 2 & 3 & 3 \\ 4 & 4 & 0 \\ 0 & 0 & 4 \end{vmatrix} = 2(16) - 3(16) + 3(0) = -16$

39. $\mathbf{u} \cdot (\mathbf{v} \times \mathbf{w}) = \begin{vmatrix} 2 & 3 & 1 \\ 1 & -1 & 0 \\ 4 & 3 & 1 \end{vmatrix} = 2(-1) - 3(1) + 1(7) = 2$

41. $\mathbf{u} \cdot (\mathbf{v} \times \mathbf{w}) = \begin{vmatrix} 1 & 1 & 0 \\ 0 & 1 & 1 \\ 1 & 0 & 1 \end{vmatrix} = 1 + 1 = 2$

Volume $= |\mathbf{u} \cdot (\mathbf{v} \times \mathbf{w})| = 2$ cubic units

43. $\mathbf{u} \cdot (\mathbf{v} \times \mathbf{w}) = \begin{vmatrix} 0 & 2 & 2 \\ 0 & 0 & -2 \\ 3 & 0 & 2 \end{vmatrix} = 0 - 2(6) + 2(0) = -12$

Volume $= |\mathbf{u} \cdot (\mathbf{v} \times \mathbf{w})| = 12$ cubic units

45. $\mathbf{u} = \langle 4, 0, 0\rangle$, $\mathbf{v} = \langle 0, -2, 3\rangle$, $\mathbf{w} = \langle 0, 5, 3\rangle$

$$\mathbf{u} \cdot (\mathbf{v} \times \mathbf{w}) = \begin{vmatrix} 4 & 0 & 0 \\ 0 & -2 & 3 \\ 0 & 5 & 3 \end{vmatrix} = 4(-21) = -84$$

Volume $= |-84| = 84$ cubic units

47. $$\mathbf{V} \times \mathbf{F} = \begin{vmatrix} \mathbf{i} & \mathbf{j} & \mathbf{k} \\ 0 & \frac{1}{2}\cos 40^\circ & \frac{1}{2}\sin 40^\circ \\ 0 & 0 & -20 \end{vmatrix} = -10\cos 40^\circ\,\mathbf{i}$$

$\|\mathbf{V} \times \mathbf{F}\| = 10\cos 40^\circ \approx 7.66$ foot-pounds

49. True. The cross product is defined for vectors in three-dimensional space.

51. $$\mathbf{u} \times \mathbf{u} = \begin{vmatrix} \mathbf{i} & \mathbf{j} & \mathbf{k} \\ u_1 & u_2 & u_3 \\ u_1 & u_2 & u_3 \end{vmatrix} = (u_2u_3 - u_2u_3)\mathbf{i} - (u_1u_3 - u_1u_3)\mathbf{j} - (u_1u_2 - u_1u_2)\mathbf{k} = \mathbf{0}$$

53. $$\mathbf{u} \times \mathbf{v} = \begin{vmatrix} \mathbf{i} & \mathbf{j} & \mathbf{k} \\ \cos\alpha & \sin\alpha & 0 \\ \cos\beta & \sin\beta & 0 \end{vmatrix} = (\cos\alpha\sin\beta - \sin\alpha\cos\beta)\mathbf{k}$$

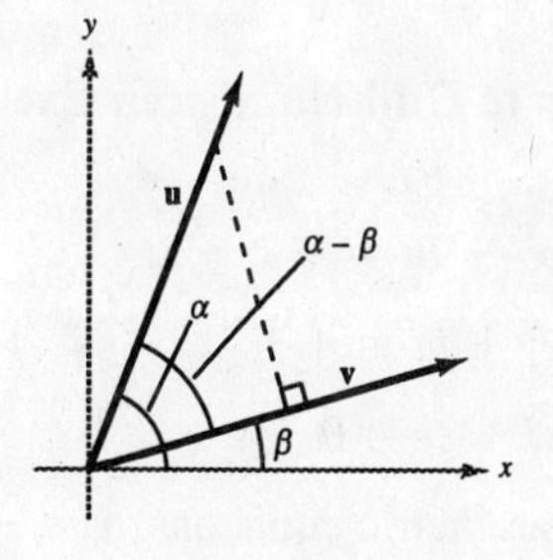

Area of triangle formed by the unit vectors **u** and **v** is

$\frac{1}{2}(\text{base})(\text{height}) = \frac{1}{2}(1)\sin(\alpha - \beta)$.

The area is also given by

$\frac{1}{2}\|\mathbf{u} \times \mathbf{v}\| = \frac{1}{2}|\cos\alpha\sin\beta - \sin\alpha\cos\beta| = \sin\alpha\cos\beta - \cos\alpha\sin\beta$.

Thus, $\sin(\alpha - \beta) = \sin\alpha\cos\beta - \cos\alpha\sin\beta$.

55. $\tan 300^\circ = -\sqrt{3}$

57. $\cos 930^\circ = \cos 210^\circ = -\dfrac{\sqrt{3}}{2}$

59. $\cos\dfrac{17\pi}{6} = \cos\dfrac{5\pi}{6} = -\dfrac{\sqrt{3}}{2}$

61. $\tan\dfrac{10\pi}{3} = \tan\dfrac{4\pi}{3} = \sqrt{3}$

63. $$\begin{aligned} 4x + 3y &= 24 \\ x + 3y &= 15 \\ \hline 3x &= 9 \\ x &= 3 \end{aligned}$$

$(x, y) = (3, 4)$

Vertex	z
(0, 0)	0
(6, 0)	36
(0, 5)	35
(3, 4)	46

Maximum value is 46 at (3, 4)

Minimum value is 0 at (0, 0)

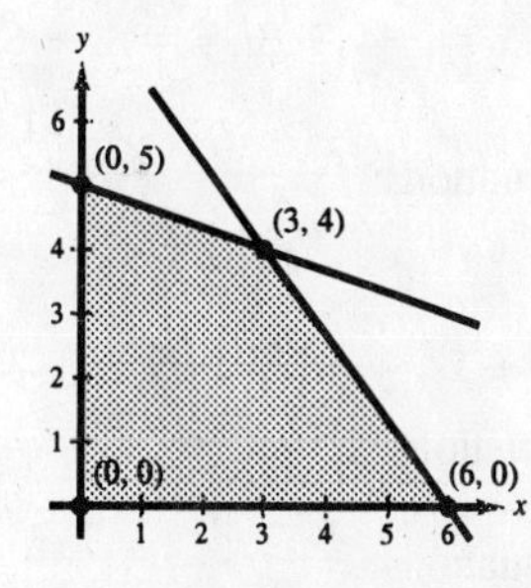

Section 11.4 Lines and Planes in Space

- The parametric equations of the line in space parallel to the vector $\langle a, b, c\rangle$ and passing through the point (x_1, y_2, z_3) are

  $$x = x_1 + at, \quad y = y_1 + bt, \quad z = z_1 + ct.$$

- The standard equation of the plane in space containing the point (x_1, y_1, z_1) and having normal vector $\langle a, b, c\rangle$ is

  $$a(x - x_1) + b(y - y_1) + c(z - z_1) = 0.$$

- You should be able to find the angle between two planes by calculating the angle between their normal vectors.
- You should be able to sketch a plane in space.
- The distance between a point Q and a plane having normal $\mathbf{n}$ is

  $$D = \|\text{proj}_{\mathbf{n}}\overrightarrow{PQ}\| = \frac{|\overrightarrow{PQ} \cdot \mathbf{n}|}{\|\mathbf{n}\|}$$

  where P is a point in the plane.

Solutions to Odd-Numbered Exercises

1. $x = x_1 + at = -1 - 2t$

$y = y_1 + bt = 4 + 4t$

$z = z_1 + ct = 0 + t$

(a) Parametric equations: $x = -1 - 2t, y = 4 + 4t, z = t$

(b) Symmetric equations: $\dfrac{x + 1}{-2} = \dfrac{y - 4}{4} = z$

3. $x = x_1 + at = -4 + \dfrac{1}{2}t,\ y = y_1 + bt = 1 + \dfrac{4}{3}t,\ z = z_1 + ct = 0 - t$

(a) Parametric equations: $x = -4 + \dfrac{1}{2}t, y = 1 + \dfrac{4}{3}t, z = -t$

Equivalently: $x = -4 + 3t, y = 1 + 8t, z = -6t$

(b) Symmetric equations: $\dfrac{x + 4}{3} = \dfrac{y - 1}{8} = \dfrac{z}{-6}$

5. $x = x_1 + at = 2 + 2t,\ y = y_1 + bt = -3 - 3t,\ z = z_1 + ct = 5 + t$

(a) Parametric equations: $x = 2 + 2t, y = -3 - 3t, z = 5 + t$

(b) Symmetric equations: $\dfrac{x - 2}{2} = \dfrac{y + 3}{-3} = z - 5$

7. (a) $\mathbf{v} = \langle 2 - 6, 1 - 0, 8 - 3\rangle = \langle 4, 1, 5\rangle$ Point $(6, 0, 3)$

$x = 6 - 4t, y = t, z = 3 + 5t$

(b) $\dfrac{x - 6}{-4} = y = \dfrac{z - 3}{5}$

9. (a) $\mathbf{v} = \langle 1 - (-3), -2 - 8, 16 - 15 \rangle = \langle 4, -10, 1 \rangle$ Point $(-3, 8, 15)$

$x = -3 + 4t, y = 8 - 10t, z = 15 + t$

(b) $\dfrac{x+3}{4} = \dfrac{y-8}{-10} = \dfrac{z-15}{1}$

11. The line is $x = -4 + 3t$, $y = -1 - t$, $z = 7$, or $(x + 4)/3 = (y + 1)/-1$, $z = 7$.

Only (b) and (c) satisfy the equation.

13.

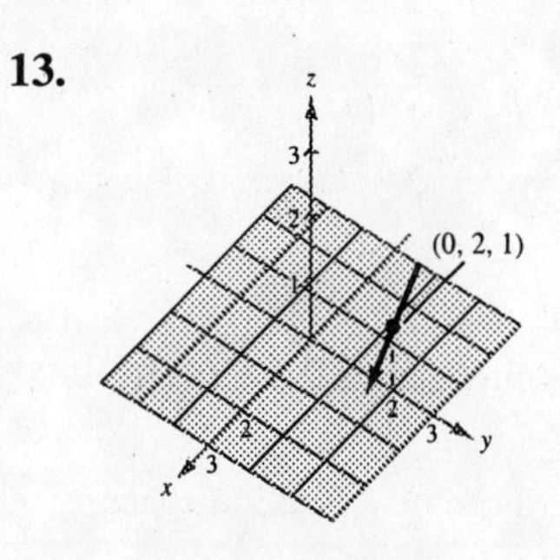

15. $a(x - x_1) + b(y - y_1) + c(z - z_1) = 0$

$0(x - 3) + 1(y - 4) + 0(z + 2) = 0$

$y - 4 = 0$

17. $-2(x - 5) + 1(y - 6) - 2(z - 3) = 0$

$-2x + y - 2z + 10 = 0$

19. $\mathbf{n} = \langle -1, -2, 1 \rangle \Rightarrow -1(x - 2) - 2(y - 0) + 1(z - 0) = 0$

$-x - 2y + z + 2 = 0$

21. $\mathbf{u} = \langle 2 - 0, 1 - 0, 3 - 0 \rangle = \langle 2, 1, 3 \rangle$

$\mathbf{v} = \langle -2 - 0, 1 - 0, 3 - 0 \rangle = \langle -2, 1, 3 \rangle$

$$\mathbf{n} = \mathbf{u} \times \mathbf{v} = \begin{vmatrix} \mathbf{i} & \mathbf{j} & \mathbf{k} \\ 2 & 1 & 3 \\ -2 & 1 & 3 \end{vmatrix} = -12\mathbf{j} + 4\mathbf{k}$$

$0(x - 0) - 12(y - 0) + 4(z - 0) = 0$

$-12y + 4z = 0$

$-3y + z = 0$

23. $\mathbf{u} = \langle 4 - 0, 1 - (-1), 6 - (-2) \rangle = \langle 4, 2, 8 \rangle$

$\mathbf{v} = \langle 1 - 0, 0 - (-1), -3 - (-2) \rangle = \langle 1, 1, -1 \rangle$

$$\mathbf{n} = \mathbf{u} \times \mathbf{v} = \begin{vmatrix} \mathbf{i} & \mathbf{j} & \mathbf{k} \\ 4 & 2 & 8 \\ 1 & 1 & -1 \end{vmatrix} = -10\mathbf{i} + 12\mathbf{j} + 2\mathbf{k}$$

$-10(x - 0) + 12(y + 1) + 2(z + 2) = 0$

$-10x + 12y + 2z + 16 = 0$

$-5x + 6y + z + 8 = 0$

25. $\mathbf{n} = \mathbf{j}$: $0(x - 2) + 1(y - 5) + 0(z - 3) = 0$

$y - 5 = 0$

27. The vectors $\langle 1, 2, 2 \rangle$ and $\langle 4 - 0, 0 - 2, 0 - 0 \rangle = \langle 4, -2, 0 \rangle$ are parallel to the plane.

$$\mathbf{n} = \begin{vmatrix} \mathbf{i} & \mathbf{j} & \mathbf{k} \\ 1 & 2 & 2 \\ 4 & -2 & 0 \end{vmatrix} = 4\mathbf{i} + 8\mathbf{j} - 10\mathbf{k}$$

is perpendicular to the plane.

Equation: $4(x - 4) + 8(y - 0) - 10(z - 0) = 0$

$4x + 8y - 10z - 16 = 0$

$2x + 4y - 5z - 8 = 0$

29. $\mathbf{n}_1 = \langle 3, 1, -4 \rangle$, $\mathbf{n}_2 = \langle -9, -3, 12 \rangle = -3\mathbf{n}_1$

Parallel

31. $\mathbf{n}_1 = \langle 2, 0, -1 \rangle$, $\mathbf{n}_2 = \langle 4, 1, 8 \rangle$

$\mathbf{n}_1 \cdot \mathbf{n}_2 = 8 - 8 = 0$ Orthogonal

33.

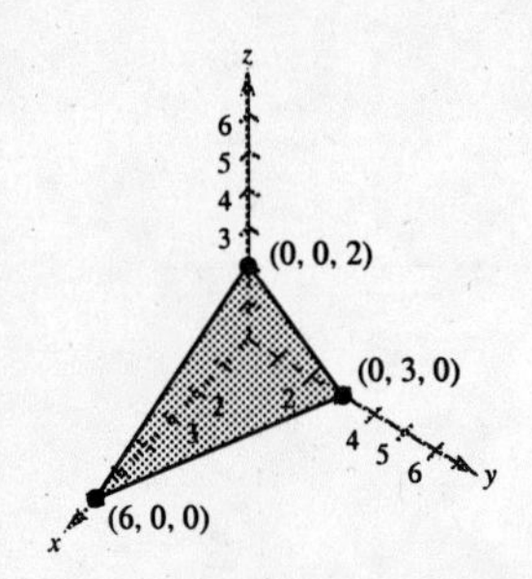

35.

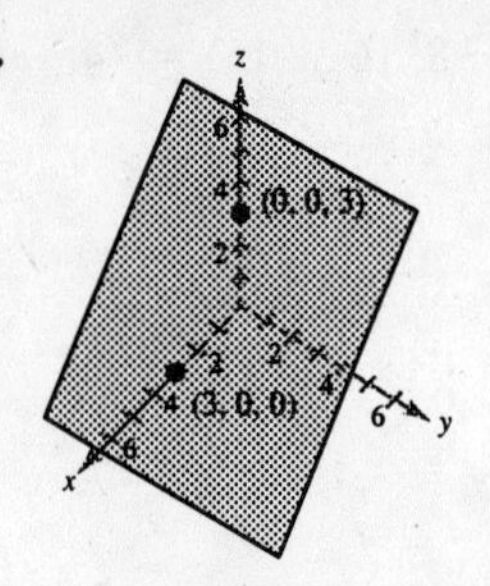

37.

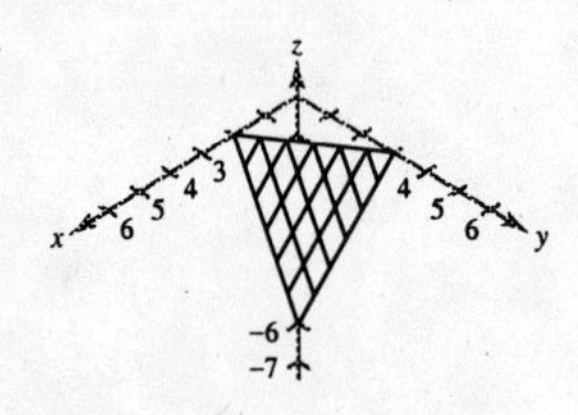

39.

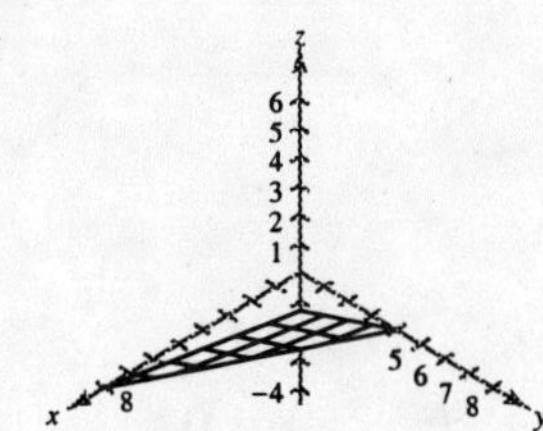

41. $D = \dfrac{|\overrightarrow{PQ} \cdot \mathbf{n}|}{\|\mathbf{n}\|}$

$P = (4, 0, 0)$ on plane, $Q = (0, 0, 0)$, $\mathbf{n} = \langle 3, 2, 1\rangle$, $\overrightarrow{PQ} = \langle -4, 0, 0\rangle$:

$$D = \frac{|\langle -4, 0, 0\rangle \cdot \langle 3, 2, 1\rangle|}{\sqrt{14}} = \frac{12}{\sqrt{14}} = \frac{6\sqrt{14}}{7}$$

43. $D = \dfrac{|\overrightarrow{PQ} \cdot \mathbf{n}|}{\|\mathbf{n}\|}$

$P = (2, 0, 0)$ on plane, $Q = (4, -2, -2)$, $\mathbf{n} = \langle 2, -1, 1\rangle$, $\overrightarrow{PQ} = \langle 2, -2, -2\rangle$

$$D = \frac{|\langle 2, -2, -2\rangle \cdot \langle 2, -1, 1\rangle|}{\sqrt{6}} = \frac{4}{\sqrt{6}} = \frac{2\sqrt{6}}{3}$$

45. (a) $\mathbf{n}_1 = \langle 3, -4, 5\rangle$, $\mathbf{n}_2 = \langle 1, 1, -1\rangle$ normal vectors to planes

$$\cos\theta = \frac{|\mathbf{n}_1 \cdot \mathbf{n}_2|}{\|\mathbf{n}_1\|\,\|\mathbf{n}_2\|} = \frac{|-6|}{\sqrt{50}\sqrt{3}} = \frac{6}{\sqrt{150}} \implies \theta \approx 60.67°$$

(b) $3x - 4y + 5z = 6$ Equation 1

$x + y - z = 2$ Equation 2

(-3) times Equation 2 added to Equation 1 gives

$$-7y + 8z = 0$$
$$y = \tfrac{8}{7}z$$

Substituting back into Equation 2, $x = 2 - y + z = 2 - \frac{8}{7}z + z$

$= 2 - \frac{1}{7}z.$

Letting $t = \dfrac{z}{7}$, we obtain

$x = 2 - t, y = 8t, z = 7t.$

47. (a) $\mathbf{n}_1 = \langle 1, 1, -1\rangle, \mathbf{n}_2 = \langle 2, -5, -1\rangle$ normal vectors to planes

$$\cos\theta = \frac{|\mathbf{n}_1 \cdot \mathbf{n}_2|}{\|\mathbf{n}_1\|\,\|\mathbf{n}_2\|} = \frac{|-2|}{\sqrt{3}\sqrt{30}} = \frac{2}{\sqrt{90}} \implies \theta \approx 77.83°$$

(b) $x + y - z = 0$ Equation 1

$2x - 5y - z = 1$ Equation 2

(-2) times Equation 1 added to Equation 2 gives

$$-7y + z = 1$$

$$y = \frac{z-1}{7}$$

Substituting back into Equation 1,

$$x = z - y = z - \frac{z-1}{7} = \frac{6z}{7} + \frac{1}{7} = \frac{1}{7}(6z + 1)$$

Letting $z = t, x = \dfrac{6t+1}{7}, y = \dfrac{t-1}{7}$.

Equivalently, let $y = t, z = 7t + 1$ and $x = 6t + 1$.

49. The normal vector to plane containing $(0, 0, 0), (2, 2, 12)$ and $(10, 0, 0)$ is obtained as follows.

$\mathbf{v}_1 = \langle 2, 2, 12\rangle, \mathbf{v}_2 = \langle 10, 0, 0\rangle$

$$\mathbf{v}_1 \times \mathbf{v}_2 = \begin{vmatrix} \mathbf{i} & \mathbf{j} & \mathbf{k} \\ 2 & 2 & 12 \\ 10 & 0 & 0 \end{vmatrix} = \langle 0, 120, -20\rangle.$$

$\mathbf{n}_1 = \langle 0, 6, -1\rangle$

The normal vector to the plane containing $(0, 0, 0), (2, 2, 12)$ and $(0, 10, 0)$ is obtained as follows.

$\mathbf{u}_1 = \langle 2, 2, 12\rangle, \mathbf{u}_2 = \langle 0, 10, 0\rangle$

$$\mathbf{u}_1 \times \mathbf{u}_2 = \begin{vmatrix} \mathbf{i} & \mathbf{j} & \mathbf{k} \\ 2 & 2 & 12 \\ 0 & 10 & 0 \end{vmatrix} = \langle -120, 0, 20\rangle.$$

$\mathbf{n}_2 = \langle -6, 0, 1\rangle$

The angle θ between 2 adjacent sides is given by

$$\cos\theta = \frac{|\mathbf{n}_1 \cdot \mathbf{n}_2|}{\|\mathbf{n}_1\|\,\|\mathbf{n}_2\|} = \frac{|-1|}{\sqrt{37}\sqrt{37}} = \frac{1}{37} \implies \theta \approx 88.45°.$$

51. False. They might be skew lines, such as

L_1: $x = t, y = 0, z = 0$ (x-axis) and

L_2: $x = 0, y = 0, z = t + 1$

53. True

55. (a) Sphere: $(x - 4)^2 + (y + 1)^2 + (z - 1)^2 = 4$

(b) Two planes parallel to given plane. Let $Q = (x, y, z)$ be a point on one of these planes, and pick $P = (0, 0, 10)$ on the given plane. By the distance formula,

$$2 = \frac{|\overrightarrow{PQ} \cdot \mathbf{n}|}{\|\mathbf{n}\|} = \frac{|\langle x, y, z - 10\rangle \cdot \langle 4, -3, 1\rangle|}{\sqrt{26}}$$

$$\pm 2\sqrt{26} = 4x - 3y + z - 10$$

$4x - 3y + z = 10 \pm 2\sqrt{26}$ (Two planes parallel to given plane)

57. $\theta = \frac{3\pi}{4} \implies \tan\theta = -1 = \frac{y}{x} \implies y = -x$ (line)

59. $r = \frac{1}{2 - \cos\theta} \implies 2r - r\cos\theta = 1 \implies 2\sqrt{x^2 + y^2} - x = 1$

$\implies 2\sqrt{x^2 + y^2} = x + 1 \implies 4(x^2 + y^2) = x^2 + 2x + 1 \implies 3x^2 + 4y^2 = 2x + 1$

61. $x^2 + y^2 - 4x = 0$

$r^2 - 4r\cos\theta = 0$

$r - 4\cos\theta = 0 \implies r = 4\cos\theta$

63. $x = 3$

$r\cos\theta = 3$

$r = 3\sec\theta$

65. $5x - 6y + 4 = 0$

$5r\cos\theta - 6r\sin\theta = -4$

$r(5\cos\theta - 6\sin\theta) = -4$

$r = \frac{4}{6\sin\theta - 5\cos\theta}$

Review Exercises for Chapter 11

Solutions to Odd-Numbered Exercises

1. (a) and (b)

3. $(-5, 4, 0)$

5. $d = \sqrt{(5-4)^2 + (2-0)^2 + (1-7)^2}$

$= \sqrt{1 + 4 + 36}$

$= \sqrt{41}$

7. $d_1 = \sqrt{(3-0)^2 + (-2-3)^2 + (0-2)^2} = \sqrt{9 + 25 + 4} = \sqrt{38}$

$d_2 = \sqrt{(0-0)^2 + (5-3)^2 + (-3-2)^2} = \sqrt{4 + 25} = \sqrt{29}$

$d_3 = \sqrt{(0-3)^2 + (5-(-2))^2 + (-3-0)^2} = \sqrt{9 + 49 + 9} = \sqrt{67}$

$d_1^2 + d_2^2 = 38 + 29 = 67 = d_3^2$

9. Midpoint: $\left(\frac{8+5}{2}, \frac{-2+6}{2}, \frac{3+7}{2}\right) = \left(\frac{13}{2}, 2, 5\right)$

11. Midpoint: $\left(\frac{10-8}{2}, \frac{6-2}{2}, \frac{-12-6}{2}\right) = (1, 2, -9)$

13. $(x-2)^2 + (y-3)^2 + (z-5)^2 = 1$

15. Radius $= 6$

$(x-1)^2 + (y-5)^2 + (z-2)^2 = 36$

17. $(x^2 - 4x + 4) + (y^2 - 6y + 9) + z^2 = -4 + 4 + 9$

$$(x - 2)^2 + (y - 3)^2 + z^2 = 9$$

Center: $(2, 3, 0)$

Radius: 3

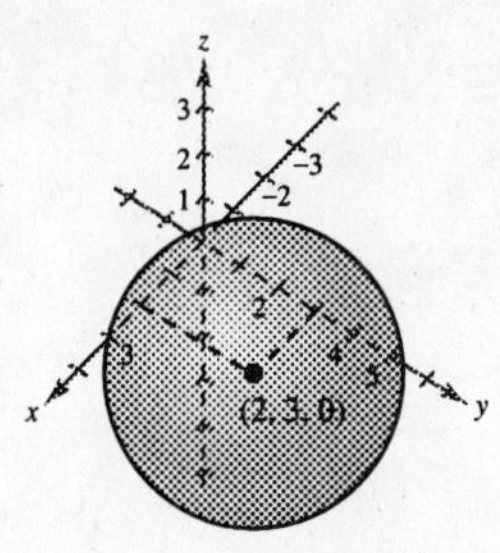

19. (a) xz-trace $(y = 0)$: $x^2 + z^2 = 7$ circle

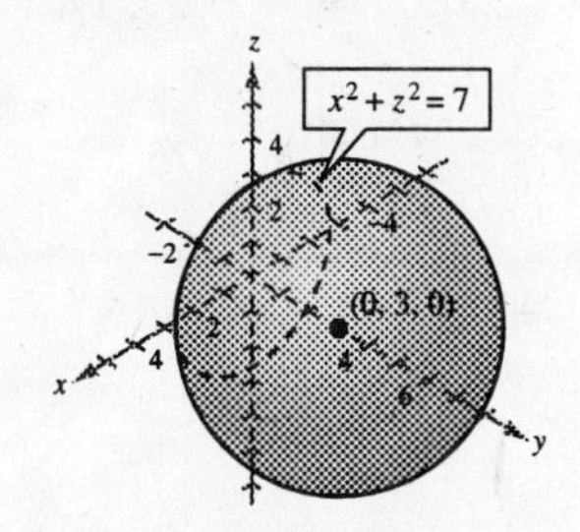

(b) yz-trace $(x = 0)$: $(y - 3)^2 + z^2 = 16$ circle

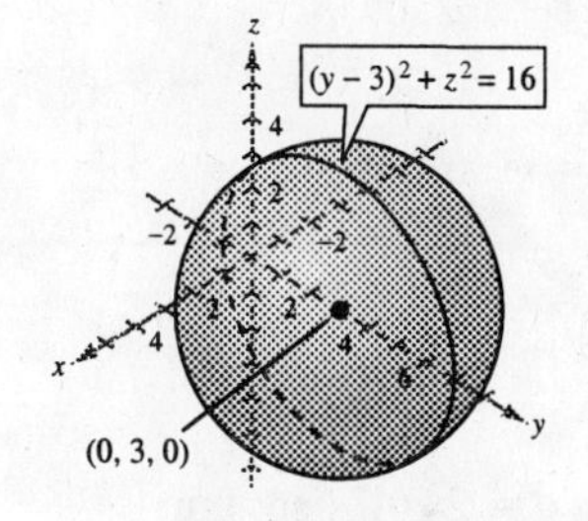

21. $\overrightarrow{PQ} = \langle 3 - 2, 3 - (-1), 0 - 4\rangle = \langle 1, 4, -4\rangle$

$\|\overrightarrow{PQ}\| = \sqrt{1^2 + 4^2 + (-4)^2} = \sqrt{33}$

23. $\overrightarrow{PQ} = \langle -3 - 7, 2 - (-4), 10 - 3\rangle = \langle -10, 6, 7\rangle$

$\|PQ\| = \sqrt{(-10)^2 + 6^2 + 7^2} = \sqrt{185}$

25. $\mathbf{u} \cdot \mathbf{v} = 2(0) - 3(6) + 4(5) = 2$

27. $\mathbf{u} \cdot \mathbf{v} = 2(1) - 1(0) + 1(-1) = 1$

29. Since $\mathbf{u} \cdot \mathbf{v} = 0$, the angle is 90°.

31. Since $-\frac{2}{3}\langle 39, -12, 21\rangle = \langle -26, 8, -14\rangle$, the vectors are parallel.

33. The vector determined by the first 2 points is $\langle -3, 4, 1\rangle$.
The vector determined by the last 2 points is $\langle 3, -4, -1\rangle$.
These vectors are parallel and equal in length.
Thus, the 4 points form a parallelogram.

35. Let $\mathbf{a}$, $\mathbf{b}$, and $\mathbf{c}$ be the three force vectors determined by $A(0, 10, 10)$, $B(-4, -6, 10)$ and $C(4, -6, 10)$.

$$\mathbf{a} = \|\mathbf{a}\|\,\langle 0, 10, 10\rangle/10\sqrt{2} = \|\mathbf{a}\|\left\langle 0, \frac{1}{\sqrt{2}}, \frac{1}{\sqrt{2}}\right\rangle$$

$$\mathbf{b} = \|\mathbf{b}\|\,\langle -4, -6, 10\rangle/\sqrt{152} = \|\mathbf{b}\|\left\langle \frac{-2}{\sqrt{38}}, \frac{-3}{\sqrt{38}}, \frac{5}{\sqrt{38}}\right\rangle$$

$$\mathbf{c} = \|\mathbf{c}\|\,\langle 4, -6, 10\rangle/\sqrt{152} = \|\mathbf{c}\|\left\langle \frac{2}{\sqrt{38}}, \frac{-3}{\sqrt{38}}, \frac{5}{\sqrt{38}}\right\rangle$$

Must have $\mathbf{a} + \mathbf{b} + \mathbf{c} = 300\mathbf{k}$. Thus,

$$\frac{-2}{\sqrt{38}}\|\mathbf{b}\| + \frac{2}{\sqrt{38}}\|\mathbf{c}\| = 0$$

$$\frac{1}{\sqrt{2}}\|\mathbf{a}\| - \frac{3}{\sqrt{38}}\|\mathbf{b}\| - \frac{3}{\sqrt{38}}\|\mathbf{c}\| = 0$$

$$\frac{1}{\sqrt{2}}\|\mathbf{a}\| + \frac{5}{\sqrt{38}}\|\mathbf{b}\| + \frac{5}{\sqrt{38}}\|\mathbf{c}\| = 300$$

From the first equation $\|\mathbf{b}\| = \|\mathbf{c}\|$.

From the second equation,

$$\frac{1}{\sqrt{2}}\|\mathbf{a}\| = \frac{6}{\sqrt{38}}\|\mathbf{b}\|$$

From the third equation,

$$\frac{1}{\sqrt{2}}\|\mathbf{a}\| = 300 - \frac{10}{\sqrt{38}}\|\mathbf{b}\|$$

Thus, $\frac{6}{\sqrt{38}}\|\mathbf{b}\| = 300 - \frac{10}{\sqrt{38}}\|\mathbf{b}\| \Rightarrow$

$\frac{16}{\sqrt{38}}\|\mathbf{b}\| = 300$ and $\|\mathbf{b}\| = \|\mathbf{c}\| = \frac{75\sqrt{38}}{4} \approx 115.58$

Finally, $\|\mathbf{a}\| = \sqrt{2}\left(\frac{6}{\sqrt{38}}\right)\left(\frac{75\sqrt{38}}{4}\right) = \frac{225\sqrt{2}}{2}$

≈ 159.10

37. $\mathbf{u} \times \mathbf{v} = \begin{vmatrix} \mathbf{i} & \mathbf{j} & \mathbf{k} \\ -2 & 8 & 2 \\ 1 & 1 & -1 \end{vmatrix} = \langle -10, 0, -10 \rangle$

39. $\mathbf{u} \times \mathbf{v} = \begin{vmatrix} \mathbf{i} & \mathbf{j} & \mathbf{k} \\ -3 & 2 & -5 \\ 10 & -15 & 2 \end{vmatrix} = \langle -71, -44, 25 \rangle$

$\|\mathbf{u} \times \mathbf{v}\| = \sqrt{7602}$

unit vector: $\dfrac{1}{\sqrt{7602}} \langle -71, -44, 25 \rangle$

41. First two points: $\langle 3, 2, 3 \rangle$

Last two points: $\langle 3, 2, 3 \rangle$

First and third points: $\langle -2, 2, 0 \rangle$

$$\begin{vmatrix} \mathbf{i} & \mathbf{j} & \mathbf{k} \\ 3 & 2 & 3 \\ -2 & 2 & 0 \end{vmatrix} = \langle -6, -6, 10 \rangle$$

$$\begin{aligned} \text{Area} = |\langle -6, -6, 10 \rangle| &= \sqrt{36 + 36 + 100} \\ &= \sqrt{172} \\ &= 2\sqrt{43} \text{ sq. units} \end{aligned}$$

43. The parallelogram is determined by the three vectors with initial point $(0, 0, 0)$.

$\mathbf{u} = \langle 3, 0, 0 \rangle$, $\mathbf{v} = \langle 2, 0, 5 \rangle$, $\mathbf{w} = \langle 0, 5, 1 \rangle$.

$$\mathbf{u} \cdot (\mathbf{v} \times \mathbf{w}) = \begin{vmatrix} 3 & 0 & 0 \\ 2 & 0 & 5 \\ 0 & 5 & 1 \end{vmatrix} = -75$$

Volume $= |-75| = 75$ cubic units

45. (a) $\mathbf{v} = \langle 5, 20, -3 \rangle$

$x = 5t,\ y = -10 + 20t,\ z = 3 - 3t$

(b) $\dfrac{x}{5} = \dfrac{y + 10}{20} = \dfrac{z - 3}{-3}$

47. (a) $\mathbf{v} = \langle 1, 1, 1 \rangle$

$x = 3 + t, y = 2 + t, z = 1 + t$

(b) $\dfrac{x - 3}{1} = \dfrac{y - 2}{1} = \dfrac{z - 1}{1}$ or

$x - 3 = y - 2 = z - 1$

49. $\mathbf{u} = \langle 5, -5, -2 \rangle$, $\mathbf{v} = \langle 3, 5, 2 \rangle$

$$\mathbf{n} = \mathbf{u} \times \mathbf{v} = \begin{vmatrix} \mathbf{i} & \mathbf{j} & \mathbf{k} \\ 5 & -5 & -2 \\ 3 & 5 & 2 \end{vmatrix} = \langle 0, -16, 40 \rangle$$

$$\begin{aligned} \text{Plane: } 0(x + 1) - 16(y - 3) + 40(z - 4) &= 0 \\ -2(y - 3) + 5(z - 4) &= 0 \\ -2y + 5z - 14 &= 0 \end{aligned}$$

51. $\mathbf{n} = \langle 1, 1, 1 \rangle$ normal vector

$$\begin{aligned} \text{Plane: } 1(x - 3) + 1(y - 1) + 1(z - 2) &= 0 \\ x + y + z - 6 &= 0 \end{aligned}$$

53.

55.

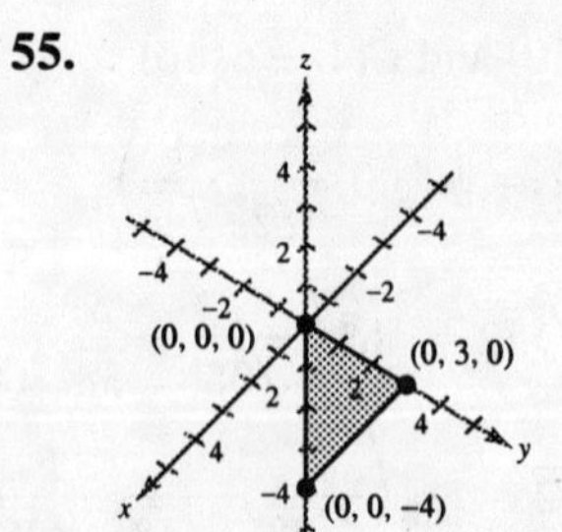

57. $D = \dfrac{|\overrightarrow{PQ} \cdot \mathbf{n}|}{\|\mathbf{n}\|}$

$Q = (1, 2, 3)$, $P = (2, 0, 0)$ on plane. $\overrightarrow{PQ} = \langle -1, 2, 3 \rangle$, $\mathbf{n} = \langle 2, -1, 1 \rangle$

$$D = \frac{|\langle -1, 2, 3 \rangle \cdot \langle 2, -1, 1 \rangle|}{\sqrt{6}} = \frac{1}{\sqrt{6}} = \frac{\sqrt{6}}{6}$$

59. $D = \dfrac{|\overrightarrow{PQ} \cdot \mathbf{n}|}{\|\mathbf{n}\|}$

$Q = (0, 0, 0), P = (0, 0, 12)$ on plane. $\overrightarrow{PQ} = \langle 0, 0, -12 \rangle, \mathbf{n} = \langle 2, 3, 1 \rangle$

$$D = \frac{|\langle 0, 0, -12 \rangle \cdot \langle 2, 3, 1 \rangle|}{\sqrt{14}} = \frac{12}{\sqrt{14}} = \frac{6\sqrt{14}}{7}$$

61. True. See page 788.

63. $\mathbf{u} \times \mathbf{v} = \begin{vmatrix} \mathbf{i} & \mathbf{j} & \mathbf{k} \\ 3 & -2 & 1 \\ 2 & -4 & -3 \end{vmatrix} = \langle 10, 11, -8 \rangle$

$\mathbf{v} \times \mathbf{u} = \begin{vmatrix} \mathbf{i} & \mathbf{j} & \mathbf{k} \\ 2 & -4 & -3 \\ 3 & -2 & 1 \end{vmatrix} = \langle -10, -11, 8 \rangle$

Thus, $\mathbf{u} \times \mathbf{v} = -(\mathbf{v} \times \mathbf{u})$.

65. $\mathbf{u} \times (\mathbf{v} + \mathbf{w}) = \mathbf{u} \times \langle 1, -2, -1 \rangle = \begin{vmatrix} \mathbf{i} & \mathbf{j} & \mathbf{k} \\ 3 & -2 & 1 \\ 1 & -2 & -1 \end{vmatrix} = \langle 4, 4, -4 \rangle$

$\mathbf{u} \times \mathbf{v} = \langle 10, 11, -8 \rangle$ (Exercise 63)

$\mathbf{u} \times \mathbf{w} = \begin{vmatrix} \mathbf{i} & \mathbf{j} & \mathbf{k} \\ 3 & -2 & 1 \\ -1 & 2 & 2 \end{vmatrix} = \langle -6, -7, 4 \rangle$

$$(\mathbf{u} \times \mathbf{v}) + (\mathbf{u} \times \mathbf{w}) = \langle 10, 11, -8 \rangle + \langle -6, -7, 4 \rangle = \langle 4, 4, -4 \rangle$$
$$= \mathbf{u} \times (\mathbf{v} + \mathbf{w})$$

Chapter 11 Practice Test

1. Find the lengths of the sides of the triangle with vertices $(0, 0, 0)$, $(1, 2, -4)$, and $(0, -2, -1)$. Show that the triangle is a right triangle.

2. Find the standard form of the equation of a sphere having center $(0, 4, 1)$ and radius 5.

3. Find the center and radius of the sphere $x^2 + y^2 + z^2 + 2x - 4z - 11 = 0$.

4. Find the vector $\mathbf{u} - 3\mathbf{v}$ given $\mathbf{u} = \langle 1, 0, -1 \rangle$ and $\mathbf{v} = \langle 4, 3, -6 \rangle$.

5. Find the length of $\frac{1}{2}\mathbf{v}$ if $\mathbf{v} = \langle 2, 4, -6 \rangle$.

6. Find the dot product of $\mathbf{u} = \langle 2, 1, -3 \rangle$ and $\mathbf{v} = \langle 1, 1, -2 \rangle$.

7. Determine whether $\mathbf{u} = \langle 1, 1, -1 \rangle$ and $\mathbf{v} = \langle -3, -3, 3 \rangle$ are orthogonal, parallel, or neither.

8. Find the cross product of $\mathbf{u} = \langle -1, 0, 2 \rangle$ and $\mathbf{v} = \langle 1, -1, 3 \rangle$. What is $\mathbf{v} \times \mathbf{u}$?

9. Use the triple scalar product to find the volume of the parallelepiped having adjacent edges $\mathbf{u} = \langle 1, 1, 1 \rangle$, $\mathbf{v} = \langle 0, -1, 1 \rangle$, and $\mathbf{w} = \langle 1, 0, 4 \rangle$.

10. Find a set of parametric equations for the line through the points $(0, -3, 3)$ and $(2, -3, 4)$.

11. Find an equation of the plane passing through $(1, 2, 3)$ and perpendicular to the vector $\mathbf{n} = \langle 1, -1, 0 \rangle$.

12. Find an equation of the plane passing through the three points $A = (0, 0, 0)$, $B = (1, 1, 1)$, and $C = (1, 2, 3)$.

13. Determine whether the planes $x + y - z = 12$ and $3x - 4y - z = 9$ are parallel, orthogonal or neither.

14. Find the distance between the point $(1, 1, 1)$ and the plane $x + 2y + z = 6$.

CHAPTER 12
Limits and an Introduction to Calculus

CHAPTER 12
Limits and an Introduction to Calculus

Section 12.1 Introduction to Limits

- If $f(x)$ becomes arbitrarily close to a unique number L as x approaches c from either side, then the limit of $f(x)$ as x approaches c is L:

  $$\lim_{x \to c} f(x) = L.$$
- You should be able to use a calculator to find a limit.
- You should be able to use a graph to find a limit.
- You should understand how limits can fail to exist:
  - (a) $f(x)$ approaches a different number from the right of c than it approaches from the left of c.
  - (b) $f(x)$ increases or decreases without bound as x approaches c.
  - (c) $f(x)$ oscillates between two fixed values as x approaches c.
- You should know and be able to use the elementary properties of limits.

Solutions to Odd-Numbered Exercises

1. (a)

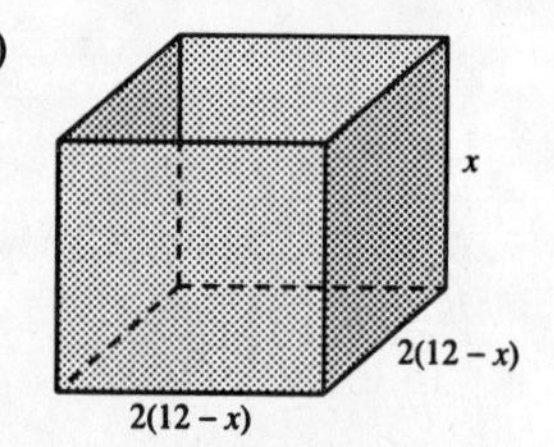

(b) $V = (\text{base})\text{height} = (24 - 2x)^2 x = 4x(12 - x)^2$

(c) $\lim_{x \to 4} V = 1024$

x	3	3.5	3.9	4	4.1	4.5	5
V	972.0	1011.5	1023.5	1024.0	1023.5	1012.5	980.0

(d)

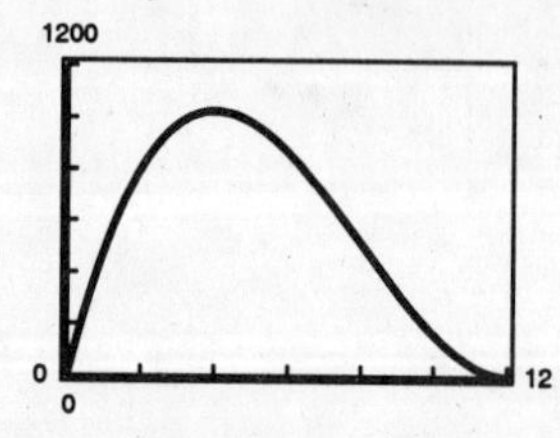

maximum at $x = 4$

3. $\lim_{x \to 3} (4 - 3x) = -5$

x	2.9	2.99	2.999	3	3.001	3.01	3.1
$f(x)$	-4.7	-4.97	-4.997	-5	-5.003	-5.03	-5.3

The limit is reached.

5. $\lim_{x \to 3} \frac{x - 3}{x^2 - 9} = \frac{1}{6}$

x	2.9	2.99	2.999	3	3.001	3.01	3.1
$f(x)$	0.1695	0.1669	0.16669	?	0.16664	0.1664	0.1639

The limit is not reached.

7. $\lim_{x \to 1} \frac{x - 1}{x^2 + 2x - 3} = \frac{1}{4}$

x	0.9	0.99	0.999	1.0	1.001	1.01	1.1
$f(x)$	0.2564	0.2506	0.2501	?	0.2499	0.2464	0.2439

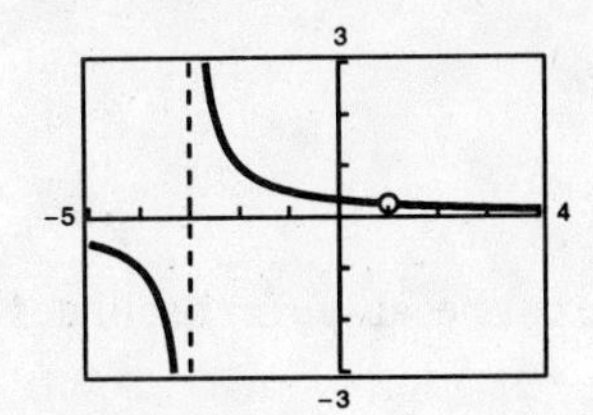

9. $\lim_{x \to 0} \frac{\sqrt{x + 5} - \sqrt{5}}{x} \approx 0.2236 \left(\text{Actual limit is } \frac{1}{2\sqrt{5}}\right)$

x	−0.1	−0.01	−0.001	0	0.001	0.01	0.1
$f(x)$	0.2247	0.2237	0.2236	?	0.2236	0.2235	0.2225

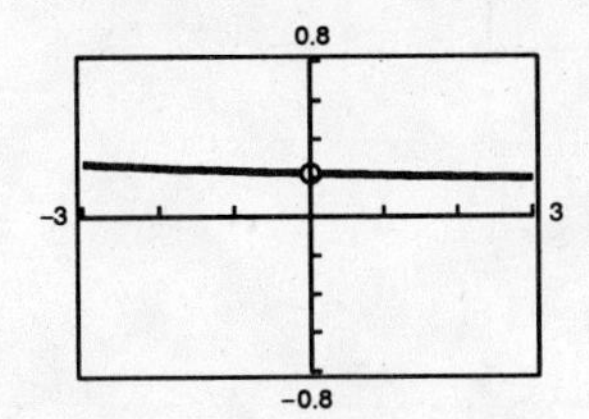

11. $\lim_{x \to -4} \frac{[x/(x + 2)] - 2}{x + 4} = \frac{1}{2}$

x	−4.1	−4.01	−4.001	−4.0	−3.999	−3.99	−3.9
$f(x)$	0.4762	0.4975	0.4998	?	0.5003	0.5025	0.5263

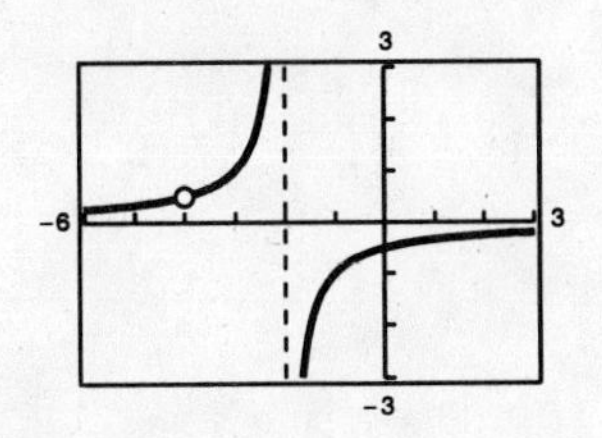

13. Make sure your calculator is set in radian mode.

$$\lim_{x\to 0} \frac{\sin x}{x} = 1$$

x	-0.1	-0.01	-0.001	0	0.001	0.01	0.1
$f(x)$	0.9983	0.99998	0.9999998	?	0.9999998	0.99998	0.9983

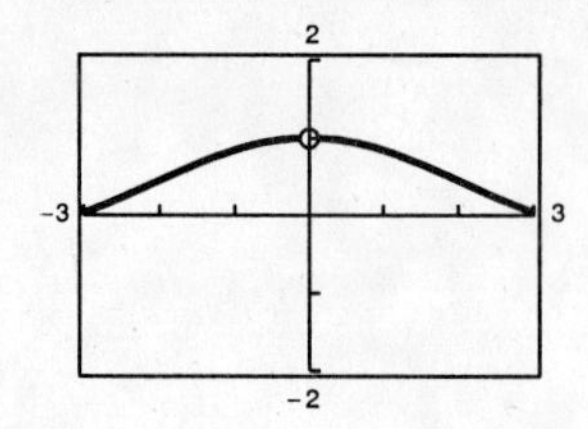

15. $\lim_{x\to 2} (3 - x) = 1$

17. $\lim_{x\to -1} \sin \frac{\pi x}{2} = -1$

19. The limit does not exist because $f(x)$ approaches different values from the left of $x = -2$ and the right of $x = -2$.

21. The limit does not exist because $f(x)$ oscillates between 2 and -2.

23.

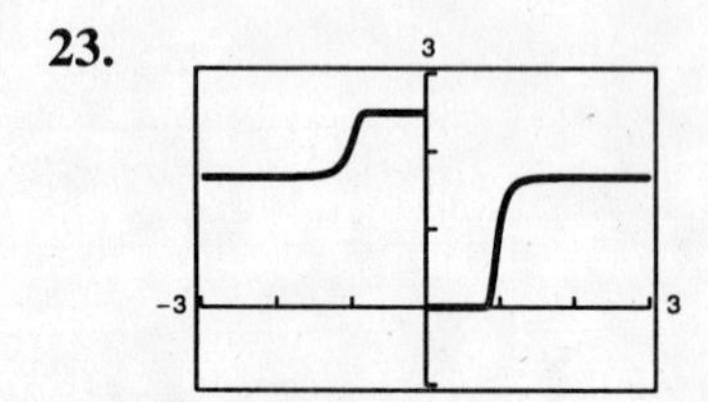

The limit does not exist.

25.

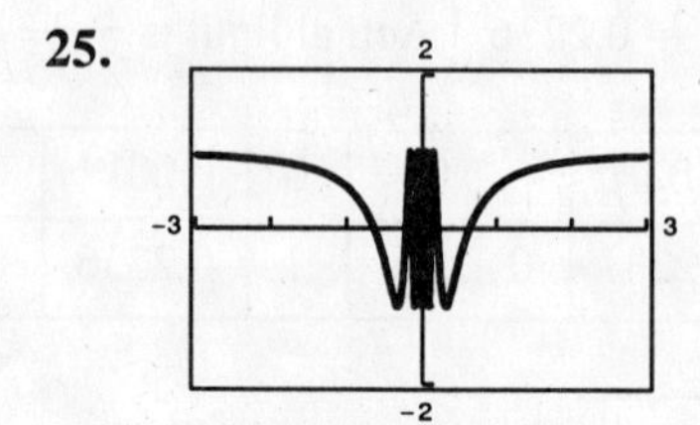

$\lim_{x\to 0} \cos \frac{1}{x}$ does not exist.

27.

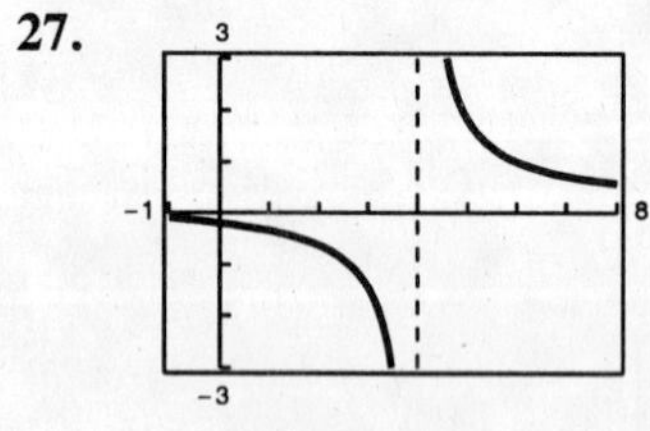

$\lim_{x\to 4} \frac{\sqrt{x+3} - 1}{x - 4}$ does not exist.

29.

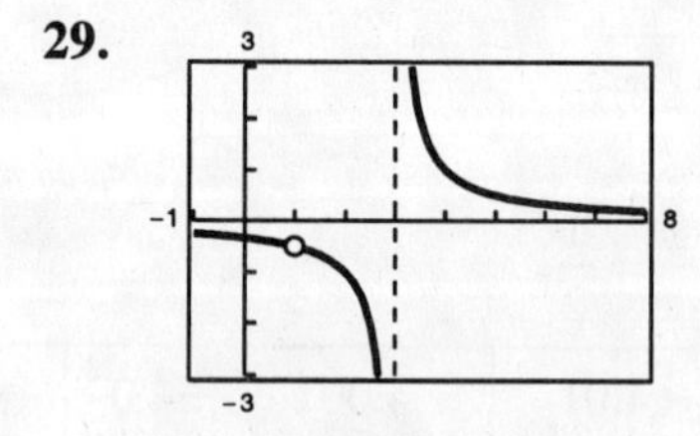

$\lim_{x\to 1} \frac{x - 1}{x^2 - 4x + 3} = -\frac{1}{2}$

31.

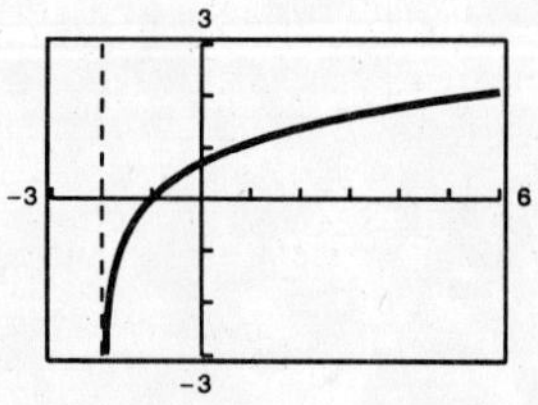

$\lim_{x\to 4} \ln(x + 2) \approx 1.7918$ (Exact limit is $\ln 6$)

33. $\lim_{x\to 1} (x^2 + 3x - 4) = 1^2 + 3(1) - 4 = 0$

35. $\lim_{x\to 2} \sqrt[3]{10x + 7} = \sqrt[3]{10(2) + 7} = \sqrt[3]{27} = 3$

37. $\lim_{x\to 3} \frac{15}{x} = \frac{15}{3} = 5$

39. $\lim_{x \to -1} \dfrac{\sqrt{x^2 - 1}}{x} = \dfrac{\sqrt{(-1)^2 - 1}}{-1} = \dfrac{0}{-1} = 0$

41. $\lim_{x \to 3} e^x = e^3 \approx 20.0855$

43. $\lim_{x \to \pi} \sin 2x = \sin 2\pi = 0$

45. $\lim_{x \to 1/2} \arcsin x = \arcsin \dfrac{1}{2} = \dfrac{\pi}{6} \approx 0.5236$

47. The limit does not exist. As x approaches 2 from the left, $f(x)$ approaches 5. As x approaches 2 from the right, $f(x)$ approaches 6.

49. (a) $\lim_{x \to c} [-2g(x)] = -2(6) = -12$

(b) $\lim_{x \to c} [f(x) + g(x)] = 3 + 6 = 9$

(c) $\lim_{x \to c} \dfrac{f(x)}{g(x)} = \dfrac{3}{6} = \dfrac{1}{2}$

(d) $\lim_{x \to c} \sqrt{f(x)} = \sqrt{3}$

51. (a) $\lim_{x \to 2} f(x) = 2^3 = 8$

(b) $\lim_{x \to 2} g(x) = \dfrac{\sqrt{2^2 + 5}}{2(2^2)} = \dfrac{3}{8}$

(c) $\lim_{x \to 2} f(x)g(x) = 8\left(\dfrac{3}{8}\right) = 3$

(d) $\lim_{x \to 2} [g(x) - f(x)] = \dfrac{3}{8} - 8 = -\dfrac{61}{8}$

53. True

55. Answers will vary.

57. No. The limit may or may not exist. And if it exists, it may not be equal to 4.

59. $\lim_{x \to 5} f(x) = 12$ means that as x gets close to 5, $f(x)$ gets close to 12.

61.

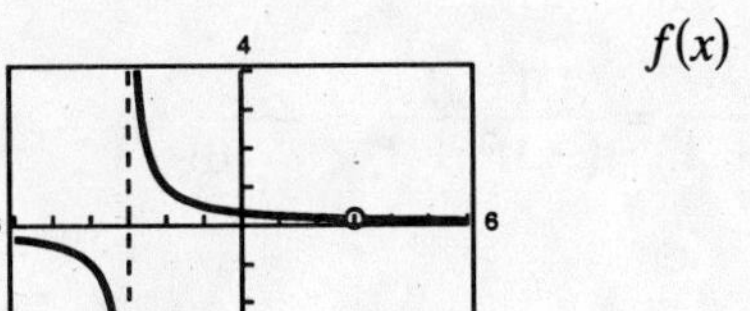

$f(x) = \dfrac{x - 3}{x^2 - 9}$

$\lim_{x \to 3} f(x) = \dfrac{1}{6}$

Domain: all $x \neq \pm 3$.

It is difficult to determine the domain solely by the graph because it is not obvious that the function is undefined at $x = 3$.

63. $\dfrac{5 - x}{3x - 15} = \dfrac{5 - x}{-3(5 - x)} = -\dfrac{1}{3}, x \neq 5$

65. $\dfrac{15x^2 + 7x - 4}{15x^2 + x - 2} = \dfrac{(3x - 1)(5x + 4)}{(3x - 1)(5x + 2)}$

$= \dfrac{5x + 4}{5x + 2}, \; x \neq \dfrac{1}{3}$

67. $\dfrac{x^3 + 27}{x^2 + x - 6} = \dfrac{(x + 3)(x^2 - 3x + 9)}{(x + 3)(x - 2)}$

$= \dfrac{x^2 - 3x + 9}{x - 2}, \; x \neq -3$

69. $d = \sqrt{(3 - 3)^2 + (2 - 2)^2 + (8 - 7)^2} = 1$

71. $d = \sqrt{(-3 - 2)^2 + (-4 - 2)^2 + (-8 + 5)^2}$

$= \sqrt{25 + 36 + 9}$

$= \sqrt{70}$

73. $d = \sqrt{(0-3)^2 + (5+3)^2 + (-5-0)^2}$
$= \sqrt{9 + 64 + 25}$
$= \sqrt{98}$
$= 7\sqrt{2}$

Section 12.2 Techniques for Evaluating Limits

- You can use direct substitution to find the limit of a polynomial function $p(x)$:
  $\lim_{x \to c} p(x) = p(c)$.
- You can use direct substitution to find the limit of a rational function $r(x) = \dfrac{p(x)}{q(x)}$, as long as $q(c) \neq 0$:
  $\lim_{x \to c} r(x) = r(c) = \dfrac{p(c)}{q(c)}, q(c) \neq 0$.
- You should be able to use cancellation techniques to find a limit.
- You should know how to use rationalization techniques to find a limit.
- You should know how to use technology to find a limit.
- You should be able to calculate one-sided limits.

Solutions to Odd-Numbered Exercises

1. $\lim_{x \to 5} (10 - x^2) = 10 - 5^2 = 10 - 25 = -15$

3. $\lim_{x \to -3} \dfrac{3x}{x^2 + 1} = \dfrac{3(-3)}{(-3)^2 + 1} = -\dfrac{9}{10}$

5. $\lim_{x \to -2} \dfrac{5x + 3}{2x - 9} = \dfrac{5(-2) + 3}{2(-2) - 9} = \dfrac{-7}{-13} = \dfrac{7}{13}$

7. $\lim_{x \to -1} \sqrt{x + 2} = \sqrt{-1 + 2} = \sqrt{1} = 1$

9. $g(x) = \dfrac{-2x^2 + x}{x}$

$g_2(x) = -2x + 1$

(a) $\lim_{x \to 0} g(x) = 1$

(b) $\lim_{x \to -1} g(x) = 3$

(c) $\lim_{x \to -2} g(x) = 5$

11. $g(x) = \dfrac{x^3 - x}{x - 1}$

$g_2(x) = x^2 + x = x(x + 1)$

(a) $\lim_{x \to 1} g(x) = 2$

(b) $\lim_{x \to -1} g(x) = 0$

(c) $\lim_{x \to 0} g(x) = 0$

13. $\lim_{x \to 6} \dfrac{x - 6}{x^2 - 36} = \lim_{x \to 6} \dfrac{x - 6}{(x - 6)(x + 6)}$

$= \lim_{x \to 6} \dfrac{1}{x + 6}$

$= \dfrac{1}{12}$

15. $\lim_{x \to -1} \dfrac{1 - 2x - 3x^2}{1 + x} = \lim_{x \to -1} \dfrac{(1 + x)(1 - 3x)}{1 + x}$

$= \lim_{x \to -1} (1 - 3x) = 4$

17. $\displaystyle \lim_{y\to 0}\frac{\sqrt{5+y}-\sqrt{5}}{y} = \lim_{y\to 0}\frac{\sqrt{5+y}-\sqrt{5}}{y}\cdot\frac{\sqrt{5+y}+\sqrt{5}}{\sqrt{5+y}+\sqrt{5}}$

$$= \lim_{y\to 0}\frac{(5+y)-5}{y\left(\sqrt{5+y}+\sqrt{5}\right)}$$

$$= \lim_{y\to 0}\frac{1}{\sqrt{5+y}+\sqrt{5}}$$

$$= \frac{1}{2\sqrt{5}}$$

19. $\displaystyle \lim_{x\to -3}\frac{\sqrt{x+7}-2}{x+3} = \lim_{x\to -3}\frac{\sqrt{x+7}-2}{x+3}\cdot\frac{\sqrt{x+7}+2}{\sqrt{x+7}+2}$

$$= \lim_{x\to -3}\frac{(x+7)-4}{(x+3)\left(\sqrt{x+7}+2\right)}$$

$$= \lim_{x\to -3}\frac{1}{\sqrt{x+7}+2}$$

$$= \frac{1}{4}$$

21. $\displaystyle \lim_{x\to 0}\frac{\dfrac{1}{1+x}-1}{x} = \lim_{x\to 0}\frac{1-(1+x)}{(1+x)x} = \lim_{x\to 0}\frac{-1}{1+x} = -1$

23. $\displaystyle \lim_{x\to 0}\frac{\sec x}{\tan x} = \lim_{x\to 0}\frac{1}{\cos x}\cdot\frac{\cos x}{\sin x} = \lim_{x\to 0}\frac{1}{\sin x}$, does not exist.

25. $\displaystyle f(x) = \frac{\sqrt{x+3}-\sqrt{3}}{x}$

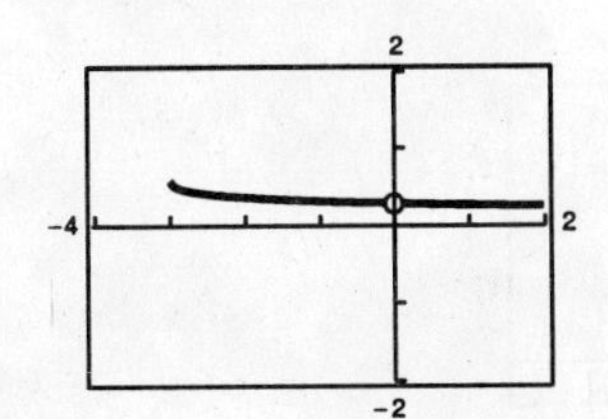

$\displaystyle \lim_{x\to 0} f(x) \approx 0.2887 \quad \left(\text{Exact limit: } \frac{1}{2\sqrt{3}}\right)$

27.

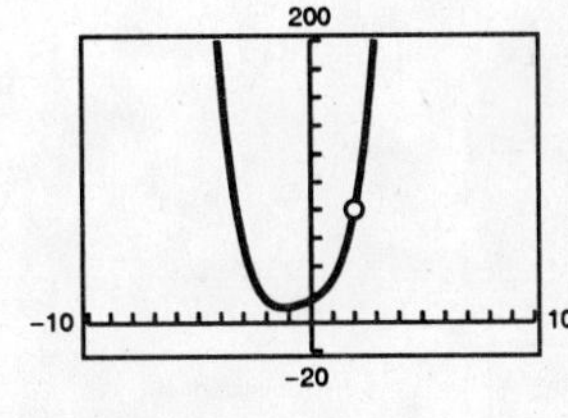

$$\lim_{x\to 2}\frac{x^5-32}{x-2} = 80$$

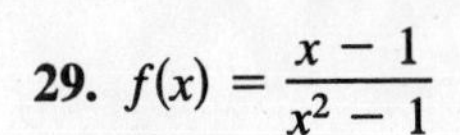

29. $\displaystyle f(x) = \frac{x-1}{x^2-1}$

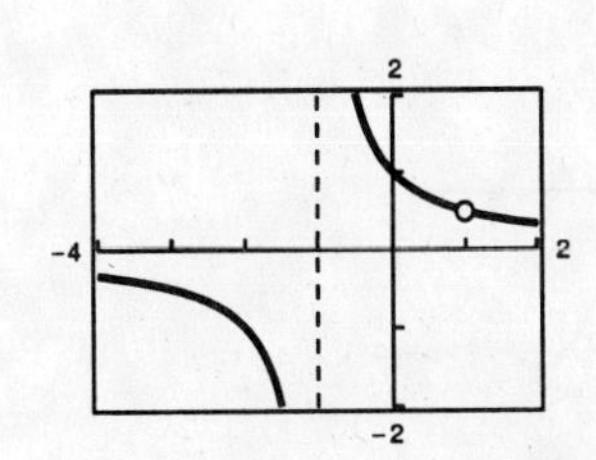

Graphically, $\displaystyle \lim_{x\to 1^-}\frac{x-1}{x^2-1} = \frac{1}{2}$

x	0.5	0.9	0.99	0.999	1
$f(x)$	0.6667	0.5263	0.5025	0.5003	0.5

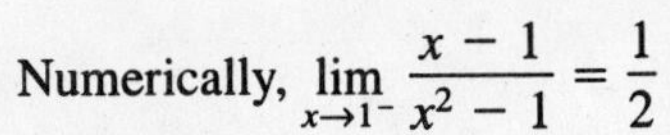

Numerically, $\displaystyle \lim_{x\to 1^-}\frac{x-1}{x^2-1} = \frac{1}{2}$

Algebraically, $\displaystyle \lim_{x\to 1^-}\frac{x-1}{x^2-1} = \lim_{x\to 1^-}\frac{x-1}{(x-1)(x+1)} = \lim_{x\to 1^-}\frac{1}{x+1} = \frac{1}{2}$

31. $f(x) = \dfrac{4 - \sqrt{x}}{x - 16}$

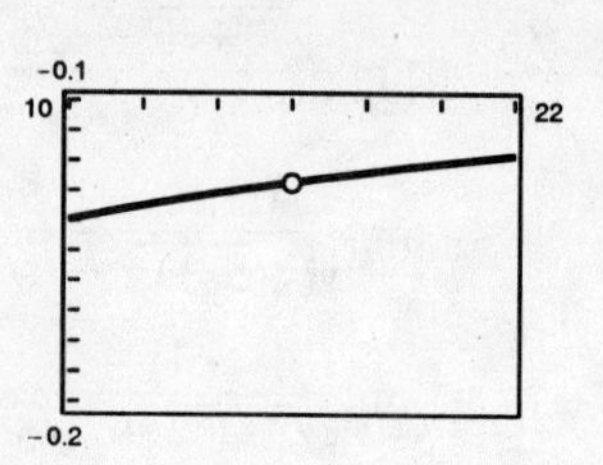

Graphically, $\displaystyle\lim_{x\to16^+} \frac{4 - \sqrt{x}}{x - 16} = -\frac{1}{8}$

x	16	16.001	16.01	16.1	16.5
$f(x)$	-0.1250	-0.1250	-0.1250	-0.1248	-0.1240

Numerically, $\displaystyle\lim_{x\to16^+} \frac{4 - \sqrt{x}}{x - 16} = -0.125$

Algebraically, $\displaystyle\lim_{x\to16^+} \frac{4 - \sqrt{x}}{x - 16} = \lim_{x\to16^+} \frac{4 - \sqrt{x}}{\left(\sqrt{x} - 4\right)\left(\sqrt{x} + 4\right)}$

$$= \lim_{x\to16^+} \frac{-1}{\sqrt{x} + 4}$$

$$= \frac{-1}{4 + 4} = -\frac{1}{8}$$

33. 

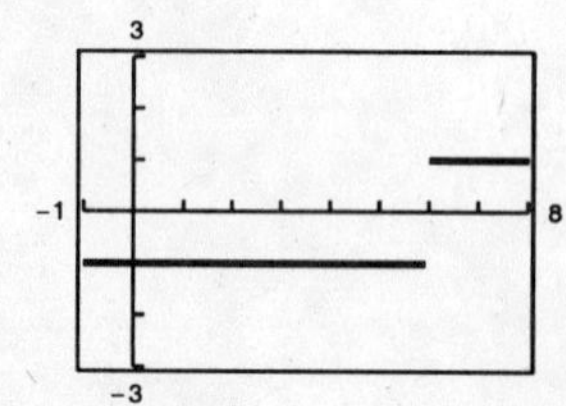

$f(x) = \dfrac{|x - 6|}{x - 6}$

$\displaystyle\lim_{x\to6^+} f(x) = 1,\ \lim_{x\to6^-} f(x) = -1$

Limit does not exist.

35.

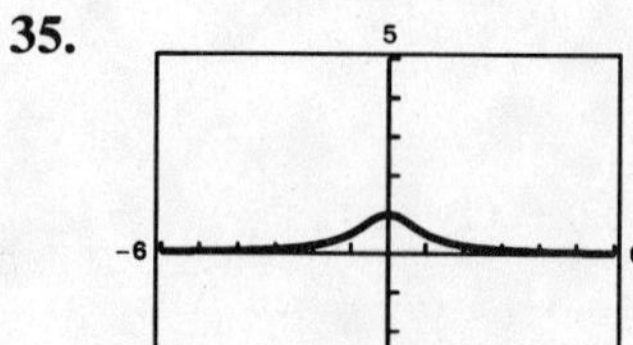

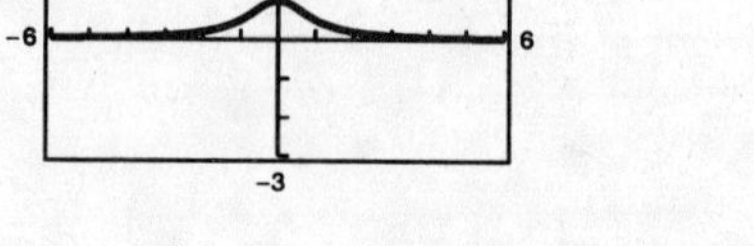

$\displaystyle\lim_{x\to1^-} \frac{1}{x^2 + 1} = \lim_{x\to1^+} = \frac{1}{x^2 + 1} = \lim_{x\to1} \frac{1}{x^2 + 1} = \frac{1}{2}$

37. $\displaystyle\lim_{x\to2^-} f(x) = 2 - 1 = 1$

$\displaystyle\lim_{x\to2^+} f(x) = 2(2) - 3 = 1$

$\displaystyle\lim_{x\to2} f(x) = 1$

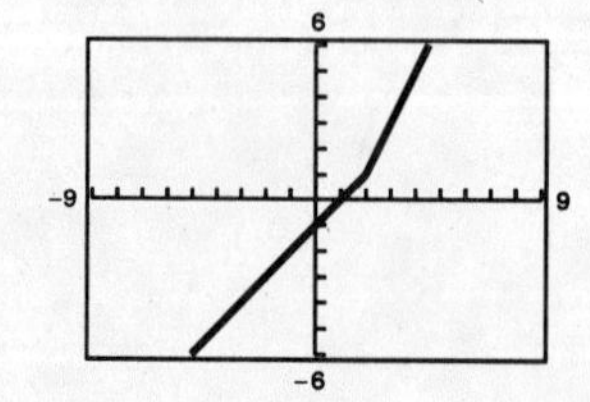

39. $\displaystyle\lim_{x\to0^+} x \ln x = 0$

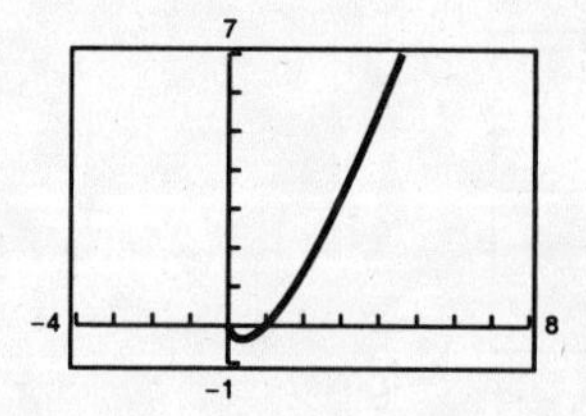

41. $\displaystyle\lim_{x\to0} \frac{\sin 2x}{x} = 2$

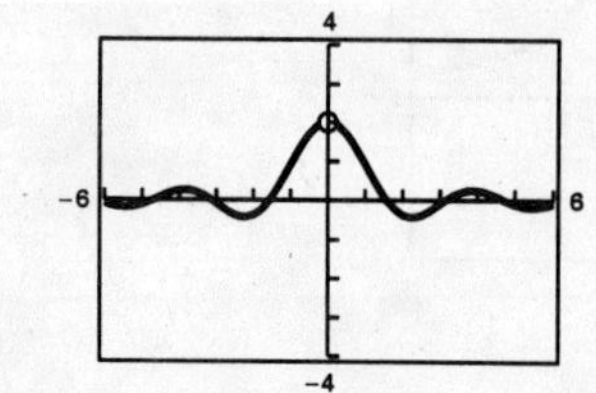

43. $\displaystyle\lim_{x\to0} \frac{\tan x}{x} = 1$

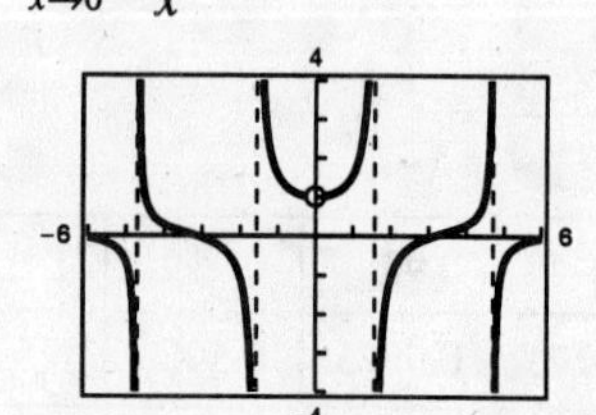

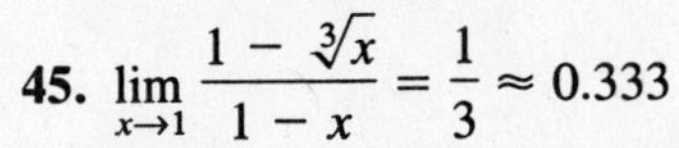

45. $\lim_{x\to 1} \dfrac{1 - \sqrt[3]{x}}{1 - x} = \dfrac{1}{3} \approx 0.333$

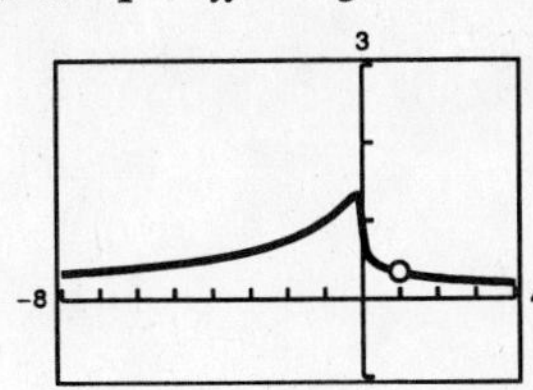

47.

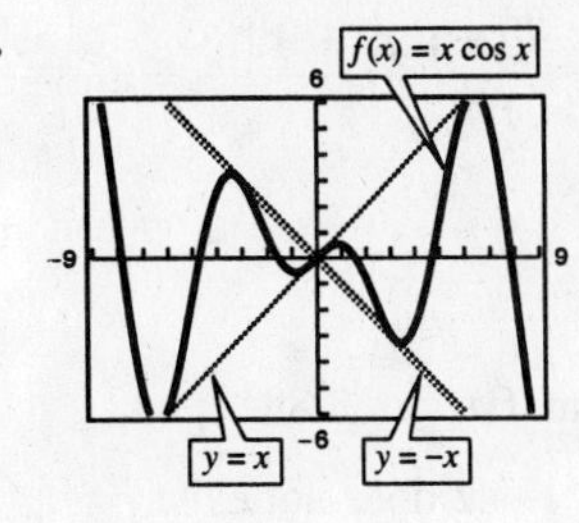

$\lim_{x\to 0} f(x) = 0$

49.

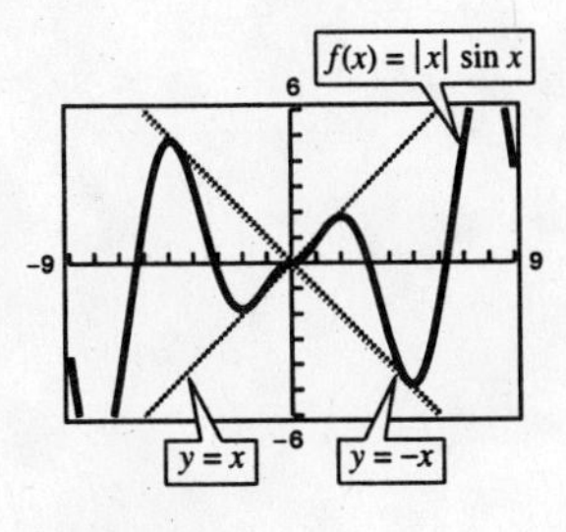

$\lim_{x\to 0} f(x) = 0$

51.

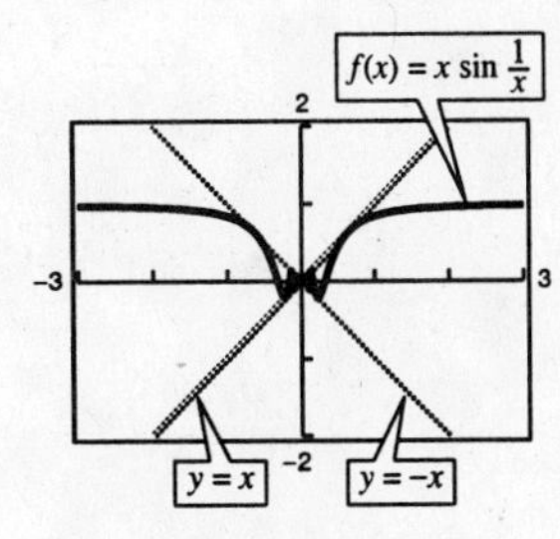

$\lim_{x\to 0} f(x) = 0$

53. (a) Can be evaluated by direct substitution:

$\lim_{x\to 0} x^2 \sin x^2 = 0^2 \sin 0^2 = 0$

(b) Cannot be evaluated by direct substitution:

$\lim_{x\to 0} \dfrac{\sin x^2}{x^2} = 1$

55.
$$\lim_{h\to 0} \frac{f(x+h) - f(x)}{h} = \lim_{h\to 0} \frac{3(x+h) - 1 - (3x - 1)}{h}$$
$$= \lim_{h\to 0} \frac{3x + 3h - 1 - 3x + 1}{h}$$
$$= \lim_{h\to 0} \frac{3h}{h} = 3$$

57.
$$\lim_{h\to 0} \frac{f(x+h) - f(x)}{h} = \lim_{h\to 0} \frac{\sqrt{x+h} - \sqrt{x}}{h} \cdot \left(\frac{\sqrt{x+h} + \sqrt{x}}{\sqrt{x+h} + \sqrt{x}}\right)$$
$$= \lim_{h\to 0} \frac{(x+h) - x}{h\left(\sqrt{x+h} + \sqrt{x}\right)}$$
$$= \lim_{h\to 0} \frac{1}{\sqrt{x+h} + \sqrt{x}} = \frac{1}{2\sqrt{x}}$$

59.
$$\lim_{h\to 0} \frac{f(x+h) - f(x)}{h} = \lim_{h\to 0} \frac{((x+h)^2 - 3(x+h)) - (x^2 - 3x)}{h}$$
$$= \lim_{h\to 0} \frac{x^2 + 2xh + h^2 - 3x - 3h - x^2 + 3x}{h}$$
$$= \lim_{h\to 0} \frac{2xh + h^2 - 3h}{h} = \lim_{h\to 0} (2x + h - 3) = 2x - 3$$

61. $\lim_{t\to 1} \frac{(-16(1)+128)-(-16t^2+128)}{1-t} = \lim_{t\to 1} \frac{16t^2-16}{1-t} = \lim_{t\to 1} \frac{16(t-1)(t+1)}{1-t}$

$= \lim_{t\to 1} -16(t+1) = -32 \frac{\text{ft}}{\text{sec}}$

63. $\lim_{t\to 2^-} f(t) = 30.80$, $\lim_{t\to 2^+} f(t) = 33.88$

Thus, the limit of f as $t \to 2$ does not exist.

65. True

67. Many answers possible.

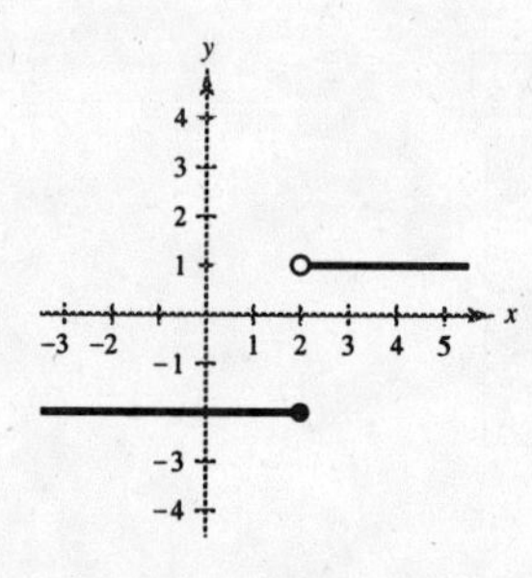

69. Answers will vary.

71. Slope between $(3, -3)$ and $(5, -2)$ is

$$\frac{-2-(-3)}{5-3} = \frac{1}{2}.$$

Line: $y + 1 = \frac{1}{2}(x - 1)$

$2y + 2 = x - 1$

$2y - x + 3 = 0$

73. $\langle 5, 5, 0\rangle \cdot \langle 0, 5, 1\rangle = 25 \neq 0.$

Not multiples of each other.

Neither parallel nor orthogonal.

75. $\langle 2, -3, 1\rangle \cdot \langle -2, 2, 2\rangle = -6 \neq 0.$

Not multiples of each other.

Neither parallel nor orthogonal.

Section 12.3 The Tangent Line Problem

- You should be able to visually approximate the slope of a graph.
- The slope m of the graph of f at the point $(x, f(x))$ is given by

$$m = \lim_{h\to 0} \frac{f(x+h) - f(x)}{h}$$

provided this limit exists.

- You should be able to use the limit definition to find the slope of a graph.
- The derivative of f at x is given by

$$f'(x) = \lim_{h\to 0} \frac{f(x+h) - f(x)}{h}$$

provided this limit exists. Notice that this is the same limit as that for the tangent line slope.

- You should be able to use the limit definition to find the derivative of a function.

Solutions to Odd-Numbered Exercises

1. Slope is 0 at (x, y).

3. Slope is $\frac{1}{2}$ at (x, y).

5.

slope ≈ 2

7.

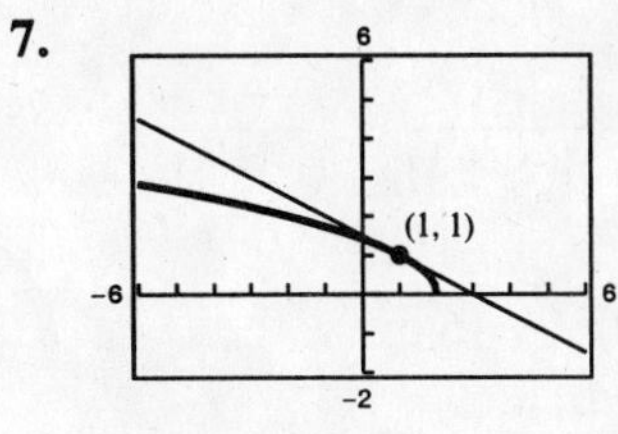

slope $\approx -\frac{1}{2}$

9. $m_{\text{sec}} = \dfrac{g(1 + h) - g(1)}{h} = \dfrac{5 - 2(1 + h) - 3}{h} = \dfrac{-2h}{h}$

$$m = \lim_{h \to 0} \frac{-2h}{h} = -2$$

11. $m_{\text{sec}} = \dfrac{g(3 + h) - g(3)}{h} = \dfrac{(3 + h)^2 - 4(3 + h) - (-3)}{h} = \dfrac{h^2 + 2h}{h}$

$$m = \lim_{h \to 0} \frac{h^2 + 2h}{h} = \lim_{h \to 0} \frac{h(h + 2)}{h} = \lim_{h \to 0} (h + 2) = 2$$

13. $m_{\text{sec}} = \dfrac{g(2 + h) - g(2)}{h} = \dfrac{\dfrac{4}{2 + h} - 2}{h} = \dfrac{4 - 2(2 + h)}{(2 + h)h} = \dfrac{-2}{2 + h}, \ h \neq 0$

$$m = \lim_{h \to 0} \left(\frac{-2}{2 + h}\right) = -1$$

15. $m_{\text{sec}} = \dfrac{h(9 + h) - h(9)}{h} = \dfrac{\sqrt{9 + h} - 3}{h} \cdot \dfrac{\sqrt{9 + h} + 3}{\sqrt{9 + h} + 3}$

$$= \frac{(9 + h) - 9}{h\left[\sqrt{9 + h} + 3\right]} = \frac{1}{\sqrt{9 + h} + 3}, \ h \neq 0$$

$$m = \lim_{h \to 0} \frac{1}{\sqrt{9 + h} + 3} = \frac{1}{6}$$

17. $m_{\text{sec}} = \dfrac{g(x + h) - g(x)}{h} = \dfrac{4 - (x + h)^2 - (4 - x^2)}{h}$

$$= \frac{-2xh - h^2}{h} = -2x - h, \ h \neq 0$$

$$m = \lim_{h \to 0} (-2x - h) = -2x$$

At $(0, 4)$, $m = -2(0) = 0$.

At $(-1, 3)$, $m = -2(-1) = 2$.

19. $m_{sec} = \dfrac{g(x+h) - g(x)}{h} = \dfrac{\dfrac{1}{x+h+4} - \dfrac{1}{x+4}}{h} = \dfrac{(x+4) - (x+4+h)}{(x+h+4)(x+4)(h)}$

$= \dfrac{-h}{(x+h+4)(x+4)h} = \dfrac{-1}{(x+h+4)(x+4)}, \ h \neq 0$

$m = \lim_{h\to 0} \dfrac{-1}{(x+h+4)(x+4)} = \dfrac{-1}{(x+4)^2}$

At $\left(0, \frac{1}{4}\right)$, $m = \dfrac{-1}{(0+4)^2} = \dfrac{-1}{16}$.

At $\left(-2, \frac{1}{2}\right)$, $m = \dfrac{-1}{(-2+4)^2} = \dfrac{-1}{4}$.

21. $f'(x) = \lim_{h\to 0} \dfrac{f(x+h) - f(x)}{h} = \lim_{h\to 0} \dfrac{5-5}{h} = 0$

23. $g'(x) = \lim_{h\to 0} \dfrac{g(x+h) - g(x)}{h} = \lim_{h\to 0} \dfrac{\left[9 - \frac{1}{3}(x+h)\right] - \left[9 - \frac{1}{3}x\right]}{h} = \lim_{h\to 0} \dfrac{-\frac{1}{3}h}{h} = -\dfrac{1}{3}$

25. $f'(x) = \lim_{h\to 0} \dfrac{f(x+h) - f(x)}{h} = \lim_{h\to 0} \dfrac{\dfrac{1}{(x+h)^2} - \dfrac{1}{x^2}}{h} = \lim_{h\to 0} \dfrac{x^2 - (x^2 + 2xh + h^2)}{(x+h)^2x^2h} = \lim_{h\to 0} -\dfrac{2x - h}{(x+h)^2x^2}$

$= -\dfrac{2x}{x^4} = -\dfrac{2}{x^3}$

27. $m_{sec} = \dfrac{f(2+h) - f(2)}{h} = \dfrac{(2+h)^2 - 1 - 3}{h} = \dfrac{4h + h^2}{h} = 4 + h, \ h \neq 0$

$m = \lim_{h\to 0} (4 + h) = 4$

Tangent line: $y - 3 = 4(x - 2)$

$y = 4x - 5$

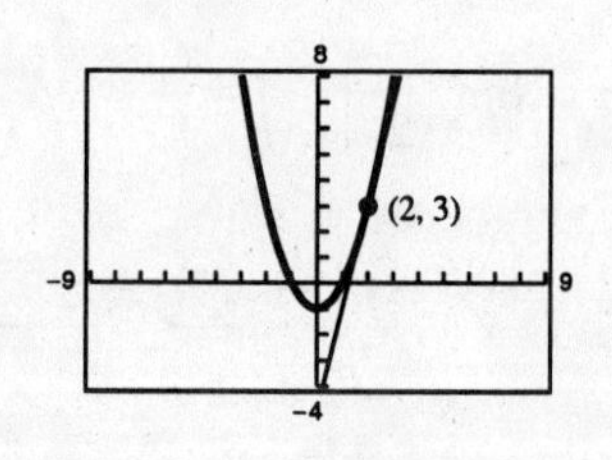

29. $m_{sec} = \dfrac{f(3+h) - f(3)}{h} = \dfrac{\sqrt{3+h+1} - 2}{h} \cdot \dfrac{\sqrt{4+h} + 2}{\sqrt{4+h} + 2} = \dfrac{(4+h) - 4}{h\left[\sqrt{4+h} + 2\right]} = \dfrac{1}{\sqrt{4+h} + 2}$

$m = \lim_{h\to 0} \dfrac{1}{\sqrt{4+h} + 2} = \dfrac{1}{4}$

Tangent line: $y - 2 = \dfrac{1}{4}(x - 3)$

$4y = x + 5$

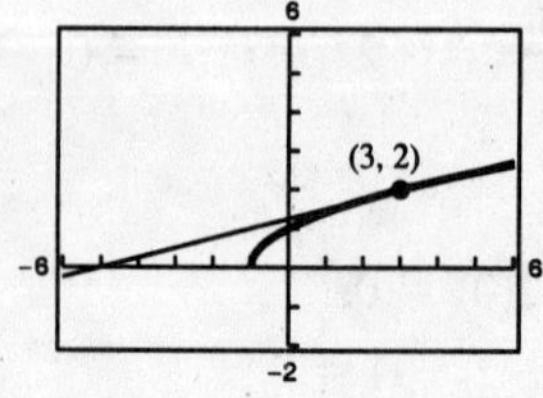

31.

x	-2	-1.5	-1	-0.5	0	0.5	1	1.5	2
$f(x)$	2	1.125	0.5	0.125	0	0.125	0.5	0.125	2
$f'(x)$	-2	-1.5	-1	-0.5	0	0.5	1	1.5	2

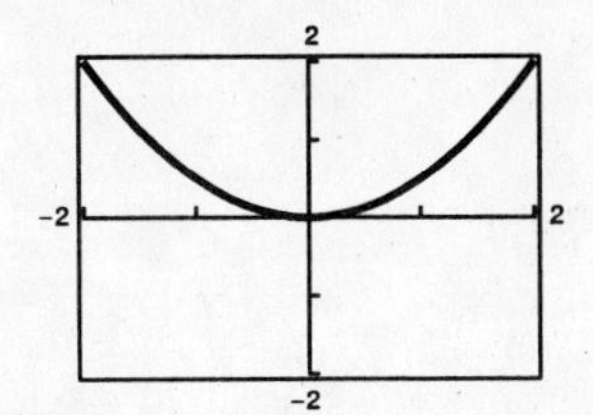

33.

x	-2	-1.5	-1	-0.5	0	0.5	1	1.5	2
$f(x)$	1	1.225	1.414	1.581	1.732	1.871	2	2.121	2.236
$f(x)$	0.5	0.408	0.354	0.316	0.289	0.267	0.25	0.236	0.224

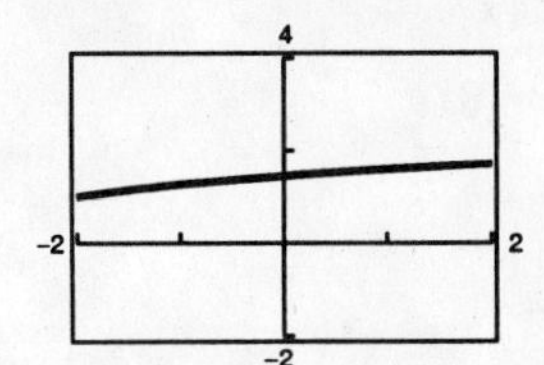

$$f(x) = \sqrt{x+3}$$

$$f'(x) = \frac{1}{2\sqrt{x+3}}$$

35. $f'(x) = \lim_{h\to 0} \frac{f(x+h) - f(x)}{h} = \lim_{h\to 0} \frac{[(x+h)^2 - 4(x+h) + 3] - [x^2 - 4x + 3]}{h}$

$= \lim_{h\to 0} \frac{(x^2 + 2xh + h^2 - 4x - 4h + 3) - (x^2 - 4x + 3)}{h}$

$= \lim_{h\to 0} \frac{2xh + h^2 - 4h}{h} = \lim_{h\to 0} 2x + h - 4 = 2x - 4$

$f'(x) = 0 = 2x - 4 \Rightarrow x = 2$

f has a horizontal tangent at $(2, -1)$.

37. $f'(x) = \lim_{h\to 0} \frac{f(x+h) - f(x)}{h} = \lim_{h\to 0} \frac{3(x+h)^3 - 9(x+h) - (3x^3 - 9x)}{h}$

$= \lim_{h\to 0} \frac{9x^2h + 9xh^2 + 3h^3 - 9h}{h} = 9x^2 - 9$

$f'(x) = 0 = 9x^2 - 9 \Rightarrow x = \pm 1$

f has horizontal tangents at $(1, -6)$ and $(-1, 6)$.

39. (a) $y = 0.0729t^3 + 7.8919t^2 - 192.3508t + 1955.1904$

(b)

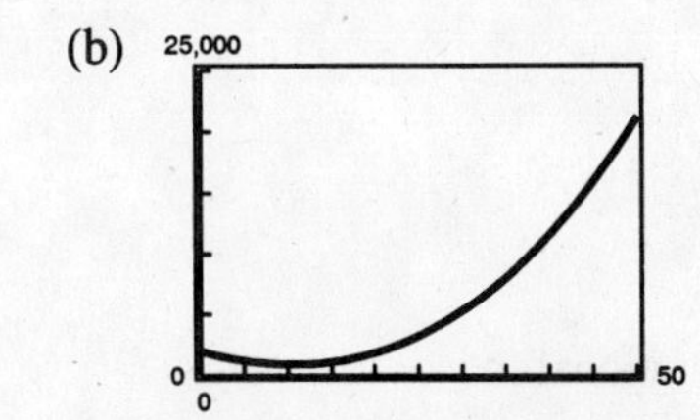

When $t = 30$, slope $\approx$ 478 billion dollars/year.

The debt is growing at 478 billion dollars per year in 1980.

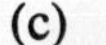
(c)

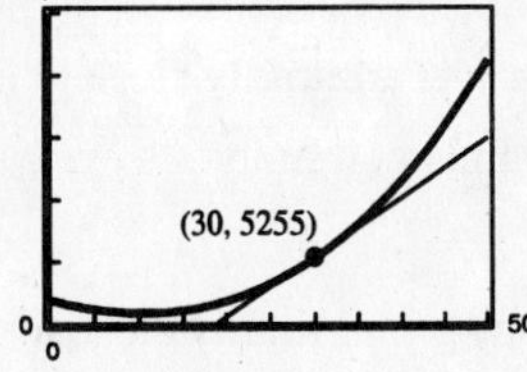

41. True. The slope is $2x$, which is different for all x.

43. Matches (b). (derivative is always positive, but decreasing)

45. Matches (d). (derivative is -1 for $x < 0$, 1 for $x > 0$.)

47. Answers will vary.

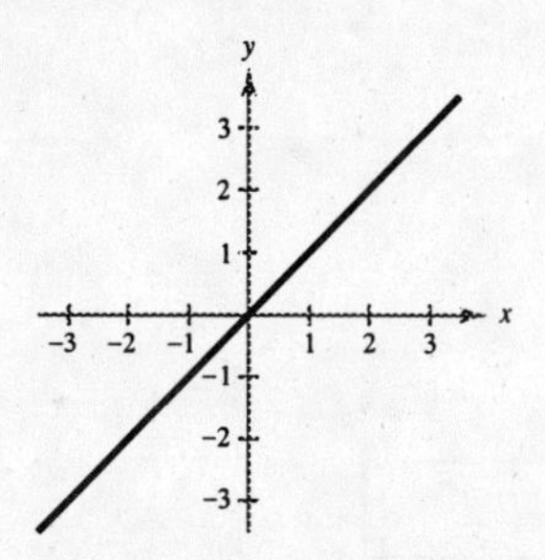

49. Answers will vary.

51. $f(x) = \dfrac{x - 2}{x^2 - 4x + 3} = \dfrac{x - 2}{(x - 3)(x - 1)}$

Vertical asymptotes: $x = 1, 3$

Horizontal asymptote: $y = 0$

Intercepts: $(2, 0)$, $\left(0, -\frac{2}{3}\right)$

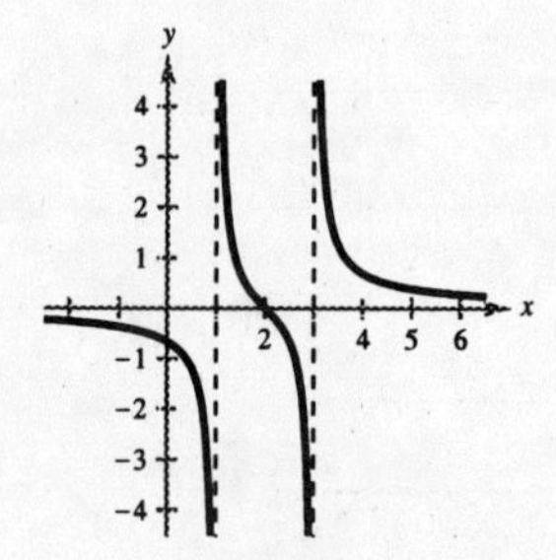

53. $f(x) = \dfrac{x^2 - 16}{x + 4} = \dfrac{(x - 4)(x + 4)}{x + 4} = x - 4, x \neq -4$

Line with hole at $(-4, -8)$

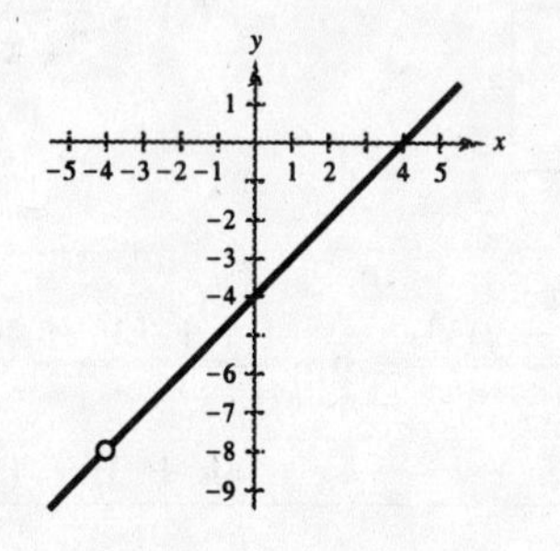

55. $\mathbf{u} \times \mathbf{v} = \begin{vmatrix} \mathbf{i} & \mathbf{j} & \mathbf{k} \\ -10 & 0 & 6 \\ 7 & 0 & 0 \end{vmatrix} = \langle 0, 42, 0 \rangle$

57. $\mathbf{u} \times \mathbf{v} = \begin{vmatrix} \mathbf{i} & \mathbf{j} & \mathbf{k} \\ 8 & -7 & 14 \\ -1 & 8 & 4 \end{vmatrix} = \langle -140, -46, 57 \rangle$

Section 12.4 Limits at Infinity and Limits of Sequences

- The limit at infinity

  $$\lim_{x \to \infty} f(x) = L$$

  means that $f(x)$ get arbitrarily close to L as x increases without bound.
- Similarly, the limit at infinity

  $$\lim_{x \to -\infty} f(x) = L$$

  means that $F(x)$ get arbitrarily close to L as x decreases without bound.
- You should be able to calculate limits at infinity, especially those arising from rational functions.
- Limits of functions can be used to evaluate limits of sequences. If f is a function such that $\lim_{x \to \infty} f(x) = L$ and if a_n is a sequence such that $f(n) = a_n$, then $\lim_{n \to \infty} a_n = L$.

Solutions to Odd-Numbered Exercises

1. Intercept: $(0, 0)$

Horizontal asymptote: $y = 4$

Matches (c).

3. Horizontal asymptote: $y = 4$

Vertical asymptote: $x = 0$

Matches (d).

5. $\lim\limits_{x\to\infty} \dfrac{3}{x^2} = 0$

7. $\lim\limits_{x\to\infty} \dfrac{3 + x}{3 - x} = -1$

9. $\lim\limits_{x\to-\infty} \dfrac{4x - 3}{2x + 1} = 2$

11. $\lim\limits_{t\to\infty} \dfrac{t^2}{t + 3}$ does not exist

13. $\lim\limits_{x\to-\infty} \dfrac{-(x^2 + 3)}{(2 - x)^2} = \lim\limits_{x\to-\infty} \dfrac{-x^2 - 3}{x^2 - 4x + 4} = -1$

15. $\lim\limits_{x\to-\infty} \left[\dfrac{x}{(x + 1)^2} - 4\right] = 0 - 4 = -4$

17. $\lim\limits_{t\to\infty}\left(\dfrac{1}{3t^2} - \dfrac{5t}{t + 2}\right) = 0 - 5 = -5$

19. 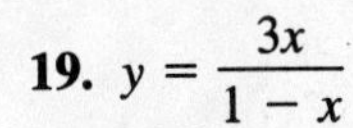
$y = \dfrac{3x}{1 - x}$

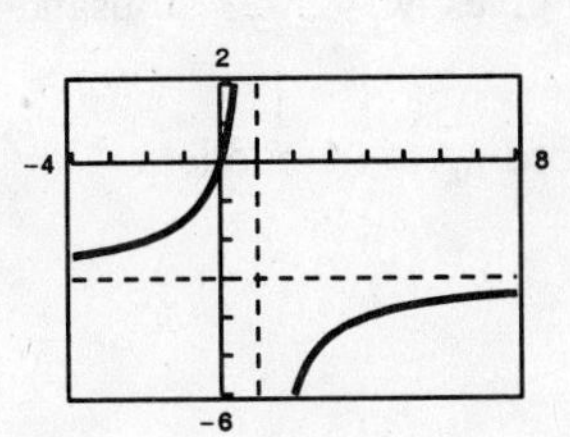

Horizontal asymptote: $y = -3$

21.

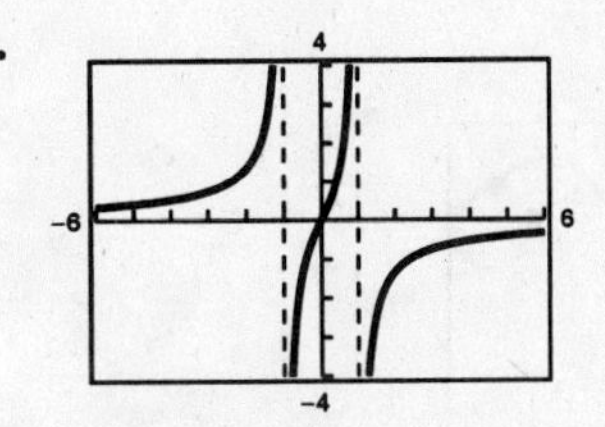

Horizontal asymptote: $y = 0$

23.

x	10^0	10^1	10^2	10^3	10^4	10^5	10^6
$f(x)$	-0.7321	-0.0995	-0.00999	-0.001	-1×10^{-4}	-1×10^{-5}	-1×10^{-6}

$\lim\limits_{x\to\infty}\left(x - \sqrt{x^2 + 2}\right) = 0$

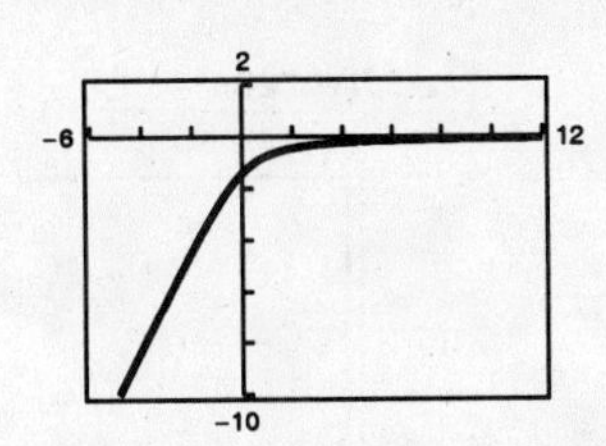

25.

x	10^0	10^1	10^2	10^3	10^4	10^5	10^6
$f(x)$	$-.7082$	$-.7454$	$-.7495$	$-.74995$	$-.749995$	$-.75$	$-.75$

$\lim\limits_{x\to\infty} 3\left(2x - \sqrt{4x^2 + x}\right) = -\frac{3}{4}$

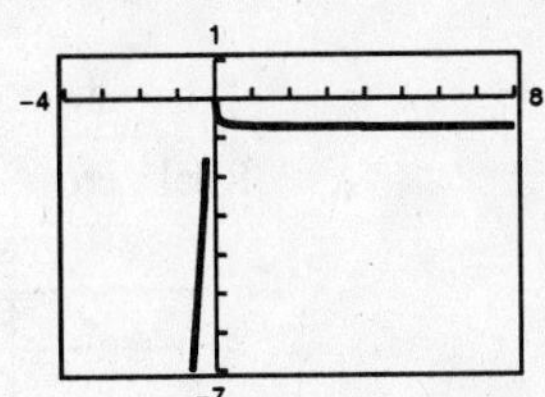

27. $1, \dfrac{3}{5}, \dfrac{2}{5}, \dfrac{5}{17}, \dfrac{3}{13} \cdot \lim\limits_{n\to\infty} \dfrac{n + 2}{n^2 + 1} = 0$

29. $\dfrac{1}{5}, \dfrac{1}{2}, \dfrac{9}{11}, \dfrac{8}{7}, \dfrac{25}{17} \cdot \lim\limits_{n\to\infty} \dfrac{n^2}{3n + 1}$ does not exist.

31. $2, 3, 4, 5, 6 \cdot \lim_{n\to\infty} \dfrac{(n+1)!}{n!} = \lim_{n\to\infty} (n+1)$ does not exist

33. $-1, \dfrac{1}{2}, -\dfrac{1}{3}, \dfrac{1}{4}, -\dfrac{1}{5} \cdot \lim_{n\to\infty} \dfrac{(-1)^n}{n} = 0$

35. $\lim_{n\to\infty} a_n = \frac{3}{2}$

n	10^0	10^1	10^2	10^3	10^4	10^5	10^6
a_n	2	1.55	1.505	1.5005	1.50005	1.500005	1.5000005

37. $\lim_{n\to\infty} a_n = \frac{16}{3}$

n	10^0	10^1	10^2	10^3	10^4	10^5	10^6
a_n	16	6.16	5.4136	5.341336	5.3341	5.33341	5.333341

39. (a)

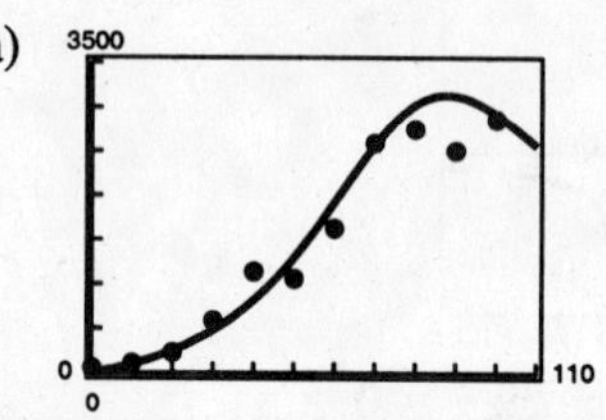

(b) For $t = 80$, the model gives $N \approx 3023$ thousand graduates.

(c) As $t \to \infty$, $N \to 0$, according to the model.

41. False. $f(x) = \dfrac{x^2+1}{1}$ does not have a horizontal asymptote.

43. For example, let $f(x) = \dfrac{1}{x^2}$ and $g(x) = \dfrac{1}{x^2}$.

Then, $\lim_{x\to 0} \dfrac{1}{x^2} = \infty$, but $\lim_{x\to 0} [f(x) - g(x)] = 0$.

45. Converges to 0

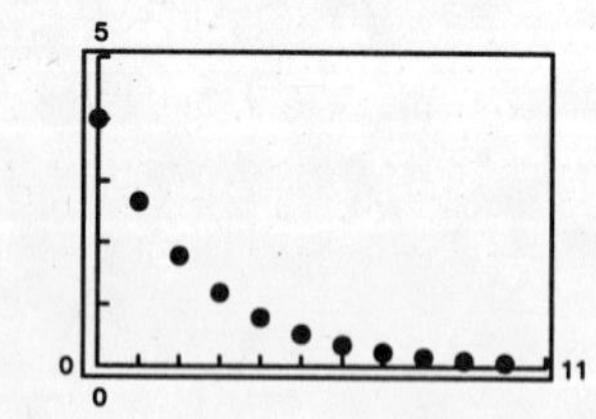

47. Diverges

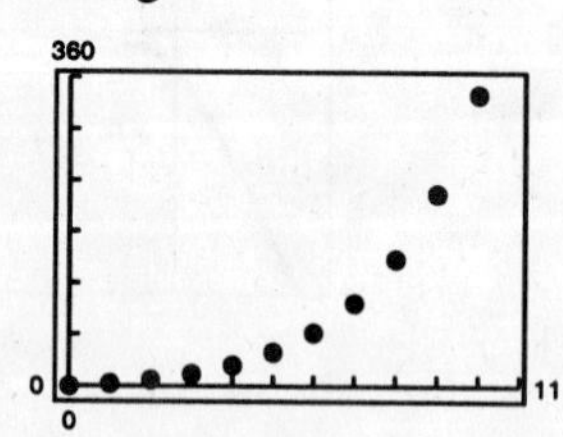

49. $f(x) = x^4 - x^3 - 20x^2$

$= x^2(x^2 - x - 20)$

$= x^2(x-5)(x+4)$

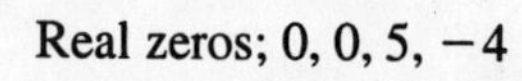

Real zeros; 0, 0, 5, −4

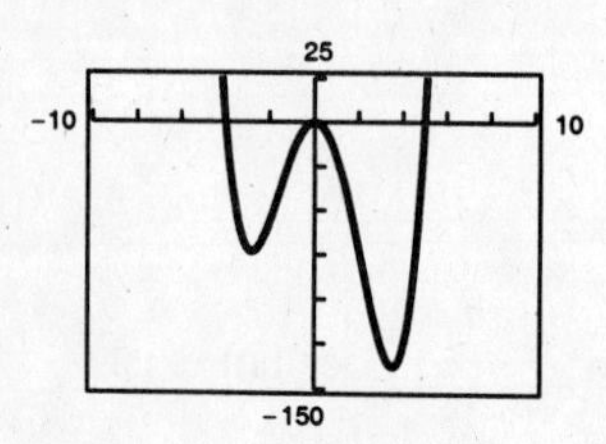

51. $f(x) = x^3 - 3x^2 + 2x - 6$

$= x^2(x-3) + 2(x-3)$

$= (x-3)(x^2+2)$

Real zero: 3

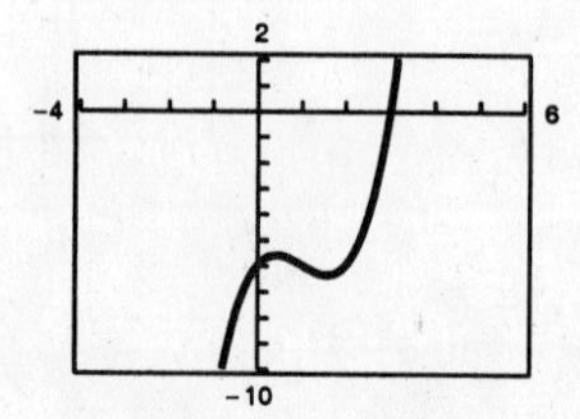

53. $\sum_{i=1}^{6} (2i + 3) = 5 + 7 + 9 + 11 + 13 + 15 = 60$

55. $\sum_{k=1}^{10} 15 = 10(15) = 150$

Section 12.5 The Area Problem

- You should know the following summation formulas and properties.

  (a) $\sum_{i=1}^{n} c = cn$ (b) $\sum_{i=1}^{n} i = \frac{n(n+1)}{2}$ (c) $\sum_{i=1}^{n} i^2 = \frac{n(n+1)(2n+1)}{6}$

  (d) $\sum_{i=1}^{n} i^3 = \frac{n^2(n+1)^2}{4}$ (e) $\sum_{i=1}^{n} (a_i \pm b_i) = \sum_{i=1}^{n} a_i \pm \sum_{i=1}^{n} b_i$ (f) $\sum_{i=1}^{n} ka_i = k\sum_{i=1}^{n} a_i$
- You should be able to evaluate a limit of a summation, $\lim_{n\to\infty} S(n)$.
- You should be able to approximate the area of a region using rectangles. By increasing the number of rectangles, the approximation improves.
- The area of a plane region above the x-axis bounded by f between $x = a$ and $x = b$ is the limit of the sum of the approximating rectangles:

  $$A = \lim_{n\to\infty} \sum_{i=1}^{n} f\left(a + \frac{(b-a)i}{n}\right)\left(\frac{b-a}{n}\right)$$
- You should be able to use the limit definition of area to find the area bounded by simple functions in the plane.

Solutions to Odd-Numbered Exercises

1. $\sum_{i=1}^{60} i = \frac{n(n+1)}{2} = \frac{60(61)}{2} = 1830$

3. $\sum_{k=1}^{20} k^3 = \frac{n^2(n+1)^2}{4} = \frac{20^2(21)^2}{4} = 44{,}100$

5. $\sum_{j=1}^{25} (j^2 + j) = \frac{25(26)(51)}{6} + \frac{25(26)}{2} = 5850$

7. $S(n) = \sum_{i=1}^{n} \frac{i^3}{n^4} = \frac{1}{n^4}\left[\frac{n^2(n+1)^2}{4}\right] = \frac{n^2 + 2n + 1}{4n^2}$

n	10^0	10^1	10^2	10^3
$S(n)$	1	0.3025	0.255025	0.25050025

$\lim_{n\to\infty} S(n) = \frac{1}{4}$

9. $S(n) = \sum_{i=1}^{n} \frac{3}{n^3}(1 + i^2) = \frac{3}{n^3}\left[n + \frac{n(n+1)(2n+1)}{6}\right] = \frac{3}{n^2} + \frac{6n^2 + 9n + 3}{6n^2}$

n	10^0	10^1	10^2	10^3
$S(n)$	6	1.185	1.0154	1.0015

$\lim_{n\to\infty} S(n) = 1$

11. $S(n) = \sum_{i=1}^{n} \left(\frac{i^2}{n^3} + \frac{2}{n}\right)\left(\frac{1}{n}\right) = \frac{1}{n}\left[\frac{n(n+1)(2n+1)}{6n^3} + \frac{2n}{n}\right] = \frac{1}{6n^3}(2n^2 + 3n + 1) + \frac{2}{n}$

n	10^0	10^1	10^2	10^3
$S(n)$	3	0.2385	0.02338	0.00233

$\lim_{n\to\infty} S(n) = 0$

13. $S(n) = \sum_{i=1}^{n}\left[1 - \left(\frac{i}{n}\right)^2\right]\left(\frac{1}{n}\right) = \frac{1}{n}\left[n - \frac{1}{n^2}\left(\frac{n(n+1)(2n+1)}{6}\right)\right] = 1 - \frac{2n^2 + 3n + 1}{6n^2}$

n	10^0	10^1	10^2	10^3
$S(n)$	0	0.615	0.66165	0.66617

$\lim_{n\to\infty} S(n) = \frac{2}{3}$

15. $f(x) = x + 4, [-1, 2], n = 6$, width $= \frac{1}{2}$

Area $\approx \frac{1}{2}[3.5 + 4 + 4.5 + 5 + 5.5 + 6] = 14.25$ sq. units

17. The width of each rectangle is $\frac{1}{4}$. The height is obtained by evaluating f at the right hand endpoint of each interval.

$$A \approx \sum_{i=1}^{8} f\left(\frac{i}{4}\right)\left(\frac{1}{4}\right) = \sum_{i=1}^{8} \frac{1}{4}\left(\frac{i}{4}\right)^3\left(\frac{1}{4}\right) = 1.265625$$

19. Width of each rectangle is $\frac{12}{n}$. The height is

$f\left(\frac{12}{n}i\right) = -\frac{1}{3}\left(\frac{12}{n}i\right) + 4.$

$$A = \sum_{i=1}^{n}\left[-\frac{1}{3}\left(\frac{12i}{n}\right) + 4\right]\left(\frac{12}{n}\right)$$

n	4	8	20	50
Approximate area	18	21	22.8	23.52

Note: exact area is 24.

21. The width of each rectangle is $\frac{3}{n}$. The height is $\frac{1}{9}\left(\frac{3i}{n}\right)^3$.

$$A \approx \sum_{i=1}^{n}\frac{1}{9}\left(\frac{3i}{n}\right)^3\left(\frac{3}{n}\right)$$

n	4	8	20	50
Approximate area	3.52	2.85	2.45	2.34

23.

$$A \approx \sum_{i=1}^{n} f\left(\frac{i}{n}\right)\left(\frac{1}{n}\right) = \sum_{i=1}^{n} 4\left(\frac{i}{n}\right)\left(\frac{1}{n}\right) = \frac{4}{n^2}\sum_{i=1}^{n} i = \frac{4}{n^2}\,\frac{n(n+1)}{2} = \frac{4n^2 + 4n}{2n^2}$$

$$A = \lim_{n\to\infty}\frac{4n^2 + 4n}{2n^2} = 2$$

25.

$$A \approx \sum_{i=1}^{n} f\left(-1 + \frac{2i}{n}\right)\left(\frac{2}{n}\right) = \sum_{i=1}^{n}\left[2 - \left(-1 + \frac{2i}{n}\right)^2\right]\frac{2}{n}$$

$$= \sum_{i=1}^{n}\left[2 - 1 + \frac{4i}{n} - \frac{4i^2}{n^2}\right]\left(\frac{2}{n}\right) = \frac{2}{n}\sum_{i=1}^{n} 1 + \frac{8}{n^2}\sum_{i=1}^{n} i - \frac{8}{n^3}\sum_{i=1}^{n} i^2$$

$$= \frac{2}{n}(n) + \frac{8}{n^2}\,\frac{n(n+1)}{2} - \frac{8}{n^3}\,\frac{n(n+1)(2n+1)}{6}$$

$$A = \lim_{n\to\infty}\left[2 + 4\,\frac{n(n+1)}{n^2} - \frac{4}{3}\,\frac{n(n+1)(2n+1)}{n^3}\right] = 2 + 4 - \frac{8}{3} = \frac{10}{3}$$

27. $A \approx \sum_{i=1}^{n} g\left(1 + \frac{i}{n}\right)\left(\frac{1}{n}\right)$

$$= \sum_{i=1}^{n} \left[8 - \left(1 + \frac{i}{n}\right)^3\right]\frac{1}{n}$$

$$= \sum_{i=1}^{n} \left[7 - \frac{3i}{n} + \frac{3i^2}{n^2} + \frac{i^3}{n^3}\right]\frac{1}{n}$$

$$= \frac{7}{n}\sum_{i=1}^{n} 1 - \frac{3}{n^2}\sum_{i=1}^{n} i - \frac{3}{n^3}\sum_{i=1}^{n} i^2 - \frac{1}{n^4}\sum_{i=1}^{n} i^3$$

$$= \frac{7}{n}(n) - \frac{3}{n^2}\frac{n(n+1)}{2} - \frac{3}{n^3}\frac{n(n+1)(2n+1)}{6} - \frac{1}{n^4}\frac{n^2(n+1)^2}{4}$$

$$A = \lim_{n\to\infty}\left[7 - \frac{3}{2}\frac{n(n+1)}{n^2} - \frac{1}{2n^3}n(n+1)(2n+1) - \frac{1}{n^4}\frac{n^2(n+1)^2}{4}\right]$$

$$= 7 - \frac{3}{2} - 1 - \frac{1}{4} = \frac{17}{4}$$

29. $A \approx \sum_{i=1}^{n} f\left(1 + \frac{3i}{n}\right)\left(\frac{3}{n}\right)$

$$= \sum_{i=1}^{n} \left[\frac{1}{4}\left(1 + \frac{3i}{n}\right)^2 + \left(1 + \frac{3i}{n}\right)\right]\left(\frac{3}{n}\right)$$

$$= \sum_{i=1}^{4} \left(\frac{1}{4} + \frac{3}{2}\frac{i}{n} + \frac{9}{4}\frac{i^2}{n^2} + 1 + \frac{3i}{n}\right)\left(\frac{3}{n}\right)$$

$$= \frac{15}{4n}\sum_{i=1}^{n} 1 + \frac{27}{2n^2}\sum_{i=1}^{n} i + \frac{27}{4n^3}\frac{n(n+1)(2n+1)}{6}$$

$$= \frac{15}{4n}(n) + \frac{27}{2n^2}\left(\frac{n(n+1)}{2}\right) + \frac{27}{4n^3}\sum_{i=1}^{n} i^2$$

$$A = \lim_{n\to\infty}\left[\frac{15}{4} + \frac{27}{4}\frac{n(n+1)}{n^2} + \frac{9}{8n^3}n(n+1)(2n+1)\right] = \frac{15}{4} + \frac{27}{4} + \frac{9}{4} = \frac{51}{4}$$

31. Area $\approx 0.167 \;\left(= \frac{1}{6}\right)$

33. Area ≈ 1.0

35. $y = (-3.0 \cdot 10^{-6})x^3 + 0.002x^2 - 1.05x + 400$

Note that $y = 0$ when $x = 500$.

Area $\approx$ 105,208.33 square feet

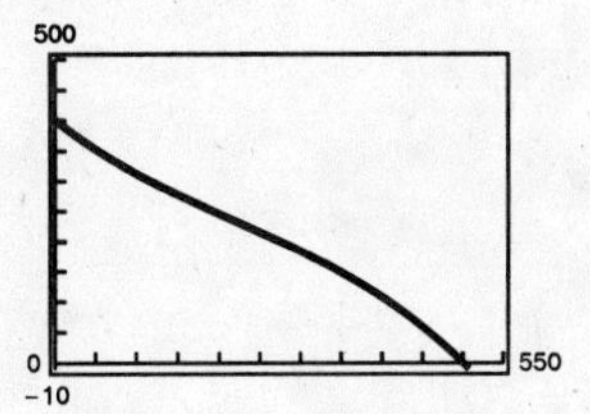

37. True. See Formula 2, page 843.

39. Answers will vary.

41. (a)

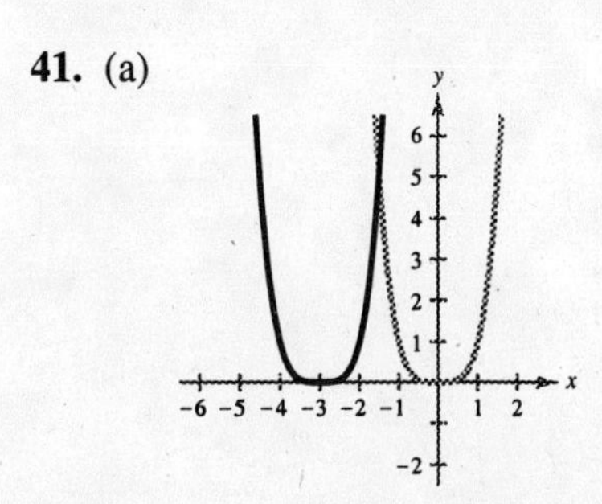

(b)

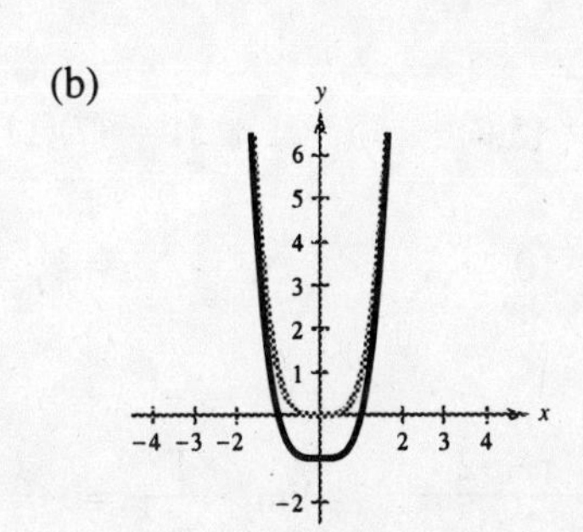

—CONTINUED—

41. —CONTINUED—

(c)

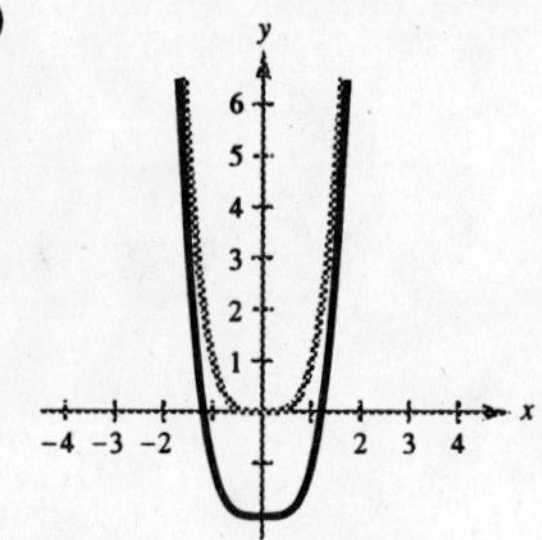

(d)

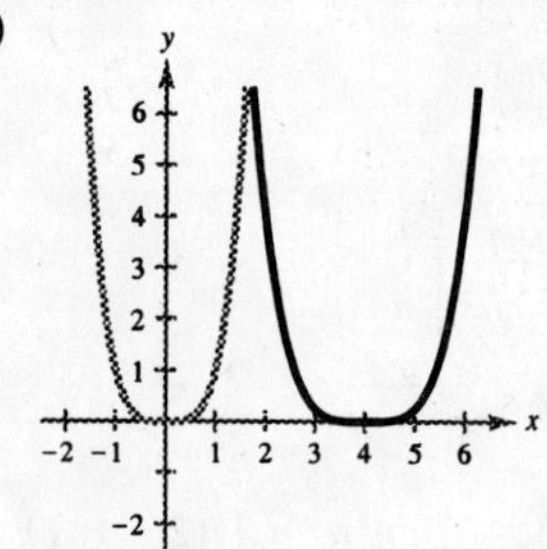

43. $(\mathbf{u} \cdot \mathbf{v})\mathbf{u} = (\langle 4, -5\rangle \cdot \langle -1, -2\rangle)\langle 4, -5\rangle$
$= 6\langle 4, -5\rangle$
$= \langle 24, -30\rangle$

45. $3\mathbf{u} \cdot \mathbf{v} = \langle 12, -15\rangle \cdot \langle -1, -2\rangle$
$= 18$

Review Exercises for Chapter 12

Solutions to Odd-Numbered Exercises

1. $\lim\limits_{x\to 2} \dfrac{x-2}{3x^2-4x-4} = \dfrac{1}{8}$

x	1.9	1.99	1.999	2	2.001	2.01	2.1
$f(x)$	.1299	.1255	.1250	?	.1250	.1245	.1205

3. Yes, $\lim\limits_{x\to 2} f(x) = 1$

5. No, the limit does not exist.

7. $\lim\limits_{x\to 4}\left(\dfrac{1}{2}x + 3\right) = \dfrac{1}{2}(4) + 3 = 5$

9. $\lim\limits_{x\to 2} \dfrac{x^2-1}{x^3+2} = \dfrac{2^2-1}{2^3+2} = \dfrac{3}{10}$

11. $\lim\limits_{x\to \pi} \sin 3x = \sin 3\pi = 0$

13. (a) $\lim\limits_{x\to c} [f(x)]^3 = 4^3 = 64$

(b) $\lim\limits_{x\to c} [3f(x) - g(x)] = 3(4) - 5 = 7$

(c) $\lim\limits_{x\to c} [f(x)\,g(x)] = (4)(5) = 20$

(d) $\lim\limits_{x\to c} \dfrac{f(x)}{g(x)} = \dfrac{4}{5}$

15. $\lim\limits_{x\to 3} (5x-4) = 5(3) - 4 = 11$

17. $\lim\limits_{x\to 2} (5x-3)(3x+5) = (5(2)-3)(3(2)+5) = (7)(11) = 77$

19. $\lim\limits_{t\to 3} \dfrac{t^2+1}{t} = \dfrac{9+1}{3} = \dfrac{10}{3}$

21. $\lim\limits_{t\to -2} \dfrac{t+2}{t^2-4} = \lim\limits_{t\to -2} \dfrac{t+2}{(t+2)(t-2)} = \lim\limits_{t\to -2} \dfrac{1}{t-2} = -\dfrac{1}{4}$

23. $\lim_{x\to 5} \frac{x-5}{x^2+5x-50} = \lim_{x\to 5} \frac{x-5}{(x-5)(x+10)} = \lim_{x\to 5} \frac{1}{x+10} = \frac{1}{15}$

25. $\lim_{x\to -2} \frac{x^2-4}{x^3+8} = \lim_{x\to -2} \frac{(x+2)(x-2)}{(x+2)(x^2-2x+4)} = \lim_{x\to -2} \frac{x-2}{x^2-2x+4}$

$$= \frac{-4}{12} = \frac{-1}{3}$$

27. $\lim_{x\to -1} \frac{\frac{1}{x+2}-1}{x+1} = \lim_{x\to -1} \frac{1-(x-2)}{(x+2)(x+1)} = \lim_{x\to -1} \frac{-(x+1)}{(x+2)(x+1)}$

$$= \lim_{x\to -1} \frac{-1}{(x+2)} = -1$$

29. $\lim_{u\to 0} \frac{\sqrt{4+u}-2}{u} = \lim_{u\to 0} \frac{\sqrt{4+u}-2}{u} \cdot \frac{\sqrt{4+u}+2}{\sqrt{4+u}+2}$

$$= \lim_{u\to 0} \frac{(4+u)-4}{u\left(\sqrt{4+u}+2\right)} = \lim_{u\to 0} \frac{1}{\sqrt{4+u}+2} = \frac{1}{4}$$

31. $\lim_{x\to 5} \frac{\sqrt{x-1}-2}{x-5} = \lim_{x\to 5} \frac{\sqrt{x-1}-2}{x-5} \cdot \frac{\sqrt{x-1}+2}{\sqrt{x-1}+2}$

$$= \lim_{x\to 5} \frac{(x-1)-4}{(x-5)\left(\sqrt{x-1}+2\right)}$$

$$= \lim_{x\to 5} \frac{1}{\sqrt{x-1}+2} = \frac{1}{2+2} = \frac{1}{4}$$

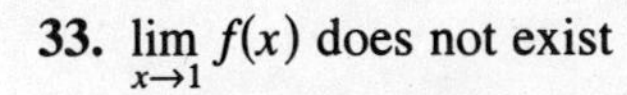

33. $\lim_{x\to 1} f(x)$ does not exist

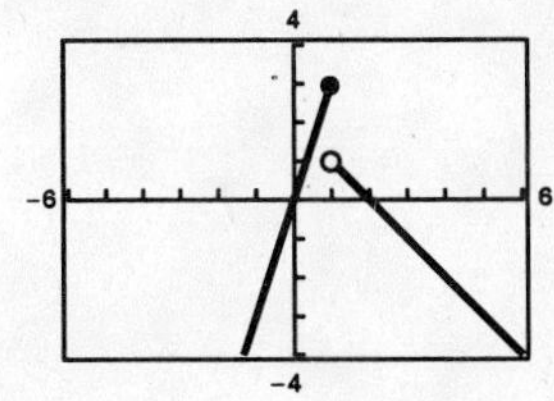

35. $\lim_{x\to 0} e^{-2/x}$ does not exist

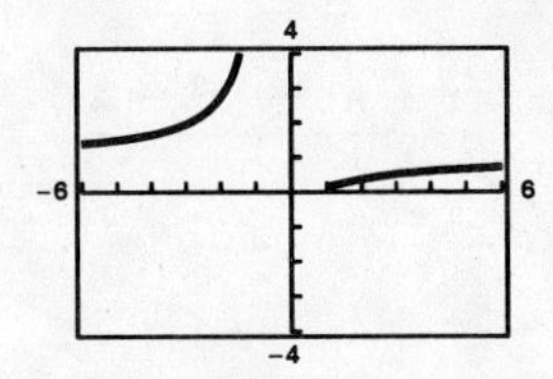

37. $\lim_{x\to 0} g(x) = 2$

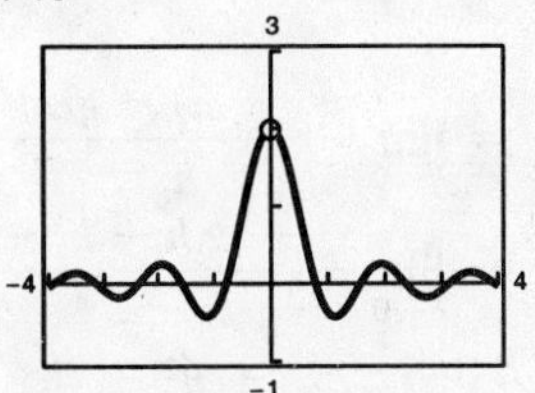

39. (a)

x	1.1	1.01	1.001	1.0001
$f(x)$	0.5680	0.5764	0.5772	0.5773

$\lim_{x\to 1^+} f(x) \approx 0.577$

(b) $\lim_{x\to 1^+} \frac{\sqrt{2x+1}-\sqrt{3}}{x-1} \cdot \frac{\sqrt{2x+1}+\sqrt{3}}{\sqrt{2x+1}+\sqrt{3}}$

$$= \lim_{x\to 1^+} \frac{(2x+1)-3}{(x-1)\left(\sqrt{2x+1}+\sqrt{3}\right)}$$

$$= \lim_{x\to 1^+} \frac{2(x-1)}{(x-1)\left(\sqrt{2x+1}+\sqrt{3}\right)}$$

$$\lim_{x\to 1^+} \frac{2}{\sqrt{2x+1}+\sqrt{3}} = \frac{2}{2\sqrt{3}} = \frac{1}{\sqrt{3}} = \frac{\sqrt{3}}{3}$$

41. $\lim_{x\to 5^-} \frac{|x-5|}{x-5} = -1$

43. $\lim_{x\to 2^+} f(x) = 2^2 - 3 = 1$

45. $f(x) = \dfrac{|x - 3|}{x - 3}$

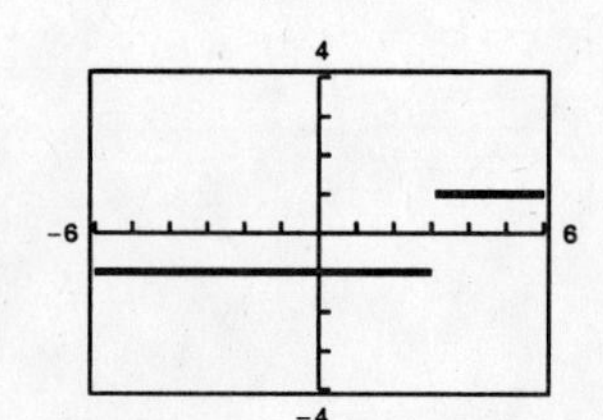

Limit does not exist because $\lim_{x\to 3^+} f(x) = 1$ and $\lim_{x\to 3^-} f(x) = -1$.

47. $f(x) = \dfrac{2}{x^2 - 4}$

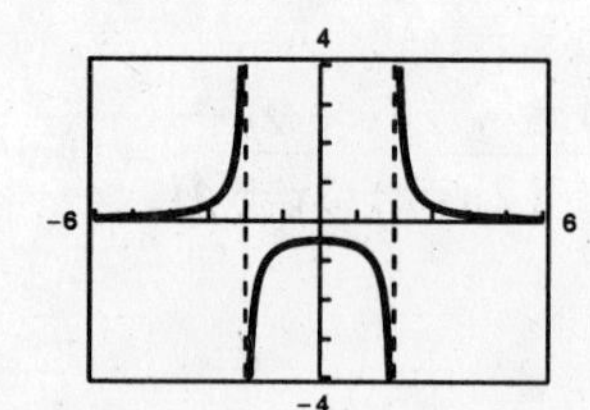

Limit does not exist.

49.
$$\lim_{h\to 0} \frac{f(x+h) - f(x)}{h} = \lim_{h\to 0} \frac{3(x+h) - (x+h)^2 - (3x - x^2)}{h}$$
$$= \lim_{h\to 0} \frac{3x + 3h - x^2 - 2xh - h^2 - 3x + x^2}{h} = \lim_{h\to 0} \frac{3h - 2xh - h^2}{h}$$
$$= \lim_{h\to 0} (3 - 2x - h) = 3 - 2x$$

51. slope $\approx \frac{5}{3}$

53.

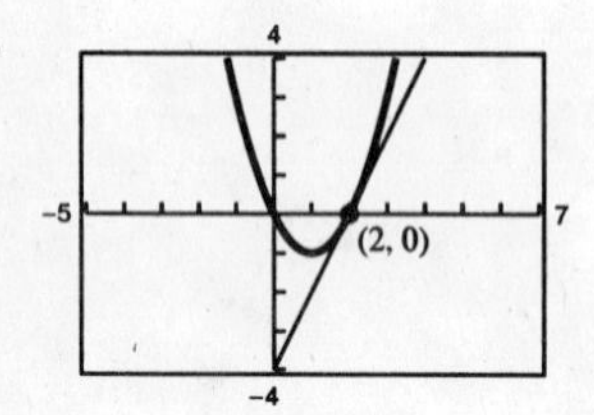

Slope at $(2, f(2))$ is approximately 2.

55.

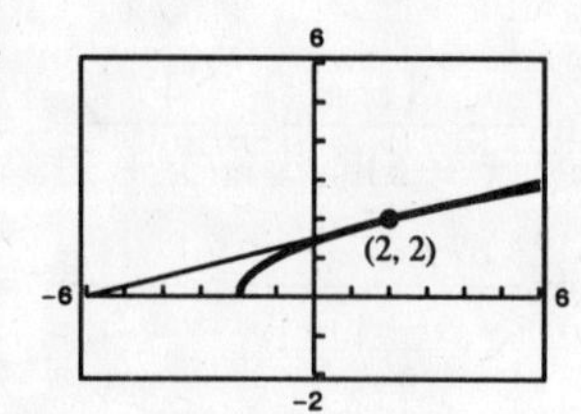

Slope is $\frac{1}{4}$ at $(2, 2)$.

57.
$$m = \lim_{h\to 0} \frac{g(x+h) - g(x)}{h} = \lim_{h\to 0} \frac{(x+h)^2 - 4(x+h) - (x^2 - 4x)}{h}$$
$$= \lim_{h\to 0} \frac{x^2 + 2xh + h^2 - 4x - 4h - x^2 - 4x}{h}$$
$$= \lim_{h\to 0} \frac{2xh + h^2 - 4h}{h} = \lim_{h\to 0} (2x + h - 4) = 2x - 4$$

(a) At $(0, 0)$, $m = 2(0) - 4 = -4$.

(b) At $(5, 5)$, $m = 2(5) - 4 = 6$.

59.
$$m = \lim_{h\to 0} \frac{f(x+h) - f(x)}{h} = \lim_{h\to 0} \frac{\dfrac{4}{x+h-6} - \dfrac{4}{x-6}}{h}$$
$$= \lim_{h\to 0} \frac{4(x-6) - 4(x+h-6)}{(x+h-6)(x-6)h} = \lim_{h\to 0} \frac{-4h}{(x+h-6)(x-6)h}$$
$$= \lim_{h\to 0} \frac{-4}{(x+h-6)(x-6)} = \frac{-4}{(x-6)^2}$$

(a) At $(7, 4)$, $m = \dfrac{-4}{(7-6)^2} = -4$

(b) At $(8, 2)$, $m = \dfrac{-4}{(8-6)^2} = -1$.

61. $f'(x) = \lim_{h\to 0} \frac{f(x+h) - f(x)}{h} = \lim_{h\to 0} \frac{5-5}{h} = 0$

63. $g'(x) = \lim_{h\to 0} \frac{g(x+h) - g(x)}{h} = \lim_{h\to 0} \frac{-4-(-4)}{h} = 0$

65. $h'(x) = \lim_{h\to 0} \frac{h(x+h) - h(x)}{h} = \lim_{h\to 0} \frac{\left[5 - \frac{1}{2}(x+h)\right] - \left[5 - \frac{1}{2}x\right]}{h}$

$= \lim_{h\to 0} \frac{-\frac{1}{2}h}{h} = -\frac{1}{2}$

67. $f'(t) = \lim_{h\to 0} \frac{f(t+h) - f(t)}{h} = \lim_{h\to 0} \frac{\sqrt{t+h+5} - \sqrt{t+5}}{h} \cdot \frac{\sqrt{t+h+5} + \sqrt{t+5}}{\sqrt{t+h+5} + \sqrt{t+5}}$

$= \lim_{h\to 0} \frac{(t+h+5) - (t-5)}{h\left(\sqrt{t+h+5} + \sqrt{t+5}\right)} = \lim_{h\to 0} \frac{1}{\sqrt{t+h+5} + \sqrt{t+5}}$

$= \frac{1}{2\sqrt{t+5}}$

69. $g'(s) = \frac{g(s+h) - g(s)}{h} = \lim_{h\to 0} \frac{\frac{4}{s+h+5} - \frac{4}{s+5}}{h}$

$= \lim_{h\to 0} \frac{4s + 20 - 4s - 4h - 20}{(s+h+5)(s+5)h}$

$= \lim_{h\to 0} \frac{-4h}{(s+h+5)(s+5)h}$

$= \lim_{h\to 0} \frac{-4}{(s+h+5)(s+5)} = \frac{-4}{(s+5)^2}$

71. $\lim_{x\to\infty} \frac{4x}{2x-3} = \frac{4}{2} = 2$

73. $\lim_{x\to\infty} \frac{2x}{x^2-25} = 0$

75. $\lim_{x\to\infty} \left(4 - \frac{7}{x^3}\right) = 4 - 0 = 4$

77. (a) Average cost per unit is

$\frac{C}{x} = \frac{22.50x + 12{,}200}{x} = \overline{C}(x)$

$\overline{C}(100) = \$144.50$

$\overline{C}(1000) = \$34.70$

(b) As $x \to \infty$, $\overline{C} \to \$22.50$.

79. $\frac{1}{2}, \frac{2}{5}, \frac{3}{10}, \frac{4}{17}, \frac{5}{26}$

$\lim_{n\to\infty} a_n = 0$

81. $a_n = 2 + \frac{2}{n}(n-1) - \frac{4}{n} = 4 - \frac{6}{n}$: $-2, 1, 2, \frac{5}{2}, \frac{14}{5}$

$\lim_{n\to\infty} a_n = \lim_{n\to\infty} \left(4 - \frac{6}{n}\right) = 4$

83. $\sum_{i=1}^{n}\left[4-\left(\frac{3i}{n}\right)^2\right]\left(\frac{3i}{n^2}\right) = \frac{12}{n^2}\sum_{i=1}^{n} i - \frac{27}{n^4}\sum_{i=1}^{n} i^3 = \frac{12}{n^2}\frac{n(n+1)}{2} - \frac{27}{n^4}\frac{n^2(n+1)^2}{4}$

$$= \frac{24n^2 + 24n - 27(n^2 + 2n + 1)}{4n^2}$$

$$= \frac{-3n^2 - 30n - 27}{4n^2} = \frac{-3}{4n^2}(n^2 + 10n + 9)$$

$$= \frac{-3(n+1)(n+9)}{4n^2}$$

n	10^0	10^1	10^2	10^3
$S(n)$	-15	-1.5675	-0.8257	-0.7575

$\lim_{n\to\infty} Sn = -\frac{3}{4}$

85. Width of rectangle: $\frac{1}{4}$; Height is f evaluated at right endpoint.

Area $\approx \frac{1}{4}\left[f\left(\frac{1}{4}\right) + f\left(\frac{1}{2}\right) + f\left(\frac{3}{4}\right) + f(1)\right]$

$= \frac{1}{4}\left[4 - \left(\frac{1}{4}\right)^2 + 4 - \left(\frac{1}{2}\right)^2 + 4 - \left(\frac{3}{4}\right)^2 + 4 - 1\right]$

$= \frac{1}{4}\left[15 - \frac{14}{16}\right] = \frac{113}{32} = 3.53125$

87. $f(x) = 4x - x^2$

n	4	8	20	50
Approximate Area	10	10.5	10.64	10.6624

(Exact area is $10\frac{2}{3}$)

89. $A = \lim_{n\to\infty}\sum_{i=1}^{n}\left[2\left(3+\frac{3i}{n}\right) - 6\right]\left(\frac{3}{n}\right)$

$= \lim_{n\to\infty}\sum_{i=1}^{n}\frac{18i}{n^2} = \lim_{n\to\infty}\frac{18}{n^2}\sum_{i=1}^{n} i$

$= \lim_{n\to\infty}\frac{18}{n^2}\frac{n(n+1)}{2} = 9$ exact area

91. $A = \lim_{n\to\infty}\sum_{i=1}^{n} 8\left(\left(\frac{i}{n}\right) - \left(\frac{i}{n}\right)^2\right)\frac{i}{n}$

$= \lim_{n\to\infty}\left[\frac{8}{n^2}\sum_{i=1}^{n} i - \frac{8}{n^3}\sum i^2\right]$

$= \lim_{n\to\infty}\left[\frac{8}{n^2}\frac{n(n+1)}{2} - \frac{8}{n^3}\frac{n(n+1)(2n+1)}{6}\right]$

$= 4 - \frac{8}{3} = \frac{4}{3}$ exact area

93. $A = \lim_{n\to\infty}\sum_{i=1}^{n}\left[4 - \left[\left(1+\frac{3i}{n}\right) - 2\right]^2\right]\left(\frac{3}{n}\right)$

$= \lim_{n\to\infty}\sum_{i=1}^{n}\left[4 - \left(\frac{3i}{n} - 1\right)^2\right]\left(\frac{3}{n}\right)$

$= \lim_{n\to\infty}\frac{3}{n}\sum_{i=1}^{n}\left(3 + \frac{6i}{n} - \frac{9i^2}{n^2}\right)$

$= \lim_{n\to\infty}\left[\frac{9}{n}\sum_{i=1}^{n} 1 + \frac{18}{n^2}\sum_{i=1}^{n} i - \frac{27}{n^3}\sum_{i=1}^{n} i^2\right]$

$= \lim_{n\to\infty}\left[\frac{9}{n}(n) + \frac{18}{n^2}\frac{n(n+1)}{2} - \frac{27}{n^3}\frac{n(n+1)(2n+1)}{6}\right]$

$= 9 + 9 - 9 = 9$ exact area

95. True (assuming all the limits exist)

Chapter 12 Practice Test

1. Use a graphing utility to complete the table and use the result to estimate the limit

$\lim_{x \to 3} \frac{x-3}{x^2-9}$.

x	2.9	2.99	3	3.01	3.1
$f(x)$			?		

2. Graph the function $f(x) = \frac{\sqrt{x+4}-2}{x}$ and estimate the limit

$$\lim_{x \to 0} \frac{\sqrt{x+4}-2}{x}.$$

3. Find the limit $\lim_{x \to 2} e^{x-2}$ by direct substitution.

4. Find the limit $\lim_{x \to 1} \frac{x^3-1}{x-1}$ analytically.

5. Use a graphing utility to estimate the limit

$$\lim_{x \to 0} \frac{\sin 5x}{2x}.$$

6. Find the limit $\lim_{x \to -2} \frac{|x+2|}{x+2}$.

7. Use the limit process to find the slope of the graph of $f(x) = \sqrt{x}$ at the point (4, 2).

8. Find the derivative of the function $f(x) = 3x - 1$.

9. Find the limits.

(a) $\lim_{x \to \infty} \frac{3}{x^4}$ (b) $\lim_{x \to -\infty} \frac{x^2}{x^2+3}$ (c) $\lim_{x \to \infty} \frac{|x|}{1-x}$

10. Write the first four terms of the sequence $a_n = \frac{1-n^2}{2n^2+1}$ and find the limit of the sequence.

11. Find the sum $\sum_{i=1}^{25} (i^2 + i)$.

12. Write the sum $\sum_{i=1}^{n} \frac{i^2}{n^3}$ as a rational function $s(n)$, and find $\lim_{n \to \infty} s(n)$.

13. Find the area of the region bounded by $f(x) = 1 - x^2$ over the interval $0 \le x \le 1$.

Chapter 1 Practice Test Solutions

1.

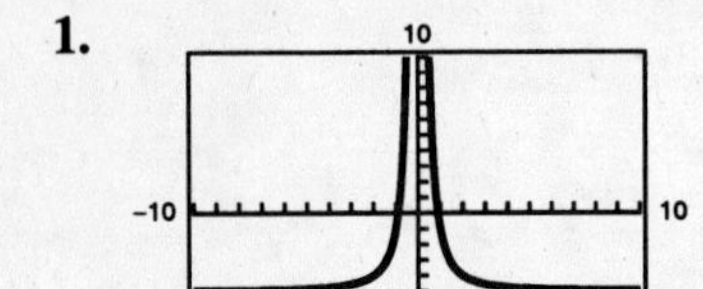

x-intercepts: ± 0.894

2.

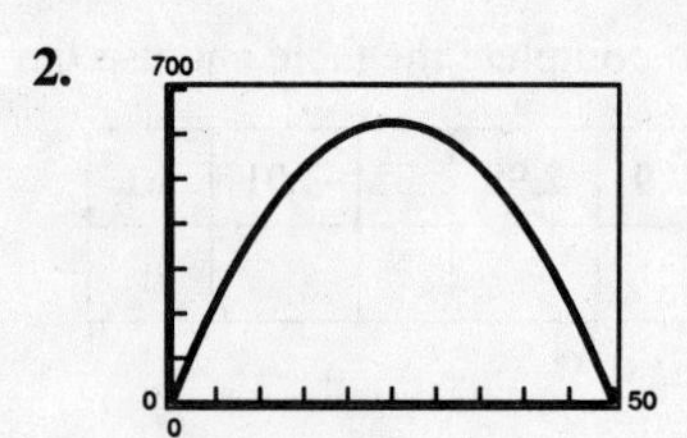

No x-intercepts

3. $3x - 5y = 15$

Line

x-intercept: $(5, 0)$

y-intercept: $(0, -3)$

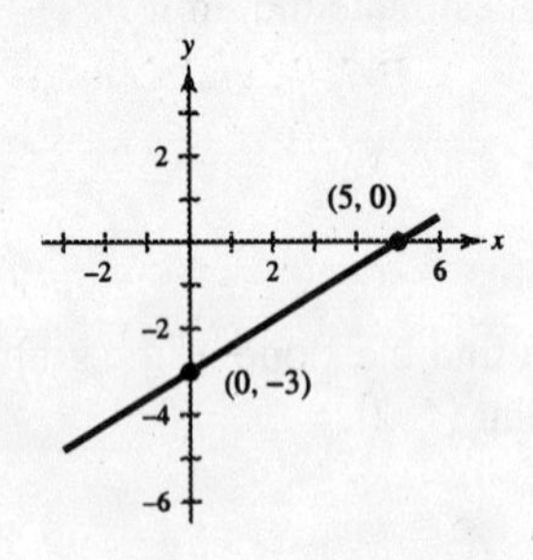

4. $y = \sqrt{9 - x}$

Domain: $(-\infty, 9]$

x-intercept: $(9, 0)$

y-intercept: $(0, 3)$

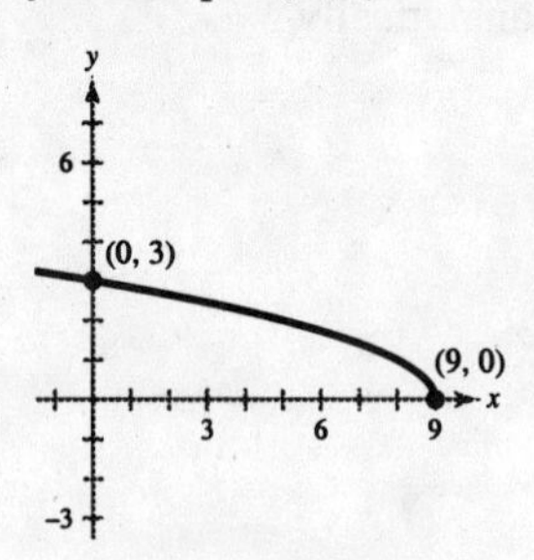

5.
$$\begin{aligned} 5x + 4 &= 7x - 8 \\ 4 + 8 &= 7x - 5x \\ 12 &= 2x \\ x &= 6 \end{aligned}$$

6.
$$\begin{aligned} \frac{x}{3} - 5 &= \frac{x}{5} + 1 \\ 15\left(\frac{x}{3} - 5\right) &= 15\left(\frac{x}{5} + 1\right) \\ 5x - 75 &= 3x + 15 \\ 2x &= 90 \\ x &= 45 \end{aligned}$$

7.
$$\begin{aligned} \frac{3x + 1}{6x - 7} &= \frac{2}{5} \\ 5(3x + 1) &= 2(6x - 7) \\ 15x + 5 &= 12x - 14 \\ 3x &= -19 \\ x &= -\frac{19}{3} \end{aligned}$$

8.
$$\begin{aligned} (x - 3)^2 + 4 &= (x + 1)^2 \\ x^2 - 6x + 9 + 4 &= x^2 + 2x + 1 \\ -8x &= -12 \\ x &= \frac{-12}{-8} \\ x &= \frac{3}{2} \end{aligned}$$

9. Slope $= \dfrac{-2 - (-5)}{3 - 4} = \dfrac{3}{-1} = -3$

$y + 2 = -3(x - 3)$

$y + 2 = -3x - 9$

$y + 3x = 7$ or $y = -3x + 7$

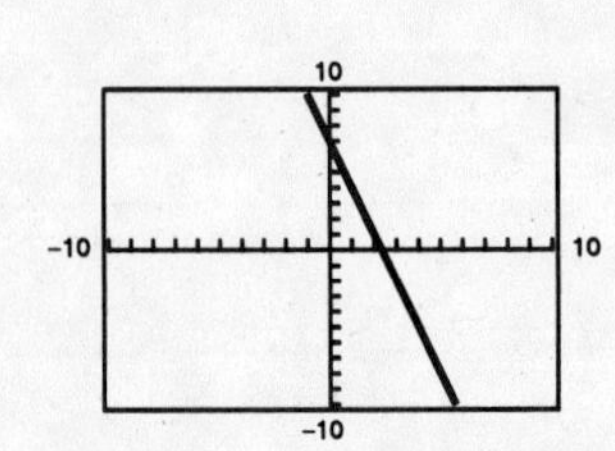

10. $y - 5 = -3(x + 1)$

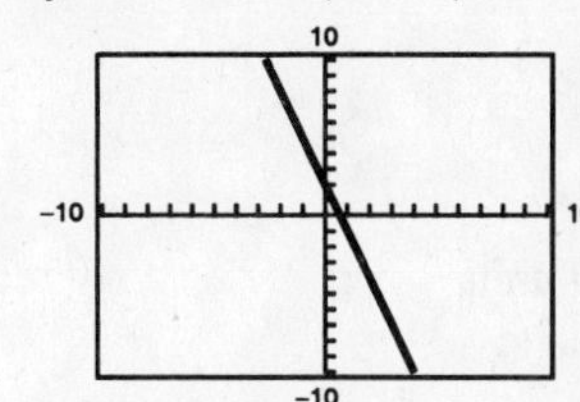

$$y - 5 = -3x - 3$$

$$y + 3x = 2 \quad \text{or} \quad y = -3x + 2$$

11. No, y is not a function of x. For example, $(0, 2)$ and $(0, -2)$ both satisfy the equation.

12. $f(0) = \dfrac{|0 - 2|}{(0 - 2)} = \dfrac{2}{-2} = -1$

$f(2)$ is not defined.

$f(4) = \dfrac{|4 - 2|}{(4 - 2)} = \dfrac{2}{2} = 1$

13. The domain of

$$f(x) = \frac{5}{x^2 - 16}$$

is all $x \neq \pm 4$.

14. The domain of $g(t) = \sqrt{4 - t}$ consists of all t satisfying $4 - t \geq 0$ or $t \leq 4$.

15.

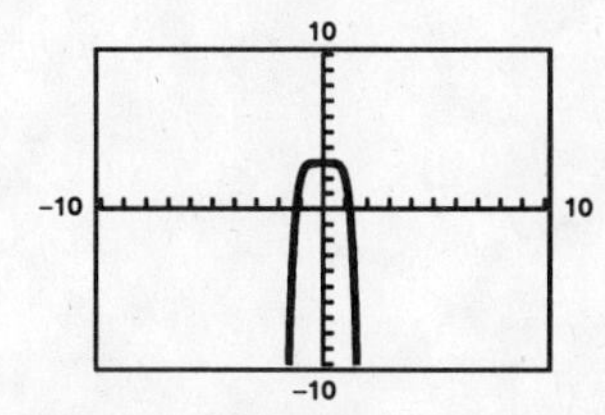

$f(x) = 3 - x^6$ is even.

16.

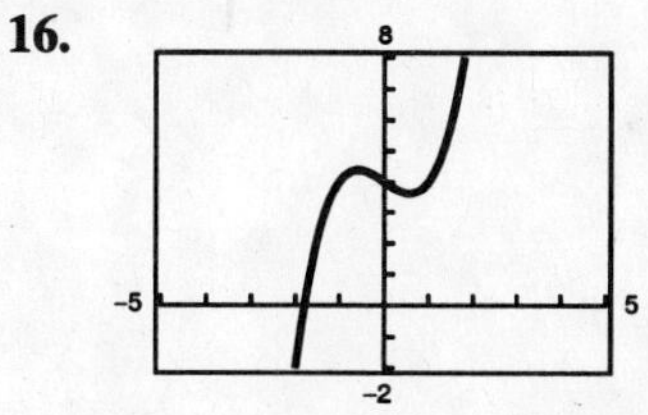

Relative minimum: $(0.577, 3.615)$

Relative maximum: $(-0.577, 4.385)$

17. $f(x) = x^3 - 3$ is a vertical shift of 3 units downward of $y = x^3$.

18. $f(x) = \sqrt{x - 6}$ is a horizontal shift 6 units to the right of $y = \sqrt{x}$.

19. $(g \circ f)(x) = g(f(x)) = g\left(\sqrt{x}\right) = \left(\sqrt{x}\right)^2 - 2 = x - 2$

Domain: $x \geq 0$

20. $\left(\dfrac{f}{g}\right)(x) = \dfrac{f(x)}{g(x)} = \dfrac{3x^2}{16 - x^4}$

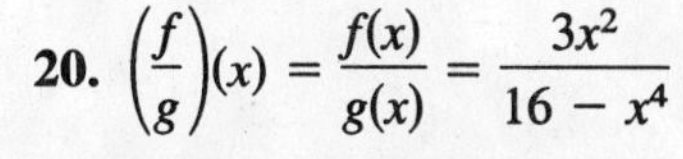

The domain is all $x \neq \pm 2$.

21. $(f \circ g)(x) = f\left(\dfrac{x - 1}{3}\right) = 3\left(\dfrac{x - 1}{3}\right) + 1 = (x - 1) + 1 = x$

$(g \circ f)(x) = g(3x + 1) = \dfrac{(3x + 1) - 1}{3} = \dfrac{3x}{3} = x$

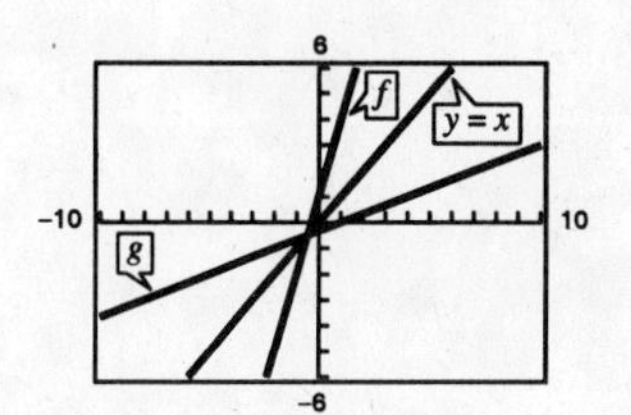

22. $y = \sqrt{9 - x^2}, \; 0 \leq x \leq 3$

$x = \sqrt{9 - y^2}$

$x^2 = 9 - y^2$

$y^2 = 9 - x^2$

$y = \sqrt{9 - x^2}$

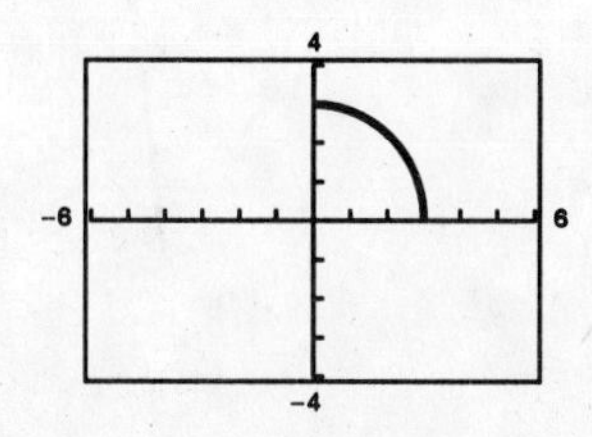

Chapter 2 Practice Test Solutions

1. x-intercepts: $(1, 0)$, $(5, 0)$

y-intercept: $(0, 5)$

Vertex: $(3, -4)$

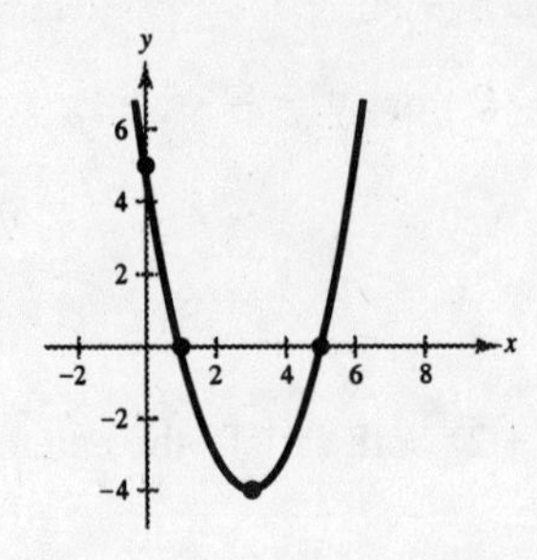

2. $a = 0.01, b = -90$

$$\frac{-b}{2a} = \frac{90}{2(.01)} = 4500 \text{ units}$$

3. Vertex: $(1, 7)$

Opening downward through $(2, 5)$

$y = a(x - 1)^2 + 7$ Standard form

$$5 = a(2 - 1)^2 + 7$$
$$5 = a + 7$$
$$a = -2$$
$$y = -2(x - 1)^2 + 7$$
$$= -2(x^2 - 2x + 1) + 7$$
$$= -2x^2 + 4x + 5$$

4. $y = \pm a(x - 2)(3x - 4)$ where a is any real number

$y = \pm(3x^2 - 10x + 8)$

5. Leading coefficient: -3

Degree: 5

Moves down to the right and up to the left.

6.
$$0 = x^5 - 5x^3 + 4x$$
$$= x(x^4 - 5x^2 + 4)$$
$$= x(x^2 - 1)(x^2 - 4)$$
$$= x(x + 1)(x - 1)(x + 2)(x - 2)$$
$$x = 0, x = \pm 1, x = \pm 2$$

7.
$$f(x) = x(x - 3)(x + 2)$$
$$= x(x^2 - x - 6)$$
$$= x^3 - x^2 - 6x$$

8. Intercepts: $(0, 0)$, $(\pm 2\sqrt{3}, 0)$

Moves up to the right.

Moves down to the left.

x	-2	-1	0	1	2
y	16	11	0	-11	-16

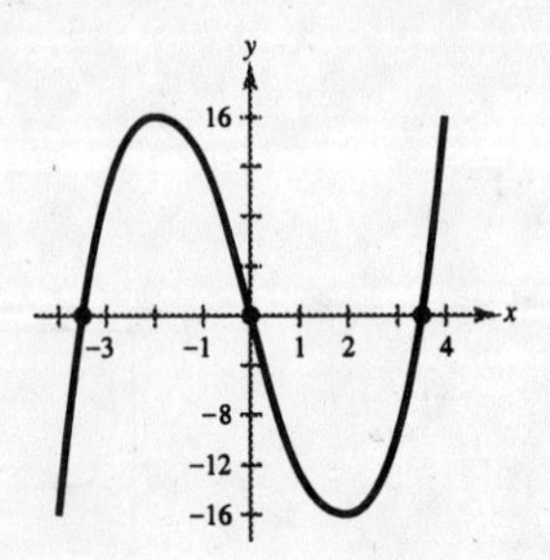

9.

$$
\begin{array}{r}
3x^3 + 9x^2 + 20x \ + 62 + \dfrac{176}{x-3} \\
x - 3 \overline{)\,3x^4 + 0x^3 - 7x^2 \ + \ 2x - 10} \\
\underline{3x^4 - 9x^3} \qquad\qquad\qquad\qquad \\
9x^3 - 7x^2 \qquad\qquad\quad \\
\underline{9x^3 - 27x^2} \qquad\qquad\quad \\
20x^2 + \ 2x \qquad\quad \\
\underline{20x^2 - 60x} \qquad\quad \\
62x - 10 \\
\underline{62x - 186} \\
176
\end{array}
$$

10.

$$
\begin{array}{r}
x - 2 + \dfrac{5x - 13}{x^2 + 2x - 1} \\
x^2 + 2x - 1 \overline{)\,x^3 + \ 0x^2 + 0x - 11} \\
\underline{x^3 + \ 2x^2 - \ x} \qquad\quad \\
-2x^2 + \ x - 11 \\
\underline{-2x^2 - 4x + \ 2} \\
5x - 13
\end{array}
$$

11.

$$
\begin{array}{r|rrrrrr}
-5 & 3 & 13 & 0 & 0 & 12 & -1 \\
 & & -15 & 10 & -50 & 250 & -1310 \\
\hline
 & 3 & -2 & 10 & -50 & 262 & -1311
\end{array}
$$

$$\frac{3x^5 + 13x^4 + 12x - 1}{x + 5} = 3x^4 - 2x^3 + 10x^2 - 50x + 262 - \frac{1311}{x + 5}$$

12.

$$
\begin{array}{r|rrrr}
-6 & 7 & 40 & -12 & 15 \\
 & & -42 & 12 & 0 \\
\hline
 & 7 & -2 & 0 & 15
\end{array}
$$

$f(-6) = 15$

13. $0 = x^3 - 19x - 30$

Possible rational roots:

$\pm 1, \pm 2, \pm 3, \pm 5, \pm 6, \pm 10, \pm 15, \pm 30$

$$
\begin{array}{r|rrrr}
-2 & 1 & 0 & -19 & -30 \\
 & & -2 & 4 & 30 \\
\hline
 & 1 & -2 & -15 & 0
\end{array}
\qquad -2 \text{ is a zero.}
$$

$0 = (x + 2)(x^2 - 2x - 15)$

$0 = (x + 2)(x + 3)(x - 5)$

Zeros: $x = -2, x = -3, x = 5$

14. $0 = x^4 + x^3 - 8x^2 - 9x - 9$

Possible rational roots: $\pm 1, \pm 3, \pm 9$

$$\begin{array}{r|rrrrr} 3 & 1 & 1 & -8 & -9 & -9 \\ & & 3 & 12 & 12 & 9 \\ \hline & 1 & 4 & 4 & 3 & 0 \end{array} \quad x = 3 \text{ is a zero.}$$

$0 = (x - 3)(x^3 + 4x^2 + 4x + 3)$

Possible rational roots of $x^3 + 4x^2 + 4x + 3$: $\pm 1, \pm 3$

$$\begin{array}{r|rrrr} -3 & 1 & 4 & 4 & 3 \\ & & -3 & -3 & -3 \\ \hline & 1 & 1 & 1 & 0 \end{array} \quad x = -3 \text{ is a zero.}$$

$0 = (x - 3)(x + 3)(x^2 + x + 1)$

The zeros of $x^2 + x + 1$ are $x = \dfrac{-1 \pm \sqrt{3}i}{2}$.

Zeros: $x = 3, x = -3, x = -\dfrac{1}{2} + \dfrac{\sqrt{3}}{2}i, x = -\dfrac{1}{2} - \dfrac{\sqrt{3}}{2}i$

15. $0 = 6x^3 - 5x^2 + 4x - 15$

Possible rational roots: $\pm 1, \pm 3, \pm 5, \pm 15, \pm\frac{1}{2}, \pm\frac{3}{2}, \pm\frac{5}{2}, \pm\frac{15}{2}, \pm\frac{1}{3}, \pm\frac{5}{3}, \pm\frac{1}{6}, \pm\frac{5}{6}$

16. $0 = x^3 - \frac{20}{3}x^2 + 9x - \frac{10}{3}$

$0 = 3x^3 - 20x^2 + 27x - 10$

Possible rational roots: $\pm 1, \pm 2, \pm 5, \pm 10, \pm\frac{1}{3}, \pm\frac{2}{3}, \pm\frac{5}{3}, \pm\frac{10}{3}$

$$\begin{array}{r|rrrr} 1 & 3 & -20 & 27 & -10 \\ & & 3 & -17 & 10 \\ \hline & 3 & -17 & 10 & 0 \end{array}$$

$0 = (x - 1)(3x^2 - 17x + 10)$

$0 = (x - 1)(3x - 2)(x - 5)$

Zeros: $x = 1, x = \frac{2}{3}, x = 5$

17. Possible Rational Roots: $\pm 1, \pm 2, \pm 5, \pm 10$

$$\begin{array}{r|rrrrr} 1 & 1 & 1 & 3 & 5 & -10 \\ & & 1 & 2 & 5 & 10 \\ \hline & 1 & 2 & 5 & 10 & 0 \end{array} \quad x = 1 \text{ is a zero.}$$

$$\begin{array}{r|rrrr} -2 & 1 & 2 & 5 & 10 \\ & & -2 & 0 & -10 \\ \hline & 1 & 0 & 5 & 0 \end{array} \quad x = -2 \text{ is a zero.}$$

$f(x) = (x - 1)(x + 2)(x^2 + 5)$

$\quad = (x - 1)(x + 2)(x + 5i)(x - 5i)$

18. $f(x) = (x - 2)[x - (3 + i)][x - (3 - i)]$

$= (x - 2)[x^2 - x(3 - i) - x(3 + i) + (3 + i)(3 - i)]$

$= (x - 2)[x^2 - 6x + 10]$

$= x^3 - 8x^2 + 22x - 20$

19.
$$\begin{array}{r|rrrr} 3i & 1 & 4 & 9 & 36 \\ & & 3i & 12i - 9 & -36 \\ \hline & 1 & 4 + 3i & 12i & 0 \end{array}$$

20. $z = \dfrac{kx^2}{\sqrt{y}}$

21. Vertical asymptote: $x = 0$
Horizontal asymptote: $y = \frac{1}{2}$
x-intercept: $(1, 0)$

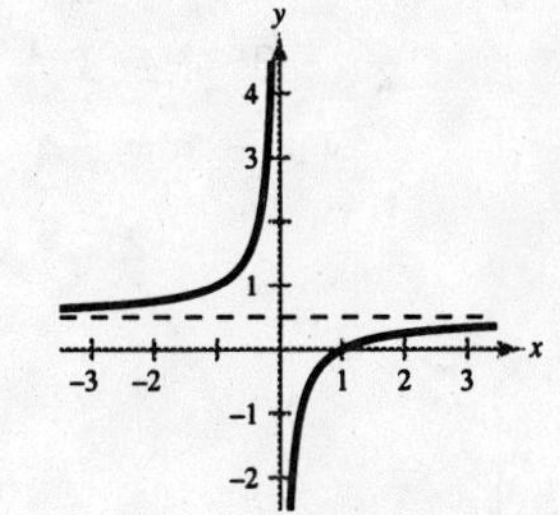

22. Vertical asymptote: $x = 0$
Horizontal asymptote: $y = 3x$
x-intercept: $\left(\pm\dfrac{2}{\sqrt{3}}, 0\right)$

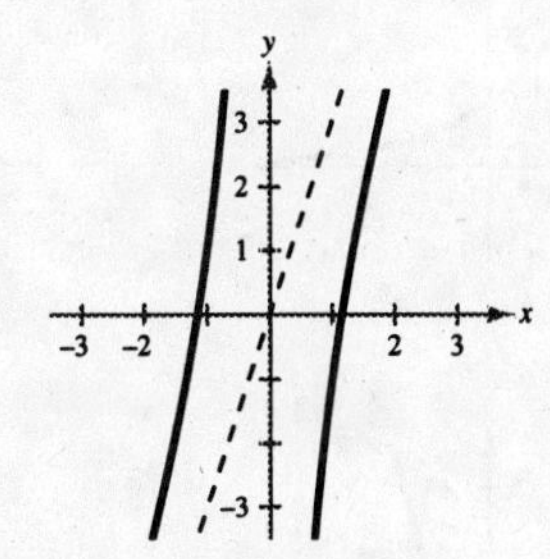

23. $y = 8$ is a horizontal asymptote since the degree on the numerator equals the degree of the denominator. There are no vertical asymptotes.

24. $x = 1$ is a vertical asymptote.

$$\frac{4x^2 - 2x + 7}{x - 1} = 4x + 2 + \frac{9}{x - 1}$$

so $y = 4x + 2$ is a slant asymptote.

25. $f(x) = \dfrac{x-5}{(x-5)^2} = \dfrac{1}{x-5}$

Vertical asymptote: $x = 5$

Horizontal asymptote: $y = 0$

y-intercept: $\left(0, -\dfrac{1}{5}\right)$

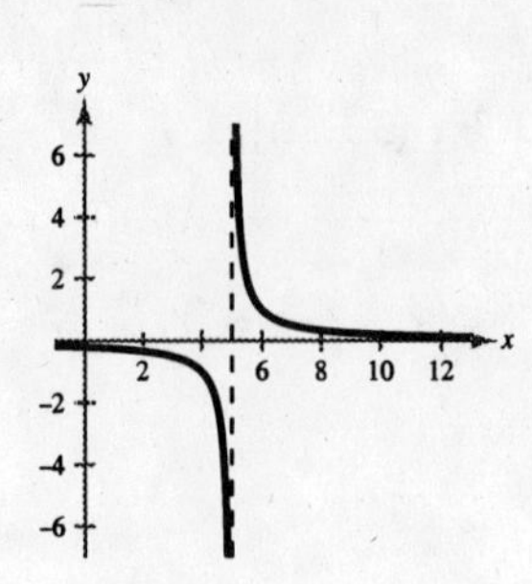

Chapter 3 Practice Test Solutions

1. $x^{3/5} = 8$

$x = 8^{5/3} = \left(\sqrt[3]{8}\right)^5 = 2^5 = 32$

2. $3^{x-1} = \frac{1}{81}$

$3^{x-1} = 3^{-4}$

$x - 1 = -4$

$x = -3$

3. $f(x) = 2^{-x} = \left(\frac{1}{2}\right)^x$

x	-2	-1	0	1	2
$f(x)$	4	2	1	$\frac{1}{2}$	$\frac{1}{4}$

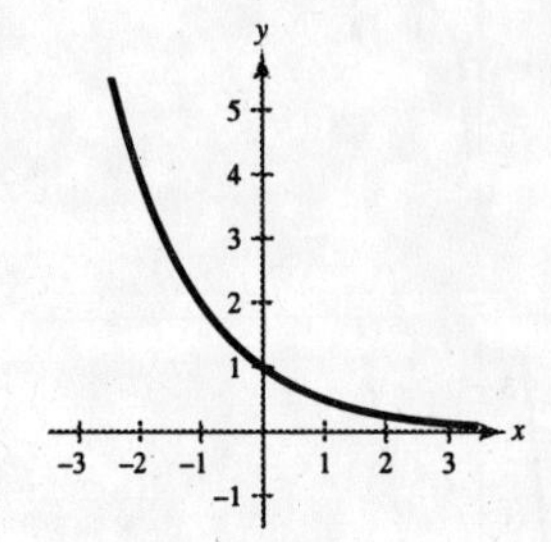

4. $g(x) = e^x + 1$

x	-2	-1	0	1	2
$g(x)$	1.14	1.37	2	3.72	8.39

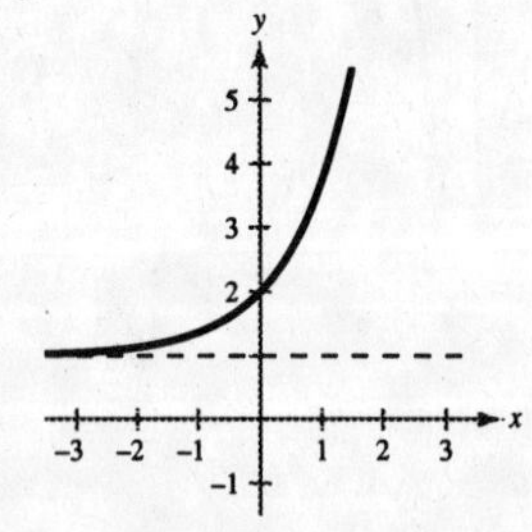

5. $A = P\left(1 + \frac{r}{n}\right)^{nt}$

(a) $A = 5000\left(1 + \frac{0.09}{12}\right)^{12(3)} \approx \$6543.23$

(b) $A = 5000\left(1 + \frac{0.09}{4}\right)^{4(3)} \approx \$6530.25$

(c) $A = 5000e^{(0.09)(3)} \approx \$6549.82$

6. $7^{-2} = \frac{1}{49}$

$\log_7 \frac{1}{49} = -2$

7. $x = 4 = \log_2 \frac{1}{64}$

$2^{x-4} = \frac{1}{64}$

$2^{x-4} = 2^{-6}$

$x - 4 = -6$

$x = -2$

8. $\log \sqrt[4]{\frac{8}{25}} = \frac{1}{4} \log_b \frac{8}{25}$

$= \frac{1}{4}[\log_b 8 - \log_b 25]$

$= \frac{1}{4}[\log_b 2^3 - \log_b 5^2]$

$= \frac{1}{4}[3 \log_b 2 - 2 \log_b 5]$

$= \frac{1}{4}[3(0.3562) - 2(0.8271)]$

$= -0.1464$

9. $5 \ln x - \frac{1}{2} \ln y + 6 \ln z = \ln x^5 - \ln \sqrt{y} + \ln z^6 = \ln\left(\frac{x^5 z^6}{\sqrt{y}}\right)$

10. $\log_9 28 = \frac{\log 28}{\log 9} \approx 1.5166$

11. $\log N = 0.6646$

$N = 10^{0.6646} \approx 4.62$

12.

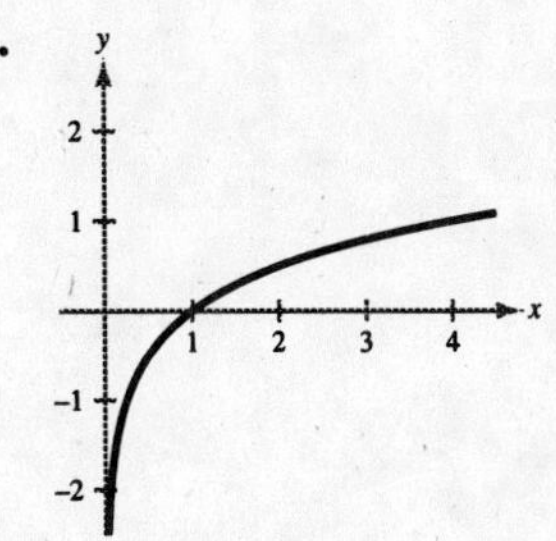

13. Domain:

$x^2 - 9 > 0$

$(x + 3)(x - 3) > 0$

$x < -3 \text{ or } x > 3$

14.

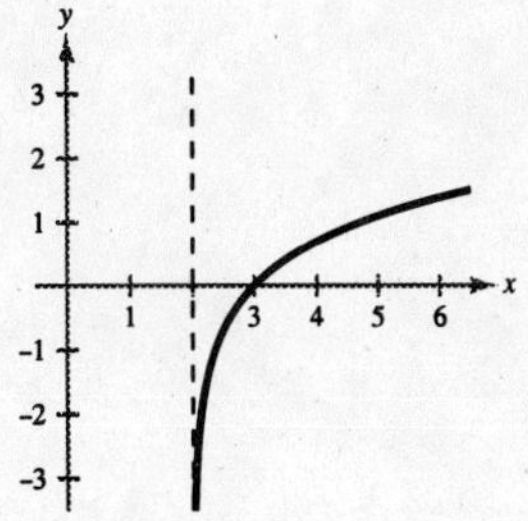

15. $\frac{\ln x}{\ln y} \neq \ln(x - y)$ since $\frac{\ln x}{\ln y} = \log_y x$.

16. $5^3 = 41$

$$x = \log_5 41 = \frac{\ln 41}{\ln 5} \approx 2.3074$$

17. $x - x^2 = \log_5 \frac{1}{25}$

$$5^{x-x^2} = \frac{1}{25}$$
$$5^{x-x^2} = 5^{-2}$$
$$x - x^2 = -2$$
$$0 = x^2 - x - 2$$
$$0 = (x + 1)(x - 2)$$
$$x = -1 \text{ or } x = 2$$

18. $\log_2 x + \log_2(x - 3) = 2$

$$\log_2[x(x - 3)] = 2$$
$$x(x - 3) = 2^2$$
$$x^2 - 3x = 4$$
$$x^2 - 3x - 4 = 0$$
$$(x + 1)(x - 4) = 0$$
$$x = 4$$
$$x = -1$$

No solution (extraneous solution)

19.
$$\frac{e^x + e^{-x}}{3} = 4$$
$$e^x(e^x + e^{-x}) = 12e^x$$
$$e^{2x} + 1 = 12e^x$$
$$e^{2x} - 12e^x + 1 = 0$$
$$e^x = \frac{12 \pm \sqrt{144 - 4}}{2}$$

$e^x = 11.9161$ or $e^x = 0.0839$

$x = \ln 11.9161$ $x = \ln 0.0839$

$x \approx 2.4779$ $x \approx -2.4779$

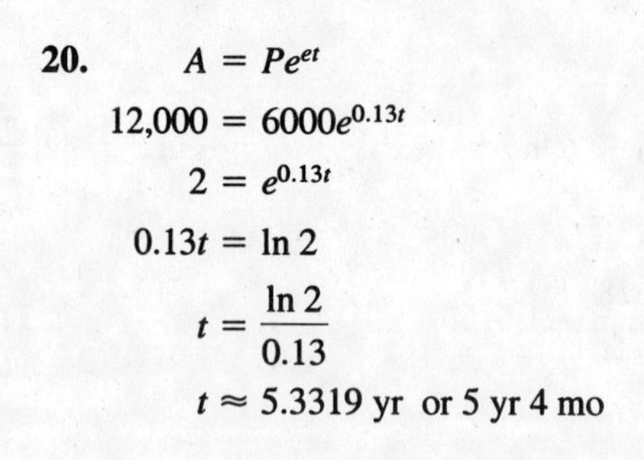

20.
$$A = Pe^{et}$$
$$12{,}000 = 6000e^{0.13t}$$
$$2 = e^{0.13t}$$
$$0.13t = \ln 2$$
$$t = \frac{\ln 2}{0.13}$$
$$t \approx 5.3319 \text{ yr or } 5 \text{ yr } 4 \text{ mo}$$

21. There are 2 points of intersection:

(0.0169, −2.983),

(1.731, 1.647)

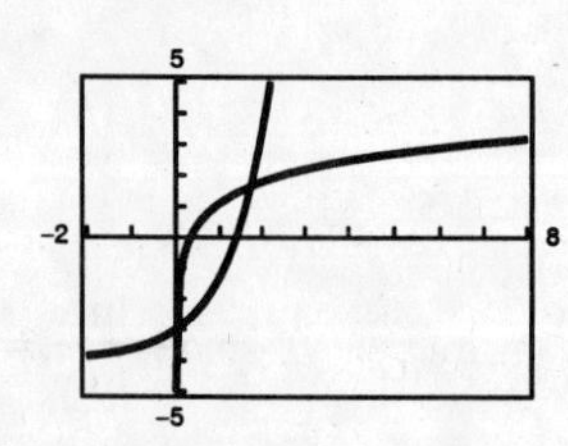

22. $y = 1.0597x^{1.9792}$

Chapter 4 Practice Test Solutions

1. $350° = 350\left(\frac{\pi}{180}\right) = \frac{35\pi}{18}$

2. $\frac{5\pi}{9} = \frac{5\pi}{9} \cdot \frac{180}{\pi} = 100°$

3. $135° \ 14' \ 12'' = \left(135 + \frac{14}{60} + \frac{12}{3600}\right)^{\circ}$

$\approx 135.2367°$

4. $-22.569° = -(22° + 0.569(60)')$

$= -22° \ 34.14'$

$= -(22° \ 34' + 0.14(60)'')$

$\approx -22° \ 34' \ 8''$

5. $\cos\theta = \frac{2}{3}$

$x = 2, \ r = 3, \ y = \pm\sqrt{9-4} = \pm\sqrt{5}$

$\tan\theta = \frac{y}{x} = \pm\frac{\sqrt{5}}{2}$

6. $\sin\theta = 0.9063$

$\theta = \arcsin(0.9063)$

$\theta = 65°$ or $\frac{13\pi}{36}$

7. $\tan 20° = \frac{35}{x}$

$x = \frac{35}{\tan 20°} \approx 96.1617$

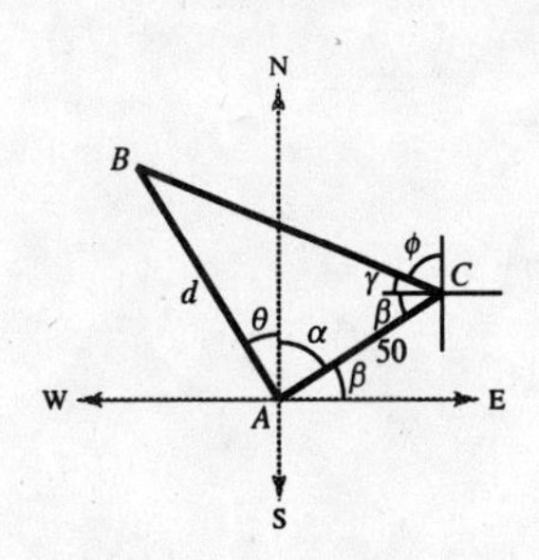

8. $\theta = \frac{6\pi}{5}$, θ is in Quadrant III.

Reference angle: $\frac{6\pi}{5} - \pi = \frac{\pi}{5}$ or $36°$

9. $\csc 3.92 = \frac{1}{\sin 3.92} \approx -1.4242$

10. $\tan\theta = 6 = \frac{6}{1}$, θ lies in Quadrant III.

$y = -6, \ x = -1, \ r = \sqrt{36+1} = \sqrt{37}$,

so $\sec\theta = \frac{\sqrt{37}}{-1} \approx -6.0828$.

11. Period: 4π

Amplitude: 3

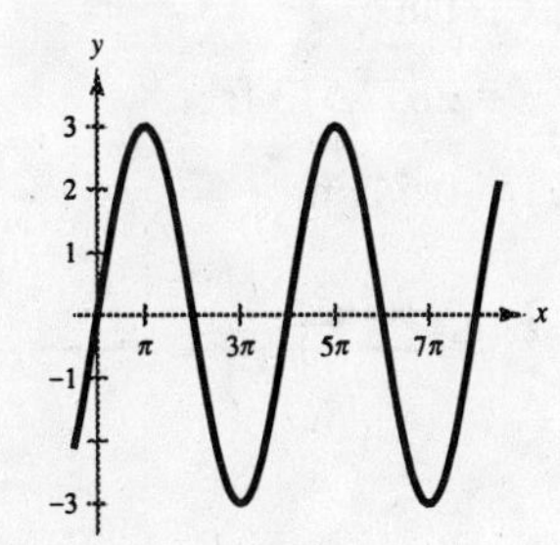

12. Period: 2π

Amplitude: 2

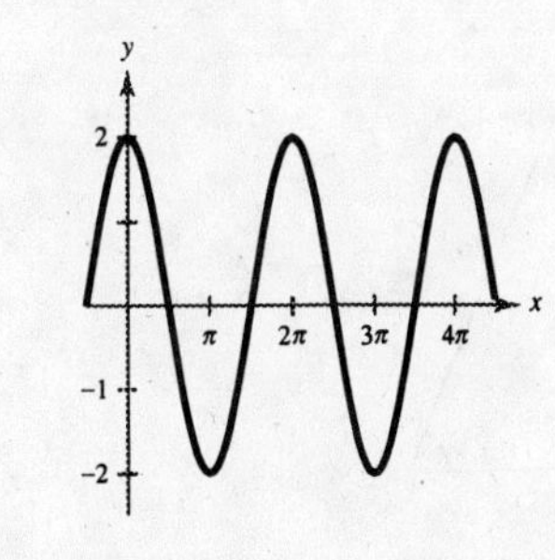

13. Period: $\frac{\pi}{2}$

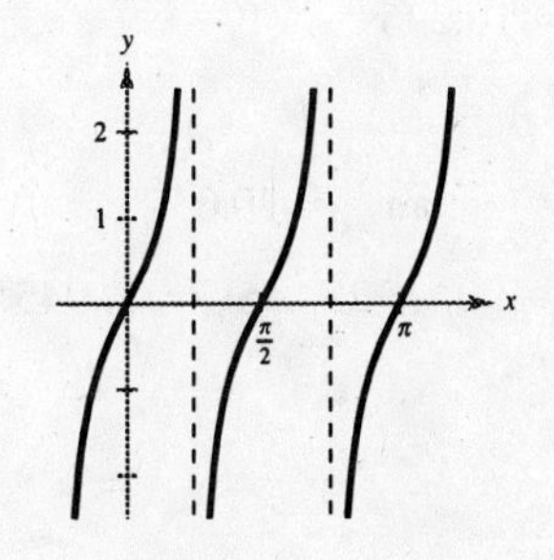

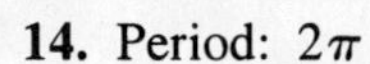

14. Period: 2π

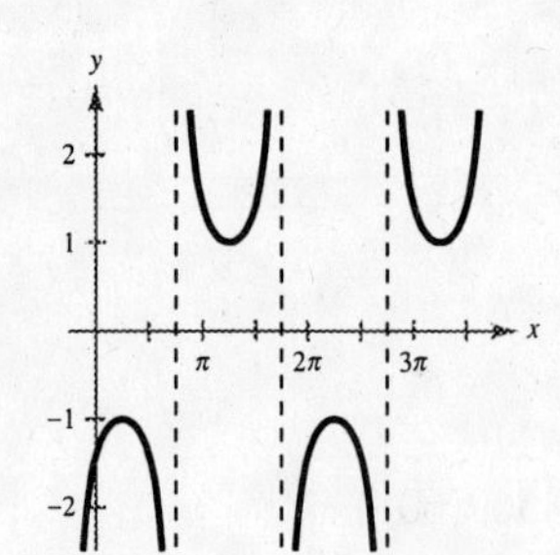

15.

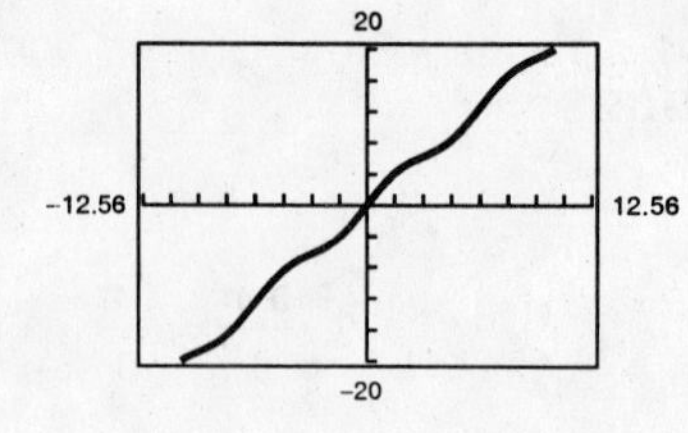

16.

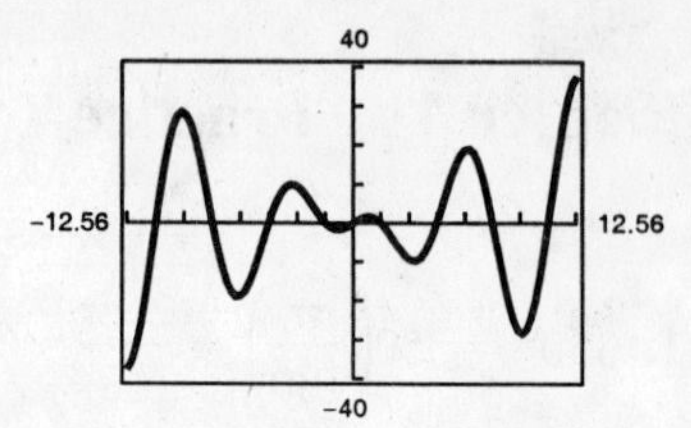

17. $\theta = \arcsin 1$

$\sin\theta = 1$

$\theta = \dfrac{\pi}{2}$

18. $\theta = \arctan(-3)$

$\tan\theta = -3$

$\theta \approx -1.249$ or $-71.565°$

19. $\sin\left(\arccos \dfrac{4}{\sqrt{35}}\right)$

$\sin\theta = \dfrac{\sqrt{19}}{\sqrt{35}} \approx 0.7368$

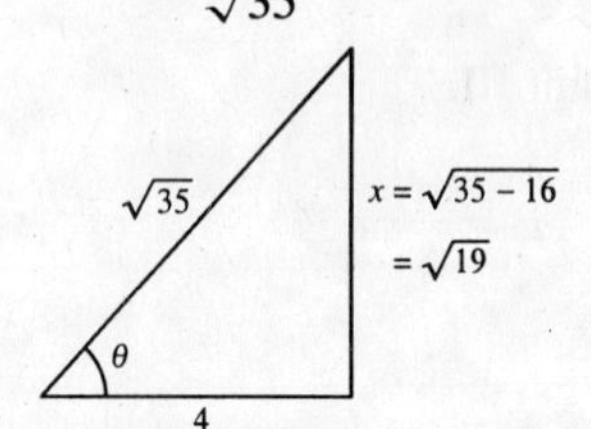

20. $\cos\left(\arcsin \dfrac{x}{4}\right)$

$\cos\theta = \dfrac{\sqrt{16 - x^2}}{4}$

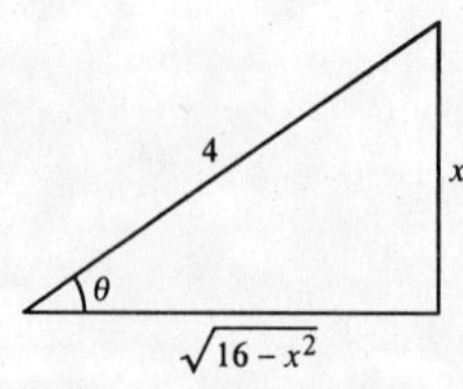

21. Given $A = 40°,\ c = 12$

$B = 90° - 40° = 50°$

$\sin 40° = \dfrac{a}{12}$

$a = 12 \sin 40° \approx 7.713$

$\cos 40° = \dfrac{b}{12}$

$b = 12 \cos 40° \approx 9.193$

22. Given $B = 6.84°,\ a = 21.3$

$A = 90° - 6.84° = 83.16°$

$\sin 83.16° = \dfrac{21.3}{c}$

$c = \dfrac{21.3}{\sin 83.16°} \approx 21.453$

$\tan 83.16° = \dfrac{21.3}{b}$

$b = \dfrac{21.3}{\tan 83.16°} \approx 2.555$

23. Given $a = 5,\ b = 9$

$c = \sqrt{25 + 81} = \sqrt{106}$

≈ 10.296

$\tan A = \frac{5}{9}$

$A = \arctan \frac{5}{9} \approx 29.055°$

$B = 90° - 29.055° = 60.945°$

24. $\sin 67° = \dfrac{x}{20}$

$x = 20 \sin 67° \approx 18.41$ feet

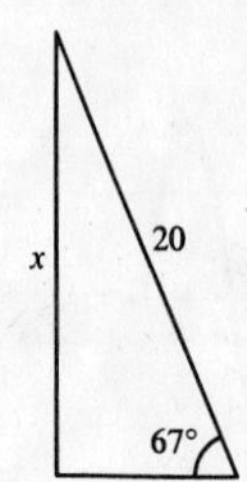

25. $\tan 5° = \dfrac{250}{x}$

$x = \dfrac{250}{\tan 5°}$

≈ 2857.513 feet

≈ 0.541 mi

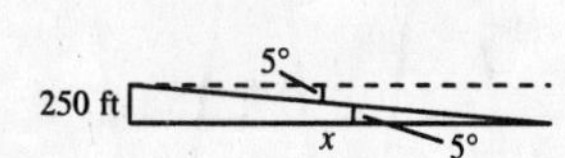

Chapter 5 Practice Test Solutions

1. $\tan x = \frac{4}{11}$, $\sec x < 0 \implies x$ is in Quadrant III.

$y = -4,\ x = -11,\ r = \sqrt{16 + 121} = \sqrt{137}$

$\sin x = -\frac{4}{\sqrt{137}} = -\frac{4\sqrt{137}}{137}$ $\qquad \csc x = -\frac{\sqrt{137}}{4}$

$\cos x = -\frac{11}{\sqrt{137}} = -\frac{11\sqrt{137}}{137}$ $\qquad \sec x = -\frac{\sqrt{137}}{11}$

$\tan x = \frac{4}{11}$ $\qquad \cot x = \frac{11}{4}$

2.
$$\frac{\sec^2 x + \csc^2 x}{\csc^2 x(1 + \tan^2 x)} = \frac{\sec^2 x + \csc^2 x}{\csc^2 x + (\csc^2 x)\tan^2 x} = \frac{\sec^2 x + \csc^2 x}{\csc^2 x + \frac{1}{\sin^2 x} \cdot \frac{\sin^2 x}{\cos^2 x}}$$
$$= \frac{\sec^2 x + \csc^2 x}{\csc^2 x + \frac{1}{\cos^2 x}} = \frac{\sec^2 x + \csc^2 x}{\csc^2 x + \sec^2 x} = 1$$

3. $\ln|\tan \theta| - \ln|\cot \theta| = \ln\frac{|\tan \theta|}{|\cot \theta|} = \ln\left|\frac{\sin \theta/\cos \theta}{\cos \theta/\sin \theta}\right| = \ln\left|\frac{\sin^2 \theta}{\cos^2 \theta}\right| = \ln|\tan^2 \theta| = 2\ln|\tan \theta|$

4. $\cos\left(\frac{\pi}{2} - x\right) = \frac{1}{\csc x}$ is true since $\cos\left(\frac{\pi}{2} - x\right) = \sin x = \frac{1}{\csc x}$.

5. $\sin^4 x + (\sin^2 x)\cos^2 x = \sin^2 x(\sin^2 x + \cos^2 x) = \sin^2 x(1) = \sin^2 x$

6. $(\csc x + 1)(\csc x - 1) = \csc^2 x - 1 = \cot^2 x$

7. $\frac{\cos^2 x}{1 - \sin x} \cdot \frac{1 + \sin x}{1 + \sin x} = \frac{\cos^2 x(1 + \sin x)}{1 - \sin^2 x} = \frac{\cos^2 x(1 + \sin x)}{\cos^2 x} = 1 + \sin x$

8.
$$\frac{1 + \cos \theta}{\sin \theta} + \frac{\sin \theta}{1 + \cos \theta} = \frac{(1 + \cos \theta)^2 + \sin^2 \theta}{\sin \theta(1 + \cos \theta)}$$
$$= \frac{1 + 2\cos \theta + \cos^2 \theta + \sin^2 \theta}{\sin \theta(1 + \cos \theta)} = \frac{2 + 2\cos \theta}{\sin \theta(1 + \cos \theta)} = \frac{2}{\sin \theta} = 2\csc \theta$$

9. $\tan^4 x + 2\tan^2 x + 1 = (\tan^2 x + 1)^2 = (\sec^2 x)^2 = \sec^2 x) = \sec^{4\,x}$

10. (a) $\sin 105° = \sin(60° + 45°) = \sin 60° \cos 45° + \cos 60° \sin 45°$
$$= \frac{\sqrt{3}}{2} \cdot \frac{\sqrt{2}}{2} + \frac{1}{2} \cdot \frac{\sqrt{2}}{2} = \frac{\sqrt{2}}{4}(\sqrt{3} + 1)$$

(b) $\tan 15° = \tan(60° - 45°) = \frac{\tan 60° - \tan 45°}{1 + \tan 60° \tan 45°}$
$$= \frac{\sqrt{3} - 1}{1 + \sqrt{3}} \cdot \frac{1 - \sqrt{3}}{1 - \sqrt{3}} = \frac{2\sqrt{3} - 1 - 3}{1 - 3} = \frac{2\sqrt{3} - 4}{-2} = 2 - \sqrt{3}$$

11. $(\sin 42°)\cos 38° - (\cos 42°)\sin 38° = \sin(42° - 38°) = \sin 4°$

12. $\tan\left(\theta + \frac{\pi}{4}\right) = \frac{\tan\theta + \tan(\pi(4)}{1 - (\tan\theta)\tan(\pi(4)} = \frac{\tan\theta + 1}{1 - \tan\theta(1)} = \frac{1 + \tan\theta}{1 - \tan\theta}$

13. $\sin(\arcsin x - \arccos x) = \sin(\arcsin x)\cos(\arccos x) - \cos(\arcsin x)\sin(\arccos x)$

$$= (x)(x) - \left(\sqrt{1 - x^2}\right)\left(\sqrt{1 - x^2}\right) = x^2 - (1 - x^2) = 2x^2 - 1$$

14. (a) $\cos(120°) = \cos)2(60°)] = 2\cos^2 60° - 1 = 2\left(\frac{1}{2}\right)^2 - 1 = -\frac{1}{2}$

(b) $\tan(300°) = \tan)2(150°)] = \frac{2\tan 150°}{1 - \tan^2 150°} = \frac{-2\sqrt{3}(3}{1 - (1(3)} = -\sqrt{3}$

15. (a) $\sin 22.5° = \sin\frac{45°}{2} = \sqrt{\frac{1 - \cos 45°}{2}} = \sqrt{\frac{1 - \sqrt{2}(2}{2}} = \frac{\sqrt{2 - \sqrt{2}}}{2}$

(b) $\tan\frac{\pi}{12} = \tan\frac{\pi(6}{2} = \frac{\sin(\pi(6)}{1 + \cos(\pi(6)} = \frac{1(2}{1 + \sqrt{3}(2} = \frac{1}{2 + \sqrt{3}} = 2 - \sqrt{3}$

16. $\sin\theta = \frac{4}{5}$, θ lies in Quadrant II $\Rightarrow$ $\cos\theta = -\frac{3}{5}$.

$$\cos\frac{\theta}{2} = \sqrt{\frac{1 + \cos\theta}{2}} = \sqrt{\frac{1 - 3(5}{2}} = \sqrt{\frac{2}{10}} = \frac{1}{\sqrt{5}} = \frac{\sqrt{5}}{5}$$

17. $(\sin^2 x)\cos^2 x = \frac{1 - \cos^2 x}{2} \cdot \frac{1 + \cos^2 x}{2} = \frac{1}{4})1 - \cos^2 2x] = \frac{1}{4}\left[1 - \frac{1 + \cos 4x}{2}\right]$

$$= \frac{1}{8})2 - (1 + \cos 4x)] = \frac{1}{8})1 - \cos 4x]$$

18. $6(\sin 5\theta)\cos 2\theta = 6\{\frac{1}{2})\sin(5\theta + 2\theta) + \sin(5\theta - 2\theta)]\} = 3)\sin 7\theta + \sin 3\theta]$

19. $\sin(x + \pi) + \sin(x - \pi) = 2\left(\sin\frac{)(x + \pi) + (x - \pi)]}{2}\right)\cos\frac{)(x + \pi) - (x - \pi)]}{2}$

20. $\frac{\sin 9x + \sin 5x}{\cos 9x - \cos 5x} = \frac{2\sin 7x\cos 2x}{-2\sin 7x\sin 2x} = -\frac{\cos 2x}{\sin 2x} = -\cot 2x$

21. $\frac{1}{2})\sin(u + v) - \sin(u - v)] = \frac{1}{2}\{(\sin u)\cos v + (\cos u)\sin v -)(\sin u)\cos v - (\cos u)\sin v]\}$

$$= \frac{1}{2})2(\cos u)\sin v] = (\cos u)\sin v$$

22. $4\sin^2 x = 1$

$\sin^2 x = \frac{1}{4}$

$\sin x = \pm\frac{1}{2}$

$\sin x = \frac{1}{2}$ or $\sin x = -\frac{1}{2}$

$x = \frac{\pi}{6}$ or $\frac{5\pi}{6}$ $\quad$ $x = \frac{7\pi}{6}$ or $\frac{11\pi}{6}$

23. $\tan^2\theta + \left(\sqrt{3} - 1\right)\tan\theta - \sqrt{3} = 0$

$(\tan\theta - 1)\left(\tan\theta + \sqrt{3}\right) = 0$

$\tan\theta = 1$ or $\tan\theta = -\sqrt{3}$

$\theta = \frac{\pi}{4}$ or $\frac{5\pi}{4}$ $\quad$ $\theta = \frac{2\pi}{3}$ or $\frac{5\pi}{3}$

24.

$$\sin 2x = \cos x$$
$$2(\sin x)\cos x - \cos x = 0$$
$$\cos x(2\sin x - 1) = 0$$

$\cos x = 0$ or $\sin x = \frac{1}{2}$

$x = \frac{\pi}{2}$ or $\frac{3\pi}{2}$ $\qquad$ $x = \frac{\pi}{6}$ or $\frac{5\pi}{6}$

25. $\tan^2 x - 6\tan x + 4 = 0$

$$\tan x = \frac{-(-6) \pm \sqrt{(-6)^2 - 4(1)(4)}}{2(1)}$$
$$\tan x = \frac{6 \pm \sqrt{20}}{2} = 3 \pm \sqrt{5}$$

$\tan x = 3 + \sqrt{5}$ $\qquad$ or $\quad$ $\tan x = 3 - \sqrt{5}$

$x \approx 1.3821$ or 4.5237 $\qquad$ $x = 0.6524$ or 3.7940

Chapter 6 Practice Test Solutions

1. $C = 180° - (40° + 12°) = 128°$

$a = \sin 40°\left(\frac{100}{\sin 12°}\right) \approx 309.164$

$c = \sin 128°\left(\frac{100}{\sin 12°}\right) \approx 379.012$

2. $\sin A = 5\left(\frac{\sin 150°}{20}\right) = 0.125$

$A \approx 7.181°$

$B \approx 180° - (150° + 7.181°) = 22.819°$

$b = \sin 22.819°\left(\frac{20}{\sin 150°}\right) \approx 15.513$

3. $\text{Area} = \frac{1}{2}ab \sin C$

$= \frac{1}{2}(3)(5) \sin 130°$

≈ 5.745 square units

4. $h = b \sin A$

$= 35 \sin 22.5°$

≈ 13.394

$a = 10$

Since $a < h$ and A is acute, the triangle has no solution.

5. $\cos A = \frac{(53)^2 + (38)^2 - (49)^2}{2(53)(38)} \approx 0.4598$

$A \approx 62.627°$

$\cos B = \frac{(49)^2 + (38)^2 - (53)^2}{2(49)(38)} \approx 0.2782$

$B \approx 73.847°$

$C \approx 180° - (42.627° + 73.847°)$

$= 43.526°$

6. $c^2 = (100)^2 + (300)^2 - 2(100)(300) \cos 29°$

$\approx 47{,}522.8176$

$c \approx 218$

$\cos A = \frac{(300)^2 + (218)^2 - (100)^2}{2(300)(218)} \approx 0.97495$

$A \approx 12.85°$

$B \approx 180° - (12.85° + 29°) = 138.15°$

7. $s = \frac{a + b + c}{2} = \frac{4.1 + 6.8 + 5.5}{2} = 8.2$

$\text{Area} = \sqrt{s(s - a)(s - b)(s - c)}$

$= \sqrt{8.2(8.2 - 4.1)(8.2 - 6.8)(8.2 - 5.5)}$

$= 11.273$ square units

8. $x^2 = (40)^2 + (70)^2 - 2(40)(70) \cos 168°$

$\approx 11{,}977.6266$

$x \approx 190.442$ miles

9. $\mathbf{w} = 4(3\mathbf{i} + \mathbf{j}) - 7(-\mathbf{i} + 2\mathbf{j})$

$= 19\mathbf{i} - 10\mathbf{j}$

10. $\frac{\mathbf{v}}{\|\mathbf{v}\|} = \frac{5\mathbf{i} + 3\mathbf{j}}{\sqrt{25 + 9}} = \frac{5}{\sqrt{34}}\mathbf{i} - \frac{3}{\sqrt{34}}\mathbf{j}$

$= \frac{5\sqrt{34}}{34}\mathbf{i} - \frac{3\sqrt{34}}{34}\mathbf{j}$

11. $\mathbf{u} = 6\mathbf{i} + 5\mathbf{j} \qquad \mathbf{v} = 2\mathbf{i} - 3\mathbf{j}$

$$\mathbf{u} \cdot \mathbf{v} = 6(2) + 5(-3) = -3$$

$$\|\mathbf{u}\| = \sqrt{61} \qquad \|\mathbf{v}\| = \sqrt{13}$$

$$\cos\theta = \frac{-3}{\sqrt{61}\sqrt{13}}$$

$$\theta \approx 96.116°$$

12. $4(\mathbf{i}\cos 30° + \mathbf{j}\sin 30°) = 4\left(\frac{\sqrt{3}}{2}\mathbf{i} + \frac{1}{2}\mathbf{j}\right)$

$$= \langle 4\sqrt{3}, 2\rangle$$

y

v

30°

x

13. $\text{proj}_{\mathbf{v}}\mathbf{u} = \left(\frac{\mathbf{u}\cdot\mathbf{v}}{\|\mathbf{v}\|^2}\right)\mathbf{v} = \frac{-10}{20}\langle -2, 4\rangle = \langle 1, -2\rangle$

14. $r = \sqrt{25 + 25} = \sqrt{50} = 5\sqrt{2}$

$$\tan\theta = \frac{-5}{5} = -1$$

Since z is in Quadrant IV,

$$\theta = 315°$$

$$z = 5\sqrt{2}(\cos 315° + i\sin 315°).$$

15. $\cos 225° = -\frac{\sqrt{2}}{2} \quad \sin 225° = -\frac{\sqrt{2}}{2}$

$$z = 6\left(-\frac{\sqrt{2}}{2} - i\frac{\sqrt{2}}{2}\right)$$

$$= -3\sqrt{2} - 3\sqrt{2}i$$

16. $[7(\cos 23° + i\sin 23°)][4(\cos 7° + i\sin 7°)] = 7(4)[\cos(23° + 7°) + i\sin(23° + 7°)]$

$$= 28(\cos 30° + i\sin 30°)$$

17. $\dfrac{9\left(\cos\frac{5\pi}{4} + i\sin\frac{5\pi}{4}\right)}{3(\cos\pi + i\sin\pi)} = \frac{9}{3}\left[\cos\left(\frac{5\pi}{4} - \pi\right) + i\sin\left(\frac{5\pi}{4} - \pi\right)\right] = 3\left(\cos\frac{\pi}{4} + i\sin\frac{\pi}{4}\right)$

18. $(2 + 2i)^8 = [2\sqrt{2}(\cos 45° + i\sin 45°)]^8 = \left(2\sqrt{2}\right)^8[\cos(8)(45°) + i\sin(8)(45°)]$

$$= 4096[\cos 360° + i\sin 360°] = 4096$$

19. $z = 8\left(\cos\frac{\pi}{3} + i\sin\frac{\pi}{3}\right),\ n = 3$

The cube roots of z are:

For $k = 0$, $\sqrt[3]{8}\left[\cos\frac{\pi/3}{3} + i\sin\frac{\pi/3}{3}\right] = 2\left(\cos\frac{\pi}{9} + i\sin\frac{\pi}{9}\right).$

For $k = 1$, $\sqrt[3]{8}\left[\cos\frac{\pi/3 + 2\pi}{3} + i\sin\frac{\pi/3 + 2\pi}{3}\right] = 2\left(\cos\frac{7\pi}{9} + i\sin\frac{7\pi}{9}\right).$

For $k = 2$, $\sqrt[3]{8}\left[\cos\frac{\pi/3 + 4\pi}{3} + i\sin\frac{\pi/3 + 4\pi}{3}\right] = 2\left(\cos\frac{13\pi}{9} + i\sin\frac{13\pi}{9}\right).$

20. $x^4 = -i = 1\left(\cos\frac{3\pi}{2} + i\sin\frac{3\pi}{2}\right)$

For $k = 0$, $\cos\frac{3\pi/2}{4} + i\sin\frac{3\pi/2}{4} = \cos\frac{3\pi}{8} + i\sin\frac{3\pi}{8}$.

For $k = 1$, $\cos\frac{3\pi/2 + 2\pi}{4} + i\sin\frac{3\pi/2 + 2\pi}{4} = \cos\frac{7\pi}{8} + i\sin\frac{7\pi}{8}$.

For $k = 2$, $\cos\frac{3\pi/2 + 4\pi}{4} + i\sin\frac{3\pi/2 + 4\pi}{4} = \cos\frac{11\pi}{8} + i\sin\frac{11\pi}{8}$.

For $k = 3$, $\cos\frac{3\pi/2 + 6\pi}{4} + i\sin\frac{3\pi/2 + 6\pi}{4} = \cos\frac{15\pi}{8} + i\sin\frac{15\pi}{8}$.

Chapter 7 Practice Test Solutions

1.
$$\begin{aligned} x + y &= 1 \\ 3x - y &= 15 \implies y = 3x - 15 \\ x + (3x - 15) &= 1 \\ 4x &= 16 \\ x &= 4 \\ y &= -3 \end{aligned}$$

2.
$$\begin{aligned} x - 3y &= -3 \implies x = 3y - 3 \\ x^2 + 5y &= 5 \\ (3y - 3)^2 + 6y &= 5 \\ 9y^2 - 18y + 9 + 6y &= 5 \\ 9y^2 - 12y + 4 &= 0 \\ (3y - 2)^2 &= 0 \\ y &= \tfrac{2}{3} \\ x &= -1 \end{aligned}$$

3.
$$\begin{aligned} x + y + z &= 6 \implies z = 6 - x - y \\ 2x - y + 3z &= 0 \qquad 2x - y + 3(6 - x - y) = 0 \implies -x - 4y = -18 \\ 5x + 2y - z &= -3 \qquad 5x + 2y - (6 - x - y) = -3 \implies 6x + 3y = 3 \\ x &= 18 - 4y \\ 6(18 - 4y) + 3y &= 3 \\ -21y &= -105 \\ y &= 5 \\ x &= 18 - 4y = -2 \\ z &= 6 - x - y = 3 \end{aligned}$$

4.
$$\begin{aligned} x + y &= 110 \implies y = 110 - x \\ xy &= 2800 \\ x(110 - x) &= 2800 \\ 0 &= x^2 - 110x + 2800 \\ 0 &= (x - 40)(x - 70) \\ x &= 40 \text{ or } x = 70 \\ y &= 70 \qquad y = 40 \end{aligned}$$

5.
$$\begin{aligned} 2x + 2y &= 170 \implies y = \frac{170 - 2x}{2} = 85 - x \\ xy &= 2800 \\ x(85 - x) &= 2800 \\ 0 &= x^2 - 85x + 2800 \\ 0 &= (x - 25)(x - 60) \\ x &= 25 \text{ or } x = 60 \\ y &= 60 \qquad y = 25 \end{aligned}$$

Dimensions: $60' \times 25'$

6.
$$\begin{aligned} 2x + 15y = 4 &\implies 2x + 15y = 4 \\ x - 3y = 23 &\implies \underline{5x - 15y = 115} \\ &\qquad 7x \qquad\;\; = 119 \\ &\qquad x = 17 \\ &\qquad y = \frac{x - 23}{3} = -2 \end{aligned}$$

7.
$$\begin{aligned} x + y = 2 &\implies 19x + 19y = 38 \\ 38x - 19y = 7 &\implies \underline{38x - 19y = 7} \\ &\qquad 57x \qquad = 45 \\ &\qquad x = \frac{45}{57} = \frac{15}{19} \\ &\qquad y = 2 - x = \frac{38}{19} - \frac{15}{19} = \frac{23}{19} \end{aligned}$$

8. $y_1 = 2(0.112 - 0.4x)$

$$y_2 = \frac{(0.13 + 0.3x)}{0.7}$$

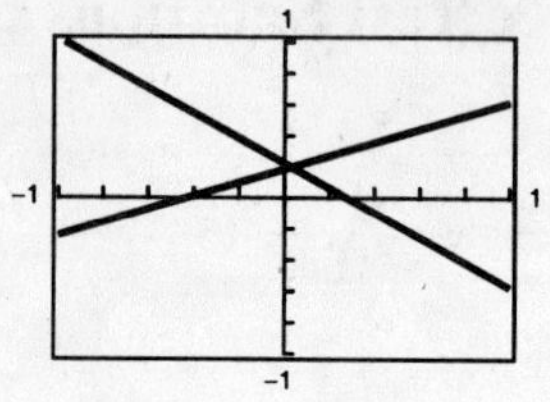

$$\begin{aligned} 0.4x + 0.5y &= 0.112 \Rightarrow & 0.28x + 0.35y &= 0.0784 \\ 0.3x - 0.7y &= -0.131 \Rightarrow & 0.15x - 0.35y &= -0.0655 \\ & & 0.43x &= 0.0129 \end{aligned}$$

$$x = \frac{0.0129}{0.43} = 0.03$$

$$y = \frac{0.112 - 0.4x}{0.5} = 0.20$$

9. Let x = amount in 11% fund and y = amount in 13% fund.

$$\begin{aligned} x + y &= 17000 \Rightarrow y = 17000 - x \\ 0.11x + 0.13y &= 2080 \\ 0.11x + 0.13(17000 - x) &= 2080 \\ -0.02x &= -130 \\ x &= \$6500 \\ y &= \$10500 \end{aligned}$$

10. Using a graphing utility, you obtain $y = 0.7857x - 0.1429$. Analytically, $(4, 3), (1, 1), (-1, -2), (-2, -1)$.

$$n = 4, \sum_{i=1}^{4} x_i = 2, \sum_{i=1}^{4} y_i = 1, \sum_{i=1}^{4} x_i^2 = 22, \sum_{i=1}^{4} x_i y_i = 17$$

$$\begin{aligned} 4b + 2a &= 1 \Rightarrow & 4b + 2a &= 1 \\ 2b + 22a &= 17 \Rightarrow & -4b - 44a &= -34 \\ & & -42a &= -33 \end{aligned}$$

$$a = \tfrac{33}{42} = \tfrac{11}{14}$$

$$b = \tfrac{1}{4}\left(1 - 2\left(\tfrac{33}{42}\right)\right) = -\tfrac{1}{7}$$

$$y = ax + b = \tfrac{11}{14}x - \tfrac{1}{7}$$

11.

$$\begin{aligned} x + y \quad &= -2 \Rightarrow & -2x - 2y \quad &= 4 & -9y + 3z &= 45 \\ 2x - y + z &= 11 \Rightarrow & 2x - y + z &= 11 & 4y - 3z &= -20 \\ 4y - 3z &= -20 & -3y + z &= 15 & -5y \quad &= 25 \end{aligned}$$

$$y = -5$$
$$x = 3$$
$$z = 0$$

12.
$$\begin{aligned} 4x - y + 5z &= 4 \Rightarrow & 4x - y + 5z &= 4 \\ 2x + y - z &= 0 \Rightarrow & -4x - 2y + 2z &= 0 \\ 2x + 4y + 8z &= 0 & \hline -3y + 7z &= 4 \end{aligned}$$

$$\begin{aligned} 2x + 4y + 8z &= 0 \\ -2x - y + z &= 0 \\ \hline 3y + 9z &= 0 \\ -3y + 7z &= 4 \\ \hline 16z &= 4 \\ z &= \tfrac{1}{4} \\ y &= -\tfrac{3}{4} \\ x &= \tfrac{1}{2} \end{aligned}$$

13.
$$\begin{aligned} 3x + 2y - z &= 5 \Rightarrow & 6x + 4y - 2z &= 10 \\ 6x - y + 5z &= 2 \Rightarrow & -6x + y - 5z &= -2 \\ & & \hline 5y - 7z &= 8 \\ & & y &= \frac{8 + 7z}{5} \end{aligned}$$

$$\begin{aligned} 3x + 2y - z &= 5 \\ 12x - 2y + 10z &= 4 \\ \hline 15x + 9z &= 9 \\ x = \frac{9 - 9z}{15} &= \frac{3 - 3z}{5} \end{aligned}$$

Let $z = a$, then $x = \dfrac{3 - 3a}{5}$ and $y = \dfrac{8 + 7a}{5}$.

14. $y = ax^2 + bx + c$ passes through $(0, -1)$, $(1, 4)$, and $(2, 13)$.

At $(0, -1)$: $-1 = a(0)^2 + b(0) + c \Rightarrow c = -1$

At $(1, 4)$: $4 = a(1)^2 + b(1) - 1 \Rightarrow 5 = a + b \Rightarrow 5 = a + b$

At $(2, 13)$: $13 = a(2)^2 + b(2) - 1 \Rightarrow 14a + 2b \Rightarrow -7 = -2a - b$

$$\begin{aligned} -2 &= -a \\ a &= 2 \\ b &= 3 \end{aligned}$$

Thus, $y = 2x^2 + 3x - 1$.

15. $s = \frac{1}{2}at^2 + v_0t + s_0$ passes through (1, 12), (2, 5), and (3, 4).

At (1, 12): $12 = \frac{1}{2}a + v_0 + s_0 \Rightarrow 24 = a + 2v_0 + 2s_0$

At (2, 5): $5 = 2a + 2v_0 + s_0 \Rightarrow -5 = -2a - 2v_0 - s_0$

At (3, 4): $4 = \frac{9}{2}a + 3v_0 + s_0 \Rightarrow 19 = -a + s_0$

$$15 = 6a + 6v_0 + 3s_0$$
$$-8 = -9a - 6v_0 - 2s_0$$
$$7 = -3a + s_0$$
$$-19 = a - s_0$$
$$-12 = -2a$$
$$a = 6$$
$$s_0 = 25$$
$$v_0 = -16$$

Thus, $s = \frac{1}{2}(6)t^2 - 16t + 25 = 3t^2 - 16t + 25$.

16. $x^2 + y^2 \geq 9$

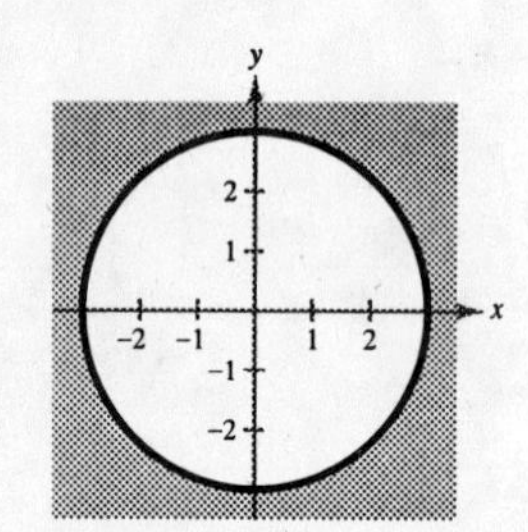

17. $x + y \leq 6$

$x \geq 2$

$y \geq 0$

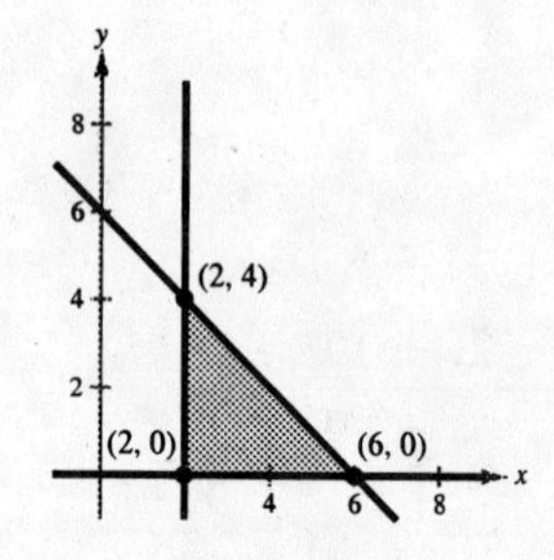

18. Line through (0, 0) and (0, 7): $x = 0$

Line through (0, 0) and (2, 3): $y = \frac{3}{2}x$ or $3x - 2y = 0$

Line through (0, 7) and (2, 3): $y = -2x + 7$ or $2x + y = 7$

Inequalities: $x \geq 0$

$3x - 2y \leq 0$

$2x + y \leq 7$

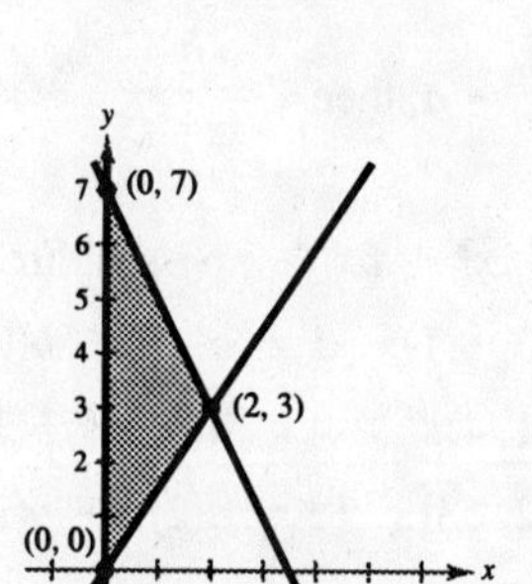

19. Vertices: (0, 0), (0, 7), (6, 0), (3, 5)

$z = 30x + 26y$

At (0, 0): $z = 0$

At (0, 7): $z = 182$

At (6, 0): $z = 180$

At (3, 5): $z = 220$

The maximum value is z is 220.

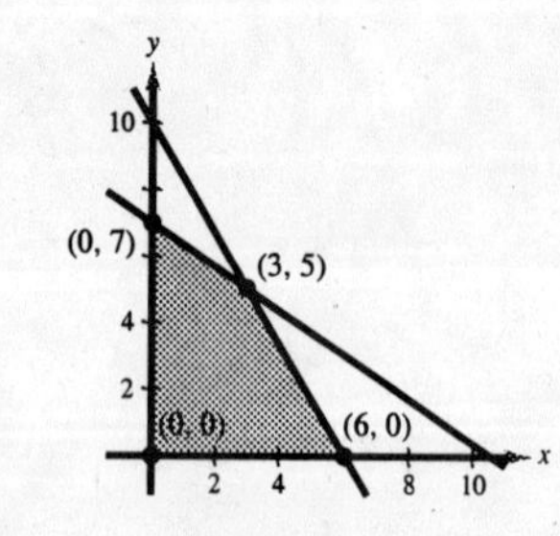

20. $x^2 + y^2 \le 4$

$(x - 2)^2 + y^2 \ge 4$

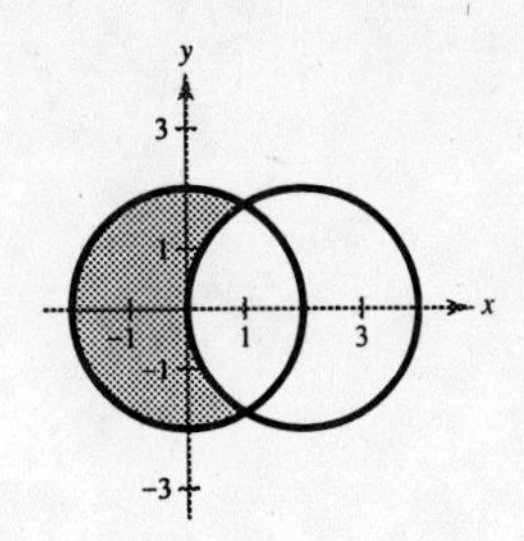

Chapter 8 Practice Test Solutions

1.

$$\begin{bmatrix} 1 & -2 & 4 \\ 3 & -5 & 9 \end{bmatrix}$$

$$-3R_1 + R_2 \rightarrow \begin{bmatrix} 1 & -2 & 4 \\ 0 & 1 & -3 \end{bmatrix}$$

$$2R_2 + R_1 \rightarrow \begin{bmatrix} 1 & 0 & -2 \\ 0 & 1 & -3 \end{bmatrix}$$

2. $3x + 5y = 3$

$2x - y = -11$

$$\begin{bmatrix} 3 & 5 & \vdots & 3 \\ 2 & -1 & \vdots & -11 \end{bmatrix}$$

$$-R_2 + R_1 \rightarrow \begin{bmatrix} 1 & 6 & \vdots & 14 \\ 2 & -1 & \vdots & -11 \end{bmatrix}$$

$$-2R_1 + R_2 \rightarrow \begin{bmatrix} 1 & 6 & \vdots & 14 \\ 0 & -13 & \vdots & -39 \end{bmatrix}$$

$$-\tfrac{1}{13}R_2 \rightarrow \begin{bmatrix} 1 & 6 & \vdots & 14 \\ 0 & 1 & \vdots & 3 \end{bmatrix}$$

$$-6R_2 + R_1 \rightarrow \begin{bmatrix} 1 & 0 & \vdots & -4 \\ 0 & 1 & \vdots & 3 \end{bmatrix}$$

Answer: $x = -4, y = 3$

3. $2x + 3y = -3$

$3x - 2y = 8$

$x + y = 1$

$$\begin{bmatrix} 2 & 3 & \vdots & -3 \\ 3 & 2 & \vdots & 8 \\ 1 & 1 & \vdots & 1 \end{bmatrix}$$

$$\begin{matrix} R_1 \rightarrow \\ \\ R_3 \rightarrow \end{matrix} \begin{bmatrix} 1 & 1 & \vdots & 1 \\ 3 & 2 & \vdots & 8 \\ 2 & 3 & \vdots & -3 \end{bmatrix}$$

$$\begin{matrix} \\ -3R_1 + R_2 \rightarrow \\ -2R_1 + R_3 \rightarrow \end{matrix} \begin{bmatrix} 1 & 1 & \vdots & 1 \\ 0 & -1 & \vdots & 5 \\ 0 & 1 & \vdots & -5 \end{bmatrix}$$

$$\begin{matrix} R_2 + R_1 \rightarrow \\ -R_2 \rightarrow \\ -R_2 + R_3 \rightarrow \end{matrix} \begin{bmatrix} 1 & 0 & \vdots & 6 \\ 0 & 1 & \vdots & -5 \\ 0 & 0 & \vdots & 0 \end{bmatrix}$$

Answer: $x = 6, y = -5$

4. $x + 3z = -5$

$2x + y = 0$

$3x + y - z = -3$

$$\begin{bmatrix} 1 & 0 & 3 & \vdots & -5 \\ 2 & 1 & 0 & \vdots & 0 \\ 3 & 1 & -1 & \vdots & 3 \end{bmatrix}$$

$$\begin{matrix} \\ -2R_1 + R_2 \rightarrow \\ -3R_1 + R_3 \rightarrow \end{matrix} \begin{bmatrix} 1 & 0 & 3 & \vdots & -5 \\ 0 & 1 & -6 & \vdots & 10 \\ 0 & 1 & -10 & \vdots & 18 \end{bmatrix}$$

$$\begin{matrix} \\ \\ -R_2 + R_3 \rightarrow \end{matrix} \begin{bmatrix} 1 & 0 & 3 & \vdots & -5 \\ 0 & 1 & -6 & \vdots & 10 \\ 0 & 0 & -4 & \vdots & 8 \end{bmatrix}$$

$$\begin{matrix} -3R_3 + R_1 \rightarrow \\ 6R_3 + R_2 \rightarrow \\ -\frac{1}{4}R_4 \rightarrow \end{matrix} \begin{bmatrix} 1 & 0 & 0 & \vdots & 1 \\ 0 & 1 & 0 & \vdots & -2 \\ 0 & 0 & 1 & \vdots & -2 \end{bmatrix}$$

Answer: $x = 1, y = -2, z = -2$

5. $\begin{bmatrix} 1 & 4 & 5 \\ 2 & 0 & -3 \end{bmatrix} \begin{bmatrix} 1 & 6 \\ 0 & -7 \\ -1 & 2 \end{bmatrix} = \begin{bmatrix} -4 & -12 \\ 5 & 6 \end{bmatrix}$

6. $3A - 5B = 3\begin{bmatrix} 9 & 1 \\ -4 & 8 \end{bmatrix} - 5\begin{bmatrix} 6 & -2 \\ 3 & 5 \end{bmatrix}$

$= \begin{bmatrix} 27 & 3 \\ -12 & 24 \end{bmatrix} - \begin{bmatrix} 30 & -10 \\ 15 & 25 \end{bmatrix}$

$= \begin{bmatrix} -3 & 13 \\ -27 & -1 \end{bmatrix}$

7. $f(A) = \begin{bmatrix} 3 & 0 \\ 7 & 1 \end{bmatrix}^2 - 7\begin{bmatrix} 3 & 0 \\ 7 & 1 \end{bmatrix} + 8\begin{bmatrix} 1 & 0 \\ 0 & 1 \end{bmatrix}$

$= \begin{bmatrix} 3 & 0 \\ 7 & 1 \end{bmatrix}\begin{bmatrix} 3 & 0 \\ 7 & 1 \end{bmatrix} - \begin{bmatrix} 21 & 0 \\ 49 & 7 \end{bmatrix} + \begin{bmatrix} 8 & 0 \\ 0 & 8 \end{bmatrix}$

$= \begin{bmatrix} 9 & 0 \\ 28 & 1 \end{bmatrix} - \begin{bmatrix} 21 & 0 \\ 49 & 7 \end{bmatrix} + \begin{bmatrix} 8 & 0 \\ 0 & 8 \end{bmatrix}$

$= \begin{bmatrix} -4 & 0 \\ -21 & 2 \end{bmatrix}$

8. False since

$(A + B)(A + 3B) = A(A + 3B) + B(A + 3B)$

$= A^2 + 3AB + BA + 3B^2.$

9.

$$\left[\begin{array}{cc:cc} 1 & 2 & 1 & 0 \\ 3 & 5 & 0 & 1 \end{array}\right]$$

$$\begin{array}{r} \\ -3R_1 + R_2 \rightarrow \end{array}\left[\begin{array}{cc:cc} 1 & 2 & 1 & 0 \\ 0 & -1 & -3 & 1 \end{array}\right]$$

$$\begin{array}{r} 2R_2 + R_1 \rightarrow \\ -R_2 \rightarrow \end{array}\left[\begin{array}{cc:cc} 1 & 0 & -5 & 2 \\ 0 & 1 & 3 & -1 \end{array}\right]$$

$$A^{-1} = \begin{bmatrix} -5 & 2 \\ 3 & -1 \end{bmatrix}$$

10.

$$\left[\begin{array}{ccc:ccc} 1 & 1 & 1 & 1 & 0 & 0 \\ 3 & 6 & 5 & 0 & 1 & 0 \\ 6 & 10 & 8 & 0 & 0 & 1 \end{array}\right]$$

$$\begin{array}{r} \\ -3R_1 + R_2 \rightarrow \\ -6R_1 + R_3 \rightarrow \end{array}\left[\begin{array}{ccc:ccc} 1 & 1 & 1 & 1 & 0 & 0 \\ 0 & 3 & 2 & -3 & 1 & 0 \\ 0 & 4 & 2 & -6 & 0 & 1 \end{array}\right]$$

$$\begin{array}{r} -R_2 + R_1 \rightarrow \\ \frac{1}{3}R_2 \rightarrow \\ -4R_2 + R_3 \rightarrow \end{array}\left[\begin{array}{ccc:ccc} 1 & 0 & \frac{1}{3} & 2 & -\frac{1}{3} & 0 \\ 0 & 1 & \frac{2}{3} & -1 & \frac{1}{3} & 0 \\ 0 & 0 & -\frac{2}{3} & -2 & -\frac{4}{3} & 1 \end{array}\right]$$

$$\begin{array}{r} \frac{1}{2}R_3 + R_1 \rightarrow \\ R_3 + R_2 \rightarrow \\ -\frac{3}{2}R_3 \rightarrow \end{array}\left[\begin{array}{ccc:ccc} 1 & 0 & 0 & 1 & -1 & \frac{1}{2} \\ 0 & 1 & 0 & -3 & -1 & 1 \\ 0 & 0 & 1 & 3 & 2 & -\frac{3}{2} \end{array}\right]$$

$$A^{-1} = \begin{bmatrix} 1 & -1 & \frac{1}{2} \\ -3 & -1 & 1 \\ 3 & 2 & -\frac{3}{2} \end{bmatrix}$$

11. (a) $x + 2y = 4$

$3x + 5y = 1$

$$\left[\begin{array}{cc:cc} 1 & 2 & 1 & 0 \\ 3 & 5 & 0 & 1 \end{array}\right]$$

$$\begin{array}{r} \\ -3R_1 + R_2 \rightarrow \end{array}\left[\begin{array}{cc:cc} 1 & 2 & 1 & 0 \\ 0 & -1 & -3 & 1 \end{array}\right]$$

$$\begin{array}{r} -2R_2 + R_1 \rightarrow \\ -R_2 \rightarrow \end{array}\left[\begin{array}{cc:cc} 1 & 0 & -5 & 2 \\ 0 & 1 & 3 & -1 \end{array}\right]$$

$$X = A^{-1}B = \begin{bmatrix} -5 & 2 \\ 3 & -1 \end{bmatrix}\begin{bmatrix} 4 \\ 1 \end{bmatrix} = \begin{bmatrix} -18 \\ 11 \end{bmatrix}$$

$x = -18, y = 11$

(b) $x + 2y = 3$

$3x + 5y = -2$

$$X = A^{-1}B - \begin{bmatrix} -5 & 2 \\ 3 & -1 \end{bmatrix}\begin{bmatrix} 3 \\ -2 \end{bmatrix} = \begin{bmatrix} -19 \\ 11 \end{bmatrix}$$

$x = -19, y = 11$

12. $\begin{vmatrix} 6 & -1 \\ 3 & 4 \end{vmatrix} = 24 - (-3) = 27$

13. $\begin{vmatrix} 1 & 3 & -1 \\ 5 & 9 & 0 \\ 6 & 2 & -5 \end{vmatrix} = 1(-45) + 3(25) + (-1)(-44) = 74$

14. $\begin{vmatrix} 1 & 4 & 2 & 3 \\ 0 & 1 & -2 & 0 \\ 3 & 5 & -2 & 1 \\ 2 & 0 & 6 & 1 \end{vmatrix} = -7$

15. $\begin{matrix} 6 & 4 & 3 & 0 & 6 \\ 0 & 5 & 1 & 4 & 8 \\ 0 & 0 & 2 & 7 & 3 \\ 0 & 0 & 0 & 9 & 2 \\ 0 & 0 & 0 & 0 & 1 \end{matrix} = 6(5)(2)(9)(1) = 540$

16. Area $= \dfrac{1}{2}\begin{vmatrix} 0 & 7 & 1 \\ 5 & 0 & 1 \\ 3 & 9 & 1 \end{vmatrix} = \dfrac{1}{2}(31)$

$= 15.5$ square units

17. $\begin{vmatrix} x & y & 1 \\ 2 & 7 & 1 \\ -1 & 4 & 1 \end{vmatrix} = 3x - 3y + 15 = 0$

OR $= x - y + 5 = 0$

18. $x = \dfrac{\begin{vmatrix} 4 & -7 \\ 11 & 5 \end{vmatrix}}{\begin{vmatrix} 6 & -7 \\ 2 & 5 \end{vmatrix}} = \dfrac{97}{44}$

19. $z = \dfrac{\begin{vmatrix} 3 & 0 & 1 \\ 0 & 1 & 3 \\ 1 & -1 & 2 \end{vmatrix}}{\begin{vmatrix} 3 & 0 & 1 \\ 0 & 1 & 4 \\ 1 & -1 & 0 \end{vmatrix}} = \dfrac{14}{11}$

20. $y = \dfrac{\begin{vmatrix} 721.4 & 33.77 \\ 45.9 & 19.85 \end{vmatrix}}{\begin{vmatrix} 721.4 & -29.1 \\ 45.9 & 105.6 \end{vmatrix}} = \dfrac{12{,}769.747}{77{,}515.530} \approx 0.1647$

Chapter 9 Practice Test Solutions

1. $a_n = \dfrac{2n}{(n+1)!}$

$a_1 = \dfrac{2(1)}{3!} = \dfrac{2}{6} = \dfrac{1}{3}$

$a_2 = \dfrac{2(2)}{4!} = \dfrac{4}{24} = \dfrac{1}{6}$

$a_3 = \dfrac{2(3)}{5!} = \dfrac{6}{120} = \dfrac{1}{20}$

$a_4 = \dfrac{2(4)}{6!} = \dfrac{8}{720} = \dfrac{1}{90}$

$a_5 = \dfrac{2(5)}{7!} = \dfrac{10}{5040} = \dfrac{1}{504}$

Terms: $\dfrac{1}{3}, \dfrac{1}{6}, \dfrac{1}{20}, \dfrac{1}{90}, \dfrac{1}{504}$

2. $a_n = \dfrac{n+3}{3^n}$

3. $\sum_{i=1}^{6}(2i - 1) = 1 + 3 + 5 + 7 + 9 + 11 = 36$

4. $a_1 = 23,\ d = -2$

$a_2 = a_1 + d = 21$

$a_3 = a_2 + d = 19$

$a_4 = a_3 + d = 17$

$a_5 = a_4 + d = 15$

Terms: 23, 21, 19, 17, 15

5. $a_1 = 12, d = 3, n = 50$

$a_n = a_1 + (n - 1)d$

$a_{50} = 12 + (50 - 1)3 = 159$

6. $a_1 = 1$

$a_{200} = 200$

$S_n = \dfrac{n}{2}(a_1 + a_n)$

$S_{200} = \dfrac{200}{2}(1 + 200) = 20{,}100$

7. $a_1 = 7, r = 2$

$a_2 = a_1 r = 14$

$a_3 = a_1 r^3 = 56$

$a_4 = a_1 r^3 = 56$

$a_5 = a_1 r^4 = 112$

Terms: 7, 14, 28, 56, 112

8. $\sum_{n=0}^{9} 6\left(\dfrac{2}{3}\right)^n, a_1 = 6, r = \dfrac{2}{3}, n = 10$

$$S_n = \frac{a_1(1 - r^n)}{1 - r} = \frac{6\left(1 - (2/3)^{10}\right)}{1 - (2/3)} \approx 17.6879$$

9. $\sum_{n=0}^{\infty}(0.03)^n, a_1 = 1, r = 0.03$

$$S = \frac{a_1}{1 - r} = \frac{1}{1 - 0.03} = \frac{1}{0.97} = \frac{100}{97} \approx 1.0309$$

10. For $n = 1, 1 = \dfrac{1(1 + 1)}{2}$.

Assume that $1 + 2 + 3 = 4 + \cdots + k = \dfrac{k(k+1)}{2}$.

Now for $n = k + 1$,

$$\begin{aligned} 1 + 2 + 3 + 4 + \cdots + k + (k + 1) &= \frac{k(k+1)}{2} + k + 1 \\ &= \frac{k(k+1)}{2} + \frac{2(k+1)}{2} \\ &= \frac{(k+1)(k+2)}{2}. \end{aligned}$$

Thus, $1 + 2 + 3 + 4 + \cdots + n = \dfrac{n(n+1)}{2}$ for all integers $n \ge 1$.

11. For $n = 4, 4! > 2^4$. Assume that $k! > 2^k$. Then

$(k + 1)! = (k + 1)(k!) > (k + 1)2^k > 2 \cdot 2^k = 2^{k+1}$.

Thus, $n! > 2^n$ for all integers $n \ge 4$.

12. ${}_{13}C_4 = \dfrac{13!}{(13 - 4)!4!} = 715$

13.
$$\begin{aligned} (x + 3)^5 &= x^5 + 5x^4(3) + 10x^3(3)^2 + 10x^2(3)^3 + 5x(3)^4 + (3)^5 \\ &= x^5 + 15x^4 + 90x^3 + 270x^2 + 405x + 243 \end{aligned}$$

14. ${}_{12}C_5x^7(-2)^5 = -25{,}344x^7$

15. ${}_{30}P_4 = \dfrac{30!}{(30-4)!} = 657{,}720$

16. $6! = 720$ ways

17. ${}_{12}P_3 = 1320$

18. $P(2) + P(3) + P(4) = \frac{1}{36} + \frac{2}{36} + \frac{3}{36}$

$= \frac{6}{36} = \frac{1}{6}$

19. $P(K, B10) = \frac{4}{52} \cdot \frac{2}{51} = \frac{2}{663}$

20. Let A = probability of no faulty units.

$P(A) = \left(\frac{997}{1000}\right)^{50} \approx 0.8605$

$P(A') = 1 - P(A) \approx 0.1395$

Chapter 10 Practice Test Solutions

1. $x^2 - 6x - 4y + 1 = 0$

$$x^2 - 6x + 9 = 4y - 1 + 9$$
$$(x-3)^2 = 4y + 8$$
$$(x-3)^2 = 4(1)(y+2) \Rightarrow p = 1$$

Vertex: $(3, -2)$
Focus: $(3, -1)$
Directrix: $y = -3$

2. Vertex: $(2, -5)$
Focus: $(2, -6)$
Vertical axis; opens downward with $p = -1$

$$(x-h)^2 = 4p(y-k)$$
$$(x-2)^2 = 4(-1)(y+5)$$
$$x^2 - 4x + 4 = -4y - 20$$
$$x^2 - 4x + 4y + 24 = 0$$

3. $x^2 + 4y^2 - 2x + 32y + 61 = 0$

$$(x^2 - 2x + 1) + 4(y^2 + 8y + 16) = -61 + 1 + 64$$
$$(x-1)^2 + 4(y+4)^2 = 4$$
$$\frac{(x-1)^2}{4} + \frac{(y+4)^2}{1} = 1$$

$a = 2, b = 1, c = \sqrt{3}$
Horizontal major axis
Center: $(1, -4)$
Foci: $(1 \pm \sqrt{3}, -4)$
Vertices: $(3, -4), (-1, -4)$
Eccentricity: $e = \dfrac{\sqrt{3}}{2}$

4. Vertices: $(0, \pm 6)$
Eccentricity: $e = \dfrac{1}{2}$
Center: $(0, 0)$
Vertical major axis
$a = 6, e = \dfrac{c}{a} = \dfrac{c}{6} = \dfrac{1}{2} \Rightarrow c = 3$
$b^2 = (6)^2 - (3)^2 = 27$
$$\frac{x^2}{27} + \frac{y^2}{36} = 1$$

5. $16y^2 - x^2 - 6x - 128y + 231 = 0$

$$16(y^2 - 8y + 16) - (x^2 + 6x + 9) = -231 + 256 - 9$$
$$16(y-4)^2 - (x+3)^2 = 16$$
$$\frac{(y-4)^2}{1} - \frac{(x+3)^2}{16} = 1$$

$a = 1, b = 4, c = \sqrt{17}$
Center: $(-3, 4)$
Vertical transverse axis
Vertices: $(-3, 5), (-3, 3)$
Foci: $(-3, 4 \pm \sqrt{17})$
Asymptotes: $y = 4 \pm \dfrac{1}{4}(x+3)$

6. Vertices: $(\pm 3, 2)$
Foci: $(\pm 5, 2)$
Center: $(0, 2)$
Horizontal transverse axis
$a = 3, c = 5, b = 4$

$$\frac{(x-0)^2}{9} - \frac{(y-2)^2}{16} = 1$$
$$\frac{x^2}{9} - \frac{(y-2)^2}{16} = 1$$

7. $5x^2 + 2xy + 5y^2 - 10 = 0$

$A = 5, B = 2, C = 5$

$$\cot 2\theta = \frac{5 - 5}{2} = 0$$

$$2\theta = \frac{\pi}{2} \Rightarrow \theta = \frac{\pi}{4}$$

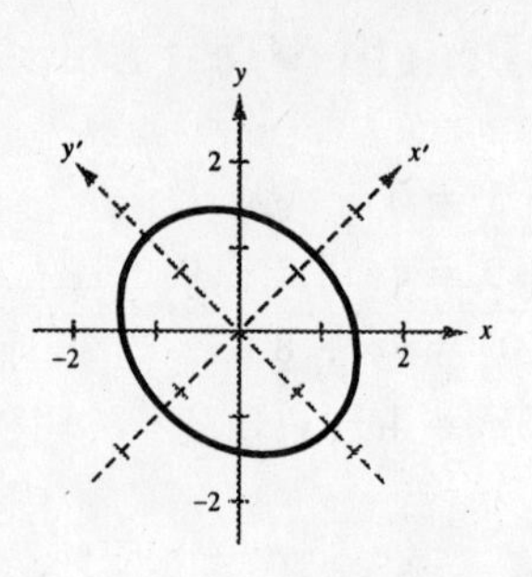

$$x = x' \cos\frac{\pi}{4} - y' \sin\frac{\pi}{4} = \frac{x' - y'}{\sqrt{2}}$$

$$x = x' \cos\frac{\pi}{4} + y' \sin\frac{\pi}{4} = \frac{x' + y'}{\sqrt{2}}$$

$$5\left(\frac{x' - y'}{\sqrt{2}}\right)^2 + 2\left(\frac{x' - y'}{\sqrt{2}}\right)\left(\frac{x' + y'}{\sqrt{2}}\right) + 5\left(\frac{x' + y'}{\sqrt{2}}\right)^2 - 10 = 0$$

$$\frac{5(x')^2}{2} - \frac{10x'y'}{2} + \frac{5(y')^2}{2} + (x')^2 - (y')^2 + \frac{5(x')^2}{2} + \frac{10x'y'}{2} + \frac{5(y')^2}{2} - 10 = 0$$

$$6(x')^2 + 4(y')^2 - 10 = 0$$

$$\frac{3(x')^2}{5} + \frac{2(y')^2}{5} = 1$$

$$\frac{(x')^2}{5/3} + \frac{(y')^2}{5/2} = 1$$

Ellipse centered at the origin

8. (a) $6x^2 - 2xy + y^2 = 0$

$A = 6, B = -2, C = 1$

$B^2 - 4AC = (-2)^2 - 4(6)(1) = -20 < 0$

Ellipse

(b) $x^2 + 4xy + 4y^2 - x - y + 17 = 0$

$A = 1, B = 4, C = 4$

$B^2 - 4AC = (4)^2 - 4(1)(4) = 0$

Parabola

9. $x = 3 - 2\sin\theta, y = 1 + 5\cos\theta$

$$\frac{x - 3}{-2} = \sin\theta, \frac{y - 1}{5} = \cos\theta$$

$$\left(\frac{x - 3}{-2}\right)^2 + \left(\frac{y - 1}{5}\right)^2 = 1$$

$$\frac{(x - 3)^2}{4} + \frac{(y - 1)^2}{25} = 1$$

10. $x = e^{2t}, y = e^{4t}$

$x > 0, y > 0$

$y = (e^{2t})^2 = (x)^2 = x^2, x < 0, y > 0$

11. Polar: $\left(\sqrt{2}, \frac{3\pi}{4}\right)$

$$x = \sqrt{2}\cos\frac{3\pi}{4} = \sqrt{2}\left(-\frac{1}{\sqrt{2}}\right) = -1$$

$$y = \sqrt{2}\sin\frac{3\pi}{4} = \sqrt{2}\left(\frac{1}{\sqrt{2}}\right) = 1$$

Rectangular: $(-1, 1)$

12. Rectangular: $(\sqrt{3}, -1)$

$$r = \pm\sqrt{(\sqrt{3})^2 + (-1)^2} = \pm 2$$

$$\tan\theta = \frac{\sqrt{3}}{-1} = -\sqrt{3}$$

$$\theta = \frac{2\pi}{3} \text{ or } \theta = \frac{5\pi}{3}$$

Polar: $\left(-2, \frac{2\pi}{3}\right)$ or $\left(2, \frac{5\pi}{3}\right)$

13. Rectangular: $4x - 3y = 12$

Polar: $4r\cos\theta - 3r\sin\theta = 12$

$r(4\cos\theta - 3\sin\theta) = 12$

$$r = \frac{12}{4\cos\theta - 3\sin\theta}$$

14. Polar: $r = 5\cos\theta$

$r^2 = 5r\cos\theta$

Rectangular: $x^2 + y^2 = 5x$

$x^2 + y^2 - 5x = 0$

15. $r = 1 - \cos\theta$

Cardioid

Symmetry: Polar axis

Maximum value of $|r|$: $r = 2$ when $\theta = \pi$.

Zero of r: $r = 0$ when $\theta = 0$

θ	0	$\frac{\pi}{2}$	π	$\frac{3\pi}{2}$
r	0	1	2	1

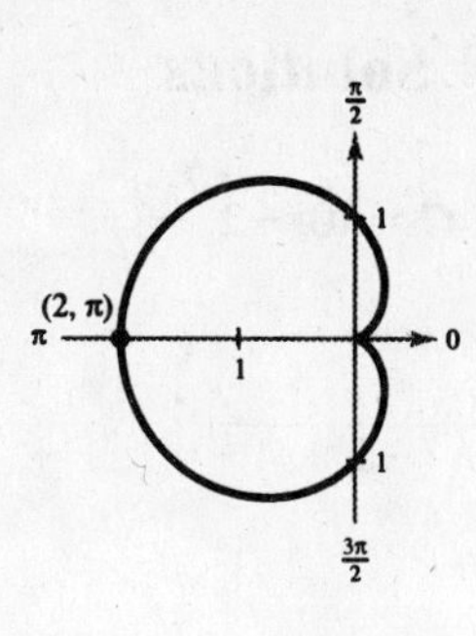

16. $r = 5\sin 2\theta$

Rose curve with four petals

Symmetry: Polar axis, $\theta = \frac{\pi}{2}$, and pole

Maximum value of $|r|$: $|r| = 5$ when $\theta = \frac{\pi}{4}, \frac{3\pi}{4}, \frac{5\pi}{4}, \frac{7\pi}{4}$

Zeros of r: $r = 0$ when $\theta = 0, \frac{\pi}{2}, \pi, \frac{3\pi}{2}$

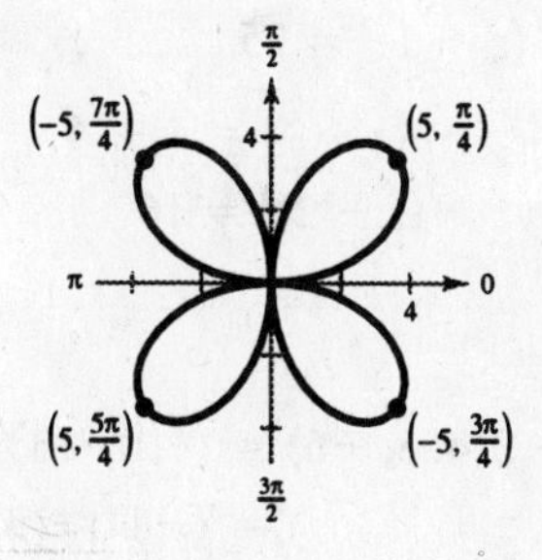

17. $r = \dfrac{3}{6 - \cos\theta}$

$$r = \frac{\frac{1}{2}}{1 - \frac{1}{6}\cos\theta}$$

$e = \frac{1}{6} < 1$, so the graph is an ellipse.

θ	0	$\frac{\pi}{2}$	π	$\frac{3\pi}{2}$
r	$\frac{3}{5}$	$\frac{1}{2}$	$\frac{3}{7}$	$\frac{1}{2}$

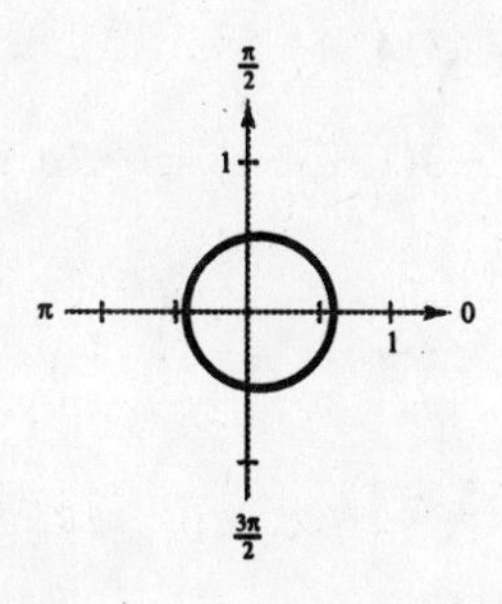

18. Parabola

Vertex: $\left(6, \frac{\pi}{2}\right)$

Focus: $(0, 0)$

$e = 1$

$$r = \frac{ep}{1 + e\sin\theta}$$

$$r = \frac{p}{1 + \sin\theta}$$

$$6 = \frac{p}{1 + \sin(\pi/2)}$$

$$6 = \frac{p}{2}$$

$$12 = p$$

$$r = \frac{12}{1 + \sin\theta}$$

Chapter 11 Practice Test Solutions

1. Let $A = (0, 0, 0)$, $B = (1, 2, -4)$, $C = (0, -2, -1)$

Side AB: $\sqrt{1^2 + 2^2 + 4^2} = \sqrt{21}$

Side AC: $\sqrt{0^2 + 2^2 + 1^2} = \sqrt{5}$

Side BC: $\sqrt{(-1)^2 + (-2-2)^2 + (-1+4)^2} = \sqrt{1 + 16 + 9} = \sqrt{26}$

$BC^2 = AB^2 + AC^2$

$26 = 21 + 5$

2. $(x - 0)^2 + (y - 4)^2 + (z - 1)^2 = 5^2$

$x^2 + (y - 4)^2 + (z - 1)^2 = 25$

3. $(x^2 + 2x + 1) + y^2 + (z^2 - 4z + 4) = 1 + 4 + 11$

$(x + 1)^2 + y^2 + (z - 2)^2 = 16$

Center: $(-1, 0, 2)$

Radius: 4

4. $\mathbf{u} - 3\mathbf{v} = \langle 1, 0, -1\rangle - 3\langle 4, 3, -6\rangle = \langle 1, 0, -1\rangle - \langle 12, 9, -18\rangle$

$= \langle -11, -9, 17\rangle$

5. $\frac{1}{2}\mathbf{v} = \frac{1}{2}\langle 2, 4, -6\rangle = \langle 1, 2, -3\rangle$

$\|\frac{1}{2}\mathbf{v}\| = \sqrt{1^2 + 2^2 + (-3)^2} = \sqrt{14}$

6. $\mathbf{u} \cdot \mathbf{v} = \langle 2, 1, -3\rangle \cdot \langle 1, 1, -2\rangle$

$= 2 + 1 + 6 = 9$

7. Because $\mathbf{v} = \langle -3, -3, 3\rangle = -3\langle 1, 1, -1\rangle = -3\mathbf{u}$, $\mathbf{u}$ and $\mathbf{v}$ are parallel.

8. $\mathbf{u} \times \mathbf{v} = \begin{vmatrix} \mathbf{i} & \mathbf{j} & \mathbf{k} \\ -1 & 0 & 2 \\ 1 & -1 & 3 \end{vmatrix} = \langle 2, 5, 1\rangle$

$\mathbf{v} \times \mathbf{u} = -(\mathbf{u} \times \mathbf{v}) = \langle -2, -5, -1\rangle$

9. $\mathbf{u} \cdot (\mathbf{v} \times \mathbf{w}) = \begin{vmatrix} 1 & 1 & 1 \\ 0 & -1 & 1 \\ 1 & 0 & 4 \end{vmatrix}$

$= 1(-4) - 1(-1) + 1(1)$

$= -4 + 1 + 1 = -2$

Volume $= |\mathbf{u} \cdot (\mathbf{v} \times \mathbf{w})| = |-2| = 2$

10. $\mathbf{v} = \langle (2 - 0), -3 - (-3), 4 - 3\rangle = \langle 2, 0, 1\rangle$

$x = 2 + 2t, y = -3, z = 4 + t$

11. $1(x - 1) - 1(y - 2) + 0(z - 3) = 0$

$x - 1 - y + 2 = 0$

$x - y + 1 = 0$

12. $\overrightarrow{AB} = \langle 1, 1, 1\rangle, \overrightarrow{AC} = \langle 1, 2, 3\rangle$

$\mathbf{n} = \overrightarrow{AB} \times \overrightarrow{AC} = \begin{vmatrix} \mathbf{i} & \mathbf{j} & \mathbf{k} \\ 1 & 1 & 1 \\ 1 & 2 & 3 \end{vmatrix} = \langle 1, -2, 1\rangle$

Plane: $1(x - 0) - 2(y - 0) + (z - 0) = 0$

$x - 2y + z = 0$

13. $\mathbf{n}_1 = \langle 1, 1, -1\rangle, \mathbf{n}_2 = \langle 3, -4, -1\rangle$

$\mathbf{n}_1 \cdot \mathbf{n}_2 = 3 - 4 + 1 = 0 \Rightarrow$ Orthogonal planes

14. $\mathbf{n} = \langle 1, 2, 1\rangle$, $Q = (1, 1, 1)$, $P = (0, 0, 6)$ on plane.

$\overrightarrow{PQ} = \langle 1, 1, -5\rangle$

$$D = \frac{|\overrightarrow{PQ} \cdot \mathbf{n}|}{\|\mathbf{n}\|} = \frac{|1 + 2 - 5|}{\sqrt{1 + 4 + 1}} = \frac{2}{\sqrt{6}} = \frac{\sqrt{6}}{3}$$

Chapter 12 Practice Test Solutions

1.

x	2.9	2.99	3	3.01	3.1
$f(x)$	0.1695	0.1669	?	0.1664	0.1639

$$\lim_{x\to 3} \frac{x-3}{x^2-9} \approx 0.1667$$

2. $$\lim_{x\to 0} \frac{\sqrt{x+4}-2}{x} \approx \frac{1}{4}$$

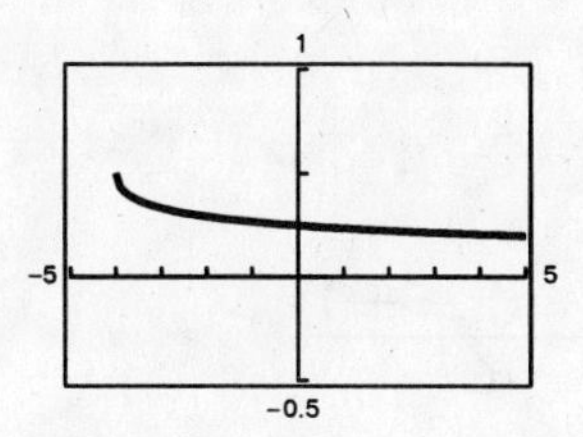

3. $$\lim_{x\to 2} e^{x-2} = e^{2-2} = e^0 = 1$$

4. $$\lim_{x\to 1} \frac{x^3-1}{x-1} = \lim_{x\to 1} \frac{(x-1)(x^2+x+1)}{x-1} = \lim_{x\to 1} (x^2+x+1) = 3$$

5. $$\lim_{x\to 0} \frac{\sin 5x}{2x} \approx 2.5$$

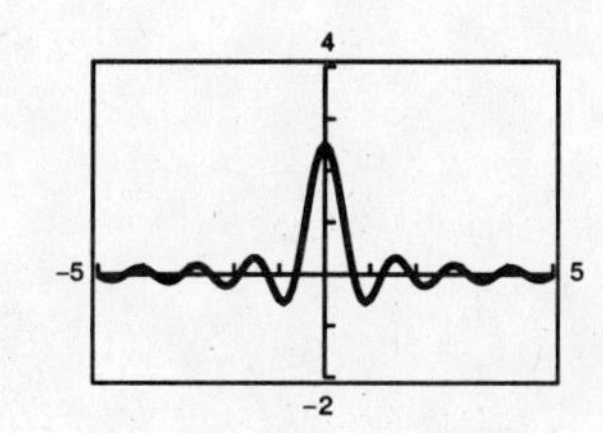

6. The limit does not exist. If $f(x) = \frac{|x+2|}{x+2}$, then $f(x) = 1$ for $x > -2$, and $f(x) = -1$ for $x < -2$.

7. $$m_{\text{sec}} = \frac{f(4+h)-f(4)}{h}$$
$$= \frac{\sqrt{4+h}-2}{h}$$
$$= \frac{\sqrt{4+h}-2}{h} \cdot \frac{\sqrt{4+h}+2}{\sqrt{4+h}+2}$$
$$= \frac{(4+h)-4}{h\left[\sqrt{4+h}+2\right]}$$
$$= \frac{h}{h\left[\sqrt{4+h}+2\right]}$$
$$= \frac{1}{\sqrt{4+h}+2},\ h \neq 0$$
$$m = \lim_{h\to 0} \frac{1}{\sqrt{4+h}+2} = \frac{1}{\sqrt{4}+2} = \frac{1}{4}$$

8. $$f^1(x) = \lim_{h\to 0} \frac{f(x+h)-f(x)}{h}$$
$$= \lim_{h\to 0} \frac{[3(x+h)-1]-[3x-1]}{h}$$
$$= \lim_{h\to 0} \frac{3x+3h-1-3x+1}{h}$$
$$= \lim_{h\to 0} \frac{3h}{h} = \lim_{h\to 0} 3 = 3$$

9. (a) $$\lim_{x\to\infty} \frac{3}{x^4} = 0$$

(b) $$\lim_{x\to-\infty} \frac{x^2}{x^2+3} = 1$$

(c) $$\lim_{x\to\infty} \frac{|x|}{1-x} = -1$$

10. $$a_1 = 0,\ a_2 = \frac{1-4}{8+1} = -\frac{1}{3},\ a_3 = \frac{1-9}{18+1} = -\frac{8}{19}$$
$$a_4 = \frac{1-16}{33} = -\frac{15}{33}$$
$$\lim_{n\to\infty} a_n = \lim_{n\to\infty} \frac{1-n^2}{2n^2+1} = -\frac{1}{2}$$

11. $\sum_{i=1}^{25} i^2 + \sum_{i=1}^{25} i = \frac{25(26)(51)}{6} + \frac{25(26)}{2} = \frac{25(26)}{6}[51 + 3] = \frac{25(26)(54)}{6} = 5850$

12. $\sum_{i=1}^{n} \frac{i^2}{n^3} = \frac{1}{n^3}\sum_{i=1}^{n} i^2 = \frac{1}{n^3}\left[\frac{n(n + 1)(2n + 1)}{6}\right] = \frac{2n^2 + 3n + 1}{6n^2} = s(n)$

$\lim_{n\to\infty} s(n) = \frac{1}{3}$

13. Width of rectangles: $\frac{b - a}{n} = \frac{1}{n}$

Height: $f\left(a + \frac{(b - a)i}{n}\right) = f\left(\frac{i}{n}\right) = 1 - \left(\frac{i}{n}\right)^2$

$a_n \approx \sum_{i=1}^{n}\left[1 - \frac{i^2}{n^2}\right]\frac{1}{n} = \sum_{i=1}^{n}\frac{1}{n} - \sum_{i=1}^{n}\frac{i^2}{n^3} = 1 - \frac{1}{n^2}\frac{n(n + 1)(2n + 1)}{6}$

$A = \lim_{n\to\infty} A_n = 1 - \frac{1}{3} = \frac{2}{3}$

PART II

Chapter P Chapter Test

1. Midpoint $= \left(\dfrac{-2+6}{2}, \dfrac{5+0}{2}\right) = \left(2, \dfrac{5}{2}\right)$

Distance $= \sqrt{(-2-6)^2 + (5-0)^2}$

$= \sqrt{64+25} = \sqrt{89} \approx 9.43$

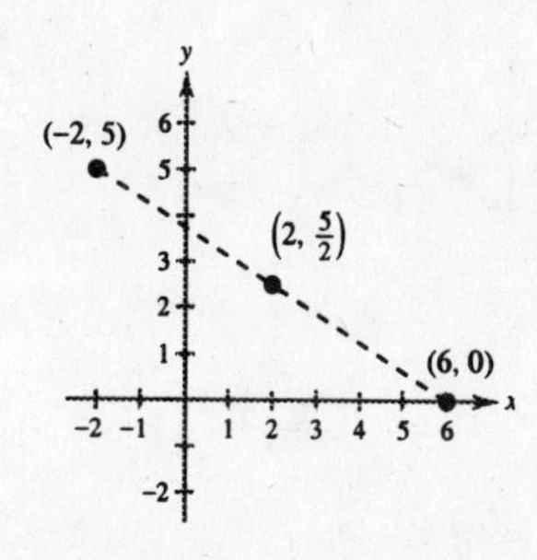

2.

3. $r = \sqrt{(4-1)^2 + (-1-4)^2} = \sqrt{9+25} = \sqrt{34}$

$(x-4)^2 + (y+1)^2 = 34$

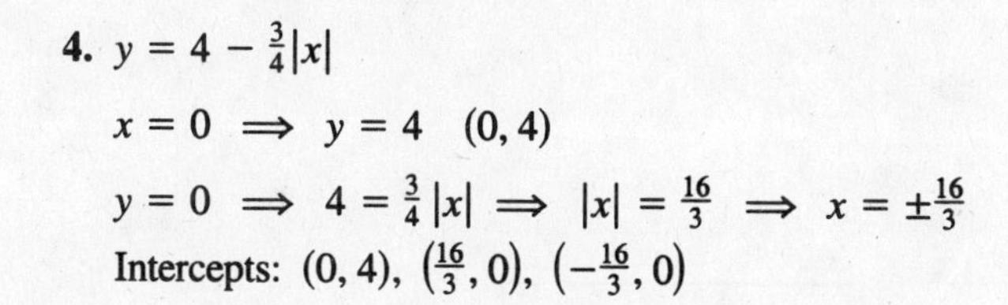

4. $y = 4 - \frac{3}{4}|x|$

$x = 0 \implies y = 4 \quad (0, 4)$

$y = 0 \implies 4 = \frac{3}{4}|x| \implies |x| = \frac{16}{3} \implies x = \pm\frac{16}{3}$

Intercepts: $(0, 4), \left(\frac{16}{3}, 0\right), \left(-\frac{16}{3}, 0\right)$

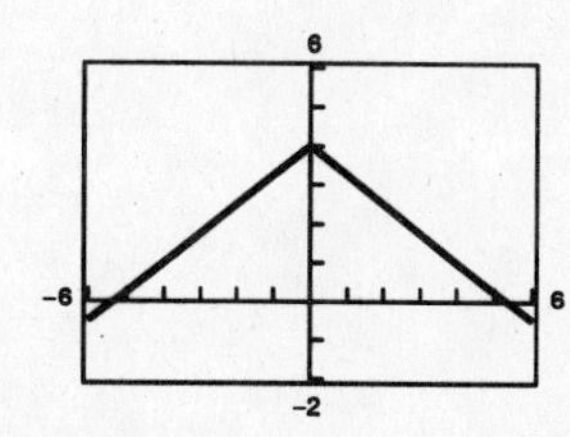

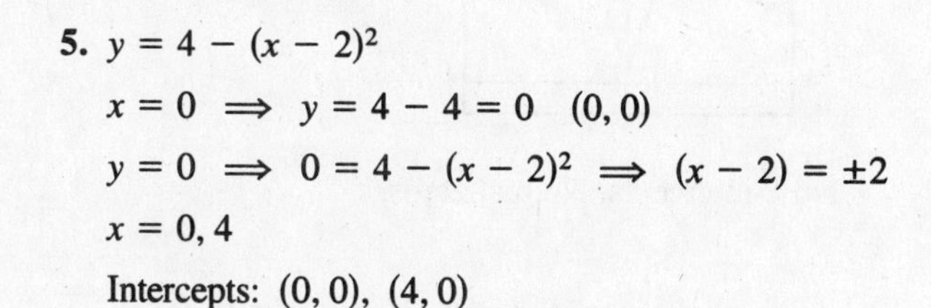

5. $y = 4 - (x-2)^2$

$x = 0 \implies y = 4 - 4 = 0 \quad (0, 0)$

$y = 0 \implies 0 = 4 - (x-2)^2 \implies (x-2) = \pm 2$

$x = 0, 4$

Intercepts: $(0, 0), (4, 0)$

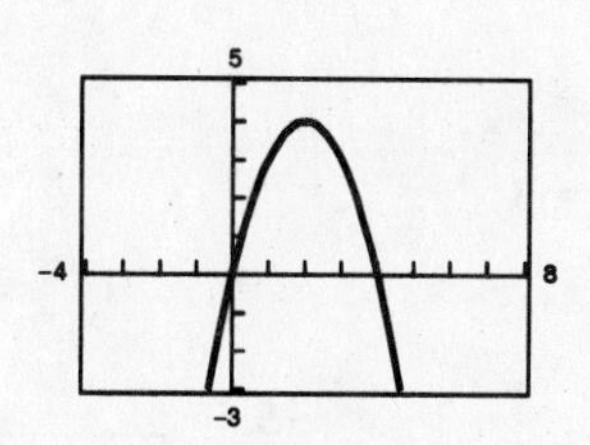

6. $y = x - x^3$

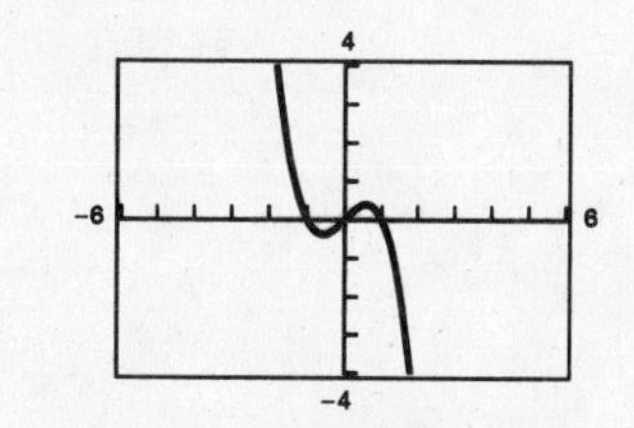

$x = 0 \implies y = 0 \quad (0, 0)$

$y = 0 \implies 0 = x(1-x)(1+x)$

Intercepts: $(0, 0), (1, 0), (-1, 0)$

7. $y = -x^3 + 2x - 4$

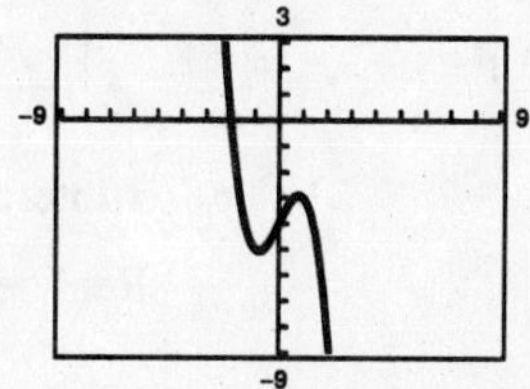

$x = 0 \implies y = -4 \quad (0, -4)$

$y = 0 \implies x = -2 \quad (-2, 0)$

Intercepts: $(0, -4), (-2, 0)$

8. $y = \sqrt{3 - x}$

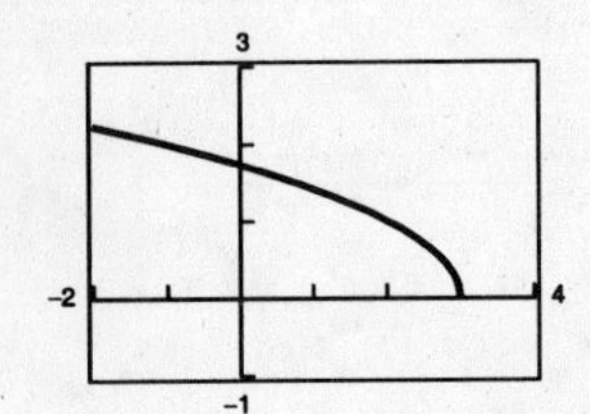

$x = 0 \Rightarrow y = \sqrt{3}$

$y = 0 \Rightarrow 0 = \sqrt{3 - x} \Rightarrow x = 3$

Intercepts: $(0, \sqrt{3})$, $(3, 0)$

9. $y = \frac{1}{2}x\sqrt{x + 3}$

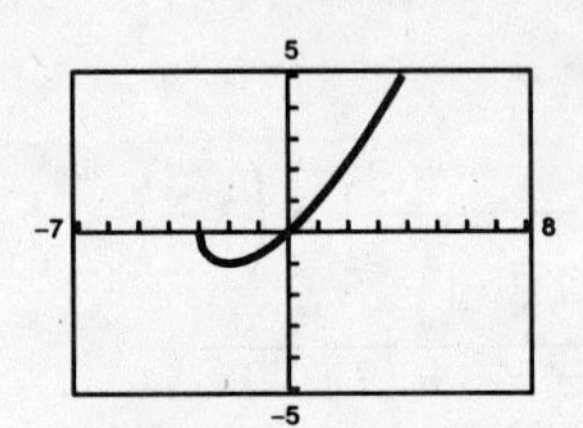

$x = 0 \Rightarrow y = 0 \quad (0, 0)$

$y = 0 \Rightarrow x = 0, -3 \quad (0, 0), (-3, 0)$

Intercepts: $(0, 0)$, $(-3, 0)$

10. $y - (-1) = \frac{3}{2}(x - 3) = \frac{3}{2}x - \frac{9}{2}$

$y = \frac{3}{2}x - \frac{11}{2}$

Additional points: $(1, -4)$, $\left(4, \frac{1}{2}\right)$, $(5, 2)$

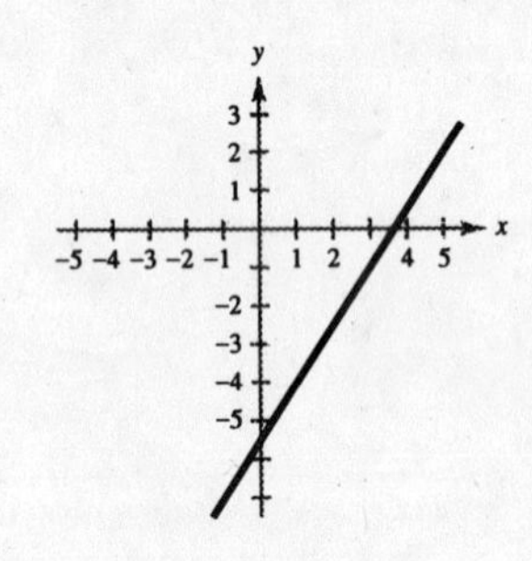

11. $5x + 2y = 3$

$2y = 3 - 5x$

$y = \frac{3}{2} - \frac{5}{2}x$

Slope of the perpendicular line is $\frac{2}{5}$.

$y - 4 = \frac{2}{5}(x - 0)$

$5y - 20 = 2x$

$2x - 5y + 20 = 0$

12 $\dfrac{12}{x} - 7 = -\dfrac{27}{x} + 6$

$\dfrac{39}{x} = 13$

$39 = 13x$

$3 = x \Rightarrow x = 3$

13.

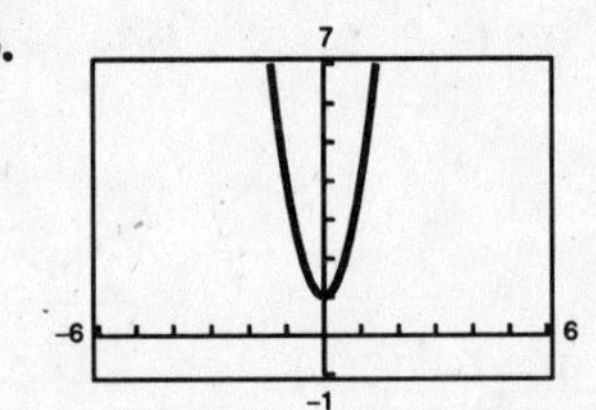

No x-intercepts. No real zeros.

14.

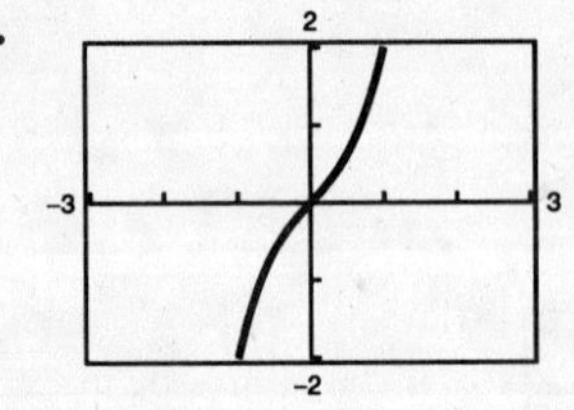

x-intercept: $(0, 0)$

Real zero: $x = 0$

15.

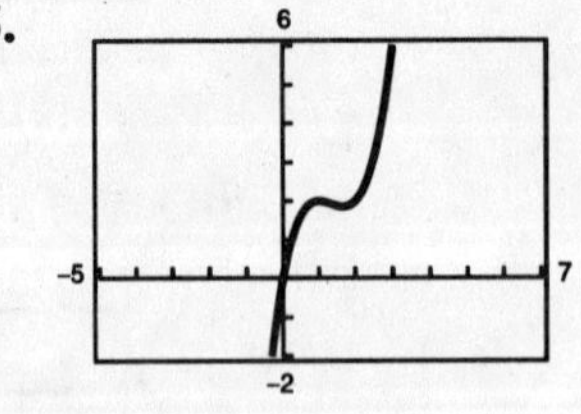

x-intercept: $(0, 0)$

Real zero: $x = 0$

16. $x^2 - 10x + 9 = 0$

$(x - 1)(x - 9) = 0$

$x = 1, 9$

17. $4x^2 - 81 = 0$

$4x^2 = 81$

$x^2 = \frac{81}{4}$

$x = \pm\frac{9}{2}$

18. $3x^3 - 4x^2 - 12x + 16 = 0$

$x^2(3x - 4) - 4(3x - 4) = 0$

$(x^2 - 4)(3x - 4) = 0$

$x = 2, -2, \frac{4}{3}$

19. $x + \sqrt{22 - 3x} = 6$

$\sqrt{22 - 3x} = 6 - x$

$22 - 3x = (6 - x)^2$

$22 - 3x = 36 - 12x + x^2$

$x^2 - 9x + 14 = 0$

$(x - 2)(x - 7) = 0$

$x = 2 \quad (x = 7 \text{ is extraneous})$

20. $(x^2 + 6)^{2/3} = 16$

$x^2 + 6 = 16^{3/2} = 64$

$x^2 = 58$

$x = \pm\sqrt{58} \approx \pm 7.616$

21. $|8x - 1| = 21$

$8x - 1 = 21$ or $-(8x - 1) = 21$

$8x = 22$ or $-8x = 20$

$x = \frac{11}{4}$ $\qquad x = -\frac{5}{2}$

22. $-\frac{5}{6} < x - 2 < \frac{1}{8}$

$2 - \frac{5}{6} < x < 2 + \frac{1}{8}$

$\frac{7}{6} < x < \frac{17}{8}$

$\frac{7}{6}$ $\frac{17}{8}$

0 1 2 3 x

23. $2|x - 8| < 10$

$|x - 8| < 5$

$-5 < x - 8 < 5$

$3 < x < 13$

3 4 5 6 7 8 9 10 11 12 13 x

24. $\dfrac{3 - 5x}{2 + 3x} < -2$

$\dfrac{3 - 5x}{2 + 3x} + 2 < 0$

$\dfrac{3 - 5x + 2(2 + 3x)}{2 + 3x} < 0$

$\dfrac{x + 7}{2 + 3x} < 0$

Critical numbers: $-7, -\frac{2}{3}$. Checking the three intervals, we obtain $-7 < x < -\frac{2}{3}$

$-\frac{2}{3}$

-8 -7 -6 -5 -4 -3 -2 -1 0 x

25. $C = 16.369t + 401.702$

$C = 16.369t + 401.702 = 600$

$16.369t = 198.298$

$t \approx 12.1$ or year 2002

Chapter 1 Chapter Test

1. No, for some x there corresponds more than one value of y. For instance, if $x = 1$, $y = \pm\frac{1}{\sqrt{3}}$.

2. $f(-6) = 10 - \sqrt{3-(-6)} = 10 - 3 = 7$

3. $f(t-3) = 10 - \sqrt{3-(t-3)} = 10 - \sqrt{6-t}$

4. $\frac{f(x)-f(2)}{x-2} = \frac{10-\sqrt{3-x}-9}{x-2} = \frac{\sqrt{3-x}-1}{2-x}$

5. $3 - x \geq 0 \Rightarrow$ domain is all $x \leq 3$.

6. $C = 5.60x + 24{,}000$

$$P = R - C = 9.20x - (5.60x + 24{,}000) = 3.60x - 24{,}000$$

7. (a)

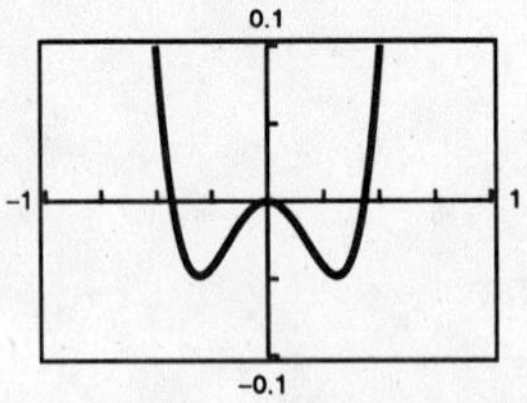

(b) Increasing: $(-0.308, 0), (0.308, \infty)$

Decreasing: $(-\infty, -0.308), (0, 0.308)$

(c) Even function: $f(x) = f(-x)$

8. (a)

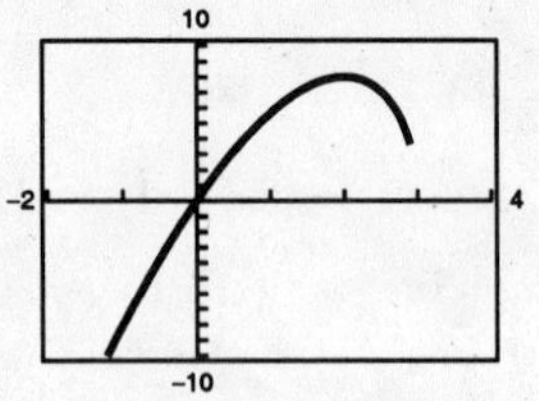

(b) Increasing: $(-\infty, 2.0)$

Decreasing: $(2.0, 3)$

(c) Neither even nor odd

9. (a)

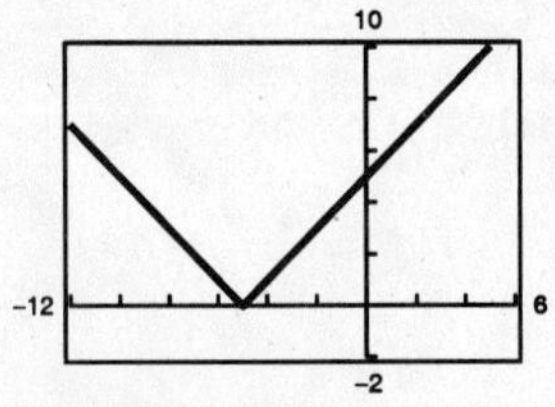

(b) Increasing: $(-5, \infty)$

Decreasing: $(-\infty, -5)$

(c) Neither even nor odd

10. (a)

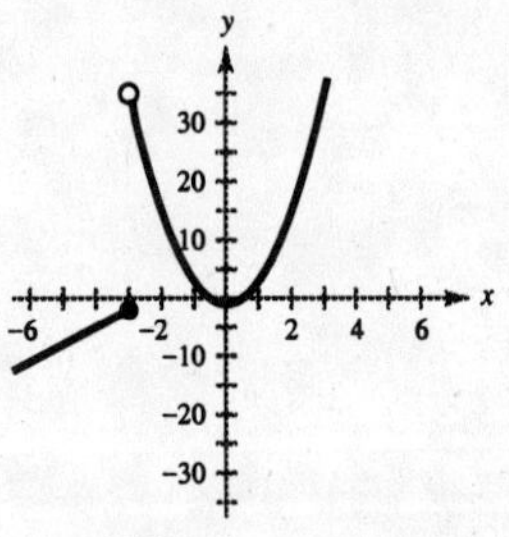

11. Relative minimum: $(-3.33, -6.52)$. Relative maximum: $(0, 12)$

12. Relative minimum: $(0.77, 1.81)$. Relative maximum: $(-0.77, 2.19)$

13. (a) Common function $g(x) = x^3$

(b) f is obtained from g by a horizontal shift 5 units to the right, a vertical stretch of 2, a reflection in the x-axis, and a vertical shift 3 units upward.

(c)

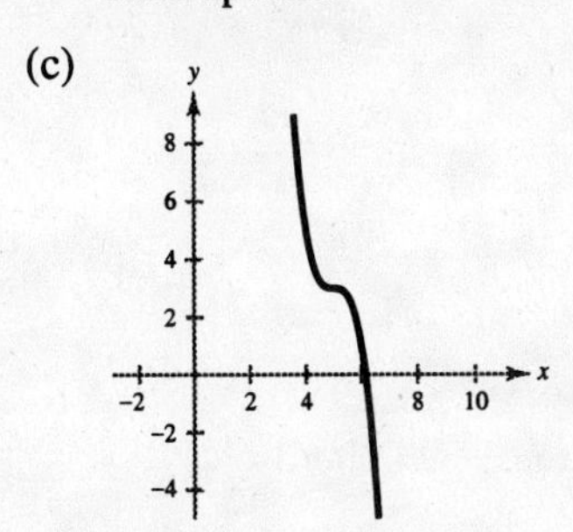

14. (a) Common function $g(x) = \sqrt{x}$.

(b) f is obtained from g by a reflection in the y-axis, and a horizontal shift 7 units to the left.

(c)

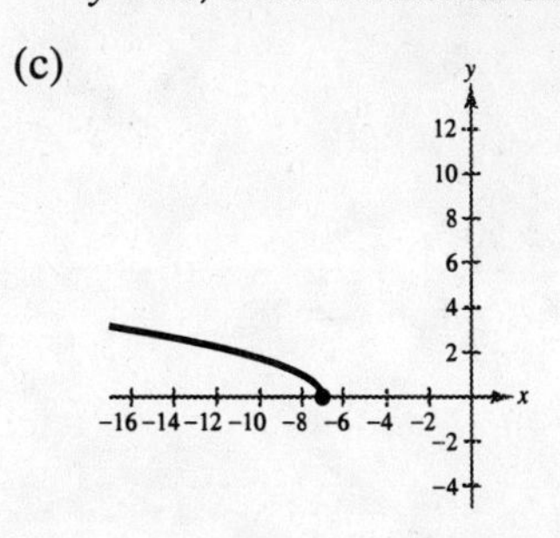

15. (a) Common function $g(x) = |x|$.

(b) $f(x) = 4|-x| - 7 = 4|x| - 7$ is obtained from g by a vertical stretch of 4 followed by a vertical shift 7 units downward.

(c)

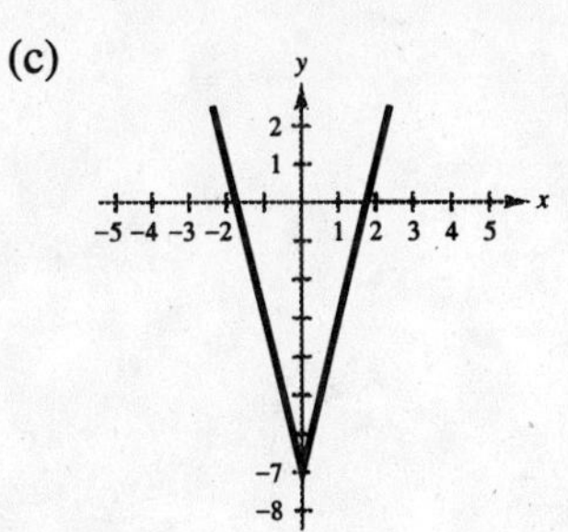

16. (a) $y = x^2$

(b) $f(x) = -(x - 3)^2 + 5$ consists of a horizontal shift 3 units to the right, a reflection in the x-axis, and a vertical shift 5 units upward.

(c)

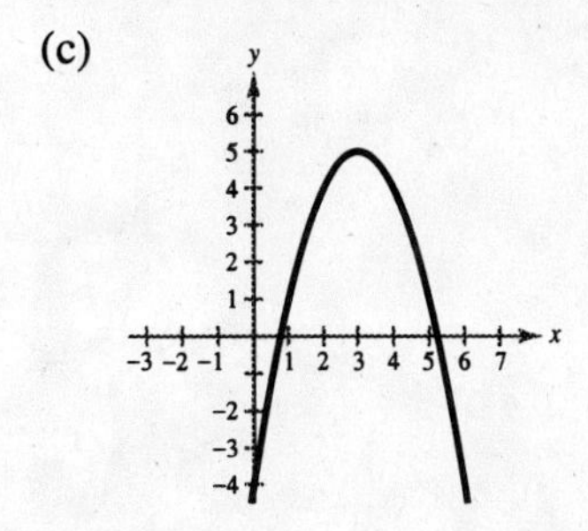

17. (a) $(f - g)(x) = x^2 - \sqrt{2 - x}$, Domain: $x \le 2$

(b) $\left(\frac{f}{g}\right)(x) = \frac{x^2}{\sqrt{2 - x}}$, Domain: $x < 2$

(c) $(f \circ g)(x) = f(\sqrt{2 - x}) = 2 - x$, Domain $x \le 2$

(d) $y = \sqrt{2 - x}$

$x = \sqrt{2 - y}$ Interchange x and y

$x^2 = 2 - y$

$y = g^{-1}(x) = 2 - x^2, \quad x \ge 0$

18. $y = x^3 + 8$

$x = y^3 + 8$

$x - 8 = y^3$

$f^{-1}(x) = \sqrt[3]{x - 8}$

19. Inverse does not exist

20. $y = \frac{3x\sqrt{x}}{8} = \frac{3}{8}x^{3/2}, x \ge 0$

$x = \frac{3}{8}y^{3/2}, y \ge 0$

$y^{3/2} = \frac{8}{3}x$

$y = \left(\frac{8}{3}x\right)^{2/3}, x \ge 0$

21. 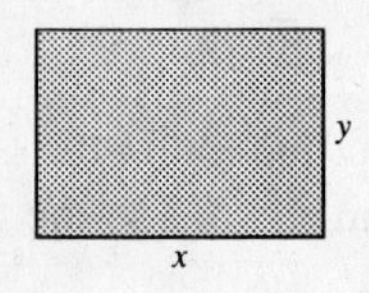

$$2x + 2y = 100$$
$$y = 50 - x$$

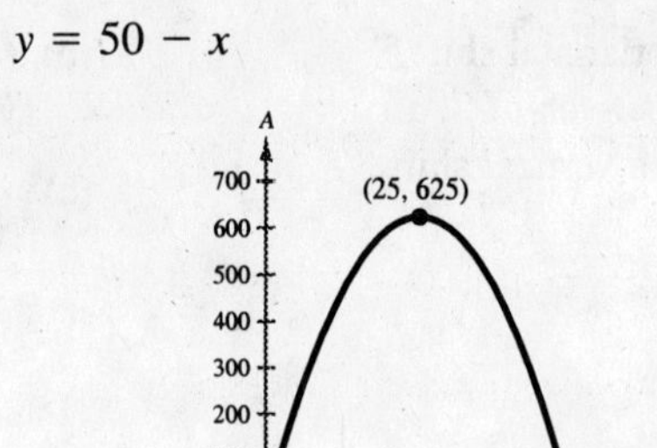

$A = xy = x(50 - x)$

Domain: $0 < x < 50$

Maximum area: 625 when $x = 25$.

Chapter 2 Chapter Test

1. (a) $g(x) = 2 - x^2$ is a reflection in the x-axis followed by a vertical shift 2 units upward.

(b) $g(x) = \left(x - \frac{3}{2}\right)^2$ is a horizontal shift $\frac{3}{2}$ units to the right.

2. $y = x^2 + 4x + 3 = x^2 + 4x + 4 - 1 = (x + 2)^2 - 1$

Vertex: $(-2, -1)$

$x = 0 \implies y = 3$

$y = 0 \implies x^2 + 4x + 3 = 0 \implies (x + 3)(x + 1) = 0 \implies x = -1, -3$

Intercepts: $(0, 3)$, $(-1, 0)$, $(-3, 0)$

3. Let $y = a(x - h)^2 + k$. The vertex $(3, -6)$ implies that $y = a(x - 3)^2 - 6$. For $(0, 3)$ you obtain

$3 = a(0 - 3)^2 - 6 = 9a - 6 \implies a = 1.$

Thus, $y = (x - 3)^2 - 6 = x^2 - 6x + 3$.

4. (a) $y = -\frac{1}{20}x^2 + 3x + 5 = -\frac{1}{20}(x^2 - 60x + 900) + 5 + 45$

$= -\frac{1}{20}(x - 30)^2 + 50$

Maximum height: $y = 50$ feet

(b) The term 5 determines the height at which the ball was thrown. Changing the constant term results in a vertical shift of the graph and therefore changes the maximum height.

5.

$$\begin{array}{r} 3x \\ x^2 + 1 \overline{) \; 3x^3 + 0x^2 + 4x - 1} \\ \underline{3x^3 + 3x } \\ x - 1 \end{array}$$

$3x + \dfrac{x - 1}{x^2 + 1}$

6.

$$\begin{array}{r|rrrrr} 2 & 2 & 0 & -5 & 0 & -3 \\ & & 4 & 8 & 6 & 12 \\ \hline & 2 & 4 & 3 & 6 & 9 \end{array}$$

$2x^3 + 4x^2 + 3x + 6 + \dfrac{9}{x - 2}$

7. Possible rational zeros:

$\pm 24, \pm 12, \pm 8, \pm 6, \pm 4, \pm 3, \pm 2, \pm 1, \pm\frac{3}{2}, \pm\frac{1}{2}$

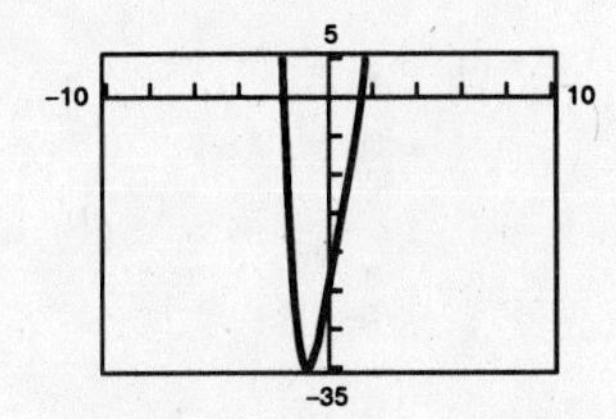

Rational zeros: $-2, \frac{3}{2}$

8. Possible rational zeros:

$\pm 2, \pm 1, \pm\frac{2}{3}, \pm\frac{1}{3}$

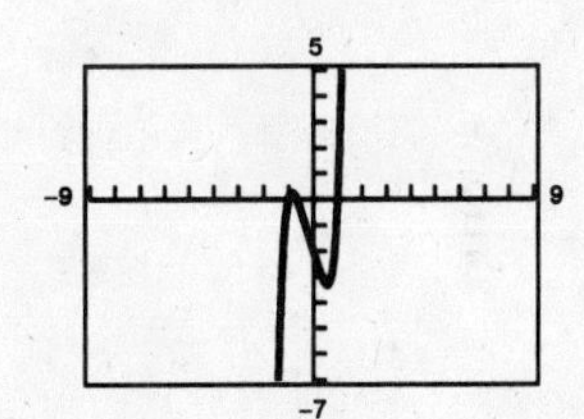

Rational zeros: $\pm 1, -\frac{2}{3}$

9. $(-8 - 3i) + (-1 - 15i) = -9 - 18i$

10. $\left(10 + \sqrt{-20}\right) - \left(4 - \sqrt{-14}\right) = 6 + 2\sqrt{5}\,i + \sqrt{14}\,i = 6 + \left(2\sqrt{5} + \sqrt{14}\right)i$

11. $(2 + i)(6 - i) = 12 + 6i - 2i + 1 = 13 + 4i$

12. $\frac{8 + 5i}{6 - i} \cdot \frac{6 + i}{6 + i} = \frac{48 + 30i + 8i - 5}{36 + 1} = \frac{43}{37} + \frac{38}{37}i$

13. Real zeros: 1.380, -0.819

14. Real zeros: -1.414, -0.667, 1.414

15. $(x - 0)(x - 3)(x - (3 + i))(x - (3 - i))$
$x(x - 3)(x^2 - 6x + 10)$
$x^4 - 9x^3 + 28x^2 - 30x$

16. $\left(x - (1 + \sqrt{3}i)\right)\left(x - (1 - \sqrt{3}i)\right)(x - 2)(x - 2)$
$(x^2 - 2x + 4)(x^2 - 4x + 4)$
$x^4 - 6x^3 + 16x^2 - 24x + 16$

17. $(x - 0)(x + 5)(x - 1 - i)(x - 1 + i)$
$(x^2 + 5x)((x - 1)^2 + 1)$
$(x^2 + 5x)(x^2 - 2x + 2)$
$x^4 + 3x^3 - 8x^2 + 10x$

18.

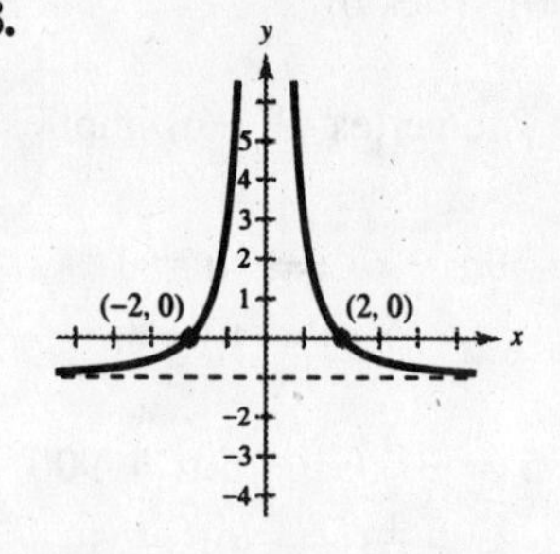

Vertical asymptote: $x = 0$
Intercepts: $(2, 0), (-2, 0)$
Symmetry: y-axis
Horizontal asymptote: $y = -1$

19. $g(x) = \frac{x^2 + 2}{x - 1} = x + 1 + \frac{3}{x - 1}$

Vertical asymptote: $x = 1$
Intercept: $(0, -2)$
Slant asymptote: $y = x + 1$

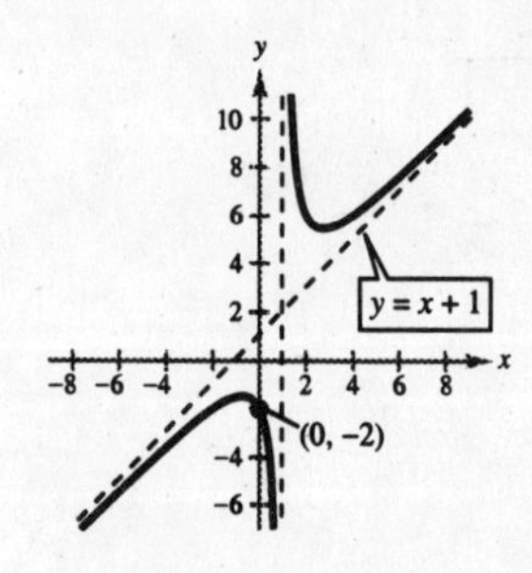

20. $f(x) = \frac{2x^2 + 9}{5x^2 + 2}$

Horizontal asymptote: $y = \frac{2}{5}$

y-axis symmetry

Intercept: $\left(0, \frac{9}{2}\right)$

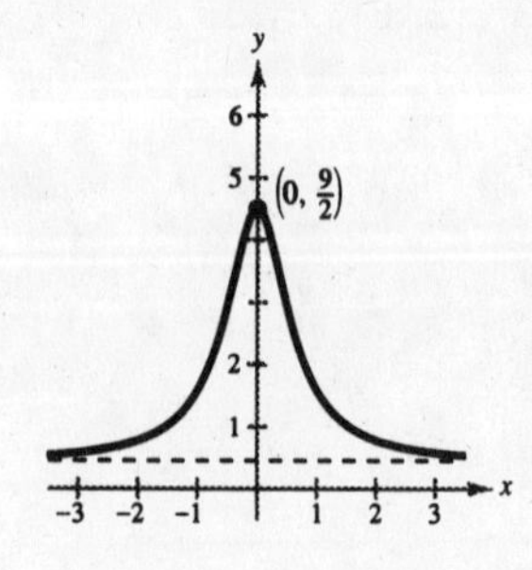

21. $30 = -0.00428x^2 + 1.442x - 3.136$
Solving for x, $x \approx 24.8$ years($x = 312$ is extraneous)

Chapter 3 Chapter Test

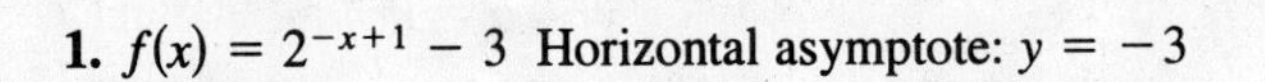

1. $f(x) = 2^{-x+1} - 3$ Horizontal asymptote: $y = -3$

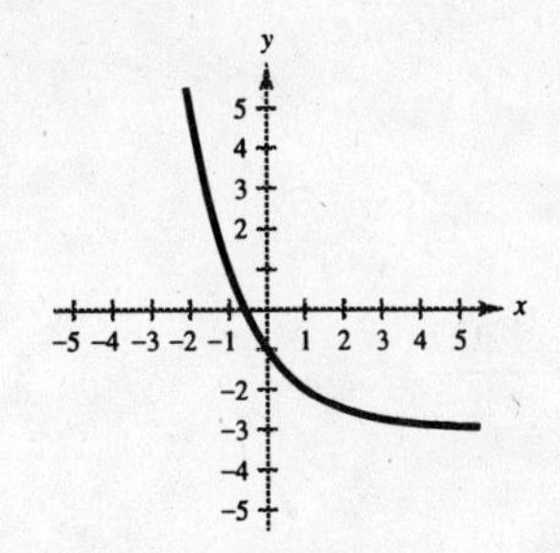

2. $f(x) = \dfrac{1000}{1 + 4e^{-0.2x}}$

These are two horizontal asymptotes.

$y = 1000$ (to the right)

$y = 0$ (to the left)

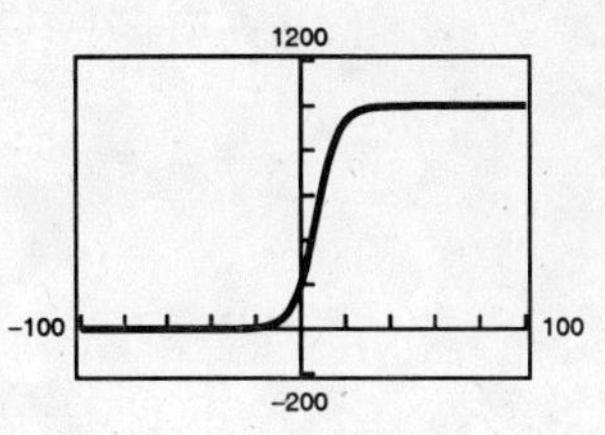

3. $200{,}000 = P\left(1 + \dfrac{0.08}{365}\right)^{(365)20}$

$200{,}000 = P(4.95216) \implies P = \$40{,}386.38$

4. $\log_4 64 = 3 \iff 4^3 = 64$

5. $g(x) = \log_3(x - 2) = \dfrac{\ln(x - 2)}{\ln 3}$

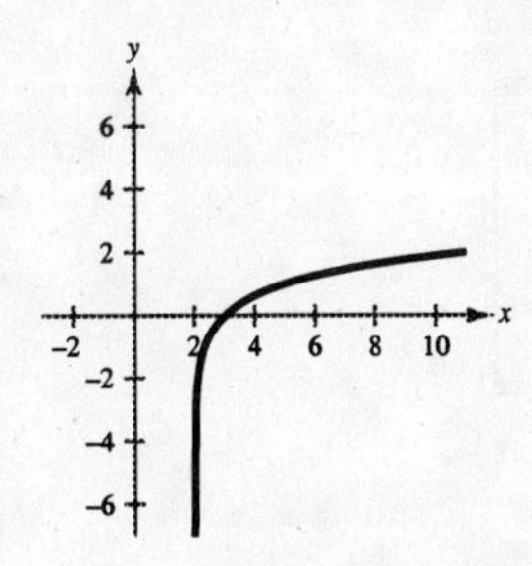

6.
$$\ln\left(\frac{6x^2}{\sqrt{x^2 + 1}}\right) = \ln 6x^2 - \ln\sqrt{x^2 + 1}$$
$$= \ln 6 + \ln x^2 - \ln(x^2 + 1)^{1/2}$$
$$= \ln 6 + 2\ln x - \frac{1}{2}\ln(x^2 + 1)$$

7. $\log_5 25 = \log_5 5^2 = 2\log_5 5 = 2$

8. $-2\ln e^2 + 1 = (-2)(2) + 1 = -3$

9.
$$8 + \frac{1}{4}e^{x/2} = 450$$
$$\frac{1}{4}e^{x/2} = 442$$
$$e^{x/2} = 1768$$
$$\frac{x}{2} = \ln 1768$$
$$x = 2\ln 1768 \approx 14.955$$

10.
$$\left(1 + \frac{0.06}{4}\right)^{4t} = 3$$
$$4t\ln\left(1 + \frac{0.06}{4}\right) = \ln 3$$
$$t = \frac{\ln 3}{4\ln\left(1 + \frac{0.06}{4}\right)} \approx 18.447$$

11. $3.6 - 5\ln(x+4) = 22$

$$-5\ln(x+4) = 18.4$$

$$\ln(x+4) = \frac{-18.4}{5} = -3.68$$

$$x + 4 = e^{-3.68}$$

$$x = -4 + e^{-3.68} \approx -3.975$$

12. $y = ae^{bx}$ $a = \$28{,}000$ $(x = 0)$

$$20{,}000 = 28{,}000e^{b(1)} \Rightarrow \tfrac{5}{7} = e^b$$

$$\Rightarrow b = \ln\tfrac{5}{7} \approx -0.33647$$

$$y = 28{,}000e^{(-0.33647)3} \approx \$10{,}204$$

13. $F(t) = 1 - e^{-t/3}$

(a) $F\left(\frac{1}{2}\right) = 1 - e^{(-1/2)/3} \approx 0.154$

(b) $F(2) = 1 - e^{-2/3} \approx 0.487$

(c) $F(5) = 1 - e^{-5/3} \approx 0.811$

14. (a) If $t = 0$, $p(0) = \dfrac{1200}{4} = 300$

(b) $p(5) = \dfrac{1200}{(1 + 3e^{-1})} \approx 570$

(c) Solving $800 = \dfrac{1200}{1 + 3e^{-t/5}}$ graphically, you obtain $t \approx 9$ years. Hence, at the end of 8 years.

15. Since the graph is symmetric about the y-axis, the function must be even. Since the graph passes through (0, 0), it must be (c). Also, $y = 6$ is a horizontal asymptote.

16. $y = 6.775(1.361)^x$, $3 \le x \le 7$

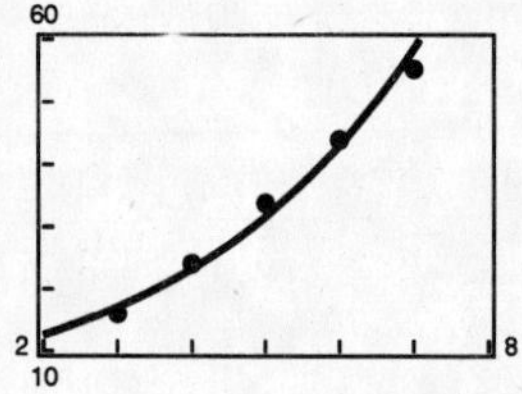

Chapter 1–3 Cumulative Test

1. No, for some x there corresponds two values of y.

2. $f(6) = \dfrac{6}{6-2} = \dfrac{3}{2}$

   $f(2)$ is undefined (division by zero).

   $f(s+2) = \dfrac{s+2}{(s+2)-2} = \dfrac{s+2}{s}$

3. (a) $f\left(-\frac{5}{3}\right) = 3\left(-\frac{5}{3}\right) - 8 = -13$

   (b) $f(-1) = 3(-1)^2 + 9(-1) - 8 = 3 - 9 - 8 = -14$

   (c) $f(0) = 3(0)^2 + 9(0) - 8 = -8$

4. 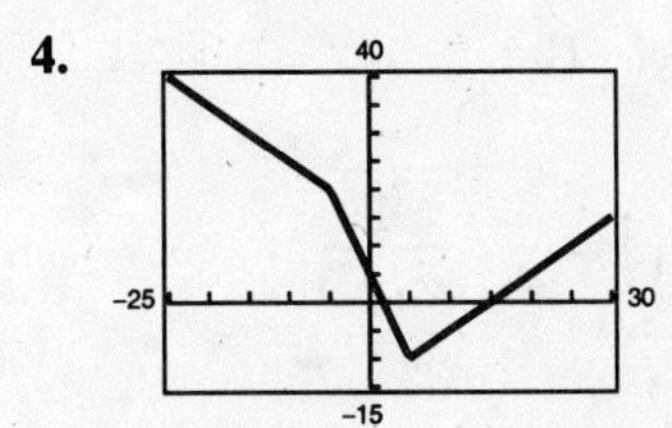

   Decreasing on $(-\infty, 5)$, increasing on $(5, \infty)$

5. 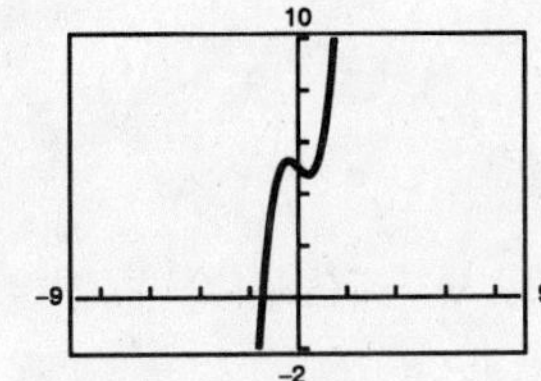

   Relative maximum: $(-0.408, 5.272)$

   Relative minimum: $(0.408, 4.728)$

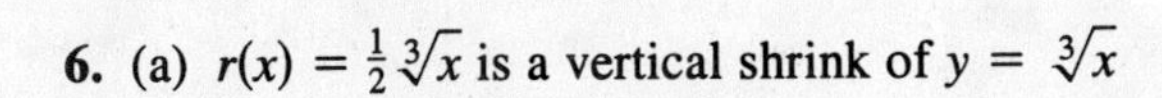

6. (a) $r(x) = \frac{1}{2}\sqrt[3]{x}$ is a vertical shrink of $y = \sqrt[3]{x}$

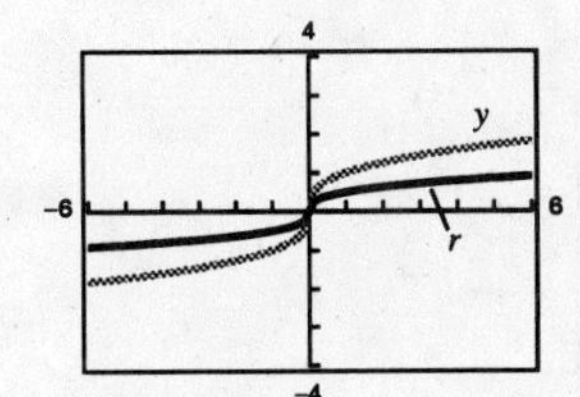

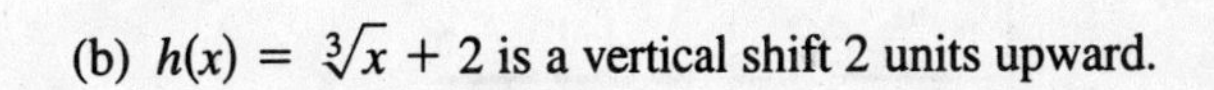

   (b) $h(x) = \sqrt[3]{x} + 2$ is a vertical shift 2 units upward.

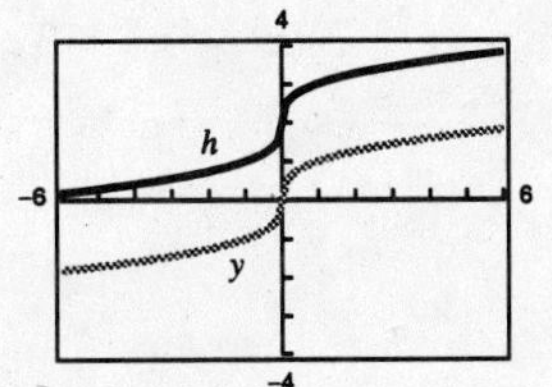

   (c) $g(x) = \sqrt[3]{x+2}$ is a horizontal shift 2 units to the left.

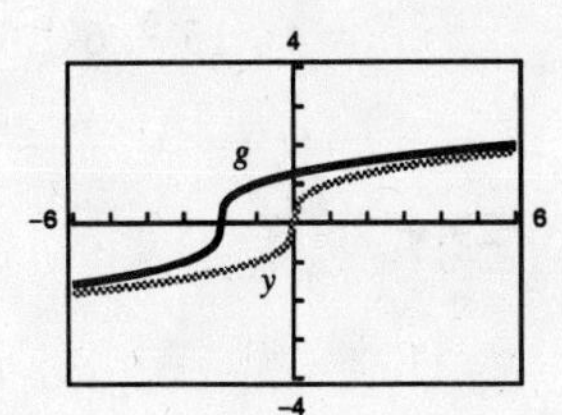

7. $(f+g)(-4) = f(-4) + g(-4) = [-(-4)^2 + 3(-4) - 10] + [4(-4) + 1]$
$= -38 - 15 = -53$

8. $(g - f)(\frac{3}{4}) = \left[4(\frac{3}{4}) + 1\right] - \left[-(\frac{3}{4})^2 + 3(\frac{3}{4}) - 10\right]$
$= 4 - (-8.3125) = 12.3125 = \frac{197}{16}$

9. $(g \circ f)(-2) = g(f(-2)) = g(-20) = 4(-20) + 1 = -79$

10. $(fg)(-1) = f(-1)g(-1) = (-14)(-3) = 42$

11. $f(g(x)) = f\left(\frac{1}{x^2 - 1}\right) = \sqrt{\frac{1}{x^2 - 1}}$.

Domain: $x^2 - 1 > 0$

$x^2 > 1$

$x > 1$ and $x < -1$

12. $y = \sqrt{x - 2}, x \geq 2, y \geq 0$

$x = \sqrt{y - 2}, x \geq 0, y \geq 2$

$x^2 = y - 2 \Rightarrow f^{-1}(x) = x^2 + 2, x \geq 0$

13. $y = 2x^2 - 3x - 5$

Vertex at $x = -\frac{b}{2a} = \frac{3}{4} \Rightarrow$ Vertex $= \left(\frac{3}{4}, -\frac{49}{8}\right)$

Intercepts: $(0, -5), \left(\frac{5}{2}, 0\right), (-1, 0)$

14. $y - 3 = a(x + 2)^2$

$9 - 3 = a(-4 + 2)^2$

$6 = 4a \Rightarrow a = \frac{3}{2} \Rightarrow y = \frac{3}{2}(x + 2)^2 + 3$

$= \frac{3}{2}x^2 + 6x + 9$

15.
$$\begin{array}{r|rrrrr} -3 & 2 & 5 & 0 & 7 & -6 \\ & & -6 & 3 & -9 & 6 \\ \hline & 2 & -1 & 3 & -2 & 0 \end{array}$$

$2x^4 + 5x^3 + 7x - 6 = (x + 3)(x^3 - x^2 + 3x - 2)$

16.

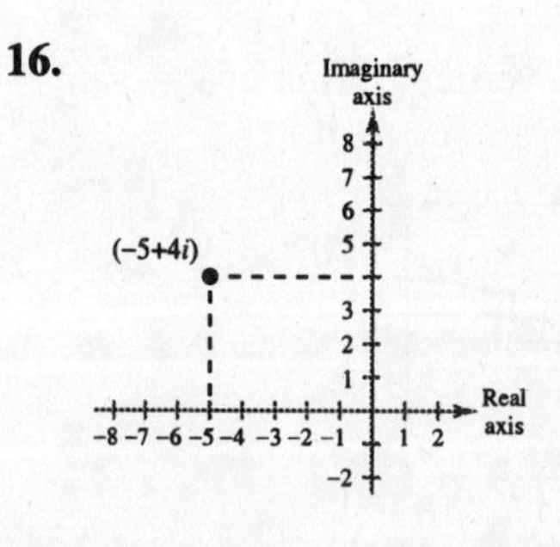

17. $(1 - 3i) - (2i - 3) = 4 - 5i$

18. $\sqrt{-20}\left(\sqrt{18} - \sqrt{-12}\right) = 2\sqrt{5}i\left(3\sqrt{2} - 2\sqrt{3}i\right)$
$= 4\sqrt{15} + 6\sqrt{10}i$

19. $\frac{3i + 7}{1 - 3i} \cdot \frac{1 + 3i}{1 + 3i} = -\frac{2 + 24i}{10} = -\frac{1}{5} + \frac{12}{5}i$

20. $(6 - i)(5i + 4) = 29 + 26i$

21. $(x + 1)(x - 3)(x - 3)(x - 4i)(x + 4i) = (x + 1)(x^2 - 6x + 9)(x^2 + 16)$
$= x^5 - 5x^4 + 19x^3 - 71x^2 + 48x + 144$

22. By synthetic division, -3 and $\frac{1}{2}$ are zeros. Thus, the zeros are $-3, \frac{1}{2}, \pm 2i$

$f(x) = (x + 3)(2x - 1)(x + 2i)(x - 2i)$

23. $f(x) = \dfrac{2x - 1}{x^2 + 3x - 10} = \dfrac{2x - 1}{(x + 5)(x - 2)}$

Asymptotes: $x = -5, x = 2, y = 0$

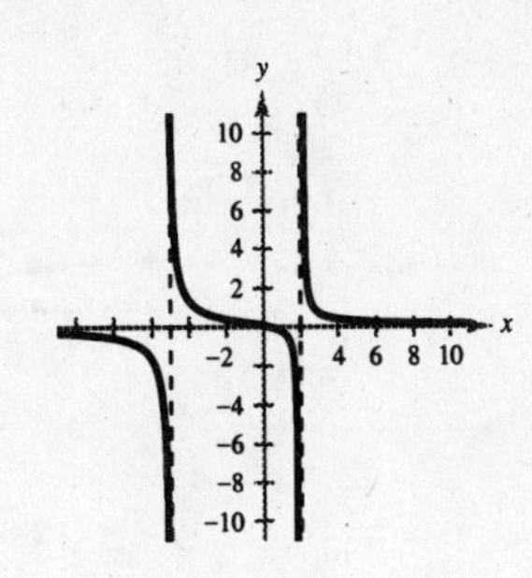

24. $\log_2 64 = 6$ because $2^6 = 64$

25. $\log_2\left(\dfrac{1}{16}\right) = -4$ because $2^{-4} = \dfrac{1}{16}$

26. $\ln e^{10} = 10 \ln e = 10$

27. $\ln\left(\dfrac{1}{e^3}\right) = \ln(e^{-3}) = -3 \ln e = -3$

28. $(1.85)^{3.1} \approx 6.733$

29. $58^{\sqrt{5}} \approx 8772.934$

30. $e^{-20/11} \approx 0.162$

31. $4e^{2.56} \approx 51.743$

32. $f(x) = -3^{x+4} - 5$

Intercept: $(0, -86)$

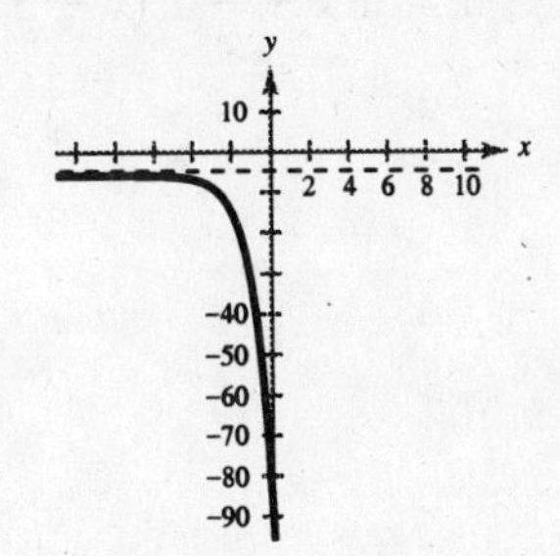

33. $f(x) = -\left(\dfrac{1}{2}\right)^{-x} - 3$

Intercept: $(0, -4)$

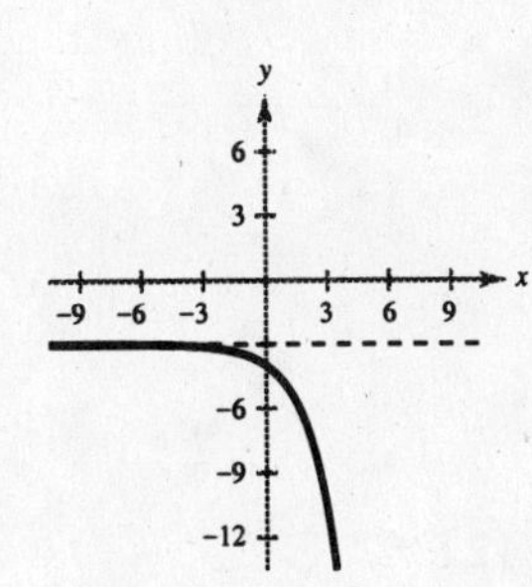

34. $f(x) = 4 + \log_{10}(x - 3)$

Domain: $x > 3$

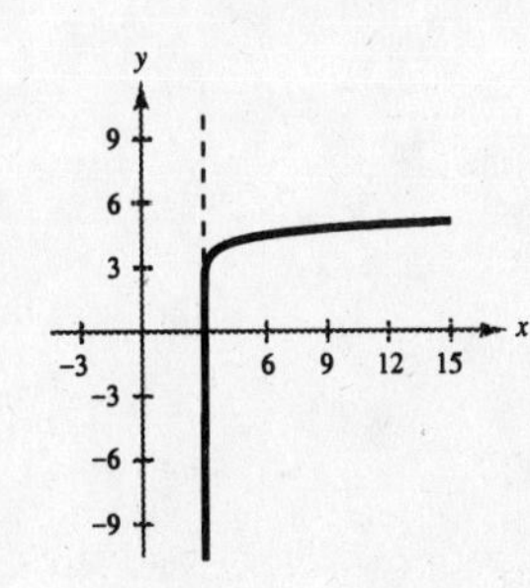

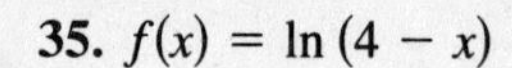

35. $f(x) = \ln(4 - x)$

Domain: $4 - x > 0$ or $x < 4$

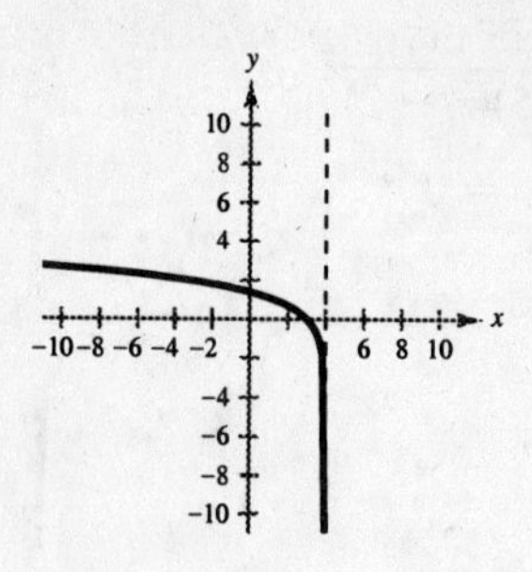

36. $\log_5 21 = \dfrac{\ln 21}{\ln 5} \approx 1.892$

37. $\log_9 6.8 = \dfrac{\ln 6.8}{\ln 9} \approx 0.872$

38. $\log_{3/4}(8.61) = \dfrac{\ln 8.61}{\ln\left(\frac{3}{4}\right)} \approx -7.484$

39. $\log_{7/8}\left(\dfrac{3}{2}\right) = \dfrac{\ln\left(\frac{3}{2}\right)}{\ln\left(\frac{7}{8}\right)} \approx -3.036$

40. $2\ln x - \dfrac{1}{2}\ln(x+5) = \ln x^2 - \ln(x+5)^{1/2} = \ln\left(\dfrac{x^2}{\sqrt{x+5}}\right)$

41. $\ln\left(x\sqrt[3]{x-5}\right) = \ln x + \ln\sqrt[3]{x-5} = \ln x + \dfrac{1}{3}\ln(x-5)$

42. $6e^{2x} = 72$

$$e^{2x} = 12$$
$$2x = \ln 12$$
$$x = \frac{1}{2}\ln 12 \approx 1.242$$

43. $4e^{x-3} + 21 = 30$

$$4e^{x-3} = 9$$
$$e^{x-3} = \frac{9}{4}$$
$$x - 3 = \ln\left(\frac{9}{4}\right)$$
$$x = 3 + \ln\left(\frac{9}{4}\right) \approx 3.811$$

44. $\log_2 x + \log_2 5 = 6$

$$\log_2(5x) = 6$$
$$5x = 2^6$$
$$x = \frac{1}{5}2^6 = \frac{64}{5} = 12.8$$

45. $-3 + \ln 4x = 0$

$$\ln 4x = 3$$
$$4x = e^3$$
$$x = \frac{1}{4}e^3 \approx 5.021$$

46. $\ln\sqrt{x+2} = 3$

$$\frac{1}{2}\ln(x+2) = 3$$
$$\ln(x+2) = 6$$
$$e^6 = x + 2$$
$$x = e^6 - 2 \approx 401.429$$

47. $\ln 4x^2 = 7$

$$4x^2 = e^7$$
$$x^2 = \frac{1}{4}e^7$$
$$x = \pm\frac{1}{2}e^{3.5} \approx \pm 16.558$$

48. (a) Let x and y be the lengths of the sides $2x + 2y = 546 \Longrightarrow y = 273 - x$

$A = xy = x(273 - x)$

(b)

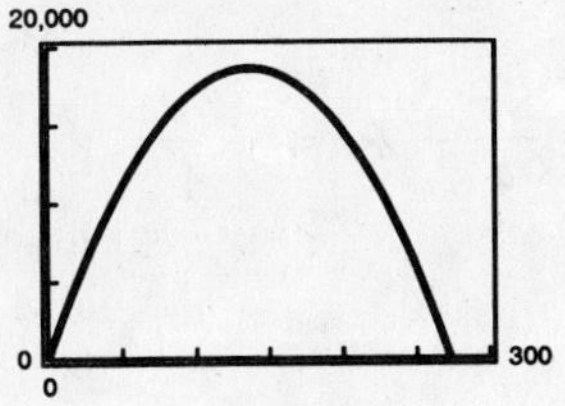

Domain: $0 < x < 273$

(c) If $A = 15000$, then $x = 76.23$ or 196.77

Dimensions in feet:

76.23×196.77 or 196.77×76.23

49. $y = 5.55(1.41)^x = 5.55e^{0.344x}$

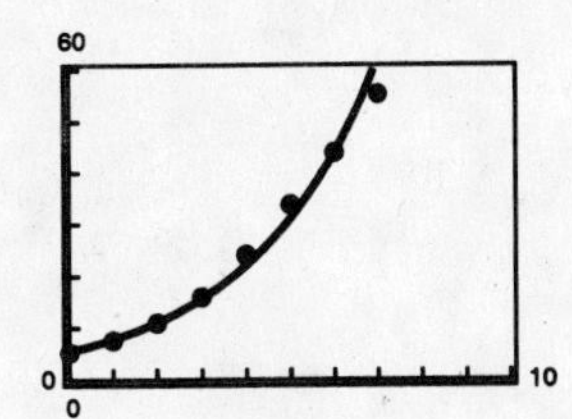

Chapter 4 Chapter Test

1. 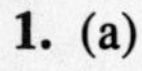
(a)

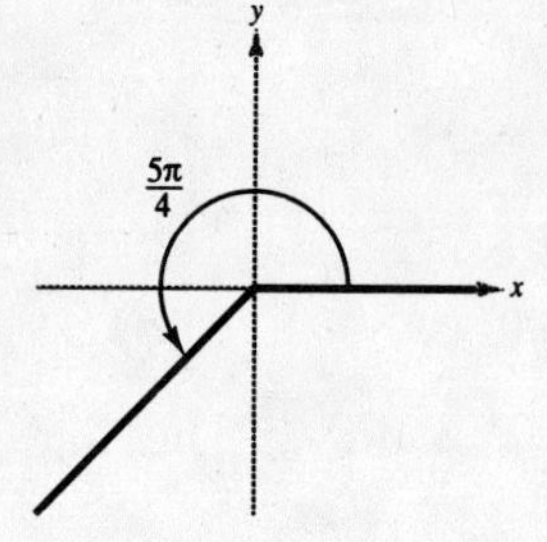

(b) $\frac{5\pi}{4} + 2\pi = \frac{13\pi}{4}$; $\frac{5\pi}{4} - 2\pi = -\frac{3\pi}{4}$

(c) $\frac{5\pi}{4} \cdot \frac{180}{\pi} = 225°$

2. $(90{,}000 \text{ meters/hr})\left(\frac{1}{60} \text{ hr/min}\right)\left(\frac{2\pi \text{ rad}}{2\pi\left(\frac{1}{2}\right) \text{ meters}}\right) = 3000 \text{ rad/min}$

3. $\sin\theta = \frac{4}{\sqrt{17}} = \frac{4\sqrt{17}}{17}$ $\quad$ $\csc\theta = \frac{\sqrt{17}}{4}$

$\cos\theta = -\frac{1}{\sqrt{17}} = -\frac{\sqrt{17}}{17}$ $\quad$ $\sec\theta = -\sqrt{17}$

$\tan\theta = -4$ $\quad$ $\cot\theta = -\frac{1}{4}$

4. $\tan\theta = \frac{11}{6} > 0 \implies \theta$ is in Quadrant I or III.

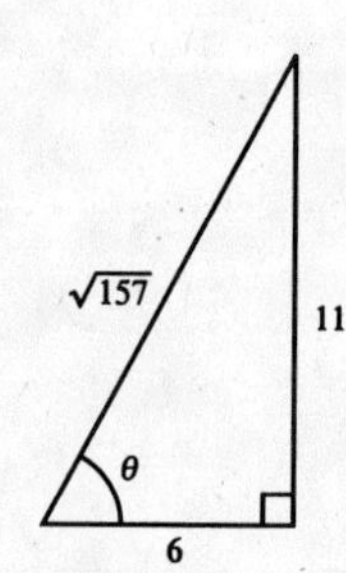

$\cot\theta = \frac{6}{11}$

$\sin\theta = \pm\frac{11\sqrt{157}}{157}$ $\quad$ $\cos\theta = \pm\frac{6\sqrt{157}}{157}$

$\csc\theta = \pm\frac{\sqrt{157}}{11}$ $\quad$ $\sec\theta = \pm\frac{\sqrt{157}}{6}$

5. $\theta = 290° \implies \theta' = 70°$

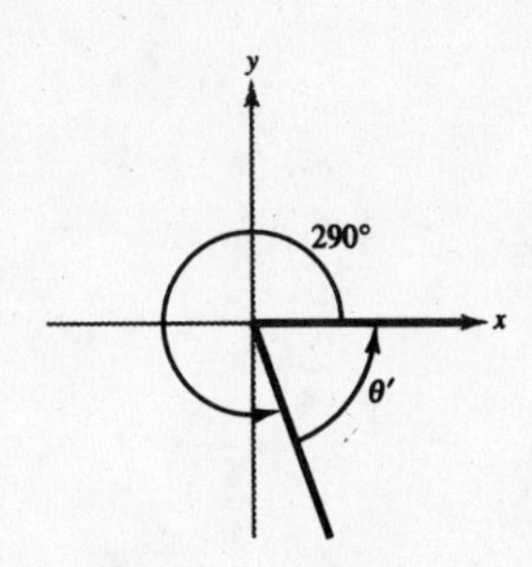

6. $\sec\theta = \dfrac{1}{\cos\theta} < 0 \implies$ Quadrants II or III

$\tan\theta > 0 \implies$ Quadrants I or III

Hence, Quadrant III.

7. If $\cos\theta = -\dfrac{\sqrt{3}}{2}$, then θ is in Quadrant II or III. $\theta = 150°,\ 210°$

8. $\csc\theta = \dfrac{1}{\sin\theta} = 1.030 \implies \sin\theta = \dfrac{1}{1.030}$ and θ in Quadrant I or II. Using a calculator, $\theta = 1.33$, 1.81 radians.

9. $\sec\theta = \dfrac{12}{10} = \dfrac{6}{5}$ and $\tan\theta < 0$ θ in Quadrant IV.

$$\cos\theta = \frac{5}{6}$$

$$\sin\theta = -\frac{\sqrt{11}}{6}$$

$$\tan\theta = -\frac{\sqrt{11}}{5}$$

$$\cot\theta = -\frac{5}{\sqrt{11}} = -\frac{5\sqrt{11}}{11}$$

$$\csc\theta = -\frac{6}{\sqrt{11}} = -\frac{6\sqrt{11}}{11}$$

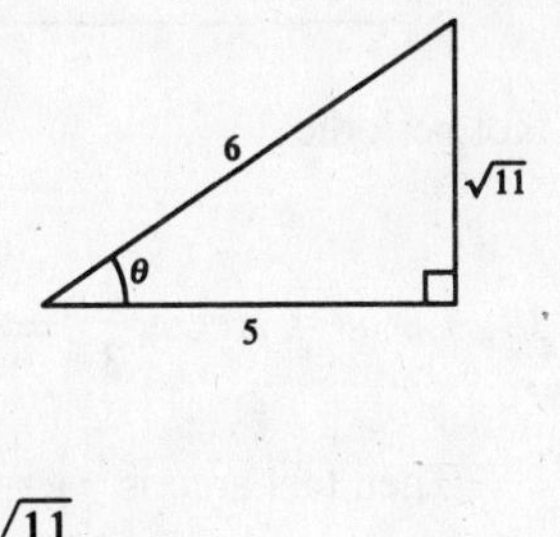

10. Amplitude: 2, shifted $\dfrac{\pi}{4}$ to the right.

11.

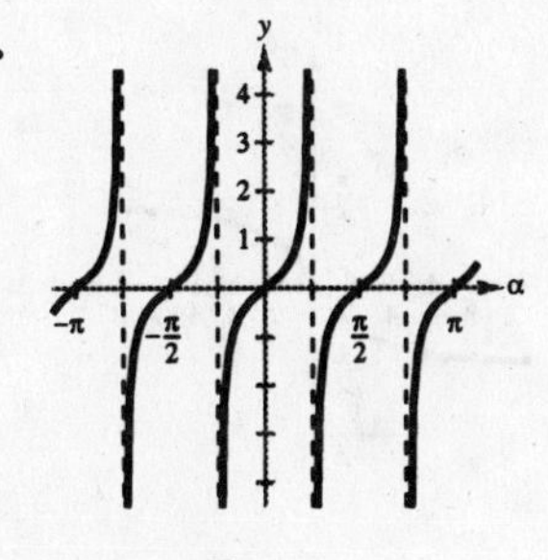

Period: $\dfrac{\pi}{2}$

12. Shifted π to the right.

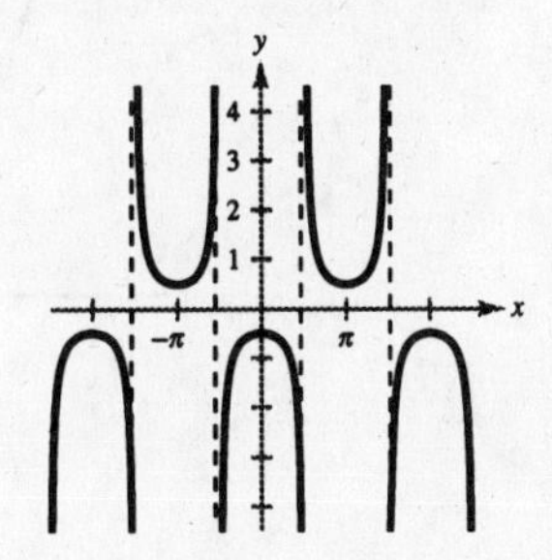

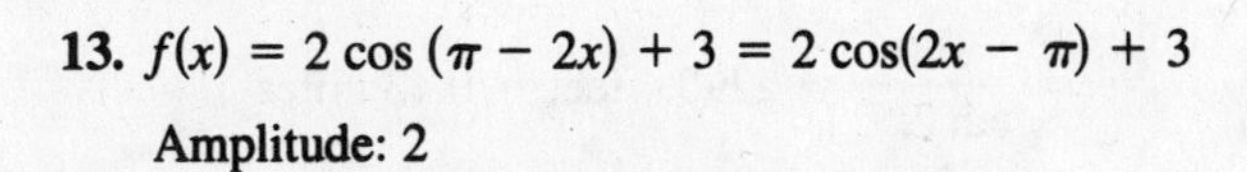

13. $f(x) = 2\cos(\pi - 2x) + 3 = 2\cos(2x - \pi) + 3$

Amplitude: 2

Shifted $\dfrac{\pi}{2}$ to the right, period π

Shifted vertically upward 3

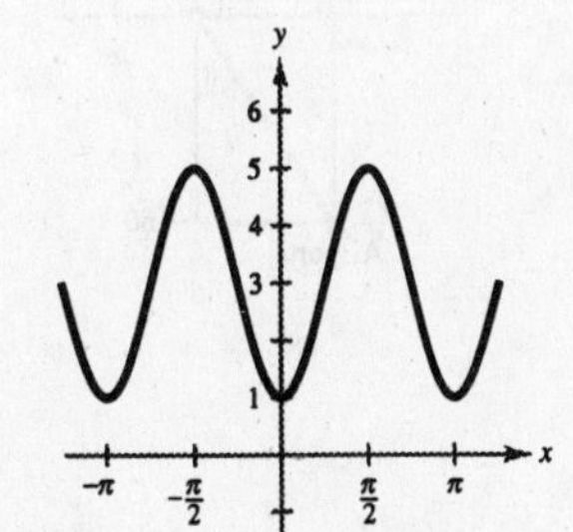

14. $f(x) = 2\csc\left(x + \dfrac{\pi}{2}\right)$

Shifted $\dfrac{\pi}{2}$ to the left

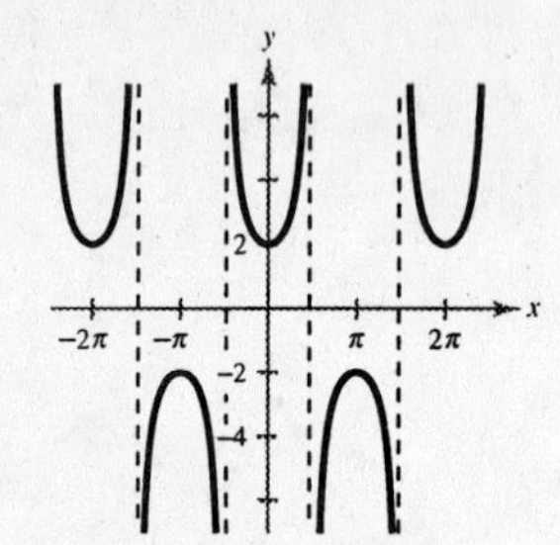

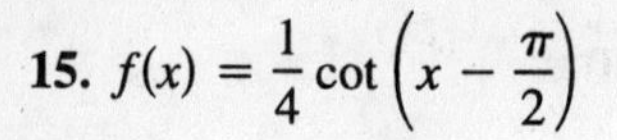

15. $f(x) = \dfrac{1}{4}\cot\left(x - \dfrac{\pi}{2}\right)$

Shifted $\dfrac{\pi}{2}$ to the right

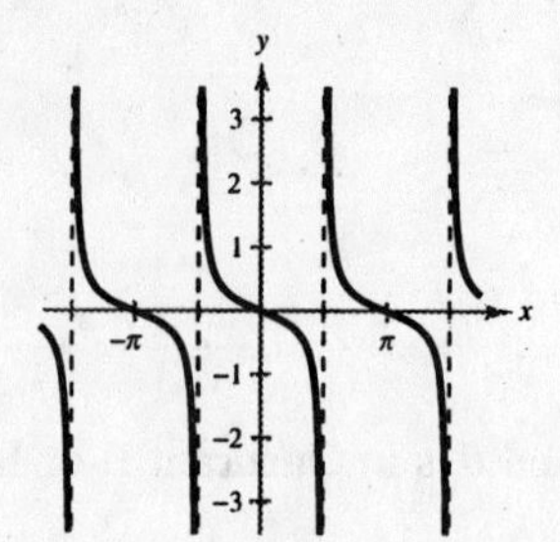

16.

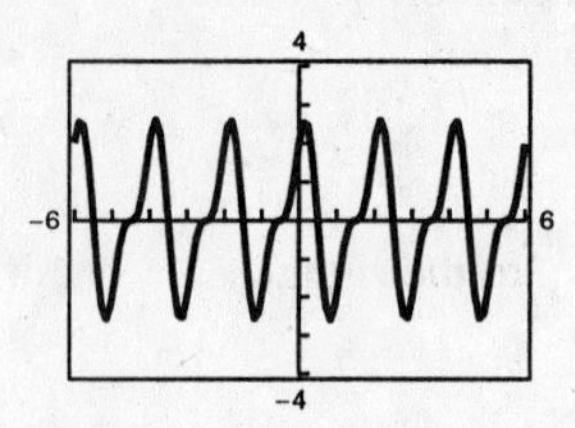

Period is 2

17.

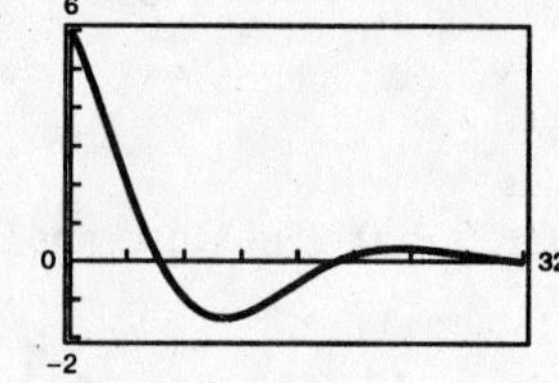

Not periodic

18. Amplitude = 2, reflected in x-axis $\Rightarrow$ $a = -2$.

Period 4π and shifted to the right: $y = -2\sin\left(\dfrac{x}{2} - \dfrac{\pi}{4}\right)$

19. Let $u = \arccos\dfrac{2}{3} \Rightarrow \cos u = \dfrac{2}{3}$.

Then $\tan\left(\arccos\dfrac{2}{3}\right) = \tan u = \dfrac{\sqrt{5}}{2}$.

20.

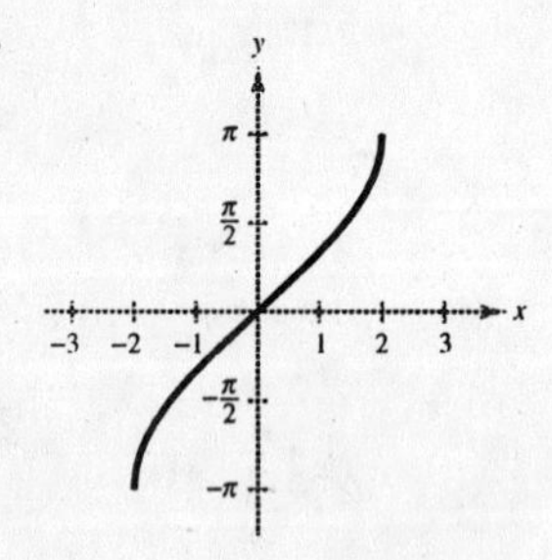

21. $f(x) = 2\arccos x$

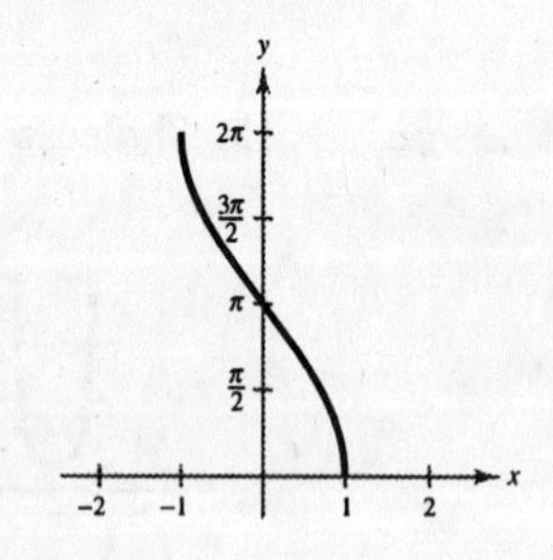

22. $f(x) = \arctan\left(\dfrac{x}{2}\right)$

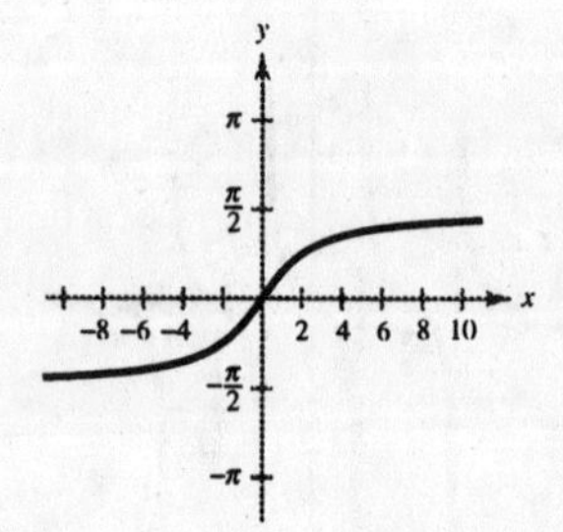

23. $\tan\theta = \dfrac{110}{160} \Rightarrow \theta \approx 34.5°$

Bearing: S 34.5° W

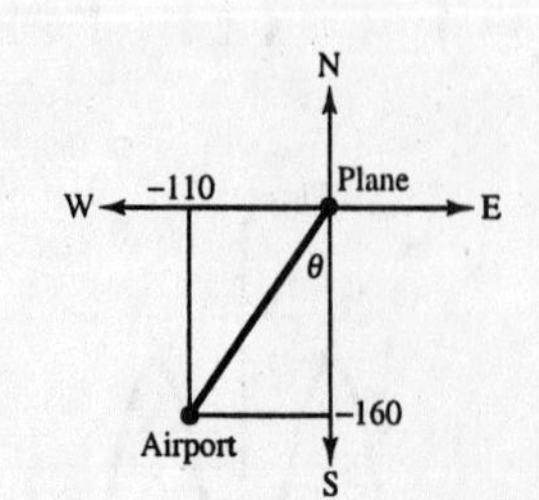

24. $\tan 2.5° = \dfrac{100}{x}$

$x = \dfrac{100}{\tan 2.5°} \approx 2290.4$ feet ≈ 0.43 miles

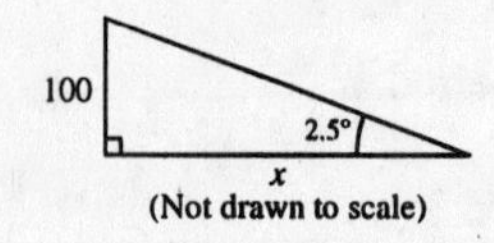

(Not drawn to scale)

Chapter 5 Chapter Test

1. $\tan\theta = \frac{6}{5}$, $\cos\theta < 0 \Rightarrow$ Quadrant III

$\tan^2\theta + 1 = \sec^2\theta \Rightarrow \frac{36}{25} + 1 = \frac{61}{25} = \sec^2\theta \Rightarrow \sec\theta = -\frac{\sqrt{61}}{5}$

$\cos\theta = \frac{-5}{\sqrt{61}} = \frac{-5\sqrt{61}}{61}$

$\sin\theta = \tan\theta\cos\theta = \frac{6}{5}\left(\frac{-5}{\sqrt{61}}\right) = \frac{-6}{\sqrt{61}} = \frac{-6\sqrt{61}}{61}$

$\csc\theta = \frac{\sqrt{61}}{-6}$

$\cot\theta = \frac{5}{6}$

2. $\csc^2\beta(1 - \cos^2\beta) = \frac{1}{\sin^2\beta} \cdot \sin^2\beta = 1$

3. $\frac{\sec^4 x - \tan^4 x}{\sec^2 x + \tan^2 x} = \frac{[(\sec^2 x) + (\tan^2 x)][\sec^2 x - \tan^2 x]}{\sec^2 x + \tan^2 x} = \sec^2 x - \tan^2 x = 1$

4. $\frac{\cos\theta}{\sin\theta} + \frac{\sin\theta}{\cos\theta} = \frac{\cos^2\theta + \sin^2\theta}{\sin\theta\cos\theta} = \frac{1}{\sin\theta\cos\theta} = \csc\theta\sec\theta$

5. Since $\tan^2\theta = \sec^2\theta - 1$ for all θ, then $\tan\theta = -\sqrt{\sec^2\theta - 1}$ in Quadrants II and IV. Thus, $\pi/2 < \theta \leq \pi$ and $3\pi/2 < \theta < 2\pi$.

6.

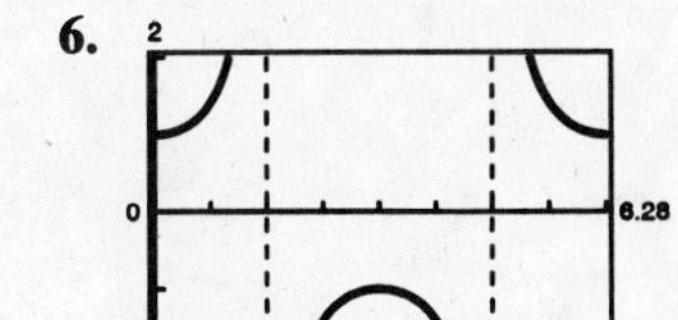

The graph appears equal.

Analytically,
$$\begin{aligned} y_1 &= \cos x + \sin x \tan x \\ &= \cos x + \sin x(\sin x/\cos x) \\ &= \frac{\cos^2 x + \sin^2 x}{\cos x} \\ &= \frac{1}{\cos x} \\ &= \sec x = y_2 \end{aligned}$$

7. $\sin\theta \cdot \sec\theta = \sin\theta \frac{1}{\cos\theta} = \tan\theta$

8. $\sec^2 x\tan^2 x + \sec^2 x = \sec^2 x(\tan^2 x + 1) = \sec^4 x$

9. $$\frac{\csc\alpha + \sec\alpha}{\sin\alpha + \cos\alpha} = \frac{\dfrac{1}{\sin\alpha} + \dfrac{1}{\cos\alpha}}{\sin\alpha + \cos\alpha} = \frac{\dfrac{\cos\alpha + \sin\alpha}{\sin\alpha \cdot \cos\alpha}}{(\sin\alpha + \cos\alpha)}$$
$$= \frac{1}{\sin\alpha\cos\alpha} = \frac{\cos^2\alpha + \sin^2\alpha}{\sin\alpha\cos\alpha}$$
$$= \frac{\cos\alpha}{\sin\alpha} + \frac{\sin\alpha}{\cos\alpha} = \cot\alpha + \tan\alpha$$

10. $$\cos\left(x + \frac{\pi}{2}\right) = \cos x \cos\frac{\pi}{2} - \sin x \sin\frac{\pi}{2}$$
$$= 0 - \sin x = -\sin x$$

11. $$\sin(n\pi + \theta) = \sin n\pi \cdot \cos\theta + \sin\theta \cdot \cos n\pi$$
$$= \cos n\pi \cdot \sin\theta$$
$$= (-1)^n \sin\theta$$

12. $$(\sin x + \cos x)^2 = \sin^2 x + \cos^2 x + 2\sin x\cos x$$
$$= 1 + \sin 2x$$

13. $$\sin\left(-\frac{7\pi}{12}\right) = -\sin\left(\frac{1}{2}\cdot\frac{7\pi}{6}\right) = -\sqrt{\frac{1 - \cos 7\pi/6}{2}} = -\sqrt{\frac{1 + \sqrt{3}/2}{2}}$$
$$= -\frac{\sqrt{2 + \sqrt{3}}}{2} = \frac{-\sqrt{6} - \sqrt{2}}{4}$$
$$\cos\left(\frac{-7\pi}{12}\right) = \cos\left(\frac{1}{2}\cdot\frac{7\pi}{6}\right) = \sqrt{\frac{1 + \cos 7\pi/6}{2}} = -\sqrt{\frac{1 - \sqrt{3}/2}{2}} = \frac{-\sqrt{2 - \sqrt{3}}}{2} = \frac{-\sqrt{6} + \sqrt{2}}{4}$$
$$\tan\left(\frac{-7\pi}{12}\right) = \frac{\sqrt{2 + \sqrt{3}}}{\sqrt{2 - \sqrt{3}}} = \sqrt{\frac{2 + \sqrt{3}}{2 - \sqrt{3}}} = \sqrt{4\sqrt{3} + 7} = \frac{1}{2 - \sqrt{3}} = 2 + \sqrt{3}$$

14. $$\frac{\sin^4 x}{\tan^2 x} = \sin^2 x \cdot \cos^2 x = \frac{1 - \cos 2x}{2} \cdot \frac{1 + \cos 2x}{2}$$
$$= \frac{1}{4}(1 - \cos^2 2x) = \frac{1}{4} - \frac{1}{4}\left(\frac{1 + \cos 4x}{2}\right) = \frac{1}{8} - \frac{1}{8}\cos 4x$$

15. $$3\sin 2\theta \sin 6\theta = 3\tfrac{1}{2}(\cos(2\theta - 6\theta) - \cos(2\theta + 6\theta))$$
$$= \tfrac{3}{2}(\cos 4\theta - \cos 8\theta)$$

16. $$\cos 5\theta + \cos 3\theta = 2\left(\cos\frac{5\theta + 3\theta}{2}\cos\frac{5\theta - 3\theta}{2}\right) = 2\cos 4\theta\cos\theta$$

17. $\tan^2 x + \tan x = 0$

$\tan x(\tan x + 1) = 0$

$\tan x = 0 \Rightarrow x = 0, \pi$

$\tan x + 1 = 0 \Rightarrow \tan x = -1 \Rightarrow x = \frac{3\pi}{4}, \frac{7\pi}{4}$

18. $\sin 2\alpha - \cos \alpha = 0$

$2 \sin \alpha \cos \alpha - \cos \alpha = 0$

$\cos \alpha(2 \sin \alpha - 1) = 0$

$\cos \alpha = 0 \Rightarrow \alpha = \frac{\pi}{2}, \frac{3\pi}{2}$

$2 \sin \alpha - 1 = 0 \Rightarrow \sin \alpha = \frac{1}{2} \Rightarrow \alpha = \frac{\pi}{6}, \frac{5\pi}{6}$

19. $4 \cos^2 x - 3 = 0$

$\cos^2 x = \frac{3}{4}$

$\cos x = \pm\frac{\sqrt{3}}{2}$

$x = \frac{\pi}{6}, \frac{5\pi}{6}, \frac{7\pi}{6}, \frac{11\pi}{6}$

20. $\csc^2 x - \csc x - 2 = 0$

$(\csc x - 2)(\csc x + 1) = 0$

$\csc x - 2 = 0 \Rightarrow \csc x = 2 \Rightarrow \sin x = \frac{1}{2} \Rightarrow x = \frac{\pi}{6}, \frac{5\pi}{6}$

$\csc x + 1 = 0 \Rightarrow \csc x = -1 \Rightarrow \sin x = -1 \Rightarrow x = \frac{3\pi}{2}$

21. Let $y = 5 \cos x - x$ on $[0, 2\pi)$.

The zero is $x \approx 1.306$

22. $\sin 2u = 2 \sin u \cos u = 2\frac{2}{\sqrt{5}} \cdot \frac{1}{\sqrt{5}} = \frac{4}{5}$

$\tan 2u = \frac{2 \tan u}{1 - \tan^2 u} = \frac{2(2)}{1 - 2^2} = -\frac{4}{3}$

23. Since $|\cos x| \leq 1$, $|\cos^2 x + \cos x| \leq 2$ for all x.

24. $\tan 105° = \tan(135° - 30°) = \frac{\tan 135° - \tan 30°}{1 + \tan 135° \tan 30°}$

$$= \frac{-1 - \frac{1}{\sqrt{3}}}{1 + (-1)\left(\frac{1}{\sqrt{3}}\right)} = \frac{-\sqrt{3} - 1}{\sqrt{3} - 1} = \frac{1 + \sqrt{3}}{1 - \sqrt{3}}$$

$$= -2 - \sqrt{3}$$

25. $n = \dfrac{\sin\left(\dfrac{\theta}{2} + \dfrac{\alpha}{2}\right)}{\sin\left(\dfrac{\theta}{2}\right)}$

$$\frac{3}{2} = \frac{\sin\left(\dfrac{\theta}{2} + 30^\circ\right)}{\sin\left(\dfrac{\theta}{2}\right)}$$

$$3\sin\frac{\theta}{2} = 2\left[\sin\frac{\theta}{2}\cos 30^\circ + \cos\frac{\theta}{2}\sin 30^\circ\right]$$

$$3\sin\frac{\theta}{2} = \sqrt{3}\sin\frac{\theta}{2} + \cos\frac{\theta}{2}$$

$$(3 - \sqrt{3})\sin\frac{\theta}{2} = \cos\frac{\theta}{2}$$

$$\tan\frac{\theta}{2} = \frac{1}{3 - \sqrt{3}}$$

$$\frac{\theta}{2} = \arctan\left(\frac{1}{3 - \sqrt{3}}\right) \approx 38.26^\circ$$

$$\theta \approx 76.52^\circ$$

Chapter 6 Chapter Test

1. $C = 180° - A - B = 180° - 37.6 - 98.4 = 44°$

$a = \frac{c}{\sin C} \sin A = \frac{18.2}{\sin 44°} \sin 37.6° \approx 15.99$

$b = \frac{c}{\sin C} \sin B = \frac{18.2}{\sin 44°} \sin 98.4° \approx 25.92$

2. $\cos A = \frac{b^2 + c^2 - a^2}{2bc} = \frac{23.2^2 + 10^2 - 14.9^2}{2(23.2)(10)} = 0.8970 \Longrightarrow A \approx 26.2°$

$\sin C = \frac{\sin A}{a} c = \frac{\sin 26.2°}{14.9}(10) = 0.2963 \Longrightarrow C \approx 17.2°$

$B = 180° - A° - C° = 180° - 26.2° - 17.2° = 136.6°$

3. $a^2 = b^2 + c^2 - 2bc \cos A = 8.4^2 + 11.2^2 - 2(8.4)(11.2)\cos 58° \approx 96.29 \Longrightarrow a \approx 9.8$

$\sin B = \frac{\sin A}{a} b = \frac{\sin 58°}{9.8} 8.4 \approx 0.7269 \Longrightarrow B \approx 46.6°$

$\sin C = \frac{\sin A}{a} c = \frac{\sin 58°}{9.8} 11.2 \approx 0.9692 \Longrightarrow C \approx 75.7°$

4. $h = 28 \sin 24.6° \approx 11.7 \Longrightarrow$ Two solutions

$\sin B = \frac{\sin A}{a} b = \frac{\sin 24.6}{15.6} 28 \approx 0.7472 \Longrightarrow B \approx 48.3°$ or $131.7°$

For $B_1 = 48.3°$, $C = 180° - 48.3° - 24.6° = 107.1$

and $c = \frac{a}{\sin A} \sin C = \frac{15.6}{\sin 24.6} \sin 107.1° \approx 35.8$

For $B_2 = 131.7°$, $C = 180° - 131.7° - 24.6° = 23.7°$

and $c = \frac{a}{\sin A} \sin C = \frac{15.6}{\sin 24.6} \sin 23.7° \approx 15.1$

5. No triangle possible $(5.2 \leq 10.1)$

6. $\sin B = \frac{\sin A}{a} b = \frac{\sin 150°}{9.4} 4.8 \approx 0.2553 \Longrightarrow B \approx 14.8°$

$C = 180° - A - B = 15.2$

$c = \frac{a}{\sin A} \sin C \approx 4.9$

7. $A = 56°$, $a = 1070$, $c = 650$

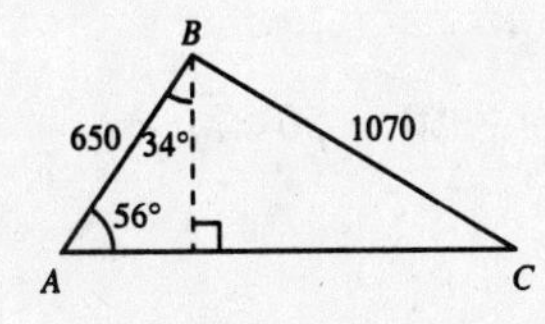

$\sin C = \frac{\sin A}{a} c = \frac{\sin 56°}{1070} 650 \Longrightarrow C \approx 30.2°$

$B = 180 - A - C = 93.8°$

$93.8 - 34° = 59.8°$

Bearing $S\ 59.8°$ E

8. Law of Cosines:

$a^2 = b^2 + c^2 - 2bc\cos\theta = 565^2 + 480^2 - 2(565)(480)\cos 80° = 455,438.2 \implies a \approx 674.9$ ft

9. $\mathbf{w} = \langle 4 - (-8), 1 - (-12)\rangle = \langle 12, 13\rangle$

$\|\mathbf{w}\| = \sqrt{12^2 + 13^2} = \sqrt{313} \approx 17.7$

10. Unit vector $= \dfrac{\mathbf{v}}{\|\mathbf{v}\|} = \dfrac{1}{\sqrt{49 + 16}}\langle 7, 4\rangle = \left\langle \dfrac{7}{\sqrt{65}}, \dfrac{4}{\sqrt{65}}\right\rangle$

11. (a) $2\mathbf{v} + \mathbf{u} = 2\langle -2, 4\rangle + \langle 0, -4\rangle = \langle -4, 4\rangle$

(b) $\mathbf{u} - 3\mathbf{v} = \langle 0, -4\rangle - 3\langle -2, 4\rangle = \langle 6, -16\rangle$

(c) $5\mathbf{u} - \mathbf{v} = 5\langle 0, -4\rangle - \langle -2, 4\rangle = \langle 2, -24\rangle$

12. (a) $2\mathbf{v} + \mathbf{u} = 2\langle 1, 5\rangle + \langle -2, -3\rangle = \langle 0, 7\rangle$

(b) $\mathbf{u} - 3\mathbf{v} = \langle -2, -3\rangle - 3\langle 1, 5\rangle = \langle -5, -18\rangle$

(c) $5\mathbf{u} - \mathbf{v} = 5\langle -2, -3\rangle - \langle 1, 5\rangle = \langle -11, -20\rangle$

13. (a) $2\mathbf{v} + \mathbf{u} = 2(6\mathbf{i} + 9\mathbf{j}) + (\mathbf{i} - \mathbf{j}) = 13\mathbf{i} + 17\mathbf{j}$

(b) $\mathbf{u} - 3\mathbf{v} = (\mathbf{i} - \mathbf{j}) - 3(6\mathbf{i} + 9\mathbf{j}) = -17\mathbf{i} - 28\mathbf{j}$

(c) $5\mathbf{u} - \mathbf{v} = 5(\mathbf{i} - \mathbf{j}) - (6\mathbf{i} + 9\mathbf{j}) = -\mathbf{i} - 14\mathbf{j}$

14. (a) $2\mathbf{v} + \mathbf{u} = 2(-\mathbf{i} - 2\mathbf{j}) + (2\mathbf{i} + 3\mathbf{j}) = -\mathbf{j}$

(b) $\mathbf{u} - 3\mathbf{v} = (2\mathbf{i} + 3\mathbf{j}) - 3(-\mathbf{i} - 2\mathbf{j}) = 5\mathbf{i} + 9\mathbf{j}$

(c) $5\mathbf{u} - \mathbf{v} = 5(2\mathbf{i} + 3\mathbf{j}) - (-\mathbf{i} - 2\mathbf{j}) = 11\mathbf{i} + 17\mathbf{j}$

15. $12\dfrac{\langle 3, -5\rangle}{\|\langle 3, -5\rangle\|} = \dfrac{12}{\sqrt{34}}\langle 3, -5\rangle = \left\langle \dfrac{36}{\sqrt{34}}, \dfrac{-60}{\sqrt{34}}\right\rangle$

16. $250(\cos 45°\mathbf{i} + \sin 45°\mathbf{j})$ first force

$130(\cos(-60°)\mathbf{i} + \sin(-60°)\mathbf{j})$ second force

Resultant: $\left[250\left(\dfrac{\sqrt{2}}{2}\right) + 130\left(\dfrac{1}{2}\right)\right]\mathbf{i} + \left[250\left(\dfrac{\sqrt{2}}{2}\right) + 130\left(-\dfrac{\sqrt{3}}{2}\right)\right]\mathbf{j}$

$= (125\sqrt{2} + 65)\mathbf{i} + (125\sqrt{2} - 65\sqrt{3})\mathbf{j}$

Magnitude: $\sqrt{(125\sqrt{2} + 65)^2 + (125\sqrt{2} - 65\sqrt{3})^2} \approx 250.15$

Direction: $\theta = \arctan\left(\dfrac{125\sqrt{2} - 65\sqrt{3}}{125\sqrt{2} + 65}\right) \implies \theta \approx 14.9°$

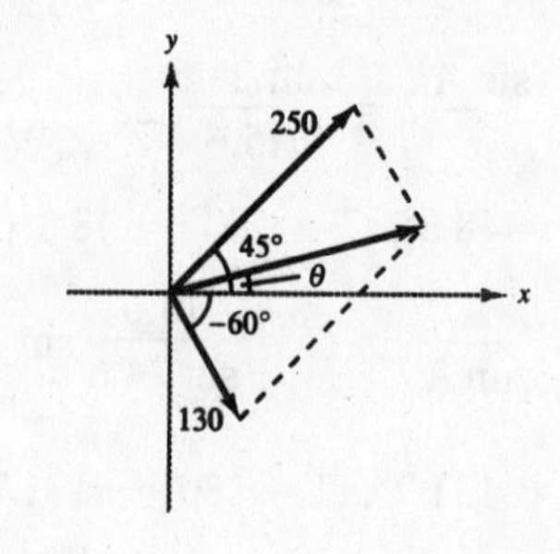

17. $\cos\theta = \dfrac{\mathbf{u}\cdot\mathbf{v}}{\|\mathbf{u}\|\,\|\mathbf{v}\|} = \dfrac{-8}{\sqrt{53}(4)} \implies \theta \approx 105.9°$

18. No, the dot product is 24, not 0.

19. $\text{proj}_{\mathbf{v}}\mathbf{u} = \dfrac{-37}{26}\langle -5, -1\rangle = \left\langle \dfrac{185}{26}, \dfrac{37}{26}\right\rangle = \mathbf{w}_1$

$\mathbf{w}_2 = \mathbf{u} - \mathbf{w}_1 = \langle 6, 7\rangle - \left\langle \dfrac{185}{26}, \dfrac{37}{26}\right\rangle = \left\langle \dfrac{-29}{26}, \dfrac{145}{26}\right\rangle$

$\mathbf{u} = \mathbf{w}_1 + \mathbf{w}_2$

20. $|z| = 2\sqrt{2}$. $z = 2\sqrt{2}\left(\cos\dfrac{3\pi}{4} + i\sin\dfrac{3\pi}{4}\right)$

21. $100(\cos 240° + i\sin 240°) = -50 - 50\sqrt{3}i$

22. $24(\cos 330° + i\sin 330°) = 12\sqrt{3} - 12i$

23. $\left[3\left(\cos\frac{5\pi}{6} + i\sin\frac{5\pi}{6}\right)\right]^8 = 3^8\left(\cos\frac{40\pi}{6} + i\sin\frac{40\pi}{6}\right)$

$= 3^8\left(-\frac{1}{2} + \frac{\sqrt{3}}{2}i\right)$

$= -3280.5 + 3280.5\sqrt{3}i$

24. $(3 - 3i)^6 = \left[3\sqrt{2}\left(\cos\frac{7\pi}{4} + i\sin\frac{7\pi}{4}\right)\right]^6$

$= 5832\left(\cos\frac{42\pi}{4} + i\sin\frac{42\pi}{4}\right) = 5832i$

25. $128(1 + \sqrt{3}i) = 256\left(\frac{1}{2} + \frac{\sqrt{3}}{2}i\right) = 256\left(\cos\frac{\pi}{3} + i\sin\frac{\pi}{3}\right).$

4th roots: $\sqrt[4]{256}\left(\cos\frac{\frac{\pi}{3} + 2\pi k}{4} + i\sin\frac{\frac{\pi}{3} + 2k\pi}{4}\right), k = 0, 1, 2, 3$

4 roots are: $4\left(\cos\frac{\pi}{12} + i\sin\frac{\pi}{12}\right)$

$4\left(\cos\frac{7\pi}{12} + i\sin\frac{7\pi}{12}\right)$

$4\left(\cos\frac{13\pi}{12} + \text{i}\sin\frac{13\pi}{12}\right)$

$4\left(\cos\frac{19\pi}{12} + i\sin\frac{19\pi}{12}\right)$

26. $x^4 = 625i$. Fourth roots of $625i = 625\left(\cos\frac{\pi}{2} + i\sin\frac{\pi}{2}\right)$

$\sqrt[4]{625}\left(\cos\left(\frac{\frac{\pi}{2} + 2\pi k}{4}\right) + i\sin\left(\frac{\frac{\pi}{2} + 2\pi k}{4}\right)\right), k = 0, 1, 2, 3$

4 roots are $5\left(\cos\frac{\pi}{8} + i\sin\frac{\pi}{8}\right)$

$5\left(\cos\frac{5\pi}{8} + i\sin\frac{5\pi}{8}\right)$

$5\left(\cos\frac{9\pi}{8} + i\sin\frac{9\pi}{8}\right)$

$5\left(\cos\frac{13\pi}{8} + i\sin\frac{13\pi}{8}\right)$

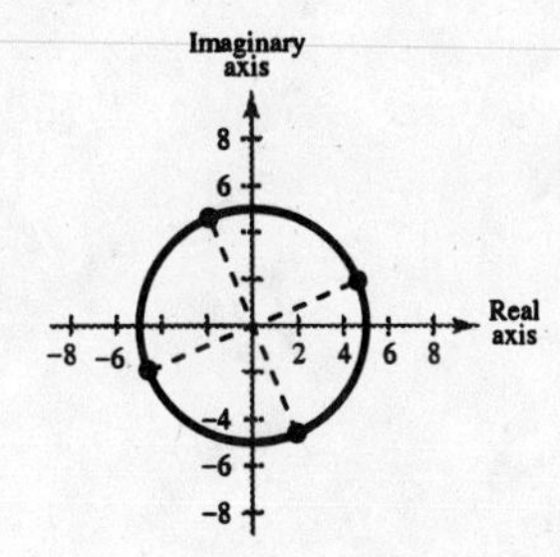

Chapters 4–6 Cumulative Test

1. (a)

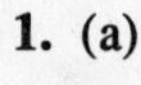

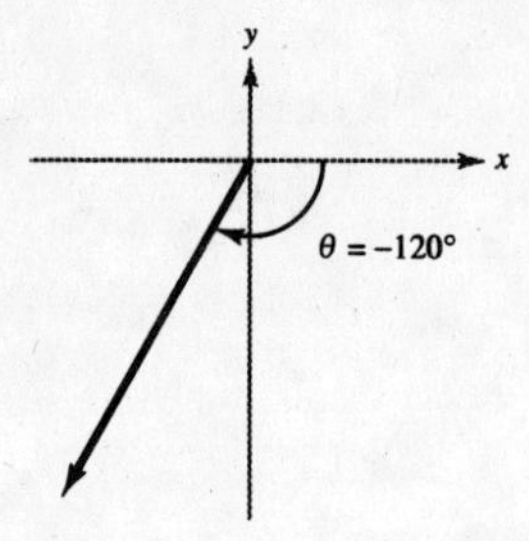

(b) $-120° + 360° = 240°$

(c) $-120° \cdot \dfrac{\pi \text{ rad}}{180 \text{ deg}} = -\dfrac{2\pi}{3}$

(d) $\theta' = 60°$

(e) $\sin\theta = -\dfrac{\sqrt{3}}{2}, \quad \cos\theta = -\dfrac{1}{2}, \quad \tan\theta = \sqrt{3}$

$\csc\theta = -\dfrac{2}{\sqrt{3}} = -\dfrac{2\sqrt{3}}{3}, \quad \sec\theta = -2,$

$\cot\theta = \dfrac{1}{\sqrt{3}} = \dfrac{\sqrt{3}}{3}$

2. 2.35 radians $\left(\dfrac{180}{\pi}\right) \approx 134.6°$

3. $\tan\theta = -\frac{4}{3} \implies \sec^2\theta = \tan^2\theta + 1 = \frac{16}{9} + 1 = \frac{25}{9}$

$\implies \sec\theta = \frac{5}{3}$ (Quadrant IV) $\implies \cos\theta = \frac{3}{5}$

4. $f(x) = 3 - 2\sin\pi x$

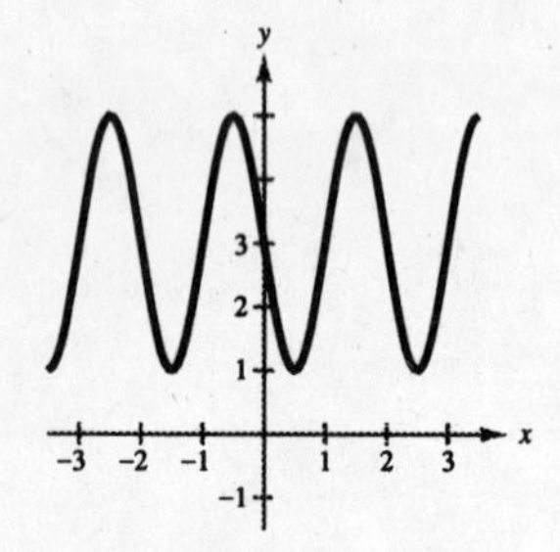

5. $f(x) = \dfrac{1}{2}\tan\left(x - \dfrac{\pi}{2}\right)$

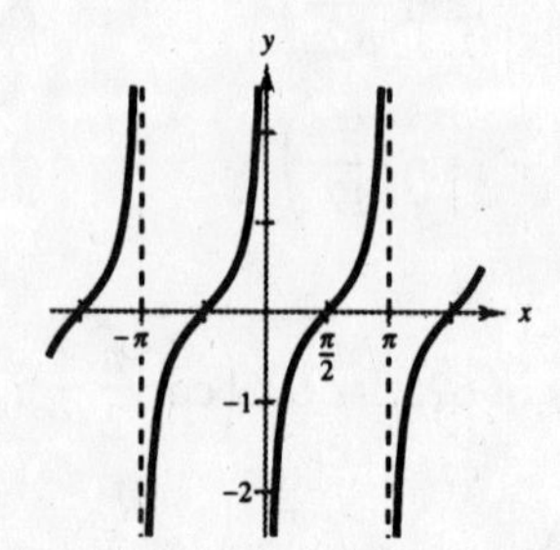

6. $f(x) = \frac{1}{2}\sec(x + \pi)$

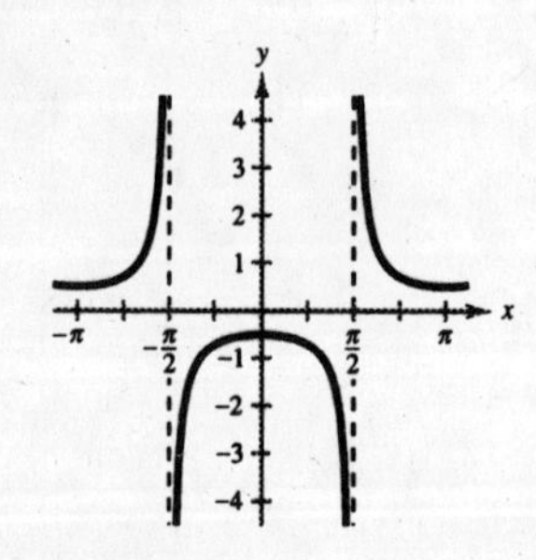

7. Amplitude: 3
Cosine curve reflected about the x-axis.
Period: 2 $\implies h(x) = -3\cos(\pi x)$
Answer: $a = -3, b = \pi, c = 0$

8. $\tan(\arctan 6.7) = 6.7$

9. $\tan\left(\arcsin\frac{3}{5}\right) = \tan y = \frac{3}{4}$

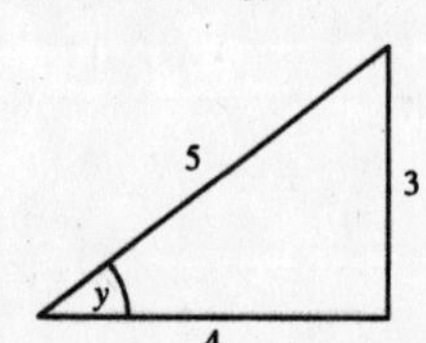

10. Let $u = \arccos 2x \implies \cos u = 2x$. Then:

$$\sin(\arccos 2x) = \sin u$$
$$= \sqrt{1 - 4x^2}$$

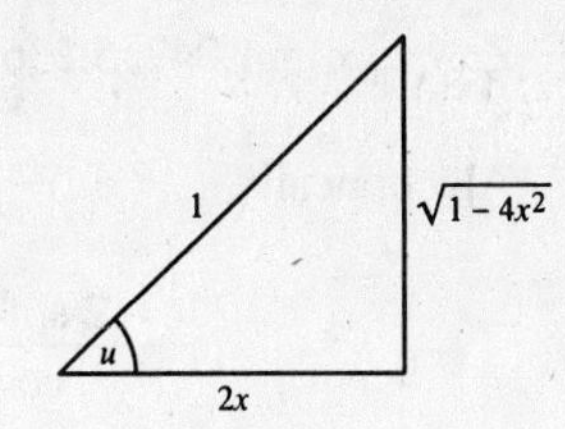

11. $h = 22 \sin 74° \approx 21.1$ feet

12. $$\frac{\sin\theta - 1}{\cos\theta} - \frac{\cos\theta}{\sin\theta - 1} = \frac{\sin^2\theta - 2\sin\theta + 1 - \cos^2\theta}{\cos\theta(\sin\theta - 1)}$$
$$= \frac{\sin^2\theta - 2\sin\theta + \sin^2\theta}{\cos\theta(\sin\theta - 1)}$$
$$= \frac{2\sin\theta(\sin\theta - 1)}{\cos\theta(\sin\theta - 1)} = 2\tan\theta$$

13. $\cot^2\alpha(\sec^2\alpha - 1) = \cot^2\alpha(\tan^2\alpha) = 1$

14. $$\sin(x + y)\sin(x - y) = [\sin x\cos y + \cos x\sin y][\sin x\cos y - \sin y\cos x]$$
$$= \sin^2 x\cos^2 y - \sin^2 y\cos^2 x$$
$$= \sin^2 x(1 - \sin^2 y) - \sin^2 y(1 - \sin^2 x)$$
$$= \sin^2 x - \sin^2 x\sin^2 y - \sin^2 y + \sin^2 y\sin^2 x$$
$$= \sin^2 x - \sin^2 y$$

15. $$2\cos^2\beta - \cos\beta = 0$$
$$\cos\beta(2\cos\beta - 1) = 0$$
$$\cos\beta = 0 \implies \beta = \frac{\pi}{2}, \frac{3\pi}{2}$$
$$2\cos\beta - 1 = 0 \implies \cos\beta = \frac{1}{2} \implies \beta = \frac{\pi}{3}, \frac{5\pi}{3}$$

16. $3\tan^2\theta - \cot\theta = 0$
$$3\tan\theta - \frac{1}{\tan\theta} = 0$$
$$3\tan^2\theta - 1 = 0$$
$$\tan\theta = \pm\frac{1}{\sqrt{3}} \implies \theta = \frac{\pi}{6}, \frac{5\pi}{6}, \frac{7\pi}{6}, \frac{11\pi}{6}$$

17. Graph $y = \cos^2 x - 5\cos x - 1$ on $[0, 2\pi)$.

Roots are $x \approx 1.765, 4.519$

18.

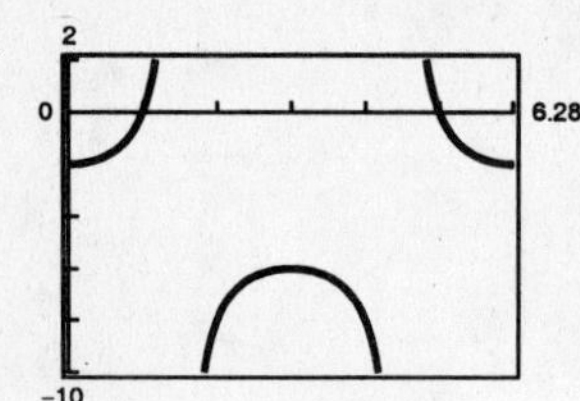

Zeros: $x \approx 1.047, 5.236$

Algebraically,

$$\frac{1 + \sin x}{\cos x} + \frac{\cos x}{1 + \sin x} = 4$$

$$\frac{1 + 2\sin x + \sin^2 x + \cos^2 x}{\cos x(1 + \sin x)} = 4$$

$$\frac{2 + 2\sin x}{\cos x(1 + \sin x)} = 4$$

$$\frac{2}{\cos x} = 4$$

$$\cos x = \frac{1}{2}$$

$$x = \frac{\pi}{3}, \frac{5\pi}{3}$$

19.

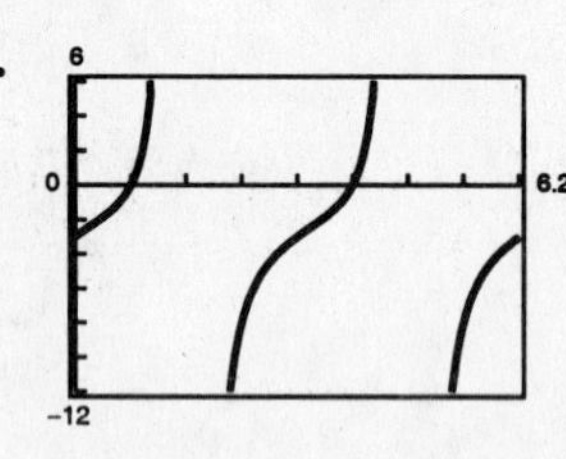

Zeros: $x \approx 0.785, 3.927$

Algebraically,

$$\tan^3 x - \tan^2 x + 3\tan x - 3 = 0$$

$$\tan^2 x(\tan x - 1) + 3(\tan x - 1) = 0$$

$$(\tan^2 x + 3)(\tan x - 1) = 0$$

$$\tan x = 1 \implies x = \frac{\pi}{4}, \frac{5\pi}{4}$$

20. Let $u = \arccos 2x \implies \cos u = 2x$

$$\cos(2\arccos 2x) = \cos 2u = \cos^2 u - \sin^2 u = 4x^2 - (1 - 4x^2) = 8x^2 - 1$$

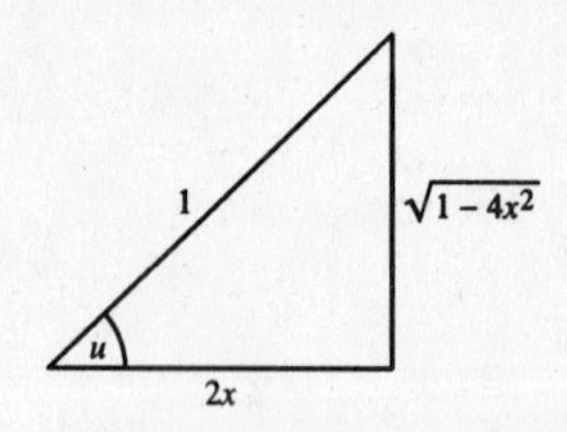

21. Let $u = \arctan x \implies \tan u = x$

$$\sin(2\arctan x) = \sin 2u$$

$$= 2\sin u \cos u$$

$$= 2\frac{x}{\sqrt{x^2+1}} \cdot \frac{1}{\sqrt{x^2+1}} = \frac{2x}{x^2+1}$$

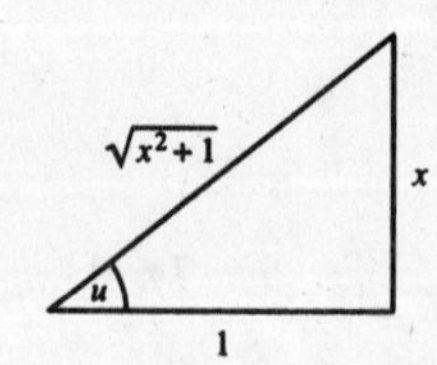

22. $\theta = 67° \, 30' = 67.5° = \frac{1}{2}(135)$. Quadrant I

$$\sin\theta = \sin\left(\frac{1}{2}(135)\right) = \sqrt{\frac{1-\cos 135°}{2}} = \sqrt{\frac{1+\sqrt{2}/2}{2}} = \frac{1}{2}\sqrt{2+\sqrt{2}}$$

$$\cos\theta = \cos\left(\frac{1}{2}(135)\right) = \sqrt{\frac{1+\cos 135°}{2}} = \sqrt{\frac{1-\sqrt{2}/2}{2}} = \frac{1}{2}\sqrt{2-\sqrt{2}}$$

$$\tan\theta = \frac{\sin\theta}{\cos\theta} = \frac{\sqrt{2+\sqrt{2}}}{\sqrt{2-\sqrt{2}}} = \sqrt{2}+1$$

23. $\cos 8x + \cos 4x = 2\cos\left(\frac{8x+4x}{2}\right)\cos\left(\frac{8x-4x}{2}\right)$

$= 2\cos 6x \cos 2x$

24. $\tan x(1-\sin^2 x) = \frac{\sin x}{\cos x}\cos^2 x = \sin x \cos x$

$= \frac{1}{2}(2\sin x\cos x)$

$= \frac{1}{2}\sin 2x$

25. $\sin 3\theta \sin\theta = \frac{1}{2}[\cos(3\theta-\theta) - \cos[3\theta+\theta]]$

$= \frac{1}{2}(\cos 2\theta - \cos 4\theta)$

26. $\sin 3x \cos 2x = \frac{1}{2}(\sin(3x+2x) + \sin(3x-2x))$

$= \frac{1}{2}(\sin 5x + \sin x)$

27. $\frac{2\cos 3x}{\sin 4x - \sin 2x} = \frac{2\cos 3x}{2\cos 3x \cdot \sin x} = \frac{1}{\sin x} = \csc x$

28. $\sin B = \frac{\sin A}{a}b = 0.2569 \Rightarrow B \approx 14.9°$

$C = 180° - 46° - 14.9° = 119.1°$

$c = \frac{a}{\sin A}(\sin C) \approx 17.0$

[Note: answers for B and C could be switched]

29. $a^2 = b^2 + c^2 - 2bc\cos A = 25.436 \Rightarrow a \approx 5.04$

$\sin B = \frac{\sin A}{a} \cdot b = 0.8 \Rightarrow B \approx 52.5°$

$C = 180° - 52.5 - 30° = 97.5°$

30. $B = 180 - 24° - 101° = 55°$

$b = \frac{a}{\sin A}\sin B \approx 20.1$

$c = \frac{a}{\sin A}\sin C \approx 24.1$

31. $\cos A = \frac{b^2+c^2-a^2}{2bc} = 0.8982 \Rightarrow A \approx 26.1°$

$\cos B = \frac{a^2+c^2-b^2}{2ac} = 0.8355 \Rightarrow B \approx 33.3°$

$C = 180° - 26.1° - 33.3° = 120.6°$

32. $A = \frac{1}{2}bh = \frac{1}{2}19 \cdot 14\sin 82° \approx 131.7$ sq inches

33. $s = \dfrac{a + b + c}{2} = 22$

$$\text{Area} = \sqrt{s(s-a)(s-b)(s-c)}$$
$$= \sqrt{22(11)(6)(5)}$$
$$\approx 85.2 \text{ sq inches}$$

34. $\mathbf{u} = \langle 3, 5 \rangle = 3\mathbf{i} + 5\mathbf{j}$

35. $\|\mathbf{v}\| = \sqrt{2}$. Unit vector: $\left\langle \dfrac{1}{\sqrt{2}}, \dfrac{1}{\sqrt{2}} \right\rangle$

36. $\mathbf{u} \cdot \mathbf{v} = 3(1) + 4(-2) = -5$

37. $\text{proj}_{\mathbf{v}} \mathbf{u} = \dfrac{8 - 10}{26}\langle 1, 5 \rangle = \left\langle -\dfrac{1}{13}, -\dfrac{5}{13} \right\rangle = \mathbf{w}_1$

$$\mathbf{w}_2 = \mathbf{u} - \mathbf{w}_1 = \langle 8, -2 \rangle - \left\langle -\frac{1}{13}, -\frac{5}{13} \right\rangle = \left\langle \frac{105}{13}, -\frac{21}{13} \right\rangle$$

38. $|z| = 3\sqrt{2}, \theta = \dfrac{3\pi}{4}$: $3\sqrt{2}\left(\cos\dfrac{3\pi}{4} + i \sin\dfrac{3\pi}{4}\right)$

39. $8\left(-\dfrac{\sqrt{3}}{2} + \dfrac{1}{2}i\right) = -4\sqrt{3} + 4i$

40. $[4(\cos 30° + i \sin 30°)][6(\cos 120° + i \sin 120°)] = 24(\cos(30 + 120°) + i \sin(30° + 120°))$

$$= 24(\cos 150° + i \sin 150°)$$
$$= 24\left(-\frac{\sqrt{3}}{2} + i\frac{1}{2}\right) = -12\sqrt{3} + 12i$$

41. $1 = 1(\cos 0 + i \sin 0)$

$$\cos\left(\frac{0 + 2\pi k}{3}\right) + i \sin\left(\frac{0 + 2\pi k}{3}\right), k = 0, 1, 2$$

$k = 0$: $\cos 0 + i \sin 0 = 1$

$k = 1$: $\cos\dfrac{2\pi}{3} + i \sin\dfrac{2\pi}{3} = -\dfrac{1}{2} + \dfrac{\sqrt{3}}{2}i$

$k = 2$: $\cos\dfrac{4\pi}{3} + i \sin\dfrac{4\pi}{3} = -\dfrac{1}{2} - \dfrac{\sqrt{3}}{2}i$

42. $x^5 = -243$. 5 fifth rots of $-243 = 243(\cos \pi + i \sin \pi)$ are

$$\sqrt[5]{243}\left(\cos\frac{\pi + 2\pi k}{5} + i \sin\frac{\pi + 2\pi k}{5}\right); k = 0, 1, 2, 3, 4$$

$$3\left(\cos\frac{\pi}{5} + i \sin\frac{\pi}{5}\right)$$

$$3\left(\cos\frac{3\pi}{5} + i \sin\frac{3\pi}{5}\right)$$

$$3\left(\cos\frac{5\pi}{5} + i \sin\frac{5\pi}{5}\right) = -3$$

$$3\left(\cos\frac{7\pi}{5} + i \sin\frac{7\pi}{5}\right)$$

$$3\left(\cos\frac{9\pi}{5} + i \sin\frac{9\pi}{5}\right)$$

43. $\tan 18° = \dfrac{h}{200}$

$\tan 16° 45' = \dfrac{k}{200}$

Hence, $f = h - k = 200 \tan 18° - 200 \tan 16° 45'$

$\approx 4.8 \approx 5$ feet.

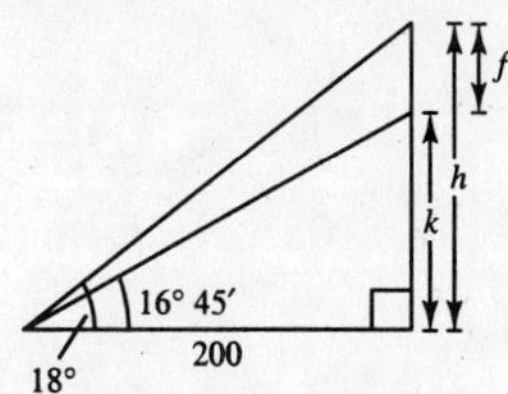

44. $y = 4\cos\left(\dfrac{\pi}{4}t\right)$ or $y = 4\sin\left(\dfrac{\pi}{4}t\right)$. Amplitude 4. Period $\dfrac{2\pi}{\frac{\pi}{4}} = 8$

45. Add the two vectors:

$500(\cos 60\mathbf{i} + \sin 60\mathbf{j}) + 50(\cos 30\mathbf{i} + \sin 30\mathbf{j}) = (250 + 25\sqrt{3})\mathbf{i} + (250\sqrt{3} + 25)\mathbf{j}$

$\tan\theta = \dfrac{250\sqrt{3} + 25}{250 + 25\sqrt{3}} \approx 1.56 \Longrightarrow \theta \approx 57.4$

Direction: N 32.6° E

Speed $= \sqrt{(250 + 25\sqrt{3})^2 + (250\sqrt{3} + 25)^2} \approx 543.9$ km/hr

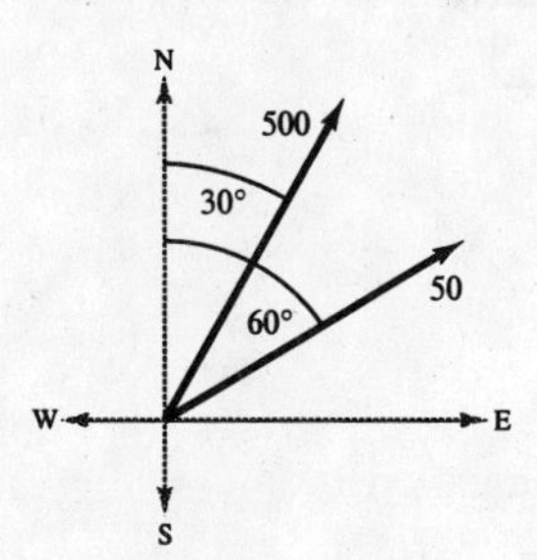

47. $\cos A = \dfrac{60^2 + 125^2 - 100^2}{2(60)(125)} = 0.615 \Longrightarrow A \approx 52.05°$

$\cos B = \dfrac{100^2 + 125^2 - 60^2}{2(100)(125)} = 0.881 \Longrightarrow B \approx 28.24$

Angle between vectors $= A + B \approx 80.3°$

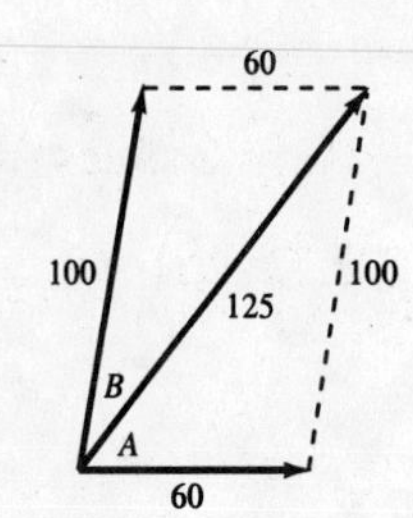

Chapter 7 Chapter Test

1. $x - y = 6 \Rightarrow y = x - 6$. Then $3x + 5(x - 6) = 2 \Rightarrow 8x = 32 \Rightarrow x = 4, y = 4 - 6 = -2$.

Answer: $(4, -2)$

2. $y = x - 1 = (x - 1)^3 \Rightarrow x = 1$ or $1 = (x - 1)^2 = x^2 - 2x + 1 \Rightarrow x^2 - 2x = 0$.

Thus, $x = 1$ or $x(x - 2) = 0 \Rightarrow x = 0, 1, 2$. *Answer:* $(0, -1), (1, 0), (2, 1)$

3. $x - y = 3 \Rightarrow y = x - 3 \Rightarrow$

$$4x - (x - 3)^2 = 7$$
$$4x - (x^2 - 6x + 9) = 7$$
$$x^2 - 10x + 16 = 0$$
$$(x - 2)(x - 8) = 0$$
$$x = 2, 8$$

Answer: $(2, -1), (8, 5)$

4. $4x - 3y = -15 \Rightarrow y = \dfrac{4x + 15}{3}$

$4x + 3y = -9 \Rightarrow y = \dfrac{-4x - 9}{3}$

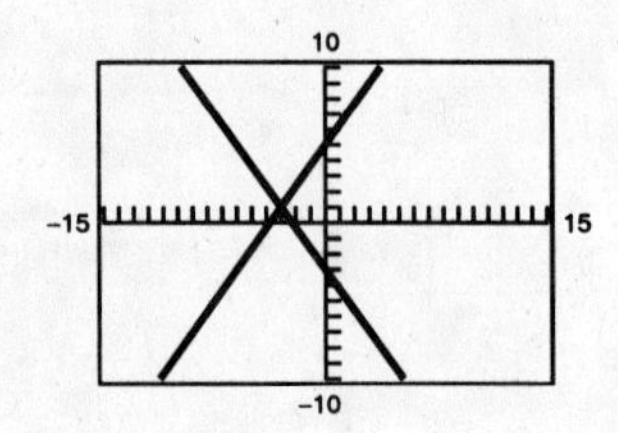

Answer: $(-3, 1)$

5. $y = 16 - x^2$

$y = x + 4$

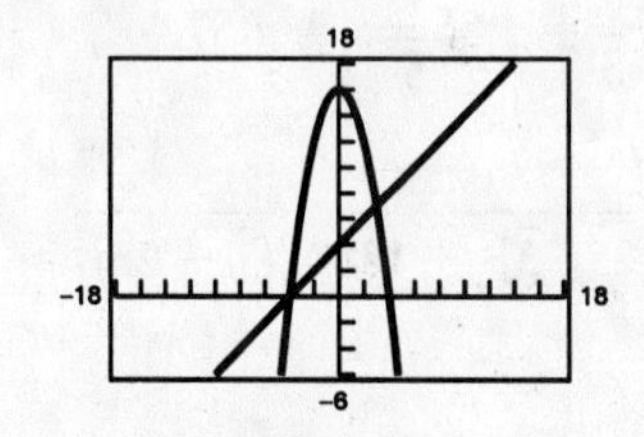

Answer: $(-4, 0), (3, 7)$

6. $y - \ln x = 8 \Rightarrow y = \ln x + 8$

$3x + y + 10 = 21 \Rightarrow y = -3x + 11$

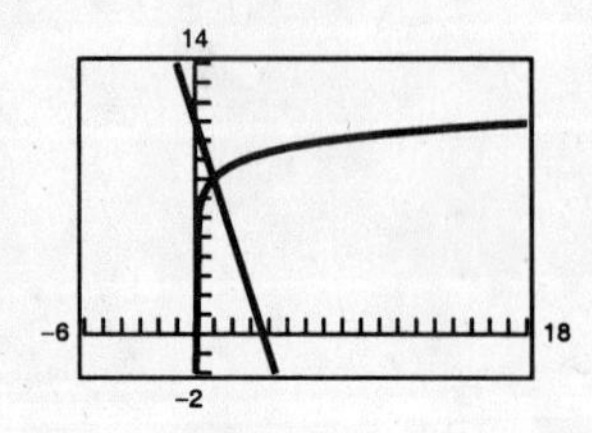

Answer: $(1, 8)$

7. $2x + 5y = -11$ Equation 1

$5x - y = 19$ Equation 2

$-\frac{5}{2}$ times Eq. 1 added to Eq. 2 produces

$-\frac{27}{2}y = \frac{93}{2} \Rightarrow y = -\frac{31}{9}$.

Then $2x + 5\left(-\frac{31}{9}\right) = -11 \Rightarrow x = \frac{28}{9}$.

Answer: $\left(\frac{28}{9}, -\frac{31}{9}\right)$

8. $x - 2y + 3z = -5$ Equation 1

$2x - z = -4$ Equation 2

$3y + z = 17$ Equation 3

$x - 2y + 3z = -5$

$4y - 7z = 6$ (-2) Eq. 1 + Eq. 2

$3y + z = 17$

$x - 2y + 3z = -5$

$4y - 7z = 6$

$\frac{25}{4}z = \frac{50}{4}$ $\left(-\frac{3}{4}\right)$ Eq. 2 + Eq. 3

$z = 2 \Rightarrow 4y - 7(2) = 6 \Rightarrow y = 5$

$z = 2, y = 5 \Rightarrow x - 2(5) + 3(2) = -5 \Rightarrow x = -1$

Answer: $(-1, 5, 2)$

9. $5x + 5y - z = 0$ Equation 1

$10x + 5y + 2z = 0$ Equation 2

$5x + 15y - 9z = 0$ Equation 3

$5x + 5y - z = 0$

$-5y + 4z = 0 \quad (-2)\text{Eq. } 1 + \text{Eq. } 2$

$10y - 8z = 0 \quad (-1)\text{Eq. } 1 + \text{Eq. } 3$

$5x + 5y - z = 0$

$-5y + 4z = 0$

$0 = 0 \quad (2)\text{Eq. } 2 + \text{Eq. } 3$

Infinite number of solutions. They are all of the form $(-3a, 4a, 5a)$ or $(-\frac{3}{5}a, \frac{4}{5}a, a)$ where a is any real number.

10. There are many correct answers. One system is:

$\frac{4}{3}(3) + (-8)(1) = -4 \quad \Rightarrow \quad 3x + y = -4$

$\frac{4}{3}(0) + (-8)(2) = -16 \quad \Rightarrow \quad 2y = -16$

11. There are many correct answers. One system is:

$-\frac{1}{2}(2) + 5(1) + (-\frac{9}{4})4 = -5 \quad \Rightarrow \quad 2x + y + 4z = -5$

$5(2) + (-\frac{9}{4})0 = 10 \quad \Rightarrow \quad 2y = 10$

$(-\frac{9}{4})4 = -9 \quad \Rightarrow \quad 4x = -9$

12. $6 = a(0)^2 + b(0) + c \Rightarrow c = 6$

$2 = a(-2)^2 + b(-2) + c$

$\frac{9}{2} = a(3)^2 + b(3) + c$

Hence, $4a - 2b + 6 = 2$ or $2a - b = -2$

$9a + 3b + 6 = \frac{9}{2}$ or $9a + 3b = -\frac{3}{2}$.

Solving this system for a and b, you obtain $a = -\frac{1}{2}, b = 1$. Thus, $y = -\frac{1}{2}x^2 + x + 6$.

13. $\dfrac{5x - 2}{(x - 1)^2} = \dfrac{A}{x - 1} + \dfrac{B}{(x - 1)^2}$

$5x - 2 = A(x - 1) + B$

$x = 1: \quad 3 = B$

$x = 0: \quad -2 = -A + 3 \Rightarrow A = 5$

$\dfrac{5x - 2}{(x - 1)^2} = \dfrac{5}{x - 1} + \dfrac{3}{(x - 1)^2}$

14. $x \geq -7$

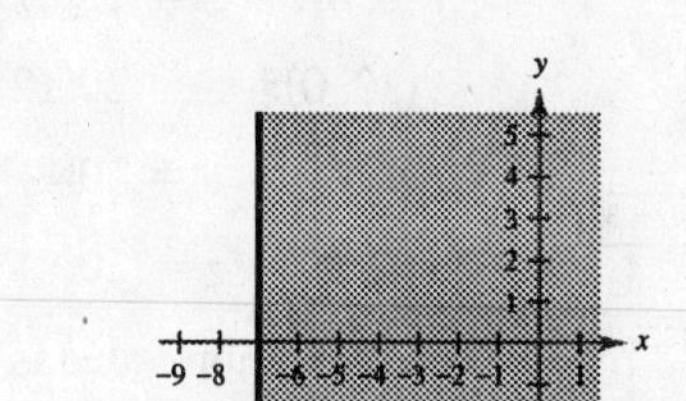

15. $-4x + 6y < -11$

$y < \frac{1}{6}(4x - 11)$

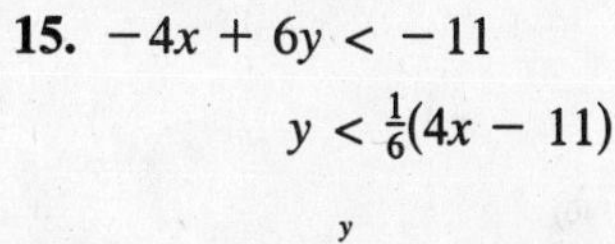

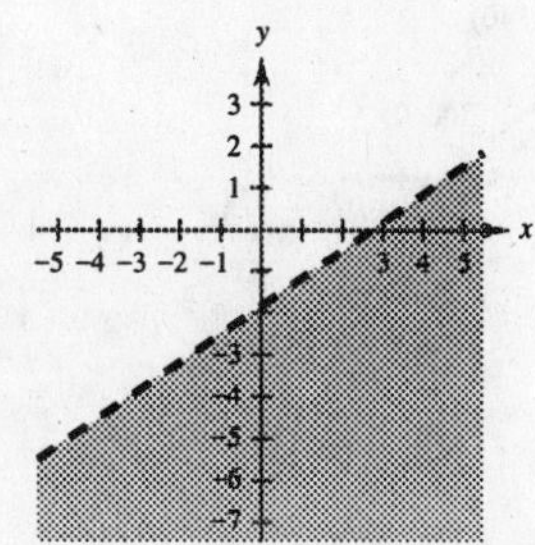

16. $(x - 4)^2 + y^2 < 16$ Interior of circle

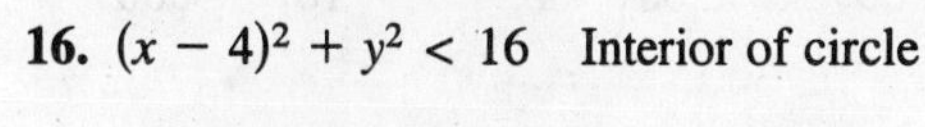

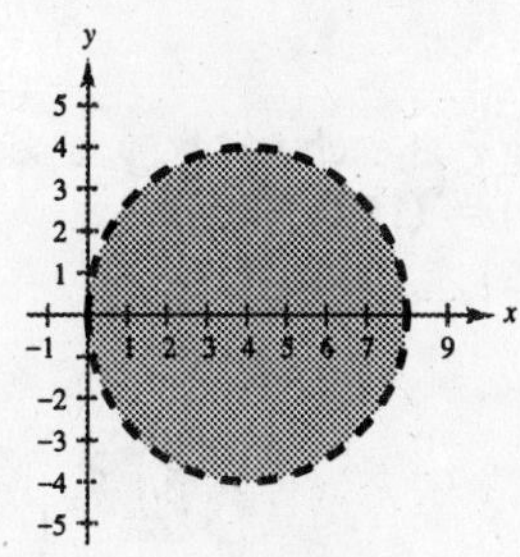

17. $2x + y \le 6$

$2x - y \le 0$

$x \ge -5$

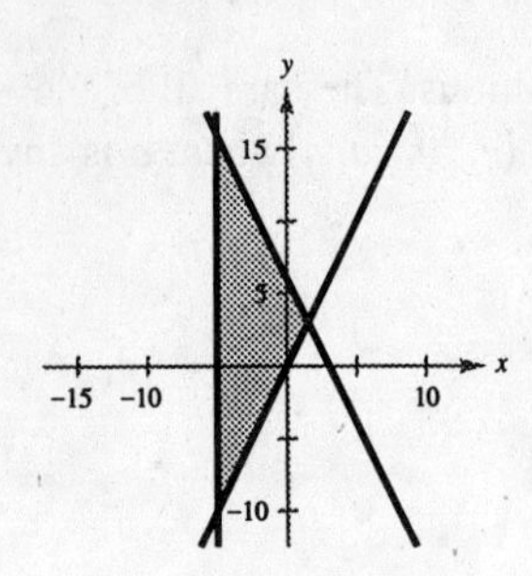

18.

$$-x^4 + x^2 + 4 = 4x$$

$$x^4 - x^2 + 4x - 4 = 0$$

$$(x + 2)(x - 1)(x^2 - x + 2) = 0$$

Intersection points: $(-2, -6)$, $(1, 3)$

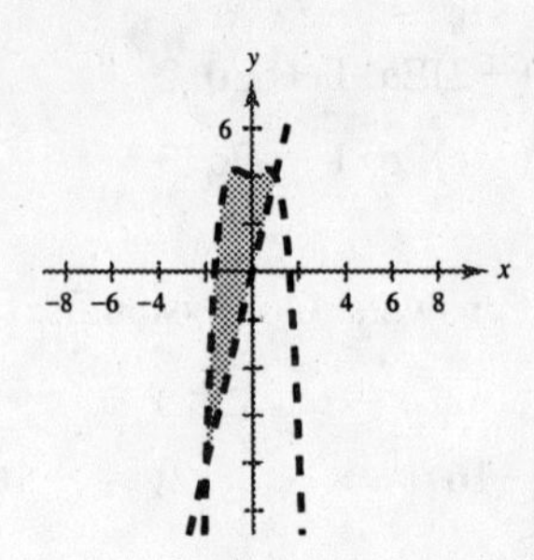

19. $x^2 + y^2 \le 16$ circle

$x \ge 1$ vertical line

$y \ge -3$ horizontal line

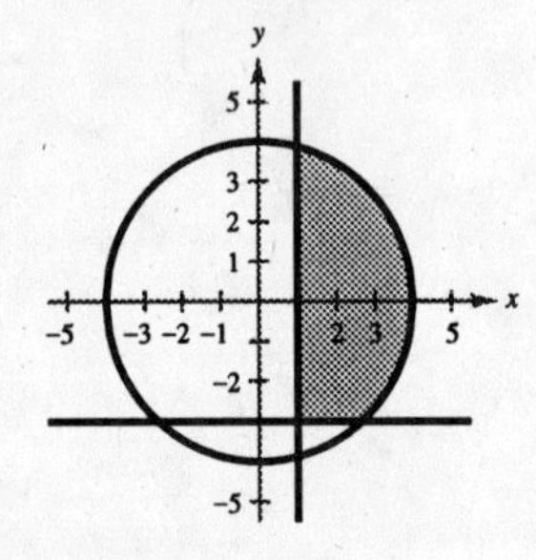

20. The line joining (0, 15) and (9, 12) is $y = -\frac{1}{3}x + 15$. The line joining (9, 12) and (12, 5) is $y = -\frac{7}{3}x + 33$.

Hence,

$$x + 3y \le 45$$

$$7x + 3y \le 99$$

$$x \le 12$$

$$x \ge 0$$

$$y \ge 0.$$

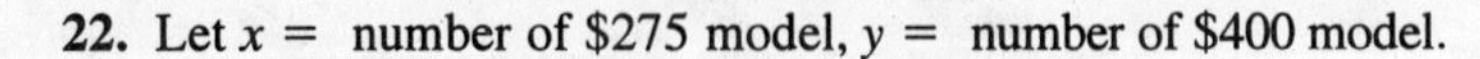

21.

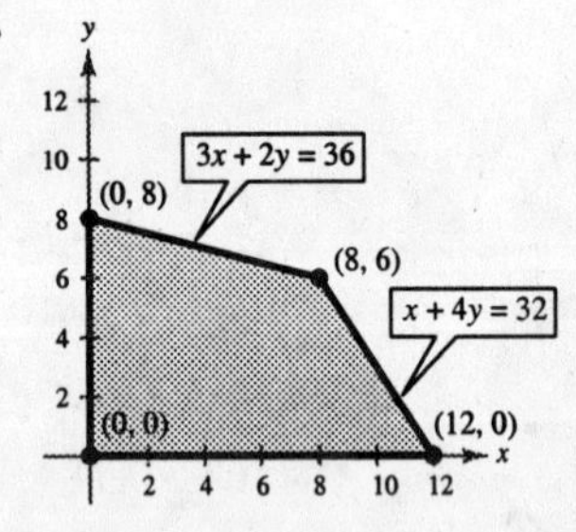

$z = 20x + 12y$

At (0, 0): $z = 20(0) + 12(0) = 0$

At (12, 0): $z = 20(12) + 12(0) = 240$

At (8, 6): $z = 20(8) + 12(6) = 232$

At (0, 8): $z = 20(0) + 12(8) = 96$

The maximum value is $z = 240$ at (12, 0).

22. Let $x =$ number of \$275 model, $y =$ number of \$400 model.

$275x + 400y \le 100{,}000$ or $11x + 16y \le 4000$

$x + y \le 300$

$x, y \ge 0$

$P = 55x + 75y$. Testing P at each vertex, you see P is a maximum at $(x, y) = (160, 140)$.

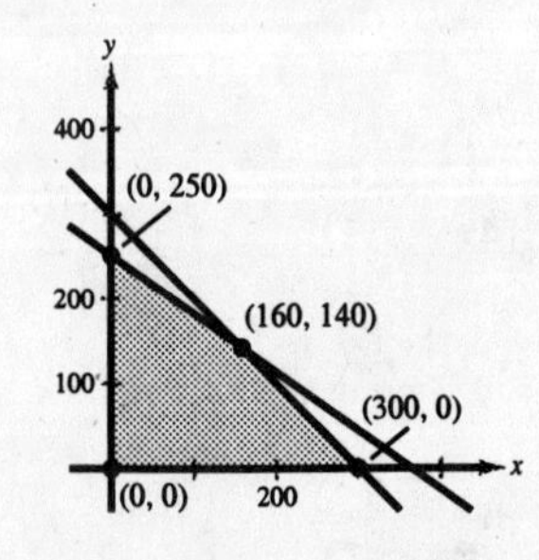

Chapter 8 Chapter Test

1. $\begin{bmatrix} 1 & -1 & 5 \\ 6 & 2 & 3 \\ 5 & 3 & -3 \end{bmatrix} \Rightarrow \begin{bmatrix} 1 & -1 & 5 \\ 0 & 8 & -27 \\ 0 & 8 & -28 \end{bmatrix}$

$\Rightarrow \begin{bmatrix} 1 & -1 & 5 \\ 0 & 8 & -27 \\ 0 & 0 & -1 \end{bmatrix}$

$\Rightarrow \begin{bmatrix} 1 & -1 & 0 \\ 0 & 8 & 0 \\ 0 & 0 & 1 \end{bmatrix}$

$\Rightarrow \begin{bmatrix} 1 & 0 & 0 \\ 0 & 1 & 0 \\ 0 & 0 & 1 \end{bmatrix}$

2. $\begin{bmatrix} 1 & 0 & -1 & 2 \\ -1 & 1 & 1 & -3 \\ 1 & 1 & -1 & 1 \\ 3 & 2 & -3 & 4 \end{bmatrix} \Rightarrow \begin{bmatrix} 1 & 0 & -1 & 2 \\ 0 & 1 & 0 & -1 \\ 0 & 1 & 0 & -1 \\ 0 & 2 & 0 & -2 \end{bmatrix}$

$\Rightarrow \begin{bmatrix} 1 & 0 & -1 & 2 \\ 0 & 1 & 0 & -1 \\ 0 & 0 & 0 & 0 \\ 0 & 0 & 0 & 0 \end{bmatrix}$

3. $\begin{bmatrix} 2 & 1 & 2 & \vdots & 4 \\ 2 & 2 & 0 & \vdots & 5 \\ 2 & -1 & 6 & \vdots & 2 \end{bmatrix}$ row reduces to $\begin{bmatrix} 1 & 0 & 2 & \vdots & 1.5 \\ 0 & 1 & -2 & \vdots & 1 \\ 0 & 0 & 0 & \vdots & 0 \end{bmatrix}$

Infinite number of solutions. Let $z = a$, $y = 2a + 1$, $x = 1.5 - 2a$.

Answer: $(1.5 - 2a, 1 + 2a, a)$, where a is any real number

4. $\begin{bmatrix} 2 & 3 & 1 & \vdots & 10 \\ 2 & -3 & -3 & \vdots & 22 \\ 4 & -2 & 3 & \vdots & -2 \end{bmatrix}$ row reduces to $\begin{bmatrix} 1 & 0 & 0 & \vdots & 5 \\ 0 & 1 & 0 & \vdots & 2 \\ 0 & 0 & 1 & \vdots & -6 \end{bmatrix}$

Answer: $(5, 2, -6)$

5. (a) $A - B = \begin{bmatrix} 5 & 4 & 4 \\ -4 & -4 & 0 \end{bmatrix} - \begin{bmatrix} 4 & -1 & 6 \\ -4 & 0 & -3 \end{bmatrix} = \begin{bmatrix} 1 & 5 & -2 \\ 0 & -4 & 3 \end{bmatrix}$

(b) $3A = 3\begin{bmatrix} 5 & 4 & 4 \\ -4 & -4 & 0 \end{bmatrix} = \begin{bmatrix} 15 & 12 & 12 \\ -12 & -12 & 0 \end{bmatrix}$

(c) $3A - 2B = \begin{bmatrix} 15 & 12 & 12 \\ -12 & -12 & 0 \end{bmatrix} - 2\begin{bmatrix} 4 & -1 & 6 \\ -4 & 0 & -3 \end{bmatrix} = \begin{bmatrix} 7 & 14 & 0 \\ -4 & -12 & 6 \end{bmatrix}$

6. $AB = \begin{bmatrix} 2 & -2 & 6 \\ 3 & -1 & 7 \\ 2 & 0 & -2 \end{bmatrix}\begin{bmatrix} 4 & 4 \\ 3 & 2 \\ 1 & -2 \end{bmatrix} = \begin{bmatrix} 8 & -8 \\ 16 & -4 \\ 6 & 12 \end{bmatrix}$

7. $A^{-1} = \dfrac{1}{ad - bc}\begin{bmatrix} d & -b \\ -c & a \end{bmatrix} = \dfrac{1}{30 - 40}\begin{bmatrix} -5 & -4 \\ -10 & -6 \end{bmatrix} = \dfrac{1}{10}\begin{bmatrix} 5 & 4 \\ 10 & 6 \end{bmatrix} = \begin{bmatrix} \frac{1}{2} & \frac{2}{5} \\ 1 & \frac{3}{5} \end{bmatrix}$

$X = A^{-1}B = \begin{bmatrix} \frac{1}{2} & \frac{2}{5} \\ 1 & \frac{3}{5} \end{bmatrix}\begin{bmatrix} 10 \\ 20 \end{bmatrix} = \begin{bmatrix} 13 \\ 22 \end{bmatrix} \Rightarrow (x, y) = (13, 22)$

8. $\begin{vmatrix} -25 & 18 \\ 6 & -7 \end{vmatrix} = (-25)(-7) - 6(18) = 67$

9. $\det(A) = \begin{vmatrix} 4 & 0 & 3 \\ 1 & -8 & 2 \\ 3 & 2 & 2 \end{vmatrix} = 4(-16 - 4) - 0 + 3(2 + 24)$

$= -80 + 78 = -2$

10. determinant $= (-10)(2)(5)(-3) = 300$ (Upper Triangular)

11. $\begin{vmatrix} -5 & 0 & 1 \\ 3 & 2 & 1 \\ 4 & 4 & 1 \end{vmatrix} = -5(2 - 4) - 0 + 1(12 - 8) = 10 + 4 = 14$

Area $= \frac{1}{2}(14) = 7$

12. $x = \dfrac{\begin{vmatrix} 11 & 8 \\ 21 & -24 \end{vmatrix}}{\begin{vmatrix} 20 & 8 \\ 12 & -24 \end{vmatrix}} = \dfrac{-432}{-576} = \dfrac{3}{4} \qquad y = \dfrac{\begin{vmatrix} 20 & 11 \\ 12 & 21 \end{vmatrix}}{\begin{vmatrix} 20 & 8 \\ 12 & -24 \end{vmatrix}} = \dfrac{288}{-576} = -\dfrac{1}{2}$

Answer: $\left(\frac{3}{4}, -\frac{1}{2}\right)$

13. $\text{Det}\begin{bmatrix} 13 & -6 \\ 26 & -12 \end{bmatrix} = 0 \Rightarrow$ Cramer's Rule not applicable.

[Note: system is inconsistent.]

14. $-2 = a(-2)^2 + b(-2) + c$

$-2 = a(2)^2 + b(2) + c$

$-2 = a(4)^2 + b(4) + c$

$14a - 2b + c = -2$

$4a + 2b + c = 2$

$16a + 4b + c = -2$

Row-reducing the augmented matrix yields

$$\begin{bmatrix} 4 & -2 & 1 & \vdots & -2 \\ 4 & 2 & 1 & \vdots & 2 \\ 16 & 4 & 1 & \vdots & -2 \end{bmatrix} \Rightarrow \begin{bmatrix} 1 & 0 & 0 & \vdots & -\frac{1}{2} \\ 0 & 1 & 0 & \vdots & 1 \\ 0 & 0 & 1 & \vdots & 2 \end{bmatrix}.$$

Thus, $y = -\frac{1}{2}x^2 + x + 2$.

15. Upper left: $400 + x_2 = x_1$

Upper right: $x_1 + x_3 = x_4 + 600$

Lower left: $300 = x_2 + x_3 + x_5$

Lower right: $x_5 + x_4 = 100$

$$\begin{aligned} x_1 - x_2 \qquad\qquad\qquad &= 400 \\ x_1 \qquad + x_3 - x_4 \qquad &= 600 \\ x_2 + x_3 \qquad + x_5 &= 300 \\ x_4 + x_5 &= 100 \end{aligned}$$

Solving this system,

$$\begin{bmatrix} 1 & -1 & 0 & 0 & 0 & \vdots & 400 \\ 1 & 0 & 1 & -1 & 0 & \vdots & 600 \\ 0 & 1 & 1 & 0 & 1 & \vdots & 300 \\ 0 & 0 & 0 & 1 & 1 & \vdots & 100 \end{bmatrix} \rightarrow \begin{bmatrix} 1 & 0 & 1 & 0 & 1 & \vdots & 700 \\ 0 & 1 & 1 & 0 & 1 & \vdots & 300 \\ 0 & 0 & 0 & 1 & 1 & \vdots & 100 \\ 0 & 0 & 0 & 0 & 0 & \vdots & 0 \end{bmatrix}$$

Letting $x_3 = a$ and $x_5 = b$ be real numbers, we have

$x_5 = b$

$x_4 = 100 - b$

$x_3 = a$

$x_2 = 300 - a - b$

$x_1 = 700 - b - a$

Chapter 9 Chapter Test

1. $a_n = \left(-\frac{2}{3}\right)^{n-1}$

$a_1 = \left(-\frac{2}{3}\right)^{1-1} = \left(-\frac{2}{3}\right)^0 = 1$

$a_2 = -\frac{2}{3}$

$a_3 = \left(-\frac{2}{3}\right)^2 = \frac{4}{9}$

$a_4 = \left(-\frac{2}{3}\right)^3 = -\frac{8}{27}$

$a_5 = \left(-\frac{2}{3}\right)^4 = \frac{16}{81}$

2. $a_1 = 12,\ a_{k+1} = a_k + 4$

$a_2 = 12 + 4 = 16$

$a_3 = 16 + 4 = 20$

$a_4 = 20 + 4 = 24$

$a_5 = 24 + 4 = 28$

3. $\dfrac{11!4!}{4!7!} = \dfrac{11!}{7!} = \dfrac{11 \cdot 10 \cdot 9 \cdot 8 \cdot 7!}{7!} = 11 \cdot 10 \cdot 9 \cdot 8 = 7920$

4. $a_n = dn + c,\ c = a_1 - d = 5000 - (-100) = 5100$

$\Rightarrow a_n = -100n + 5100 = 5000 - 100(n - 1)$

5. $a_n = a_1 r^{n-1},\ a_1 = 4,\ r = \dfrac{1}{2} \Rightarrow a_n = 4\left(\dfrac{1}{2}\right)^{n-1}$

6. $\displaystyle\sum_{n=1}^{12} \frac{2}{3n + 1}$

7. $3 + 6 + \cdots + 150 = 3(1 + 2 + \cdots + 50) = 3 \cdot \dfrac{50(51)}{2} = 3825$

8. $\displaystyle\sum_{n=1}^{7} (8n - 5) = 8\left(\frac{7(8)}{2}\right) - 5(7) = 224 - 35 = 189$

9. $\sum_{n=1}^{8} 24\left(\frac{1}{6}\right)^{n-1} = 24\left(\frac{1-\left(\frac{1}{6}\right)^8}{1-\frac{1}{6}}\right) = 24\left(\frac{6}{5}\right)\left(1-\left(\frac{1}{6}\right)^8\right) \approx 28.79998$

10. $\sum_{n=1}^{\infty} 5\left(\frac{1}{10}\right)^{n-1} = 5[1 + 0.1 + 0.01 + \ldots] = 5[1.1111\ldots] = 5.555\ldots = \frac{50}{9}$

11. $A_1 = 50\left(1 + \frac{0.08}{12}\right)^1 = 50(1.00667)$

$A_{300} = 50\left(1 + \frac{0.08}{12}\right)^{300} 50(1.00667)^{300}$

$S_n = 50(1.00667)\left[\frac{1-(1.00667)^{300}}{1-1.00667}\right] \approx \$47,868.33$

12. (1) For $n = 1, 3 = \frac{3(1)(1+1)}{2}$

(2) Assume $S_k = 3 + 6 + \ldots + 3k = \frac{3k(k+1)}{2}$.

Then $S_{k+1} = 3 + 6 + \ldots + 3k + 3(k+1)$

$= S_k + 3(k+1)$

$= \frac{3k(k+1)}{2} + 3(k+1)$

$= \frac{k+1}{2}[3k+6]$

$= \frac{3(k+1)(k+2)}{2}$

Therefore, the formula is true for all positive integers n.

13. ${}_9C_3 = 84$

14. ${}_{20}C_3 = 1140$

15. ${}_{18}C_5 = 8568$

16. ${}_{40}C_{38} = 780$

17. ${}_8C_3 = \frac{8!}{5!\,3!} = \frac{8 \cdot 7 \cdot 6}{3 \cdot 2} = 56$

18. $26 \cdot 10 \cdot 10 \cdot 10 = 26,000$ ways

19. ${}_{25}C_4 = \frac{25!}{21!\,4!} = \frac{25 \cdot 24 \cdot 23 \cdot 22}{24} = 12,650$ ways

20. There are 6 red face cards $\Rightarrow$ probability $= \frac{6}{52} = \frac{3}{26}$.

21. There are ${}_2C_4 = 6$ ways to select 2 spark plugs. Only one way corresponds to both being selected. Probability $= \frac{1}{6}$.

22. (a) $\left(\frac{30}{60}\right)\left(\frac{30}{60}\right) = \frac{1}{2} \cdot \frac{1}{2} = \frac{1}{4}$

(b) $\frac{11}{60} \cdot \frac{11}{60} = \frac{121}{3600} \approx 0.0336$

(c) $\frac{1}{60} \approx 0.0167$

Chapter 10 Chapter Test

1. $y^2 = 8x = 4(2)x$

$p = 2$

Vertex: $(0, 0)$

Focus: $(2, 0)$

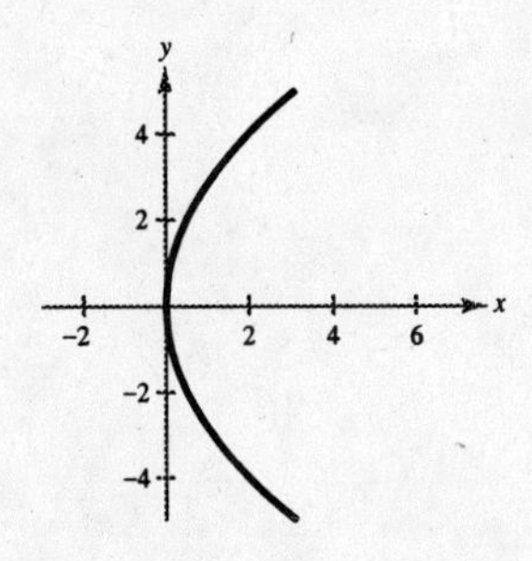

2. Center: $(3, 2)$

$a = 10, b = 9, c = \sqrt{19}$

Vertices: $(3, 12), (3, -8)$

Foci: $\left(3, 2 \pm \sqrt{19}\right)$

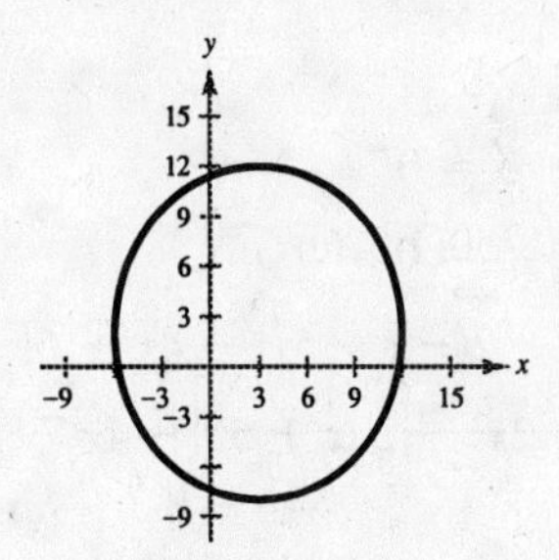

3. $(x^2 - 6x + 9) = -2y - 9 + 9$

$$(x - 3)^2 = -2y = 4\left(-\frac{1}{2}\right)y, p = -\frac{1}{2}$$

Vertex: $(3, 0)$

Focus: $\left(3, -\frac{1}{2}\right)$

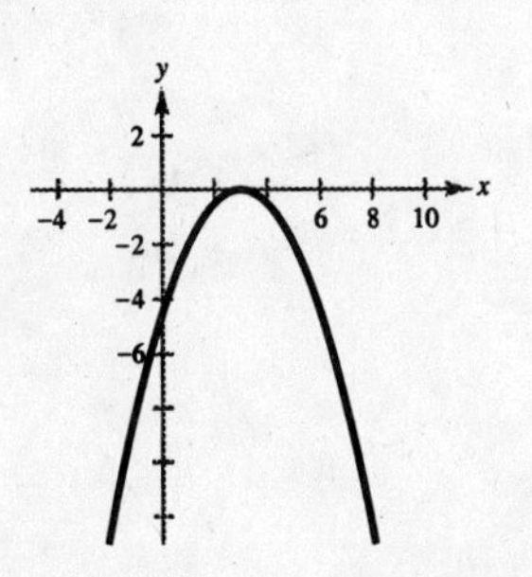

4. $(x^2 - 4x + 4) - 4y^2 = 4$

$$(x - 2)^2 - 4y^2 = 4$$

$$\frac{(x - 2)^2}{4} - \frac{y^2}{1} = 1$$

Center: $(2, 0)$

Vertices: $(0, 0), (4, 0)$

$a = 2, b = 1, c = \sqrt{5}$

Foci: $\left(2 \pm \sqrt{5}, 0\right)$

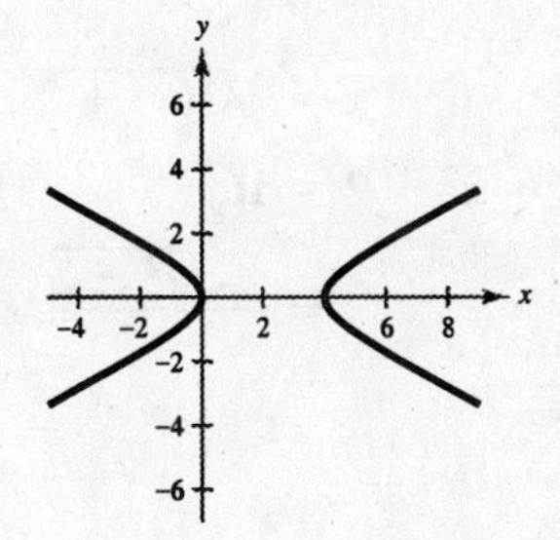

5. Vertex: $(6, -2), p = 2$

$(y + 2)^2 = 4(2)(x - 6)$

$(y + 2)^2 = 8(x - 6)$

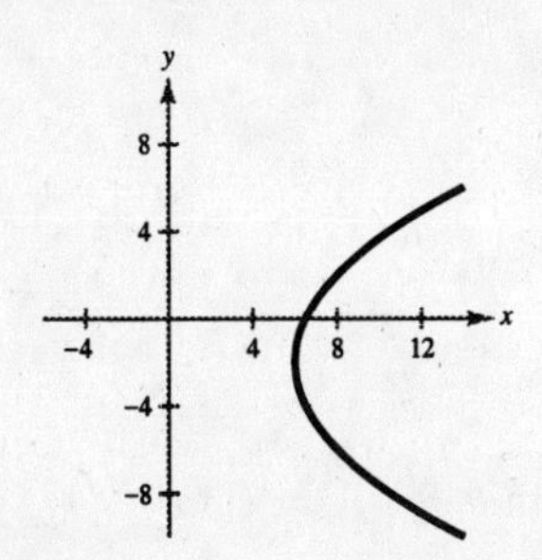

6. $a = 7, b = 4$, center: $(-6, 3)$

$$\frac{(x + 6)^2}{16} + \frac{(y - 3)^2}{49} = 1$$

7. Center: $(0, 0)$, $a = 3, \frac{3}{2}x = \frac{a}{b} \Rightarrow b = 2$

$$\frac{y^2}{9} - \frac{x^2}{4} = 1$$

8. (a) $\cot 2\theta = \dfrac{A - C}{B} = \dfrac{1 - 1}{6} = 0 \implies \theta = \dfrac{\pi}{4}$ or $45°$

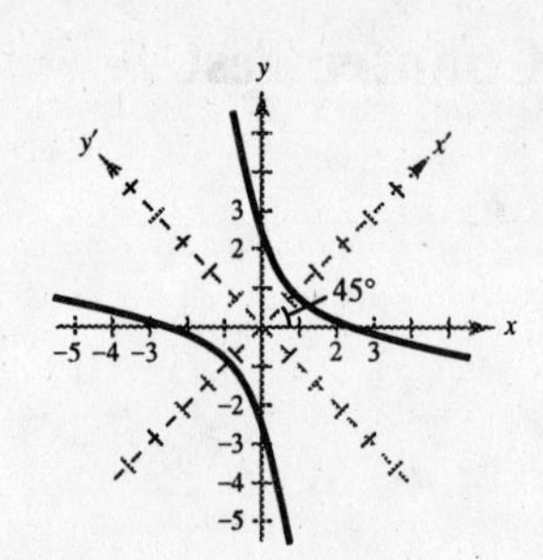

(b) $B^2 - 4AC = 36 - 4 = 32 > 0 \implies$ Hyperbola

$$y^2 + 6xy + (x^2 - 6) = 0$$

$$y = \frac{-6x \pm \sqrt{36x^2 - 4(x^2 - 6)}}{2}$$

9. $x^2 + 2y^2 - 4x + 6y - 5 = 0$

$$x + y + 5 = 0$$

$y = -x - 5$: $\quad x^2 + 2(-x - 5)^2 - 4x + 6(-x - 5) - 5 = 0$

$$x^2 + 2x^2 + 20x + 50 - 4x - 6x - 30 - 5 = 0$$

$$3x^2 + 10x + 15 = 0$$

This quadratic has no real solutions.

Therefore, no solution.

10. $x = \sqrt{t^2 + 2}$

$y = \dfrac{t}{4} \implies t = 4y$

$x = \sqrt{16y^2 + 2}$

$x^2 - 2 = 16y^2$

$y = \dfrac{\pm\sqrt{x^2 - 2}}{4}, x \ge \sqrt{2}$

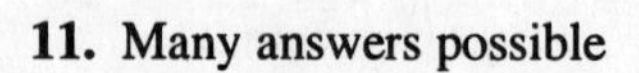

11. Many answers possible

$x = t \qquad\qquad x = 2t$

$y = \dfrac{1}{4}t - 5 \qquad y = \dfrac{1}{2}t - 5$

12. $(r, \theta) = \left(-14, \dfrac{5\pi}{3}\right) \implies (x, y) = \left(-14\cos\dfrac{5\pi}{3}, -14\sin\dfrac{5\pi}{3}\right)$

$$= \left(-7, 7\sqrt{3}\right)$$

13. $(x, y) = (-2, -2) = (r, \theta) = \left(2\sqrt{2}, \dfrac{5\pi}{4}\right)$,

$\left(-2\sqrt{2}, \dfrac{\pi}{4}\right), \left(2\sqrt{2}, -\dfrac{3\pi}{4}\right)$

14. $r^2 - 12r\sin\theta = 0$

$$r = 12\sin\theta$$

15. $r = 2 + 3\sin\theta$

Limaçon

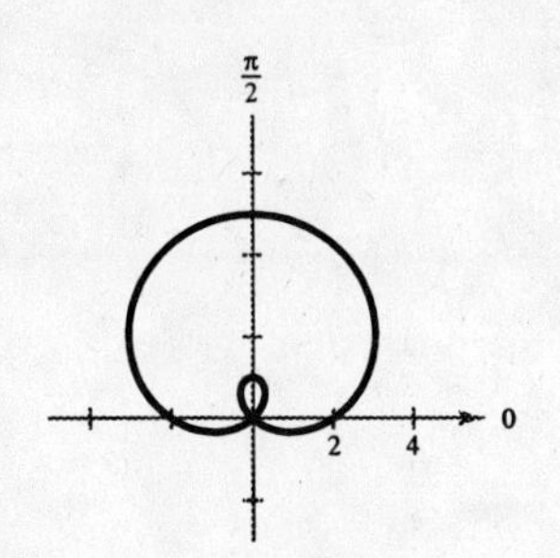

16. $r = 8\cos 3\theta$

Rose curve

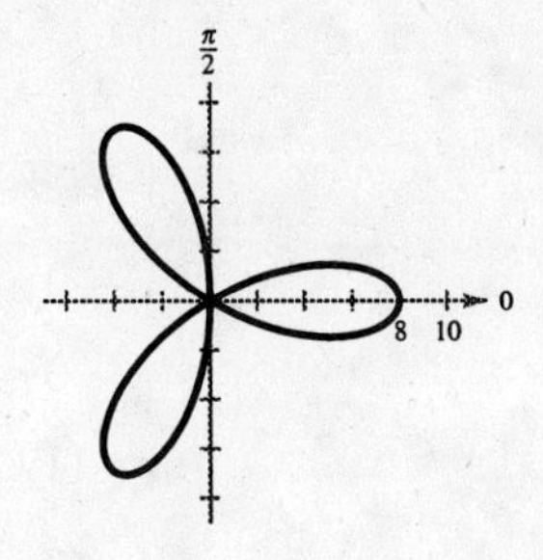

17. $r = \dfrac{8}{4 + 6\sin\theta}$

$= \dfrac{2}{1 + \frac{3}{2}\sin\theta}$

Hyperbola

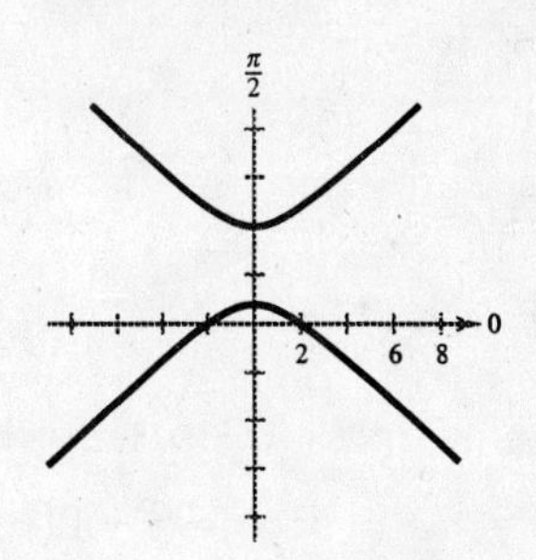

18. $r = \dfrac{ep}{1 + e\sin\theta}, e = \dfrac{1}{4}, p = 4$

$r = \dfrac{1}{1 + \frac{1}{4}\sin\theta}$

19. $r = \dfrac{ep}{1 + e\sin\theta}, e = 0.2056$

$72{,}000{,}000 = 2a = \dfrac{ep}{1 + e} + \dfrac{ep}{1 - e} = \left(\dfrac{0.2056}{1.2056} + \dfrac{0.2056}{0.7944}\right)p$

$\Rightarrow \quad p = 167{,}695{,}676.5$ and $ep \approx 34{,}478{,}231$

$r = \dfrac{34{,}478{,}231}{1 + 0.2056\sin\theta}$

Chapters 7–10 Cumulative Test

1. $-x - 3y = 5$

$4x + 2y = 10$

$x + 3y = -5$

$-10y = 30$

$y = -3, x = -3y - 5 = 9 - 5 = 4$

Answer: $(4, -3)$

2. $x - 3y + 3z = 12$

$2x - 3y + z = 13$

$-4x + y - 2z = -6$

$x - 3y + 3z = 12$

$3y - 5z = -11$

$-11y + 10z = 42$

$x - 3y + 3z = 12$

$y - \frac{5}{3}z = -\frac{11}{3}$

$-\frac{25}{3}z = \frac{5}{3}$

$z = -\frac{1}{5}, y = \frac{5}{3}\left(-\frac{1}{5}\right) - \frac{11}{3} = -4,$

$x = 3(-4) - 3\left(-\frac{1}{5}\right) + 12 = \frac{3}{5}$

Answer: $\left(\frac{3}{5}, -4, -\frac{1}{5}\right)$

3. $3A - 2B = \begin{bmatrix} -7 & -10 & -16 \\ -6 & 18 & 9 \\ -12 & 16 & 7 \end{bmatrix}$

4. $5A + 3B = \begin{bmatrix} -18 & 15 & -14 \\ 28 & 11 & 34 \\ -20 & 52 & -1 \end{bmatrix}$

5. $AB = \begin{bmatrix} 3 & -31 & 2 \\ 22 & 18 & 6 \\ 52 & -40 & 14 \end{bmatrix}$

6. $BA = \begin{bmatrix} 5 & 36 & 31 \\ -36 & 12 & -36 \\ 16 & 0 & 18 \end{bmatrix}$

7. (a) $A^{-1} = \begin{bmatrix} -175 & 37 & -13 \\ 95 & -20 & 7 \\ 14 & -3 & 1 \end{bmatrix}$

(b) $\det(A) = 1$

8. $A^{-1} = \begin{bmatrix} -18 & -7 & 5 \\ -3 & -1 & 1 \\ -5 & -2 & 1 \end{bmatrix}, B = \begin{bmatrix} 8 \\ -19 \\ 3 \end{bmatrix}$

$A^{-1}B = \begin{bmatrix} 4 \\ -2 \\ 1 \end{bmatrix}$

Answer: $(4, -2, 1)$

9. $\begin{vmatrix} 0 & 0 & 1 \\ 6 & 2 & 1 \\ 8 & 10 & 1 \end{vmatrix} = 60 - 16 = 44$

Area $= \frac{1}{2}(44) = 22$ sq. units

10. $a_1 = 8$

$a_{20} = 8 + 19(5) = 103$

Sum $= \frac{20}{2}(8 + 103) = 1110$

11. $\sum_{k=1}^{6} (7k - 2) = 135$

12. $\sum_{k=1}^{4} \frac{2}{k^2 + 4} = \frac{47}{52} \approx 0.9038$

13. $\sum_{n=0}^{10} 9\left(\frac{3}{4}\right)^n \approx 34.4795$

14. $\sum_{n=1}^{\infty} 8(0.9)^{n-1} = \frac{8}{1 - 0.9} = 80$

15. $\sum_{k=40}^{94} k = 3685$

16. For $n = 1$, $3 = 1[2(1) + 1]$

Assume true for k, and consider

$$\begin{aligned} 3 + 7 + \cdots + (4k - 1) + [4(k + 1) - 1] &= k(2k + 1) + (4k + 3) \\ &= 2k^2 + 5k + 3 \\ &= (2k + 3)(k + 1) \\ &= (k + 1)[2(k + 1) + 1] \end{aligned}$$

which shows that the formula is valid for $n = k + 1$.

17. $(x - 2y)^6 = x^6 - 12x^5y + 60x^4y^2 - 160x^3y^3 + 240x^2y^4 - 192xy^5 + 64y^6$

18. $\dfrac{10!}{3!2!2!} = 151{,}200$

19. Hyperbola

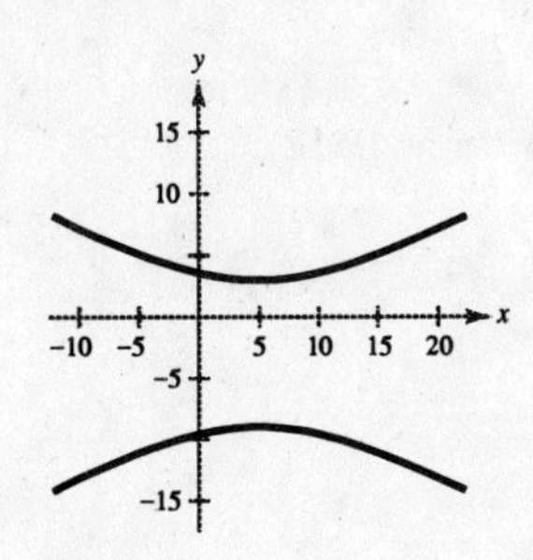

20. Ellipse

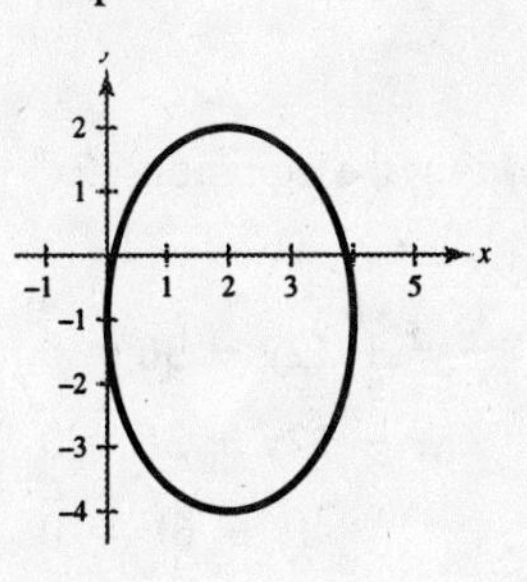

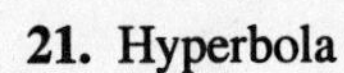

21. Hyperbola

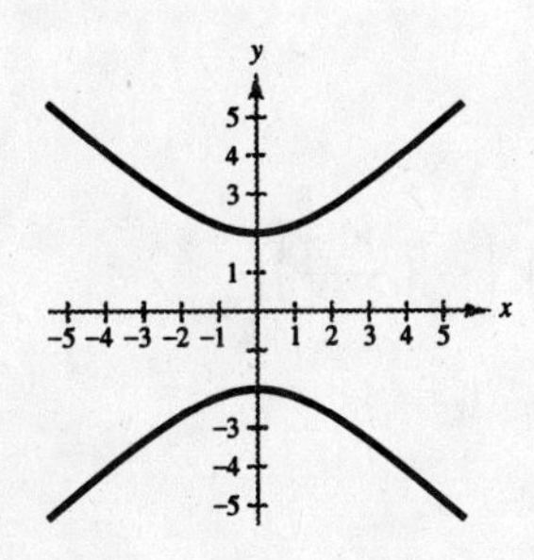

22. Circle

$(x - 1)^2 + (y - 2)^2 = 9$

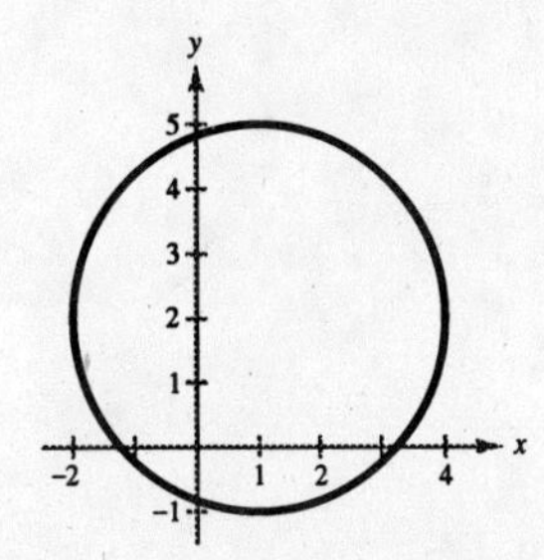

23. Vertex: $(2, 3)$: $(x - 2)^2 = c(y - 3)$

Passes through $(0, 0)$: $4 = -3c \Rightarrow c = -\dfrac{4}{3}$

$(x - 2)^2 = -\dfrac{4}{3}(y - 3)$

24. Center: $(1, 4)$ $a = 5, b = 2$

$$\frac{(x - 1)^2}{25} + \frac{(y - 4)^2}{4} = 1$$

25. Center: $(0, -4)$, $a = 2$

$$\frac{(y + 4)^2}{4} - \frac{x^2}{c} = 1$$

$(4, 0)$ on curve: $4 - \dfrac{16}{c} = 1 \Rightarrow 3 = \dfrac{16}{c} \Rightarrow c = \dfrac{16}{3}$

Answer: $\dfrac{(y + 4)^2}{4} - \dfrac{x^2}{\left(\frac{16}{3}\right)} = 1$

26. $c = 2$, center: $(0, 2)$ Vertical transverse axis

$$\frac{a}{b} = \frac{1}{2} \Rightarrow b = 2a \qquad c^2 = 4 = a^2 + b^2 = a^2 + 4a^2 \Rightarrow a^2 = \frac{4}{5} \Rightarrow b^2 = \frac{16}{5}$$

$$\frac{(y-2)^2}{\left(\frac{4}{5}\right)} - \frac{x^2}{\left(\frac{16}{5}\right)} = 1$$

27. $B^2 - 4AC = 16 - 8 = 8 \Rightarrow$ Hyperbola

$$\cot 2\theta = \frac{2-1}{-4} = -\frac{1}{4} \Rightarrow \theta \approx 38°$$

Graph as: $2y^2 - 4xy + (x^2 - 6) = 0$

$$y = \frac{4x \pm \sqrt{16x^2 - 8(x^2 - 6)}}{4}$$

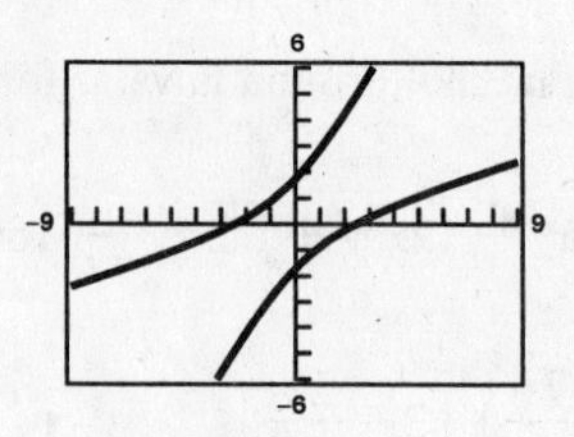

28. Adding the equations, $2x^2 - 24x = 0 \Rightarrow x = 0, 12$

For $x = 0$,

$$-y^2 + 12y - 36 = 0$$
$$y^2 - 12y + 36 = 0$$
$$(y-6)^2 = 0 \Rightarrow y = 6$$

For $x = 12$,

$$144 - y^2 - 144 + 12y - 36 = 0$$
$$y^2 - 12y + 36 = 0 \Rightarrow y = 6$$

Answer: $(0, 6), (12, 6)$

29. $x = 4 \ln t \Rightarrow t = e^{x/4}$

$$y = \frac{1}{2}t^2$$

$$y = \frac{1}{2}(e^{x/4})^2 = \frac{1}{2}e^{x/2}$$

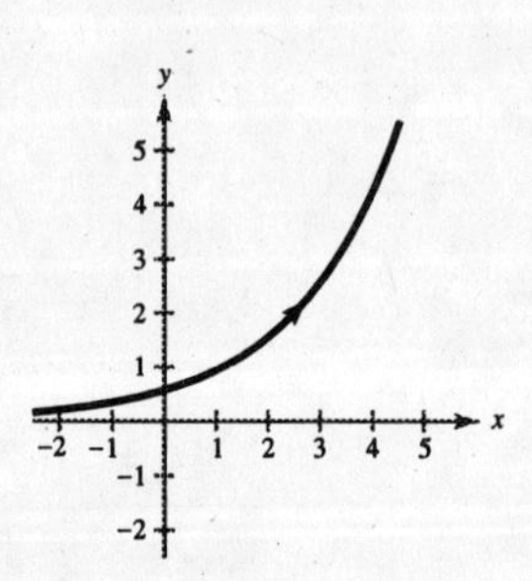

30. $\left(8, -\frac{7\pi}{6}\right), \left(-8, -\frac{\pi}{6}\right), \left(-8, \frac{11\pi}{6}\right)$

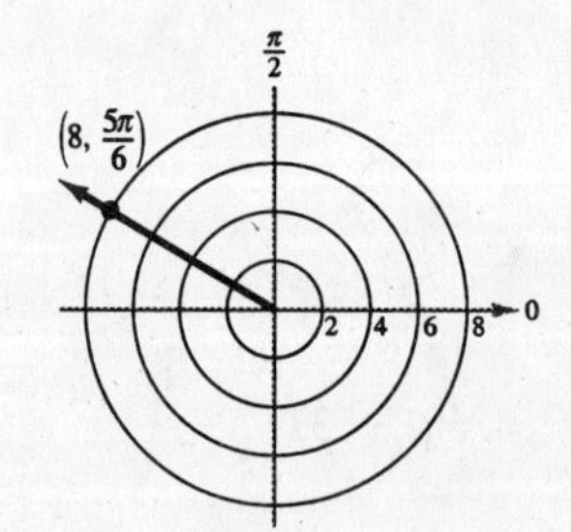

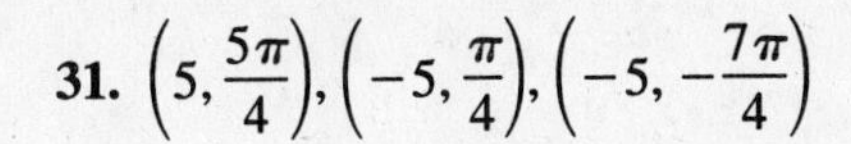

31. $\left(5, \frac{5\pi}{4}\right), \left(-5, \frac{\pi}{4}\right), \left(-5, -\frac{7\pi}{4}\right)$

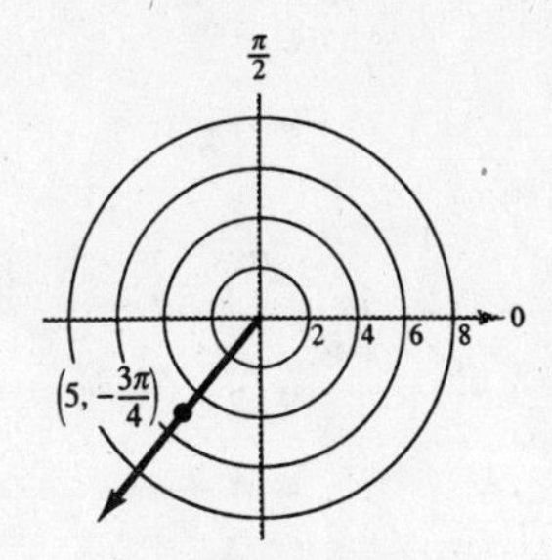

32. $\left(-2, -\frac{3\pi}{4}\right), \left(2, \frac{\pi}{4}\right), \left(2, -\frac{7\pi}{4}\right)$

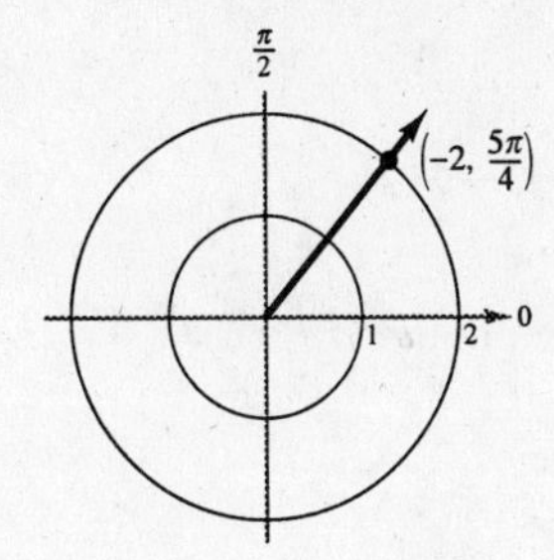

33. $\left(-3, \frac{\pi}{6}\right), \left(3, \frac{7\pi}{6}\right), \left(3, -\frac{5\pi}{6}\right)$

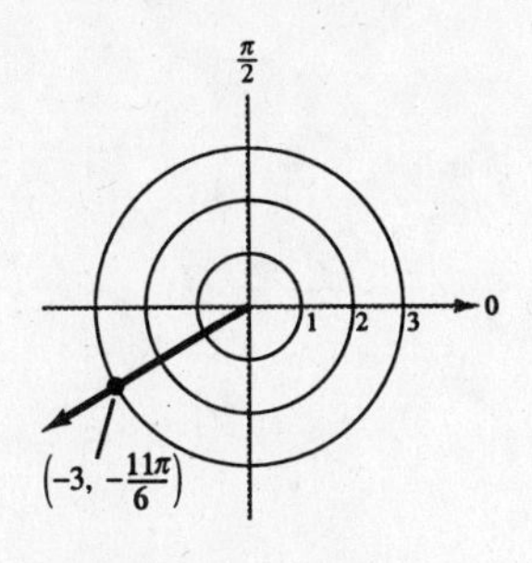

34. $-8r\cos\theta - 3r\sin\theta = -5$

$$r = \frac{5}{8\cos\theta + 3\sin\theta}$$

35.

$$\begin{aligned} 4r - 5r\cos\theta &= 2 \\ 4(x^2 + y^2)^{1/2} - 5x &= 2 \\ 4\sqrt{x^2 + y^2} &= 5x + 2 \\ 16(x^2 + y^2) &= 25x^2 + 20x + 4 \\ 9x^2 - 16y^2 + 20x + 4 &= 0 \end{aligned}$$

36. $r = -\frac{\pi}{6}$

Circle

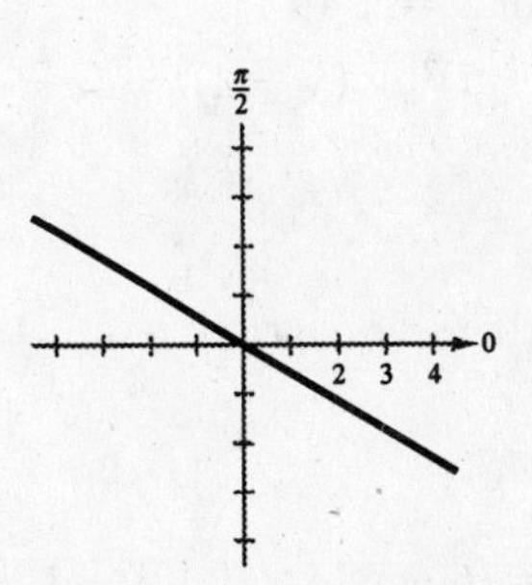

37. $r = 3 - 2\sin\theta$

Limaçon

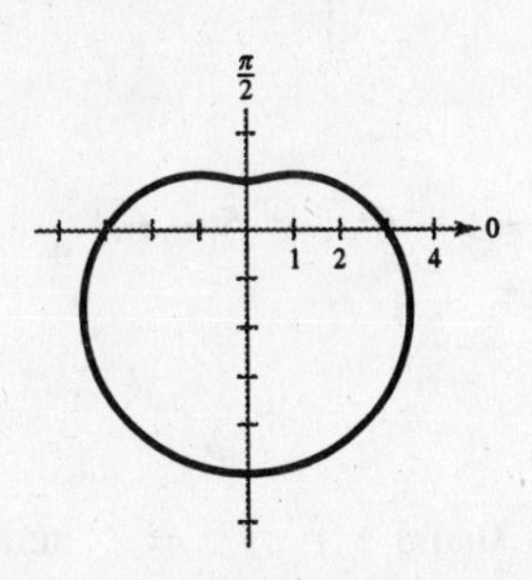

38. $r = 2 + 5\cos\theta$

Limaçon

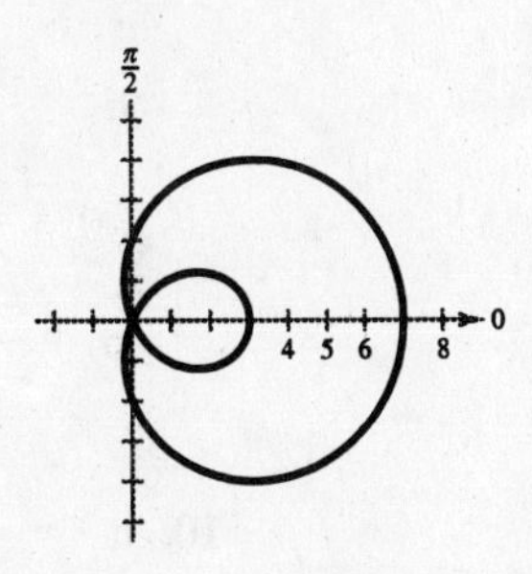

39. $r = \frac{14}{4 - 3\cos\theta}$

40. Since the positions are different, use

${}_{15}P_4 = 32{,}760$ ways

41. ${}_{72}C_6 = 156{,}238{,}908$ subsets

42. (2)(2)(1) choices.
Probability $= \frac{1}{4}$.

Chapter 11 Chapter Test

1.

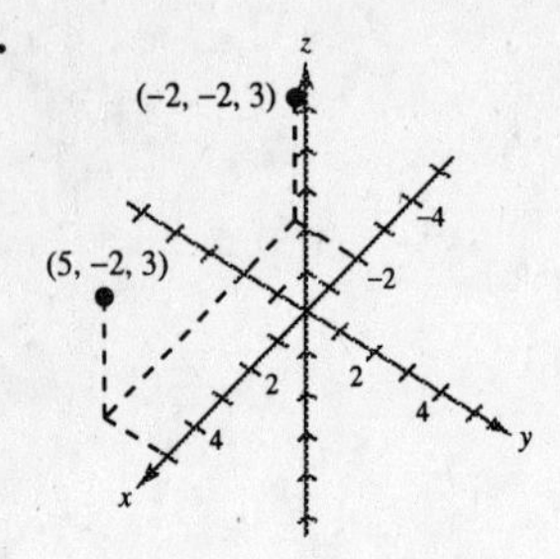

2. $AB = \sqrt{(8-6)^2 + (-2-4)^2 + (5+1)^2} = \sqrt{76}$

$AC = \sqrt{(8+4)^2 + (-2-3)^2 + (5-0)^2} = \sqrt{144+25+25} = \sqrt{194}$

$BC = \sqrt{(6+4)^2 + (4-3)^2 + (-1-0)^2} = \sqrt{100+1+1} = \sqrt{102}$

No. $\left(\sqrt{76}\right)^2 + \left(\sqrt{102}\right)^2 \neq \left(\sqrt{194}\right)^2$

3. Midpoint $= \left(\dfrac{8+6}{2}, \dfrac{-2+4}{2}, \dfrac{5-1}{2}\right) = (7, 1, 2)$

4. Diameter $= \sqrt{(8-6)^2 + (-2-4)^2 + (5+1)^2} = \sqrt{4+36+36} = \sqrt{76}$

Radius $= \sqrt{19}$

$(x-7)^2 + (y-1)^2 + (z-2)^2 = 19$

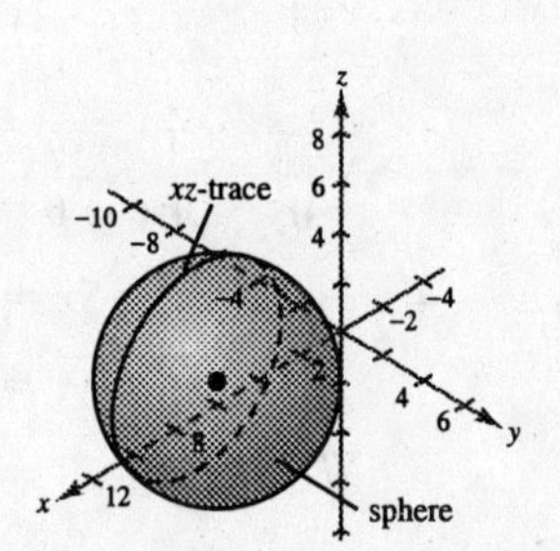

5. $\mathbf{u} = \langle 6-8, 4-(-2), -1-5\rangle = \langle -2, 6, -6\rangle$

$\mathbf{v} = \langle -4-8, 3-(-2), 0-5\rangle = \langle -12, 5, -5\rangle$

6. (a) $\|\mathbf{v}\| = \sqrt{(-12)^2 + 5^2 + (-5)^2} = \sqrt{194}$

(b) $\mathbf{u} \cdot \mathbf{v} = (-2)(-12) + 6(5) + (-6)(-5) = 84$

(c) $\mathbf{u} \times \mathbf{v} = \begin{vmatrix} \mathbf{i} & \mathbf{j} & \mathbf{k} \\ -2 & 6 & -6 \\ -12 & 5 & -5 \end{vmatrix} = \langle 0, 62, 62\rangle$

7. $\cos\theta = \dfrac{\mathbf{u} \cdot \mathbf{v}}{\|\mathbf{u}\|\,\|\mathbf{v}\|} = \dfrac{84}{\sqrt{76}\sqrt{194}} \approx 0.6918$

$\Rightarrow \theta \approx 46.23°$ or 0.8068 radians

8. (a) $x = 8 - 2t, y = -2 + 6t, z = 5 - 6t$

(b) $\dfrac{x-8}{-2} = \dfrac{y+2}{6} = \dfrac{z-5}{-6}$

9. Normal vector: $\mathbf{n} = \mathbf{u} \times \mathbf{v} = \langle 0, 62, 62\rangle$ or $\langle 0, 1, 1\rangle$.

$0(x-8) + 1(y+2) + 1(z-5) = 0$

$y + z - 3 = 0$

10. $\mathbf{u} \cdot \mathbf{v} = 0 - 2 - 6 \neq 0$ and $\mathbf{u} \neq c\mathbf{v} \Rightarrow$ neither

11. $\mathbf{u} \cdot \mathbf{v} = -2 + 3 - 1 = 0 \implies$ orthogonal

12. Let $A(0, 0, 5)$ be the vertex.

$\mathbf{u} = \overrightarrow{AD} = \langle 4, 0, 0 \rangle$, $\mathbf{v} = \overrightarrow{AB} = \langle 0, 10, 0 \rangle$, $\mathbf{w} = \overrightarrow{AE} = \langle 0, 1, -5 \rangle$

$$\mathbf{u} \cdot (\mathbf{v} \times \mathbf{w}) = \begin{vmatrix} 4 & 0 & 0 \\ 0 & 10 & 0 \\ 0 & 1 & -5 \end{vmatrix} = 4(-50) = -200$$

Volume $= |-200| = 200$ cubic units

13.

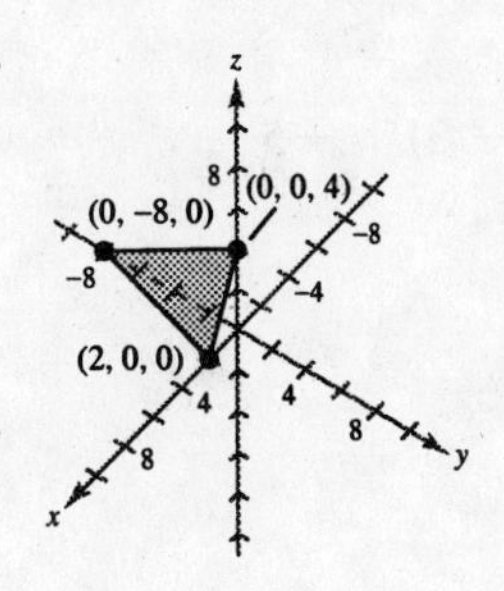

14.

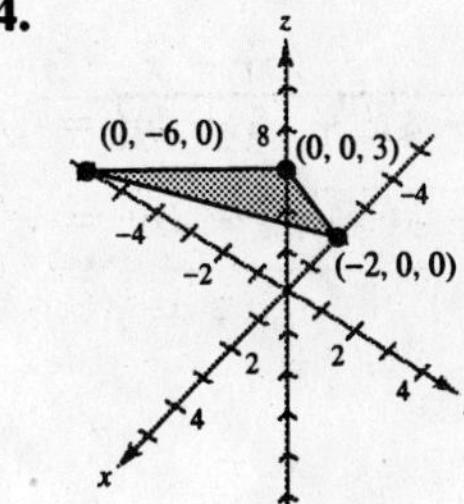

15. $\mathbf{n} = \langle 3, -2, 1 \rangle$, $Q = (4, 3, 8)$, $P = (0, 0, 6)$ in plane, $\overrightarrow{PQ} = \langle 4, 3, 2 \rangle$.

$$D = \frac{|\overrightarrow{PQ} \cdot \mathbf{n}|}{\|\mathbf{n}\|} = \frac{8}{\sqrt{14}} = \frac{4\sqrt{14}}{7}$$

16.

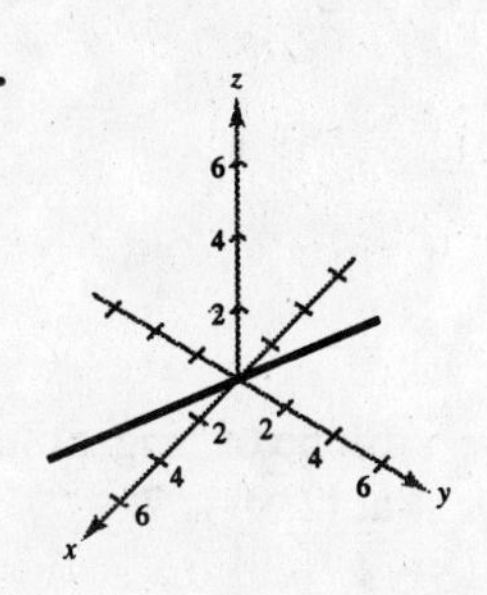

17. diagonal length $= \sqrt{(6-0)^2 + (0-6)^2 + (0-6)^2}$

$= \sqrt{36 + 36 + 36}$

$= 6\sqrt{3}$

Chapter 12 Chapter Test

1.

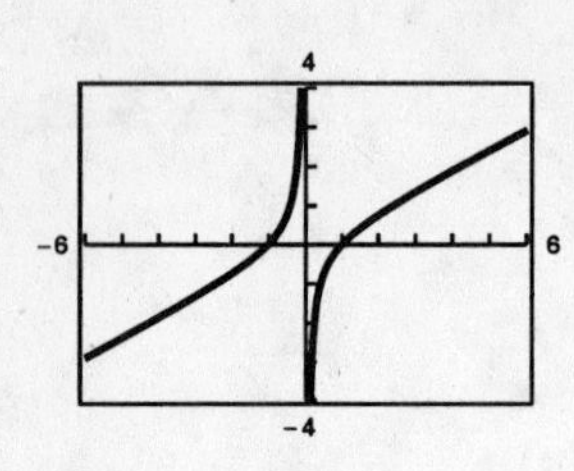

$$\lim_{x\to -2} \frac{x^2 - 1}{2x} = \frac{(-2)^2 - 1}{2(-2)} = -\frac{3}{4}$$

limit is -0.75

2.

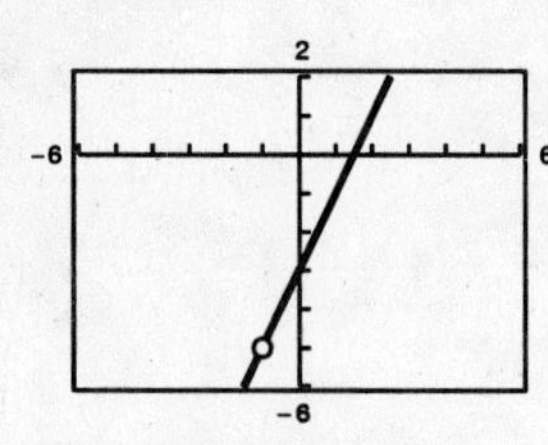

$$\lim_{x\to -1} \frac{2x^2 - x - 3}{x + 1} = \lim_{x\to -1} \frac{(x + 1)(2x - 3)}{x + 1} = \lim_{x\to -1} (2x - 3) = -5$$

limit is -5

3.

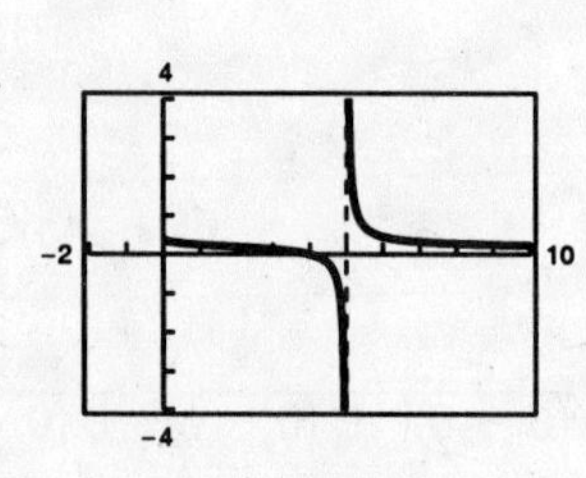

$$\lim_{x\to 5} \frac{\sqrt{x} - 2}{x - 5}$$

does not exist

4.

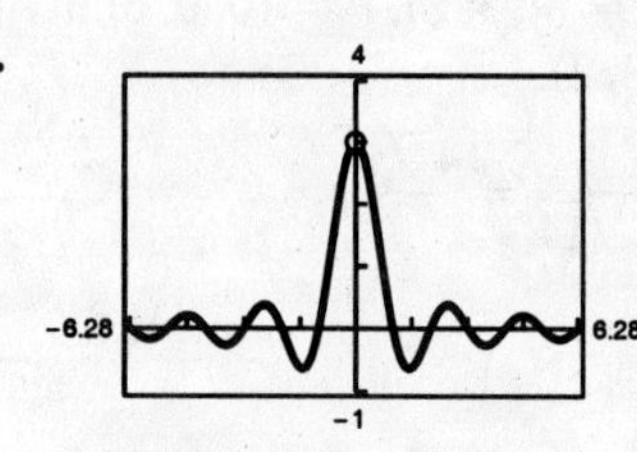

$$f(x) = \frac{\sin 3x}{x}$$

$$\lim_{x\to 0} \frac{\sin 3x}{x} = 3$$

5.

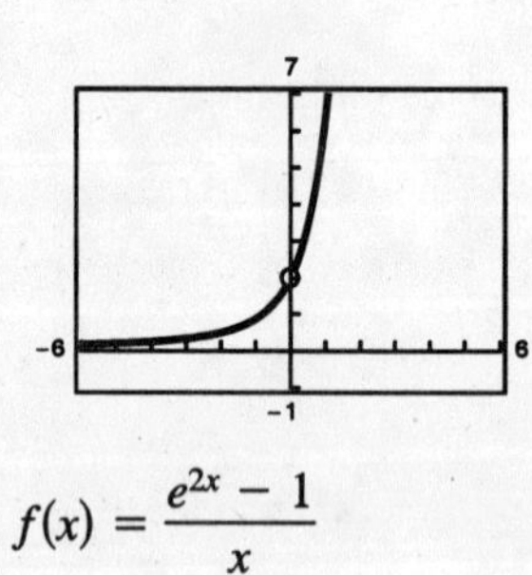

$$f(x) = \frac{e^{2x} - 1}{x}$$

$$\lim_{x\to 0} \frac{e^{2x} - 1}{x} = 2$$

6.

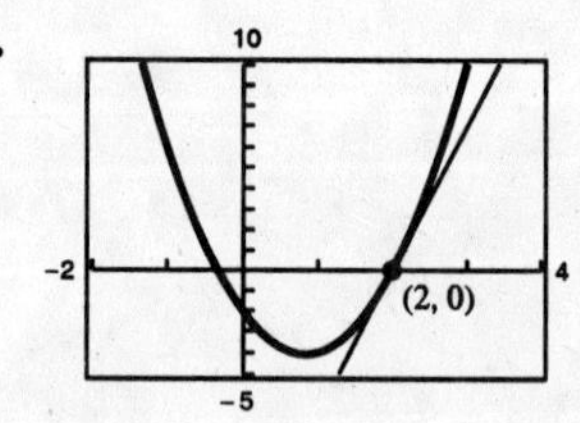

Slope of tangent line at $(2, 0)$ is 7.

7.

$$\frac{f(x + h) - f(x)}{h} = \frac{[2(x + h)^3 + 6(x + h)] - [2x^3 + 6x]}{h}$$

$$= \frac{2x^3 + 6x^2h + 6xh^2 + 2h^3 + 6x + 6h - 2x^3 - 6x}{h}$$

$$= \frac{6x^2h + 6xh^2 + 2h^3 + 6h}{h}$$

$$= 6x^2 + 6xh + 2h^2 + 6, \; h \neq 0$$

$$f'(x) = \lim_{h\to 0} [6x^2 + 6xh + 2h^2 + 6] = 6x^2 + 6$$

$$f'(-1) = 6(-1)^2 + 6 = 12$$

8. $f'(x) = \lim_{h\to 0} \frac{f(x+h) - f(x)}{h}$

$= \lim_{h\to 0} \frac{6 - \frac{2}{3}(x+h) - \left(6 - \frac{2}{3}x\right)}{h}$

$= \lim_{h\to 0} \frac{-\frac{2}{3}h}{h} = -\frac{2}{3}$

9. $f'(x) = \lim_{h\to 0} \frac{f(x+h) - f(x)}{h} = \lim_{h\to 0} \frac{2(x+h)^2 - 7(x+h) + 3 - (2x^2 - 7x + 3)}{h}$

$= \lim_{h\to 0} \frac{2x^2 + 4xh + 2h^2 - 7x - 7h + 3 - 2x^2 + 7x - 3}{h}$

$= \lim_{h\to 0} \frac{4xh + 2h^2 - 7h}{h}$

$= \lim_{h\to 0} 4x + 2h - 7 = 4x - 7$

10. $f'(x) = \lim_{h\to 0} \frac{f(x+h) - f(x)}{h} = \lim_{h\to 0} \frac{\frac{1}{x+h+7} - \frac{1}{x+7}}{h}$

$= \lim_{h\to 0} \frac{x + 7 - x - h - 7}{(x+h+7)(x+7)h}$

$= \lim_{h\to 0} \frac{-h}{(x+h+7)(x+7)h}$

$= \lim_{h\to 0} \frac{-1}{(x+h+7)(x+7)} = \frac{-1}{(x+7)^2}$

11. $\lim_{x\to\infty} \frac{6}{5x - 1} = 0$

12. $\lim_{x\to\infty} \frac{1 - 3x^2}{x^2 - 5} = -3$

13. $\lim_{x\to -\infty} \frac{3x^3}{x + 2}$ does not exist

14. $0, \frac{3}{4}, \frac{14}{19}, \frac{12}{17}, \frac{36}{53}$

$\lim_{n\to\infty} a_n = \frac{1}{2}$

15. $0, 1, 0, \frac{1}{2}, 0$

$\lim_{n\to\infty} a_n = 0$

16. Width of each rectangle: $\frac{1}{2}$

Heights: $8, \frac{15}{2}, 6, \frac{7}{2}$

Area $\approx \frac{1}{2}\left[8 + \frac{15}{2} + 6 + \frac{7}{2}\right] = \frac{25}{2}$

17. width: $\frac{4}{n}$ height: $f\left(-2 + \frac{4i}{n}\right) = \left(-2 + \frac{4i}{n}\right) + 2 = \frac{4i}{n}$

$A \approx \sum_{i=1}^{n} \left(\frac{4i}{n}\right)\left(\frac{4}{n}\right) = \frac{16}{n^2}\sum_{i=1}^{n} i = \frac{16}{n^2}\frac{n(n+1)}{2}$

$A = \lim_{n\to\infty} \frac{16}{n^2} \cdot \frac{n(n+1)}{2} = 8$

18. width: $\frac{1}{n}$, height: $f\left(\frac{i}{n}\right) = 1 - \frac{i^3}{n^3}$

$$A \approx \sum_{i=1}^{n}\left(1 - \frac{i^3}{n^3}\right)\left(\frac{1}{n}\right) = \sum_{i=1}^{n}\left(\frac{1}{n} - \frac{i^3}{n^4}\right)$$

$$= \frac{1}{n}\sum_{i=1}^{n} 1 - \frac{1}{n^4}\sum_{i=1}^{n} i^3$$

$$= \frac{1}{n}(n) - \frac{1}{n^4}\left(\frac{n^2(n+1)^2}{4}\right)$$

$$= 1 - \frac{(n+1)^2}{4n^2}$$

$$A = \lim_{n\to\infty}\left(1 - \frac{(n+1)^2}{4n^2}\right) = 1 - \frac{1}{4} = \frac{3}{4}$$

19. (a) $y = 8.786x^2 - 6.243x - 0.429$

(b) velocity = derivative = $17.572x - 6.243$

At $x = 5$, velocity ≈ 81.6 ft/sec

Chapters 11–12 Cumulative Test

1. $(-4, 2, 3)$

2. $(0, -4, 0)$

3. $d = \sqrt{(4-(-2))^2 + (-5-3)^2 + (1-(-6))^2}$
$= \sqrt{36 + 64 + 49}$
$= \sqrt{149}$

4. $d_1 = 3, d_2 = 4, d_3 = \sqrt{4^2 + 3^2} = 5$
$d_1^2 + d_2^2 = d_3^2$

5. midpoint: $\left(\frac{3-5}{2}, \frac{4+0}{2}, \frac{-1+2}{2}\right) = \left(-1, 2, \frac{1}{2}\right)$

6. Center $= (2, 2, 4)$
Radius $= \sqrt{2^2 + 2^2 + 4^2} = \sqrt{24}$
$(x-2)^2 + (y-2)^2 + (z-4)^2 = 24$

7. *xy*-trace: $(z = 0)$ $(x-2)^2 + (y+1)^2 = 4$ circle
yz-trace: $(x = 0)$ $4 + (y+1)^2 + z^2 = 4$ or $(y+1)^2 + z^2 = 0$ Point

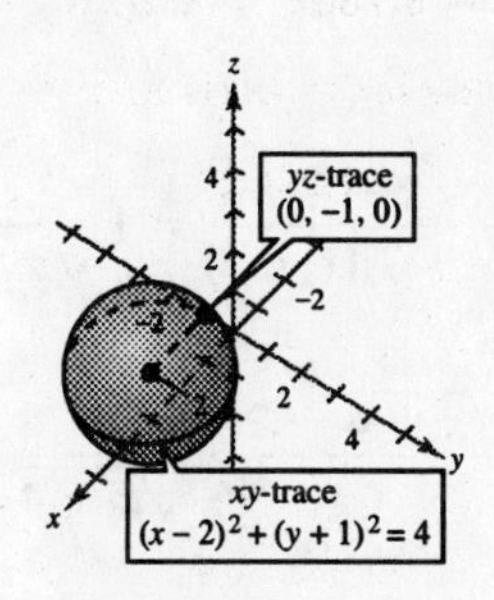

8. $\langle -3, 4, 1\rangle \cdot \langle 5, 0, 2\rangle = -15 + 2 = -13$
$$\mathbf{u} \times \mathbf{v} = \begin{vmatrix} \mathbf{i} & \mathbf{j} & \mathbf{k} \\ -3 & 4 & 1 \\ 5 & 0 & 2 \end{vmatrix} = \langle 8, 11, -20\rangle$$

9. $\mathbf{u} \cdot \mathbf{v} \neq 0, \mathbf{u} \neq c\mathbf{v} \implies$ neither

10. $\mathbf{u} \cdot \mathbf{v} = -8 - 12 + 20 = 0 \implies$ orthogonal

11. $3\mathbf{u} = \langle -3, 18, 9\rangle = \mathbf{v} \implies$ parallel

12. $\overrightarrow{DA} = \langle 0, -2, 0\rangle, \overrightarrow{DC} = \langle 2, 1, 0\rangle, \overrightarrow{DH} = \langle 0, 0, 3\rangle$
$$\begin{vmatrix} 0 & -2 & 0 \\ 2 & 1 & 0 \\ 0 & 0 & 3 \end{vmatrix} = 12 \text{ cubic units}$$

13. (a) Vector is $\langle 5+2, 8-3, 25-0\rangle = \langle 7, 5, 25\rangle$
$x = -2 + 7t, y = 3 + 5t, z = 25t$

(b) $\frac{x+2}{7} = \frac{y-3}{5} = \frac{z}{25}$

14. $\mathbf{u} = \langle -2, 3, 0\rangle, \mathbf{v} = \langle 5, 8, 25\rangle$
$$\mathbf{u} \times \mathbf{v} = \begin{vmatrix} \mathbf{i} & \mathbf{j} & \mathbf{k} \\ -2 & 3 & 0 \\ 5 & 8 & 25 \end{vmatrix} = \langle 75, 50, -31\rangle$$
Normal to plane.
Plane: $75x + 50y - 31z = 0$

15.

16. $\mathbf{n} = \langle 2, -5, 1\rangle, Q = (0, 0, 25), P = (0, 0, 10)$ in plane $\overrightarrow{PQ} = \langle 0, 0, 15\rangle$
$$D = \frac{|\overrightarrow{PQ} \cdot \mathbf{n}|}{\|\mathbf{n}\|} = \frac{15}{\sqrt{30}} = \frac{\sqrt{30}}{2} \approx 2.74$$

17. Normal to plane containing $(-1, -1, 3)$, $(0, 0, 0)$ and $(2, 0, 0)$ is

$$\langle -1, -1, 3\rangle \times \langle 2, 0, 0\rangle = \begin{vmatrix} \mathbf{i} & \mathbf{j} & \mathbf{k} \\ -1 & -1 & 3 \\ 2 & 0 & 0 \end{vmatrix} = \langle 0, 6, 2\rangle \text{ or } \mathbf{n}_1 = \langle 0, 3, 1\rangle$$

Normal to front face is

$$\langle 1, -1, 3\rangle \times \langle 0, 2, 0\rangle = \begin{vmatrix} \mathbf{i} & \mathbf{j} & \mathbf{k} \\ 1 & -1 & 3 \\ 0 & 2 & 0 \end{vmatrix} = \langle -6, 0, 2\rangle \text{ or } \mathbf{n}_2 = \langle -3, 0, 1\rangle$$

Angle between sides:

$$\cos\theta = \frac{|\mathbf{n}_1 \cdot \mathbf{n}_2|}{\|\mathbf{n}_1\|\,\|\mathbf{n}_2\|} = \frac{1}{\sqrt{10}\sqrt{10}} = \frac{1}{10} \implies \theta \approx 84.26°$$

18. $\lim_{x\to 4} (5x - x^2) = 5(4) - 4^2 = 4$

19. $\lim_{x\to -2} \frac{x+2}{(x+2)(x-1)} = \lim_{x\to -2} \frac{1}{x-1} = -\frac{1}{3}$

20. $\lim_{x\to 7} \frac{x-7}{(x-7)(x+7)} = \lim_{x\to 7} \frac{1}{x+7} = \frac{1}{14}$

21. $\lim_{x\to 0} \frac{\sqrt{x+4}-2}{x} \cdot \frac{\sqrt{x+4}+2}{\sqrt{x+4}+2} = \lim_{x\to 0} \frac{(x+4)-4}{x\left(\sqrt{x+4}+2\right)} = \lim_{x\to 0} \frac{1}{\sqrt{x+4}+2} = \frac{1}{2+2} = \frac{1}{4}$

22. $\lim_{x\to 4^-} \frac{|x-4|}{x-4} = -1$

23. $\lim_{x\to 2} f(x) = 2$

24. $m_{\text{sec}} = \frac{f(1+h)-f(1)}{h} = \frac{3-(1+h)^2-2}{h} = \frac{3-(1+2h+h^2)-2}{h} = \frac{-2h-h^2}{h}$

Slope $= \lim_{h\to 0} \frac{-h^2-2h}{h} = \lim_{h\to 0} (-h-2) = -2$

25. $m_{\text{sec}} = \frac{f(-2+h)-f(-2)}{h} = \frac{\sqrt{-2+h+3}-1}{h} = \frac{\sqrt{h+1}-1}{h}$

$= \frac{\sqrt{h+1}-1}{h} \cdot \frac{\sqrt{h+1}+1}{\sqrt{h+1}+1} = \frac{(h+1)-1}{h\left(\sqrt{h+1}+1\right)} = \frac{h}{h\left(\sqrt{h+1}+1\right)}$

Slope $= \lim_{h\to 0} \frac{h}{h\left(\sqrt{h+1}+1\right)} = \frac{1}{2}$

26. $m_{\text{sec}} = \frac{f(1+h)-f(1)}{h} = \frac{\frac{1}{1+h+3}-\frac{1}{4}}{h} = \frac{4-(h+4)}{(h+4)4h} = \frac{-1}{4(h+4)}$

Slope $= \lim_{h\to 0} \frac{-1}{4(h+4)} = -\frac{1}{16}$

27. $m_{\sec} = \dfrac{f(-1+h) - f(-1)}{h} = \dfrac{(-1+h)^4 - 1}{h} = \dfrac{1 - 4h + 6h^2 - 4h^3 + h^4 - 1}{h}$

$= \dfrac{-4h + 6h^2 - 4h^3 + h^4}{h}$

Slope $= \lim_{h \to 0} \dfrac{-4h + 6h^2 - 4h^3 + h^4}{h} = -4$

28. $\lim_{x \to \infty} \dfrac{2x^4 - x^3 + 4}{x^2 - 9}$

does not exist

29. $\lim_{x \to \infty} \dfrac{2 - 5x}{1 - 2x} = \dfrac{5}{2}$

30. $\lim_{x \to \infty} \dfrac{3x^2 + 1}{x^2 + 4} = 3$

31. $\lim_{x \to \infty} \dfrac{2x}{x^2 + 3x - 2} = 0$

32. $\sum_{i=1}^{50} \dfrac{1}{4} i^2 = \dfrac{1}{4} \dfrac{50(51)(101)}{6} = 10731.25$

33. $\sum_{k=1}^{20} (3k^2 - 2k) = 3\dfrac{20(21)(41)}{6} - 2\dfrac{20(21)}{2}$

$= 8610 - 420 = 8190$

34. $\sum_{i=1}^{40} (12 + i^3) = 12(40) + \dfrac{40^2(41)^2}{4}$

$= 480 + 672{,}400 = 672{,}880$

35. Area $\approx \frac{1}{2}[4.875 + 4.5 + 3.875 + 3] = 8.125$ sq. units

36. Area $\approx \dfrac{1}{4}\left[\dfrac{1}{1 + (-3/4)^2} + \dfrac{1}{1 + (-1/2)^2} + \dfrac{1}{1 + (-1/4)^2} + \dfrac{1}{1 + 0} + \dfrac{1}{1 + (1/4)^2} + \dfrac{1}{1 + (1/2)^2} + \dfrac{1}{1 + (3/4)^2} + \dfrac{1}{1 + 1^2}\right]$

$= \dfrac{1}{4}\left[2(0.64) + 2(0.8) + 2(0.941176) + 1 + \dfrac{1}{2}\right]$

≈ 1.566 sq. units

37. Width: $\dfrac{1}{n}$

Height: $f\left(\dfrac{i}{n}\right) = 1 - \left(\dfrac{i}{n}\right)^3$

$A \approx \sum_{i=1}^{n} \left(1 - \left(\dfrac{i}{n}\right)^3\right)\left(\dfrac{1}{n}\right) = \dfrac{1}{n}\sum_{i=1}^{n} - \dfrac{1}{n^4}\sum_{i=1}^{n} i^3$

$= \dfrac{1}{n}(n) - \dfrac{1}{n^4}\left[\dfrac{n^2(n+1)^2}{4}\right]$

$A = \lim_{n \to \infty} \left[1 - \dfrac{1}{n^4}\left(\dfrac{n^2(n+1)^2}{4}\right)\right] = 1 - \dfrac{1}{4} = \dfrac{3}{4}$

38. Width: $\dfrac{6}{n}$

Height: $f\left(-3 + \dfrac{6i}{n}\right) = \left(-3 + \dfrac{6i}{n}\right) + 3 = \dfrac{6i}{n}$

$A \approx \sum_{i=1}^{n} \left(\dfrac{6i}{n}\right)\left(\dfrac{6}{n}\right) = \dfrac{36}{n^2}\sum_{i=1}^{n} i$

$= \dfrac{36}{n^2}\dfrac{n(n+1)}{2}$

$= \dfrac{18(n+1)}{n}$

$A = \lim_{n \to \infty} \left[\dfrac{18(n+1)}{n}\right] = 18$

39. Width: $\frac{2}{n}$, height: $f\left(-1+\frac{2i}{n}\right)=\left(-1+\frac{2i}{n}\right)^2=1-\frac{4i}{n}+\frac{4i^2}{n^2}$

$$A \approx \sum_{i=1}^{n}\left[1-\frac{4i}{n}+\frac{4i^2}{n^2}\right]\left(\frac{2}{n}\right)$$

$$= \frac{2}{n}\sum_{i=1}^{n}1-\frac{8}{n^2}\sum_{i=1}^{n}i+\frac{8}{n^3}\sum_{i=1}^{n}i^2$$

$$= \frac{2}{n}(n)-\frac{8}{n^2}\frac{n(n+1)}{2}+\frac{8}{n^3}\frac{n(n+1)(2n+1)}{6}$$

$$= 2-\frac{4(n+1)}{n}+\frac{4(n+1)(2n+1)}{3n^2}$$

$$A = \lim_{n\to\infty}\left[2-\frac{4(n+1)}{n}+\frac{4(n+1)(2n+1)}{3n^2}\right]$$

$$= 2-4+\frac{8}{3}=\frac{2}{3}$$